Table B
Long Form of the Periodic Table[a,b]

D0022981

s Block / d Block / p Block

Group	1	2		3	4	5	6	7	8	9	10	11	12		13/III	14/IV	15/V	16/VI	17/VII	18/VIII
NVE	1	2		3	4	5	6	7	8	9	10	11	12		3	4	5	6	7	8
Per. 1 Z =	1 H	2 He[c]																		
Per. 2 Z =	3 Li	4 Be													5 B	6 C	7 N	8 O	9 F	10 Ne
Per. 3 Z =	11 Na	12 Mg													13 Al	14 Si	15 P	16 S	17 Cl	18 Ar
Per. 4 Z =	19 K	20 Ca		21 Sc	22 Ti	23 V	24 Cr	25 Mn	26 Fe	27 Co	28 Ni	29 Cu	30 Zn		31 Ga	32 Ge	33 As	34 Se	35 Br	36 Kr
Per. 5 Z =	37 Rb	38 Sr		39 Y	40 Zr	41 Nb	42 Mo	43 Tc	44 Ru	45 Rh	46 Pd	47 Ag	48 Cd		49 In	50 Sn	51 Sb	52 Te	53 I	54 Xe
Per. 6 Z =	55 Cs	56 Ba		71 Lu	72 Hf	73 Ta	74 W	75 Re	76 Os	77 Ir	78 Pt	79 Au	80 Hg		81 Tl	82 Pb	83 Bi	84 Po	85 At	86 Rn
Per. 7 Z =	87 Fr	88 Ra		103 Lr	104 Rf	105 Db	106 Sg	107 Bh	108 Hs	109 Mt	110 Ds	111 Rg	112 Cn		113 Nh	114 Fl	115 Mc	116 Lv	117 Ts	118 Og

f Block

	3	4	5	6	7	8	9	10	11	12	13	14	15	16
Per. 6 Z =	57 La	58 Ce	59 Pr	60 Nd	61 Pm	62 Sm	63 Eu	64 Gd	65 Tb	66 Dy	67 Ho	68 Er	69 Tm	70 Yb
Per. 7 Z =	89 Ac	90 Th	91 Pa	92 U	93 Np	94 Pu	95 Am	96 Cm	97 Bk	98 Cf	99 Es	100 Fm	101 Md	102 No

[a] Rows are labeled with the period numbers.

[b] Columns are labeled with the number of valence electrons. These are also the group numbers in the *s* and *d* blocks. In the *p* block these are the group numbers in the old style using Roman numerals. For example, 3 valence electrons in the *p* block = Group 13/III.

[c] Although its chemical inertness justifies positioning He in Group 18/VIII, it does not have 8 valence electrons or any *p* electrons, so for purposes of figuring electron configurations, it is included in the *s* block in this table.

FOUNDATIONS OF
Inorganic Chemistry

Gary Wulfsberg

Middle Tennessee State University

UNIVERSITY SCIENCE BOOKS
MILL VALLEY, CALIFORNIA

University Science Books
Mill Valley, California
www.uscibooks.com

EDITOR Jane Ellis
PRODUCTION MANAGER Julianna Scott Fein
MANUSCRIPT EDITOR John Murdzek
ILLUSTRATOR Laurel Muller
TEXT DESIGN Yvonne Tsang
COVER DESIGN Genette Itoko McGrew
COMPOSITOR Michael Starkman at Wilsted & Taylor
PRINTER AND BINDER Bang Printing

This book is printed on acid-free paper.

Print ISBN 978-1-891389-95-5
eBook ISBN 978-1-938787-94-2

LIBRARY OF CONGRESS CATALOGING-IN-PUBLICATION DATA

Names: Wulfsberg, Gary, 1944– author.
Title: Foundations of inorganic chemistry / Gary Wulfsberg,
    Middle Tennessee State University.
Description: Mill Valley, California : University Science Books,
    [2017] | Includes bibliographical references.
Identifiers: LCCN 2017010601| ISBN 9781891389955 (print : alk.
    paper) | ISBN 9781938787942 (ebook)
Subjects: LCSH: Chemistry, Inorganic.
Classification: LCC QD151.3 .W85 2017 | DDC 546—dc23
LC record available at https://lccn.loc.gov/2017010601

Printed in the United States of America
10   9   8   7   6   5   4   3   2   1

To my wife, children, and grandchildren,
who have given me so much personal support—
Marlys, Joanna, Paul, Reema, Kuzey, Zade—
and in memory of my father, Paul,
and parents-in-law, Herbert and Muriel.
To my graduate research supervisor,
Robert C. West, Jr., who awakened
my interest in teaching that actively
engages the interest of students.

CONTENTS IN BRIEF

CONTENTS

CONNECTIONS AND AMPLIFICATIONS

The following Connection and Amplification sections are interspersed throughout the text.

Chapter 7

Chapter 8

Chapter 12 *(cont.)*

PREFACE

To the Instructor

This textbook is principally designed for a one-semester Foundations of Inorganic Chemistry course with no physical chemistry prerequisite.[1] This textbook will have wide appeal, not only to inorganic chemistry students, but also to other chemists and to professionals in other fields of science and in the health professions in which inorganic chemistry has become important. Consequently, such a course may well have great growth in enrollment. In addition, the text has much of the material needed for a second-semester In-Depth/Advanced Inorganic Chemistry course.

Chemistry majors. A Foundations of Inorganic Chemistry course is currently required by the American Chemical Society Committee on Professional Training for professional chemistry majors in departments that they certify.[2] It covers about 98% of the topics included in the ACS Exams Institute 2016 exam for the Foundations of Inorganic Chemistry course.

Chemists and professionals in other fields. The inorganic chemistry that is most appealing and useful to the students in the most populous branches of chemistry and in allied professional fields [e.g., preprofessional (premedical, etc.) studies, medicinal chemistry, analytical chemistry, biochemistry, chemical education, environmental chemistry, and geochemistry] is the chemistry of ions of the elements as it takes place in water, in solid-state salts, and in complexes, along with the chemistry of a selection of representative inorganic and organometallic molecules. This chemistry is presented as *periodic trends* in Chapters 1–8 of this textbook, organized in the categories that students used in general chemistry—namely, acid–base chemistry (Chapter 2), coordination (complex-ion) chemistry (Chapters 3, 5, and 7), precipitation chemistry and the solid-state chemistry of salts (Chapters 4 and 8), and oxidation–reduction chemistry (Chapter 6). Chapter 9 covers the theoretical reasons for the periodic trends in fundamental atomic properties covered in Chapter 1, and can be integrated with the first chapter if desired. Numerous twenty-first century developments in inorganic chemistry have been incorporated.

Chapters 10, 11, and 12 cover the major topics needed to begin the study of in-depth/advanced inorganic chemistry. Most other texts cover these topics (especially molecular orbital theory) at the beginning of the text, when the students actually know very little factual inorganic chemistry to which the theory can be applied. Using this text, students will know much more about inorganic ions and compounds and their reactions, and can apply molecular orbital theory much more fruitfully than they could using a traditional advanced inorganic chemistry text. Students can therefore more fully see how the central idea of inorganic chemistry—periodicity—develops.

Many potential users of this text come from a background of teaching a one-semester Advanced Inorganic Chemistry course and may feel uncomfortable with the idea of teaching an up-to-date course emphasizing the aqueous chemistry of the elements and the connections of inorganic chemistry to other fields. To assist such instructors (and their students), we include a "Background Reading" section at the end of each chapter.

For colleges and universities that have only one inorganic Foundations course per year, we suggest that both future inorganic chemists and the rest of chemists and allied science students can be served if the university *offers alternate-year versions of the Foundations course*. We have described the structure that we use to achieve this at MTSU.[3]

Aids to learning. In response to user feedback on our previous (Year 2000) textbook, *Inorganic Chemistry*, extensive improvements have been incorporated.

1. We have increased, not only the number of exercises (at least half of which were tested on MTSU students by putting them on exams), but also the number with answers. The book contains over 900 exercises, ranging from elementary ones at the beginning of each subsection to some very challenging ones later. The odd-numbered exercises are answered in the back of the book. There are over 180 worked examples—at least one per subsection.

2. Since students learn the topics better when doing (graded) homework, the even-numbered exercises are *not* answered in the back of the book. The answers are included in an Instructors' Manual available only to professors who have officially adopted the textbook.

3. Since our experience (as well as that of the ACS Committee on Professional Training) is that student retention of the numerous concepts introduced in general chemistry is not optimal, we have added a number of general chemistry review subsections, which can be assigned or made available for students who need review. These sections should be particularly useful in schools in which the Foundations course includes freshman students.

4. To make students' review before exams less daunting, we have included study objectives for each section and a quick synopsis of the main ideas in the boxed overview at the beginning of each section.

5. *All* types of chemistry students want to know "What is this material good for?" In common with most modern texts, we have boxed optional sections that we call "Connections" to 18 other fields of chemistry and allied sciences. Unlike some textbooks, however, these 70 Connection sections are closely connected to inorganic concepts and to specific other fields of chemistry and

allied sciences. Students can see how interrelated modern inorganic chemistry and the other fields of chemical science have become. Indeed, more than once students have told me that they felt, after using this text, that inorganic chemistry was the *central* division of chemistry, being the most connected to the other fields. Some have told me that the 2000 book is the *only* Foundations textbook that they have felt was worth keeping.

6. We have tried to restrict the inorganic concepts covered to the most broadly useful ones. About 70 concepts that would mainly be useful in a more advanced inorganic chemistry course are in boxes entitled "Amplifications."

7. The readability of the text has been improved, with a reading level that is two or more grade levels below that of the 2000 text. We have not, however, removed clear explanations of concepts. (Students have made online comments such as, "If you actually want to understand this concept, you need to consult Wulfsberg's [2000] text.")

8. We have added two-color illustrations and flow charts outlining the most complex concepts.

9. The best first step in understanding inorganic reactivity is to see and think about examples. In Experiments 1–10 (found on the University Science Books website), we present several optional simple inorganic experiments or computational exercises in which students not only observe trends in inorganic reactivity, but are challenged to reason inductively from them to discover for themselves many of the main principles of the text. These technically simple experiments include asking the student to design parts of the procedure, in which they must understand how to control multiple variables—these experiments are practice in the scientific method of a sort not normally encountered until students initiate research.[4] In courses that do not have laboratories, these can be done as demonstrations followed by small-group discussions to try to design the procedure and reason inductively to the principles that they discover. To make these more easily performed and more widely available, videos of three of these demonstration/discussions have been prepared.[5]

To the Student

With the help of careful research, including input from such techniques as functional magnetic resonance imaging of brains in action, cognitive scientists have changed some of our ideas about how to write a book and how to study in a course.

1. Although students must begin with factual knowledge (i.e., memorize some things), such knowledge is much more likely to remain in your long-term memory if you have thought about the meaning of the concepts. We not only state principles and give equations, but also take the time to explain them. However, it also helps if you work a variety of problem types during the same study session. Hence, the problems given with each section are not only direct applications of the equations or principles given in the study

objectives, but also include problems that may require you to deepen your grasp of the principle and see how it applies in different situations. This should help you remember the fundamentals when it comes time in your career (e.g., when taking the Graduate Record Exam) to show that you know your chemistry.

2. If material is covered and used only once, it is unlikely to be stored in long-term memory. Over time, numerous principles and equations have been introduced in inorganic chemistry. However, many of them apply only in limited circumstances, while others are useful in many situations. We have chosen to remove or de-emphasize the concepts of limited usefulness that you might not use again, and so would not keep in long-term memory. Instead, we focus on broadly useful principles that you will use again in later chapters.

3. Scientific papers routinely begin with a summary of what will be discussed and conclude with a summary of what has been discussed. Each section begins with an overview that includes the study objectives and a summary of the concepts of that section, and references to the previous sections that contain the concepts that you need to recall. We also mention future sections in which you will again use the concepts developed in this section.

4. The concepts and categories we use in this text are based on concepts introduced in a General Chemistry course or, in some cases later in the book, in a Foundations of Organic Chemistry course. It may be that you have not used them since that date, so that you may have forgotten them. Since these may come back to haunt you when taking exams such as the GRE or MCAT, we have included a review of these ideas and some homework assignments in which to apply them. This course can be a very helpful part of reviewing inorganic aspects of general chemistry when preparing for exams such as the MCAT. However, we caution you that concepts introduced in general chemistry are often simplified, whereas they are refined in this text.

5. Doing homework assignments or taking frequent quizzes is more than a way of assessing students. Each such assignment or quiz is also a powerful chance for you to learn. Hence, we have increased the number of exercises for each concept to at least six (with at least one worked example), and the number per chapter to around 60 to 100. The odd-numbered exercises have answers in the back of the book to help you assess your own learning. However, we have *not* answered the other half, so that you can have the learning opportunity of doing them as homework assignments or in-class quizzes.

6. Boldfaced terms have their definitions nearby in the textbook; they are also included in the index so that you can find their definitions later. Italicized words are important in your understanding of the procedure or concept.

Acknowledgments

I want to thank my students at MTSU who not only caught errors but put up with my annoying habit of first trying out new end-of-chapter exercises as exam questions. Bill Ilsley also class-tested and error-checked my manuscript and offered valuable suggestions. After my retirement (to emeritus status), my replacement, Keying Ding, also did the same.

The external reviewers of the manuscript were also very helpful, especially Sabrina G. Sobel and Daniel R. Talham, who reviewed the entire manuscript. Others who provided valuable comments on the manuscript include Lothar Stahl, David Marx, Mike Heinekey, Dale Ensor, Jared Paul, Marc Walters, and Susan K. VanderKam. Rick Nelson helped with pedagogical ideas, John DiVincenzo reviewed the environmental chemistry of the oxides, and Dave Finster helped with suggestions for the coverage of safety. Preston MacDougall arranged for the digitization of the supplemental video labs, and Rebecca Jones provided valuable corroboration from her teaching at Austin Peay State University that a Foundations course that included nonchemists could work and generate much larger enrollments.

I am much indebted to my publisher, Bruce Armbruster; my editor, Jane Ellis; my copy editor, Dr. John Murdzek; and the artists and others at University Science Books for their enthusiastic support of this long project.

Notes

1 Co-registration in an Organic Chemistry course is helpful later in the book.

2 American Chemical Society Committee on Professional Education. Undergraduate Professional Education in Chemistry: ACS Guidelines and Evaluation Procedures for Bachelor's Degree Programs. 2015. http://www.acs.org/content/dam/acsorg/about/governance/committees/training/2015-acs-guidelines-for-bachelors-degree-programs.pdf. Accessed June 16, 2015.

3 G. Wulfsberg, *J. Chem. Educ.* 89, 1220 (2012).

4 We have found that our students, even the seniors, usually do not understand how to design an experiment in which two variables must be controlled.

5 L. H. Laroche, G. Wulfsberg, and B. Young, *J. Chem. Educ.* 80, 962 (2003). Available at www.uscibooks.com/Wulfsberg_supplements.htm.

FOUNDATIONS OF INORGANIC CHEMISTRY

MARIE SKLODOWSKA CURIE (1867–1934) was born in Warsaw when that region of Poland was still part of the Russian Empire. Women were not allowed to study at the University of Warsaw, so she attended a "floating university" that met at night in different locations to avoid detection by the czar's authorities. In 1891, she joined her older sister Bronislawa at the University of Paris, and within three years she had earned masters degrees in physics and in math. During this time, she met and married Pierre Curie, a member of the University's physics faculty. Continuing her studies at the University, in 1903 she became the first woman in France to earn a doctorate degree for her pioneering research on the radioactivity first observed in uranium salts by Henri Becquerel in 1896. Marie and Pierre found that the mineral pitchblende was more radioactive than uranium, and they reasoned that it contained other radioactive substances. The substances turned out to be two previously unknown elements, polonium and radium, which they painstakingly extracted from tons of ore. Marie, Pierre, and Bequerrel were awarded the 1903 Nobel Prize in Physics for their studies of radioactivity, and Marie was awarded the 1911 Nobel Prize in Chemistry for her discovery and isolation of polonium and radium. Despite her obvious contributions to science, she was never elected to the French Academy of Sciences because of the prevailing prejudice against women. During World War I, Marie developed and operated 20 mobile X-ray labs to assist battlefield surgeons. She died in 1934 of aplastic anemia, no doubt due to a lifelong exposure to radiation.

CHAPTER 1

Periodic Trends in Fundamental Properties of Atoms and Simple Ions

1.1. The Connections of Inorganic Chemistry to Other Fields of Chemical Science

The central concept of inorganic chemistry is *periodicity*. We intend to show the amazing breadth of this concept as it unfurls in inorganic chemistry, in other branches of chemistry and science, and in modern life itself. Inorganic chemistry plays a crucial role in modern interdisciplinary research. Interdisciplinary discoveries may originate in the research laboratories of inorganic chemistry, in the environmental field, or in outer space. The discoveries may involve exotic new compounds of types no one had ever imagined, or familiar old inorganic compounds [such as Prussian blue (Section 5.3A) and Egyptian blue (Section 8.6A)] that were discovered hundreds or thousands of years ago. Chemists and scientists such as you who develop a good fundamental understanding of the foundations of inorganic chemistry are more likely to contribute to the most exciting interdisciplinary discoveries of the future.

As is true for any of the major fields of chemistry, it takes at least a year to cover the foundations of inorganic chemistry. An undergraduate four-year chemistry curriculum does not give time for all students to take this many courses in each field, however. Recognizing this, the 2008 guidelines of the Committee on Professional Training (CPT) of the American Chemical Society (ACS) feature required one-semester Foundations courses in five fundamental areas of chemistry, including inorganic chemistry, with an elective second semester In-Depth (i.e., Advanced) Inorganic Chemistry course to be available as an elective. In a one-semester course, one must select which foundation concepts to cover. We have selected for this book the concepts of the greatest importance for perhaps 80% of inorganic students—those enrolled in 18 of perhaps 30 branches of chemistry and allied fields.[1] This chemistry tends to be aqueous and bio-inorganic chemistry: It is the chemistry of ions of the elements as it takes place in water, in salts, and in complexes, along with the chemistry of a selection of representative inorganic and organometallic molecules.

In Chapters 1–9 we develop inorganic chemistry as *periodic trends* across the *entire* periodic table in the categories of chemical reactions that you used in general chemistry—namely, acid–base chemistry (Chapter 2), coordination (complex-ion) chemistry (Chapters 3, 5, and 7), precipitation chemistry and the solid-state chemistry of salts (Chapters 4 and 8), and oxidation–reduction chemistry (Chapter 6). Principles that help explain these trends in chemical reactivity are given in Chapter 9. To understand this chemistry, the bonding concepts introduced in General Chemistry courses generally suffice. The topics of symmetry and molecular orbital theory are foundational for students going on to In-Depth (Advanced) Inorganic Chemistry, so these are introduced in Chapters 10 and 11 and applied in Chapter 12.

To emphasize how extensively inorganic chemistry is interconnected with other chemical sciences, we introduce boxed text sections describing specifically the connections to the 18 selected areas of chemical science. For example, in this introductory Chapter 1, we introduce connections to biochemistry (Section 1.2B) and industrial chemistry and geopolitics through lithium-ion batteries, conflict minerals, and rare earth elements (Sections 1.2C). Knowing the details of these (often rapidly changing) connections is not vital, but we recommend reading them to gain new insights into the degree of interconnectedness found in modern chemistry.

This text has many other features designed to make it more accessible and helpful to you, the student. These are summarized in the Preface, which we encourage you to read. One that is important in Chapter 1 is that we include many subsections that review concepts that were probably introduced to you in general chemistry. However, concepts that were introduced in the past but not used again tend to be lost. For these subsections you could test yourself by trying the recommended odd-numbered (starred) Exercises, which are answered in the back. If the concepts have become too rusty, you then have a review subsection in which to brush up your abilities. These abilities, after all, are going to be expected in exams you may be taking in the future, such as admissions exams for graduate or professional school, or the standard final exam for an American Foundations of Inorganic Chemistry exam.

1.2. The Periodic Table and Ions of the Elements

OVERVIEW. Know the blocks of elements found in the periodic table (Section 1.2A). Know the difference between the atoms, elemental forms, and the ions of an element, and be able to write balanced half-reactions for interconversions among them (Section 1.2B). Know the common patterns of monatomic ion charges and names in different blocks of the periodic table. Know in which blocks colored ions are commonly formed (Tables 1.3 and 1.4, Section 1.2C). You may practice these concepts by trying Exercises 1.1–1.8). (Tables 1.2–1.6 contain many bits of information that are illustrative and which you are not expected to know, such as the colors and biological roles of specific ions.) Given the names or formulas of a cation and an anion, deduce the formula of the salt that they form. Given the name or formula of an ionic salt, deduce the charges of its cation and anion. You may review these concepts of Section 1.2D by trying Exercises 1.9–1.13. The concepts of this section are fundamental for understanding and applying inorganic chemistry, and will be used again in almost all future chapters.

1.2A. Periods, Groups, and Blocks of the Periodic Table (Review). The invention of the periodic table, which preceded any knowledge of subatomic structure or quantum mechanics, was an attempt to systematize the known physical properties and chemical reactions of the elements and their ions. The periodic table was organized such that, if the elements were listed in order of an increasing fundamental property (originally atomic weight, now atomic number Z, the number of protons in the atom), very similar chemical properties would recur periodically. When the properties first recurred, the horizontal listing of the elements would be disrupted, and a new **period** of elements would be begun, with the chemically similar elements and ions listed above and below each other in a **group**.

To do this with the known elements, it is necessary to set up seven horizontal periods of elements. As it turns out, however, these periods become considerably longer as atomic numbers get larger, so that the periodic table becomes inconveniently wide (32 groups at the bottom). Hence, the more common form of the periodic table separately lists the last groups to appear (the "lanthanides" and "actinides") at the bottom. We will usually resort to this shorter form, too, but in Table B (inside the front cover) we present the long form of the periodic table.

The periodic table has four main regions, each of which has more than one name in common use. Both the two groups at the far left and the six groups at the far right are often called the representative elements. To distinguish them, we call the two groups at the far left the **s-block elements** and the six groups at the far right the **p-block elements**. Between 2000 and 2010, new elements 113–118 were added to the p block in the seventh period (Chapter 9). The International Union of Pure and Applied Chemistry (IUPAC) approved the names flerovium (symbol Fl) for element 114 and livermorium (symbol Lv) for element 116 in 2012. Then, in 2016, their respective discoverers proposed the names nihonium (Nh) for element 113, moscovium (Mc) for element 115, tennessine (Ts) for element 117, and oganesson (Og) for element 118.

The long block of elements next to the s-block elements in Table B has a variety of names: "rare-earth elements" (although many of them are not rare), "inner transition elements," or "lanthanides and actinides." For simplicity, we call these elements the **f-block elements**.

Between the f- and p-block elements in Table B are the "transition elements," which for consistency we will call the **d-block elements**. Since 2000, three new d-block elements in the seventh period have been confirmed and named: Element 110 is darmstadtium (Ds), 111 is roentgenium (Rg), and 112 is copernicium (Cn).

Note the general shape of the periodic table: Every *two* periods a new block of elements is introduced, just after the s block. Moreover, each new block of elements is wider than the ones introduced before: The s block is two elements wide, the p block is six elements wide, the d block is 10 elements wide, and the f block is 14 elements wide.

Numbering of the Groups. Group numbers serve multiple purposes. Primarily, they allow access to information about the elements in that group. When this is done by computer searching, groups have to have unambiguous numbers. This was not the case with the old numbering based on Roman numerals I through VIII, followed by the letters A and B, because A and B were used in different blocks in different parts of the world. Another useful function of group numbers is in determining the numbers of valence electrons (NVE) and valence electron configurations in atoms (Section 1.3B).

According to the current IUPAC recommendations, the s-block groups should be numbered 1 and 2. The IUPAC recommends (for clarity in literature searching) that the p-block groups be numbered 13 through 18, but for counting valence electrons the older Roman numerals work better (without the A or B). We will attempt to compromise these numbering traditions in the p block—that is, the boron group will be numbered 13/III. We follow the current IUPAC practice of numbering the d-block groups from 3 to 12, which is more unambiguous and more useful for counting valence electrons than the old system of numbering from III to VIII and then back to I and II. The f-block groups are not usually numbered at all, but they do have characteristic numbers of valence electrons. Hence, in Table B we also label the groups with their blocks and numbers of valence electrons.

EXAMPLE 1.1

New elements are currently being created (Chapter 9), in part with the goal of extending the periodic table to see whether it continues the familiar pattern. If the pattern continues, the as-yet-unknown period 8 should include a new block of elements. How many elements wide should this block be? With what element number should it begin and end?

SOLUTION: Table B shows the following pattern to the structure of the periodic table: Every other period a new block of elements is introduced after the s block, and each new block is four elements wider than the previous one. In period 8, a new block is expected after the s block (which should end with element 120). The new block should be $14 + 4 = 18$ elements wide, so it should begin with element 121 and end with element 138.

1.2B. Ions, the Common Forms of Most Elements. Inorganic chemistry includes the study of the chemistry of the **elements**, their **compounds** with other elements, and their **ions** (with the exception of most of the compounds and ions of carbon). We use the periodic table (Section 1.2A) to organize this chemistry.

Most of the known elements do not appear in nature as neutral atoms. Instead, we encounter most of them as ions in ionic compounds. **Ionic compounds** consist of **cations** (positively charged ions) and **anions** (negatively charged ions) in solid **salts** or in solutions—for example, we may find sodium ion in the solid NaCl(s) or in aqueous solution as NaCl(aq).

Atoms of most elements are almost never encountered alone, but are nearly always bonded with other atoms in the form of **polyatomic ions** or **molecules**. If the other atoms are of the same element, then we have **elemental substances**, which include not only the diatomic molecules, such as those of Group 17/VII, but also larger molecules, such as C_{60}, or macromolecular forms such as metals. Because the bonding in many elemental substances requires more advanced concepts of bonding, we postpone coverage of elemental substances to Chapter 12.

Our language blurs the difference between cations and elemental substances. Cations are named by appending the word "ion" to the name of the element. Unfortunately,

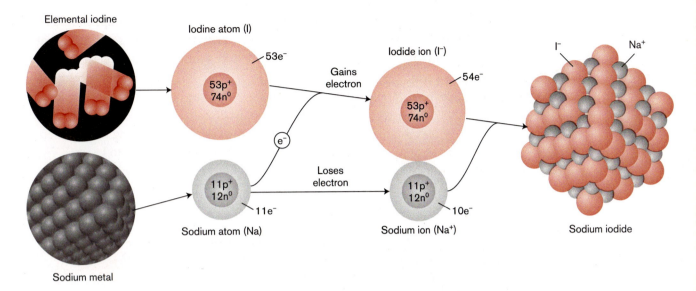

Figure 1.1. Reaction of sodium metal with elemental iodine to give sodium and iodide ions, then solid sodium iodide. [Adapted from M. S. Silberberg, *Principles of General Chemistry*, 2nd ed., McGraw-Hill: New York, 2010, p. 48.]

in common usage the word "ion" is omitted, leaving no distinction between the ion, the atom, and the elemental substance. When a nutritional article says, "Calcium is essential in the diet," is it referring to calcium atoms, Ca(*s*), a silvery metal that reacts violently with water, or the calcium ion, Ca^{2+}, which is found in solution or in various salts such as limestone (calcium carbonate)?

Experiment 1 may help you visualize and understand the difference between elemental substances and ions in salts or solution. We often talk about sodium being very common in nature and in our diet, but which is sprinkled on your French fries: Na(*s*) or Na^+? What happens if the cook adds the wrong form? Figure 1.1 shows a reaction connected with this Experiment, conducted in the absence of French fries.

EXPERIMENT 1 MAY BE ASSIGNED NOW.
(See www.uscibooks.com/foundations.htm.)

At the particulate and symbolic levels, atoms have the same number of protons and electrons, so they are uncharged. Ions come in two types: positively charged ions called cations and negatively charged ions called anions. Cations are formed from atoms by the *loss* of electrons. The following half-reaction describes this process for the formation of the sodium ion:

$$Na \rightarrow Na^+ + e^- \tag{1.1}$$

Anions are formed from atoms by the *gain* of electrons:

$$I + e^- \rightarrow I^- \tag{1.2}$$

Half-reactions are chemical equations that include electrons as either reactants or products. They must be balanced for charge as well as for the elements involved.

EXAMPLE 1.2

Experiment 1 includes a "Zinc Cycle" in which the elemental and ionic forms of zinc and iodine are interconverted. Write balanced half-reactions for the reactions of the two elemental substances to form their respective aqueous monatomic ions. (Note that zinc is a solid metal that forms ions with +2 charges, whereas elemental iodine occurs as diatomic molecules that form monatomic ions with –1 charges.)

SOLUTION: For zinc, the two forms differ by two electrons:

$$Zn(s) \rightarrow Zn^{2+}(aq) + 2e^- \tag{1.3}$$

Equation 1.3 is balanced both for atoms (one Zn on the left, one on the right) and for charge [no charge on the left, $+2 + (-2) = 0$ net charge on the right]. For iodine, we start with a diatomic molecule and end up with two monatomic ions:

$$I_2(s) + 2e^- \rightarrow 2I^-(aq) \tag{1.4}$$

Equation 1.4 is balanced for iodine atoms (two I on the left, two on the right) and for charge (–2 on the left, –2 on the right).

If you did this experiment, you would also note the difference in appearance of these forms: $Zn(s)$ is a gray solid and $I_2(s)$ is a dark purple solid with purple vapors when warm, whereas the solution containing $Zn^{2+}(aq)$ and $I^-(aq)$ is colorless (so each ion is also colorless); the solid containing Zn^{2+} and I^-, $ZnI_2(s)$, is white. Which form of zinc do you think is in your multivitamin pill? Which form of iodine is in the iodized salt you probably consume?

A CONNECTION TO BIOCHEMISTRY

Preview of the Biological Functions of Metal Ions

As we proceed through the next chapters, you will develop the background to better understand these functions, so we will return to these biological functions as indicated.

1. In Chapter 2 we will see that the acidity of a metal ion polarizes the water molecule in a hydrated ion $[M(H_2O)_x]^{y+}$ and causes it to release hydrogen ions (Figure 2.1). In biochemistry an acidic metal ion also enhances the acidity of other molecules attached (coordinated) to metal ions, making them more reactive than they would be otherwise. In Chapter 3 we will see that this acidity originates in the metal ion, not the water, so that metal ions act as Lewis acids and can catalyze Lewis acid–base reactions.

2. The ions K^+, Na^+, Ca^{2+}, and Mg^{2+} trigger and control certain biochemical mechanisms; the passage of Na^+ ions across nerve cell walls constitutes an electrical current involved in nerve impulse transmission (Figure 4.16).

3. Metal ions can change the conformation of an enzyme or biomolecule containing negatively charged functional groups that are close together (such as triphosphate ion groups in DNA and RNA). In the absence of metal ions with a +2 charge, DNA and RNA double helices tend to unwind due to the repulsion of the triphosphate groups. The mutual attraction of the negatively charged triphosphate groups to the Mg^{2+} ion helps retain the double helix. This is an application of Coulomb's law, which governs the attraction of opposite charges and the repulsion of like charges (Chapters 2 and 4).

4. If two molecules that are to react with each other are each attached to the same metal ion (Figure 5.13), the statistical odds of their finding each other and colliding may be greatly enhanced. This tendency is called the **template effect**.

5. The *d*-block metals with several oxidation states are particularly useful in catalyzing biological electron-transfer and oxidation–reduction reactions (Chapter 6), and the transfer of groups or atoms such as CH_3, S, and O.

6. Inorganic salts deposit in life forms as shells, bones, and teeth (Chapters 4 and 8). Table 1.1 shows a recent compilation of the elements that have major or minor biological roles.

TABLE 1.1. The Biological Periodic Table of the Elements

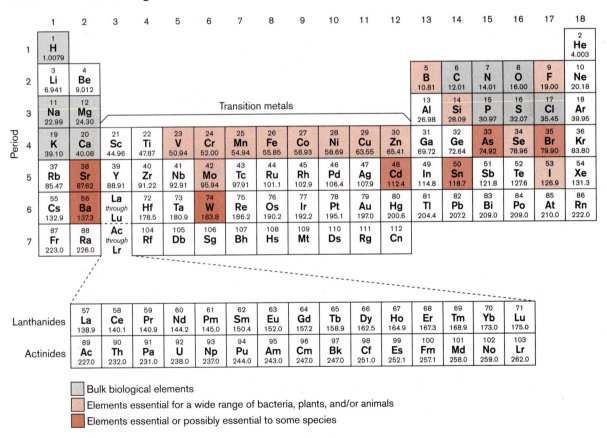

Bulk biological elements
Elements essential for a wide range of bacteria, plants, and/or animals
Elements essential or possibly essential to some species

Source: I. Bertini, H. B. Gray, E. I. Steifel, and J. S. Valentine, *Biological Inorganic Chemistry: Structure and Reactivity*, University Science Books: Mill Valley, CA, 2007, p. 2. Reprinted with permission.

1.2C. Cations Characteristic of the Blocks of the Periodic Table. Cations of the *s* block form colorless ions with charges equal to their group numbers, 1 or 2. These cations are formed by the loss of *n* electrons from the atoms of the element (where *n* = group number). Most of the *s*-block ions have important biological functions (Table 1.2):

TABLE 1.2. Cations of the *s*-Block Metals and Their Biological Functions

Period	Group 1 Cation	Biological Function[a]	Group 2 Cation	Biological Function[a]
2	Li^+	Used to treat bipolar disorder (2)	Be^{2+}	Rare; extremely toxic (2, 3)
3	Na^+	Essential for nerve transmission in animals (2)	Mg^{2+}	Essential; in chlorophyll; involved in enzyme functions and with DNA (2, 3)
4	K^+	Essential in cytoplasm, nerve function, cardiac function (2)	Ca^{2+}	Essential; acts as a messenger; involved in blood clotting; found in bones (2, 6)
5	Rb^+	Rare	Sr^{2+}	In skeletons of *acantharia* (6)
6	Cs^+	Rare	Ba^{2+}	Moderately toxic (2)

[a] The numbers in parentheses correspond to the numbers of the types of functions described in the preceding Preview (Connection to Biochemistry).

A CONNECTION TO INDUSTRIAL CHEMISTRY AND GEOPOLITICS

Lithium-Ion Batteries

Lithium is an *s*-block element that's been in the news lately. In Sections 6.5B and 8.3B we will describe the new technology being used for an enormous number of electronic applications, including electric cars, energy storage, batteries for small, light electronic devices such as laptop computers and cell phones, and energy storage. Each new plug-in hybrid or all-electric car may contain about 7 kg of lithium (involved in batteries as both the cation and the metal). Where will the lithium for "the new lithium economy"[2] come from? Current sources are concentrated in Chile and Argentina, with large reserves awaiting development in the impoverished nations of Bolivia and Afghanistan.[3]

Cations Characteristic of the f Block. The charges on the cations of the f block are not quite as predictable as those of the s block: Most of the sixth-period elements form mainly or exclusively +3 ions. None of them is known to have positive biological functions. They are much more attractive to the eye than the s-block cations: Most of them have pale, pastel colors in aqueous solution (Table 1.3).

As can be seen from Table 1.3, there are more than one cation for some of these elements. In such cases, the ions are distinguished by indicating the charge in parentheses after the name of the atom but before the word "ion." The most common way of doing this is to give the oxidation number (the Roman numeral equivalent of the charge) in parentheses. Thus, the pink ion of Am is named americium(III) ion and the red ion is americium(IV) ion.

TABLE 1.3. Major Monatomic Cations of the f Block and Their Colors (Aqueous)

1st Half	Period 6	2nd Half	Period 6		Period 7
La^{3+}	Colorless			Ac^{3+}	Colorless
Ce^{3+}	Colorless	Yb^{3+}	Colorless	Pa^{4+}	Colorless
Pr^{3+}	Green	Tm^{3+}	Green	U^{3+}	Red-brown
Nd^{3+}	Lilac	Er^{3+}	Lilac	Np^{4+}	Yellow-green
Pm^{3+}	Pink	Ho^{3+}	Yellow	Np^{3+}	Purple
Sm^{3+}	Yellow	Dy^{3+}	Yellow	Pu^{4+}	Tan
Sm^{2+}	Red			Pu^{3+}	Blue-violet
Eu^{3+}	Pink	Tb^{3+}	Pink	Am^{4+}	Red
Eu^{2+}	Colorless	Gd^{3+}	Colorless	Am^{3+}	Pink
				Cm^{4+}	Yellow

A CONNECTION TO INDUSTRIAL CHEMISTRY AND GEOPOLITICS

Chemistry of the Rare Earths

The sixth-period f-block elements are similar to the Group 3 elements and are often included together as "rare earths." Their importance has been growing very rapidly in recent years. Their oxides and related compounds have many uses—in ceramics, yttrium-aluminum garnets (YAG) in lasers, catalytic zeolites, and high-temperature superconductors (Section 8.4), and in phosphors for color television screens.[4] The most rapidly growing and crucial uses of the metallic forms of the elements are in hydrogen-storage alloys such as $LaNi_5$ and high-flux magnets (samarium-cobalt and neodymium-iron-boron alloys, Chapter 12), which play crucial roles in computer memory, hybrid car production, the production of renewable energy from windmills, and room-temperature magnetic refrigeration. Traces of dysprosium allow magnets in electric motors to be 90% lighter, while the use of terbium in lighting cuts electricity use by 80%.[5] The U.S. Department of Energy has determined that the most critical shortages are of the five metals Y, Nd, Eu, Tb, and Dy.[6]

Most of these elements are in fact not rare, but are widely distributed as salts in low concentration in various minerals. When they are found in high concentration, they are often found with the seventh-period actinide elements, which are radioactive and therefore add hazard to the mining process. Ores that are free from radioactive actinides are not common and are found, for example, at Mountain Pass, CA, in the USA, but most abundantly in Inner Mongolia, China. Around the turn of the century, Chinese rare earths were marketed at cheaper prices and the U.S. source was driven out of business, so that in 2010 95% of the world's production came from China. The Chinese leadership recognized the geopolitical advantage that this gave them: "There is oil in the Middle East and rare earths in China."[7] Although this was not officially acknowledged, China limited quantities and imposed heavy tariffs on the export of many rare earths, especially erbium, terbium, and dysprosium. As a result, the cost of cerium oxide jumped from $3100 per metric ton in 2009 to $110,000 per metric ton outside of China (but only 25% of this inside China).[5] In a 2010 conflict with Japan over an imprisoned fisherman, the vital exports of rare earths to Japan suddenly ceased, and Japan was forced to release the fishermen. Because of their dependence on rare earths, Japan began recycling electronic and computer waste to recover these valuable metals. Efforts were made to reopen the U.S. mine that was previously driven out of business. Consequently, by 2014 the Chinese share of the world market for rare earths had fallen to 70%. In 2014 the World Trade Organization ruled that China's export quotas violated international trade rules, so China cancelled those quotas.

Cations Characteristic of the *d* Block. The charges on the cations of the *d* block of the periodic table are even less predictable (Table 1.4): The most common charges are +2 and +3. Most of the fourth-period elements have positive biological functions as ions or in complexes (Chapter 7). Most of the fourth-period ions are colored, too—they are not pale, but of intermediate intensity (Section 7.3). The colors depend sensitively on the specific complex ion they are in (Chapter 7). Some of these metal ions can show virtually any color in the rainbow. The colors shown in Table 1.4 are for the ions surrounded by water molecules, in aqueous solution. Because of the variability of charge, the charges should be indicated in the names of these ions (if only one ion exists, this is not necessary—e.g., Sc^{3+} and Zn^{2+}).

TABLE 1.4. Important Fourth-Period *d*-Block Monatomic Cations

Group	Biological Function[a]	Ion	Color[b]	Ion	Color[b]
3	None; extremely rare	Sc^{3+}	Colorless		
4	None	Ti^{4+}	Colorless	Ti^{3+}	Violet
5	Anti-diabetes; essential for sea squirts (5)	V^{3+}	Blue	V^{2+}	Violet
6	Anti-diabetes; moderately toxic (5)	Cr^{3+}	Violet	Cr^{2+}	Blue
7	Essential; in numerous enzymes (1, 5)	Mn^{2+}	Pale pink		
8	Essential; in electron-transfer processes and hemoglobin (5)	Fe^{3+}	Violet	Fe^{2+}	Pale green
9	Essential; in vitamin B_{12}; prevents pernicious anemia (1, 5)	Co^{3+}	Green	Co^{2+}	Pink
10	Essential in a few enzymes; moderately toxic (1, 5)	Ni^{2+}	Green		
11	Essential in electron-transfer processes (1, 5, 6); moderately toxic	Cu^{2+}	Blue	Cu^+	Colorless
12	Essential; in enzymes; moderately toxic (1, 3)	Zn^{2+}	Colorless		

Source: I. Bertini, H. B. Gray, E. I. Steifel, and J. S. Valentine, *Biological Inorganic Chemistry: Structure and Reactivity*, University Science Books: Mill Valley, CA, 2007. Reprinted with permission.

[a] The numbers in parentheses correspond to the numbers of the types of functions given in the preceding Connection to Biochemistry.

[b] Colors are for hydrated ions in aqueous solution.

EXAMPLE 1.3

Before fleeing the Nazis, the Nobel Prize winners Max von Laue and James Frank had left their gold Nobel Prize medals with Neils Bohr for safekeeping. When the Nazis overran Denmark, Bohr and the chemist George de Hevesy were confronted with the question of how to hide their gold Nobel Prizes in such a way that they would not be found, even though they were in plain sight. How could they do this?

SOLUTION: People recognize gold when it is in the familiar elemental form, but most do not realize that it can also exists as an ion, with a totally different appearance. The Nobel Prize winners converted their gold medals to a solution of gold ions, which they left in bottles in their lab. The Nazis paid no attention to this brown liquid. After the war, the prize winners returned their solution of gold ions to elemental gold, which they returned to the Nobel Prize committee in Stockholm, which re-made their Nobel Prize medals.[8]

A CONNECTION TO INDUSTRIAL CHEMISTRY AND GEOPOLITICS

Conflict Minerals

Lithium-ion batteries (Chapters 6 and 8) currently contain d-block metal ions such as those of cobalt (found in the war-torn eastern part of the Congo) and manganese. Having resources such as this can potentially make a nation rich, although unfortunately it usually makes the rulers rich and uninterested in having a democracy. In the eastern Congo, fighting for control of the wealth of these minerals has fueled the repeated savage fighting. A recent U.S. law on "conflict minerals" has attempted to target this source of carnage by requiring companies to certify that imports of tin, tantalum, tungsten, and gold do not come from rebel-controlled mines. But this law could deprive these regions of their main livelihood.[9] It has been reported that a World War I "battle" occurred in central Colorado over control of the molybdenum mined there, which was needed by Germany to strengthen the steel used in building its Big Bertha siege guns.[10] Afghanistan also has an extraordinary variety of critical minerals worth perhaps a trillion dollars, with seven world-class deposits of copper, iron, tin, gold, and rare earths.[3,11]

Platinum, palladium, and the related metals come from more stable countries (South Africa, Russia, and the USA), but their supplies are limited and their prices very high;[12] likewise, vital niobium is entirely obtained from abroad (mainly Brazil and Canada). As we shall see in Chapter 6, some of the valuable d-block metals are considerably more abundant in places on the ocean floor or in asteroids, so proposals have been made to mine underseas or in outer space.[13]

Cations Characteristic of Metallic Elements of the p Block. The metallic (and to some degree metalloid) elements of the p block of the periodic table are the ones to the lower left of the block and are the ones that form cations (Table 1.5). Many of them form two cations with charges that can be identified from the old-style group number (e.g., III rather than 13). The two possible ions have charges equal to (a) the old-style group number, and (b) the old-style group number minus two. These cations are colorless. Because of the variability of charge, the charges should be indicated in the names of these ions (this is not done for Al^{3+}).

TABLE 1.5. Major Cations of the p Block

Period	Group 13/III	Biological Function	Group 14/IV	Biological Function	Group 15/V	Biological Function
3	Al^{3+}					
4	Ga^{3+}	Rare	Ge^{4+}; Ge^{2+}	Rare		
5	In^{3+}; In^+	Rare	Sn^{4+}; Sn^{2+}		Sb^{3+}	
6	Tl^{3+}; Tl^+	Highly toxic (1)	Pb^{4+}; Pb^{2+}	Highly toxic (1)	Bi^{3+}	

Monatomic Anions Characteristic of Nonmetallic Elements of the *p* Block. The common monatomic anions all arise from the nonmetal atoms of the upper right part of the *p* block; each such element forms only one monatomic anion (Table 1.6). These colorless anions form by adding electrons. Since electrons have negative charges, these anions have charges equal to the old-style group number minus eight. The anions have names clearly different from the parent atoms or elements; these names replace the last syllable (or two) of the element name with the suffix –ide. Collectively, the anions of Group 17/VII are often called the halide ions, those of Group 16/VI the chalcogenide ions, and those of Group 15/V the pnicogenide ions.

TABLE 1.6. Monatomic Anions of the *p* Block

Period	Group 16/VI	Name	Biological Function	Group 17/VII	Name	Biological Function
2	O^{2-}	Oxide	Reacts with water	F^-	Fluoride	Essential but moderately toxic
3	S^{2-}	Sulfide	Quite toxic; molecular derivatives essential	Cl^-	Chloride	Essential; involved in cystic fibrosis
4	Se^{2-}	Selenide	Quite toxic; some molecular derivatives essential	Br^-	Bromide	May be essential; sedative
5	Te^{2-}	Telluride	Quite toxic; users smell like garlic	I^-	Iodide	Essential; goiter from deficiency
6	Po^{2-}	Polonide	None (radioactive)	At^-	Astatide	None (radioactive)

EXAMPLE 1.4

Hydrogen is a unique element that poses some difficulties for placement in the blocks of the periodic table. Its elemental substance is H_2, and it forms two ions, H^+ and H^-. (a) Which of these properties is a better fit for H in the *s* block? In the *p* block? (b) Name these two ions. (c) Write a balanced half-reaction for the conversion of H^+ to H^-.

SOLUTION: (a) A diatomic elemental substance that forms an anion is more characteristic of Group 17/VII *p*-block elements, whereas forming a +1 cation is characteristic of a Group 1 *s*-block element. (b) H^+ is hydrogen ion while H^- is hydride ion. (c) $H^+ + 2e^- \rightarrow H^-$.

1.2D. Naming and Writing Formulas of Ionic Salts (Review).

When given the name of a salt and asked to write its formula:

1. Write the formula of the cation.

2. Write the formula of the anion.

3. Find the electrically neutral combination.
   Conventionally, the ion charges are then omitted.

EXAMPLE 1.5

Write the formula of calcium arsenide.

SOLUTION: Calcium arsenide contains the Ca^{2+} cation and the As^{3-} anion. For the salt to be electrically neutral, the total charges of all cations and anions must add up to zero. The simplest formula results when the smallest possible number of each is used (three Ca^{2+} ions and two As^{3-} anions). Dropping charges gives us the formula of Ca_3As_2.

When given the formula of a salt and asked to name it:

1. From the formula of the salt, identify the anion and its charge, and write its name.

2. Using the fact that the salt must be electrically neutral, derive the charge of the cation. Name the cation, using its oxidation number in Roman numerals if needed.

3. List the name of the cation before the name of the anion.

EXAMPLE 1.6

Name the salt Cr_3N_2.

SOLUTION: The salt Cr_3N_2 contains two N^{3-} ions, named nitride, with six negative units of charge. The three Cr ions must total six positive units of charge, so each is a Cr^{2+} ion. The cation is named the chromium(II) ion, so the salt is chromium(II) nitride.

1.3. Core and Valence Electrons; Characteristic Valence Electron Configurations of Atoms and Ions

OVERVIEW. Given a short-form periodic table and the symbol (or atomic number) of an element, you should be able to write its characteristic full electron configuration (Section 1.3A; Exercises 1.14 and 1.15). Valence orbitals are those involved in chemical change and are of two types: the last *s* orbital and the last orbital filled. Valence electron configurations of atoms show the occupation of valence orbitals by the group's number of valence electrons; Lewis dot symbols show the number of valence electrons present as dots (Section 1.3B). You may practice this concept with Exercises 1.16–1.19. Valence electron configurations of ions (Section 1.3C) are derived from the full configurations and are even more useful: s-Block cations have no valence electrons; f-block cations have only f valence electrons; d-block cations have only d valence electrons. p-Block cations have either no valence electrons or two valence s electrons. p-Block anions have ns^2np^6 valence electron configurations. You may practice these concepts by trying Exercises 1.20–1.27.

You will employ valence electron configurations in Sections 3.4, 4.2, and 8.4, and in Chapters 7 and 10. The reasons for the distinction between full and valence electron configurations will be explored in Section 9.2C and expanded further in Section 9.6.

1.3A. Characteristic Full Electron Configurations of the Elements (Review). More than one way of deriving full electron configurations of elements is valid and is taught in general chemistry. We will review using the periodic table to write out the **characteristic full electron configurations** of the elements.

Step 1. If necessary, locate the element using Table D (inside the back cover). This table gives you the symbol, period number, block type, group number, and atomic number of the element in question.

Step 2. Locate the element in Table B (inside the front cover, but you should be able to do this in any standard periodic table). Read off the order of filling of the orbitals from the periodic table, going horizontally by period through the various blocks that are encountered in each period, until you encounter the element you want. The last orbital filled is of the type listed in Table 1.7.

TABLE 1.7. Blocks of the Long Form of the Periodic Table, Showing the Last Orbitals Filled

Period Starts with Atomic No. $Z=$	s Block Groups 1 and 2	f Block, 14 Elements Wide	d Block, 10 Elements Wide Groups 3–12	p Block, 6 Elements Wide Groups 13/III–18/VIII Ends with $Z=$
1 ($Z = 1$)	$1s$			
2 ($Z = 3$)	$2s$			$2p$ ($Z = 10$)
3 ($Z = 11$)	$3s$			$3p$ ($Z = 18$)
4 ($Z = 19$)	$4s$		$3d$	$4p$ ($Z = 36$)
5 ($Z = 37$)	$5s$		$4d$	$5p$ ($Z = 54$)
6 ($Z = 55$)	$6s$	$4f$	$5d$	$6p$ ($Z = 86$)
7 ($Z = 87$)	$7s$	$5f$	$6d$	$7p$ ($Z = 118$)

Step 3. Put in superscripts to indicate the number of electrons in each set (subshell) of orbitals. This number generally equals the width of the block, so it is *two* for the *s* block, *six* for the *p* block, *10* for the *d* block, and *14* for the *f* block. However, when you are coming up to the element in question, the last set of orbitals (which is of the block type) does not fill completely. The number of electrons that occupy the last set of orbitals equals *one for each element your element is from the left side of the block*. For this to work in the *f* block, though, you must start with La and Ac, not Ce and Th. You should end up with a characteristic full electron configuration as found in Table 1.8, in which the electron configuration of the previous Group 18/VIII element is abbreviated—that is, [Xe] for the configuration of the 54 electrons of xenon.

AN AMPLIFICATION

Exceptional Electron Configurations

For various reasons, the electron configurations of gaseous atoms of the elements, as predicted by the above methods, are not totally correct for some of the elements in the *d* and *f* blocks. Fortunately for us, gaseous atoms of the *d*- and *f*-block atoms are only encountered in high vacuum (such as interstellar space). Most chemists encounter the *ions* of these elements far more frequently. As we shall see shortly, there are *no* exceptional electron configurations among the ions. Therefore, we will not concern ourselves with exceptional electron configurations of atoms.

TABLE 1.8. Characteristic Full Electron Configurations of the Elements[a–d]

Period	*s* Block 2 Elements Wide	*f* Block 14 Elements Wide	*d* Block 10 Elements Wide	*p* Block 6 Elements Wide
1	$1s^n$			
2	$[He]2s^n$			$[He]2s^2 2p^{n-2}$
3	$[Ne]3s^n$			$[Ne]3s^2 3p^{n-2}$
4	$[Ar]4s^n$		$[Ar]4s^2 3d^{n-2}$	$[Ar]4s^2 3d^{10} 4p^{n-2}$
5	$[Kr]5s^n$		$[Kr]5s^2 4d^{n-2}$	$[Kr]5s^2 4d^{10} 5p^{n-2}$
6	$[Xe]6s^n$	$[Xe]6s^2 4f^{n-2}$	$[Xe]6s^2 4f^{14} 5d^{n-2}$	$[Xe]6s^2 4f^{14} 5d^{10} 6p^{n-2}$
7	$[Rn]7s^n$	$[Rn]7s^2 5f^{n-2}$	$[Ar]7s^2 5f^{14} 6d^{n-2}$	

[a] n = Number of valence electrons = group number (taken as 3 through 8 in the *p* block).

[b] [He], etc. is the noble-gas core part of the electron configuration.

[c] **$3d^{10}$**, etc. (in **brown**) is the pseudo-noble-gas core part of the electron configuration.

[d] $1s^n$, etc. (in black) is the valence part of the electron configuration (Section 1.3B).

AN AMPLIFICATION

Which Elements Belong in the *f* Block?

A few elements have such similar chemical and electronic properties that their placement into groups of the periodic table is somewhat arbitrary. Examples include the pairs of elements lanthanum (La) and lutetium (Lu), and actinium (Ac) and lawrencium (Lr), each of which resembles the early elements in Group 3, scandium (Sc) and yttrium (Y). As pointed out by Jensen,[14] the metallurgical resemblance to *d*-block metals is much stronger for lutetium than for lanthanum, so we have adopted the metallurgist's convention of listing Lu (and by extension Lr) below Sc and Y. An important additional advantage of this is that the periodic table becomes more symmetrical when the *f* block begins with La and Ac and ends with Yb and No, and it becomes easier to predict electron configurations. Scerri[15] points out that recent determinations of the electron configurations of most of the *f*-block elements[16] show that they do not include *d* electrons, as you would infer by starting the *f* block with Ce and Th with characteristic d^1 electron configurations.

EXAMPLE 1.7

Write the characteristic full electron configuration of element number 76, Os.

SOLUTION: Step 1. Reading across Table B, we encounter the orbitals in the following order: *1s 2s 2p 3s 3p 4s 3d 4p 5s 4d 5p 6s 4f 5d*. We stop at *5d*, because this is the last orbital filled in the element Os.

Step 2. Fill in 2, 6, 10, and 14 as superscripts to indicate the numbers of electrons in the filled *s*, *p*, *d*, and *f* sets of orbitals, and count across the *d* block six elements until we encounter Os, the sixth element in the *5d* set of elements. We thus obtain the following characteristic electron configuration: $1s^2 2s^2 2p^6 3s^2 3p^6 4s^2 3d^{10} 4p^6 5s^2 4d^{10} 5p^6 6s^2 4f^{14} 5d^6$.

1.3B. Core and Valence Electrons in the Electron Configurations of Atoms
(Review). Although we will be needing to use the full electron configuration of an element in Chapter 9, for most purposes we are more interested in orbitals and electrons that are involved in chemical change: the gain, loss, or sharing of electrons. These are known as **valence orbitals** and **valence electrons**, and are the ones shown in black type in Table 1.8.

Core Electrons

Although core electrons indirectly influence chemical reactivity, they are not gained, lost, or shared. They remain, for example, in all ions. Core electrons can be subdivided into two types, called noble-gas core electrons and pseudo-noble-gas core electrons. **Noble-gas core electrons**, which make up most of the core electrons, are those that are found in the preceding noble gas. These are often abbreviated in full electron configurations with the noble-gas symbol in brackets—for example, $[Ne] = 1s^22s^22p^6$. However, for many elements with atomic numbers of 31 and above, there is an additional category of core electrons, those from already-filled d and f subshells, which are not gained, lost, or shared by elements outside that block. These additional core electrons are sometimes called **pseudo-noble-gas core electrons**, and are indicated in **brown** type in Table 1.8. As-yet-unoccupied orbitals in the period in question are sometimes called **post-valence orbitals**.

One of the justifications for grouping elements as we do in the periodic table is that *elements in a group of the periodic table have the same number of valence electrons in the same shapes of orbitals*. A given element has two types of valence electrons: those in the last-filled ns orbital, and those in the last-filled orbital characteristic of the block—namely, np in the p block, $(n-1)d$ in the d block, and $(n-2)f$ in the f block (Table 1.7). The **characteristic valence electron configuration** shows the expected occupancy of just these one or two types of orbitals.

These valence electrons are the ones symbolized by dots in Lewis (electron-dot) symbols used in general and organic chemistry (e.g., the Group 13/III atoms are ·B:, ·Al:, ·Ga:, ·In:, and ·Tl:). In general, *the number of valence electrons of an atom* (as shown in Table B) *equals its group number*. However, with the new IUPAC system for numbering p-block groups from 13 to 18, the number of valence electrons equals the *group number minus 10*. The groups of f-block elements are not numbered at all in a conventional periodic table, so we must count over from La and Ac as 3 or Ce and Th as 4.

When we draw Lewis electron-dot symbols for atoms, we represent each valence electron with one dot. Due to the large number of dots that may result with the d- and f-block elements, Lewis symbols are used mainly with s- and p-block elements.

EXAMPLE 1.8

How many valence electrons do atoms of Se, Fe, Co, and Pr have? How many dots should be included in the Lewis dot symbol of each atom?

SOLUTION: Se is located in Group 16/VI of the periodic table, so it has six valence electrons and six dots in its Lewis symbol. Fe is located in Group 8; it has eight valence electrons (and eight dots if we were to draw its Lewis symbol). Co is located in Group 9 of the modern periodic table, but also in Group VIII or VIIIB of an older one. It has to have one more electron than Fe, and because this is a valence electron, we assign nine valence electrons to cobalt. Pr is a Group $5F$ f-block element (located two elements beyond the first element of the f block, La, which has three valence electrons), so it has five valence electrons.

With these principles, it is possible to read valence electron configurations of atoms from the periodic table. We need only read off valence electrons from the period in which the element is located. Only the last-filled *s* orbital remains a valence orbital outside of its block, the *s* block; it remains a valence orbital across the entire period before becoming a core orbital in the next period. All other types of orbitals remain valence orbitals only in their characteristic block of the periodic table.

Having determined the number of valence electrons and the types of orbitals in which they go, to obtain the valence electron configuration we simply assign the valence electrons to the valence orbitals, starting with the *s* orbital.

EXAMPLE 1.9

Write the characteristic valence electron configurations of atoms of Ba, As, Zn, and Tm.

SOLUTION: First, we note the blocks and periods in which the elements are located. Ba is in the *s* block in Period 6; As is in the *p* block in Period 4; Zn is in the *d* block in Period 4; and Tm is in the *f* block in Period 6. All of the elements will use the *ns* orbital as a valence orbital, in which *n* is the period number of the element. In addition, As will use the *np* orbital, Zn will use the $(n - 1)d$ orbital, and Tm will use the $(n - 2)f$ orbital. Thus, the valence orbitals are as follows: Ba, only the 6*s* orbital; As, the 4*s* and 4*p* orbitals; Zn, the 4*s* and 3*d* orbitals; and Tm, the 6*s* and 4*f* orbitals.

Second, we find the group numbers. Ba is in Group 2; As is in Group 15/V; Zn is in Group 12; and Tm is in Group 15*F*. Hence, these are the number of valence electrons (for As, there are five). Filling in these numbers of valence electrons into the *ns* orbitals first, we obtain: Ba, $6s^2$; As, $4s^2 4p^3$; Zn, $4s^2 3d^{10}$; and Tm, $6s^2 4f^{13}$.

Alternately, since the elements of a group have the same number of valence electrons in the same shapes of orbitals, we can obtain the valence electron configuration of a later element in a group from the configuration of the first member of the group. The only difference is that the principal quantum number of each orbital increases by one for each additional period in the periodic table. For example, if we recall that the valence electron configuration of N is $2s^2 2p^3$, then counting down two periods to As means that its electron configuration is $4s^2 4p^3$.

1.3C. Valence Electron Configurations of Ions (Review). The basis for the classification of electrons as core electrons is most easily seen in the formation of ions (which are the most chemically important form of most of these elements). In Table 1.9 we show the full and (in **boldface**) the valence electron configurations of the common monatomic ions of the elements.

TABLE 1.9. Characteristic Electron Configurations of Monatomic Ions of the Elements[a–d]

Period	s-Block Cations	f-Block Cations	d-Block Cations	p-Block Cations	p-Block Anions
1	$1s^0$				
2	$[He]2s^0$			$[He]2s^0$	$[He]2s^22p^6$
3	$[Ne]3s^0$			$[Ne]3s^0$	$[Ne]3s^63p^6$
4	$[Ar]4s^0$		$[Ar]3d^{g-x}$	$[Ar]4s^23d^{10}$; $[Ar]4s^03d^{10}$	$[Ar]4s^23d^{10}4p^6$
5	$[Kr]5s^0$		$[Kr]4d^{g-x}$	$[Kr]5s^24d^{10}$; $[Kr]5s^04d^{10}$	$[Kr]5s^24d^{10}5p^6$
6	$[Xe]6s^0$	$[Xe]4f^{g-x}$	$[Xe]4f^{14}5d^{g-x}$	$[Xe]6s^24f^{14}5d^{10}$; $[Xe]6s^04f^{14}5d^{10}$	$[Xe]6s^24f^{14}5d^{10}6p^6$
7	$[Rn]7s^0$	$[Rn]5f^{g-x}$	$[Ar]5f^{14}6d^{g-x}$		

[a] g = Group number; x = charge of cation.

[b] [He], etc. is the noble-gas core part of the electron configuration.

[c] $3d^{10}$, etc. (in **brown**) is the pseudo-noble-gas core part of the electron configuration.

[d] $1s^2$, etc. (in **boldface**) is the valence part of the electron configuration.

We note the following generalities in Table 1.9:

1. Cations of the s block have lost a number of valence electrons from their ns valence orbital (where n is the period number) equivalent to their group number g and have ns^0 valence electron configurations. The Lewis symbols of these ions have *no* dots.

2. Cations of the f-block metals have lost the number of electrons equal to their charge, first from their ns valence orbital, then (if required by their charge) from their $(n-2)f$ valence orbital. All valence electrons are in their $(n-2)f$ orbitals.

3. Cations of the d-block metals have lost their valence ns electrons and (if necessary) additional electrons from their $(n-1)d$ valence orbital. All of their valence electrons are in the $(n-1)d$ valence orbitals.

4. The lower-charged cations of the heavier elements of the p block have lost *all* of their valence np electrons and have ns^2 valence electron configurations and *two* dots in their Lewis symbols. The higher-charged cations of these elements have also lost *all* of their valence ns electrons and have no valence electrons and *no* dots in their Lewis symbols.

5. Monatomic anions of the p block have *gained* electrons in their np valence orbital, filling it to give a total of eight valence electrons, with the valence electron configuration ns^2np^6. The Lewis symbols of these ions have *eight* electrons (an octet).

EXAMPLE 1.10

Write the valence electron configurations and Lewis symbols (if appropriate) of the following ions: Ba^{2+}, As^{3+}, Zn^{2+}, and Tm^{3+}.

SOLUTION: First, we write the valence electron configurations of the neutral atoms (we have already done this in Example 1.9). Next, we remove one electron for each unit of positive charge. For Ba^{2+} this means removing two s electrons to give a final configuration of $6s^0$; its Lewis symbol has no dots and is just Ba^{2+}. Removing three p electrons gives As^{3+} the configuration of $4s^2$ (i.e., $4s^2 4p^0$); its Lewis symbol has two dots and can be drawn as $:As^{3+}$. Loss of two s electrons leaves Zn^{2+} with a $3d^{10}$ configuration. Tm must lose two s electrons followed by one f electron, which results in a valence electron configuration of $4f^{12}$ for Tm^{3+}. The Lewis symbol for this ion would, in principle, have 12 dots, but Lewis symbols are not usually drawn for ions with more than eight valence electrons.

1.4. Identifying "Ions" in Chemical Formulas; Assigning Oxidation Numbers

OVERVIEW. Oxidation numbers are ionic charges obtained from a chemical formula on the assumption that the bonding is ionic. We can calculate these using simple rules (Section 1.4A). You may practice these calculations by trying Exercises 1.28–1.37. At the left sides of the d, f, and p blocks, the maximum oxidation numbers obtainable are equal to the group number (Section 1.4B). You may apply these trends to specific elements in Exercises 1.38–1.43.

Reasons for the trends in oxidation numbers and ion charges will be developed in Section 9.3 and amplified in Sections 9.4 and 9.6. Oxidation numbers and maximum oxidation numbers will be used again in Sections 3.1, 3.2, 3.3, and 3.9, and in Chapters 5, 6, and 7.

1.4A. Deducing Charges or Oxidation Numbers of Possible Cations from Chemical Formulas (Review).

Chemical formulas as normally written do not tell whether the compound is molecular (with covalent bonding) or ionic (containing cations and anions). Oxidation numbers are formalisms, based on the assumption of ionic bonding, and give real ion charges seen before in Section 1.2 if the compound is indeed a binary ionic compound. If the compound is covalent or contains polyatomic ions (Chapter 3), the oxidation numbers are assigned for each atom as if it were a monatomic ion. So, oxidation numbers can be calculated for ionic or covalent compounds, or even for polyatomic ions. We will be able to use them to supplement real ionic charges in predicting chemical reactivity in future chapters.

The following set of rules work for calculating oxidation numbers as long as the specified elements do not have unusual bonding features (e.g., metal carbonyls in Section 11.6A) and provided the molecule or ion does not have more than three elements in it (e.g., in complex ions in Section 3.2B). If the compound is ionic and has more than three elements, the rules should be applied separately to each ion.

1. Consult Table A (inside front cover) and identify the *least* electronegative element in the species; it is probably present as a cation of predictable charge. If it is in Group 1, assign it an oxidation number of +1; if it is in Group 2, assign it +2; if it is in Group 3 or Group 3 of the *f* block, assign it +3. (In Group 13/III, Al, Ga, and In can normally be assigned oxidation numbers of +3.)

2. If only one element remains, assign it an unknown oxidation number, x. More commonly, two elements remain to be assigned. Identify the *most* electronegative element. If the most electronegative element is in Group 17/VII, assign it an oxidation number of –1; if it is in Group 16/VI, assign it –2; if it is nitrogen from Group 15/V, assign it –3. Only one element should now remain unassigned; assign it an unknown oxidation number of x.

3. Multiply each element's oxidation number by the number of atoms of that type present in the molecule or ion. Add all of these terms up. The sum *must* equal the *charge* on the ion, or zero if this is a neutral molecule. Solve this equation for the unknown oxidation number x.

This procedure gives the average oxidation number of all atoms of a given element in a given species, but it will not differentiate atoms of an element in different bonding environments. Consequently, it can give fractional oxidation numbers.

When we apply this procedure often enough, we can eventually find the highest positive oxidation number/state in known compounds of any element (Table 1.10). For some of our predictions, these oxidation numbers/states can supplement the lists of positive ions shown in Table 1.3 for the *f* block, Table 1.4 for the *d* block, and Table 1.5 for the *p* block.

EXAMPLE 1.11

Compute the oxidation numbers of each element in the ions (a) $H_2IO_6^{3-}$; (b) $S_3O_6^{2-}$; (c) Na_2O_2.

SOLUTION: (a) The sum of all oxidation numbers in this ion must equal –3, the charge on the ion. First, we assign the oxidation number of the least electronegative atom, H, as +1. Then we assign oxygen (the most electronegative element) the oxidation number –2 and I the unknown oxidation number x. Taking into account the number of atoms of each type, we have $2(+1) + x + 6(-2) = -3$. When we solve the equation for x, the oxidation number of I turns out to be +7.

(b) The sum of all oxidation numbers in this ion must be –2. Rule 2 allows us to assign –2 as the oxidation number of oxygen, so sulfur is assigned the unknown oxidation number x. Since there are three S atoms and six O atoms, $3x + 6(-2) = -2$. Solving this equation gives $x = 10/3$.

(c) The sum of all oxidation numbers in this neutral compound must be zero. Rule 1 allows us to assign the oxidation number of +1 to each Na atom. By Rule 2, the oxidation number of the last element, oxygen, must be x. Since there are two of each type of atom, we have $2(+1) + 2x = 0$. Thus, x, the oxidation number of oxygen, must be –1.

1.4B. Periodic Trends in Oxidation States. Among cations of the elements, the oxidation state (often expressed as a Roman numeral in parentheses) is also equal to the ionic charge. A necessary trend is that the *highest oxidation number of an element cannot exceed its group number*, since it cannot, by the definition of valence electrons, lose or share anything beyond its valence electrons. In ions that have lost all of their valence electrons, the Lewis symbol will be without dots; the valence electron configuration shows the valence orbitals holding zero electrons. Table 1.10 shows that the highest[17] or *group oxidation state* is achievable at the *left side* of each block, but is most often not achievable at the right side, where it would be +8 (*p* block), +12 (*d* block), or +16 (*f* block).

TABLE 1.10. **Highest Positive Oxidation Number for Each Element**[a,b]

[a] Shaded boxes enclose elements for which the maximum oxidation number equals the group number.

[b] Oxidation numbers in *italics* either are found only by low-temperature matrix isolation or have been computed but not observed.

High Oxidation States: To Be or Not to Be?

Modern chemical techniques complicate the assembly of Table 1.10 by posing the question: When can we say that the compound or ion showing that oxidation state *exists*? Some high oxidation numbers have been attributed to compounds that can exist only at ultra-low temperatures, such as 77 K, while surrounded by inert-gas atoms in a condition known as *matrix isolation*. Two notable examples are the first known argon compound, HArF, formed by photolysis of HF in a solid argon matrix,[18] and the (disputed[19]) production of HgF_4 in neon or argon matrices.[20] In other cases there may be only convincing computations of the properties of molecules such as RgF_4.[21] The oxidation state +9 has been characterized in the vapor phase in IrO_4^+, but this cation has not yet been isolated in a solid.[22] Oxidation states assigned by these methods are indicated *in italics* in Table 1.10.

In the *p* block, the second likely cationic electron configuration is that in which only the *np* electrons have been lost. The valence electron configurations of these ions are ns^2. The oxidation state of such a *p*-block cation is then *two less than the (old-style) group number*.

In the *d* block, the group oxidation state is found for less than half of the elements, those being at the left of the *d* block. Otherwise there are no uniquely favored electron configurations—instead, we often find oxidation states and ionic charges equal to +2, +3, or +4. These lower oxidation states are also preferred more frequently at the *top* of the *d* block.

The group oxidation number is found only at the far left of the *f* block, but more extensively at the bottom of the block. The most common oxidation number for most of the *f*-block elements is +3. The topic of most common oxidation states will be explored in more depth in Section 6.2C, but for now these oxidation states are summarized in Table 1.11.

TABLE 1.11. Most Common Positive Oxidation States for Each Element

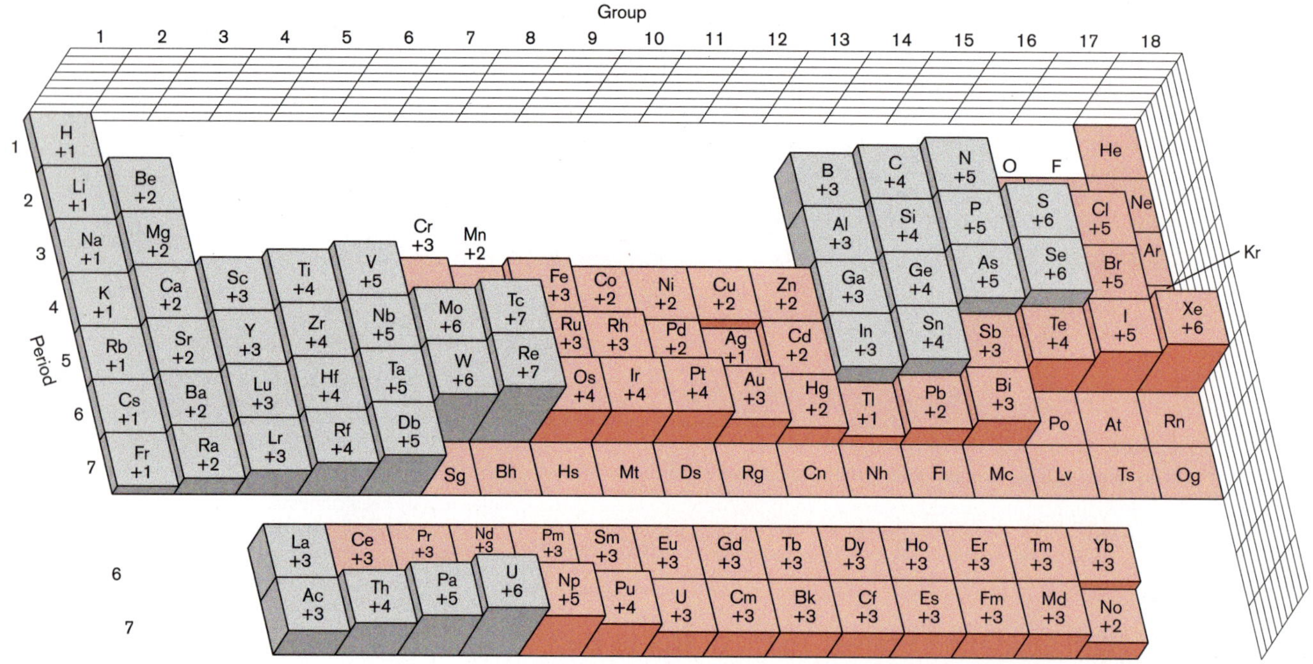

EXAMPLE 1.12

Which of these statements are true? If they are not true in all blocks, identify in which blocks they are true. (a) All atoms with two valence electrons can have the +2 oxidation state. (b) All atoms with four valence electrons can have the +4 oxidation state. (c) All atoms with six valence electrons can have the +6 oxidation state. (d) All atoms with eight valence electrons can have the +8 oxidation state.

SOLUTION: If necessary, check Table B to identify the groups that have the specified numbers of valence electrons.

(a) Atoms with two valence electrons are found only in the *s* block; these Group 2 atoms have two valence electrons.

(b) Atoms with four valence electrons are found in the *f* block and in Groups 4 and 14/IV. These groups are far enough left in their blocks that all such atoms can have +4 oxidation states (Table 1.10).

(c) Atoms with six valence electrons can also be found in the *f* block and in Groups 6 and 16/VI. The +6 oxidation state is found in all Group 6 elements, but not in Nd in the *d* block or O in the *p* block.

(d) +8 is a high oxidation state and is not found for all atoms in any block with eight valence electrons. It is found for three Group 8 elements (Fe, Ru, and Os) and one Group 18/VIII element (Xe).

1.5. Types of Atomic Radii and Their Periodic Trends

OVERVIEW. There are four types of atomic radii to be distinguished and compared in magnitude for a given atom—namely, covalent radius, cationic radius, anionic radius, and van der Waals radius. Cationic radii are roughly 60 pm smaller than the covalent radii of the same atoms, which are roughly 60 pm smaller than the (roughly equal) anionic and van der Waals radii of the same atom (Section 1.5A). You may practice explaining and predicting these results by trying Exercises 1.44–1.50. Without referring to Table C or Table 1.13, you should be able to describe and apply the main horizontal and vertical periodic trends in radii of atoms (Section 1.5B) and ions (Section 1.5C). You should know when the scandide and lanthanide contractions come into play to alter these trends. You may practice applying these trends by trying Exercises 1.51–1.59.

Cationic and anionic radii (e.g., ionic radii) will be employed extensively in Chapters 2, 4, and 5, and in Sections 8.2 and 8.3. Covalent radii will be used in Sections 9.4, 9.5, and 9.6. Van der Waals radii will be used in Sections 11.3C and 12.4C. Reasons for the trends in atomic and ionic radii will be explored in Section 9.4 and amplified in Section 9.6.

1.5A. Types of Atomic Radii. In the previous sections, we have examined periodic trends in two fundamental properties of atoms and ions—their *charges* and *electron configurations*. In this section, we add the property of *size*, as measured by the *radius* of an atom or ion. The radius of an atom conceptually is the distance from the nucleus of an atom to the outer edge of the outermost electron in its electron configuration. However, as was discussed in your Introductory Chemistry course and as is reviewed in Chapter 9, electrons in orbitals do *not* have outer edges. Instead, they are described by *wave functions*, in which the probability of finding an electron at a certain distance from the nucleus can be described. That probability gradually diminishes with increasing distance, but never goes to zero. Hence, there is no practical way of measuring the radii of isolated gaseous atoms.

As a result, we normally measure the distances of atomic nuclei from other nuclei in a solid, or sometimes a liquid or even a gaseous *molecule*. That internuclear distance then is considered to be the sum of the radii of the two interacting atoms. These interactions can be of different types: (1) covalent interactions of overlapping atoms in either a *covalent* or *metallic* material, (2) ionic (Coulombic) interactions of cations and anions, or (3) intermolecular (London or van der Waals, dipole–dipole, or induced dipole–induced dipole) interactions of distinct molecules. We commonly distinguish five different types of atomic radii: (1a) covalent radii, (1b) metallic radii, (2a) ionic radii of cations, (2b) ionic radii of anions, and (3) van der Waals radii.

The radii are obtained by measuring distances between neighboring atomic nuclei, then taking the distances as being the sums of the appropriate radii. Figure 1.2 illustrates how covalent radii (1a) and van der Waals radii (3) are obtained. [Metallic radii (1b) are obtained in a similar manner to covalent radii, except that bonding is the same in all directions—there are no van der Waals distances or radii.] In a molecule such as Cl_2 in which both atoms are identical, we define the **covalent radius** to be one-half the internuclear distance—that is, one-half the covalent bond length. We take one-half the internuclear distance to the nearest chlorine nucleus in the next molecule to be the **nonbonded** or **van der Waals (VDW) radius** of the chlorine atom.

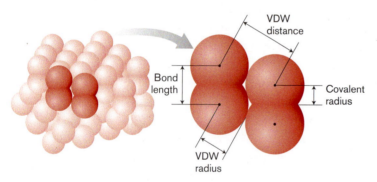

Figure 1.2. Covalent bond length and radius (1a), and van der Waals (VDW) distance and radius (3) in a crystal of the solid form of a diatomic molecule such as $Cl_2(s)$.

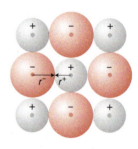

Figure 1.3. Interionic distances in ionic compounds are the sums of the ionic radii of (2a) cations (r^+) and (2b) anions (r^-).

Figure 1.3 shows how the ionic radii of cations (2a) and anions (2b) are determined in ionic solids such as $(Na^+)(Cl^-)(s)$.

To probe more deeply into the differences in these types of radii, we have drawn a diatomic species in Figure 1.4. The inner circle for each atom represents the outermost extent of its *core* electrons; the outer circle for each atom represents the limit of its outer *valence* electrons.

Suppose that the diatomic species at the top of Figure 1.4a is an **ion pair**—that is, a cation touching an anion. For many elements, the characteristic cation has lost all of its valence electrons. We may then suppose that the cationic radius equals the radii of the core electrons. Monatomic anions characteristically fill their valence orbitals, so we may suppose that the anionic radius equals the radius of the valence electrons.

Suppose next that the species at the bottom of Figure 1.4a is a *covalent molecule* of an element, such as Cl_2. In a covalent molecule, the valence orbitals overlap so that a valence electron of each atom can be shared with the other atom. Overlap is improved by bringing the two atoms closer, but ultimately we would expect the overlapping valence electrons of one atom to "bump into" and be repelled by the core electrons of the other atom. The covalent radius is half the covalent bond distance, so it extends through the core electrons and halfway through the valence electrons. Thus, *the cationic radius is less than the covalent radius, which is less than the anionic radius.*

Imagine the two Cl_2 molecules of Figure 1.4a are in the solid state, in a crystal in which

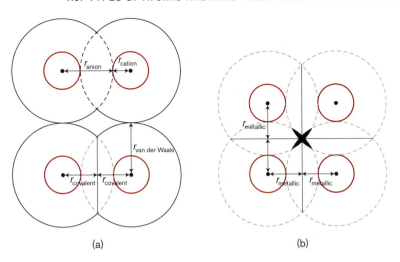

Figure 1.4. (a) Drawing of two adjacent molecules of a diatomic species; (b) drawing of four adjacent atoms in a metallic solid. The inner circles represent the outer extent of the core electrons of each atom, whereas the outer circles represent the extent of the valence electrons. For explanations of the different lines and radii, see the text. The blackened areas of Figure 1.4b represent regions of overlap of nonadjacent orbitals that are important in metallic bonding (Chapter 12).

each just touches other molecules above it and below it. Since this happens when core electrons "bump into" each other, the van der Waals radius of an atom should exceed its covalent radius and be approximately *equal to its anionic radius.*

In Figure 1.4b we represent four atoms of a solid metallic element. In solid metals, each atom characteristically has 8 or 12 nearest-neighbor atoms, all at the same distance. Hence, we have pushed the two molecules of Figure 1.4a together until they merged, with all distances becoming equal. This merger results in a **metallic radius** for each atom that somewhat exceeds the covalent radius of the "unmerged" atoms of Figure 1.4a, due to the effect of the large number of atoms being brought into close proximity. Since the main applications of metallic radii are in Chapter 12, we do not emphasize this type of radius now.

Table 1.12 compares the values of these different types of radii for selected atoms, and Figure 1.5 compares the values for S, Se, and Te graphically. The units of radius used in this text are picometers (pm, where 1 pm = 10^{-12} m). (Also commonly used are angstrom units, Å, where 1 Å = 100 pm.)

These data show the following general trends:

1. The smallest radius for a given element is its *cationic radius.*

2. The *covalent radius* for a given element is about 55 pm larger than its cationic radius.

3. The *anionic radius* and the *van der Waals* radius are approximately equal to (within 10 pm of) each other and are about 64 pm larger than the covalent radius of the element. For many of the Exercises it will suffice to treat the large differences described here in (2) and (3) as being about 60 pm each.

Unfortunately, choosing the proper type of bonding and type of structure for many compounds is more complicated than we might wish. We will gradually

TABLE 1.12. Different Types of Atomic Radii for Selected Atoms[a]

Element	Cation	Covalent	Anion	Van der Waals	Cov – Cat[b]	An – Cov[c]
Li	90	128		181	38	
Na	116	166		227	50	
K	152	203		275	51	
Be	59	96		153	37	
Mg	86	141		173	55	
B	41	84		192	43	
Al	67	121		184	54	
Ga	76	122		187	46	
In	94	142		193	48	
Tl	102	145		196	43	
N		71	132	155		61
P		107		180		
As	60	119		185	59	
Sb	74	139		206	65	
O		66	126	152		60
S	43	105	170	180	62	65
Se	56	120	184	190	64	64
Te	70	138	207	206	68	69
F		57	119	147		62
Cl	41	102	167	175	61	65
Br	53	120	182	183	67	62
I	67	139	206	198	72	67

Sources: Ionic radii are from Table C of this text. Covalent radii are from B. Cordero, V. Gomez, A. E. Platero-Prats, M. Revés, J. Echeverría, E. Cremades, F. Barragán, and S. Alvarez, *Dalton Trans.*, 21, 2832 (2008). Van der Waals radii are from M. Mantina, A. C. Chamberlain, R. Valero, C. J. Cramer, and D. G. Truhlar, *J. Phys. Chem. A* 113, 5806 (2009).

[a] All radii are in units of picometers (pm).

[b] The Cov – Cat column gives the difference between the covalent (cov) and cationic (cat) radii of the element. This difference averages 55 pm and ranges from 45 to 65 pm.

[c] The An – Cov column gives the difference between the anionic (an) and covalent (cov) radii of the element. This difference averages 64 pm and ranges from 61 to 67 pm.

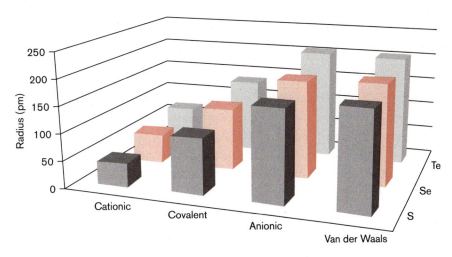

Figure 1.5. Comparison of four different types of atomic radii for the Group 16/VI elements S, Se, and Te. (The data come from Table 1.12.)

develop ways of doing this, ending in Chapter 12. These methods, however, depend in part on the sizes of the atoms/ions. As a result, we will start with the assumption of *ionic* bonding. Less electronegative elements will be presumed to be cation-like, and we will use their cationic radii (Table C), while more electronegative elements will be characterized using their anionic radii (left side of Table C). Later we can improve on this starting assumption.

Computationally, a given bond length is roughly the same if it is computed by adding the cationic radius to the anionic radius, or if it is computed by adding the two covalent radii. This is only a first approximation, though, because the bond length *does* vary by some picometers, depending on whether the bonding is ionic or covalent (Example 1.13). Therefore, comparing observed and predicted bond lengths can help in more advanced work to choose the bonding types.

EXAMPLE 1.13

Which is longer—the covalent bond between two elements or the ionic bond between the same two elements? How much is the difference?

SOLUTION: Using the three general trends that we observed in the data in Table 1.12, the average differences between cationic and covalent radii and between anionic and covalent radii can be put into equation form as follows:

Cationic radius + 55 pm = covalent radius   (1.5)

Anionic radius – 64 pm = covalent radius   (1.6)

Adding these, we obtain:

Cationic radius + anionic radius – 9 pm = 2(covalent radius)   (1.7)

Thus, an ionic bond is 9 pm longer on average than a covalent bond between the same two elements.

1.5B. Trends in Covalent Radii. The covalent radii of the elements are given in Table 1.13. Note that the *main horizontal trend* among neutral atoms (as found in the elements, metals, or in covalent compounds) is to become *smaller* from *left to right* in the periodic table. Figure 1.6 compares covalent radii for atoms of Groups 2, 4, and 13. In a given period, the covalent radius of the Group 2 atom is largest, and the Group 13 atom is smallest. In Periods 2 and 3 the decrease from Group 2 to Group 13 is relatively small. In Periods 4, 5, and 6, however, the decrease is larger. This corresponds with the introduction of the *d*-block elements, and the period crosses 10 additional groups. Finally, the decrease from Group 2 to Group 4 is relatively small in Periods 4 and 5, but it is larger in Period 6. This corresponds with the introduction of the *f*-block elements, and the period crosses 14 additional groups.

TABLE 1.13. Covalent Radii of the Elements[a]

Group	1	2	3	4	5	6	7	8	9	10	11	12	13	14	15	16	17	18
Per. 1	H																	He
	31																	28
Per. 2	Li	Be											B	C	N	O	F	Ne
	128	96											84	76	71	66	57	58
Per. 3	Na	Mg											Al	Si	P	S	Cl	Ar
	166	141											121	111	107	105	102	106
Per. 4	K	Ca	Sc	Ti	V	Cr	Mn	Fe	Co	Ni	Cu	Zn	Ga	Ge	As	Se	Br	Kr
	203	176	170	160	153	139	139	132	126	124	132	122	122	120	119	120	120	116
Per. 5	Rb	Sr	Y	Zr	Nb	Mo	Tc	Ru	Rh	Pd	Ag	Cd	In	Sn	Sb	Te	I	Xe
	220	195	190	175	164	154	147	146	142	139	145	144	142	139	139	138	139	140
Per. 6	Cs	Ba	Lu	Hf	Ta	W	Re	Os	Ir	Pt	Au	Hg	Tl	Pb	Bi	Po	At	Rn
	244	215	187	175	170	162	151	144	141	136	136	132	145	146	148	140	150	150
Per. 7	Fr	Ra	Lr	Rf	Db	Sg	Bh	Hs	Mt	Ds	Rg	Cn	Nh	Fl	Mc	Lv	Ts	Og
	260	221																

Group			3F	4F	5F	6F	7F	8F	9F	10F	11F	12F	13F	14F	15F	16F
Per. 6			La	Ce	Pr	Nd	Pm	Sm	Eu	Gd	Tb	Dy	Ho	Er	Tm	Yb
			207	204	203	201	199	198	198	196	194	192	192	189	190	187
Per. 7			Ac	Th	Pa	U	Np	Pu	Am	Cm	Bk	Cf	Es	Fm	Md	No
			215	206	200	196	190	187	180	169						

Source: B. Cordero, V. Gomez, A. E. Platero-Prats, M. Reves, J. Echeverria, E. Cremades, F. Barragan, and S. Alvarez, *Dalton Trans.* 21, 2832 (2008).

[a] All values are in picometers.

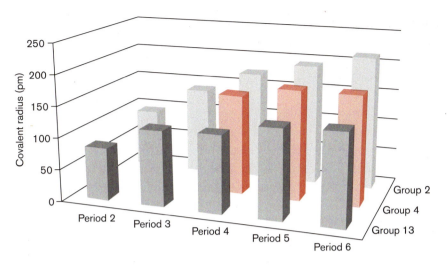

Figure 1.6. Covalent radii of atoms from Groups 2, 4, and 13. The data come from Table 1.13.

The main *vertical* trend within a group is for atoms to get *larger* as we go to *higher* period numbers. This trend is well established in the *s* block (e.g., Group 2 in Figure 1.6). In the *d* block (e.g., Group 4 in Figure 1.6), the atoms start getting larger as we go to higher period numbers, but the expansion fails to occur in Period 6. In the *p* block (e.g., Group 12 in Figure 1.6), the expansion fails to occur in Periods 4 and 6. The reasons for these anomalous countertrends in the *p* and *d* blocks are fundamentally the same as the reasons for the anomalous countertrends in the horizontal trends just discussed. These countertrends are consequences that follow the first introduction of a new block of elements in the periodic table. (There is a new decrease for each element in the new block.)

In Period 4, the *d* block is first introduced with the element scandium, Sc. The early fourth-period elements that follow the introduction of the *d* block then experience a **scandide contraction** (or nonexpansion). By the time we reach Ga, this scandide contraction almost exactly counterbalances the normal vertical expansion that we expect on going from third-period Al to fourth-period Ga.

In Period 6, the *f* block is first introduced with the element lanthanum, La. The sixth-period elements that follow the introduction of the *f* block then experience a **lanthanide contraction** (or nonexpansion). The early *d*-block elements, such as Hf, have the same size as the corresponding fifth-period elements, such as Zr. Furthermore, the sixth-period *p*-block elements Tl and Pb are almost the same size as the fifth-period elements In and Sn. (As shall be seen in Section 9.5, relativistic contraction is also involved in this decrease.)

EXAMPLE 1.14

Each of the following pairs of atoms consists of neighbors in the same group: Na and K; Y and La. (a) Which pair of atoms should be closest in size and why? (b) In the remaining pair, which atom should be the largest?

SOLUTION: (a) Y and La are closest in size, because La follows the first introduction of the *f* block and therefore experiences the lanthanide contraction (nonexpansion).

(b) K is larger than Na, because atomic radii increase going down a group when no lanthanide or scandide contraction first comes into play.

1.5C. Trends in Ionic Radii (Review). The best available measurements of ionic radii, the Shannon–Prewitt values, are listed in Table C, inside the back cover of the book. Ions can differ from each other in a number of respects, so to see clear trends, we must control the number of variables among the ions.

1. The main *horizontal* trend also shows up among cations, provided we keep their charges constant. The most extensive horizontal series of cations of common charge is in the *f* block of elements, among the 3+ ions of the lanthanides (elements of Period 6). Quite analogously to the neutral atoms, their +3 cations decrease slightly in radii from left to right (Table C). The same trend may be seen in a given period among the *d*-block elements of constant charge (i.e., +2, +3, or +4), although some anomalies may be noted (see Chapter 7).

2. A series of ions of differing atomic number is **isoelectronic** if they have the same electron configuration. To examine such a trend, we may start with a Group 1 cation (with one more proton than its total number of electrons), then add protons, producing cations in later groups with increased charge due to increased numbers of protons but with the same number of electrons. For example, we may compare the cationic radii of the following isoelectronic ions in Period 5: Rb^+, Sr^{2+}, Y^{3+}, Zr^{4+}, Nb^{5+}, Mo^{6+}, and Tc^{7+}. As the positive charge on the nucleus increases, the attraction for the 36 electrons becomes stronger, resulting in a strong *contraction* of the ion with *increasing* positive charge. We may also extend this series to the left of Group 1 back to the preceding isoelectronic anions, Br^- and Se^{2-}; these ions increase in size as their negative charges increase. (We have put isoelectronic anions on the left side of Table C, before cations. This unusual choice makes it easier to see a series of isoelectronic anions and cations.)

3. Among metals that form more than one cation, the radius of the cation *decreases* as the positive charge *increases* (e.g., compare the data for V, Cr, or Mn in Table C). By removing electrons, we increase the surplus of protons over electrons and reduce the tendency of electron clouds to expand due to repulsion of their like charges.

4. Most importantly for our purposes, going down a group such as Group 2, the main *vertical* trend is that the radius of the ion *increases* in size. Since the principal quantum number of the valence orbital steadily increases down a group, this result matches what we saw with neutral atoms (Figure 1.6). Just as for neutral atoms, there are anomalous countertrends in the early *p* block between the third and the fourth periods due to the scandide contraction, and between the fifth and the sixth periods in the early *d* and *f* blocks due to the lanthanide contraction (Figure 1.7).

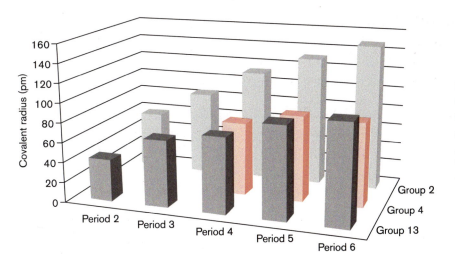

Figure 1.7. Radii of the +2 cations of Group 2, the +4 cations of Group 4, and the +3 cations of Group 13/III. The data come from Table C.

EXAMPLE 1.15

Assume that the recently discovered elements (a) Db (dubnium), (b) Sg (seaborgium), and (c) Bh (bohrium) all form +5 ions. Extrapolate reasonable radii for these ions.

SOLUTION: (a) Dubnium is in Group 5, and Table C lists the radii for all of the +5 ions in Group 5: V^{5+} = 68 pm, Nb^{5+} = 78 pm, and Ta^{5+} = 78 pm (Ta^{5+} is not larger than Nb^{5+} due to the lanthanide contraction). The normal trend to larger ions down the group should resume, so the radius of Db^{5+} should be > 78 pm. (b) The radius of Sg^{5+} should be larger than that of W^{5+}, so Sg^{5+} should be greater than 76 pm, but less than its neighbor to the left, Db^{5+}. (c) The radius of Bh^{5+} should be larger than that of Re^{5+} (72 pm), but less than its neighbor to the left, Sg^{5+}.

1.6. Periodic Trends in Pauling Electronegativities of Atoms

OVERVIEW. Without referring to Table A, describe the characteristic main horizontal and vertical trends in Pauling electronegativities of atoms, which are roughly the inverse of the corresponding trends in atomic radii (Section 1.6A). You may practice these concepts by trying Exercises 1.60–1.68. Describe how and where in the periodic table the scandide and lanthanide contractions affect the Pauling electronegativities of atoms (Section 1.6B). You may practice these concepts by trying Exercises 1.69–1.73.

The concept of electronegativity is widely used throughout this text, including in Sections 2.1–2.3, 3.3, 3.6, 4.3, 5.3–5.5, 5.8, 6.2, 6.4, 8.2, 9.5, and 9.6. Reasons for the trends in electronegativity values will be explored in Section 9.5 and amplified in Section 9.6.

When atoms of different kinds form a covalent bond, the attraction of the two nuclei for the shared electron pair will not in general be the same and the electron pair will, on the average, be closer to one of the nuclei. Pauling sought a measure of the relative abilities of the different atoms to attract such bond electrons to themselves, which he called the atom's **electronegativity**. It was not clear how such a property of atoms was to be measured; different people have proposed different measures of electronegativity. Pauling himself examined bond energies of molecules (which we shall look at in Section 3.3D) and obtained a table of what we now call **Pauling electronegativities**. An updated version of these values is included inside the front cover as Table A.

1.6A. Main Periodic Trends (Review). The overall trends in any scale of electronegativity are: (1) electronegativities of atoms generally *increase* from *left to right* across the periodic table, as atoms decrease in size; (2) electronegativities generally *decrease* as we go down a column of the periodic table, as atoms increase in size. In part, increases in electronegativities result from decreases in size. That is, by Coulomb's law, the attraction of a positively charged nucleus for shared negatively charged electron pairs should decrease with increasing distance of the shared electron pair from the nucleus. (A more complete explanation of the trends is developed in Section 9.5.)

The general horizontal trend holds well in Periods 2 and 3, with large increases in electronegativity from one element to the next in the *s* and *p* blocks. Electronegativities rise more slowly and sometimes decrease across the *d*-block elements in Periods 4 to 6. Electronegativities scarcely vary at all across the *f*-block elements.

EXAMPLE 1.16

Based on where they often sat during class, I could have constructed a partial periodic table of my previous inorganic chemistry students, which is shown below. Assume that these students show the same periodic properties as elements.

Period/Group	3	4	5	6	7
2	Standley	Scott	Cox	Mayfield	Klasek
3			Kim		
4			Chandler		
5			Williams		

(a) Who is the largest student in Group 5? (b) Who is the smallest student in Period 2? (c) Who is the least electronegative student in Period 2? (d) Who is the most electronegative student in Group 5?

SOLUTION: (a) The largest student in Group 5 is Williams, the one lowest in the group. (b) The smallest in Period 2 is Klasek, who is furthest to the right in the row. (c) Standley is the least electronegative in Period 2 because Standley is the furthest to the left in the row. (d) Cox is the highest student in Group 5, so Cox is the most electronegative. Disclaimer: These descriptions bear no relation to the actual sizes or negative natures of these students.

1.6B. Anomalies. Just as with trends in size, there are parts of the periodic table in which the overall trends disappear or even reverse. If Pauling electronegativities strongly influence chemical properties, we would expect to find some anomalous chemical trends in these areas. Going *down groups* of the *s*-block elements, we see normal trends, but going down groups of the *f*- and *d*-block elements, we often encounter significant *increases* in electronegativities. Going down Groups 13/III and 14/IV in the *p* block, the electronegativity values first decrease, then increase, then decrease again.

In Section 1.5B we noted how the scandide and lanthanide contractions resulted in anomalous trends in atomic size in these same regions of the periodic table. Since the sizes of atoms affect their electronegativities, we invoke the scandide and lanthanide contractions in explaining the anomalous increases in electronegativities low in the *f*, *d*, and *p* blocks. (The anomalously high Pauling electronegativities of gold and its near neighbors are also strongly influenced by relativistic effects (Section 9.5). In the next chapters, we may watch for consequences of these anomalies in unusual trends in chemical reactivity in the lower parts of these blocks of the periodic table.

EXAMPLE 1.17

Extrapolate reasonable values of the Pauling electronegativity for (a) Lr, (b) Rf, and (c) Db.

SOLUTION: The lanthanide contraction presumably does not apply when comparing Period 6 and Period 7 elements in the same group; it applies only after the first filling of the *f* block (in Period 6 but not Period 5). Hence, we expect lower electronegativities in Period 7 (the normal trend). (a) Lr < 1.27 (value for Lu); (b) Rf < 1.30 (value for Hf), but greater than the value for Lr; (c) Db < 1.50 (value for Ta), but greater than the value for Rf.

1.7. Background Reading for Chapter 1

Connections of Inorganic Chemistry to Other Fields
G. Wulfsberg, "What Are the 'Foundations of Inorganic Chemistry'? Two Answers," *J. Chem. Educ.* 89, 1220–1223 (2012).

Biological Functions of Ions of Elements: Connections
I. Bertini, H. B. Gray, E. I. Steifel, and J. S. Valentine, *Biological Inorganic Chemistry: Structure and Reactivity*, University Science Books: Mill Valley, CA, 2007, Chapter 1: Introduction and Text Overview.

Lithium-Ion Batteries: Connections
S. Fletcher, *Bottled Lightning: Superbatteries, Electric Cars, and the New Lithium Economy*, Hill and Wang: New York, 2011.

Rare-Earth Chemistry: Connections
T. Folger, "The Secret (Chinese) Ingredients of (almost) Everything," *Natl. Geogr.*, June 2011, pp. 136–145.

Conflict Minerals: Connections
S. Simpson, "Afghanistan's Buried Riches," *Sci. Amer.*, 305(4), 58–65 (Sept. 2011).

Where Does the *f* Block Begin? Amplification
W. B. Jensen, "The Positions of Lanthanum (Actinium) and Lutetium (Lawrencium) in the Periodic Table: An Update," *Found. Chem.* 17, 23–31 (2015); ibid., *J. Chem. Educ.* 59, 634–636 (1982).
E. R. Scerri, "Which Elements Belong in Group 3?" *J. Chem. Educ.* 86, 1188 (2009); ibid., *J. Chem. Educ.* 68, 122–127 (1991).

Periodic Trends in Oxidation States

S. Riedel and M. Kaupp, "The Highest Oxidation States of the Transition Metal Elements," *Coord. Chem. Rev.* 253, 606–624 (2009).

History of the Elements

H. Aldersey-Williams, *Periodic Tales: A Cultural History of the Elements from Arsenic to Zinc*, Harper-Collins: New York, 2011.

S. Kean, *The Disappearing Spoon and Other True Tales of Madness, Love, and the History of the World from the Periodic Table of the Elements*, Little, Brown: New York, 2010.

Various Authors, "It's Elemental: The Periodic Table," *Chem. Eng. News* 81(36), 27–190 (2003).

Numbered references from this chapter may be viewed online at www.uscibooks.com /foundations.htm.

1.8. Exercises

Odd-numbered exercises (those preceded by a star, *) are answered at the end of the book.

1.1. *If iodized salt is used in their preparation, French fries may be a source of iodine in the diet. (a) In that case, in which form is iodine found in French fries: I, I_2, or I^-? (b) If your French fries turn blue-black after adding iodine, which form of iodine has been added: I_2 or I^-? (c) Write the balanced half-reaction for the conversion of iodine atoms to iodide ions. (d) Write the balanced half-reaction for the conversion of (diatomic) iodine molecules to iodide ions.

1.2. French fries are commonly said to be high in sodium. (a) In which form is sodium found in French fries: $Na(g)$, $Na(s)$, or Na^+? (b) If your French fries burst into flame after adding sodium, which form of sodium has been added: $Na(s)$ or Na^+? (c) Write the balanced half-reaction for the conversion of sodium atoms to sodium ions.

1.3. *Which of these colors—colorless, pale green, or silvery—applies to (a) praseodymium(III) ion; (b) elemental praseodymium?

1.4. Which of these colors—colorless, blue, or silvery—applies to (a) lead(II) ion; (b) elemental lead?

1.5. *Write the balanced half-reaction for the conversion of lead(II) ion to lead(IV) ion.

1.6. Which is essential in your diet and is used in electron-transfer processes: copper metal or copper(II) ion?

1.7. *(a) The fudge bar I just ate contained 200 mg of potassium. Does this refer to K^+ or K? (b) Spinach is reputed to be high in iron. Does this mean Fe, Fe^{2+}, Fe^{3+}, or either Fe^{2+} or Fe^{3+}?

1.8. Name each of the following anions: C^{4-}, Ge^{4-}, O^{2-}, Se^{2-}, F^-, and Br^-.

1.9. *Name these ions and salts, indicating ion charges if appropriate: (a) Cr^{2+}; (b) Cr^{3+}; (c) Se^{2-}; (d) CrSe; (e) $CrCl_3$; (f) CrN.

1.10. Write the formulas of the following ions or salts: (a) selenide ion; (b) titanium(IV) selenide; (c) titanium(II) nitride; (d) titanium(IV) carbide.

1.11. *Divide each set of ions and salts into smaller sets, each containing the metal ion with the same ionic charge. Name the ions and salts, indicating ionic charges if appropriate. (a) Fe^{2+}, Fe^{3+}, FeS, $FeCl_3$, and FeN; (b) Sn^{2+}, Sn^{4+}, $SnCl_2$, SnO_2, Sn_3N_2, and Sn_3N_4; (c) Tl^+, Tl^{3+}, TlCl, TlN, Tl_2O, and Tl_2O_3.

1.12. Write the formulas of the following ions or salts: (a) gold(I) ion; (b) gold(III) ion; (c) arsenide ion; (d) sulfide ion; (e) gold(III) oxide; (f) gold(I) iodide; (g) gold(III) arsenide; (h) gold(I) sulfide.

1.13. *Write the formulas of the following ions or salts: (a) copper(I) ion, copper(II) ion; copper(II) oxide, copper(I) chloride, and copper(II) nitride; (b) chromium(III) ion, chromium(III) oxide, chromium(III) fluoride, chromium(VI) ion, chromium(VI) oxide, and chromium(VI) fluoride.

1.14. Write the characteristic full electron configurations of the following atoms: (a) Pd; (b) Ce; (c) Co; (d) Au; (e) Am; (f) Bi; (g) Nd.

1.15. *Write the characteristic full electron configurations of the following atoms: (a) Fr; (b) As; (c) Pt; (d) Dy; (e) Ge.

1.16. Write the characteristic valence electron configurations and Lewis symbols (if appropriate) of the following atoms: (a) Pd; (b) Ce; (c) Co; (d) Au; (e) Am; (f) Bi; (g) Nd.

1.17. *Write the characteristic valence electron configurations and Lewis symbols (if appropriate) of the following atoms: (a) Fr; (b) As; (c) Pt; (d) Dy; (e) Ge.

1.18. Write the characteristic valence electron configurations and Lewis symbols (if appropriate) of the following atoms: (a) F; (b) Os; (c) U; (d) Mo; (e) Pu; (f) Se.

1.19. *Write the characteristic valence electron configurations and Lewis symbols (if appropriate) of the following atoms: (a) Sr; (b) At; (c) W; (d) Bk; (e) Mn; (f) Pa.

1.20. Write the characteristic valence electron configurations and Lewis symbols (if appropriate) of the following ions: (a) Pd^{2+}; (b) Sb^{3-}; (c) Nd^{3+}; (d) Am^{4+}; (e) Pb^{2+}; (f) Co^{3+}.

1.21. *Write the characteristic valence electron configurations and Lewis symbols (if appropriate) of the following ions: (a) Bi^{3-}; (b) Bi^{3+}; (c) Ba^{2+}; (d) Pt^{2+}; (e) Dy^{3+}; (f) Ge^{2+}; (g) Ge^{4-}.

1.22. Write the characteristic valence electron configurations and Lewis symbols (if appropriate) of the following ions: (a) S^{2-}; (b) Tl^{+}; (c) Cs^{+}; (d) W^{4+}; (e) Bk^{3+}; (f) Mn^{4+}.

1.23. *Write the characteristic valence electron configurations and Lewis symbols (if appropriate) of the following ions: (a) Cl^-; (b) At^{3+}; (c) Os^{2+}; (d) Mo^{4+}; (e) Se^{4+}; (f) Se^{2-}.

1.24. Several chemical educators, including E. R. Scerri [*J. Chem. Educ.* 68, 122 (1991)] have suggested an alternate form of the periodic table in which the *s* block is placed at the right, rather than the left, of the table. Discuss the advantages and disadvantages of this type of table in presenting the periodic trends of this chapter.

1.25. *The characteristic valence electron configurations listed first are found in which of the species listed afterwards? (a) $5s^2 5p^6$ is found in Xe, Te^{2+}, Te^{2-}, or none of these. (b) $4s^2 3d^5$ is found in V, Mn, Co^{2+}, or none of these. (c) $4f^2$ is found in Ce, Pr, Pr^{3+}, Nd^{3+}, Pm^{3+}, or none of these. (d) $6s^2 5d^{10}$ is found in Hg, Pt, Au^+, Au^-, Pb^{2+}, or none of these.

1.26. List all elements in the periodic table that (a) have five valence electrons; (b) have a Lewis symbol :X: (include elements for which we do not normally draw Lewis symbols); (c) have eight valence electrons; (d) can form ions with −2 charges.

1.27. *Write the characteristic *valence* electron configurations of the following atoms or ions, each of which has a grand total of 80 valence plus core electrons: (a) Hg; (b) Bi^{3+}; (c) Au^-; (d) Tl^+.

1.28. Assign oxidation numbers to each atom in (a) HCN; (b) SO_3^{2-}; (c) NO_2^+; (d) XeF_4; (e) ICl_4^-.

1.29. *Assign oxidation numbers to each atom in (a) CO_2; (b) NO_2^-; (c) NO^+; (d) H_2O_2; (e) CH_4; (f) IF_5; (g) OsO_4.

1.30. Assign oxidation numbers to each kind of atom in (a) BF_4^-; (b) NaH; (c) $H_2S_5O_6$; (d) As_7^{3-}; (e) O_2F_2.

1.31. *Write the oxidation numbers of each type of atom in (a) $S_4O_6^{2-}$; (b) P_4S_7.

1.32. Assign oxidation numbers to the boron (B) atoms in (a) B_8Br_8; (b) B_2O_3; (c) $Na_2B_4O_7$; (d) $B_6Br_6^{2-}$.

1.33. *Using numerical rules, calculate the oxidation number of the imaginary element Branchium (Bc) in (a) $Na_2Bc_2F_8$; (b) BcO_6^{7-}.

1.34. Using numerical rules, calculate the (average) oxidation number of nitrogen in (a) $HNNH$; (b) N_3^-; (c) NF_4^+.

1.35. *Write the (average) oxidation numbers of each type of atom in (a) H_2O_2; (b) $S_3O_6^{2-}$.

1.36. Using numerical rules, calculate the oxidation number of the imaginary element Mnirajdine (Mj) in (a) $Ca_3Mj_2O_8$; (b) Mj_2; (c) MjO_6^{7-}.

1.37. *Assign oxidation numbers to the silicon (Si) or aluminum (Al) atoms in (a) Si_9^{4-}; (b) Si_2F_6; (c) Na_3AlF_6.

1.38. Assuming the following species consist of ions, write the valence electron configurations and (if appropriate) the Lewis symbols of the cations in (a) UF_4; (b) WF_8^{3-}; (c) SCl_3^+; (d) ClO_2^-.

1.39. *Assuming the given species consist of ions, write the valence electron configurations and (if appropriate) the Lewis symbols of the cations in (a) $FeCl_2$; (b) SnO_2; (c) PCl_3; (d) $SbCl_6^{3-}$.

1.40. Considering the location of each of the following elements within the *s*, *p*, *d*, or *f* blocks of elements, predict which should show the group oxidation number and what that number should be: (a) Th; (b) Cl; (c) Zn; (d) W; (e) Ra; (f) Si.

1.41. *List all of the known chemical elements that have six valence electrons (and no more). Give the atomic numbers of the next four elements that can be expected to have just six valence electrons.

1.42. In which of the following parts of the periodic table are you most likely to find the group oxidation states: (a) *s* block; (b) left side of the *p* block; (c) right side of the *p* block; (d) left side of the *d* block; (e) right side of the *d* block; (f) left side of the *f* block; and/or (g) right side of the *f* block?

1.43. *Which of the following elements show maximum positive oxidation numbers equal to their group numbers? Take group numbers to be 3 through 8 in the *p* block: S, Ne, Ba, Ni, Ta, Er, and U.

1.44. Arrange the following types of atomic radii in order of increasing size for a given element: anionic radius, cationic radius, covalent radius, and van der Waals radius. Be sure to indicate when two types of radii are approximately equal.

1.45. *The covalent radius of Na is 166 pm. Give a reasonable estimate of the following radii, assuming that the increment between radii of different magnitudes is 60 pm: (a) the van der Waals radius of Na; (b) the cationic radius of Na; (c) the anionic radius of Na.

1.46. The cationic radius of Au (in Au^{3+}) is 99 pm. Give a reasonable estimate of the following radii for Au (assume the increment between radii of different magnitudes is 60 pm): (a) its covalent radius; (b) its anionic radius; (c) its van der Waals radius.

1.47. *The anionic radius of I^- is 206 pm. Give a reasonable estimate of the following radii, assuming that the increment between radii of different magnitudes is 60 pm: (a) covalent radius of I; (b) cationic radius of I^{7+}; (c) van der Waals radius of I. Then find a measured radius in the tables to check your calculations.

1.48. The covalent radius of Pt is 136 pm. Give a reasonable estimate of the following radii, assuming that the increment between radii of different magnitudes is 60 pm: (a) the van der Waals radius of Pt; (b) the cationic radius of Pt^{6+}; (c) the anionic radius of Pt^{2-}.

1.49. *(a) Which has the largest radius: Pr or Pr^{3+}? Is it about 60 pm or 120 pm larger? (b) Which has the largest radius: As^{5+} or As^{3-}? Is it about 60 pm or 120 pm larger?

1.50. Recently solids containing anions of the Group 1 metals Na^-, K^-, Rb^-, and Cs^- have been isolated. (a) Compare and contrast the valence electron configurations of these anions with those of the p-block anions and of the d-block anion Au^-. (b) Predict the radii of these ions and compare with the radii reported in R. H. Huang, D. L. Ward, and J. L. Dye, *J. Am. Chem. Soc.*, 111, 5707 (1989); and R. H. Huang, D. L. Ward, M. E. Kuchenmeister, and J. L. Dye, *J. Am. Chem. Soc.*, 109, 5561 (1987).

1.51. *Suppose that you have a metallic atom bonded to a nonmetallic atom, and you are uncertain whether to classify the bonding as ionic or covalent. Using appropriate tabulated radii, calculate the internuclear distances you would expect to find for each type of bonding for the following cases: (a) Al and Cl; (b) Tl and I in TlI_3; (c) Na and F. Which type of bonding appears to give longer bonds?

1.52. From each of the following sets of atoms or ions, select the *largest* and the *smallest*. As you go through these sets, you will find two atoms or ions that should be of very similar size; identify these two. (a) I, Rb, Sn, and Zr; (b) B, In, Al, and Ga; (c) I^{7+}, I^-, I^+, and I^{3+}.

1.53. *Without the use of tables, arrange each of the following sets in order of increasing size: (a) Li, C, F, and Ne; (b) Be, Ca, Ba, and Ra; (c) B, Al, Ga, In, and Tl; (d) V, Nb, Ta, and Db (element 105).

1.54. Without the use of tables, arrange each of the following sets of atoms in order of increasing size, noting any cases in which two atoms should be of very similar size: (a) As, Br, Ca, and Ga; (b) B, In, Al, and Ga; (c) C, Si, Ge, and Sn; (d) Ge, Se, K, and Zn.

1.55. *Without referring to Table C, arrange the following sets of ions in order of increasing radii: (a) Cr^{6+}, Cr^{4+}, Cr^{2+}, and Cr^{3+}; (b) Ra^{2+}, Mg^{2+}, Be^{2+}, and Sr^{2+}; (c) the ions of valence electron configuration $6s^0$ between Cs and Re; (d) the f block +3 ions of Period 7.

1.56. Put each of the following series of cations in order of increasing size: (a) La^{3+}, Lu^{3+}, Eu^{3+}, and Gd^{3+}; (b) Tl^{3+}, Ga^{3+}, and B^{3+}; (c) Mn^{7+}, K^+, Sc^{3+}, and V^{5+}; (d) V^{5+}, V^{2+}, V^{4+}, and V^{3+}; (e) No^{3+}, No^{2+}, Md^{2+}, and Fm^{2+}.

1.57. *Put each of the following series of cations in order of increasing size: (a) Cl^+, At^+, I^+, F^+, and Br^+; (b) Au^+, Au^{7+}, Au^{3+}, and Au^{5+}; (c) Pb^{2+}, Au^-, Tl^+, and Bi^{3+}; (d) H^+, Fr^+, Rb^+, and Li^+; (e) Te^{4+}, In^+, Sb^{3+}, and Sn^{2+}.

1.58. (a) The *scandide* contraction contributes to making which two Group 14/IV atoms nearly equal in radius? (b) The *lanthanide* contraction contributes to making which two Group 4 atoms nearly equal in size?

1.59. *Given the following pairs of atoms that are next to each other in the periodic table, identify the pair in each set that are most nearly identical in size: (a) Li, Be or Mn, Fe or Nd, Pm; (b) B, Al or Al, Ga; (c) Ti, Zr or Zr, Hf; (d) Ga, In or In, Tl.

1.60. Fill in each blank with one of the following terms or phrases: increase, decrease, stay the same, more rapidly, or less rapidly. (a) Electronegativities of atoms change _____ when crossing the *f* block of elements than when crossing the *d* block. (b) Radii change _____ when crossing the *f* block than when crossing the *p* block.

1.61. *Fill in each blank with one of the following terms or phrases: increase, decrease, stay the same, more rapidly, or less rapidly. (a) Electronegativities of atoms ___ from left to right across a period. (b) Electronegativities of atoms ___ from top to bottom down a group. (c) Electronegativities of atoms change ___ when crossing the *d* block of elements than when crossing the *p* block. (d) The radii of atoms ___ from left to right across the *f* block in Period 6, but the radii ___ across the *f* block ___ than they do across the *p* block in Period 6. (e) Radii of atoms ____ from top to bottom going down Group 2; radii ___ ___ from top to bottom going down Group 13/III.

1.62. Use the symbols > for "greater than," < for "less than," and ≈ for "approximately equal to" to indicate the relative electronegativity values for the following elements: (a) Be ____ B ____ C; (b) C ____ Si ____ Ge in the absence of the scandide contraction; (c) C ____ Si ____ Ge in the presence of the scandide contraction; (d) Ti ____ Zr ____ Hf in the presence of the lanthanide contraction.

1.63. *Describe the main periodic trends of the following properties from top to bottom in the periodic table: (a) electronegativity; (b) radius; (c) common oxidation numbers.

1.64. Select the *most* electronegative and the *least* electronegative atom from each of the following sets: (a) I, Rb, Sn, and Zr; (b) B, In, Al, and Ga.

1.65. *Without the use of tables, arrange each of the following sets of atoms in order of increasing Pauling electronegativity: (a) Ba, Be, Ca, Mg, and Sr; (b) Al, Cl, Na, and P; (c) C, Si, and Ge; (d) Ag, Au, and Rg.

1.66. The following is a partial periodic table of elements with disguised symbols:

	Group 13/III	Group 14/IV	Group 15/V
Period 2	Jo	Tr	Vo
Period 3	St	Ka	Gr
Period 4	Tn	Em	Da

(a) In Group 15/V, should the element Vo have the largest or the smallest covalent radius? The highest or the lowest electronegativity? (b) In Group 13/III, which two elements will have approximately the same radii?

1.67. *Use the symbols > for "greater than," < for "less than," and ≈ for "approximately equal to" to indicate the relative electronegativity values of the following elements: (a) Mg _____ Si _____ S; (b) Mg _____ Sr _____ Ra; (c) B _____ Al _____ Ga; (d) difference between B and C _____ difference between Sc and Ti _____ difference between Ce and Pr.

1.68. The electronegativities of the second-period elements increase quite steadily from left to right: Li 0.98, Be 1.57, B 2.04, ..., O 3.44, F 3.98. (a) From Table 1.10, what are the highest positive oxidation numbers for each of the second-period elements? (b) Generally the electronegativities and the highest oxidation numbers both increase from left to right. Why does this relationship fail at the far right of the period?

1.69. *Describe the main periodic trends of the following properties from left to right in the periodic table: (a) electronegativity; (b) radius; (c) common oxidation numbers. In which blocks of the periodic table (if any) are there anomalous countertrends?

1.70. In which blocks of the periodic table are there anomalous vertical trends in Pauling electronegativities? Are there corresponding anomalous vertical trends in these blocks in ionic radii? Are there corresponding anomalous vertical trends in these blocks in common positive oxidation numbers?

1.71. *(a) Which atom in Period 4 is the largest? (b) Which atom in Group 14/IV (the carbon group) is the smallest? (c) Which atom in Period 2 is the least electronegative? (d) Which atom in Group 11 (the copper group) is the most electronegative? (e) Which two Group 14/IV atoms are nearly equal in electronegativity due to the scandide contraction? (f) Which two Group 4 atoms are nearly equal in size due to the lanthanide contraction?

1.72. Use five pairs of atoms—B and Al, Al and Ga, Zr and Hf, K and Kr, and Cu and Au— to answer the following questions (each pair is from the same group or the same period): (a) Which pair is nearly equal in electronegativity due to the scandide contraction? (b) Which pair is nearly equal in electronegativity due to the lanthanide contraction? (c) In each of the other three pairs, which atom has the higher electronegativity?

1.73. *Use the given pairs of neighboring atoms to answer the following questions: Which one is substantially *higher* in electronegativity than the other? Which pairs are *nearly identical* in electronegativity (within 0.05 of each other)? (a) Pm and Sm; (b) Li and Be; (c) O and S; (d) Zr and Hf.

CHARLES-AUGUSTIN DE COULOMB (1736–1806) was born in Angoulême, France, to parents who came from aristocratic families. He entered military school in 1759 and graduated in 1761 from the Royal Engineering School of Mézières. His early work for the military involved structural design and soil mechanics. Beginning in 1764 he was stationed in Martinique, West Indies, where he was responsible for building Fort Bourbon. He returned to France in 1773 ill with fever. He continued to work as an engineer, but he also began studies on applied mechanics and presented his findings to the Académie des Sciences in Paris. He invented the torsion balance, an apparatus that measures very weak forces, and used it to study the torsional forces on metal wires. He eventually used it to measure the attractive force between oppositely charged spheres. His discovery that the force was proportional to the product of the charges and inversely proportional to the square of the distance between the spheres has come to be called Coulomb's law. The SI unit of electric charge, the coulomb, is also named after him. The French Revolution, which began in 1789, forced many aristocrats from government, Coulomb included. He retired from the Corps of Engineers in 1791, but continued his research at his small estate at Blois. He eventually returned to Paris to help the Revolutionary government develop the weights and measures that became the metric system, but years of declining health eventually led to his death in 1806.

CHAPTER 2

Monatomic Ions and Their Acid–Base Reactivity

Overview of the Chapter

Most of the elements are commonly encountered in nature, in our bodies, and even in our laboratories as ions, rather than as free elements. Here in Chapter 2, we investigate the interaction of some common cations and monatomic anions of the elements with water and see how the periodic trends in these reaction tendencies can be related to the properties of atoms and ions (especially charge, size, and electronegativity) surveyed in Chapter 1. We develop methods of categorizing ions that are especially useful, and can be used again for the other types of chemistry included in Chapters 3 through 6.

In Section 2.1 your instructor may assign Experiment 2 as a laboratory experiment, a classroom demonstration/discussion, a computer video discussion, or a YouTube video discussion. Observe the process of dissolving the cations of the elements (in the form of their chlorides) in water. This may sound trivial, but you will find some unexpected drama in the process. You will find, upon analyzing the results, that even so simple a reaction involves some important chemistry. You may draw your own conclusions as to what happens when a chloride comes into contact with water, and how the extent of this reaction depends on the charge, size, and electronegativity of the cation. After this is discussed in Section 2.2, you can predict whether other chlorides you have never handled will react violently and be a hazard to you, your employer, and the environment! Is it always a good idea to flood a chemical spill with water? In Section 2.3 we construct a physical model of what is happening during these reactions. When and why do the expected reactions sometimes fail to occur? Which chlorides, fluorides, etc., are very unreactive, and is this entirely a good thing or a bad thing?

In Section 2.5 we similarly analyze the process of dissolving anions of the elements (in the form of their sodium salts) in water. In Sections 2.4 and 2.6 we develop a useful graphical way of summarizing and predicting acid–base reactivity by means of acid–base predominance diagrams.

You will find that the principles you derive, and the classification scheme that is developed in this chapter, apply to far more than just the reaction of a cation or an anion with the humble water molecule; ions react similarly with many other molecules and biomolecules (Section 1.2B). Your understanding of positively and negatively charged species from this chapter will pay off in subsequent chapters.

2.1. Acidity of Cations

OVERVIEW. Most metal ions, when placed in water, give solutions that are acidic. The first step of the reaction of the hydrated ion with water, Equation 2.3 and Figure 2.1, can be characterized by an acid ionization equilibrium expression similar to Equation 2.5 and has associated K_a and pK_a values (Table 2.1, Section 2.2A). You may practice these concepts by trying problems from the equilibrium chapters of your general chemistry text or by trying Exercises 2.1–2.4. From its pK_a value, you can classify the acidity of a cation as nonacidic, feebly acidic, weakly acidic, moderately acidic, strongly acidic, or very strongly acidic. Then you can use this classification to tell whether the cation will react with water, roughly what the pH of the resulting solution will be, and whether this reaction with water can be reversed (Section 2.1B). You may practice applying these concepts by trying Exercises 2.5–2.13. These are fundamental concepts, which will be applied again in several sections in Chapters 3, 4, and 5, and in Sections 6.5, 7.7, and 8.5–8.7.

Metal chlorides such as NaCl and $TiCl_4$ are important commercial materials that are used and transported in the environment all the time. What happens if there is a spill of such a chloride on a highway? The answers are dramatically different for these two chlorides. Rock salt (NaCl) is deliberately spread on icy highways to melt the ice. Although the practice has some negative environmental consequences, it improves highway safety. $TiCl_4$ is *not* used for this purpose! A tank car containing nearly 3200 gallons of this metal chloride broke open on Interstate 40 in Tennessee in July 2008. This accident required the closing of the interstate for nearly two days,

EXPERIMENT 2 MAY BE ASSIGNED NOW.
(See www.uscibooks.com/foundations.htm.)

with massive multi-hour traffic jams and traffic diversion onto secondary routes that also jammed up. Ultimately a 60-mile detour via two other interstates through a distant city was required. Why is there such a huge difference in reactivity between NaCl and $TiCl_4$?

If you watched the video version of Experiment 2, you will have seen the author of this book experiencing (on a vastly smaller scale) what the drivers on Interstate 40 would have experienced had they been allowed through. $TiCl_4$ reacts violently with water and even with the moisture in the air, generating strongly acidic fumes and/or a strongly acidic solution:

$$TiCl_4(l) + 6\ H_2O \rightarrow [Ti(H_2O)_6]^{4+} + 4\ Cl^- \tag{2.1}$$

$$[Ti(H_2O)_6]^{4+} + H_2O \rightleftharpoons [Ti(H_2O)_5(OH)]^{3+} + H_3O^+ \tag{2.2}$$

On the left side of Figure 2.1 we show cross-sections of the structures of the reactant and product metal ions of Equation 2.2 (i.e., when M^+ is Ti^{4+}).

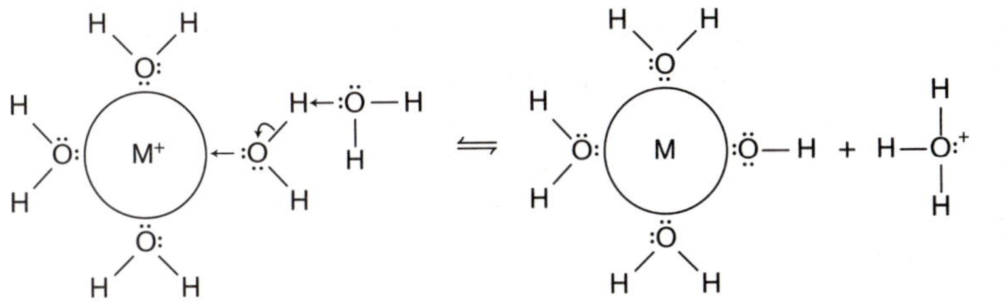

Figure 2.1.
Reaction of a hydrated cation with water to generate H_3O^+.

2.1A. Acid Dissociation Equilibria and Constants K_a (Review).

Often equations such as Equation 2.2 are simplified by representing the intact water molecules as (*aq*):

$$M^{z+}(aq) + H_2O \rightleftharpoons M(OH)^{(z-1)+}(aq) + H^+(aq) \text{ [or } H_3O^+(aq)] \qquad (2.3)$$

Note the similarity of this equation to the equation for the equilibrium process of dissociation of a weak acid such as acetic acid:

$$HC_2H_3O_2 + H_2O \rightleftharpoons C_2H_3O_2^- + H_3O^+ \qquad (2.4)$$

The equilibrium constant expression for this acid-producing reaction of metal ions (Eq. 2.2) can be written in the same manner as that for acetic acid:

$$K_a = \frac{[Ti(OH)^{3+}][H^+]}{[Ti^{4+}]} \approx 1 \times 10^4 \qquad (2.5)$$

Note, however, that whereas the K_a for acetic acid is a moderately small number, 1.7×10^{-5}, the K_a for Ti^{4+} is a large number—so large that it cannot be accurately measured.

Just as we commonly use pH to more clearly represent the small numbers that commonly are found for hydrogen-ion concentrations, we can represent K_a values more clearly by taking their negative logarithms:

$$pK_a = -\log K_a = 4.74 \text{ for acetic acid; } pK_a = -4 \text{ for } TiCl_4 \qquad (2.6)$$

An important consequence of taking *negative* logarithms, as in pH values, is that *lower* values of pK_a and pH correspond to *higher* degrees of reaction and *higher* acidities of solutions.

Of most practical importance to us is the hydrogen-ion concentration $[H^+]$ and hence the pH that results upon adding the metal halide to water. You may recall from your general chemistry course the procedure for calculating $[H^+]$ for a weak acid such as acetic acid. This depends on the initial concentration of the weak acid. *In this course, we want to emphasize broad periodic trends rather than exact calculations.* We assume for simplicity an initial concentration of 1.0 M for the weak acid or the metal ion. Then we take the equilibrium concentration (or activity) of the hydrogen ion and of the conjugate base of the weak acid [i.e., the acetate ion, or the $Ti(OH)^{3+}$ ion] as being x, and the equilibrium concentration of the acid (acetic or metal ion) as being $1 - x \approx 1$ M:

$$K_a = \frac{[\text{conjugate base}][H^+]}{[\text{conjugate acid}]} = \frac{x^2}{(1-x)} \approx x^2 = [H^+]^2 = K_a \qquad (2.7)$$

From Equation 2.7 we can then derive (after taking negative logarithms of each side of the equation) the approximate pH of a solution of an (initially) 1.0 M weak acid:

$$pH \approx pK_a/2 \qquad (2.8)$$

Thus, the pH of a 1.0 M acetic acid solution is approximately 2.37, while the pH of a 1.0 M solution of $TiCl_4$, although not well described by Equation 2.8, *is less than zero.* (Natural water with a pH of –3.6 has been found in mine waters from Iron Mountain, California.[1]) The pH of a 1.0 M solution of $TiCl_4$ is comparable to the pH of a >1.0 M solution of $HCl(aq)$. Consequently, it makes sense to categorize $TiCl_4(aq)$ or $Ti^{4+}(aq)$ as a strong acid.

EXAMPLE 2.1

In Experiment 2 the pH values given for the solutions of each cation are (a) 6 for Li^+, (b) 5 for Zn^{2+}, (c) 2 for Al^{3+}, and (d) 0 for Ti^{4+}. Assuming (very dubiously) that the solutions shown are 1.0 M and that Equation 2.8 is quite accurate, calculate the approximate pK_a values of each of these ions and compare them with the pK_a values given in Table 2.1 in Section 2.2A. Draw conclusions on the limitations of the measurements and on the accuracy of these assumptions.

SOLUTION: Rearranging Equation 2.8 gives us $pK_a \approx 2$ pH. When this equation is applied to our four cations, (a) the calculated pK_a = (2)(6) = 12; the tabulated pK_a = 13.6 (note that effects of dissolved CO_2 on the measurement are ignored). (b) The calculated pK_a = 10; the tabulated pK_a = 9.0. (c) The calculated pK_a = 4; the tabulated pK_a = 5.0. (d) The calculated pK_a = 0; the tabulated pK_a = –4.0. Note that pHydrion paper will not record a pH below zero, and that Equation 2.8 is not valid in this case.

2.1B. Acidity Categories. There are 43 cations listed in Table 2.1 (Section 2.2A), but these do not show 43 different types of reactivity as acids that we need to memorize or calculate! Instead, we define *six categories of acidity* into which these cations can be grouped, so that we can easily anticipate their reactivity (Figure 2.2). We find experimentally that the ions in a given category of acidity share several important chemical properties. These properties include not only the degree to which the ions react with water but also their ability to form oxo anions (Chapter 3), the solubility or insolubility of salts formed by these ions (Chapter 4), and the properties of compounds formed by these ions (such as their oxides, Chapter 8) in the absence of water.

Cations with pK_a values of 14 or greater show such negligible reaction with water by Equation 2.3 that they do not measurably alter the pH of their solutions; we call these cations **nonacidic** (or *neutral*) **cations**. Examples include Cs^+ and Rb^+, whose acidic reactions with water are too slight to be measured.

Cations with pK_a values between 11.5 and 14 (such values can be measured in sensitive experiments) have acidities that are also not normally directly detectable, but they have significant consequences for their mineral solubilities (Chapter 4). We refer to these ions as **feebly acidic cations**. By Equation 2.8, the pH of such 1 M solutions should be between 5.75 and 7. Such slight acidity is normally masked by the weak acidity due to dissolved carbon dioxide (carbonic acid) in the solution. An example of a feebly acidic cation from Experiment 2 is Li^+.

pK_a	Acidity category	Approx. pH of 1 M solution
−4	Very strongly acidic	
−2	Strongly acidic	Eq. (2.8) is invalid
0		0
2	Moderately acidic	1
4		2
6		3
8	Weakly acidic	4
10		5
12	Feebly acidic	6
14		7
16	Nonacidic	7

Figure 2.2. Acidity categories defined for metal cations based on their pK_a values, and the approximate pH of their 1 M solutions (based on Eq. 2.8). Note that the term "weakly acid" from general chemistry has been divided into three categories in this text: feebly acidic, weakly acidic, and moderately acidic.

Cations with pK_a values between 6 and 11.5 we call **weakly acidic cations**. When 1 M solutions of these cations (such as Zn^{2+} in Experiment 2) are made up in water, the pH should end up between 3 and 5.75. Acidity is an important part of their chemistry. Included in this group are many important +2 ions, such as Mg^{2+} and the +2-charged d-block ions whose biological functions revolve around their acidity.

Moderately acidic cations such as Al^{3+} and the +3-charged d-block ions have pK_a values of 1 to 6 and are comparable in acid strength to carboxylic acids such as acetic acid. Their 1 M solutions are likewise unmistakably acidic, with pH values between 0.5 and 3. As noted in Experiment 2, the reactions of the chlorides with water may be exothermic.

Strongly acidic cations have pK_a values between −4 and 1. These ions (e.g., Ti^{4+}) react violently and nearly completely with water, giving strongly acidic solutions, because the products of Equation 2.3 are favored except in very concentrated strong acids (in which $[H^+] > 1$ M so that pH < 0). Reactions of these chlorides with water are quite exothermic.

For completeness, we define a category of **very strongly acidic** cations with pK_a values below −4. These "cations" react irreversibly and very exothermically with water. Thus, this category of "cation" is strictly hypothetical in water. Very strongly acidic cations cannot exist in aqueous solution in measurable concentrations because of their very high reactivity, and their pK_a values cannot be measured directly. Examples include carbonium ions, such as CH_3^+ or C^{4+}, which organic chemists know cannot exist in water—instead, they react completely with water to generate $CH_3OH + H^+$, and $CO_2 + 4\,H^+$, respectively.

Further Reactions of Multiply Charged Metal Cations. Equation 2.2 is only the first of several reactions that may occur to a multiply charged metal cation such as Ti^{4+}. A second, then a third water molecule in the hydrated ion may react, producing more hydronium ions and giving rise to hydrated hydroxy cations containing more than one hydroxy group:

$$[Ti(H_2O)_5(OH)]^{3+} + H_2O \rightleftharpoons [Ti(H_2O)_4(OH)_2]^{2+} + H_3O^+ \tag{2.9}$$

$$[Ti(H_2O)_4(OH)_2]^{2+} + H_2O \rightleftharpoons [Ti(H_2O)_3(OH)_3]^+ + H_3O^+ \tag{2.10}$$

$$[Ti(H_2O)_3(OH)_3]^+ + H_2O \rightleftharpoons [Ti(H_2O)_2(OH)_4](s) + H_3O^+ \tag{2.11}$$

As with polyprotic acids in general chemistry, each of these reactions has its own equilibrium-constant expression and pK_a value (designated as pK_{a2} for Equation 2.9, pK_{a3} for Equation 2.10, and pK_{a4} for Equation 2.11). For most metal ions these pK_a values are close to each other, which will allow simplifications in Section 2.5A.

When the process has continued to the point at which the original charge of the metal ion is completely neutralized, a *metal hydroxide* results. This species is no longer an ion, does not hydrate as readily, and is usually insoluble—hence a precipitate appeared in the solution of $TiCl_4$. Often the metal hydroxides will subsequently lose molecules of water to give insoluble hydrous or anhydrous *metal oxides*, as represented by Equation 2.12:

$$[Ti(H_2O)_2(OH)_4](s) \rightarrow Ti(OH)_4(s) \rightarrow TiO_2 \cdot nH_2O(s) \rightarrow TiO_2(s) \tag{2.12}$$

Since it is difficult for us to tell whether this has happened, we will not attempt to distinguish metal hydroxide and metal oxide precipitates.[2]

Reactions similar to Equations 2.9–2.11 are reversible for strongly acidic cations; their chlorides can be dissolved in strongly acidified water (e.g., concentrated HCl). Otherwise, strongly acidic cations in water give precipitated hydroxides, hydrous oxides, or oxides. These reactions *cannot* be reversed for chlorides of very strongly acidic cations; the hydroxides, hydrous oxides, or oxides *do* precipitate.

EXAMPLE 2.2

Choose an acidity classification—nonacidic, very strongly acidic, or weakly acidic—and one of the following pK_a values—less than –4, over 14, or from 6 to 11—for each of the following imaginary element chlorides: (a) chanslerium chloride, which reacts violently with water to give a cloudy solution of pH 0; (b) goochium chloride, which dissolves in water with no change in pH and no temperature change to give a clear solution; and (c) sullinium chloride, which dissolves in water with some heat evolution to give a clear solution of pH 3.5.

SOLUTION: (a) Because of its very strong reaction with water and low pH value, this cation is (strongly or) very strongly acidic with a pK_a less than –4. (b) Because of the absence of a reaction with water and the absence of a pH change, this cation is (feebly acidic or) nonacidic with a pK_a over 14. (c) Because of its moderate heat evolution and moderate pH change, this cation is (moderately or) weakly acidic with a pK_a from 6 to 11.

2.2. Predicting the Acidity Classifications of Cations

OVERVIEW. The acidity of a cation is directly proportional to its charge and electronegativity and is inversely proportional to its size. Using these properties, the acidity classification of a metal ion can be predicted based on its Z^2/r ratio (Section 2.2A and Table 2.2) and modified if its electronegativity equals or exceeds 1.8 (Section 2.2B). You can practice these predictions by trying Exercises 2.17–2.24. As in the last part of Experiment 2, you should be able to control variables in designing a scientific experiment; you can practice this by trying Exercises 2.25–2.27.

In this section you will be applying many concepts from Chapter 1, such as recognizing ions and their charges (Section 1.2 and 1.4), using the proper radii (Table C and Section 1.5), and using their electronegativities (Table A and Section 1.6). You will use these calculations in Chapters 3, 4, and 5, and in Sections 6.5, 7.7, and 8.5–8.7.

There are important practical consequences of the acidic properties and acid dissociation reactions of cations. The suggested laboratory experiment hints at some of these—namely, a compound with an innocent-looking formula such as $TiCl_4$ may react quite violently with water. It may react with the water vapor in the air and fill the laboratory with choking fumes of the acid HCl. You may be trying to prepare a solution of a metal ion for some experiment in biology, only to find a precipitate forming. You may need to know the form a metal-ion pollutant takes in a lake, to know whether it will end up as an insoluble sludge of oxide at the bottom of the lake or whether it will remain in solution as a cation or as an oxo anion (Chapter 3)—which may be taken up or rejected by quite different mechanisms by living organisms. If you worked for your state's Department of Transportation and heard that a tank car full of $GaCl_3$ had spilled on the interstate, should you close the interstate?

Thus, it is important for us to be able to gauge the approximate acidity of a given cation so that we can anticipate (even in absence of a pK_a value in Table 2.1) how violently a new compound will react with water or atmospheric humidity and whether its hydroxide or oxide will precipitate or whether an oxo anion will be produced. There are over 150 cations listed in Table C; hence, it is very useful to assign these ions to the six categories of acidity, so that we can easily anticipate their reactivity.

2.2A. Predictions Based on Cation Charge and Radius.
If you performed Experiment 2, you discovered that the acidity of a solution of a metal cation in water depends on three variables: its charge, its size, and its electronegativity. To sort out the relationships, you had to control all but one of the variables, then determine how the acidity depended on the remaining variable. In Figure 2.3 we show typical pK_a values for varying charges among sets of cations of similar radii and (low) electronegativities. The graph plunges strongly from left to right, indicating that acidity *increases strongly* as charge increases.

If we examine Figure 2.3 from front to back for two cations of constant charge, we see that pK_a increases and acidity decreases (less dramatically) as the radius increases.

TABLE 2.1. pK_a Values for Metal Cations

Electronegativity < 1.5				Electronegativity > 1.5				
Ion	Radius (pm)	Z^2/r	pK_a	Ion	Radius (pm)	Z^2/r	(a)	pK_a
+1 Ions								
K	152	0.007	14.5	Tl	164	0.006	0.016	13.2
Na	116	0.009	14.2	Ag	129	0.008	0.049	12.0
Li	90	0.011	13.6					
+2 Ions								
Ba	149	0.027	13.5					
Sr	132	0.030	13.3	Pb	133	0.030	0.066	7.7
				Sn				3.4
Ca	114	0.035	12.8	Hg	116	0.034	0.082	3.4
				Cd	109	0.037	0.055	10.1
				Cr	94	0.043	0.043	10.0
				Mn	97	0.041	0.046	10.6
				Fe	92	0.043	0.075	9.5
				Co	88	0.045	0.082	9.6
				Ni	83	0.048	0.088	9.9
Mg	86	0.047	11.4	Zn	88	0.045	0.060	9.0
				Be	59	0.068	0.074	6.2
+3 Ions								
Pu	114	0.079	7.0					
La	117	0.077	8.5	Bi	117	0.077	0.127	1.1
Lu	100	0.090	7.6	Tl	102	0.088	0.140	6.6
Y	104	0.086	7.7	Au	99	0.091	0.191	−1.5
Sc	88	0.102	4.3	In	94	0.096	0.123	4.0
				Ti	81	0.111	0.115	2.2
				Ga	76	0.118	0.148	2.6
				Fe	78	0.115	0.147	2.2
				Cr	75	0.120	0.135	4.0
				Al	67	0.134	0.145	5.0
+4 Ions								
Th	108	0.148	3.2					
Pa	104	0.154	−0.8					
U	103	0.155	0.6					
Np	101	0.158	1.5					
Pu	100	0.160	0.5					
Ce	101	0.158	−1.1					
Hf	85	0.188	0.2	Sn	83	0.193	0.222	−0.6
Zr	86	0.186	−0.3	Ti	74	0.216	0.220	−4.0

Sources: Values of acid dissociation constants (pK_a) taken from C. F. Baes and R. E. Mesmer, *The Hydrolysis of Cations*, Wiley-Interscience: New York, 1976; and from J. Burgess, *Metal Ions in Solution*, Ellis Horwood: Chichester, UK, 1978, pp. 264–267. Radii are cationic radii in pm taken from Table C, and Pauling electronegativities are from Table A.

[a] The $Z^2/r + 0.096(\chi_P - 1.50)$ for the cation, as is introduced below in Equation 2.14.

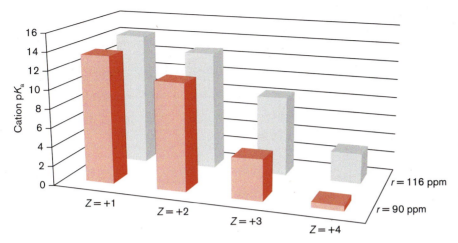

Figure 2.3. Dependence of cation pK_a on charge for metals with Pauling electronegativity values (χ_P) < 1.50 and (a) cationic radius of about 90 pm (e.g., Li$^+$, Mg^{2+}, Al^{3+}, and Ti^{4+}, as in Experiment 2); (b) cationic radius of about 116 pm (e.g., Na$^+$, Ca^{2+}, La^{3+}, and Th^{4+}, as in Table 2.1). Note that acidity increases from top to bottom of this figure.

This suggests that we try finding a relationship between pK_a of a cation and its Z^2/r ratio. Using data for a number of (low-electronegativity) cations from the left side of Table 2.1, we find the following empirical relationship:

$$pK_a = 15.14 - 88.16\, Z^2/r \qquad (2.13)$$

A graphical plot of these data is shown (using closed circles) in Figure 2.4. Data points that correspond to cations that fall in the same acidity category are enclosed by labeled boxes.

Normally it is inconveniently slow to solve Equation 2.13 to obtain pK_a values for each of the 150 or so cations. Instead we evaluate Z^2/r in order to place it in a one of the six ranges or categories of acidity enclosed by boxes in Figure 2.4. If the Pauling electronegativity (χ_P) of the element in the cation is below 1.50, we can use the Z^2/r ratio of the cation to assign the acidic category, as indicated on the left side of Table 2.2.

Figure 2.4. pK_a of cations as a function of charge, size, and electronegativity (χ_P). Black circles = metals of electronegativity < 1.5, plotted as a function of Z^2/r. **Brown** circles = metals of electronegativity > 1.5, plotted as a function of $Z^2/r + 0.096(\chi_P - 1.50)$.

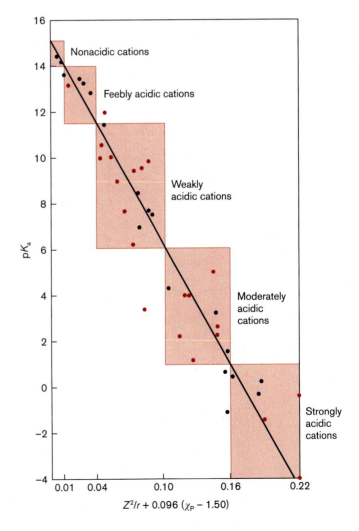

TABLE 2.2. **Relationship Between Cationic** Z^2/r **Ratios and Acidities**

Examples	Category if $\chi_P < 1.8$	Z^2/r Ratio	Category if $\chi_P > 1.8$	Examples
Most +1 s-block ions	Nonacidic	0.00–0.01	Feebly acidic	Tl$^+$
Most +2 s, f-block ions	Feebly acidic	0.01–0.04	Weakly acidic	Most +2 d-block ions
All +3 f-block ions	Weakly acidic	0.04–0.10	Moderately acidic	Most +3 d-block ions
Most +4 f-block ions	Moderately acidic	0.10–0.16	Strongly acidic	Most +4 d-block ions
	Strongly acidic	0.16–0.22	Very strongly acidic	Most p-block cations
	Very strongly acidic	0.22 and up	Very strongly acidic	Most p-block cations

EXAMPLE 2.3

How large would a +5-charged cation have to be in order to exist in water at some pH?

SOLUTION: The cation would have to have a Z^2/r ratio below 0.22. If $Z = 5$, then this is $25/r = 0.22$, or $r = 25/0.22 = 114$ pm. Table C contains no +5-charged ions this large.

2.2B. Effects of Electronegativity. So far we have not attempted to take into account the effects of high electronegativity on the cation, which you may have discovered if you performed Experiment 2. Figure 2.5 shows that the cations of metals with Pauling electronegativities (χ_P) over 1.5 are more acidic than other metal ions of similar charge and size.

We have derived a rough relationship between the "excess" Pauling electronegativity of a cation and its "excess" acidity, which allows us to modify Equation 2.13 to include the effect of electronegativity:

$$pK_a \approx 15.14 - 88.16[Z^2/r + 0.096(\chi_P - 1.50)] \tag{2.14}$$

This equation should be used *if and only if* the Pauling electronegativity of the metal exceeds 1.50. Otherwise Equation 2.13 should be used for this type of prediction.

In Figure 2.4 we have used **brown** circles to show the relationship of the pK_a of cations to their "modified" Z^2/r ratio, $Z^2/r + 0.096(\chi_P - 1.50)$. There is more scatter in the data for these more electronegative cations than there was for the "simple," less electronegative cations, which suggests that we have oversimplified the relationship to electronegativity. Our main purpose, however, is to categorize quickly the relative acidity of these cations.

The scatter in Figure 2.4 is not serious enough to prevent us from suggesting an even simpler rule of thumb: If the electronegativity of the metal ion is 1.8 or greater, move it up one category in acidity—that is, read to the right in Table 2.2.

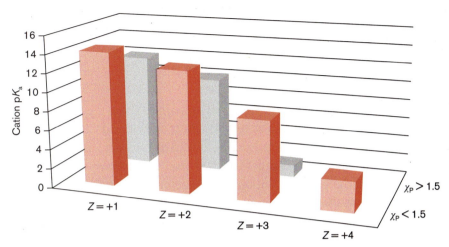

Figure 2.5. pK_a values of cations with a radius of ~116 pm versus charge for (a) metals of electronegativity < 1.5; (b) metals of electronegativity > 1.5. Data are from Table 2.1. Note that acidity increases from top to bottom in this figure.

EXAMPLE 2.4

Classify each of the following cations and describe their reactions with water: Eu^{2+}, B^{3+}, and W^{6+}.

SOLUTION: We begin by finding ionic radii in Table C; then we compute Z^2/r. For Eu^{2+} this ratio works out to be $2^2/131 = 0.031$; for B^{3+}, $3^2/41 = 0.220$; and for W^{6+}, $6^2/74 = 0.487$. Without considering electronegativities, we would classify Eu^{2+} as a feebly acidic cation, B^{3+} as strongly or very strongly acidic, and W^{6+} as very strongly acidic. Checking the table of electronegativities, we find that the latter two elements have electronegativities in excess of 1.8, so both are definitely in the class of very strongly acidic cations. Eu^{2+} remains classified as feebly acidic, and should be present largely unchanged (as a hydrated ion) in solutions of normal pH. The latter two "cations" will not actually be present at all in water. Halides of these two cations will react violently with water to generate hydrohalic acid and the oxides or hydroxides (or, as we will see in Chapter 3, oxo acids or oxo anions) of these elements.

2.3. Explaining the Acidic Tendencies of Cations: Why Halides of Some Cations Fail to React

OVERVIEW. The fact that hydration energies (Section 2.3A) and acidities (Section 2.3B) of cations depend directly on the charge and inversely on the radius of the cation is a consequence of Coulomb's law. Their direct (though less strong) relationship to the electronegativity of the cation is due to the significant covalency of the electronegative metal–oxygen bond. You can practice explaining and applying the relationship of hydration energy and acidity of a cation to its charge, radius, and electronegativity by trying Exercises 2.28–2.33. The reactions of halides with water are slow, however, if the cation has already reached its maximum coordination number. You may practice applying this Section 2.3C concept by trying Exercises 2.34–2.38.

Again you will be employing the concepts of charge, size, and electronegativity from Chapter 1. You will use the concepts of hydration energies again in Sections 4.4, 6.6, and 9.3. The concept of maximum coordination numbers will be useful to you again in Sections 3.4, 3.5, and 4.1.

2.3A. Explaining the Hydration of Cations.
In order to understand the chemistry of ions (such as the reactions of cations with water), you must understand Coulomb's law, a fundamental principle of physics—namely, opposite charges attract with forces and energies that are proportional to the charges involved, and are inversely proportional to the distances between the charges. Cations are therefore attracted to the partial negative charge present on the oxygen-atom end of the polar water molecule. As a result, a cation placed in water surrounds itself with water molecules, with the oxygen ends inward toward (coordinated to) the ion (see the left side of Figure 2.6). An anion surrounds itself with water molecules, too, but with the hydrogen ends inward (see the right side of Figure 2.6).

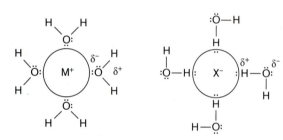

Figure 2.6. A hydrated cation (left) and a hydrated anion (right). Not shown are water molecules above and below the plane of the figure.

The attraction of opposite charges is quite exothermic. If we were to plunge one mole of gaseous cations into water, they would form hydrated ions and release a large amount of energy, which we call the **hydration enthalpy** (or energy) ΔH_{hyd} of the cation.* As illustrated in Figure 2.7, hydration energies have very large (negative) magnitudes, and they depend strongly on the charge and inversely on the radius of the cation.

* This experiment is quite impossible to perform, but the energy released can be determined indirectly, as we will see in Section 4.4. Note in Experiment 2 that both cations and anions in a liquid or solid were added to water. This makes a huge difference; consequently, there is a much more modest release of energy.

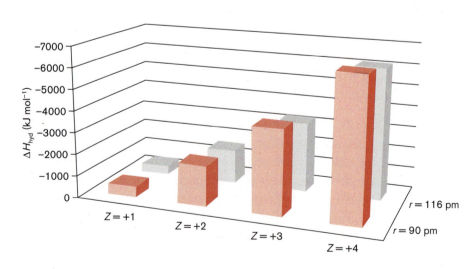

Figure 2.7. Hydration enthalpies (kJ mol⁻¹) for (a) cations of radius ~90 pm (Li⁺, Mg²⁺, Sc³⁺, and Zr⁴⁺); (b) cations of radius ~116 pm (Na⁺, Ca²⁺, La³⁺, and Th⁴⁺). [Sources: Same as Table 2.3.]

Hydration energies of a number of cations are listed in Table 2.3. Latimer[3] observed that if the electronegativity of the metal is not too great, the hydration energies of metal ions are given approximately by Equation 2.15:

$$\Delta H_{hyd} = \frac{-60,900\,Z^2}{(r+50)}\ kJ\ mol^{-1}$$ (2.15)

where Z is the charge on the cation and r is the cationic radius (in pm).

TABLE 2.3. Hydration Enthalpies of Metal Cations

Electronegativity < 1.5			Electronegativity > 1.5		
Ion	Radius (pm)	ΔH_{hyd} (kJ mol^{-1})	Ion	Radius (pm)	ΔH_{hyd} (kJ mol^{-1})
+1 Ions					
Cs	181	−263			
Rb	166	−296	Tl	164	−326
K	152	−321			
Na	116	−405	Ag	129	−475
Li	90	−515	Cu	91	−594
H		−1091			
+2 Ions					
Ra		−1259			
Ba	149	−1304			
Eu	131	−1458			
Sr	132	−1445	Pb	133	−1480
No	124	−1485	Sn		−1554
Yb	116	−1594	Hg	116	−1824
Ca	114	−1592	Cd	109	−1806
			Ag	108	−1931
			Ti	100	−1862
			V	93	−1918
			Cr	94	−1850
			Mn	97	−1845
			Fe	92	−1920
			Co	88	−2054
			Ni	83	−2106
			Cu	91	−2100
Mg	86	−1922	Zn	88	−2044
			Be	59	−2487

Table 2.3 continues on next page ▶

TABLE 2.3. *(cont.)*

Electronegativity < 1.5			Electronegativity > 1.5		
Ion	Radius (pm)	ΔH_{hyd} (kJ mol^{-1})	Ion	Radius (pm)	ΔH_{hyd} (kJ mol^{-1})
+3 Ions					
Pu	114	−3441			
La	117	−3283			
Lu	100	−3758	Tl	102	−4184
Y	104	−3620	In	94	−4109
			Ti	81	−4154
			V	78	−4375
			Cr	75	−4402
			Mn	78	−4544
			Fe	78	−4376
			Co	75	−4651
			Al	67	−4660
+4 Ions					
Th	108	−6136			
U	103	−6470			
Ce	101	−6489			
Zr	86	−6593			
Hf	85	−7120			

Sources: Ionic radii are from Table C (inside the back cover). Hydration enthalpies are taken from J. Burgess, *Metal Ions in Solution*, Ellis Horwood: Chichester, UK, 1978, pp. 182–183; D. W. Smith, *Inorganic Substances: A Prelude to the Study of Descriptive Inorganic Chemistry*, Cambridge University Press: Cambridge, UK, 1990, p. 160; and Y. Marcus, *Ion Solvation*, Wiley-Interscience, Chichester, UK, 1985, pp. 107–109. There are discrepancies of up to 2–3% among these sources.

No attempt is made in Latimer's equation to include the effects of electronegativity, but examination of the data for metals of Pauling electronegativities greater than 1.5 (on the right side of Table 2.3) shows that their hydration energies are substantially higher than those of ions of comparable radius and charge on the left side of the table. Such metals have electronegativities within about two units of that of oxygen. This suggests that for these metals there is not just a Coulombic electrostatic attraction between the metal ion and the negative end of the water molecule, but there also may be some degree of covalent bond formation, in which a lone pair of electrons on water is shared with the metal ion.

EXAMPLE 2.5

(a) Estimate the hydration energies of C^{4+}, Si^{4+}, and Sn^{4+}.
(b) Explain the trend in these values.

SOLUTION: (a) Using the Latimer equation (Eq. 2.15) and the radii of these ions from Table C, we obtain

$$\Delta H_{hyd} = - \frac{(60,900)(4^2)}{(30 + 50)} \text{ kJ mol}^{-1} = -12,200 \text{ kJ mol}^{-1}$$

for C^{4+}! For Si^{4+}, with its larger radius of 54 pm, we obtain -9370 kJ mol^{-1}, and for Sn^{4+} we calculate -7330 kJ mol^{-1}. (b) The calculated hydration energies decrease among these +4-charged ions because their radii increase down the group.

2.3B. Explaining the Acid Dissociation Reactions of Cations. If the attraction of the metal ion for the negative end of the water dipole is strong enough, the water molecule itself is affected (Figure 2.1). As the unshared electron pairs of the water molecule are pulled closer to (or even shared with) the metal ion, the electrons in the H–O bonds move closer to the oxygen to compensate some of its loss of electron density. Consequently, the hydrogen ends up with an increased positive charge, which makes it more closely resemble a hydrogen ion. Eventually the H^+ may dissociate completely, leaving a hydroxide group attached to the metal. In doing so, the H^+ attaches itself to solvent water molecules to make a hydronium ion, thus making the solution acidic, as we represented earlier by Equation 2.3.

EXAMPLE 2.6

a) Which mercury cation should have the greatest hydration energy, Hg^{2+} or the methylmercury cation, $H_3C–Hg^+$? Give two explanations.
(b) Which mercury cation should be the most acidic? Give one explanation.

SOLUTION: (a) Hg^{2+}, because it has double the charge, and because the methyl group in $H_3C–Hg^+$ blocks hydration on that side of the mercury cation.
(b) Hg^{2+}, because its greater charge results in a stronger pull on the electrons in the water molecules.

2.3C. Halides That Fail to React with Water; Maximum Coordination Numbers.
According to the rules we have developed, carbon tetrachloride should be one of the most reactive of all the chlorides of the elements, because Z^2/r is much higher for C^{4+} (0.533) than it is for Ti^{4+} (0.216). Although we saw how exothermic and extensive the reaction of $TiCl_4$ was, the predicted very violent reaction of CCl_4 and water fails to occur at or near room temperature. Indeed, CCl_4 is used as an inert solvent that is insoluble in and unreactive with water. The key to this seeming violation of our concepts lies in the *very* small size of the carbon atom, which is so crowded in by the four large chlorine atoms that the water molecule cannot get in to start the reaction—the CCl_4 molecule is *sterically hindered* from bonding a fifth atom to the carbon.

This nonreaction (at low temperatures) illustrates an aspect of the radius in the central atom in a molecule: There is a limit to the number of atoms that may be bonded to the central atom. This number is called the **maximum coordination number** of the atom, and although it depends on the size of the atoms around the central atom, it *increases* down the periodic table. Elements of Period 2 almost never have more than four atoms bonded to them in compounds; this is an important basis of the octet rule. Hence we say that the maximum coordination number for an element in Period 2 is four. But Table C shows that the ions of Periods 3 and 4 (which have similar sizes due to the scandide contraction; Section 1.5C) are substantially larger than those of Period 2; hence they can bond to as many as *six* other atoms. In Periods 5 and 6, the ions (which are similar in size to each other due to the lanthanide contraction; Section 1.5C) are larger still, and cases of atoms bonded to *more than six (often eight)* neighboring atoms are known.

Lower Coordination Numbers and Higher Cation Acidities

Actual coordination numbers below the maximum are often found. Thus, I^- is larger than F^-, so it is often the case that an element has a smaller coordination number in its iodide than in its fluoride (see Section 4.1B). For relativistic reasons (Section 9.6) some ions near the center of Periods 5 and 6 (e.g., Ag^+, Sn^{2+}, Hg^{2+}, and Au^+) attach as few as two water molecules. The polarizing power of the cationic charge is then concentrated on a smaller number of water molecules, so each H_2O molecule is polarized to a greater extent. The acidity of the hydrated Hg^{2+} and Sn^{2+} ions are therefore also much higher than expected from Equation 2.14.

Another case of "unexpected" unreactivity of a halide is that of sulfur hexafluoride, SF_6, which resists reacting with steam at 500°C and resists molten KOH. The sulfur atom in SF_6 exhibits its maximum coordination number of six; there is no room for the water molecule to attach to the sulfur to begin forming the hydrated ion (Figure 2.8). Hence the process of generating H_3O^+ as in Figure 2.1 cannot occur. Because atoms get larger as we go down Group 16/VI, SeF_6 is unreactive with water at 25°C but is hydrolyzed in the respiratory tract at 37°C (with quite undesirable consequences). Due to the higher maximum coordination number of Te, TeF_6 is able to react slowly with cold water.

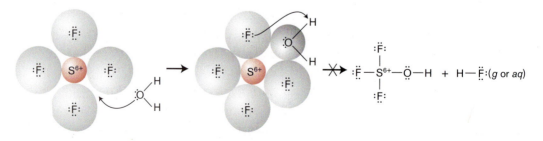

Figure 2.8. The first step in hydration and reaction of SF_6 with H_2O fails to occur. (Fluorine atoms above and below sulfur are not shown.)

A CONNECTION TO ENVIRONMENTAL CHEMISTRY

Halides in the Atmosphere

If they fail to react with water and with other atmospheric molecules and solar energy, unreactive halides such as CCl_4, SF_6, and other chlorofluorocarbons (Freons) can persist in the atmosphere for long periods—about 40 years in the case of CCl_4. In the upper atmosphere, such halides are broken apart by sunlight, leading to a series of reactions that result in the destruction of ozone, O_3 (Section 12.3). Since ozone prevents high-energy sunlight from reaching Earth's surface, its loss could lead to increased genetic damage and skin cancer. The Montreal Protocol of the 1980s led to the virtual elimination of the use of Freons. To its credit, the major manufacturer of Freons, DuPont, cooperated in this elimination and in finding substitute halides to use in air conditioning and refrigeration.

Gases such as CO_2 (Section 8.3) and the Freons have high abilities to trap infrared light (heat) being radiated into space by Earth. This causes Earth to warm up in a process known as the **greenhouse effect** (Sections 5.3, 5.7, and 8.5). A positive side effect of the Montreal Protocol was that it led to a slowing of the rate of warming of Earth since the 1990s.[4]

Among the other halide greenhouse gases is SF_6, which is valued as an inert gas for use in high-voltage transformers and has an estimated lifetime in the atmosphere of 3200 years; fortunately its concentration, although growing, is still only in the parts per trillion range. Recently another persistent halide, CF_3–SF_5, has been found to have the highest known ability to absorb infrared light of any substance—18,000 times that of an equivalent amount of CO_2.[5,6] So far its concentration is very low (0.2 parts per trillion), but it is growing at 6% per year. This gas does not occur naturally and is not manufactured for any purpose either; it may arise from side reactions of the "inert gas" SF_6 that is used in electronics.

A CONNECTION TO ORGANIC CHEMISTRY

Why Can Life Exist on Earth?

Although the rate of reaction of water and carbon tetrachloride is negligible at room temperature, it is still true that the end products of their very slow reaction (CO_2 + 4HCl) are more stable than the reactants (CCl_4 + $2H_2O$) by 52 kJ. At high temperatures the reaction does indeed occur. An intermediate product that forms in this reaction (which may happen when a carbon tetrachloride fire extinguisher is used along with water on a very hot fire) is carbonyl chloride or phosgene, $COCl_2$, a deadly poison and war gas. It has been suggested that a main reason that life can exist based on carbon chemistry rather than silicon chemistry is that the small size of carbon causes the reactions of carbon compounds to be much slower, so that many more carbon compounds that are not inherently thermodynamically stable can exist in the presence of reactive compounds such as water and oxygen.[7]

EXAMPLE 2.7

Which Group 14/IV chloride—CCl_4, $SiCl_4$, or $SnCl_4$—reacts most rapidly and exothermically with water and why? Which requires the greatest precaution in handling in moist air?

SOLUTION: Although we often expect the most exothermic reaction (that of C^{4+}; Example 3.4) to go most rapidly, this is not the case here, because C^{4+} in CCl_4 has reached its maximum coordination number of 4, so there is steric hindrance to the attack of the water molecule. Of the other two reactants, the reaction of $SiCl_4$ would be most exothermic. Neither Si^{4+} nor Sn^{4+} has reached its maximum co-ordination number in the chloride, so both $SiCl_4$ and $SnCl_4$ should react rapidly with water. Overall, the greatest caution should be employed in handling $SiCl_4$ in moist air.

2.4. Predominance Diagrams and Nonmetal Hydrides as Acids

OVERVIEW. Acid–base predominance or speciation diagrams show the species of an element at highest concentration at a given pH. The diagrams are handy for quick classification and comparison of acidity classes, because the pK_a of the acid normally equals the pH at the right boundary of the predominance range of the acid (Section 2.4A). In Section 2.4B we introduce predominance diagrams for the hydrogen compounds of the nonmetals, because these include more typical polyprotic acids and predominance diagrams. You may practice drawing and using these diagrams by trying Exercises 2.39–2.42. Comparison of the acid–base predominance diagrams of two potential reactants can be useful to show whether an acid–base reaction will occur, and if so, what the products may be (Section 2.4C; Exercises 2.43–2.45). The nomenclature of nonmetal hydrides is covered in Section 2.4D. You may practice applying these concepts by trying Exercises 2.46–2.49. You will apply predominance diagrams again in Section 2.5 to metal cations, in Section 2.6 to monatomic anions, in Chapter 3 to polyatomic anions, and in Chapter 6 to oxidation–reduction reactions.

2.4A. Predominance (or Speciation) Diagrams. Let us quickly review the process of *titration* of a soluble weak acid such as acetic acid with a strong base such as NaOH. If we monitor the pH of the solution with a pH meter during the titration, we can plot a titration curve (Figure 2.9). Before any NaOH has been added, the solution has a low pH. As more NaOH is added, more of the acetic acid is neutralized to give its conjugate base, the acetate ion. As NaOH is added, the concentration of the acid decreases and the concentration of its conjugate base increases. The pH at any point can be calculated using the **Henderson–Hasselbalch equation**:

$$pH = pK_a + \log \frac{[\text{conjugate base}]}{[\text{conjugate acid}]} \tag{2.16}$$

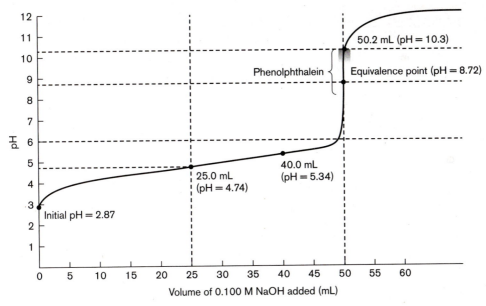

Figure 2.9. Curve for the titration of 50.0 mL of 0.100 M acetic acid with 0.100 M NaOH. [Adapted from J. W. Moore, X. L. Stanitski, and P. C. Jurs, *Chemistry: The Molecular Science*, 3rd ed., Thompson Brooks/Cole: Belmont, CA, 2008, p. 840.]

In titrations you may have performed previously, you were interested in the **equivalence point**, which is achieved when the number of moles (or equivalents) of NaOH added exactly equals the number of moles (or equivalents) of acetic acid present. At this time we are particularly interested in the point at which the titration is 50% complete (the half-equivalence point—at 25.00 mL base in the case of the titration of Figure 2.9). At this point, regardless of the initial concentrations of the acid and base, their solution concentrations are equal. Therefore,

$$\log \frac{[\text{conjugate base}]}{[\text{conjugate acid}]} = \log 1 = 0$$

and Equation 2.16 simplifies to Equation 2.17:

pH at half-equivalence point = pH at right boundary of acidic species = pK_a (2.17)

The **predominance** or **speciation diagram** of acetic acid (Figure 2.10) shows which form or species (acetic acid or acetate ion) is predominant at a given pH. To the left of the boundary at pH = pK_a, the concentration of acetic acid is higher than that of the acetate ion, and the *predominance range* is labeled with the identity of the conjugate acid. To the right of the boundary, the concentration of acetate ion is higher, so this predominance region is labeled with the identity of the conjugate base. As we move to the left or right of the boundary, the concentration of the less-predominant species diminishes. However, it is more than 5% of the total of all species within about 2 pH units of the boundary.

Figure 2.10. (Acid–base) predominance or speciation diagram for acetic acid.

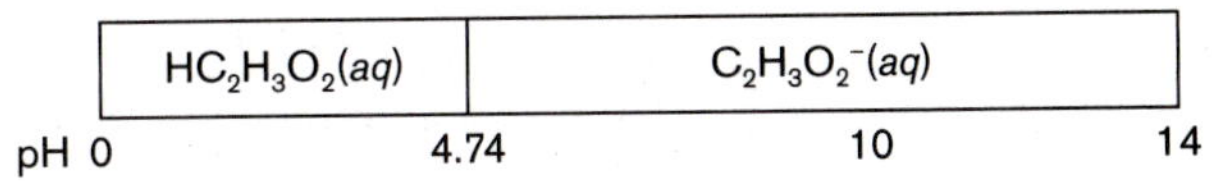

$HC_2H_3O_2(aq)$	$C_2H_3O_2{}^-(aq)$

pH 0   4.74   10   14

Predominance diagrams give handy quick identifications of the main species present at a given pH. Since they use pK_a values, they also allow us to classify the acidity of the protonated species. Thus, by inspection we can see that the pK_a of acetic acid is 4.74; therefore, acetic acid is *moderately acidic* (pK_a between 1 and 6). The relative acidity of different ions may be assessed by inspection if acid–base predominance diagrams are available. *The more acidic the species* (such as a metal ion), *the more it is confined to predominance in small ranges of extremely low pH at the left of a diagram.* Very strongly acidic species will not even be on the diagram, but they would be found off the left side, so they represent the ultimate in small predominance ranges. As an example, the predominance diagram for trifluoroacetic acid (not shown) has its boundary at pH 0.50, with $HC_2F_3O_2(aq)$ predominant at pH values < 0.50 and $C_2F_3O_2^-(aq)$ predominant at pH values > 0.50. From this boundary we can classify trifluoroacetic acid as a strong acid and a stronger acid than acetic acid.

2.4B. Nonmetal Hydrides as Acids. We introduce the hydrogen compounds of the nonmetals at this point because their predominance diagrams are more typical than those of the metal cations. The diagrams include no insoluble species, and the Group 16/VI hydrides are polyprotic acids with large differences between pK_{a1} and pK_{a2}. Figure 2.11 shows acid–base predominance diagrams for four binary weak-acid nonmetal hydrides—namely, HF ($pK_a = 3.15$), H_2Te ($pK_{a1} = 2.6$), H_2Se ($pK_{a1} = 3.6$), and H_2S ($pK_{a1} = 6.89$).

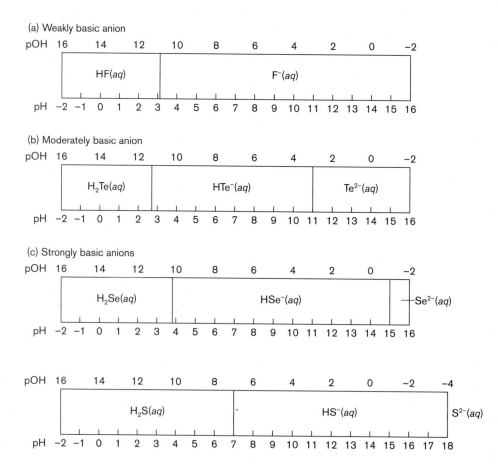

Figure 2.11. Acid–base predominance diagrams for (1 M total concentrations of) binary nonmetal hydrides HF, H_2Te, H_2Se, and H_2S, and their partially protonated and fully deprotonated forms (monatomic anions). The descriptions of the pOH scale and the basic properties of the anions are introduced in Section 2.6.

After the first hydrogens of H_2S, H_2Se, and H_2Te have been neutralized, the remaining hydrogen atoms can also ionize. For example, for H_2Te the equilibrium is:

$$HTe^-(aq) + H_2O \rightleftharpoons H_3O^+(aq) + Te^{2-}(aq) \qquad (2.18)$$

For this equilibrium, $pK_{a2} = 11.0$. In Figure 2.11 there is a separate predominance range for HTe^-, between pK_{a1} (2.6) and pK_{a2} (11.0). Finally, in the range above pH 11.0, the monatomic anion Te^{2-} predominates.

EXAMPLE 2.8

From their predominance diagrams in Figure 2.11, classify the acidic strength of (a) HF, (b) H_2S, (c) HTe^-, and (d) HSe^-. (e) Which Group 16/VI hydride is the strongest acid?

SOLUTION: We can read the pK_a values from the right boundary of the predominance range of the species in question, then classify the acidity of the species. This applies to the partially protonated anions as well as the neutral nonmetal hydrides. (a) The right boundary for HF is at a pH of approximately 3, which is therefore its pK_a. HF is moderately acidic. (b) The right boundary for H_2S is in the next pH range (6–11.5), which allows us to classify it as weakly acidic. Its pK_a is about 7. (c) The right boundary for HTe^- is likewise in the 6–11.5 range, which also makes it weakly acidic (although it is close to being feebly acidic). Its pK_a appears to be 11. (d) The right boundary for HSe^- is beyond 14, so it is nonacidic. Its pK_{a2} appears to be 15. (e) H_2Te is confined to a smaller predominance range than either H_2Se or H_2S, so is the strongest acid of the three Group 16/VI hydrides.

2.4C. Acid–Base Reactivity from Predominance Diagrams. At this point we are ready to predict a limited amount of reactivity of different acids and bases with each other. Predominance diagrams will help make these predictions possible by inspection.

Should a solution of a strong acid such as HCl (the predominance range of which is off scale to the left in Figure 2.11) react with a solution of a salt of the anion F^- (the predominance range of which lies above pH 3 in Figure 2.11a)? If HCl transfers a proton, it becomes the anion Cl^-, which predominates over the entire pH range of water. The predominance range of the anion F^- does not overlap the predominance range of HCl (Figure 2.12a, b). If two related reactants (e.g., HCl and F^-) have *nonoverlapping* predominance ranges, they cannot co-exist as predominant forms at the one pH that will result after mixing, so reaction tends to occur to give products that have *overlapping* predominance ranges and thus can co-exist at the same pH. In this reaction mixture, the two products with overlapping predominance ranges are Cl^- (pH –2 to +16) and HF (pH –2 to +3) (Figure 2.12c, d). If the two related reactants start with *overlapping* predominance ranges, *no* acid–base reaction between them is expected.

2.4D. Nomenclature. The Group 17/VII compounds of hydrogen—HF, HCl, HBr, and HI—when pure are the hydrogen halides. In solution, they undergo acid dissociation reactions, and are named as the hydrohalic acids. HCl, HBr, and HI are very strongly

Figure 2.12.
Acid–base reactivity predicted using predominance diagrams.
(a) Nonoverlapping shaded predominance ranges of the reactants HCl and F⁻.
(b) Overlapping shaded predominance ranges of the products Cl⁻ and HF.

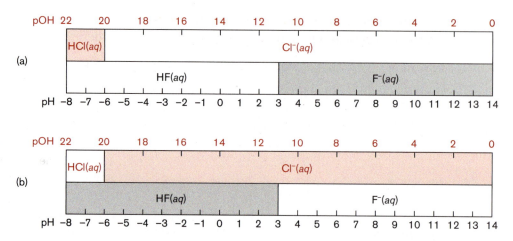

acidic; unlike HF, they fall off the left side of acid–base predominance diagrams. The Group 16/VI hydrides—H_2S, H_2Se, and H_2Te—when pure are named hydrogen sulfide, hydrogen selenide, and hydrogen telluride, respectively. The partially protonated anions sometimes have special names. Although HS^- and SeH^- are just called the hydrogen sulfide ion and the hydrogen selenide ion, respectively, OH^- is called the hydroxide ion. The names of the other nonmetal hydrides end in –ane if they are in Group 14/IV [plumbane, stannane, germane, silane, and (exceptionally) methane] and in –ine if in Group 15/V [bismuthine, stibine, arsine, phosphine, and (exceptionally) ammonia]. It is even customary to write the formulas of the acidic hydrogen compounds of Groups 17/VII and 16/VI with hydrogen first, as is common for acids, while the hydrogen compounds in Groups 14/IV and 15/V are written with hydrogen last to emphasize that they are *not* predominantly acidic (CH_4, NH_3, etc.). The most commonly encountered partially protonated forms are NH_2^- (the amide ion) and CH_3^- (the methide ion or methyl anion).

EXAMPLE 2.9

The imaginary Group 16/VI element Quinnium (Qn) forms an acidic hydride. Its incomplete predominance diagram is as follows:

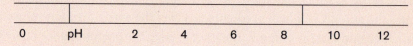

(a) Fill in the compounds and ions of Quinnium in the diagram. (b) What are the pK_a values of these compounds and ions? (c) What are the acidity classifications of these compounds and ions? (d) Name these compounds and ions.

SOLUTION: (a) The completed predominance diagram is as follows:

H_2Qn	HQn^-		Qn^{2-}
0	pH 2 4	6 8	10 12

(b) For H_2Qn, $pK_{a1} = 1$. For HQn^-, $pK_{a2} = 9$. Qn^{2-} is not an acid, so it does not have a pK_a value. (c) H_2Qn is on the border between moderately and strongly acidic, whereas HQn^- is weakly acidic. (d) Proceeding from left to right, the names are hydrogen quinnide, hydrogen quinnide ion, and quinnide ion.

2.5. Predominance Diagrams and the Precipitation of Metal Cations as Oxides or Hydroxides

OVERVIEW. In the predominance diagram for a metal cation that has an insoluble oxide or hydroxide, the right pH boundary and the pH of precipitation, at which the metal hydroxide or oxide precipitates, is *approximately* equal to pK_a (Section 2.5A). As the acidity of a metal cation increases, its predominance range decreases, and it becomes confined to the far left of the diagram (or may even be off scale at the left) (Section 2.5B). You may practice applying these concepts by trying Exercises 2.48–2.59. These concepts will be used again in Sections 3.7, 3.8, 5.6, 5.7, 6.5, and 8.5.

2.5A. Predominance Diagrams and pH Values of Precipitation for Metal Cations and Their Insoluble Oxides or Hydroxides.

Before presenting these predominance diagrams in Figure 2.15, we need to define the **pH of precipitation** for a metal cation. This is the pH above which more than half of the metal ion is precipitated as the hydroxide or oxide. As described in the following Amplification, the pH of precipitation of the metal ion (the pH at its right boundary) is located *approximately* at the pK_a.*

2.5B. Trends in the pH of Precipitation of Metal Oxides and Hydroxides.

The predominance diagrams of nine representative metal cations are shown in Figure 2.15 on page 68.

The predominance diagrams of *nonacidic cations* ($pK_a \geq 14$; Figure 2.15.a) are very simple: The cation is predominant at all practical pH values. The hydroxide will not precipitate from a 1 M solution at any pH below about 14. This illustrates a general principle for predominance diagrams—namely, unreactive, stable species have *broad* predominance ranges.

Feebly acidic cations precipitate hydroxides from 1 M solutions at pH values above approximately 11.5 and 14 (that is, in strongly basic solutions).

Weakly acidic cations (pK_a values between 6 and 11.5) show enough acidity to precipitate insoluble metal hydroxides in neutral or just slightly basic solutions—but if the solution is not made basic, the hydrated metal ion will be the predominant form (Figure 2.15.c).

Moderately acidic cations (pK_a values of 1 to 6) can be predominant in solution only at quite low pH values (Figure 2.15d). If the solutions of these ions are not kept acidic, metal hydroxides will precipitate (giving at least a cloudy appearance to the solution).

Strongly acidic cations (pK_a values between –4 and 1) can only predominate in very concentrated acids; at most practical pH values these cations are found as hydroxide or oxide precipitates (Figure 2.15e). Very reactive species have *narrow* predominance ranges, because there are few pH values at which they are stable. As we noted in Section 2.4A, strongly acidic species are confined to the *left* side of a predominance diagram.

Very strongly acidic "cations" (pK_a values below –4) cannot exist as such in water, because pH values this low are not realistically achievable in water (although some can exist as weakly or moderately acidic *oxo cations* such as VO_2^+ and UO_2^{2+}). The "cations" of the purely nonmetallic elements of the periodic table (e.g., cations of carbon such as C^{4+} and CH_3^+) have pK_a ratios in this range, so they cannot exist as aqueous cations.

* Note the difference between the pH of precipitation and the pH from Equation 2.8. Equation 2.8 gives the pH when the metal cation (as a chloride) is added to pure water. In the current case, we can think of the metal cation as being added to a solution, the pH of which is fixed at the right boundary (i.e., a buffered solution).

Predominance Diagrams for Metal Cations and Their Insoluble Oxides or Hydroxides

Predominance diagrams that involve ions in solution *and* precipitates (such as insoluble oxides or hydroxides) differ somewhat from diagrams such as those in which all species are soluble (e.g., Figure 2.10). In Section 2.1 we indicated that a multiply charged cation such as Ti^{4+} reacts with water in steps according to Equations 2.2, 2.9, 2.10, and 2.11, until the metal oxide or hydroxide finally precipitates. Each of these reaction steps has its own K_a and pK_a, which we could label pK_{a1} through pK_{a4}. Unfortunately, the pK_a values beyond that for the first step (pK_{a1}) are extraordinarily difficult to measure and have been measured in only a few cases. (For example, the three pK_a values for Al^{3+} are 4.89, 5.43, and 5.86.[8]) For Cr^{3+}, pK_{a1} = 4.10, which gives the boundary shown in Figure 2.13a. The pK_{a2} is 5.55, so the right boundary for $Cr(OH)^{2+}$ is 5.55. The predominance range for $Cr(OH)^{2+}$ is therefore quite narrow (Figure 2.13b). For simplicity, therefore, we ignore the intermediate predominance ranges and focus on only the original metal cation in solution and its insoluble hydroxide or oxide, as shown for chromium(III) in Figure 2.13d.

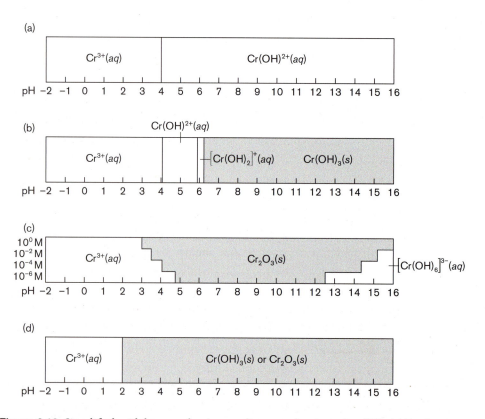

Figure 2.13. Simplified acid–base predominance diagrams for chromium(III). (a) Predominance diagram showing only $Cr^{3+}(aq)$ and $Cr(OH)^{2+}(aq)$. (b) Predominance diagram also showing $Cr(OH)_2^+(aq)$ and the precipitate $Cr(OH)_3(s)$. (c) Predominance diagram showing the effects of changing the total concentration of chromium(III) from 10^0 M (top) down to 10^{-6} M (bottom). (d) Predominance diagram considering only $Cr^{3+}(aq)$ and $Cr_2O_3(s)$ at a total concentration of 1.0 M. Shaded regions of pH are those in which precipitates (insoluble substances) predominate.

If we add more water to a precipitated insoluble metal hydroxide or oxide, more of it dissolves. If we add enough water, all of it dissolves, and the predominance region of the insoluble metal hydroxide or oxide disappears. As the $Cr(OH)_3$ dissolves, it releases $OH^-(aq)$, which shifts the right boundary of the $Cr^{3+}(aq)$ predominance range to a higher pH. Figure 2.13c shows the effects on the predominance ranges of the different chromium(III) species as the total concentration of all chromium(III) species is changed in steps from 1 M to 10^{-2} M to 10^{-4} M to 10^{-6} M. In general the predominance ranges of soluble species *expand* with increasing dilution, while those of insoluble species *contract*. Because the standard concentration (activity) is 1.0 M, we will use that total concentration in our predominance diagrams. Geochemists and environmental chemists generally encounter much lower total metal-ion concentrations, so their predominance diagrams have broader ranges for soluble species such as metal cations.

It can be shown[9] that 50% precipitation or dissolution of a metal hydroxide or oxide occurs at a pH that is fairly close to the pK_a value of the metal ion:

$$\text{pH of precipitation} = pK_a \text{ for } M^{z+} - \left(\frac{1}{z}\right)\log[M^{z+}] - \frac{5.6}{z} \tag{2.19}$$

The pH dependence on charge is significant for +1-charged ions, but is less serious for the more common multiply charged ion, because the last two terms in Equation 2.19 shrink.

EXAMPLE 2.10

Figure 2.14 shows the predominance diagrams for 1 M solutions of three +1-charged cations—namely, $Na^+(aq)$, $Tl^+(aq)$, and $Ag^+(aq)$.

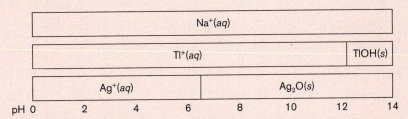

Figure 2.14. Acid–base predominance diagrams for 1 M solutions of the +1-charged cations $Na^+(aq)$, $Tl^+(aq)$, and $Ag^+(aq)$.

Test the validity of the approximation that $pK_a \approx$ pH of precipitation (a) for the +1-charged ions in Figure 2.14; and (b) for the Ca^{2+} cation, which precipitates as $Ca(OH)_2$ from 1 M solution at a pH of 12.8.

SOLUTION: For Na^+, the pH of precipitation is off range (above pH 14); for Tl^+, the pH is read as 12.0; for Ag^+, the pH is 6.0; and for Ca^{2+}, the pH is given as 11.4. These pH values are approximately equal to the pK_a values of the cations, which we find in Table 2.1: 14.2 for Na^+, 13.2 for Tl^+, 12.0 for Ag^+, and 12.8 for Ca^{2+}. Let us take the agreement as satisfactory if the agreement is within 2 units, and if the two values give the same acidity classification of the cation. This is the case except for the +1-charged Ag^+, for which the discrepancy is well above +2. (Application of Eq. 2.19 gives a more refined prediction of the pH of precipitation of Ag^+ as 12.0 – 5.6/1 = 6.4, which is in much better agreement with the predominance diagram in Figure 2.14.)

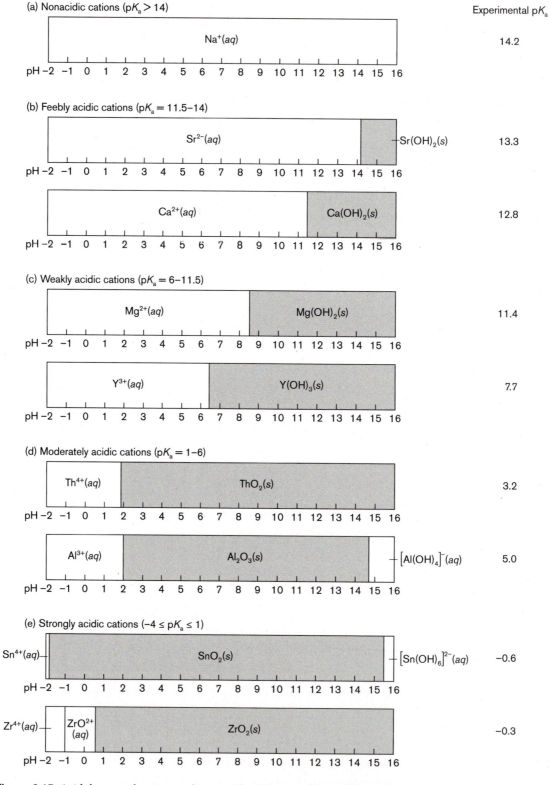

Figure 2.15. Acid–base predominance diagrams for (a) a typical nonacidic cation, (b) two typical feebly acidic cations, (c) two typical weakly acidic cations, (d) two typical moderately acidic cations, and (e) two typical strongly acidic cations. For comparison, experimental pK_a values (from Table 2.1) are shown at the right side of each diagram. Note: The oxo anions on the right side of the predominance diagrams for Al^{3+} and Sn^{4+} have been included (see Chapter 3). [Source: Data from M. Pourbaix, *Atlas of Electrochemical Equilibria in Aqueous Solutions*, National Association of Corrosion Engineers: Houston, 1974.]

EXAMPLE 2.11

The following predominance diagrams are for newly discovered elements, which you have established by titration experiments (to determine pK_a values) and conductance experiments (to determine ionic charges).

(a) Which of the cations shown is most acidic?
(b) Roughly speaking, what is the pK_a of each cation?
(c) Give the acidity classification of each cation. (d) On each predominance diagram, identify the pH range in which you expect to see a precipitate.

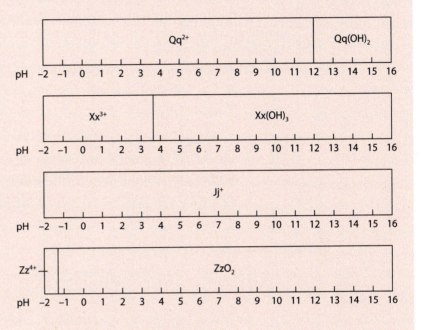

SOLUTION: (a) The most acidic cation is the one most confined to the left side of its predominance diagram: Zz^{4+}.
(b) Very approximately, the pK_a of each cation is the pH at which it predominantly converts to the hydroxide or oxide: 12 for Qq^{2+}, 4 for Xx^{3+}, >16 for Jj^+, and −1 for Zz^{4+}. (c) Based on these estimated pK_a values, we would classify Qq^{2+} as feebly acidic, Xx^{3+} as moderately acidic, Jj^+ as nonacidic, and Zz^{4+} as strongly acidic. (d) Precipitates of hydroxides or oxides predominate to the right of (on the high-pH or basic side of) the bars in these predominance diagrams.

2.6. Monatomic Anions as Bases

OVERVIEW. The basicities of monatomic anions are also proportional to Z^2/r and can be classified into groups based on their pK_b values (Section 2.6A). You can practice explaining the relationship of the basicities of these anions to their charges and anionic radii by trying Exercises 2.60–2.62. Periodic trends in basicities of monatomic and partially protonated anions are covered in Section 2.6B and can be practiced by trying Exercises 2.63–2.69. Given pK_b data, you can draw the predominance diagram of a monatomic anion and locate on it the partially and fully protonated (neutral) forms of the anion (Section 2.6C). You can practice these concepts by trying Exercises 2.70–2.74.

These concepts generally will be used in the same sections that the concepts of cation acidity are used (see Overviews for Sections 2.1 and 2.2). Predominance diagrams for anions generally will be employed in the same sections as predominance diagrams for cations (see the Overview for Section 2.5).

2.6A. Base Ionization Constants pK_b and Basicity Classifications. Just as metal cations do not exist "bare" in aqueous solutions, neither do anions. They also attract water molecules to form hydrated ions, although in this case the *positive* end of the dipolar water molecule—that is, one of the hydrogen atoms—is attracted to each of the unshared electron pairs on the anion, forming a hydrogen bond to the anion (Figure 2.6). Once again the interaction between an ion (in this case, an anion) and its waters of hydration frequently is great enough that one or more of the water molecules may be pulled apart. In this case the partially positively charged hydrogen atom of the water bonds to the anion, forming a (partially) protonated anion and releasing the remainder of the water molecule—a hydroxide ion—thereby producing a basic solution:

$$X^{y-} + H_2O \rightleftharpoons HX^{(y-1)-} + OH^- \tag{2.20}$$

TABLE 2.4. Aqueous Basicity of Monatomic Anions and Their Partially Protonated Forms

A. Monatomic Ions			
(C^{4-} very strong)[a]	N^{3-} very strong	O^{2-} very strong $pK_{b1} = -22$	F^- weak $pK_{b1} = 10.85$
(Si^{4-} very strong)[a]	P^{3-} very strong	S^{2-} strong $pK_{b1} = -4$[b]	Cl^- nonbasic $pK_{b1} = 20.3$
(Ge^{4-} very strong)[a]	As^{3-} very strong	Se^{2-} strong $pK_{b1} = -1$[b]	Br^- nonbasic $pK_{b1} = 22.7$
		Te^{2-} moderate $pK_{b1} = 3.0$[b]	I^- nonbasic $pK_{b1} = 23.3$
B. Partially Protonated Anions			
CH_3^- very strong $pK_{b4} = -30$	NH_2^- very strong $pK_{b3} = -25$	OH^- strong $pK_{b2} = -1.74$	
SiH_3^- very strong $pK_{b4} = -21$	PH_2^- very strong $pK_{b3} = -13$	SH^- moderate $pK_{b2} = 7.11$	
GeH_3^- very strong $pK_{b4} = -11$	AsH_2^- very strong $pK_{b3} = -9$	SeH^- weak $pK_{b2} = 10.3$	
		TeH^- feeble $pK_{b2} = 11.4$	

Sources: The pK_b values are calculated from pK_a values given in R. J. Myers, *J. Chem. Educ.* 63, 687 (1986); S. Licht, F. Forouzan, and K. Longo, *Anal. Chem.* 62, 1356 (1990); W. L. Jolly, *Modern Inorganic Chemistry*, McGraw-Hill: New York, 1984, p. 177; R. V. Dilts, *Analytical Chemistry*, Van Nostrand: New York, 1974, p. 553; and W. H. Nebergall, H. H. Holtzclaw, Jr., and W. R. Robinson, *General Chemistry*, 6th ed., Heath: Lexington, MA, 1980.

[a] Some "anions" such as Si^{4-} are listed in parentheses, because these cannot exist in any solution or the solid state. Many pK_b values in this table are too negative to be measurable in water, so they are estimated from values in other types of solutions.

[b] The pK_b values for S^{2-}, Se^{2-}, and Te^{2-} are subject to a lot of uncertainty because their equilibrium concentration is very low in water, and these anions are easily oxidized.

An equilibrium constant can be written for Equation 2.20, which is often called a base ionization constant and is symbolized by K_b. Table 2.4A lists the negative logarithms of these equilibrium constants, which are symbolized by pK_b or pK_{b1}.

We place *anions* in categories of *basicity* that match those used for the acidity of cations—we can think of these as categories of *acid–base reactivity*. Thus, we classify the simple anions that have no detectable basicity (pK_b values over 14) as **nonbasic anions**. Examples, which include Cl^-, Br^-, and I^-, have no tendency to combine with hydrogen ions in aqueous solution. This means that their basicity is so much less than that of water molecules that they cannot compete with water for hydrogen ions. (Later we will see other circumstances in which Lewis basic properties of the halide ions are evident.)

Anions that have pK_b values between 11.5 and 14 we call **feebly basic anions**; although there are no examples immediately at hand we will soon encounter some. The first anions with some commonly noticeable basicity, such as fluoride ion, have pK_b values between 6 and 11.5; we call these **weakly basic anions**. **Moderately basic anions** (e.g., Te^{2-}) have pK_b values between 1 and 6, whereas **strongly basic anions** (e.g., Se^{2-} and S^{2-}) have pK_b values between −4 and 1. **Very strongly basic anions**, such as the oxide ion, have pK_b values below −4. These anions react irreversibly with water, as shown in Equation 2.21.

$$O^{2-} + H_2O \rightarrow 2OH^- \tag{2.21}$$

AN AMPLIFICATION

Leveling Effect

Equation 2.21 demonstrates the *leveling* properties of very strong bases—that is, any (very strong) base that is stronger than the characteristic base of the solvent, OH^-, will react with the solvent (in this case, water) to generate its characteristic base (i.e., hydroxide ion). No base stronger than OH^- can persist in water. Likewise any (very strong) acid that is stronger than the characteristic acid of the solvent, H_3O^+, will react with water to generate its characteristic acid, hydronium ion. No acid stronger than H_3O^+ can persist in water.

Analogous to the acidities of cations, the basicities of anions increase strongly with increasing negative charge (i.e., from right to left in a period); the anions with charges of −3 or −4 are very strong bases that cannot persist in water. Additionally, the basicities of these anions decrease with increasing size (down a group). Both of these periodic trends match those found with cations. Thus, acid–base reactivity increases with increasing charge and decreasing radius. A plot of the pK_{b1} values in Table 2.4A versus their Z^2/r ratios is shown in Figure 2.16.

These data show a reasonable fit to Equation 2.22:

$$pK_{b1} \approx 29 - 1200Z^2/r \tag{2.22}$$

Comparison of Equation 2.22 with Equation 2.13 or comparison of Figure 2.16 with Figure 2.4 shows that the basicity of anions increases much more rapidly with

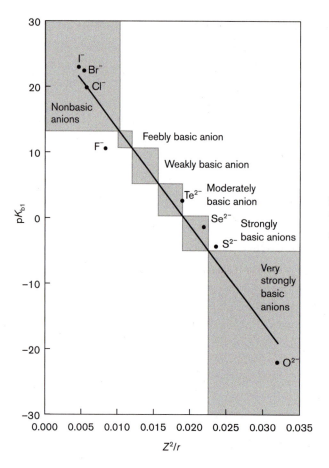

Figure 2.16. The pK_{b1} values (from Table 2.4) of monatomic nonmetal anions versus their Z^2/r ratios.

increasing Z^2/r than does the acidity of cations. As a result, the Z^2/r ranges used to classify the acidity of cations (e.g., < 0.01 = nonacidic) cannot be carried over without modification to classify the basicity of anions. We could select new Z^2/r ranges for the anions, but there are really too few monatomic anions to justify doing this. However, we can compare Z^2/r ratios of different monatomic anions to rank their relative basicities.

EXAMPLE 2.12

If possible, estimate the values of pK_{b1} and the basicity classifications of (a) Po^{2-} and (b) H^-.

SOLUTION: (a) Based on the data in Table C (going down Group 16/VI and also going to the left in Period 6), a reasonable radius for Po^{2-} would be about 230 pm. Then, Z^2/r would be 4/230 = 0.0174, and Equation 2.24 would yield pK_{b1} = 29 – (1200)(0.018) = 8.2. This anion should be weakly basic. (b) No ionic radius is given for hydride ion, H^-. We can only extrapolate that it should be smaller than F^- (119 pm) and therefore more basic (i.e., moderately basic or beyond).

2.6B. Periodic Trends in Basicity. There are corresponding periodic trends in acidity for cations and basicity for anions. In Figure 2.17 we indicate these reactivity classifications for a few anions and cations (the latter with charges equal to their group numbers). Notice that the most *nonreactive* (nonacidic, nonbasic) ions are found at the *outside bottom* of the periodic table. Thus, the further up and into the periodic table you go, the more reactive the ion is, as long as the charges on the ion increase with movement into the table.

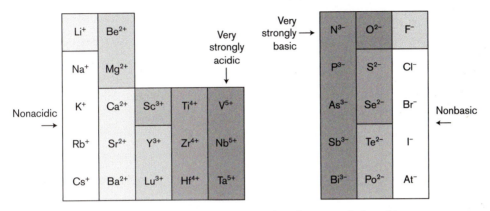

Figure 2.17. Reactivity (acidity, basicity) classifications of selected ions as a function of periodic table position. Unshaded = nonacidic or nonbasic; feebly shaded = feebly acidic or basic; shaded = weakly acidic or basic; moderately shaded = moderately acidic or basic; strongly shaded = strongly acidic or basic; very strongly shaded = very strongly acidic or basic.

EXAMPLE 2.13

Rank "C^{4-}," "Sn^{4-}," O^{2-}, and Te^{2-} in order of increasing basicity, and discuss periodic trends in their basicity.

SOLUTION: The Z^2/r ratios for O^{2-} and Te^{2-} are 0.031 and 0.019, respectively. Table C does not give the (hypothetical) anionic radii of "C^{4-}" or "Sn^{4-}," but from periodic trends we might extrapolate radii that are a little larger than those of O^{2-} and Te^{2-}, respectively. With $r = 138$ pm, Z^2/r for C^{4-} becomes 0.116; with $r = 188$ pm, Z^2/r for Sn^{4-} becomes 0.085. The predicted order of basicity is $Te^{2-} < O^{2-} \ll Sn^{4-} < C^{4-}$.

This order makes sense in terms of periodic trends. We expect dramatically greater basicity with greater charge, which puts the –4-charged anions at the head of the list. Of lesser significance is the effect of decreasing size to give increasing basicity, which puts "C^{4-}" ahead of "Sn^{4-}" and O^{2-} ahead of Te^{2-}. This reasoning could have allowed us to *rank* these anions without extrapolating the unknown radii.

2.6C. Partially Protonated Anions; Predominance Diagrams for Anions. The product of Equation 2.20 is a partially protonated anion, which can in turn undergo subsequent steps of reaction with water (see Eqs. 2.23 and 2.24). These are distinguished by the symbols pK_{b2}, pK_{b3}, and so on. Some measured or estimated values of these equilibrium constants are listed in Table 2.4B.

$$SH^- + H_2O \rightleftharpoons H_2S + OH^- \quad pK_{b2} = 7.11 \tag{2.23}$$

$$NH_2^- + H_2O \rightleftharpoons NH_3 + OH^- \quad pK_{b3} = -25 \text{ (estimated)} \tag{2.24}$$

The values listed in Table 2.4B are all for the last stage of base dissociation of the –1-charged partially protonated anions. These are less basic than the parent monatomic ions with multiple negative charges; even so, these anions cannot persist in water for the elements of Groups 14/IV and 15/V. Even with charges held constant, the same periodic trends suggested by Figure 2.17 still hold. Basicity increases from right to left and from bottom to top, with the partially protonated anions as well as the monatomic anions.

When constructing predominance diagrams for anions, we label them with a pH scale and a pOH scale, because the pOH scale is more directly related to the strength of bases.

$$pOH = -\log[OH^-] = 14 - pH \tag{2.25}$$

$$pOH = pK_b + \log \frac{[\text{conjugate acid}]}{[\text{conjugate base}]} \tag{2.26}$$

When the anion and its partially protonated form (conjugate base and its conjugate acid) are present in equal concentration, this equation simplifies to Equation 2.27.

$$pOH \text{ at left boundary of predominance range of basic species} = pK_b \tag{2.27}$$

Figure 2.11 (Section 2.4A) shows predominance diagrams for several of the monatomic anions, including their (partially) protonated forms. As illustrated by these anions, the *more strongly basic* the species, the more it is confined to an *extreme right* position in the predominance diagram. Because the nonbasic (neutral) anions Cl^-, Br^-, and I^- predominate at all pH values, they are not shown in Figure 2.11; neither are the very strongly basic anions, which do not predominate at any pH. Note that this *left* boundary for the *basic species* is also the *right* boundary of its *conjugate acid* species, for which Equation 2.17 applies.

Some of the partially protonated anions have predominance ranges that are confined to the *middle* of their predominance diagrams. These anions are *both* acidic and basic—they are **amphiprotic**—but they are not strong in either sense. The basic reaction of HS^-, for example, is given by Equation 2.23, whereas the acidic reaction is given by Equation 2.28.

$$HS^- + H_2O \rightleftharpoons H_3O^+ + S^{2-} \tag{2.28}$$

The relative positions of the predominance ranges suggest which property is more significant in aqueous solution. For example, dissolving the salt Na^+SH^- (which contains a nonacidic cation that will not affect the pH) will produce a solution that is *basic*, because the predominance range of SH^- (bottom of Figure 2.11) is mainly on the basic (right) side of the diagram.

In acid–base neutralization reactions, stronger acids predominate at the low pH side (left side) of the diagram, while stronger bases predominate at the high pH (right) side of the diagram. They tend to react to give species with overlapping predominance ranges closer to the center of the diagram. Consequently, *stronger acids* (such as HCl) and *stronger bases* (such as F⁻) tend to react to give *weaker acids* (such as HF) and *weaker bases* (such as Cl⁻) (Figure 2.12, Section 2.4D).

EXAMPLE 2.14

Some anions of new nonmetallic elements have been discovered. For each one, sketch a likely predominance diagram. (a) Mm⁻, which is nonbasic. (b) Nn⁻, which is moderately basic. (c) Pp²⁻, which is moderately basic. (d) QqH⁻, which is weakly basic and also weakly acidic.

SOLUTION: (a) A nonbasic species such as Mm⁻ will not be protonated at any pH, so it will predominate over the entire pH range (Figure 2.18a).

(b) Because Nn⁻ is moderately basic, its pK_b will be in the range of 1 to 6—let us say 3. The pOH at which it will stop predominating is therefore about 3—this is a pH of 11. Below pH 11 we will find the once-protonated species, HNn (Figure 2.18b).

(c) Because Pp²⁻ is also moderately basic, we can apply the same reasoning as for Nn⁻ and set up the left edge of its predominance range at a pH of about 11. To the left of this will be the once-protonated species, PpH⁻. Because PpH⁻ is still an anion, somewhere to the left of it there will be another predominance range for H₂Pp (Figure 2.18c).

(d) QqH⁻ is a partially protonated anion, so there must be a predominance range for Qq²⁻ somewhere to the right of its predominance range and a predominance range for H₂Qq somewhere to the left. Because QqH⁻ is a weak base, its pK_b is between 6 and 11.5—say 9. Thus, its left boundary is at pOH 9, which is pH 5. Because QqH⁻ is also a weak acid, its pK_a is also between 6 and 11.5—say 9. This is the approximate right pH boundary for its predominance range (Figure 2.18d).

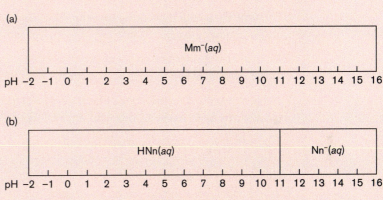

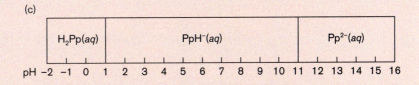

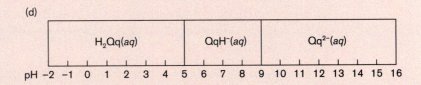

Figure 2.18. Answers for Example 2.14.

2.7. Background Reading for Chapter 2

Ion Hydration; Acidity of Cations; Basicity of Anions

Jack Barrett, *Inorganic Chemistry in Aqueous Solution*, Royal Society of Chemistry: Cambridge, UK, 2003, pp. 13–57 (Chapter 2: Liquid Water and the Hydration of Ions, and Chapter 3: Acids and Bases; Forms of Ions in Aqueous Solution; Ion Hydrolysis and Compound Solubility). Note that the Shannon–Prewitt ionic radii from Table C are not used, so conclusions can differ in detail.

John Burgess, *Ions in Solution: Basic Principles of Chemical Interactions*, 2nd ed., Horwood Publishing: Chichester, UK, 1999, pp. 62–73 (Chapter 5: Acid-Base Behavior: Hydrolysis and Polymerization).

Stephen Hawkes, "All Positive Ions Give Acid Solutions in Water," *J. Chem. Educ.* 73, 516–517 (1996).

M. Henry, J. P. Jolivet, and J. Livage, "Aqueous Chemistry of Metal Cations: Hydrolysis, Condensation, and Complexation," *Struct. Bond.* 77, 153–206 (1992). Advanced discussion of conditions governing the formation of solid hydroxides versus oxides. Discusses sol-gel synthesis of metal oxides.

Numbered references from this chapter may be viewed online at www.uscibooks.com /foundations.htm.

D. K. Nordstrom, C. N. Alpers, C. J. Ptacek, and D. W. Blowes, "Negative pH and Extremely Acidic Mine Waters from Iron Mountain, California," *Environ. Sci. Technol.* 34, 254–258 (2000).

2.8. Exercises

2.1. *Write the equilibrium-constant expressions and give the K_a values for the following cations: (a) oberlium cation (Ob^{2+}, $pK_a = 3$); (b) cathium cation (Ct^{4+}; $pK_a = -5.5$).

2.2. Write the equilibrium-constant expressions and give the pK_a values for the following cations: (a) hallium cation (Hl^{3+}, $K_a = 4.4 \times 10^{-5}$); (b) patelium cation (Pl^+; $K_a = 12.2$).

2.3. *Give two other metal cations that should react quite similarly to the each of the following cations in Experiment 2: (a) Li^+; (b) Zn^{2+}; (c) Al^{3+}; (d) Ti^{4+}.

2.4. A leak develops in an industrial tank of liquid standing above ground in an industrial district. Clouds of white, corrosive smoke pour from around the leak. (a) Suggest the possible contents of the tank, and explain what is happening to generate the smoke. (b) If you are the first responder, what should you do about this?

2.5. *Which category of metal (or nonmetal) cation: (a) gives a neutral solution in water; (b) gives a faintly acidic solution in water, but the acidity is masked by that due to dissolved carbon dioxide; (c) gives a weakly acidic solution (comparable in acidity to vinegar); (d) reacts reversibly with water to give a strongly acidic solution; (e) reacts irreversibly with water.

2.6. The pK_{a1} values of some new metal cations have been measured. Classify the acidity of each cation, and tell the approximate pH of the resulting solution. (a) Heltonium ion, $pK_a = 18.0$; (b) azharium, $pK_a = -18.0$; (c) lawhornium, $pK_a = 8.0$; (d) oyegbium, $pK_a = -3.0$; (e) robertsium, $pK_a = 3.0$.

2.7. *You have found three new elements, which you name in honor of your classmates: cochranium (Cc), nogium (Ng), and rominium (Ro). In the quantitative analysis laboratory you titrate their cations in order to determine their pK_a values: Cc cation, +12; Ng cation, –8; and Ro cation, +3. (a) Which of these three cations is the most acidic? (b) Which is the least acidic? (c) Assign each cation to an acidity category (very strongly acidic, etc.).

2.8. What charge is found on the majority of ions in Table 2.1 that fall into each of the following acidity categories: (a) nonacidic; (b) feebly acidic; (c) weakly acidic; (d) moderately acidic; (e) strongly acidic?

2.9. *Choose one of the acidity classifications (i.e., nonacidic, strongly acidic, or weakly acidic) and choose one of the pK_a values (i.e., between –4 and 1, over 14, or from 6 to 11.5) for *each* of the following element chlorides: (a) oxsherium chloride, which dissolves in water with some heat evolution to give a clear solution of pH 4; (b) chamberlainium chloride, which reacts violently with water to give a cloudy solution of pH 0; (c) mcphersium chloride, which dissolves in water with no change in pH and no temperature change to give a clear solution.

2.10. Choose one of the acidity classifications (i.e., nonacidic, strongly acidic, or weakly acidic) and choose one of the pK_a values (i.e., between –4 and 1, over 14, or from 6 to 11) for *each* of the chlorides of the following imaginary elements: (a) haltomium chloride, which dissolves in water with some heat evolution to give a clear solution of pH 3; (b) schwerium chloride, which reacts violently with water to give a cloudy solution of pH 0; (c) etheridgium chloride, which dissolves in water with no change in pH and no temperature change to give a clear solution.

2.11. *A 1 M solution containing just the neutral Cl^- anion and the patelium cation has a pH of 2.0. (a) What is the approximate pK_a of the cation? (b) What is the K_a of the cation? (c) What acidity category applies to this patelium cation?

2.12. You have found a new element, which you name lankfordium (Lf) in honor of your classmate. In the quantitative analysis laboratory you titrate the lankfordium cation in order to determine that its $pK_a = +8.1$. (a) Is lankfordium cation more or less acidic than the stoltzium cation, which has $pK_a = -8.1$? (b) Assign the lankfordium cation to an acidity category (very strongly acidic, etc.). (c) A 1 M solution of lankfordium cation will have roughly which pH: 1, 4, 8, or 13?

2.13. *A 1 M solution containing just the neutral Cl^- anion and the hawkinsium cation has a pH of 4.5. (a) What is the approximate pK_a of the cation? (b) What is the K_a of the cation? (c) What acidity category applies to this hawkinsium cation?

2.14. Titanium forms three cations, which are shown in Table C. Calculate Z^2/r and assign the acidity category of each.

2.15. *Assuming that the effects of their electronegativities are negligible, use charge/size ratios to classify the acidity of each of the cations listed in Table C for (a) chromium and (b) neptunium. Tell which fluorides of these cations would react violently and irreversibly with water for (c) chromium and (d) neptunium.

2.16. Which metal probably shows the *greatest* range of pK_a values and acidities among its known cations? Calculate these values and assign the acidities of these cations.

2.17. *Consider the following cations: U^{3+}, Ag^+, Pa^{5+}, C^{4+}, As^{3+}, Tl^+, and Th^{4+}. (a) Classify the acidity of each of these cations and describe the reactions of their chlorides with water. (b) Which of these would give "cloudiness" or precipitation upon dissolving in water? What could you do to rectify this if it occurred?

2.18. (a) Use Equations 2.13 and 2.14 to calculate pK_a values for each of the cations in Exercise 2.17, and determine whether any of the ions could be shifted to another category of acidity. (b) When possible, compare these values with the experimental pK_a values to determine whether any of the ions should be shifted to another category of acidity.

2.19. *(a) Calculate Z^2/r for each of the cations B^{3+}, Ba^{2+}, Ag^+, Th^{4+}, and As^{5+}. (b) Without considering electronegativity, classify the acidity (feebly acidic, etc.) of each of the cations. (c) Now reclassify each, including the effects of electronegativity. (d) Which two ions will have the most nearly neutral solutions? (e) Which ion(s) do not exist in this form in water under any conditions of pH?

2.20. (a) Calculate Z^2/r for each of the cations As^{3+}, Pb^{2+}, Rb^+, and N^{3+} (assume 40 pm radius for N^{3+}). (b) Without considering electronegativity, classify the acidity of these cations. (c) Now reclassify each, including the effects of electronegativity. (d) Which one of these will have the most nearly neutral solution? (e) Which cation does not exist in this form in water under any conditions of pH?

2.21. *(a) Classify the acidity of each of the cations Rb^+, La^{3+}, and P^{5+}. (b) The pK_a values of these ions are –33.1, 8.5, and 14.6, not necessarily in that order. Without actually calculating pK_a, show that you understand pK_a by assigning each pK_a value to the metal ion most likely to show it. (c) The pH values of some solutions of these ions are 0, 5.5, and 6.5. Without actually calculating pH, show that you understand pH by assigning each pH value to the appropriate metal ion. (d) Describe briefly what will happen when you try to dissolve soluble salts of each of these cations in water.

2.22. (a) Calculate Z^2/r for the cations Rg^+ and Sg^{4+}. Assume that, as a result of relativistic effects, the metallic radius of Rg is 155 pm, its cationic radius is 144 pm, and its Pauling electronegativity is 2.77; assume, too, that the metallic radius of Sg is 132 pm, its cationic radius is 81 pm, and its Pauling electronegativity is 1.68. (b) Without considering electronegativity, classify the acidity of these cations (weakly acidic, etc.). (c) Now reclassify each, including the effects of electronegativity.

2.23. *Referring to Tables A and C, and using approximate rules of classification rather than Equation 2.14, derive a list of (a) all nonacidic cations and (b) all feebly acidic cations.

2.24. (a) Referring to Table 2.3, compute the smallest and the largest radius that will allow a +1-charged ion to fit into each category of cation acidity. Then, referring to Table C (and the electronegativity table), list all +1 ions from Table C that fit into each of the categories. Describe the location in the periodic table of the ions of each category. (b) Do the same for +2-charged cations.

2.25. *Measuring the pH values of which of the following pairs of chlorides represents the best-designed test of the effects of size on acidity? Explain your choice. (a) $AlCl_3$ versus $TlCl_3$, or (b) $AlCl_3$ versus $PaCl_3$, or (c) $AlCl_3$ versus VCl_2.

2.26. Measuring the pH values of which of the following sets of ions represents the best-designed test of the effects of electronegativity on acidity? Explain your choice. (a) Ca^{2+} versus Fe^{2+} versus Mn^{2+} versus Zn^{2+}, or (b) Pr^{3+} versus Bi^{3+}, or (c) Tl^+ versus Ba^{2+}, or (d) Pr^{3+} versus P^{3+}.

2.27. *Measuring the pH values of which of the following sets of chlorides represents the best-designed test of the effects of charge on acidity? Explain your choice. (a) NaCl versus $CaCl_2$ versus $LaCl_3$, or (b) NaCl versus $HgCl_2$ versus $LaCl_3$.

2.28. What are the three main factors that influence the acidity and the hydration energy of a cation? Explain (in terms of Coulombic and/or covalent interactions) why each factor influences the acidity of a cation the way that it does.

2.29. *(a) Use the Latimer equation to estimate the hydration energies of the cations Na^+, Ca^{2+}, La^{3+}, and Hg^{2+}. (b) Compare these values with the experimental values listed in Table 2.3. (c) For which of these cations is the discrepancy the greatest? Why?

2.30. (a) Calculate Z^2/r for each of the cations Ga^{3+}, Zr^{4+}, Rg^+, and Sg^{4+}. (Assume the following radii for Rg^+: metallic 155 pm; cationic 144 pm. Assume the following radii for Sg^{4+}: metallic 132 pm; cationic 81 pm.) (b) Without considering electronegativity, classify the acidity of these cations (weakly acidic, etc.) (c) Now reclassify each, including the effects of electronegativity. (d) Which of these (gaseous) cations releases the least energy when it forms a hydrated ion? (e) Which of these (gaseous) cations releases the most energy when it forms a hydrated ion?

2.31. *What +5-charged "cation" listed in Table C would be expected to have the lowest hydration energy? Estimate this energy. Can this hydrated cation exist in water?

2.32. The earlier f-block cations (La^{3+} through Nd^{3+}) give nonahydrated cations $[M(H_2O)_9]^{3+}$, while the later f-block cations (Eu^{3+} through Yb^{3+}) give octahydrated cations $[M(H_2O)_8]^{3+}$. Explain why this trend holds rather than the reverse.

2.33. *Suppose that we have two new cations of about equal electronegativity. The grecium ion has twice the positive charge of the lokitium ion, and also has twice the radius of the lokitium ion. Which ion should have the highest hydration energy? Which ion should have the highest acidity? Why?

2.34. Explain briefly what the concept of the maximum coordination number of an element is. Tell what this number is for each period of the periodic table (starting with Period 2). (b) Briefly explain why BCl_3 and $TiCl_4$ are more likely to fume in tropical air than is CCl_4.

2.35. *Describe briefly what will happen when you try to dissolve the following compounds in water: (a) KCl; (b) $NbCl_5$; (c) $AlBr_3$; (d) CBr_4; (e) BaI_2; (f) IF_7. Which of these would be likely to fume in air and why?

2.36. Explain briefly why WCl_6 fumes in humid air, but SF_6 does not.

2.37. *In each of the following pairs of compounds, choose the one which would react with water much more rapidly (or tell if both would react at similar rates). (a) UF_6 or SF_6; (b) $COCl_2$ or CCl_4; (c) SO_2F_2 or SF_6. Give the name of the concept we use to predict these results; use it to explain your answers.

2.38. The halides $AlBr_3$, CBr_4, PCl_5, and $ClSF_5$ (S is central atom) are all calculated to contain very strongly acidic cations. (a) Which two of them do *not* react with water to give an acidic solution? (b) Explain why it is these two.

2.39. *What form (species) of H_2Te is predominant at the following pH values: (a) 1; (b) 8; (c) 13?

2.40. You dissolve $H_2Se(g)$ in a solution buffered to pH 5.00. (a) What is the predominant form of $H_2Se(aq)$ that forms in the solution? (b) Are the concentrations of the other two forms negligible?

2.41. *The imaginary Group 15/V element hoffmium (Hm) forms an acidic hydride. Its complete predominance diagram is as follows:

(a) Fill in the compounds and ions of hoffmium in the diagram. (a) What are the pK_a values of these compounds and ions? (b) What are the acidity classifications of these compounds and ions? (c) Name these compounds and ions.

2.42. Use the given predominance diagrams of two hypothetical hydrides, H_2By and H_2Ks, to answer the following questions:

$H_2By(aq)$	$HBy^-(aq)$	$By^{2-}(aq)$

$H_2Ks(aq)$	$HKs^-(aq)$

pH 0   2   4   6   8   10   12   14

(a) What is the pK_a of H_2By? (b) What is the pK_a of HKs^-? (c) What is the acidity classification of H_2By? (d) What is the acidity classification of HKs^-? (e) If By is bydekium and Ks is klassium, name these species.

2.43. *List a plausible product when 1 mol of solids containing each of the following ions or substances are stirred together in 1 L of water: (a) H_2S and Te^{2-}; (b) HCl and S^{2-}; (c) HTe^- and Se^{2-}.

2.44. Use the predominance diagrams in the text to predict whether the following molecules and ions will react with each other. If so, predict the predominant products when the mixing results in the final pH mentioned. (a) H_2Te and HS^-, pH of 6.0; (b) HTe^- and HS^-, pH of 10.0.

2.45. *Based on Figure 2.7 and your answers to Exercise 2.42, predict whether the following molecules and ions will react with each other. If so, predict plausible products. (a) H_2By + HKs^-; (b) HBy^- + HKs^-; (c) $HC_2H_3O_2$ + HBy^-.

2.46. (a) Give the names of the hydrides SiH_4, PH_3, and HBr. (b) Which of these hydrides is most acidic?

2.47. *(a) Name the anions S^{2-}, NH_2^-, SH^-, and N^{3-}. (b) Write the formulas of the salts of these anions with the La^{3+} cation.

2.48. Write the formulas of and name the –1-charged partially protonated forms of the anions C^{4-}, Ge^{4-}, O^{2-}, and Se^{2-}.

2.49. *Write the formulas of the Ca^{2+} salts of the anions NH_2^-, SeH^-, P^{3-}, and I^-. (b) Write the names of the Fe^{3+} salts of the same anions.

2.50. Use the given acid–base predominance diagrams of two cations to answer the following questions:

Zhaoium ion	Zhaoium hydroxide

Weatherlium ion	Weatherlium hydroxide

pH 0   2   4   6   8   10   12   14

What is the pK_a of zhaoium ion? (b) Which of the two cations is the most acidic? (c) What is the acidity classification of the weatherlium ion? (d) What is the acidity classification of the zhaoium ion? (e) Adjusting the pH to 11 will cause which solution or solutions to turn opaque (cloudy)? (f) The pH of a 1.0 M solution of zhaoium chloride should be about what? (g) If you dissolve some zhaoium chloride in water and the solution turns cloudy, which of the following treatments is most likely to give a clear solution? (1) Add more water; (2) add HCl; (3) add NaOH; or (4) nothing, because the reaction with water is irreversible.

2.51. *Choose the acidity category to which each cation belongs, and give its approximate pK_{a1} value.

(a)

| Cation(*aq*) | Oxide(*s*) |

(b)

| Cation(*aq*) | Hydroxide(*s*) |

(c)

| Cation(*aq*) |

(d)

| Oxide(*s*) |

pH   −4   −2   0   2   4   6   8   10   12   14   16

2.52. Surprisingly, the Bi^+ ion has been found in a solid compound. (a) Estimate its radius by extrapolation from known ionic radii for bismuth. Use this to compute its Z^2/r value, its acidity classification, and its pK_a value. (b) If suggested by your instructor, refine your pK_a value by using Equation 2.14. (c) Draw the acid–base predominance diagram you would expect for this cation. Label the pH range and the ranges in the diagram.

2.53. *The following cations have the approximate pK_a values listed here: Pb^{4+}, −7.4; Pb^{2+}, 7.7; and Yb^{2+}, 12.1. Use the unlabeled predominance diagrams provided to answer the following questions:

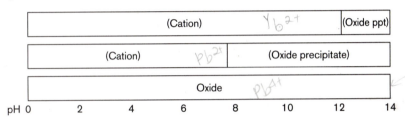

(a) Assign each predominance diagram to one of the above ions.
(b) Which cation does not exist as a cation in water under any conditions of pH?

2.54. The following cations have the approximate pK_a values listed here: Bi^{3+}, 1.1; Bi^{5+}, −13.8; K^+, 14.5; Eu^{2+}, 12.5. Use the unlabeled predominance diagrams provided to answer the following questions.

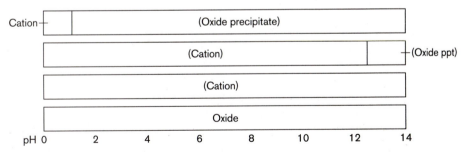

(a) Assign each predominance diagram to one of the above ions.
(b) Which two of these will have the most nearly neutral solutions? (c) Which cation does not exist in this form in water under any conditions of pH?

2.55. *Assign each of the unlabeled predominance diagrams provided to one of the following elements in the specified oxidation state: As(III), Pb(II), Rb(I), and N(III).

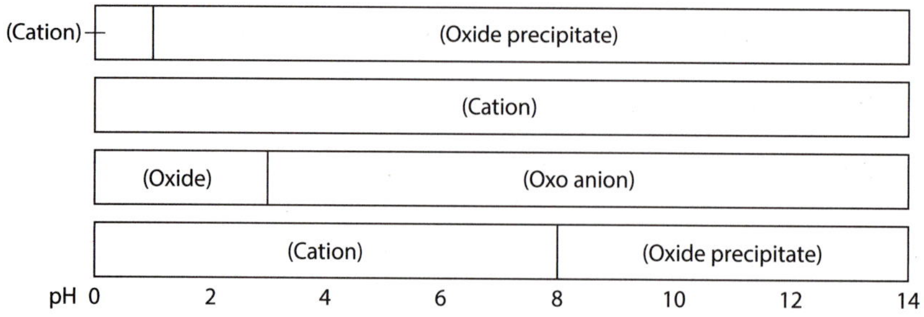

2.56. Use the pK_a data in Table 2.1 for the ions Au^{3+}, Sc^{3+}, Tl^+, and Co^{2+} to (a) draw predominance diagrams including these cations; (b) estimate the pH above which their hydroxides would precipitate; and (c) select the most acidic of these cations.

2.57. *You are following up your work in Exercise 2.7 on the three new elements, Cc, Ng, and Ro. (a) Assign three of the four unlabeled predominance diagrams provided to the appropriate new elements. (b) Tell how you can dissolve the oxide precipitate shown in the first diagram. (c) Tell how you can dissolve the oxide precipitate shown in the fourth diagram.

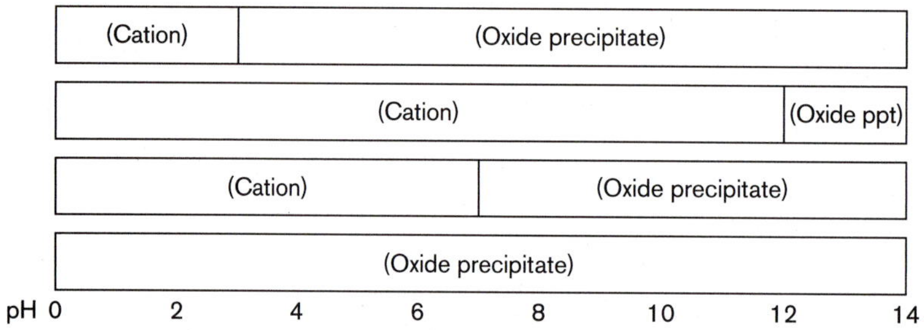

2.58. Classify the acidity (feebly acidic, etc.) of each of the cations U^{4+}, Pd^{4+}, K^+, and Ag^+. Show the calculations you use to carry out the classifications. (b) Which one of these ions would most likely react with water to give a solution of pH about 0? (c) Which one of these ions would most likely have a pK_a of about 14? (d) The hydroxides of which (one or more) of these ions would be insoluble in a solution of pH 7? (e) If you ordered and received sealed bottles of the chlorides of these cations, which should you be most reluctant to open in humid air?

2.59. *For each of the six categories of metal-ion acidity, tell whether the hydroxides or oxides of metal cations of that category would be soluble or insoluble in buffered solutions of pH: (a) 0; (b) 7; (c) 14.

2.60. These elements form anions, the basicity of which has not been studied. Predict the basicity classification of each of the following anions: (a) the Au^- ion (radius = 188 pm); (b) the anion of Po; and (c) the anion of Bi. For the last two anions, use periodic trends to estimate expected charges and radii.

2.61. *Rank the following imaginary nonmetal anions in order of increasing basicity: Aa^{3-} (radius = 140 pm), Bb^- (radius = 330 pm), Cc^{2-} (radius = 220 pm), Dd^- (radius = 170 pm), and Ee^{3-} (radius = 355 pm).

2.62. Fill in each blank with one of the following words: INCREASES, DECREASES, POSITIVE, NEGATIVE, OXYGEN, HYDROGEN, METAL. (a) The acidity of a cation _____ as its charge increases, because the _____ charge of the cation attracts the _____ end of the water molecule. (b) The basicity of an anion _____ as its charge increases, because the _____ charge of the anion attracts the _____ end of the water molecule. (c) The acidity of a cation _____ as its radius increases, because the charge of the cation is then further away from the _____ atom of the water molecule. (d) The acidity of a cation _____ as its electronegativity increases. When this happens, the covalency of the cation–oxygen bond _____, which allows more electrons to flow from the _____ atom to the _____ atom.

2.63. *(a) Which one of the four anions CH_3^-, C^{4-}, F^-, and I^- is the least basic? (b) Which is the most strongly basic?

2.64. Which of the anions S^{2-}, NH_2^-, SH^-, N^{3-}, and I^- are very strongly basic? (b) Which are nonbasic?

2.65. *What do a *feebly acidic* cation and a *feebly basic* anion have in common?

2.66. The two reactions provided here cannot be run in water because the anions are very strongly basic, but they could perhaps be run in some other solvent. Using basic-strength trends, predict whether reactants or products are favored in each of these equilibria:
(a) $CH_3^- + NH_3 \rightleftharpoons CH_4 + NH_2^-$; (b) $CH_3^- + SnH_4 \rightleftharpoons SnH_3^- + CH_4$.

2.67. *Use the anions C^{4-}, Ge^{4-}, O^{2-}, Se^{2-}, F^-, and Br^- to answer the following questions: (a) Which one is the weakest base? (b) Which of these are too strongly basic to exist in water? (c) Two of these anions can give partially protonated anions that can exist in water. Write the formulas (including charges) of the two water-stable partially protonated anions.

2.68. Use the predominance diagrams of two monatomic cations and two monatomic anions to answer the following questions.

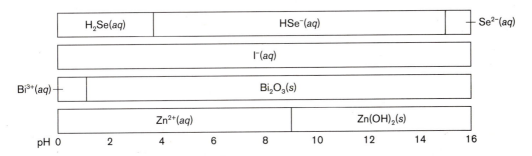

(a) What is the pK_b of $Se^{2-}(aq)$? (b) What is the pK_a of $Bi^{3+}(aq)$? (c) What is the basicity classification of $I^-(aq)$? (d) What is the acidity classification of $Zn^{2+}(aq)$? (e) Which is the strongest acid, H_2Se or HI or Bi^{3+} or Zn^{2+}? (f) Which is the strongest base, Se^{2-} or I^- or Bi_2O_3 or Zn^{2+}?

2.69. *Use the periodic table of nonmetal atoms with disguised symbols to answer the following questions. These elements form monoatomic anions.

	Group 15/V	Group 16/VI	Group 17/VII
Period 2	Vo	Su	Sh
Period 3	Gr	Go	Gb
Period 4	Da	Ch	Bw

(a) What are the charges expected for each anion? (b) Which is the most basic of these anions? (c) Which is the least basic of these anions?

2.70. Assign each of the unlabeled predominance diagrams provided to one of the following elements in the specified oxidation state: As(–III), Te(–II), F (–I), and I(–I).

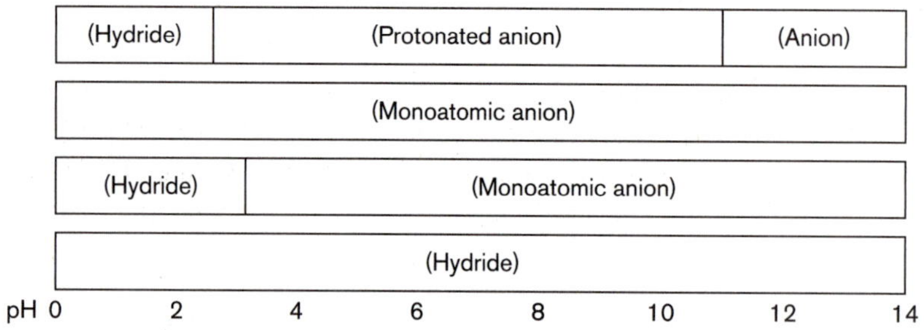

(Hydride)	(Protonated anion)	(Anion)

(Monoatomic anion)

(Hydride)	(Monoatomic anion)

(Hydride)

pH 0   2   4   6   8   10   12   14

2.71. *In the quantitative analysis laboratory you are titrating four unidentified anions in order to find their pK_{b1} values. The ion names and their pK_{b1} values are beckwithide ion = –6, campbellide ion = +10, gallianide ion = 0, and udezide ion = +18. (a) Which of these four anions is the most basic? Which is the least basic? (b) Assign each anion to a basicity category. (c) Which of these four (gaseous) anions will release the most energy when it forms a hydrated ion? Which will release the least? (d) If it later turns out that these four ions are in the same group of the periodic table, which one would you expect to find at the top of the group? Which would you expect to find at the bottom? (e) Match each of the four anions to the correct predominance diagram:

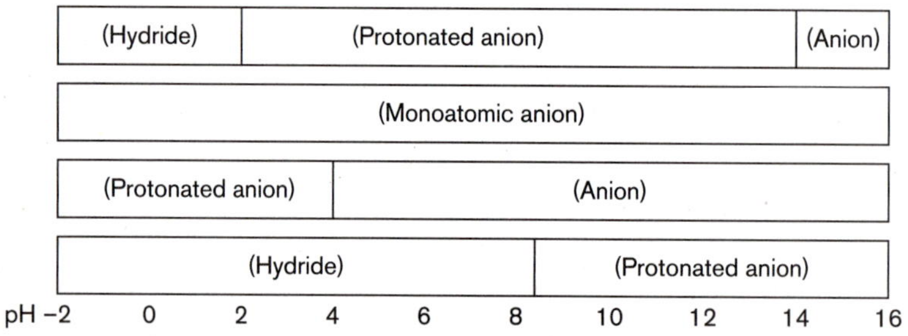

(Hydride)	(Protonated anion)	(Anion)

(Monoatomic anion)

(Protonated anion)	(Anion)

(Hydride)	(Protonated anion)

pH –2   0   2   4   6   8   10   12   14   16

2.72. In the quantitative analysis laboratory you are titrating three unidentified anions in order to find their pK_{b1} values. As their discoverer, you tentatively name these ions in honor of your classmates. The ion names and their pK_{b1} values are pfalmide ion, a –1-charged anion with pK_{b1} = +12; whealide ion, a –2-charged anion with pK_{b1} = +22; and hooperide ion, a –3-charged anion with pK_{b1} = –9. (a) Assign each anion to a basicity category (very strongly basic, etc.). (b) Assign each anion to one of the four acid–base predominance diagrams provided.

(Hydride)	(Protonated anion)	(Anion)

(Monoatomic anion)

(Prot. anion)	(Anion)

(Hydride)	(Protonated anion)

pH –2   0   2   4   6   8   10   12   14   16

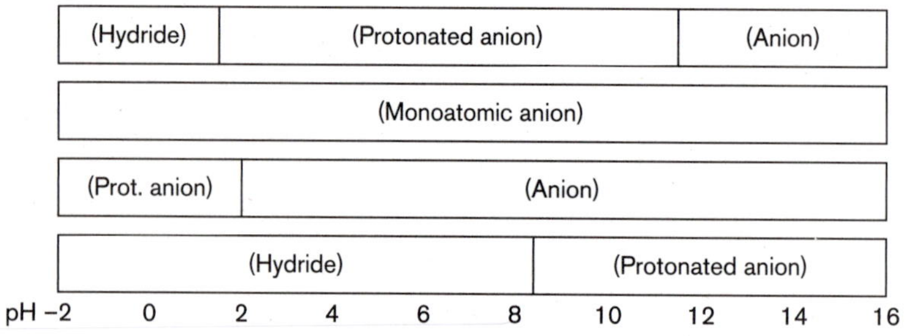

2.73. *Some anions of new nonmetallic elements have been discovered. For each one, sketch a likely predominance diagram. (a) Ff^{2-}, which is weakly basic; (b) GgH^-, which is feebly basic and also feebly acidic; (c) Jj^{3-}, which is very strongly basic; (d) Kk^-, which is moderately basic.

2.74. Use the predominance diagrams of two hypothetical monatomic anions, bydalikide (By^{2-}) and klasside (Kl^{2-}), to answer the following questions: (a) What is the pK_b of $By^{2-}(aq)$? (b) What is the pK_b of $HKs^-(aq)$? (c) What is the basicity classification of $By^{2-}(aq)$? (d) What is the basicity classification of $Ks^{2-}(aq)$? (e) Which of these anions has the greatest *hydration energy*?

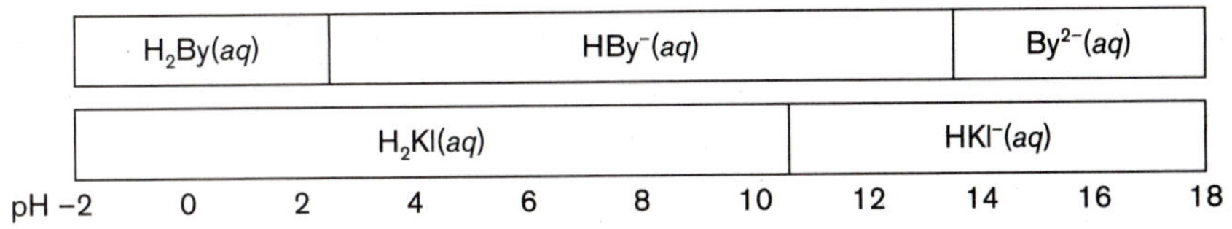

$H_2By(aq)$	$HBy^-(aq)$	$By^{2-}(aq)$

$H_2Kl(aq)$	$HKl^-(aq)$

pH −2   0   2   4   6   8   10   12   14   16   18

GILBERT NEWTON LEWIS (1875–1946) was born in Weymouth, Massachusetts. He received a bachelor's degree (1896), a master's degree (1898), and a doctorate (1899), all in chemistry and all from Harvard University. His dissertation was on the electrochemistry of zinc and cadmium amalgams. After graduation, he spent a year in the laboratories of Wilhelm Ostwald and Walther Nernst in Germany, three years as an instructor at Harvard, another year in the Philippines as the Superintendent of Weights and Measures, then seven years on the faculty at the Massachusetts Institute of Technology. In 1912, Lewis was made Dean of the College of Chemistry at the University of California at Berkeley, which he developed into one of the finest teaching and research departments in the world. Most undergraduate chemistry students know Lewis from the electron dot diagrams that he developed to symbolize the electronic structures of atoms and molecules, from his description of the covalent bond as a sharing of electrons, and from his description of acids as electron-pair acceptors and bases as electron-pair donors. Additionally, he made enormous contributions to the field of chemical thermodynamics, including the concepts of fugacity, activity coefficients, and ionic strength. Despite these and many other contributions to chemistry, he never won the Nobel Prize.

CHAPTER 3

Polyatomic Ions
Their Structures and Acid–Base Properties

Overview of the Chapter

Here in Chapter 3, we focus on the structures and acid–base chemistry of polyatomic anions and complexes, with special emphasis on the **oxo anions**. In this book we seek periodic trends and their applications. Where is the periodicity among the structures and basicities of oxo anions? In Group 15/V, for example, nitrate (NO_3^-) is nonbasic (neutral), while phosphate (PO_4^{3-}) has a different structure and basicity. There is periodicity behind the structures and charges of oxo anions (Section 3.5) and in their nomenclature (Section 3.9), too. These oxo anions are categorized into familiar basicity classes in Section 3.6. In Section 3.8 we draw together information on the acidity of cations from Chapter 2 and basicity of oxo anions and their protonated forms, and construct predominance diagrams for elements in specified oxidation states in water. From these diagrams we can make preliminary predictions of the soluble or insoluble form that is found for that element in water at a specified pH—for example, corresponding to a pristine or polluted natural water, or the body. In Chapter 4 we begin applying these concepts to the patterns of solubility of salts of oxo anions (which are very different for phosphates and nitrates). We stand upon the results of these patterns: Phosphates, carbonates, and silicates are found in important minerals in Earth's crust.

3.1. Drawing Lewis Structures (Review); Homopolyatomic Ions

OVERVIEW. The formulas of polyatomic ions such as the homopolyatomic ions of Section 3.1A are often enclosed in parentheses to distinguish them from monatomic ions. This is not the usual practice when only one polyatomic ion is present, so their presence or absence must be deduced using the principles of Section 1.4. You may practice this concept by trying Exercises 3.1–3.5. The Lewis structure of a molecule or polyatomic ion shows an arrangement of bonding and nonbonding (unshared) electron pairs. It is built using each atom's valence electrons, and adding one electron for each negative charge of an anion, or deleting one electron for each positive charge of a cation. You may review this concept of Section 3.1B by trying Exercises 3.7–3.11. Drawing Lewis structures is vital later in this chapter, in organic chemistry, and in Sections 5.1, 5.2, 5.3, 5.4, 7.1, 7.9, 8.5, and 8.6.

3.1A. Homopolyatomic Ions. Ions that contain more than one atom, all of the same element, are called **homopolyatomic ions**. The most common example from Group 17/VII is I_3^-, the triiodide ion.

Homopolyatomic anions are more common in Group 16/VI. Sulfur forms a number of them, the most common of which is S_2^{2-}, the disulfide ion. The tetrasulfide ion S_4^{2-} is moderately basic, with $pK_{b1} = 3.8$ and $pK_{b2} = 6.3$. Two important homopolyatomic anions of oxygen are formed during the chemical or biochemical reduction of elemental O_2. The first one formed is the **superoxide ion** O_2^-, a **radical anion**. It has an unpaired electron, which causes it to be very reactive. It causes such serious cell damage that an enzyme, superoxide dismutase, is needed to catalyze its decomposition. Addition of another electron gives rise to O_2^{2-}, the **peroxide ion**, a moderately basic anion with $pK_{b1} = 3.38$. (Upon protonation, O_2^{2-} becomes hydrogen peroxide, H_2O_2.)

In Group 15/V the most common polyatomic anion is the **azide** ion, N_3^-. The azide ion is weakly basic ($pK_b = 9.4$), very toxic (Section 5.5), and very reactive in oxidation–reduction reactions (Section 6.3). Upon protonation it becomes the very explosive compound hydrazoic acid, HN_3.

In Group 14/IV the monatomic carbide ion, C^{4-}, is so strongly basic that it is only found in a few quite covalent derivatives (WC, Be_2C, and Al_4C_3). More common is the **acetylide** ion, C_2^{2-}, which on protonation gives acetylene or ethyne, $H-C\equiv C-H$. Many salts of this anion are known, some of which are explosive (Section 6.3). The best-known salt is calcium "carbide," CaC_2. There is also a triatomic anion, C_3^{4-}, which on protonation gives propyne, $H_3C-C\equiv CH$.

Finally, the most common polyatomic cation, Hg_2^{2+}, has a Hg–Hg single bond. The oxidation number of mercury in Hg_2^{2+} is +1, so it is most commonly named the mercury(I) cation.

The nomenclature of these anions is embarrassingly unsystematic. A more significant problem is distinguishing monatomic and homopolyatomic ions in chemical formulas. If more than one of them is present in the compound, the polyatomic ion is enclosed in parentheses—for example, $(Hg_2)_2SiO_4$ [mercury(I) silicate]. If only *one* polyatomic ion is present, however, then the parentheses or brackets are usually omitted—for example, Hg_2SO_4 [mercury(I) sulfate]. In some cases you can use the principles of Section 1.4 to deduce whether the nonmetal anion is monatomic or homopolyatomic.

EXAMPLE 3.1

Identify the "carbide" ions present in the Group 2 carbides Be_2C, Mg_2C_3, and CaC_2.

SOLUTION: In each case the Group 2 metal ion is first assigned a +2 charge. Because the compounds are neutral overall, the anion in Be_2C must be C^{4-}. In Mg_2C_3 the three carbons must together have a −4 total charge. This is impossible with three anions of −4 charge, or 1.5 anions of −2 charge; the anion must instead be one C_3^{4-} ion. Similarly, the only possible identity for the carbon anion in CaC_2 is C_2^{2-}.

AN AMPLIFICATION

Other Salts of Homopolyatomic Anions

If the cation is not from Group 1, Group 2, or (usually) a type of Group 3, we cannot apply this procedure, and the anion may be difficult or impossible to identify from the formula. Consider the two compounds FeS_2 (pyrite or fool's gold) and MoS_2 (the solid-state lubricant found in WD-40™). The first compound could be iron(IV) sulfide, $(Fe^{4+})(S^{2-})_2$, or it could be iron(II) disulfide, $(Fe^{2+})(S_2^{2-})$. The same two possibilities exist for the molybdenum sulfides. According to Table C, molybdenum and iron both have +2 and +4 oxidation states, so we cannot rule out either formulation. It then becomes necessary to obtain physical evidence to determine which structure is present. The most obvious method is to determine the crystal structure to determine whether S–S bonds are present or absent. Such a determination shows that S–S bonds are present in fool's gold, which is therefore iron(II) disulfide, $(Fe^{2+})(S_2^{2-})$. But normal-length S–S bonds are absent in MoS_2, which is therefore molybdenum(IV) sulfide, $(Mo^{4+})(S^{2-})_2$.

3.1B. Drawing Lewis Structures of *p*-Block Molecules and Ions (Review). The rules for drawing Lewis structures given here should be similar or at least equivalent to those you learned in general chemistry. However, we apply them to some more complexes and ions. The flow chart in Figure 3.1 summarizes the steps.

1. Count the total number of valence electrons in the molecule or ion (Section 1.3B). To do this, add up the group numbers for each atom in the molecule. Then add one extra electron for each unit of negative charge, or deduct one electron for each unit of positive charge.

2. Draw a skeleton structure of the molecule or ion, joining the **central atom** to the **outer** or **terminal** atoms with single bonds. There are some conventions in writing formulas that help in identifying central and outer atoms. (a) Hydrogen atoms normally form only one bond, so must be outer atoms. (b) If the species has a formula AB_n, the unique A atom is normally

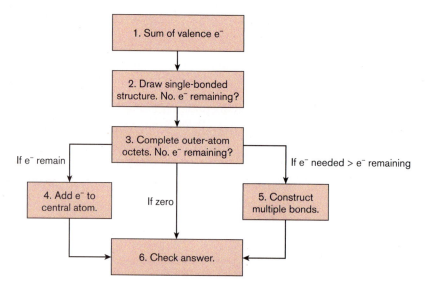

Figure 3.1. Flow chart for drawing Lewis structures.

the central atom; to it are bonded *n* B outer atoms. (c) Often, but not always, if three atoms are written in order, for example SCN⁻, the middle atom can be taken as the central atom.

Now deduct two electrons for each bond drawn from the total number of valence electrons (Step 1) to obtain the number of electrons still available to complete the Lewis structure.

3. Complete the octets of non-hydrogen outer atoms with unshared electron pairs. Deduct two for each such pair to see how many of the available electrons (Step 2) still remain to be located. If *no* electrons remain and *no* electrons are needed to complete central-atom octets, proceed to step 6.

4. If electrons remain (Step 3), use them to complete the octet of the central atom. If electrons *still* remain, put the extra electrons on the central atom, which will then exceed an octet of electrons. (In this case, the central atom must be from the third period of the periodic table or below; for explanations on how this can be, see "Postvalence Orbitals in Bonding?—An Amplification" in Section 3.3B.)

5. If the central atom still needs electrons to complete its octet, but no electrons remain (Step 3), move unshared electron pair(s) from outer atom(s) so as to connect the central and the outer atom(s) with double (or triple) bond(s).

6. Finally, check your answer! Be sure that (a) the total number of electrons shown agrees with the number calculated in Step 1; (b) each atom has an octet of electrons (except hydrogen atoms with two, and, if necessary, central atoms from the third period or below, which may have more than eight).

Lewis structures can be refined using the concepts of resonance structures and formal charges, but for most of our uses of Lewis structures these refinements are not mandatory. The concepts of resonance structures and formal charges are given as Amplifications at the end of this section.

We begin with an example in which we draw the Lewis structures of some homopolyatomic anions.

EXAMPLE 3.2

Draw Lewis structures of the following homopolyatomic anions:
(a) O_2^{2-}; (b) O_2^-; (c) linear N_3^-; (d) linear I_3^-.

SOLUTION: N_3^- and I_3^- are linear triatomic molecules with central atoms, which require Steps 4 and 5. O_2^{2-} and O_2^- are diatomics, which do not need Steps 4 and 5, so we draw their Lewis structures first.

In Step 1 we count the total number of valence electrons as follows. (a) Two oxygen atoms plus two negative charges gives us $2(6) + 2 = 14$ valence electrons. (b) $2(6) + 1 = 13$ valence electrons.

In Step 2 we draw single bonds connecting the atoms: (a) O–O; (b) O–O. We deduct two electrons for each bond drawn to obtain the number of electrons still available to complete the Lewis structures. In (a) this is $14 - 2 = 12$; in (b) this is $13 - 2 = 11$.

In Step 3 for (a), each outer atom needs six electrons to complete an octet, which requires 12 electrons total. (See the top row of Figure 3.2.) In (b) 12 electrons are needed, but only 11 are available. We cannot draw electrons we don't have, so we place only five electrons on one of the oxygen atoms, resulting in a radical anion.

Ions (a) and (b) have no central atoms, so Steps 4 and 5 do not apply; these ions are ready for Step 6, verifying the structures, which we leave to you.

Turning now to N_3^- and I_3^-, in Step 1 we count (c) $3(5) + 1 = 16$ valence electrons and (d) $3(7) + 1 = 22$ valence electrons. In Step 2 we draw single bonds connecting the atoms: (c) N–N–N and (d) I–I–I. We calculate the numbers of valence electrons still available to complete the Lewis structures; in (c) this is $16 - 4 = 12$; in (d) this is $22 - 4 = 18$.

In Step 3 for (c) four electrons are needed to complete the octet of the central atom, but none remain, so we proceed to Step 5. In (d) six electrons remain, so we proceed to Step 4.

In Step 4 for (d), we place the six remaining electrons on the central I atom (which then exceeds an octet—but this is okay because it is from the third period or below). This structure is complete (bottom row of Figure 3.2).

In Step 5 for (c), the central N atom needs four electrons but none are available. We use two lone pairs on two outer nitrogen atoms to make two N=N double bonds. (We could have used two lone pairs from one outer nitrogen atom to make one N≡N triple bond, which is an alternate resonance structure—see the Amplification.)

We leave it to you to verify the structures by applying Step 6.

Example 3.2 continues on next page ▶

EXAMPLE 3.2 *(cont.)*

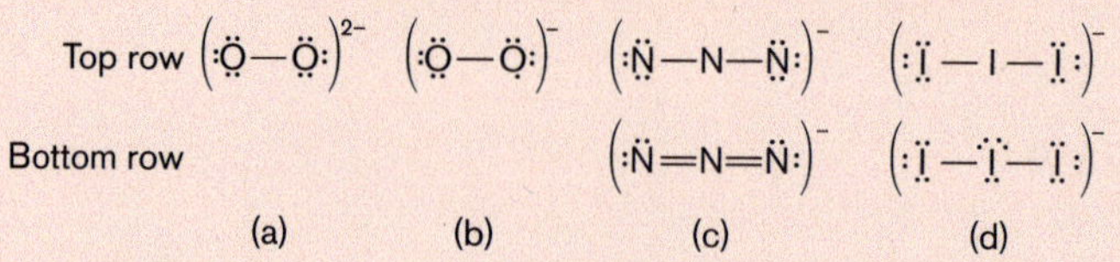

Top row $\left(:\!\ddot{O}\!-\!\ddot{O}\!:\right)^{2-}$ $\left(:\!\ddot{O}\!-\!\ddot{O}\!:\right)^{-}$ $\left(:\!\ddot{N}\!-\!N\!-\!\ddot{N}\!:\right)^{-}$ $\left(:\!\ddot{I}\!-\!I\!-\!\ddot{I}\!:\right)^{-}$

Bottom row $\left(:\!\ddot{N}\!=\!N\!=\!\ddot{N}\!:\right)^{-}$ $\left(:\!\ddot{I}\!-\!\ddot{I}\!-\!\ddot{I}\!:\right)^{-}$

(a)   (b)   (c)   (d)

Figure 3.2. Steps in drawing Lewis structures of polyatomic anions in Example 3.2. Note that we generally enclose the Lewis structures of ions in parentheses or brackets, with the charge falling outside the right parenthesis or bracket.

In Example 3.3, more complex molecules and ions are considered. For these we suggest skeleton formulas.

EXAMPLE 3.3

Draw the Lewis structures of the following polyatomic molecules or ions: (a) cyclo-$(HBNH)_3$, (b) $[O_3S\text{–}S\text{–}S\text{–}SO_3]^{2-}$, and (c) NSF_3.

SOLUTION: (a) Step 1: There are $3(1 + 3 + 5 + 1) = 30$ valence electrons. Step 2: The skeleton structures that result from this step are shown at the top of Figure 3.3. The prefix "cyclo–" implies that the three HBNH units form a ring, but hydrogen atoms form only one bond, so they must be external to the ring, which involves alternate B and N atoms. $30 - 24 = 6$ electrons are still available to complete the Lewis structure. Step 3: There is no change, because hydrogen atoms do not take octets. Step 5: Three more pairs are needed than are available; these can be shared between adjacent B and N atoms as shown in the middle of Figure 3.3a. Boron is not a good double-bonding atom (Section 3.3C), however, so we may prefer the structure without double bonds, in which the electron pairs are only on the more electronegative N atoms (bottom row of Figure 3.3a).

(b) and (c) Step 1: In (b) there are 10 Group 16/VI atoms and two negative charges for a total of 62 valence electrons; in (c) there are $5 + 6 + 3(7) = 32$ valence electrons. Step 2: In (b), the unique atom in each SO_3 group, S, is taken as the central atom of that group. The S–S connections are given, so the four S atoms act as central atoms. In (c), the S atom in the middle of the formula is taken as the central atom, and the N and three F atoms are taken as outer atoms. For (b), $62 - 18 = 44$ electrons are still available. For (c), $32 - 8 = 24$ electrons are available.

Step 3: For (b) we supply six lone pairs to each of the six outer oxygen atoms, so $44 - 36 = 8$ electrons remain. For (c) we supply 12 lone pairs to the outer N and F atoms, so $24 - 24 = 0$ electrons remain. The S atom has an octet, so no further operations are required. The structures in the middle of Figure 3.3 result.

Step 4: In (b) the eight remaining electrons are given to the two middle S atoms as lone pairs, completing octets for each.

Example 3.2 continues on next page ▶

EXAMPLE 3.3 *(cont.)*

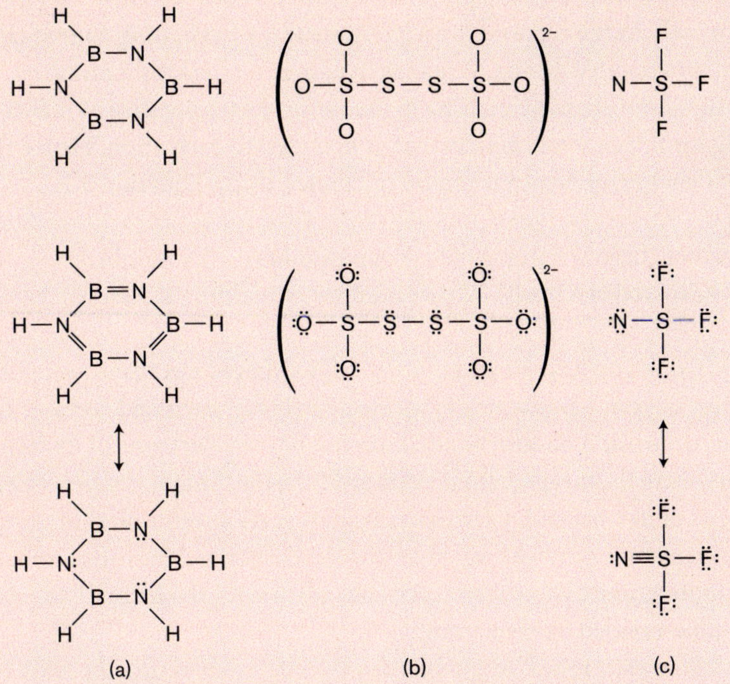

Figure 3.3. Drawing Lewis structures for Example 3.3. (Top row) Skeleton formulas from Step 2; (middle and bottom rows) alternate resonance Lewis structures.

Resonance Structures

AN AMPLIFICATION

Often it is possible to draw more than one acceptable structure. There are other ways we could have drawn the structure of the azide ion (bottom row of Figure 3.2c)—that is, instead of using an unshared electron pair from each outer N atom to form two double bonds, we could have taken two unshared electron pairs from the left N atom to form one triple bond to the central atom. Or we could have done the same with the right N atom. These are called **resonance structures** of the azide ion. We say that the true structure of the azide ion is a composite to which each resonance structure contributes.

Resonance structures for cyclo-$(HBNH)_3$ and NSF_3 are shown in the middle and bottom rows of Figure 3.3. (There is also another resonance structure for cyclo-$(HBNH)_3$ in which the single and double bonds are interchanged.)

Formal Charge

Formal charge represents an alternate way of assigning shared electron pairs to bonded atoms. Whereas in assigning oxidation numbers, we assign both shared electrons to the more electronegative atom, in assigning formal charges one electron is assigned to each atom. Hence the formal charge for an atom can be calculated using Equation 3.1:

Formal charge of atom = (its group number) – (its number
of unshared electrons) – ½(its number of bonding electrons)   (3.1)

Counting its own and one-half of its shared electrons, a single-bonded outer oxygen has seven valence electrons of its own and has a –1 formal charge, while a double-bonded oxygen has six electrons of its own and a zero formal charge. A single-bonded fluorine atom has seven valence electrons of its own and a zero formal charge. As a result, the single-bonded oxygen atom is more basic than the single-bonded fluorine atom, and is more prone to favor the resonance form with the formally uncharged double-bonded oxygen atom.

Likewise, in the single-bonded resonance structure for NSF_3 (middle row of Figure 3.3c), the N atom has a formal charge of $5 – 6 – ½(2) = –2$, while the central S atom has a formal charge of $6 – 0 – ½(4) = +2$. In contrast, in the triple-bonded resonance structure for NSF_3 (bottom row of Figure 3.3c), both the N and the S atoms have formal charges of zero. Usually the *lowest-energy* structure is the one with the *smallest* formal charges on the atoms—thus, the triple-bonded structure for NSF_3 is preferred, even though it requires the S atom to exceed an octet.

Another part of the concept of formal charges works better for nonmetal–nonmetal bonds (as in organic chemistry) than for metal–nonmetal bonds (as in coordination compounds). There is a preference for structures in which negative formal charges are placed on more electronegative atoms while positive formal charges are placed on more electropositive charges. Unfortunately this is violated severely even in simple hydrated ions such as $[Al(H_2O)_6]^{3+}$, in which Al has a formal charge of –3, while each electronegative oxygen has a formal charge of +1. In view of difficulties such as these, formal charges have not been used as frequently in inorganic chemistry as they have in organic chemistry. For most of our purposes, the most important aspect of Lewis structures is to find atoms that have unshared electron pairs, which is apparent from any of the resonance structures we have considered. As we proceed in this chapter, we will tend to illustrate using the acceptable resonance structure with single bonds rather than the alternatives with double or triple bonds.

3.2. The Lewis Acid–Base Concept and Complex Ions

OVERVIEW. In Section 3.2A we compare terms used in inorganic, organic, and biochemistry for Lewis acid–base (coordination) chemistry. You will learn to identify examples for these terms in the formulas or structures of given coordination compounds. Potential Lewis bases or ligands have the important feature in their Lewis structures of unshared (lone) pairs of electrons on donor atoms. You may practice identifying potential ligands by trying Exercises 3.11–3.16. In Section 3.2B you use the principle of charge balance to help deduce the charges of complex ions, metal ions, and ligands in a complex salt. You may practice the use of these terms and the underlying concepts by trying Exercises 3.17–3.26. Given or presuming a reasonable coordination number in the product, you can predict possible products of Lewis acid–base reactions like those of Experiment 3. In Section 3.2C we briefly develop some acid–base chemistry of complex ions, apply these ideas to one function of metal ions in biochemistry and identify some that might be nonacidic. We also provide a Connection of these concepts to analytical and physical chemistry. You may practice these concepts by trying Exercises 3.27– 30.

You may want to review the principle of charge balance, which you used previously in Section 1.4A to find oxidation numbers. Coordination chemistry terms will often be used in the remainder of the text, particularly in Chapters 5 and 7. Identifying charges of complex ions and of their central atoms will prove particularly useful later in this chapter and in Sections 5.7, 5.8, 8.2, 8.3, 8.6, and 8.7.

3.2A. Lewis Acids and Bases. *Acid* and *base* are collective terms defined by chemists in such a way as to include all compounds having similar chemical properties. Thus a statement that "acids do thus and so" economically describes hundreds or thousands of reactions of the compounds included in the definition. Several times in the history of chemistry, chemists have noted additional groups of compounds that show many properties or reactions analogous to those of recognized acids and bases; hence, several new, broader definitions of acids and bases have appeared. In order to allow you to discover one of these broadened concepts of acidity and basicity, and to give you more concrete experience with this type of chemistry, your instructor may now assign Experiment 3.

Among his many contributions to chemistry, G. N. Lewis noted that many reactions and properties that resembled acid–base reactions and properties occurred in compounds in the complete absence of water and even

EXPERIMENT 3 MAY BE ASSIGNED NOW.
(See www.uscibooks.com/foundations.htm.)

occurred in compounds that did not contain the elements hydrogen and oxygen. Organic amines such as pyridine (C_5H_5N; left side of Figure 3.4) have unshared electron pairs on their nitrogen atoms and act like bases in that they react with (neutralize) acids, yielding solid salts such as pyridinium chloride, $[C_5H_5NH]^+Cl^-$ (Figure 3.4a). But Lewis thought it significant that bases such as pyridine also react with numerous metal ions and metal compounds to give neutralized solids and solid salts such as $[Cd(C_5H_5N)_2Cl_2]$ (Figure 3.4b) and $[Ni(C_5H_5N)_4]^{2+}(ClO_4^-)_2$ (Figure 3.4c). In these reactions, the metal ions and salts seem to be reacting very much like familiar, proton-containing acids. Lewis applied his work on the octet rule and on the drawing of dot structures to these reactions, which he felt should also be classified as acid–base reactions. In such reactions, the two electrons in this bond both came from the same reactant, which had been designated as the Lewis base. The resulting bond is called a **coordinate covalent** or dative bond.

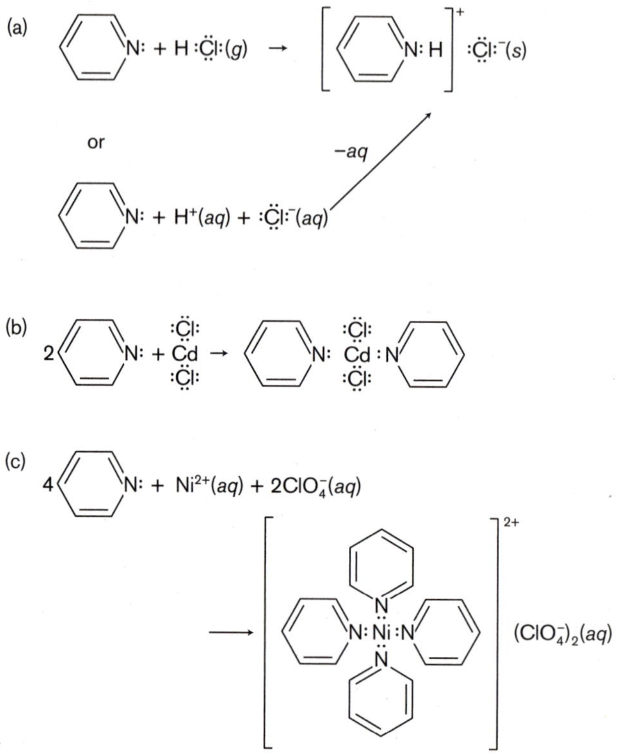

Figure 3.4. Representation of different Lewis acid–base neutralization reactions, all involving the same Lewis base, pyridine. (a) Reaction with gaseous or aqueous HCl; (b) reaction with $CdCl_2$ to give a molecular coordination compound; and (c) reaction with $Ni(ClO_4)_2$ to give a product containing a complex ion and unattached perchlorate ions.

AN AMPLIFICATION

Coordinate Covalent or Covalent Bond?

There are many cases in which covalent bonds can be formed either using two electrons from one atom and none from the other, or using one electron from each atom, and the bonds have the same properties regardless of which way they form. Haaland[1] has proposed that bonds can be unambiguously classified as covalent or coordinate covalent based on how they break apart in the gas phase. With this distinction, a number of differences in the two types of bond emerge, which are beyond the scope of the present discussion.

Thus, Lewis proposed broadening the definition of bases to include any species that can donate a pair of electrons. Such species are now commonly called **Lewis bases, donors**, or **ligands**; organic chemists commonly call these **nucleophiles**. Correspondingly, Lewis proposed broadening the definition of acids to include not only hydrogen-ion producing materials, but also such species as metal ions, because they can all accept a pair of electrons. These species are now often called **Lewis acids** or **acceptors** or just central atoms; organic chemists commonly call these **electrophiles** (electron lovers). In the product (usually a solid), the two atoms are held together by sharing of the pair of electrons that came from the donor atom in the Lewis base. The species in which the Lewis acid and base are joined together in chemical matrimony is known by various names—namely, **acid–base adduct, coordination complex** or **coordination compound**, or (if it is charged) **complex ion**. Figure 3.5 shows two examples of the assembly or disassembly of two different complex ions of cobalt. The synonymous terms used for the reactants and the product are included as well.

Any atom in the Lewis base that possesses an unshared electron pair may be capable of donating it to a Lewis acid, so this atom is a potential **donor atom**. There are compounds and ions in which any nonmetal atom other than the lighter noble gases may have unshared pairs of electrons and thus be a potential donor atom—even metal atoms occasionally can be donors. When carbon atoms (such as the C atom in the methide ion, $:CH_3^-$) act as donor atoms, metal–carbon bonds are formed; such compounds are known as **organometallic compounds**.

Not only inorganic anions, and many inorganic molecules, but all classes of organic compounds containing oxygen, nitrogen, sulfur, or sometimes even halogen

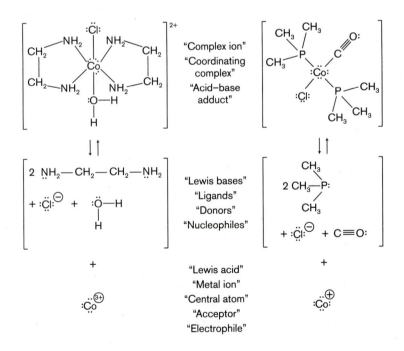

Figure 3.5. Pictorial representation of the assembly or disassembly of two complex ions from their component ligands and metal ions. (The Lewis structures shown include electrons in *d* orbitals—these are usually omitted.)

atoms[2] with unshared electron pairs can function as ligands. Thus, we may see that the number of potential Lewis acid–base reactions in the world, or even in our bodies, is enormous. (From an inorganic chemist's point of view, cell fluids are "soups full of ligands'" waiting to attach to metal ion Lewis acids.) If we can apply the acid–base concepts we already have to this greatly expanded list of acids and bases, our power to predict chemical reactions and properties certainly will be enhanced (e.g., in upcoming Chapters 5 and 6).

EXAMPLE 3.4

Identify the donor atoms in each of the following molecules or ions:
(a) HNNH; (b) (linear) N_3^-; (c) NF_4^+; (d) OCS.

SOLUTION: The Lewis structures of these molecules are shown in the Appendix as solutions for Exercise 3.7. In (a) the two nitrogen atoms each have a lone pair of electrons, so the nitrogen atoms are potential donor atoms. In (b) only the two outer N atoms have lone pairs, so they are donor atoms. (c) The fluorine atoms have lone pairs, so they might function as donor atoms (although this is unusual in a cationic species). (d) The oxygen and sulfur atoms have lone pairs and are potential donor atoms.

When writing a formula of a complex ion, it is customary to enclose it in brackets. Also, each polyatomic ligand (anion or neutral molecule) is enclosed in parentheses, even if there is only one of them. This allows us to distinguish between the ligands O_2^{2-} and O^{2-} (or even the neutral molecule O_2, on the one hand) from two oxide ligands, on the other hand, within the formula of the complex ion; two oxide ions will appear as O_2 without any parentheses. Although each ligand may have more than one donor atom, for the time being we will assume that each ligand uses only one of its potential donor atoms. In Section 5.1 we will see the conditions under which this may not be the case.

The term *coordination number*, which we introduced in Chapter 2 to refer to the number of nearest-neighbor atoms or ions to a given atom or ion, is especially useful here. The coordination number of the Lewis acid or acceptor atom or ion in coordination compounds is an important property and is in most circumstances equal to the number of donor atoms attached to the acceptor atom (for now, one per ligand). Provided that there is no multiple bonding to the acceptor atom, it is also equal to the number of "ordinary" and coordinate covalent bonds to the acceptor atom.

3.2B. Charge Balance in Complex Ions and Salts. A complex ion, when isolated from solution, must be accompanied by ion(s) of opposite charge (**counterions**). To show that there is no sharing of electrons between these ions and the Lewis acid or base in the complex, the complex ion itself is traditionally enclosed in brackets. In Figure 3.4, for example, we showed formulas and structures of two coordination complexes, $[Cd(C_5H_5N)_2Cl_2]$ and $[Ni(C_5H_5N)_4]^{2+}(ClO_4^-)_2$. The cadmium compound is one uncharged molecule, enclosed by the brackets, with no charge. But the nickel compound contains a complex ion enclosed by brackets, the +2 charge of which is balanced by two perchlorate ions not joined to the nickel ion by any coordinate covalent bond. If the charge(s) of the counterions are known, then the charge of the complex ion can be computed by the principle of charge balance:

Charge of complex ion = –sum of charges of counterion(s) $\qquad$ (3.2)

Once we have computed or been given the charge of the complex ion, the principle of charge balance can also be used to help identify the charges on the metal ion and ligands:

Charge of complex ion = charge of metal ion + sum of charges of all ligands $\quad$ (3.3)

Three other general guidelines for identifying ligands are as follows:

1. Most ligands obey octet rules, so Cl_2 (without parentheses) in a formula refers to two eight-electron chloride ions, not two seven-electron chlorine atoms, and not a Cl_2 molecule. Also, particularly in this chapter, the ligands will be familiar ions or molecules (H_2O or NH_3, not H_2O^- or NH_3^-, which are unfamiliar species that do not obey octet rules).

2. The Lewis structure of your ligand will include *both* electrons that were originally in its coordinate covalent bond to the central atom.

3. Cations very rarely act as ligands or Lewis bases (because cations tend to be acidic).

EXAMPLE 3.5

Consider the following complexes: (a) $[Au(CH_3)_2(NH_3)_2]^+$, (b) $[SnI_6]^{2-}$, (c) $HgCl_2$, and (d) $[Hg(CH_3)_2]$. Treat all the bonds to the metal as coordinate covalent bonds. Identify the ligands and metal ions in these complexes. Identify the donor atoms in the ligands found in each complex. Give the coordination number of the metal atoms in each of these complexes.

SOLUTION: (a) The ligands are $:NH_3$ and $(:CH_3)^-$, with C and N donor atoms; if we alternately assumed that the ligands were $:CH_3$ and $(:NH_3)^-$, neither one would have a Lewis structure that obeys the octet rule. Applying Equation 3.3, $+1$ = (charge on Au) $+ 2(-1) + 2(0)$, so the charge on Au is $+3$. The Au^{3+} has four ligands attached via one donor atom each, so its coordination number is 4.

(b) The six iodines are six different I^- anions—if they were two triiodide ions, they would have been written as $(I_3)_2$ in the formula, and neutral iodine atoms would disobey the octet rule. Thus, there are six I donor atoms, the charge on Sn is $+4$, and Sn^{4+} has coordination number 6.

(c) Hg^{2+} is coordinated by two chloride ions (two Cl donor atoms), so it has coordination number 2.

(d) Hg^{2+} is coordinated by two $(:CH_3)^-$ ligands, with two C donor atoms; it has coordination number 2.

EXAMPLE 3.6

Deduce the charge (oxidation number) of the central metal ion in each of the following complex ions or coordination compounds: (a) $[Al(H_2O)_6]^{3+}$, (b) $[Co(NH_3)_5Cl]Cl_2$, and (c) $K_3[FeF_4(OH)_2]$.

SOLUTION: The brackets in these formulas separate the complex ion from the counterions, which are two Cl^- ions in (b) and three potassium ions (K^+) in (c). After applying Equation 3.2, we obtain the charges on the complex ions: +2 in (b) and –3 in (c).

Next, we apply the principle of charge balance (Eq. 3.3) to deduce the charge on the central metal ion (after the ligands have been removed). (a) For $[Al(H_2O)_6]^{3+}$, removal of six neutral water molecules still leaves a +3 charge; the Lewis acid is Al^{3+}. (b) Removal of some neutral ammonia molecules and a Cl^- from $[Co(NH_3)_5Cl]^{2+}$ leaves a Co^{3+} ion as the Lewis acid. (c) Removal of six –1-charged ligands from the –3-charged iron complex ion leaves a Fe^{3+} ion as the Lewis acid.

AN AMPLIFICATION

The First Nobel Prize in Inorganic Chemistry

The distinction between ligands and counterions is real: Among the complexes shown in Figure 3.4, only $[Ni(C_5H_5N)_4]^{2+}(ClO_4^-)_2$, when dissolved in an appropriate nonreacting solvent, conducts an electrical current. The distinction did not come easily to chemists, however: Alfred Werner, a Swiss chemist, won the first Nobel Prize awarded in inorganic chemistry (1913) for his study of the chemistry of coordination complexes. He prepared numerous such complexes, among them a series of complexes of the Co^{3+} ion with the ligand NH_3 (ammonia), which differed in composition, color, and reactivity. He found that aqueous solutions of the yellow compound $CoCl_3 \cdot 6NH_3$ conducted electricity as well in solution as, say, $LaCl_3$, and reacted with 3 mol $AgNO_3$ rapidly to precipitate 3 mol $AgCl$. The purple compound $CoCl_3 \cdot 5NH_3$ had a lesser conductivity, comparable to $BaCl_2$, and yielded only 2 mol $AgCl$ immediately on treatment with excess $AgNO_3$. He also found two isomers (green and violet, respectively) of $CoCl_3 \cdot 4NH_3$, each of which had a conductivity comparable to that of $NaCl$, and each of which released 1 mol $AgCl$. Later, he prepared the compound $CoCl_3 \cdot 3NH_3$, the solution of which did not conduct electricity and did not react immediately with $AgNO_3$. Werner saw that these data could be explained by assuming that there were two kinds of chloride in these compounds. The chloride not present as free chloride ions could not conduct an electric current or react with Ag^+ because it was attached to the Co^{3+} (i.e., was a ligand). He rewrote the above formulas to make this distinction: $[Co(NH_3)_6]Cl_3$, $[Co(NH_3)_5Cl]Cl_2$, $[Co(NH_3)_4Cl_2]Cl$, and $[Co(NH_3)_3Cl_3]$, respectively (Figure 3.6). Werner noticed that in each of these cobalt complexes and in many others he studied, the experimental data could best be explained by assuming a constant coordination number of six for the Co^{3+} ion.

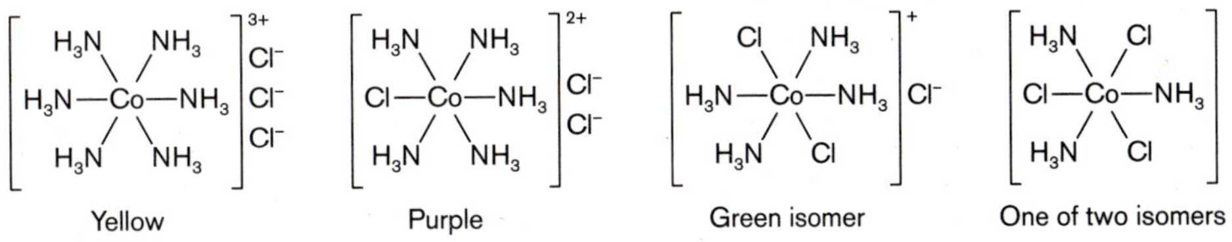

Figure 3.6. Structures of some key complexes discovered by Alfred Werner: yellow $[Co(NH_3)_6]Cl_3$, purple $[Co(NH_3)_5Cl]Cl_2$, the green isomer of $[Co(NH_3)_4Cl_2]Cl$, and one of two isomers of $[Co(NH_3)_3Cl_3]$. (Isomers are discussed in more detail in Section 10.2.) Note that the complex ions are enclosed in the square brackets, and counterions are shown outside the brackets. Note, too, that only the chloride ions within the brackets act as ligands.

3.2C. Acid–Base Strength and Complex Ions. In Chapter 2, we observed that metal cations can distort the electron distribution in water to the point that hydrogen (hydronium) ions are released, so we classified metal cations as acids. Similarly, we classified anions as bases due to their ability to accept protons from the solvent, water, thus releasing hydroxide ions. A characteristic of concentrated solutions of ions is that there are not enough water molecules present to act as solvent molecules or even to complete the hydration of the ions. For example, in a dilute aqueous solution of Zn^{2+}, we expect the predominant species to be $[Zn(H_2O)_6]^{2+}$. When we melt the hydrated salt $ZnCl_2 \cdot 2H_2O$, there is insufficient water to serve as free solvent or to complete the formation of $[Zn(H_2O)_6]^{2+}$.

Acid–base properties are extraordinary in very concentrated solutions. A concentrated (19 M) solution of KOH behaves in many respects in a manner that we would expect from a 10,000 M solution of hydrated hydroxide ions! The molten salt $ZnCl_2 \cdot 2H_2O$ behaves, not as a weak acid containing $[Zn(H_2O)_6]^{2+}$ ($pK_a = 9.0$), but as a strong acid ($pK_a < -4$), comparable to concentrated $HCl(aq)$. In the important enzyme carboxypeptidase A (Section 5.4D), in which the Zn^{2+} is five-coordinate, it has a pK_a of 7. Thus, we have the first biological function of metal ions (previewed earlier in Section 1.2)—that of enhancing the acidity of coordinated ligands.

A CONNECTION TO ANALYTICAL CHEMISTRY AND PHYSICAL CHEMISTRY

Activity Coefficients of Ions

Analytical and physical chemists are often concerned with this phenomenon. Concentrated solutions of ions do not show ideal colligative properties (freezing-point depression, etc.) or ideal compliance to the law of chemical equilibrium unless account is taken of the enhanced or reduced activity such ions have. This activity is often measured in terms of the activity coefficient of a dissolved ion, which is the ratio of its apparent or effective concentration to its actual concentration. In sufficiently dilute solutions, in which nearly all the ions are completely hydrated, the activity coefficients of ions approach 1.00. In solutions of moderate concentrations, activity coefficients are less than 1.00 due to the greater tendency for pairing of cations and anions to occur. But in highly concentrated solutions, activity coefficients become much greater than 1.00.

TABLE 3.1. The pK_a Values of Free and Coordinated Ligands[a]

	No Metal	M = Ca^{2+}	M = Ni^{2+}	M = Cu^{2+}
$[M(NH_3)]^{2+} \rightleftharpoons H^+ + [M(NH_2)]^+$	35.0		32.2	30.7
$[M(H_2O)]^{2+} \rightleftharpoons H^+ + [M(OH)]^+$	14.0	12.8	9.9	7.3
$[M(HImid)]^{3+} \rightleftharpoons H^+ +]M(Imid)]^{2+}$	7.0		4.0	3.8
$[M(CH_3CO_2H)]^{2+} \rightleftharpoons H^+ + [M(CH_3CO_2)]^+$	4.7	4.2	4.0	3.0

Source: Data from S. J. Lippard and J. M. Berg, *Principles of Bioinorganic Chemistry*, University Science Books: Mill Valley, CA, 1994, p. 25, and from Table 2.2 and references cited therein.

[a]Imid = Imidazole =

Such findings suggest the ultimate source of the acidity and basicity of the type of complex ions known as hydrated ions. The acidity or basicity is inherent not in the water, but in the cation or the anion—the water actually serves to dampen out the activity of the metal ion or the anion. Consequently, inorganic chemists nowadays do many of their reactions in the absence of water and employ the language of the Lewis acid–base concept to describe their results.

Just as the acidity of a metal ion polarizes the water molecule in the complex ion $[M(H_2O)_x]^{y+}$ and causes it to release hydrogen ions, an acidic metal ion will enhance the acidity of other ligands (Table 3.1).[3] Then, for example, bonds other than O–H may also be attacked by water or other Lewis bases, producing products more readily than would otherwise be the case.

Nonacidic Complex Cations. As can be seen from Table 3.1, ammonia (NH_3), even when coordinated to metal ions, has such a high pK_a that it is nonacidic. Thus, complex cations provide important additional members of the class of nonacidic cations. Complex cations are, strictly speaking, not spherical, so it is impossible to rigorously define a radius for them, but we can readily imagine some kind of "effective" radius that is likely to be substantially larger than that of Cs$^+$ or Fr$^+$, so that a Z^2/r calculation, if possible, would usually support this classification. The hydroxide of the nonacidic $[Co(NH_3)_6]^{3+}$ cation, $[Co(NH_3)_6](OH)_3$, is a soluble strong base.

The simplest "complex" cation is that of ammonia with the smallest cation, H$^+$: the ammonium ion, NH_4^+. In size and charge, it closely resembles the K$^+$ ion, so in many respects it behaves as a nonacidic cation. However, its solution is weakly acidic due to the very strong acidity of the extremely tiny H$^+$ ion acting on the coordinated ammonia molecule (pK_a = 9.25). If we replace the hydrogens on this ion by organic R groups [where R = methyl (CH_3), ethyl (C_2H_5), etc.], the resulting cations are unambiguously nonacidic and, being even larger than Cs$^+$, show the properties of nonacidic cations even more clearly. Hence, chemists frequently use the tetramethylammonium ion, $(CH_3)_4N^+$, and the tetraethylammonium ion, $(C_2H_5)_4N^+$, when they need a very large nonacidic cation (Section 4.3C).

3.3. Orbital Shapes, Covalent Bond Types, and the Periodicity of Bond Energies

> **OVERVIEW.** You can review how to draw the shapes of s, p, and d orbitals, and add the ability to identify nodal planes and changes in sign of the lobes of these orbitals (Section 3.3A). You may practice this concept by trying Exercises 3.31 and 3.32. You may also identify cases of positive, negative, and zero overlap between specified s, p, or d orbitals on two different atoms, and identify cases of σ, π, and δ bonds resulting from such overlap. You may practice this concept in Section 3.3B by trying Exercises 3.33–3.40. You should be able to describe and explain the main periodic trends in covalent σ, π, and δ bond energies, and practice this concept in Section 3.3C by trying Exercises 3.41–3.43. Polar bonds between unlike atoms are stronger than those between identical atoms. From this difference in bond energies Pauling derived his scale of electronegativities. You can calculate Pauling electronegativities from bond energies or vice versa (Section 3.3D); try this with Exercises 3.44–3.49.
>
> To understand bond energy trends, you may want to review covalent radii (Section 1.5B) and Pauling electronegativities (Section 1.6). The ability to identify σ, π, and δ bonds will help you in Chapters 7 and 10 and in Sections 8.4, 9.1, 9.2, and 9.6. Bond energies and their trends will be used again in Sections 6.3, 6.6, 9.5, and 10.2. A Connection of weak bonds with explosives and therefore chemical safety is included.

3.3A. Shapes and Signs of Orbital Wave Functions.

This concept is a review of a topic from quantum chemistry in general chemistry, but it adds to those concepts. To begin, recall that electrons move in **orbitals** that are expressed mathematically by wave functions. Similar to water waves, orbitals cannot be said to be present only at certain distances and directions from the nucleus; but there are distances and directions at which the electron waves have maximum probabilities. Hence the orbitals we draw in Figure 3.7 do not actually end suddenly at the edge of the shaded areas, but instead decrease in amplitude beyond the "edge."

Water and electron waves have crests (regions in which they are above the mean sea level, signified by a + sign for the wave function and lobe) and troughs (regions in which they are below the mean sea level, signified by a – sign for the wave function and lobe). Between each crest and the adjacent trough the wave function passes through a **nodal plane**, in which the wave is at sea level, and is neither + nor –.

The s orbitals are spherical. In terms of a particle model of the electron, it is as likely to be found in one direction from the nucleus as another. The wave function does not change sign as it changes angle around the nucleus.

The p orbitals have the familiar dumbbell shape (Figure 3.7), with two lobes, one on each side of the nucleus at the center. The mathematical angular wave function for a p orbital involves some constants times $\cos \theta$ or $\sin \theta$. The angular wave function for the p_z orbital contains $\cos \theta$, which goes from +1 (a crest) to zero (at a nodal plane) to –1 (a trough) as we move from the north pole to the equator to the south pole, so the *sign* of the wave function changes. The *amplitude* of the wave is proportional to the square of the wave function, so it is equally positive in both lobes. This amplitude corresponds to the probability of finding the electron at that spot. The shaded plane

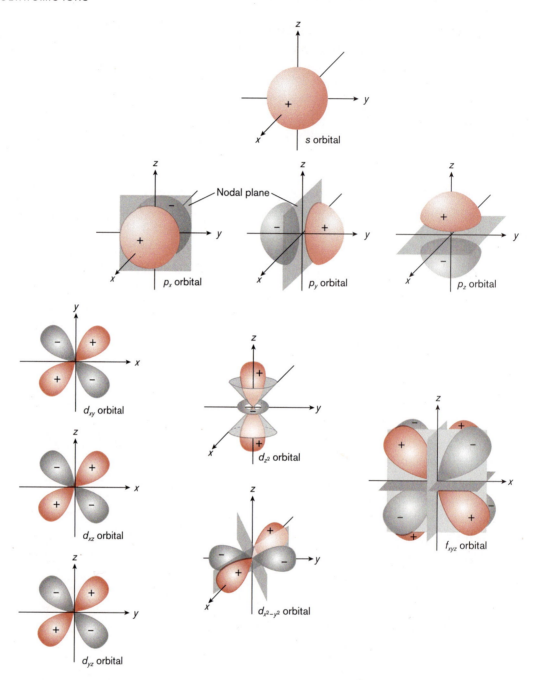

Figure 3.7. Angular wave amplitudes of *s*, *p*, *d*, and an *f* orbital showing signs of the wave functions in different lobes and indicating nodal planes with dark planes or solid lines. [Source: R. L. DeKock and H. B. Gray, *Chemical Structure and Bonding*, University Science Books: Mill Valley, CA, 1989.]

in Figure 3.7 through the nucleus and the "equator" of the p_z orbital is the *one* nodal plane of the p_z orbital, the plane in which the wave amplitude is zero and the electron cannot be found.

In a given atom, there are five otherwise-identical *d* orbitals and seven otherwise-identical *f* orbitals that differ in their orientations. A *d* orbital generally has the double-dumbbell shape indicated in Figure 3.7, with four lobes and *two* nodal planes. The four lobes alternately correspond to crests and troughs, but all have positive wave amplitudes and correspond to regions in which there are positive probabilities of finding the electron. The difference in appearance of the d_{z^2} orbital is an artifact of mathematical operations; like the others it has two nodal planes through the nuclei, but these surfaces happen to be planes wrapped around to form cones. In preparation for discussing bonding in d-block complexes (Chapter 7), you will need to know the designations, shapes, and orientations of the five d orbitals. Note that three of the orbitals—d_{xy}, d_{xz}, and d_{yz}—have lobes between the named axes, while the other two—$d_{x^2-y^2}$ and d_{z^2}—have lobes along the named axes.

A typical f orbital has the quadruple-dumbbell shape shown in Figure 3.7. Other f orbitals in a given atom have a somewhat different appearance due to the different mathematical operations used to eliminate the imaginary numbers from the wave functions, but all of them share the characteristic of having *three* nodal planar or conic surfaces passing through the nucleus.

The p, d, and f orbitals not only have characteristic shapes but also are oriented in certain directions. The p_z orbital we described earlier has its crest at its north pole and its trough at its south pole and is thus oriented along this polar axis, which we label the z axis. A given atom also has two otherwise-identical p orbitals aligned at right angles to the first one and to each other along an x and a y axis; we conventionally label these three orbitals as the p_x, p_y, and p_z orbitals (Figure 3.7).

EXAMPLE 3.7

Give the spectroscopic shorthand designation (name) for an orbital that has (a) only one nodal plane, the xy plane; (b) two positive lobes, both along the z axis; (c) two negative lobes, both along the x axis.

SOLUTION: (a) The p orbitals have only one nodal plane; if this is the xy plane, then the p orbital is *not* in this plane, so the orbital is a p_z orbital. b. Two positive lobes would be balanced by two negative lobes, which is characteristic of a d orbital. Only the d_{z^2} orbital has lobes along the z axis. (c) Again, this is a d orbital. Only the $d_{x^2-y^2}$ orbital has lobes along the x (and y) axis.

3.3B. Sigma, Pi, and Delta Orbital Overlap. For a covalent bond to form by sharing electrons between two orbitals on two different atoms, the shapes and wave function signs must be such that the two atomic orbitals have positive overlap. **Positive overlap** means that the signs of the two orbitals *match* in the region of overlap, as in Figure 3.8c. (One negative lobe overlapping another negative lobe gives positive overlap, because the signs of the lobes match!) Under these conditions, the crest of one atomic orbital is in phase with (enhances) the crest of the other atomic orbital (or the two troughs are in phase with and enhance each other). Then the two electrons that are allowed to be in such an orbital have an enhanced probability of being present in the region between the two nuclei. In this region, the two electrons are attracted to both nuclei and serve to bond the two (otherwise mutually repulsive) nuclei to each other.

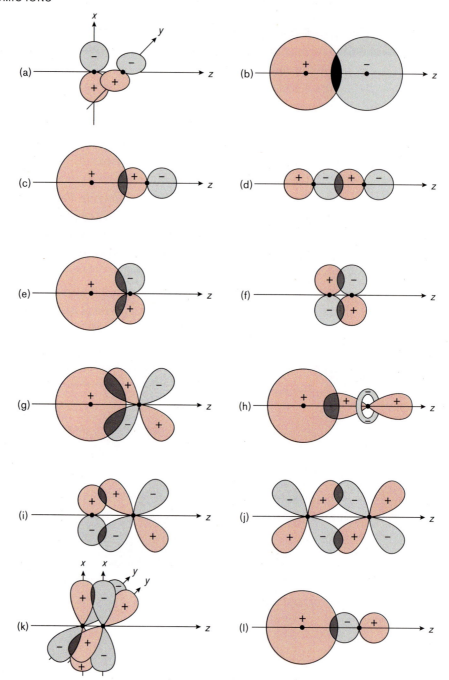

Figure 3.8. Combinations of atomic orbitals to be tested for positive, negative, and zero overlap (see the Exercises). In every case, x is the vertical axis and z is the horizontal axis.

No bonding interaction or molecular orbital can result if equal amounts of positive and negative overlap occur. Suppose that we were to bring two atoms together along their mutual z axes, and try to overlap the p_x orbital of one atom with the p_y orbital of the other (Figure 3.8a). No bond would result, because the positive lobe of the p_x orbital on the one atom would have equal overlap with the positive and the negative lobes of the p_y orbital on the other atoms, and the same would hold for the negative lobe of the p_x orbital on the first atom.

If the signs of two overlapping orbitals oppose each other in the region of overlap, the crest of one atomic orbital overlaps with the trough of the other, resulting in destructive interference and loss of wave amplitude in the desirable region between the nuclei. The electrons must then spend most of their time beyond the nuclei and do not serve to attract the nuclei together. The mutual repulsions of the like-charged nuclei are not overcome, and a repulsive or antibonding interaction results, as in Figure 3.8b.

EXAMPLE 3.8

Consider two atoms in a diatomic molecule, using the two orbitals specified. (In every case x is the vertical axis and z is the horizontal axis.) Draw the orbitals and fill in the signs of the lobes so that they achieve the type of overlap (positive, zero, or negative) specified. (a) Atom A = p_x, B = s, zero overlap; (b) Atom C = d_{xz}, D = p_x, positive overlap; (c) Atom E = s, F = $d_z{}^2$, positive overlap; (d) Atom G = s, H = p_x, zero overlap; (e) Atom I = d_{xz}, J = d_{xz}, positive overlap; (f) Atom K = $d_z{}^2$, L = p_z; positive overlap.

SOLUTION: Positive overlap (as in CD, EF, IJ, and KL) means that the signs of the overlapping orbitals match. Hence both can have positive signs, or both can have negative lobes. Zero overlap (as in AB and GH) means that every region of positive overlap (top halves of AB and GH) is cancelled out by a corresponding region of negative overlap, where orbital signs do not match (bottom halves of AB and GH). Choices of signs that give the desired types of overlap are shown in Figure 3.9.

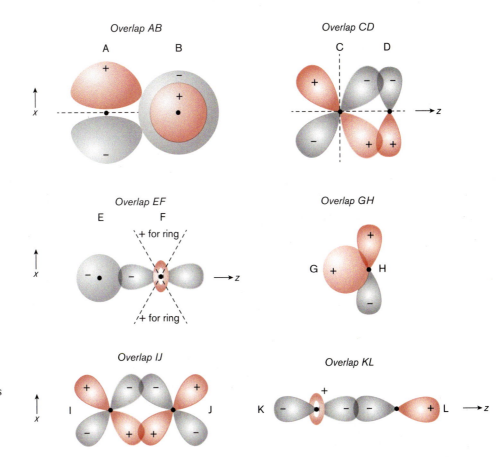

Figure 3.9. Drawings of pairs of orbitals for use in Examples 3.7, 3.8, and 3.9.

Bonding molecular orbitals that result from positive overlap of two atomic orbitals are commonly classified as **sigma (σ)** bonds, **pi (π)** bonds, and **delta (δ)** bonds. (The next type, not shown, is a **φ** bond involving two f orbitals.) If a bonding molecular orbital has no nodal plane through both atomic nuclei, it is called a σ bonding orbital. As shown in the first column on the left of Figure 3.10, σ bonds can be formed from two atomic s, p_z, or d_{z^2} orbitals; each combination overlaps "head-on" in one shaded region to give a σ bond. Although not pictured, a σ bond can also arise from the overlap of an s orbital on one atom with a p_z orbital on the other atom, and so on. If two atoms have just one covalent bonding orbital linking them (i.e., if they are joined by a single bond), this bonding is nearly always σ bonding.

If the molecular orbital has "sideways" overlap of its atomic orbitals and has *one* internuclear nodal plane passing through both atomic nuclei, it is called a π bond (center of Figure 3.10). Two p-block atoms can form two such π bonds, one using their valence p_x orbitals, and one using their valence p_y orbitals. Two d-block atoms can also form two π bonds, one with their d_{xz} orbitals and one with their d_{yz} orbitals. Mixed π bonds involving a valence d_{xz} or d_{yz} orbital on one atom and a valence p_x or p_y orbital on the other are also formed between atoms of these two blocks of the periodic table. Double

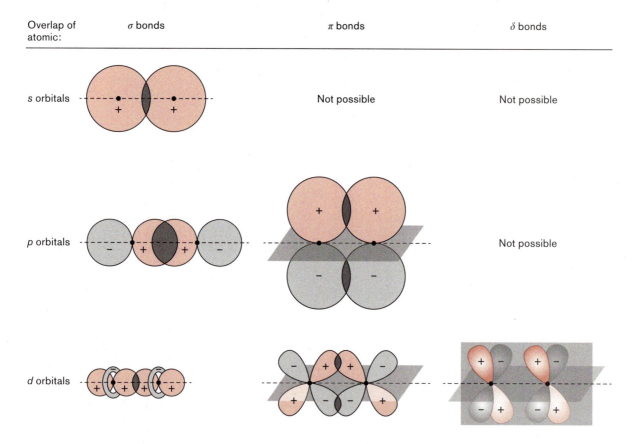

Figure 3.10. Overlap of atomic orbitals of two atoms, the nuclei of which are represented by dots, to give covalent bonds. Regions of positive overlap are indicated by shading. The z axis for each atom is taken as the internuclear axis, and it is shown by the dashed line (- - - -). Internuclear nodal planes are sketched and shaded—namely, one for each π bond, and two for the δ bond shown.

bonds nearly always consist of one σ plus one π bond, whereas triple bonds consist of one σ plus two π bonds.

If the molecular orbital is formed by "face-to-face" overlap of two atomic orbitals that have *two* nodal planes passing through both atomic nuclei, then it is a δ bond (third column of Figure 3.10). Two δ bonds can be formed per pair of *d*-block atoms, one using two d_{xy} and the other using two $d_{x^2-y^2}$ orbitals that overlap in four regions (one hidden from view). A *p*-block element cannot form a δ bond, and hence cannot participate in a quadruple or quintuple bond, but these are possible in the *d* block, and they are well known in a series of acetates and other derivatives of quadruple bonded diatomic ions of early *d*-block metals: Cr_2^{4+}, Mo_2^{4+}, W_2^{4+}, Re_2^{6+}, and so on. More recently, a series of *quintuply* bonded complexes formed by the diatomic ions Cr_2^{2+} and Mo_2^{2+} have been synthesized.[4,5]

EXAMPLE 3.9

Consider again the diatomic molecules described in Example 3.8, with the two orbitals overlapping as indicated in Figure 3.9. For each case in which there is positive overlap, identify the bond that is formed as σ, δ, or π. Then identify the two cases that probably give the *strongest* bonds, and the two cases that give the *weakest* bonds.

SOLUTION: Positive overlap occurs in cases CD, EF, IJ, and KL (Figure 3.9). Bond CD = π (one nodal plane passing through the C and D nuclei). Bond EF = σ. Bond IJ = π. Bond KL = σ. The strongest bonds are probably the σ bonds, EF and KL, and the weakest are probably the π bonds, CD and IJ.

AN AMPLIFICATION

Postvalence Orbitals in Bonding?

In the second period, each of the four available valence orbitals ($2s$, $2p_x$, $2p_y$, and $2p_z$) (or some hybridized version of them; Section 3.3B) can form one covalent bond or hold one unshared electron pair. In the third and fourth periods, there are still only four available valence orbitals, but the central atom can have more than an octet of electrons (Step 4 of the Figure 3.1 flow chart). Two possibilities have been suggested to accommodate the extra electrons. In the valence bond picture, six covalent bonds or unshared electron pairs require the use of six central-atom atomic orbitals, thereby using some of the next-highest-energy set of orbitals, the $3d$ or $4d$ orbitals (or hybrids thereof, such as d^2sp^3 hybrid orbitals. The preferred explanation, however, is based on the molecular orbital theory of bonding (Chapter 11). Four atomic orbitals on a central atom may suffice to form six bonds because one central-atom orbital can overlap the atomic orbitals of more than one outer atom. This is multicentered covalent bonding (Section 11.5C).

In the *d* block, there is no doubt that unoccupied orbitals beyond the valence ns and $(n-1)d$ orbitals—namely, the post-valence np orbitals—do become involved in the bonding. This gives a total of nine available orbitals to form covalent bonds to other atoms, so the high numbers of covalent bonds present in compounds such as the one shown in Figure 3.13 are not surprising.

Hybrid Orbitals

It is possible to construct mathematical orbital types that show better overlap properties than do the pure *s*, *p*, and *d* orbitals, which have lobes (or hemispheres for *s* orbitals) that are aimed away from the region of overlap and hence are "wasted." For example, an *s* and a *p* orbital on the same atom can be mixed or hybridized with each other to produce two new *sp* hybrid orbitals (Figure 3.11). Combining the crest (+ lobe) of an *s* orbital with a *p* orbital with its crest to the right produces an *sp* hybrid orbital with most of its amplitude at the right (Figure 3.11a), where it can overlap well with an orbital of a neighboring atom. (Of the original trough part of the *p* orbital, only a vestigial tail remains.) The other possible combination (which must also be made) is identical in shape but is oriented so as to give maximal overlap with an orbital of an atom on the left (Figure 3.11b). Although it requires energy to mix different orbitals in this manner, the improved overlap that results afterward often make this feasible for second-period elements. For third-period and later atoms, the greater size difference between valence *s* and *p* orbitals makes hybridization less effective (Section 9.4). Consequently, this concept is much more successful in the realm of organic chemistry than in inorganic chemistry. In inorganic chemistry the concept is also found to have low predictive power—it is better for rationalizing trends after the fact. To be sure, hybrid orbitals are part of the valence bond theory description of chemical bonding on which this text is based; the main alternative approach, the molecular orbital theory, is introduced in Chapter 11.

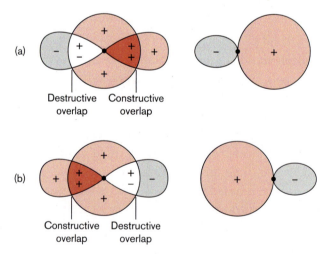

Figure 3.11. Hybridization of an *s* and a *p* orbital to produce two *sp* hybrid orbitals.

3.3C. Periodic Trends in Bond Energies. σ-Bond Energies. Periodic trends in the strengths of σ bonds show up most clearly among the bond (dissociation) energies of homoatomic single bonds of the elements (Table 3.2)—these are the energies required to break single covalent bonds. We see that single-bond energies tend to increase as we proceed to the right in a given period. However, they reach a maximum in Group 14/IV, then drop as Group 15/V is entered and electrons are added to already half-occupied orbitals. We may also note the vertical trend manifested in the *s* and *p* blocks—bond dissociation energies *usually* increase as we go up a group.

These trends are related to the trends in covalent radii (Section 1.5), where small atoms generally form stronger covalent bonds. In bonds between smaller atoms, the shared electrons are closer to (and therefore more strongly attracted to) both atomic nuclei.

π-Bond Energies. Because double and triple covalent bond energies are available for many of the *p*-block elements, we can subtract σ- (single-) bond energies from these to obtain estimates for π-bond dissociation energies, shown in Table 3.3. In the case of triple-bond energies, after subtraction the result is divided by two, because a triple bond contains two π bonds. The π-bond energies show the same basic periodic trends as σ-bond energies in the *p* block: They strengthen going to the right or up the periodic table. The data also show that π bonds are generally weaker than σ bonds between the same elements. Although Figure 3.10 does not give a computationally precise representation of the overlap of atomic orbitals, it suggests that this trend arises from the smaller total degree of overlap involved in π bonding than in σ bonding between orbitals of like shape and size.

TABLE 3.2. Element–Element Covalent Single-Bond Dissociation Energies

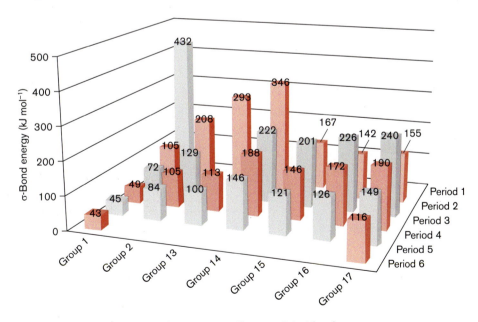

Source: The data come from J. E. Huheey, *Inorganic Chemistry: Principles of Structure and Reactivity*, 3rd ed., Harper and Row: New York, 1983, Table E-1.

**AN AMPLIFICATION AND CONNECTION TO
ORGANIC CHEMISTRY AND CHEMICAL SAFETY**

Anomalously Weak σ Bonds

An anomaly in the main trends in σ-bond energies occurs in the upper right of the p block. The N–N, O–O, and F–F single bonds are substantially weaker than the P–P, S–S, and Cl–Cl bonds, even though the second-period atoms are much smaller. This weakness will have explosive consequences in Section 6.3! This anomaly is another case in which too small a size leads to diminishing returns. Molecules containing these single bonds (such as $H_2N–NH_2$, HO–OH, and F_2) also have unshared pairs of electrons. These electrons are brought so close to each other that substantial electron-pair–electron-pair repulsion results, weakening the net bonding (Figure 3.12). Consequently, fluorine shows a much larger covalent radius (71 pm) in F_2 than it does in most of its compounds (57 pm; Table 1.11).

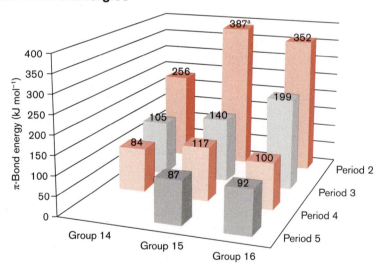

Figure 3.12. (a) Electron-pair–electron-pair repulsion between two small second-period atoms, illustrated for one conformation of $H_2N–NH_2$; (b) reduction of the repulsion when a double bond is present, illustrated for HN=NH; (c) absence of the repulsion when a triple bond is present, illustrated for N≡N.

TABLE 3.3. π-Bond Energies

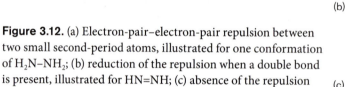

Source: The data come from M. W. Schmidt, P. N. Truong, and M. S. Gordon, *J. Amer. Chem. Soc.* 109, 5217 (1987).

[a] Schmidt et al. give a value for the π-bond energy of the N–N bond, 251 kJ mol⁻¹, which is not calculated as described above, and which therefore does not also include the effect of reducing unshared electron-pair repulsion on going from the single bond to the double bond. The 387 kJ mol⁻¹ value should be used in calculating double- or triple-bond energies, but the 251 kJ mol⁻¹ value is a better representation of the intrinsic strength of the π bond itself, and so is better to compare with neighboring values.

The data in Table 3.3 also suggest that π-bond energies tend to drop off with increasing distance even faster than σ bonds. Consequently, the tendency of *p*-block elements to form double or triple bonds drops off in the order O > N = C >> S > P > Si.[6] As Figure 3.12 also suggests, the repulsion of electron pairs on neighboring N, O, or F atoms is greatly diminished or absent in double or triple bonds, so the π- (hence, double- and triple-) bond energies of N and O are not anomalously weak.

Because the π-bond energies are added to the already existing σ-bond energies, double bonds are stronger than single bonds (although not twice as strong) and triple bonds are stronger than double bonds. As a result of the additional orbital overlaps and electron-pair sharing in multiple bonds, the greater the bond order (multiple nature) of the bond between a given pair of atoms, the shorter it is. Figure 3.13 shows a molecule that features a single, a double, and a triple bond between tungsten and carbon, all in the same molecule. Note that the usual trend in bond lengths is shown: the triple bond is shorter than the double bond, which is shorter than the single bond between the same two elements. The shortness of the bond (as compared to known bonds of lower order between the same pair of elements) is the feature most frequently used to identify multiple bonds in X-ray structural determinations on crystalline substances.

In the *d* block, π-bonding strength and ability is greater in the fifth and sixth periods than in the fourth period. The π-bonding tendency reaches a maximum at Mo and W, but is also strong in Tc and Re. Not only do these elements (and to a lesser extent their neighbors) form multiple bonds to other like atoms, but also to *p*-block elements such as C (Figure 3.13), N, and O. We will see in Section 3.5 that this affects the type of oxo anions that these fourth- and fifth-period elements form.

δ-Bond Energies. From chemical experience, the periodic trends among δ-bond energies are much like those of π bonds in the *d* block. As suggested qualitatively by the drawings of Figure 3.10, overlap is even weaker in the δ-bonding situation than in the π-bonding situation, so δ bonds are weaker than π bonds. But the tendency to form δ bonds increases down the *d* block: The quadruple-bonded species Mo_2^{4+} and W_2^{4+} have proportionately shorter and more robust bonds than Cr_2^{4+}, with the δ bonds of W_2^{4+} being slightly stronger than those of Mo_2^{4+}.[7]

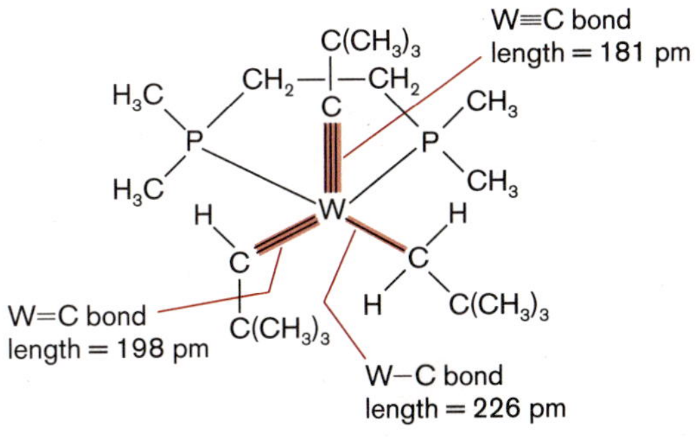

Figure 3.13. Structure of a molecule containing W–C, W=C, and W≡C bonds. [Source: The data come from M. R. Churchill and W. J. Youngs, *Inorg. Chem.* 18, 2454 (1979). Adapted from D. E. Ebbing, *General Chemistry*, 3rd ed., Houghton Mifflin: Boston, 1990, p. 338.]

EXAMPLE 3.10

Consider the compound with the structure shown in Figure 3.13. (a) How many σ and how many π bonds are present in each W–C bond? (b) Which one is the strongest of these three W–C bonds? Is it three times as strong as the weakest one? Why or why not? (c) Suppose that your friend wanted to prepare a similar but better compound that not only contains a W–C single bond, a W=C double bond, and a W≡C bond bond, but also a W≡C quadruple bond. Can your friend succeed? Why or why not?

SOLUTION: (a) The W–C single bond includes one σ and no π bond; the W=C double bond includes one σ and one π bond; the W≡C triple bond includes one σ and two π bonds. (b) The triple bond is strongest, but it is not three times as strong as the single bond, because π bonds are weaker than σ bonds. (c) Your friend will *not* succeed because a quadruple bond needs a $d_{x^2-y^2}$ or d_{xy} orbital on each atom to form its δ component, and carbon does not have these valence orbitals. Also, W does not have enough valence electrons to form this bond in addition to a single, double, and triple bond.

3.3D. Polar Covalent Bond Energies and Pauling Electronegativities.
Table 3.4 lists and Figure 3.14 summarizes the bond dissociation energies measured in gaseous fluorides EF_n of s- and p-block elements. (Bond energies for chlorides, bromides, and iodides are given in G. Wulfsberg, *Inorganic Chemistry*, University Science Books: Mill Valley, CA, 2000, pp. 332–333.)

Linus Pauling noticed the following interesting periodic trends, which are displayed in Figure 3.14, in these bond dissociation energies:

1. The polar bonds of elements to halogens such as fluorine (Table 3.4) are also almost always stronger than the nonpolar bonds of elements to themselves (Table 3.2). However, such covalent bonds do contribute to the total bond energies.

2. The strongest bonds in the gaseous state are for the halides of the metals at the far left of the periodic table. Gaseous metal halides consist of pairs or larger groups of ions; their ionic cation–anion attractions also contribute to the total bond energies.

3. At the left of Figure 3.14 the bond energies seem to level off, as if the bond can only become so ionic.

4. Between the nonpolar F–F bond energy at the upper right of Figure 3.14 and the ionic bond energies on the left of Figure 3.14, the bond energies of gaseous polar covalent fluorides increase as the distance of the element from fluorine increases. In summary, ion pairs show higher bond energies than polar covalent bonds, which show higher bond energies than nonpolar covalent bonds.

TABLE 3.4. Element–Fluorine Bond Dissociation Energies[a,b]

HF	565														
LiF	573	BeF_2	632	BF_3	613	CF_4	485	NF_3	283	OF_2	189	F_2	155		
NaF	477	MgF_2	513	AlF_3	583	SiF_4	565	PF_3	490	SF_6	284	ClF_5	142		
												ClF_3	172		
												ClF	249		
KF	490	CaF_2	550	GaF_3	469	GeF_4	452	AsF_5	406	SeF_6	285	BrF_5	187	KrF_2	50
						GeF_2	481	AsF_3	484	SeF_4	310	BrF_3	201		
										SeF_2	351	BrF	249		
RbF	490	SrF_2	553	InF_3	444	SnF_4	414	SbF_5	402	TeF_6	330	IF_7	231	XeF_6	126
				InF	523	SnF_2	481	SbF_3	440	TeF_4	335	IF_5	268	XeF_4	130
										TeF_2	393	IF_3	272	XeF_2	131
CsF	502	BaF_2	578	TlF	439	PbF_4	331	BiF_5	297						
						PbF_2	394	BiF_3	393						

Source: Data are from J. E. Huheey, E. A. Keiter, and R. L. Keiter, *Inorganic Chemistry: Principles of Structure and Reactivity*, 4th ed., Harper-Collins: New York, 1993, pp. A-25–A-33.

[a]Units are in kilojoules per mole (kJ mol⁻¹).

[b]Compounds are shown for which each bond dissociation energy applies.

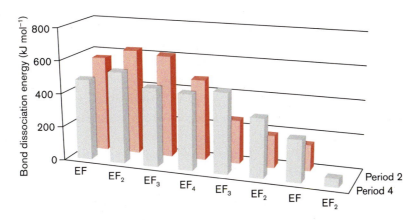

Figure 3.14. Element–fluorine bond dissociation energies (in kJ mol⁻¹) for second- and fourth-period fluorides of the same stoichiometry EF_n.

Pauling thus suggested that *bond energies* $\Delta H(E-X)$ could be analyzed as a *sum* of contributions from *covalent bonding* and *ionic bonding*:

$$\Delta H(E-X) = \text{Covalent bond contribution} + \text{ionic bond contribution} \qquad (3.4)$$

Pauling took the covalent contribution to the bond energy as the average of the halogen–halogen (X–X) and E–E covalent bond energies (Table 3.2). We may suppose that the ionic contribution should depend on the magnitude of the partial positive charge of the element times the magnitude of the partial negative charge of the halide divided by the bond distance (i.e., Z^2/r). If we neglect the variability of the bond distance, an equation for the element(E)–halogen(X) bond energy, $\Delta H(E-X)$, begins to emerge:

$$\Delta H(E-X) = \frac{1}{2}[\Delta H(X-X) + \Delta H(E-E)] + k\left(\begin{array}{c}\text{partial postive}\\ \text{charge of E}\end{array}\right)\left(\begin{array}{c}\text{partial negative}\\ \text{charge of X}\end{array}\right) \qquad (3.5)$$

The partial positive and negative charges in polar covalent bonds each should depend on the difference between the relative electron-attracting abilities of the two elements: The more unequal this ability, the more charge will build up at each end of the bond. Pauling called this electron-attracting ability the **electronegativity (χ_P)** of that element; he defined it as the attraction of an atom in a molecule (or polyatomic ion) for the electrons in its covalent bonds. Thus, each of the partial charges would be expected to build up in proportion to the difference between the electronegativities, $\Delta\chi_P = \chi_P(X) - \chi_P(E)$, of the two atoms, until the charge separation became as complete as it could be in the ion pairs, at a $\Delta\chi_P$ of about 1.8:

$$\Delta H(E-X) = \frac{1}{2}[\Delta H(X-X) + \Delta H(E-E)] + 96.5[\chi_P(X) - \chi_P(E)]^2 \qquad (3.6)$$

To use Equation 3.6, we must limit the value of $\chi_P(X) - \chi_P(E)$ to 1.8.

To obtain values of the Pauling electronegativities of the *s*- and *p*-block elements, $\chi_P(E)$, we rearrange Equation 3.6 to place the unknown on the left side and not in a squared expression:

$$\chi_P(E) = \chi_P(X) \pm \sqrt{\left(\frac{1}{96.5}\right)\{\Delta H(E-X) - \frac{1}{2}[\Delta H(X-X) + \Delta H(E-E)]\}} \qquad (3.7)$$

(Since extracting square roots from Eq. 3.6 gives positive and negative square roots, Eq. 3.7 gives two electronegativity values, one of which is nonsensical—higher than that of F, for example). Once we arbitrarily set one electronegativity value (3.98 for F), Equation 3.6 can be used, along with modern experimental gaseous bond dissociation energies, to obtain $\chi_P(E)$ for the *s*- and *p*-block elements. (The method of obtaining χ_P for *d*- and *f*-block elements is discussed in Section 7.5B.) These values were obtained separately for each kind of gaseous halide and for other single-bonded compounds of the elements, such as hydrides. The values in Table A represent mean values for these different types of compounds, and have standard deviations of ±0.05.

Changing oxidation numbers (Section 1.4) can change the Pauling electronegativity of atoms, as shown in Example 3.11.

EXAMPLE 3.11

The I–F bond energy differs in the different iodine fluorides; it is 231 kJ mol^{-1} in IF_7, 268 kJ mol^{-1} in IF_5, 272 kJ mol^{-1} in IF_3, and 278 kJ mol^{-1} in IF (Table 3.4). Compute the Pauling electronegativity of iodine in each of these compounds.

SOLUTION: We take the electronegativity of fluorine to be 3.98 and look up the I–I and F–F bond energies in Table 3.2; they are 149 and 155 kJ mol^{-1}, respectively. Now, substituting these values in Equation 3.7 and using the negative square root, we obtain:

$$\chi_P(I) = 3.98 - \sqrt{\left(\frac{1}{96.5}\right)\left\{\Delta H(I\text{–}F) - \frac{1}{2}[155 + 149]\right\}}$$

$$\chi_P(I) = 3.98 - \sqrt{\left(\frac{1}{96.5}\right)[\Delta H(I\text{–}F) - 152]}$$

Now substitute in the I–F bond energy for IF_7, which is 231 kJ mol^{-1}:

$$\chi_P(I) = 3.98 - \sqrt{\left(\frac{1}{96.5}\right)(231 - 152)}$$

$$\chi_P(I) = 3.98 - \sqrt{\left(\frac{1}{96.5}\right)(79)}$$

$$\chi_P(I) = 3.98 - \sqrt{(0.818)}$$

$$\chi_P(I) = 3.98 - 0.91 = 3.07 \text{ for I(VII) in } IF_7.$$

To do the other examples, we use the other three I–F bond energies, and obtain 2.88 for I(V) in IF_5, 2.87 for I(III) in IF_3, and 2.84 for I(I) in IF.

Example 3.11 shows that the Pauling electronegativity of an element increases with increasing oxidation number, which makes sense, because a more electron-poor atom should have a greater attraction for shared electrons in its covalent bonds. The example also shows that the change in electronegativity with changing oxidation state is usually not great enough to justify tabulating separate values, especially since the values have inherent uncertainties of ±0.05. In the few cases where such data are available, the variation is not much greater than 0.20, even over a range of several oxidation numbers. There are two major exceptions—as discussed in Section 9.6A, the electronegativities of thallium and lead are very sensitive to their oxidation number (changes of over 0.40 for a change of only 2 in their oxidation states). Hence, Table A contains separate values for the two oxidation states indicated for these two elements.

It is common to use increasing Pauling electronegativity differences to classify the bond type of a given bond as being less covalent or more ionic, up to a limit at roughly $\Delta\chi = 1.8$. Recall from Chapter 2 that we dealt with metal–oxygen bonds in hydrated ions, which only have appreciable covalent character if the electronegativity of the metal exceeds 1.50 (Eq. 2.14, Table 2.2) or 1.8, as in the approximation at the end of Section 2.2B. In Section 12.6, we refine this common way of distinguishing ionic and covalent bonding.

3.4. Lewis Structures, Maximum Total Coordination Numbers, and Fluoro Anions

OVERVIEW. After drawing the Lewis symbol of a central ion, you can add the outer atoms as ligand anions by forming coordinate covalent bonds to the central atom. This is an alternate way of drawing (assembling) the Lewis dot structure of a specified molecule or ion (Section 3.4A). You may practice this skill by trying Exercises 3.50–3.55. Using the concept of the maximum coordination number of a cation, you can add that number of ligands to the central cation to predict the formula and draw the Lewis structure of the central ion's highest fluoride, fluoro cation, or fluoro anion (Section 3.4B). You may practice constructing these species by trying Exercises 3.56–3.62.

This section again employs concepts from Chapter 1—namely, the Lewis symbols of atoms (Section 1.3), and oxidation numbers and charge balance (Section 1.4). The maximum coordination number of central atoms was covered in Section 2.3C. Assembling Lewis structures of complex ions from metal ions and a given number of ligands will be useful again in Sections 5.4 and 8.5.

3.4A. Drawing Lewis Structures Using the Lewis Acid–Base Concept. Some central-atom–outer-atom bonds are usually formed by the sharing of one electron from each atom (forming normal covalent bonds), and others are formed by the donation of electron pairs from the outer atom (forming coordinate covalent bonds). For example, in part 5 of Experiment 3 you may have reacted CsCl with $SnCl_4$ to form $Cs_2[SnCl_6]$. In this reaction two Sn–Cl coordinate covalent bonds are formed:

$$SnCl_4 + 2Cl:^- \rightarrow [SnCl_6]^{2-} \tag{3.8}$$

However, the $SnCl_4$ you used was probably formed by the reaction of $Sn + 2Cl_2$, during which we can envision the formation of four normal covalent bonds:

$$\cdot \overset{\cdot}{\underset{\cdot}{Sn}} \cdot + 4Cl\cdot \rightarrow Sn(-Cl)_4 \tag{3.9}$$

We could then conclude that the product, $[SnCl_6]^{2-}$, contained four covalent and two coordinate covalent bonds. However, there is no experimental difference among the six Sn–Cl bonds in $[SnCl_6]^{2-}$, so it is not very profitable to try to judge which bonds are of which type. In this section we will explore the advantages of assuming that all six of the Sn–Cl bonds are formed as coordinate covalent bonds, by the reaction of a Lewis acid and six Lewis bases:

$$Sn^{4+} + 6Cl^- \rightarrow [SnCl_6]^{2-} \tag{3.10}$$

This gives us an alternate procedure for assembling Lewis structures, based on coordinate covalent bond formation between A^{x+} cations (Lewis acids) and X^{y-} anions. This method offers some advantages: It not only gives a satisfactory Lewis structure (as the method of Section 3.1 also does), but it automatically locates unshared electron pairs on the central atom, and it predicts the overall charge of the $AX_n^{z\pm}$ product. The process is outlined in Figure 3.15, which is useful (with minor modifications) in this section and the next.

When we know the formula $AX_n^{z\pm}$ of the product, its Lewis structure can be assembled as follows:

1. Obtain the cation charge by calculating oxidation numbers according to the simple method in Section 1.4.

2. Treat each element with a positive oxidation number as a cation; write its Lewis symbol. If it has unshared s or p electron pairs, these will still be on the same atom in the Lewis structure of the final product.

3. Treat each element with a negative oxidation number as an anion; write its Lewis symbol (as in Section 1.4).

4. The coordination number of A in $AX_n^{z\pm}$ is just the value of n.

5. Attach each Lewis base (anion) to the Lewis acid (cation) with a coordinate covalent bond, normally represented by a line. Calculate the charge of the complex ion using Equation 3.3, the principle of charge balance.

6. If the central atom remains short of an octet of electrons (i.e., if less than four Lewis bases are attached) and the central and Lewis base donor atoms are capable of forming π bonds, convert unshared electron pair(s) on the donor atom to π bond(s).

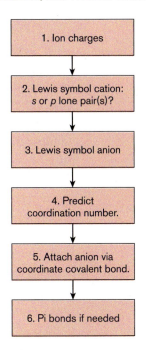

Figure 3.15. Flow chart for predicting formulas and assembling Lewis structures of $AX_n^{z\pm}$.

EXAMPLE 3.12

Draw the Lewis structures of (a) PCl_3, (b) SO_3, (c) SO_3^{2-}, (d) XeO_4^{2-}, (e) $SbCl_5$, and (f) $SbCl_5^{2-}$, treating each as being formed from the appropriate Lewis acids and bases.

SOLUTION: 1. In each of the molecules and ions in question, you should assign the oxidation number of –1 to chlorine and the oxidation number of –2 to oxygen. The oxidation numbers of the other elements are (a) +3 for P, (b) +6 for S, (c) +4 for S, (d) +6 for Xe, (e) +5 for Sb, and (f) +3 for Sb.

2. (a) The P^{3+} ion has an oxidation number two less than its group number. Hence, it retains two sp-hybridizable electrons and its Lewis symbol is $:P^{3+}$. Similarly, for (c) we obtain $:S^{4+}$, for (d) we obtain $:Xe^{6+}$, and for (f) we obtain $:Sb^{3+}$. In the other two cases, the elements are in the group oxidation states, so their Lewis symbols show no valence electrons: (b) S^{6+} and (e) Sb^{5+}.

3. Cl^- and O^{2-} are each drawn with an octet of electrons.

4. The value of n (the coordination number of A) is already given in the questions.

5. One electron pair on each Cl^- or O^{2-} is used to form a coordinate covalent bond to the —central atom, shown as –. This completes all of the Lewis structures except that of SO_3, which lacks an octet on S. (We have drawn the single-bonded resonance structures when possible.)

6. Oxygen is good at forming π bonds; it can act as a π-donor ligand, forming one of these bonds from O to S completes the Lewis structure of SO_3. The completed Lewis structures are to the right.

3.4B. Predicting the Formulas and Charges of Complex Ions. By using the Lewis acid–base concept, it is even possible to predict the number of ligands to include in the formulas and the charges of these $AB_n^{z\pm}$ species, if the cation has a predictable coordination number. Most cations reach their maximum coordination number (Section 2.4) when the donor atom is small (hence, sterically undemanding) and not too negatively charged. Adding this number of ligands gives us the "n" in the formula of the $AB_n^{z\pm}$; adding the charges of these ligands to the charge of the cation gives us the charge of the $AB_n^{z\pm}$.

Fluoride ion is probably the ideal ligand in this respect, because it does not readily form π bonds due to its high electronegativity, so Step 6 in Figure 3.15 does not apply. Second-period cations of moderate, strong, or very strong acidity characteristically form coordinate covalent bonds to *four* fluoride ions; this results in a neutral covalent fluoride, a fluoro cation, or a fluoro anion. Larger acidic third- and fourth-period cations characteristically show a maximum coordination number of *six* fluoride ions. The cations from the fifth and sixth periods are less completely predictable, but the maximum coordination number *eight* is found in UF_8^{2-}, WF_8^{2-}, ReF_8^{2-}, ReF_8^{-}, TeF_8^{2-}, and IF_8^{-}.[8,9]

EXAMPLE 3.13

Given as many fluoride ions as they can accommodate, what complex species will the second-period "cations" from Be^{2+} onward (each in their group oxidation state) form? What will be the Lewis structures of these species?

SOLUTION: Cation charges (Step 1) are equal to group numbers, so their Lewis symbols have no unshared electron pairs (Step 2). In Step 4, we assume a coordination number of 4 for each cation. In Step 5, the formulas and Lewis structures are obtained by attaching four fluoride ions to each "cation," being certain also to add one negative charge for each fluoride ion to the charge of the cation. The resulting formulas are BeF_4^{2-}, BF_4^{-}, CF_4, and NF_4^{+}. (Note that this particular series includes one fluoro cation, one fluoride, and two fluoro anions. All of these are known species.) Because the group oxidation number is not known for O, F, or Ne, the series terminates at N. Feebly acidic Li^+ does not form a complex ion with F^-.

The Lewis structures each contain four coordinate covalent M–F bonds; these are isoelectronic species:

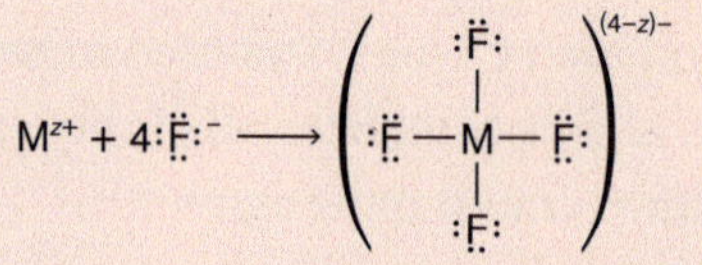

$$M^{z+} + 4\,:\!\ddot{F}\!:^{-} \longrightarrow \left(:\!\ddot{F}\!-\!\overset{\displaystyle :\!\ddot{F}\!:}{\underset{\displaystyle :\!\ddot{F}\!:}{M}}\!-\!\ddot{F}\!: \right)^{(4-z)-}$$

Here M represents the second-period element and z is its group number. Only the charge on the complex species varies across the period.

Now, what happens when the Lewis acid is a *p*-block "cation" with one (or more) electron pairs already on it. If only one electron pair is present, we know that pair is located in a valence *ns* orbital in a true cation. But that orbital, being full, cannot be used to accept electrons from a donor atom of a Lewis base. The spherical *ns* electron pair would repel all the donor atoms in all directions from the Lewis acid, preventing the formation of any very good coordinate covalent (or covalent) bonds. Rather than give up this advantage, the Lewis acid will usually confine the electron pair to some type of *sp* hybrid orbital, as in Figure 3.11. Then the repulsive effect is confined to one direction, so the Lewis bases can approach from all of the directions except that one. Thus, the total number of fluoride ions (ideal Lewis bases) expected to coordinate to the Lewis acid is *one less than the maximum coordination number.* [It has been suggested[10] that lone pairs take twice as much space in the third and fourth periods, and three times as much space in the fifth period. Sometimes, therefore, the number of ions coordinated may be two (or even three) less than the maximum coordination number.]

To render the preceding discussion more precise, it is useful to introduce the concept of the **total coordination number (TCN)**.[11] This is the sum of the number of other atoms attached to the central atom (Lewis acid) in question plus the number of unshared *p*-block valence electron pairs (*sp* hybrids, etc.) about it. In terms of the VSEPR model of structure of *p*-block compounds of general formula AX_nE_m (reviewed in Section 5.1A), with *n* ligands and *m* unshared electron pairs, $TCN = n + m$. We may take as a working hypothesis that there is a *maximum total coordination number* for Lewis-acidic cations in each period—namely, four in the second period, six in the third and fourth periods, and in excess of six in the fifth and sixth periods. (We will be able to refine our predictions when we cover radius ratios in Section 4.1B.)

Why Ignore Unshared Metal Electron Pairs in the *d* and *f* Blocks?

AN AMPLIFICATION

Any unshared electrons present on *d*- or *f*-block cations are not in *sp* hybrid orbitals pointing at incoming ligands, but in the $(n - 1)d$ or $(n - 2)f$ orbitals, which frequently fall *between* the ligands. Thus, in complexes with six ligands, the ligands fall along the *x*, *y*, and *z* axes, while the d_{xy}, d_{xz}, and d_{yz} orbitals fall between the *x*, *y*, and *z* axes (Figure 3.7). Frequently, these orbitals are also inside the valence or post-valence orbitals used to accept the donated electrons. Therefore, they have no effect on the number of donor atoms coordinated to the Lewis acid. Hence, in the case of *d*- or *f*-block Lewis acids, we ignore any remaining valence electrons in predicting the formula of the species.

EXAMPLE 3.14

Predict the formulas and draw the Lewis structures (if appropriate) of the fluoride, fluoro anion, or fluoro cation formed by these Lewis acids when an excess of F^- is available: (a) Cl^{5+}, (b) I^{5+}, (c) Cl^{3+}, and (d) Fe^{3+}.

SOLUTION: (a) The Lewis formula of the Lewis acid is $:Cl^{5+}$, which leaves room for one less than the maximum total coordination number (6) of fluoride ions to coordinate: We expect the highest fluoro species to be neutral ClF_5. (Unexpectedly, very unstable ClF_6^- and more stable BrF_6^- ions have been prepared.[12]) (b) The isoelectronic $:I^{5+}$ cation is from the fifth period, so it has a maximum total coordination number over six. If the maximum TCN is seven, then the species $:IF_6^-$ should be formed. (c) The Lewis formula of the Lewis acid is $:Cl:^{3+}$, with two unshared pairs of electrons; these should block the approach of two fluoride ions, allowing four to enter, giving ClF_4^-. (d) The valence electron configuration of the fourth-period Lewis acid Fe^{3+} is $3d^5$. But because these $(n-1)d$ electrons tend not to block access of donor atoms (see the preceding Amplification), we ignore them and predict the formula FeF_6^{3-}. Lewis structures are customarily not drawn for d-block complexes; the Lewis structures of the other examples are as follows:

(a)   (b)   (c)

Nomenclature of Fluoro Anions

The nomenclature of fluoro anions is based on the general pattern used for the nomenclature in coordination compounds in general. Three elements are involved in naming a fluoro cation or anion.

1. The number of fluoro groups is indicated by a prefix (di– for two, tri– for three, tetra– for four, penta– for five, hexa– for six, hepta– for seven, and octa– for eight) placed in front of fluoro–, which is placed in front of the name of the central atom.

2. In a fluoro anion, the ending of –ate is added after the name of the central atom, and the Latin root name of the element is used if it has one (Table D).

3. If the central atom can vary its oxidation number, this is indicated in either of two ways within parentheses at the end of the name of the species. In the Stock convention, the oxidation number of the central atom is indicated in Roman numerals in the parentheses; in the Ewens–Basset convention, the charge on the ion is indicated in Arabic numerals in the parentheses. (The Stock convention is probably more common, but the Ewens–Bassett convention is used in Chemical Abstracts.)

As examples, the ion SiF_6^{2-} may be named hexafluorosilicate(IV) ion (Stock convention) or hexafluorosilicate(2–) ion (Ewens–Bassett convention); similarly, FeF_6^{3-} may be named hexafluoroferrate(III) ion or hexafluoroferrate(3–) ion.

AN AMPLIFICATION

3.5. The Formulas of Oxo Anions

OVERVIEW. The central atoms of oxo anions usually achieve a lower (penultimate) total coordination number than the maximum total coordination numbers shown by central atoms of fluoro anions; this number is three in the second period, four in the third and fourth periods, and six in the fifth and sixth periods. You may practice using this concept to predict the formulas of oxo anions (and acids) of elements in specified oxidation states by trying Exercises 3.63–3.67. You can also suggest an explanation of why oxo anions tend to achieve the penultimate rather than the maximum total coordination number of the central atom; try Exercise 3.68. Fifth- and sixth-period metal atoms in oxidation states of five or higher can be so good at ϖ bonding that they may limit themselves to achieving an even lower (antepenultimate) total coordination number of four; apply this concept by trying Exercises 3.69 and 3.70.

This section expands the concept of maximum total coordination numbers that you learned in Section 2.3C and applied to fluoro anions in Section 3.4. These concepts are extensions of those used in Section 3.4 and will find utility in the same chapters and sections.

Fluoro anions are useful for illustrating the assembly of a complex anion from a Lewis acid plus some Lewis bases, and they are useful in the laboratory, but fluorine is a relatively rare element in nature, so fluoro anions have comparatively few practical applications. **Oxo anions**, MO_x^{y-}, which can be assembled from Lewis acids plus smaller numbers of oxide ions, O^{2-}, as Lewis bases, are far more important in ordinary life; in fact, most of Earth's crust consists of the salts of oxo anions. (We will explore the geochemistry of these anions in Sections 4.3, 8.6, and 8.7.) The resulting terminal oxygen atoms, which are bonded to M and to no other atom (shown in **brown** in Figure 3.16), are known as **oxo groups**.

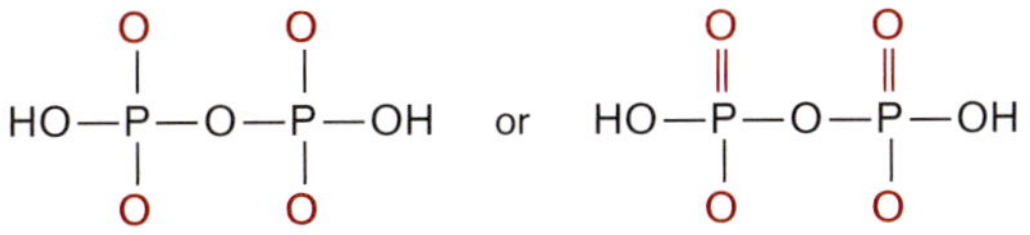

Figure 3.16. Diphosphoric acid (two resonance structures), showing oxo groups in **brown**. (The other oxygen atoms are parts of hydroxo groups or are bridging oxygens.)

Oxo anions are among the several classes of inorganic compounds in which the central Lewis acid does not normally achieve the maximum total coordination number, but instead achieves the next lower common coordination number, which we will call the **penultimate total coordination number**. (Sometimes a still lower common coordination number is reached, which we call the **antepenultimate total coordination number**.) In the second period, the penultimate TCN is three; in the similarly sized third and fourth periods, this number is four; and in the fifth, sixth, and seventh periods, this number is six (Table 3.5). When applicable, the antepenultimate TCNs are two in the second period, three in the third and fourth periods, and four in the fifth, sixth, and seventh periods.

TABLE 3.5. Maximum and Lower Coordination Numbers in the Periodic Table

Periods	Maximum Coordination Number (Example)	Penultimate Coordination Number (Example)	Antepenultimate[a] Coordination Number (Example)
2	4 (CF_4)	3 (CO_3^{2-})	2 (CO_2, CN_2^{2-})
3, 4	6 (SF_6, PF_6^-)	4 (SiO_4^{4-})	3 (SO_3, GaN_3^{6-})
5, 6	8 (IF_8^-)	6 [$Sb(OH)_6^-$]	4 (OsO_4, MoO_4^{2-})

[a] "Second from the highest." Note that coordination number 5 is not commonly favored in non-d-block complexes.

Why is the lower total coordination number realized in oxo anions? One difference of the oxide ion is that it has a –2 charge, double that of the fluoride ion, so attaching a large number of these ions builds up oxo anions with extremely high negative charges. Those negative charges would be expected to repel each other, perhaps enough to destabilize the anion.

The main reason for the low total coordination number is probably the one advanced by Mayer[13]—namely, the oxide ion (but not the fluoride ion) tends to form π bonds. Unlike F^-, O^{2-} is a very strong base, so it is more likely to continue donating unshared electron pairs by forming π bonds. For this to happen there must be empty acceptor orbitals on the Lewis acid. In a hypothetical second-period oxo anion showing the maximum total coordination number, such as CO_4^{4-}, there would be no empty valence orbital. For carbon to accept a second pair of electrons from oxygen, the oxide ion must be expelled (Figure 3.17). Once this is done, a strong second-period π bond is formed, excessive negative charge is dispelled, and some relief of overcrowding may be achieved. Thus π-bonded oxo anions are preferred in the second period for carbon and nitrogen central atoms (although not for the poorly π-bonding boron atom).

In some cases the process of Figure 3.17 can be repeated again to form either one triple bond or two double bonds, and result in a still lower total coordination number. This can happen when the central atom is highly acidic (charge of +5 or higher, especially Mo^{6+}, W^{6+}, Tc^{7+}, and Re^{7+}) and the outer atoms are nitrogen atoms in **nitrido anions** such as CN_2^{2-}, or when a lot of negative charge can be expelled from the oxo anion (e.g., $MoO_6^{6-} \rightarrow MoO_4^{2-} + 2O^{2-}$).

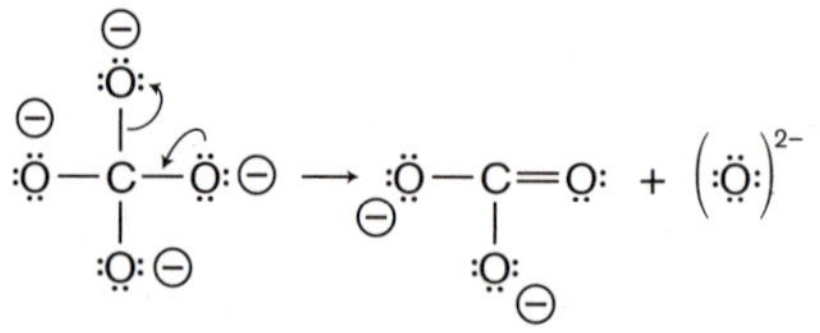

Figure 3.17. Expulsion of an oxide ion from a hypothetical CO_4^{4-} ion to give a stable π-bonded CO_3^{2-} ion.

Nitrido Anions

Nitrido anions are complexes in which N^{3-} outer atoms may form additional π bonds to the central atom. This can allow the antepenultimate coordination number to occur in anions (Table 3.4). Period 2 examples include BN_3^{3-}, CN_2^{2-} (cyanamide ion), and NN_2^- or N_3^- (azide ion; Figure 5.15). In Periods 3 and 4 an example is GaN_3^{6-}; examples in Periods 5 and 6 include MN_4^{6-} for Group 6 metals and MN_4^{7-} for Group 7 metals. Note the decreasing coordination numbers in the series GaF_6^{3-}, GaO_4^{5-}, and GaN_3^{6-}.[14]

EXAMPLE 3.15

Predict the formulas of the oxo anions formed by the following elements, each with the specified oxidation number: N(III), P(V), As(III), Se(IV), I(VII), I(V), and Mn(VI).

SOLUTION: 1. The Roman numerals indicate the charges on the central "cations."

2. Some of these have sp lone pairs. Their Lewis symbols are $:N^{3+}$, P^{5+}, $:As^{3+}$, $:Se^{4+}$, I^{7+}, $:I^{5+}$, and Mn^{6+} (in Mn^{6+}, the lone electron is not in an sp-hybridizable orbital).

3. In all cases the anion is O^{2-}, with an octet of electrons.

4. For oxo anions, we predict antepenultimate TCNs: 3 for $:N^{3+}$; 4 for P^{5+}, $:As^{3+}$, $:Se^{4+}$, and Mn^{6+}; and 6 for I^{7+} and $:I^{5+}$.

5. The number of oxide ions we attach is one less than the antepenultimate TCN for each unshared sp-hybridizable electron pair on the central atom. We determine the charge of the oxo anion by adding charges of the central Lewis acid and the oxide ions:

$$:N^{3+} + 2O^{2-} \rightarrow :NO_2^- \qquad\qquad P^{5+} + 4O^{2-} \rightarrow PO_4^{3-}$$
$$:As^{3+} + 3O^{2-} \rightarrow :AsO_3^{3-} \qquad\qquad :Se^{4+} + 3O^{2-} \rightarrow :SeO_3^{2-}$$
$$I^{7+} + 6O^{2-} \rightarrow IO_6^{5-}$$
$$:I^{5+} + 5O^{2-} \rightarrow :IO_5^{5-} \text{ (in fact, only three oxides attach, to give } :IO_3^-)$$
$$\cdot Mn^{6+} + 4O^{2-} \rightarrow MnO_4^{2-}.$$

6. In NO_2^- the central N atom lacks an octet, so a double bond to one oxygen is required. The Lewis structures of the oxo anions are as follows, where single-bonded resonance structures are shown unless a double bond is required:

Predicted   Actual   (Has one *d* electron)

Tables 3.6 and 3.7 list the major oxo anions of the elements. As can be seen, our first five predictions are all confirmed; MnO_4^{2-} also has the predicted formula. The iodine(V) oxo anion is in fact IO_3^-. In some cases the expulsion of oxide ions is reversible, so there are two known oxo anions for the element (e.g., IO_6^{5-} and IO_4^-).

When the central atom contains an unshared pair of sp-hybridizable electrons, its oxo anion contains (usually one, sometimes two or more) fewer oxo groups (Table 3.7).

There are also a few cases in which the central atom oxidation number is four or six less than the group number, such as NO^-, ClO_2^-, BrO_2^-, IO_2^-, ClO^-, BrO^-, and IO^-. (In the d block, the difference can be an odd number, as in MnO_4^{2-}.)

EXAMPLE 3.16

Write the formulas of the oxo anions that would be formed by each of the Group 14/IV elements (carbon group) in the +2 oxidation state (do not worry that some of these do not actually exist). Then explain briefly why the formulas change as you go down the group.

SOLUTION: In the +2 oxidation state, each Group 14/IV ion has an unshared electron pair. Therefore, to achieve the penultimate total coordination number for each element, we reserve one space for an sp hybrid electron pair. Thus, $:C^{2+}$, with an expected antepenultimate total coordination number of 3 (last column of Table 3.7), adds (3 − 1) oxide ions to give CO_2^{2-}. $:Si^{2+}$ and $:Ge^{2+}$ add (4 − 1) oxide ions to give SiO_3^{4-} and GeO_3^{4-}, respectively. (Note that formulas as commonly written do not show the unshared electron pairs.) In principle, $:Sn^{2+}$ and $:Pb^{2+}$ add (6 − 1) oxide ions to give SnO_5^{8-} and PbO_5^{8-}, but in practice SnO_3^{4-} and PbO_3^{5-} are formed. The number of oxo groups increases down the group as the penultimate total coordination number increases—that is, as the central atom increases in size and ability to attach more oxo groups.

EXAMPLE 3.17

Which of the following Group 6 cations are expected to form predominantly oxo anions of formula (Element)O_4^{2-}, and why: S^{6+}, Se^{6+}, Te^{6+}, Cr^{6+}, Mo^{6+}, W^{6+}, Cr^{3+}, Mo^{3+}, and W^{3+}?

SOLUTION: S^{6+}, Se^{6+}, and Cr^{6+} do this because of moderate π-bonding ability, which allows them to achieve their penultimate total coordination number (4). Mo^{6+} and W^{6+} also do this because their high acidity gives them strong π-bonding ability, so they usually achieve their antepenultimate total coordination number (4). Cr^{3+}, Mo^{3+}, and W^{3+} are not highly charged enough to be good π bonders, so they do not form oxo anions with a total coordination number as low as four.

TABLE 3.6. Formulas and Names of the Important Oxo Anions in Which the Central Atom Oxidation Number Matches the Group Number

Period	III	IV	5 or V	6 or VI	7 or VII	8 or VIII	Penultimate CN
2	BO_3^{3-a}	CO_3^{2-}	NO_3^-				3
3	AlO_4^{5-}	SiO_4^{4-}	PO_4^{3-}	SO_4^{2-}	ClO_4^-		4
4	GaO_4^{5-}	GeO_4^{4-}	AsO_4^{3-}	SeO_4^{2-}	BrO_4^-		4
4			VO_4^{3-}	CrO_4^{2-}	MnO_4^-		4
5		SnO_6^{8-}	SbO_6^{7-}	TeO_6^{6-}	IO_6^{5-}	XeO_6^{4-}	6
5						RuO_6^{4-}	6
5				TeO_4^{2-}	IO_4^-		4 (Antepenultimate)
5				MoO_4^{2-}	TcO_4^-		4 (Antepenultimate)
6		PbO_6^{8-}	BiO_6^{7-}		ReO_6^{5-}	OsO_6^{4-}	6
6				WO_4^{2-}	ReO_4^-		4 (Antepenultimate)
Names	–ate	–ate	–ate	–ate	Per– –ate	Per– –ate	

[a] The larger boron atom is less prone to π bonding and does not have as clear a preference. Boron oxo anions can show either coordination number three or four in different circumstances.

TABLE 3.7. Formulas and Names of the Important Oxo Anions in Which the Central Atom Oxidation Number Is Two Less Than the Group Number

Period	III	IV	5 or V	6 or VI	7 or VII	8 or VIII	Penultimate TCN
2			$:NO_2^-$				3
3				$:SO_3^{2-}$	$:ClO_3^-$		4
4			$:AsO_3^{3-}$	$:SeO_3^{2-}$	$:BrO_3^-$		4
4						$:FeO_4^{2-}$	4
5		$:SnO_3^{4-}$	$:SbO_3^{3-}$	$:TeO_3^{2-}$	$:IO_3^-$	$:XeO_4^{2-}(TCN=5)$	4 (Antepenultimate)
5						$:RuO_4^{2-}$	4 (Antepenultimate)
6						$:OsO_4^{2-}$	4 (Antepenultimate)
Names		–ite	–ite	–ite	–ate	–ate	

Oxo Cations

The elements with good π-bonding ability can also form fewer numbers of π bonds, producing **oxo cations**. As an example, U^{6+} is a very strongly acidic cation, and we can imagine it reacting as in Section 2.2 in four stages to produce a cation with four hydroxy groups (Eq. 3.11). But (at least in cations) U^{6+} also has strong π-bonding ability, and it can eliminate water so as to replace four U–O single bonds with two U=O double bonds in the linear hydrated *uranyl ion*:

$$[U(H_2O)_6]^{2+} \rightarrow [U(OH)_4(H_2O)_2]^{2+} \rightarrow [O=U=O]^{2+}(aq) + 4H_2O \qquad (3.11)$$

Similar oxo cations with one or two oxo groups are known, not only for the aforementioned Period 5 and 6 elements, but also in the fourth period—for example, with attached chloride ligands in $O=VCl_3$ and $(O=)_2CrCl_2$, which are vanadyl chloride and chromyl chloride, respectively. Similar chlorides from the *p* block include $O=PCl_3$ and $(O=)_2SCl_2$, which are phosphoryl chloride and sulfuryl chloride, respectively. ($O=SCl_2$ is known as thionyl chloride.)

The acidity of oxo cations is intermediate between the acidity of the same bare central atom and the acidity of monatomic cations bearing the same net charge. For example, the pK_{a1} value of PuO_2^+ is 9.6,[15] making it a weak acid—much less acidic than Pu^{5+} but more acidic than any +1 monatomic cation in Table 2.1. The pK_{a1} value of the more highly charged PuO_2^{2+} ion is 3.34, making it moderately acidic (more acidic than PuO_2^+ or any monatomic +2 cation).

3.6. The Basicity of Oxo Anions

OVERVIEW. The pK_b of an oxo anion can be calculated with an uncertainty of ±1.0 based on its charge and number of oxo groups, then classified into one of the basicity categories previously defined in Section 2.6 (Section 3.6A). You may practice the calculation and classification by trying Exercises 3.71–3.78. You can use one of several concepts to explain why some oxo anions are more basic than others (Section 3.6B). You can apply your explanations to these simple or some important larger oxo anions by trying Exercises 3.79–3.82.

This classification scheme is the same one that you previously learned in Section 2.1 for cations and Section 2.6 for monatomic anions, and it again employs charges, sizes, and electronegativities from Chapter 1. You will use the ability to calculate and classify the basicity of oxo anions again in Sections 4.3, 4.4, 4.5, 8.6, and 8.7.

Oxo anions are hydrated in the same manner as monatomic nonmetal ions (Section 2.5). Hydrated oxo anions also undergo reactions with water to give basic solutions containing OH^- and partially protonated anions:

EXPERIMENT 4 MAY BE ASSIGNED NOW.
(See www.uscibooks.com/foundations.htm.)

$$MO_x^{y-} + H_2O \rightleftharpoons [MO_{(x-1)}OH]^{(y-1)-} + OH^- \tag{3.12}$$

A familiar example of this process is the protonation of carbonate ion to give bicarbonate:

$$CO_3^{2-} + H_2O \rightleftharpoons HCO_3^- + OH^- \tag{3.13}$$

Simple oxo anions span the whole range of basicity found in monatomic anions (Section 2.6B). We will use the same terms to classify the degrees of basicity. (As it happens, however, there are no practical consequences of subdividing the weakly basic, moderately basic, and strongly basic categories.) As before, *nonbasic* oxo anions show no detectable tendencies to undergo reaction with water and do not alter the pH of their aqueous solution. Their salts tend to be water soluble (Section 4.3B), but the insoluble ones are useful in syntheses (Section 4.3C). *Feebly basic* oxo anions have distinct solubility patterns and mineral forms (Section 4.3D), but in aqueous solution also give almost no pH change to the solution. The *weakly, moderately,* and *strongly basic* oxo anions include the ones on which equilibrium calculations were done in general chemistry, probably under the heading of "salts of weak acids." Their solutions are distinctly basic, and they show yet another solubility pattern and produce a large category of important minerals (Section 4.3A). The *very strongly basic* anions are normally found only in insoluble minerals (Section 4.3A) or anhydrous solids, because they react completely or nearly completely in water according to Equation 3.12. Table 3.8 gives examples of these categories of oxo anions, along with their measured pK_b values when these are available.

TABLE 3.8. pK_b Values and Basicity Classifications of Oxo Anions

Calculated pK_{b1} Range	Classification	Type	Examples with Known pK_{b1}
>14.0	Nonbasic anions	MO_4^-	M = Cl, Br, Mn, Tc, Re
		MO_3^-	M = N (13.6), Cl, Br (13.0), I (13.2)
14 to ~11.5	Feebly basic anions	MO_4^{2-}	M = S (12.1), Se (12.0), Xe, Ru, Os, Mo (9.9), W (9.4), Cr (7.5), Fe (6.2), Mn (3.8)
		MO_2^-	M = N (10.7), Cl (12.1)
~11.5 to ~(−4)	Weakly, moderately, and strongly basic anions	MO_3^{2-}	M = C, S (6.8), Se (7.4), Te (6.3)
		MO^-	M = Cl (6.5), Br (5.3), I (3.4)
		MO_6^{4-}	M = Xe, Os
		MO_4^{3-}	M = P (2.0), As (1.5), V (1.0), Fe (2.9)
< ~(−4)	Very strongly basic anions	MO_3^{3-}	M = As (0.01), Sb
		MO_6^{5-}	M = I, Np
		MO_4^{4-}	M = Si, Ge
		MO_3^{4-}	M = Sn
		MO_6^{6-}	M = Te
		MO_3^{4-}	M = B, Al, Ga

Sources: Known pK_b values are calculated from the appropriate pK_a values given in F. A. Cotton and G. Wilkinson, *Advanced Inorganic Chemistry: A Comprehensive Text*, 5th ed., Wiley-Interscience: New York, 1988, p. 105; R. C. Weast, Ed., *Handbook of Physics and Chemistry*, 84th ed., Chemical Rubber Publishing Co.: Cleveland, 1998; J. A. Dean, Ed., *Lange's Handbook of Chemistry*, 13th ed., McGraw-Hill: New York, 1985; B. H. J. Bieleski, *Free Radical Res. Commun.* 12–13, 469 (1991); and D. D. Perrin, *Ionization Constants of Inorganic Acids and Bases in Aqueous Solution*, 2nd ed., Pergamon: Oxford, UK, 1982.

3.6A. Effects of Charge, Number of Oxo Groups, and Electronegativity.

Again we will try to develop simple rules that will enable us to predict just how basic the solution of an oxo anion should be, and that will enable us to predict some practical consequences. For example, any solution that is highly basic is also corrosive to many materials, including human tissue! With cations (Section 2.2) and monoatomic anions (Section 2.6) we looked at three variables: charge, size, and electronegativity of the metal ion. The size of oxo anions is determined less by the radius of the central atom than by the number of oxo groups present. Thus, the basicity of oxo anion of formula MO_x^{y-} depends on the effects of the charge of the oxo anion, $-y$, its number of oxo groups, x, and (secondarily) the electronegativity of the central atom M.

Effect of Charge. Increasing negative charge on an anion increases its tendency to give basic solutions. Referring to the (partial) lists of oxo anions in Tables 3.6 and 3.7, we see that the negative charges of oxo anions can be quite substantial indeed. Therefore, it is not surprising that many of them are very strongly basic. Comparison of the pK_{b1} values of otherwise-similar oxo anions with increasing negative charges leads to the conclusion that the pK_{b1} of an oxo anion *decreases* by 10.2 units for *each (additional)* negative charge on it. This increase in basicity is enough, in favorable cases, to change the basicity by two categories with a change of only one unit of charge. Thus, –3-charged PO_4^{3-}, with $pK_{b1} = 2.0$, is moderately basic, while –2-charged SO_4^{2-}, with $pK_{b1} = 12.1$, is feebly basic, not weakly basic.

Effect of Number of Oxo Groups. Most of the nonmetals show more than one oxidation number and can form oxo anions that differ in the number of oxo groups that are attached to the nonmetal atom. These different oxo anions differ substantially in basicity. The most complete series of oxo anions is that of chlorine, which forms four different oxo anions with these pK_{b1} values: ClO^-, 6.5; ClO_2^-, 12.1; ClO_3^- and ClO_4^-, too nonbasic to be measured in water. Examination of the pK_{b1} values for a number of such sets of oxo anions (Table 3.8) shows that, on the average, *each additional* oxo group in an oxo anion *increases* its pK_{b1} by 5.7 units.

We may incorporate the effects of the number of oxo groups and of the negative charge of an oxo anion into one equation that allows a reasonably accurate calculation of the constant for the first hydrolysis of an oxo anion:

$$pK_{b1} = 10.0 + 5.7x - 10.2y \pm 1.0 \qquad (3.14)$$

Here x is the number of oxo groups and y is the number of units of negative charge in the oxo anion of formula MO_x^{y-}. Keep in mind the physical reasoning behind this equation: (1) higher pK_{b1} values correspond to weaker basicity (lower pH); (2) additional oxo groups weaken basicity, hence add to pK_{b1}; and (3) additional negative charges increases basicity, hence subtract from pK_{b1}.

additional oxo groups INCREASE acidity

EXAMPLE 3.18

Compute the pK_{b1} values and classify the basicity of each of the following oxo anions (some of which are hypothetical or quite unstable): BrO_3^{3-}, MnO_4^{3-}, and UO_4^{2-}.

SOLUTION: Using Equation 3.14 results in the following calculations of pK_{b1}: For BrO_3^{3-}, $pK_{b1} = 10.0 + 5.7(3) - 10.2(3) = -3.5$, which with an error of ±1.0 may fit in either the category of strongly basic or very strongly basic. For MnO_4^{3-}, $pK_{b1} = 10.0 + 5.7(4) - 10.2(3) = 2.2$, which fits the defined range for a moderately basic anion. For UO_4^{2-}, $pK_{b1} = 10.0 + 5.7(4) - 10.2(2) = 12.4$, which falls in the range defined for a feebly basic anion.

Effect of Electronegativity. The electronegativity of the nonmetal atom also influences the basicity of an oxo anion. Comparing the pK_{b1} values of ClO^- (6.50), BrO^- (5.3), and IO^- (3.4), we find that reducing the electronegativity of the halogen atom increases the basicity of the oxo anion. For *d*-block metals of much lower electronegativity than *p*-block elements, this effect may be greater. The pK_{b1} for MO_4^{2-} is 12.1 for M = S and 12.0 for M = Se; for the somewhat less electronegative metals Mo and W, the pK_{b1} values are 9.9 and 9.4, respectively, while for the still less electronegative metals

Cr, Fe, and Mn, the pK_{b1} values are even lower: 7.5, 6.2, and 2.9, respectively. (But no such trend is noted for the MO_4^{3-} ions shown in Table 3.8.) Qualitatively it appears that electronegativity has a somewhat smaller effect than the charge or the number of oxo groups in the oxo anion. Therefore, we generally apply Equation 3.14 and ignore the influence of electronegativity.

Fluoro Anions. Most fluoro anions have charges from –1 to –3, and have at least six fluoro groups, so their ratio of Z^2/size should generally be quite small. The conjugate acid of a fluoro anion would be protonated on fluorine, giving it a positive formal charge, which is quite unfavorable for the extremely electronegative fluorine atom. Therefore, most or all fluoro anions should probably be *nonbasic*. It is difficult to measure pK_b values for nonbasic anions, so there is little experimental data: pK_b for BF_4^- is either 0.5 or –4.9, based on indirect measurements. The main use of fluoro anions in inorganic chemistry is as valuable nonbasic anions (Section 4.3C). (Conversely, most nitrido anions are extremely basic, so they react with water and are found only in the solid state.)

EXAMPLE 3.19

Boron in its group oxidation state of +3 forms a fluoro anion BF_4^- but an oxo anion of formula BO_3^{3-}. (a) Does the reduction in charge on going from BO_3^{3-} to BF_4^- make BF_4^- more basic than BO_3^{3-} or less basic? (b) Does the change in size on going from BO_3^{3-} to BF_4^- make BF_4^- more basic than BO_3^{3-} or less basic?

SOLUTION: (a) BF_4^- has less charge, so it is less basic. (b) BF_4^- is larger, so it is less basic.

3.6B. Why Should the Number of Oxo Groups Have an Effect on the Basicity of an Oxo Anion? Students contemplating this question in the past have devised a number of explanations that fit the facts and are chemically reasonable.

1. *Resonance structures.* Those who have already studied organic chemistry recognize that the more oxo groups there are in an oxo anion, the more resonance structures can be drawn by which the negative charge of the anion is delocalized (compare the three plausible double-bonded resonance structures possible for SO_3^{2-} with the 10 for SO_4^{2-} drawn in Figure 3.18). There is greater resonance stabilization in anions with more oxo groups.

2. *Charge to size ratio/formal charges.* Another explanation looks at the Lewis dot structures of oxo anions and asks the question, "On which atoms of this oxo anion are the negative charges formally located?" Charges are not really on any one atom, because oxo anions are (internally) covalently bonded species, but knowing that oxygen is more electronegative than the other atom in the oxo anion, we can assume that the charges formally are on oxygen. But in an oxo anion such as ClO_4^-, there is no reason to assume that the negative charge is on one particular oxygen any more than on any of the others, so we may suppose that, overall, each oxygen atom bears a partial charge

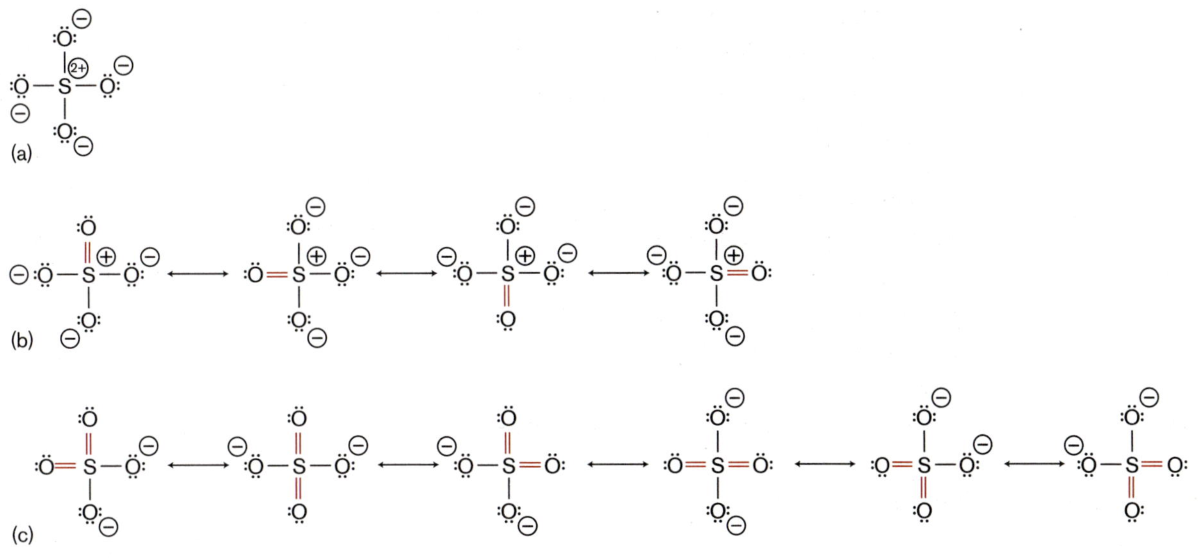

Figure 3.18. Resonance structures and formal charges in the sulfate ion, SO_4^{2-}. The circled numbers are formal charges located on the indicated atoms of each structure. (a) Structures with only single bonds; these have the advantage of not using post-valence *d* orbitals, but have the disadvantage of requiring the separation of many formal charges. (b) Structures with one S=O double bond. (c) Structures with two S=O double bonds, which minimize formal charge buildup but require maximal utilization of postvalence orbitals on S.

of (formally) −¼. Likewise in ClO_3^-, we suppose a greater partial charge of −⅓, while the oxygen atom in ClO^- is stuck with the whole charge of −1. The oxygen atom in ClO^- therefore should be more basic than any other oxygen atom in any of the other oxo anions of chlorine. Although these numbers are approximations, there should indeed be partial negative charges on oxygen, which should decrease as the number of oxo groups increases, resulting in reduced basicity. From a slightly different perspective, the more oxo groups are present, the larger the oxo anion is, so the more the charge is dispersed over space.

3. *Oxidation effects.* The conversion of ClO^- to ClO_2^- is actually an oxidation process (Chapter 6), which can be achieved by attaching an O atom to an unshared pair of electrons that was on the Cl atom of ClO^-. Oxygen is a very electronegative atom, so when we add an oxo group to an anion such as ClO^-, the new oxygen atom strongly withdraws these electrons from the Cl. (In effect, the Cl now acts as a "cation" with a positive "charge" increased by 2.) The Cl now compensates its electron loss by attracting electrons more strongly from the other oxygen atom(s); hence, there is less effective negative charge at each oxo group for attracting the hydrogen atom of the water of hydration.

It is not infrequent in science to have more than one explanation of a phenomenon such as this. If the explanations are all satisfactory, we may continue to use more than one, depending on the situation or on personal preference. If you think about the preceding explanations, you will see that they are far from being completely different from one another—to a large degree they are just different ways of saying the same thing.

EXAMPLE 3.20

In 12 words or less, explain why (a) NO_2^- is more basic than NO_3^-; (b) PO_4^{3-} is more basic than SO_4^{2-}; (c) NO_2^- and $C_2O_4^{2-}$ have about the same values of pK_{b1} (10.7 and 10.19, respectively).

SOLUTION: (a) Each oxo group in NO_2^- has more negative charge than in NO_3^- ($-\frac{1}{2}$ vs. $-\frac{1}{3}$). (b) PO_4^{3-} has a greater negative charge than SO_4^{2-}. (c) Each ion has $-\frac{1}{2}$ charge per oxo group.

Larger Oxo Anions. An additional group of oxo anions of importance are those with *two* or *three* identical central atoms connected to each other, either by element–element bonds or by element–oxygen–element bonds. An example of the former is the **oxalate** ion, $[O_2C-CO_2]^{2-}$ ($pK_{b1} = 10.19$). A familiar example of the latter is the formation of the orange **dichromate** ion, $Cr_2O_7^{2-}$, upon partial protonation of the yellow chromate ion:

$$2HCrO_4^- \rightleftharpoons [O_3Cr-O-CrO_3]^{2-} + H_2O \quad K_{eq} = 10^{2.2} \tag{3.15}$$

A particularly important set of these biochemically are the **diphosphate** or **pyrophosphate** and the **triphosphate** ions ($P_2O_7^{4-}$ and $P_3O_{10}^{5-}$, respectively, each with a pK_{b1} of 4.75). Upon heating, partially protonated PO_4^{3-} ions can eliminate water and form these ions:

$$2HPO_4^{2-} + heat \rightarrow [O_3P-O-PO_3]^{4-} + H_2O \tag{3.16}$$

$$H_2PO_4^- + 2HPO_4^{2-} + heat \rightarrow [O_3P-O-PO_2-O-PO_3]^{5-} + 2H_2O \tag{3.17}$$

Despite the large negative charges of the latter two, they are not very strongly basic. You may also be able to offer multiple reasons why this is so.

3.7. Predominance Diagrams for Oxo Anions and Oxo Acids

OVERVIEW. In this section, we will construct predominance diagrams involving oxo anions and oxo acids (this is similar to what you did earlier in Section 2.5). We will also identify the perhaps partially protonated anions or oxo acids that occur in such diagrams, and the form that predominates at a given pH. You may practice these concepts found in Section 3.7A by trying Exercises 3.83–3.89. Oxo acids can be classified as weakly, moderately, strongly, or very strongly acidic, based on the number of oxo groups they contain. You may apply and develop this idea (presented in Section 3.7B) by trying Exercises 3.90–3.94.

Strengths of oxo acids and their positions in predominance diagrams will come into play again in Sections 3.8, 6.5, 8.5, 8.6, and 8.7. A Connection is given to some esters of inorganic oxo acids that are important in organic chemistry, biochemistry, and agricultural chemistry.

3.7A. Predominance Diagrams for Oxo Anions. The product of Equation 3.12, the once-protonated oxo anion $[MO_{(x-1)}OH]^{(y-1)-}$, is still a base, but it now has one less negative charge and one less oxo group, because one oxo group has been converted by protonation to a hydroxo group. If we calculate the basicity of this monoprotonated oxo anion using Equation 3.14, we find that its pK_{b2} is *4.5 units more* than its pK_{b1}. Likewise, the resulting diprotonated oxo anion is still a base, but of further diminished basicity due to a pK_{b3} that is yet another 4.5 units higher.

We previously constructed predominance diagrams in Section 2.5. These included predominance regions of variable width for partially protonated monatomic anions. The process is easier for oxo anions, because the predominance region of each partially protonated oxo anion is about 4.5 pH (or pOH) units wide. (This occurs because all oxo anions have the same donor atom, oxygen.)

Four basic steps are involved, as outlined in Figure 3.19.

1. Calculate the pK_b values of the oxo anions and their partially protonated forms. This can be done most easily by applying Equation 3.14 to calculate pK_{b1} for the oxo anion, then adding 4.5 to the answer to calculate pK_{b2} of the once-protonated anion, then repeating through the last anion. There will be one pK_b value for each unit of negative charge on the original anion.

2. Set up a blank predominance or speciation diagram in the manner of Figure 2.10, with a pH scale below. Add a pOH scale above the diagram, which runs in descending order from left to right from the pOH that matches the highest pK_b to the lowest, as described in Section 2.6C and specifically by Equation 2.25.

3. Draw vertical lines as boundaries at the pOH value matching each pK_b value.

4. Write in the formulas of each oxo anion, starting at the right with the unprotonated oxo anion. Fill the others in from right to left, with the last oxo anion falling to the right of the last (leftmost) boundary. To the left of the leftmost boundary write in the formula of the fully protonated *oxo acid*.

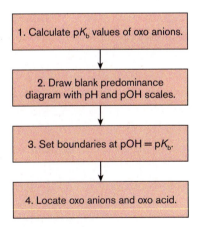

Figure 3.19. Flow chart for constructing predominance diagrams for oxo anions.

EXAMPLE 3.21

Sketch the expected predominance diagram for AlO_4^{5-}.

SOLUTION: (1) From Equation 3.14, we calculate the pK_{b1} of AlO_4^{5-} to be $10.0 + 5.7(4) - 10.2(5)$ = 18.2. The pK_{b2} is obtained by adding 4.5 to −18.2, so it is −13.7. The $pK_{b3} = -13.7 + 4.5 = -9.2$; $pK_{b4} = -9.2 + 4.5 = -4.7$; $pK_{b5} = -4.7 + 4.5 = -0.2$. We stop here, because at this point all negative charges have been neutralized. (2) The blank predominance diagram is set up with a pOH range from (at least) −0.2 on the left to −18.2 on the right. The corresponding pH range is set up on the bottom of the diagram by applying Equation 2.25 (pOH = 14 − pH) from (at least) 14.2 on the left to −32.3 on the right. (We could expand on the left so that the pH range starts at 0). (3) Vertical lines are drawn at the five pOH values matching the pK_{bn} values calculated in Step 1. (4) We fill in the six forms of AlO_4^{5-}, each having one more proton than the form on the right, until finally the fully protonated neutral oxo acid is obtained at the left (Figure 3.20).

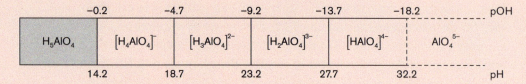

Figure 3.20. Expanded predominance diagram for AlO_4^{5-}.

Much of Figure 3.20 is meaningless in aqueous solution. In Section 3.2C, we mentioned that a concentrated KOH solution behaves as if it were 10,000 M (because not enough water molecules are present to fully hydrate the hydroxide ions). Such a concentration corresponds to a pOH of −4 or a pH of 18, which is about the highest effective pH value that can be achieved in water. Hence, the only two species in the above diagram that can exist in water are $H_4AlO_4^-$ and H_5AlO_4, which can be rewritten as $Al(OH)_3 \cdot H_2O$ or as $Al(OH)_3(s)$ or even as $Al_2O_3(s)$.

The portion of this diagram that involves possible pH values is included at the bottom of Figure 3.21, which shows examples of predominance diagrams for oxo anions of the four most important basicity classifications.

Figure 3.21 is arranged by increasing anion negative charge from top to bottom. Given that basicity depends on the square of charge, it is not surprising that the oxo anions have larger predominance regions at the top of Figure 3.21. As we move down, the oxo anion gradually requires more basic conditions to predominate in water. If the starting oxo anion has a more negative charge than −3, it is not predominant in water at any realistic pH (one exception is MO_6^{4-}), although we will see in Chapter 4 that such very strongly basic anions can exist in solids such as minerals. Thus, although the silicate anion SiO_4^{4-} is found in several minerals, it is not really present in a solution of sodium silicate due to its very strong basicity (calculated $pK_{b1} = -8.0$). Instead, the main species in solution is apparently the diprotonated ion $[H_2SiO_4]^{2-}$ or $[SiO_2(OH)_2]^{2-}$, with a calculated pK_{b3} of 1.0.

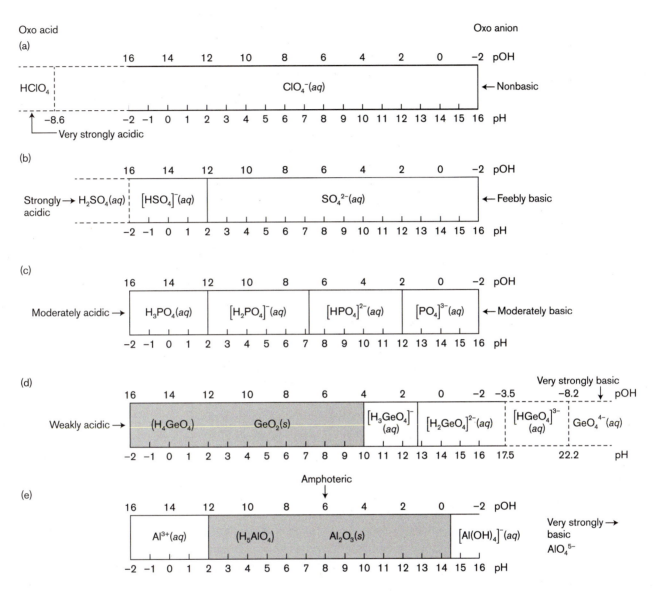

Figure 3.21. Predominance diagrams for several oxo anions having four oxo groups, MO_4^{y-}, and their partially protonated forms and fully protonated oxo acids. The dashed lines represent theoretical boundaries computed for pH ranges that are not achievable in water.

If the charge on the original oxo anion equals or exceeds the number of oxo groups ($x \geq y$), protonation eventually results in, for example, H_xMO_x, which can be rewritten as $M(OH)_x$. This hydroxide is usually insoluble, as described in Section 2.5B. It can often readily and unpredictably lose water to give a metal oxide:

$$M(OH)_x(s) \rightleftharpoons MO_{x/2} + \left(\frac{x}{2}\right)H_2O \qquad (3.18)$$

EXAMPLE 3.22

Sketch the expected predominance diagram for PO_4^{3-}. From the diagram, read off the predominant phosphate species found in solution at a pH of 3.0 (the approximate pH of a cola beverage).

SOLUTION: Applying Equation 3.14, we calculate the pK_b of PO_4^{3-} to be 2.2. This number equals the pOH of the left boundary of this moderately basic oxo anion (Eq. 2.28 and Figure 3.21c). To the left of this boundary, the once-protonated anion, $[(HO)PO_3]^{2-}$, is predominant; its pK_b is calculated to be 6.7—that is, this pK_{b2} is 4.5 units more than the previously calculated pK_{b1}. This pOH value is at the left boundary of $[(HO)PO_3]^{2-}$. To the left of this, $[(HO)_2PO_2]^-$ predominates, with a calculated pK_{b3} of 11.2 and an experimental pK_{b3} of 11.84. To the left of this pOH value, phosphoric acid, $(HO)_3PO$ or (more familiarly) H_3PO_4, predominates.

   With the pH scale at the bottom, we can read from Figure 3.21c that, at pH = 3.0, the predominant species in a phosphate solution is $H_2PO_4^-$. (Because this pH is close to the boundary, there is also be a substantial amount of H_3PO_4 present.)

3.7B. Oxo Acids. Protonation of an oxo anion MO_x^{y-} in which $x > y$ may ultimately give us an **oxo acid** of formula H_yMO_x. This is expected to have a structure corresponding to $MO_{x-y}(OH)_y$, still possessing $(x - y)$ oxo groups. Similarly, protonation of a hydroxo anion, followed by the loss of some water molecules, leads to a metal hydroxide $M(OH)_y$, which can be considered to be an oxo acid with no oxo groups. These oxo acids have various tendencies to ionize:

$$H_yMO_x + H_2O \rightleftharpoons H_3O^+ + H_{y-1}MO_x^- \tag{3.19}$$

We can characterize the strength of these oxo acids by their pK_a values. If necessary, these can be calculated from the pK_b of the final protonated oxo anion (conjugate base) on the right side of Equation 3.19 by using the relationship in Equation 3.20:

$$pK_a = 14 - pK_b \tag{3.20}$$

From the calculations, the Ricci equation[16] emerges, in which *strength* of the oxo acid H_yMO_x depends only on the *number* of oxo groups present:

$$pK_a = 8.5 - 5.7(x - y) \tag{3.21}$$

Writing the alternate formula for the oxo acid, $(HO)_yMO_{x-y}$, shows the number of oxo groups directly. Because existing oxo acids of this type have from one to three oxo groups, this allows us to categorize the oxo acids as follows:

1. Oxo acids with *three* oxo groups (in practice, HMO_4) are expected to have pK_a values of about –8.6 (i.e., to be *very strongly acidic*).

2. Oxo acids with *two* oxo groups (HMO_3, H_2MO_4, and H_4MO_6) are expected to have pK_a values of about –2.9 (i.e., to be *strongly acidic*).

3. Oxo acids with *one* oxo group (HMO_2, H_2MO_3, H_3MO_4, and H_5MO_6) are expected to have pK_a values of about 2.8 (i.e., to be *moderately acidic*). The pK_a values of a number of these acids have been measured and are within a standard deviation of ±0.9 of this value.

4. "Oxo acids" *without* oxo groups (hydroxides of the nonmetals) are expected to have pK_a values of about 8.5 (i.e., to be *weakly acidic*). The measured values for these acids fall within ±1.0 of this value. Generally speaking, these hydroxides differ from the true oxo acids (that have oxo groups) by being insoluble in water.

5. Fluoro acids (such as $HAsF_6$) and HCl, HBr, and HI, the conjugate acids of nonbasic anions, are also very strong acids.

EXAMPLE 3.23

Classify the acidity of the oxo acids derived from the four oxo anions of Exercise 3.77: MO_3^-, MO_3^{2-}, MO_3^{3-}, and MO_3^{4-}.

SOLUTION: HMO_3 has one hydroxo group, but more importantly, has two oxo groups and is strongly acidic. H_2MO_3 has one oxo group and is moderately acidic. H_3MO_3 and H_4MO_3 have no oxo groups (just three hydroxo oxo groups and one coordinated H_2O ligand) and are weakly acidic or even nonacidic.

EXAMPLE 3.24

The partially protonated anion $H_4IO_6^-$ is in equilibrium in solution with the anion IO_4^-:

$$H_4IO_6^- \rightleftharpoons IO_4^- + 2H_2O$$

(a) Classify the basicity of these two ions. (b) Would there be much of a change in pH as this equilibrium was reached? (c) Classify the basicities of the partially protonated anion $H_4TeO_6^{2-}$ and of the alternate Te(VI) oxo anion TeO_4^{2-}. (d) How could you simply determine which form of the Te(VI) oxo anion is favored?

SOLUTION: (a) IO_4^- is nonbasic (calculated $pK_b = 22.6$); $H_4IO_6^-$ is near the boundary between weakly and feebly basic [$pK_b = 10.0 + 5.7(2) -10.2(1) = 11.2$].

(b) No, there would not be much of a change in H, becaues neither one has appreciable basicity (the solution of IO_4^- would be neutral, while that of feebly basic $H_4IO_6^-$ would have a pH only slightly different from neutral.

(c) TeO_4^{2-} would be feebly basic ($pK_b = 12.4$); $H_4TeO_6^{2-}$ is at the boundary between moderately and strongly basic [$pK_b = 10.0 + 5.7(2) - 11.2(2) = -1$].

(d) Measure the pH of the solution: If it is close to neutral (pH ≈ 7), then the oxo anion TeO_4^{2-} predominates; if it has a pH close to 14 (a pOH close to –1), then $H_4TeO_6^{2-}$ should be predominant.

Anomalous Oxo Acids

There are some oxo acids that have anomalous acidities. Notable among these are two oxo acids of phosphorus, H_3PO_3 and H_3PO_2, which have about the same pK_a (~2) as H_3PO_4, suggesting that all have one oxo group. In fact, there is independent evidence that this is the case, and that the structures of H_3PO_3 and H_3PO_2 do not have three O–H groups, but have two O–H groups, one oxo group, and one P–H group instead (Figure 3.22). In contrast, H_3AsO_3 has the normal structure with no H–As bond and no oxo groups.

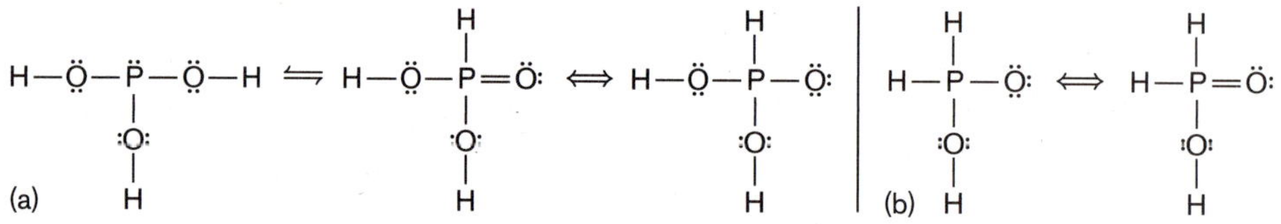

Figure 3.22. Proton-rearrangement and resonance structures of (a) H_3PO_3 and (b) H_3PO_2

The measured pK_{a1} for carbonic acid (6.38) also seems to be out of line for an oxo acid with one oxo group. But it is found that, in a solution of carbonic acid, most of the carbon is in the form of hydrated CO_2, not $(HO)_2C=O$.

$$CO_2(aq) + H_2O(aq) \rightleftharpoons H_2CO_3(aq); K_{eq} = 2 \times 10^{-3} \tag{3.22}$$

Formation of H_2CO_3 is not favored because it requires breaking one of the two very strong π bonds in CO_2. If a correction is made for this, it is found that the pK_a for the H_2CO_3 that is actually present in the solution is 3.58, which is in the expected range for an oxo acid with one oxo group. Similar equilibrium shifts that reduce acidity somewhat have been noted for some other simple acidic oxides—namely, XeO_4, XeO_3, RuO_4, OsO_4, and SO_2 (Section 8.5A).[17]

AN AMPLIFICATION

AN AMPLIFICATION AND CONNECTION TO ORGANIC CHEMISTRY AND BIOCHEMISTRY

Inorganic Esters

Just as oxo acids may ultimately be formed by the reaction of chlorides of strongly or very strongly acidic cations with water, **inorganic esters** can often be formed by the reaction of these chlorides with alcohols, especially in the presence of a basic amine to remove the proton from the alcohol ROH:

$$PCl_3 + 3ROH + 3R_3N \rightarrow P(OR)_3 + 3R_3NH^+Cl^- \qquad (3.23)$$

This product is known as a trialkyl phosphite and is a useful ligand that provides a phosphorus donor atom (Section 7.4). In contrast, trialkyl phosphates have only oxygen donor atoms. Tributyl phosphate, $(CH_3\text{–}CH_2\text{–}CH_2\text{–}CH_2\text{–}O)_3PO$, is an important solvent for separating compounds of the lanthanide elements (Section 1.2B). These phosphates, and especially the corresponding sulfates such as dimethyl sulfate, $(MeO)_2SO_2$, still have polarizable C–O bonds. They act as electrophiles—sources of the Lewis acid CH_3^+—which they can attach to nitrogen (or other) donor atoms, for example in DNA. Thus, they are powerful *alkylating agents*, and may be powerful *mutagens* and/or *carcinogens* and can be quite toxic.

Organophosphate (and thiophosphate) pesticides are very important in modern agriculture. These interfere with nerve transmission in insects (and mammals) by inhibiting the enzyme cholinesterase. An especially toxic ester is tetraethyl pyrophosphate, $(EtO)_2PO\text{–}O\text{–}PO(OEt)_2$. Some common pesticides contain a thio (S–) group in place of the oxo group: for example, parathion, $S\text{–}P(OEt)_2(OC_6H_4NO_2)$. Also related are some chemical warfare agents, the nerve gases, such as Sarin or GB, $MePOF(OCHMe_2)$, the lethal dose of which may be 1 mg. Although very toxic, these pesticides have the advantage that they undergo slow reaction with water to give the starting alcohols and phosphoric acid, so they do not persist in the environment. Not all organophosphates are toxic: The biochemical compounds DNA, RNA, ATP (adenosine triphosphate), and ADP (adenosine diphosphate) are all organophosphates, and are why phosphorus is essential for life. The interconversion of ATP and ADP is of crucial importance in storing and transferring energy in biochemistry:

$$[\text{Adenosine–O–PO(OH)–O–PO(OH)–O–PO}_2\text{OH}]^- + H_2O \rightleftharpoons H_2PO_4^- + [\text{Adenosine–O–PO(OH)–O–PO}_2\text{OH}] + 20\text{ kJ} \qquad (3.24)$$

Can Arsenate Replace Phosphate in DNA?

Fourth-period arsenate strongly resembles third-period phosphate in size, number of oxo groups, basicity, and so on. A very controversial paper [F. Wolfe-Simon et al., *Science* 332, 1163 (2011)] recently reported that a bacterium grown in the very alkaline, phosphorus-poor but arsenic-rich environment of Mono Lake in California was able to substitute arsenate groups for phosphate groups in its biomolecules. A summary of the objections [D. S. Tawfik and R. E. Viola, *Biochemistry* 50, 1128 (2011)] emphasized two points of difference between the fundamentally very similar phosphate and arsenate ions and esters. First, arsenate is much more susceptible to reduction than is phosphate (Figures 6.2 and 6.3). Second, while phosphate esters are fairly stable in water, arsenate esters undergo rapid reaction with water. More recent work seems to establish that arsenate groups do *not* replace phosphate groups in DNA, but instead that arsenate bonds noncovalently to the outside of the DNA (summarized in C. Drahl, *Chem. Eng. News*, Jan. 30, 2012, p. 42).

3.8. Most Common Forms of the Elements in Natural Waters

OVERVIEW. We can apply the concepts of Chapter 2 involving the species formed by reactions of cations and oxo anions with water to give a preliminary prediction of the forms in which an element in a given oxidation state should be found in natural waters of various pH values. The predominance diagrams of cations developed in Section 2.5 and of oxo anions developed in Section 3.7 can be combined, then used to predict the form in which it would likely be found (hydrated cation, insoluble oxide or hydroxide, or oxo anion) in unpolluted or (acid–base) polluted natural waters. This Section 3.8A concept can first be practiced using existing predominance diagrams by trying Exercises 3.95–3.100. Then you can start from the most likely oxidation state and apply the concepts of Chapters 2 and 3 to draw up a reasonable predominance diagram, and then use it to predict the speciation of the element (Exercises 3.101–3.106). This section is a good chance to integrate, review, and then apply the concepts of charge, size, and electronegativity from Chapter 1, acidity, basicity, predominance diagrams and insolubility of oxides and hydroxides from Chapter 2, as well as the concepts earlier in this chapter.

In Section 3.8B no new concepts are introduced, but instead this subsection consists entirely of Connections of predominance diagrams and speciation to the aqueous environmental chemistry, biological effects, and toxicology of selected elements (Tables 1.1, 1.2, 1.4, and 1.5). The concept of speciation is developed further in Sections 4.4D, 5.7, 5.8, 7.9, and 8.1.

3.8A. Predicting the Speciation (Predominant Forms) of an Element. In this section, we summarize the concepts in Chapters 2 and 3 by using them to make preliminary predictions of the forms in which the different chemical elements are likely to be found (the **speciation** of the elements) in natural waters such as lakes. Unpolluted natural waters usually have pH values between 6 and 9; moderately polluted waters can span a range from about 3 to 10. If an element is found in a soluble form at the pH in question, it will be much more available for (essential or toxic) biological functions than if it is found in an insoluble form such as an insoluble oxide or hydroxide.

The chemistry of the elements in natural waters is actually quite complex and involves types of chemistry that we have not yet dealt with, such as precipitation of salts (Chapter 4), complexation by ligands (Chapter 5), and oxidation by air or reduction by pollutants (Chapter 6). At this time, we will confine ourselves to discussing a rather unnatural body of natural water, one that contains no oxidizing or reducing agents or complexing agents, and in which no salts (except hydroxides) are allowed to precipitate. Since no oxidation–reduction reactions are allowed to occur, we assume that each element is in the common positive oxidation state specified in Table 1.11. We then construct an overall predominance diagram combining the features of the two types of predominance diagrams we have previously constructed.

1. Identify the most common positive oxidation state of the element from Table 1.11 and evaluate the acidity of the corresponding cation. Recall that to do this, we need to compute Z^2/r, adjust for χ_P, and classify the acidity of the cation (Section 2.2).

2. Draw the cation's predominance diagram as in Section 2.5A. If the cation is nonacidic or feebly acidic, the next two steps are unnecessary, since the cation does not form an oxo anion.

3. If the cation is sufficiently acidic, construct the diagram for the potential oxo anion of the element as in Section 3.7A, showing the soluble oxo anion, protonated oxo anions, and insoluble hydroxides and oxides of that element. (Recall that the left pOH boundaries for the anions are equal to their pK_b values.)

4. Once the two diagrams have been drawn, they can be merged, because both may feature an insoluble oxide or hydroxide in the central pH range.

5. From the combined predominance diagram, we find the predominant species at the pH range in question.

The steps of this process are summarized in Figure 3.23.

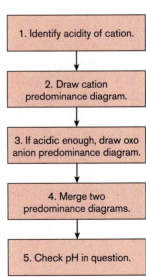

Figure 3.23. Flow chart for predicting the speciation of an element.

1. Identify acidity of cation.

2. Draw cation predominance diagram.

3. If acidic enough, draw oxo anion predominance diagram.

4. Merge two predominance diagrams.

5. Check pH in question.

EXAMPLE 3.25

Draw a combined predominance diagram for aluminum and use it to predict the predominant form that aluminum should take in (a) an unpolluted natural water of pH 5.5; (b) a severely acid-rain-polluted water of pH 0.

SOLUTION: 1. The expected oxidation state of Al is +3 (Section 1.2 or Table 1.11). The radius of Al^{3+} is 67 pm (Table C), so Z^2/r for Al^{3+} is $9/67 = 0.13$. Because the electronegativity of Al is below 1.8 (Table A), the cation is moderately acidic.

2. Such cations have pK_a values between 1 and 6 (Section 2.2). This is the *approximate* pH value at the right boundary of the predominance region for Al^{3+} (Section 2.5A). Beyond this right boundary we place the insoluble $Al(OH)_3$ or Al_2O_3. Such a diagram (using pH = 2 for the boundary) was shown in Figure 2.15d (below pH 14).

3. Because Al^{3+} is neither nonacidic nor feebly acidic, we need to consider whether Al(III) forms a stable oxo anion or protonated oxo anion. Al^{3+}, being in the third period, is expected to adopt the penultimate total coordination number of four in its oxo anion (Section 3.5A), so that its oxo anion would be AlO_4^{5-}. Next, we calculate pK_b for this oxo anion and its partially protonated forms, and include them in a predominance diagram. We already did this in Example 3.21, obtaining the diagram of Figure 3.20.

4. The overall predominance diagram is a merger of the one from Step 2 (Figure 2.15d) below pH 14 and the one from Step 3 (Figure 3.20) above pH 14. The merged predominance diagram is the complete one in Figure 2.15d, which includes achievable pH values up to 16.

5. On the overall diagram, we find the predominant form of aluminum(III) to be (a) at pH 5.5, $Al(OH)_3$ or Al_2O_3; (b) at pH 0, Al^{3+}.

EXAMPLE 3.26

The following radioisotopes (as fallout from an atomic bomb blast) are being deposited in a lake of pH 5.5–7: (a) ^{99}Tc; (b) ^{90}Sr; and (c) ^{244}Pu. Predict the speciation of these elements in this lake.

SOLUTION: 1. Table 1.11 lists the most common oxidation states: Tc(VII), Sr(II), and Pu(IV), respectively. Of these, only Sr^{2+} is (nonacidic or) feebly acidic. The Tc^{7+} ion ($Z^2/r = 49/70$) is very strongly acidic with $pK_a < -4$. Pu^{4+} ($Z^2/r = 16/100$) is on the line between moderately and strongly acidic with a pK_a of about 1.

2. The predominance diagram of Sr^{2+} was previously shown in Figure 2.15b; strontium is therefore present in the lake as the hydrated ion $^{90}Sr^{2+}(aq)$. The Tc^{7+} cation cannot exist at any pH; it does not appear on the predominance diagram. The Pu^{4+} cation is predominant below pH ≈ 1; to the right of this, $PuO_2(s)$ is expected.

3. Next, we determine whether Tc and Pu can form stable oxo anions (or partially protonated oxo anions) in solution at achievable pH values. (a) Technetium is in the fifth period and in an oxo anion might show a penultimate total coordination number of six. Because it is in the region of good π-bonding elements (Section 3.5), we are more likely to find the antepenultimate total coordination number of four. Therefore, we predict its oxo anion to have the formula TcO_4^-.

Example 3.26 continues on next page ▶

EXAMPLE 3.26 (cont.)

We calculate the pK_{b1} of TcO_4^- to be 22.6. This ion is nonbasic and is expected to persist in water at all achievable pH values. (b) For seventh-period Pu(IV) we predict that an oxo anion, if formed, would be PuO_6^{8-}, which would be very strongly basic and could not persist in water. Detailed calculations, if carried out, would also show that none of the partially protonated forms could persist at neutral pH values either.

4. In neither of these cases are there two predominance diagrams to merge. Tc(VII) has only an oxo-anion predominance diagram, with TcO_4^- predominating at all reasonable pH values. Pu(IV) has only a cationic predominance diagram:

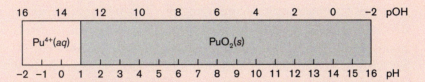

5. (a) At any pH, the predominant form of Tc(VII) is TcO_4^-. (c) At the specified pH of 5.5 to 7, Pu(IV) is expected to be the insoluble oxide, PuO_2, or the hydroxide, $Pu(OH)_4$. (Pu^{4+} is so exceedingly toxic that the fact its concentration is well below 1 M at this pH is of little consolation, because even the tiny amount present in solution may be harmful.)

EXAMPLE 3.27

Which ion in each of the following pairs should show the greater availability as a toxin or nutrient in unpolluted natural waters of pH 5.5: (a) Be^{2+} or Ba^{2+}; (b) iron(II) or iron(III); (c) chromium(III) or chromium(VI)?

SOLUTION: (a) Be^{2+} is weakly acidic [$Z^2/r = 4/59 = 0.068$; $pK_a = 6.2$ (Table 2.1)], so its hydroxide, unavailable $Be(OH)_2(s)$, would predominate in natural waters of pH 6–9. Ba^{2+} is feebly acidic, so it predominates at pH 6–9, and as a soluble hydrated cation it is more available. (b) Fe^{2+} is weakly acidic [$Z^2/r = 4/92 = 0.043$; $pK_a = 9.5$ (Table 2.1), so it predominates at pH 6–9. Fe^{3+} is moderately acidic [$Z^2/r = 9/78 = 0.11$; $pK_a = 2.2$ (Table 2.1)], so its insoluble hydroxide $Fe(OH)_3$ or oxide Fe_2O_3 predominates at pH 6–9. Therefore, Fe(II) is more available. (c) Cr^{3+} is very similar to Fe^{3+} [$Z^2/r = 9/75 = 0.12$; $pK_a = 4.0$ (Table 2.1)], so its insoluble hydroxide $Cr(OH)_3$ or oxide Cr_2O_3 predominates at pH 6–9. Its predicted oxo anion, $[CrO_4]^{5-}$, is very strongly basic; even when protonated to the level of $[Cr(OH)_4]^{2-}$ its pK_b by Equation 3.14 is out of range for persistence at pH 6–9. Because the insoluble hydroxide predominates, Cr(III) is unavailable at the normal pH ranges. [Note: Due to crystal field effects (Chapter 7), the actual oxo anion is six-coordinate, as seen in Figure 2.13d. Nonetheless this oxo anion, even when protonated, does not persist at normal pH ranges.] In contrast, Cr(VI) is much too acidic to exist in water, while its oxo anion, $[CrO_4]^{2-}$, is expected to be feebly basic and therefore available at pH 6–9.

3.8B. Availability of Nutrient or Toxic Elements—A Series of Connections to Environmental Water Chemistry, Biochemistry, and Toxicology. *d*-Block Micronutrients (Tables 1.1 and 1.4).

Many essential micronutrients are weakly acidic +2-charged cations: Zn^{2+}, Cu^{2+}, Co^{2+}, Fe^{2+}, and Mn^{2+} (Table 1.4). As micronutrients, they are not needed in anywhere near 1 M concentration, so they may be available enough to suffice. But they may precipitate from dilute solutions at pH values above 5.3–8.5, so liming (adding CaO to neutralize) an acidic soil or acid-rain-polluted lake may remove essential micronutrients.

Iron. One exception is the element iron, which is biochemically very important even though its most common form, Fe^{3+}, is moderately acidic and therefore unavailable in natural waters at pH values above about 2.0. Organisms have very elaborate strategies for obtaining and conserving this common yet inaccessible nutrient (iron-deficiency anemia may be the most prevalent dietary deficiency disease among humans). In one-third of the world's ocean—especially the Antarctic Ocean—this nutrient is not even common, because iron is often brought to ocean waters in the form of dust blowing from dry soils or deserts, which are scarce in Antarctica. The Antarctic Ocean is incredibly fertile in other nutrients, however, and is quite productive of plant and animal life. The productivity of the Antarctic is limited by the low availability of iron. In 1990 J. H. Martin proposed that this productivity could be multiplied 10-fold if enough iron were supplied.[18] It has been estimated that enough photosynthesis would result to remove a significant fraction (perhaps 15%) of the excess CO_2 that humans have put into the atmosphere; this CO_2 is a major contributor to the greenhouse effect (see Section 8.5C). It has been confirmed experimentally several times that adding tons of $FeSO_4$ to regions of the ocean results in spectacular blooms of algae, with an increase in photosynthesis and therefore CO_2 fixation.[19] Normally, when the phytoplankton are eaten, the CO_2 is then returned to the air—unless the plankton sink to the deep ocean. (After several hundred years, nearly all carbon that sinks to the deep ocean returns to the surface by upwelling. A small amount remains and eventually becomes petroleum and natural gas.) Whether this would suffice to remove enough CO_2 from the atmosphere to ocean depths to reverse climate change is less certain.[20,21] As will be discussed in Chapter 5, a major factor in survival of the fittest organisms in the oceans is based on their ability to form complexes with iron(III), so such a plan would have major ecological effects.

Aluminum. If a body of water becomes highly polluted with acidic pollutants, it is possible for moderately acidic cations to go into solution. The adverse consequences of acid rain for fish seem to involve just this process. Lakes that are poorly buffered can undergo quite substantial changes in pH when acid rainfall runs off into them. The pH of such a lake can drop to the point that normally insoluble aluminum is converted to $Al^{3+}(aq)$. The toxic effects on fish seem to be due not to the enhanced concentration of hydrogen ion but to the presence of aluminum ion, which is normally not encountered in natural waters.

A rise in the pH of an acidic body of water by dilution or neutralization can cause the weakly and moderately acidic ions to precipitate as hydroxides. This difference in pH between the acid water and the more neutral environment of the gills of fish apparently causes gelatinous aluminum hydroxide to precipitate there, coating the gills. The fish seem to sneeze themselves to death in an attempt to get rid of this gelatinous precipitate.

Acid Mine Drainage. In the acid mine drainage problem of old coal mines, the mineral pyrite, FeS_2, is slowly oxidized by air to concentrated aqueous iron(III) sulfate, $Fe_2(SO_4)_3$. The Fe^{3+} reacts with the water to give hydrogen ions and a very low pH. As the pH of the acid water is raised by dilution with uncontaminated streams of water outside of the mine, the Fe^{3+} converts to yellow insoluble iron(III) hydroxide, which precipitates as unsightly "yellow boy" along the stream banks.

Titanium and Silicon. The elements with moderately and strongly acidic cations are the ones most likely to be insoluble (as oxides or hydroxides) in unpolluted natural waters. Consequently, these elements are seldom available for biological activity, either in natural waters or in the digestive system or in body fluids, unless chemical processes other than acid–base reactions can somehow make them available. These unavailable elements thus are seldom either essential for life or toxic to life. Despite the abundance of titanium and silicon and the relatively high abundance of many of the "rare" earth elements (Table 1.4), few or no forms of life make use of them.

Some or many of the very strongly acidic "cations" are in fact found in natural waters as oxo anions or partly protonated oxo anions, which then may be the available forms for these cations. Although this is beyond the scope of this text, the charge of the anion can affect how readily it is taken up by plant roots, for example, so speciation (degree of protonation) of oxo anions may also affect availability. In Chapter 8 we will see how the availability of cations is also affected further by their binding in soil polysilicates (clays), which will refine our ability to predict availability.

A series of predictions for the elements in 1 M solution in their most common positive oxidation states (Table 1.11) is embodied in Figure 3.24. Because environmental chemists seldom encounter any species in a 1 M concentration, these are rough approximations to environmental reality!

Figure 3.24. Main forms of the elements in moderately aerated water of pH 5.5–7. Shaded areas represent insoluble compounds. [Source: Based on data given in M. Pourbaix, *Atlas of Electrochemical Equilibria in Aqueous Solutions*, National Association of Corrosion Engineers: Houston, 1974, with preference given to positive oxidation states when possible.]

1	2	3	4	5	6	7	8	9	10	11	12	13/III	14/IV	15/V	16/VI	17/VII
H_2O																
Li^+	$Be(OH)_2$											$B(OH)_3$	CO_2 / HCO_3^-	NO_3^-	H_2O	F^-
Na^+	Mg^{2+}											$Al(OH)_3$	SiO_2	$H_2PO_4^-$ / HPO_4^{2-}	SO_4^{2-}	Cl^-
K^+	Ca^{2+}	$Sc(OH)_3$	TiO_2	$H_3V_2O_7^-$ / H_3VO_4	$Cr(OH)_3$	Mn^{2+} / MnO_2	$Fe(OH)_3$	Co^{2+}	Ni^{2+}	Cu^{2+}	Zn^{2+}	$Ga(OH)_3$	GeO_2	$H_2AsO_4^-$ / $HAsO_4^{2-}$	SeO_4^{2-}	Br^-
Rb^+	Sr^{2+}	Y^{3+} / $Y(OH)_3$	ZrO_2	Nb_2O_5	MoO_4^{2-}	TcO_4^-	$Ru(OH)_3$	Rh_2O_3	$Pd(OH)_2$	Ag^+	Cd^{2+}	$In(OH)_3$	SnO_2	Sb_2O_3	$HTeO_3^-$	IO_3^-
Cs^+	Ba^{2+}	Lu^{3+}	HfO_2	Ta_2O_5	WO_3 / WO_4^{2-}	ReO_4^-	OsO_2	IrO_2	PtO_2	Au metal	HgO	Tl^+	Pb^{2+}	Bi_2O_3	$HPoO_3^-$?	

Stable oxidation number equals group number

Lower oxidation number more stable

Stable oxidation number is 2 less than group number

La^{3+}	Ce^{3+}	Pr^{3+}	Nd^{3+}	Pm^{3+}	Sm^{3+}	Eu^{3+}	Gd^{3+}	Tb^{3+}	Dy^{3+}	Ho^{3+}	Er^{3+}	Tm^{3+}	
Ac^{3+}	ThO_2	Pa_2O_5	UO_2^{2+}	NpO_2^+	PuO_2	Am^{3+} / $Am(OH)_4$							

3.9. Nomenclature of Oxo Anions and Acids

OVERVIEW. Given the formula of an oxo acid, an oxo anion, or the salt of an oxo anion, apply nomenclature principles to name the oxo acid or anion (or its salt). You may practice this concept found in Section 3.9A by trying Exercises 3.107–3.110. Similarly to this, given the name of an oxo acid, oxo anion, or its salt, you should be able to write its formula, as in Section 3.9B and Exercises 3.111–3.118. The Exercises for this chapter conclude with some general review (see Exercises 3.119–3.124) that may help you review the chapter in preparation for an exam.

This nomenclature is based on oxidation numbers of the central atom, which you first studied in Section 1.4. Because oxo acids, salts, and oxo anions are often referred to by name without giving their formulas, this is a generally useful skill, which applies particularly in Sections 4.3, 5.4, 6.2, and 8.3.

Learning the formulas and nomenclature of polyatomic ions is often assigned in general chemistry as an exercise in memorization. Unfortunately, memorized facts are normally promptly forgotten. The alternative is to understand and apply the relevant principles of inorganic nomenclature, which we present here. Applying the principles requires you to use several principles from the first three chapters, so it is not as easy as people think, and it is a good way of using and applying those principles so as to review them and increase your competence with them.

3.9A. Naming Oxo Anions and Acids. If oxo anions were named by the system described for fluoro anions in Section 3.4C, which is sanctioned by the International Union of Pure and Applied Chemistry (IUPAC), their names would be easier to derive but would be longer and (above all) unfamiliar: NO_2^- would be named dioxonitrate(III) ion or dioxonitrate(–1) ion. In contrast to fluoro anions, oxo anions have been extremely important in many fields for so many centuries that an older, more difficult naming system is entrenched in usage in most fields connected to chemistry (such as commerce and agriculture).

This older but far more common naming system uses suffixes (and sometimes prefixes) to indicate relative oxidation numbers (Table 3.9). All names of oxo anions replace the final suffix of the name of the element (e.g., –ium) with either the suffix –ate or the suffix –ite. For oxo acids, the suffixes –ic acid and –ous acid are used. In addition, a prefix (per– or hypo–) may be added.

- The oxo anion/acid involving the *most common* oxidation state of the central atom takes the name ending in –ate/–ic acid (e.g., arsenate for AsO_4^{3-}).

- The oxo anion/acid in which the central atom has the *next lower* oxidation number (two lower if in the *p* block) is given the name ending in –ite/–ous acid (e.g., arsenite for AsO_3^{3-}).

- Any oxo anion/acid in which the central atom has a still lower oxidation number (four lower if in the *p* block) is given the prefix hypo– and the suffix –ite/–ous acid.

- If there is a central-atom oxidation number higher (two higher if in the *p* block) than the most common one, the prefix per– and the suffix –ate/–ic acid are used.

- If the element has a Latin name (shown in Table D in parentheses), this Latin name is used in the nomenclature.

Because oxidation states of +7 and +8 are generally uncommon, even in Groups 17/VII, 18/VIII, 7, and 8, the prefixes and suffixes used differ in these groups (last column of Table 3.9).

TABLE 3.9. **Relationship Between Names of Oxo Anions and Acids and the Oxidation Number of the Central Atom**

Prefix	Suffixes	In Most Groups	In Groups 17/VII, 18/VIII, 7, and 8
per–	–ate, –ic acid	Does not occur	Oxidation number = Group number
	–ate, –ic acid	Oxidation number = Group number	Oxidation number = Group number – 2
	–ite, –ous acid	Oxidation number = Group number – 2	Oxidation number = Group number – 4
hypo–	–ite, –ous acid	Oxidation number = Group number – 4	Oxidation number = Group number – 6

AN AMPLIFICATION

Exceptions in This Naming System

Complications occur in the *d* block, where the oxidation number can be one less than the group number. Here we often see MnO_4^{2-} called "manganate" and RuO_4^- called "perruthenate." Furthermore, sometimes the prefix "per–" refers to the presence of a *peroxo* group $-O_2$ in place of an oxo group!

Change in Total Coordination Number

One other set of prefixes is sometimes encountered. In the fifth and sixth periods, in which the expected total coordination number is variable (Table 3.5), either type of oxo anion may sometimes be encountered. In such cases, the oxo anion with the lower total coordination number is given a pre-prefix of meta– (IO_4^- is metaperiodate); the one with the higher total coordination number has the pre-prefix of ortho– [$H_4IO_6^-$ is the (tetrahydrogen) orthoperiodate ion].

EXAMPLE 3.28

Give the names of the following oxo anions: NO_2^-, PO_4^{3-}, AsO_3^{3-}, SeO_3^{2-}, FeO_4^{2-}, and IO_3^-.

SOLUTION: First, it is necessary to check two things. Is the group number 17/VII, 18/VIII, 7, or 8? If so, the right column of Table 3.9 is used; if not, the column to its left is used. Second, how does the oxidation number of the central atom compare with the group number? From this, select the row from Table 3.9 that applies. For NO_2^-, PO_4^{3-}, AsO_3^{3-}, and SeO_3^{2-}, the element is in a group numbered 6 or less, so if the central-atom oxidation number equals the group number, the suffix –ate is used: PO_4^{3-} is phosphate. For the other three, the oxidation number equals the group number minus two, so NO_2^- is nitrite, AsO_3^{3-} is arsenite, and SeO_3^{2-} is selenite.

For the iodine and iron oxo anions, the group numbers are 17/VII and 8, so the right column of Table 3.9 is used. Because the iodine and iron oxidation numbers are two less than the group numbers, IO_3^- is iodate and FeO_4^{2-} is ferrate.

EXAMPLE 3.29

What are the names of the acids obtained by the neutralization of (a) ClO^-; (b) Cl^-; (c) ClO_3^-; (d) ClO_4^-?

SOLUTION: Chlorine is in group 17/VII, so we use the right column of Table 3.9. (a) The Cl oxidation number is +1, six less than the group number, so the acid is *hypochlorous acid*. (b) This is not an oxo anion—referring to Section 2.5E, the acid is *hydrochloric acid*. (c) The Cl oxidation number is +5, two less than the group number, so $HClO_3$ is *chloric acid*. (d) The Cl oxidation number is +7, equal to the group number, so the acid is *perchloric acid*.

3.9B. Formulas of Oxo Acids and Anions from Their Names. To go from the name of an oxo anion to its formula, the reverse process is suggested.

- First, determine whether the group number of the element is a "7" or an "8" and note the prefixes and suffixes in the name of the oxo anion. This determines how the oxidation number of the central atom relates to the group number (Table 3.9).

- Once the oxidation number of the central atom has been determined, the problem is just like Example 3.15, and the same three steps are used.

EXAMPLE 3.30

Give the formulas of the vanadate, perxenate, and chlorite ions.

SOLUTION: First, vanadium is in Group 5, so it is named normally. The suffix –ate tells us that the oxidation number of vanadium equals its group number, so it is +5. Both Xe and Cl are in Groups 17/VII and 18/VIII, so their names come from the right column of Table 3.9. The per– and –ate with xenon tells us, in this case, that the oxidation number equals the group number, +8. The –ite suffix for chlorine is in the third row of Table 3.9, so the oxidation number is four less than the group number of Cl, making it +3.

Second, the penultimate total coordination numbers for third-period chlorine and fourth-period vanadium are each four, while that of fifth period xenon is six (Table 3.5).

Third, we examine oxidation numbers to see whether p-block elements have unshared pairs. These occur only for Cl(III) in chlorite, which (because the oxidation number is four below the group number) has two unshared pairs of electrons.

Finally, we mentally merge cations with enough oxide ions to generate oxo anions with the expected penultimate total coordination numbers. Combining V^{5+} with four O^{2-}, we obtain VO_4^{3-}. Combining $:Cl:^{3+}$ with two O^{2-}, we obtain ClO_2^-, which has a total coordination number of four. Combining Xe^{8+} with six O^{2-}, we obtain XeO_6^{4-}.

3.10. Background Reading for Chapter 3

Quintuple Bonds

Y.-C. Tsai and C.-C. Chang, "Recent Progress in the Chemistry of Quintuple Bonds," *Chem. Lett.* 38, 1122–1129 (2009).

Maximum Total Coordination Numbers

W. L. Jolly, *Modern Inorganic Chemistry*, McGraw-Hill: New York, 1984, pp. 77–90 ("Valence-Shell Electron Repulsion").

Multiple Bonding in Oxo Anions

J. M. Mayer, "Why Are There No Terminal Oxo Complexes of the Late Transition Metals? Or The Importance of Metal–Ligand π Antibonding Interactions," *Comments Inorg. Chem.* 8, 125–135 (1988).

Life Based on Arsenate in Place of Phosphate?–
A Connection and Scientific Controversy

F. Wolfe-Simon et al., "A Bacterium That Can Grow by Using Arsenic Instead of Phosphorus," *Science* 332, 1163–1166 (2011). A very controversial paper!

R. S. Goody, "How Bacteria Choose Phosphate," *Angew. Chem. Intl. Ed.* 52, 2406–2407 (2013). The choice, despite the very close similarities between third-period phosphate and fourth-period arsenate.

S. Silver, "Beyond the Fringe: When Science Moves from Innovative to Nonsense," *FEMS Microbiol. Lett.* 350, 2–8 (2014). Discusses sources of error in scientific publications and science journalism.

Iron and the Productivity of the Antarctic Ocean–A Connection

J. H. Martin, "Glacial-Interglacial CO_2 Change: The Iron Hypothesis," *Paleoceanogr.* 5, 1–13 (1990); J. H. Martin, "Iron in Antarctic Waters," *Nature* 345, 156–158 (1990).

P. W. Boyd et al., "Mesoscale Iron Enrichment Experiments 1993–2005: Synthesis and Future Directions," *Science* 315, 612–617 (2007).

P. Williamson et al., "Ocean Fertilization for Geoengineering: A Review of Effectiveness, Environmental Impacts and Emerging Governance," *Process Saf. Environ. Prot.* 90, 475–488 (2012). An Institute of Chemical Engineers summary of conclusions on proposals for ocean fertilization with iron.

Numbered references from this chapter may be viewed online at www.uscibooks.com /foundations.htm.

3.11. Exercises

3.1. *Which oxide ion is present in the following compounds: (a) KO_2; (b) BaO_2; (c) TiO_2?

3.2. Which iodide ion is present in the following compounds: (a) AlI_3; (b) $Ra(I_3)_2$; (c) TlI_3? Name each compound.

3.3. *The CrO_8^{3-} ion is known. What is the apparent oxidation number of chromium? Is this a possible oxidation number for Cr? Suggest an alternative interpretation of this preposterous-looking chemical formula.

3.4. Assign oxidation numbers in the homopolyatomic ions/molecules: (a) O_2, O_2^-, O_2^{2-}, O_3^- (ozonide ion); (b) N_2, N_3^-, N^{3-}; (c) C^{4-}, C_2^{2-}, C_3^{4-}, C_{60}^{3-} (fullerene trianion). (d) Why don't we automatically assign oxidation number –2 to oxygen in the numerical rules of Section 1.4?

3.5. *Discuss the advantages and disadvantages of using parentheses around the formulas of polyatomic ions even if only one is present. Discuss the advantages and disadvantages of always showing ion charges in ionic compounds.

3.6. Draw the Lewis structures of each of the following molecules or ions: (a) H_2N-OH (bonded in that order); (b) (linear) Br_3^-; (c) $[NSN]^{2-}$ (S is central atom).

3.7. *Draw the Lewis structures of each of the following molecules or ions: (a) HNNH (bonded in that order); (b) (linear) N_3^-; (c) NF_4^+; (d) OCS (C is central atom).

3.8. Draw Lewis structures of (a) XeF_4; (b) $[N-C-N]^{2-}$; (c) TeF_5^-; (d) $[C(NH_2)_3]^+$.

3.9. *Draw Lewis structures of the following molecules or ions: (a) $HCCl_3$; (b) NSF; (c) C_2H_6; (d) C_3H_7Cl; (e) $Fe(CO)_5$; (f) $Cl-CH_2-SiH_2-SiH_2-CH_2Cl$.

• **3.10.** Draw Lewis structures of the following molecules or ions: (a) H_2O_2; (b) HCN; (c) H_2N-NH_2; (d) H_2N-OH; (e) H_3C-SiH_3; (f) $[H_3C-C(=O)(-O-O)]^-$.

3.11. *Draw Lewis structures of the following molecules or ions: (a) HOF; (b) Si_2H_6; (c) Si_3H_8; (d) CSe_2; (e) $[V(CO)_6]^-$; (f) $ClCH_2-CH_2Cl$.

3.12. Examine the Lewis structures you drew in Exercise 3.8, and identify the potential donor atoms in (a) XeF_4; (b) $[NCN]^{2-}$; (c) TeF_5^-; (d) $[C(NH_2)_3]^+$.

3.13. *Draw the Lewis dot structures of the following molecules or ions, and identify the potential donor atoms in each: (a) NH_3; (b) NH_2-CH_2-COOH (an amino acid); (c) SO_4^{2-}; (d) SO_3^{2-}.

3.14. Examine the Lewis structures you drew in Exercise 3.10, and identify the potential donor atoms in (a) H_2O_2; (b) HCN; (c) H_2N-NH_2; (d) H_2N-OH; (e) H_3C-SiH_3; (f) $[H_3C-C(=O)(-O-O)]^-$.

3.15. *Examine the Lewis structures you drew in Exercise 3.9, and identify the potential donor atoms in (a) $HCCl_3$; (b) NSF; (c) C_2H_6; (d) C_3H_7Cl; (e) $Fe(CO)_5$; (f) $Cl-CH_2-SiH_2-SiH_2-CH_2Cl$.

3.16. Consider the following coordination compound, where the three dots near the center represent three d electrons on the Cr atom:

(a) What are the formulas and charges of the two different types of ligands? (b) What is the coordination number of Cr? (c) The Cr atom/ion may be referred to as being which of the following in this complex? Select all that apply: counterion; Lewis acid; electrophile; Lewis base; nucleophile; acceptor atom. (d) What is the charge on the Cr ion? (e) What is the donor atom of the organic ligand? (f) What is the counterion in this compound?

3.17. *Examine the Lewis structures you drew in Exercise 3.11, and identify the potential donor atoms in (a) HOF; (b) Si_2H_6; (c) Si_3H_8; (d) CSe_2; (e) $[V(CO)_6]^-$; (f) $ClCH_2-CH_2Cl$.

3.18. Consider the compound $[AlF_2(NH_3)_4]Cl$. Write the formula (including charge, if any) of (a) the counterion, (b) the Lewis acid, and (c) the two different types of ligands. (d) What is the coordination number of the Lewis acid? (e) What is the donor atom of the molecular ligand?

3.19. *Some metalloid elements such as As can act either as Lewis acids or Lewis bases in different situations. Write a likely reaction forming the indicated bonds to As from likely starting materials for each of the following compounds, and tell in each case whether the As acted as a Lewis acid or a Lewis base: (a) As–(–F)$_5$; (b) $(CH_3)_2O$–AsCl$_3$; (c) [Cl–AsCl$_3$]$^-$; (d) Ni–(–AsF$_3$)$_4$$^{2+}$.

3.20. Give the coordination number of the metal atoms in each of the following complexes: (a) [Hg(SCH$_3$)$_2$]; (b) [Pt(NH$_2$)$_3$(NH$_3$)$_3$](NO$_3$); (c) K[NpF$_8$]. Then give the symbols of each different type of donor atom found in the ligands in each complex.

3.21. *Two common organic ions or molecules found in coordination compounds are methylamine or methaneamine, CH$_3$NH$_2$, and methoxide ion, CH$_3$O$^-$. (a) Give one term used in coordination chemistry to describe the role of these organic species in complexes. (b) Identify the donor atom found in each organic species. (c) Give one term (other than metal ion) used in coordination chemistry to describe the role of metal ions in coordination compounds of these organic species.

3.22. Given [CoF$_6$]$^{3-}$, determine (a) the charge on each fluorine atom and (b) the charge on the cobalt atom.

3.23. *Given the complex ion [V(CO)$_3$(NH$_3$)$_3$]$^-$, determine (a) the charge on each CO, (b) the charge on each NH$_3$, and (c) the charge on the vanadium atom.

3.24. Assign charges to the central metal atom and the outer molecules or anions in (a) [UO$_2$F(CH$_3$)] and (b) [Os(NH$_3$)$_4$I$_2$]F$_2$.

3.25. *Assign charges (oxidation numbers) to the central metal (or nonmetal) atom in each of the following complex ions or compounds: (a) [Ru(NH$_3$)$_5$I]Cl$_3$; (b) [SbCl$_5$(OH$_2$)]; (c) UO$_2$$^{2+}$; (d) Ba$_3$[AlF$_6$]$_2$.

3.26. Consider the compound Rb$_2$[MoO$_2$(OCH$_3$)$_4$]. Write the formula (including charge, if any) of (a) the counterion, (b) the Lewis acid, and (c) the two different types of ligands. (d) What is the coordination number of the Lewis acid? (e) What is the donor atom of the polyatomic ligand?

3.27. *Give plausible products (or just one product) for each of the following reactions, or tell if no reaction is expected: (a) CsCl + AlCl$_3$ → ; (b) SiCl$_4$ + 4C$_5$H$_5$N → (a conductor of electricity in solution); (c) PbCl$_4$ + 2C$_5$H$_5$N → (a nonconductor of electricity in solution).

3.28. For each set of reactants, label the Lewis acid and the Lewis base. Then write a likely formula for the product formed. If it is a complex ion, enclose it in brackets and label the counterion that accompanies the complex ion. (a) H$^+$(aq) + Cl$^-$(aq) + C$_5$H$_5$N →; (b) 2C$_5$H$_5$N + SnCl$_4$ →; (c) 2Cs$^+$ + 2Cl$^-$ + SnCl$_4$ →.

3.29. *Pretend you are Alfred Werner and that you have isolated the following five compounds:

PdCl$_2$·4(CH$_3$)$_3$P	Contains 3 ions; 2 Cl$^-$ ions precipitated by Ag$^+$
PdCl$_2$·3(CH$_3$)$_3$P	Contains 2 ions; 1 Cl$^-$ ion precipitated by Ag$^+$
PdCl$_2$·2(CH$_3$)$_3$P	Contains 0 ions; 0 Cl$^-$ ions precipitated by Ag$^+$
PdCl$_2$·(CH$_3$)$_3$P·KCl	Contains 2 ions; 0 Cl$^-$ ions precipitated by Ag$^+$
PdCl$_2$·2KCl	Contains 3 ions; 0 Cl$^-$ ions precipitated by Ag$^+$

(a) Write the formulas of each of these compounds in the modern form, with complex ions enclosed in brackets. (b) Identify the counterions in each compound. (c) What is the coordination number of palladium in this series of compounds? (It is constant.)

3.30. Write formulas of plausible nonacidic cations with the following central atoms and coordination numbers: (a) N^{5+} and 4; (b) Br^{7+} and 6; (c) Pt^{4+} and 6; (d) Pt^{2+} and 4.

3.31. *Draw each of the following orbitals, showing their shapes, + and – lobes, and nodal planes: (a) the p_y orbital; (b) the $d_{x^2-y^2}$ orbital; (c) the d_{xz} orbital.

3.32. Draw and name an orbital that has (a) four lobes falling between the x and y axes; (b) four lobes falling along the x and y axes; (c) no nodal planes; (d) three nodal planes (an incomplete name is called for); (e) two nodal planes, which are the xz and yz planes.

3.33. *Figure 3.8a–f shows different pairs of atoms with orbitals overlapping. (a) Identify the specific atomic orbital (e.g., d_{xy}) used by each atom in each case. (b) In each case, tell whether there is positive, negative, or zero overlap between the orbitals of the two atoms. (c) For each case in which there is positive overlap, identify the bond that is formed as σ, π, or δ.

3.34. Figure 3.8g–l shows different pairs of atoms with orbitals overlapping. (a) Identify the specific atomic orbital used by each atom in each case. (b) In each case, tell whether there is positive, negative, or zero overlap between the orbitals of A and of B. (c) For each case in which there is positive overlap, identify the bond that is formed as σ, π, or δ.

3.35. *(a) How many nodal planes through the two nuclei would you expect to find in a φ bond? (b) Referring to the f_{xyz} orbital pictured in Figure 3.7, imagine another such orbital approaching this orbital along the z axis so as to give positive overlap. Would the resulting bond be σ, π, δ, or φ?

3.36. Which of the choices given are impossible for two atoms having positive overlap along the z axis? (a) A δ bond is formed from an s orbital on the first atom overlapping a d_{z^2} orbital on the second. (b) A π bond is formed from a p_x orbital on the first atom overlapping a d_{xz} orbital on the second. (c) A σ bond is formed from a p_x orbital on the first atom overlapping a p_y orbital on the second.

3.37. *Give the name (σ, etc.) of each of the kinds of bonds drawn here, and arrange them in order of increasing bond dissociation energy. Fill in the signs of the wave functions necessary to give bonding interactions in each case.

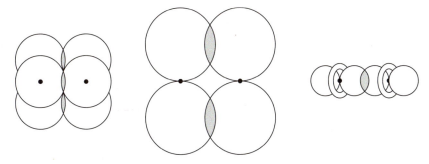

3.38. Which element in each of the following groups shows the greatest tendency to form double or triple bonds: (a) C, Si, Ge, Sn, or Pb; (b) Mn, Tc, or Re; (c) Lu, W, or Pt?

3.39. *Which of the following is a better conclusion? (a) Pi bonds are especially stable among the second period elements, or (b) σ bonds between certain second-period elements are especially unstable. Explain your conclusion, using data from Tables 3.2 and 3.3.

3.40. Which type of overlap—σ, π, δ, or none—is possible between the specified pairs of orbitals overlapping along the z axis? (a) s and $d_{x^2-y^2}$; (b) s and d_{z^2}; (c) p_z and d_{xz}; (d) p_z and d_{xy}.

3.41. *Without the use of tables, arrange each of the following sets in order of increasing covalent-bond energies (increasing stability of covalent bonds): (a) single bonds: C–C, Si–Si, Ge–Ge, Sn–Sn, Pb–Pb; (b) single bonds: Li–Li, B–B, C–C; (c) Mo–Mo bonds: σ, δ, π.

3.42. The C–halogen bond energy is 327 kJ mol^{-1} for C–Cl, 285 kJ mol^{-1} for C–Br, and 213 kJ mol^{-1} for C–I. Use this information and data from Table 3.2 to obtain four independent calculations of the electronegativity of carbon. Do these results fall within ±0.05 of the value tabulated in Table A?

3.43. *From considerations of their Lewis structures, which homopolyatomic (atom or ion) has the strongest element–element bond? (a) O_2 or O_2^{2-}? (b) N_2 or N_3^-? (c) C_2^{2-} or C_3^{4-}?

3.44. Some new elements have just been discovered in the author's laboratory. The following atomic parameters have been obtained:

Element	Sheltonium	Bennerine	Coatsium	Kamelogen
Electronegativity	1.06	3.82	1.97	2.34
Element–element single-bond energy (kJ mol^{-1})	78	166	88	183

Calculate the covalent bond energies expected for (a) the coatsium–kamelogen bond and (b) the sheltonium–bennerine bond.

3.45. *The electronegativity of carbon in CCl_4 is estimated to be 2.64; the electronegativity of carbon in $(CH_3)_3CCl$ is estimated to be 2.29. Compute the expected C–Cl bond energy in each compound.

3.46. Consider an element, udezium (Ud), in the middle of the p block. The Ud–Ud bond energy is 193 kJ mol^{-1}, the Ud–F bond energy is 543 kJ mol^{-1}, and the F–F bond energy is 155 kJ mol^{-1}. From this information, calculate the Pauling electronegativity of Ud.

3.47. *Consider an element, taylorium (Ty), in the middle of the p block. The Ty–Ty bond energy is 114 kJ mol^{-1}, the Ty–F bond energy is 503 kJ mol^{-1}, and the F–F bond energy is 155 kJ mol^{-1}. From this information, calculate the Pauling electronegativity of Ty.

3.48. Consider an element, przybylskium (Pz), in the middle of the p block. The Pz–Pz bond energy is 293 kJ mol^{-1}, the Pz–F bond energy is 613 kJ mol^{-1}, the F–F bond energy is 155 kJ mol^{-1}, and the I–I bond energy is 149 kJ mol^{-1}. (a) From this information, calculate the Pauling electronegativity of Pz. (b) From your Pauling electronegativity of Pz, calculate the expected bond energy of the Pz–I bond.

3.49. *You are given the following bond energies: Ge–Ge = 188 kJ mol^{-1}, Cl–Cl = 240 kJ mol^{-1}, and Ge–Cl (in $GeCl_2$) = 385 kJ mol^{-1}. Calculate the electronegativity of Ge in $GeCl_2$. Does your answer differ from the Pauling electronegativity of Ge in $GeCl_4$ (2.01) in the expected manner? Explain.

3.50. Based on the calculated oxidation state of each central atom, draw Lewis dot structures of the following molecules and ions: (a) SO_2^{2+}; (b) XeF_6; (c) SiS; (d) IF_4^+; (e) $TeCl_6^{2-}$.

3.51. *Based on the calculated oxidation state of each central atom, draw the Lewis dot structures of the following molecules or ions (treat hydrogen as the hydride anion, H$^-$): (a) OsO_4; (b) NO_2^+; (c) XeF_4; (d) SiH_4; (e) ICl_4^-.

3.52. Based on the calculated oxidation state of each central atom, draw Lewis dot structures of the following molecules and ions: (a) IO_3^-; (b) IF_4^-; (c) XeF_5^-; (d) BCl_4^-; (e) TeF_4.

3.53. *Based on the calculated oxidation state of each central atom, draw the Lewis dot structures of each of the following molecules or ions: (a) NO_2^+; (b) XeF_4; (c) $SbCl_5$; (d) IF_5; (e) ClNO.

3.54. Based on the calculated oxidation state of each central atom, draw the Lewis dot structures of each of the following molecules or ions: (a) $[Te(CH_3)_2Cl_3]^-$ (central atom is Te in oxidation state +4); (b) BrF_4^-; (c) BO_3^{3-}; (d) $XeOF_4$; (e) ClO_2^-; (f) SCl_4.

3.55. *Based on the calculated oxidation state of each central atom, draw the Lewis dot structures of each of the following molecules or ions: (a) XeF_3^+; (b) $SbCl_5^{2-}$; (c) XeO_4; (d) XeF_5^+; (e) NCN^{2-}.

3.56. Predict the formulas of the highest fluoro anions of the following elements in the specified oxidation states: (a) Si^{4+}; (b) B^{3+}; (c) As^{5+}; (d) Cr^{3+}.

3.57. *Predict the formulas of the highest fluoro anions of the following elements in the specified oxidation states: (a) Mo^{5+}; (b) Se^{4+}; (c) Xe^{6+}; (d) Xe^{8+}.

3.58. Predict the formulas of the highest fluoro anions of the following elements in the specified oxidation states: (a) Ge^{4+}; (b) Se^{4+}; (c) Xe^{4+}.

3.59. *Predict the formulas (including charges) of the fluoro anions formed by the following +6-charged Period 6 cations: (a) Te^{6+}; (b) Xe^{6+}; (c) W^{6+}; (d) Os^{6+}.

3.60. Assuming that an excess of F^- is available and referring to Tables 1.11, 1.10, and C, predict the formulas of the fluoro species expected for (a) each third-period element from Al to Cl in each of its positive oxidation states; (b) the fourth-period d-block elements in their highest oxidation states; (c) the fifth-period p-block elements in their highest oxidation states; (d) the Group 15/V elements in their +5 oxidation states; (e) the Group 15/V elements in their +3 oxidation states.

3.61. *Predict the formulas of the highest fluoro anion (or cation, or neutral compound) of each of the following elements in the specified oxidation state: (a) C^{4+}; (b) Si^{4+}; (c) Cl^{3+}; (d) Cl^{7+}; (e) I^{7+}.

3.62. Draw the Lewis structures of the highest fluoro anion (or cation, or neutral compound) of each of the following elements in the specified oxidation state: (a) Xe^{4+}; (b) S^{4+}; (c) N^{5+}.

3.63. *Write the formula (including charge) of the oxo anion of (a) boron with oxidation number +3; (b) bromine with oxidation number +3; (c) arsenic with oxidation number +5; (d) sulfur with oxidation number +4.

3.64. Predict the formulas of the highest oxo anion of each of the following elements in the specified oxidation state: (a) C^{4+}; (b) Si^{4+}; (c) Cl^{3+}; (d) Cl^{7+}; (e) I^{7+}.

3.65. *Write the formulas of the oxo anions that the following elements would have if they had the oxidation number and the (hypothetical) total coordination number listed: (a) boron with oxidation number +3 and total coordination number 4; (b) boron with oxidation number +3 and total coordination number 3; (c) arsenic with oxidation number +3 and total coordination number 5; (d) ruthenium with oxidation number +8 and total coordination number 5; (e) iodine with oxidation number +5 and total coordination number 5.

3.66. In addition to the bromate ion formed by bromine with the common oxidation number +5, bromine also forms another oxo anion with an oxidation number higher than +5, and two other oxo anions with oxidation numbers lower than +5. Predict the formulas of each of these three oxo anions.

3.67. *Draw the Lewis structures of the highest oxo anion of each of the following elements in the specified oxidation state: (a) Xe^{4+}; (b) S^{4+}; (c) N^{5+}.

3.68. Tell whether this statement is true or false, and explain your reasons: "The second-period oxo anions CO_3^{2-}, NO_3^-, NO_2^-, and NO^- (which is important in some biochemical reactions) do not obey the octet rule, because they do not have four oxygen atoms bonded to their central atoms."

3.69. *Explain why the oxides of the Group 14/IV elements in their group oxidation state all have the formula MO_2, while the oxo anions of the same elements change their formulas, including charges, down the group.

3.70. The oxo anions of some of the fifth- and sixth-period d-block elements have the same formulas (including charges) as their fourth-period counterparts in the same group, but this is not true of the fluoro anions. Illustrate with examples and explain.

3.71. *(a) Write the formula of the oxo anion likely to be formed by the following elements with the oxidation states shown in parentheses: Re(VII); As(III); S(II) (imaginary); Cl(V); Sb(V); P(V). (b) Calculate the pK_b for the formula you have written. (c) Classify the basicity of this anion (weakly basic, etc.).

3.72. The elements N, Se, and I each exhibit two common positive oxidation numbers; each occurs in an oxo anion. Predict the formulas (including charges) of those oxo anions. Classify each of the resulting six oxo anions as (a) nonbasic, (b) feebly basic, (c) moderately basic, or (d) very strongly basic.

3.73. *Calculate pK_b values and select the appropriate category of basicity for each of the following hypothetical oxo anions: (a) MO_5^{2-}; (b) MO_5^{5-}; (c) MO_7^{6-}.

3.74. (a) Write the formula of the oxo anions likely to be formed by the following elements with the oxidation states shown in parentheses: P(III); Br(V); Br(III); Si(IV); C(IV); S(IV). (b) Calculate the pK_b for each formula you have written. (c) Classify the basicity of each anion (weakly basic, etc.).

3.75. *Consider the three oxo anions ReO_4^-, AlO_4^{5-}, and AsO_4^{3-}. (a) Classify the basicity of each oxo anion. (b) The pK_{b1} values of these three anions are −18.2, 2.2 and 22.6 (not necessarily in that order). Without actually calculating pK_b values, associate each value with the oxo anion most likely to show it. (c) Solutions of the three anions are found to have pH values of 6.5, 12.0, and 14.0 (not necessarily in that order). List each pH with the ion most likely to have a solution of that pH. (d) Are any of these oxo anions incapable of existing in water?

3.76. When the use of triphosphates in detergents was under attack, sodium carbonate and sodium silicate were tried as replacements for use in high concentration in detergents. However, concern was expressed that such detergents would be very dangerous if infants should swallow them; the detergents might also be quite corrosive to the washing machines. Explain why.

3.77. *Arsenic in the +3 oxidation state forms the AsO_3^{3-} oxo anion. (a) Calculate the pK_{b1} for this oxo anion. (b) Classify the strength of this oxo anion. (c) Write the formula of the calcium salt of this oxo anion. (d) Draw the Lewis structure of this anion.

3.78. Arsenic in the +5 oxidation state forms the AsO_4^{3-} oxo anion. (a) Calculate the pK_{b1} for this oxo anion. (b) Classify the strength of this oxo anion. (c) Write the formula of the thorium(IV) salt of this oxo anion. (d) Draw the Lewis structure of this anion.

3.79. *Explain how and why the basicity of an oxo anion depends on its charge, the number of oxo groups present, and the electronegativity. Use your reasoning to try to compare the basicity of the PO_4^{3-} anion with that of $[O_3P-O-PO_3]^{4-}$ (pyrophosphate ion) and $[O_3P-O-PO_2-O-PO_3]^{5-}$ (triphosphate ion).

3.80. (a) Explain briefly why a given cation might have a maximum total coordination number. Tell what this number is for each period of the periodic table (starting with the second period). (b) Give one possible reason why, in another compound, the same cation might only achieve its penultimate total coordination number. Tell what this number is for each period of the periodic table (starting with the second period). (c) Which total coordination number (maximum or penultimate) is likely to apply in determining the formulas of fluoro anions? Apply the appropriate number to the case of Si^{4+} and predict the formula of the fluoro anion of Si^{4+}. (d) Which total coordination number (maximum or penultimate) is likely to apply in determining the formulas of oxo anions? Apply the appropriate number to the case of Si^{4+} and predict the formula of the oxo anion of Si^{4+}. Apply the appropriate number to the case of S^{4+} and predict the formula of the oxo anion of S^{4+}. (e) Explain the difference between the coordination number and the total coordination number of an element. Illustrate it by giving the actual values of these two numbers for the sulfur atom in the oxo anion of S^{4+}.

3.81. *Assuming that the basicity of fluoro anions can be qualitatively evaluated using the same concepts as those used for anions, select the least basic in each set of known fluoro anions, and give your reason for your choice: (a) BeF_4^{2-} or IF_4^-; (b) BeF_4^{2-}, PbF_6^{2-}, or WF_8^{2-}; (c) TeF_7^-, TaF_7^{2-}, or ZrF_7^{3-}.

3.82. (a) Draw a reasonable structure of a hydrated oxo anion, and use it to explain what effect increasing the negative charge of the oxo anion has on the basicity of the oxo anion. (b) Although the $[O_2C-CH_2-CH_2-CO_2]^{2-}$ ion has twice the negative charge of the $CH_3CO_2^-$ ion, its basicity is not much different (the pK_{b1} values are 9.24 and 9.75 respectively). Explain why not.

3.83. *The computed pK_b of the AsO_3^{3-} anion is –3.5. Sketch the predominance diagram of this anion and its protonated forms over the pH range of 0–14. Draw in the boundary lines and write the pH values of the boundaries on this graph.

3.84. (a) Calculate the pK_b values for and give the basicity classifications of the anions CrO_4^{2-} and VO_4^{3-}. (b) A predominance diagram is shown below. To which one of these two anions does this diagram belong? (c) Write the formulas (including charges) of species A, B, and C in the diagram.

3.85. *(a) Sketch predominance diagrams over the pH range 0–14 for each of the following oxo anions and their protonated forms: MO_3^-, MO_3^{2-}, MO_3^{3-}, and MO_3^{4-}. (b) What anionic form of each is expected to predominate at a pH of 14?

3.86. (a) Sketch predominance diagrams over the pH range 0–14 for each of the following oxo anions and their protonated forms: MO_6^{4-}, MO_6^{5-}, MO_6^{6-}, and MO_6^{7-}. (b) Which form of each anion is expected to predominate at a pH of 14? (c) Classify the acidity of the oxo acids derived from the four anions.

3.87. *Two predominance diagrams of arsenic oxo anions are provided. (a) Which diagram (the first or the second) is for arsenic in the +3 oxidation state? Which is for arsenic in the +5 oxidation state? (b) Fill in the formulas of the two oxo anions, and of their partially and fully protonated forms, in the appropriate regions of the two predominance diagrams. (c) What is the formula of the (hydr)oxo acid of arsenic(III)? What is its acidity category? (d) What is the formula of the oxo acid of arsenic(V)? What is its acidity classification? (e) What form of each oxo anion predominates at pH 6? (f) What is the pH of a buffer that is equimolar in the arsenic(V) oxo acid and its –1-charged anion?

3.88. (a) A predominance diagram is provided. It belongs to which of the following sulfur oxo anions: SO_4^{2-}, SO_3^{2-}, or (hypothetical) SO_2^{2-}? (b) Fill in the formulas of the various forms of this oxo anion in the appropriate region of the predominance diagram. (c) Which two forms of this oxo anion could you use to produce a buffer solution with pH about 2?

3.89. *Calculations, presented in D. Himmel, C. Knapp, M. Patzsche, and S. Riedel, *Chem. Phys. Chem.* 11, 865 (2010), suggested that iridium might exhibit an oxidation state of +9. (a) Suggest two likely formulas for the oxo (cation or) anion of Ir(IX). (b) Draw a likely predominance diagrams for the oxo anion of Ir(IX). (c) What anionic form of the oxo anion of Ir(IX) is likely to be predominant at pH 4? (d) What species does the calculation consider as the most likely form of Ir(IX)? [This form was later detected in the vapor phase—see G. Wang, M. Zhou, J. T. Goettel, G. J. Schrobilgen, J. Su, J. Li, T. Schlöder, and S. Riedel, *Nature* 514, 475 (2014).]

3.90. Choose the most acidic of each pair of the following oxo acids (if possible), then give a reason for your choice: (a) $HClO_4$ or $HClO$; (b) H_3BO_3 or H_6MoO_6; (c) H_2CO_3 or H_4SiO_4.

3.91. *Classify the acidity of each of the following oxo acids: (a) H_4XeO_6; (b) H_6TeO_6; (c) a hypothetical oxo acid formed by Te(IV) in which Te has a total coordination number of 6; (d) the oxo acid that may someday be formed by iridium in the +9 oxidation state.

3.92. Molybdic acid is a substantially weaker acid than chromic acid, sulfuric acid, or selenic acid. How might this be explained, given that Mo is in the fifth period, in which the penultimate and antepenultimate total coordination numbers can both be achieved? [See J. J. Cruywagen and J. B. B. Heyns, *J. Chem. Educ.* 66, 861 (1989).]

3.93. *Sodium triphosphate is a major component of detergents; the acid itself has the formula $H_5P_3O_{10}$ and can be written out as $(HO)_2P(=O)-O-P(=O)(OH)-O-P(=O)(OH)_2$. (a) How many oxo groups are present in this acid? Given the number of oxo groups, how would you classify its acidity using the steps in Section 3.7? (b) Actually, its acid strength is very similar to that of H_3PO_4. How can you rationalize this, and what caution would you recommend in applying the steps in Section 3.7 to the case of very large oxo acids?

3.94. In saturated KOH solution, the acid H_3AsO_3 can be fully neutralized to give the AsO_3^{3-} ion, but the third hydrogen of the acid H_3PO_3 and the second and third hydrogens of the acid H_3PO_2 cannot. Explain. (Hint: See an Amplification in Section 3.7B.)

3.95. *Referring to the predominance diagrams in Figure 2.15, tell which of the elements—Na, Sr, Ca, Mg, Y, Th, Al, Sn, Ze—are available in the illustrated oxidation state in waters of pH (a) 5.5; (b) 0; (c) 16.

3.96. Referring to the predominance diagrams in Example 2.5B, tell which of the unknown elements—Mm, Nn, Pp, Qq—should be available in the illustrated oxidation state in waters of pH (a) 5.5; (b) 0; (c) 16.

3.97. *Referring to the predominance diagrams in Figure 3.21, which of the elements—Cl, S, P, Ge, Al—(in the illustrated oxidation states) should be available, as what species, in solutions of pH (a) 1; (b) 12?

3.98. The oxalate ion, $C_2O_4^{2-}$, has $pK_{b1} = 10.19$ and $pK_{b2} = 12.75$. Which unprotonated or protonated species of the oxalate ion should be predominant in (a) neutral unpolluted natural waters; (b) acid mine-drainage waters (pH = 0); (c) the streams of Corner Brook, Newfoundland, Canada (pH = 11)?

3.99. *Which two categories of cations are likely to be found in unpolluted natural waters: (a) nonacidic; (b) feebly acidic; (c) strongly acidic; (d) very strongly acidic?

3.100. Which two categories of anions are *least* likely to be found in acid-rain-polluted natural waters (pH = 2): (a) nonbasic; (b) feebly basic; (c) strongly basic; (d) very strongly basic?

3.101. *For each of the following elements, give its expected positive oxidation number, and give the formula of the form (species) in which you would expect to find it in water of pH 5.5–7: (a) Li; (b) Al; (c) W; (d) Hs.

3.102. Assume that the following elements in the specified oxidation states form oxo anions: N(III), Mn(VII), and Xe(IV). (a) Predict the formula of the oxo anion formed by each element in that oxidation state. (b) Classify the basicity of each oxo anion. (c) Predict whether each element would be found in solution as an oxo anion in a lake of normal pH. (d) Tell whether any oxo anion could never be found in water at any pH, but would only exist in water in a partially protonated form.

3.103. *A lake near the Oak Ridge National Laboratory in Tennessee has become contaminated with plutonium from the reprocessing of spent fuel from nuclear reactors. Predict whether most of the plutonium in this lake is likely to be dissolved in the water or would be found in the sediments at the bottom of the lake. Also predict whether this might be altered if the lake were strongly subject to the effects of acid rain.

3.104. Two radioactive elements, technetium (Tc, $Z = 43$) and promethium (Pm, $Z = 61$) do not occur naturally on Earth but are found in the fallout from atomic bomb explosions. Predict the formulas of the forms (cation, anion, or oxide) in which each would likely be found in a lake of pH 5.5–7.

3.105. *The recent concern about the presence of radon is actually mainly concerned with its radioactive decay product, polonium. Assume that polonium in aqueous solution would be present with an oxidation state of +4. (a) Predict whether Po would more likely be present in a lake of pH 5.5–7 as a cation or as an oxide. (b) Predict two possible formulas for the oxo anion of polonium. Tell whether each could be present in a lake of pH 5.5–7.

3.106. Consider the cations Gd^{3+}, Au^{3+}, B^{3+}, Pa^{5+}, Ac^{3+}, and Tl^+. (a) Classify the acidity of each of the cations. (b) Describe briefly what will happen when you try to dissolve each of these cations in water. (c) Write the formula of the predominant form that each cation would take in a lake highly polluted with acid rain (pH 3).

3.107. *Name each of the following salts: (a) $TiCl_3$; (b) $FePO_4$; (c) Ag_5IO_6; (d) Hg_3TeO_6.

3.108. Name each of the following oxo anions, calculate its approximate pK_{b1}, and tell whether in aqueous solution it will be nonbasic, feebly basic, moderately basic, or very strongly basic: (a) CO_3^{2-}; (b) BrO_4^-; (c) IO_6^{5-}; (d) XeO_6^{4-}; (e) AsO_3^{3-}; (f) IO^-.

3.109. *Name each of the following salts: (a) $Sr(BrO_3)_2$; (b) $Au_2(SeO_4)_3$; (c) K_4SnO_3; (d) $Hg_5(IO_6)_2$; (e) $TlIO_3$; (f) $Zn_5(NpO_6)_2$.

3.110. Name each of the following salts: (a) $TlClO$; (b) Sn_3N_2; (c) $Al(RuO_4)_3$; (d) $HgSO_3$; (e) K_8SnO_6; (f) $Sr(ClO)_2$.

3.111. *Write the formula of each of the following oxo acids and tell whether in aqueous solution it will be very strongly acidic, strongly acidic, moderately acidic, or weakly acidic: (a) permanganic acid; (b) selenic acid; (c) arsenious acid; (d) selenous acid; (e) telluric acid.

3.112. Write the formula of each of the following oxo acids and tell whether in aqueous solution it will be very strongly acidic, strongly acidic, moderately acidic, or weakly acidic: (a) molybdic acid; (b) arsenic acid; (c) ferric acid; (d) hypobromous acid; (e) hyposelenous acid (hypothetical).

3.113. *Write the formulas of each of the following oxo anions and classify each as nonbasic, feebly basic, moderately basic, or very strongly basic: (a) silicate; (b) tellurate; (c) perbromate; (d) sulfite.

3.114. Write the formulas of the following salts: (a) thorium(IV) phosphate; (b) barium perxenate; (c) aluminum(III) sulfite; (d) magnesium sulfite; (e) cerium(III) pertechnetate; (f) molybdenum(IV) sulfide.

3.115. *Write the formulas of each of the following oxo anions and classify each as nonbasic, feebly basic, moderately basic, or very strongly basic: (a) hypochlorite; (b) perneptunate; (c) nitrite; (d) ferrate.

3.116. Write the formulas of the following salts: (a) calcium phosphate; (b) calcium nitrite; (c) silver(I) perxenate; (d) thorium(IV) chlorite; (e) iron(III) silicate; (f) chromium(III) silicate.

3.117. *Give the formulas of the following salts: (a) calcium borate; (b) strontium perchlorate; (c) europium(II) sulfite; (d) iron(II) phosphate; (e) chromium(II) carbonate.

3.118. Give the formulas of the following salts: (a) cesium phosphate; (b) zinc(II) perbromate; (c) potassium perbromate; (d) calcium bromite.

3.119. *Consider the oxo anions of the following elements in their +5 oxidation states: nitrogen (N), bromine (Br), phosphorus (P), antimony (Sb), and neptunium (Np). (a) Write the total coordination number (or numbers) you expect each element to show in its oxo anions. (b) Write the expected formula, including charge, of each oxo anion. (If more than one total coordination number is possible, use the most likely one.) (c) Write the name of each oxo anion. (d) List the basicity classification of each oxo anion. (e) Tell whether that oxo anion is likely to be dissolved in the water of a lake of normal pH, or whether it will be found in the sediments of the lake as an oxide or hydroxide instead. (f) Which oxo anion that you mentioned in part (e) as being soluble in natural waters is actually present in natural waters in partially protonated form (i.e., as a type of hydroxo anion)? Write the formula (including charge) of a partially protonated form of this oxo anion that you would actually expect to find in natural waters.

3.120. Among the elements near uranium in the periodic table are neptunium (Np), pluto-nium (Pu), and protactinium(Pa). (a) Carry out the calculations necessary to predict whether the +3 cations of these elements could appear in natural waters of normal pH as hydrated cations. (b) Which one oxidation state listed for each one of these elements in Table C would give rise to the least basic oxo anion of that element? (c) Assuming that these three oxo anions can exist, predict their formulas (including charges) and give their names. (d) Now determine whether these oxo anions could in fact exist in natural waters. Show your calculations or rea-soning. (e) Write the formulas of the calcium salts of each of these three oxo anions.

3.121. *(a) Name each of the following anions: MnO_4^-, NpO_6^{5-}, WF_8^{2-}, and IO^-. (b) Classify each of these anions as nonbasic, feebly basic, moderately basic, or very strongly basic. (c) Calculate pK_{b1} values for each oxo anion. (d) Calculate the pK_b for the partially protonated oxo anion $H_4NpO_6^-$. (e) Which of these five ions could be present in a lake of pH 3–10?

3.122. Consider an oxo anion with the formula MO_3^{2-}. (a) Classify the basicity of this ion (as weakly basic, etc.) (b) This oxo anion will also form a partially protonated (hydroxo) anion of formula HMO_3^-. Classify its basicity. (c) Which (if any) Group 14/IV elements would give an oxo anion with this formula? Give the name of this oxo anion or these oxo anions (if any). (d) Which Group 16/VI elements would give an oxo anion with this formula? Give the name(s) of this oxo anion or these oxo anions. (e) If the element Xe formed such an oxo anion (i.e., XeO_3^{2-}), what would this ion be named?

3.123. *If you did not do Experiments 3 and 4, go back and try to apply the principles of this chapter to predict what would have happened in each experiment. Also answer the questions included in each experiment.

3.124. This question concerns the following cations and oxo anions: Rb^+, P^{5+}, La^{3+}, ReO_4^-, AlO_4^{5-}, and AsO_4^{3-}. (a) Draw reasonable structures for the hydrated ions that would be formed by the above cations and anions upon their being plunged into water. Use the draw-ings to explain why some of these cations or anions could not actually exist in water; tell what products would be produced by their reaction with water. (b) The three cations in the above list have pK_a values of –33.1, 8.5, and 14.6. Make the most likely match of each cation with its pK_a. (c) Solutions of the three cations (of similar and reasonable concentrations) have pH values of 0, 5.5, and 6.5. Make the most likely match of each cation with the pH of its solu-tion. (d) The three anions in the above list have pK_b values of –18.2, 2.2, and 22.6. Make the most likely match of each anion with its pK_b. (e) Solutions of the three anions (of similar and reasonable concentrations) have pH values of 14.0, 12.0, and 6.5. Make the most likely match of each cation with the pH of its solution. (f) Which of the above cations and anions would you expect to find biochemically available in solution in a lake of normal pH? Which (if any) would only be rendered soluble in a lake highly contaminated by acid rain? By "basic rain"? (g) Suppose that the three anions in the above list were transported to a moon of a planet on which the relative abundance of the elements oxygen and fluorine were reversed. On this moon the lakes would consist of liquid HF, and the oxo anions would be converted to fluoro anions. Assume that this happens unaccompanied by any redox chemistry. Write likely for-mulas for the fluoro anions of Re, Al, and As. Which one of these is actually found in a real mineral on Earth? Compare the basicity of each fluoro anion (in aqueous solution) to the oxo anion of the same element (in aqueous solution).

MAX BORN (1882–1970) was born in Breslau, German Empire—what is now Wroclaw, Poland. He studied mathematics and physics at universities in Breslau, Heidelberg, Zurich, and Göttingen. He received his doctorate from the University of Göttingen in 1907 for his study of the stability of elastic wires and tapes. In 1918, Born discussed with Fritz Haber the way in which ionic compounds are formed from a metal and a halogen. The result was the Born–Haber cycle that countless undergraduates have used to analyze reaction energies (especially lattice energies). After serving in the German army during World War I, Born returned to Göttingen as professor in 1921. His research focused on crystal lattices and quantum theory—including the matrix mechanics representation of quantum mechanics with Werner Heisenberg. Other collaborators (some of whom received their doctorates under Born) included Enrico Fermi, Gerhard Herzberg, Friedrich Hund, Pascual Jordan, Wolfgang Pauli, Edward Teller, Max Delbrück, Robert Oppenheimer, and Maria Goeppert-Mayer. Born, along with many other German intellectuals, was forced to emigrate in 1933 when the Nazis came to power. He moved to the United Kingdom, where he took a position at the University of Edinburgh in 1936. He shared the 1954 Nobel Prize in Physics with Walther Bothe "for his fundamental research in quantum mechanics, especially for his statistical interpretation of the wavefunction."

CHAPTER 4

Ionic Compounds in the Solid State, in Minerals, and in Solution

Overview of the Chapter

In this chapter we will move our attention from ions in solution (Chapters 2 and 3) to ions in the solid state. In Section 4.1 we look at the role of the charges and relative *sizes* of the ions in determining their coordination numbers; these computed coordination numbers often but not always agree with maximum or penultimate coordination numbers. From the coordination numbers and the stoichiometries of the salts, we select likely lattice types for the salts. We also explore how these are affected by covalency (electronegativity) and high pressures. In Section 4.2 we examine the consequences of Coulomb's law on the attractive and repulsive interactions of cations and anions to generate the extremely large lattice energies of ionic solids.

In Sections 4.3–4.5 we turn our attention to the solubility properties of ionic solids in water (and, at the end, in some organic solvents). In place of the complex set of specific solubility rules presented in general chemistry, in Section 4.3 we develop a more general set of four solubility rules, stated in terms of the acidity categories of the cations and the basicity categories of the oxo or fluoro anions. We also see the connections of these rules to the occurrence of minerals in nature, and the process of isolating natural products by organic chemists. In Sections 4.4 and 4.5 we see how the Coulombic attractions of cations for anions balance out closely against the Coulombic attractions of either type of ion for the oppositely charged end of water molecules. We analyze the general solubility rules using thermodynamics, and find that both enthalpy changes and entropy changes can be dominant. We present connections of these solubility rules to the processes of solvent extraction in analytical chemistry, bioamplification in biology, and nerve transmission in biochemistry. In what biological situations are smaller cations larger than large cations, and why? How do ions migrate through cellwall membranes? The concepts here in Chapter 4 pave the way for understanding the concepts of materials science that are presented in Chapters 8 and 12.

4.1. Ionic Lattice Types and Radius Ratios

OVERVIEW. The lattice structures of ionic solids are the result of optimizing the Coulombic attractions between oppositely charged ions and the repulsions of like-charged ions. The lattice type chosen must correspond to the cation-to-anion stoichiometry of the ionic salt. To verify that the structural drawings of unit cells of ionic solids show the correct stoichiometry, it is useful to know how to count the pictured ions depending on their location in the unit cell (Section 4.1A). You may practice this in Exercises 4.1–4.6.

Having established a lattice with the correct stoichiometry, you can calculate the radius ratio of the two monatomic ions. Usually it is possible to predict the coordination numbers of the cations and anions using Table 4.2 and Equation 4.1. [Your calculated coordination numbers will sometimes be an improvement on the expected maximum coordination numbers (Section 2.3), which do not consider the full range of variability in ion radius across a period.] Often it is possible to identify a specific lattice type for the salt (Section 4.1B). You may master such calculations by trying Exercises 4.7–4.16. In Section 4.1C some limitations of radius–ratio calculations are noted. (1) The distance between ions depends somewhat on their coordination numbers. (2) A given salt can often show polymorphism or polymerization isomerism and crystallize in more than one lattice type. (3) The effects of covalency in the cation–anion bonding are not covered in radius–ratio calculations. (4) The results do not apply at very high pressures. Nonetheless, these calculations, although far from infallible, are generally useful in predicting periodic trends. These Section 4.1C ideas are useful for solving Exercises 4.17 and 4.18.

You will employ the stoichiometry of ionic salts (as determined in Section 1.2D) and ionic radii (Section 1.5C and Table C). The radius ratios and lattice types you obtain and their limitations will be revisited and refined in Chapters 8 and 12, which deal with materials science.

4.1A. Ionic Lattice Types and Stoichiometry. Besides being found as hydrated ions in water solutions, ions are characteristically found in solids constructed from **ionic lattices**. In ionic lattices, enormous numbers of cations are surrounded by similar numbers of anions and vice versa. These lattices are held together by the strong Coulombic attractions of oppositely charged cations and anions. However, Coulombic repulsions exist between like-charged ions. These ionic lattices avoid close cation–cation and anion–anion contacts, but the nature of the ionic solid is such that cations will be surrounded by other cations at longer distances, so that (weaker) repulsions are inevitable. These strong attractions and repulsions give characteristic physical properties to ionic solids—for example, melting points in the thousands of degrees (above those of any organic compound) in ceramics and rather high densities.

For a number of salts of 1:1 cation:anion stoichiometry (MX), the optimal balance of attractions and repulsions is found when the cations and anions are arranged in the lattice type found in $Na^+Cl^-(s)$ (Figure 4.1). In this structure both Na^+ and Cl^- have their maximum coordination numbers of six of the oppositely charged ion. The

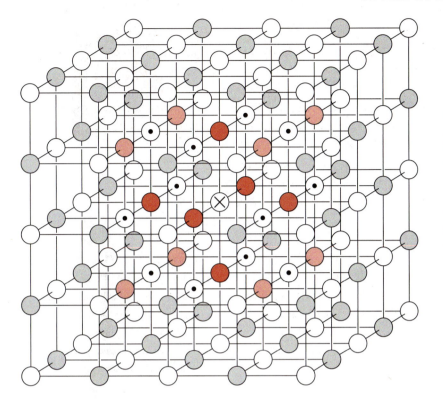

Figure 4.1. Crystal structure of NaCl. Starting with the Na$^+$ ion marked $\otimes$, there are six nearest-neighbor anions (●), 12 next nearest-neighbor cations (⊙), 8 next, next nearest-neighbor anions (●), etc. [Adapted from J. E. Huheey, E. A. Keiter, and R. L. Keiter, *Inorganic Chemistry: Principles of Structure and Reactivity*, 4th ed., Harper-Collins: New York, 1993.]

equilibrium Na–Cl distance r_0 in this structure is 283 pm, which is the sum of the ionic radii of Na$^+$ and of Cl$^-$ taken from Table C.

Lattices such as this are difficult to visualize unless we focus on a smaller unit of the lattice, the **unit cell**. This is the smallest unit of volume that contains all of the information necessary to build up the large-scale structure of the lattice. The unit cell of Na$^+$Cl$^-$ is shown in Figure 4.2a.

For other salts, other arrangements of ions (lattice types) are more satisfactory in that they result in lower energies (lattice energies, calculated in the next section). Unit cells of some of the other lattice types that are important for binary ionic substances are shown in Figure 4.2.

The NaCl lattice* has sites for an equal number of monatomic (Chapter 2) or polyatomic (Chapter 3) cations and anions.* If the *stoichiometry* of the compound is other than 1:1, another lattice type that provides the required ratio of cation and anion sites is needed. In Figure 4.2, equal numbers of cations and anions are found in the rock salt (halite), cesium chloride, zinc blende (also known as sphalerite*), and wurtzite structures, while twice as many anions as cations are found in the fluorite* and rutile* structures. Because of its 1:1 cation:anion stoichiometry, another salt such as CaO cannot possibly adopt the CaF$_2$ or TiO$_2$ lattice type; it *might* take one of the other lattice types shown.

* Structures indicated with asterisks (*) can be built using the Polyhedral Model Kit produced by the Materials Research Science and Engineering Center at the University of Wisconsin-Madison, which also provides rotatable views of these structures at http://mrsec.wisc.edu/Edetc/index.html. Other useful solid-state model kits are available from the Institute for Chemical Education at http://ice.chem.wisc.eduicatalogitems/ScienceKits.htm#SolidState. The kits are distributed by the Institute of Chemical Education, http//ice.chem.wisc.edu.

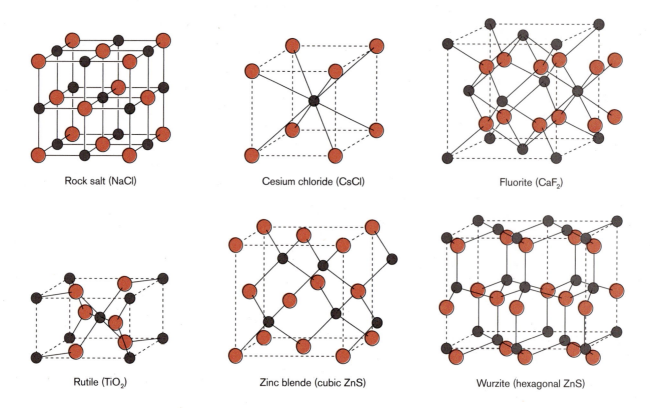

Rock salt (NaCl) Cesium chloride (CsCl) Fluorite (CaF$_2$)

Rutile (TiO$_2$) Zinc blende (cubic ZnS) Wurzite (hexagonal ZnS)

Figure 4.2. Unit cells for important lattice types for binary ionic compounds. Small black circles denote metal cations, whereas large **brown** circles denote anions. [Adapted from F. A. Cotton and G. Wilkinson, *Advanced Inorganic Chemistry*, 5th ed., Wiley-Interscience: New York, 1988, p. 5.]

Part of interpreting drawings of unit cells of solids such as those in Figure 4.2 is knowing how to confirm that the drawing indeed shows the proper stoichiometry of ions. In the structure of a crystalline solid, unit cells are repeated in all three dimensions. Consequently, any atom or ion:

- appearing on the *surface (face)* of the unit cell also appears in the neighboring unit cell(s),

- appearing in the *interior* of a unit cell is *completely* in that unit cell,

- appearing along the *edge* of a unit cell is shared among the three or four other unit cells touching that edge, and

- appearing at the *corner* of a unit cell is also shared among the six or eight adjacent unit cells touching that corner.

Consequently, in unit-cell drawings such as those of Figure 4.2, many or most atoms or ions are only fractionally in that particular unit cell (Table 4.1).

Doing the count from two-dimensional drawings such as those of Figure 4.2 requires development of a sense of perspective. More satisfactorily, solid-state models can be constructed and the number location of atoms or ions directly verified by inspecting the model (including its interior).

TABLE 4.1. Fraction of Each Atom Contained in the Unit Cell

Unit Cell Corner Angles	Interior	Face	Edge	Corner
90° (cubic, tetragonal, orthorhombic)	1	½	¼	⅛
120° (hexagonal; e.g., wurzite)	1	½	⅓	⅙

EXAMPLE 4.1

The crystal structure of a rhenium oxide* is shown in Figure 4.3. Count the atoms in the unit cell to determine the stoichiometry of this oxide.

SOLUTION: The unit cell in this structure is the cube and does not include the three open (oxygen) atoms at the upper left. This structure has no atom entirely inside the cube; all atoms are on the surface. None of the atoms is in the centers of the face. The 12 open (oxygen) atoms are on edges that are shared with three neighboring cubes, so count as 12 × ¼ = 3 oxygen atoms. The eight solid (rhenium) atoms are each on a corner and are shared with seven neighboring cubes, so count as 8 × ⅛ = 1 rhenium atom. This crystal structure is of ReO_3.

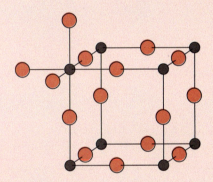

Figure 4.3. Structure of an oxide (**brown** circles) of rhenium (black circles), ReO_3.* [Adapted from F. A. Cotton and G. Wilkinson, *Advanced Inorganic Chemistry*, 5th ed., Wiley-Interscience: New York, 1988, p. 851.]

Because the ReO_3 lattice type has one cation per every three anions, it can potentially also be adopted by a salt such as AlF_3, which also has one cation per every three anions.

Normally we have more anions than cations, or equal numbers of each, but some salts such as Na_2O have the inverse: more cations than anions. Often we then find that one of the known lattice types is adopted, but the positions of the cations and the anions are reversed—the cation, which is now the more abundant ion, takes the position held by the usually more abundant anion. The lattice type is then named with the prefix anti-. As an example, the salt Cu_3N has a 3:1 ratio of cations to anions. It takes an anti-ReO_3 lattice type, in which the Cu atoms occupy the positions of the **brown** circles (O atoms) in Figure 4.3, and the N atoms occupy the black-circle (Re atom) sites.

4.1B. Importance of the Relative Sizes of the Ions (Radius Ratio). The second major reason why most ionic solids adopt a lattice type other than the NaCl type is that, while the NaCl lattice incorporates coordination numbers of six for the cation and six for the anion, this coordination number is not optimal for many combinations of cations and anions. If the cation is too small relative to the anion, it may not touch the anions, while the large anions still touch each other (Figure 4.4a). This reduces cation–anion attractive forces while keeping the anion–anion repulsive forces strong—a situation to be avoided, if possible, by going to a lattice in which the coordination number of the cation is lower so that a smaller number of anions can touch the cation without touching each other (Figure 4.4b). If the cation is too large relative to the anion (Figure 4.4c), it may be possible to fit additional anions around the cation without introducing excessive anion–anion repulsion; hence the compound will be expected to adopt a different lattice in which the cation has a higher coordination number (Figure 4.4d). Thus, the relative sizes of the monoatomic cation and anion (the **radius ratio** of the cation to the anion) is an important element in determining the type of crystal lattice that an ionic compound will adopt.

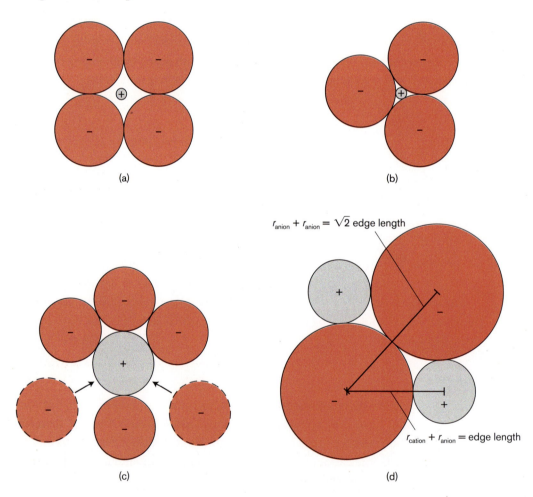

Figure 4.4. Radius–ratio limitations for a cation. (a) The cation is too small for four anions (**brown**), so (b) the cation touches three anions. (c) The cation is too large for four anions, so additional anions coordinate. (d) Relationships of edge and diagonal lengths that apply when the six-coordinate cations and anions are just touching.

If a cation and an anion have a radius ratio between 0.414 and 0.732, the NaCl lattice is expected to be chosen, because enlarging the cation in a NaCl lattice spreads the touching anions apart, reducing repulsions, while shrinking the cation in the CsCl structure reduces cation–anion contact, reducing attractions. Similar ranges for the other common binary lattice types for different stoichiometries are shown in Table 4.2.

AN AMPLIFICATION

Derivation of a Radius Ratio

We can use solid geometry to work out the ideal radius ratio for "perfect packing" (i.e., cations and anions just touching each other; anions and anions just touching each other) of many lattices. For example, for the NaCl lattice type (Figure 4.1 or Figure 4.4d), we can consider a square of two Na^+ ions at opposite corners and two Cl^- ions at the other corners. The Na^+ and Cl^- ions touch along the edges, and the larger Cl^- ions touch each other across the diagonal of the square. Consequently, we have the following two relationships:

$$r_{cation} + r_{anion} = a \text{ (the edge length)}$$

$$2r_{anion} = \sqrt{2}\, a \text{ (the diagonal length)}$$

By eliminating a between these two equations and rearranging, we can solve for the "ideal" radius ratio, $r_{cation}/r_{anion} = \sqrt{2} - 1 = 0.414$.

Likewise, starting with the cube of the CsCl lattice (Figure 4.2) and using the fact that the cube diagonal is equal to $\sqrt{2}$ times the edge, we can calculate the ideal radius ratio for this lattice to be $\sqrt{2} - 1 = 0.732$.

TABLE 4.2. Lattice Types and Madelung Constants for Different Stoichiometries and Radius Ratios of Cations and Anion

Radius Ratio	Lattice Type	Coordination Number of Cation	Coordination Number of Anion	Madelung Constant[a]
Never favored	Ion pair	1	1	1.0000
0.000–0.155	Depends on stoichiometry	2		
0.155–0.225	Depends on stoichiometry	3		
0.225–0.414	Depends on stoichiometry	4		
0.414–0.732	Depends on stoichiometry	6		
0.732–1.000	Depends on stoichiometry	8		
1.000		12		

Table 4.2 continues on next page ▶

TABLE 4.2. *(cont.)*

Radius Ratio	Lattice Type	Coordination Number of Cation	Coordination Number of Anion	Madelung Constant[a]
1:1 Stoichiometry of Salt (MX)				
0.225–0.414	Wurtzite (ZnS)*[b]	4	4	1.63805
	Zinc blende (ZnS)	4	4	1.64132
0.414–0.732	Rock salt (NaCl)*	6	6	1.74756
0.732–1.000	Cesium chloride (CsCl)	8	8	1.76267
1:2 Stoichiometry of Salt (MX$_2$)				
0.225–0.414	Beta-quartz (SiO$_2$)	4	2	2.201
0.414–0.732	Rutile (TiO$_2$)*	6	3	2.408
	Cadmium chloride (CdCl$_2$)*	6	3	2.244
	Cadmium iodide (CdI$_2$)*	6	3	2.355
0.732–1.000	Fluorite (CaF$_2$)*	8	4	2.51939
2:3 Stoichiometry of Salt (M$_2$X$_3$)				
0.414–0.732	Corundum (Al$_2$O$_3$)*	6	4	4.1719

[a] The Madelung constant is discussed in Section 4.2A.

[b] Structures indicated with an asterisk (*) can be built with the appropriate model kit. See footnote, p. 167.

It is possible to predict coordination numbers for salts not corresponding to any of the lattice types listed in Table 4.1 or shown in Figure 4.2 (e.g., ReO$_3$ or AlF$_3$).

1. Calculate the radius ratio, r_{cation}/r_{anion}. (When the cation is larger than the anion, calculate the inverse radius ratio, r_{anion}/r_{cation}.)

2. The radius ratio should predict the coordination number for the less abundant ion in any lattice type and for any stoichiometry. Except in some compounds such as K$_2$O, the less abundant ion is the cation; hence from the radius ratios given at the top of Table 4.2, the expected coordination number of the cation can be predicted.

3. The average coordination number of the anion (the more abundant ion) can then be calculated from the stoichiometry of the salt of M and X:

(Coordination number of M)(Number of M in formula) =
(Average coordination number of X)(Number of X in formula)　　　　　(4.1)

If in fact the lower part of Table 4.2 lists a lattice type with the calculated radius ratio and coordination number of the cation, this can simply be read from the table (as a substitute for step 3 in Figure 4.5).

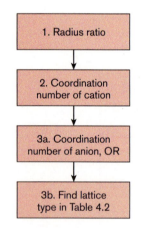

Figure 4.5. Flow chart for characterizing lattice types.

1. Radius ratio

2. Coordination number of cation

3a. Coordination number of anion, OR

3b. Find lattice type in Table 4.2

EXAMPLE 4.2

Use radius ratios to predict the coordination number of Al^{3+} in its halides. Then predict the average coordination numbers of the halide ions.

SOLUTION: The radius of Al^{3+} is 67 pm; the ratio to the halide ion radii (left side of Table C) are 0.563 for F^-, 0.401 for Cl^-, 0.368 for Br^-, and 0.325 for I^-. Therefore, the predicted coordination numbers for Al^{3+} are six in AlF_3, and four in $AlCl_3$, $AlBr_3$, and AlI_3. These coordination numbers (CN) are in fact found in the trihalides except for $AlCl_3$, in which $CN(Al) = 6$. Note that the $AlCl_3$ radius ratio is the closest to the dividing line for coordination numbers in Table 4.1.

When the coordination number of Al^{3+} is six, applying Equation 4.1, we find that the (average coordination number of X) = (coordination number of M)(number of M in formula)/(number of X in formula) = (6)⅓ = 2. When the coordination number of Al^{3+} is four, the (average coordination number of X) = (coordination number of M)(number of M in formula)/(number of X in formula) = (4)(⅓) = 1⅓. This fractional average coordination number of X means that not all of the X^- ions have the same coordination number (some have a coordination number of two, while others have a coordination number of one).

EXAMPLE 4.3

Predict the lattice types that are adopted by the oxides MgO, ThO_2, and Na_2O.

SOLUTION: First, we obtain the appropriate radii from Table C and then calculate the radius ratios, r_{cation}/r_{anion}: 86/126 = 0.68, 108/126 = 0.86, and 116/126 = 0.920, respectively.

Second, we refer to the top of Table 4.1 to obtain the expected coordination number of the most abundant ion: 6 for Mg^{2+} in MgO, 8 for Th^{4+} in ThO_2, and 8 for O^{2-} in Ag_2O.

Third, if these stoichiometries and sets of coordination numbers match any of the entries in the remainder of Table 4.1, we can read the lattice type from the table. We thus predict that MgO ought to adopt the NaCl structure, and ThO_2 ought to adopt the fluorite structure. The radius ratio for Na_2O suggests the fluorite structure; however, its stoichiometry is 2:1 (M_2X) instead of 1:2 (MX_2). The coordination number of eight applies to the oxide ions instead of the less abundant sodium ions. Na_2O should adopt the anti-fluorite structure, in which each Na^+ is surrounded by four O^{2-} ions, and each O^{2-} ion is surrounded by eight Na^+ ions. These structures are in fact found according to Wells.[1]

AN AMPLIFICATION

Effect of Coordination Number on Ionic Radius

One factor that comes into play near the radius-ratio dividing line is that the radii of ions change when their coordination numbers change: as more anions crowd around a cation, they repel each other more and cannot approach as closely to a cation. In the gaseous state, Na–Cl exists as gaseous *ion pairs* with one-coordinate ions and a Na–Cl distance of 236 pm, while in solid NaCl, six-coordinate ions are present and the interionic distance is 283 pm. The radius of Na^+ varies with its coordination number—113 pm when four-coordinate, 116 pm when six-coordinate, and 132 pm when eight-coordinate. Therefore, its radius ratio to a given anion also changes somewhat, which would produce erroneous predictions when we are close to the boundary ratios. The radii tabulated in Table C are for six-coordinate ions, so they give a correct prediction of both lattice type and interatomic distance for solid NaCl.

4.1C. Other Factors Influencing the Prediction of Lattice Types Using the Radius Ratio Rule. The radius-ratio rule succeeds in predicting lattice types about two-thirds of the time.[2] The large-scale periodic trends are predicted correctly—for a given anion, larger cations further left or down in the periodic table adopt lattices in which the cations have larger coordination numbers—but the radius ratio at which the transition from one lattice type to another occurs is often somewhat different than the boundary ratios given in Table 4.2. The radius ratio Si^{4+}/O^{2-}, for example, is 0.428, close to the boundary ratio predicted between coordination numbers four and six, 0.414; radius ratios that are further from the boundary ratios are more likely to give correct predictions.

Role of Covalency. The choice of lattice type can also be influenced by the ionic or covalent nature of the cation–anion bonding. If the anion is two- or three-coordinate, different bond angles at the anion are optimal for ionic or covalent bonding. If an oxide or fluoride ion is two-coordinated to two genuine cations (as in ReO_3 or AlF_3), the repulsions between the two cations are minimized if the cation–anion–cation bond angle is 180°. However, a (less-electronegative) sulfide or heavier halide ion is more likely to be sharing valence *np* (or hybrid *nsp³*) electron pairs with the "cations." Because these electron pairs are oriented at 90° or 109° to each other, a cation–anion–cation bond angle in this range is preferred. As examples, the cation–anion–cation bond angles are in the 90°–109° range in $BeCl_2$ and AuCl (Figure 4.6a, b).

Similarly, a three-coordinate oxide in a truly ionic solid would tend to have minimal cation–cation repulsion if the bond angles are about 120°, as in rutile (Figure 4.2).

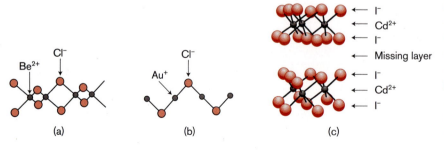

Figure 4.6. Polymeric structures of dihalides and monohalides having some covalent character in their bonding. Darkened circles represent metal atoms; **brown** circles represent halogens. (a) The chain structure of $[BeCl_2]$. (b) The chain structure of [AuCl]. (c) The CdI_2 layer structure.

But the three-coordinate chloride and iodide ions in $CdCl_2$ and CdI_2 lattice types achieve smaller Cd–X–Cd bond angles by sandwiching all of the Cd^{2+} ions between every other pair of layers of halide ions (Figure 4.6a, b). Between the alternate pairs of layers there is an *absence* of cations. Such structures, known as *layer structures*, seem to have direct contacts of anions in adjacent layers, contradicting expectations based on the Coulombic repulsion of these ions, but they have the bond angles needed for covalent bonding. In general, layer structures are common among *d*-block di- and trichlorides, bromides, iodides, and sulfides, whereas the corresponding fluorides and oxides tend to adopt the expected ionic lattice types.

Polymorphs and Polymerization Isomers: Alternate Lattice Types for a Compound. The radius-ratio rule cannot differentiate between **polymorphs** of a given salt that have different atomic arrangements involving the same coordination numbers, such as the two polymorphs of zinc sulfide, wurtzite, and zinc blende (Figure 4.2). Sometimes the different atomic arrangements involve different coordination numbers (CN), in which case we will refer to different **polymerization isomers**.[3]

Silicon dioxide is a very important compound in geochemistry, and shows numerous polymorphs and polymerization isomers. Table 4.3 lists some of these, and Figure 4.7 illustrates them. Many of these structures feature rings of many alternating Si and O atoms that have fairly large Si–O–Si bond angles such as 144°. These large rings have large pores (voids) in their centers, which give rise to low densities.

Role of Pressure. Polymerization isomers with unusually high densities and higher coordination numbers can result from application of high *pressure*, which can compress additional anions closer to cations to give the cations higher coordination numbers.[4] Thus, high pressure favors the higher-density forms of SiO_2. Pressures of 35,000 to 120,000 atm enable the formation of the polymorph coesite and the polymerization isomer stishovite, which are found in meteor craters. The even more dense seifertite was found in a meteor ejected from Mars.

TABLE 4.3. Some Polymorphs and Polymerization Isomers of SiO_2

Mineral	Figure 4.5 Reference	CN(Si)	CN(O)	Ring Size	Density, ρ (g cm^{-3})	Occurrence, Structure Reference
ITQ-4	f	4	2	Up to 12 Si + 12 O	1.70	Synthetic
Tridymite	b	4	2	6Si + 6O	2.27	Mineral
β-Cristobalite	c	4	2	6Si + 6O	2.33	Mineral
β-Quartz	a	4	2	3Si + 3O; 6Si + 6O	2.65	Mineral
Coesite	d	4	2	2Si + 4O; 4Si + 8O	2.92	Meteor craters
Stishovite	e	6	3	3Si + 3O	4.29	Meteor craters
Seifertite		6	3		4.30	Mars

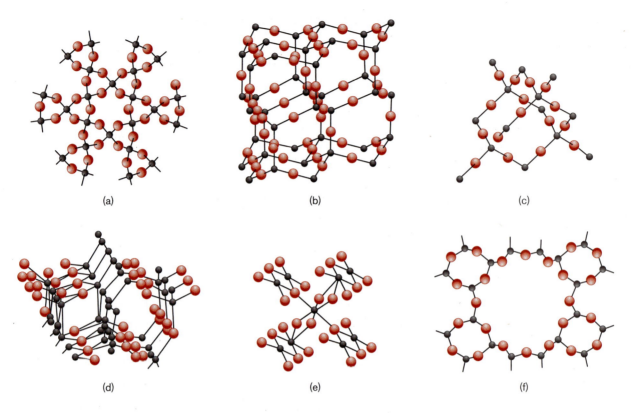

(a)   (b)   (c)

(d)   (e)   (f)

Figure 4.7. Polymorphs and polymerization isomers of SiO_2 and their densities (ρ). Silicon atoms are shown in black and oxygen atoms in **brown**. (a) β-Quartz (ρ = 2.533 g cm⁻³); (b) β-tridymite (ρ = 2.265 g cm⁻³); (c) β-crystobalite (ρ = 2.334 g cm⁻³); (d) coesite (ρ = 2.921 g cm⁻³); (e) stishovite (ρ = 4.290 g cm⁻³); (f) ITQ-4 (ρ = 1.70 g cm⁻³). [*Sources:* Adapted from: (a)–(c) P. A. Barrett, M. A. Camblor, A. Corma, R. H. Jones, and L. A. Villaescusa, *Chem. Mater.* 9, 1713 (1997); ibid., *J. Phys. Chem. B* 102, 4147 (1998). (d) K. L. Geisinger, M. A. Spackman, and G. V. Gibbs, *J. Phys. Chem.* 91, 3237 (1987). (e) A. Kirfel, H.-G. Krane, P. Blahs, K. Schwarz, and T. Lippman, *Acta Crystallogr., Sect. A* A57, 663 (2001). (f) Y. Kuwayama, K. Hirose, and Y. Ohishi, *Science* 309, 923 (2005).]

EXAMPLE 4.4

(a) Using our distinction between polymorphs and polymerization isomers, which of the minerals listed in Table 4.3 are polymerization isomers of β-quartz and which are polymorphs of it? (b) The nitride Na_3N, which was only fairly recently synthesized,[5] was found to have a coordination number of six for N^{3-}. A more recent paper[6] has described the effects of increasing pressure on this structure. What would you expect high pressure to do to the coordination numbers of N^{3-}? Consult the paper to describe the details. (c) The Si–O–Si bond angles in the SiO_2 polymorphs of Figure 4.7 are often in the neighborhood of 144°. What does this angle suggest about the nature of the bonding?

SOLUTION: (a) The minerals in Table 4.3 with different coordination numbers than β-quartz are polymerization isomers of β-quartz; these are stishovite and seifertite. (b) Higher pressure should give denser polymerization isomers and increase the coordination numbers; it is reported to increase CN(N^{3-}) from 6 to 8 to 9 to 11 to 14. (c) Note that the Si–O–Si angle is much larger than the typical 109° M–O–M angle for covalent tetrahedral sp^3-hybridized two-coordinate oxygen, but it is also much smaller than the 180° angle found for two-coordinate oxygen in ionic ReO_3, suggesting an intermediate nature of the bonding.

4.2. Ionic Solids: Coulombic Attractions and Lattice Energies

OVERVIEW. We can calculate the lattice energy of an ionic salt using Equation 4.9 after finding its lattice type (as in Section 4.1), ionic radii, Madelung constant (Table 4.2), and average Born exponent of the ions (Table 4.4). You may master such calculations by trying Exercises 4.19–4.30. It is remarkable how close these calculations come to experimental lattice energies for truly ionic salts, and gives us confidence that the Coulombic model of ionic attractions and repulsions is a good model for ionic bonding.

Born exponents are obtained from the full electron configurations of ions (Section 1.3A). Lattice energies will prove useful in Sections 5.5, 6.6 and 7.2, while Coulombic attractions and lattice energies play roles in Chapters 5, 7, 8, and 12.

4.2A. Coulomb's Law and the Energetics of Ion-Pair Formation.

Despite the modest success of quantitative predictions with the radius-ratio rule, the concept gives us insight into the consequences of the strong attractions of the oppositely charged cations and anions, and the strong repulsions of like-charged ions (Coulombic attractions and repulsions) in ionic compounds. The energy of the Coulombic attraction of one mole of gaseous cations and anions can be derived from Coulomb's law in physics:

$$E_{attraction} = \frac{Z_+ Z_- e^2 N}{4\pi\varepsilon_0 d} \tag{4.2}$$

In this equation, the variables are the charge on the cation, Z_+, the charge on the anion, Z_-, and the distance between the two, d. Thus, it has dimensions of charge squared over distance and is another example of this familiar type of ratio in chemistry.

Incorporating the appropriate conversion factors and collecting all constant terms into one term, we can simplify Equation 4.2 to Equation 4.3:

$$E_{attraction} = \frac{138,900 \; Z_+ Z_-}{d} \; kJ \; mol^{-1} \tag{4.3}$$

This is an exothermic interaction (E is negative), because the charge of the anion is negative.

Constants in Equation 4.2

This equation includes certain constants needed to give the answers in SI units: e is the charge on the electron, 1.602×10^{-19} C (coulombs); N is Avogadro's number (so that the result will be expressed per mole of ionic compound); and ε_0 is the dielectric constant (permittivity) of a vacuum, 8.854×10^{-12} C^2 m^{-1} J^{-1}. We prefer, however, to use pm instead of SI units of length (m), and to have our answer in kJ mol^{-1}, not in J mol^{-1}.

AN AMPLIFICATION

In developing Coulombic relationships further, it is desirable to use spherical cations and anions—that is, not oxo anions. Hence, let us consider the attraction of 1 mol of sodium ions and 1 mol of chloride ions at opposite ends of the universe (i.e., $d =$ infinity), so that the energy of attraction is zero. We now allow the cations and anions to approach until each cation just touches one other anion, but no other interactions occur; now we have a mole of gaseous **ion pairs** of NaCl. We could start by assuming that the distance d in Equation 4.3 is 283 pm, the sum of the ionic radii in Table C:

$$d = r_{cation} + r_{anion} = 116 \text{ pm} + 167 \text{ pm} = 283 \text{ pm} \tag{4.4}$$

However, this would give too low an attraction energy, because the radii in Table C are for six-coordinate ions while the ions are one-coordinate in an ion pair. Experimentally the NaCl distance in such an ion pair is much shorter—236 pm. Substituting 236 pm in Equation 4.3, we estimate an attractive energy of -589 kJ mol^{-1}.

Although bringing the ions closer would bring the energy still lower, this cannot happen because the like-charged electrons of the oppositely charged ions repel each other (Figure 4.8).

The energy of the repulsion is given by Equation 4.5:

$$E_{repulsion} = \frac{NB}{d^n} \tag{4.5}$$

where B is a constant and n is known as the *Born exponent*. Its value depends on the principal quantum numbers of the electrons involved in the repulsion (Table 4.4) and can be determined by measuring the compressibility of an ionic compound.

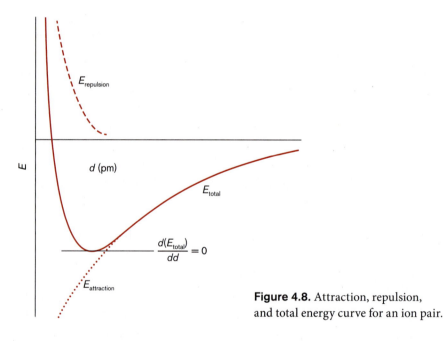

Figure 4.8. Attraction, repulsion, and total energy curve for an ion pair.

TABLE 4.4. **Born Exponents and Electron Configurations of Ions**

Born Exponent	Principal Quantum Number of Outermost Electrons of Ion
5	1 ($1s^2$: H^-, Li^+)
7	2 ($2s^2 2p^6$: F^-, Na^+)
9	3 ($3s^2 3p^6$, perhaps $3d^{10}$: Cl^-, K^+, Zn^{2+}, Ga^{3+})
10	4 ($4s^2 4p^6$, perhaps $4d^{10}$: Br^-, Rb^+, Cd^{2+}, In^{3+})
12	5 ($5s^2 5p^6$, perhaps $5d^{10}$: I^-, Cs^+, Au^+, Tl^{3+})

The total energy of the ion pairs is the sum of the attraction represented by Equation 4.3 and the repulsion represented by Equation 4.5.

$$E_{total} = E_{attraction} + E_{repulsion} \tag{4.6}$$

In the general case, the minimum in total energy is found (using calculus) by taking the derivative of the total energy with respect to distance and setting that derivative equal to zero:

$$0 = \frac{\partial E_{total}}{\partial d} = \frac{-138,900 \, Z_+ Z_-}{d_0^2} - \frac{nNB}{d_0^{n+1}} \tag{4.7}$$

Equation 4.7 can be solved for the constant B, and the new value for B can be substituted back into the total energy (Eq. 4.6) to give the final form for the total energy of 1 mol of ion pairs:

$$E_{total} = \frac{138,900 \, Z_+ Z_-}{d_0} \left(1 - \frac{1}{n} \right) \tag{4.8}$$

To evaluate this expression for the NaCl ion pair, we note that the Born exponent for Na^+ is 7 and that for Cl^- is 9, so we use an *average* Born exponent of 8. The last term of Equation 4.7 becomes $(1 - \frac{1}{8})$, and the *total* energy of formation of the NaCl ion pair from the infinitely separated ions is $\frac{7}{8}$ of -589 kJ mol^{-1}, or -515 kJ mol^{-1}.

We know that solid NaCl does not consist of isolated ion pairs. The maximum coordination numbers of Na^+ and of Cl^- are by no means met in the ion pair. Additional chloride ions from other ion pairs can still be attracted to the sodium ion in one ion pair and vice versa. This attraction leads ultimately to the *structure* of solid NaCl (Figure 4.1), in which each Na^+ surrounds itself with as many Cl^- ions as can fit (six), and each Cl^- likewise is surrounded by six Na^+ ions. The equilibrium Na–Cl distance d_0 in this structure is the sum of the six-coordinate ionic radii from Table C.

This crystal lattice has many more Coulombic attractions between oppositely charged ions than does the ion pair. In addition there are many Coulombic repulsions between like-charged ions. Each one can be evaluated using Equation 4.7. All attractions and repulsions must then be counted up, the distance involved in each interaction noted, and all the energy terms added over the entire crystal! This summation gives rise to the **Madelung constant M. M** is characteristic of the geometry of this crystal lattice type.

AN AMPLIFICATION

Derivation of the Madelung Constant

The Madelung constant M allows for the calculation of the electric potential of all ions of the lattice. To begin, we note from Figure 4.1 that each Na^+ ion has six nearest-neighbor Cl^- ions at a distance d; these contribute six times the energy term of Equation 4.8 to the total energy. However, the next-nearest neighbors to each Na^+ ion are 12 other Na^+ ions, each across the diagonal of a square from the original Na^+ and thus at a distance from it of $\sqrt{2}$ times d; these *subtract* a total of $12/\sqrt{2}$ times the energy term of Equation 4.8 from the total energy. Likewise the next, next-nearest neighbors to each Na^+ ion are the eight Cl^- ions across the body diagonal of a cube (at a distance of $\sqrt{3}d_0$) from the Na^+; this adds $8/\sqrt{3}$ energy terms to the total energy. And so the process continues, giving rise to the infinite series in Equation 4.9:

$$M = \left(\frac{6}{1} - \frac{12}{\sqrt{2}} + \frac{8}{\sqrt{3}} - \frac{6}{\sqrt{4}} + \frac{24}{\sqrt{5}} - \cdots \right) \tag{4.9}$$

The infinite series in Equation 4.9 converges rather slowly to the Madelung constant for the NaCl lattice type. It has the value 1.74756, and it applies to any of the dozens of ionic compounds in which the ions are arranged in this manner.

4.2B. Born–Landé Equation for the Lattice Energy U. When we take the total energy of 1 mol of ion pairs (Eq. 4.8), and multiply it by M, we obtain the *Born–Landé equation* for the **lattice energy U**, the enthalpy change for producing one mole of an ionic solid from its component gaseous ions:

$$U = \frac{138,900 \, M \, Z_+ Z_-}{d_0} \left(1 - \frac{1}{n} \right) \tag{4.10}$$

As outlined in Figure 4.9, we can substitute into Equation 4.10 the values of the Madelung constant, ionic charges, average Born exponent, and interionic distance to obtain the theoretical lattice energy of NaCl: $U = -751$ kJ mol^{-1}. This is quite close to the experimental lattice energy of -774 kJ mol^{-1}. This calculation shows that the major factor responsible for the stability of an ionic lattice is the *strong Coulombic attraction of the oppositely charged ions*, which extends in all directions from a given ion and throughout space. This attraction is strong enough to overcome the Coulombic repulsions that exist between like-charged ions and the repulsions that exist between outer electrons of adjacent ions.

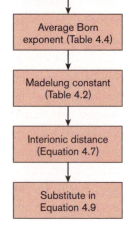

Figure 4.9. Flow chart for calculating lattice energies.

Lattice energies are very large enthalpy changes that can amount to many thousands of kJ mol^{-1} in magnitude. Some typical values are shown in Figures 4.10 and 4.11, and are included later in the Examples and in Table 4.6. Figures 4.10 and 4.11 emphasize the key variables of charge (of cation and of anion) and distance (size) in determining the magnitudes of the lattice energies.

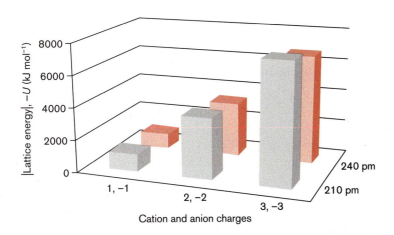

Figure 4.10. Absolute values of lattice energies ($-U$, kJ mol^{-1}) versus cation and anion charges in 1:1 binary salts MX. Front row: M–X distances are about 210 pm, in salts LiF, MgO, and ScN. Back row (in **brown**): M–X distances are about 240 pm, in salts NaF, CaO, and LaN. [Based on data from *Handbook of Chemistry and Physics*, 84th ed., Chemical Rubber Publishing Co.: Cleveland, 2006, pp. 12.22–12.30.]

Figure 4.11. Absolute values of lattice energies ($-U$, kJ mol^{-1}) versus d (see Eq. 4.7) in selected Group 1 halides LiF, NaCl, KBr, RbI. [Data from *Handbook of Chemistry and Physics*, 84th ed., Chemical Rubber Publishing Co.: Cleveland, 2006, pp. 12.22–12.30.]

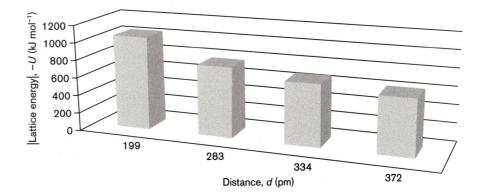

EXAMPLE 4.5

Compute the lattice energy of ThO$_2$. Explain why this lattice energy is about four times the magnitude of the lattice energy of CaF$_2$, –2651 kJ mol^{-1}.

SOLUTION: Equation 4.10 requires us to provide the cation and anion charges (+4 and –2), the Madelung constant, the interionic distance, and the average Born exponent for the two ions. In Example 4.3 we determined that ThO$_2$ should and does adopt the fluorite lattice, which from Table 4.2 has a Madelung constant of 2.51939. The interionic distance is presumed to be the sum of the ionic radii of Th^{4+} and O^{2-}, 108 + 126 = 234 pm. From Table 4.3 we find the Born exponent for O^{2-} (electron configuration $2s^2\,2p^6$) to be 7. Th^{4+} has the valence electron configuration $7s^0$ or, considering its outermost electrons, $6s^2 6p^6$. The Born exponent for this configuration is not listed, so we extrapolate a value of 14.

Example 4.5 continues on next page ▶

EXAMPLE 4.5 *(cont.)*

Averaging the two Born exponents, 7 and 14, gives us $n = 10.5$ to use in Equation 4.10. Substituting these numbers gives us

$$\frac{(138,900)(2.51939)(4)(-2)}{234}\left(1-\frac{1}{10.5}\right) = -10,824 \text{ kJ mol}^{-1}$$

This lattice energy is about four times that of CaF_2 because the cation and the anion charges in ThO_2 each are double the corresponding charges in CaF_2 (see Figure 4.10).

Examination of the Madelung constants in Table 4.2 shows that polymorphs often have very small differences in Madelung constants, which translates into small differences in lattice energies and densities. For example, M for wurtzite (1.63805) is almost the same as M for zinc blende (1.64132); the two polymorphs of TiO_2, anatase and rutile, have almost the same Madelung constants (2.400 and 2.408, respectively). However, a change from an ionic to a layer lattice gives a much larger change: M for rutile is very different from the Madelung constants for $CdCl_2$ (2.244) and CdI_2 (2.355). Hence the lattice energies in the layer structures would be computed to be much less than in the corresponding ionic lattice. Counteracting this, covalent bond energies contribute to the total stability of $CdCl_2$ and CdI_2.

EXAMPLE 4.6

The experimental lattice energy for CdI_2 is –2455 kJ mol^{-1}. Calculate the theoretical lattice energy of CdI_2 (a) with its actual CdI_2 lattice type and (b) assuming that it took the ionic rutile lattice type.

SOLUTION: For both (a) and (b), we apply Equation 4.7, assuming that the interionic distance is the sum of the ionic radii of Cd^{2+} and I^-, 109 + 206 = 315 pm. The Born exponent for I^- (valence electron configuration $5s^25p^6$) is 12; the Born exponent n for Cd^{2+} (valence electron configuration $5s^04d^{10}$) is 10; so the average $n = 11$. For (a) we take the Madelung constant of CdI_2, 2.355, so Equation 4.10 gives us

$$\frac{(138,900)(2.355)(2)(-1)}{315}\left(1-\frac{1}{11}\right) = 1888 \text{ kJ mol}^{-1}$$

For (b) we take the Madelung constant of TiO_2, 2.408, and obtain

$$\frac{(138,900)(2.408)(2)(-1)}{315}\left(1-\frac{1}{11}\right) = 1930 \text{ kJ mol}^{-1}$$

Both calculations fall well short of the experimental value; the covalent Cd–I bond energy contributes to making up the difference. (We explore this later in Section 5.5.)

4.3. Solubility Rules and Tendencies I–IV for Salts of Oxo and Fluoro Anions: Their Geochemical Consequences

OVERVIEW. Know and apply the four generalized solubility rules and tendencies to predict the solubility or insolubility of salts of oxo and fluoro anions, oxides, and fluorides. Know which mineral classes are covered by each solubility rule or tendency. You may practice this concept by trying Exercises 4.31–4.37. Select a suitable counterion to precipitate a given ion, or write specific solubility rules for additional anions or cations. You may practice this concept by trying Exercises 4.38–4.44. This section is rich in Connections to geochemistry, as well as biochemistry and medicinal chemistry.

If necessary, you should first review the methods of classifying the acidity of cations (Sections 2.1 and 2.2) and the basicity of oxo anions (Section 3.6). These solubility rules and tendencies will come into play again in Sections 5.6, 6.4, and 8.5. Solubility Rules V and VI will be added in Section 5.6A.

It is important to be able to anticipate when an ionic salt will dissolve in water, or conversely, when a precipitation reaction will occur in a solution containing given cations and anions:

$$y M^{m+}(aq) + m X^{y-}(aq) \;\rightleftharpoons\; M_y X_m(s) \tag{4.11}$$

Precipitation happens, for example, in the body when the products of the concentrations of certain cations and anions come to exceed the solubility products of their salts, as when calcium ions and oxalate ions precipitate as calcium oxalate, one form of kidney stone.

Precipitation happens in nature when natural waters rich in an appropriate cation mix with another rich in an appropriate anion at high temperatures, generating mineral deposits by *hydrothermal* reactions. If these are rich enough to be mined, they are known as **ores** and are sources of the cations (and polyatomic anions) of the elements that we introduced in Chapter 1. These minerals are named after scientists of many nationalities, often with the ending –ite. Often their crystals are beautiful, especially when photographically magnified in a book such as "The Magic of Minerals."[7] Minerals pictured in this book are indicated with a dagger (†); however, photos of these and others can readily be found online.

The solubility properties of ionic salts in water present certain complexities not found in the solubilities of other classes of materials. As a generalization, "like dissolves like," which means that nonpolar covalent materials generally dissolve in nonpolar covalent solvents, and polar hydrogen-bonding molecules generally dissolve in hydrogen-bonding solvents. For ionic solids in the most important polar covalent solvent, water, no such simple blanket statement can be made—many ionic salts are soluble in water, and many are not.

The need to be able to anticipate the solubility properties of salts leads many general chemistry textbooks to present a very limited series of solubility rules, arranged by anion, such as all sulfates are soluble except those of Ca^{2+}, Sr^{2+}, Ba^{2+}, Hg^{2+}, Pb^{2+}, and Ag^+, and so on. Such rules can cover only a fraction of all the possible or known inorganic salts. Thus, there is a need for *generalized* solubility rules that can be applied easily to predict the solubility of a wide variety of inorganic salts.

EXPERIMENT 5 MAY BE ASSIGNED NOW.
(See www.uscibooks.com/foundations.htm.)

Experiment 5 gives observations from which you may be able to devise generalized *solubility rules for salts of polyatomic oxo and fluoro anions* (e.g., SO_4^{2-} and PF_6^-). Solubility rules for salts of other anions such as chloride and sulfide will follow in Section 5.6A.

The results of experiments such as Experiment 5 are best organized by the acidity classifications of the cations and the basicity classifications of the anions being combined to form a given (soluble or insoluble) salt. When this is done, patterns emerge (Figure 4.12) among salts of fluoro and oxo anions: Precipitation of insoluble salts occurs with cation–anion combinations that fall in the *shaded* regions of Figure 4.12.

Precipitation occurs along the diagonal line representing matching of the reactivity classifications of the cation and anion. *No precipitation* occurs (salts are soluble) in the clear regions at the lower left and upper right edges, which represent mismatching of the reactivity classifications of the cation and the anion. These reliable patterns we will describe as **Solubility Rules I and II**.

Less consistent results (some precipitation, some cases of salts remaining in solution) occur in the remaining (partially shaded) regions. Because there are exceptions to the patterns, we describe them with **Solubility Tendencies III and IV**. When dealing with salts in these regions, you are well advised to confirm their solubilities in a chemistry handbook, such as the *Handbook of Chemistry and Physics*,[8] in its extensive tabulation of "Physical Constants of Inorganic Compounds."

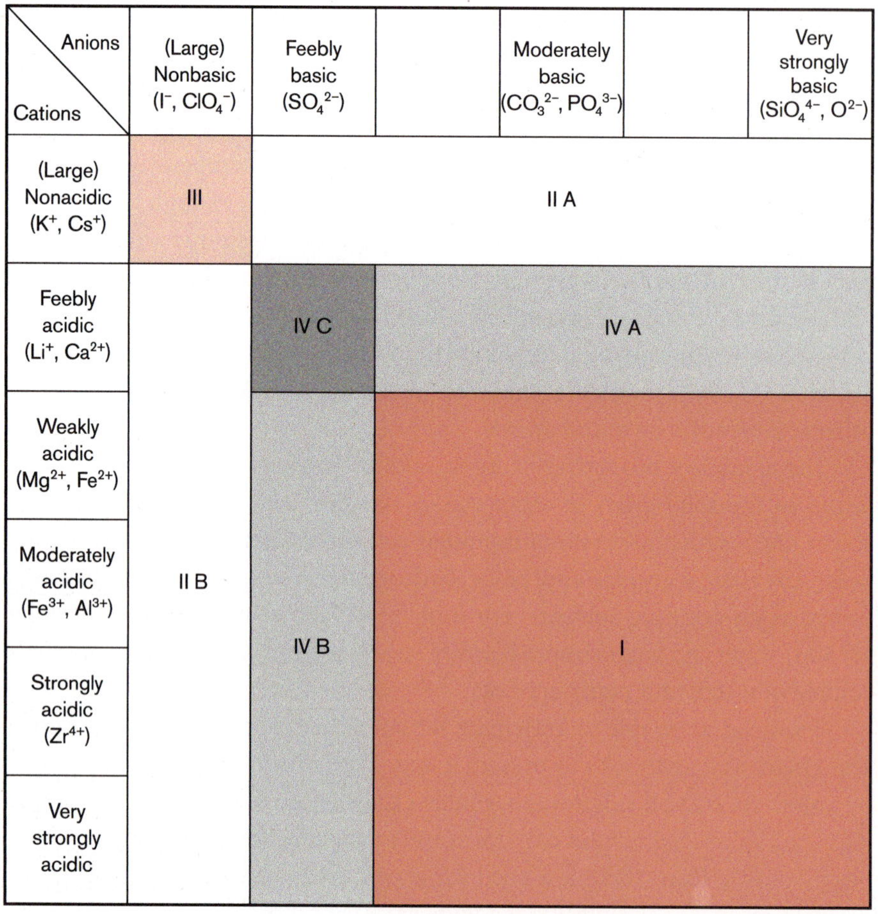

Cations \ Anions	(Large) Nonbasic (I^-, ClO_4^-)	Feebly basic (SO_4^{2-})		Moderately basic (CO_3^{2-}, PO_4^{3-})		Very strongly basic (SiO_4^{4-}, O^{2-})
(Large) Nonacidic (K^+, Cs^+)	III	II A				
Feebly acidic (Li^+, Ca^{2+})		IV C	IV A			
Weakly acidic (Mg^{2+}, Fe^{2+})			I			
Moderately acidic (Fe^{3+}, Al^{3+})	II B	IV B				
Strongly acidic (Zr^{4+})						
Very strongly acidic						

Figure 4.12. Characteristic pattern of precipitation on mixing 1 N (0.33–1.0 M) °solutions of cations with 1 N solutions of oxo anions. Shaded regions represent combinations that give precipitates; in unshaded regions, no precipitation occurs; in partially shaded regions, some combinations give precipitates and others do not. The Roman numerals correspond to the numbers of the generalized solubility rules and tendencies discussed here in Section 4.3B.

Solubility Rule I applies in Region I of Figure 4.12 (**dark brown**) to combinations of acidic cations plus basic anions. Because most metal cations found in aqueous solution are at least weakly acidic, and most oxo anions are at least weakly or moderately basic, the reaction of *acidic cations and basic anions to give insoluble salts* is an important general property. By *acidic cations* we mean any weakly, moderately, strongly, or very strongly acidic cations. *Basic anions* refer to any anions that are weakly, moderately, strongly, or very strongly basic.

This generalized solubility rule sums up the main parts of some commonly given specific insolubility rules: (1) All carbonates, phosphates, arsenates, silicates (Chapter 8), oxalates, etc., are insoluble except those of the Group 1 elements and NH_4^+; (2) all hydroxides are insoluble except those of the Group 1 elements, Sr^{2+}, and Ba^{2+}. (Oxide and hydroxide solubility is discussed in more detail in Section 8.5A.)

A CONNECTION TO AGRICULTURAL CHEMISTRY AND GEOCHEMISTRY

Phosphate Fertilizers

Phosphorus is one of the six elements needed in large quantities in the biochemistry of life, and it is often the key element needed to increase agricultural yields and food production. Phosphorus is found in nature in various insoluble phosphate minerals. The most common of these in nature are the calcium phosphates $Ca_5(PO_4)_3(OH)$ (hydroxylapatite) and $Ca_5(PO_4)_3F$ (apatite)[†]. Mineral deposits of apatites are found particularly in Morocco, the Western Sahara, China, the United States (in particular, in Bone Valley, near Tampa, Florida), and South Africa. Hydroxylapatite is deposited in the body as the main inorganic component of bones and teeth.

Because of their insolubility, apatites are unsuitable for use as fertilizers. They may instead be converted to soluble, moderately acidic phosphoric acid. This is produced by the reaction of a strong acid, sulfuric acid, with the moderately basic phosphate ion in apatite:

$$Ca_5(PO_4)_3F + 5H_2SO_4 \rightarrow 3H_3PO_4 + 5CaSO_4 + HF$$

The reaction mixture is diluted with water and the insoluble calcium sulfate is filtered off; the solution may then be concentrated. This reaction also produces a serious air pollutant, gaseous HF.

For use in fertilizers, it is unnecessary to protonate the PO_4^{3-} ion completely. Partial protonation of the phosphate ion reduces its basicity, so the salt $Ca(HPO_4)$ becomes soluble enough to be used (mixed with calcium sulfate) as "superphosphate" fertilizer.

$$Ca_5(PO_4)_3OH + 2H_2SO_4 \rightarrow 3CaHPO_4 + 2CaSO_4 + H_2O$$

The agricultural use of phosphate fertilizers is so great that it is projected that the world's supplies of economically accessible phosphate rock may be depleted in a few decades.[9] If this were to happen, billions of people would starve. There is a phosphate cycle in nature by which phosphate is geologically recycled, but it is much too slow. Our bodies excrete phosphate ions in urine. In wastewater treatment plants this phosphate precipitates as a "troublesome" mineral, struvite, $(NH_4)MgPO_4$. Struvite contains three important nutrient elements, and it is soluble enough to be used as a slow-release fertilizer. People do not excrete enough urine, however, to meet the agricultural need. Cows excrete much more phosphate than people do, so developing ways of catching agricultural runoff are also being investigated.[10]

Phosphates as Mineral Sources of Metal Ions

As implied by Solubility Rule I (and Tendency IVA), most metal phosphates, arsenates, and vanadates are insoluble. Hence, there are a number of minerals in this category, including the iron(II) phosphate vivianite[†] $[Fe_3(PO_4)_2 \cdot 8H_2O]$; the cobalt(II) arsenate erythrite[†] $[Co_3(AsO_4)_2 \cdot 8H_2O]$; the copper(II) phosphate turquoise $[CuAl_6(OH)_8(PO_4)_4 \cdot 4H_2O]$; pharmacolite[†] $[Ca(HVO_4)]$; and the lead(II) salts pyromorphite[†] $[Pb_5(Cl)(PO_4)_3]$, mimetite[†] $[Pb_5(Cl)(AsO_4)_3]$, and vanadinite[†] $[Pb_5(Cl)(VO_4)_3]$. Of greatest importance now are the phosphates that serve as the main mineral sources[†] of the rare earth and actinide elements, including carnotite [hydrated $K(UO_2)(VO_4)$], monazite $[(RE)PO_4]$, and xenotime $[(RE)PO_4]$, where RE is a rare earth ion. The crystal structures of monazite and xenotime differ in the coordination number of the rare earth ion: Monazite accommodates larger rare earth ions such as La^{3+}, Ce^{3+}, Pr^{3+}, and Nd^{3+}, and also actinide ions, by allowing a coordination number up to 10. Xenotime does not allow a coordination number over eight, so it accommodates smaller rare earth ions and Y^{3+}. Extraction of rare earth elements (Section 1.2C) from monazite is complicated by the radioactivity of the mineral; it is also an important source of Th^{4+}. Xenotime, in contrast, is sometimes not radioactive, and it contains a higher proportion of Y^{3+} and the more valuable smaller, later rare earth ions, which are in shortest supply, including Dy^{3+}, Tb^{3+}, Er^{3+}, and Yb^{3+}.

Carbonate Minerals and Ocean Acidity

As implied by Solubility Rule I (and Tendency IVA), there are also many insoluble metal carbonates. These include the pink rhodochrosite[†] $(MnCO_3)$, siderite[†] $(FeCO_3)$, smithsonite[†] $(ZnCO_3)$, cerussite[†] $(PbCO_3)$, strontianite[†] $(SrCO_3)$, the two polymorphs of $CaCO_3$ (calcite[†] and aragonite[†] as well as impure forms such as limestone and coral), dolomite $[CaMg(CO_3)_2]$, the rare earth mineral bastnasite $[LaF(CO_3)]$, the blue-green malachite[†] $[CuCO_3 \cdot Cu(OH)_2]$, and the blue azurite[†] $[2CuCO_3 \cdot Cu(OH)_2]$. We do not find carbonates of moderately acidic cations such as Fe^{3+} or Al^{3+} or stronger cations, because these cations polarize the carbonate ion, decomposing it to the oxide of the cation and gaseous CO_2 (which escapes). Even copper(II) carbonate decomposes partially to azurite or malachite.

Hydrogen ion also decomposes carbonates to evolve gaseous CO_2 and a solution of the metal ion, so carbonates readily dissolve in acids with bubbling. Even the moderately acidic carbonic acid, H_2CO_3 (formed from CO_2 and water), slowly dissolves $CaCO_3$ to form the more soluble hydrogen carbonate ("bicarbonates"):

$$CaCO_3(s) + H_2O + CO_2 \rightleftharpoons Ca^{2+}(aq) + 2\ HCO_3^-(aq)$$

This process creates caves in deposits of limestone. After the solution has formed, the reaction may reverse as $CO_2(g)$ is lost, and $CaCO_3(s)$ is deposited as stalactites and stalagmites.

Many aquatic species such as corals in reefs, some types of phytoplankton, formaminifera, and marine snails depend on maintaining $CaCO_3$ (especially the more soluble polymorph, aragonite) in their skeleton, shells, and cells, and cannot thrive or even survive if the $CaCO_3$ cannot be produced or maintained. Reducing the pH of seawater from its normal 8.0–8.3 (slightly alkaline) by just a few tenths of a pH unit would imperil these species and the higher organisms that feed on them.[11,12] This could happen by the end of the century as a result of increasing CO_2 levels in the air (Section 8.5C). Even greater pH changes in the ocean occurred during the end-Permian mass extinction 251 million years ago (Section 5.7B), when about 79–95% of all oceanic species were wiped out, and more recently during the Paleocene–Eocene Thermal Maximum 55 million years ago. The current rate of ocean acidification is at least 10 times as fast as occurred 55 million years ago, or at any other time during the past 300 million years. Although ocean species can slowly evolve in response to such environmental disasters, they have not been tested by this rapid a rate of change in the last 250 million years.[13]

Solubility Rule II applies in the white Regions IIA and IIB of Figure 4.12 to *cross-combinations* of ions—namely, nonacidic cations plus all other basic anions (including even feebly basic anions) in Region IIA and acidic cations (including even feebly acidic cations) plus nonbasic anions in region IIB. These *cross-combinations give soluble* salts. Two common solubility rules are related to this rule. The first applies to Region IIA and states that all Group IA and ammonium salts are soluble. This rule includes the exceptions of the common solubility rules for phosphates, carbonates, etc., just discussed. If you need to prepare a solution of a given oxo or fluoro anion, you would dissolve its salt with a nonacidic cation (generally Na^+ for economic reasons).

The second common solubility rule applies to Region IIB and is that for nitrate: All nitrates are soluble. Because most cations are acidic, their soluble oxo salts will be of nonbasic anions such as nitrate and perchlorate. Also soluble are salts of the nonbasic fluoro anions and (usually) of the nonbasic chloride, bromide, and iodide ions. Salts of these anions (especially inexpensive chloride or nitrate) are the ones you would choose in order to make up solutions of most of the metal cations for use in experiments.

CONNECTIONS TO GEOCHEMISTRY AND GEOPOLITICS

Soluble Minerals

The sources of cations and anions that mainly fall in Regions IIA and IIB include seawater, brines, and salt flats and very dry deserts (such as the Atacama Desert). Seawater contains many ions in low to moderate concentration, so it is an economical source of only a handful of ions of elements (e.g., Mg^{2+} and Br^-). Brines have higher concentrations of ions than seawater, so they serve as sources of some additional ions (e.g., Li^+ for lithium-ion batteries from evaporated brines in Bolivia and Nevada, and brines used as a source of geothermal energy in California). When seawater evaporates in tidal flats such as the Rann of Kutch in India, sea salt can readily be gathered. Because Na^+ is necessary for animal life but is not used or accumulated by plants, animals may have to go long distances in moist environments to find *salt licks* to obtain their Na^+. We now take access to salt for granted, but it has been a valuable commodity in the past and has played an important part in world history.[14]

Oxo anions that are historically or currently mined from dry or desert regions include borates (as salts such as borax, $Na_2B_4O_7 \cdot 10\ H_2O$, mined in Death Valley, California), and nitrates such as saltpeter (KNO_3) and soda niter ($NaNO_3$) from the Atacama Desert. Because Germany and its allies did not have desert territories, in World War I the Allies prevented their trade with Chile and other sources of the nitrates needed to make explosives and gunpowder (Section 6.2). World War I would have ground to a halt had not the brilliant German chemist, Fritz Haber, discovered the Haber process for converting atmospheric nitrogen to ammonia (which can be oxidized to nitrate, Section 8.5C).

Solubility Tendency III applies in the light brown Region III of Figure 4.12 to combinations of nonacidic cations plus nonbasic anions. The generalized solubility tendency in this region is that *large nonacidic cations and large nonbasic anions give* relatively *insoluble* salts. This is a tendency rather than a rule, and the insolubility is much less pronounced than in Region I. If you performed Experiment 5, you may have also noticed this qualitative difference: In Region I, due to their highly insoluble nature, the precipitates form immediately upon mixing, which does not allow time for visibly crystalline material to grow. (Only very tiny crystals are formed.) But in Region III, the salt is only moderately insoluble, so often the precipitation does not immediately begin. This allows time for the growth of larger crystals.

This tendency to precipitate or crystallize out of solution is almost completely absent for the smallest of the nonacidic cations, Na^+ (hence the specific solubility rule that all sodium salts are soluble), and the smallest of the nonbasic anions, NO_3^-, Cl^-, Br^-, and I^-. This tendency becomes more pronounced with Cs^+ salts and perchlorates ($CsClO_4$ itself precipitates readily), and it becomes most pronounced (and most useful) with even larger cations and anions. Other things being equal, it helps minimize the solubility (and maximize the yield) to match, not only the *size* of the cation and anion, but also to match the *charges*—for example, to precipitate a large complex −3-charged anion with a large complex +3-charged cation.[15]

Summarizing our categories of nonacidic cations, we have:

1. the larger group 1 cations (Section 2.2);

2. complex cations such as $[Co(NH_3)_6]^{3+}$ (Section 3.2C); and

3. tetraalkylammonium ions (Section 3.2C).

Our categories of nonbasic anions include:

1. larger Group 17/VII (halide) anions (Section 2.5D);

2. most or all fluoro anions (Section 3.5);

3. nonbasic oxo anions (Section 3.6); and

4. complex anions such as $[Cr(NH_3)_2Cl_4]^-$.

A CONNECTION TO BIOCHEMISTRY AND ORGANIC CHEMISTRY

Isolating Nitrogen-Based Natural Products

When inorganic chemists synthesize a new complex cation in solution, they generally crystallize it out of the solution by adding a solution of a large nonbasic anion such as ClO_4^- (see Section 6.2 for a reason why this choice often is hazardous) or BF_4^- or PF_6^- (much safer choices). Even nitrate or iodide may work well if used in excess (so that the solubility is reduced by the common ion effect).

The same principle is used by organic natural-products chemists when isolating organic nitrogen compounds (amines, alkaloids, etc.): They extract their organic nitrogen compound into an aqueous strong acid, which converts the compound into a large organic *ammonium* cation, which is then precipitated by a suitable large nonbasic anion. Conversely, new complex or large organic anions are best crystallized out of solution by adding a large nonacidic cation such as Cs^+ or $(CH_3)_4N^+$, $(C_2H_5)_4N^+$, and so on.

Solubility Tendency IV applies to salts containing feebly acidic cations (in Region IVA of Figure 4.12) and/or feebly basic anions (in Region IVB). The most common feebly basic anion is sulfate; the traditional solubility rule for sulfates is that most sulfates are soluble, except salts such as $CaSO_4$, $SrSO_4$, $BaSO_4$, and $PbSO_4$.

The driving forces (to be discussed in the next section) favoring insolubility in Region I and solubility in Region II overlap with each other and come into conflict in Region IV. The solubility of salts of these ions is particularly complex and difficult to predict. In Region IVA, precipitation usually prevails, but oxides and hydroxides of feebly acidic cations are (at least somewhat) soluble. In Region IVB, solubility usually prevails, but (for example) sulfates of some large, weakly acidic cations also show low solubility.

The clearest result occurs at the intersection of these two regions, Region IVC, where large, feebly acidic cations (such as Ba^{2+}) give *precipitates* with large, feebly basic anions (such as SO_4^{2-} and SeO_4^{2-}). Thus, in gravimetric determinations in a quantitative analysis course, barium and sulfate ions are always precipitated with each other.

Solubility Tendency IV applies somewhat to the fluoride ion itself, although with $pK_b = 10.85$, it is actually just barely weakly basic: LiF is insoluble, and the most insoluble fluoride is CaF_2, which is widely found in nature as the mineral fluorite. The tendency also applies somewhat to oxalates ($pK_b = 10.15$). Calcium oxalate (CaC_2O_4) precipitates and accumulates in mineral deposits, in kidney stones, and in numerous toxic plants, such as the leaves of rhubarb. The antifreeze ethylene glycol (HO–CH_2–CH_2–OH) is toxic because it is oxidized in the body to oxalate, which then precipitates as calcium oxalate. Fluoride and oxalate also manifest their weak basicity in that the fluorides and oxalates of most acidic cations are relatively insoluble.

CONNECTIONS TO GEOCHEMISTRY AND MEDICINAL CHEMISTRY

Sulfate, Selenate, Chromate, Molybdate, and Tungstate Minerals and Medicines

Several related minerals form from the combination of feebly acidic cations and feebly basic anions—namely, the sulfate minerals gypsum[†] or selenite ($CaSO_4 \cdot 2H_2O$), anhydrite ($CaSO_4$), celestite ($SrSO_4$), barite[†] ($BaSO_4$), and anglesite ($PbSO_4$); chromates such as the red crocoite[†] ($PbCrO_4$); molybdates such as the orange wulfenite[†] ($PbMoO_4$); and tungstates such as scheelite ($CaWO_4$). These tend to have lower solubility as one goes down Group 2, which has allowed the use of $CaWO_4$ to remove radioactive $^{90}Sr^{2+}$ from water or milk.

Some readers may be familiar with $BaSO_4$ as the "barium milkshake" in medicine. The digestive system does not diffract X rays well. Heavy barium ions do, but these are toxic (Section 5.7B). To obtain good X rays of the digestive system without poisoning the patient, a very insoluble barium salt is needed—a milky suspension of $BaSO_4$. (Students have assured me that this milkshake has a remarkably bad taste.)

Selenite is also an interesting mineral—first, it contains no selenium, let alone selenite ion! It also provides an example of slow crystal growth—an incredible case of crystallization of $CaSO_4 \cdot 2H_2O$ at virtually constant temperatures over many thousands of years to give single crystals much larger than human beings is illustrated in N. Shea, *Natl. Geogr.*, 214(5), 64 (2008).

EXAMPLE 4.7

The most highly symmetrical polyatomic ion is the icosahedral ion $B_{12}H_{12}^{2-}$. Estimate the basicity of this ion using the principles you would for an ion of formula $B_{12}F_{12}^{2-}$. If you prepared this ion in high concentration in solution, which one of the following ions would be the best to add to form a crystalline precipitate with this anion: Li^+, Be^{2+}, Al^{3+}, $[(CH_3)_4N]^+$, $[C_2H_4N_2(CH_3)_6]^{2+}$, $[Co(NH_3)_6]^{3+}$, ClO_4^-, $[SiF_6]^{2-}$, or $[AlF_6]^{3-}$?

SOLUTION: A large fluoro anion of formula $B_{12}F_{12}^{2-}$ would be quite nonbasic; even though H is not as electronegative as F, we might suppose that $B_{12}H_{12}^{2-}$ might also be nonbasic. This combination falls in Region III of Figure 4.12. To precipitate this large nonbasic anion, we need a large nonacidic cation, of which we have three choices: $[(CH_3)_4N]^+$, $[C_2H_4N_2(CH_3)_6]^{2+}$, and $[Co(NH_3)_6]^{3+}$. Of these, the best would likely be the one with the charge that matches that of our non-basic anion, $[C_2H_4N_2(CH_3)_6]^{2+}$.

EXAMPLE 4.8

Without reference to a table of solubilities, classify each of the following salts as soluble or insoluble in water: Ag_2SeO_4, K_3PO_4, $ZnSO_4$, and $Th_3(PO_4)_4$.

SOLUTION: To locate the regions of these salts in Figure 4.12, we need to categorize (1) the acidities of the cations, as in Section 2.2, and (2) the basicities of the oxo anions, as in Section 3.6.

1. The charges of the cations and anions in each salt must be identified so that their acidity and basicity can be classified. The cations are Ag^+, K^+, Zn^{2+}, and Th^{4+}. Referring to Table C for radii, we compute Z^2/r ratios for these four cations: 0.008, 0.007, 0.046, and 0.145, respectively, which allows the first two to be categorized as nonacidic, the third as weakly acidic, and the fourth as moderately acidic. The table of electronegativities, however, shows that Ag has an electronegativity exceeding 1.8, so it should be reclassified as feebly acidic.

2. The anions are SeO_4^{2-}, PO_4^{3-}, and SO_4^{2-}. Any oxo anion with a –3 or higher charge (such as phosphate) is automatically at least moderately basic. A calculation using Equation 3.14 (or reference to Table 3.8) classifies sulfate and selenate as feebly basic.

Hence the salt Ag_2SeO_4 is a combination of a feebly acidic cation and a feebly basic anion and should be insoluble (Region IVC). K_3PO_4 is a combination of a nonacidic cation and a basic anion and should be soluble (Region IIB). $ZnSO_4$ is a combination of a (weakly) acidic cation and a feebly basic anion (Region IVB) and should probably be soluble. $Th_3(PO_4)_4$ is a combination of a (moderately) acidic cation and a (moderately) basic anion and should be insoluble (Region I). A check of the *Handbook of Chemistry and Physics* confirms all of these predictions.

4.4. The Contrasting Effects of Lattice and Hydration Energies on Salt Solubilities

OVERVIEW. You can calculate enthalpies of precipitation of salts from Equation 4.12 given their lattice energies (Table 4.6) and the hydration-energy data from Tables 2.3 and 4.5. You may practice this calculation in Section 4.4A by trying Exercises 4.45–4.48. (Reasons for Solubility Rules II and III are also explored in this section.) In Section 4.4B, we see that the solubility of a given inorganic salt, coordination compound, or organometallic compound depends on its speciation. Its solubility can be predicted: (a) in water, using the solubility rules from the last section; (b) in nonpolar solvents, based on the presence of organic groups in the ligands; (c) in polar solvents, based in part on the possibility of Lewis acid–base reactions between the salt and the solvent. You may practice making these predictions by trying Exercises 4.49 and 4.50. These concepts have connections to concepts in analytical chemistry, organic chemistry, biochemistry, and green chemistry.

The concept of speciation was introduced earlier in Sections 3.8 and 3.9. Solubility in organic liquids has important health implications in Section 5.7 and 5.8.

4.4A. Analyzing ΔH for Precipitation Reactions of Salts.
The enthalpy change ΔH for precipitation of an ionic salt (i.e., for Eq. 4.10) can be related to the properties of the cation and anion involved if we break down the relatively complex reaction (Eq. 4.10) into three simpler steps: (1) The hydrated cation is dehydrated and converted to a gaseous bare cation. (2) The same is done to the anion. (3) The gaseous cations and anions are then allowed to come together to form the ionic solid. This thermodynamic analysis is shown schematically in Figure 4.13.

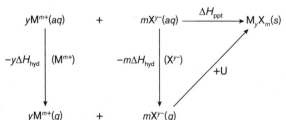

Figure 4.13. Thermodynamic cycle for analysis of the precipitation of a solid.

By Hess's law, the enthalpy change for the overall reaction (Eq. 4.10) must equal the sum of the enthalpy changes for each of the component steps:

$$\Delta H_{\text{pptn}}(M_yX_m) = -y\Delta H_{\text{hyd}}(M^{m+}) - m\Delta H_{\text{hyd}}(X^{y-}) + U(M_yX_m) \qquad (4.12)$$

Steps 1 and 2 are simply the reverse of the processes of formation of the hydrated cations and hydrated anions, so their enthalpy changes are readily obtained by reversing the signs of the hydration enthalpies given in Tables 2.3 and 4.5.

TABLE 4.5. Hydration Enthalpies of Some Anions

Anion	Radius (pm)	Hydration Energy (kJ mol^{-1})	Anion	Radius (pm)	Hydration Energy (kJ mol^{-1})
F$^-$	119	−513	OH$^-$	119	−460
Cl$^-$	167	−370	NO$_3^-$	165	−314
			CN$^-$	177	−342
Br$^-$	182	−339	N$_3^-$	181	−298
I$^-$	206	−294	SH$^-$	193	−336
			BF$_4^-$	215	−223
			ClO$_4^-$	226	−235
S^{2-}	170	−1372	CO$_3^{2-}$	164	−1314
			SO$_4^{2-}$	244	−1059

Sources: Radii for monatomic anions are from Table C; "radii" for polyatomic anions, which are not truly spherical, are thermochemical radii, taken from J. E. Huheey, E. A. Keiter, and R. L. Keiter, *Inorganic Chemistry: Principles of Structure and Reactivity*, 4th ed., Harper-Collins: New York, 1993, p. 118. Hydration enthalpies are from A. G. Sharpe, *J. Chem. Educ.* 67, 309 (1990); D. W. Smith, *Inorganic Substances*, Cambridge University Press: Cambridge, UK, 1990, p. 160; and M. C. Ball and Norbury, *Physical Data for Inorganic Chemists*, Longman: London, 1974.

The enthalpy change for Step 3 is the lattice energy U of the ionic solid; experimental values of U are found in the second column of Table 4.6. These three energy terms add up to give the overall enthalpy of precipitation (experimental values of ΔH_{pptn} are given in the third column of Table 4.6). Thus, we can evaluate the lattice energies of several salts to obtain enthalpies of precipitation. When we take the difference of the two types of very large (negative) energy terms, only a relatively small enthalpy of precipitation results. When doing precipitation reactions in Experiment 5, ΔH_{pptn} or its converse, ΔH_{soln}, was usually so small that you did not notice it during the reaction.

From Equation 4.11 we deduce that the enthalpies of precipitation of ionic salts are a result of a balance between lattice energies and hydration energies. Checking Column 2 of Table 4.6, we see that ΔH_{pptn} can either be positive or negative, depending largely on which solubility rule or tendency is involved.

EXAMPLE 4.9

Calculate the enthalpy of precipitation for the reaction of nonacidic K$^+$ with nonbasic ClO$_4^-$.

SOLUTION: Equation 4.11 becomes, in this case, $\Delta H_{pptn}(KClO_4) = -\Delta H_{hyd}(K^+) - \Delta H_{hyd}(ClO_4)^- + U(KClO_4)$. Taking data from Tables 2.3, 4.5, and 4.6, we compute $\Delta H_{pptn} = -(-321 \text{ kJ}) - (-235 \text{ kJ}) + (-595 \text{ kJ}) = -39 \text{ kJ}$.

TABLE 4.6. Thermodynamic Data for Precipitation Reactions of Ionic Salts[a]

Salt	U	ΔH_{pptn}	$-T\Delta S_{pptn}$ (at 298 K)[b]	ΔG_{pptn}[b]	Solubility (moles/kg H_2O)
	**Rule I. Acidic Cations + Basic Anions → Precipitates (ΔG_{pptn} is negative)**				
$Be(OH)_2$	–3620	–31	**–90**	–121	0.000008
$Mg(OH)_2$	–2998	–3	**–61**	–63	0.0002
$MgCO_3$	–3122	28	**–74**	–45	0.0093
$FePO_4$	–7300	78	**–180**	–102	slight
	Rule II. Cross Combinations → Soluble Salts (ΔG_{pptn} is positive)				
	IIA. Nonacidic Cations + Basic Anions				
KOH	–796	**55[c]**	7	62	19.1
RbOH	–765	**63**	11	74	17.6
CsOH	–732	**71**	12	83	26.4
K_2CO_3	–1846	**35**	1	36	8.12
Rb_2CO_3	–1783	**41**	9	50	19.5
Cs_2CO_3	–1722	**62**	10	73	8.0
	IIB. Acidic Cations + Nonbasic Anions				
$Mg(NO_3)_2$	–2521	**85**	4	89	1.65
$Ca(NO_3)_2$	–2247	**20**	13	32	2.08
$Sr(NO_3)_2$	–2151	**–18**	14	–3	1.89
$Ba(NO_3)_2$	–2035	**–40**	27	–13	0.33
$Mg(ClO_4)_2$		**141**	4	144	2.24
$Ba(ClO_4)_2$	–1769	**12**	34	46	5.91
	Tendency III. Nonacidic Cations + Nonbasic Anions → Precipitates (ΔG_{pptn} is negative)				
$KClO_4$	–595	**–51**	39	–12	0.054
$RbClO_4$	–576	**–57**	43	–14	0.027
$CsClO_4$	–550	**–55**	44	–12	0.034
$NaNO_3$	–763	**–21**	27	6	8.59
	Tendency IVA. Feebly Acidic Cations + Basic Anions → Usually Precipitates				
$Ca(OH)_2$	–2637	16	**–44**	–28	0.025
Li_2CO_3	–2254	18	**–34**	–17	0.18
$CaCO_3$	–2811	10	**–57**	–48	0.0002
$SrCO_3$	–2688	3	**–56**	–52	0.00007
$BaCO_3$	–2554	–4	**–43**	–47	0.00011

Table 4.6 continues on next page ▶

TABLE 4.6. *(cont.)*

Salt	U	ΔH_{pptn}	$-T\Delta S_{pptn}$ (at 298 K)[b]	ΔG_{pptn}[b]	Solubility (mol/kg H_2O)
Tendency IVB. Acidic Cations + Feebly Basic Anions → Usually Soluble Salts					
$Al_2(SO_4)_3$		**338**	–241	96	0.92
$BeSO_4$		**123**	–64	59	2.40
$MgSO_4$		**91**	–61	30	2.88
Tendency IVC. Feebly Acidic Cations + Feebly Basic Anions → Precipitates					
$CaSO_4$	–2480	18	**–45**	–27	0.014
$SrSO_4$	–2484	9	**–43**	–34	0.006
$BaSO_4$	–2374	–19	**–31**	–50	0.00001

Sources: C. S. G. Phillips and R. J. P. Williams, *Inorganic Chemistry*, Oxford University Press: Oxford, UK, 1965, p. 254; D. A. Johnson, *Some Thermodynamic Aspects of Inorganic Chemistry*, Cambridge University Press: Cambridge, UK, 1968, p. 107; B. G. Cox and A. J. Parker, *J. Am. Chem. Soc.* 95, 6879 (1973); *Handbook of Chemistry and Physics*, 84th ed., Chemical Rubber Publishing Co.: Cleveland, 2006, pp. 12.22–12.30. Solubilities listed are often for the hydrated salts and are for temperatures between 0 and 30°C.

[a] Energies are in kJ mol⁻¹. The dominant term (either ΔH or $-T\Delta S$) is shown in **boldface**.

[b] Columns of data for use in Section 4.5.

[c] If you have ever made up a concentrated solution of NaOH or KOH, you will have noticed that the solution becomes quite warm or hot.

EXAMPLE 4.10

Calculate the enthalpies of precipitation of $Ba(ClO_4)_2$ and $Mg(OH)_2$.

SOLUTION: In doing this summation, it is necessary to multiply each energy (in kJ mol⁻¹) from the tables by the number of moles of that ion involved. The hydration enthalpies are given in kilojoules per mole of ion; for each of these salts we are using 2 mol of anions. Thus, for $Ba(ClO_4)_2$ we have

$$\Delta H_{pptn} = -(-1304 \text{ kJ mol}^{-1})(1 \text{ mol}) - (-235 \text{ kJ mol}^{-1})(2 \text{ mol}) + (-1769 \text{ kJ mol}^{-1})(1 \text{ mol}) = +5 \text{ kJ}$$

For $Mg(OH)_2$ we calculate

$$\Delta H_{pptn} = -(-1922 \text{ kJ mol}^{-1})(1 \text{ mol}) - (-460 \text{ kJ mol}^{-1})(2 \text{ mol}) + (-2998 \text{ kJ mol}^{-1})(1 \text{ mol}) = -86 \text{ kJ}$$

The Reason for Solubility Rule II. Cross-combination precipitation reactions usually have positive values of ΔH_{pptn} (Table 4.6)—that is, such reactions are endothermic (they absorb heat). Rule 2, however, states that precipitation reactions are *not* to be expected; instead, we expect these salts to be soluble to give 1 M solutions. Dissolution reactions of these salts are exothermic (they give off heat). For these salts, the sums of hydration energies are larger in magnitude than their lattice energies. Acidic

Hydrated Salts

Not only are Solubility Rule 2 salts quite soluble in water, but if the solution is evaporated until these salts do crystallize, they often crystallize as **hydrated salts**, in which the hydration sphere of the cation (usually) is retained. This is good in Region IIA because the radius ratio of the *hydrated* cation to the anion is much larger than that of the anhydrous cation to the anion. For example, most *d*-block metals form hydrated sulfates, nitrates, chlorides, etc., which are often written as $Cu(NO_3)_2 \cdot 6\,H_2O$, and so on, but which could more accurately be written as including hydrated cations, such as $[Cu(H_2O)_6]^{2+}$. When such salts are obtained anhydrous, they are often very good **desiccants** or drying agents, capable of removing water from solvents or from the air. $Mg(ClO_4)_2$, $CaCl_2$, and $CaSO_4$ are used in this manner.

Implicit Assumptions (a Caveat)

The solubility rules and tendencies implicitly depend on the ions involved forming relatively normal hydrated ions and lattices. They may not apply well to salts of ions that are very irregularly shaped, such as long-chain organic carboxylate anions,[16] $C_nH_{2n+1}COO^-$, which can only hydrate on one end, and which form atypical layer lattices as well.[17]

cations and nonbasic anions give rise to soluble salts because such ions are quite different in size and give poorer lattice energies than hydration energies. In extreme cases the cation may be so much smaller than the anion that there may be poor contact between cations and anions. Hence, a low Madelung constant must be used in Equation 4.10. In solution, however, good contact may be achieved between cations and small water molecules, and separately by anions and small water molecules. Each is free to hydrate with as many waters as it can.

The Reason for Solubility Tendency III. In this part of Table 4.6, ΔH_{pptn} is modestly negative. The ultimate reason that nonacidic cations and nonbasic anions give rise to insoluble salts is that such cations and anions are both large (i.e., similar in size), so they form especially stable crystal lattices with large Madelung constants and large (negative) lattice energies. Large ions, however, do not give especially large hydration energies (in comparison to the lattice energy U).

4.4B. Solubilities and Speciation of Ionic Salts and Metal–Organic Compounds.

Nonpolar solvents include the saturated hydrocarbons such as pentane, hexane, heptane, "petroleum ether" (a mixture of saturated hydrocarbons), and cyclohexane. Also included in this group are weakly solvating aromatic hydrocarbons such as benzene and toluene, the very slightly polar and weakly Lewis-basic chlorocarbons such as CH_2Cl_2 and $CHCl_3$, and the fats and lipids in our bodies.

Such solvents cannot generally interact effectively either with the cation or the anion of an ionic salt, and thus there is virtually no energy of solvation to offset the lattice energy. Therefore, most ionic salts are insoluble in such solvents. But these can be good solvents for nonpolar *hydrophobic* solutes such as nonpolar organic compounds and *uncharged* complexes and organometallic compounds that have carbon- and hydrogen-rich groups on their outer surfaces. The organic groups may be (among others) *alkyl groups* C_nH_{2n+1} (collectively symbolized by R) or *aryl groups* (collectively symbolized by Ar or a hexagon with an inscribed circle and based on substitution within the cyclic *phenyl group* C_6H_5). The hydrophobic compounds may have these groups directly attached to the metal, in which case they are organometallic compounds such as tetramethylsilane (Figure 4.14a) or dimethylmercury (Figure 4.14b). Or the hydrophobic compounds may be neutral complexes of organic-substituted anions such as the methoxide ion (CH_3O^-), the methylthiolate ion (CH_3S^-), the dimethylamide ion [$(CH_3)_2N^-$], or collectively the alkoxide ions (RO^-), the alkylthiolate or mercaptide ions (RS^-), and the dialkylamide ions, (R_2N^-). An example of one such complex is tetraethyl silicate (Figure 4.14c). These uncharged complexes and organometallic compounds are generally *insoluble in water* because they cannot form hydrated ions. However, they also have very small lattice energies (intermolecular attractions) because they are uncharged; hence, they often dissolve in nonpolar solvents. (These provide additional examples of the *speciation* of metals in modifying their properties.)

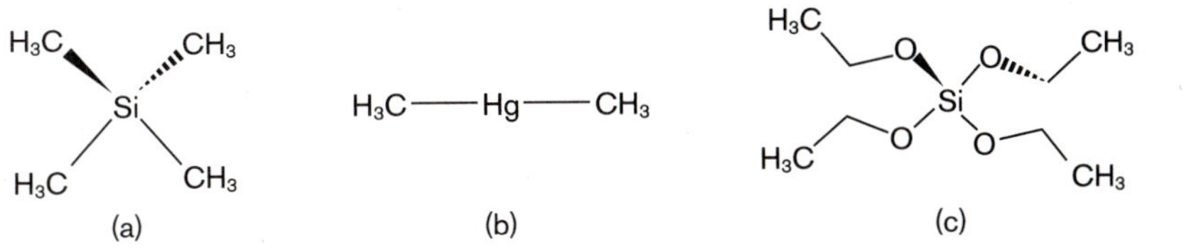

Figure 4.14. The metal–organic compounds (a) tetramethylsilane, (b) dimethylmercury, and (c) tetraethyl silicate.

CONNECTIONS TO ANALYTICAL AND BIOCHEMISTRY

Solvent Extraction and Bioamplification

This property is made use of very often in *solvent extraction* in analytical chemistry. A metal ion present in very low concentration in a natural water is complexed with an organic ligand to give an uncharged complex. This complex is then extracted into a much smaller volume of a nonpolar solvent and is thus separated from many impurities and concentrated at the same time. Complexes of this type tend to accumulate in fatty tissue and lipids in the body and are not easily excreted in the watery urine; hence, toxic effects of metals in these complexes persist longer. When an animal dies and is eaten by another animal higher up the food chain, the uncharged complex is passed on to it. Because these animals eat many times their own weight of the lower-chain animal, the concentration of the uncharged metal complex builds up; this process is known as **bioamplification** of the toxin.

Phase-Transfer Catalysis

Even charged complexes can be rendered soluble in nonpolar solvents if either the cation or the anion can be converted to the form of large complex ions with hydrophobic organic groups on the outside. The *tetraalkylammonium* cations, such as $(CH_3)_4N^+$ (Section 3.2C), but especially those with even longer organic groups, fit this description and can carry a polar counterion from water into nonpolar organic solvents. For example, potassium permanganate ($KMnO_4$) is a very useful oxidizing agent in organic chemistry, but it has some disadvantages. It is insoluble in the nonpolar organic solvents that best dissolve the organic compound to be oxidized. It is not even very soluble in water (Solubility Tendency III). When it does dissolve, the MnO_4^- ion hydrates, which reduces its reactivity. These problems can be overcome by adding a small, catalytic quantity of a tetraalkylammonium cation, which will form some ion pairs with MnO_4^-. These ion pairs are then very soluble in unhydrated, highly reactive forms in the organic solvent. Such tetraalkylammonium cations function as **phase-transfer catalysts**.

A particularly favored nonpolar solvent is **supercritical carbon dioxide**. Carbon dioxide is a gas at room temperature and atmospheric pressure, but is converted to a **supercritical fluid** above its critical temperature of 304 K and its critical pressure of 72.8 atm (where its density is 0.45 g cm^{-3}, intermediate between gas-like and liquid-like densities). Unlike other nonpolar solvents, it is obtained from the atmosphere; on releasing the high pressure, it vaporizes and again returns to the atmosphere. Hence, it is a nonpolluting **green solvent**, which can be used to extract relatively nonpolar organic products from reaction mixtures and then be allowed to vaporize, leaving the organic product behind without introducing any air pollution from escaping vapors of the organic solvent.

EXAMPLE 4.11

Tell whether the following salts or complexes are likely to be soluble in water or in nonpolar organic solvents: (a) $[PtCl_2(NH_3)_2]$; (b) PtO; (c) $[PtCl(NH_2C_{10}H_{21})_3]Cl$; (d) $[PtCl(NH_2C_{10}H_{21})_3](TcO_4)$.

SOLUTION: (a) $[PtCl_2(NH_3)_2]$ is an uncharged complex, so it has no ions to hydrate exothermically; it is not likely to be very water soluble. The ligands do not have C–H or other nonpolar bonds, so the complex will also not be soluble in nonpolar organic solvents.

(b) PtO is a salt of a (moderately) acidic cation and a (very strongly) basic anion, so it is insoluble in water. There is nothing organic in its structure to make it soluble in nonpolar organic solvents.

(c) $[PtCl(NH_2C_{10}H_{21})_3]Cl$ is a complex that contains a large nonacidic cation with a long organic chain and a small nonbasic anion, Cl^-. This salt *would probably* be soluble in water (as an exception to Solubility Tendency III) and should be soluble in nonpolar organic solvents. It could be a good (but expensive) phase-transfer catalyst.

(d) $[PtCl(NH_2C_{10}H_{21})_3](TcO_4)$ differs only from the complex in (c) in that its anion is a larger nonbasic anion. It probably will still be (at least somewhat) soluble in the nonpolar organic solvent, but will be insoluble in water (Solubility Tendency III).

A full consideration of the solubility of a salt or complex consists of two questions. Its solubility in water is predicted using Solubility Rules and Tendencies I–IV. Its solubility in nonpolar solvents depends on its speciation (substituents) and is predicted using the concepts of this subsection.

AN AMPLIFICATION

Solubilities of Ionic Salts in Polar Solvents

Polar protic solvents such as methanol (CH_3OH), ethanol (CH_3CH_2OH), or glycerol [$C_3H_5(OH)_3$] can also solvate metal cations and simple or polyatomic anions (via hydrogen bonding), but less exothermically than water can hydrate them. Nonetheless this may be sufficient to dissolve salts with low lattice energies. Because these solvents are also good at dissolving organic compounds, the use of alcohols as solvents can be good for promoting reactions between inorganic salts and organic molecules.

There are also polar aprotic solvents such as acetone [$(CH_3)_2C{=}O$] and dimethyl sulfoxide [$(CH_3)_2S({=}O)$ or DMSO], which can solvate metal cations but not anions. Some of these have quite polar bonds to terminal oxygen atoms (which bear a formal negative charge), so they give substantial energies of solvation of cations. Hence, they may dissolve some inorganic salts with low lattice energies:

$$2CoCl_2(s) + 6(CH_3)_2CO \rightarrow Co[OC(CH_3)_2]_6^{2+} + [CoCl_4]^{2-}$$

Because organics are also soluble in these solvents, they are also useful in promoting metal-ion reactions with organics. DMSO achieved some notoriety a couple of decades ago when it was discovered that it dissolves itself in the lipid layer of the skin, so it penetrates the skin, taking in with it whatever solutes it contains. It proved a good way of introducing poisonous inorganic contaminants into the body!

4.5. The Role of Entropy Changes in Precipitation Reactions

OVERVIEW. You may want to review the thermodynamic terms entropy and enthalpy, and review the Gibbs free energy equation (Section 4.5A); to assist, try Exercises 4.51–4.54. By analyzing the Gibbs free energy ΔG for the precipitation reaction in terms of the enthalpy and entropy changes for precipitation, you can explain why Solubility Rule I works (Section 4.5B). You should be able to describe multiple hydration spheres of ions and tell which charges and sizes of ions, or which acidity/basicity classes of ions, are likely to have the largest hydration spheres. You may practice this concept with Exercises 4.55–4.59. A Connection to nerve transmission (biochemistry) is included.

Acidity/basicity classes of ions were introduced in Sections 2.1, 2.2, and 3.6. The concept of hydration spheres will be revisited in Section 8.7. The concept of entropy will again be useful in Section 6.3.

4.5A. Thermodynamic Measures of Tendencies to React (Review). In this section we attempt to analyze why ionic solids show such complex solubility patterns. To understand this we need to review some thermodynamic concepts. What causes some reactions to favor products (to be "spontaneous") while others favor reactants (are "nonspontaneous")? So far we have been using one principle, based on the *enthalpy change* ΔH for a reaction. A reaction tends to favor *products* (be spontaneous) if it is *exothermic* (ΔH is negative); it tends to favor its *reactants* if it is *endothermic* (ΔH is positive). As examples, in Experiment 2 the dissolution and acid dissociation of compounds of strongly acidic cations such as $TiCl_4$ or PCl_3 gave off notable amounts of heat (were exothermic). In Section 4.2 we saw that gaseous cations and anions combine to give ionic solids, releasing an enormous amount of (heat) energy known as the lattice energy.

But this cannot be the only factor determining whether a reaction will favor products. In Experiment 2 we noted that other ionic solids such as LiCl and $ZnCl_2$ tend to dissolve even though little or no heat is liberated. Some ionic solids dissolve even though heat is absorbed in the process (the dissolution process is endothermic). The dissolution of organic liquids in other like organic liquids (e.g., methanol in ethanol) releases little or no heat. Ice cubes melt at room temperature, but they do not release heat to their surroundings (i.e., your hand)!

The second principle is that a reaction tends to favor *products* (be spontaneous) if it increases the dispersal of molecules or ions over a *larger* volume of space; it tends to favor its *reactants* if it is concentrates the molecules or ions in a *smaller* volume of space. The measure of this dispersal or disorder is known as the *entropy S* of the substance. A *positive* entropy change ΔS for a reaction indicates increasing dispersal or disorder, and tends to favor the formation of *products*. Four common examples of this principle are as follows:

- Entropies of solids are usually less than those of the corresponding liquids, which are in turn usually much less than those of the corresponding gases.

- Entropies usually increase when a pure liquid or non-ionic (molecular) solid dissolves in a liquid solvent (ΔS is positive).

- Entropies decrease when a gas dissolves in a liquid (ΔS is negative).

- Entropies usually increase when chemical reactions produce more moles of gas than they consume.

But products are not always formed when ΔH is negative or when ΔS is positive. Because we have two principles operating, it is not always the case that they will operate in harmony with each other. If both ΔH and ΔS have matching signs, they operate in opposition to each other. There is also a third factor involved: the *temperature* of the system. Ice cubes melt at room or body temperature, but they do not melt when the temperature is below 0°C (e.g., in the freezer).

The resolution of the competing principles lies in combining them in the **(Gibbs) free energy equation**. A reaction goes to *products* (is spontaneous) if its Gibbs free energy change ΔG is *negative*. Under standard conditions,

$$\Delta G° = \Delta H° - T\Delta S° \tag{4.13}$$

As consequences of the inclusion of the temperature term, we find that:

- At very low temperatures, the $-T\Delta S$ term is negligible, so exothermic reactions are always favored.

- At very high temperatures (above 5000 or 10,000 K) the $-T\Delta S$ term dominates, so reactions that increase dispersal or disorder are always favored. Thus, at such temperatures, ionic salts and covalent molecules always dissociate into gaseous atoms. At much higher temperatures even atoms cannot exist, but dissociate into subatomic particles.

- At or near 298 K the $-T\Delta S$ term is usually (but not always) smaller in magnitude than about ± 40 kJ mol^{-1}. Consequently, if ΔH exceeds ± 40 kJ mol^{-1}, it usually dominates. But not always! The dissolution of ionic solids includes one of the most spectacular examples of when this is not the case.

AN AMPLIFICATION

Standard Conditions

Standard conditions mean that the substance is in the state (solid, liquid, gas, or polymorph) most commonly found at 298 K, and that gases and solutes are at unit activity (approximately 1 atm pressure and, most importantly for us, 1 M concentration for solutes). The "°" appended to ΔG, ΔH, and ΔS indicates that standard conditions are being applied.

AN AMPLIFICATION

What Does "Spontaneous" Mean?

The word "spontaneous" is not being used in its common meaning in this discussion. We are saying nothing about how *quickly* the reaction proceeds to products. It often happens that a product-favored "spontaneous" reaction is hindered by a very high activation energy that must first be provided. In the absence of this initiating energy, the reaction may take a very long time to occur. In Section 2.3C we discussed halides that are expected to react very exothermically with water. When this reaction occurs (at high temperature), it is very exothermic, but the rate at ordinary temperatures is vanishingly small because the activation energy needed to force water molecules into the vicinity of the cation (which already has its maximum coordination number) or to remove a chloride ion (to make room for the water) is unavailable. Hence, these reactions are thermodynamically favored, but are kinetically slow. As another example, the reaction of you personally with oxygen to yield CO_2 and H_2O (along with dust and ashes) is thermodynamically favored but is not observed at room temperature.

4.5B. Analyzing Solubility Rule I: Entropy Changes and the Structure of Hydrated Ions.

The Gibbs free energy equation (Eq. 4.12) may be applied specifically to the process of the precipitation of an ionic salt:

$$\Delta G_{pptn} = \Delta H_{pptn} - T\Delta S_{pptn} \tag{4.14}$$

Equation 4.14 tells us that formation of a precipitate is favored (ΔG_{pptn} is negative) if ΔH_{pptn} is negative (as discussed in Section 4.5A) and/or $-T\Delta S_{pptn}$ is negative (disorder is increased upon precipitation). Data for $-T\Delta S_{pptn}$ and ΔG_{pptn} have been gathered in Table 4.6 along with the previously used data for ΔH_{pptn} and U, the lattice energy. As previously discussed, a positive ΔH_{pptn} (which is to say a negative ΔH_{soln}) results in Solubility Rule 2; a negative ΔH_{pptn} results in Solubility Tendency III. However, the very important Solubility Rule I is a consequence of the entropy term $-T\Delta S_{pptn}$ being negative—that is, acidic cations and basic anions give precipitates because *disorder is increased* during the process.

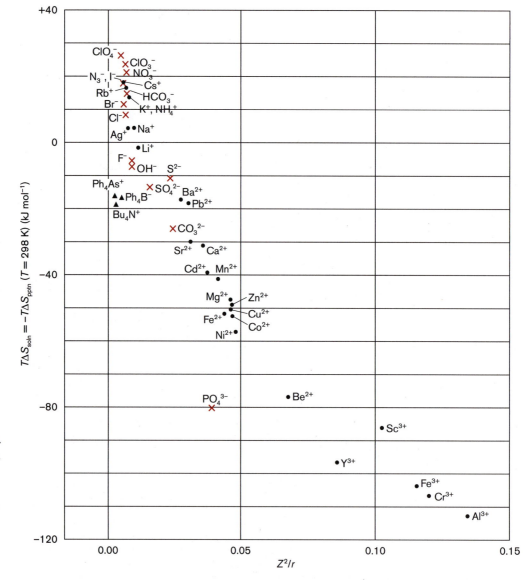

Figure 4.15. The 298 K entropy terms ($T\Delta S_{soln} = -T\Delta S_{pptn}$) for cations and anions as a function of Z^2/r. Crosses (✗) are for anions; dots (•) are for metal cations; triangles (▲) are for organic ions. [Thermodynamic data come from B. G. Cox and A. J. Parker, *J. Am. Chem. Soc.* 95, 6879 (1973); thermochemical radii of oxo anions come from J. E. Huheey, E. A. Keiter, and R. L. Keiter, *Inorganic Chemistry: Principles of Structure and Reactivity*, 4th ed., Harper-Collins: New York, 1993, p. 118.]

It may seem surprising at first that precipitation reactions of acidic cations and basic anions should be due to a negative entropy term, $-T\Delta S$, because this term corresponds to increasing randomness or disorder in the system, which is not what we expect in a reaction that produces a crystalline solid precipitate. But the reacting cations and anions exist as hydrated ions, and upon formation of the precipitate, water molecules are released. If enough water molecules are released, then the resulting positive entropy change (negative $-T\Delta S$) may exceed the ordering effect of producing a crystalline precipitate.

Experimentally, the convention has been to consider the reverse of Equation 4.10—the dissolution of an ionic solid—and to tabulate entropies of solution of the cations and anions. Figure 4.15 shows the $T\Delta S_{soln}$ (that is to say, the $-T\Delta S_{pptn}$) terms at 298 K for the hydration of the separated cations and anions, and their relationship to the familiar Z^2/r ratio of the ions.

It can be seen from Figure 4.15 that *all but the nonacidic cations* and *all but the nonbasic anions* make negative contributions to the $-T\Delta S_{pptn}$ term in Equation 4.14. Acidic cations and basic anions, upon dissolving and hydrating in water, coordinate enough water molecules to give a net ordering effect.

The more acidic or basic the ion (i.e., the higher its charge or the smaller its radius), the greater is the magnitude of the ion's ordering effect. The inverse relationship to size may seem surprising, because larger cations have greater maximum coordination numbers, so we might expect them to attach more water molecules. Evidently the smaller ion attaches more water molecules (or at least orders them more effectively). Similarly, although increasing the charge of a cation does not increase its maximum coordination number, it does increase the ordering of water molecules in the hydrated ion. One measure of this can be found in Table 4.7, which lists the hydration numbers of several common cations.

TABLE 4.7. Hydration Numbers and Hydrated Radii of Some Hydrated Ions

Ion	Z^2/r	Hydration Number	Hydrated Radius (pm)
Cs^+	0.0055	6	228
K^+	0.0066	7	232
Na^+	0.0088	13	276
Li^+	0.0111	22	340
Ba^{2+}	0.0268	28	
Sr^{2+}	0.0303	29	
Ca^{2+}	0.0351	29	
Mg^{2+}	0.0465	36	
Cd^{2+}	0.0549	39	
Zn^{2+}	0.0599	44	

Sources: Hydration numbers come from A. T. Rutgers and Y. Hendrikx, *Trans. Faraday Soc.* 58, 2184 (1962). Hydrated radii come from R. P. Hanzlik, *Inorganic Aspects of Biological and Organic Chemistry*, Academic Press: New York, 1976, p. 31.

It can be seen in Table 4.7 that larger Group 1 cations have smaller hydration numbers (attach fewer water molecules). The same is true in Group 2. To explain this unexpected relationship, we theorize that hydrated cations and anions have more complex structures than we have assumed up to now. All hydrated metal ions and nonmetal or oxo anions have a layer of water molecules surrounding the ion, as shown in Figure 2.6. This layer is referred to as the **primary hydration sphere** or **inner sphere** of the hydrated ion, and the larger the "bare" cation or anion, the more water molecules can be accommodated in this primary hydration sphere. But ions that are at all acidic or basic exert such a strong attraction for the water molecules in their primary hydration sphere. Ions pull electron pairs towards a cation or hydrogens toward an anion. Then these water molecules exert an enhanced hydrogen-bonding attraction for other water molecules and organize them into a **secondary hydration sphere** or **outer sphere** around the first layer (Figure 4.16).

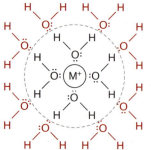

Figure 4.16. Primary and secondary hydration spheres.

In summary, the reason that most acidic cations and basic anions react to give precipitates is due to the disorder resulting from the release of numerous water molecules from the multiple hydration spheres of the cations and anions. This is apparent when we include reasonable numbers of water molecules in the seemingly simple precipitation reaction of, say, magnesium and carbonate ions:

$$Mg(H_2O)_{36}^{2+} + CO_3(H_2O)_{28}^{2-} \rightarrow MgCO_3(s) + 64H_2O \qquad (4.15)$$

Secondary Hydration Spheres

This secondary hydration sphere may consist of more than one layer of water molecules—the stronger the attraction of the bare ion for water molecules (i.e., the greater its acidity or basicity), the more water molecules that may be attached in additional layers. Each subsequent layer would be expected to involve weaker attractions, however, so water molecules would stay immobilized by hydrogen bonds for shorter and shorter periods of time. Many methods have been used to study this phenomenon of multiple hydration layers, but these diverse methods of measuring the number of waters attached or the radius of the hydrated ion[18] give inconsistent answers, because each method requires a different amount of time for the water molecules to remain attached in order to be counted. But nearly all measurements give consistent *trends* in hydration numbers among series of ions, such as the results for the ions in Table 4.5, obtained from transference measurements by way of the *Stokes radii* or *hydrated radii* of the hydrated ions, also listed in Table 4.5.

AN AMPLIFICATION

A CONNECTION TO BIOCHEMISTRY

Nerve Transmission

Because the *hydrated* ions of smaller bare ions may be larger than those of larger bare ions, we sometimes see unexpected chemical results. One of the important functions of metal ions in biochemistry given in Section 1.2B was the "trigger and control" function: The passage of the ions K^+, Na^+, Ca^{2+}, and Mg^{2+} across nerve cell walls constitutes an electrical current that triggers and controls certain biochemical mechanisms. An important question in biochemistry has to do with how metal ions can migrate across cell membranes, which consist largely of lipids in which metal ions are quite insoluble. In 1998 MacKinnon published Nobel-Prize-winning work using X-ray crystallography to show the high-resolution structure of an ion channel (Figure 4.17) that allows the passage of K^+ ion, but not the "smaller" Na^+ ion, through the cell wall.[19] Subsequent work[20] has shown that the K^+ ion approaches as a $K(H_2O)_8^+$ ion, but this is too large to fit through the channel. However, the channel has eight oxygen donor atoms arranged to mimic the geometry of the $K(H_2O)_8^+$ ion, which assist the dehydration of this ion and allow it to pass through the channel. Interestingly, large nonacidic quaternary ammonium ions such as $(C_2H_5)_4N^+$ are also attracted to the pore, but cannot become smaller by dehydration to pass through; they are potent blockers of the channel.

PORE HELIX

Figure 4.17. Ribbon representation of the view from outside the cell of the tetrameric ion-channel protein KcsA. [From D. A. Doyle, J. M. Cabral, R. A. Pfuetzner, A. Kuo, J. M. Gulbis, S. L. Cohen, B. T. Chait, and R. MacKinnon, *Science* 280, 69 (1998). Reprinted by permission.]

EXAMPLE 4.12

How does the partial protonation of a multiply charged oxo anion alter its $T\Delta S_{soln}$? Describe what happens to the ordering properties of such an ion upon partial protonation. Does this entropy effect favor the partial protonation of such ions? (The hydronium ion itself is only a weakly ordering ion.)

SOLUTION: Adding a proton to a multiply charged oxo anion reduces its negative charge, and hence reduces its ability to organize multiple hydration spheres around it, so its $T\Delta S_{soln}$ is made more negative. (Compare the $+25.5$ kJ mol^{-1} value for CO_3^{2-} in Figure 4.15 with the -13.8 kJ mol^{-1} value for HCO_3^-.) Because the disappearance of the hydrated proton itself has little effect on the overall order in the solution, this change has reduced the order in the solution. Hence, protonation of the anion is favored by this entropy effect.

4.6. Background Reading for Chapter 4

Photographs of Crystals

O. Medenbach and H. Wilk, *The Magic of Minerals*, Springer-Verlag: Berlin, 1986. An electronic version of this book is available.

Structures of Water and Hydrated Ions

John Burgess, *Ions in Solution: Basic Principles of Chemical Interactions*, 2nd ed., Horwood Publishing: Chichester, UK, 1999, pp. 28–35 (Chapter 2: Solvation Numbers) and pp. 51–54 (Section 4.3: Thermochemistry of Ion Solvation).

H. Ohtaki and T. Radnai, "Structure and Dynamics of Hydrated Ions," *Chem. Rev.* 93, 1157–1203 (1993).

E. K. Wilson, "A Renaissance for Hofmeister: Flurry of New Research Overturns Long-Held Ideas about Ions, Water, and Macromolecules," *Chem. Eng. News*, Nov. 26, 2007, pp. 47–50.

Y. Marcus, "Effects of Ions on the Structure of Water," *Pure Appl. Chem.* 82, 1889–1899 (2010). Contends that some of the articles cited in the previous reference have misleading titles and over-extended conclusions.

Phosphates for Fertilizers

D. A. Vaccari, "Phosphorus: A Looming Crisis," *Sci. Amer.* 300(6), 54–59 (2009). A vital component of fertilizers, which is too often flushed away.

Carbonates and Ocean Acidification

S. C. Doney, "The Dangers of Ocean Acidification," *Sci. Amer.* 294(3), 58–65 (2006).

M. J. Hardt and C. Safina, "Threatening Ocean Life from the Inside Out," *Sci. Amer.* 303(2), 66–73 (2010).

B. Hönisch et al., "The Geological Record of Ocean Acidification," *Science* 335, 1058–1063 (2012).

Soluble Minerals

M. Kurlansky, *Salt: A World History*, Walker and Co.: New York, 2002.

Numbered references from this chapter may be viewed online at www.uscibooks.com /foundations.htm.

4.7. Exercises

4.1. *Which ionic salts have the correct stoichiometry to adopt the fluorite structure (Figure 4.2)? Which have the correct stoichiometry to adopt the anti-fluorite structure? (a) GeO_2; (b) GeF_2; (c) GeF_4; (d) Rb_2O; (e) $Na_2[SiF_6]$; (f) $Ba(ClO_4)_2$.

4.2. The salt Na_3N was only synthesized recently; it has a structure in which N^{3-} has a coordination number of six. (a) What is the average coordination number of the Na^+ ion? Is this coordination number expected for Na^+? (b) Look up the structure in the literature to find its lattice type. To which lattice type in Figures 4.2 or 4.3 is this most closely related?

4.3. *The salt Li_3N has a structure in which N^{3-} has a coordination number of eight. (a) What is the average coordination number of the Li^+ ion? How can such a coordination number be achieved? (b) An anion coordination number of eight is found in the CsCl structure. Does Li_3N take the CsCl structure? Why or why not?

4.4. Consider a cubic unit cell. In counting atoms to determine the simplest formula, each atom in a cubic unit cell counts as what fraction of an atom in calculating the formula? (Fill in fractions, or the term "a whole.") (a) An atom in the center of a face counts as _____ atom; (b) An atom on a corner counts as _____ atom; (c) An atom in the interior of the unit call counts as _____ atom; (d) An atom in the center of an edge counts as _____ atom.

4.5. *Calculate the total numbers of cations and anions shown in the unit cell (Figure 4.2) of (a) rock salt and (b) cesium chloride.

4.6. Calculate the total numbers of cations and anions shown in the unit cell (Figure 4.2) of (a) fluorite; (b) rutile; (c) zinc blende.

4.7. *Using Tables C and 4.1, calculate the radius ratio for, and predict the cation and anion coordination numbers of and the type of lattice that would be formed by (a) VSe; (b) V_2Se_3; (c) VSe_2.

4.8. Select the most likely lattice types for each of the following salts: (a) BeF_2; (b) CaO; (c) BeI_2; (d) BeTe; (e) HgF_2.

4.9. *Assume that the rare, radioactive elements astatine and polonium form anions At^- and Po^{2-}, each of which has a radius of about 226 pm. Which of the lattice types (ZnS, NaCl, CsCl, SiO_2, TiO_2, or CaF_2) would be expected to be adopted by (a) the astatide of each Group 1 metal (Na through Cs); (b) the astatide of each Group 2 metal (Mg through Ba); (c) the polonide of each Group 2 metal (Mg through Ba).

4.10. Do radius-ratio calculations to determine the most likely lattice type (or if this cannot be identified, the cation and anion coordination numbers) for (a) UN; (b) UN_2; (c) U_3N_4.

4.11. *Using Tables C and 4.2, calculate the radius ratio for, and predict the cation and anion coordination numbers of and the type of lattice that would be formed by (a) CeO_2; (b) $SiTe_2$; (c) Tl_2O_3; (d) BN.

4.12. Cesium and gold form an ionic compound, Cs^+Au^-, with a cesium–gold distance of 369 pm. What type of lattice will CsAu adopt? [See U. Zachwieja, *Z. Anorg. Allg. Chem.* 619, 1095 (1993) for the actual lattice type.]

4.13. *The elements radium and astatine both arise during the radioactive decay of uranium. Assume that they get together in the form of the salt $RaAt_2$; assume, too, that the radius of the Ra^{2+} ion is 162 pm and the radius of the At^- ion is 226 pm. Predict the coordination numbers of Ra and At and the type of lattice that $RaAt_2$ will adopt.

4.14. The elements polonium and thorium both arise during the radioactive decay of uranium. Assume that they get together in the form of the salt $ThPo_2$; assume, too, that the radius of the Th^{4+} ion is 108 pm and the radius of the Po^{2-} ion is 226 pm. Predict the coordination numbers of Th and Po and the type of lattice that $ThPo_2$ will adopt.

4.15. *The separation of the *f*-block elements of the sixth period (the "lanthanides" or "rare earths") and the associated Group 3 ions Sc^{3+} and Y^{3+} is a very tricky business due to the identical +3 charges they normally show in their cations, and the close similarity of their ionic radii. In the early days the separation was attempted by fractional crystallization, which would work if the ions would crystallize in different lattice types. Assume for simplicity that you have only the largest and the smallest of these ions, La^{3+} and Sc^{3+}, and that you want to crystallize their nitrides. (a) Predict the coordination numbers that will be found in LaN. Which lattice type is expected for LaN? (b) Predict the coordination numbers that will be found in ScN. Which lattice type is expected for ScN?

4.16. (a) The ionic radius of Re^{6+} is not listed in Table C, so interpolate a reasonable value for it. (b) Compute the radius ratio for ReO_3. (c) Compute the coordination numbers of the cation and anion in ReO_3. (d) Does your answer correspond to the observed coordination numbers (Figure 4.3)? What are the observed bond angles at the two ions?

4.17. *Can the coordination numbers and the β-quartz lattice type given in Table 4.2 be used for all polymerization isomers of SiO_2 (Figure 4.7)? Which lattice types given in Table 4.2 for 1:2 stoichiometry would probably always be inappropriate, and why?

4.18. Which three lattice types given in Table 4.2 show the appropriate stoichiometry and radius ratio for the compound $MoTe_2$? Is there any way to make a choice among these three lattice types—if so, how would you choose?

4.19. *The Ag^+ ion has a coordination number of six in both AgF(s) and AgBr(s). (a) Is this coordination number expected for these halides by radius-ratio rules? (b) Using the observed coordination number of six for the two halides, calculate the lattice energies of AgF(s) and AgBr(s). (c) The measured lattice energies of the two halides are –974 kJ mol^{-1} for AgF(s) and –905 kJ mol^{-1} for AgBr(s). Are your calculated lattice energies close to the measured values? (d) Which calculation is further from the experimental value? How might you explain this?

4.20. The Ag–X distances are shorter in the gaseous ion pairs $(Ag^+)(X^-)(g)$ than in the six-coordinate solids $(Ag^+)(X^-)(s)$. The observed Ag–X(g) distances are 198 pm in AgF(g), 228 pm in AgCl(g), 239 pm in AgBr(g), and 254 pm in AgI(g). (a) Calculate the total energies of these gaseous ion pairs (E_{total}) according to Equation 4.8. (b) Calculate the distances expected in the corresponding solids $(Ag^+)(X^-)(s)$. Why are these larger? (c) Calculate the total energies the gaseous ion pairs would have if they had the same Ag–X distances as is found in the solids.

4.21. *Calculate the expected lattice energy for the salts (a) CaF_2 and (b) CeO_2.

4.22. In honor of the 60th anniversary of the United Nations, let us consider the hypothetical ionic salt uranium(III) nitride. (a) Write the formula of uranium(III) nitride. (b) Assuming that it adopts the CsCl lattice type, compute the expected lattice energy for uranium(III) nitride, extrapolating a reasonable value for the Born exponent for uranium.

4.23. *The ionic salt ThS_2 adopts the TiO_2 lattice type. Estimate a suitable Born exponent for the thorium ion, then calculate the lattice energy of ThS_2.

4.24. Cesium and gold form an ionic compound, Cs^+Au^-, with a cesium–gold distance of 369 pm. Using your predicted lattice type for this compound from Exercise 4.12a, and estimating a suitable Born exponent for the Au^- ion, calculate the lattice energy of CsAu.

4.25. *Using your results from Exercise 4.9b and Table 4.1, calculate the lattice energies of (a) barium astatide and (b) $RaAt_2$.

4.26. Using your lattice type from Exercise 4.9c, calculate the lattice energy of $ThPo_2$.

4.27. *Calculate the lattice energy of lanthanum nitride, LaN, using your prediction of its lattice type from Exercise 4.15a.

4.28. The ionic salt Sc_2S_3 adopts the Al_2O_3 (corundum) lattice type. Calculate the lattice energy of Sc_2S_3.

4.29. *The Th^{4+} cation forms a salt with the Te^{2-} ion that crystallizes in the rutile crystal lattice type. (a) Calculate the lattice energy of this salt. (b) The Ti^{2+} cation has virtually the same radius as the Th^{4+} cation, and the I^- anion has virtually the same radius as the Te^{2-} anion. Can titanium(II) iodide crystallize in the rutile crystal lattice? (c) Which salt, thorium(IV) telluride or titanium(II) iodide, will have the lattice energy of the greatest magnitude?

4.30. The mineral TiN(s), osbornite, is said to be one of the first dozen minerals formed in the universe.[21] It is in interstellar space, forming from the heavy elements being emitted by exploding supernovas. Its lattice energy is –8033 kJ mol^{-1}. The measured lattice energies of some other binary titanium salts are TiO, –3811 kJ mol^{-1}; TiO_2, –12150 kJ mol^{-1} (calculated);

and Ti_2O_3, –14149 kJ mol^{-1}. Explain why TiN(s) has a greater or lesser lattice energy than (a) TiO; (b) TiO_2; (c) Ti_2O_3.

4.31. *Which of the following salts will be insoluble in water: $Cd(NO_3)_2$, $CsBrO_4$, $CePO_4$, Cs_3AsO_4, $BaSeO_4$, $Cd_5(IO_6)_2$, and TiO_2?

4.32. Which of the following compounds will be insoluble in water: $ZnMoO_4$, Cs_2MoO_4, Al_2TeO_6, $Al(NO_3)_3$, CdO, NaOH, $RbTcO_4$, KIO_4, and K_5IO_6?

4.33. *Which of the following compounds will be insoluble in water: $Zn(ReO_4)_2$, $CsReO_4$, K_2FeO_4, Al_2TeO_6, EuO, and Ga_2O_3?

4.34. You are preparing solutions for the organic and analytical chemistry labs, and you are told to make a 1 M solution of potassium permanganate, $KMnO_4$. Can you do this?

4.35. *Consider each of the cations U^{3+}, Rb^+, Pb^{4+}, and Al^{3+}. (a) Classify the acidity of each of these cations. (b) Which of these cations give a precipitate with the ClO_4^- ion? (c) Which of these cations give a precipitate with the PO_4^{3-} ion?

4.36. Assume that the given elements form oxo anions when they are in the indicated oxidation states: Tc(VII); Pu(VIII); As(III). (a) Predict the formula of the oxo anion formed by each element in that oxidation state. (b) Name each oxo anion. (c) Classify the basicity of each oxo anion. (d) Predict whether the K^+ salt of each will be insoluble. (e) Predict whether the Fe^{3+} salt of each will be insoluble.

4.37. *What are the possible insoluble salts that can arise on mixing solutions of the following cations—Pu^{4+}, $(CH_3)_4N^+$, and Eu^{2+}—with solutions of the following anions—PF_6^-, PO_4^{3-}, and SO_4^{2-}?4.

4.38. The *Handbook of Chemistry and Physics* has solubility data for hexafluorosilicate $[SiF_6]^{2-}$ salts of the following cations. Solubility of 390 g L^{-1} water or more: Zn^{2+}, Cu^{2+}, Pb^{2+}, Co^{2+}, Mg^{2+}, Li^+. Solubility from 0.3 g L^{-1} water to 32 g L^{-1}: Ca^{2+}, Sr^{2+}, Ba^{2+}, Na^+, K^+, Rb^+. Based on this information, how would you attempt to classify the basicity of this anion?

4.39. *Write general chemistry-type specific solubility rules for the salts of the oxo anions: (a) chromate; (b) ferrate; (c) pertechnetate; (d) silicate; (e) tellurate. These rules should state whether most salts of this anion are soluble or insoluble, then list some ions that are exceptions.

4.40. Liming (adding CaO to) a lake to neutralize acid rain in it could seriously affect the availability of anion nutrients such as phosphate and molybdate, but not nitrate. Explain.

4.41. *When you mix an aqueous solution of the Hg^{2+} ion with an aqueous solution of the SiO_4^{4-} ion, you obtain two precipitates, one yellow and one white. (a) What are the logical formulas for those two precipitates? (b) Design a simple mixing experiment using Hg^{2+} and another anion to determine which precipitate is which compound. (c) When you mix Hg^{2+} with PO_4^{3-} instead of with SiO_4^{4-}, you obtain only one precipitate. Why would PO_4^{3-} be less likely to give the second precipitate?

4.42. In order to get tenure and a promotion at your school, you have to publish a new laboratory experiment. You decide to devise a qualitative analysis scheme of the anions that uses cations to separate the following anions by precipitation: ReO_4^-; SeO_4^{2-}; SeO_3^{2-}; PO_4^{3-}; SiF_6^{2-}; OH^-. (a) Your Anion Group I is precipitated with the $(CH_3)_4N^+$ ion. Which anions will precipitate in this group? Write the formulas of the salt that will precipitate. (b) After removing the Group I precipitates, your Anion Group II is precipitated with the Sr^{2+} ion. Write the formulas of the salts that will precipitate in this group.

4.43. *In an organic chemistry lab you have just synthesized the listed new ions in solution. Tell how you would isolate each from solution to get a stable crystalline solid. (a) Tropylium ion, $C_7H_7^+$; (b) squarate ion, $C_4O_4^{2-}$; (c) cyclopentadienide ion, $C_5H_5^-$.

4.44. The generalized solubility rules and tendencies implicitly treat the phenomenon of precipitation as a type of acid–base reactivity. However, uncharacteristically one of the solubility rules and tendencies shows that there is (precipitation) reactivity between two types of acid–base species that have overlapping predominance areas (see Section 2.4). Which two types of species are these, and which solubility rule or tendency is involved?

4.45. *Enthalpies of precipitation are known for the following halides: LiF, –5 kJ mol^{-1}; NaCl, –4 kJ mol^{-1}; AgCl, –65 kJ mol^{-1}; CaCl$_2$, +83 kJ mol^{-1}. Use data on hydration enthalpies (Tables 2.1 and 4.5) to calculate the lattice energies of these compounds.

4.46. (a) Use the known lattice energies for $(CH_3)_4NCl$ (–566 kJ mol^{-1}), $(CH_3)_4NBr$ (–553 kJ mol^{-1}), and $(CH_3)_4NI$ (–544 kJ mol^{-1}), along with data from Table 4.5 and precipitation enthalpies (–4, –24, and –42 kJ mol^{-1}, respectively), to compute the hydration energy of the tetramethylammonium ion. (b) Compare your result with the hydration energy of the cesium ion and explain any differences between the two.

4.47. *(a) Use the known lattice energies for $NaNO_2$ (–772 kJ mol^{-1}) and KNO_2 (–687 kJ mol^{-1}), along with data from Table 2.3 and precipitation enthalpies (–14 and –13 kJ mol^{-1}, respectively), to compute the hydration energy of the nitrite ion. (b) Compare your result with the hydration energy of the nitrate ion and explain any differences between the two.

4.48. Usually, $\Delta H°$ for the precipitation of an ionic salt is small—roughly zero (Table 4.6). Use this approximate value and the hydration energies for Al^{3+} and for ClO_4^- (Tables 2.3 and 4.5) to estimate the lattice energy of aluminum perchlorate.

4.49. *Tell whether the following salts or complexes are likely to be soluble in water, in nonpolar solvents, in both, or in neither: (a) FeO; (b) $KClO_4$; (c) $(C_3H_7)_4Cl$; (d) $(C_3H_7)_4PF_6$; (e) $Ba(NO_3)_2$; (f) $[Cr(NH_3)_3Cl_3]$; (g) $[Cr(NH_2C_8H_{17})_3Cl_3]$.

4.50. Of the compounds $Fe(BrO_4)_3$, $Zn_5(IO_6)_2$, $[(C_2H_5)_4N][ClO_4]$, and $BaFeO_4$, (a) which one is the best desiccant and is obtained as a hydrate upon evaporating its aqueous solution? (b) Which one is most likely to be soluble in body fat? (c) Which one is most likely to be involved in phase-transfer catalysis?

4.51. *Predict whether the entropy change for the system (reaction) is positive or negative, and whether it is favorable or unfavorable for forming products, when (a) Humpty Dumpty (an egg) falls from the wall and breaks, spilling his insides; (b) all the king's horses and all the king's men succeed in putting Humpty Dumpty together again; (c) liquid water → steam; (d) liquid water → ice; (e) 50 coins in roll, heads all up → 50 coins in roll, but some heads up, some down.

4.52. Predict whether the entropy change for the system (reaction) is positive or negative when (a) a jigsaw puzzle is assembled; (b) this class ends and the students scatter across campus; (c) sugar dissolves in water; (d) nitroglycerine explodes, $C_3H_5(NO_3)_3(l) \rightarrow 3CO_2(g) + 2\frac{1}{2}H_2O(g) + \frac{1}{2}N_2O(g) + N_2(g)$; (e) crystalline solid → gaseous cations + gaseous anions.

4.53. *(a) If the $-T\Delta S_{pptn}$ term in Table 4.6 is negative at 298 K, will it be more negative, less negative, or the same at the boiling point of water (373 K)? (b) Will this cause the ΔG_{pptn} term to be more negative, less negative, or the same at the boiling point of water? (c) Will this cause the solubility of the salt to increase, decrease, or be the same at the boiling point of water? (d) Consulting Table 4.6, to which solubility rules and tendencies do the suppositions in parts (a)–(c) of this question generally apply?

4.54. Tell which of the following changes favor products and which favor reactants (left side). In each case tell whether your answer is due mainly to a favorable entropy term (in the equation for ΔG, the Gibbs free energy), an unfavorable entropy term, a favorable enthalpy term, or an unfavorable enthalpy term (in some cases more than one answer is possible). Briefly explain why each answer is correct. (a) Broken glass beaker → reassembled intact glass beaker; (b) coal (C) + $O_2(g)$ → $CO_2(g)$; (c) unexploded nitroglycerine → exploded gaseous products; (d) Humpty Dumpty (an egg) on top of the wall → Humpty Dumpty after the fall; (e) an unassembled jigsaw puzzle on the table → an assembled jigsaw puzzle on the same table.

4.55. *Select one of the following ions [Al^{3+}, Li^+, Rb^+, or $(C_4H_9)_4N^+$] as an example of each of the following properties: (a) the ion that attaches the fewest water molecules when it dissolves in water; (b) the ion that forms the largest hydrated ion; (c) the ion with the largest primary hydration sphere.

4.56. Select one of the following ions (Fr^+, Fe^{3+}, Li^+, or Ac^{3+}) as an example of each of the following properties: (a) the ion that attaches the greatest number of water molecules when it dissolves in water; (b) the ion that forms the smallest hydrated ion; (c) the ion with the largest primary hydration sphere.

4.57. *Consider the ions K^+, Zn^{2+}, Al^{3+}, SiO_4^{4-}, ClO_4^-, and SeO_4^{2-}. (a) Which cation will form the largest hydrated ion? (b) Which cation(s) will have the smallest secondary hydration spheres? (c) Which cation(s) will give insoluble salt(s) with the ClO_4^- ion? (d) Which cation(s) will give insoluble salt(s) with the SiO_4^{4-} ion? (e) Write the formula of any salt containing just these ions that will precipitate for reasons connected with an entropy change. (f) Write the formula of any soluble salt containing just these ions and explain (in terms of lattices, etc.) why it is soluble.

4.58. Consider the cations Cs^+, Li^+, and Ga^{3+}. (a) Which cation will form the largest hydrated ion? (b) Which cation(s) will have the largest secondary hydration spheres? (c) Which cation(s) will give insoluble salt(s) with the MnO_4^- ion? (d) Which cation(s) will give insoluble salt(s) with the CO_3^{2-} ion?

4.59. *Fill in the blanks with the appropriate term from the following list (soluble salts, precipitates, ΔH, $-T\Delta S$), then complete the explanation in your own words. (a) Acidic cations and basic anions tend to give _____. Thermodynamically this is favored because of the _____ term in ΔG. (b) Cross combinations (such as nonacidic cations plus basic anions) tend to give _____. Thermodynamically this is favored because of the _____ term in ΔG. The reason that this happens is:

CHARLES J. PEDERSEN (1904–1989) was born in Pusan, Korea (now South Korea). His father was Norwegian and his mother was Japanese. After growing up in Japan, he moved to the United States in 1922 to study chemical engineering at the University of Dayton in Ohio, where he had family and friends. After getting his bachelor's degree, he obtained a master's degree in organic chemistry from MIT. Instead of pursuing a doctorate, he got a job in 1927 as a research chemist at the du Pont Company, where he worked until he retired in 1969. His research interests ranged widely and included the catalytic properties of transition elements (particularly copper), the oxidative degradation of petroleum products and rubber, polymerization initiators, and photochemistry. In 1960, while studying the effects of bidentate and multidentate phenolic ligands on the catalytic properties of the VO group, he obtained a small amount of dibenzo 18-crown-6, the first crown ether. Eventually, Pedersen was able to vary the size of the polyether ring, making it possible to form stable structures with a variety of alkali metal ions. Donald J. Cram (University of California, Los Angeles) and Jean-Marie Lehn (Université Louis Pasteur and Collège de France) subsequently and individually expanded upon Pedersen's work on crown ethers, and all three were awarded the 1987 Nobel Prize in Chemistry. Charles Pedersen is one of the few Nobel Prize winners in science to lack a Ph.D.

CHAPTER 5

Trends in Coordination Equilibria

In Section 3.2 we introduced the topics of complex-ion formation and reviewed the concepts of Lewis acids such as metal ions and Lewis bases (*ligands*) such as simple anions and water and other small molecules. Ions and molecules need not be small to act as ligands, however. Many or most of the biological functions of metal ions (Section 1.2) are filled, not by hydrated ions, but by complex ions involving very large molecules as ligands. Indeed, the large majority of organic compounds and inorganic anions possess *donor atoms* and can act as *ligands*. You may recall that in Figure 3.5 you identified ligands in complex metal ions. Two very important biomolecules—namely, chlorophyll (Figure 5.1a) and hemoglobin (Figure 5.1b)—have long been known to have metal ions at the centers of their large ligands. Chlorophyll is the key to photosynthesis in plants, whereas the heme unit in hemoglobin transports the oxygen molecule in the body. The O_2 molecule acts as a ligand and is coordinated to the iron atom of heme during its transport. One complex vitamin, vitamin B_{12} or cobalamin (Figure 5.1c), incorporates a cobalt atom that, in the various forms of this vitamin, is identifiable as a Co^{3+}, Co^{2+}, or Co^+ ion.

Countless complexes could conceivably be formed by a given metal in natural waters, in biological fluids, or in wastewaters. Important properties such as solubilities in fat or water (Section 4.6) and toxicity depend on the chemical form in which the element is present (the speciation of the element; Section 3.8A). Most of these chemical forms are complexes.

Fortunately, there are two very important qualitative generalizations that we will emphasize here in Chapter 5 that will enable us to predict some of the most likely complexes in a complicated mixture of metal ions and ligands. The first of these is based on the geometry of the ligands (which is based on their Lewis structures; Section 3.4A) and the spacing of the donor atoms. We explore the geometrical factors in Sections 5.1–5.3.

The second generalization is based not on the positions of the donor atoms but on their identities. In the previous chapters we dealt mainly with anions having oxygen and fluorine donor atoms. Now we will broaden our perspective to include other

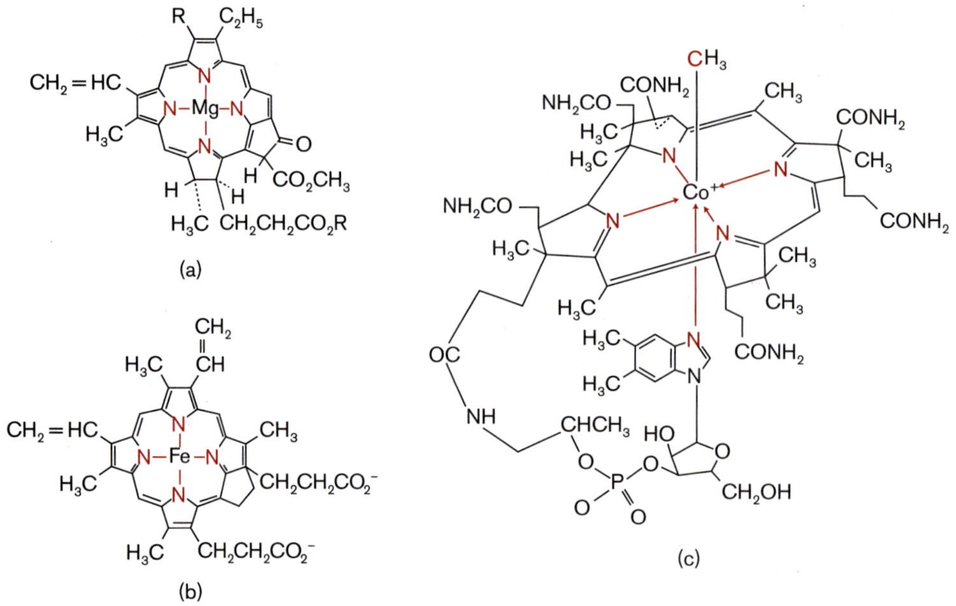

Figure 5.1. Structures of some biologically significant macrocyclic complexes. (a) Chlorophyll contains a Mg^{2+} ion coordinated to four nitrogen donor atoms of an organic (porphyrin) ligand. (b) Heme contains a Fe^{2+} ion coordinated to four nitrogen donor atoms of a porphyrin ligand and a nitrogen donor atom from a protein. (c) Methylcobalamin is one derivative of vitamin B_{12}. [Adapted from F. A. Cotton and G. Wilkinson, *Basic Inorganic Chemistry*, John Wiley and Sons: New York, 1976.]

donor atoms. In Sections 5.4–5.8 we will develop the hard and soft acid–base (HSAB) principle. The HSAB principle allows us to predict rather successfully what kinds of metal ions prefer to form coordinate covalent bonds with which kinds of donor atoms. In Section 5.6 we add the last two solubility rules, to cover the solubilities of salts of anions having donor atoms other than O and F.

The HSAB principle has many important practical applications. In Section 5.6 we examine the geochemical applications of the principle to produce many important minerals beyond those covered in Chapter 4. Then, in Sections 5.7, we apply this principle to the biological functions and toxicology of the elements. Finally, in Section 5.8, we bring the two principles of this chapter together to survey some of the many recent developments in the field of medicinal inorganic chemistry. The Connections in this chapter are the subject of many new developments. These Connections should serve to highlight your awareness of current research; however, by the time you read this textbook, they may not be completely up-to-date.

5.1. Structural Analysis of Ligands: Monodentate and Chelating Ligands

OVERVIEW. One way of classifying molecules and ions as ligands is based on their Lewis structures and the numbers and positions of their donor atoms. You will name geometries and obtain approximate bond angles from valence-shell electron-pair repulsion (VSEPR) theory (Section 5.1A; Exercises 5.1–5.4). If the molecule (or ion) has only one usable donor atom, it is classified as a monodentate ligand. Molecules with two or more donor atoms geometrically positioned so that they can attach to the *same* metal ion are potential chelating ligands

(Section 5.1B). You may practice identifying potential chelating ligands by trying Exercises 5.5–5.11. Chelating ligands form more stable complexes than similar monodentate ligands (Section 5.1C). You can practice applying and explaining this effect by trying Exercises 5.12–5.14. Connections to environmental chemistry and biochemistry are included.

Please note that this is the first chapter that assumes that you have some knowledge of organic chemistry (you should at least be simultaneously enrolled in the first semester of organic chemistry). You will also use Lewis structures of larger molecules and ions. If necessary, you can review drawing Lewis structures and identifying donor atoms by rereading Sections 3.1 and 3.2, and re-trying Exercises 3.1–3.5 and 3.11–3.15. Chelating ligands will appear again in Sections 7.4 and 7.8. Bond angles and geometries will play an important part in Chapter 7.

5.1A. Bond Angles and Geometries (Review of VSEPR from General and Organic Chemistry).

The geometric shapes of molecules, ions, and coordination compounds of the p-block elements, and the approximate angles between their covalent bonds, may be predicted remarkably well using a simple scheme known as the valence-shell electron-pair repulsion (VSEPR) model. This model assumes that the shapes of AX_mE_n molecules and ions are determined primarily by the electrostatic repulsions of the negatively charged A–X covalent bond electron pairs and unshared pairs of sp-hybridized valence electrons E about the central atom. The geometry and bond angles adopted are those that allow these pairs of electrons to move as far apart as possible.

We begin the determination of molecular geometry with a correct Lewis dot structure of the species (Section 3.1). If it is at all large, there will be more than one central atom, and each central atom may have its own geometry and bond angles that must be determined separately. For each central atom in this structure, we determine the total coordination number (Section 3.4). Whether the atoms are attached by single, double, or triple bonds does not matter in this count, because all the electrons in double or triple bonds must occupy the same general region of space—between the two atoms in the double or triple bond. For each total coordination number, there is ideally one bond angle (or set of angles) that minimizes the repulsions of the electron pairs. There is then a term for the geometry that describes the locations of the atoms in a molecule having these bond angles.

Compounds with No Unshared s or sp-Hybridized Electron Pairs on the Central Atom (AX_mE_0). We begin with this, the simpler case; in coordination chemistry, these apply to the central (metal) atoms/ions of the complex. For these complexes, the total coordination number of each central atom equals its coordination number m in AX_mE_0. The least repulsive arrangements of electron pairs for total coordination numbers between two and nine are shown in Figure 5.2. Also given are the names of the geometries and (when it is fixed) the donor atom–metal–donor atom bond angles.

A total coordination number of two about a given atom (as in HgF_2 or O=C=O) means that there are two independent pairs of electrons repelling each other. Repulsion is minimized with the two electron pairs at a bond angle of 180°, giving a **linear** geometry about that central atom. The total coordination number of three gives rise to bond angles of 120° between electron pairs and an **equilateral triangular** (or **trigonal planar**) arrangement of bonds.

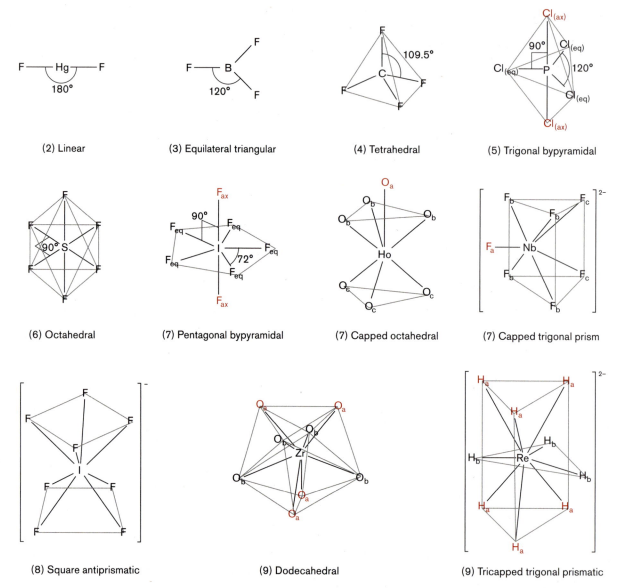

Figure 5.2. Idealized arrangements of m electron pairs that minimize repulsions among from two to nine electron pairs (m = 2–9) about a central atom in $AX_m E_0$. The examples shown are complete molecules or complex ions except for the capped octahedron (7), which is part of the structure of $[Ho(H_2O)(C_{15}H_{11}O_2)_3]$; the square antiprism (8); and the dodecahedron (8), which is part of the complex $[Zr(C_5H_7O_2)_4]$. Equivalent sets of outer atoms are indicated by the same subscripts: either a, b, or c, or (for pyramidal structures) "ax" for axial and "eq" for equatorial.

For a total coordination number of four or higher, three-dimensional structures give less repulsion than structures that confine all the atoms and bond electrons to a single plane. Thus, a square array of four electron pairs about a central atom separates the pairs by 90°, but a **tetrahedral** array gives a greater separation of 109.5°.

For a total coordination number of five, it is impossible for all of the electron pairs to be equivalently situated about the central atom, even if all of the outer atoms are identical. Instead, there is a three-layered **trigonal bipyramidal** arrangement of outer atoms or unshared electron pairs into two different types. Those above and below the central atom are said to be in **axial** positions, whereas those in the same plane as the

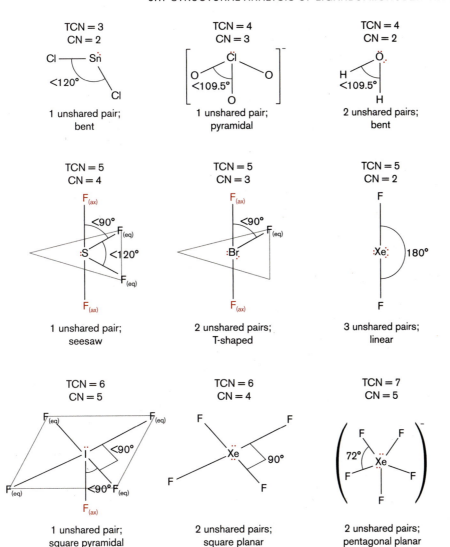

Figure 5.3. Bond angles and geometries described by the central and outer atoms in p-block molecules and ions AX_mE_n having n unshared electron pairs about the central atom. TCN = total coordination number $m + n$, and CN = coordination number m of central atom.

central atom are said to be in **equatorial** positions. The angles between equatorial electron pairs are 120°, whereas the angles between any of these pairs and any axial electron pair are 90°.

The preferred arrangement of six electron pairs involves 90° bond angles. This geometric shape, the **octahedron**, is highly symmetrical: No two pairs are any more axial than any other two pairs.

For a total coordination number of seven, there are three geometries of nearly equal efficiency in separating electron pairs—namely, **pentagonal bipyramidal**, **capped octahedral**, and **capped trigonal prismatic**. Except in the case of the pentagonal bipyramid, the bond angles are variable. A total coordination number of eight gives rise to two arrangements of minimal repulsion, the square antiprism and the dodecahedron. For total coordination number nine, a unique geometry, the **tricapped trigonal prism**, is predicted to minimize the electron-pair repulsions.

p-**Block Compounds AX_mE_n with Unshared Electron Pairs on the Central (Donor) Atom ($n > 0$).** The approximate bond angles are determined by the total coordination number $m + n$ (Figure 5.3). The choice of the term for the **molecular geometry** is based on the coordination number m, which excludes electron pairs on p-block central atoms. This results in some new names for geometries.

To obtain the geometric shape and the approximate bond angles of a ligand, we follow these four steps: (1) Draw a correct Lewis structure of the species AX_mE_n. (2) Determine the total coordination number $m + n$ about each central atom. (3) Determine the coordination number m of the central atom. (4) Choose the resulting geometry and identify the approximate bond angles using Figure 5.2 if $n = 0$ or Figure 5.3 if $n > 0$.

EXAMPLE 5.1

Predict the geometric shapes and the bond angles of (a) NO_2^-; (b) ClO_2^-; (c) CO_3^{2-}; (d) SO_3^{2-}; (e) PCl_2F_3; and (f) TeF_7^-.

SOLUTION: After drawing the Lewis structures (Step 1) and applying Steps 2 and 3, we obtain the total coordination number $m + n$ and coordination number m as follows: (a) For NO_2^-, $m + n = 3$ and $m = 2$. (b) For ClO_2^-, $m + n = 4$ and $m = 2$. (c) For CO_3^{2-}, $m + n = 3$ and $m = 3$. (d) For SO_3^{2-}, $m + n = 4$ and $m = 3$. (e) For PCl_2F_3, $m + n = 5$ and $m = 5$. (f) For TeF_7^-, $m + n = 7$ and $m = 7$. From these, the application of Step 4 and examination of Figure 5.2 or Figure 5.3 gives the following geometries and bond angles: (a) bent and ~120° for NO_2^-; (b) bent and ~109.5° for ClO_2^-; (c) equilateral triangular and ~120° for CO_3^{2-}; (d) pyramidal and ~109.5° for SO_3^{2-}; (e) trigonal bipyramidal and 90° and 120° for PCl_2F_3; (f) no defined bond angle and one of three possible geometries for TeF_7^-. (Of the three possible geometries, this ion happens to take the pentagonal bipyramidal geometry, with 90° and 72° bond angles.)

5.1B. Monodentate and Chelating Ligands. Monodentate Ligands. Monodentate ligands have only *one usable donor atom* with which they can bond to Lewis acids (such as metal ions). Figure 5.4 shows a number of monodentate ligands, with their atoms having unshared electron pairs (potential donor atoms) shown in **brown**.

The simplest monodentate ligands have only one donor atom with only one unshared pair of electrons; you should be able to spot five of these in Figure 5.4.

Other monodentate ligands (such as water and the halide and oxide ions) have more than one electron pair on the one donor atom. When they coordinate to just one metal ion, they are also monodentate. Many such ligands may also use the second electron pair to coordinate to a second Lewis acid, which (if this happens) leads them to be classified as bridging ligands (Section 5.3). For example, the various forms of SiO_2 shown in Figure 4.7 contain bridging oxide ions.

Some monodentate ligands have more than one donor atom, but are unable (for geometric reasons to be illustrated in Figure 5.7) to use them both to bond to the same metal atom. Examples include most of the homopolyatomic anions discussed in Section 3.1A, most simple oxo anions, and many simple organic ligands such as the acetate ion, $CH_3CO_2^-$ (Figure 5.4). Indeed, many organic functional groups contain more than one atom with unshared electron pairs, but cannot donate these electron pairs simultaneously to the same metal ion. Molecules with only one functional group nearly always can function only as monodentate ligands or bridging ligands.

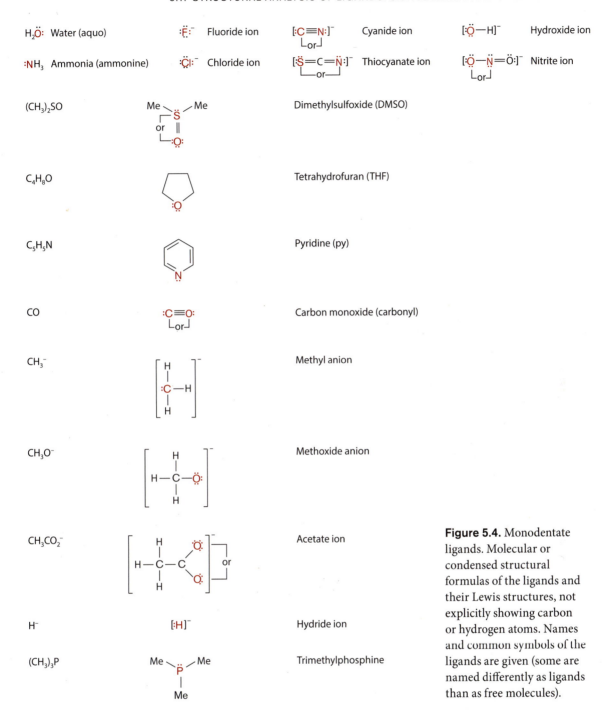

Figure 5.4. Monodentate ligands. Molecular or condensed structural formulas of the ligands and their Lewis structures, not explicitly showing carbon or hydrogen atoms. Names and common symbols of the ligands are given (some are named differently as ligands than as free molecules).

Recall from Section 3.2A a special class of monodentate (or other) ligands—namely, those in which the donor atom is carbon. The complexes formed by these ligands contain direct carbon-to-metal bonds and are organometallic compounds. The three biologically significant complexes in Figure 5.1 all contain metal ions and carbon, but only the third one (methylcobalamin) contains a metal–carbon bond and is therefore an organometallic compound. Historically, some hydrogen-free carbon-donor ligands such as CO, CN⁻, and the carbide ions (Section 3.1B) have been regarded as inorganic, so compounds containing these ligands bonded to metals are not listed as organometallic.

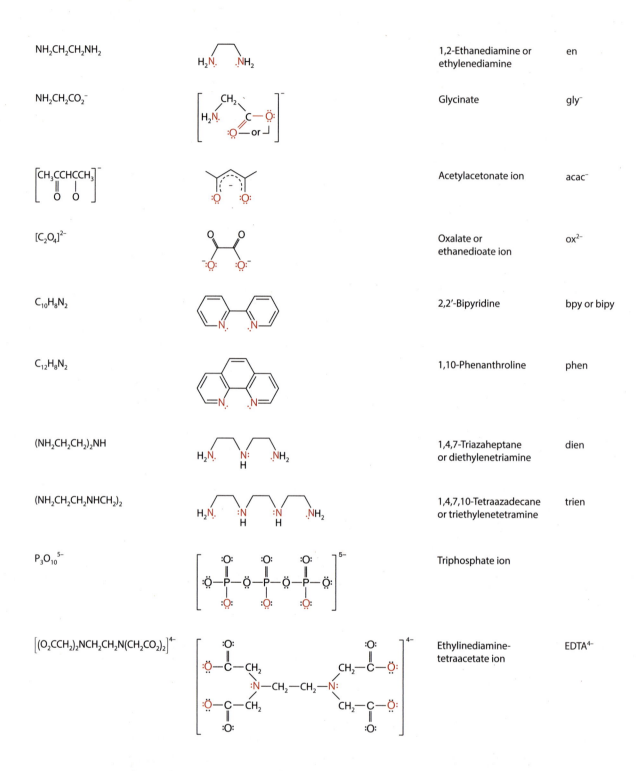

Figure 5.5. Some bidentate, tridentate, tetradentate, and hexadentate chelating ligands. Left column: molecular or condensed structural formulas of the ligands. Second column: Lewis structures, not explicitly showing carbon or hydrogen atoms. Third column: names of the ligands (common and IUPAC-approved). Fourth column: common symbols for the ligands.

EXAMPLE 5.2

As currently defined, which of the following can be monodentate ligands? Which, if attached to a metal ion, give organometallic compounds? (a) Methyl anion, CH_3^-; (b) methane, CH_4; (c) methoxide ion, CH_3-O^-.

SOLUTION: (a) Methyl anion (CH_3^-) has an unshared electron pair on carbon, so it is a monodentate ligand. On attaching to a metal ion, a carbon–metal bond results, and hence an organometallic compound (or ion) is formed.

(b) Methane (CH_4) has no unshared electron pairs, so it does not fit our current definition of a monodentate ligand.

(c) Methoxide ion (CH_3-O^-) has an unshared electron pair on oxygen, so it is a monodentate ligand. However, on attaching to a metal ion, an oxygen–metal bond is formed, not a carbon–metal bond, so the product is *not* an organometallic compound.

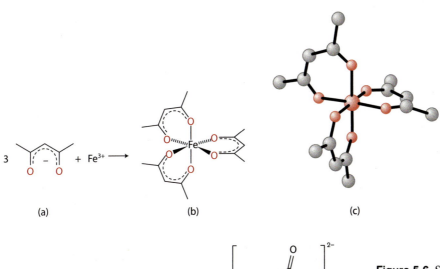

(a)   (b)   (c)

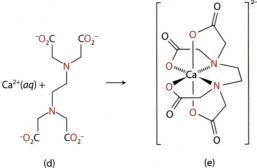

(d)   (e)

Figure 5.6. Six-coordinate octahedral complexes formed by two chelating ligands. (a) Bidentate acac⁻ reacts with Fe^{3+} to give the neutral complex $Fe(acac)_3$. The structure of $Fe(acac)_3$ is shown in (b) and is depicted as a ball-and-stick model in (c). (d) Hexadentate $EDTA^{4-}$ reacts with Ca^{2+} to give (e) the complex ion $[Ca(EDTA)]^{2-}$.

Chelating Ligands. Many nonlinear ligands with more than one donor atom have geometries such that they can donate electrons from two or more donor atoms (in two or more functional groups) to the same metal ion. Such ligands are known as **chelating ligands** (Figure 5.5). They form cyclic (ring) complexes (e.g., Figure 5.6) with bonds to the metal or central atom having reasonable (unstrained) bond angles (Figure 5.7c, d).

You may recall from organic chemistry that to have a planar ring involving *n* atoms, the bond angles must add up to (*n* – 2) straight angles—that is, to 180° in a

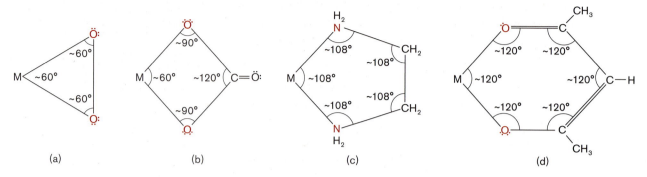

Figure 5.7. Geometric requirements for chelation by a ligand with all bond lengths being equal. (a) and (b) Bond angles at the metal are unfavorable if chelation is attempted with O_2^{2-} or simple oxo anions such as CO_3^{2-}, because some very acute (60°) bond angles result. (c) and (d) Bond angles are favorable upon chelation with ethylenediamine or the acetylacetonate anion, giving five- and six-membered rings with relatively favorable bond angles.

three-membered ring, 360° in a four-membered ring, 540° in a five-membered ring, and 720° in a six-membered ring. If the bond distances in the ring are all equal, the bond angles should be as illustrated in Figure 5.7. Figures 5.7a and 5.7b show that three- and four-membered rings lead to very strained ~60° donor atom–metal–donor atom bond angles. (Such chelation, although unfavorable, sometimes happens.) Figures 5.7c and 5.7d show the formation of more stable five- and six-membered rings, counting the metal ion. For this to happen, there should be two or three atoms intervening between the two donor atoms.

Chelating ligands are classified according to the number of donor atoms they possess that are suitably positioned to be simultaneously donated to the same metal ion. Ligands with two suitably positioned donor atoms (such as most of those shown in Figure 5.5) are **bidentate chelating ligands**; those with three (such as diethylenetriamine and triphosphate ion) are called **tridentate chelating ligands**, and so on. In general, these kinds of ligands are said to be **polydentate**. Note that the prefix (i.e., bi-, tri-, etc.) gives the **denticity** (number of "teeth") of the ligand.

Identification of the chelating potential of a ligand is more challenging when (as is usually the case) you are given only a condensed structural formula of the ligand, with no indication of the presence of unshared electron pairs. The following procedure will help identify chelating ligands.

1. Expand the condensed structural formula into a full structural formula that shows all bonds in the organic species. Add unshared electron pairs to complete the octets of the non-hydrogen atoms.

2. Identify the potential donor atoms (Sections 3.1 and 3.2) and functional groups.

3. To function as a chelating ligand, the donor atoms must be adequately far apart, in separate functional groups. If the compound is polyfunctional, pick one donor atom in one functional group, and label it "No. 1." If one of the donor atoms is clearly at the center of the whole molecule, it is advantageous to start the numbering from this donor atom [e.g., the N atom of $N(CH_2CH_2OH)_3$]. Proceed along the backbone of the organic molecule

toward the next functional group, numbering upward until another functional group is reached. If the number of this donor atom is No. 4 or 5, the two donor atoms are far enough apart to chelate.

4. Donor atoms 1 and 4 or 5 must also be free to form a closed ring along with the metal ion. Circumstances that can prevent this include the two donor atoms being tied back in a small (under ~12-atom) ring compound, a bond angle of 180° rather than the usual 109.5° or 120° occurring along the backbone, or the functional groups being in the meta or para positions of an aromatic ring.

5. If these geometric conditions allow chelation, the ligand is a bidentate chelating ligand.

6. To determine whether the ligand could be tridentate, repeat Steps 3 and 4, starting from either of the identified donor atoms toward some other functional group in the molecule. Repeat until all functional groups have been checked.

These steps are summarized in Figure 5.8.

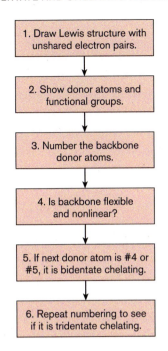

Figure 5.8. Flow chart for identifying chelating ligands.

EXAMPLE 5.3

Which of the following are potential chelating ligands: (a) malonate ion, $^-OC(=O)CH_2C(=O)O^-$; (b) 2,5,8-nonanetrione, $CH_3C(=O)CH_2CH_2C(=O)CH_2CH_2C(=O)CH_3$?

SOLUTION: First, draw the full structural formulas of these species, completing all octets. Second, the atoms with unshared electron pairs are potential donor atoms (drawn in **brown** in the full structural formulas shown in Figure 5.9a). Ion A contains two carboxylate functional groups and molecule B contains three carbonyl (ketone) functional groups (shaded boxes of Figure 5.9b).

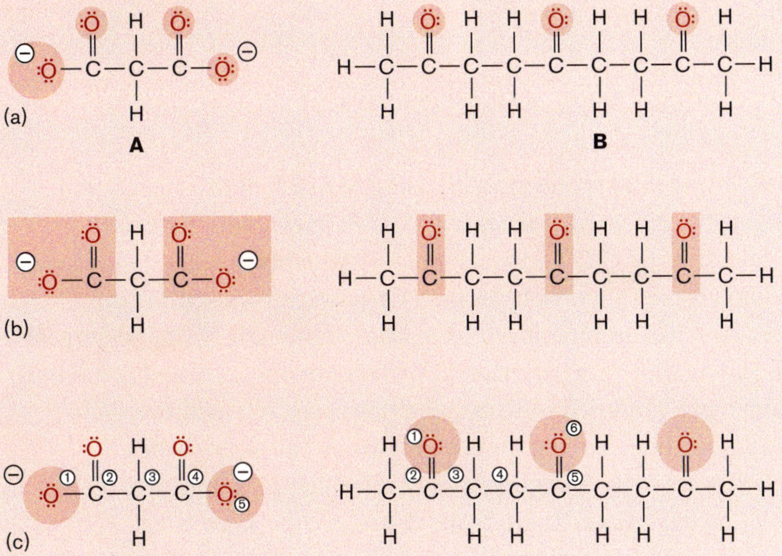

Figure 5.9. Analyses of two organic structures for Example 5.3.

EXAMPLE 5.3 *(cont.)*

Third, beginning at the leftmost donor atom of each species, the backbone atoms are numbered as shown in Figure 5.9c until a donor atom in the next functional group is reached. In ion A, either O in the second carboxylate functional group could be No. 5, but in molecule B the O in the second carbonyl group is No. 6.

Fourth, no geometric factor that could prevent chelation is present: In neither case are the functional groups attached to an aromatic ring, nor are the two donor atoms already part of a small ring. Applying VSEPR theory, we find that no backbone carbon atoms have 180° bond angles (all are 109° or 120°).

Fifth, we conclude that ion A has the proper spacing and bond angles to be a potential bidentate chelating ligand. Because there are no further functional groups, that is its final classification. Molecule B has its donor atoms spaced too far apart, so it fails this test. Sixth, because there is an additional carbonyl functional group in B, the third and fourth steps should be repeated, by renumbering the oxygen in the middle functional group as atom No. 1, and counting out to the oxygen in the right functional group, which again is atom No. 6. Hence, 2,5,8-nonanetrione (molecule B) is *not* a chelating ligand.

A CONNECTION TO ENVIRONMENTAL CHEMISTRY

Inorganic Polyphosphate Chelating Ligands

In Section 3.6B we introduced the *diphosphate* or *pyrophosphate* ion, $(O_3P-O-PO_3)^{4-}$, and the *triphosphate* ion, $(O-PO_2-O-PO_2-O-PO_2-O)^{5-}$. The $P_3O_{10}^{5-}$ ion (Figure 5.5) has been of great importance in phosphate detergents. This chelating anion complexes the +2 ions (especially Ca^{2+} and Mg^{2+}) found in "hard" water, preventing them from precipitating the insoluble salts ("bathtub scum") of long-chain carboxylate and alkylsulfonate ions responsible for cleaning:

$$Ca^{2+} + 2C_{17}H_{35}COO^- \rightleftharpoons Ca(OOCC_{17}H_{35})_2(s)$$

The environmental problems with using so much of this material in detergents stem from its hydrolysis to give the simple phosphate ions (e.g., the reverse of Eq. 3.17), which are plant nutrients in lakes. Many lakes that have high inputs of phosphates have been overgrown with algae. Consequently, it has been removed from home automatic washer detergents sold in the USA. Unfortunately, the $P_3O_{10}^{5-}$ ion not only chelated metal ions, it maintained pH, inhibited washing machine corrosion, and suspended insoluble dirt, without leaving spots and films on the clean dishes. Manufacturers are now struggling to find more complex mixtures of other chelating ligands and ingredients to achieve the same effect.

A N   A M P L I F I C A T I O N

Designating the Number of Donor Atoms in Use by a Ligand

Ligands such as O_2 or O_2^{2-} sometimes act as monodentate ligands as expected, but they also can bond to metals using electron pairs on both oxygen atoms in bidentate fashion to form three-membered rings (Figure 5.7a). To clarify whether these ligands are attached to the metal ion via one or more than one donor atoms, we can insert a *hapto* number, h^1 or η^1, as a prefix to indicate that the following ligand is monodentate, or h^2 or η^2 to indicate that the following ligand is using two donor atoms. Thus, η^1-O_2 in a formula tells us that oxygen is acting as a monodentate ligand in the complex, while η^2-O_2 tells us that two donor atoms are attached, so that the ligand is acting as a bidentate chelating ligand. We do this in some of the upcoming Examples and Exercises to help in analyzing the structures of some complexes.

5.1C. The Chelate Effect. Three generalizations can be used to predict which ligands will be complexed with which metal ions:

1. Chelate ligands form more stable complexes than analogous monodentate ligands. Consequently, ligands that can chelate will chelate.

2. Chelates of higher denticity are expected to give more stable complexes than chelates of lower denticity.

3. Chelate and other ligands are complexed more strongly by ions of higher charge.

To see the **chelate effect**, we need to set up a competition (for a metal ion) between 1 mol of an *n*-dentate chelating ligand and *n* mol of a very similar monodentate ligand. For example, if we mix 1 mol Ni^{2+} ion with 3 mol ethylenediamine and 6 mol ammonia, the equilibrium will favor the formation of the complex with ethylenediamine:

$$[NiNH_3)_6]^{2+} + 3NH_2CH_2CH_2NH_2 \rightleftharpoons$$
$$[Ni(NH_2CH_2CH_2NH_2)_3]^{2+} + 6NH_3 \quad K_{eq} = 10^{7.35} \qquad (5.1)$$

The tendency for polydentate ligands to chelate is so prevalent that if we see a line formula for a complex written as $[Ni(NH_2CH_2CH_2NH_2)]$, and the two potential donor atoms can attach to the nickel, then we conclude that they will. We classify this as a bidentate complex and draw its structure with two Ni–N bonds.

As an illustration of the effect of higher denticity, we note the much larger equilibrium constant for complexation with 1 mol of the hexadentate amine ligand:

$$[Ni(NH_3)_6]^{2+} + NH_2(CH_2CH_2NH)_4CH_2CH_2NH_2 \rightleftharpoons$$
$$[Ni\{NH_2(CH_2CH_2NH)_4CH_2CH_2NH_2\}]^{2+} + 6NH_3 \quad K_{eq} = 10^{19.1} \qquad (5.2)$$

The analytical reagent, medicinal agent, and food preservative $EDTA^{4-}$ (ethylene-diaminetetraacetate ion; Figures 5.5 and 5.6) is a very powerful hexadentate chelating ligand indeed, and chelates metal ion impurities in foods that would otherwise catalyze air oxidation (spoiling) of the food.

$$[M(OH_2)_6]^{z+} + EDTA^{4-} \rightleftharpoons [M(EDTA)]^{-4+z} + 6H_2O \qquad (5.3)$$
$$K_{eq} = 10^{2.79} \text{ for } Li^+; 10^{8.64} \text{ for } Mg^{2+}; 10^{25.1} \text{ for } Fe^{3+}$$

In Equation 5.3, all three metal ions are of similar size, but increasing their charge strongly drives the reaction further to the right.

EXAMPLE 5.4

Predict whether reactants or products are favored in the following equilibria:

(a) $[Ni\{NH(CH_3)_2\}_6]^{2+} + 2$ [dimethylamino-ethyl triamine ligand structure] $\rightleftharpoons$

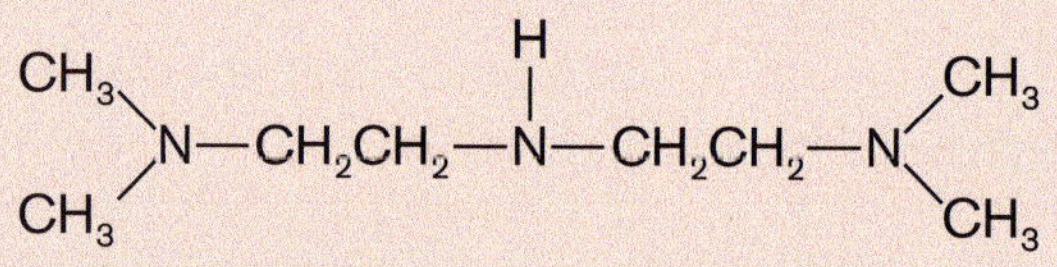

$$Ni\{[(CH_3)_2NCH_2CH_2]_2NH\}_2^{2+} + 6NH(CH_3)_2$$

(b) $[Tl(SCH_2CH_2SCH_2CH_2S)]^+ + 3CH_3S^- \rightleftharpoons$
$$[Tl(SCH_3)_3] + (SCH_2CH_2SCH_2CH_2S)^{2-}$$

(c) $[Ni(monodentate\text{-}NH_2CH_2CH_2NH_2)_3(NH_3)_3]^{2+} \rightleftharpoons$
$$3NH_3 + [Ni(bidentate\text{-}NH_2CH_2CH_2NH_2)_3]^{2+}$$

SOLUTION: First, note the controlled situations: Both products and reactant complexes contain the same number of Ni–N or Tl–S bonds.

(a) The ligand $NH(CH_3)_2$, dimethylamine, has only the one (N) donor atom, while the pictured ligand has three N donor atoms. If the center N donor atom is No. 1, then the left and the right donor atoms are No. 4. Hence, the pictured ligand is a tridentate chelating ligand, and it forms a more stable complex than the monodentate ligand. The more stable complex ion is on the right side, among the *products*.

(b) The ligand CH_3S^- has only one (S) donor atom, so it is monodentate. The ligand $(SCH_2CH_2SCH_2CH_2S)^{2-}$ has three (S) donor atoms. If the center S atom is No. 1, then each of the outer S atoms is a No. 4. This is a tridentate chelating ligand, and it forms the more stable complex with Tl^{3+}. This complex is among the *reactants*.

(c) In the reactant complex, the 1,2-ethanediamine is attached by only one N donor atoms, so it is monodentate, while in the product complex the 1,2-ethanediamine is bidentate. *Products* are favored.

AN AMPLIFICATION

Thermodynamics of the Chelate Effect

The free energy change driving Equation 5.1 to the right is -53 kJ mol^{-1}, which breaks down into a very small enthalpy change of -17 kJ mol^{-1} and a larger entropy change term ($-T\Delta S = -36$ kJ mol^{-1}). The enthalpy change is small because each complex ion in Equation 5.1 involves six very similar Ni–N coordinate covalent bonds. The entropy effect is responsible for the shift of this and similar equilibria in favor of the complex with the chelated ligand attached to the metal ion. The major reason for the favorable entropy change is that there are 4 mol of solute particles on the left side of Equation 5.1 and 7 mol on the right; the increasing degrees of freedom (disorder) among the products favor complex formation by the chelate ligand. The imbalance in moles is even larger in Equations 5.2 and 5.3. [1,2]

A CONNECTION TO BIOCHEMISTRY

Silicon Bioavailability via Chelation

As indicated in Table 1.1, silicon is essential for life for aquatic species such as diatoms, radiolaria, horsetail rushes, and some sponges. The availability of silicon is a problem, however. Si^{4+} is very strongly acidic and is never found in water; moreover, SiO_2 and its oxo acid, H_4SiO_4, are of very low solubility in water. Chelation by bidentate chelating ligands such as *ortho*-$[C_6H_4O_2]^{2-}$ to give five- or six-coordinate organosilicate complexes such as $[Si(O_2C_6H_4)_3]^{2-}$ can occur in strongly basic solutions, but the pH values required are too high to be relevant to biological processes. However, certain sugars (which are bidentate chelating ligands) can form similar complexes in aqueous solutions at relevant pH values. Such complexes are formed by diatoms. [3]

5.2. Coordination by Macrocyclic Ligands

OVERVIEW. Chelating ligands that are already large ring compounds even before attaching to a metal ion are potential macrocyclic ligands (Section 5.2A). You may practice identifying macrocyclic ligands by trying Exercises 5.15–5.19. Macrocyclic ligands form more stable complexes than analogous chelating or monodentate ligands, and macrobicyclic (crypt) ligands form more stable complexes than analogous macrocyclic ligands (Section 5.2B). You may practice applying these concepts to predicting the positions of equilibria by trying Exercises 5.20–5.22.

Connections to organic chemistry, biochemistry, and medicinal chemistry appear in this section. Medicinal macrocyclic ligands appear in Section 5.8. An important macrocyclic ligand, heme, will appear again in Section 7.8.

5.2A. Macrocyclic Ligands. A special class of chelating ligand includes those that are large (~15-membered to ~21-membered) ring compounds even without a metal atom present and that can position several donor atoms *inside* their ring to donate to one central metal ion. These kinds of ligands are said to be **macrocyclic**. Chlorophyll, heme, and vitamin B_{12} all contain tetradentate macrocyclic ligands (Figure 5.1). The organic ligands of chlorophyll and heme contain a 16-membered *porphyrin* ring that includes four nitrogen donor atoms, whereas vitamin B_{12} contains a 15-membered *corrin* ring that also includes four nitrogen atoms. Because of the conjugated double bonds in these macrocyclic ligands, the ligands are quite rigidly planar—the nitrogen–metal bond distance in a planar porphyrin complex is constrained to fall in a narrow range between 193 and 210 pm.

The crown ethers (Figure 5.10) are also commonly used macrocyclic ligands that are more flexible. The crown ether shown on the left in Figure 5.10, *cyclo*-$(CH_2CH_2O)_6$, is commonly called 18-crown-6, because it has an 18-membered ring with six oxygen atoms. Related important crown ethers include 21-crown-7 [*cyclo*-$(CH_2CH_2O)_7$] and 15-crown-5 [*cyclo*-$(CH_2CH_2O)_5$]. The 1987 Nobel Prize in Chemistry recognized the importance of the discovery of these compounds and their usefulness.

In the conformation shown in Figure 5.10, 18-crown-6 has a cavity of a size that is governed by the van der Waals radius of the oxygen atoms, 152 pm (Table 1.12). The six oxygen donor atoms touch each other at this radius, forming a hexagon with sides of 300 pm. By trigonometry, the optimal metal–oxygen distance is calculated to be 300 pm, which results (after subtracting the van der Waals radius of oxygen) in an optimal metal-ion radius of 148 pm. Consistent with this calculation, 18-crown-6 forms a more stable complex with K^+ (radius = 152 pm) than with other Group 1 ions that are smaller or larger than K^+. In contrast, the smaller 15-crown-5 forms its most stable Group 1 complexes of almost equal stability with Na^+ and K^+.

Cryptands differ from crown ethers in being large three-dimensional *bicyclic* (or polycyclic) ligands. In order to achieve three dimensions, the cryptand ligands normally replace (at least) two of the ether oxygen atoms with tertiary amine nitrogen

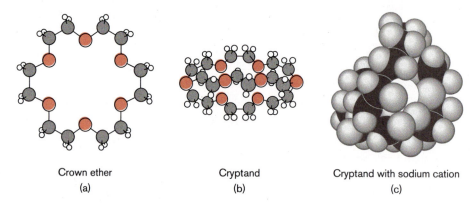

Crown ether	Cryptand	Cryptand with sodium cation
(a)	(b)	(c)

Figure 5.10. Molecular models of a crown ether, a "cryptand" or crypt polyether *bicyclo*-$N(CH_2CH_2OCH_2CH_2OCH_2CH_2)_3N$ (which also contains two nitrogen atoms), and a Group 1 cation (in white) encapsulated by a crypt polyether ligand. [Adapted from J. L. Dye, *Sci. Amer.* 237(1), 92 (1977).]

atoms. Thus, the cryptand in Figure 5.10b has nitrogen atoms at the left and right sides, so its structural formula can be written as *bicyclo*-N(CH$_2$CH$_2$OCH$_2$CH$_2$OCH$_2$CH$_2$)$_3$N. These encapsulate a metal ion in three dimensions, so the metal ion is "buried" inside (Figure 5.10c), and outside reactants cannot reach it. As an example, the very reactive Ge^{2+} cation has been stabilized inside this cryptand.[4]

Note that crown ethers and cryptands form stable complexes with nonacidic Group 1 ions. Hence, the term "nonacidic" for the Group 1 cation is relative to the complexing ability of the ligand: Group 1 ions can function as Lewis acids under very favorable circumstances. These complex cations have nonpolar CH$_2$ groups on the outside, so they impart solubility to their salts in nonpolar solvents. Therefore crown-ether- and cryptand-complexed cations are useful as phase-transfer catalysts (Section 4.4B).

A CONNECTION TO BIOCHEMISTRY

Antibiotics and the Question of Nerve Impulse Transmission

The trigger, control, and nerve transmission functions of the Na$^+$ and K$^+$ ions involve their flow across nonpolar cell membranes. At equilibrium, the concentrations of these ions are unequal within the cell and outside the cell (Table 5.1). Such unequal concentrations generate an electrochemical potential according to the Nernst equation (Eq. 5.4), in which R is the gas constant, z is the charge on the ion, and F is the Faraday:

$$E = 2.303 \left(\frac{RT}{zF} \right) \log \left[\frac{(\text{Extracellular concentration})}{(\text{Intracellular concentration})} \right] \tag{5.4}$$

TABLE 5.1. Free Ionic Concentrations and Equilibrium Potentials for Mammalian Skeletal Muscle

Ion	Extracellular Concentration (mM)	Intracellular Concentration (mM)	Equilibrium Potential (mV)
Na$^+$	145	12	+68
K$^+$	4	155	−99
Ca^{2+}	1.5	<10^{-4}	>+128
Cl$^-$	123	4.2	−90

Source: Adapted from S. J. Lippard and J. M. Berg, *Principles of Bioinorganic Chemistry*, University Science Books: Mill Valley, CA, 1994, p. 154.

A flow of charged particles constitutes an **electric current**. The flow must be rapid for nerve transmission, and it must be selective for the passage of the proper ion or ions. This requires both an active pumping mechanism and some sophisticated ion selectivity to distinguish the two ions involved in nerve transmission, Na$^+$ and K$^+$.

In early work on the mechanism of this selectivity and transport, two models were developed—namely, the **pore/channel model** (channels are proteins that form pores in the cell membrane) and the **carrier model** (in which a lipid-soluble complex is formed that diffuses through the cell membrane). Both models were based on the mechanisms of action of certain antibiotics that kill cells by causing catastrophic leakage of ions across the cell membranes.

The carrier model was based on the structures and behavior of antibiotics such as valinomycin and nonactin (Figure 5.11), which resemble crown ethers in structure and ion selectivity (nonactin selects for K^+ over Na^+ and Cs^+).

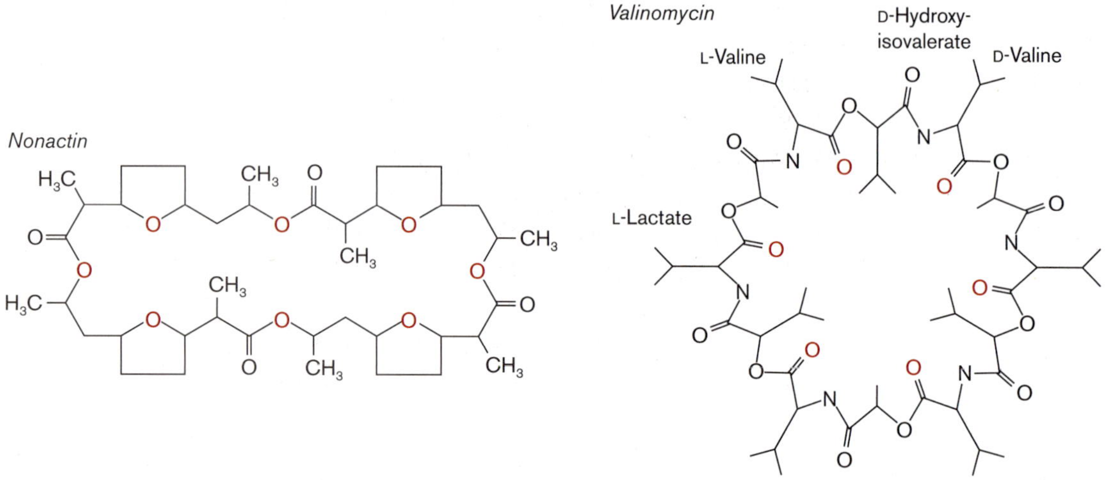

Valinomycin

D-Hydroxy-isovalerate

L-Valine   D-Valine

L-Lactate

Nonactin

Figure 5.11. Nonactin (left) and valinomycin (right) are macrocyclic antibiotics. [Adapted from M. N. Hughes, *The Inorganic Chemistry of Biological Processes*, 2nd ed., John Wiley and Sons Ltd.: Chichester, UK, 1981, p. 267. Copyright © 1981 by John Wiley and Sons Ltd.]

Figure 5.12. A model of the structure of gramicidin A, showing the pore at the center.

The pore/channel model was based on the behavior of the antibiotic gramicidin A (Figure 5.12), which inserts in the cell wall and has a 400-pm diameter channel through which hydrated ions may pass. The relative rate of passage of ions through this pore is $Cs^+ > Rb^+ > K^+ > Na^+ > Li^+$. This makes sense if the hydrated cesium ion is indeed the form in which the ion is carried across the membrane, because Cs^+ is the smallest hydrated ion in this series (Table 4.7). As mentioned in Section 4.5B, this is the model that has prevailed. There are different types of channels, such as the voltage-gated sodium channel, which is deduced to have a narrower pore that allows $Na(H_2O)_1^+$ to pass 11 times more easily than the larger unhydrated K^+ ion.[5] The malfunctions of the channels for the passage of Na^+ and K^+ are involved in the disease epilepsy, and the absence of the channel for Cl^- is implicated in cystic fibrosis.[6]

A CONNECTION TO ORGANIC CHEMISTRY

The Template Effect

If two molecules that are to react with each other can each act as a ligand and be coordinated next to each other on the same metal ion, they are more likely to collide and react with each other, and the rate of reaction is greatly enhanced. This is the template effect that is often an important factor in biochemistry (see the Connection entitled "Preview of the Biological Functions of Metal Ions" in Section 1.2B).

One example of the template effect is the syntheses of 18-crown-6 from the reaction of chelated η^4-$HO(CH_2CH_2O)_2CH_2CH_2OH$ and η^2-$Br(CH_2CH_2O)_2CH_2CH_2Br$ with 2 mol of MOR. If MOR is $K^+{}^-OC_4H_9$, the two reactants presumably first coordinate to the K^+, which brings their ends close together and allows the formation of the product, 18-crown-6 coordinated to K^+, in good yield (Figure 5.13):

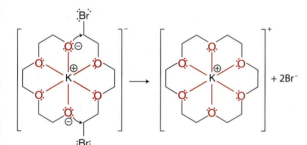

Figure 5.13. Template effect in the synthesis of the $[K(18\text{-crown-}6)]^+$ complex ion from two already coordinated chelate ligands.

However, if MOR is $(C_4H_9)_4N^+OH^-$, the cation cannot complex the reactants, the reaction is slow, and the main product is a long polymer $(CH_2CH_2O)_x$. In this case it is improbable that the ends can get together and form the large ring present in 18-crown-6.

Another example of the chelate effect is the reaction of 4 mol of *ortho*-$C_6H_4(C\equiv N)_2$ at 300°C with metals such as copper to generate very stable, inert complexes of a macrocyclic dianion (Figure 5.14). These **phthalocyanines**[7] have rigid ring systems and quite inflexible hole sizes.

Figure 5.14. Structure of a phthalocyanine. M can be most of the metals in the periodic table.

EXAMPLE 5.5

Determine which of ligands **A–E** are (a) ligands that can chelate, (b) macrocyclic ligands, and (c) tridentate ligands.

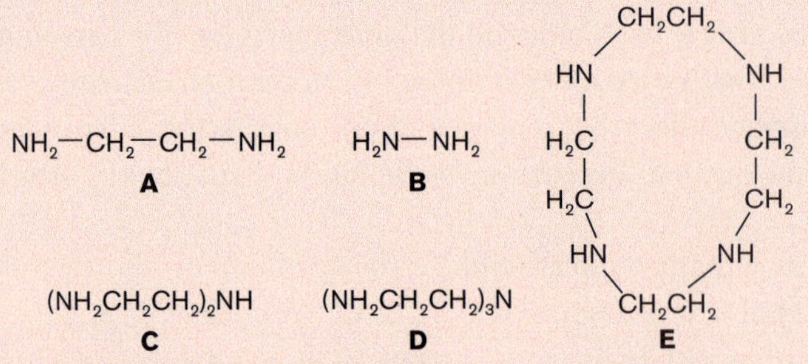

$NH_2-CH_2-CH_2-NH_2$
A

H_2N-NH_2
B

$(NH_2CH_2CH_2)_2NH$
C

$(NH_2CH_2CH_2)_3N$
D

E

SOLUTION: (a) We use the procedure outlined in Example 5.3 to identify chelate ligands. When the full structural formulas of **A–E** are drawn (showing octets), each of the nitrogen atoms in these molecules has an unshared electron pair and is therefore a potential donor atom. All functional groups are simple amine functional groups. None of the ligands is linear or inflexible. Designating an end or central nitrogen atom on each molecule as No.1, the other nitrogen atoms are numbered as follows:

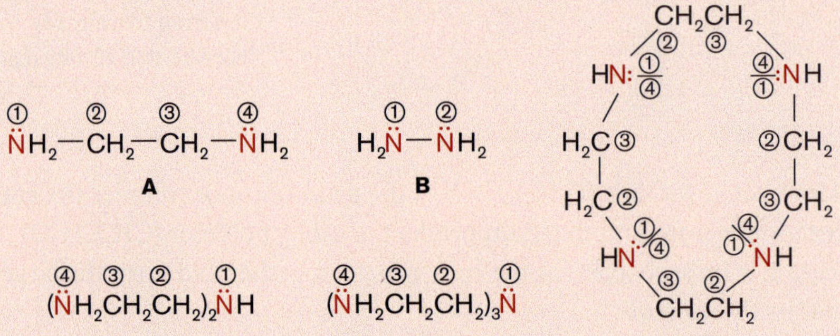

Hence, the chelating ligands are **A, C, D,** and **E**.

(b) To be a macrocyclic ligand, the chelate ligand must already be a large ring compound before it is attached to a metal atom. Only **E** qualifies.

(c) In **E**, numbering to the next N donor atom makes it No. 4. If we then renumber this donor atom as No. 1 and then number from this second donor N atom to the next N, the next N is No. 4. This means that **E** is multidentate, together with **C** and **D**. Tridentate ligands have three and only three usable donor atoms, so only **C** is tridentate. Ligands **D** and **E** have four donor atoms, so they are tetradentate.

5.2B. Macrocyclic Effect. A competition in equilibrium between either monodentate or noncyclic chelating ligands and a macrocyclic (chelating) ligand having the same number and type of donor atoms favors complex formation by the macrocyclic ligand. This is known as the **macrocyclic effect.**

$$K[\eta^6\text{-}CH_3O(CH_2CH_2O)_5CH_3]^+ + cyclo\text{-}(CH_2CH_2O)_6 \rightleftharpoons$$
$$K[\eta^6\text{-}cyclo\text{-}(CH_2CH_2O)_6]^+ + CH_3O(CH_2CH_2O)_5CH_3 \quad K_{eq} = 10^4 \quad (5.5)$$

$$K_{eq} = 10^{5.2} \quad (5.6)$$

Macrobicyclic cryptand ligands form more stable complexes more selectively even than the crown ethers. The macrobicyclic cryptand shown in Figure 5.10b displaces the corresponding crown ether:

$$Ba[\eta^6\text{-}cyclo\text{-}HN(CH_2CH_2OCH_2CH_2OCH_2CH_2)_2NH]^{2+}$$
$$+ bicyclo\text{-}N(CH_2CH_2OCH_2CH_2OCH_2CH_2)_3N \rightleftharpoons$$
$$Ba[\eta^6\text{-}bicyclo\text{-}N(CH_2CH_2OCH_2CH_2OCH_2CH_2)_3N]^{2+}$$
$$+ cyclo\text{-}HN(CH_2CH_2OCH_2CH_2OCH_2CH_2)_2NH \quad K_{eq} = 10^1 \quad (5.7)$$

This cryptand removes radioactive strontium and radium ions, as well as lead ions, from the body without disturbing calcium ions. It selects Cd^{2+} over Ca^{2+} and Zn^{2+} by a factor of one million and thus can be used for the selective removal of toxic cadmium ion (Section 5.8B).

EXAMPLE 5.6

Predict whether reactants (left side) or products (right side) will be favored in each of the following equilibria:

(a) $[Li(21\text{-crown-7})]^+ + [Rb(15\text{-crown-5})]^+ \rightleftharpoons [Rb(21\text{-crown-7})]^+ + [Li(15\text{-crown-5})]^+$

(b) $[K(cyclo\text{-}OC_4H_8)_6]^+ + 18\text{-crown-6} \rightleftharpoons [K(18\text{-crown-6})]^+ + 6 cyclo\text{-}OC_4H_8$

(c) $[K(18\text{-crown-6})]^+ + bicyclo\text{-}N(CH_2CH_2OCH_2CH_2OCH_2CH_2)_3N \rightleftharpoons$
$K[bicyclo\text{-}N(CH_2CH_2OCH_2CH_2OCH_2CH_2)_3N]^+ + 18\text{-crown-6}$

SOLUTION: (a) Products (right side) are favored, because the smaller Li^+ is a better fit for the smaller 15-crown-5 while the larger Rb^+ is a better fit for the larger 21-crown-7.

(b) $Cyclo\text{-}OC_4H_8$ is a monodentate ligand (commonly known as tetrahydrofuran or THF; Figure 5.4), so it cannot compete in complexation with the macrocyclic 18-crown-6. Products (right side) are favored.

(c) Products are favored. $Bicyclo\text{-}N(CH_2CH_2OCH_2CH_2OCH_2CH_2)_3N$ is the "cryptand" ligand shown in Figure 5.10b, and it forms even more stable complexes than the macrocyclic 18-crown-6.

5.3. Bridging Ligands and Metal–Organic Framework Compounds

OVERVIEW. When molecules or ions have more than one unshared pair of electrons, but are unable to donate these to the same metal ion, they often donate them to *different* metal ions, thus acting as bridging ligands (Section 5.3A). You may practice identifying bridging ligands by trying Exercises 5.23– 5.26. When linear bridging ligands are combined with octahedrally coordinating metal ions, Prussian blues and metal–organic framework (MOF) compounds may result, which have low densities and pore volumes that in some cases are high enough that important gases such as H_2 and CH_4 may be stored in them (Section 5.3B). You may practice evaluating pore diameters, pore volumes, and densities by trying Exercises 5.27–5.33.

Connections to environmental chemistry are included. Some of the concepts involved with metal–organic framework compounds will appear again in a more completely inorganic context in Section 8.7C on zeolites.

What will the fuels of the future be? Will automobiles run on hydrogen—a fuel whose combustion produces only water (not normally considered to be an environmental pollutant)? Utilizing any new source of energy requires solving a host of problems. In this section we will consider one large problem for either hydrogen or methane (CH_4) as a fuel—they are gases under standard conditions. To store them at room temperature they are normally compressed in very heavy steel cylinders that would add prohibitively to the weight of an automobile, lowering the fuel efficiency tremendously. Have you ever had to change a gas cylinder on an instrument? If so, you know what we are talking about. Hopefully you have never had the valve on the cylinder break off, and the cylinder take off like a rocket, breaking through concrete walls! In this section we will consider how bridging ligands might help us construct lighter containers for fuels such as these, and more.

5.3A. Bridging Ligands; Prussian Blues. Most of the monodentate ligands shown in Figure 5.4 have more than one electron pair on their donor atom or atoms. In several of these ligands, the second (and third) electron pair may be donated to a second (and even a third) metal ion. A ligand that donates more than one electron pair to more than one metal ion or other Lewis acid is known as a **bridging ligand**. Among the most important ligands in Figure 5.4 that can act as bridging ligands are the oxide, halide, cyanide, and hydroxide ions. (Sulfide is another important bridging ligand, but water itself is not, except in hydrogen-bonding situations.)

Bridging oxide and halide ions are essential components of the crystal lattices of metal oxides and halides (Section 4.1). In Section 4.1C we discussed the effects of electronegativity difference on details of the bridging interaction. If the metal–anion difference is large (as in AlF_3 with the ReO_3 structure; Figure 4.3), the metal–anion–metal bond angle is large (in this case 180°), which separates the mutually repelling cations. If the electronegativity difference is small (as in AuCl or CdX_2; Figure 4.6) the metal–anion–metal bond angles are smaller, which allows good covalent overlap of orbitals.

Organic dicarboxylates form an important class of bridging ligands. A commonly used example of such a bridging ligand is the dianion of 1,4-benzenedioic acid

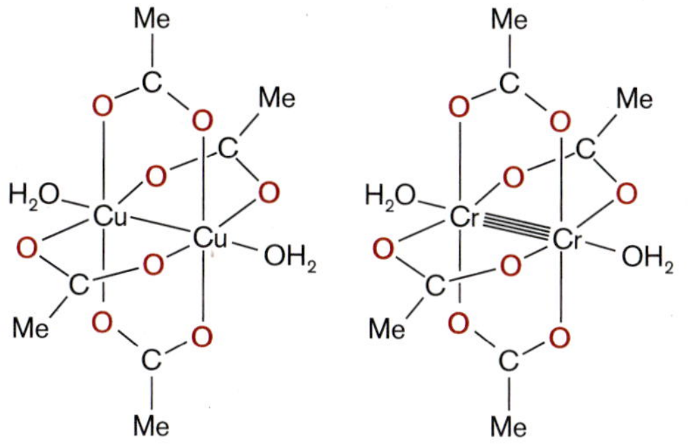

Figure 5.15. Structures of two *d*-block acetate-bridged dimetal(II) tetraacetates. The copper compound has a metal–metal single bond, whereas the chromium compound has a metal–metal quadruple bond.

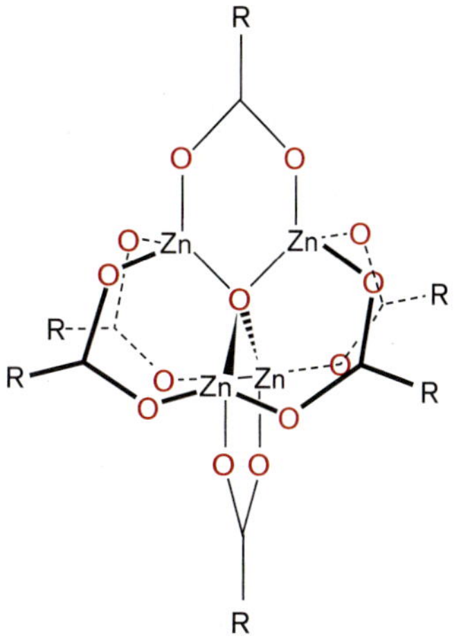

Figure 5.16. A Zn_4O^{6+} metal-ion cluster bridged by six carboxylate bridging ligands. When R = CH_3, the compound is basic zinc acetate.[8]

(terephthalic acid), $(OOC–C_6H_4–COO)^{2-}$. Organic carboxylates can form roughly linear bridges if the bond angle at the bridging oxygen is altered somewhat. Carboxylate groups can also bridge *two* metal ions, which are then linked directly or indirectly to form five- or six-membered chelate rings.

Many *d*-block dimetal tetraacetates such as copper(II) acetate, chromium(II) acetate, and so on (Figure 5.15) contain five-membered rings in which the metal ions bond to each other not only by σ bonds, but also may form π and δ bonds, giving double, triple, quadruple, or even quintuple bonds (Section 3.3B). This bridging mode stabilizes these multiple bonds, forming neutral, fairly insoluble complexes (Figure 5.15):

$$2Cr^{2+}(aq) + 4CH_3COO^-(aq) \rightleftharpoons Cr_2(\eta^2\text{-}OOCCH_3)_4(s) \qquad (5.8)$$

Figure 5.16 illustrates another bridging mode, in which a six-membered ring is formed by two metal ions and a multiply bridging oxide ion. The resulting Zn_4O^{6+} polyatomic ion called a **metal-ion cluster**.

Many important bridging ligands are linear, with donor atoms at the two ends of the ligand. Among the most important of these are some of the **pseudohalide ions** (Figure 5.17). (Pseudohalides got their name because they resemble the halide ions in their relatively weak basicity and in their oxidation–reduction chemistry.)

One of the most important monodentate/bridging ligands is the **cyanide ion**, $(:C\equiv N:)^-$. This ion is often monodentate, as in the nonbasic cyano anions $[Co(CN)_6]^{3-}$ and $[Fe(CN)_6]^{3-}$ [ferricyanide or hexacyanoferrate(III) ion] and the weakly basic $[Fe(CN)_6]^{4-}$ ion ($pK_{b1} = 9.5$).[9] The $[Fe(CN)_6]^{4-}$ [ferrocyanide or hexacyanoferrate(II)]

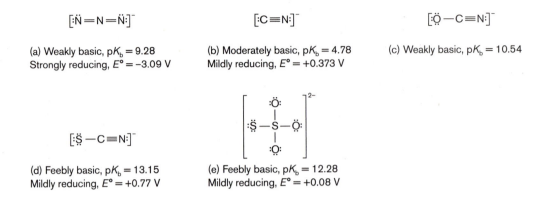

(a) Weakly basic, $pK_b = 9.28$
Strongly reducing, $E° = -3.09$ V

(b) Moderately basic, $pK_b = 4.78$
Mildly reducing, $E° = +0.373$ V

(c) Weakly basic, $pK_b = 10.54$

(d) Feebly basic, $pK_b = 13.15$
Mildly reducing, $E° = +0.77$ V

(e) Feebly basic, $pK_b = 12.28$
Mildly reducing, $E° = +0.08$ V

Figure 5.17. Lewis dot structures and basic properties of some pseudohalide ions: (a) azide ion (N_3^-); (b) cyanide ion (CN^-); (c) cyanate ion (CNO^-); (d) thiocyanate ion (CNS^-); (e) thiosulfate ion ($S_2O_3^{2-}$). [The pK_b values are taken from R. V. Dilts, *Analytical Chemistry*, Van Nostrand: New York, 1974, p. 533. Standard reduction potentials are also shown for later use in Chapter 6 and are taken from J. E. Huheey, *Inorganic Chemistry: Principles of Structure and Reactivity*, 3rd ed., Harper and Row: New York, 1983.]

ion is so extraordinarily stable that it is used as an additive to table salt, which means that almost none of the intensely toxic cyanide ion is released. Many other (less stable) d-block cyano anions exist, such as $[M(CN)_6]^{4-}$ for M = V^{2+} and Mn^{2+}; $[M(CN)_6]^{3-}$ for M = Ti^{3+}, Cr^{3+}, Mn^{3+}, and Co^{3+}; and $[Mn(CN)_6]^{2-}$.[10]

Because the cyanide ion has unshared electron pairs at both ends, it can also form strictly linear bridges, M(1)–C≡N–M(2). When these cyano anions are mixed with solutions of d-block metal ions, some very intensely colored insoluble compounds such as $KFe[Fe(CN)_6]$, **Prussian blue**, are formed. Prussian blue is one of the first coordination compounds ever synthesized, and it is the first one containing metal–carbon bonds. It was obtained accidentally in 1704 by a Berlin manufacturer of artists' color named Diesback, from heating together horses' hooves and potash (a mixture including potassium hydroxide and potassium carbonate) in an iron kettle.[11] It is currently used in black and bluish ink and laundry bluing, and was involved in the cyanotype form of blueprint.

$$[Fe(CN)_6]^{4-} + Fe^{3+} + K^+ \rightarrow KFe[Fe(CN)_6] \tag{5.9}$$

$$[Fe(CN)_6]^{4-} + Fe^{2+} + 2K^+ \rightarrow K_2Fe[Fe(CN)_6] \text{ (a white insoluble salt)} \tag{5.10}$$

Because these salts are three-dimensional coordination polymers, they are very insoluble, precipitate rapidly, often incorporate impurities and vacancies, and are very difficult to crystallize for crystal structure determination. Of most importance to us are the two salts $KFe[Fe(CN)_6]$ (Prussian blue) and the white $K_2Fe[Fe(CN)_6]$. The cyanide ions bridge different iron ions to form three-dimensional linked cubes, as shown in Figure 5.18.

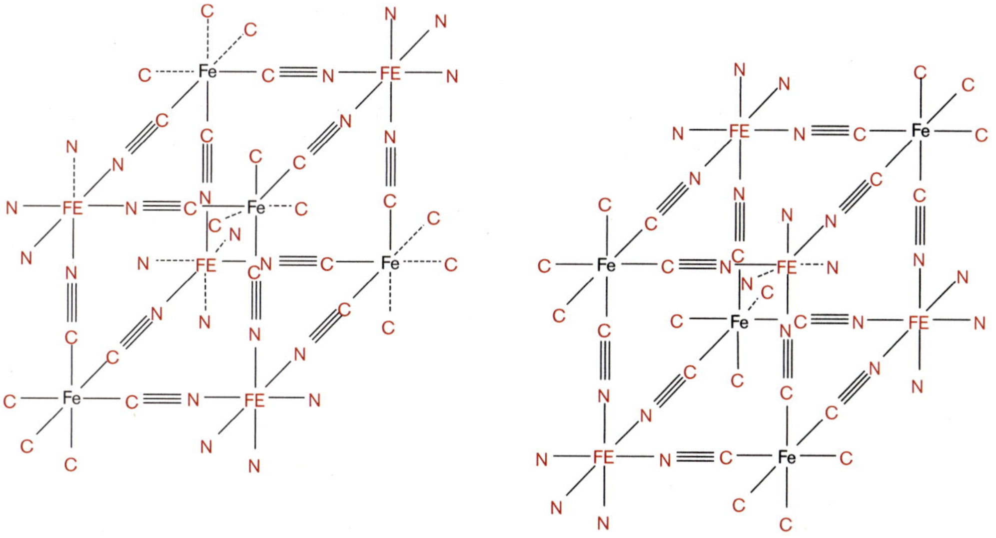

Figure 5.18. Structures of two adjacent linked cubes in soluble Prussian blue. For clarity in viewing, the two cubes are shown separately, although they share the same face of four cyanide ions, two "Fe" ions (which are Fe^{2+} ions) and two "FE" ions (which are Fe^{3+} ions).

In Prussian blue, the ions labeled "Fe" are Fe^{2+} ions, and each of them is octahedrally coordinated inside six C donor atoms, while the ions labeled "FE" are Fe^{3+} ions, and each of them is octahedrally coordinated inside six N donor atoms. Inside each cube is a pore. Half of these pores are occupied by K^+ ions. In white $K_2Fe[Fe(CN)_6]$, both "Fe" and "FE" are Fe^{2+} ions, and *all* of the pores are occupied by K^+ ions. The Prussian blues are useful ion-exchange materials—K^+ ions are preferentially replaced in their pores by larger aqueous cations, such as the very toxic Tl^+ and Cs^+ ions. This is particularly dangerous if Cs^+ is the radioactive isotope Cs-137, one of the most dangerous products of nuclear fission (bombs and power plants). Prussian blue is insoluble and nontoxic enough that it is a preferred antidote[12] for poisoning by Tl^+ and $(Cs-137)^+$.

EXAMPLE 5.7

Which of the homopolyatomic anions discussed in Section 3.1B would be most suitable for creating salts having three-dimensional linked cube structures similar to that of soluble Prussian blue? Why would the others be unsuitable?

SOLUTION: The anions should have lone pairs at 180° angles to the element–element bonds, which likely means one unshared electron pair on each terminal atom. Drawing Lewis structures of most of the anions discussed reveals the presence of three unshared electron pairs on the terminal atoms of triiodide ion (I_3^-) and peroxide ion (O_2^{2-}), and two unshared electron pairs on each terminal atom of the azide ion (N_3^-). The most suitable might be the acetylide ion, (C_2^{2-}), which has one unshared electron pair on each carbon atom.

To obtain the same crystal structure as soluble Prussian blue, this acetylide should have the same stoichiometry—namely, three different cations with the opposite total charge as the six bridging anions (–6 for six cyanides, but –12 for six acetylides). One such possibility might be a potassium tungsten(V) tungsten(VI) acetylide, $KW[W(C_2)_6]$.

5.3B. Metal–Organic Framework Compounds. We have previously seen that crown ethers and cryptands (Section 5.2A) and Prussian blues (Section 5.3A) have internal holes or pores that can be used to bury or store ions. If we can create larger pores in stable compounds, we can store important molecules. In this section we explore the use of larger bridging organic ligands to create porous **metal–organic framework (MOF)** coordination oligomers or polymers.

Construction of MOF compounds is normally done by reacting bridging ligands having at least two donor atoms with metal ions or clusters having at least two available coordination sites. Figure 5.19 shows some smaller bridging organic ligands that are useful in creating MOF compounds.

The MOF structures can be made even larger if, in place of a simple M^{2+} metal ion, a secondary building unit (a cluster of metal ions around a central bridging oxide ion) is used.[13] One such example is the Zn_4O^{6+} cluster (Figure 5.16). A variety of MOFs (Figure 5.20 on page 240) have been produced by Yaghi et al.[14] that incorporate this secondary building unit. It is constructed during the synthesis of the MOFs by the reaction of $Zn(NO_3)4\cdot4H_2O$ and the acid form of the bridging ligand in an organic solvent at an elevated temperature:

$$4Zn(NO_3)_2\cdot4H_2O + 3HOOC–C_6H_4–COOH \rightarrow$$
$$Zn_4O(OOC–C_6H_4–COO)_3 + 3H_2O + 8HNO_3 \quad (5.11)$$

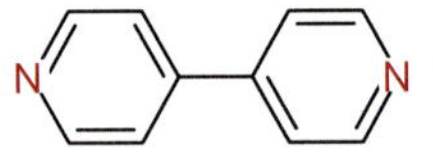

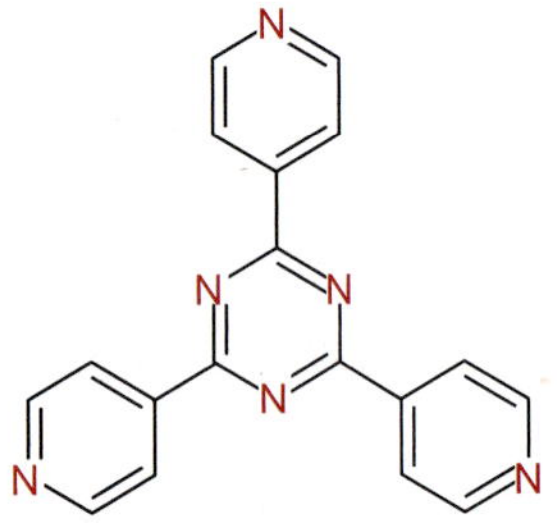

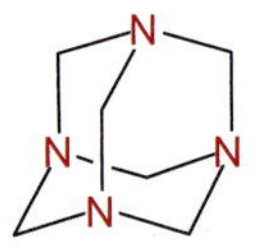

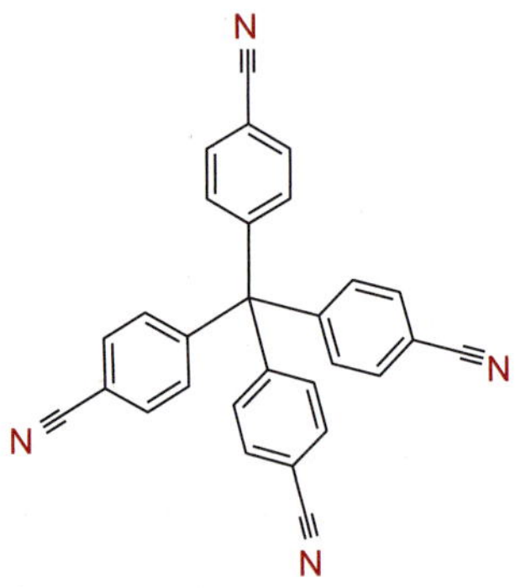

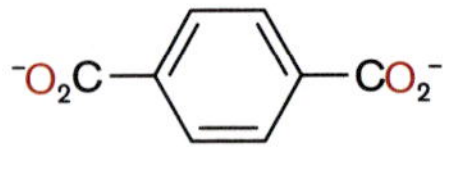

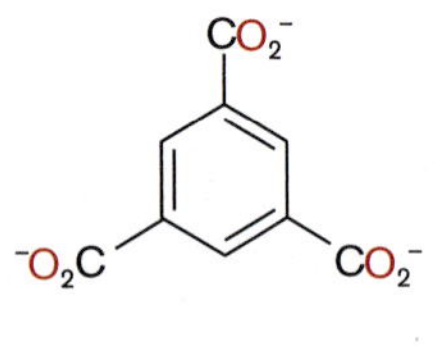

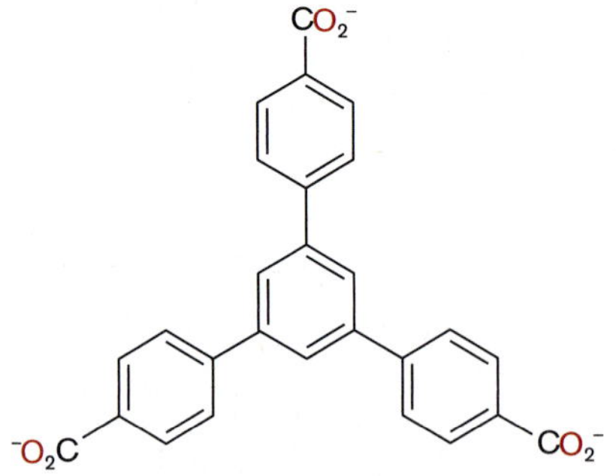

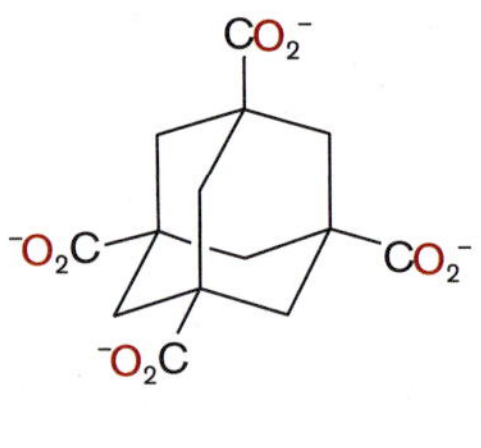

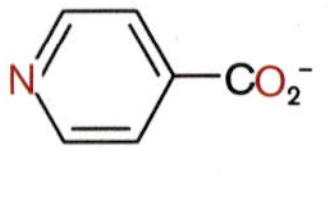

Figure 5.19. Smaller rigid bridging organic ligands used in the synthesis of metal–organic framework compounds. [Adapted from S. L. James, *Chem. Soc. Rev.* 32, 276 (2003).]

EXAMPLE 5.8

(a) Which of the ligands shown in Figure 5.19 should be most suitable, upon reacting with an $M(H_2O)_6^{2+}$ solution, for forming three-dimensional cubic-structured metal–organic framework compounds similar to Prussian blues? (b) Propose an additional suitable organic molecule that is more closely related to the cyanide ion.

SOLUTION: (a) Ligand **1**, 4,4'-bipyridine, is the most suitable because the lone pairs of its donor atoms are collinear. Ligands **5** and **9** may also give linear coordination if the para-carboxylate groups can coordinate to metal ions to give an overall linear framework.

   (b) Placing the linear CN groups at para positions on a benzene ring (as $:N\equiv C-C_6H_4-C\equiv N:$; that is, 1,4-dicyanobenzene or terephthalonitrile) would also allow linear linkage, as would using the cyanoacetylide anion, $(:N\equiv C-C\equiv C:)^-$.

EXAMPLE 5.9

Carefully examine the MOF crystal structures shown in Figure 5.20. (a) Which MOF contains the bridging ligand **A**? (b) Which one contains ligand **B**? (c) Which one contains ligand **C**? (d) Of what order of magnitude are the pore volumes in these MOFs—20%, 50%, or 90%?

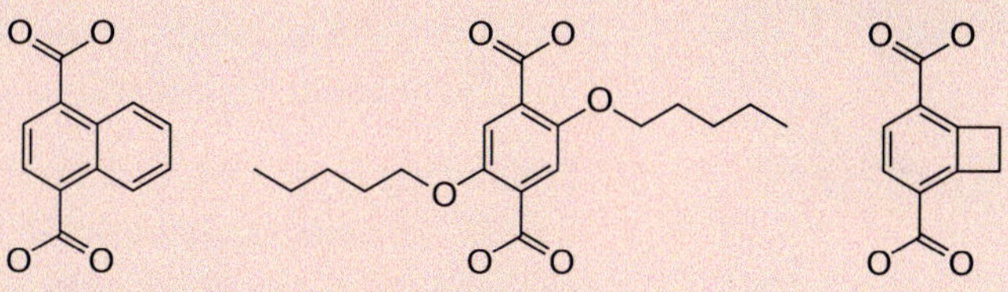

A   **B**   **C**

SOLUTION: (a) Ligand **A** is based on a naphthalene ring (it is the 1,4-naphthalenedioate ion). Structure **7** in Figure 5.20 contains this ligand.

   (b) Ligand **B** has two C_5H_9O ether side chains attached to the benzene ring, which is found in structure **5**.

   (c) Ligand **C** has a cyclobutane ring fused onto the benzene ring, which is found in structure **6**.

   (d) Although the MOF is not shown in its full volume, in these three MOFs the pore volumes are probably about 50% of the total volume of the MOFs.

The pore volumes of the MOFs (suggested in Figure 5.20 by the shaded spheres) must be large enough to accommodate the molecule that is to be stored there. There are several ways of estimating the porosity of these materials.

1. *Low density* is a crude indicator of high porosity. For example, returning to the polymorphs of SiO_2 in Figure 4.7, it is visually apparent that the highest porosity is found in ITQ-4, which has the lowest density (1.70 g cm^{-3}) of those polymorphs. MOFs such as those illustrated in Figure 5.20 have densities that are among the lowest ever recorded for crystalline materials, 0.41–0.21 g cm^{-3};[15] the density of MOF-5 in Figure 5.20 is 0.59 g cm^{-3}.

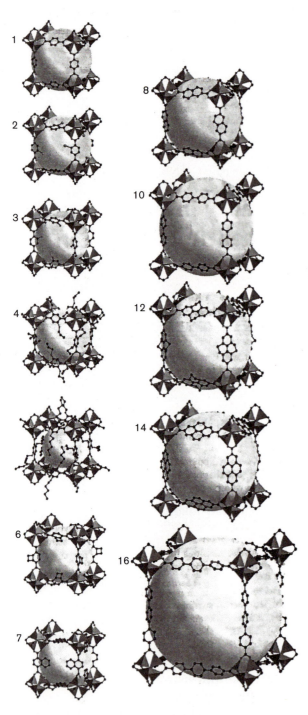

Figure 5.20. Crystal structures of some Zn$_4$O(dicarboxylate)$_6$ metal–organic framework compounds. The shaded spheres represent the largest pores that are within van der Waals distances of the framework atoms.

2. *Pore diameter*, *pore volume*, and the *surface area* of the material are all interrelated, but they can be measured, calculated, or expressed in different ways. The pore diameter should exclude the van der Waals radii of the inner atoms of the MOF. When calculated in this manner, a pore diameter of about 280–330 pm seems to be ideal for enclosing H$_2$ gas, which has a kinetic diameter of about 280 pm.[16] The pore volume of MOF-5 is calculated to be 0.61 cm^3 cm^{-3} (i.e., the pore occupies 61% of the unit cell). The surface area of the pore in MOF-5 can be estimated as 2900 m^2 g^{-1}. The more recently synthesized MOF-210 has a pore surface area of 6420 m^2 g^{-1}, which corresponds to nearly the area of an official soccer field in one gram of material![17]

3. For a given gas being stored, the storage capacity can be expressed in *weight percent* or *weight of gas per unit volume*. The U.S. Department of Energy set targets for H$_2$ storage of 4.5% (45 g H$_2$ per kg of MOF) or 28 g H$_2$ per L volume of the MOF. Some MOFs exceed one or both of these values— but it matters whether the measurement is made at 77 K or room temperature.

4. For storing H$_2$ gas,[18] it turns out that larger pore volume is not necessarily optimal. Of the MOFs shown in Figure 5.20, the best results are obtained with MOF-5, while those with larger pores do not perform as well. It seems that attractive intermolecular forces between the MOF and the stored gas are important. These are strongest when there is close contact between the MOF and the stored gas. This attraction can be quantified by measuring the *enthalpy of adsorption*, ΔH_{ads}. If the ΔH_{ads} is too small, the gas may be adsorbed well at 77 K under high pressure but not as well at room temperature and pressure, which is the more practical temperature of operation. On the other hand, we want to not only store the gas but also get it out for use. Hence, too *great* a value of ΔH_{ads} may also be undesirable.

Methane is another fuel gas that is desirable to store in MOFs. Progress has been less spectacular for this purpose— MOF-6 adsorbs 155 L of methane gas in 1 L of MOF-6 at 298 K and 36 atm, as compared to the Department of Energy goal of 180 liters per liter.[16]

Storage of Methane in Ice

Nature has a different way of building a framework of *ice* around methane to store it as a **clathrate compound** inside the pores of a different structure of ice, an *icosahedral* structure in which water molecules hydrogen-bond to each other in pentagonal rings rather than the hexagonal rings found in normal ice. This *methane hydrate* (Figure 5.21) decomposes if it is not under mild positive pressure, and it melts at slightly above 0°C (this form of ice burns!).[19] Massive deposits of methane hydrate, containing twice as much carbon as in all other oil and gas reserves, have been located 300–1000 m beneath the sea and under permafrost. Energy-poor Japan has succeeded in extracting methane from offshore deposits. However, warming could suddenly release vast quantities of methane gas to the atmosphere. Because it is 21 times as potent a greenhouse gas as CO_2, it is believed that methane could trigger, and actually has triggered, sudden rapid global warming episodes (four times in the last 60,000 years),[20] and could do this again.

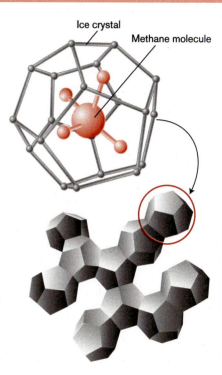

Figure 5.21. The structure of methane hydrate. [Adapted from E. Suess, G. Bohrmann, J. Greinert, and E. Lausch, *Sci. Amer.* 281(5), 76 (1999).]

Capture of CO_2

For purposes of reducing the climate-change effect of increasing atmospheric concentrations of CO_2, it would be useful to have an MOF that could capture CO_2 from smokestack gases and that could then release it for storage or other uses, thus allowing the catalyst to be regenerated. A magnesium-based MOF, Mg-MOF-74, captures only CO_2 from a gaseous mixture of CO_2 and CH_4, retaining 89 g CO_2 per kilogram of MOF, and then releases it all at the relatively mild temperature of 80°C. A crystal structural study of Ni-MOF-74 revealed that, upon preliminary dehydration, a five-coordinate Ni^{2+} ion absorbs η^1-CO_2 as a coordinated Lewis base.[21] More recently, a secondary building unit that incorporates an amine side group has also been used to capture CO_2; a crystal structure shows bonding of the NH_2 group to the CO_2.[22]

5.4. The Hard and Soft Acid–Base Principle

OVERVIEW. Based on charge, size, and electronegativity (or position in the periodic table), we classify Lewis acids and Lewis bases as hard, soft, or borderline. We then apply the hard and soft acid–base (HSAB) principle to predict whether a given reaction will go to the left or to the right (Section 5.4A; try Exercises 5.34–5.40). The hardness or softness of a donor or acceptor atom can be modified by changing its oxidation state (charge) or by changing the hardness or softness of the substituents on the atom (Section 5.4B; Exercises 5.41–5.53). Ambidentate ligands have different donor atoms, which normally differ in softness. If the ambidentate ligand acts as a monodentate ligand, you can use the HSAB principle to predict which donor atom will preferentially bond to a given Lewis acid (in some cases, two linkage isomers result; Section 5.4C; Exercises 5.54–5.57). A Connection to biochemistry is included.

The HSAB principle is an application of what you have already learned about the Pauling electronegativities (χ_P; Section 1.4), charges (Section 1.1), and sizes (Section 1.5C) of donor atoms. For classifying oxo anions, you may want to review the construction of their formulas and Lewis structures (Sections 3.3 and 3.4). You will see additional applications of the HSAB principle in Sections 6.2, 6.3, 6.4, 7.5, 7.8, 9.5, and 11.3.

In Experiment 6, you can discover the principle that enables us to make qualitative predictions of the predominant products that result when Lewis acids and Lewis bases (of the same denticity, such as all monodentate or all bidentate) compete with each other to form precipitates or complex ions. After we develop this principle, we will use it to organize the chemistry of such diverse phenomena as the following:

1. additional solubility rules for inorganic compounds (Section 5.6);

2. the geochemistry of the elements and why they occur in the ores in which they are found (Section 5.6);

3. the nutritional and toxic effects of many metals and nonmetals in the body (e.g., micronutrients and toxic effects of heavy metals) (Section 5.7); and

4. the ways in which medicinal chemists devise drugs to counteract the effects of heavy metal poisoning (Section 5.8).

EXPERIMENT 6 MAY BE ASSIGNED NOW.
(See www.uscibooks.com/foundations.htm.)

In Experiment 6, you may have observed that in mixtures of several Lewis acids and Lewis bases, a certain set of metal ions tend to combine with the iodide ion in preference to the fluoride ion. This same set of metal ions prefers the sulfide ion to the hydroxide ion, thiourea to urea, and the sulfide ion to the silicate ion.

This last experiment is related to the *differentiation* of the elements in nature. Berzelius first noted in 1796 that certain metal ions tend to occur in nature as sulfides, while others tend to occur as oxides, carbonates, sulfates, or silicates. Only about 50 years ago was it realized just how broadly these observations could be generalized and how many thousands of chemical reactions and phenomena could be (reasonably well) predicted using the generalized observation. The general answer to the question "Which metal ion will tend preferentially to form a complex ion with which ligand?" was best summarized by Pearson[23a]: *Hard (Lewis) acids tend to combine with hard (Lewis) bases, whereas soft acids prefer soft bases.*

This statement is now known as the **hard and soft acid–base (HSAB) principle**. It is simply a summary of the observed results of thousands of chemical reactions in inorganic chemistry, organic chemistry, biochemistry, aquatic chemistry, medicinal chemistry, geochemistry, and so on. Put into equation form, for a chemical reaction among complex ions, the equilibrium will tend to favor the products on the right side of Equation 5.12:

$$[\text{Hard acid}(:\text{soft base})_n] + [\text{soft acid}(:\text{hard base})_n] \rightarrow$$
$$[\text{hard acid}(:\text{hard base})_n] + [\text{soft acid}(:\text{soft base})_n] \quad (5.12)$$

To be able to use the HSAB principle, we must find a way to classify metal ions as either hard or soft. Next, we must classify all ligands as either hard or soft. We can do this reasonably well, although some acids and bases will not fall clearly into either class; we will call them borderline. The classification scheme that is usually devised is shown in Table 5.2, where it is superimposed upon a table of Pauling electronegativity values.

TABLE 5.2. Hard and Soft Acids and Bases[a]

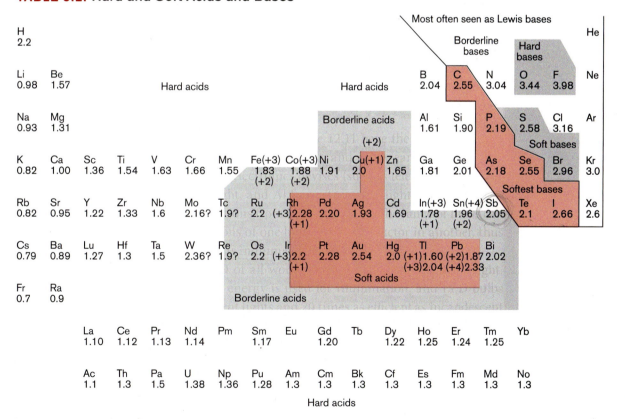

[a] Numbers in parentheses are oxidation numbers. The number below each atomic symbol is the Pauling electronegativity of that element. When not listed, the oxidation state of metals is taken to be the most common positive oxidation state (Table 1.11). However, changing the oxidation states of these metals is not usually expected to alter the HSAB classification.

5.4A. Hard and Soft Acids and Bases. **Soft Acids.** Despite their presence in different groups of the periodic table, these metal ions, popularly known as the heavy metal ions, have a number of chemical similarities. They have insoluble chlorides, bromides, and iodides (Section 5.6). They occur in nature as the free elements or in sulfide and related minerals (Section 5.6). In terms of fundamental atomic properties (as you may have noted in the lab experiment), these metals are characterized by quite high Pauling electronegativities for metals, generally in the range of 1.9–2.54. Other characteristics that they share to a lesser extent are large size (ionic radii in excess of 90 pm) and low charge (usually +1 or +2). Simply put, they form a triangle of metals around the element gold, which is the most electronegative of all metals.

Soft Bases. According to the HSAB principle, the group of ligands with which these metal ions tend to associate are called soft bases. The soft-acid metals are found in nature either in the form of the native (free) elements or as chlorides, bromides, iodides, sulfides, selenides, tellurides, arsenides, and so on. They are not found as oxides or fluorides or as the salts of various oxo anions such as sulfate, silicate, carbonate, and so on (Section 4.3). If we have the misfortune to swallow the ions of the soft-acid metals and live through the experience, we will find the ions bonded to certain donor atoms in the proteins and enzymes of our body—to sulfur and to selenium, but not to oxygen, fluorine, or (probably) nitrogen (Section 5.7). Table 5.2 indicates that the following donor atoms are characteristic of soft bases: C, P, As, S, Se, Te, Br, I, and Xe.[24] (We may also add H when it occurs as a donor atom in the hydride ion, H^-.) Fundamentally, these are nonmetals with moderate Pauling electronegativities of 2.1–2.96. These are also the largest nonmetal atoms, with anionic radii in excess of 170 pm.

Hard Bases. Contrasting most strongly with the soft nonmetal donor atoms are oxygen and fluorine, which are seldom found associated in nature with the soft acids. These donor atoms are of very high electronegativity (3.44 and 3.98) and are the smallest of the nonmetal atoms (anionic radii of ~120 pm). Although this seems like a small group of atoms, they (oxygen in particular) are the donor atoms in countless ligands—namely, the oxo anions such as sulfate, carbonate, and silicate; the anions of organic acids such as acetate; and several classes of organic molecules such as alcohols and ketones.

Hard Acids. Although there are only two unambiguously hard donor atoms, in nature a large number of metal ions occur associated with ligands containing O and F donor atoms. These hard acids occupy a good share of the periodic table (Table 5.2); they have in common low electronegativities of (usually) 0.7 to 1.6. They frequently (but not invariably) have relatively small cationic radii (< perhaps 90 pm) and may often have high charges (+3 or higher). (Note that some atoms may function as "cations" or Lewis acidic sites in some compounds and as donor atoms in others. For example, arsenic as As^{3+} is a hard acid, but as a donor (As^{3-}, etc.) it is a soft base. Similarly, H^+ is a hard acid but H^- is a soft base.

As you may have noticed in the lab experiment, the most important atomic variable in classifying hard and soft acids and bases is the Pauling electronegativity of the donor or acceptor atom (Table 5.3). Sometimes the other variables (charge and size) override considerations of electronegativity. Thus, H^+ is classified as a hard acid despite its high electronegativity on the basis of its extremely small size. Both B^{3+} and C^{4+} are considered hard acids despite their electronegativities both because of their very small size and their high charges.

Once we have classified the two Lewis acids and the two Lewis bases, we apply the HSAB principle to determine whether the equilibrium in question favors products or reactants.

TABLE 5.3. **Characteristic Properties of Hard and Soft Acids and Bases**

Property	Hard Acids	Soft Acids	Soft Bases	Hard Bases
Electronegativity	0.7–1.6	1.9–2.5	2.1–3.0	3.4–4.0
Ionic Radius (pm)	< 90	> 90	> 170	~120
Ionic Charge	≥ +3	≤ +2	—	—

EXAMPLE 5.10

Predict whether reactants or products are favored in each of the following equilibria:
(a) $Nb_2S_5 + 5HgO \rightleftharpoons Nb_2O_5 + 5HgS$
(b) $La_2(CO_3)_3 + Tl_2S_3 \rightleftharpoons La_2S_3 + Tl_2(CO_3)_3$
(c) $2CH_3MgF + HgF_2 \rightleftharpoons (CH_3)_2Hg + 2MgF_2$
(d) $Ca^{2+}(g) + S^{2-}(g) \rightleftharpoons CaS(s)$

SOLUTION: (a) The two cations are Nb^{5+} and Hg^{2+}. On the basis of its electronegativity and charge, Nb^{5+} is a hard acid, whereas Hg^{2+} is a soft acid. The two anions are O^{2-} and S^{2-}, which are hard and soft bases, respectively. It is convenient to write "HA," "HB," "SA," and "SB" under the species as they occur in the equation:

$$Nb_2S_5 + 5HgO \rightleftharpoons Nb_2O_5 + 5HgS$$
$$\text{HA:SB} \quad\quad \text{SA:HB} \quad\quad\quad \text{HA:HB} \quad\quad \text{SA:SB}$$

The side of the equation under which the combinations HA:HB and SA:SB occur will be favored. Thus, products are favored in this reaction.

(b) Based on electronegativity considerations, we classify La^{3+} as a hard acid and Tl^{3+} as a soft acid. We classify S^{2-} as a soft base. In CO_3^{2-}, we must identify the donor atom, which is oxygen; hence, CO_3^{2-} is a hard base. The combinations HA:HB and SA:SB occur under the reactants in this case.

(c) Mg^{2+} is a hard acid and Hg^{2+} is a soft acid; F^- is a hard base and CH_3^-, with a carbon donor atom, is a soft base. As a result, the species CH_3MgF represents SB:HA:HB. Notice on the right side, though, that F^- is still coordinated to Mg^{2+}. It has not participated in the reaction, so the relevant exchange is of the soft base CH_3^- away from the hard acid Mg^{2+} to the soft acid Hg^{2+}. This equilibrium favors products.

(d) This is a trick question! There is only one Lewis acid and one Lewis base shown, so the HSAB principle cannot be applied. In this case, the lattice energy of $CaS(s)$ drives the reaction to the product.

5.4B. Relative and Borderline Softness. Suppose that we are considering a reaction, such as the one in Equation 5.13, in which both Lewis acids are soft and both Lewis bases are soft:

$$CdSe + HgS \rightleftharpoons CdS + HgSe \qquad (5.13)$$

The HSAB principle can also be used to predict this result (the reaction goes to the right), because it is possible to assign relative softnesses to different soft acids and bases. Among soft donor atoms, the following orders of softness have been observed:

- Group 17/VII   I > Br > Cl > F
- Group 16/VI   Te = Se > S >> O
- Group 15/V   Sb < As = P > N

EXAMPLE 5.11

Students in an inorganic lab have just discovered some new elements, which have the following symbols and properties:

Symbol	Fx	Mq	Mr	Az	Za	Mm
Electronegativity	2.22	1.33	2.08	2.60	3.33	1.91
Ion formed	Fx^{2-}	Mq^{3+}	Mr^+	Az^{2-}	Za^{2-}	Mm^{2+}
Ion radius (pm)	188	62	125	155	117	103

(a) Based on the information given, classify each of the ions as a hard acid, soft acid, hard base, or soft base. (b) Rank the soft acids and the soft bases in order of increasing softness. (c) Predict whether products or reactants are favored in the following equilibrium:

$$Mr_2Az + MmFx \rightleftharpoons Mr_2Fx + MmAz$$

SOLUTION: (a) and (b) The anions are expected to be bases, and those with electronegativities below 2.96 (Fx^{2-} and Az^{2-}) likely will be soft bases. The radius of Fx^{2-} is consistent with this while the radius of Az^{2-} is a little small. Because its electronegativity is also higher, Az^{2-} will be less soft than Fx^{2-}. The Za^{2-} ion is small and has a high electronegativity, so it is a hard base.

The cations are expected to be Lewis acids, and those with electronegativities above 1.9 (Mr^+ and Mm^{2+}) likely will be soft acids. Their low charges and radii in excess of 90 pm are consistent with this. Because Mr^+ has a lower charge, larger radius, and a higher electronegativity than Mm^{2+}, it will be the softer of the two. Based on its high charge, small radius, and low electronegativity, Mq^{3+} is expected to be a hard acid.

(c) Both acids are soft, but Mr^+ is softer than Mm^{2+}. Both bases are soft, but Fx^{2-} is softer than Az^{2-}. We can enter this information below the equation:

Mr_2Az	+	$MmFx$	$\rightleftharpoons$	Mr_2Fx	+	$MmAz$
Softer acid:	+	Less-soft acid:		Softer acid:	+	Less-soft acid:
less-soft base		softer base		softer base		less-soft base

The salt containing the softer acid and the softer base, Mr_2Fx, is favored, as is the salt containing the less-soft acid and the less-soft base, $MmAz$. Both are products, so products are favored.

As we see in Table 5.2, the softest donor atoms are those of the lowest electronegativity values, and they form a "ridge" along the metal–nonmetal boundary. We find, therefore, in a competitive precipitation experiment involving all the halide ions, that AgI will precipitate preferentially (it is the least soluble). Silver bromide (AgBr) is not quite as insoluble as AgI; AgCl is not quite as insoluble as AgBr; and AgF, which involves a hard base, is much more soluble than any of the others.

Evaluating the relative softness of the metal ions has proved somewhat harder. There are some minor discrepancies among different competitive equilibria involving different soft metal ions with different soft bases. For our purposes, we can say that the closer the metal is to gold, the softer it is.

For use with competitive equilibria in which softness is present in more than one Lewis acid and more than one Lewis base, we may restate the HSAB principle as follows: Less soft acids tend to combine with less soft bases, whereas softer acids prefer softer bases. Thus, in Equation 5.13, mercury, being closer to gold in the periodic table, should be softer, and should prefer the softer base, the selenide ion.

Borderline Acids and Bases. Given that there are degrees of softness, it is not surprising that there are some metal and nonmetal atoms that show such a small degree of softness that softness will not consistently govern the results of their competitive equilibria. As Table 5.2 shows, there are several such metal ions, which are labeled **borderline acids**. (Many of these have insoluble sulfides but soluble halides, for example.) Likewise, two of the nonmetal atoms, chlorine and nitrogen, are classified as **borderline bases**. In these cases, the electronegativities of the atoms or their sizes may fall between the ranges that are typical for soft and for hard species. Although the chemistry of borderline acids and bases is a little harder to predict, we can still use the HSAB principle: A borderline base (such as Cl) will be softer than F but not as soft as I.

Modifying the Softness of an Atom. Because high oxidation numbers are one characteristic of a hard acid, and low oxidation numbers are a characteristic of a soft acid, it is possible to alter the softness of a metal ion by changing its oxidation number. This is a particularly important characteristic of the later d-block metals of the fourth period, which are often borderline acids to begin with. Among the metals Fe to Zn, the +1 oxidation state, which occurs only for Cu, is definitely soft, whereas the +2 oxidation state is borderline, and the +3 ion (such as for Fe) is characteristically hard. In Figure 5.22, we see this effect in a chelate complex of iron in which three strands of the ligand offer a total of six hard oxygen donor atoms low in the strands and six borderline nitrogen donor atoms higher in the strands. When the iron is oxidized to the hard Fe^{3+}, it prefers the six hard donor atoms; when it is reduced to Fe^{2+}, it jumps along the strands in order to be chelated by the six borderline donor atoms.[25]

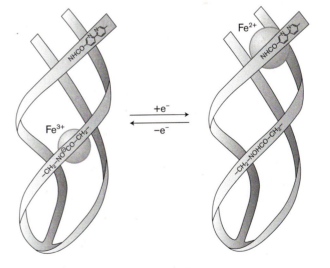

Figure 5.22. Schematic representation of the iron complex of a triple-stranded chelate ligand offering a choice of six hard oxygen or six borderline nitrogen donor atoms. [Adapted from L. Zelikovich, J. Libman, and A. Shanzer, *Nature* 374, 790 (1995).]

Several other examples of changes in categories of a metal with change in its oxidation state can also be seen in Table 5.2. We also note that, in terms of their catalytic effects, metallic forms of elements (oxidation number = 0) are generally soft acids. The relative stability of compounds such as $Cr(CO)_5Xe^{26}$ can thus be rationalized: Although Cr^{3+} is indubitably a hard acid, the oxidation number of Cr in this compound is zero. Here it coordinates a soft base, a xenon atom, consistent with Cr(0) being classified as a soft acid.

The softness of a donor or acceptor atom can also be altered by changing the substituents on the atom. Attaching soft bases to a Lewis acid generally softens it as a Lewis acid in any further reactions with additional Lewis bases; the converse holds true for the attachment of hard bases. For example, in the form of hydrated ions (water is a hard base), all of the Group 13/III +3 metal ions except Tl^{3+} are hard acids; in the form of their methyl compounds $M(CH_3)_3$, all are soft acids except $B(CH_3)_3$ and $Al(CH_3)_3$.

EXAMPLE 5.12

Predict whether reactants or products are favored in the following equilibria:

(a) $ZnI_2 + HgCl_2 \rightleftharpoons ZnCl_2 + HgI_2$

(b) $(CH_3)_2O{:}BF_3 + (CH_3)_2S{:}BH_3 \rightleftharpoons (CH_3)_2S{:}BF_3 + (CH_3)_2O{:}BH_3$

(c) $CuI_2 + Cu_2O \rightleftharpoons 2CuI + CuO$

SOLUTION: (a) The I^- ion is a soft base and Hg^{2+} is a soft acid. Although Zn^{2+} is a borderline acid (Table 5.2) and Cl^- is a borderline base, relative to Hg^{2+} and I^-, they are each the harder species. Thus, the SA:SB combination is HgI_2 and the less-soft acid:less-soft base combination is $ZnCl_2$; *products* are favored.

(b) In the four coordination compounds chosen, the change that occurs is a switch of B:O and B:S coordinate covalent bonds. Breaking apart the compounds at these bonds, we identify the hard base as $(CH_3)_2O$ and the soft base as $(CH_3)_2S$. The two Lewis acids BH_3 and BF_3 differ in their substituents; BH_3 should be the softer because F withdraws more charge from boron. Hence, the *reactants* should be preferred.

(c) Of the bases, I^- is soft and O^{2-} is hard. The copper occurs in two oxidation states—namely, Cu^+ in Cu_2O and CuI, and Cu^{2+} in CuO and CuI_2. Because Cu^+ is softer, the SA:SB combination is CuI, and *products* are favored.

5.4C. Alternate Binding Sites on Ligands. Amino Acids as Ligands.
Enzymes are nature's catalysts for speeding up chemical reactions in the body to rates as much as 100,000 times as fast as they would occur without catalysis. It is estimated that about 40% of all enzyme-catalyzed reactions involve metals. Many enzymes contain from one to several metal ions firmly incorporated into the protein structure of the enzyme; these are known as **metalloenzymes**. Other enzymes, although not incorporating the metals irreversibly into their structures, do require the reversible coordination of metal ions in order to become active; these are called **metal-activated enzymes**. Most of the metal ions of the third and fourth periods of the periodic table function in one or more metalloenzymes or in activating numerous enzymes (Table 5.4).

TABLE 5.4. Functions and Preferred Ligand Binding Groups for Essential Metal Ions

Metal Ion	Function	Ligand Groups, with Donor Atoms in Parentheses
A. Hard Acids		
Na^+	As charged ion	Hydrated ions (O)
K^+	As charged ion	Singly charged oxygen donor atoms or neutral oxygen ligands (O)
Mg^{2+}	As charged ion, structural	Carboxylate (O), phosphate (O), nitrogen donors (N)
Ca^{2+}	As charged ion, structural	Like Mg^{2+} but less affinity for nitrogen donors, phosphate, and other multidentate anions
Fe^{3+}	Redox reactions	Carboxylate (O), tyrosine (O), $-NH_2$ (N), porphyrin ("hard" N)
Co^{3+}	Redox reactions	Similar to Fe^{3+}
V^{n+}	Essential to sea squirts; mimics insulin	
Cr^{3+}	Glucose tolerance factor	
Al^{3+}	May activate two enzymes	
B. Borderline Acids		
Mn^{2+}	Lewis acid	Similar to Mg^{2+}
Fe^{2+}	Redox reactions	$-SH$ (S), $-NH_2$ (N) > carboxylates (O)
Zn^{2+}	Lewis acid	Imidazole (N), cysteine (S)
Cu^{2+}	Redox reactions	Amines (N) >> carboxylates (O)
Ni^{2+}	In urease; stabilizes coiled ribosomes	
Mo^{2+}	Redox reactions	$-SH$ (S)
W^{n+}	Essential for high-temperature microbes near black smokers	
C. Soft Acids		
Cu^+	Redox reactions	Cysteine (S)

Data from M. N. Hughes, *Inorganic Chemistry of Biological Processes*, 2nd ed., John Wiley and Sons Ltd.: Chichester, UK, 1981; J. E. Huheey, E. A. Keiter, and R. L. Keiter, *Inorganic Chemistry: Principles of Structure and Reactivity*, 4th ed., Harper-Collins: New York, 1993, pp. 943–948; *Chem. Eng. News*, April 12, 1993, p. 39; ibid., Feb. 21, 1994, p. 35; and ibid., May 22, 1995, p.23.

Side Groups of Metalloenzymes and Metal-Activated Enzymes. Many of the most important ligands in biochemistry are derived from **amino acids**, $R-CH(NH_2)-COOH$. Any free amino acid anion, $R-CH(NH_2)-COO^-$, can function as a chelating agent using its borderline nitrogen and one of its hard oxygen donor atoms. In biochemistry, the amino acids are polymerized into polypeptides and enzymes, $(NH-CHR-CO)_x$. In polypeptides and enzymes, the NH nitrogen and the CO oxygen atoms are in a peptide linkage, where they are both very weak bases and sterically protected from coordination to metal ions. Hence, in metalloenzymes and metal-activated enzymes, we are interested in amino acids that have donor atoms in their side groups. These are shown in Figure 5.23.

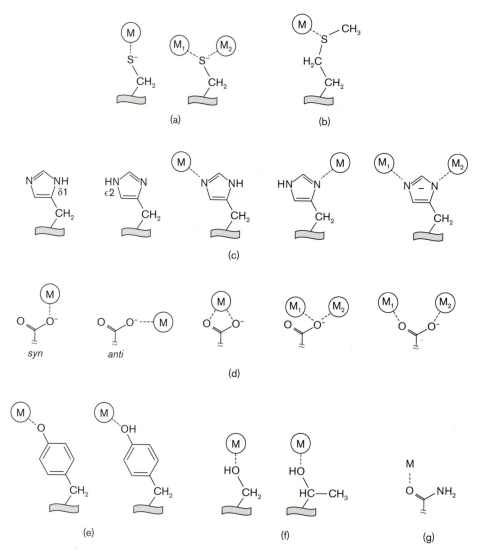

Figure 5.23. Amino acid side-chain coordination to metal ions: (a) cysteinate anion (selenocysteinate anion is similar); (b) methionine; (c) histidine (at left, two forms of the imidazole functional group from proton rearrangement; in the center, coordination by each form; at right, coordination by the histidine anion); (d) carboxylate groups of glutamate anion and aspartate anion; (e) tyrosinate anion and tyrosine; (f) serine and threonine; and (g) asparagine and glutamine. [Adapted from I. Bertini, H. B. Gray, E. I. Steifel, and J. S. Valentine, *Biological Inorganic Chemistry: Structure and Reactivity*, University Science Books: Mill Valley, CA, 2007.]

The Metalloenzyme Zinc Carboxypeptidase A

One of the best-understood metalloenzymes is carboxypeptidase A, shown in Figure 5.24, which helps cleave protein molecules by hydrolyzing the amide group linking the carboxyl-terminal amino acid to the polypeptide or protein:

$$-C(=O)-NH-CHR-C(=O)-O^- + H_2O \rightarrow -C(=O)-O^- + {}^+NH_3-CHR-C(=O)-O^-$$

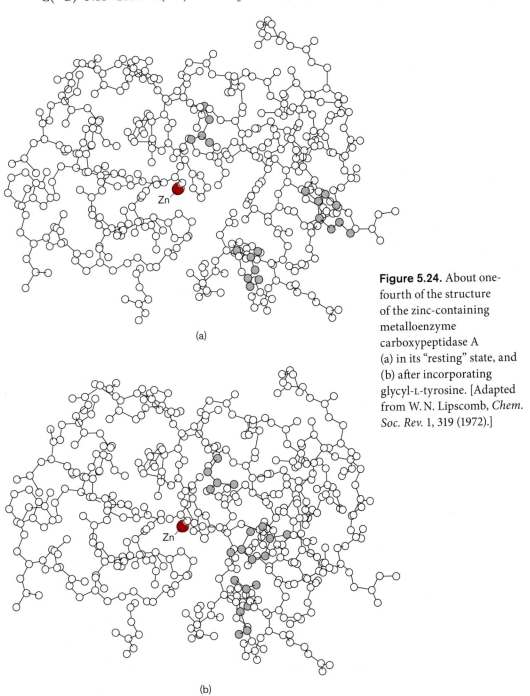

(a)

(b)

Figure 5.24. About one-fourth of the structure of the zinc-containing metalloenzyme carboxypeptidase A (a) in its "resting" state, and (b) after incorporating glycyl-L-tyrosine. [Adapted from W. N. Lipscomb, *Chem. Soc. Rev.* 1, 319 (1972).]

In the center of the 307 amino acids linked together to form this enzyme is one Zn^{2+} ion, which is intimately involved in the mechanism of action of this enzyme.[27] Before the polypeptide enters, the Zn^{2+} is coordinated to a H_2O molecule and to the enzyme at three sites (two N-donor histidine groups and one bidentate O-donor glutamate anion; Figure 5.25).

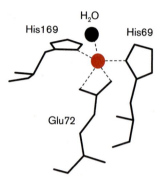

Figure 5.25. The coordination environment of zinc in carboxypeptidase. Glu = glutamate ion and His = histidine. [Adapted from I. Bertini, H. B. Gray, E. I. Stiefel, and J. S. Valentine, *Biological Inorganic Chemistry: Structure and Reactivity*, University Science Books: Mill Valley, CA, 2007, p. 37.]

The Zn^{2+} ion then coordinates the polypeptide (Figure 5.24b). The oxygen of the polypeptide carbonyl group (C=O) next to the N–C bond is also attached to zinc, which enhances the carbonyl group's acidity. Then the coordinated water is deprotonated to a hydroxide ion, which is more basic than water and is better able to attack the peptide N–C bond. The proximity of the Lewis-acidic polypeptide carbonyl group and the Lewis-basic hydroxide ion also facilitates the reaction (the template effect). Nature does not recognize our distinction of fields of chemistry, however: The mechanism also depends critically on the contribution of hydrogen bonding to nearby polypeptide functional groups.

The search for simpler coordination complexes that mimic the behavior of complicated biochemical complexes is one of the activities of the field of research called **bioinorganic chemistry**. With simpler ligands present, there is more hope of being able to sort out or even calculate the reasons for the functions of the catalyst. For example, possible mechanisms for the activity of the Zn^{2+} in carboxypeptidase A in promoting the hydrolysis of polypeptides were modeled in much simpler enzyme-free Co^{3+} complexes of the polypeptide.[28]

Ligands such as the pseudohalide ions (Figure 5.17) are capable of donating electron pairs to different atoms and thus are potential bridging ligands. In many complexes, however, this potential is unfulfilled, and only one donor atom is used to coordinate one metal ion. If a ligand has two different types of donor atoms, but uses only one, a choice as to which donor atom is used must be made. Ligands that can use either of two (or more) different donor atoms are called **ambidentate ligands**.[29] Note that all of the pseudohalide ions except azide are ambidentate ligands. Other ambidentate ligands include some others from Figure 5.4—namely, nitrite ion (NO_2^-) and dimethylsulfoxide [$(CH_3)_2SO$]. Still other ambidentate ligands include the structural isomer of cyanate ion, called the *fulminate* ion (CNO^-), and the sulfite ion (SO_3^{2-}).

When the two donor atoms are sulfur and oxygen (as in thiocyanate and dimethylsulfoxide), we expect the ligand to coordinate via its sulfur atom to a soft acid but via its oxygen atom to a hard acid. In some cases with less clear-cut differences in softness (e.g., N and O donor atoms in nitrite), different complexes are known in which either donor atom is chosen; these kinds of complexes are known as **linkage isomers**. As an example, when the ambidentate nitrite ion coordinates to the $[Co(NH_3)_5]^{3+}$ Lewis acid, the product is yellow when the nitrogen atom is used as the donor atom of nitrite (Figure 5.26a), and it is red when the oxygen atom is used (Figure 5.26b).

Figure 5.26. Linkage isomers of the ambidentate NO_2^- ion with the $[Co(NH_3)_5]^{3+}$ Lewis acid. (a) The yellow complex cation that results when NO_2^- uses its nitrogen donor atom and (b) the red complex cation that results when NO_2^- uses an oxygen donor atom.

EXAMPLE 5.13

Given alkoxide (RO^-) and related ions **A–C**, (a) which one is ambidentate? (b) If the ambidentate ligand uses only one donor atom, which donor atom is expected to bond to mercury(II)? Which donor atom is expected to bond to chromium(III)? (c) Is there a reason that both donor atoms could bond?

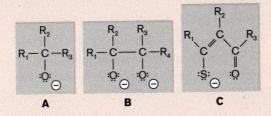

SOLUTION: (a) Ligand **C** is ambidentate because it has a choice of two different donor atoms, S and O.

(b) The sulfur atom is a soft-base donor atom, so it preferentially bonds to the soft acid, Hg^{2+}, as in **D** below. The oxygen atom is a hard-base donor atom, so it preferentially bonds to the hard acid, Cr^{3+}, as in **E**.

(c) The ligand **C** is bidentate and can function as a chelating ligand, as in **F**. Chelated complexes tend to be more stable than monodentate complexes. As we shall see in the next section, when two principles such as HSAB and the chlelate effect conflict, the choice does not always go one way or the other.

5.5. Bonding Principles Behind the HSAB Principle and Polar Covalent Bonding

OVERVIEW. We explain why (in terms of thermodynamics and bonding types) the HSAB principle works, and how covalency and softness affects lattice-energy calculations (Section 5.5A; Exercises 5.58–5.63). We show that the HSAB principle may not work well if there is a great difference in strength or chelating ability in the acids or bases involved (Section 5.5B; Exercises 5.64–5.66).

Lattice-energy calculations were covered in Section 4.2. Section 12.6 will amplify on the need to consider more than one parameter in evaluating bonding.

5.5A. The HSAB Principle and Polar Covalent Bonding.
The HSAB principle is empirical—that is, it is based on observation. Your conclusions upon doing the lab experiments have also been empirical. Although empirical relationships are very valuable, science also seeks to find theoretical reasons for the relationships. For the HSAB principle, this has been more difficult than usual, in part because it is a qualitative relationship: We do not have good numerical values for the softness of species. Sophisticated methods and arguments (such as perturbation molecular orbital theory[30]) have been applied to the question of why the principle works, but these are beyond the level of this text.

For now let us analyze the principle at a very simple level, looking at the fundamental properties of hard and soft acids and bases (Table 5.3). Of these properties, the most important in determining hardness or softness is the Pauling electronegativity.[23b] In Section 3.3D, we saw that Pauling suggested that *bond energies* can be analyzed as a sum of contributions from covalent bonding and ionic bonding. We can now apply Equation 3.6 to the analysis of the archetypical HSAB equilibrium reaction occurring in the gas phase. This is Equation 5.12 with $n = 1$:

$$\text{HA:SB}(g) + \text{SA:HB}(g) \rightarrow \text{HA:HB}(g) + \text{SA:SB}(g) \tag{5.14}$$

As shown in the following Amplification, ΔH for this typical reaction to give the products HA–HB and SA–SB is very favorable ($\Delta H \approx -150$ kJ mol^{-1}).

AN AMPLIFICATION

Thermochemical Analysis of the HSAB Principle

We evaluate the bond energies of each of the four reactants and products in Equation 5.14 using Equation 3.6. We then take the sum of the bond dissociation energies of the reactants minus the sum of the bond dissociation energies of the products (in which bonds are formed, not dissociated). When we do this, all the terms involving element–element bond energies [ΔH(HA–HA), etc.] drop out, leaving four terms:

$$\Delta H = 96.5\{[\chi_P(\text{HA}) - \chi_P(\text{SB})]^2 + [\chi_P(\text{SA}) - \chi_P(\text{HB})]^2 - [\chi_P(\text{HA}) - \chi_P(\text{HB})]^2 - [\chi_P(\text{SA}) - \chi_P(\text{SB})]^2\} \tag{5.15}$$

Let us now take some typical electronegativity values χ_P for the donor and acceptor atoms of hard and soft acids and bases: $\chi_P(HA) = 1.6$; $\chi_P(HB) = 3.4$; $\chi_P(SA) = \chi_P(SB) = 2.5$. (These values are consistent with the ranges found in Table 5.3 and the requirement that the electronegativity difference in Equation 3.6 not exceed 1.8.) Substituting these values in Equation 5.15 gives

$$\Delta H = 96.5(0.81 + 0.81 - 3.24 - 0) = -156 \text{ kJ mol}^{-1}$$

Thermodynamics of HA:HB and SA:SB Interactions. There is a thermodynamic difference between the hard-acid–hard-base and the soft-acid–soft-base interactions. We noted in Section 4.5B that the most common hard-acid–hard-base precipitation reaction, that of an acidic cation and a basic anion, is accompanied by increasing disorder as many solvent water molecules are released. The typical hard-acid–hard-base reaction (Eq. 5.16) involves a small, unfavorable ΔH of $+51$ kJ mol^{-1} and is driven by its $-T\Delta S$ term of -121 kJ mol^{-1}.

$$H^+(OH_2) + F^-(H_2O) \rightleftharpoons HF(aq) + (H_2O:H_2O) \qquad \Delta G = -70 \text{ kJ} \qquad (5.16)$$

Reactions of the large, low-charged soft acids and bases generally do not involve much of an entropy change but are associated with a favorable enthalpy change. Equation 5.17, a typical soft-acid–soft-base reaction, involves a negligible $-T\Delta S$ term of $+10$ kJ mol^{-1} and is driven by the magnitude of its ΔH term, which is -315 kJ mol^{-1}.

$$Hg^{2+}(OH_2) + I^-(H_2O) \rightleftharpoons HgI^+(aq) + (H_2O:H_2O) \qquad \Delta G = -305 \text{ kJ} \qquad (5.17)$$

A hard and soft acid–base reaction includes both a HA:HB reaction similar to Equation 5.16 and a SA:SB reaction similar to Equation 5.17. Although Equation 5.16 is entropy-driven, the magnitude of its $-T\Delta S$ term is *much smaller* than the magnitude of the ΔH term of Equation 5.17. Therefore, the HSAB reaction is strongly *enthalpy-driven* (as suggested in the previous Amplification).

In contrast to Equation 5.17, many SA:SB products are insoluble solids, so they have very large lattice energies (Section 4.2). Following the thinking of Pauling, we would expect experimental lattice energies (which are enthalpies) to include contributions from soft acid:soft base covalency as well as from ionic bonding.

Polar Covalent Bonds Are Most Reactive. On examining the combination of a hard base and a soft acid (or a hard acid and a soft base), we find that the relative electronegativities and sizes are not optimal for either ionic or covalent bonding. The electronegativity differences of about 1 are characteristic of **polar covalent bonds**. The HSAB principle, in a sense, is restating the observation (common in organic chemistry) that polar covalent compounds tend to be rather reactive if they can react to give an ionic product and a nonpolar covalent product. Many of the more reactive species in organic chemistry (such as Grignard reagents and alkyl halides) have polar covalent bonds (C–Mg and C–halogen, respectively), and they often react according to the HSAB principle to generate new covalent and new ionic bonds:

$$
\begin{array}{ll}
R_2C(=O) & \text{(polar covalent)} \\
+ \\
H_3C\text{–}MgBr & \text{(polar covalent)}
\end{array}
\;\rightarrow\;
\begin{array}{ll}
R_2C\text{–}O^+MgBr & \text{(ionic)} \\
\;\vert \\
H_3C & \text{(covalent)}
\end{array}
\tag{5.18}
$$

EXAMPLE 5.14

AgBr(s) crystallizes in the NaCl lattice type. (a) Calculate the lattice energy U expected for AgBr(s). (b) The experimental lattice energy of AgBr(s) is -905 kJ mol^{-1}. Explain the discrepancy.

SOLUTION: (a) We apply Equation 4.9:

$$U = \frac{138{,}900\, Z_+Z_-}{d_0}\left(1 - \frac{1}{n}\right)$$

From Table 4.4 we find that the Born exponents n for both Ag$^+$ and Br$^-$ are 10. From Table 4.2 we find that the Madelung constant M for the NaCl lattice type is 1.74756. Table C gives us the radii of Ag$^+$ and Br$^-$; their sum $d_0 = (129 + 182)$ pm = 311 pm. Substituting these in Equation 4.9, we obtain

$$U = \frac{(138{,}900)(1.74756)(+1)(-1)}{311}\left(1 - \frac{1}{10}\right) = -702 \text{ kJ mol}^{-1}$$

(b) The formation of the AgBr(s) crystal lattice is more than 200 kJ mol^{-1} more exothermic than calculated—a very substantial amount. Equation 4.9 assumes that only ionic (electrostatic) bonding contributes to the lattice energy. The extra energy is due to the addition of covalent bonding energy to the lattice energy.

Visualizing Differences Among Ionic, Covalent, and Polar Covalent Bonding

AN AMPLIFICATION

Overlap between orbitals of the Lewis acid and the Lewis base is expected to be considerable in a covalent bond formed by a soft acid and a soft base. This overlap is expected to be absent in an ionic salt formed by a hard acid and a soft base. A polar covalent bond should be intermediate in nature. Pictures of such bonds can be generated using molecular orbital calculations, and experimental confirmation of overlap can be deduced using photoelectron spectroscopy (Section 11.3B). Results from these two methods have been presented by Lai-Shing Wang and co-workers,[31,32] showing the absence of overlap in the HA:HB ionic salt complex CsI$_2^-$, and strong overlap in the SA:SB complexes AuI$_2^-$ and Au(CN)$_2^-$. Intermediate overlap suggesting polar covalent bonding was shown for CuI$_2^-$, AgI$_2^-$, Cu(CN)$_2^-$, and Ag(CN)$_2^-$.

5.5B. The Need for More than One Parameter for Predicting Reaction Tendencies.
Despite its many successful predictions (many of which will be illustrated in the remainder of the chapter), we are by no means ready to replace all of the principles from earlier in the book with the HSAB principle. In general, it is impossible to characterize the Lewis acid–Lewis base interaction by a *single* parameter. Softness is just one parameter, which must be considered in addition to other parameters or concepts, such as the *strength* of the acid or base, and sometimes the *chelating ability* of the base.

When we evaluate the Z^2/r ratio for an acid or base, or count its oxo groups, we are evaluating its strength. When we have reactions involving stronger and weaker acids and bases, as in Section 2.4C and Figure 2.12 involving $HCl(aq)$, $HF(aq)$, $Cl^-(aq)$, and $F^-(aq)$, we might say that *stronger acids tend to react with stronger bases, and weaker acids prefer weaker bases.* (This also *approximately* summarizes the solubility rules and tendencies in Chapter 4.)

If a competitive equilibrium involves only one species with any degree of softness, we cannot apply the HSAB principle. Instead, we use the concepts of strength from the first few chapters of this book. For example, the aqueous equilibrium

$$Hg(ClO_4)_2 + 2KOH \rightleftharpoons Hg(OH)_2 + 2KClO_4 \qquad (5.19)$$

involves only one soft acid and no soft base. We can predict that products will be favored and $Hg(OH)_2$ and $KClO_4$ will precipitate by evaluating the strengths of the Lewis acids and Lewis bases. The K^+ ion is a nonacidic cation and should give a precipitate with the nonbasic ClO_4^- anion; Hg^{2+} is a weakly to moderately acidic cation and should give a precipitate with the strongly basic hydroxide anion.

Because a competitive Lewis acid–base reaction can be governed by two or three independent principles, we may expect to find cases in which the two principles come into conflict and in which the HSAB principle by itself fails to give a correct prediction. Equation 5.20 is an example:

$$[CH_3HgSO_3]^- \ + \ [OH_2{:}OH]^- \ \rightleftharpoons \ CH_3HgOH \ + \ [OH_2{:}SO_3]^{2-} \qquad (5.20)$$

SA:SB	HA:HB	SA:HB	HA:SB
Strong A–Weak B	Weak A–Strong B	Strong A–Strong B	Weak A–Weak B

Note that we have included the solvent, water, as a hard acid; it has a hard acceptor atom, H, engaged in hydrogen bonding to the anion. The analyses beneath the equation show that the HSAB principle favors reactants, while considerations of acid and base strengths favor products. The observed equilibrium constant of only 10 indicates that neither reactants nor products are strongly preferred over the other.

Fortunately, in everyday experience the HSAB principle usually overrides the tendency of stronger acids and bases to react, or the tendency of reactions to give chelation. Typical Gibbs free energy and enthalpy changes connected with HSAB covalent-bond formation are relatively large (e.g., $\Delta G = -305$ kJ mol^{-1} for Eq. 5.17). Typical Gibbs free energy or entropy changes driving chelation (ΔG for Fe^{3+} in Eq. 5.3 is -143 kJ mol^{-1}) or the aqueous reactions of acids and bases ($\Delta G = -70$ kJ mol^{-1} for Eq. 5.16) are smaller.

Although the HSAB principle clearly is not infallible, in real life (especially in aqueous solutions) it is highly useful and can help us understand and predict a number of phenomena in chemistry and related fields. Examples of these will be given in the remainder of this chapter.

EXAMPLE 5.15

The equilibrium constant for the reaction

$$HgI_2 + 2KOH \rightleftharpoons Hg(OH)_2 + 2KI$$

is relatively close to 1. Briefly explain why this reaction does not strongly tend to go to one side or the other.

SOLUTION: The HSAB principle favors the reactants, because the softest acid–softest base combination is HgI_2. But acid–base strength considerations favor products, because the strongest acid–strongest base combination is $Hg(OH)_2$. The two effects evidently approximately cancel each other out.

5.6. HSAB and the Solubilities of Halides and Chalcogenides

OVERVIEW. Two last solubility rules, V and VI, are based on HSAB and acidity classifications and allow you to predict whether a given halide, pseudohalide, sulfide, selenide, or telluride will be soluble or insoluble in water (Section 5.6A; Exercises 5.67–5.72). The geochemical classification of elements as atmophiles, lithophiles, or chalcophiles or siderophiles, is connected to the HSAB principle. They will help you identify likely mineral sources of elements. Soft-acid–soft-base salts often have structures that are quite different from those of the purely ionic salts that we examined in Section 4.1 (Section 5.6B; Exercises 5.73–5.81).

Before continuing, you may want to review Solubility Rules I–IV from Section 4.3, which were based on acidity classifications (Section 2.2). Other related properties of halides and sulfides appear in Sections 6.4 and 12.5. The topic of differentiation of the elements in the history of Earth is continued in Section 6.4.

5.6A. Solubility Rules V and VI for Halides and Chalcogenides.
In Section 4.3, we developed solubility rules, but these were only for oxo and fluoro anions, which are hard bases. Using the HSAB principle, we can now supplement these rules with solubility rules for salts of soft bases.

For the precipitation reaction involving an anion X^- from Group 17/VII (a halide ion), or a pseudohalide ion such as cyanide, thiocyanate, or azide,

$$KX(aq) + AgNO_3(aq) \rightleftharpoons AgX(s) + KNO_3(aq) \tag{5.21}$$

we are, in aqueous solution, actually dealing with hydrated ions. Water is a hard Lewis base (oxygen donor atom) and a hard Lewis acid (hydrogen acceptor atom). In net ionic form in aqueous solution, the precipitation reaction in Equation 5.21 may be represented as

$$X^-{:}H_2O \quad + \quad Ag^+{:}OH_2 \rightleftharpoons \quad AgX(s) \quad + \quad H_2O{:}H_2O \tag{5.22}$$

?B:HA   SA:HB   SA:?B   HB:HA

Thus, the HSAB principle favors products *if X^- is a soft base.*

TABLE 5.5. Insoluble Chlorides of the Elements[a]

Group 8	Group 9	Group 10	Group 11	Group 12	Group 13/III	Group 14/IV
			CuCl			
		$PdCl_2$	AgCl			
$OsCl_2$	$IrCl_2$	$PtCl_2$	AuCl	Hg_2Cl_2	TlCl	$PbCl_2$

[a] The lists of insoluble bromides and of insoluble iodides are similar but slightly larger. Note that in terms of solubility (and many other properties), chloride and azide behave like the soft bases, bromide and iodide.

Solubility Rule V, then, is the following: The soft halides and pseudohalides (chlorides, bromides, iodides, cyanides, thiocyanates, and azides) of the soft acids are insoluble. This applies to the +1- and +2-charged soft-acid metal ions (except Hg^{2+}). These metal ions (Table 5.5) are approximately the group of low-charged soft-acid metal ions identified in Table 5.2; note that they form (roughly) a triangle around the softest acid, Au^+.

A similar solubility rule applies to sulfides, selenides, and tellurides. **Solubility Rule VI** states that the **chalcogenides** (sulfides, selenides, and tellurides) and the salts of some homopolyatomic anions from Groups 15/VI and 16/VI (such as As_2^{4-} and S_2^{2-}) of the soft and borderline acids are insoluble. We may rationalize this solubility rule by noting the following:

1. The S^{2-}, Se^{2-}, and Te^{2-} ions are stronger bases than the Cl^-, Br^-, and I^- ions.

2. The borderline-acid metal ions tend to be somewhat stronger acids than the soft-acid metal ions, because the borderline acids generally have higher charges or smaller sizes than the soft-acid metal ions.

This combination of softness and some strength enables these borderline acids to combine with the more strongly basic (–2-charged) soft bases. The closer in electronegativity the soft and borderline acids are to these soft bases, the lower the solubility and K_{sp} are of the products—HgS, for example, has an extraordinarily low K_{sp}.

Soluble Salts of the Soft Halide and Chalcogenide Ions. The HSAB principle cannot be used to predict the solubility of the halides and chalcogenides of the hard acids, because *no soft acid* is involved in their solubility equilibria. The concept of strength can be used, however. The halide ions are nonbasic anions, so they give insoluble salts only with large nonacidic (hard) acids (Solubility Principle III).

The sulfides, selenides, and telluride ions are basic anions, so they give insoluble salts with (hard) acidic cations, such as La^{3+}, Ti^{4+}, or Al^{3+} (Solubility Principle I). However, the oxide ion is more strongly basic than the sulfide, selenide, or telluride ions, so the oxides of hard-acidic cations are even more insoluble than the sulfides, selenides, or tellurides. Consequently, the sulfides, selenides, and tellurides of the more acidic hard acids, although insoluble, decompose in water to give the even more insoluble oxides or hydroxides.

$$TiS_2(s) + 2H_2O \rightarrow TiO_2(s) + 2H_2S(aq \text{ or } g) \tag{5.23}$$

EXAMPLE 5.16

Suppose that we add a solution of a mixture of sulfide and thiocyanate ions to solutions of (a) Sb^{3+} and (b) Ag^+. Write the formula of the salt that is likely to precipitate (if any) in each case.

SOLUTION: (a) Sb^{3+} is a borderline acid, and sulfide is a soft base, so by Solubility Rule VI, Sb_2S_3 precipitates, but by Solubility Rule V, the thiocyanate does not.

(b) Ag^+ is a soft acid and sulfide is a soft base, so by Solubility Rule VI, Ag_2S precipitates. By Solubility Rule V, AgSCN also precipitates. Thus, a mixture of precipitates is expected.

EXAMPLE 5.17

Using the hypothetical elements classified in Example 5.11, determine whether their bromides, thiocyanates, cyanates, and sulfides will be soluble. Write the formulas of the insoluble salts.

SOLUTION: Let us begin by classifying the four anions and three cations by softness and strength.

Ion	Hardness or Softness	Strength
Br^-	Soft base (Table 5.2)	Nonbasic (Section 2.6)
SCN^-	Soft base (S donor)	Feebly basic (Figure 5.17)
OCN^-	Borderline/hard base	Weakly basic (Figure 5.17)
S^{2-}	Soft base (Table 5.2)	Strongly basic (Section 2.6)
Mr^+	Soft acid (Example 5.11)	Feebly acidic (see calculations below)
Mm^{2+}	Soft acid (Example 5.11)	Weakly acidic (see calculations below)
Mq^{3+}	Hard acid (Example 5.11)	Moderately acidic (see calculations below)

Based on their Z^2/r ratios and electronegativities (Section 2.2), Mr^+ has $Z^2/r = 1/125 = 0.008$, but has an electronegativity over 1.8, so it is feebly acidic. Mm^{2+} has $Z^2/r = 4/103 = 0.039$, but has an electronegativity over 1.8, so it is weakly acidic. Because Mq^{3+} has $Z^2/r = 9/62 = 0.145$, it is moderately acidic.

Because the HSAB principle, when applicable, usually trumps acid/base strength, we employ it first with the soft bases Br^- and SCN^- and the soft (or borderline) acids Mr^+ and Mm^{2+}. Solubility Rule V applies to bromide and thiocyanate ion, and states that MrBr, MrSCN, $MmBr_2$, and $Mm(SCN)_2$ should be insoluble. Solubility Rule VI applies to sulfide ion, and states that Mr_2S and MmS should be insoluble.

The cyanate ion is hard, so HSAB does not apply. As a weak base, it should give insoluble salts with acidic cations (Solubility Rule I), and it may give precipitates with feebly acidic cations (Solubility Rule IVA). Therefore, $Mm(OCN)_2$ and $Mq(OCN)_3$ would probably be insoluble, while MrOCN might be insoluble.

Because sulfide is also a strong base, by Solubility Rule I it could still give insoluble salts with acidic cations—for example, MmS and Mq_2S_3. However, these are probably going to react with the water present to give the HA:HB precipitates MmO or $Mm(OH)_2$, and $Mq(OH)_3$.

5.6B. The Geochemical Classification of the Elements—Chalcophile Minerals.

While studying the types of minerals in which different elements are found, Berzelius noted in the first half of the nineteenth century that certain metals tend to occur as sulfides and others as carbonates or oxides. Victor Goldschmidt, the father of modern geochemistry, classified the elements into four classes—namely, lithophiles, chalcophiles, atmophiles, and siderophiles—according to their predominant geological pattern of behavior[33] (Table 5.6).

1. **Lithophiles** are the metals and nonmetals that tend to occur as cations in oxides, fluorides, silicates, sulfates, or carbonates. Because these anions (except F⁻) all possess oxygen as a donor atom, the metal ions involved are the hard acids. The nonmetals that are classified as lithophiles either are inherently hard bases or have been oxidized by atmospheric oxygen to oxo anions, which are hard bases. Because these oxo anions and fluoride anion are hard bases, the insoluble minerals mentioned in Chapter 4 exemplify minerals formed by lithophiles.

2. **Chalcophiles** occur in nature as cations in sulfides (and less commonly with other, scarcer soft bases such as telluride and arsenide). The chalcophiles are mostly the borderline and some soft acids. Logically, many of the soft bases are also listed as chalcophiles, because they are found as the anions in such minerals.

We will pay less attention here in Chapter 5 to the other two categories, the atmophiles and siderophiles, which are based on behavior in oxidation–reduction chemistry (Section 6.4B).

3. **Atmophiles** are the chemically unreactive nonmetals that occur in the atmosphere in elemental form (N_2 and the noble gases).

4. **Siderophiles** are the metals that tend to occur native (in the elemental form). When they are found in compounds, these tend to be sulfides, telluride, or arsenides. Here in Chapter 5 the siderophiles are grouped with the chalcophiles.

TABLE 5.6. Geochemical Classification of the Elements[a]

					Lithophiles										Atmophiles		
																	He
Li	Be											B	C	N	O	F	Ne
Na	Mg											Al	Si	P	S	Cl	Ar
K	Ca	Sc	Ti	V	Cr	Mn	Fe	Co	Ni	Cu	Zn	Ga	Ge	As	Se	Br	Kr
Rb	Cr	Y	Zr	Nb	Mo		Ru	Rh	Pd	Ag	Cd	In	Sn	Sb	Te	I	Xe
Cs	Ba	Lu	Hf	Ta	W	Re	Os	Ir	Pt	Au	Hg	Tl	Pb	Bi			
								Siderophiles					Chalcophiles				

[a] This classification emphasizes behavior under conditions at the surface of Earth; many elements can display other classes of behavior as well.

EXAMPLE 5.18

Classify each element as either a lithophile or either a chalcophile or a siderophile. Then choose its most likely mineral source. (a) Element: La; source: $LaPO_4$, LaAs, LaI_3, or $LaCl_3$. (b) Element: Pt; source: $PtAs_2$, PtN_2, $PtSiO_4$, or PtF_2. (c) Element: Zr; source: $ZrSiO_4$, $ZrPbS_4$, or $ZrCl_4$. (d) Element: Sb; source: $Sb_2(SiO_3)_3$, Sb_2S_3, SbF_3, or $SbCl_3$. (e) Element: Te; source: TeF_4, Na_2Te, PbTe, or TeI_4.

SOLUTION: (a) La is a hard acid, so it is likely a lithophile; mineral source: $LaPO_4$.

(b) Pt is a soft acid, so it is likely a chalcophile or siderophile; mineral source: $PtAs_2$.

(c) Zr is a hard acid, so it is likely a lithophile; mineral source: $ZrSiO_4$.

(d) Sb is a borderline acid, so it is likely a chalcophile; mineral source: Sb_2S_3.

(e) Te is a soft base, so it is likely a chalcophile; mineral source: PbTe.

Distinctive Structures for Chalcogenide Minerals. As previewed in Section 4.1C, covalency in these compounds usually results in changes in their crystal structures. Referring back to Table 4.2, salts of 1:1 stoichiometry and radius ratios between 0.225 and 0.414 commonly take one of the ZnS structures, whether they are predominantly ionic or covalent. When a higher coordination number is selected, the rock salt (NaCl) lattice type is often bypassed in favor of the nickel arsenide (NiAs) lattice type. This structure has six-coordination around the As atom. The coordination around the Ni atoms involves six As atoms but also two other Ni atoms within bonding distance. Because close approach of Ni cations to each other is unexpected, the bonding in NiAs and the two dozen similar 1:1 late d-block chalcogenides, arsenides, etc., is better treated as *metallic covalent* (Section 12.4). (Note also that Eq. 4.1 cannot be applied to these compounds, because it is based on counting cations and anions.)

Soft-base compounds of apparent 1:2 stoichiometry may in fact have, not monatomic anions, but diatomic or larger anions. For example, FeS_2 (pyrite or fool's gold) has equal numbers of Fe^{2+} and S_2^{2-} anions, and adopts a structure similar to the NaCl type. Similar anions account for the strange formulas of FeAsS, $CoAs_2$, and $PtAs_2$.

As mentioned in Section 4.1.C, soft-acid–soft-base compounds of true 1:2 stoichiometry often adopt *layered* structures, such as the CdI_2 structure (Figure 4.6c) or the molybdenum disulfide structure (Figure 5.27). The molybdenum disulfide structure differs from the cadmium halide structures in having trigonal prismatic coordination around the Mo atoms. This geometry brings the sulfur atoms fairly close together, which is unexpected for S^{2-} ions repelling each other; some degree of S–S bonding is suggested in this structure. (Further important properties of MoS_2 and related layered compounds are discussed in Section 12.5.)

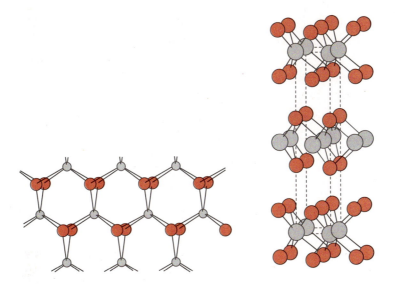

Figure 5.27. The molybdenum disulfide (MoS_2) structure from different perspectives. Gray circles = Mo atoms; **brown** circles = S atoms. [Adapted from A. F. Wells, *Structural Inorganic Chemistry*, 3rd ed., Clarendon Press: Oxford, UK, 1962; and C. E. Stanton, S. B. T. Nguyen, J. M. Kesselman, P. E. Laibinis, and N. S. Lewis, "Semiconductors," in *Encyclopedia of Inorganic Chemistry*, R. B. King, Ed., John Wiley and Sons Ltd.: Chichester, UK, 1994, p. 3727.]

A CONNECTION TO GEOCHEMISTRY

Chalcophile Minerals

By far the most common of these are the numerous *sulfides* of the soft- and borderline-acid minerals. Chalcophile minerals* are major sources of their metals. For soft acids we mention black *argentite* (Ag_2S) and red *cinnabar* (HgS). Many borderline-acid metals are found in or obtained from sulfides; those with asterisks (*) are also found in minerals with various hard bases. Arsenic* is found in yellow *orpiment* (As_2S_3) and red *realgar*[†] (molecular As_4S_4), antimony is found in black *stibnite*[†] (Sb_2S_3), bismuth* in brown-black *bismuthinite* (Bi_2S_3), lead in black *galena*[†] (PbS), zinc* as *zinc blende* or *wurzite*[†] (ZnS), cadmium as yellow-orange *greenockite* (CdS), copper* as black *chalcocite* (Cu_2S) and *chalcopyrite*[†] ($CuFeS_2$), nickel* as *pentlandite* [$(Ni,Fe)_9S_8$], cobalt as dark gray *linneaite* (Co_3S_4), iron* as golden *pyrite*[†] or *fool's gold* (FeS_2), and molybdenum as black *molybdenite* (MoS_2).

Some of the soft acids can be found associated with softer bases than sulfide—namely *telluride* and *arsenide*. Many of these minerals (indicated with the symbol §) are more frequently found as pure elements, so they are also classified by geochemists as siderophiles. Among these are gold[§] in *calaverite* ($AuTe_2$) and *sylvanite* [$(Ag,Au)_2Te_4$], cobalt in gray *smaltite* [$(Co,Fe,Ni)As_2$], nickel in red *niccolite* (NiAs), and platinum[§] in *sperrylite* ($PtAs_2$).

* Often their crystals are beautiful, especially when photographically magnified in a book such as O. Medenbach and H. Wilk, *The Magic of Minerals*, Springer-Verlag: Berlin, 1986. Minerals pictured in this book are indicated with a dagger (†); however, photos of these and others can readily be found online.

Halide minerals are far less common than sulfide, selenide, and arsenide minerals, because halides are not as insoluble. Although insoluble soft halide minerals such as AgX, CuX, and *calomel* [mercury(I) chloride, Hg_2Cl_2] are known, the only one that is an important mineral source is *cerargerite* or *horn silver* (AgCl).

In contrast, F^- is a hard, feebly-to-weakly basic anion, so insoluble fluoride minerals are mainly known for hard feebly acidic cations—namely, *villiaumite* (NaF), *sellaite* (MgF_2), and much more importantly *fluorite*† (CaF_2). Because most halides are soluble, halide mineral sources of some elements are found in desert areas in which water solutions have evaporated—namely, *carnallite* ($KMgCl_3 \cdot 6H_2O$), *sylvite* (KCl), and *halite*† or *rock salt* (NaCl). Lithium salts are often extracted from *brines* (salty waters) that have not evaporated.

5.7. HSAB and the Biological Functions and Toxicology of the Elements

OVERVIEW. In this section we hope to develop your ability to avoid poisoning yourself and others! When using chemicals, it is valuable to be able to read GHS (Globally Harmonized System) pictograms symbolizing their hazards (Section 5.7A; Exercises 5.82–5.84). The toxicities and biological functions of Lewis acid and base forms of elements can be discussed and to some extent predicted with reference to the HSAB principle (Section 5.7B; Exercises 5.85–5.92). Different forms or "species" of elements can differ in their toxicity and methods of entry into the body (Section 5.7C; Exercises 5.93–5.97). This section is generally connected to the fields of chemical safety and toxicology; a connection to environmental chemistry is also included.

Note that we previewed the biological functions of the elements in Section 1.2. The concept of speciation was introduced in Sections 3.8, 3.9, and 4.4. The biological properties of hemoglobin will be amplified in Section 7.9. We return to GHS pictograms in Section 6.2.

5.7A. Reading Safety and Toxicity Symbols on Commercial Chemicals. The objective of this subsection is to extend your life expectancy as a person who uses chemicals. In the space available we cannot hope to duplicate the excellent coverage given in the book *Laboratory Safety for Chemistry Students*,[34] but we will highlight a few useful things to know. Safety information on chemicals is listed in the U.S. more or less usefully in the Materials Science Data Sheets (MSDS) provided by manufacturers. We hope to help you anticipate possible problems so that you are more likely to refer to MSDS sheets when it is clearly important.

In order to facilitate safety alerts across national and linguistic boundaries, the United Nations has devised a set of pictograms and an associated classification system, known as the Globally Harmonized System for Classification and Harmonization of Chemicals, or GHS. These have been adopted by the U.S. Occupational Health and Safety Administration and are now being implemented by chemical suppliers in the U.S. and elsewhere. The nine basic pictograms are shown in Figure 5.28.

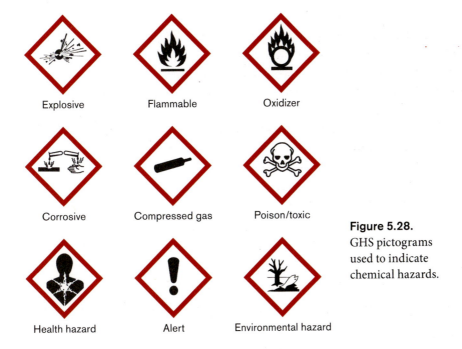

Figure 5.28.
GHS pictograms used to indicate chemical hazards.

These nine pictograms identify four categories of hazards:

1. "Explosive," "Flammable," and "Oxidizer" refer to oxidation–reduction hazards that we cover in Section 6.2.

2. "Corrosives" most often have acid–base reactivities that you should now be able to broadly anticipate from what you learned in Chapters 2 and 3.

3. "Compressed Gases" are hazardous, not for their chemical properties, but because their cylinder containers contain (often liquefied) gases under high pressure. If the valve at the top of the cylinder should break off, the pressure might be released so rapidly that the cylinder would become a rocket, possibly able to shoot itself through concrete walls (or you!). Such cylinders should *always* be restrained by belts that will prevent them from falling over and damaging their valves!

4. Toxicity issues are indicated when any of the last four pictograms in Figure 5.28 are present on the label of the chemical bottle. We focus on the acute toxicities of the chemicals by oral ingestion. The material is considered to be **poisonous** by oral ingestion (swallowing) and to be fatal if swallowed if its lethal dose for half of test animals (LD_{50}) is < 5 mg per kg of body weight (level 1 hazard), or if its LD_{50} is between 5 and 50 mg kg^{-1} (level 2 hazard). The material is considered **toxic** if its LD_{50} is between 50 and 300 mg kg^{-1} (level 3 hazard). If any of these apply (or corresponding levels for skin or inhalation exposure), the Poison/Toxic <Skull and Crossbones> pictogram is placed on the bottle. If the LD_{50} is between 300 and 2000 mg kg^{-1}, it is considered to be "Harmful if Swallowed" and the Alert <!> pictogram is present. Other types of toxicity issues are symbolized either by the Alert, Health Hazard <head and chest>, and/or Environmental Hazard <fish and tree> pictograms; consult a text for details.[34]

5.7B. Essential and Toxic Effects of the Elements and the HSAB Principle.

In Chapter 1 we gave a preview of the biological functions of the elements, indicating essential elements in Table 1.1 and illustrating some biological functions by blocks of the periodic table in Tables 1.2, 1.4, 1.5, and 1.6. Table 5.4 again lists some of these functions and indicates the donor atoms of biochemical ligands to which the essential metal ions prefer to bind. In this section we will review and regroup these functions using the HSAB principle, then add additional details which that principle facilitates. The choices of donor atoms follow the general pattern we would expect from the HSAB principle.

Hard Acids. Toxicities of metal cations and simple and polyatomic anions tend to be related to the HSAB principle. As we see from Tables 5.1, 1.1, 1.2, and the left groups of Table 1.4, several hard-acid metal ions are essential for life. The metal ions Na^+, K^+, Mg^{2+}, and Ca^{2+} often function in the form of their hydrated ions, which the body allows to pass selectively across various barriers. This passage of charged ions constitutes an electric current, which can transmit a nerve impulse or trigger some response (Sections 4.5B and 5.2). Both Mg^{2+} and Ca^{2+} also serve structural purposes—Mg^{2+} in preserving the double helical structure of DNA and Ca^{2+} in the form of calcium phosphate or calcium carbonate as bones, teeth, and shells.

Most other hard-acid metal ions are small and/or highly charged. Thus, they are unavailable in natural waters (Section 3.8B), because they precipitate as hydroxides. Hence, most hard-acid metal ions are *neither essential nor toxic*. Beryllium is one exception—it is very dangerously poisonous—by inhalation it causes a fatal or disabling disease called berylliosis. Apparently, it substitutes for Mg^{2+} in certain enzymes in the body, but being much more acidic, it disrupts the functions of those enzymes. Barium is also toxic if soluble, because its size is so similar to that of K^+ that it can pass through the same channels into cells. There it behaves differently and is toxic, because its acidity is higher than that of K^+. In the form of insoluble $BaSO_4$ (Solubility Tendency IVC), it is safely given as "barium milkshakes" to make the stomach opaque to X-rays.

Borderline Acids. Many of the lighter borderline-acid metal ions, although needed only in small quantities, have critical biochemical functions. Their appreciable Lewis acidity enables them to bind certain ligands (e.g., O_2) or to enhance the acidity and reactivity of organic ligands as a step in synthesizing or metabolizing them. Many of these metal ions also have two or more accessible oxidation states differing by only one electron, which allows them (in appropriate enzymes) to catalyze important redox reactions.

In excessive quantities, these metal ions can be toxic; hence, borderline acids tend to be both *essential* and *toxic*. For example, excess iron causes siderosis and may result in an increased risk of heart attacks. High exposure to manganese in the workplace gives rise to symptoms similar to Parkinson's disease.[35] In Wilson's disease, Cu^{2+} cannot be eliminated, producing toxic concentrations.

Soft Acids. In order to function properly, enzymes must maintain their correct three-dimensional conformational structure. This is maintained by comparatively weak hydrogen bonds and the sulfur–sulfur single bonds found in the "dimerized" amino acid cystine. Its S–S bond, which is formed by oxidation of the S–H bonds of two cysteine amino acid components, is easily ruptured by the action of soft-acid metal ions (giving S–metal bonds instead); this deactivates or *denatures* the enzyme. Critical

EXAMPLE 5.19

Check the labels of chemicals in the stockroom, or (more conveniently) check in a chemical company's catalog to determine (using just GHS pictograms) the relative toxicity hazards of beryllium chloride, barium acetate, and barium sulfate.

SOLUTION: We consulted the 2012–2014 Sigma-Aldrich™ *Handbook of Fine Chemicals*. As expected from the above discussion, there were no GHS pictograms shown for barium sulfate. Barium acetate showed the Alert pictogram, which indicates less than maximal skin or eye irritation, toxicity, or specific organ toxicity. Beryllium chloride shows three pictograms: poison/toxic, health hazard, and environmental hazard. $BeCl_2$ is the most dangerous of the three (its LD_{50} value is 4.4 mg kg^{-1}). What would the lethal dose be for you?

enzymes, being very efficient at catalyzing biochemical reactions, are often present in quite low concentrations and are quite susceptible to deactivation by soft-metal ions. Hence, all soft-metal ions are *toxic* when present in the body at concentrations of tens of milligrams per kilogram of body weight (tens of parts per million, ppm); none is essential (other than protein-bound complexed Cu^+ found in cells and involved in re-dox reactions of Cu^{2+}). This protein-bound Cu^+ allows us to detect the smell of toxic organosulfur compounds in air at extraordinarily low concentrations.[36] (See Section 12.4D for details on Ag^+.)

Lead is a notorious soft-acid poison, occurring in the modern environment be-cause of the former practice of using a lead hydroxy-carbonate as a white paint pig-ment and more recently from the use of tetraethyllead $[(CH_3CH_2)_4Pb]$ as an antiknock agent in gasoline. As expected, the lead ion reacts with enzyme SH groups. Its most serious effects arise from the consequent deactivation of two enzymes needed for the biosynthesis of heme and from adverse effects upon the function of the brain. There are suspicions that childhood exposure to lead ions is linked to increased frequency of criminal activity in adults.[37] Too much soil is contaminated with excessive lead levels to be removed and replaced by clean topsoil. Precipitating the Pb^{2+} with S^{2-}, which is itself very toxic, is similarly unworkable. However, Pb^{2+} is not only a soft acid, but it is also weakly acidic and forms a sufficiently insoluble phosphate, $Pb_5Cl(PO_4)_3$. It has proven practical and safe to precipitate Pb^{2+} in soils by treating the soil with dry fish bones (a good source of phosphate).[38]

Biochemically, cadmium behaves as a soft acid and is quite toxic. In excess, it causes itai-itai disease, in which Cd^{2+} substitutes for Ca^{2+} in the bones; the bones be-come brittle and break easily and painfully. Its levels also correlate with increased lev-els in the population of diseases related to high blood pressure. (However, it has been found as a cofactor for a carbonic anhydrase used in marine microbes.[39])

Tl^+ is very toxic but is seldom encountered. Because of its similar charge and size, Tl^+ accumulates in tissues along with K^+. It is soft, however, so it behaves differently once in cells, reacting with –SH groups of vital enzymes.

Hard Bases. Toxicity is *not* a general property of the most common hard bases, the oxo anions. The use of fluoride in drinking water for preventing tooth decay resulted in the oldest environmental health controversy. It converts the mineral hydroxylapatite [$Ca_5(PO_3)_3(OH)$; Section 4.3A] in the surface of teeth to the less basic and more acid-resistant $Ca_5(PO_3)_3F$ (apatite). Although fluoridation of water supplies has reduced the incidence of tooth decay, fluoride is toxic at slightly higher levels.[40] Its conjugate acid, HF, penetrates the skin and causes extreme irritation to that part of the body.

Another recent controversy involves the potential toxicity of the perchlorate ion (ClO_4^-), which is used in rocket fuels (Section 6.2) but also found naturally in small amounts. Iodide and perchlorate are both nonbasic and have similar radii (I^-, 206 pm; ClO_4^-, effectively 226 pm) Perchlorate is a competitive inhibitor of the uptake of iodide ion, which is essential in the production of thyroid hormones.[41] Iodide deficiency was formerly a major problem, producing the disease goiter and leading to mental retardation and possibly other problems in infants. This problem was solved by the simple expedient of adding sodium iodide to salt to give iodized salt. Unfortunately, in recent decades the iodide level in pregnant women and in breast milk has declined substantially. Perchlorate was suspected as a cause of this problem, but the problem may be due to a deficiency of iodine in modern diets. This may be because, although purchased salt is iodized, 80% of the salt used in preprocessed and fast foods is *not* iodized.[42]

Soft Bases. Many of the soft-base nonmetals are beneficial in small doses and may even be essential. Sulfur and phosphorus are major components of amino acids, nucleic acids, and so on. Chlorine is essential as Cl^-, whereas iodine is essential in the form of the hormone thyroxin. Of particular interest are the very soft arsenic and selenium, which are both *essential* and *toxic*. Selenium is a component of glutathione peroxidase, and because of its reducing properties, it protects against free radicals, which may be implicated in aging and causing cancer. Selenocysteine is incorporated in peptides synthesized in cells as the 21st amino acid.[43] Because selenium is softer than sulfur, it attracts soft metal ions preferentially and thus protects against their toxic effects (such as carcinogenesis). Thus, despite the fact that tuna have been found with elevated mercury levels, no one has been found to suffer mercury poisoning from high consumption of tuna. Tuna is also high in selenium, which may tie up the mercury in the form of insoluble mercury selenide species. Recent studies have confirmed that mercury binds Se more strongly than S. Because Se is essential, it may be that part of mercury's toxicity is due to its binding to selenoenzymes and precipitating HgSe.[44] An application of the strong Hg–Se affinity is in the use of a cloth that is impregnated with nanoparticles of elemental selenium to capture the metallic mercury vapor that is released by a broken compact fluorescent light bulb.[45]

On the other hand, both of these elements are very toxic. Selenium occurs in nature at high levels in certain soils in New Zealand and the western United States, where it is concentrated in certain plants such as locoweed. Cattle eating this weed develop a selenium-poisoning disease known as the blind staggers. Based on the HSAB principle, the accumulation of Se much in excess of that needed to tie up soft-acid metals may result in binding to borderline-acid metals that are essential in enzyme functions. Whether or not this happens, it is not a contradiction to say that selenium (or other substances) can be both essential and (in slightly higher doses) also highly toxic. ("The dose makes the poison.")

A CONNECTION TO ENVIRONMENTAL CHEMISTRY

H_2S, Hibernation, and Mass Extinction

Surprisingly, despite its toxicity, H_2S is now known to have important biological roles in cell signaling, where it mediates metabolic rate, inflammation, and blood pressure.[46] It was recently found, at lower than toxic levels, to drastically to reduce the O_2 consumption, body temperature, and rate of CO_2 production of mice—in effect, putting the mice in suspended animation or "hibernation."[47]

The "mother" of all mass extinctions occurred 251 million years ago between the Permian and Triassic eras. The current theory is that massive and prolonged volcanic activity in the Siberian Traps (lasting as long as 200,000 years) released enormous volumes of the greenhouse gas CO_2. The atmospheric level of CO_2 jumped tenfold from levels similar to those we are currently experiencing, which caused rapid global warming and acidification of the rain and the ocean.

At this time about 79% of marine animals went extinct,[48] due mainly to the enhanced acidity of the ocean water making it impossible to make shells out of calcium carbonate (Section 4.3A). Acid rain also dissolved more phosphate minerals from the land, causing (apparently) eutrophication of the ocean with the growth of algae, which deprived the ocean of oxygen. Anaerobic H_2S-rich deep waters reached the surface, releasing 2000 times as much H_2S to the atmosphere as occurs in current times. This suffocated and poisoned nearly all oxygen-breathing life and deprived Earth of its protective ozone layer.[49] Over a few million years the composition of the atmosphere dropped from about 30% O_2 down to 15% O_2,[50] which is not enough to sustain many forms of life, particularly in utero. It is thought that around 90% of all terrestrial species became extinct at this time. Only one medium-sized animal survived the horrendous conditions: *Lystrosaurus* dominated the nearly barren Earth for a few million years, and then gave rise to all dinosaurs and birds on the future Earth. This may have been the time when life most nearly became extinct on Earth.[51]

Other soft bases are also well-known toxins. The soft carbon donor atoms of carbon monoxide (:CO) and the cyanide ion (:CN⁻) bind strongly to borderline-acid metal ions such as Fe^{2+} in heme. Thus, the hard base O_2 is unable to compete equitably with :CO for the iron ion in hemoglobin, and cyanide prevents the heme-based enzyme cytochrome c oxidase from binding O_2. Many nonmetal hydrogen (Section 2.4D) compounds, such as phosphine, arsine, hydrogen sulfide, hydrogen selenide, and their methylated forms, such as trimethylphospine [$(CH_3)_3P$] and so on, have soft donor atoms and are very poisonous. Being molecular and of fairly low molecular weights, these compounds are volatile and pass readily through skin. They generally have very revolting odors (rotten egg or garlic smells), and our bodies have learned to sense them and warn us.

5.7C. Speciation and Toxicity. The toxic effects of many elements depend on the chemical form or "species" in which they are found.

Speciation and Toxicity of Mercury. For a complete evaluation of the toxicity of environmental mercury, we need to know its chemical form or "species."[52] The

---

EXAMPLE 5.20

(a) Which of the hypothetical ions of Example 5.11 are most likely to be toxic?
(b) Suppose two of the toxic ions occurred together in one (binary) compound. Would this compound likely be more or less toxic than the separate ions ingested at different times?

SOLUTION: (a) The soft acids Mr^+ and Mm^{2+} will be toxic; probably the two soft bases, Fl^{2-} and Az^{2-}, will be, too. (b) Binary compounds involving these ions would have one soft acid and one soft base. The resulting compounds (Mr_2Az, Mr_2Fl, $MmAz$, or $MmFl$) would be insoluble, so they likely would be less available and less toxic than the separate hydrated ions.

---

forms of importance include (1) elemental or metallic mercury, (2) soluble inorganic mercury salts (mostly salts of hard bases), (3) insoluble mercury salts (mostly of soft bases), (4) salts of the methylmercury cation (CH_3Hg^+), and (5) dimethylmercury [$(CH_3)_2Hg$]. [Confusingly, both CH_3Hg^+ and $(CH_3)_2Hg$ are commonly referred to as "methylmercury."]

Metallic mercury is insoluble and passes through the digestive system unchanged (but this is not true of surface contaminants of inorganic mercury). However, it is volatile, and the vapors are exceedingly toxic. Spills of metallic mercury need to be cleaned up carefully, because scientists working for years in laboratories containing spilled metallic mercury have been poisoned. Mercury is also encountered in the form of *amalgams*, in which other metals are dissolved in the mercury. The most familiar amalgam is the silver amalgam used in dental fillings. Over 20 years, most of the mercury evaporates from such a filling, and what remains is converted to HgS (see below). Metallic mercury is also used to dissolve metallic gold from river beds, but much is then lost to the environment.

Inorganic mercury as Hg^{2+} is toxic and, even when it is not directly placed in water as a pollutant, is formed by oxidation of metallic mercury vapors emitted into the upper atmosphere during volcanic eruptions and the combustion of coal. This Hg^{2+} dissolves in rainwater as soluble mercury(II) salts such as $HgCl_2$, which are then absorbed in the gastrointestinal tract. If a single dose of a soluble or stomach acid-soluble inorganic mercury salt is ingested and does not cause death, the body can excrete the Hg^{2+} in the urine over a period of months. In contrast, very insoluble mercury salts of soft bases such as HgSe and HgS are relatively nontoxic to begin with, because they are nonvolatile and insoluble and cannot be absorbed by the body.

Insoluble mercury salts such as Hg_2Cl_2 (calomel), HgS, and HgSe are insoluble by Solubility Rule V—HgS and HgSe are so insoluble that they are virtually nontoxic. HgS has been used for millennia as the red paint and cosmetic pigment *cinnabar*.

The *methylmercury cation* (CH_3Hg^+) is formed from soluble inorganic mercury salts by biomethylation. Cases of methylmercury poisoning have occurred from eating fish that came from waters into which only Hg^{2+} had been dumped. Certain aquatic microorganisms are able to methylate Hg^{2+} using the natural methylating agent methylcobalamin (vitamin B_{12}; Figure 5.1c). Methylcobalamin acts as "nature's Grignard agent" and transfers the soft base CH_3^- to Hg^{2+} or other soft-acid metal cations (such as those of arsenic, selenium, and tellurium), thus generating species such as CH_3Hg^+:

$$CH_3\text{–cobalamin} + H_2O + Hg^{2+} \rightarrow CH_3Hg^+ + [(H_2O)\text{–cobalamin}]^+ \qquad (5.24)$$

In most cases, biomethylation increases the solubility of the metal or metalloid in fats and lipids (Section 4.4D), which increases toxicity. CH_3Hg^+ is perhaps the softest of all cations, and it coordinates to sulfur and selenium donor atoms (e.g., in SH groups in the side chains of cysteine, selenocysteine, and methionine; Figure 5.23a, b) in the body, readily moving from one such donor atom to another. It crosses the blood–brain barrier as a water-soluble complex with L-cysteine, $CH_3Hg\text{–S–}CH_2\text{–CH}(NH_2)\text{–}$ COOH. Methylmercury thiol complexes are difficult to excrete and tend to accumulate in plankton and then increase in concentration in fish feeding on them in the process of bioamplification (Section 4.4D). Methylmercury complexes cause brain damage and birth defects.

Dimethylmercury [$(CH_3)_2Hg$] is both volatile and absorbable through the skin and digestive system. It is much more soluble in fats and lipids than in water, and it cannot readily be excreted. Due to this lipid solubility, it can also penetrate the blood–brain and the placental barriers, the skin, and even rubber gloves. $(CH_3)_2Hg$ was adopted as the mercury NMR standard. In 1997 a chemist named Karen Wetterhahn was preparing a sample of this, when one to a few drops fell on her latex glove. It went through the glove, through her skin, and into her bloodstream. The poisoning did not show up for several months, but it was untreatable; tragically she died.[53]

Speciation and Toxicity of Arsenic. Arsenic poisoning has been well known for millennia. Its toxicity is also affected by speciation. The most common environmental form, arsenate (AsO_4^{3-}), passes through skin slowly. Arsenic in the +3 oxidation state is more toxic than arsenate, and it poisons by forming As–S bonds with critical SH-containing enzymes; it passes through skin 29 times more rapidly. As we will be able to deduce in Chapter 6, arsenite is the dominant species in flooded rice paddies. Methylated arsenic species such as methylarsonic acid [$CH_3\text{–As}(O)(OH)_2$] and dimethylarsinic acid [$(CH_3)_2AsOOH$] are formed bacterially[54] by the action of another biological methylating agent, S-adenosylmethionine (Figure 5.29).[55] With its positively charged sulfur atom, S-adenosylmethionine transfers CH_3^+ to metalloids having an unshared electron pair on the metalloid atom. (This is followed by acid ionization of one As–OH group.)

$$CH_3^+ + \text{:As}(OH)_3 \rightarrow H^+ + CH_3\text{–As}(O)(OH)_2 \qquad (5.25)$$

Processes similar to Equation 5.25 can produce dimethylarsinic acid, which passes through the skin 59 times faster than arsenate.[56]

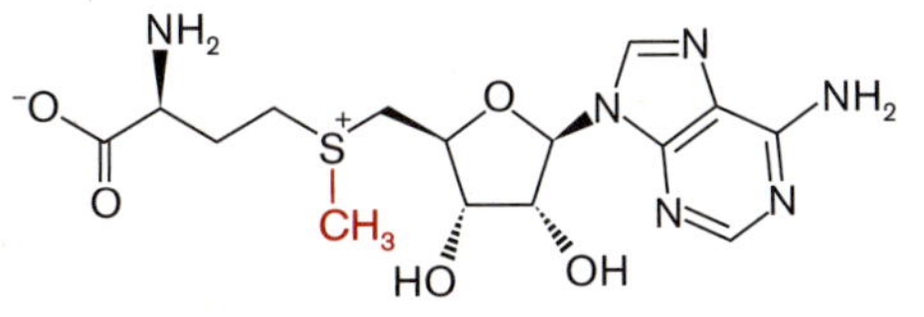

Figure 5.29. The biological methylating agent S-adenylmethionine. The methyl group attached to the S atom is transferred as CH_3^+.

Arsenic Poisoning in Bangladesh and the United States

The world's most serious environmental poisoning problem is present in Bangladesh. The well water on which 70 million people depend is contaminated with toxic levels of arsenic. This comes from subsurface rocks, and up to 200,000 people have been diagnosed with arsenic poisoning. The arsenic seems to come from oxygenation of arsenide-contaminated FeS_2 to give FeOOH and arsenate, which binds strongly to the FeOOH. However, in the mangrove swamps, the FeOOH is reduced and solubilized as Fe^{2+}, releasing arsenate to the groundwater that goes into the wells.[57]

The standard method of purifying drinking water from arsenate involves the addition of Fe^{3+}, followed by adjusting the pH to between about 5 and 7.2.[58] To some extent this may precipitate $FeAsO_4$ by Solubility Rule 1, but probably the main route is by precipitating FeOOH or $Fe(OH)_3$, followed by adsorption of arsenate by hydrogen bonding to surface hydroxyl groups:

$$[Fe]-OH + H_2AsO_4^- \; \rightleftharpoons \; [Fe]-OH_2^+ \text{----} O_3As(OH)^{2-} \qquad (5.26)$$

This treatment is unfeasible for village use, though, so an international competition for a $1 million prize was held to find a way of detoxifying Bangladeshi well water, which would be within the very limited financial resources of the villagers. This prize was won by Prof. Abdul Hassam of George Mason University, who devised a two-step filter costing less than $40.00. The first layer uses sand to remove sediment, and processed cast iron containing some Mn to catalyze the oxidation of $As(OH)_3$ to $H_2AsO_4^-$ and $HAsO_4^{2-}$. These bond to the surface of the cast iron perhaps according to Equation 5.26 or perhaps more strongly by direct $Fe-O-AsO_3$ bonding. The second layer uses wood charcoal to adsorb organic arsenic compounds. The arsenic species are bound so irreversibly to the iron that they do not come off with continued use, and there is little or no hazard associated with disposal of the spent filter.

In the U.S. inorganic arsenic is a contaminant of concern in rice grown in fields that were previously used to grow cotton. Although the practice of using lead arsenate as an insecticide in cotton fields was stopped in the 1980s, levels in the soil persist decades later when the fields have been converted to growing rice.[59]

5.8. Medicinal Inorganic Chemistry

> **OVERVIEW.** This section does not involve new concepts but rather provides an opportunity to review and re-apply old concepts from this chapter and from Section 2.1 (on acidity), Sections 3.2 and 3.3 (on bonding types and charges in complex ions), and speciation and solubility (Sections 3.8 and 4.4). Connections to medicinal chemistry and biochemistry are included. You can use HSAB and the chelate principles to select likely ligands for use in obtaining metal nutrients or detoxifying and eliminating toxic metals. Practice these Section 5.8A concepts by trying Exercises 5.98–5.103. Metal-based medicinals are also widely used to treat various diseases such as cancer (Section 5.8B). Given structures of these medicinals, you can review the application of concepts from this and the earlier chapters by trying Exercises 5.104–5.109.

5.8A. Chelation and HSAB Treatment of Toxic Inorganics. Given the potency and variety of metal ions in biochemical systems, the removal of excess or unwanted metal ions selectively from the body (or even from analytical samples or from wastewaters) is no simple task. Nor is the acquisition of essential but normally unavailable metal ions (Section 3.8) a trivial matter. Any number of ligands can be imagined that may form a complex with a given metal ion, but a number of additional constraints are usually imposed. Concentrations of the offending metal ion and the reagent to be used may both need to be low, which does not favor complex formation unless a very stable complex results. At the same time, the ligand should not form complexes that are too stable with other metal ions, because these may need to be undisturbed. In analytical chemistry, it may be desirable for the complex to be soluble in nonpolar solvents so that it can be extracted and concentrated. To some extent, this is necessary in medicinal chemistry, too, so that the ligand and metal complex can cross nonpolar cell membranes, but if the complex is too water insoluble, it cannot be excreted in the urine.

Recent review articles give more details on chelating medicinals that are used in treating a dozen or more metal intoxicants.[60,61] In designing reagents for these tasks, the medicinal, analytical, or environmental chemist has two main principles to guide him or her: the HSAB principle and the principle that chelating ligands form more stable complexes than do monodentate ones (Section 5.1B). Figure 5.30 shows some drugs used to remove excess or toxic metal ions from the body.

Some of the *f*-block elements (such as plutonium) of the seventh period are exceedingly toxic due to their radioactivity. Plutonium tends to concentrate in the bones, where even a few millionths of a gram emit enough dangerous alpha particles to irradiate the bone marrow and cause leukemia. DPTA (Figure 5.30) is the treatment of choice for plutonium poisoning, being one of the few ligands capable of dislodging the Pu^{4+} ion from its precipitated form in the bones. The problem lies in its nonselectivity—it prefers highly charged metal ions but will form very stable complexes with almost any metal ion except a +1-charged one. Excess DPTA can remove calcium ions from the hydroxyapatite (Section 4.3) in bones. This problem can be overcome by administering DPTA with calcium ions already chelated in it.

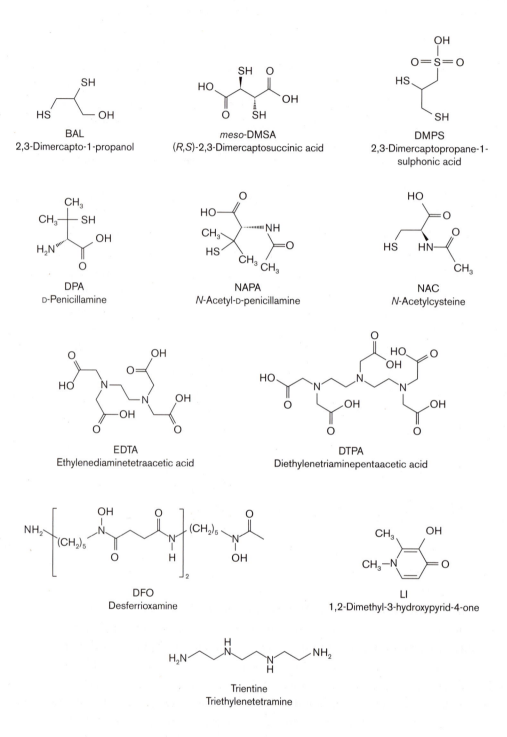

Figure 5.30. Some chelating ligands used as drugs to remove metal ions from the body. [Adapted from M. Blanuša, V. M. Varnai, M. Piasek, and K. Kostial, *Curr. Med. Chem.* 12, 2771 (2005).]

Among the numerous chelating ligands, selectivity is enhanced using the HSAB principle. The first such drug, British anti-Lewisite (BAL; Figure 5.30), when deprotonated at the –SH groups, chelates with soft-base sulfur donor atoms, and thus selectively chelates soft acids in the body. Originally developed during World War I as an antidote for the war gas Lewisite, $ClCH=CH_2AsCl_2$, it was subsequently used to treat the poisoning by other soft acids such as mercury and thallium ions. BAL is quite toxic and water insoluble, so it has been supplanted by newer drugs that contain the –CO_2H and –SO_3H functional groups needed to enhance water solubility and reduce toxicity. EDTA (Section 5.1C and Figure 5.30) was an early chelating medicinal with predominantly hard-base donor atoms. The more recently developed DPTA ligand is a variation on EDTA.

Copper—A Toxin and a Nutrient. As an example, we consider the treatment of two diseases rooted in opposite difficulties with copper-ion metabolism—namely, Wilson's disease (in which the body cannot excrete excesses of copper ion, which accumulates to toxic levels) and Menke's disease (in which the body cannot absorb needed copper ions from the intestine). One specific antidote for Wilson's disease is the inorganic salt ammonium tetrathiomolybdate, $(NH_4)_2MoS_4$, which was discovered in the sulfide-rich stomachs of sheep and cows feeding on molybdate-rich soils (they developed copper deficiency).[62] This soft-base anion is thought to reduce free Cu^{2+} to the soft acid Cu^+, which is then complexed. Also useful in treating Wilson's disease and other types of soft-acid poisoning are D-penicillamine and NAPA (Figure 5.30); these were discovered as metabolism products of the antibiotic penicillin. In contrast, Menke's disease is treated by providing the patient with a nontoxic complex of Cu^{2+} with the amino acid histidine (Figure 5.31).

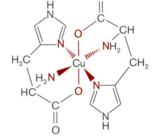

Figure 5.31. Main form of the complex of copper(II) with histidine.

Iron—Another Toxin and Nutrient. Iron is another metal whose ions are both essential and toxic (Section 5.7B). Iron-deficiency problem (anemia) is common, because Fe^{2+} is normally unstable to oxidation and Fe^{3+} is too acidic to be available at the pH of the intestine, where absorption must take place—it precipitates as $Fe(OH)_3$. Hence, vitamins usually contain, not simple iron salts, but salts of chelating organic anions. In Section 3.8B we discussed the low availability of iron in ocean waters, where it is often the nutrient limiting the growth of plankton and other organisms on which fish depend. Survival of these organisms is then favored by developing the most efficient way of extracting the limited iron from water. Bacteria achieve this by secreting hard-base chelating ligands known as siderophores. The fungus *Streptomyces pilosus* produces the chelating ligand desferrioxamine (Figure 5.30) to obtain its iron. Desferrioxamine contains three of the unusual hydroxamic acid functional groups, any one of which [when in the deprotonated form –C(=O)–N(–O^-)–] is already able to chelate iron.

EXAMPLE 5.21

You are trying to design drugs to treat poisoning by Hg^{2+} and Be^{2+} ions by forming very stable, excretable complexes of these metals. (a) Given the following mercury-containing complexes, choose the one that would give the best excretion. Explain your choice in a few words.

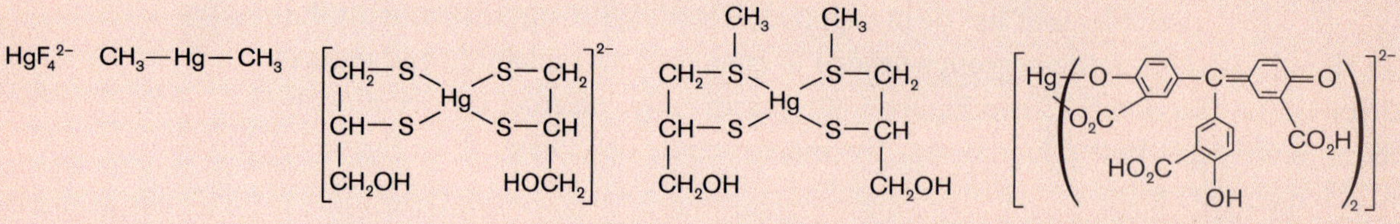

(b) Likewise choose the best beryllium-containing complex. Explain your choice in a few words.

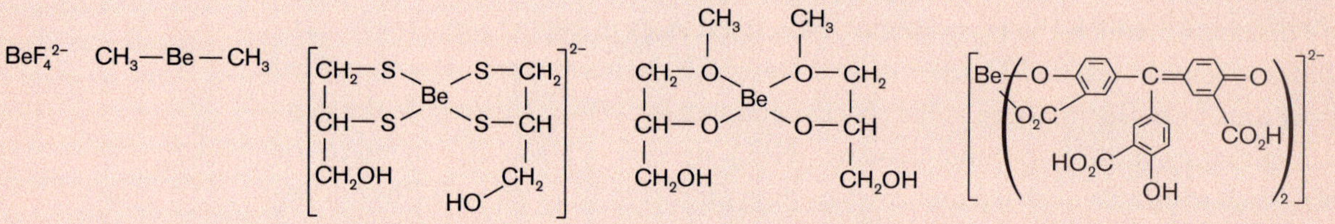

SOLUTION: (a) $Hg[(SCH_2)SCH(CH_2OH)]_2^{2-}$ has soft-base donor atoms, chelate rings, and is charged. Hence, it is more likely to be water soluble and excretable (Section 4.4D).

(b) The last (aromatic) Be complex has hard-base donor atoms, chelate rings, and is charged. Hence, it is more likely to be water soluble and excretable.

In the opposite situation, iron overload[63] then has toxic effects when iron builds up in organs, because the body has evolved no mechanism (other than bleeding) for excreting excess iron. Iron poisoning (siderosis) is treated with desferrioxamine, which has much better complexing ability for iron and aluminum ions than for other essential cations such as those of Ca, Mg, Cu, and Zn.

A CONNECTION TO BIOCHEMISTRY

Biological Responses to Excess Soft-Acid Metal Ions

Many biological organisms have developed their own methods (not always known to man) to deal with excesses of toxic elements or deficiencies of essential elements. In the body, cadmium concentrates in the kidneys, where it is bound in the protein metallothionein. This protein contains 20 cysteines binding Zn^{2+}, Cu^+, and Cu^{2+} ions, among its 60–70 amino acid groups. When Cd^{2+} is taken up, it displaces Zn^{2+} and is tetrahedrally coordinated by four sulfur atoms. Metallothionein also binds other soft acids such as Cu^+, and has other biological functions.[64] Bacteria confronted with mercury can also respond by activating a mercury-detoxification enzyme that reduces the metal ion to elemental mercury.

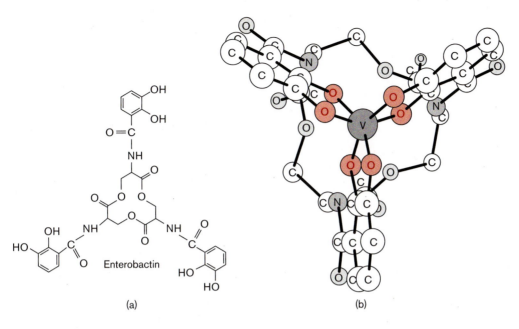

Figure 5.32. The structures of (a) enterobactin, and (b) enterobactin chelating a V^{4+} cation. [Part (b) is adapted from S. J. Lippard and J. M. Berg, *Principles of Bioinorganic Chemistry*, University Science Books: Mill Valley, CA, 1994, p. 109.]

Enterobactin (Figure 5.32), another siderophore, is a hexadentate chelating ligand (binding with six deprotonated phenolic oxygen donor atoms), which has a formation constant for complexation of one Fe^{3+} ion to one enterobactin hexaanion of 10^{49}! The bacterial siderophore alcaligin also contains three hydroxamic acid groups on a more preorganized organic backbone, which is therefore very efficient at binding iron.

Hyperaccumulator plants[65] have evolved in areas with naturally high concentrations of certain normally toxic metal or nonmetal ions such as Zn^{2+}, Cd^{2+}, Ni^{2+}, Pb^{2+}, and selenium, and have developed ways of storing them in startlingly high concentrations without harm to themselves. At an extreme, a tree from the Pacific island of New Caledonia exudes a green latex that is 25% nickel by weight; a single tree may contain about 37 kg of Ni. Other plants hyperaccumulate essential/toxic metal ions such as those of Mn, Fe, Cu, and Zn, as well as mainly toxic elements such as Cd, Hg, Pb, Se, and As. The ash from such plants can have a very high concentration of the specific metal accumulated, so it can be a useful ore. These plants are being investigated for use in a natural process of cleaning up soils or waste dumps that contain hazardous concentrations of Ni, Zn, Cd, As, and Se;[66] some are even suitable for beautifying dumps. Crop plants have been genetically modified to produce excess citrate ion to chelate Al^{3+}, which can occur at toxic levels in acidic soils.[67]

Often the hyperaccumulated metals are stored in vacuoles of the plants as complexes of the citrate ion (Figure 5.33), but citrate does not form stable enough complexes to extract these ions from soils in which their concentrations are low. Specific ligands may be secreted to complex these metal ions and transport them to the vacuoles. One simple ligand that is involved is histidine (as in Figure 5.31). Another series of (probably hexadentate) chelating ligands that are secreted include nicotianamine (Figure 5.34) and its derivatives.

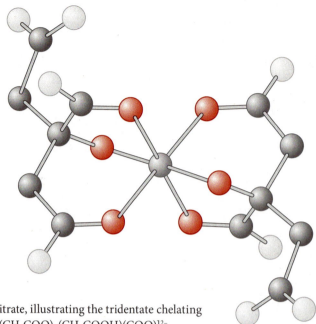

Figure 5.33. Structure of gallium citrate, illustrating the tridentate chelating function of the citrate anion [HOC(CH$_2$COO)-(CH$_2$COOH)(COO)]$^{2-}$. The central **gray** atom is Ga; the **brown** atoms are coordinated O atoms, the white atoms are other O atoms, and the black atoms are C. Note that one carboxylate group in each citrate ligand is not chelated; this carboxylate group is also not deprotonated. [Adapted from M. Claussén, L.-O. Öhman, and P. Persson, *J. Inorg. Biochem.* 99, 716 (2005).]

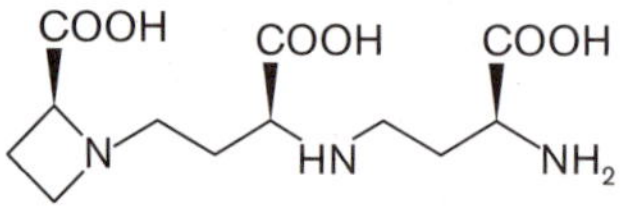

Figure 5.34. Structure of nicotianamine, a plant product that strongly chelates metals.

5.8B. Inorganics in Medicinal Chemistry.

Inorganic elements and compounds have been used for medicinal purposes for 5000 years. Near the beginning of the twentieth century, Paul Ehrlich discovered the first cure for syphilis. The 606th compound that he tried, it was an organoarsenic compound, known as salvarsan or arsphenamine or just "606." It is now known to be a mixture of cyclic organoarsenic compounds, *cyclo*-As$_n$-[(4-HO)(3-H$_2$N)C$_6$H$_3$]$_n$ (n = 3, 5).[68] This discovery was followed by the successful application of other soft acids or soft bases to the treatment of syphilis, including various gold, antimony, and mercury compounds. However, the side effects and toxicities of these early drugs were severe (the LD$_{50}$ for Ehrlich's compound is 100 mg kg^{-1}), and within a couple of decades they fell out of use, being superseded by less dangerous organic medicine. The loss of this application led research in inorganic chemistry to have a couple of "lost decades" during which it was thought of as a dead field. Medicinal inorganic chemistry itself was reborn in 1969 with the serendipitous discovery by Barnett Rosenberg of the anticancer properties of the complex *cisplatin*, *cis*-[Pt(NH$_3$)$_2$Cl$_2$]. During the current renaissance of the field, the discovery of new inorganic drugs has characteristically been followed by studies of the mechanism of their action, as well as studies of ways to modify their structures to improve their therapeutic properties. We have selected a few examples from two book-length reviews or textbooks of medicinal inorganic chemistry.[69,70]

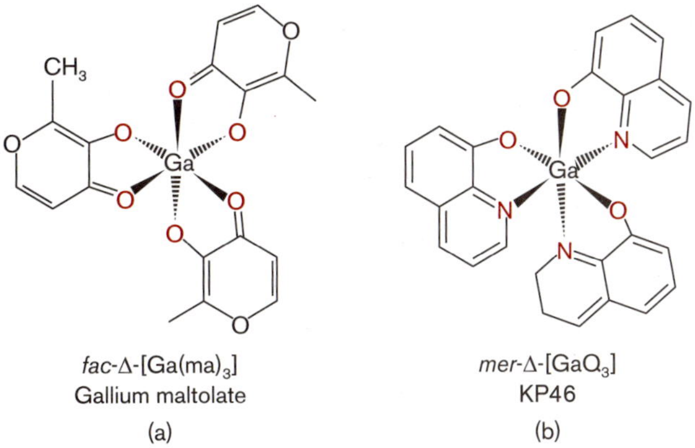

fac-Δ-[Ga(ma)$_3$]
Gallium maltolate
(a)

mer-Δ-[GaQ$_3$]
KP46
(b)

Figure 5.35. Gallium complexes of medicinal interest: (a) One isomer of the complex of Ga^{3+} with three maltolate anions; (b) one isomer of the complex of Ga^{3+} with three 8-hydroxyquinolate anions. [Adapted from J.C. Dabrowiak, *Metals in Medicine*, John Wiley and Sons Ltd.: Chichester, UK, 2009, p. 179.]

Lithium and Gallium Ions. The use of Li$^+$ in treating bipolar disorder and preventing suicides was noted in Table 1.3. Its action is not well understood, but may involve competition with weakly acidic Mg^{2+} and feebly acidic Ca^{2+} in their binding sites. It is thought to be exchanged for Na$^+$ and even K$^+$ when crossing cell membranes.[71] Ga^{3+} has been used for various medicinal reasons for 80 years, but is now of interest for its anticancer properties. Due to its acidity, it is unavailable (Section 3.8) at physiological pH values, so it is administered in a chelated form, with the bidentate chelating ligands maltol or 8-hydroxyquinoline (Figure 5.35). Because its charge, size, and electronegativity are extremely good matches for those of Fe^{3+}, it follows iron biochemical pathways in the body, often substituting for Fe^{3+} in metalloproteins.

Technetium Imaging Agents. ^{99m}Tc is an artificially produced γ-ray emitter with a half-life of 6 hours that is administered in medicine as a 10^{-10} M solution to obtain images of the heart, bones, brain, and kidneys. The γ rays are energetic and easily detected, but the short half-life means that the patient is not exposed to these rays for long. It is concentrated in the living, growing parts of the target organ, and allows contrast with the organ parts that are not growing.

Technetium is produced originally as the ^{99m}TcO$_4^-$ ion, which is not readily taken up by any organ. However, it can be reduced with SnCl$_2$ to lower oxidation states. The oxidation state that is stabilized is determined by the ligand that is provided to complex it. The charge and nature of the product then determines which organ selectively takes it up. Figure 5.36 shows three of the commonly used technetium complexes. The +1-charged "Cardiolite" cation (Figure 5.36a) accumulates in living heart muscle via the path that takes in K$^+$. Then γ rays are detected coming from living heart tissue, but not from tissue that has its blood supply interrupted due to a clot. The technetium complex of the methylenediphosphonate anion (Figure 5.36b) accumulates in growing bones, which also involve phosphate groups. Bones grow particularly after breaks or in bone cancer, so ^{99m}Tc γ-emission shows where such growth is occurring. The neutral Tc-HMPAO complex (Figure 5.36c) has many methyl groups around its periphery, so it readily crosses the blood–brain barrier and can be used in diagnosing various brain disorders such as stroke and dementia.

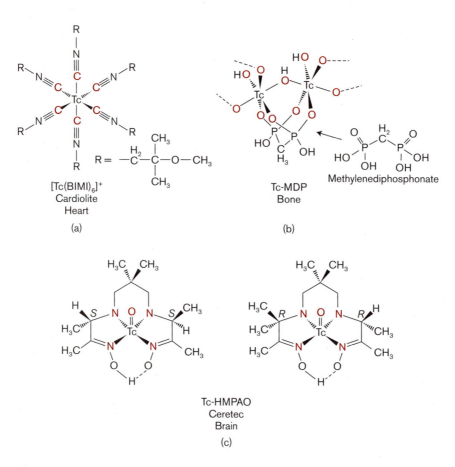

Figure 5.36. Some commonly used ^{99m}Tc imaging agents. (a) The $[Tc(BIMI)_6]^+$ or Cardiolite cation; (b) a portion of the Tc-methylenediphosphonate or Tc-MDP polymeric complex, with the parent acid of the methylenediphosphonate anion; (c) and the two optical enantiomers of the Tc-HMPAO or Ceretec complex. [Adapted from J. C. Dabrowiak, *Metals in Medicine*, John Wiley and Sons Ltd.: Chichester, UK, 2009, p. 255.]

Platinum Anticancer Agents. Cisplatin is a square planar complex with chloride ligands in adjacent cis positions. (Figure 5.37) Its anticancer action involves binding to the N-7 donor atoms of two adjacent guanine bases on the same strand of DNA, forming a 17-membered macrocyclic ring. This bends the strand of DNA about 40° off its normal axis and thus inhibits the process of cell division.

Before cisplatin can reach a target, it can undergo reaction with water (in the body or in the medicine itself), forming the acidic cations *cis*-$[(H_3N)_2PtCl(OH_2)]^+$ and *cis*-$[(H_3N)_2Pt(OH_2)_2]^{2+}$. This undesirable outcome is reduced by administering the cisplatin as a 0.0033 M solution that is also 0.154 M in NaCl. Cisplatin is unfortunately toxic to the kidneys. Cancer cells also build up resistance to it, as they increase concentrations of the soft-base ligands glutathione and metallothionein to trap platinum(II). As a result, second-generation platinum anticancer drugs have been developed and have largely replaced cisplatin. One of these is carboplatin (Figure 5.38), which is less susceptible to reaction with water and is much less toxic than cisplatin; it is now in widespread use. It does suppress white blood cell production, however. Oxaliplatin (Figure 5.38) is also used against colorectal cancer and cisplatin-resistant tumors, but it has other side effects.

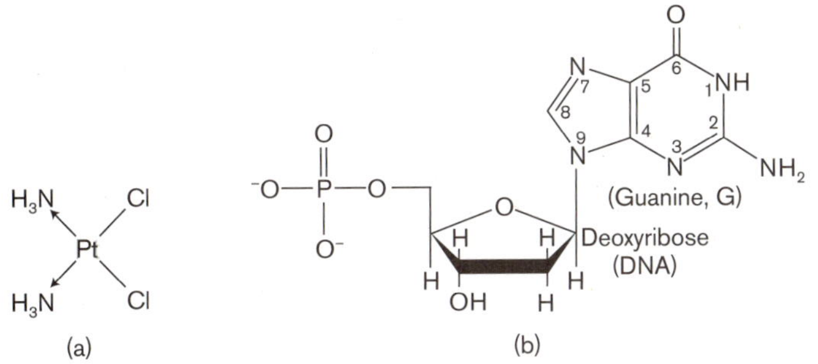

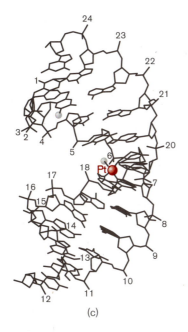

(a)

(b)

(c)

Figure 5.37. Binding of cisplatin to the N-7 donor atom of guanine in DNA. (a) The structure of cisplatin; (b) the structure of the nucleotide, deoxyguanosine 5′-monophosphate; and (c) the crystal structure of cisplatin bound to a shortened version of DNA. [Parts (a) and (b) are adapted from J. C. Dabrowiak, *Metals in Medicine*, John Wiley and Sons Ltd.: Chichester, UK, 2009, p. 97. Part (c) is adapted from P. M. Takahara et al., *Nature*, 377, 649 (1995).]

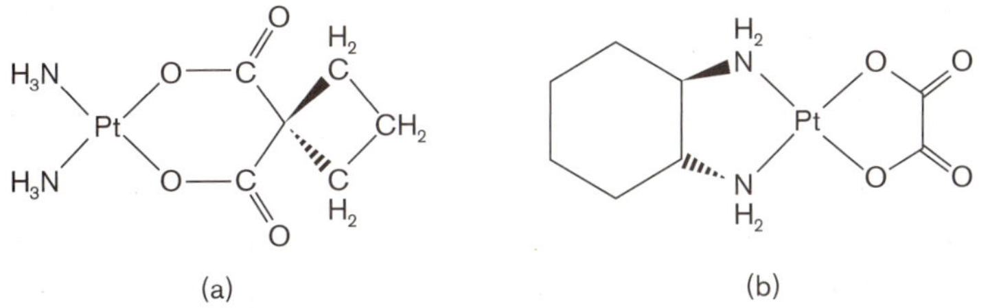

(a)

(b)

Figure 5.38. Structures of (a) carboplatin and (b) oxaliplatin.

5.9. Background Reading for Chapter 5

Chelate and Macrocyclic Ligands and Complexes

John Burgess, *Ions in Solution: Basic Principles of Chemical Interactions*, 2nd ed., Horwood Publishing: Chichester, UK, 1999, Chapter 6: Stability Constants, pp. 74–92.

Metal-Organic Framework Compounds

M. Eddadoudi and J. F. Eubank, "Insight into the Development of Metal-Organic Materials (MOMs): At Zeolite-Like Metal-Organic Frameworks (ZMOFs)," in *Metal-Organic Frameworks: Design and Application*, L. R. MacGillivray, Ed., John Wiley and Sons: Hoboken, NJ, 2010, pp. 37–83.

L. Margonelli, "An Inconvenient Ice," *Sci. Amer.* 311(4), 83–89 (2014). The energy potential and environmental hazards of massive deposits of methane hydrate.

Photographs of Crystals

O. Medenbach and H. Wilk, *The Magic of Minerals*, Springer-Verlag: Berlin, 1986. An electronic version of this book is available.

Metal Sulfides (Historical Significance)

D. Rickard, *Pyrite: A Natural History of Fool's Gold*, Oxford University Press: Oxford, UK, 2015. The importance of the most common sulfide, FeS_2, from its earliest role in starting fires by striking flint against pyrite.

Laboratory Safety

R. C. Hill, Jr. and D. C. Finster, *Laboratory Safety for Chemistry Students*, John Wiley and Sons: Hoboken, NJ, 2010. For information on GHS Symbols, see Section 3.2.1, "The Globally Harmonized System of Classification and Labelling of Chemicals (GHS), pp. 3-25 to 3-29, and Section 6.2.1, "Using the GHS to Evaluate Chemical Toxic Hazards," pp. 6-11 to 6-22.

End-Permian Extinction and the Role of H_2S

M. J. Benton, *When Life Nearly Died: The Greatest Mass Extinction of All Time*, Thames & Hudson: London, 2003.

P. D. Ward, "Impact from the Deep," *Sci. Amer.* 295(4), 64–71 (2006). Most mass extinctions are associated with global warming, oxygen depletion in the air and ocean, and the surfacing of deep-ocean anaerobic bacteria to emit H_2S into the air.

R. Wang, "Toxic Gas, Lifesaver," *Sci. Amer.* 302(3), 66–71 (2010). Key roles of H_2S in the body.

Medicinal Inorganic Chemistry

J. C. Dabrowiak, *Metals in Medicine*, John Wiley and Sons Ltd.: Chichester, UK, 2009.

P. J. Sadler, C. Muncie, and M. A. Shipman, "Metals in Medicine," in *Biological Inorganic Chemistry: Structure and Reactivity*, I. Bertini, H. B. Gray, E. I. Steifel, and J. S. Valentine, Eds., University Science Books: Mill Valley, CA, 2007, pp. 95–135.

J. Emsley, *The Elements of Murder: A History of Poison*, Oxford University Press: New York, 2005. A very readable history of poisoning with the soft acids Hg, As, Sb, Pb, and Tl. Be careful who sees you reading this!

Numbered references from this chapter may be viewed online at www.uscibooks.com /foundations.htm.

5.10. Exercises

5.1. *Identify the approximate bond and lone pair–donor atom–X angles at each potential donor atom in Exercise 3.13.

5.2. Identify the approximate bond and lone pair–donor atom–X angles at each potential donor atom in Exercise 3.14.

5.3. *Give the geometries and approximate donor atom–metal–donor atom angles at each central atom of the complexes in Exercise 3.20.

5.4. Give the geometries and approximate donor atom–metal–donor atom bond angles at each central atom of the complexes in Exercise 3.25.

5.5. *Which of the following can be a chelate ligand? (a) NH_3; (b) $NH_2-CH_2-CH_2-NH_2$; (c) NH_2-NH_2; (d) $NH_2-CH_2-CH_2-NH_3^+$.

5.6. Consider the following five molecules: (1) $H_2N-CH_2-(C=O)-OH$; (2) $CH_3-O-CH_2-O-CH_3$; (3) $H_3N-CH_2-C(=O)O$; (4) $[HO(O=)C-CH_2]_3N$; and (5) $H-C(=O)-OH$. (a) Complete the Lewis structures of these molecules by drawing in their unshared electron pairs. Underline each potential donor atom. (b) Circle the functional groups in the molecules. Which of the molecules are polyfunctional (have more than one functional group)?

(c) Which of these molecules have donor atoms appropriately spaced for them to be chelating ligands? (d) Classify each chelating ligand as bidentate, tridentate, tetradentate, and so on.

5.7. *Which of the following are chelating ligands: (a) PO_4^{3-}; (b) 9-crown-3 [*cyclo*-$(O-CH_2CH_2)_3$]; (c) $CH_3-O-CH_2-O-CH_3$; (d) $(CH_3)_3N$; (e) $N(CH_2-CO_2^-)_3$; and (f) $[NH_3CH_2CH_2NH_3]^{2+}$?

5.8. (a) Which of the ligands drawn below are chelating ligands? (b) The Pd^{2+} ion normally has a coordination number of four. Predict the formula (including overall charge) of the complex ion formed by the Pd^{2+} ion, showing a coordination number of four, with the needed number of each ligand.

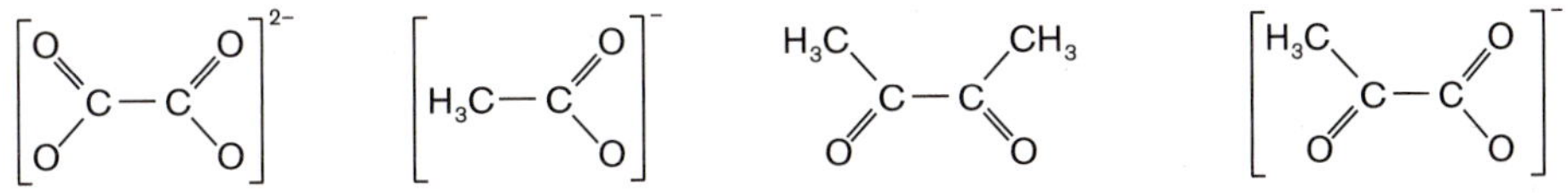

5.9. *Molecules and ions **A–E** may be capable of acting as ligands. (a) Pick out all of them that are capable of acting as chelating ligands. (b) Classify each potentially chelating ligand as (potentially) bidentate, tridentate, and so on. (Choose the highest possible denticity for each.)

A $[H_3N-CH_2-CH_2-NH_3]^{2+}$

C $[(O_2C-CH_2)_2NCH_2CH_2N(CH_2CO_2)_2]^{4-}$

B

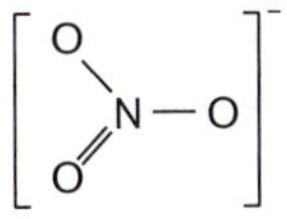

D

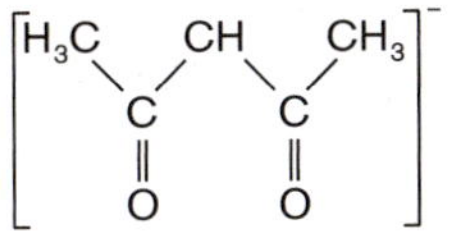

E $:N\equiv C-\ddot{S}-\ddot{S}-C\equiv N:$

5.10. Below is drawn a neutral coordination compound of the metal niobium, Nb. (a) What is the coordination number of Nb in this complex? (b) What is the likely geometric shape about the Nb atom? (c) What are the likely approximate bond angles about the Nb atom? (d) List all chelating ligands present in this complex.

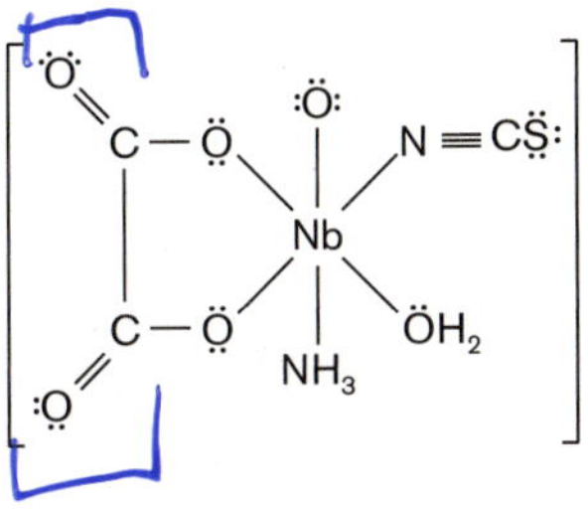

5.11. *Below are drawn the Lewis structures of chelated complexes **A–D**. (a) For each atom that is part of the actual chelate ring (i.e., the metal atom, the two donor atoms, and the atoms that connect the donor atoms) use VSEPR theory to predict the ideal bond angles within the chelate ring at that atom. (b) If the chelate ring is to be planar, these angles must add up to $(n - 2)$ times 180°, where n is the number of atoms in the chelate ring. Add up the predicted angles for each chelated complex. Which complexes, with which values of n, come closest to the VSEPR ideal? (c) For some structures that show strain due to nonideal angles in the planar structure, a nonplanar structure can be found that will allow nearly ideal angles to be achieved. Using molecular models, determine for which nonideal ring sizes (values of n) this is feasible.

A   **B**   **C**   **D**

5.12. Predict whether reactants or products are favored in each of the equilibria:

(a) $[Ni\{\eta^6\text{-}NH_2(CH_2CH_2NH)_4CH_2CH_2NH_2\}]^{2+} + 6NH_3 \rightleftharpoons$
$$NH_2(CH_2CH_2NH)_4CH_2CH_2NH_2 + [Ni(NH_3)_6]^{2+}$$

(b) $[Cu(\eta^2\text{-}CH_3OCH_2CH_2OCH_3)_2]^{2+} + 4cyclo\text{-}(CH_2CH_2O)_2 \text{ ("dioxane")} \rightleftharpoons$
$$[Cu\{\eta^1\text{-}cyclo\text{-}(CH_2CH_2O)_2\}_4]^{2+} + 2CH_3OCH_2CH_2OCH_3$$

(c) $[Ni\{\eta^6\text{-}NH_2(CH_2CH_2NH)_4CH_2CH_2NH_2\}]^{2+} + 3NH_2CH_2CH_2NH_2 \rightleftharpoons$
$$NH_2(CH_2CH_2NH)_4CH_2CH_2NH_2 + [Ni(\eta^2\text{-}NH_2CH_2CH_2NH_2)_3]^{2+}$$

5.13. *Predict whether reactants or products are favored in each of the following equilibria:

(a) $2(NH_2CH_2CH_2)_2NH + [Ni(cyclo\text{-}NC_5H_5)_6]^{2+} \rightleftharpoons$
$$[Ni\{\eta^3\text{-}(NH_2CH_2CH_2)_2NH\}_2]^{2+} + 6cyclo\text{-}NC_5H_5$$

(b) $[Ca(\eta^3\text{-}O\text{-}PO_2\text{-}O\text{-}PO_2\text{-}O\text{-}PO_2\text{-}O)_2]^{8-} + 3(O\text{-}PO_2\text{-}O\text{-}PO_2\text{-}O)^{4-} \rightleftharpoons$
$$[Ca(\eta^2\text{-}O\text{-}PO_2\text{-}O\text{-}PO_2\text{-}O)_3]^{10-} + 2(O\text{-}PO_2\text{-}O\text{-}PO_2\text{-}O\text{-}PO_2\text{-}O)^{5-}$$

5.14. Data for examining the effects of ionic radius on the complexation tendencies of EDTA^{4-} have been collected by Mitchell[72] and are tabulated below.

Cation	log K_{eq}	ΔH (kJ mol^{-1})	$-T\Delta S$ (kJ mol^{-1})
Mg^{2+}	8.64	+14.6	−62.6
Ca^{2+}	10.6	−27.4	−32.8
Sr^{2+}	8.53	−17.1	−32.8
Ba^{2+}	7.63	−20.6	−23.8

(a) Based on Chapter 2 and the results of Experiment 2, what trend in the ΔH term for complexation as a function of ionic radius would you expect? Is this found? (b) Based on the principles of Section 4.5B, what trend in the $-T\Delta S$ term would you expect? Is this found? (c) There is an anomaly in the trend of the ΔH term. How can this be explained? (You may want to consult the original article to answer this.)

5.15. *Identify donor atoms in the molecules $(CH_3)_2N\text{-}N(CH_3)_2$, $cyclo\text{-}(SCH_2CH_2)_3$, and $CH_3SCH_2CH_2SCH_2CH_2SCH_3$ by filling in the proper number of unshared electron pairs on the atoms that should have them in complete Lewis structures of each molecule. (b) Which of these molecules are chelating ligands? (c) Which are macrocyclic ligands? (d) Classify each chelating or macrocyclic ligand as bidentate, tridentate, and so on.

5.16. Below are the structures of three molecules.

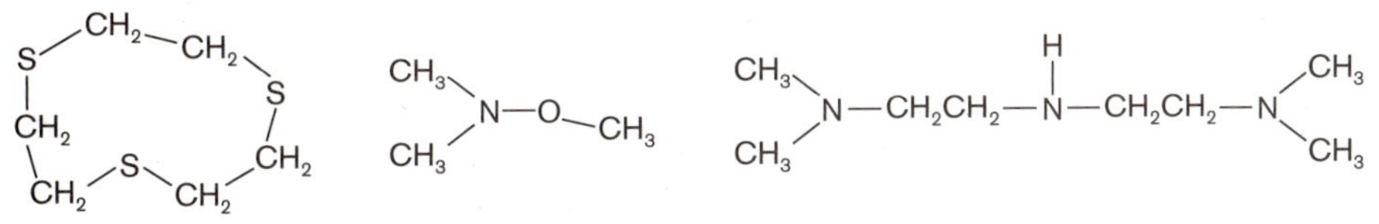

(a) Which atoms in each of the above structures are donor atoms? How many unshared electron pairs are on each? (b) Which of the three are chelating ligands? (c) Which are macrocyclic ligands? (d) Classify each ligand as monodentate, bidentate, tridentate, etc.

5.17. *The correct Lewis structures for **A**, **B**, and **C** are as follows:

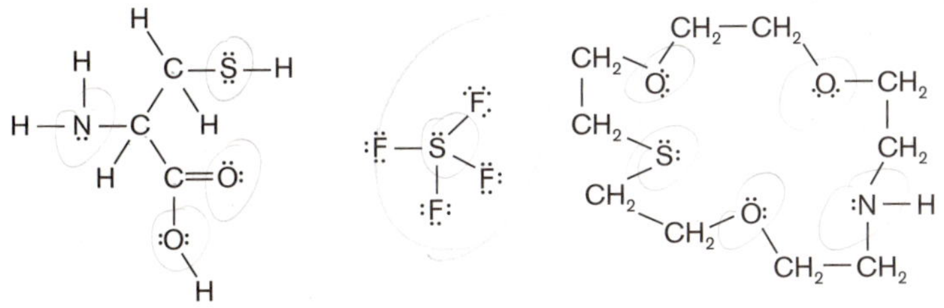

(a) Identify all possible donor atoms in molecules **A**, **B**, and **C**. (b) Which (if any) of molecules **A**, **B**, **C** can be chelating ligands; which can be macrocyclic ligands? (c) Which one of these molecules would probably form the most stable complex with a metal ion?

5.18. Draw the Lewis dot structure (including unshared electron pairs) of any example of your choice of the following: (a) a bidentate chelating ligand having two carboxylate anion functional groups; (b) a tetradentate macrocyclic ligand having two ether and two amine functional groups; and (c) an aromatic ligand having two carboxylate anion functional groups that *cannot* chelate.

5.19. *Consider the three complexes **A–C**: **A** is $[Pt(CN)_2(NH_2CH_2CH_2NH_2)_2](ClO_4)_2$, **B** is $[Pt(CN)_2(NH_3)_4]SO_4$, and **C** is drawn below. (a) List these three complexes in order of increasing stability. (b) Which complex (**A**, **B**, **C**, or none) contains a chelating ligand? (c) Which complex contains a macrocyclic ligand?

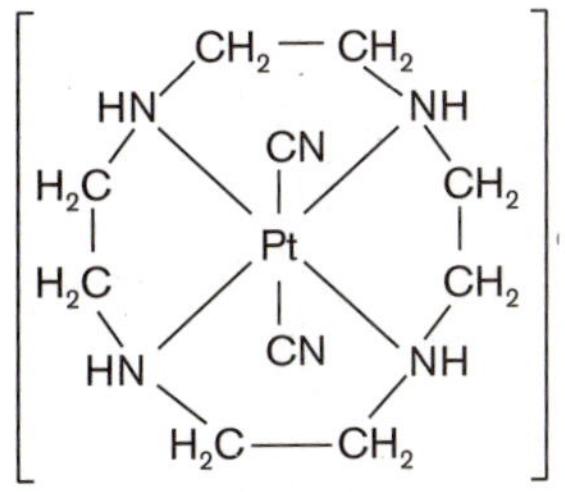

5.20. Explain why chelating ligands give more stable complexes than monodentate ligands; use a balanced chemical equation in your explanation.

5.21. *Predict whether reactants (left side) or products (right side) will be favored in each of the following equilibria.

(a) [Fe(porphyrin)(N from protein)] + 4(N from protein) $\rightleftharpoons$

(porphyrin)$^{2-}$ + [Fe(N from protein)$_5$]$^{2+}$

(b) $[ZnCl_2(15\text{-crown-}5)] + CH_3O(CH_2CH_2O)_4CH_3 \rightleftharpoons$
$$[ZnCl_2\{CH_3O(CH_2CH_2O)_4CH_3\}] + 15\text{-crown-}5$$

(c) $[HgCl_2(18\text{-crown-}6)] + CH_3O(CH_2CH_2O)_5 CH_3 \rightleftharpoons$
$$[HgCl_2\{CH_3O(CH_2CH_2O)_5CH_3\}] + 18\text{-crown-}6$$

(d) $[MgCl_2\{CH_3O(CH_2CH_2O)_4CH_3\}] + 15\text{-crown-}5 \rightleftharpoons$
$$[MgCl_2(15\text{-crown-}5)] + CH_3O(CH_2CH_2O)_4CH_3$$

5.22. Predict whether reactants (left side) or products (right side) will be favored in each of the following equilibria.

(a) $[MgCl_2(15\text{-crown-}5)] + 6CH_3OH \rightleftharpoons [MgCl_2(CH_3OH)_6] + 15\text{-crown-}5$

(b) $[Cu(porphyrin)]^{2+} + 4NHC_4H_4 \rightleftharpoons (porphyrin)^{2-} + [Cu(NHC_4H_4)_4]^{2+}$

(c) $[Rh\{(NH_2CH_2CH_2)_2NH\}]^{3+} + cyclo\text{-}(NHCH_2CH_2)_6 \rightleftharpoons$
$$[Rh\{cyclo\text{-}(NHCH_2CH_2)_6\}]^{3+} + 2(NH_2CH_2CH_2)_2NH$$

5.23. *Below is drawn the structure of a (hypothetical) truly complex complex ion.

(a) Write the formulas, including charges if any, of the ligands found in this complex. Underline the donor atoms on each ligand. (b) List all chelating ligands in the above complex, if any. Identify each as bidentate, and so on. (c) List all bridging ligands in the above complex, if any. (d) Give the coordination number of each Mo ion in the above complex. (They may or may not be the same.)

5.24. (a) Complete the Lewis dot structures of potential ligands **A–C** (all obey the octet rule), showing unshared electron pairs with dots, and circling the potential donor atoms in each:

(b) Which one of the three ligands is least likely to be a chelate ligand? (c) Which one is most likely to be a tridentate chelating ligand? (d) Which one is more likely to be a bridging (rather than a chelate) ligand? (e) Which one (if any) is a macrocyclic ligand? (f) The $:Pb^{2+}$ ion tends to have a maximum total coordination number in excess of six. Write the formula (including charge) of the complex ion it would form, having such a maximum total coordination number, with one or more moles of the highest denticity chelating ligand from the above set.

5.25. *For which of the bridging ligands shown in Figure 5.19 can you move donor atoms or functional groups so as to create ligands that are chelating rather than bridging? Explain why this cannot be done for the others.

5.26. Draw bifunctional bridging aromatic organic ligands of the following types (functional groups indicated): (a) phenolate carboxylates (⁻O– and –COO⁻); (b) amino benzoic acids (–NH₂ and –COO⁻); (c) phenolate nitriles (⁻O– and –C≡N:); (d) nitrile and isonitrile (:N≡C– and –N≡C:).

5.27. *(a) Which ligands in Figure 5.19 are most likely to give Prussian-blue-like three-dimensional metal–organic framework compounds upon coordination with metal ions? (b) Which type of metal ion—tetrahedrally coordinating $[M(OH_2)_4]^{2+}$ or octahedrally coordinating $[M(OH_2)_6]^{2+}$—is most likely to form the Prussian-blue-like MOF? (c) Which of your MOF compounds from (a), or Prussian blue itself, should have the smallest pore volume?

5.28. Assume (contrary to fact) that the unit cell for Prussian blue is either of the cubes shown in Figure 5.18. (a) Apply the counting procedures of Section 4.1A to determine how many Fe, FE, and CN groups are in each "unit cell." Show that these add up to confirm the stoichiometry of $KFe[Fe(CN)]_6$ for soluble Prussian blue, and that charge balance holds. (b) Now imagine letting each Fe and FE be Fe^{2+}. Replacing each CN^- with ligand **1** of Figure 5.19, and using counterions NO_3^- or K^+ to maintain charge neutrality, what counterion would fill what fraction of the pores? (c) Repeat the operations of part (b), this time using ligand **5**. (d) Repeat the operations of parts (b) and (c), this time using ligand **9**. (e) Assuming that the counterions were all of equal volume, which of these MOF compounds would leave enough pore volume free for small molecules to be incorporated (as in the next section)?

5.29. *Carefully examine the MOF crystal structures shown in Figure 5.20. (a) Which MOF contains the bridging ligand 2,6-NDC shown on the left? (b) Which one contains the center ligand, TPDC? (c) Which one contains the ligand on the right, BPDC? (d) The MOF based on which ligand has the largest pore volume? (e) The MOF based on which ligand has the highest density?

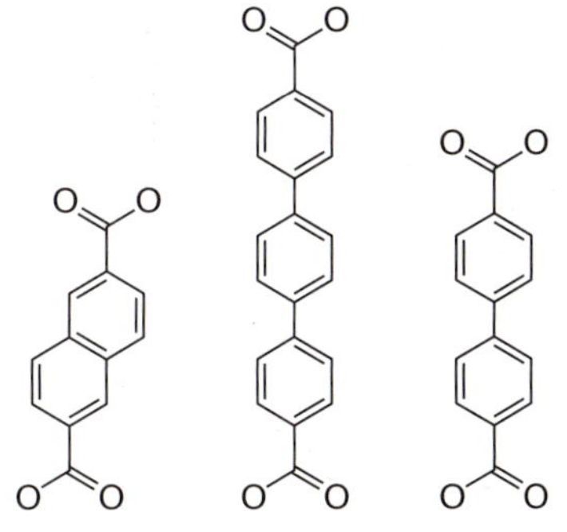

2,6-NDC   TPDC   BPDC

5.30. Suppose that you wanted to design a MOF that would be ideal for trapping radioactive radon gas found in the basements of many homes. (a) What pore diameter would you want? (b) Would MOF-5 likely be a good MOF to use? (c) What pore volume would you want?

5.31. *(a) What is the main problem with trying to use densities to compare porosities of different substances [e.g., MOF-5 and ITQ-4 (Figure 4.7g)]? (b) Why does this problem *not* apply to using densities to compare porosities of different polymorphs of SiO_2 (Figure 4.7)? (c) Irrespective of this, why might low density be inherently desirable in some applications?

5.32. Prussian blues and analogues actually occur in heavily hydrated form; on heating, 15 mol water can be removed per mole of $Ni_3[Co(CN)_6]_2 \cdot 15H_2O$. The resulting pores can take up H_2 gas [S. S. Kaye and J. R. Long, *J. Am. Chem. Soc.* 127, 6506 (2005)]. (a) Comparing Figures 5.20 and 5.18 (which is related to the structure of Prussian blue, with no cations in the cube centers), would you expect MOF-5 or anhydrous $Ni_3[Co(CN)_6]_2$ to have the larger pore volume and surface area? (b) Consult the article by Kaye and Long to obtain measured pore surface areas. (c) Consult the article to compare the weight percents of H_2 absorbed by, and the enthalpies of absorption of, the Prussian blue and by MOF-5. On which measure do the authors consider Prussian blue to have an advantage?

5.33. *Consider the bridging ligands other than **1**, **5**, and **9** shown in Figure 5.19. (a) Could they form MOFs? If so, would they resemble Prussian blues in having cubic unit cells? (b) Given secondary building units that match the general shape of the other bridging ligands, what two general shapes of pores might be expected, employing which of the other ligands?

5.34. Which of the following carbon ions or compounds are plausible soft bases: (a) CH_3^+, (b) CH_3^-, (c) CO, (d) CO_2, (e) CO_3^{2-}, (f) C_2^{2-}?

5.35. *Which of the following xenon ions or compounds are plausible examples of xenon atoms acting as a soft base: (a) XeO_3, (b) $AuXe_4^{2+}$, (c) $AuXe_2^+$, (d) $AuXe_4^{4-}$, (e) $PtXe_4^{2+}$, (f) $ThXe_4^{4+}$?

5.36. Students in inorganic lab have just discovered some new elements, which have the names, symbols, and properties given here:

Property	DeVersine (Dv)	Malium (Ml)	Derricium (Dm)	Carterogen (Ct)
Electronegativity	2.22	1.43	2.33	3.78
Ion Formed	Dv^{3-}	Ml^{4+}	Dm^+	Ct^{2-}
Ion Radius (pm)	198	52	144	125

Based on this information, classify these ions as hard acids, soft acids, hard bases, or soft bases.

5.37. *Predict whether reactants (left side) or products (right side) will be favored in each of the following equilibria:

(a) $As_2S_5 + 5HgO \rightleftharpoons As_2O_5 + 5HgS$

(b) $SrI_2 + HgF_2 \rightleftharpoons SrF_2 + HgI_2$

(c) $La_2(CO_3)_3 + Tl_2S_3 \rightleftharpoons La_2S_3 + Tl_2(CO_3)_3$

(d) $2CH_3MgF + HgF_2 \rightleftharpoons (CH_3)_2Hg + 2MgF_2$

(e) $AgF + LiI \rightleftharpoons AgI + LiF$

5.38. Predict whether reactants (left side) or products (right side) will be favored in each of the following equilibria:

(a) $Hg(CH_3)_2 + Ca(ClO_4)_2 \rightleftharpoons Hg(ClO_4)_2 + Ca(CH_3)_2$

(b) $HgSeO_4 + SrS \rightleftharpoons HgS + SrSeO_4$

(c) $CuF + LiI \rightleftharpoons CuI + LiF$

(d) $2CuF + CaBr_2 \rightleftharpoons 2CuBr + CaF_2$

5.39. *Predict whether reactants (left side) or products (right side) will be favored in each of the following equilibria:

(a) $[No(thiourea)_6]^{2+} + [Os(urea)_6]^{2+} \rightleftharpoons [Os(thiourea)_6]^{2+} + [No(urea)_6]^+$

(b) $HgSO_4 + CaBr_2 \rightleftharpoons HgBr_2 + CaSO_4$

(c) $Pt(MoS_4) + Ba(MoO_4) \rightleftharpoons Pt(MoO_4) + Ba(MoS_4)$

5.40. Give plausible products (or just one product) for each of the following reactions, or tell if no reaction is expected.

(a) $PdSO_4 + MgS \rightarrow$

(b) $(CH_3)_2Hg + CaF_2 \rightarrow$

5.41. *In each of the following complexes, classify both the metal ions and each ligand as a hard, soft, or borderline acid or base: (a) $(NH_4)_2[Pd(-SCN)_4]$ and (b) $[Co(NH_3)_4(SeO_4)]ClO_4$.

5.42. Classify each Lewis acid and each Lewis base in the following complexes as a hard, soft, or borderline acid or base: (a) $[Na(CH_3-O-CH_2-CH_2-O-CH_3)_3]^+$; (b) $[SnI_6]^{2-}$; (c) $[HgCl_2]$; and (d) $[Hg(CH_3)_2]$.

5.43. *Rewrite each of the following lists of Lewis acids or bases so that it is in increasing order of softness: (a) Cu^+, Au^+, Ag^+, K^+; (b) Br^-, I^-, F^-, Cl^-; (c) Mo^{2+}, Mo^{6+}, Mo^{4+}; (d) BF_3, $B(OCH_3)_3$, $B(CH_3)_3$; (e) FHg^+, CH_3Hg^+, Fe^{2+}, Fe^{3+}; and (f) $(CH_3)_2S$, $(CH_3)_2Se$, $(CH_3)_2O$.

5.44. Rewrite each of the following lists of Lewis acids or bases so that it is in increasing order of softness: (a) Cu^+, Cu^{2+}, Cu^{3+}; (b) $(CH_3)_2SO$, $(CH_3)_2S$, Cl_2SO (all using S as donor atom); (c) $(CH_3)_2O$, $(CH_3)_2Se$, $(CH_3)_2S$, $(CF_3)_2O$; (d) BF_3, BCl_3, $B(CH_3)_3$; (e) Sr^{2+}, Hg^{2+}, Pb^{2+}.

5.45. *Rewrite each of the following lists of Lewis acids or bases so that it is in increasing order of softness: (a) I^-, Cl^-, Br^-, F^-; (b) Ir^+, Ir^{3+}, Ir^{4+}; (c) Po^{2+}, Pt^{2+}, Pd^{2+}, Po^{4+}, Pa^{5+}; (d) Ag^{2+}, As^{3+}, Ac^{3+}, Au^+, Ag^+; (e) InF_3, BF_3, InI_3, TlI_3; and (f) $[Co(NH_3)_5]^{3+}$, $[Co(H_2O)_5]^{3+}$, $[Co(NH_3)_5]^{2+}$, $[Co(PH_3)_5]^+$.

5.46. Select the *hardest* and the *softest* from each list of Lewis acids or bases: (a) O^{2-}, Se^{2-}, S^{2-}; (b) Mn^{6+}, Mn^+, Mn^{2+}; (c) Mt^{3+}, Mn^{3+}, Md^{3+}; (d) $[PbF_3]^+$, $[Pb(CH_3)_3]^+$, $[Pb(NH_2)_3]^+$.

5.47. *Predict whether reactants (left side) or products (right side) will be favored in each of the following equilibria:

(a) $[Cu(thiourea)_4]^+ + [Cu(urea)_4]^{2+} \rightleftharpoons [Cu(thiourea)_4]^{2+} + [Cu(urea)_4]^+$

(b) $CdSO_4 + CaS \rightleftharpoons CdS + CaSO_4$

(c) $CaS + ZnSeO_4 \rightleftharpoons ZnS + CaSeO_4$

5.48. Predict whether reactants (left side) or products (right side) will be favored in each of the following equilibria:

(a) $PbSe + HgS \rightleftharpoons HgSe + PbS$

(b) $3FeO + Fe_2S_3 \rightleftharpoons Fe_2O_3 + 3FeS$

(c) $Zn(SCH_3)_2 + Hg(SeCH_3)_2 \rightleftharpoons Hg(SCH_3)_2 + Zn(SeCH_3)_2$

5.49. *Predict whether reactants (left side) or products (right side) will be favored in each of the following equilibria:

(a) $CH_3HgI + ClHgCl \rightleftharpoons CH_3HgCl + ClHgI$

(b) $PbSeO_4 + HgS \rightleftharpoons PbS + HgSeO_4$

(c) $3FeBr_2 + 2FeF_3 \rightleftharpoons 3FeF_2 + 2FeBr_3$

(d) $PtSeO_4 + PbSe \rightleftharpoons PtSe + PbSeO_4$

5.50. Classify each amino acid side-chain functional group shown in Figure 5.23 as a hard base, a borderline base, or a soft base ligand; classify each as a monodentate, chelating, or bridging ligand, too.

5.51. *Identify the potential donor atoms in each of the following molecules or ions, and classify the donor atoms as hard, soft, or borderline bases: (a) NH_3; (b) NH_2-CH_2-COOH (an amino acid); (c) SO_4^{2-}; and (d) SO_3^{2-}.

5.52. Referring to Figure 5.23, identify three of the listed metal ions that are likely to be coordinated by the side chains of (a) cysteinate anion; (b) histidine; (c) glutamate ion. Metal ions from which to choose: Ca^{2+}, Cu^+, Cu^{2+}, Fe^{2+}, Fe^{3+}, Mg^{2+}, Ni^{2+}, Zn^{2+}.

5.53. *Predict whether reactants (left side) or products (right side) will be favored in each of the following equilibria:

(a) $3Fe(CN)_2 + 2FeF_3 \rightleftharpoons 3FeF_2 + 2Fe(CN)_3$

(b) $Pb(S_2O_3) + HgSO_4 \rightleftharpoons PbSO_4 + Hg(S_2O_3)$

(c) $TlAsO_4 + AlAs \rightleftharpoons TlAs + AlAsO_4$

(d) $2Fe(SCN)_3 + 3Fe(OCN)_2 \rightleftharpoons 2Fe(OCN)_3 + 3Fe(SCN)_2$

5.54. Most oxo anions (Tables 3.6 and 3.7) are hard bases, but some are ambidentate and might also act as soft bases. Name some of these.

5.55. *For the ambidentate oxo anions you identified in Exercise 5.54, there are usually also monatomic anions of the same central atom. For each central atom, which is the softer base: the monatomic anion or the oxo anion? Why?

5.56. Draw the Lewis structures (and indicate the charges) of these pseudohalide ions. Considering that they may be ambidentate, classify them as hard, borderline, and/or soft bases: (a) the azide ion; (b) the thiocyanate ion; (c) the cyanate ion; and (d) the thiosulfate ion.

5.57. *Write the formulas of the following compounds, positioning the donor atom used by ligand immediately after the metal ion: (a) mercury(II) cyanide; (b) silver(I) thiosulfate; (c) magnesium thiosulfate; (d) mercury cyanate; and (e) mercury fulminate.

5.58. What do relative electronegativities suggest about the nature of the bond between (a) a hard acid and a hard base, (b) a soft acid and a soft base, and (c) a hard acid and a soft base?

5.59. *Hydrogen and boron do not have electronegativities characteristic of hard acids, yet the H^+ and B^{3+} ions are classified as hard acids. Which of their properties other than electronegativity might justify such a classification?

5.60. Are the ionic radii and charges of hard acids and bases also favorable to the kind of bonding you cited in part (a) of Exercise 5.58? Explain.

5.61. *In Example 4.6 we calculated the lattice energy of $CdI_2(s)$ and obtained a value of -1888 kJ mol^{-1}. This was well short of the experimental value of -2455 kJ mol^{-1}. (a) Calculate the lattice energy U expected for $CaI_2(s)$, which also crystallizes in the CdI_2 lattice type. (b) The experimental lattice energy of $CaI_2(s)$ is -2087 kJ mol^{-1}. How much more exothermic is the formation of this lattice than what you just calculated in part (b)? Explain the discrepancy. (c) Explain, using the HSAB principle, why the discrepancy is so much greater for CdI_2 than for CaI_2.

5.62. Some new elements have just been discovered in the author's laboratory. The following atomic parameters have been obtained:

Property	Sheltonium	Bennerine	Coatsium	Kamelogen
Electronegativity	1.06	3.82	1.97	2.34
Radius (pm)	(Cation) 107	(Anion) 123	(Cation) 105	(Anion) 189

Consider the equilibrium: sheltonium benneride + coatsium kamelide $\rightleftharpoons$ sheltonium kamelide + coatsium benneride. (a) Identify the type of bonding (ionic, covalent, or polar covalent) that is predominant in each product and reactant. (b) Will products or reactants be favored in this equilibrium? (c) In this equilibrium, the single most favorable enthalpy term is associated with which product or reactant?

5.63. *In each of the following equilibria, the single most favorable enthalpy term is associated with which reactant or product?

(a) $As_2S_5 + 5HgO \rightleftharpoons As_2O_5 + 5HgS$

(b) $SrI_2 + HgF_2 \rightleftharpoons SrF_2 + HgI_2$

(c) $La_2(CO_3)_3 + Tl_2S_3 \rightleftharpoons Tl_2S_3 + Bi_2(CO_3)_3$

(d) $2CH_3MgF + HgF_2 \rightleftharpoons (CH_3)_2Hg + 2MgF_2$

(e) $AgF + LiI \rightleftharpoons AgI + LiF$

5.64. In the HSAB equilibrium shown in Equation 5.12, (a) in which of the four species shown is the bonding closest to *pure covalent bonding*? (b) In which of the four species shown is the bonding closest to pure *ionic bonding*? (c) In which two of the four species shown is the bonding closest to polar covalent bonding? (d) What other factors about the acids and bases besides their softness may need to be considered?

5.65. *Which factors (hardness/softness, strength, chelating or macrocyclic effects) need to be considered in evaluating the position of each of the following equilibria?

(a) $Na_4(EDTA) + PbI_2(s) \rightleftharpoons Na_2[Pb(EDTA)] + 2NaI$

(b) $[Tl(EDTA)]^- + 3OH^- \rightleftharpoons Tl(OH)_3(s) + EDTA^{4-}$

(c) $Fe_2O_3 + 6HSCH_3 \rightleftharpoons 2Fe(SCH_3)_3 + 3H_2O$

5.66. (a) What other characteristics of acids and bases—besides their hardness and softness—determines whether a reaction will go to the left or to the right? (b) Find thermodynamically characterized equations for reactions involving these characteristics, and compare the magnitudes of the entropy or enthalpy changes with the magnitude involved in a typical HSAB reaction.

5.67. *Identify all insoluble compounds: CdTe, AgI, AgF, KI, EuSe, TiO_2, TiTe_2, and PtAs_2.

5.68. Identify all insoluble compounds in each series: (a) BaTe, TiTe_2, TlI, Tl_2S, CoAs_2 (contains Co^{2+} ions), CoI_2, CaTe, CaI_2, Li_3As; (b) BaO, CoSO_4, TiO_2, CaF_2. (c) Which of these compounds are likely to react with water to give oxides or hydroxides?

5.69. *Suppose that each of the following ions is treated with a mixture of sulfide and carbonate ions: (a) Tl^+; (b) Cu^+; (c) Rb^+; (d) Cr^{3+}; and (e) Sn^{2+}. Write the formula of the salt that is likely to precipitate (if any).

5.70. Answer the questions about each of the ions Co^{2+}, Tl^+, Pr^{3+}, and Cd^{2+}. (a) Is its bromide soluble or insoluble? (b) Is its selenide soluble, insoluble, or does it react with water to give an oxide or hydroxide?

5.71. *Answer the questions about the metal ions Zr^{4+}, Ag^+, and Sb^{3+}. (a) Is its bromide soluble or insoluble? (b) Is its selenide soluble or insoluble?

5.72. Answer the questions about each of the ions Pd^{2+}, Bk^{3+}, Sb^{3+}, and Ra^{2+}. (a) Is its bromide soluble or insoluble? (b) Is its sulfide soluble, insoluble, or does it react with water to give an oxide or hydroxide?

5.73. *Answer the questions about each of the ions Sr^{2+}, Bi^{3+}, Eu^{2+}, and Co^{2+}. (a) Is its chloride soluble or insoluble? (b) Is its telluride soluble or insoluble? (c) Is this metal a lithophile or a chalcophile? (d) Which is the more likely mineral source—a silicate or a sulfide?

5.74. Answer the questions about each of the ions Co^{2+}, Tl^+, Pr^{3+}, and Cd^{2+}. (a) Is this metal more likely a lithophile or a chalcophile? (b) Which is the more likely mineral source—a silicate or a sulfide? (c) Which is the more likely mineral source—an arsenate or an arsenide?

5.75. *Answer the questions about the metal ions Zr^{4+}, Ag^+, and Sb^{3+}. (a) Is this metal a lithophile? (b) Which is the more likely mineral source—a silicate, a sulfide, or in seawater?

5.76. Answer the questions about each of these four ions: Pd^{2+}, Bk^{3+}, Sb^{3+}, and Ra^{2+}. (a) Is this metal more likely a lithophile or a chalcophile? (b) Which is the more likely mineral source: a silicate or a telluride?

5.77. *Answer the questions about each of the ions Sr^{2+}, Bi^{3+}, Eu^{2+}, and Co^{2+}. (a) Is this metal a lithophile or a chalcophile? (b) Which is the more likely mineral source—a silicate or a sulfide?

5.78. Answer the questions about each of the ions Ag^+, Eu^{3+}, Eu^{2+}, and Bi^{3+}. (a) Classify each as either a lithophile or a chalcophile/siderophile. (b) Write the formula of a plausible mineral source of that ion.

5.79. *Classify each element as either a lithophile or a chalcophile/siderophile. Then choose its most likely mineral source. (a) Element: Be; source: $Be^{2+}(aq)$, Be_2SiO_4, or $BeSeS_2$. (b) Element: F; source: $F^-(aq)$, CaF_2, or HgF_2. (c) Element: Co; source: $CoCO_3$, $CoAs_2$, or $Co^{2+}(aq)$. (d) Element: Th; source: ThS_2, ThO_2, or $Th^{4+}(aq)$.

5.80. (a) Compare the observed Ni–Ni distance in nickel arsenide, 252 pm, with the sum of Ni covalent radii and the sum of Ni van der Waals radii, to see whether the proposal of Ni–Ni bonding is supported. (b) Why does this crystal structure seem very improbable for an ionic compound? (c) Suggest a reason why FeS, CoS, and NiS adopt the NiAs structure type, but ZnS does not.

5.81. *Consider the compounds (a) FeS_2, (b) $PtAs_2$, and (c) $PdSb_2$. If you assumed that the compounds contained monatomic S^{2-}, As^{3-}, or Sb^{3-} ions, what would the charges of the metal ions be? Compare these with Tables 1.10 and C to see how likely these charges or oxidation numbers are.

5.82. Barium sulfate, mercury(II) sulfate, and lead(II) sulfate are all insoluble. Ba^{2+}, Hg^{2+}, and Pb^{2+} have toxic properties. The chemical catalog entry for $BaSO_4$, however, contains no GHS pictograms, whereas the entry for $PbSO_4$ has three pictograms (Alert, Health Hazard, and Environmental Hazard), and the entry for $HgSO_4$ contains three pictograms (Poison/Toxic, Health Hazard, and Environmental Hazard). How can you explain the differences?

5.83. *The borderline acids of the fourth period in the d block are essential for life. Can they also be toxic? If so, which are the most toxic; which is the least toxic? To answer this, check a chemical catalog for GHS pictograms for the following: (a) iron(II) sulfate; (b) cobalt(II) sulfate; (c) nickel(II) sulfate; (d) copper(II) sulfate; (e) zinc sulfate.

5.84. The chemical form or speciation of an element may also influence its toxicity. Consider the following forms of mercury: $Hg(l)$, $Hg(NO_3)_2$, HgS, $(CH_3)_2Hg$, and HgI_2. (a) Which toxicity pictograms are present for each? (b) Which one is least toxic? Why?

5.85. *Which are more generally toxic, soft-acid metal ions or hard-acid metal ions? To which of the following types of biochemical ligand would soft-acid metal ions bind most strongly: (a) phosphate groups; (b) porphyrin groups (nitrogen donor); (c) cysteine groups (sulfur donor)?

5.86. Briefly explain why the element selenium is necessary for life and protects against mercury poisoning, but is also highly poisonous.

5.87. *Which of the following manners of disrupting protein structure would you expect to be most uniquely associated with the heavy metal ions: (a) hydrolysis of the CO–NH (peptide) linkages, (b) disruption of hydrogen bonding between different CO–NH groups, (c) disruption of S–S bond formation between different cysteine amino acid S–H groups, and (d) disruption of the electrostatic attraction of oppositely charged amino acid side chains?

5.88. Which two classes of metal ions and ligands are most likely to be both essential to life and toxic to life: hard acids, borderline acids, soft acids, hard bases, borderline bases, and soft bases?

5.89. *You are the health officer for a company that has just had a spill of the following radioactive ions: francium ion (Fr^+), actinium ion (Ac^{3+}), polonide ion (Po^{2-}), perastatate ion (AtO_6^{5-}), and meitnerium ion (Mt^{2+}). Ignore the effects of the radioactivity of these ions and think only of their possible chemical toxicity. (a) Which ion is the one most likely to bind to triphosphate groups ($P_3O_{10}^{5-}$) that are part of DNA? (b) Which ion is most likely to disrupt the quaternary structural elements of enzymes that depend on S–S bonds? (c) Which ion is most likely to bind to Fe^{2+} ions in essential enzymes? (d) Which ion is most likely to remain as an (unhydrolyzed) hydrated ion without binding to anything, and hence will be easily excreted in the urine?

5.90. You are the health officer for a company that has just had a spill of the following ions: Rb^+, Pr^{3+}, HTe^-, $H_4TeO_6^{2-}$, and Pt^{2+}. Think of their possible chemical toxicity. (a) Which ion is the one most likely to bind to triphosphate groups ($P_3O_{10}^{5-}$) that are part of DNA? (b) Which ion is most likely to disrupt the quaternary structural elements of enzymes that depend on S–S bonds? (c) Which ion is most likely to remain as an (unhydrolyzed) hydrated ion without binding to anything, and hence will be easily excreted in the urine? (d) Which ion is most likely to bind to Fe^{2+} ions in essential enzymes?

5.91. *You are the county health officer. Three new industries are moving to your area, and each will discharge a different inorganic pollutant. In each case, briefly discuss the likelihood and nature of the health effects resulting from that industry. (a) The Shahrokhi Semiconductor Co., which will discharge some hydrogen telluride vapors. (b) The Kinningham Ceramic Engines Corp. of America, which will discharge some zirconium(IV) oxide in its wastewaters. (c) The Bingham Brewery, Inc., which will discharge some cobalt(II) chloride in its wastewaters. (d) Which industry is of most concern?

5.92. Tell whether each of the following classes of metal ions and ligands is essential to life, or toxic to life, or neither, or both essential and toxic to life: (a) typical hard acids; (b) typical borderline acids; (c) typical soft acids; (d) selenium.

5.93. *You are the health officer for a company that has just had a spill of some radioactive ions. Which ion in each set is most likely to be methylated by vitamin B_{12} in the system and rendered even more toxic? (a) Francium ion (Fr^+), actinium ion (Ac^{3+}), polonide ion (Po^{2-}), perastatate ion (AtO_6^{5-}), or meitnerium ion (Mt^{2+}); (b) Rb^+, Pr^{3+}, HTe^-, $H_4TeO_6^{2-}$, or Pt^{2+}; (c) $CH_3SeO_3^-$, Na^+, Mg^{2+}, or Pb^{2+}.

5.94. Tell whether the following are likely to be methylated by methylcobalamin or by S-adenosylmethionine: a soft acid; a nonmetal atom with an unshared electron pair.

5.95. *Consider the following forms of mercury: $Hg(CH_3)_2(l)$, $HgCl_2(aq)$, and $HgS(s, insoluble)$. Which form or forms will (a) pass unabsorbed through the digestive system (absorption requires solubility in lipids or water); (b) be most easily eliminated in urine through the kidneys; (c) most readily cross the (nonpolar) blood–brain barrier; (d) most readily undergo bioamplification?

5.96. Consider the following forms of platinum: $[Pt(CH_2CH_3)_4]$, $Li_2[PtCl_6]$, and PtS. Which form will most likely (a) be soluble in water but not nonpolar organic solvents; (b) be soluble in nonpolar organic solvents but not water; (c) pass unabsorbed through the digestive system; (d) be eliminated from the body through the kidneys; (e) cross the (nonpolar) blood–brain barrier.

5.97. *Classify the following forms (species) of arsenic as a hard acid, hard base, soft acid, soft base, more than one, or none, and justify your answer: (a) AsF_5, (b) $AsCl_3$, (c) $As(C_6H_5)_3$, (d) AsO_4^{3-}, (e) AsO_3^{3-}, and (f) $FeAsO_4$. (g) Which forms are likely nontoxic or most toxic, and why?

5.98. Consider ligands **A–F**. (a) List them in order of increasing softness (note that some will be equally soft). (b) Which one of these might make the best medicine to combat poisoning by the Pt^{2+} ion? What principles did you use to choose this ligand?

5.99. *Which of the medicinal chemicals **A–E** would be most effective in combating poisoning by a soft-acid metal ion?

5.100. Answer each of the following questions for the metal ions Co^{2+}, Tl^+, Pm^{3+}, and Cd^{2+}. (a) Which two of these ions are most likely to have toxic chemical properties? (b) Which one of these ions is most likely to have nutritional value? (c) Which one of these ions would most likely be eliminated from the body by the use of the medicinal $EDTA^{4-}$ (Figure 5.5)? (d) Which one of these ions would most likely be eliminated from the body by the use of the medicinal penicillamine (Figure 5.30)?

5.101. *Three of the six compounds (**A–F**) listed below have been tried for treating poisoning by Hg^{2+}, Be^{2+}, and Cu^{2+} ions; the other three are not useful for any of these. Identify the useless compounds. For each medicinally useful ligand, identify the one metal ion of the above three that it would be used to treat.

5.102. You are a physician dealing with victims of a spill of the following radioactive ions: francium ion (Fr^+), actinium ion (Ac^{3+}), polonide ion (Po^{2-}), perastatate ion (AtO_6^{5-}), and meitnerium ion (Mt^{2+}). (a) Which ion is the one most likely to be removed from the body upon administration of the drug $EDTA^{4-}$ (Figure 5.5)? (b) Which ion is most likely to be removed from the body upon administration of the drug British anti-Lewisite (Figure 5.30)?

5.103. *Below are the structures of seven ligands (**A**–**G**). From among them find (a) the ligand that would be best suited for reacting with K^+ in the body; (b) the ligand that would be best suited for removing Hg^{2+} from the body.

NH_2— $CH_2CH_2NH_2$
A

H_2N — NH_2
B

$(NH_2CH_2CH_2)_2NH$
C

$(NH_2CH_2CH_2)_3N$
D

E

F

G

5.104. The structure of Paul Ehrlich's organoarsenic compound for treating syphilis was, until recently, not written as $cyclo\text{-}As_n\text{-}[(4\text{-}HO)(3\text{-}H_2N)C_6H_3]_n$ ($n = 3, 5$), but rather as $[(4\text{-}HO)(3\text{-}H_2N)C_6H_3]As=As[(4\text{-}HO)(3\text{-}H_2N)C_6H_3]$. What should have been suspicious about the latter formula for this compound?

5.105. *To determine the charges of the three Tc central ions in the imaging agents of Figure 5.36, we should first identify the charges on their ligands. (a) The ligand in $Tc(BIMI)^+$ is an isomer of the more familiar organonitrile ligand, $R\text{–}C{\equiv}N{:}$. What is the charge of this ligand? As a result, what is the charge of the central Tc ion? (b) The ligand labeled as "methylenediphosphonate" in Figure 5.36b is actually the protonated form of this ligand. What is the charge of the true methylenediphosphonate ligand in this complex? Considering also the two OH ligands on each Tc ion in this neutral (polymeric) complex, what charge is on the central Tc ion? (c) The pseudo-macrocyclic ligand in Tc-HMPAO has a –3 charge. Considering also the terminal oxide ligand, what charge is on this central Tc ion? (d) Rank the three Tc ions in order of increasing softness. Rank the three ligand sets in order of increasing softness. Is the softest Tc ion associated with the softest ligand set? Is the hardest Tc ion associated with the hardest ligand set? Discuss your answer.

5.106. The pK_a values of the cisplatin aquation products are $cis\text{-}[(H_3N)_2PtCl(OH_2)]^+$, 6.41; $cis\text{-}[(H_3N)_2Pt(OH_2)_2]^{2+}$, pK_{a1} 5.37 and pK_{a2} 7.21. (a) Classify the acidities of these cations. (b) The simple hydrated ion $[Pt(OH_2)_4]^{2+}$ does not form, so its pK_a value is not known. Would you expect it to be more or less acidic than $[(H_3N)_2Pt(OH_2)_2]^{2+}$? Why? (c) Based on its charge, size, and electronegativity, predict the pK_{a1} value of $[Pt(OH_2)_4]^{2+}$ and its acidity classification. How does this compare with your expectation in part (b) of this question?

5.107. *Based on the principles of this chapter, would you have predicted that cisplatin would have attacked two adjacent guanine bases in a DNA strand to form a 17-membered cyclic ring? Why or why not?

5.108. (a) What structural effect is responsible for the greater resistance of carboplatin and oxaliplatin (Figure 5.38) to reaction with water to give acidic complex ions? (b) Which of the second-generation complexes, carboplatin or oxaliplatin, is likely to react with DNA to give the same product (Figure 5.37) as cisplatin?

MARCEL POURBAIX (1904–1998) was born in Myshega, Russia, to Belgian parents. He obtained both his undergraduate (1927) and graduate degrees (1940) from the Université Libre de Bruxelles in Belgium. Throughout his career, the main focus of his research was corrosion and how to prevent it. By 1938, Pourbaix had developed the voltage potential versus pH diagrams, now called Pourbaix diagrams, which identify the stable phases of an aqueous electrochemical system. They are somewhat analogous to phase diagrams, which identify the particular combinations of temperature and pressure that give rise to the stable states of elements and compounds. Pourbaix lectured all over the world and actively collaborated with other electrochemists. He helped create the International Corrosion Council to encourage research and international cooperation in corrosion science and engineering, and in 1949 he helped found the organization that eventually became the International Society of Electrochemistry. In 1952 he founded the IUPAC's Commission of Electrochemistry, which was tasked with clarifying the sign conventions for electrode potentials. During the 1950s and 1960s, Pourbaix prepared potential–pH diagrams for all of the known elements. These data were published in the *Atlas of Electrochemical Equilibria* in 1963.

Principles of Oxidation–Reduction Reactivity

In the previous chapters, we examined the acid–base, complexation, and precipitation reactions of compounds of the elements in fixed oxidation states. Reactions in which elements change their oxidation states (such as the electron transfer reactions in Section 1.4) are known as **oxidation–reduction** or **redox** reactions and are the focus here in Chapter 6. Oxidation–reduction reactions can be extremely exothermic, giving off much more energy than is usually observed for acid–base reactions, so they can result in explosions (Figure 6.1, Section 6.3).

When the reactants are separated from each other in electrochemical cells (batteries), the release of energy can be controlled in the form of an electric current. The reactants can be in solution (Figure 6.2) or in the solid state (Figure 6.3).

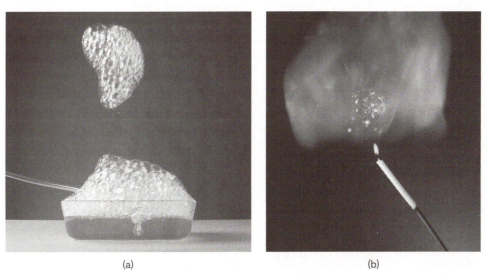

(a) (b)

Figure 6.1. The redox reaction of gaseous hydrogen and oxygen, contained in soap bubbles.
(a) The soap bubbles before ignition (to provide the activation energy).
(b) The explosion that results after ignition.

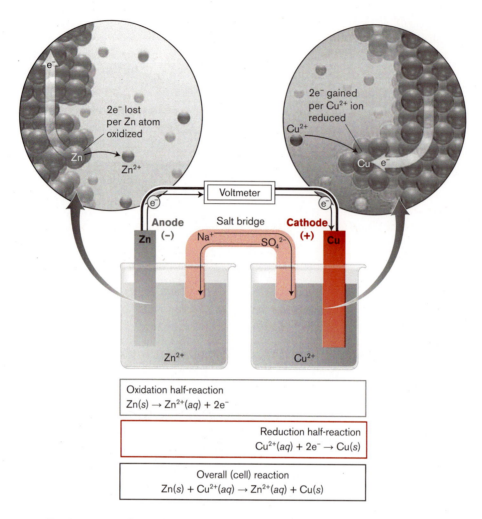

Figure 6.2. An electrochemical cell in which the reaction of zinc metal and aqueous copper ions is carried out. [Adapted from Martin S. Silberberg, *Principles of General Chemistry*, 2nd ed., McGraw-Hill: New York, 2010, p. 712.]

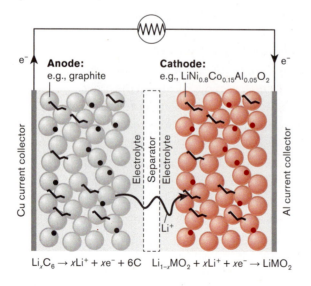

Figure 6.3. Schematic drawing of a lithium-ion battery. In the anode, small black circles represent lithium atoms and larger gray circles represent carbon atoms of graphite. In the cathode, small brown circles represent lithium ions and larger light brown circles represent other metal atoms (Ni, Co, and Al). Li$^+$ moves from left to right through the electrolyte and the separator. [Adapted from M. Jacoby, *Chem. Eng. News*, Feb. 11, 2013, pp. 33–37.]

In Section 6.1 we review the use of cell potentials (voltages) and introduce the use of redox predominance diagrams to determine whether the cell reactions go to products. In Section 6.2 we look at periodic trends of some common types of redox reactions that go to products. Section 6.3 covers cases in which these reactions can get out of hand and lead to explosions. Sections 6.4 and 6.5 cover the interrelationships of redox reactivity with the other reaction tendencies we studied earlier—acid–base, precipitation, and complexation—because many real reactions combine reaction types. Finally, in Section 6.6 we analyze why redox reactions do or do not go to products.

6.1. Standard Reduction Potentials and Their Diagrammatic Representation

OVERVIEW. We begin with a review of balancing oxidation–reduction half-reactions and combining them to give balanced equations for whole redox reactions. When carried out in an electrochemical cell, such a reaction generates a cell voltage or potential that is a combination of the potentials of the oxidation and reduction half-reactions (Section 6.1A). You may review applying these concepts by trying Exercises 6.1–6.4. Redox predominance diagrams (Section 6.1B) place thermodynamically stable species of an element, and their standard reduction potentials, in order from strongly oxidizing species of high reduction potentials, which are confined to the top, to strongly reducing species of very negative reduction potentials, confined to the bottom. You can practice identifying oxidizing and reducing agents and ranking them in order of strengths by trying Exercises 6.5–6.14. From their locations in redox predominance diagrams, you can tell whether two species (or one species plus solvent water) should undergo a redox reaction. If so, you can identify possible products and compute the standard cell voltage for the expected reaction. You can practice these concepts by trying Exercises 6.15–6.21.

You will again apply the concepts of half-reactions (Section 1.2A), oxidation numbers (Section 1.4A), and predominance diagrams (Sections 2.4, 2.5, and 3.8). Oxidation–reduction half-reactions will be revisited in Sections 8.4.

6.1A. Oxidation–Reduction Half-Reactions, Reactions, and Cell Voltages

(Review). Redox Half-Reactions. Redox reactions require the presence of two reactants. In one reactant, the oxidation number of an element is reduced, and in the other, an oxidation number of an element increases. These two processes can be treated as separate **half-reactions**. They can often be carried out in separate beakers of an **electrochemical cell** (Figure 6.2) if these two beakers are connected by (1) a wire to conduct the electrons and (2) a salt bridge or other separator to conduct the ions between these two half-reactions.

For example, the redox reaction depicted in Figure 6.2,

$$Zn(s) + Cu^{2+}(aq) \rightarrow Zn^{2+}(aq) + Cu(s) \tag{6.1}$$

can be separated into one half-reaction involving zinc and another half-reaction involving copper. The oxidation number of zinc increases from 0 in the zinc metal electrode to +2 in the zinc(II) ion. We say that the zinc is being oxidized, and that Equation 6.2 is an **oxidation half-reaction**:

$$Zn(s) \rightarrow Zn^{2+}(aq) + 2e^- \tag{6.2}$$

These electrons find their way through the wire to the copper electrode. There they reduce the aqueous Cu^{2+} ions to copper metal. Since this reduces the oxidation number of copper from +2 to 0, Equation 6.3 is a **reduction half-reaction**:

$$Cu^{2+}(aq) + 2e^- \rightarrow Cu(s) \tag{6.3}$$

In the half-reaction in Equation 6.2—the oxidation of the zinc metal—is accomplished by the copper(II) ion, which we therefore call the **oxidizing agent** in this redox reaction. The reduction of the copper from the +2 to the zero oxidation state in the half-reaction in Equation 6.3 is accomplished by the zinc metal, which we call the **reducing agent**. Thus, the oxidizing agent gets reduced, and the reducing agent gets oxidized.

Balancing Redox Half-Reactions. Half-reactions often involve more complex species, such as oxo anions. In such cases, the balancing of the half-reactions is a little less obvious, so it is summarized here in the following five-step procedure and in the flowchart in Figure 6.4:

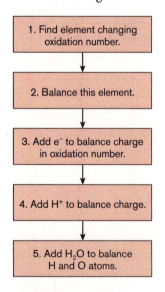

Figure 6.4. Flow chart summarizing the steps used to balance a redox half-reaction.

1. Assign oxidation numbers and identify the element being oxidized or reduced: Its oxidation number changes.

2. Balance the number of atoms of the elements being oxidized or reduced.

3. Balance the total change in oxidation number of all atoms of this element by adding the needed number of electrons (which, in effect, have oxidation numbers of –1). Reduction half-reactions involve electrons as reactants; oxidation half-reactions produce electrons.

4. Balance the charges in the half-reaction by adding the needed number of H^+ ions (OH^- ions if in basic solution).

5. Balance the hydrogen and oxygen atoms in the half-reaction by adding the appropriate number of water molecules.

EXAMPLE 6.1

Balance the half-reaction for the reduction of $Cr_2O_7^{2-}$ (the dichromate ion) to Cr^{3+}.

SOLUTION: 1. The oxidation number of each Cr atom changes from +6 in $Cr_2O_7^{2-}$ to +3 in Cr^{3+}. This is a reduction of three for each Cr atom.

2. There are two Cr atoms in $Cr_2O_7^{2-}$, so we have $Cr_2O_7^{2-} \rightarrow 2Cr^{3+}$.

3. The two Cr atoms are reduced by a total of six oxidation numbers, so they consume six electrons as reactants: $Cr_2O_7^{2-} + 6e^- \rightarrow 2Cr^{3+}$.

4. Because we have eight negative charges on the left side and six positive charges on the right, we add 14 hydrogen ions to the left side, so that it will also have six positive charges: $Cr_2O_7^{2-} + 6e^- + 14H^+ \rightarrow 2Cr^{3+}$.

5. We have seven oxygen and 14 hydrogen atoms on the left, and we need the same number on the right. These are provided by seven water molecules on the right:

$$Cr_2O_7^{2-}(aq) + 6e^- + 14H^+(aq) \rightarrow 2Cr^{3+}(aq) + 7H_2O \tag{6.4}$$

Balancing Whole Redox Reactions. The complete balanced redox reaction can be readily obtained by combining the balanced oxidation and reduction half-reactions in an appropriate manner. To do so, we add two more steps to our procedure:

6. Multiply each half-reaction by the smallest whole number needed to cause the two half-reactions being combined to have the same number of electrons.

7. Add the two half-reactions together, cancelling out the equal numbers of electrons, and anything else that may cancel (i.e., some or all of the hydrogen ions and/or water molecules).

EXAMPLE 6.2

Balance the whole redox reaction that is obtained by combining the oxidation half-reaction (Eq. 6.2) and the reduction half-reaction (Eq. 6.4).

SOLUTION: 6. The half-reaction in Equation 6.2 produces 2 electrons, and the one in Equation 6.4 consumes 6 electrons. The lowest common multiple of 2 and 6 is 6, so each half-reaction is rebalanced to include 6 electrons. This does not change Equation 6.4, but causes Equation 6.2 to become

$$3Zn(s) \rightarrow 3Zn^{2+}(aq) + 6e^-$$

7. Combining the two half-reactions gives an initial whole reaction that reads as follows:

$$3Zn(s) + Cr_2O_7^{2-}(aq) + 6e^- + 14H^+(aq) \rightarrow 3Zn^{2+}(aq) + 6e^- + 2Cr^{3+}(aq) + 7H_2O$$

Now the $6e^-$ on each side cancel out. The result reads as follows:

$$3Zn(s) + Cr_2O_7^{2-}(aq) + 14H^+(aq) \rightarrow 3Zn^{2+}(aq) + 2Cr^{3+}(aq) + 7H_2O$$

Electrochemical Cell Voltages (Electromotive Forces, emf) as Measures of Strength of Oxidizing and Reducing Agents. Thermodynamically, Equation 6.1 and other redox reactions go to products if the free energy change for the reaction, ΔG, is negative. Experimentally, it is more difficult to measure ΔG than to measure the cell electromotive force (emf, symbolized as E and measured in volts) generated when the reaction occurs spontaneously in a voltaic (galvanic) electrochemical cell. A spontaneous reaction occurring in a galvanic cell is considered to generate a *positive* emf E. If the reaction occurs under reversible conditions, the emf can be related to the free energy change by Equation 6.5,

$$\Delta G = -nFE \tag{6.5}$$

where n is the number of electrons exchanged in the whole reaction as written, and F is the conversion between electrochemical and thermodynamic units, 96.5 kJ V^{-1} mol^{-1}.

We can measure the potential E_{cell} of a whole redox reaction in an electrochemical cell such as that shown in Figure 6.2. It must be due to the sum of the contributions of the two half-reactions:

$$E_{cell} = E_{reduction} - E_{oxidation} \tag{6.6}$$

However, we cannot measure the potential generated by any half-reaction by itself. (Why not?) Therefore, (by convention) we assign a potential of 0.000 V to the half-reaction in which hydrogen ion is being reduced to hydrogen gas under standard conditions:

$$2H^+(aq, \text{activity} = 1) + 2e^- \rightarrow H_2(\text{pressure} = 1 \text{ atm}) \quad E° = 0.000 \text{ V} \tag{6.7}$$

Using Standard States Gives Us Standard Potentials $E°$

AN AMPLIFICATION

The potential generated by this and other half-cells depends on the conditions of the reactants and products, so we define conditions such that each substance is in its standard state. For gases, the standard state means a pressure of 1.000 atm for an ideal gas (or its equivalent if the gas is not ideal); for pure liquids, solvents, and solids, this is the pure liquid or solid at 1.000-atm pressure. The standard state for a solute is a concentration that gives an activity or thermodynamic concentration of 1.000 M. When H^+ is a product or reactant, its standard concentration will also be 1.000 M—hence the standard pH will be 0.00. If all reactants and products are in their standard states, we place a superscript, °, on E (and ΔG) to designate this fact.

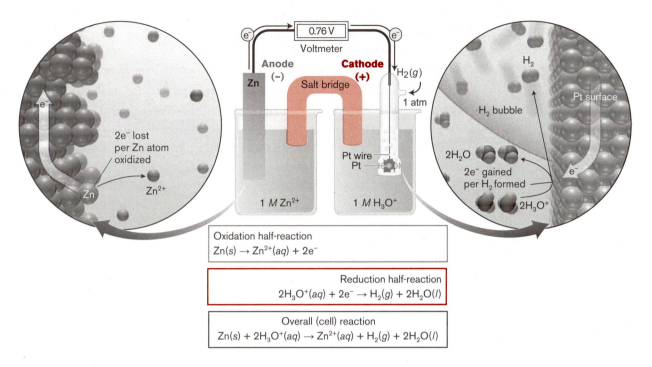

Figure 6.5. Electrochemical cell for the determination of the standard reduction potential of zinc ion using a platinum electrode at which the half-reaction of H_2 and H^+ can occur. [Adapted from Martin S. Silberberg, *Principles of General Chemistry*, 2nd ed., McGraw-Hill: New York, 2010, p. 712.]

We may now set up electrochemical cells in which all reactants and products are present in their standard state. In one half-cell, hydrogen gas is being oxidized to hydrogen ion, and in the other half-cell a chemical species is being reduced (reversibly) to a product in which an element is in a lower oxidation state. The emf generated in this cell (or required as an input if the reaction is not spontaneous) is called the standard reduction potential, $E°$, for reduction of that species.

However, the reduction of Zn^{2+} to Zn is *not* spontaneous. Instead the reverse reaction, shown in Figure 6.5, occurs. The Zn is oxidized, the H^+ is reduced, and the cell potential $E°_{cell}$ is +0.76 V. Because the reaction proceeds in the reverse direction, the standard *reduction* potential for Zn^{2+} is also reversed, making it –0.76 V. (The potentials of oxidation half-reactions are simply –1 times the potentials of the corresponding oxidation half-reactions.)

Extensive tabulations[1] have been made of standard reduction potentials of various chemical species. A selection for some important oxidizing agents is listed in Table 6.1.

The more positive the $E°$ of a reactant species is, the more easily it is reduced (and the *stronger* an oxidizing agent it is). If the species has a negative standard reduction potential, it cannot be reduced by hydrogen (under standard conditions), and it is thus a poor oxidizing agent. Thus, the strongest oxidizing agent in Table 6.1 is MnO_4^{2-}, with an $E°$ of +2.72 V.

When the whole redox reaction rather than the half-reactions are given, the process is illustrated in Example 6.3.

TABLE 6.1. **Standard Reduction Potentials of Selected Oxidizing Agents**

No.	Reduction Half-Reaction	$E°$ (volts)
H1	$2H^+(aq, \text{activity} = 1) + 2e^- \rightarrow H_2(\text{pressure} = 1\text{atm})$	0.0000
H2	$O_2(g) + 4e^- + 4H^+ \rightarrow 2H_2O$	+1.229
	Iron Half-Reactions	
Fe1	$FeO_4^{2-} + 8H^+ + 3e^- \rightarrow Fe^{3+} + 4H_2O$	+2.20
Fe2	$Fe^{3+} + e^- \rightarrow Fe^{2+}$	+0.77
Fe3	$Fe^{2+} + 2e^- \rightarrow Fe(s)$	−0.44
	Manganese Half-Reactions	
Mn1	$MnO_4^- + e^- \rightarrow MnO_4^{2-}$	+0.56
Mn2	$MnO_4^- + 4H^+ + 3e^- \rightarrow MnO_2(s) + 2H_2O$	+1.70
Mn3	$MnO_4^{2-} + 4H^+ + 2e^- \rightarrow MnO_2(s) + 2H_2O$	+2.72
Mn4	$MnO_2 + 4H^+ + 2e^- \rightarrow Mn^{2+} + 2H_2O$	+1.23
Mn5	$MnO_2(s) + 4H^+ + e^- \rightarrow Mn^{3+} + 2H_2O$	+0.95
Mn6	$Mn^{3+} + e^- \rightarrow Mn^{2+}$	+1.54
Mn7	$Mn^{2+} + 2e^- \rightarrow Mn(s)$	−1.18

Source: D. R. Lide, Ed., *Handbook of Chemistry and Physics*, 84th ed., CRC Press: Boca Raton, FL, pp. 5-4 to 5-60.

EXAMPLE 6.3

(a) Calculate $E°_{cell}$ for the redox reaction: $2MnO_4^- + 3Mn^{2+} + 2H_2O \rightarrow 5MnO_2(s) + 4H^+$. (b) Does this reaction favor products or reactants? (c) What is ΔG for the reaction?

SOLUTION: (a) We need to determine from which two half-reactions this reaction resulted. One of them converts MnO_4^-, in which Mn has oxidation number +7, to MnO_2, in which Mn has an oxidation number of +4. In Table 6.1 we find these in the "Mn2" half-reaction, $MnO_4^- + 4H^+ + 3e^- \rightarrow MnO_2(s) + 2H_2O$, which has an $E°$ of +1.70 V. The other half-reaction converts Mn^{2+} to MnO_2. This is an oxidation, so it is the reverse of the "Mn4" half-reaction. We apply Equation 6.6 to obtain the cell voltage, $E°_{cell} = +1.70$ V − (+1.23 V) = +0.47 V.

(b) Because this is a positive $E°_{cell}$, the reaction proceeds to give the product, MnO_2.

(c) In order to apply Equation 6.5, we need to find n, the number of electrons involved in the balanced whole reaction. This is the lowest common multiple (6) of the numbers of electrons in the two half-reactions ("Mn2" and "Mn4"). Therefore, $\Delta G = (-6 \text{ mol})(96.5 \text{ kJ V}^{-1} \text{ mol}^{-1})(0.47 \text{ V}) = -579$ kJ.

If this is confusing to you, join the club! The main difficulties students have with redox chemistry include remembering how the different terms relate (oxidizing agent, oxidation, etc.) and how to handle the sign conventions in applying Equation 6.6. To simplify this task, we introduce a diagrammatic way of showing standard reduction potentials that can be referenced to an easy-to-remember situation—namely, the oxygenation levels of a (stratified) lake or other body of water. These we call **redox predominance diagrams**. (Later in the chapter we will expand these diagrams to give Pourbaix diagrams, which are used in many areas of applied chemistry, such as environmental, geochemistry, and corrosion chemistry.)

6.1B. Diagrammatic Representations of Standard Reduction Potentials—Redox Predominance Diagrams. Redox predominance diagrams are modeled on the acid–base predominance diagrams of Sections 2.4A and 2.4C. In the acid–base predominance diagrams (see Figure 2.8), the predominant acid–base form of each element (in a fixed oxidation state) is shown as a function of pH. The more *acidic* a species, the more it is confined to the *left* side of its acid–base predominance diagram. The more *basic* a species, the more it is confined to the *right* side of its acid–base predominance diagram. If two different species can both persist at some common pH, they will have *overlapping predominance regions*; they will *not* tend to react with each other. If they have *nonoverlapping* predominance regions, they will tend to react via an acid–base reaction to give products that can coexist at a common pH (i.e., they will have overlapping predominance regions).

A redox predominance diagram shows the predominant (thermodynamically most stable) chemical forms of an element at different electrochemical potentials, while at a constant pH of zero and under standard conditions. Each numbered horizontal bar in a redox predominance diagram separates two chemical forms of the element in question. In Figure 6.6, we show a typical redox predominance diagram—that of iron. Above the $E°$ shown at the bar, the (more strongly oxidizing) chemical form above the bar is the predominant form; below that $E°$, the (less oxidizing) chemical form below the bar is the predominant form.

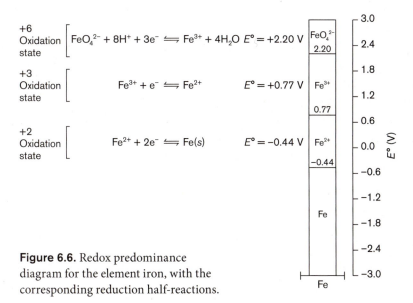

+6 Oxidation state $FeO_4^{2-} + 8H^+ + 3e^- \rightleftharpoons Fe^{3+} + 4H_2O$ $E° = +2.20$ V

+3 Oxidation state $Fe^{3+} + e^- \rightleftharpoons Fe^{2+}$ $E° = +0.77$ V

+2 Oxidation state $Fe^{2+} + 2e^- \rightleftharpoons Fe(s)$ $E° = -0.44$ V

Figure 6.6. Redox predominance diagram for the element iron, with the corresponding reduction half-reactions.

In redox predominance diagrams, *more strongly* oxidizing stable chemical forms are confined to *higher* regions of the diagram. Correspondingly, *strongly reducing* stable chemical forms are confined to *low* parts of redox predominance diagrams. Species in the middle of the diagrams that have other species above and below them in a diagram can act either as oxidizing agents or reducing agents. (Elements with no redox chemistry, such as Ar, predominate at all potentials and cover the entire diagram.)

Of the chemical forms of iron shown in Figure 6.6, the most strongly oxidizing is the ferrate ion, FeO_4^{2-}; the most strongly reducing is iron metal, Fe. Fe^{3+} and Fe^{2+} have both oxidizing and reducing properties. Since Fe^{3+} is higher in the diagram than Fe^{2+}, it is a stronger oxidizing agent than Fe^{2+}. Fe(s) has nothing below it, so it is not an oxidizing agent.

Note that this arrangement in the redox predominance diagrams corresponds to the way real bodies of water (lakes) stratify. The upper layer of a lake (the epilimnion) tends to become well aerated; hence it contains oxidized forms of elements, such as $Fe^{3+}(aq)$. The lower layer (hypolimnion) tends to accumulate reducing impurities; hence it contains reduced forms of elements, such as $Fe^{2+}(aq)$.

Note also that the thermodynamically stable forms of the elements are arranged in the diagrams so that higher oxidation states of elements occur higher in redox predominance diagrams. This finding can be confirmed by assigning oxidation states to the species shown in the iron redox predominance diagram. Thus, iron(VI), in the chemical form FeO_4^{2-}, is the most stable form in a solution of pH 0 and a total iron concentration of 1.000 M above a potential of +2.20 V. As the potential is lowered, the stable oxidation state drops, first to +3, then to +2, then to 0, in elemental iron.

Redox predominance diagrams of all of the elements in acidic solution (pH = 0) are presented in Figures 6.7–6.10.

Generating Balanced Redox Reactions from Predominance Diagrams. Unbalanced oxidation and reduction half-reactions can be read from the redox predominance diagram, but the balancing is incomplete: electrons, water, and H^+ must be added and balanced. A reduction half-reaction has the species above the boundary as reactant, being reduced to the species below the boundary as product. (It is an oxidation half-reaction when the species below the boundary is oxidized to the species above the boundary.)

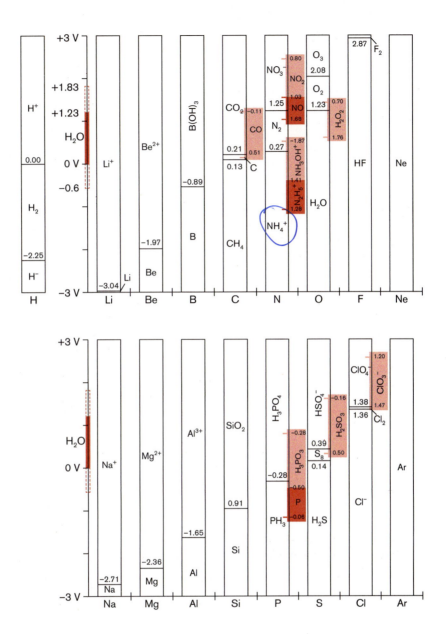

Figure 6.7. Redox predominance diagrams of the lighter *s*- and *p*-block elements, showing the thermodynamically stable form of each element at the potential indicated on the left. The numbers shown are the standard reduction potentials between the two adjacent forms of the element. Selected standard reduction potentials involving thermodynamically unstable (metastable) species are indicated in the right margins for certain elements. [Sources of the data: B. Douglas, D. McDaniel, and J. J. Alexander, *Concepts and Models of Inorganic Chemistry*, 2nd ed., John Wiley and Sons: New York, 1983, pp. 772–782; D. F. Shriver, P. W. Atkins, and C. H. Langford, *Inorganic Chemistry*, W. H. Freeman: New York, 1990, pp. 642–663; and J. P. Birk, *Predicting Inorganic Reactivity: Expert System*, Version 2.30, Project SERAPHIM, 1989 (also in J. Chem. Educ. Software). Some significant inconsistencies were noted for Ge and No.]

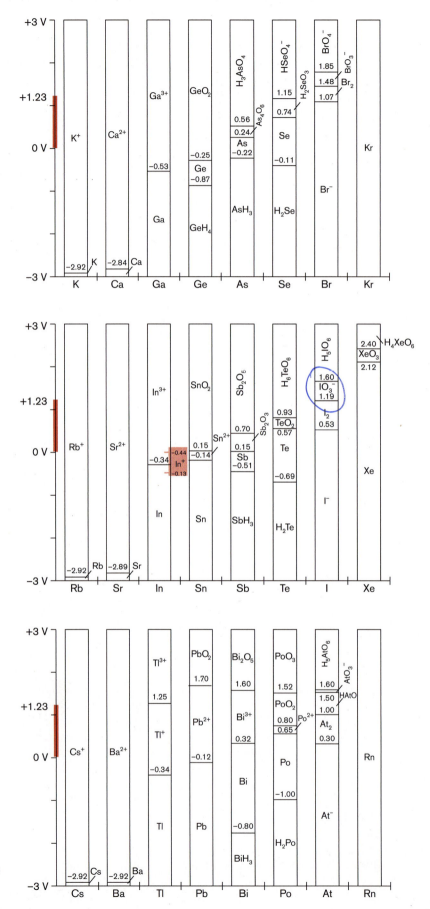

Figure 6.8. Redox predominance diagrams of the heavier *s*- and *p*-block elements. (Sources and notes as for Figure 6.7.)

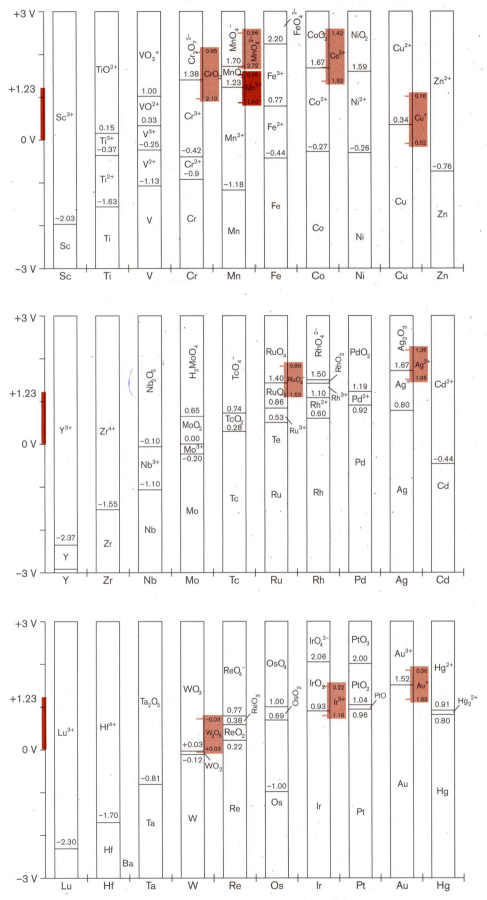

Figure 6.9. Redox predominance diagrams of the *d*-block elements. (Sources and notes as for Figure 6.7.)

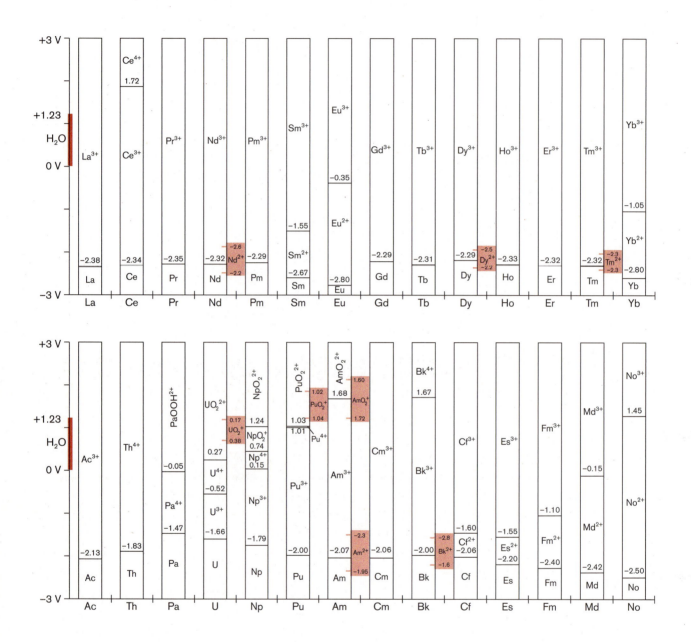

Figure 6.10. Redox predominance diagrams of the *f*-block elements. (Sources and notes as for Figure 6.7.)

EXAMPLE 6.4

From the redox predominance diagrams (Figures 6.7–6.10), find: (a) the strongest stable oxidizing agent in the p block, (b) the strongest oxidizing agent among stable oxo acids or anions in the p block, and (c) the strongest stable reducing agent in the d block. For each of these, write the balanced half-reaction by which it is oxidized or reduced, and tell the standard reduction potential of that half-reaction.

SOLUTION: (a) The strongest oxidizing agent in the p block will be the species in Figures 6.7 or 6.8 that is most severely confined to the top of its diagram; this is F_2, which is reduced to F^- at a standard reduction potential of +2.87 V. The reduction of F_2 produces $2F^-$. Each of the two fluorine atoms is reduced from oxidation state 0 to oxidation state –1; two electrons are required to give the balanced half-reaction:

$$F_2 + 2e^- \rightarrow 2F^- \quad E° = +2.87 \text{ V}$$

(b) The strongest oxidizing agent among oxo acids or anions in the p block similarly is the one most confined to the top: H_4XeO_6. It is reduced to XeO_3 at $E° = +2.40$ V. The half-reaction for this reduction requires $2e^-$ to reduce the Xe atom from an oxidation state of +8 in H_4XeO_6 to +6 in XeO_3. The charge of the $2e^-$ must be balanced with $2H^+$ as a reactant to give an equal charge (zero) on both sides of the half-reaction. Finally, water is added as a product to give the balanced half-reaction:

$$H_4XeO_6 + 2e^- + 2H^+ \rightarrow XeO_3 + 3H_2O \quad E° = +2.40 \text{ V}$$

(c) The strongest stable reducing agent in the d block is the species most severely confined to the bottom of its predominance diagram; this is Y, which has a standard reduction potential of –2.37 V. The negative sign means that a potential will not be generated, but must be applied to a solution of Y^{3+} to reduce it to Y. Three electrons are required to carry out this reduction:

$$Y^{3+} + 3e^- \rightarrow Y \quad E° = -2.37 \text{ V}$$

Predicting Redox Reactivity. Redox reactivity may be predicted from the redox predominance diagrams in much the same way as acid–base reactivity may be predicted from the acid–base diagrams. We illustrate the predictions with the help of Figure 6.11.

1. Two species that have *touching* or *overlapping* redox predominance regions have a potential at which they can coexist and will *not* tend to react with each other. Imagine that we want to know whether $H_2(g)$ and $Ca^{2+}(aq)$ will react with each other. In Figure 6.11a we show the predominance regions of H and Ca shifted so that they (partially) overlap each other. The region for $H_2(g)$ overlaps the region for $Ca^{2+}(aq)$. These two species do *not* react with each other.

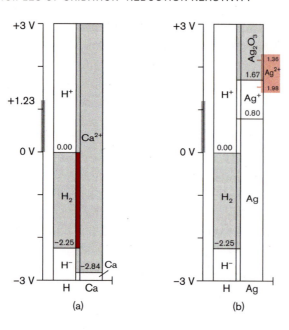

Figure 6.11. Judging the overlap of predominance regions. (a) Superposition of the redox predominance diagrams for H and Ca, showing that the (gray) predominance regions of Ca^{2+} and H_2 *do* overlap (in the **brown** region). (b) The (gray) predominance regions of Ag_2O_3 and H_2 do *not* overlap.

2. Two species that have *nonoverlapping* predominance regions are expected to react with each other to give new products that are stable in each other's presence—that is, have overlapping predominance regions. The redox reaction generates a standard cell emf $E°$ that equals the numerical *gap* between their predominance regions.

Does $H_2(g)$ react with $Ag_2O_3(s)$? In Figure 6.11b we allow the predominance diagrams of H and Ag to overlap. We see that the region for $H_2(g)$ does *not* overlap the region for Ag_2O_3—they should react with each other. They should give products with overlapping regions—$H^+(aq)$ and either $Ag^+(aq)$ or $Ag(s)$. (Which will form depends on the quantities of each present.) The cell emf $E°$ gap between the H_2 region and the Ag_2O_3 region, $(1.87 \text{ V}) - (0.00 \text{ V}) = +1.87 \text{ V}$, applies to the reactions that produce the product directly across the boundaries; in the case of Ag_2O_3, the one involving the reduction half-reaction, $Ag_2O_3 + 6H^+ + 4e^- \rightarrow 2Ag^+ + 3H_2O$.

To revisit Example 6.3, in which we calculated the $E°_{cell}$ for the redox reaction, $2MnO_4^- + 3Mn^{2+} + 2H_2O \rightarrow 5MnO_2(s) + 4H^+$, we can simply locate (in Figure 6.9) the lower boundary for MnO_4^- being reduced to MnO_2; this is at $+1.70 \text{ V}$. We then locate the upper boundary for Mn^{2+} being oxidized to MnO_2; this is at $+1.23 \text{ V}$. The gap between two is the difference between the upper boundary and the lower boundary, 0.47 V, as calculated previously in Example 6.3.

Redox Predominance Diagram for Water at pH 0. If one is carrying out a synthesis or reaction in aqueous solution, water is a potential reactant! Above an $E°$ of $+1.229 \text{ V}$, water is oxidized to oxygen:

$$2H_2O \rightarrow 4H^+(aq) + O_2 + 4e^- \quad E°_{redn} = +1.229 \text{ V} \tag{6.8}$$

Therefore, the upper boundary of the predominance region of water is $+1.229 \text{ V}$. Three important oxo anions from Figure 6.9 that fall entirely above this boundary are $Cr_2O_7^{2-}$, MnO_4^-, and FeO_4^{2-}. Aqueous solutions of these ions (1.0 M) at pH 0 are thermodynamically unstable.

If a reducing agent is introduced that has a predominance region entirely below 0.00 V (standard conditions), the hydrogen ion in water will be reduced to H_2 by the half-reaction in Equation 6.7. Therefore, the lower boundary of the predominance region of water is 0.0000 V. Water is fairly easy to oxidize and to reduce. Some other solvents, such as acetonitrile ($CH_3C\equiv N$), are more inert to redox chemistry. Most of the d-block metals from the fourth period (top of Figure 6.9) are unstable in water with a pH = 0 (i.e., acidic solution). As a reminder that water is so often a potential reactant to be considered, we have included small predominance diagrams for water along the left margins of Figures 6.7–6.10.

EXAMPLE 6.5

Predict the products (if any) of the following aqueous reactions under standard conditions: (a) Mn + H^+; (b) Yb^{2+} + H_2O; (c) IrO_4^{2-} + H_2O. Write complete balanced equations for those reactions that do occur.

SOLUTION: (a) Regions for Mn and H^+ do not overlap, so we expect them to react to give Mn^{2+} and H_2, which have overlapping predominance regions:

$$Mn + 2H^+ \rightarrow Mn^{2+} + H_2$$

(b) The predominance regions for Yb^{2+} falls below those for water and for H^+ (lower boundaries at 0.00 V), so we expect them to form Yb^{3+} and H_2, which have overlapping predominance regions. Balancing requires adding H^+:

$$2Yb^{2+} + 2H^+ \rightarrow 2Yb^{3+} + H_2$$

(c) The predominance regions of IrO_4^{2-} and H_2O do not overlap, so we expect them to form IrO_2 and O_2, which have overlapping predominance regions. Balancing requires adding H^+:

$$2IrO_4^{2-} + 4H^+ \rightarrow 2IrO_2 + O_2 + 2H_2O$$

Thermodynamically Unstable Species and Predominance Diagrams. Let us consider another Mn-based redox reaction:

$$Mn^{2+} + MnO_2 + 4H^+ \rightarrow 2H_2O + 2Mn^{3+} \tag{6.9}$$

If we turn to Table 6.1 we can find the two half-reactions that add up to give this redox reaction—namely, "Mn5" as the reduction half-reaction ($E° = +0.95$ V) and "Mn6," after reversal, as the oxidation half-reaction ($E° = +1.54$ V). Applying Equation 6.6, we calculate $E_{cell} = (+0.95$ V) $- (+1.54$ V) $= -0.59$ V. The product Mn^{3+} is *not* favored in this reaction. Instead, the reverse of the reaction in Equation 6.9 is favored: $2H_2O + 2Mn^{3+} \rightarrow Mn^{2+} + MnO_2 + 4H^+$. We say that the thermodynamically unstable Mn^{3+} **disproportionates**—it reacts with itself to generate Mn in a higher and Mn in a lower oxidation state.

If we were to try to place Mn^{3+} on the redox predominance diagram for Mn, the region for Mn^{3+} would have an *upper* boundary for its oxidation by the reverse of reaction "Mn5" in Table 6.1, at +0.95 V. It would also have a *lower* boundary for its reduction by reaction "Mn6," at +1.54 V. However, the *lower* boundary is at a *higher* $E°$ value than the *higher* boundary! Because Mn^{3+} is unstable, it does not predominate anywhere on this diagram. Its predominance region is imaginary! The only way we can locate it on the diagram is in a box off the side of the predominance diagram (shaded). By doing this, we can place the two $E°$ values on the table, albeit with the higher one being lower in voltage than the lower one.

6.2. Activity Series of Metals, Nonmetals, and Oxo Anions/Acids

OVERVIEW. Reactions that interconvert H_2 and H^+, or O_2 and H_2O, tend to proceed relatively slowly; they require overvoltages of about 0.6 V to get them to proceed rapidly. By identifying gaps of less than 0.6 V between predominance regions, you can identify some oxidizing and reducing agents that react slowly with water (Section 6.2A). You may practice this concept by trying Exercises 6.22–6.25. You can rank a group of metals into an activity series directly from their $E°$ values or make predictions based on their Pauling electronegativities. From this, you can predict the results of the reaction of any metal with H_2O, HCl, HNO_3, or another metal ion (Section 6.2B and Exercises 6.26–6.32). You can classify nonmetallic elements as to relative strength as oxidizing and reducing agents, and their anions and hydrides as reducing agents. You should know the general periodic trends in the stability and oxidizing ability of oxides, oxo anions, and oxo acids of elements in their highest oxidation states, and be able to know which GHS pictographs should apply to which reagents (Section 6.2C; Exercises 6.33–6.39).

Here in Section 6.2 there is a Connection to biochemistry. You again apply the concepts of ion charges (Section 1.2), oxidation numbers (Section 1.4), ionic radii (Section 1.5), electronegativity (Section 1.6), the HSAB principle (Section 5.4), and GHS pictographs (Section 5.7).

6.2A. Kinetic Barriers (Overpotentials) in Redox Reactions. Although oxidizing and reducing agents with increasingly nonoverlapping predominance ranges may favor formation of products more exothermically, how *quickly* they react may be another issue. Some exothermic reactions may be completed within seconds, minutes, or hours of mixing (i.e., they occur on a laboratory time scale), while others require thousands or millions of years to reach equilibrium (i.e., they occur on a geological time scale). Reactions may be slow because there are difficult steps in the mechanism producing intermediates of high energy.

Let us contrast two "simple" redox reaction of hydrogen—one with chlorine and one with oxygen. The $E°$ gaps between H_2 and either of the two oxidizing agents are rather similar (1.23 V and 1.36 V). The reaction with chlorine, $H_2(g) + Cl_2(g) \rightarrow 2HCl(g)$, proceeds rapidly. In contrast, it is possible to mix hydrogen and oxygen gases in soap bubbles, and nothing happens (Figure 6.1a) unless an activation energy is supplied (in this case, a spark). Then the reaction, $2H_2(g) + O_2(g) \rightarrow 2H_2O(g)$, proceeds explosively (Figure 6.1b).

The hydrogen oxidation half-reaction and the chlorine reduction half-reaction each involve two electrons. The hydrogen and oxygen half-reactions, on the other hand, involve different numbers of electrons, so they are said to be **noncomplementary**. The +1.229-V reduction of O_2 to $2H_2O$ (reaction "H2" in Table 6.1) requires that the reducing agent supply four electrons. Reductions generally proceed by steps that produce one or two electrons (or transfer one atom). If three or more electrons are required, a one-step mechanism is impossible, and other mechanisms are needed. This can cause the whole redox reaction to be slow.

The first step in the reduction of O_2 (Figure 6.12) endothermically produces an unstable intermediate, the high-energy, toxic superoxide ion, O_2^-, which is protonated by hydrogen superoxide at pH 0:

$$O_2 + e^- + H^+ \rightarrow HO_2 \quad E° = -0.125 \text{ V} \tag{6.10}$$

This is then reduced exothermically to another lower-energy intermediate, hydrogen peroxide (H_2O_2).

$$HO_2 + e^- + H^+ \rightarrow H_2O_2 \quad E° = +1.51 \text{ V} \tag{6.11}$$

Hydrogen peroxide then exothermically accepts two electrons to produce the lowest-energy final product, water.

$$H_2O_2 + 2e^- + 2H^+ \rightarrow 2H_2O \quad E° = +1.763 \text{ V} \tag{6.12}$$

Although H_2O is very much favored energetically over O_2, the reaction must first provide an activation energy or **overpotential**[2] to reach the unfavorable intermediate, HO_2. Insufficient activated particles may be present to provide this activation energy, so the reaction is slow.

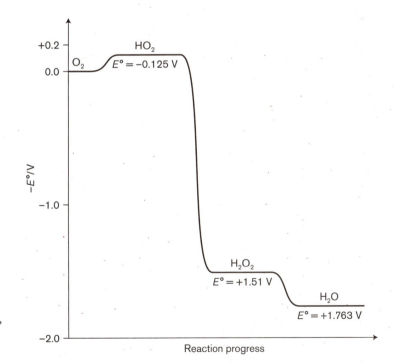

Figure 6.12.
Reaction profile for the reduction of O_2 to H_2O via the three steps—Equations 6.10, 6.11, and 6.12.

Thus, the question "Will hydrogen react with oxygen?" can have two answers. Over the long (geological) term, the answer is "yes," because hydrogen and oxygen have nonoverlapping predominance diagrams. But over the short (laboratory) time scale, the answer may be "no," because the reduction of O_2 requires an activation energy that may be absent.

Activation energies can vary tremendously and so can rates. For our purposes, a rough estimate can be made by saying that *reactions may proceed slowly unless an overpotential of an additional 0.6 volts is provided* in order to produce appreciable concentrations of activated particles.

The long-term predominance region of stability of water from 0.000 V to +1.229 V (Figure 6.13a) therefore describes *thermodynamic* stability after slow reactions have had time to occur (i.e., after geochemical periods of time). Compounds or ions with predominance regions not overlapping 0.000 V to +1.229 V (at pH = 0) are generally not found in geological or mature environmental samples (but may be important in nonaqueous or solid-state reactions). Many of the oxidizing agents in Table 6.1 cannot persist in acidic solution (pH = 0), such as those involved in reactions "Fe1," "Mn2," "Mn3," and "Mn6." Similarly, water (at pH = 0) is not stable geochemically over a long period of time in the presence of reducing agents with $E° < 0.00$ V. In the next subsection we shall see that this includes many metals, such as iron.

The short-term predominance region of stability of water is wider, from about −0.6 V to +1.8 V, and describes *kinetic* stability before slow reactions have had time to occur (i.e., on the laboratory time scale). Oxidizing agents with predominance regions confined to above about +1.8 V placed in water (at pH 0) decompose rapidly, releasing O_2, which is in fact the fate of the iron(VI) species FeO_4^{2-} ($E° = +2.20$ V) under standard conditions. In contrast, a somewhat milder oxidizing agent such as MnO_4^- ($E° = +1.70$ V, reaction "Mn2" of Table 6.1) can be used in aqueous solution without immediate decomposition. However, if you examine the solution a year later, you will notice that it has deposited brown MnO_2, the product of "Mn2." This means that MnO_4^- is useful for short-term laboratory work, but has no importance in geochemistry. Similarly, metals such as iron with $E° > -0.6$ V may react slowly with water, so they may persist for some years. As an example, consider the ship *Titanic* (Fe metal, $E° = -0.44$ V), introduced into the water in 1912. It should have been oxidized to Fe^{2+} (at pH = 0, not the conditions of the North Atlantic). The process was slow, but the oxidation of the *Titanic* is occurring (as catalyzed by microorganisms and their enzymes; Section 6.4B).

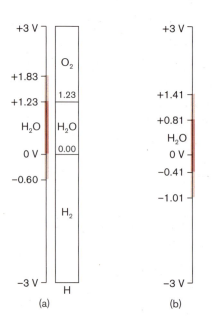

Figure 6.13. Predominance diagram for water showing the stability ranges of water at (a) pH = 0 (standard conditions) and (b) pH = 7 (neutral conditions; discussed in Section 6.5). Dark shading indicates the stability ranges of water over long periods of contact with another reactant (0 to +1.23 V at pH 0 and −0.41 to +0.81 V at pH 7). Light shading indicates the larger stability ranges for short periods of contact (−0.6 to +1.83 V at pH 0, and −1.01 to +1.41 V at pH 7).

If a *catalyst* is added, the mechanism may change, which may allow the reaction to proceed much more rapidly by another mechanism that does not involve unstable intermediates. Because *d*-block metals often have several fairly stable cations differing in charge from each other by only +1, they are especially useful as catalysts for redox reactions, such as those involving O_2. Microorganisms may catalyze many of these slow reactions using *d*-block metal ions in metalloenzymes (Section 6.4B).

A CONNECTION TO BIOCHEMISTRY

Fermentation and Disproportionation

Upon examining the redox predominance diagram for carbon (Figure 6.7), you may be struck by the fact that almost all of the tens of millions of known organic compounds are absent! This is because they are thermodynamically unstable (even though some of them have been around in oil deposits for millions of years). There are formidable kinetic barriers that slow down their conversion to the three thermodynamically favored forms (see "Why Life Exists on Earth; Section 2.3C). Life, however, has developed enzymes to catalyze some of these *disproportionation* reactions in the process of *anaerobic metabolism* or *fermentation.*

Let us consider a very common example—the fermentation of sugar to give carbon dioxide and ethanol, where we use $C_6H_{12}O_6$ to represent our sugar:

$$C_6H_{12}O_6(s) \rightarrow 2CH_3CH_2OH(l) + 2CO_2(g) \tag{6.13}$$

If we assign oxidation states of carbon (0, –2, and +4, respectively), we see that this is a disproportionation reaction from oxidation state 0 in sugar to oxidation states –2 in ethanol and +4 in CO_2). In addition, Equation 6.13 can be written as the sum of oxidation and reduction half-reactions:

$$C_6H_{12}O_6(s) + 6H_2O(l) \rightarrow 6CO_2(g) + 24e^- + 24H^+(aq)$$

$$C_6H_{12}O_6(s) + 12e^- + 12H^+(aq) \rightarrow 3CH_3CH_2OH(l) + 3H_2O(l)$$

Other than (perhaps) in a fuel cell, however, these reactions are difficult to carry out electrochemically. Since the ΔG_f values are well known, we can calculate ΔG for this reaction, which turns out to be –221.08 kJ. The process of fermentation does proceed to products, and it provides energy for the yeast carrying it out. Using Equation 6.5, we can calculate the $E°_{cell}$ that such a fuel cell would produce. Equation 6.13 involves eight electrons as written, so

$$E°_{cell} = \frac{-(-221.08 \text{ kJ})}{(8)(96.5 \text{ kJ V}^{-1})} = +0.29 \text{ V}$$

EXAMPLE 6.6

We know that the following mixtures do not react immediately. Consult the predominance diagrams and the relevant half-reactions and tell whether this is due to the mixture being thermodynamically stable or not. If the mixture is not thermodynamically stable, suggest a reason why there is no immediate reaction, and suggest how the reaction could be achieved. (a) $CH_4(g) + O_2(g)$; (b) $O_2(g) + Cl_2(g)$; (c) you personally, plus air.

SOLUTION: (a) The voltage gap between $CH_4(g)$ and $O_2(g)$ is 1.23 V – 0.13 V = 1.10 V (Figure 6.7). This mixture is unstable. However, the half-reactions to give the stable products, $CO_2(g) + H_2O(l)$, involve multiple and noncomplementary numbers of electrons (8 and 4, respectively), so a kinetic barrier is likely. A catalyst such as a burning match or cigarette will likely start the reaction!

(b) The predominance regions of O_2 and Cl_2 overlap between 1.36 V and 1.38 V, so the mixture is thermodynamically stable and will not react.

(c) You are not found personally on the predominance diagram, because you are not thermodynamically stable! However, your mixture with air is probably not so different from the mixture of methane and air, so you would be expected to react. Nevertheless, kinetic barriers exist, preserving you until your demise. After that, various microbes will catalyze your oxidation to CO_2 and H_2O (and other products such as clay). Unfortunately you will not live forever!

6.2B. Periodicity in the Activity of Metals and the Reduction of Metal Ions.

Among the most common and simplest redox reactions are those in which metals are oxidized to hydrated metal ions in common oxidation states, or in which the hydrated metal ions are reduced back to metals. In the simplest version of this reaction, $H^+(aq)$ (perhaps from the self-ionization of water) is used to oxidize the metal. We may expect that metals with predominance regions entirely below that of $H^+(aq)$ (metals with negative standard reduction potentials) will react with $H^+(aq)$ to give hydrated metal ions and $H_2(g)$. For this reaction, the rate of evolution of the product $H_2(g)$ is proportional to the cell voltage (the gap between the predominance regions of the metal ion and H^+); the kinetic barriers to this reaction are fairly constant for all metals. Hence, this **activity series of metals** is sometimes also called the electromotive series of metals. The activity of a metal increases as the standard reduction potential of the cation produced decreases (Figure 6.14).

Experiment 7 involves the interpretation of the periodicity in the activity series of metals. If you experimentally developed an activity series of metals while in general chemistry, your instructor may assign only part F of the experiment. Even if you did not do Experiment 7, you can still compare Figure 6.14 visually with graphic representations of charge/oxidation state (Table 1.11) and electronegativity (Table A). For size, you have a choice of sizes to use (Section 1.5): either covalent/metallic radii (graphically in Figure 1.6 or as tabulated in Table 1.13), or ionic radius of the cation involved (graphically

EXPERIMENT 7 MAY BE ASSIGNED NOW.
(See www.uscibooks.com/foundations.htm.)

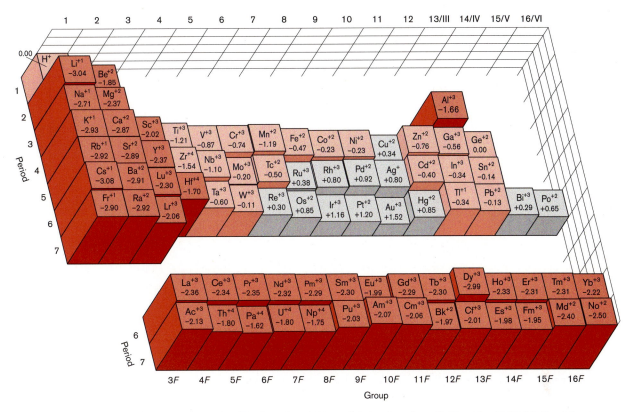

Figure 6.14. Standard reduction potentials of hydrated metal ions. [Data taken from B. Douglas, D. H. McDaniel, and J. J. Alexander, *Concepts and Models of Inorganic Chemistry*, John Wiley and Sons: New York, 1983; M. C. Ball and A. H. Norbury, *Physical Data for Inorganic Chemists*, Longman: London, 1974; and D. R. Lide, Ed., *Handbook of Chemistry and Physics*, 84th ed., CRC Press: Boca Raton, FL, pp. 8–23 to 8–28.]

in Figure 1.7 or as tabulated in Table C). The bar graph of Figure 6.14 most closely resembles (correlates with) which graph or data set—that of charge of ion involved, covalent/metallic radius of atom involved, ionic radius of ion involved, or Pauling electronegativity of the element involved?

As you may have discovered in your analysis, the activity (reducing ability) of a metal correlates fairly well with its Pauling electronegativity, as also seen in the sets of redox predominance diagrams in Figures 6.15 and 6.16a.

1. The most active metals (the left side of Figures 6.15 and 6.16a) have low electronegativities. We may include in a group of **very electropositive metals** all metals with Pauling electronegativities below 1.4. The cations of these metals generally have standard reduction potentials of –1.6 V or below. The predominance regions of the metals themselves are small and low, so these metals are very good *reducing* agents. They are so reactive that they react with the low concentration of H⁺ present in neutral water (or steam) to release hydrogen. This group includes the metals at the far left of the periodic table (Groups 1–3, and the groups of the *f* block).

   Conversely, their metal ions are not good oxidizing agents at all: They have very large predominance regions and cannot be reduced to the metal in aqueous solution, because water or H⁺(*aq*) is more easily reduced than these cations.

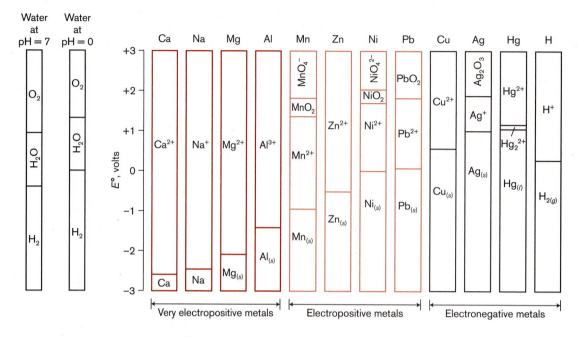

Figure 6.15. Predominance diagrams for a short activity series of metals, shown in increasing order of their Pauling electronegativity values.

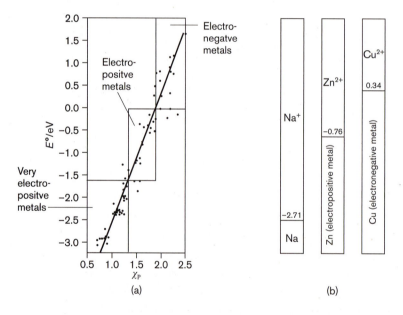

Figure 6.16. (a) The standard reduction potential ($E°$) of the most common cation of each metallic element versus its Pauling electronegativity value. (b) Redox predominance diagrams for Na, a very electropositive metal; Zn, an electropositive metal; and Cu, an electronegative metal.

2. We may identify a group of **electropositive metals** with Pauling electronegativities between 1.4 and 1.9. The cations generally have standard reduction potentials between 0.0 and –1.6 V. This group includes the d-block elements of the fourth period, and the p-group metals of the fourth and fifth periods (the middle of Figures 6.15 and 6.16a). Although these metals do not react very readily with neutral water, they do react with hydrogen ion (acids).

3. There is a group of **electronegative metals** with Pauling electronegativities between 1.9 and 2.54 (the right side of Figures 6.15 and 6.16a). Their cations generally have *positive* standard reduction potentials. These cations have

fairly small predominance regions, so they are good oxidizing agents that are easily reduced by hydrogen gas. The metals themselves have large predominance regions. This group of metals is not oxidized by hydrogen ion, and it includes most of the *d*-block elements of the fifth and sixth periods and the *p*-group metals of the sixth period.

This relative order of reactivity of metals applies not only to oxidation by water or the hydrogen ion, but (approximately) to oxidation by other oxidizing agents as well. Although H^+ cannot attack the electronegative metals, oxidizing acids such as concentrated nitric acid contain anions that can attack and dissolve most of them. The lower boundary of the predominance region of NO_3^- (Figure 6.7) is much higher than that of H^+, so it is a stronger oxidizing agent, especially when concentrated.

$$Ag(s) + 2H^+(aq) + NO_3^-(aq) \rightarrow Ag^+(aq) + NO_2(g) + H_2O(l)$$

As another example, in a *displacement reaction* a more-reactive (more electropositive) metal reacts with the cation of a less-active (more electronegative) metal to give the (positive) cation of the more electropositive metal, while the more electronegative element ends up in the zero oxidation state in the elemental form, after combining with (negative) electrons. Displacement is also readily predicted by examining the redox predominance (or Pourbaix) diagrams of the two elements.

EXAMPLE 6.7

Which of the following displacement reactions are expected to go to the products indicated? (a) $Fe^{2+} + Cu \rightarrow Fe + Cu^{2+}$; (b) $Mn + 2Ag^+ \rightarrow Mn^{2+} + 2Ag$; (c) $2Na + Cu^{2+} \rightarrow 2Na^+ + Cu$; and (d) $Bh + Hs^{2+} \rightarrow Bh^{2+} + Hs$.

SOLUTION: (a) As seen from the predominance diagram in Figure 6.9, the predominance regions of Fe^{2+} and Cu overlap, so *no* reaction is expected. This failure to react can also be predicted from electronegativities: The most electronegative metal, Cu, is expected to remain in elemental form.

(b) The predominance diagrams of Mn and Ag^+ do *not* overlap, so reaction is predicted. The most electronegative metal, Ag, is expected to be produced in elemental form, which is done by Reaction (b).

(c) As in (b), this reaction would be expected to go, and would in fact go readily in suitable nonaqueous solvents. In water, however, a complication would arise: Na, a very electropositive metal, also reacts with water to give $NaOH(aq)$ and $H_2(g)$, and the concentration of water in the solution is much higher than the concentration of Cu^{2+}. Furthermore, the NaOH byproduct of the reaction with water would tend to precipitate the Cu^{2+} as $Cu(OH)_2$. This reaction would not proceed as cleanly in water as expected from a simple examination of electronegativities or predominance diagrams.

(d) Neither predominance diagrams nor electronegativities are known for the newly discovered elements Bh and Hs. However, from periodic trends we expect Hs to be more electronegative than Bh, so elemental Hs is the expected product of this reaction, as shown.

Terminology and GHS Symbols for Strong Reducing Agents. In Section 5.7A and in particular in Figure 5.28, we introduced the Globally Harmonized System and symbols for labeling bottles of toxic chemicals. We return to that system and focus on the second pictogram (shown at left), to designate flammables. These are defined as follows:[3]

- An **inflammable** or **flammable** compound is one that is easily ignited and burns very rapidly.

- A **pyrophoric** chemical readily ignites and burns in air spontaneously, without a source of ignition.

The very electropositive metals Na, Ca, and Mg from the activity series of Experiment 7 are inflammable/flammable, and carry the "Flammable" pictogram. Beyond Mg, the flammability of the metals depends on their particle size as well as their position in the activity series: Finer surfaces allow for more rapid burning in air. Thus, Al foil and Al and Mn pellets do not carry this pictogram, but Al and Mn powders do. Granular iron has no pictogram, but finely powdered iron carries the "Flammable" pictogram.

6.2C. Redox Chemistry of Nonmetals. The redox chemistry of the nonmetals are complicated by the presence of positive, negative, and zero oxidation states for most of these elements. In addition, most of them have two positive oxidation states and some have more. Nonetheless, it will still be useful to categorize the nonmetals into two groups, based on their Pauling electronegativity values.

We identify **very electronegative nonmetals** as those with Pauling electronegativities over 2.8—F_2, Cl_2, Br_2, O_2, and N_2. Their redox predominance diagrams are shown together in Figure 6.17a, with the predominance regions of the elemental forms being shaded. Note the following overall patterns:

1. The predominance diagrams of very electronegative nonmetals are dominated by their broad predominance ranges in negative oxidation states (anions: Cl^-, Br^-, and I^-, or hydrogen compounds: HF and H_2O). These are generally the most stable oxidation states, so these species are relatively weak reducing agents.

2. Very electronegative nonmetals in positive oxidation states (oxo anions) are either absent or are strong oxidizing agents.

3. The elemental forms (shown shaded in gray) also fall rather high in their predominance diagrams, so these elements tend to be strong oxidizing agents. At the extreme, F_2 can oxidize gaseous water and cause it to catch fire! Due to its strong triple bond, N_2 is the exception to this generalization and is a poor oxidizing agent. Among Group 1 metals, only Li metal is oxidized to form a red nitride, Li_3N ($\Delta H_f = -164.56$ kJ mol^{-1}). To produce K_3N and blue Na_3N ($\Delta H_f = +64$ kJ mol^{-1}), the strong $N{\equiv}N$ bond must be broken by electrical discharge, whereupon the N atoms will react with atomic K or Na[4] or elemental Na.[5]

Terminology and GHS Symbols for Strong Oxidizing Agents. The Globally Harmonized System pictogram to designate "Oxidizers" is shown at left. An **oxidizer** or **oxidizing agent** is defined as a chemical that can rapidly bring about an oxidation

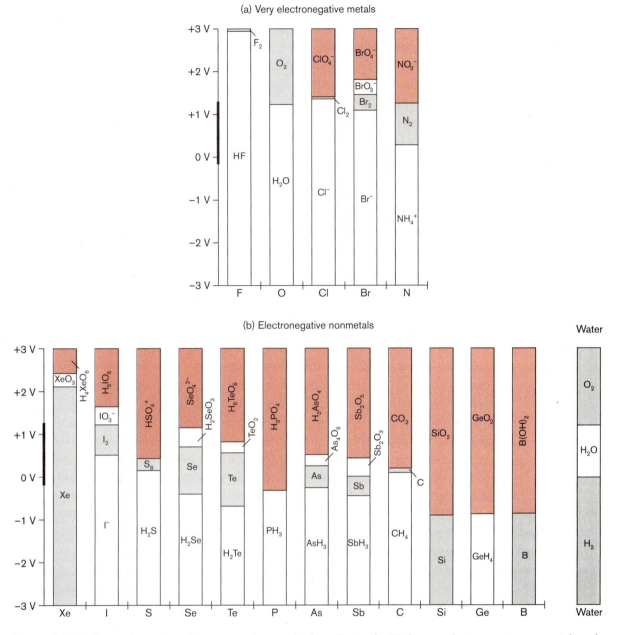

Figure 6.17. Redox predominance diagrams under standard conditions for (a) the very electronegative nonmetals and (b) the electronegative nonmetals. The elements in their group oxidation states (as oxides, oxo anions, or oxo acids) are shaded in light brown. The elemental forms themselves (atoms with oxidation state zero) are shaded in gray.

reaction by supplying oxygen or receiving electrons during oxidation. In the context of safety symbols, the adjective *rapidly* is involved.

We may be able to predict which oxidizing agents react *rapidly* and hence merit the "Oxidizer" pictogram. Strong and rapid oxidizing agents having the symbol include the salts potassium nitrate, as well as some *d*-block salts in high oxidation states, such as potassium permanganate, potassium ferrate (K_2FeO_4), and potassium manganate (K_2MnO_4) (Table 6.1).

Electronegative nonmetals include those with Pauling electronegativity values between 1.9 and 2.8; their predominance diagrams are shown in Figure 6.17b.

1. The elemental forms mostly fall in the middle of their predominance diagrams (e.g., carbon), so these elements tend to be weak oxidizing agents (see Example 6.8) *and* weak reducing agents. They can often be found in nature in elemental form (Section 6.4B). However, some of the electronegative nonmetals can be flammable in some allotropic forms (Section 12.3). Red phosphorus is sold with the "Flammable" pictogram, whereas pyrophoric white or yellow phosphorus is too flammable to be shipped. Likewise, the hydrides of the very electropositive metals and of the electronegative nonmetals tend to be flammable or even pyrophoric. Most carbon hydrides (e.g., hydrocarbons) are flammable.

2. Forms in positive oxidation states (oxides, oxo anions, and oxo acids) have fairly broad predominance ranges, so they are at best weak oxidizing agents.

3. Forms in negative oxidation states (anions or hydrogen compounds) also have fairly broad predominance ranges. However, many of them are below the predominance range of water, or penetrate its range only slightly, so they are in practice significant reducing agents.

Periodic Trends Among Nonmetals in Positive Oxidation States. Examination of the redox predominance (Figures 6.7 and 6.8) or Pourbaix diagrams of the nonmetals shows us the main horizontal trend in stability of nonmetals in positive oxidation states (oxides, oxo acids, and oxo anions). As the electronegativities of nonmetals increase from left to right across a given period, the predominance regions of the very high oxidation states constrict. The nonmetal oxo acids and anions (or oxides) become better oxidizing agents as one goes to the right in a given period. (Concentrated nitric acid and perchloric acid are labeled as oxidizers, but not concentrated sulfuric acid or phosphoric acid.) Thus, the last element of the second period that is able to achieve the group oxidation number is nitrogen (Table 1.10); nitric acid and nitrates are commonly used as strong oxidizing agents. At the end of the third period, perchloric acid exists, but perchlorates, chlorates, perchloric acid, and chloric acid are good oxidizing agents.

The vertical trends in stability among the high oxidation states of the nonmetals are perhaps a bit more surprising. Comparing the predominance diagrams for the Group 15/V elements (Exercise 6.9 and Figures 6.7 and 6.8), notice that the largest predominance region for the group oxidation state belongs to the *third*-period element, phosphorus (as H_3PO_4). Stability drops off going up to nitrogen (NO_3^-); it also drops off going down to arsenic (AsO_4^{3-}), antimony (Sb_2O_5), and bismuth (Bi_2O_5). In the *p* block, the group oxidation number is most stable in the *third* period. This periodic trend is also manifest in Groups 16/VI (Figure 6.18) and 17/VII (Exercise 6.8).

Figure 6.9 shows the corresponding trends in the stability of the group oxidation state among the *d*-block elements. The horizontal periodic trends are the same as for the *p*-block elements: As one goes to the right, the increasingly high oxidation states lose stability, disappearing after Group 7 or 8. However, the vertical periodicity is different: This time the *fourth*-period elements show the greatest reluctance to adopt the group oxidation state, which is more stable in the fifth and especially the sixth periods (Exercise 6.10).

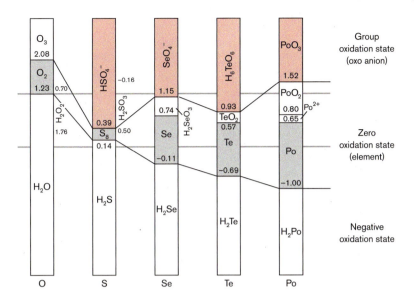

Figure 6.18. Predominance diagrams of Group 16/VI elements. The group (maximum) oxidation state is shown in light brown at the top of each diagram. The elements themselves (zero oxidation states) are in the gray shaded regions.

EXAMPLE 6.8

Based on their predominance diagrams from Figures 6.17 and 6.9, suggest additional strongly oxidizing oxides, oxo acids, and oxo anions among the (a) very electronegative nonmetals, (b) electronegative nonmetals, and (c) *d*-block metals.

SOLUTION: (a) We have already mentioned nitric acid and nitrates, perchloric acid and perchlorates, and chloric acid and chlorates. We can add species above the stability range of water from Figure 6.17a—namely, perbromate, bromate, perbromic acid, and bromic acid.

(b) From Figure 6.17b we find three species well above the stability range of water: H_4XeO_6, XeO_3, and H_5IO_6. (Others are just above the stability range when at standard concentration, but might be well above as concentrated acids: H_2SeO_4 and HIO_3.)

(c) From the *d* block, in addition to the previously mentioned Mn and Fe species from Table 6.1, there are CrO_4^{2-}, CoO_2, NiO_2, RuO_4, RhO_4^{2-}, and Ag_2O_3.

A CONNECTION TO ENVIRONMENTAL CHEMISTRY

A Naturally Occurring Strong Oxidizing Agent, Chromate

Because chromate/dichromate has a predominance region (above 1.38 V) just above that of water (1.23 V and below), we should not find it as long-lasting species in our aqueous world. It has been discovered, however, that CrO_4^{2-} is generated from Cr^{3+} by action of the manganese(III, IV) oxide mineral birnessite.[6] This is a significant example of the importance of speciation (Section 3.8): Cr^{3+} may be essential and is quite unavailable by Solubility Rule I (Section 4.4), but CrO_4^{2-} salts are not as insoluble (Solubility Tendency IV) and, due to their oxidizing ability, act as mutagens, teratogens, and carcinogens. Hence, the toxicity properties of chromium are much different in the +6 oxidation state as compared to the +3 oxidation state.

6.3. Explosions and Safety in the Laboratory

> **OVERVIEW.** You can recognize compounds and mixtures likely to react very exothermically or explosively on the basis of large $E°$ gaps between the molecules or cations and anions, and on the number of moles of gas released (Section 6.3A; Exercises 6.40–6.46). You must know appropriate safety precautions to take in handling these compounds. Other potential explosives can be recognized from the presence of oxidizing and reducing functional groups in the same molecule, or the presence of certain weak chemical bonds in the molecule (Section 6.3B; Exercises 6.47–6.53).
>
> In this section there are many Connections to chemical safety. You again use the concepts of entropy (Section 4.5A), oxidation numbers (Section 1.4), bond energies (Section 3.3), GHS pictographs (Section 5.7), and the HSAB principle (Section 5.4).

6.3A. Explosion Dangers from Mixtures or Salts Containing Oxidizing and Reducing Components.

In this section we try to develop an ability to anticipate when compounds or mixtures pose the danger of explosions. If you do well in learning the concepts involved in this section, your life expectancy may increase! In the previous section we saw ways of predicting the (in)flammability of compounds. Even more hazardous are **explosives,** which can be defined as chemicals or mixtures that can produce a sudden release of *energy* or *gas* when subjected to ignition, sudden shock, or high temperature.[3] The GHS pictograph for an explosive (left) shows materials being ejected in all directions.

The definition highlights three conditions for explosions:

1. There must be a kinetic barrier to be overcome before the reaction commences; this is accomplished by providing the needed *activation energy.* **Primary explosives** have relatively low activation energies (i.e., 105–167 kJ mol^{-1})[7] and can be initiated by heat or shock, but are generally not as powerful as *secondary* explosives. **Secondary explosives** have higher activation energies (i.e., 176–242 kJ mol^{-1}), so they require initiation by a primary explosive and are generally more powerful.

2. The reaction must be very exothermic. Reactions of strong oxidizing and strong reducing agents are more exothermic than even the most exothermic acid–base reactions (Reaction numbers AB1–AB3 in Table 6.2).

3. The third feature that contributes to explosive power is the pressure of gas emitted, which depends on the net change in number of moles of gas (and the temperature) during the reaction (third column in Table 6.2).

Very Exothermic Reactions ($E°$ Gap > 1 V, ΔG > 96.5 kJ mol^{-1} of Electrons). A classic reaction of a strong reducing agent with a strong oxidizing agent is the very exothermic reaction of solid sodium metal in chlorine gas (Reaction 1A in Table 6.2). The reaction begins very quickly, because little in the way of activation energy is required. This reaction is very exothermic, as measured by ΔH either in units of kJ mol^{-1} or in terms of kJ kg^{-1} of reactants. The $E°$ gap in Figure 6.7 is very large (4.07 V). The

TABLE 6.2. Examples of Very Exothermic (Acid–Base and) Redox Reactions

No.	Reaction	Mol gas[a]	ΔH^b (kJ mol^{-1})	ΔH^b (kJ kg^{-1})	ΔG^b (kJ mol^{-1})	$E°$ (V)
AB1	$H^+(aq) + OH^-(aq) \rightarrow H_2O(l)$	0	−55.8	−3100	−79.8	
AB2	$H_2SO_4(l) \rightarrow H^+(aq) + HSO_4^-(aq)$	0	−73.3	−748		
AB3	$P_4O_{10}(s) + 4H_2O(l) \rightarrow$ $4H_3PO_4(aq)$	0	−1002.1	−2556		
1A	$Na(s) + \frac{1}{2}Cl_2(g) \rightarrow NaCl(s)$	−½	−411.15	−7052	−384.14	3.98
1Aq	$Na(s) + \frac{1}{2}Cl_2(g) \rightarrow NaCl(aq)$	−½	−407.27	−6986	−393.13	4.07
1B	$H_2(g) + \frac{1}{2}O_2(g) \rightarrow H_2O(l)$	−1½/−½	−285.8	−15,860	−237.1	1.229
1C	$CH_4(g) + 2O_2(g) \rightarrow CO_2(g) +$ $2H_2O(l)$	−2/0	−890.8	−11,135	−817.8	1.06
1D	$C_3H_8(g) + 5O_2(g) \rightarrow 3CO_2(g) +$ $4H_2O(l)$	−3/+1	−2219.0	−10,877	−2155.7	1.117
1E	$2B_5H_9(l) + 12O_2(g) \rightarrow 5B_2O_3(s) +$ $9H_2O(g)$	−3/+9	−8617.9	−16,573	−8364.9	1.806
2A	$NH_4NO_3(s) \rightarrow 2H_2O(g) + N_2(g)$ $+ \frac{1}{2}O_2(g)$	+2½	−118.0	−1475	−299.7	1.035
2Aq	$NH_4NO_3(aq) \rightarrow 2H_2O(l) + N_2(g)$ $+ \frac{1}{2}O_2(g)$	+1½	−231.79	−2897	−283.68	0.980
2B	$NH_4ClO_4(s) \rightarrow 2H_2O(g) +$ $\frac{1}{2}N_2(g) + O_2(g) + \frac{1}{2}Cl_2(g)$	+4	−187.8	−1598	−368.6	1.27
3	$Pb(N_3)_2(s) \rightarrow Pb(s) + 3N_2(g)$	+3	−469	−1610		3.21
4	$C_3H_5(NO_3)_3(l) \rightarrow 3CO_2(g) +$ $2\frac{1}{2}H_2O(g) + N_2(g) + \frac{1}{2}O_2(g)$	+4½/+7	−1406	−6194		
5A	$H_2O_2(l) \rightarrow H_2O(l) + \frac{1}{2}O_2(g)$	+½/+1½	−98.0	−2880	−116.7	0.605
5B	$2NH_3NI_3(s) \rightarrow N_2(g) + 3I_2(s) +$ $2NH_3(g)$	+3	−176	−212		

Sources: NIST-JANAF Thermochemical Tables, http://kinetics.nist.gov/janaf/; J. A. Dean, Ed., *Lange's Handbook of Chemistry*, 13th ed., McGraw-Hill: New York, 1985, pp. 9-4 to 9-69; D. R. Lide, Ed., *Handbook of Chemistry and Physics*, 84th ed., CRC Press: Boca Raton, FL, pp. 5-4 to 5-60.

[a] The first number shows the net number of moles of gas produced if water is produced as a liquid; the second number assumes the water is produced as a gas.

[b] Explosions occur as far from equilibrium and from standard conditions (273K, 1 atm) as we will ever want to get, so the values of $\Delta H°$ and $\Delta G°$ listed in Table 6.2 are only roughly applicable. Products other than those shown may also be formed.

reaction begins rapidly and becomes very hot very fast. No gases are produced, so it does not *usually* give a very powerful explosion.

The reaction of sodium metal with $CCl_4(l)$ is also very exothermic. The reduction of $CCl_4(l)$, $CCl_4(l) + 4e^- \rightarrow C(s) + 4Cl^-(aq)$, has almost as high a value of $E°$ (i.e., +1.18 V) as does that of $Cl_2(g)$.

TABLE 6.3. Additional Oxidizing and Reducing Agents Useful in Analytical and Organic Chemistry

Reduction Half-Reaction	$E°$ (V)
A. Oxidizing agents (in boldface)	
$\mathbf{MnO_4^-}(aq) + 8H^+(aq) + 5e^-(aq) \rightarrow Mn^{2+}(aq) + 4H_2O(l)$	1.507
$\mathbf{CCl_4(l)} + 4e^-(aq) \rightarrow C(s)(aq) + 4Cl^-(aq)$	1.18
B. Reducing agents (in boldface)	
$Fe(CN)_6^{3-}(aq) + e^-(aq) \rightarrow \mathbf{Fe(CN)_6^{4-}}(aq)$	0.358
$S_4O_6^{2-}(aq) + 2e^-(aq) \rightarrow \mathbf{2S_2O_3^{2-}}(aq)$	0.08
$2CO_2(aq) + 2H^+(aq) + 2e^-(aq) \rightarrow \mathbf{H_2C_2O_4}(aq)$	−0.49
$CO_2(aq) + 2H^+(aq) + 2e^-(aq) \rightarrow \mathbf{HCOOH}(aq)$	−0.199
$HCOOH(aq) + 2H^+(aq) + 2e^-(aq) \rightarrow \mathbf{HCHO}(aq) + H_2O(l)$	0.056

Source: D. R. Lide, Ed., *Handbook of Chemistry and Physics*, 84th ed., CRC Press: Boca Raton, FL, pp. 8-23 to 8-33; and J. A. Dean, Ed., *Lange's Handbook of Chemistry*, 13th ed., McGraw-Hill: New York, 1985, pp. 6-4 to 6-19.

The predominance regions of Na(s) and $CCl_4(l)$ also have a large $E°$ gap of 3.89 V. This reaction has a very high activation energy, however, so it introduces greater dangers of explosion. Reaction with sodium metal used to be a standard method of characterizing chlorocarbons, but it sometimes led to treacherous explosions, as did attempts to extinguish burning metals with CCl_4-based fire extinguishers.

Moderately Exothermic Reactions ($E°$ Gap = 0.6 to 1.0 V, ΔG = 58 to 96.5 kJ mol^{-1} of Electrons). In contrast, redox reactions with more modest $E°$ gaps (less exothermic reactions) and *low* activation energies proceed rapidly and have been very useful in analytical chemistry for classical volumetric redox analysis. The high heat capacity of the water used as the solvent means that the heat produced in these reactions does not cause a dangerous rise in temperature. Some of the classically useful oxidizing agents for volumetric analysis include Ce^{4+}, $K_2Cr_2O_7$, I_2 [as $I_3^-(aq)$], and $KMnO_4$ (Figures 6.8 and 6.9 and Table 6.3A): Some classically useful reducing agents include Fe^{2+}, $Na_4Fe(CN)_6$, and $Na_2S_2O_3$ (sodium thiosulfate) (Figure 6.9 and Table 6.3B).

Organic oxidations such as the dichromate oxidations of alcohols to aldehydes and ketones, and of aldehydes to carboxylic acids (Table 6.3B), typically have higher activation energies, so they take longer to occur than redox titrations. Since they may be carried out in more concentrated, nonaqueous solutions, there can be more buildup of heat, which might lead to boiling of a solvent. These do not commonly result in explosions. Explosions can occur in the cases of organic compounds or oxidants with *very high* activation energies, such as those of alkanes with oxygen.

Slightly Exothermic Reactions ($E°$ Gap < 0.6 V, ΔG < 58 kJ mol^{-1} of Electrons). An example of this type of reaction having a low activation energy is the displacement reaction of copper metal with silver ion in Experiment 7,

$$Cu(s) + 2Ag^+(aq) \rightarrow Cu^{2+}(aq) + 2Ag(s)$$

Activation energies often are unknown and may depend on the presence of traces of catalysts. This makes the rates of new redox reactions difficult to predict. What we can suggest is that the *larger* the $E°$ gap (cell voltage), the *more* exothermic the reaction, and the *greater* the degree of precaution you should take in case the activation energy proves to be high. Our suggestions for anticipating hazards are as follows:

1. If the *$E°$ gap is > 1.0 V*, the reaction will be quite exothermic and might possibly be explosive. Take precautions as described below!

2. If the *$E°$ gap is between 0.6 and 1.0 V*, the reaction will be significantly exothermic, but there is less risk of an explosion. Just in case, wear safety goggles (and if possible) work in a hood (fume cupboard). Provide adequate cooling (unless being run in dilute aqueous solution).

3. If the *$E°$ gap is < 0.6 V*, you may need to add heat or a catalyst, or allow plenty of time for the reaction to occur.

As a first general lesson in **chemical safety**, always take precautions when you are carrying out a reaction of a strong oxidizing and a strong reducing agent—be sure to check for the GHS pictographs on the bottles. If the chemical compounds were not purchased and data are unavailable online, look at the $E°$ gap. Especially if it exceeds 1.0 V, keep in mind that the reaction will be highly exothermic. Add the second reactant *slowly* with *good stirring*. If an organic solvent is used, it could be heated so hot that it catches fire or boils. Allow a vent for any gases produced. Note the locations of the appropriate types of fire extinguishers. Unless you are doing this reaction in a dilute aqueous solution, provide external cooling. If an explosion should occur, the glass of the reaction vessel may be shattered into shards, which may be hurled into your skin and eyes, so *wear safety goggles* and do the reaction behind a *safety shield*, such as the window of a hood or fume cupboard.

Be especially careful if the expected exothermic reaction does *not* start upon initial cautious addition of the first small amount of the second reagent. This means that an activation energy or catalyst is needed but is (temporarily) absent. Under these conditions, heating or disturbing the system may suddenly provide the activation energy for part of the molecules. Their reaction in turn may provide the activation energy for many other molecules, resulting in a chain reaction and a violent explosion. DO NOT INCREASE THE RATE OF ADDITION OF THE REACTANT TO GET THE REACTION GOING—instead, get the safety shield down. If you have foolishly already added a lot of the second reactant, EVACUATE THE LABORATORY IMMEDIATELY!

Unfortunately, mechanisms are so diverse that the 0.6-V and 1.0-V guidelines suggested above are by no means reliable. Neither is it possible to provide a complete list of chemicals, or combinations of chemicals, that give rise to explosions—chemists are discovering new ones every month, usually unintentionally! (Such new discoveries are usually promptly reported in the Letters to the Editor section of *Chemical & Engineering News*.) However, five categories of potentially hazardous combinations[8] are suggested in the remainder of this section. The first three can be anticipated based on predominance diagrams (Figure 6.19), although these do not suggest the magnitudes of the activation energies.

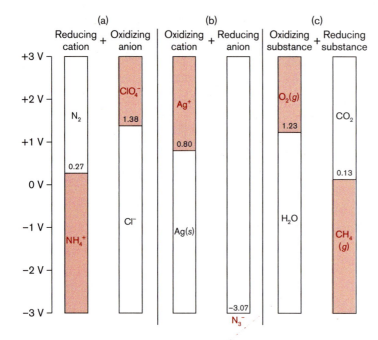

Figure 6.19. Predominance diagrams to illustrate the first three classes of explosion dangers: (a) from a reducing cation and an oxidizing anion; (b) from an oxidizing cation and a reducing anion; (c) from a mixture of oxidizing and reducing substances. The potentially explosive reactants are shown in **brown**.

Hazard Category 1. *Mixtures of strongly oxidizing and reducing substances* present the possibility of a fire or an explosion. Such substances should be stored in separate areas of the stockroom or laboratory. Mixture is particularly easy if both substances are gases; the physical chemistry that determines whether a fire or an explosion will result has been well studied.[9] Likely gaseous reducing agents include hydrogen and many of the hydrides of the nonmetals (such as the hydrocarbons). Likely gaseous oxidizing agents include F_2, Cl_2, Br_2, O_2, and the volatile oxides of the elements (Section 8.5C).

Explosions or fires may also occur, however, even if one or neither reactant is a gas. For solids, rates increase with surface areas, so finely powdered combustible solids (such as metal dusts[10] or carbohydrate dust in grain elevators) can form explosive mixtures with air. Mixtures of oxidizing and reducing solids such as gunpowder (S, C, KNO_3) can also be explosive. The thermite reaction,

$$2Al(s) + Fe_2O_3(s) \rightarrow 2Fe(s) + Al_2O_3(s) \qquad (6.14)$$

is so exothermic,

$$\Delta H = \Delta H_f(Al_2O_3) - \Delta H_f(Fe_2O_3) = -851.5 \text{ kJ mol}^{-1}$$

that the iron is produced in the molten state (Figure 6.20). No gases are produced, however, so this reaction is not normally explosive. High activation energies are required to initiate such reactions between solids. The thermite reaction is initiated with a burning Mg ribbon.

Figure 6.20. The thermite reaction.

A CONNECTION TO CHEMICAL SAFETY

Stockroom Design

Explosive mixtures can be formed unexpectedly and disastrously. This can happen in chemistry stockrooms when earthquakes strike or fires cause shelves to buckle. The old practice of placing chemicals on shelves in alphabetical order is therefore strongly discouraged! At a minimum we want to keep strong oxidizing agents in a separate area from strong reducing agents. Keeping strong acids and strong bases separate is a good idea, too. Some of the remaining classes of explosives are, in effect, pre-mixed, so they should probably be kept separate from everything else, and they should be kept in small quantities only. More detailed guides are available for the safe storage patterns for chemicals.[11]

Hazard Category 2. A salt that is the combination of a *strongly reducing cation* and a *strongly oxidizing anion* (Figure 6.19b) may persist for kinetic reasons, but it is thermodynamically unstable and is an explosion hazard. From the predominance diagram for nitrogen (Figure 6.7) it can be seen that the ammonium ion (NH_4^+) has a limited predominance range. Its salts with strongly oxidizing oxo anions are often explosive. Thus, ammonium nitrate, a common fertilizer, is a powerful explosive when ignited ($E° = 0.98$ V; see also Reaction 2A in Table 6.2). The worst industrial accident in U.S. history occurred in 1947 in Texas City, TX, when two shiploads of NH_4NO_3 detonated, killing 581 people and destroying two-thirds of the city.[12] Ammonium perchlorate (predominance gap = 1.11 V; see also Reaction 2B in Table 6.2) is used as an oxidant for solid-fueled rockets. In 1988 one of two American plants manufacturing this salt, at Henderson, NV, was destroyed by a series of explosions. There are also many organic ammonium ions containing one-to-four organic R groups in place of the ammonium hydrogen ions, and many complex cations involving ammonia or organic amines as ligands. These are large cations that form nicely insoluble perchlorates, but these perchlorates are treacherous explosives! It is much safer to precipitate these cations with nonbasic anions that are not oxidizing, such as BF_4^- or PF_6^-.

Hazard Category 3. The same considerations apply with the combination of a *strongly oxidizing cation* and a *strongly reducing anion*. The cations of the electronegative metals (Section 6.2B) are strongly oxidizing. Two important strongly reducing anions are the azide ion, N_3^- ($E° = -3.09$ V; Figure 5.17a) and the acetylide ion, C_2^{2-}. Thus, lead(II) azide (predominance gap = +2.97 V) is a primary explosive that is very shock sensitive and is used to provide the activation energy needed to detonate more stable and powerful explosives. Sodium azide mixed with potassium nitrate was long used in the automobile air bag, because the reaction of this mixture is readily initiated, and the air bag is then inflated by the large volume of N_2 released:[13]

$$10NaN_3(s) + 2KNO_3(s) \rightarrow 5Na_2O(s) + K_2O(s) + 16N_2(g)$$

However, the soft-base azides are quite poisonous, so they were replaced in air bags, most recently by cheaper but more explosive NH_4NO_3, which resulted in several deaths and millions of recalls.

Similarly, acetylides such as Ag_2C_2 and Cu_2C_2 are very explosive. Several explosive silver salts can be precipitated in common laboratory situations.[14]

EXAMPLE 6.9

Which of the following appear the most likely to show explosive properties: (a) $Eu(ClO_4)_2$, (b) $AgClO_4$, or (c) $TlClO_4$?

SOLUTION: The standard reduction potentials of the cations and anions of these salts are obtained from Figures 6.2–6.5. The Eu^{2+} ion is (mainly) a reducing agent, confined below an $E°$ of –0.35 V. Ag^+ is (mainly) an oxidizing agent, predominant above an $E°$ of +0.80 V. ClO_4^- is exclusively an oxidizing agent, found above an $E°$ of +1.38 V. Tl^+ can either be weakly oxidizing or weakly reducing, but relative to the ClO_4^- ion it would be reducing, with its upper boundary at +1.25 V being relevant.

(a) As shown in the diagram on the left, there is a gap of +1.73 V between the predominance regions of Eu^{2+} and ClO_4^-. Based on thermodynamics, then, these two ions should react with each other very exothermically. If there is no kinetic barrier, we expect this salt to be nonexistent. If the cation and anion react slowly enough to allow the salt to be isolated, the salt should be treated as a potential explosive.

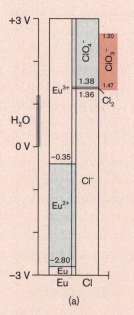

(a)

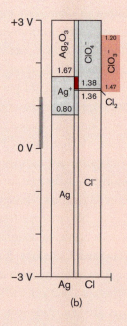

(b)

(b) The salt $AgClO_4$ contains two oxidizing ions, which have predominance regions that overlap between +1.38 and +1.67 V (in the diagram on the right). Hence, the salt should not decompose by a redox reaction between the cation and the anion.

(c) The predominance region for ClO_4^-, with a lower boundary at +1.38 V, does not overlap the region for Tl^+, which has an upper boundary at 1.25 V. The gap, however, is much less than 1.0 V. We cautiously conclude, then, that although this salt should not be thermodynamically stable, if it does decompose, it will probably not release enough energy to cause an explosion. This compound exists.

Lithium-Ion Batteries

Around 4 billion of these rechargeable batteries (Figure 6.3) are manufactured every year. They range in size from the tiny "button" batteries found in consumer electronic devices to the 1-pound batteries found in laptop computers to the 63-pound batteries found in the Boeing 787 Dreamliner plane to the 435-pound batteries found in the Chevy Volt electric automobile.

During charging, lithium ions are reduced to lithium atoms located between graphite layers (Section 12.3B) in the anode. The cathode material is a solid-state oxide or phosphate such as $LiCoO_2$, $LiMn_2O_4$, or $LiFePO_4$, often containing mixed metal ions (Sections 8.3 and 8.4). During charging, solvated lithium ions are transported through pores in the separator from the cathode to the anode (and back again during the discharge process, when the battery is in use); this conducts the current across the separator. In the process the d-block metal ion is taken to a high oxidation state [ideally, as $CoO_2(s)$, MnO_2, or $FePO_4$]. Because lithium-ion batteries involve Li metal (which is still very active even when located in graphite layers), the batteries cannot use water as a solvent. Normally the solvent is a polar organic solvent that is capable of dissolving the electrolyte, which is a lithium salt of a large nonbasic ion (Section 4.4D). In some batteries, however, there is no solvent; the ions cross from one electrode to the other while in the solid state, which must allow lithium ions to move through their lattices. (These are super-ionic solid-state conductors; Section 8.3B).

We can estimate the voltage of the solid-state reaction, $Li(s) + CoO_2(s) \rightarrow LiCoO_2(s)$, using predominance diagrams for lithium and cobalt. The species $LiCoO_2(s)$ does not appear on the cobalt predominance diagram, but let us estimate the voltage of the lithium-ion battery by replacing it with $Li(s) \rightarrow Li^+(aq)$ and $CoO_2(s) \rightarrow Co(OH)_3(s)$, which do appear:

$$CoO_2(s) + e^- + 2H_2O \rightarrow Co(OH)_3(s) + OH^-(aq) \quad E° = +0.62 \text{ V}[15]$$

The gap between this half-reaction and that of Li is 3.66 V, which is quite close to the actual cell voltage of the lithium-ion battery. This voltage is so large that an explosion is possible if the separator is pierced. Unfortunately this can happen if the battery is overcharged, producing more $Li(s)$ than the graphite can hold. In that case, sharp dendrites of Li metal are produced that can penetrate the pores of the separator, giving a short-circuit. Overcharging can take the cathode metal to an even higher oxidation state, which could react with the organic solvent. Or a crash could result in a puncture, allowing hot Li-ion-bearing organic solvent to escape and catch fire.[16] If you were the first-responding firefighter, should you attempt to put this fire out with water (Section 6.3A)? (What color would the flames be?)

Failures due to causes such as these occur rarely, about once in each 10 million cells. However, since 4 billion cells are produced each year, this does occur. Bulk shipments of large or many lithium-ion batteries pose dangers of producing high-temperature fires, which could ignite other cargo on a plane, so airlines generally ban such shipments. Another danger arises with the tiny lithium-ion batteries found so commonly in consumer electronics. The tiny "buttons" attract young children, who swallow them about 3500 times a year. If they rupture in the esophagus, the alkaline solution can burn the esophageal wall and attack the aorta. A useful summary of the theoretical and practical considerations involved in choosing anode and cathode materials for lithium-ion batteries is given by Treptow.[17]

EXAMPLE 6.10

Will the following redox reactions tend to occur? If so, what are the cell emfs?
(a) $Br_2 + 2NH_3OH^+ \rightarrow 2Br^- + N_2 + 2H_2O + 4H^+$
(b) $2MnO_4^{2-} + 6Br^- + 8H^+ \rightarrow 3Br_2 + 2MnO_2 + 4H_2O$

SOLUTION: (a) Its relatively high predominance region means that Br_2 is probably acting as an oxidizing agent. (Its oxidizing ability is confirmed by the listing of Br^- as a product.) The relevant predominance region boundary for Br_2 is thus the lower one, +1.07 V. The boundary applying to NH_3OH^+ as a reducing agent is its upper one, –1.87 V, at which it is converted to the adjacent species in the next higher oxidation state, N_2. There is a large gap between the two regions, (+1.07 V) – (–1.87 V) = +2.94 V, so this reaction will proceed, possibly explosively.

(b) The Br^- ion is a reducing agent with an upper boundary of +1.07 V. Hence, MnO_4^{2-}, if it is to react, would have to act as an oxidizing agent, so its lower boundary of +2.72 V is relevant. There is a gap of +1.65 V between this lower boundary for MnO_4^{2-} and the upper boundary for Br^-, so the reaction will be expected to proceed, perhaps explosively.

6.3B. Explosives Identified by Functional Groups and Bond Types.

Hazard Category 4. It is possible to combine an oxidizing group of atoms (functional group) and a reducing group in the same molecule; the resulting substances are often explosive. Familiar explosives such as nitroglycerine [$C_3H_5(NO_3)_3$] (Reaction 4 in Table 6.2 and Figure 6.21), trinitrotoluene [TNT, $C_7H_5(NO_2)_3$], and picric acid [$C_6H_3O(NO_2)_3$] contain oxidizing nitro (–NO_2) or nitrate ester groups and reducing hydrocarbon groups.

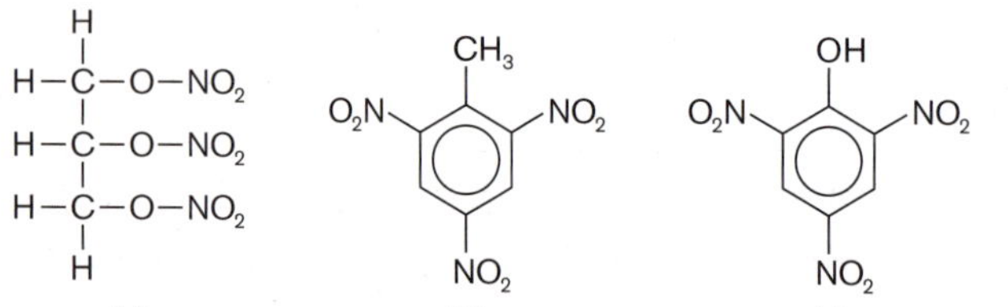

Figure 6.21.
Explosive compounds:
(a) nitroglycerine,
(b) trinitrotoluene, and
(c) picric acid.

Upon reaction, these compounds generate not only a great deal of energy but large volumes of gaseous CO, CO_2, H_2O, and N_2. Even more hazardous organic chlorate or perchlorate esters are sometimes accidentally produced by combining chloric or perchloric acid and alcohols—this should never be deliberately attempted. If the organic compound contains an unfamiliar functional group, you may be able to estimate its oxidizing or reducing properties by replacing the organic groups with H atoms, then locating the resulting species on a redox predominance diagram.

Hazard Category 5. Finally, recall from Table 3.2 and Section 3.3C that single covalent bonds between the very electronegative nonmetals (especially the second-period elements N, O, and F) are intrinsically weak. Compounds containing these bonds are therefore not only good oxidizing agents, but are often thermodynamically metastable and prone to quite exothermic disproportionation. Hence, compounds containing these bonds should be treated as potentially explosive:

- O–O [e.g., hydrogen peroxide (H_2O_2) or organic peroxides (R_2O_2)];

- N–N and –N=N– (e.g., azides, nitramines, and many nitrogen-rich compounds such as diazonium salts);

- O–Cl (perchloric acid, chlorates, chlorites, and the oxides of chlorine); O–Br; O–Xe (oxides of xenon);

- O–N [the nitro organics mentioned in the previous paragraph and compounds such as silver and mercury fulminates, AgONC and $Hg(ONC)_2$]; and

- N–Cl, N–Br, and N–I ($NI_3 \cdot NH_3$ is detonated by the force of a housefly stomping its foot).

EXAMPLE 6.11

Each of the compounds (a) ferrocenium perchlorate, (b) styphnic acid, (c) RDX, and (d) pentaerythritol dixenate is or could be an explosive. Explain what it is about each structure that makes its explosive properties plausible.

SOLUTION: (a) The ferrocenium cation has the formula $[Fe(C_5H_5)_2]^+$. It contains potentially reducing C–H and C–C bonds. The perchlorate anion is a known strong oxidizing agent; it also qualifies because of having weak Cl–O bonds.

(b) Styphnic acid contains oxidizing –NO_2 groups. In the same molecule it contains reducing C–C, C–H, and C–OH (phenolic) groups.

(c) RDX contains reducing C–H bonds, oxidizing NO_2 functional groups, and weak N–N bonds.

(d) This unknown molecule contains an unknown xenon-containing functional group that has weak Xe–O bonds. In addition, if we replace the O–C bonds with O–H bonds, the xenon group becomes H_2XeO_4 (xenic acid). Although this is not directly in Figure 6.8, it is in the same oxidation state as XeO_3, which is a strong oxidizing agent. The central part of this molecules contains many reducing C–H and C–C bonds, so whoever is brave or foolish enough to try to synthesize this molecule should observe all of the precautions described for working with powerful explosives!

A CONNECTION TO CHEMICAL SAFETY

Explosive Nitrogen-Rich Compounds

Compounds of nitrogen fit into so many of the above five categories of explosives that most research searching for new explosives focuses on nitrogen compounds, and many methods of airport security screening look for nitrogen content. Reactions are made more exothermic and moles of gas evolved are often maximized when the products of the explosion include the very stable triple-bonded N_2 molecule. For example, the oxidizing V-shaped cation, N_5^+, has been made from (explosive) reactants:[18]

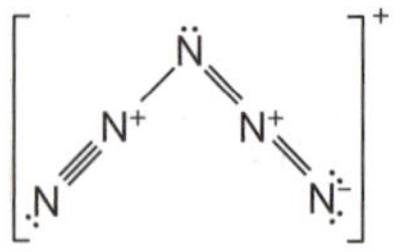

$$[N{\equiv}N{-}F]^+[AsF_6]^- + HN_3 \rightarrow [N_5]^+[AsF_6]^- + HF$$

This powerfully oxidizing cation ignites organic substances and reacts explosively with water:

$$4[N_5]^+[AsF_6]^- + 2H_2O \rightarrow O_2 + 10N_2 + 4HF + 4AsF_5$$

The salt $[N_5]^+[N_3]^-$ would decompose even more exothermically to give 4 mol of nontoxic gaseous N_2, but it has not been made—the activation energy for this decomposition is much too small. Another high-nitrogen compound of interest is the salt $[N_5]^+[B(N_3)_4]^-$ or BN_{17}.[19] A useful nitrogen-based, carbon-free oxidizing anion that has been synthesized is the dinitramide anion, $[N(NO_2)]^-$.[20] Its salts with nitrogen-based reducing cations such as ammonium and hydrazinum ($N_2H_5^+$) show explosive properties.

A CONNECTION TO CHEMICAL SAFETY AND GREEN CHEMISTRY

Pyrotechnics[21]

Pyrotechnics (fireworks) have metal ions that give colored flames: red from Sr^{2+}, yellow from Na^+, blues and greens from Cu^{2+}, and green from Ba^{2+}. The "green" qualities of the pyrotechnic are enhanced when the principal product is nontoxic N_2 rather than CO or other partially oxidized carbon compounds. (However, smoke containing barium is toxic, so the common green pyrotechnic is not "green.") If these cations are complexed with high-nitrogen anions or ligands such as tetrazole (H_2CN_4), they are more volatile, which results in less metal-ion release to the environment. Perchlorates have been the oxidizing agents of choice in pyrotechnics, but due to their teratogenic properties (Section 5.7B), replacements have been sought for them, such as high-nitrogen compounds and IO_4^- salts.[22]

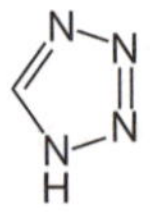

6.4. Effects of Precipitation, Complexation, and Softness on Oxidizing and Reducing Ability

OVERVIEW. The potentials of redox reactions can be influenced by precipitation or complexation of metal ions with hard or soft anions/ligands. Be able to apply the Nernst equation (Eq. 6.15) and Le Châtelier's principle in your explanations. You may practice this concept (Section 6.4A) by trying Exercises 6.54–6.59. Charge-transfer absorptions can be responsible for the intense colors in compounds in which predominance regions of oxidizing and reducing agents barely overlap. This concept can be applied to choose halides or sulfides that are most likely to be nonexistent or intensely colored (Section 6.4B; Exercises 6.60–6.64).

In this section there are many Connections to geochemistry, as well as Connections to environmental chemistry, physical chemistry, and materials science. You again apply trends in electronegativities (Section 1.6), the formation of oxo anions (Section 3.5A), the HSAB principle (Section 5.4), the solubility rules (Sections 4.3 and 5.6), and the differentiation of the elements (Section 5.6). Charge-transfer absorptions will be seen again in Sections 7.3 and 8.3B.

6.4A. Effects of Precipitation and Complexation on Redox Ability. Two common oxidizing cations that have very similar standard reduction potentials are Fe^{3+} ($E° = +0.77$ V) and Ag^+ ($E° = +0.80$ V). The predominance region of neither overlaps that of I^- (top boundary = +0.53 V), so we might expect that either one would oxidize I^- to $I_2(s)$ or $I_3^-(aq)$. [$I_2(s)$ and $I_3^-(aq)$ can be considered to have the same predominance region.] One molar solutions of Fe^{3+} and I^- do react as expected:

$$2Fe^{3+} + 3I^- \rightarrow 2Fe^{2+} + I_3^-$$

However, on mixing 1 M solutions of Ag^+ and I^-, a redox reaction does *not* happen—instead we have the precipitation of $AgI(s)$. Why is there a difference? The product, AgI, is an insoluble soft acid–soft base combination. Precipitation reduces the solution concentrations of Ag^+ and I^- below the standard 1.0 M concentration assumed in the predominance diagram. Indeed, because the solubility product (K_{sp}) of AgI is 8.52×10^{-17} and the starting concentrations of Ag^+ and I^- are equal, it is possible to calculate that, when precipitation is done, $[Ag^+] = [I^-] = \sqrt{K_{sp}} = 9.2 \times 10^{-9}$ M.

The effect of nonstandard concentrations or activities on $E°$ at room temperature is given by the **Nernst equation**:

$$E = E° - \frac{0.059}{n} \log Q \tag{6.15}$$

In the Nernst equation, n represents the number of electrons transferred and Q is the reaction quotient, which has the same form as the equilibrium expression but includes concentrations that apply when the system is *not* at equilibrium. For example, application of the Nernst equation to the reduction of Ag^+ to $Ag(s)$ gives

$$E = 0.80 \text{ V} - \frac{0.059}{1} \log \frac{1}{[Ag^+]} = 0.80 \text{ V} + 0.059 \log[Ag^+] = 0.80 \text{ V} + 0.059(-8.03) = 0.33 \text{ V}$$

In this precipitation reaction the concentration of Ag^+ is drastically reduced below the standard 1 M, so the lower predominance-region boundary for Ag^+ is lowered. At concentrations below the standard, the boundary between a soluble species and a solid species is changed so as to enlarge the predominance region of the soluble species (Figure 2.16c). When adjusted for the lower concentrations present in a saturated solution of AgI, the redox predominance regions of Ag^+ and I^- overlap, so no redox reaction occurs.

We can also look at this result using Le Châtelier's principle. The redox reaction that we anticipated was

$$2Ag^+ + 3I^- \rightleftharpoons 2Ag(s) + I_3^-$$

This reaction is shifted away from products by the removal of both Ag^+ and I^- by precipitation of $AgI(s)$. No such effect occurs with Fe^{3+} and I^-, because the hard acid–soft base combination FeI_3 (if it were stable) should be soluble (Section 5.6A).

AN AMPLIFICATION

Effect of High Concentration on Oxidizing Agents

Conversely, at concentrations above the standard, the predominance region of a soluble species contracts. For example, the predominance region for sulfuric acid (as HSO_4^-) has a lower boundary at 0.39 V, while that for iodide ion has an upper boundary at 0.53 V. The two regions overlap, so 1 M H_2SO_4 does not oxidize 1 M I^-. However, concentrated sulfuric acid is a stronger oxidizing agent because of its contracted predominance region. Sulfuric acid oxidizes iodide ion during attempts to prepare hydrogen iodide from NaI and concentrated H_2SO_4:

$$2NaI(s) + 3H_2SO_4(l) \rightarrow I_2(g) + SO_2(g) + 2H_2O + 2NaHSO_4(aq)$$

The corresponding experiment with the less easily oxidized bromide ion (as NaBr) gives some HBr and some Br_2, whereas NaCl gives exclusively HCl:

$$NaCl(s) + H_2SO_4(l) \rightarrow HCl(g) + NaHSO_4(aq)$$

The formation of complex ions is often used to increase the activity of a metal in redox reactions. For example, nitric acid is a good oxidizing agent, but it coats many electropositive metals with tough films of insoluble oxides. If the metal is a hard acid, this may often be overcome by the use of a mixture of nitric and hydrofluoric acids. The surface oxide then dissolves with the aid of the hard base F^- to give a soluble fluoro anion:

$$Nb + 5HNO_3 + 6HF \rightarrow H^+ + NbF_6^- + 5NO_2 + 5H_2O$$

Even nitric acid is not a strong enough oxidizing agent to oxidize the softest acid metal, gold. However, a mixture of hydrochloric and nitric acid, known as aqua regia, is able to dissolve gold, because the gold ion is complexed by the chloride ion:

$$2Au + 11HCl + 3HNO_3 \rightarrow 2H^+ + 2AuCl_4^- + 3NOCl + 6H_2O$$

An even softer base, the cyanide ion, forms such stable complexes with the soft acids that the parent metals may be oxidized with atmospheric oxygen. Although gold is quite inert to the oxygen in the air (the predominance diagrams of Au and O_2 overlap by 0.29 V), complexation shifts the equilibrium. This allows the extraction of gold from deposits having very low percentages of metal:

$$4Au + 8CN^- + O_2 + 2H_2O \rightarrow 4[Au(CN)_2]^- + 4OH^-$$

The water used in this process is terribly polluted with cyanide.

Related to this is the observation that, although silver is quite unreactive to the strong oxidizing agent in the air, O_2, it is readily tarnished by sulfur or by H_2S and air, to give black silver sulfide.

Because of the increased softness associated with low oxidation numbers, we generally find that low oxidation states of metals are stabilized by the presence of soft-base ligands. Oxidation states for metals of zero or below are mainly found in their organometallic compounds, in which the soft donor atom carbon is attached to the metal atom. High oxidation states of metals are usually stabilized by the presence of hard-base ligands. Thus, the maximum oxidation states of metals are usually found in their fluorides, oxides, fluoro anions, or oxo anions.

AN AMPLIFICATION

Sodide and Electride Ions as Reducing Soft Bases

The very electropositive metals (such as the Group 1 metals and Ca, Sr, Ba, Eu, and Yb) react violently with water (Experiment 7), producing H_2 and OH^-. Liquid ammonia is more resistant than water to reduction (to H_2 and the NH_2^- ion). Unless a catalyst is present, the above metals do not react with liquid ammonia, but instead dissolve to give intensely blue solutions of remarkably low density. If additional Group 1 metal is added to one of the above blue solutions, a second, even less dense liquid of bronze color forms, which is insoluble in the first and floats on it. These properties, which have been known since 1864, have fascinated chemists ever since, but the nature of the solutions proved difficult to determine, in part because evaporating either the blue or bronze solution to dryness generally gave back the metal.

It was deduced that these solutions contain the colorless ammonia equivalents of hydrated cations, $[M(NH_3)_n]^{2+}$ or $[M(NH_3)_n]^+$, which decompose upon evaporation. More stable complex cations were produced[23] using macrocyclic ligands (the crown or crypt polyethers). These complex cations did *not* decompose on removal of solvent; their larger size allowed the production of relatively stable crystals (Section 4.1B) containing the blue and bronze anions. By using a new crypt ligand with eight nitrogen donor atoms, crystalline salts stable at room temperature have finally been isolated.[24]

The bronze solutions gave golden crystals of a material containing one molecule of crown or crypt polyether and two atoms of Group 1 metal ion.

X-ray crystallography of the product with Na showed that one Na atom sits at the center of the cryptand as a typical complex cation (Figure 5.10). The other sodium atom sits outside by itself, and hence must be present as the Na^- ion (the *sodide* ion).

The combination of a sodium cation and a sodide anion is normally less stable than metallic sodium:

$$Na^+ + Na^- \rightleftharpoons 2Na(s)$$

Evidently either the extra energy of coordinate covalent bond formation with the polyether or its physical barrier to electron transfer (or both) gives marginal stability to these remarkable anions.

The blue solution gave dark blue crystals containing solvated electrons, $e^-(H_3N)_m$. The deep blue color results from excitation of the electron by light. The low density results from the fact that the ammonia molecules solvating the electron are remarkably far apart: The solvated electrons act as if they are "*electride*" ions with a surprisingly large "ionic radius" of about 150–170 pm, which is characteristic of soft bases.

The idea that the electride ion might be the ultimate soft base gives us an easy way to rationalize the results of displacement reactions: The soft-acid metal cation ends up combined with the ultimate soft base, the electron:

$$Hg^{2+}(OH_2)_n(aq) + Mn^{2+}(e^-)_2(s) \rightleftharpoons Hg^{2+}(e^-)_2(l) + Mn^{2+}(OH_2)_n(aq)$$

| SA:HB | HA:SB | $\rightleftharpoons$ | SA:SB | HA:HB |

EXAMPLE 6.12

(a) Which ligand—F^-, H_2O, NH_3, or CO—would best stabilize the –1 oxidation state of Mn? (b) Which metal—Hf, Co, or Os—is most likely to react more readily with HF(*aq*) than with HI(*aq*)? (c) Which metal—Hf, Co, or Os—is most likely to react more readily with HI(*aq*) than with HF(*aq*)?

SOLUTION: (a) The soft base $C\equiv O$ is most likely to coordinate to, hence form a stable complex with, the very soft acid Mn^-.

(b) HF, which contains the hard base F^-, can increase the activity of Hf (which forms the hard acid Hf^{4+} on dissolution) by forming a complex such as HfF_6^{2-}.

(c) Os, although not very active due to its high electronegativity, would be expected to form a soft-acid cation on dissolution. If the cation is Os^{2+}, this can precipitate as insoluble $OsI_2(s)$, shifting the equilibrium to favor dissolution. (The equilibrium may be further shifted to the right by complexation of the OsI_2 with additional I^- to form OsI_6^{4-}.)

6.4B. Charge-Transfer Absorptions and Marginal Redox Stability. As was illustrated in a Connection in Section 5.6B, chalcophile minerals (e.g., sulfides of soft acids) often have intense colors (bright yellow, orange, red, and black, with a semimetallic sheen in some cases). This is true despite the fact that both the cations and the sulfide anion are colorless in solution. In contrast, sulfides of hard acids are colorless both in solution and in the solid state.

If the oxidizing power of an ion and the reducing power of its counterion are just insufficient to produce a redox reaction, the compound can exist (although it will have a low heat of formation), and it will likely be intensely colored. A commonplace observation is that intensely colored compounds are sometimes quite reactive (in a redox sense) and should be treated with respect.

This intense color results from an electronic **charge-transfer absorption**: Light energy is capable of transferring the electron from a valence orbital of the reducing ion (e.g., an np orbital of S^{2-}) to an empty valence orbital of the oxidizing ion (e.g., a $3d$ orbital of Hg^{2+}). Because it can be moving in the same direction as the electron being transferred, the light wave has an unusually high probability of being absorbed in the charge-transfer process. Therefore, the color depletion of the reflected light is strong, and we perceive intense color.

Periodic trends in color due to charge transfer can therefore be related to periodic trends in the oxidizing ability of a cation or the reducing ability of an anion. For example, iron(III) fluoride is a white solid, while $FeCl_3$ and $FeBr_3$ are intensely colored (red-brown if anhydrous); the very difficult-to-prepare FeI_3 is intensely black.[25] In Section 6.3 we noted that compounds containing nitrogen–halogen bonds tend to be unstable and may be explosive. This tendency is particularly strong for nitrogen triiodide, which is not only a very easily detonated explosive, but is also intensely colored. Red-violet nitrogen tribromide is also easily detonated. Yellow nitrogen trichloride is not as explosive. In contrast, nitrogen trifluoride is a stable, nonexplosive, colorless compound.

EXAMPLE 6.13

(a) Why, among the third-period iodides in the group oxidation state, is only PI_5 colored (it is black)? (b) Why, among the Group 14/IV tetraiodides, is only SiI_4 colorless (CI_4 is red, GeI_4 is orange, and SnI_4 is yellow, while PbI_4 does not exist)?

SOLUTION: (a) The cations of a period increase in oxidizing ability from left to right (Section 6.2C). Iodide is easily oxidized, so evidently P^{6+} and Cl^{7+} are too oxidizing for their iodides to exist. P^{5+} is almost too oxidizing; its iodide exists but shows charge-transfer absorptions.

(b) The cations of a p-block group are least oxidizing in Period 3 (e.g., Si^{4+}) and most oxidizing in Period 6 (see Section 6.2C). Hence, SiI_4 is colorless and PbI_4 does not exist. The remaining tetraiodides are of marginal stability, so they show intense colors.

Intense color due to charge-transfer absorptions is not confined to soft-acid oxidizing agents in combination with soft-base reducing agents. Oxo anions are usually colorless; VO_4^{3-} is colorless and of low redox activity, while the good oxidizing agent CrO_4^{2-} is intensely yellow, and the very good oxidizing agent MnO_4^- is intensely purple. In Section 3.5A we envisioned forming oxo anions by coordination of the very difficult-to-oxidize oxide ion to acidic central "cations." If this "cation" is in a high enough oxidation state, its "predominance region" may barely overlap that of the oxide ion. The resulting oxo anion can then be intensely colored.

Normally, the transferred electron subsequently returns to the original atom, but sometimes the compound is permanently decomposed, so that it is photosensitive. Silver bromide is an example of a photosensitive halide; its decomposition after exposure to light is the basis of the traditional process of photography.

A CONNECTION TO PHYSICAL CHEMISTRY AND ENVIRONMENTAL CHEMISTRY

Titanium Dioxide as a Pollution-Eating Photocatalyst

The anatase polymorph of TiO_2 (Section 4.1) has the ability to harness the energy of sunlight to promote the oxidation in air of almost any organic vapor. In *nanoparticles* of $TiO_2(s)$, ultraviolet light transfers an electron from (what we can roughly describe as) an oxide ion lone pair on the surface to orbitals in the interior based on Ti $3d$ orbitals, producing a transient excited state resembling $(Ti^{4+})(e^-)(O^{2-})(O^-)$.

$TiO_2(s) + h\nu \rightarrow e^-$ in interior [i.e. $(Ti^{4+})(e^-)$] + missing valence e^- on surface (i.e., O^-). The missing electron in the surface O^- may be supplied by a hydrogen atom from water to generate a reactive *hydroxyl radical*, which is a potent oxidizer of organics.

$$\text{Ti}-\overset{\oplus}{\overset{\times\times}{\underset{\times\times}{\text{O}}}}\overset{\ominus}{^\times} + \text{(H}\overset{\cdot\cdot}{\text{O}}-\text{H} \longrightarrow \text{Ti}-\overset{\oplus}{\overset{\times\times}{\underset{\times\times}{\text{O}}}}\overset{\ominus}{^\times}\text{H} + \cdot\overset{\cdot\cdot}{\text{O}}-\text{H} \tag{6.16}$$

The interior electrons may react with atmospheric O_2 to generate superoxide ions, O_2^-, which are also potent oxidizers of organic vapors:

$$O_2(g) + e^- \text{ in interior} \rightarrow \cdot O_2^- \tag{6.17}$$

TiO_2 is of great interest to environmental chemists because of its ability, in sunlight, to oxidize almost anything (nitrogen oxides, cigarette smoke, grease, bacteria, viruses, mercury, cancer cells, SO_2, oil spills, and chlorobenzenes). It is also nontoxic, chemically inert, and inexpensive. A most dramatic use of this technology is in the insertion of the intensely white TiO_2 into the cement used to construct a visually stunning and self-cleaning church built in Rome called the Jubilee Church or Dio Padre Misericordioso Church.[26]

The principal drawback of TiO_2 is its intensely white color: It absorbs only the ultraviolet component of sunlight, not the visible light. Sunlight consists of about 5% ultraviolet light, about 48% visible light, and about 44% near-infrared (NIR) light (Section 7.3).[27] Absorption in the visible region and superior photocatalytic activity is improved by preparing carbon-containing TiO_2 by reacting $TiCl_4$ with tetrabutylammonium hydroxide, then heating at 550°C.[28] Absorption of both visible and infrared light, and photocatalytic activity, is enhanced by hydrogenating the surface of very small crystals of TiO_2 to produce disorder on the surface; the crystals are transformed from bright white to black.[29]

Solar Cells

If it were colored, TiO_2 could absorb visible light, and in an appropriate device with electrodes (since positive and negative charges are separated in the excited state), it would generate electricity from sunlight. Organic or metal-complex dyes can be absorbed on the surface of the TiO_2 to absorb visible light, but the transfer of charge to the TiO_2 is inefficient, so the conversion of sunlight to electrical energy is also inefficient. Recently it has been found that black salts of organic cations such as $CH_3NH_3^+$ with halo anions PbX_3^- efficiently absorb onto the surface of TiO_2 and generate electricity with a great improvement in efficiency.[30] Surprisingly, the efficiency is still there if these perovskite-structured salts (see Figure 8.10d and Section 8.3B) are absorbed on redox-inactive oxides such as Al_2O_3.[31] Such devices can be assembled using aqueous precipitation reactions (via Solubility Tendency III; Section 4.3C) much more cheaply than silicon solar cells can be produced.

6.5. Effects of pH on Redox Chemistry: Pourbaix Diagrams

OVERVIEW. Pourbaix diagrams combine acid–base predominance diagrams and redox predominance diagrams while allowing pH values to be varied. You will learn how to construct or interpret Pourbaix diagrams for thermodynamically stable species, and how to use them to classify and rank these species as oxidizing or reducing agents (Section 6.5A). You may practice applying these concepts by trying Exercises 6.65–6.68. The long- and short-term stability of water to redox reactivity also depends on pH and can be shown as a window of stability of water. This can be superimposed on a predominance diagram and then used to identify forms of an element that will oxidize or reduce water at high or low pH values, rapidly or slowly. Figure 6.29 can be used to find likely $E°$ and pH values of natural waters, allowing you to suggest the forms in which given elements will occur in natural waters (Section 6.5B; Exercises 6.69–6.74). You will also learn how to distinguish the primary and secondary differentiation of the elements, and how to identify siderophile elements as they occur in the core and in the crust of Earth (Section 6.5C; Exercises 6.75–6.78).

Section 6.5 includes Connections to environmental chemistry, soil and water chemistry, industrial chemistry, and materials science. You will again use the concepts of cation acidity (Section 2.2), anion basicity (Sections 2.3, 2.6, and 3.6), acid–base predominance diagrams (Sections 2.4, 2.5, and 3.7), and the speciation of elements in natural waters (Sections 3.8 and 3.9).

6.5A. Pourbaix Diagrams: Combining Acid–Base and Redox Predominance Diagrams.
In Section 2.5A, we visually represented acid–base equilibria by horizontal predominance diagrams. In Section 6.1B, we represented redox equilibria by vertical predominance diagrams. However, these two broad types of chemical equilibria are not as cleanly distinct from each other as we have made them out to be (acid–base chemistry in Chapters 2–4 and redox chemistry here in Chapter 6). The predominant form of an element in the environment (in soil or natural waters) is influenced by *both*

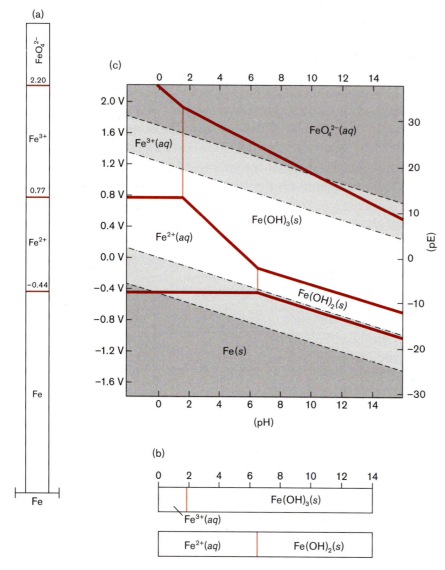

Figure 6.22. The Pourbaix diagram for iron. (a) At left, the redox predominance diagram for iron; (b) below, the acid–base predominance diagrams for Fe(III) and Fe(II); and (c) above right, the Pourbaix diagram showing the predominant form of iron at any given combination of redox potential and pH for 1 M iron solutions. Light brown vertical lines (——) separate species related by acid–base equilibria. Solid brown lines (——) separate species in different oxidation states. The long-term "window" of stability of water is unshaded (in the center of the figure). Around this is the short-term window of stability of water, which is shaded light gray. At the top and the bottom (shaded in dark gray), water is unstable to oxidation or to reduction.

types of chemical equilibria simultaneously. Hence, in applied work it is necessary to have a visual representation of the effects of *both* redox potential and pH on the predominant form of an element. Because most practical chemistry occurs at pH values that are not close to zero, and different oxidizing or reducing conditions, there is a definite need for a diagram that combines acid–base and redox predominance diagrams. Such a diagram, having these two variables as its two axes, is known variously as a **Pourbaix diagram**,[32,33] a predominance-area diagram, an $E°$–pH diagram, or a pE–pH diagram. Figure 6.22 shows the Pourbaix diagram for iron.

The diagram is flanked by the redox predominance diagram for iron (Figure 6.22a) and the acid–base predominance diagram for iron (Figure 6.22b). Note that the redox predominance diagram for iron matches exactly the Pourbaix diagram if the pH is held at zero. In both types of diagram, *redox reactions involve vertical motion crossing a horizontal or diagonal (**solid brown**) boundary*. For example, at the horizontal boundary at 0.77 V in either diagram, there are equal concentrations of Fe^{3+} and Fe^{2+}. Crossing this horizontal boundary means that most of the iron undergoes a redox reaction to the form of iron shown in that area of the diagram. Also note that *higher stable oxidation states and potentials are found higher* in either type of diagram.

The acid–base predominance diagrams for Fe(III) and Fe(II), respectively, are reproduced in Figure 6.22b. These approximately duplicate the part of the Pourbaix diagram in which pH is a variable but the potential is held constant at +0.78 V for Fe(III) and –0.43 V for Fe(II), respectively. In both types of diagram, *motion across a vertical* (light brown) *boundary* corresponds to an *acid–base reaction*. Thus, to the right of pH 2 for Fe(III), most of it precipitates as its hydroxide or oxide.

Pourbaix diagrams are more than the sum of the separate types of predominance diagrams, however. Many of the boundary lines in Figure 6.22 are *diagonal*. During a change of pH and/or potential that crosses one of these lines, a reaction occurs that is *both* an acid–base reaction and a redox reaction. For example, the reaction by which FeO_4^{2-} is converted to Fe^{3+} (top of Pourbaix diagram, between pH 0 and 2) involves *both* electrons and hydrogen ions:

$$FeO_4^{2-} + 3e^- + 8H^+ \rightleftharpoons Fe^{3+} + 4H_2O \qquad (6.18)$$

It follows from Le Châtelier's principle that the position of this equilibrium, hence its *redox potential*, depends on the pH. The appearance of H^+ (or OH^-) as a reactant or product is characteristic of most half-reactions involving hydrides, oxo anions, oxo acids, and oxides and hydroxides of the elements. The potentials for these species can be *strongly* influenced by the pH.

The effect of nonstandard concentrations and activities (as of H^+) on potential at room temperature is given by the Nernst equation (Eq. 6.15). In the case of the half-reaction in Equation 6.18, the potential is

$$E = 2.20 - \frac{0.059}{3}\left(\frac{0.059}{3}\right)\log\frac{[Fe^{3+}]}{[FeO_4^{2-}][H^+]^8}$$

$$= 2.20 - 0.157\,pH - 0.020\log\frac{[Fe^{3+}]}{[FeO_4^{2-}]} \qquad (6.19)$$

The very high half-cell potential for ferrate ion means that this ion is quite unstable under acidic conditions. Removing hydrogen ion from the equilibrium in Equation 6.18 by making the solution strongly basic strongly shifts the equilibrium to the left and increases the stability of ferrate ion. At pH 14 the potential is 0.00 V minus the last term of Equation 6.19—but because there can be very little Fe^{3+} ion in solution at pH 14, the potential has dropped all the way to 0.72 V for 1 M ferrate ion.

In a Pourbaix diagram, the separate conventions of the acid–base and the redox predominance diagrams both hold, as shown in Figure 6.23:

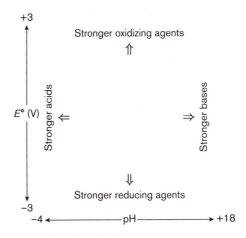

Figure 6.23. Locations of stronger acids, stronger bases, oxidizing agents, and reducing agents in a Pourbaix diagram.

Stronger *oxidizing agents* are more confined to the *top* of the diagram. In Figure 6.22, the FeO_4^{2-} ion is a strong oxidizing agent.

- Stronger *reducing agents* are more confined to the *bottom* of the diagram. In Figure 6.22, Fe is a relatively strong reducing agent. However, there are many other metals lower on the activity series (Section 6.2B) that are stronger reducing agents than Fe.

- Stronger *acids* are more confined to the *left* of the diagram. In Figure 6.22, Fe^{3+} is the strongest acid shown.

- Stronger *bases* are more confined to the *right* of the diagram. No species in Figure 6.22 is even remotely confined to the right side of the diagram. Iron forms no strong bases.

- Species that predominate over wider areas have weaker redox or acid–base properties. No species in Figure 6.22 has such a wide predominance region that it has negligible redox or acid–base properties.

- Strong oxidizing agents and reducing agents tend to react, so species that do not overlap in a vertical sense will tend to undergo redox reactions with each other. If the predominance areas fail to overlap by more than 1 V, the reaction will be very exothermic, and it may even produce an explosion. Such could be the case with a mixture of powdered Fe and a salt of FeO_4^{2-}. Even the salt $(Fe^{2+})(FeO_4^{2-})$ would be unlikely to exist unless there was a strong activation energy hindering the reaction of the two ions. If there were such an activation energy, the salt could well be explosive!

Any given (well-mixed) solution can be characterized by a particular E and a particular pH. Finding this point on the Pourbaix diagram of an element will give us the thermodynamically most stable (in principle, most abundant) form of that element at that E and pH. For example, we can see from Figure 6.22c that, at a reduction potential of +0.8 V and a pH of 14, the predominant form of iron in a solution with a total of 1 mol of iron present per liter is FeO_4^{2-}.

EXAMPLE 6.14

On the Pourbaix diagram for iron find (a) the chemical form of iron that is the strongest oxidizing agent; (b) the form of iron that is the strongest reducing agent; (c) the form of iron that would predominate in a neutral solution at a potential of 0.00 V; (d) the standard reduction potential for FeO_4^{2-} being reduced to Fe^{3+}; and (e) the standard reduction potential for Fe^{2+} being reduced to Fe metal.

SOLUTION: (a) The most strongly oxidizing form of iron shown in Figure 6.22 is the form with the highest lower boundary to its predominance area—namely, the ferrate ion (FeO_4^{2-}). This form contains iron with its highest oxidation number, +6.

(b) Iron metal has the lowest upper boundary on the Pourbaix diagram and is the most strongly reducing form of iron shown in the Pourbaix diagram.

Example 6.14 continues on next page ▶

EXAMPLE 6.14 *(cont.)*

(c) A line drawn straight up at pH 7 intersects a line drawn across at a potential of 0.00 V in the area labeled "Fe(OH)$_3$(s)." Under these conditions, this is the most abundant form of iron present. Because we are close to the regions labeled "Fe^{2+}(aq)" and "Fe(OH)$_2$(s)," at equilibrium lesser amounts of these will also be present.

(d) and (e) Standard reduction potentials are found at pH 0, because 1 M is the standard concentration of H$^+$. The standard reduction potential for the reduction of ferrate ion to iron(III) ion is found at the intersection of the pH 0 vertical line with the boundary separating FeO$_4^{2-}$ and Fe^{3+}. This is approximately at 2.2 V on the diagram. Similarly, the standard reduction potential for Fe^{2+}(aq) being reduced to metallic iron is found at about –0.5 V.

6.5B. Pourbaix Diagrams and Redox Reactions with Water. The half-reactions involving the oxidation and the reduction of water (Eqs. 6.8 and 6.7 in Section 6.1B) are both pH-dependent. After applying the Nernst equation and using the definition of pH, it is found that the potentials for these two half-reactions are as follows:

$$E \text{ for oxidation of water by Equation 6.8} = +1.229 \text{ V} - (0.059)(\text{pH}) \qquad (6.20)$$

$$E \text{ for reduction of water by Equation 6.7} = -0.059(\text{pH}) \qquad (6.21)$$

The potential for the half-reaction in Equation 6.8 gives the theoretical and long-term upper boundary for the predominance area of water, shown in Figure 6.22c as the upper diagonal dashed line. The potential for the half-reaction in Equation 6.7 gives the lower theoretical and long-term boundary for the predominance area of water, represented by the lower diagonal dashed line. The white region between these two lines is sometimes called the **window of** (redox) **stability of water**.

Since both of these half-reactions involve forming diatomic gases (O$_2$ and H$_2$, respectively), overpotentials (Section 6.2A) are significant for both boundaries of water's window of stability. Expanding each boundary by 0.6 V gives the second set of diagonal dashed lines in Figure 6.22. Species located between these sets of diagonal dashed lines, in the region shaded light gray, may be stable for relatively short periods.

Species confined above the top dashed line or below the bottom dashed line (in the regions shaded in dark gray) are so far outside the window of stability of water that they cannot exist even for short periods.

It may be seen at the upper right of Figure 6.22 that the predominance area of FeO$_4^{2-}$ falls within the light gray region above a pH of 10. This means that a 1 M solution of ferrate ion may be kinetically stable in water in strongly basic solutions for short periods. In fact, it can be prepared and used under these conditions if it is used quickly. Because it does not fall within the white window of stability of water, the ferrate ion will eventually oxidize the water to oxygen:

$$4\text{FeO}_4^{2-} + 10\text{H}_2\text{O} \rightarrow 4\text{Fe(OH)}_3 + 3\text{O}_2 + 8\text{OH}^- \qquad (6.22)$$

Pourbaix diagrams for 1 M total concentrations of the elements in the periodic table are given in Figures 6.24–6.27. In these diagrams, only the long-term (thermodynamic) stability area for water is represented, as white windows.

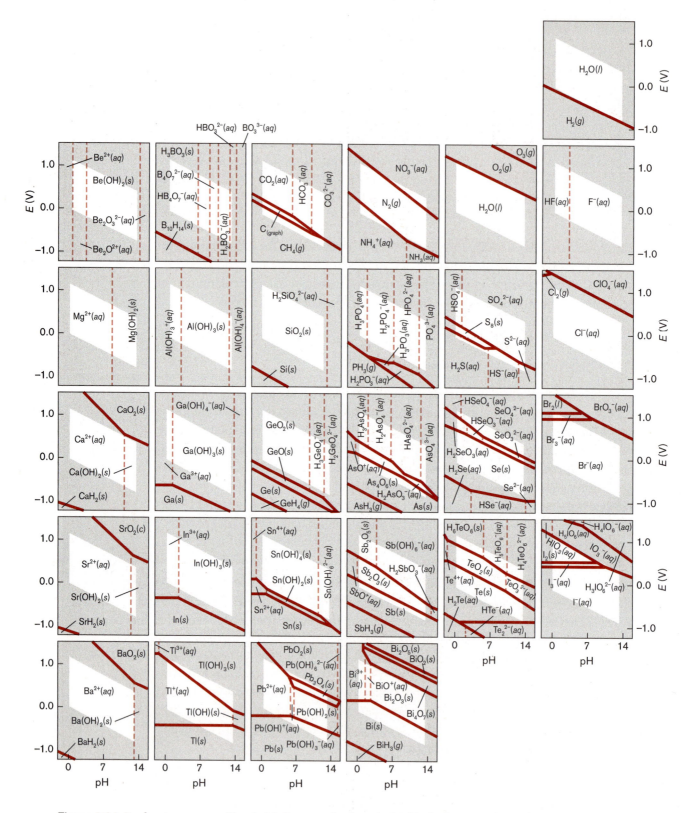

Figure 6.24. Predominance-area (Pourbaix) diagrams for the *s*- and *p*-block elements of Groups 2 through 17/VII. (The diagrams of Groups 1 and 18/VIII are not included because they have only one predominant species in each diagram.) [Adapted from J. A. Campbell and R. A. Whiteker, *J. Chem. Educ.* 46, 90 (1969).]

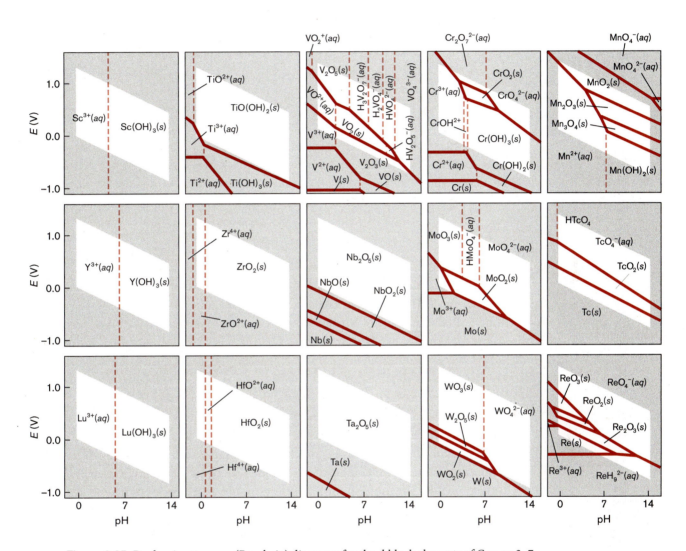

Figure 6.25. Predominance-area (Pourbaix) diagrams for the *d*-block elements of Groups 3–7. [Adapted from J. A. Campbell and R. A. Whiteker, *J. Chem. Educ.* 46, 90 (1969).]

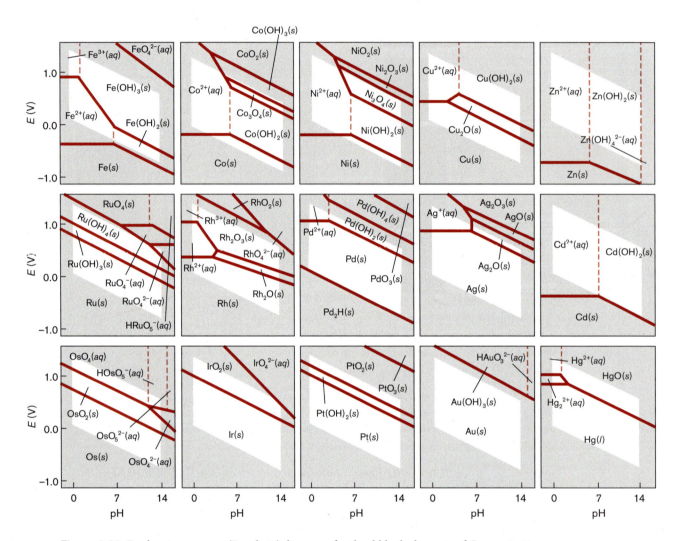

Figure 6.26. Predominance-area (Pourbaix) diagrams for the *d*-block elements of Groups 8–12.
[Adapted from J. A. Campbell and R. A. Whiteker, *J. Chem. Educ.* 46, 90 (1969).]

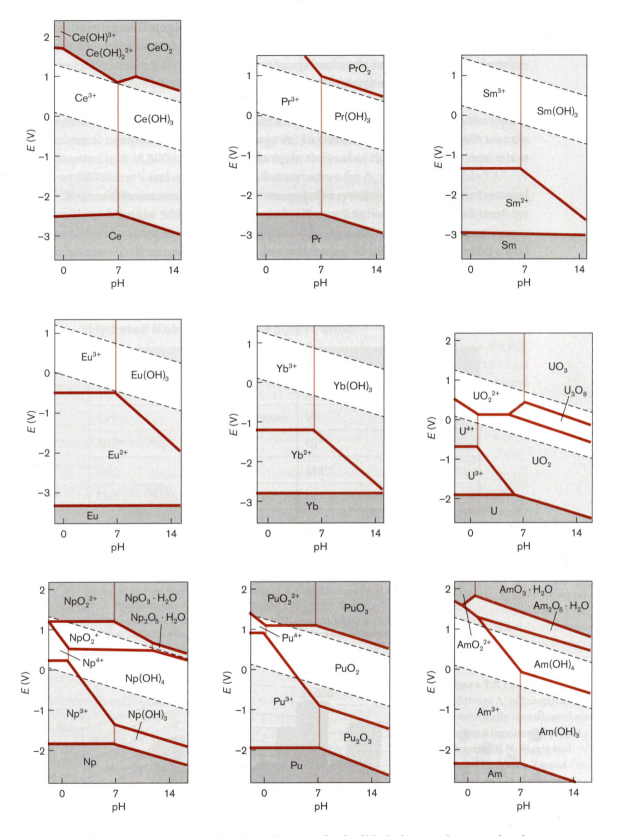

Figure 6.27. Predominance-area (Pourbaix) diagrams for the *f*-block elements showing redox chemistry. [Adapted from M. Pourbaix, *Atlas of Electrochemical Equilibria in Aqueous Solutions*, National Association of Corrosion Engineers: Houston, 1974.]

Note that the predominance area for iron(II) species [Fe^{2+} and $Fe(OH)_2$] narrows considerably at higher pH values. In some cases [e.g., chlorine(0), which appears as Cl_2 in Figure 6.24], the predominance area for a given oxidation state may disappear completely above or below a given pH. When the predominance area of a given oxidation state disappears, that oxidation state loses its thermodynamic stability and is likely to undergo disproportionation (Section 6.1B). Thus, Cl_2 is stable in water at pH < 0, but at higher pH values it undergoes disproportionation to chlorine in negative and positive oxidation states.

EXAMPLE 6.15

Without consulting the textbook diagrams:

(a) Label the areas in the Pourbaix diagram for plutonium (below) with the following species: PuO_2, PuO_3, Pu_2O_3, PuO_2^{2+}, Pu^{4+}, Pu^{3+}, and Pu.

(b) Which species is the strongest oxidizing agent? Realistically, is this species a strong enough oxidizing agent to oxidize the water it is dissolved in at a reasonable rate? (The boundaries for water shown in the diagram are the theoretical ones.)

(c) Which species is the strongest reducing agent? Realistically, is this species a strong enough reducing agent to reduce the water it is dissolved in at a reasonable rate?

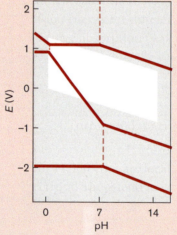

SOLUTION: (a) Plutonium in its highest oxidation state (VI, in PuO_2^{2+} and PuO_3) will go in the top predominance areas, above the top **solid brown** line. Of these two forms of Pu(VI), the more acidic one, PuO_2^{2+}, will predominate at the left, and the less acidic or more basic one, PuO_3, will predominate at the right. The diagram is then filled in downward with lower and lower oxidation states of Pu, ending with Pu metal at the bottom.

(b) The strongest oxidizing agent would probably be taken with reference to the stability window of water, so PuO_3 would be chosen, although in an absolute sense PuO_2^{2+} has a higher $E°$ at most pH values. However, neither appears to be far enough outside the stability window of water to cause the water to be oxidized rapidly.

(c) The strongest reducing agent is Pu metal. It is below even an extended window of stability of water, so it should reduce water at a rapid rate.

CONNECTIONS TO INDUSTRIAL CHEMISTRY AND MATERIALS SCIENCE

Alkaline Batteries and Super-Ion Batteries

The conventional alkaline battery uses $MnO_2(s)$ to oxidize $Zn(s)$. If we assume a pH of 14 and a 1 M concentration of the zinc product $[Zn(OH)_4]^{2-}$, then the upper boundary for the oxidation of zinc (Figure 6.26) is at -1.28 V:[33]

$$Zn(s) + 4OH^- \rightarrow [Zn(OH)_4]^{2-} + 2e^-; \quad E = 0.37 \text{ V} - 0.118(\text{pH}) = -1.28 \text{ V} \tag{6.23}$$

At this pH, the lower boundary for the reduction of $MnO_2(s)$ is at $+0.23$ V:

$$2MnO_2(s) + H_2O + 2e^- \rightarrow Mn_2O_3(s) + 2OH^-; \quad E = 1.06 \text{ V} - 0.059(\text{pH}) = +0.23 \text{ V}$$

Hence, the gap is $+0.23$ V $- (-1.28$ V$) = +1.51$ V, the normal voltage of such batteries.

A higher-voltage, longer-lasting "super-iron" alkaline battery has been developed[34] based on the ferrate ion as the oxidizing agent:

$$FeO_4^{2-} + 4H_2O + 3e^- \rightarrow Fe(OH)_3(s) + 5OH^-; \quad E = 2.03 \text{ V} - 0.118(\text{pH}) = +0.658 \text{ V} \tag{6.24}$$

With zinc as the reducing agent, as in Equation 6.23, the voltage gap at pH 14 is $+1.94$ V. Experimentally, Licht found the voltage to be close to this—namely, $+1.85$ V—using barium ferrate:

$$2BaFeO_4(s) + 3Zn(s) + 4H_2O \rightarrow Fe_2O_3(s) + ZnO(s) + 2Ba[Zn(OH)_4](s)$$

Connecting Pourbaix Diagrams to Soil and Water Chemistry. As suggested in Section 6.4A, the positions of acid–base and redox equilibria for an element also depend on its concentration. The 1 M total concentrations of the standard Pourbaix diagrams are inappropriate for use by geochemists, soil chemists, and water chemists. Figure 6.28 is a Pourbaix diagram for a 10^{-6} M total iron concentration. Comparison with Figure 6.28c shows that, for example, the right boundary for Fe^{3+} moves from a pH of about 1.7 to about 3.4. As discussed previously in Section 6.4A, increasing dilution enlarges the predominance area of soluble species. (This expansion is an application of the Nernst equation and Le Châtelier's principle.)

Pourbaix diagrams with appropriate total concentrations[35] allow us to make more realistic predictions of the forms that the different elements will take in natural waters (Section 3.8). The surface waters of a clean lake are well aerated, and they have dissolved oxygen concentrations that are high enough that their potentials are reasonably close to the oxygen standard reduction potential (Figure 6.29). In a lake that is highly polluted with organic reducing agents—in the bottom layer of a thermally stratified lake, or in a swamp—conditions may be not only quite anaerobic, but also actively reducing, so that the lower boundary of reduction to H_2 may be approached. Under these conditions, many of the most common nutrient elements are converted to unfamiliar forms. According to Figure 6.24, carbon may be converted to methane (CH_4), which is sometimes referred to as swamp gas. Nitrogen is reduced to the ammonium ion (NH_4^+) or to ammonia (NH_3). Ammonia does no good for the odor of the water and is quite toxic. Sulfur is reduced to hydrogen sulfide (H_2S), and emissions of this gas to the atmosphere are a major source of acid rain after the sulfur has been reoxidized. Given overpotentials, it is even possible for phosphorus to be reduced to low concentrations of phosphine (PH_3), a toxic, foul-smelling gas that commonly ignites upon exposure to air, giving rise to eerie light emission over swamps at night.

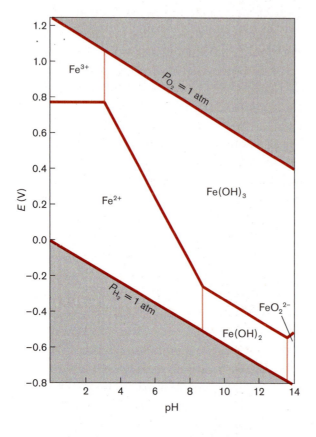

Figure 6.28. Pourbaix diagram for iron at a total activity of 10^{-6} M (only the region within the window of stability of water is shown). [Adapted from D. G. Brookins, E_h–pH Diagrams for Geochemistry, Springer-Verlag: Berlin, 1988, p. 75.]

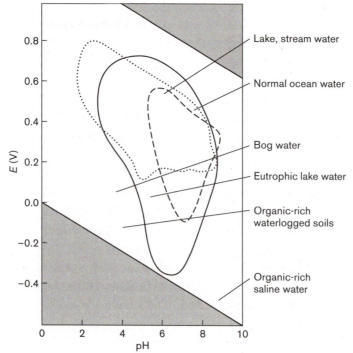

Figure 6.29. The approximate potential and pH values for some environmental waters, indicated by the lines shown or by the regions enclosed as follows:
—— soil water,
----- subsurface waters,
······ acid mine drainage waters.
[Adapted from J. E. Fergusson, Inorganic Chemistry and the Earth, Pergamon Press: Oxford, UK, 1982.]

6.5C. The Origins of Inorganic Compounds on Earth.[36] Phase 1.

The first 12 or so minerals were formed in interstellar space by condensation of the atoms emitted by exploding supernova. In order to form, these minerals had to have high lattice or covalent bond energies. They included a few oxides and silicates (Chapter 8), diamond, graphite, and the minerals *moissanite* (SiC) and *osbornite* (TiN).

Phase 2. Planetismals and chondritic meteorites formed to include about 250 minerals of many types: more oxides and silicates such as olivine and zircon (Section 8.6), some sulfides and phosphides, and iron-nickel alloys. These hot, still-molten metals and minerals coalesced into a larger planet—Earth. The **siderophiles**, the very dense electronegative metals (Section 6.2B)—iron, nickel, and most of the supply of the metals less active (Section 6.2) than iron—sank to form Earth's **core**. Hard-acid metal ions tended to associate with the silicate ions and formed the mantle and Earth's crust. The crust also contains borderline-acid and soft-acid metal sulfides, the chalcophile minerals, as well as extremely low concentrations of the seven soft-base siderophiles of electronegativity 2.20 or higher (Table 5.6). This separation of minerals into the core, mantle, and crust is known as the **primary differentiation of the elements**.

Phases 3 and 4. Following the coalescence of the planet Earth, it entered the *Hadean* geologic era (from 4.5 billion years ago to 4 billion years ago). It was frequently bombarded by other planetismals, and re-melted and slowly re-cooled over the next 2 billion years. The atmosphere did not contain oxygen, but plate tectonics were underway. Minerals were subjected to repeated heating and cooling, and changes from low pressure to high pressure (during subduction) and back. These processes led to the **secondary differentiation of the elements**, giving rise to about 1500 minerals. These slow processes often concentrated particular minerals in large crystals and ores[37]—mineral deposits that are rich enough to mine. Some oceanic volcanic sulfide deposits have been found to be significant ore deposits for copper, silver, and gold, containing, for example, Cu at a concentration of 10% as compared to surface ore concentrations of 0.5%. Such situations have focused a lot of attention on the resources that lie at the bottom of the sea and have made quite important the question of ownership of such resources.

Phase 5. The first evidence for single-celled life was found in 3- to 3.5-billion-year-old minerals from Western Australia,[38] during the *Archaean* era. These anaerobic life forms of the Archaean type produced methane as a major atmospheric component.

Phase 6. Around 2.2–2.4 billion years ago, a remarkable change occurred. Possibly as a result of declining volcanic production of nickel[39] (essential to Archaea), aerobic algae evolved, with the ability to carry out *photosynthesis*. This *Great Oxidation Event* introduced significant amounts of O_2 to the atmosphere, and changed it from anaerobic to aerobic. Soft-base anions in the crust such as sulfide, phosphide, and arsenide, and the atmospheric methane were then oxidized to form hard-base anions such as sulfate, phosphate, arsenate, and carbonate. The group of lithophile minerals expanded enormously, increasing the total number of known minerals to around 4000. In terms of weight percent composition, however, Earth's crust is nearly 94% silica or silicates. The valuable ores are included in only 6% by mass of the crust.

Some (and perhaps many) mineral deposits may be the result of the action of microorganisms, which obtain their energy by mediating redox reactions.[40,41] The more familiar *heterotrophic* bacteria oxidize organic compounds. However, there are also *autotrophic* bacteria that can thrive in a completely inorganic environment, even deep within Earth's crust,[42] oxidizing inorganic materials and using the energy to reduce CO_2 to organic compounds. Gallionella lives by oxidizing iron(II), and excretes copious quantities of Fe_2O_3. Acidthiobacillus thiooxidans also oxidizes Fe^{2+} in acidic solution to give Fe^{3+}, and oxidizes sulfur (as H_2S, S_8, FeS_2, or $S_2O_3^{2-}$) to sulfate and excretes it as 0.5 M sulfuric acid! Acidthiobacillus thiooxidans is therefore very important in producing acid mine drainage water, which can be very strongly acidic. However, the combination of Fe^{3+} and H_2SO_4 can oxidize insoluble copper sulfides, freeing Cu^{2+} from low-grade deposits and allowing it to be efficiently reduced to copper metal. Around 20% of the world's production of copper now comes from *biomining* processes such as this.[43]

Many strains of bacteria, including Gallionella and the newly discovered Halomonas titanicae, obtain energy by catalyzing the exothermic but slow air oxidation of metallic iron (in the form of the wreckage of the *Titanic*). The bacteria that oxidize Fe^{2+} to Fe^{3+} can generally also oxidize Mn^{2+} to Mn^{4+} (as MnO_2). Bacteria and some other organisms can also use oxidized inorganic materials as oxidizing agents when O_2 is unavailable. Nitrate, sulfate, arsenate, and iron(III) are commonly reduced. Some bacteria (including those in the human digestive tract) have a mercury-reducing merA gene that enables them to reduce toxic Hg^{2+} to volatile elemental mercury, which then escapes, reducing their exposure.

A CONNECTION TO GEOCHEMISTRY

Formation of Mineral Deposits at the Mid-Oceanic Ridges

Oceanographic research has revealed that the processes of secondary differentiation that generate minerals are still going on beneath the ocean, at the mid-oceanic ridges.[44] Here magma from the mantle at a temperature of 1200°C continuously rises and is cooled to form the oceanic crust. This rock cracks as it moves off the mid-oceanic ridges, allowing seawater to seep in close to the hot magma. In this superheated water (400°C, but still liquid, due to the very high pressures at the bottom of the ocean), chemical processes are speeded up. In particular, Mg^{2+} and SO_4^{2-} are removed from the ocean water. The Mg^{2+} reacts with the basaltic rock, which acts as a source of SiO_2, and water to give an insoluble magnesium hydroxysilicate and hydrogen ion. The H^+ releases *d*-block metal ions from their basalt minerals. The sulfate is reduced by iron(II) silicate found in the mantle rock to give hydrogen sulfide, dissolved H_4SiO_4, and Fe^{3+}:

$$SO_4^{2-} + 26H^+ + 4Fe_2SiO_4 \rightarrow 8Fe^{3+} + 4H_4SiO_4 + H_2S + 4H_2O \qquad (6.25)$$

The superheated water of a pH of about 3 finally emerges from vents at the top of the mid-oceanic ridge at 400°C. As this superheated solution spews into the cold nonacidic ocean water, the softer *d*-block metal ions (Fe^{2+}, Ni^{2+}, Cu^{2+}, and Zn^{2+}) precipitate as their dark sulfides. These form a *black smoker* that gives rise to a black chimney. The harder Mn^{2+} and Fe^{3+} ions, however, tend to be carried away from the smoker, where bacteria oxidize and precipitate them as FeOOH (goethite), magnetite, a magnetic oxide of iron(II) and iron(III) (Sections 8.3 and 8.4), and MnO_2 in the form of manganese-rich nodules found on the ocean floor.[45] The iron oxide particles also precipitate/absorb basic anions (HPO_4^{2-}, $HAsO_4^{2-}$, HVO_4^{2-}, and CrO_4^{2-}), and the cations of rare earth elements. This chemistry is summarized in Figure 6.30.

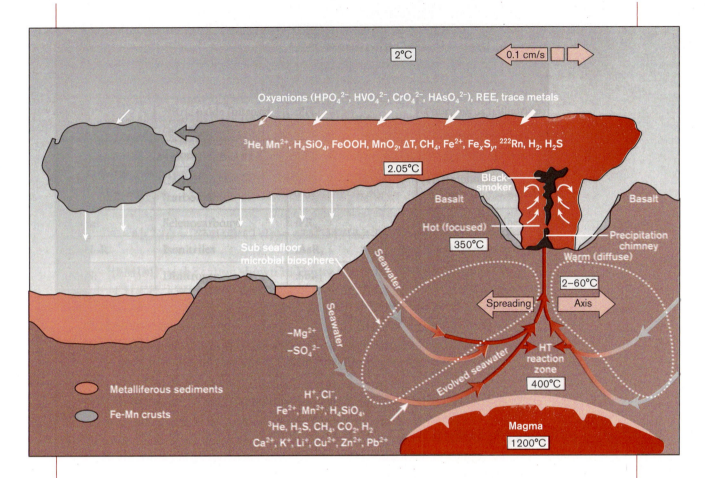

Figure 6.30. Chemistry of "black smoker" hydrothermal vents (REE = rare earth elements). [Adapted from R. L. Rawls, "Some Like It Hot," *Chem. Eng. News*, Dec. 21, 1998, pp. 235–239.]

Microorganisms live in these black smokers, despite their very high temperatures and pressures, drawing energy from the oxidation of Fe^{2+} and Mn^{2+}. These Archaea differ in many ways from more familiar organisms (e.g., in using tungsten-based enzymes[46]), and are thought to be the oldest group of organisms on Earth.

A totally different type of hydrothermal vent process has more recently been discovered. This occurs a few kilometers off the mid-oceanic ridge. The "Lost City" smoker produces *white* chimneys that are as high as 18-story buildings.[47,48] The pH of the fluid coming from these smokers is from 9 to 10, in stark contrast with that of the black smokers. These chimneys seem to consist of $CaCO_3(s)$ and $Mg(OH)_2(s)$. The temperatures of the white smokers are also much milder: 40–75°C instead of 400°C. The chimneys grow out of the mantle rock peridotite, which consists of pyroxenes (chain inosilicates; Section 8.6A) and more importantly the simple silicate olivine, $(Mg,Fe)_2SiO_4$. This contains the very strongly basic SiO_4^{4-} ion, which exothermically reacts with seawater to generate the mineral serpentine (Sections 8.6 and 8.7) and the basic $Mg(OH)_2$ found in the smoker:

$$4Mg_2SiO_4 + 6H_2O \rightarrow Mg_6(OH)_8[Si_4O_{10}] \text{ (serpentine)} + 2Mg(OH)_2$$

A CONNECTION TO BIOCHEMISTRY

Inorganic Origins of Organic Compounds on Earth

The surfaces of black and white smoker chimneys separate very different interior and exterior environments, providing gradients of very different pH values, reduction potentials, and ion concentrations. The flow of hydrogen ions, metal ions, and electrons across these membranes generate potential energy, which hypothetically might have powered the chemical evolution of life.[49] It has been suggested that the organic compounds necessary for the evolution of life on Earth may have originated at the mid-oceanic ridges in the acidic, high-temperature black smoker chimneys or (perhaps more likely) in the alkaline, low-temperature white smoker chimneys.[50,51] The production of these organic compounds depends in no way on the sun or photosynthesis.

Earlier work demonstrated that carbon–carbon bonds can be formed from reactants thought to be present in black smokers, CH_3SH and CO, under the influence of a mixture of iron and nickel sulfides (also present) at 100°C:

$$2CH_3SH + CO \rightarrow CH_3C(=O)SCH_3 + H_2S \qquad (6.26)$$

The methyl thioacetate produced by this reaction then hydrolyzes to give acetic acid. Further reactions then occur to generate small amounts of dipeptides and tripeptides.[52]

Equation 6.26 bears a remarkable similarity to the vital biosynthesis of carbon chains in more advanced organisms using vitamin B_{12} (Figure 5.1C; Sections 5.2A and 5.7C) and acetyl-coenzyme A:

$$CH_3\text{-cobalt} + CO + HS\text{-Coenzyme A} \rightarrow CH_3C(=O)S\text{-Coenzyme A}$$

The mineral olivine involved in the basic, low-temperature white smoker chimneys contains the reducing Fe^{2+} ion, which brings about the reduction of CO_2 and H_2O to give some $H_2(g)$ and some simple organic compounds, including methane (CH_4), the acetate ion (CH_3COO^-), the formate ion ($HCOO^-$), and some complex fatty acids that self-assemble to form membranes in water. The ^{13}C isotopic composition of these products indicates that they are *not* formed by life processes. More recently the iron(II,III) sulfide greigite, $Fe_3S_4(s)$, which resembles the enzyme ferredoxin, has been found to be capable of converting CO_2 and H_2 (likely present in the young Earth) at atmospheric pressure and room temperature to formic acid, methanol, acetic acid, and pyruvic acid.[53]

Extinction of the Dinosaurs

Sediments deposited around Earth about 65,000,000 years ago are a thousand times enriched in (at least) two siderophile electronegative metals, osmium and iridium, as compared to normal sediments. Because asteroids never melted and concentrated these metals in cores, this enrichment has been taken as evidence that an asteroid may have collided with Earth at that time, providing a shower of dust over Earth that (incidentally) was iridium rich and (more importantly) may have shut out a good deal of sunlight for many years. This was the period in which the dinosaurs and all other large animals of the time became extinct.[54]

Enriched amounts of such metals can also be produced by massive volcanic eruptions (e.g., the Permian extinction; Section 5.7B), so such evidence has to be evaluated carefully.

EXAMPLE 6.16

Classify each element as a lithophile, a chalcophile, a siderophile found in Earth's core only, or a siderophile also found at Earth's surface. Then choose its most likely mineral source. (a) Element: Be; source: $Be^{2+}(aq)$, Be_2SiO_4, or $BeSeS_2$. (b) Element: F; source: $F^-(aq)$, CaF_2, or HgF_2. (c) Element: Co; source: $CoCO_3$, $CoAs_2$, or $Co^{2+}(aq)$. (d) Element: Th; source: ThS_2, ThO_2, or $Th^{4+}(aq)$.

SOLUTION: (a) Be (as Be^{2+}) is a hard-acid lithophile, and it is most likely to be found in a mineral with the hard-base silicate, as Be_2SiO_4.

(b) F (as F^-) is a hard base and is therefore also a lithophile, likely to be found in an insoluble salt with a hard acid, as CaF_2.

(c) Co (as Co^{2+}) is a borderline acid chalcophile on Earth's surface (electronegativity between that of Fe and 2.20), so it is found as $CoAs_2$. Because cobalt's electronegativity is above that of iron, it would be a siderophile in the core, dissolved in the iron metal.

(d) Thorium (as Th^{4+}) is a strong hard acid, so it would be a lithophile found in an insoluble salt with a strong hard base, as ThO_2.

6.6. Thermochemical Analysis of Redox Reactions

OVERVIEW. Given thermodynamic and electrochemical data, you can set up a Born–Haber cycle and an equation for the oxidation of a metal, and calculate an energy term from it. This cycle employs atomization energies, ionization energies, and electron affinities. You may practice this by reading Section 6.6A and trying Exercises 6.79–6.83. Similarly, you can set up a thermochemical cycle for a redox reaction of the activity series of elements and calculate energy terms or explain periodic trends in activity using it. You may practice this by reading Section 6.6B and trying Exercises 6.84–6.93.

In this section you again apply the concepts of hydration energies (Sections 2.3 and 2.6), the thermochemical cycle (Section 4.2), bond energies (Section 3.3), lattice energies (Section 4.2), and entropy (Section 4.5).

6.6A. The Born–Haber Cycle. In developing an understanding of the reasons for the solubility rules in Chapter 3, we found it useful to do a thermochemical analysis of the process of precipitation of an ionic salt. In this section, we carry out additional thermochemical analyses of some of the redox reactions of the elements that we found here in Chapter 6.

Even such a simple redox reaction as the burning of sodium in chlorine (Eq. 6.27) is complex enough that it is best broken down into a sum of simpler reactions that can be easily analyzed thermochemically. Then, by Hess's law, the enthalpy change for the reaction—$\Delta H_f°$, the enthalpy of formation of NaCl—is equal to the sum of the enthalpy changes for the different simpler reactions that add up to Equation 6.27.

$$Na(s) + 0.5Cl_2(g) \rightarrow NaCl(s) \qquad \Delta H_f° = -411 \text{ kJ mol}^{-1} \tag{6.27}$$

The thermochemical cycle of simpler reactions we construct should relate this complex redox reaction to simpler gas-phase reactions; hence, we use the strategy outlined in Figure 6.31.

The thermochemical analysis of the reactions by which ionic compounds are formed from their elements in the standard state was pioneered by Born and Haber,[55] so such an analysis is now known as a **Born–Haber cycle**. The classical Born–Haber cycle for the formation of NaCl from Na metal and Cl_2 gas is shown in Figure 6.32.

In Step 1 of this thermochemical cycle, we convert sodium metal (a solid) and chlorine gas, which consists of diatomic molecules, to gaseous atoms. The energy required to produce 1 mol of gaseous atoms of an element from the form of the element that is usual

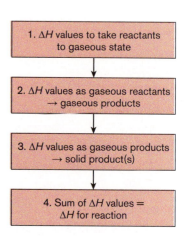

Figure 6.31. Using a thermochemical cycle to calculate ΔH for a reaction.

at room temperature and 1-atm pressure is known as the **enthalpy** (or heat) **of atomization** of that element; these are tabulated in Table 6.4. For Na, this is $+107$ kJ mol^{-1}; for Cl, this is $+122$ kJ mol^{-1}. Therefore, ΔH for Step 1 equals $+107$ kJ mol^{-1} $+ 122$ kJ mol^{-1} $= +229$ kJ mol^{-1}.

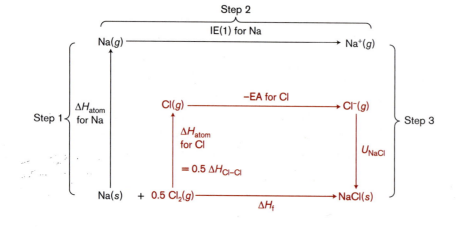

Figure 6.32. The Born–Haber cycle for the formation of NaCl from the elements in their standard states. ΔH_{Cl-Cl} is the Cl—Cl bond dissociation energy. For interpretation of the other symbols, see the text.

TABLE 6.4. Enthalpies of Atomization of the Elements[a]

Group

Period	1	2	3	4	5	6	7	8	9	10	11	12	13/III	14/IV	15/V	16/VI	17/VII	18/VIII
1	H 218																	He 0
2	Li 159	Be 324											B 563	C 717	N 473	O 249	F 79	Ne 0
3	Na 107	Mg 146											Al 326	Si 456	P 315	S 279	Cl 122	Ar 0
4	K 89	Ca 178	Sc 378	Ti 471	V 515	Cr 397	Mn 283	Fe 415	Co 426	Ni 431	Cu 338	Zn 131	Ga 277	Ge 377	As 303	Se 227	Br 112	Kr 0
5	Rb 81	Sr 165	Y 423	Zr 605	Nb 733	Mo 659	Tc 661	Ru 652	Rh 556	Pd 377	Ag 285	Cd 112	In 244	Sn 302	Sb 262	Te 197	I 107	Xe 0
6	Cs 76	Ba 182	Lu [b]414	Hf 621	Ta 782	W 860	Re 776	Os 789	Ir 671	Pt 564	Au 368	Hg 64	Tl 182	Pb 195	Bi 207	Po 142	At	Rn 0
6			La 423	Ce 419	Pr 356	Nd 328	Pm 301	Sm 207	Eu 178	Gd 398	Tb 389	Dy 291	Ho 301	Er 317	Tm 232	Yb 152		
7			Ac [b]293	Th 575	Pa [b]481	U 482	Np [b]337	Pu 352	Am [b]239	Cm	Bk	Cf	Es	Fm	Md	No		

Sources: Heats (enthalpies) of atomization of the *s*- and *d*-block elements were taken from W. L. Jolly, *Modern Inorganic Chemistry*, McGraw-Hill: New York, 1984, p. 292; those of the *d*-block elements were taken from W. W. Porterfield, *Inorganic Chemistry: A Unified Approach*, Addison-Wesley: Reading, MA, 1984, p. 84; those of the *f*-block elements are from N. N. Greenwood and A. Earnshaw, *Chemistry of the Elements*, Pergamon: Oxford, UK, 1984.

[a] Units are in kilojoules per mole (kJ mol^{-1}).

[b] These values are enthalpies of *vaporization*, which are normally slightly less than true heats of atomization, since the metals vaporize in part as diatomic or polyatomic molecules.

In Step 2, we allow the gaseous atoms to react to give the products in gaseous form. In this case, we remove an electron from Na to form the gaseous Na^+ ion, and we give the electron to the Cl atom to form the gaseous Cl^- ion. The energy required to remove the outermost (most loosely bound) electron from a gaseous atom to produce a gaseous cation of +1 charge is known as its **first ionization energy** (often called ionization potential, and abbreviated IE or IP).

$$Na(g) \rightarrow Na^+(g) + e^- \qquad \Delta H = IE(1) = +495.8 \text{ kJ mol}^{-1} \qquad (6.28)$$

Such a process is inherently endothermic, because separating positive and negative charges (cations and electrons) in the gas phase requires an input of energy. The first ionization energies of the elements are provided in Table 6.5 (see p. 363).

Additional energy terms would need to be included in the Born–Haber cycle for the formation of a metal di- or trihalide. To form a dihalide, we remove a second electron to produce a gaseous cation of +2 charge, which requires a **second ionization energy** [IE(2), Table 6.6—see p. 364]. Formation of a trihalide requires a third ionization energy [IE(3), Table 6.7—see p. 365].

The energy change involved in *adding* an electron to an atom to form an anion is known as the **electron affinity** of that element (EA). Due to an unfortunate tradition regarding the signs of electron affinities, they are better regarded as the energies required to *remove* the electron of a gaseous anion of –1 charge to produce a gaseous atom of that element. Hence we list them in Table 6.8 (p. 366) as ionization energies of the –1 ions; they could also be called *zeroth* ionization energies:

$$Cl^-(g) \rightarrow Cl(g) + e^- \qquad \Delta H = EA = IE(0) = +348.8 \text{ kJ mol}^{-1} \tag{6.29}$$

Finally in Step 2, $\Delta H = +496$ kJ mol^{-1} – 348 kJ mol^{-1} = +148 kJ mol^{-1}.

Finally, in Step 3, we allow the gaseous ions to come together to form the final solid product. Recall from Chapter 4 that a very large amount of energy, the lattice energy, U, is released at this step. Adding all of these energy terms should then give us the observed enthalpy of formation of a metal halide such as NaCl:

$$\Delta H_f(MX_n) = \Delta H_{atom}(M) + n\Delta H_{atom}(X) + \Sigma IE(n) - nEA(X) + U \tag{6.30}$$

Applying this specifically to NaCl, we have:

$$\Delta H_f(NaCl) = \Delta H_{atom}(Na) + \Delta H_{atom}(Cl) + IE(1)(Na) - EA(Cl) + U \tag{6.31}$$

$$\Delta H_f(NaCl) = +107 \text{ kJ mol}^{-1} + 122 \text{ kJ mol}^{-1} + 496 \text{ kJ mol}^{-1} - 348 \text{ kJ mol}^{-1} + U \tag{6.32}$$

In fact, one of the practical uses of the Born–Haber cycle is to obtain lattice energies from known values of ΔH_f. The enthalpy of formation of NaCl is –411 kJ mol^{-1}. Substituting this value in Equation 6.32 and solving for U gives us an experimental lattice energy of –787 kJ mol^{-1}. The calculated value of the lattice energy for NaCl (–751 kJ mol^{-1}; Section 4.2A) is close enough to this to give us confidence in the essential correctness of our model for the Coulombic attractive forces in ionic compounds (Chapter 4).

Another use of Born–Haber calculations is in predicting whether unknown compounds will be thermodynamically stable or not. When Bartlett prepared the surprising ionic compound $O_2^+PtF_6^-$, he noticed that the first ionization energy of the "noble gas" xenon was lower than the first ionization energy of the O_2 molecule. He made a rough Born–Haber calculation to predict that there might be a stable compound $Xe^+PtF_6^-$. So, (in a vacuum system) he mixed xenon gas and PtF_6 gas and obtained a solid ionic product[56] that created quite a sensation in inorganic chemistry, because at the time it was thought that noble gases could form no compounds.

TABLE 6.5. First Ionization Energies of Atoms of the Elements[a,b]

Period	s^1	s^2	s^2d^1	s^2d^2	s^2d^3	s^2d^4	s^2d^5	s^2d^6	s^2d^7	s^2d^8	s^2d^9	s^2d^{10}	s^2p^1	s^2p^2	s^2p^3	s^2p^4	s^2p^5	s^2p^6
1	H 1312																	He 2372
2	Li 520	Be 899											B 801	C 1086	N 1402	O 1314	F 1681	Ne 2081
3	Na 496	Mg 738											Al 578	Si 786	P 1012	S 1000	Cl 1251	Ar 1520
4	K 419	Ca 590	Sc 631	Ti 658	V 650	Cr 653	Mn 717	Fe 759	Co 758	Ni 737	Cu 746	Zn 906	Ga 579	Ge 762	As 944	Se 941	Br 1140	Kr 1351
5	Rb 403	Sr 550	Y 616	Zr 660	Nb 664	Mo 685	Tc 702	Ru 711	Rh 720	Pd 805	Ag 731	Cd 868	In 558	Sn 709	Sb 832	Te 869	I 1008	Xe 1170
6	Cs 376	Ba 503	Lu 524	Hf 654	Ta 761	W 770	Re 760	Os 840	Ir 880	Pt 870	Au 890	Hg 1007	Tl 589	Pb 716	Bi 703	Po 812	At 930	Rn 1037
7	Fr 400	Ra 509	Lr 479															

Period	s^2f^1	s^2f^2	s^2f^3	s^2f^4	s^2f^5	s^2f^6	s^2f^7	s^2f^8	s^2f^9	s^2f^{10}	s^2f^{11}	s^2f^{12}	s^2f^{13}	s^2f^{14}
6	La 538	Ce 528	Pr 523	Nd 530	Pm 536	Sm 543	Eu 547	Gd 592	Tb 564	Dy 572	Ho 581	Er 589	Tm 597	Yb 603
7	Ac 490	Th 590	Pa 570	U 590	Np 600	Pu 585	Am 578	Cm 581	Bk 601	Cf 608	Es 619	Fm 627	Md 635	No 642

Sources: First ionization energies, IE(1), of the elements listed are taken from J. E. Huheey, E. A. Keiter, and R. L. Keiter, *Inorganic Chemistry: Principles of Structure and Reactivity*, 4th ed., Harper-Collins: New York, 1993, pp. 36–37; and T. K. Sato et al., *Nature* 520, 209 (2015).

[a] Units are in kilojoules per mole (kJ mol⁻¹).

[b] Group headings indicate the characteristic valence electron configurations of the atoms being ionized.

TABLE 6.6. Second Ionization Energies of (+1 Ions of) the Elements[a,b]

Period	s^1	(d-block)										s^2	s^2p^1	s^2p^2	s^2p^3	s^2p^4	s^2p^5	s^2p^6
1																	He 5250	Li 7298
2	Be 1757											B 2427	C 2353	N 2856	O 3388	F 3374	Ne 3952	Na 4562
3	Mg 1451											Al 1817	Si 1577	P 1903	S 2251	Cl 2297	Ar 2666	K 3051
4	Ca 1145	Sc 1235	Ti 1310	V 1414	Cr 1496	Mn 1509	Fe 1561	Co 1646	Ni 1753	Cu 1958	Zn 1733	Ga 1979	Ge 1537	As 1798	Se 2045	Br 2100	Kr 2350	Rb 2633
5	Sr 1064	Y 1181	Zr 1267	Nb 1382	Mo 1558	Tc 1472	Ru 1617	Rh 1744	Pd 1875	Ag 2074	Cd 1631	In 1821	Sn 1412	Sb 1595	Te 1790	I 1846	Xe 2046	Cs 2230
6	Ba 965	Lu 1340	Hf 1440	Ta 1500	W 1700	Re 1260	Os 1600	Ir 1680	Pt 1791	Au 1980	Hg 1810	Tl 1971	Pb 1450	Bi 1610	Po 1800	At 1600	Rn	Fr 2100
7	Ra 979	Lr 1428																

Period														
6	La 1067	Ce 1047	Pr 1018	Nd 1034	Pm 1052	Sm 1068	Eu 1085	Gd 1170	Tb 1112	Dy 1126	Ho 1139	Er 1151	Tm 1163	Yb 1175
7	Ac 1145	Th 1147	Pa 1129	U 1438	Np 1129	Pu 1129	Am 1158	Cm 1196	Bk 1187	Cf 1206	Es 1216	Fm 1225	Md 1235	No 1254

Source: Second ionization energies, IE(2), of the +1 ions of the elements listed are taken from J. E. Huheey, E. A. Keiter, and R. L. Keiter, *Inorganic Chemistry: Principles of Structure and Reactivity*, 4th ed., Harper-Collins: New York, 1993, pp. 36–37.

[a] Units are in kilojoules per mole (kJ mol⁻¹).

[b] Group headings indicate the valence electron configurations of the ions being ionized a second time.

TABLE 6.7. Third Ionization Energies of (+2 Ions of) the Elements[a,b]

Period	d^1	d^2	d^3	d^4	d^5	d^6	d^7	d^8	d^9	d^{10}	s^1	s^2	s^1p^1	s^2p^2	s^2p^3	s^2p^4	s^2p^5	s^2p^6
1																	Li 11815	Be 14849
2											B 3660	C 4621	N 4578	O 5300	F 6050	Ne 6122	Na 6912	Mg 7733
3											Al 2745	Si 3232	P 2912	S 3361	Cl 3822	Ar 3931	K 4411	Ca 4912
4	Sc 2389	Ti 2652	V 2828	Cr 2987	Mn 3248	Fe 2957	Co 3232	Ni 3393	Cu 3554	Zn 3833	Ga 2963	Ge 3302	As 2736	Se 2973	Br 3500	Kr 3565	Rb 3900	Sr 4210
5	Y 1980	Zr 2218	Nb 2416	Mo 2621	Tc 2850	Ru 2747	Rh 2997	Pd 3177	Ag 3361	Cd 3616	In 2705	Sn 2943	Sb 2440	Te 2698	I 3200	Xe 3100	Cs 3400	Ba 3600
6	Lu 2022	Hf 2250	Ta 2100	W 2300	Re 2510	Os 2400	Ir 2600	Pt 2800	Au 2899	Hg 3300	Tl 2878	Pb 2082	Bi 2466	Po 2700	At 2900	Rn		
7																	Fr 3100	Ra 3300

Period	f^1	f^2	f^3	f^4	f^5	f^6	f^7	f^8	f^9	f^{10}	f^{11}	f^{12}	f^{13}	f^{14}
6	La 1850	Ce 1949	Pr 2086	Nd 2130	Pm 2150	Sm 2260	Eu 2400	Gd 1990	Tb 2110	Dy 2200	Ho 2200	Er 2190	Tm 2284	Yb 2415
7	Ac 1900	Th 1978	Pa 1814	U 1843	Np 1872	Pu 2103	Am 2161	Cm 2045	Bk 2152	Cf 2277	Es 2325	Fm 2354	Md 2451	No 2605

Sources: Third ionization energies, IE(3), of the +2 ions of the elements listed are taken from J. E. Huheey, E. A. Keiter, and R. L. Keiter, *Inorganic Chemistry: Principles of Structure and Reactivity*, 4th ed., Harper-Collins: New York, 1993, pp. 36–37; and D. F. Shriver, P. W. Atkins, and C. H. Langford, *Inorganic Chemistry*, W. H. Freeman: New York, 1993, pp. 638–639.

[a] Units are in kilojoules per mole (kJ mol^{-1}).

[b] Group headings indicate the valence electron configurations of the ions being ionized a third time.

TABLE 6.8. Electron Affinities of the Elements[a]

First Electron Affinities (Ionization Energies of the –1 Ions of the Elements)

Period	s^1	s^2	s^2p^1											s^2p^2	s^2p^3	s^2p^4	s^2p^5	s^2p^6
1		H 73																
2	He –48	Li 60	Be >0											B 27	C 122	N –7	O 141	F 328
3	Ne –116	Na 53	Mg >0											Al 43	Si 134	P 72	S 200	Cl 349
4	Ar –96	K 48	Ca 2	Sc 18	Ti 8	V 51	Cr 64	Mn 0	Fe 16	Co 64	Ni 112	Cu 118	Zn 0	Ga 29	Ge 130	As 78	Se 195	Br 325
5	Kr –96	Rb 47	Sr 5	Y 30	Zr 41	Nb 86	Mo 72	Tc 53	Ru 101	Rh 110	Pd 54	Ag 126	Cd 0	In 29	Sn 116	Sb 103	Te 190	I 295
6	Xe –77	Cs 46	Ba 14	Lu 48	Hf 0	Ta 31	W 79	Re 14	Os 106	Ir 151	Pt 205	Au 223	Hg 0	Tl 19	Pb 35	Bi 91	Po 183	At 270

Second and Third Electron Affinities (Ionization Energies of the –2 and –3 Ions of the Elements)

Element	2nd EA	3rd EA
N	–673	–1070
P	–468	–886
As	–435	–802
O	–744	
S	–456	
Se	–410	

Source: Electron affinities (EA) of the elements listed are taken from R. T. Myers, *J. Chem. Educ.* 67, 307 (1990); group headings indicate the characteristic valence electron configurations of the anions being ionized. Second and third electron affinities are taken from R. G. Pearson, *Inorg. Chem.* 30, 2856 (1991). Group 2 electron affinities are from J. C. Wheeler, *J. Chem. Educ.*, 74, 123 (1997).

[a] Units are in kilojoules per mole (kJ mol^{-1}).

EXAMPLE 6.17

We want to calculate the enthalpy of formation, ΔH_f, of CaSe(s), which is ΔH for the reaction Ca(s) + Se(s) → CaSe(s). (a) Fill in the empty squares in Figure 6.33 with the symbols and states of the chemical species involved in this Born–Haber cycle. (For example, the reactants and products of this reaction go in the boxes on the bottom line of this drawing.)

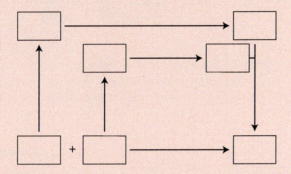

Figure 6.33. Blank drawing of a Born–Haber thermochemical cycle.

(b) Assume that the lattice energy U(CaSe) = –2946 kJ mol^{-1} and that the other Se energy terms add up to +492 kJ mol^{-1}. Adding the needed information from Tables 6.4–6.8, compute ΔH_f of CaSe(s).

SOLUTION: (a) On the left side we take the reactants to the gaseous state. On the top we take the gaseous reactants to the gaseous products. On the right side we take the gaseous products to their final, solid state.

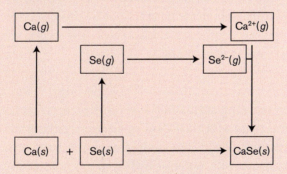

(b) $\Delta H = \Delta H_{atom}$(Ca) + IE(1)Ca + IE(2)Ca + Se terms + U = 178 + 590 + 1145 – 2946 + 492 = –541 kJ mol^{-1}.

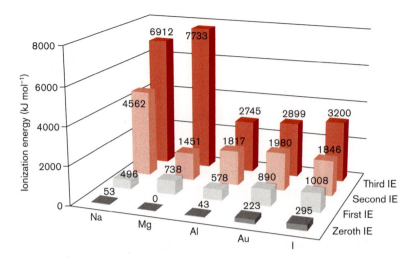

Figure 6.34. Trends in successive ionization energies among third-period elements of the first three groups (at the left), and between two elements near the lower right edge of the *d* and *p* blocks, respectively. The data come from Tables 6.5–6.8.

Trends in Successive Ionization Energies. Periodic trends within a given type of ionization energy (including within electron affinities) are explored in Section 9.3, while periodic trends in atomization energies are explored in Section 11.4. Here we review trends in successive ionization energies of a given element (Figure 6.34). The second ionization energy IE(2) is substantially larger than the first ionization energy IE(1), because an electron must be pulled away from an already positively charged ion. Likewise, IE(3) is substantially larger than IE(2). The changes are much larger if the electron being ionized is a *core* electron, rather than a valence electron—for example in Figure 6.34, IE(2)(Na) is much greater than IE(1)(Na).

Note in Figure 6.34 that the zeroth ionization energies of elements are substantially less than their first ionization energies. As we see from Table 6.8, not all gaseous anions hold on to their electrons well; many have zero or negative electron affinities. No gaseous anion holds onto a second extra electron well: The second electron affinity even of oxygen (as the O^{2-} ion) is -744 kJ mol^{-1}.

6.6B. The Activity Series of the Elements. In Experiment 7, we examined the activity of the metals as reducing agents (for water or hydrogen ion), and found a better correlation of their activity with their Pauling electronegativity values than with their (first) ionization energies. Why? This finding may well have surprised you, because the reaction of the metals with water or hydrogen ion produces metal ions and does involve ionization of the metals. Also, why is gold the least active metal? Its low activity correlates with its high Pauling electronegativity, but finding a correlation is not the same as finding an explanation. Ionization energies are more easily explained (Section 9.3) than electronegativities, and we note at the right in Figure 6.34 that Au has very high ionization energies, comparable to those of a typical electronegative nonmetal such as iodine.

One problem in correlating activity and first ionization energies is that most of the metals form, not +1, but +2 or +3 (or even +4) ions. Hence, the second and third (and perhaps fourth) ionization energies should have been involved in the correlation. However, there is still another problem: Ionization energies apply to gaseous metal atoms giving gaseous metal ions, whereas the reactions of metals with water or hydrogen ion involve solid (bulk) metals giving hydrated metal ions. To find the relationship of activity to ionization energies, we need to construct a thermochemical cycle for the relatively complicated process of dissolving a metal in standard (1 M) acid:

$$M(s) + zH^+(aq) \rightarrow M^{z+}(aq) + (z/2)H_2(g) \tag{6.33}$$

We can generalize the thermochemical cycle for the reaction of any metal with 1 M hydrogen ion; the generalized cycle is shown schematically in Figure 6.35.

Let us apply the procedure of Figure 6.33 to the (dangerously exothermic) reaction of sodium metal with 1 M aqueous hydrogen ion. In Step 1, we take the two reactants to the gaseous state. For the hydrogen ion, we must remove its water of hydration by supplying its enthalpy of hydration, $-\Delta H_{hyd} = +1091$ kJ mol^{-1} (Table 2.3). The sodium metal must be atomized, and its heat of atomization ΔH_{atom} is +107 kJ mol^{-1} (Table 6.4).

In Step 2, for the gaseous sodium atom and hydrogen ion to react, we must supply the first ionization energy of Na, IE(1) = +496 kJ mol^{-1} from Table 6.5, but we allow the reverse of the first ionization of hydrogen to occur, –IE(1) = –1312 kJ mol^{-1}.

In Step 3, we allow the gaseous sodium ion to hydrate itself, $\Delta H_{hyd} = -405$ kJ mol^{-1}. We allow the individual hydrogen atoms to come to their final state, as gaseous diatomic molecules. This energy is the negative of ΔH_{atom} of H, –218 kJ mol^{-1}.

In Step 4, we apply Hess's law and add the energy changes in all these steps. The terms involving the hydrogen are +1091 kJ mol^{-1} – 1312 kJ mol^{-1} – 218 kJ mol^{-1} = –439 kJ mol^{-1}. The sodium terms are +107 kJ mol^{-1} + 496 kJ mol^{-1} – 405 kJ mol^{-1} = +198 kJ mol^{-1}. Overall the theoretically calculated enthalpy change for the reaction of sodium metal with 1 M acid is –241 kJ mol^{-1}.

Experimentally, we know that the standard reduction potential of sodium ion is –2.71 V (Figure 6.14). This value can be converted to thermodynamic units using

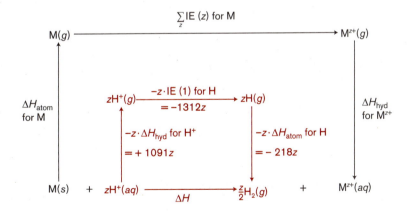

Figure 6.35. Thermochemical cycle for the reaction of a metal in its standard state with 1 M aqueous hydrogen ion to give an aqueous metal ion of charge +z and gaseous hydrogen. For interpretation of the symbols for the enthalpy changes, see the text.

Equation 6.5 (i.e., $\Delta G° = -nFE°$). If we neglect entropy changes, we can equate ΔG and ΔH and obtain ΔG for the oxidation of sodium $\approx (-2.71 \text{ V})(1 \text{ e}^-)(96.5 \text{ kJ V}^{-1} \text{ mol}^{-1}/\text{e}^-) = -262 \text{ kJ mol}^{-1}$. In view of our neglect of entropy changes, the agreement with the calculated ΔH, -241 kJ mol^{-1}, is good.

The calculations for the activity of sodium can be applied to the activity of any metal by substituting the appropriate atomization, ionization, and hydration energies. Note that the overall theoretical enthalpy changes involving hydrogen are constant at -439 kJ mol^{-1} of H^+ involved. Thus, the generalized ΔH for the reaction of M to give M^{z+} is:

$$\Delta H = \Delta H_{\text{atom}}(M) + \sum IE(z)(M) + \Delta H_{\text{hyd}}(M) - 439z \qquad (6.34)$$

Finally, assuming the equality of ΔH and ΔG, we apply Equation 6.5 to replace ΔG by $-FE°$, and, noting that $E°$ for this oxidation reaction is the reverse of the standard reduction potential, we obtain in general:

$$96.5E° \approx \frac{\Delta H_{\text{atom}}}{z} + \frac{\sum IE(z)}{z} + \frac{\Delta H_{\text{hyd}}}{z} - 439 \qquad (6.35)$$

Dividing through by ($96.5 \text{ kJ V}^{-1} \text{ mol}^{-1}$), we finally obtain:

$$E° \text{ (in V)} \approx \left(\frac{1}{96.5z}\right)[\Delta H_{\text{atom}} + \sum IE(z) + \Delta H_{\text{hyd}}] - 4.55 \qquad (6.36)$$

Inserting the values for Na, including $z = 1$, we obtain $E° \approx (1/96.5)(107 + 496 - 405) - 4.55 = 2.05 - 4.55 = -2.50 \text{ V}$, in rough agreement (considering that ΔH and ΔG are not really the same) with the experimental value of -2.71 V from Figure 6.14.

To see the chemical significance of the atomization energies of the metals, let us calculate standard reduction potentials from Equation 6.36 with the ΔH_{atom} term deleted (i.e., for the reaction of gaseous metal atoms with acids). This calculation amounts to subtracting a factor of $\Delta H_{\text{atom}}/96.5z$ from the recorded standard reduction potentials (Figure 6.14); in the case of sodium, this changes its standard reduction potential from -2.71 to -3.82 V. More interestingly, it changes the reduction potential of the least active metal, gold, from $+1.68$ to -2.13 V (i.e., it makes gold almost as active as magnesium actually is).

In fact, it is now possible to start reactions with gaseous metal atoms by providing the atomization energy separately with a high-temperature resistance heater in a high-vacuum apparatus known as a metal-atom reactor. The metal atoms produced by this means are in fact much more reactive than bulk metals as we know them. They must be handled in high vacuum at very low temperatures to prevent them from reacting with each other, with other gaseous substances, or with the walls of the reaction vessel. With the enhanced activity possessed by gaseous metal atoms, it is possible to carry out many reactions that are completely impossible with solid or liquid metals.[57] As previously mentioned in Section 6.2B, the molecular nonmetal N_2 can similarly be atomized to the more active N atom, which can oxidize Na or K atoms to give the otherwise-unknown nitrides Na_3N and K_3N.[4]

EXAMPLE 6.18

(a) If possible, calculate the standard reduction potentials of the (bulk) metals Hg, Pt, Pb, and Ba (all of which form +2 ions). (b) Compare your values with the known $E°$ values of these metals. (c) In the case of Pt, use the known $E°$ value to determine the missing hydration energy of the Pt^{2+} ion. (d) Calculate the standard reduction potentials of the gaseous atomic metals Hg, Pt, Pd, and Ba. (e) Which metal has its $E°$ value altered the most? Write an activity series of these four metals in the gaseous atomic state.

SOLUTION: (a) We use Equation 6.36 with $z = 2$. For Hg we obtain $E° = (1/193)(64 + 1007 + 1810 - 1824) - (455) = 0.93$ V. For Pt we obtain $E° = (1/193)(564 + 870 + 1791 - x) - 4.55$, which we cannot compute because we do not know x, the hydration energy of the Pt^{2+} ion. For Pb we obtain $E° = (1/193)(195 + 716 + 1450 - 1480) - 4.55 = +0.01$ V. For Ba we obtain $E° = (1/193)(182 + 503 + 965 - 1304) - 4.55 = -2.75$ V.

(b) Each of the three computed values is within 0.16 V of the experimental values from Figure 6.14: Hg = 0.85 V; Pb = -0.13 V; Ba = -0.91 V.

(c) From Figure 6.14 we obtain $E° = +1.2$ V for Pt. Using Equation 6.35 with $x/2$ for the unknown hydration energy of Pt^{2+}, we obtain $(96.5)(1.2) = 564/2 + 870/2 + 1791/2 - x/2 - 439$, so $x/2 = 116 + 439 - 282 - 435 - 845.4$), and $x = 2(-1007.5) = -2015$ kJ mol^{-1} for Pt^{2+}, which is qualitatively in line with those values given in Table 2.3.

(d) To calculate the gas-phase activities we can either use Equation 6.36, omitting the atomization term, or we can subtract the atomization energy [divided by (96.5)(2) to convert from kJ mol^{-1} to volts] from the standard reduction potentials. Taking the latter approach, we subtract the following amounts from the (experimental) standard reduction potentials: Ba, 0.94 V; Pb, 1.01 V; Hg, 0.33 V; and Pt, 1.2 V. Therefore, the reduction potentials of the gaseous atoms become Hg(g) = +0.52 V, Pt(g) = -1.7 V, Pb(g) = -1.14 V, and Ba(g) = -3.85 V.

(e) Platinum is altered the most in activity, because it has by far the largest atomization energy. The gaseous activity series reads Ba(g) (-3.85 V) >> Pt(g) (-1.7 V) > Pb(g) (-1.14 V) >> Hg(g) (+0.52 V).

6.7. Background Reading for Chapter 6

Redox Potentials for Aqueous Metal Ions

John Burgess, *Ions in Solution: Basic Principles of Chemical Interactions*, 2nd ed., Horwood Publishing: Chichester, UK, 1999, pp. 93–105 (Chapter 7: Redox Potentials).

Laboratory Safety with Strong Oxidizing and Reducing Agents

R. C. Hill, Jr. and D. C. Finster, *Laboratory Safety for Chemistry Students*, John Wiley and Sons: Hoboken, NJ, 2010. See Section 5.1.2 (Flammables—Chemicals with Burning Passion), pp. 5-13 to 5-22; Section 5.2.2 (The Chemistry of Fire and Explosions), pp. 5-31 to 5-38; Section 5.2.3 (Incompatibles—A Clash of Violent Proportions), pp. 5-39 to 5-48;

Section 5.3.2 (Peroxides—Potentially Explosive Hazards), pp. 5-61 to 5-68; Section 5.3.3 (Reactive and Unstable Laboratory Chemicals), pp. 5-69 to 5-78; and Chapter 8 (Chemical Management—Inspections, Storage, Wastes, and Security), pp. 8-1 to 8-44.

Pyrotechnics/Fireworks

G. Steinhauser and T. M. Klapotke, "Green Pyrotechnics: A Chemists' Challenge," *Angew. Chem. Intl. Ed.* 47, 3330–3347 (2008).

G. Steinhauser and T. M. Klapotke, "Using the Chemistry of Fireworks to Engage Students in Learning Basic Chemical Principles: A Lesson in Eco-Friendly Pyrotechnics," *J. Chem. Educ.* 87, 150–156 (2010).

Titanium Dioxide as a Pollution-Eating Photocatalyst

S. Kwon, M. Fan, A. Cooper, and H. Yang, "Photocatalytic Applications of Micro- and Nano-TiO_2 in Environmental Engineering," *Crit. Rev. Env. Sci. Tech.* 38, 197–226 (2008).

Solar Cell Enhancement using Organohalide "Perovskites"

Q. Lin, A. Armin, P. L. Burn, and P. Meredith, "Organohalide Perovskites for Solar Energy Conversion," *Acc. Chem. Res.* 49, 545–553 (2016). This entire issue is devoted to the topic.

Pourbaix Diagrams

G. K. Schweitzer and L. L. Pesterfield, *The Aqueous Chemistry of the Elements*, Oxford University Press: Oxford, UK, 2010. General background on Pourbaix diagrams is found in Chapters 1 and 2, pp. 3–46.

Origins of Inorganic and Organic Compounds and Life on the Earth

R. M. Hazen, "Evolution of Minerals," *Sci. Amer.* 302(3), 58–65 (2010).

E-I. Ochiai, "Global Metabolism of Elements—Principles and Applications in Bioinorganic Chemistry—XI," *J. Chem. Educ.* 74, 926–930 (1997).

G. Wächterhäuser, "On the Chemistry and Evolution of the Pioneer Organism," *Chem. Biodivers.* 4, 584–602 (2007).

D. Wacey, "Earliest Evidence for Life on Earth—an Australian Perspective," *Aust. J. Earth Sci.* 89, 153–166 (2012).

H. Lee, "Anatomy of a Mass Murderer," *Sci. Amer.* 314(3), 64–65 (2016). The role of massive volcanic eruptions in four of five mass extinctions on Earth.

Numbered references from this chapter may be viewed online at www.uscibooks.com /foundations.htm.

6.8. Exercises

Note: There are small discrepancies between the standard reduction potentials and chemical formulas given in the redox predominance and the Pourbaix diagrams for several elements, which mainly result from the fact that the electrochemical data used for the two types of diagrams were compiled from different sources in different decades of the 1900s.

6.1. *Identify the balanced reduction and oxidation half-reactions that contribute to the (unbalanced) whole redox reactions. (a) $ClO_2 + Br_2 \rightarrow BrO_3^- + Cl^-$; (b) $V_2O_5 + NO \rightarrow VO^{2+} + NO_3^-$; (c) $O_2 + HF \rightarrow OF_2 + H_2O$.

6.2. Write balanced half-reactions for: (a) the reduction of $Au(OH)_3$ to Au; (b) the oxidation of $H_2C_2O_4$ (oxalic acid) to CO_2; (c) the reduction of $S_4O_6^{2-}$ to $S_2O_3^{2-}$; (d) the oxidation of HN_3 to N_2.

6.3. *(a) Combine the following two half-reactions so as to give a balanced whole redox reaction in which U^{4+} is oxidized:

$MnO_4^- + 3e^- + 4H^+ \rightarrow MnO_2 + 2H_2O$

$UO_2^{2+} + 2e^- + 4H^+ \rightarrow U^{4+} + 2H_2O$

(b) If $E°$ for the second half-reaction is +0.33 V, what is $E°_{cell}$ for this redox reaction?

6.4. We want to reduce $PtCl_6^{2-}$ to $PtCl_4^{2-}$ ($E° = +0.73$ V) using $PdCl_4^{2-}$ as the reducing agent as it goes to $PdCl_6^{2-}$ ($E° = +1.29$ V). (a) Write the two balanced half-reactions. (b) Calculate $E°_{cell}$ for this redox reaction. Does it go to products?

6.5. *Below are three examples of reductions of an element. In each case, (a) identify the oxidation numbers of the element in the reactant and then in the product; (b) tell in which block in the given redox predominance diagrams each compound should be located; (c) complete and balance the reduction half-reactions (using H^+, in acid solution).

$$WO_3 \rightarrow WO_2 \qquad H_2MoO_4 \rightarrow MoO_2 \qquad Cr_2O_7^{2-} \rightarrow Cr^{3+}$$

$E° = -0.03$ V $\begin{array}{|c|} \hline A \\ \hline B \\ \hline \end{array}$ $E° = +0.65$ V $\begin{array}{|c|} \hline C \\ \hline D \\ \hline \end{array}$ $E° = +1.38$ V $\begin{array}{|c|} \hline E \\ \hline F \\ \hline \end{array}$

6.6. For the reductions of an element given in (a)–(g), first identify the oxidation states of the element in the reactant and then in the product; second, complete and balance the reduction half-reaction (using H^+, in acid solution); and third, find the standard reduction potential for the half-reaction from the redox predominance diagrams (Figures 6.7–6.10).

(a) $MnO_4^- \rightarrow MnO_2$

(b) $TcO_4^- \rightarrow TcO_2$

(c) $ReO_4^- \rightarrow ReO_3$

(d) $GeO_2 \rightarrow Ge$

(e) $H_3AsO_4 \rightarrow As_4O_6$

(f) $HSeO_4^- \rightarrow H_2SeO_3$

(g) $BrO_4^- \rightarrow BrO_3^-$

6.7. *According to redox predominance or Pourbaix diagrams, (a) what is the most oxidizing chemical form of cobalt? (b) What is the most strongly reducing form of selenium in acidic solutions? (c) Which +2 ion of the *f*-block metals from the sixth period is least strongly reducing? (d) What is the most strongly oxidizing form of iridium? (e) Which MO_2^{2+} ion of the *f*-block metals from the seventh period is most strongly oxidizing?

6.8. According to redox predominance or Pourbaix diagrams, for which Group 7 or 17/VII element is the +7 oxidation state (a) least oxidizing (most stable) and (b) most oxidizing (least stable)?

6.9. *According to redox predominance or Pourbaix diagrams, for which Group 5, 15/V, or 5F element is the +5 oxidation state (a) least oxidizing (most stable) and (b) most oxidizing (least stable)?

6.10. Using redox predominance or Pourbaix diagrams, choose the best oxidizing agent in each set: (a) MnO_4^-, TcO_4^-, ReO_4^-; (b) GeO_2, H_3AsO_4, SeO_4^{2-}, BrO_4^-; (c) SO_4^{2-}, SeO_4^{2-}, H_6TeO_6, (PoO_3 or H_6PoO_6, depending on diagram type); and (d) ($H_2Cr_2O_7$ or $Cr_2O_7^{2-}$, depending on diagram type), H_2MoO_4, WO_3.

6.11. *According to redox predominance or Pourbaix diagrams, for which of the following fourth-period elements is the group oxidation number found in the most strongly oxidizing (least stable) species: Zn, Ga, Ge, As, Se, or Br?

6.12. Which species in each set is the weakest oxidizing agent? (a) RuO_4, OsO_4; (b) NO_3^-, H_3PO_4, H_3AsO_4, H_7SbO_6, H_7BiO_6; (c) H_7SbO_6, H_6TeO_6, H_5IO_6, H_4XeO_6; (d) H_4GeO_4, H_3AsO_4, SeO_4^{2-}, BrO_4^-; and (e) SO_4^{2-}, SeO_4^{2-}, H_6TeO_6, H_6PoO_6.

6.13. *(a) Identify the most oxidizing chemical form of manganese. (b) Identify the most reducing form of manganese. (c) Write the balanced half-reaction for the reduction process that occurs in the manganese redox predominance diagram at -1.18 V. (d) Write the balanced half-reaction for the reduction process that occurs in the manganese redox predominance diagram at $+1.70$ V.

6.14. (a) Write the balanced half-reaction for the reduction process that occurs in the nitrogen redox predominance diagram at +1.25 V. (b) Identify the most oxidizing chemical form of nitrogen. (c) Identify the most reducing form of nitrogen.

6.15. *Using your results from Exercise 6.6, compute the cell potentials $E°$ for redox reactions (a)–(d), and tell whether they will go to give products. Support your answer either by computation or using the redox predominance diagrams.

(a) $MnO_4^- + TcO_2 \rightarrow MnO_2 + TcO_4^-$

(b) $MnO_4^- + ReO_3 \rightarrow MnO_2 + ReO_4^-$

(c) $As_4O_6 + GeO_2 \rightarrow Ge + H_3AsO_4$

(d) $H_2SeO_3 + BrO_4^- \rightarrow BrO_3^- + H_2SeO_4$

6.16. Using your results and the predominance diagrams from Exercise 6.5, balance redox reactions (a) and (b), compute their cell potentials $E°$, and tell whether they will go to give products.

(a) $H_2MoO_4 + WO_2 \rightarrow MoO_2 + WO_3$

(b) $Cr_2O_7^{2-} + WO_2 \rightarrow WO_3 + Cr^{3+}$

6.17. *Which elements—Zn, Cl_2, At_2, Au, Bi, Sc, Hg, Tm, Os—will react (a) with a solution of I^- to generate I_2; (b) with a solution of Ag^+ to generate Ag; and (c) with a solution of Na^+ to generate Na?

6.18. Which elements—O_2, Rb, Sb, Pt, Y, Sm, Mn, Rh, Sn, Li—will react (a) with a solution of I^- to generate I_2; (b) with a solution of Ag^+ to generate Ag; and (c) with a solution of Na^+ to generate Na?

6.19. *(a) List two chemical forms of nitrogen that theoretically should oxidize Mn to Mn^{2+}. (b) Balance the equation for the redox reaction that should occur in acid solution between one of these forms of nitrogen (your choice) and Mn, yielding Mn^{2+} and a likely nitrogen-containing product. (c) List two chemical forms of manganese that should oxidize NH_4^+ to N_2. (d) Balance the equations for the two redox reactions that should occur between each of these two chemical forms of Mn and NH_4^+ to give N_2 and a Mn-containing product.

6.20. (a) What are the likely thermodynamically stable products of the redox reaction between NH_4^+ and IO_3^- in aqueous acid? (b) Write a balanced redox equation for this predicted reaction. (c) Compute the emf for this predicted reaction.

6.21. *According to the redox predominance diagrams and not considering overvoltages (Section 6.2), which elements from the specified blocks of the periodic table have only one stable oxidation state in aqueous solution? (a) The s block; (b) the p block; (c) the d block; (d) the f block.

6.22. By consulting Figure 6.7, from the stable forms of the lighter s- and p-block elements, identify (a) all oxidizing agents that should rapidly oxidize water to O_2; (b) all oxidizing agents that should slowly oxidize water to O_2; (c) all reducing agents that should rapidly reduce water to H_2; (d) all reducing agents that should slowly reduce water to H_2.

6.23. *By consulting Figure 6.8, from the stable forms of the heavier s- and p-block elements, identify (a) all oxidizing agents that should rapidly oxidize water to O_2; (b) all oxidizing agents that should slowly oxidize water to O_2; (c) all reducing agents that should rapidly reduce water to H_2; (d) all reducing agents that should slowly reduce water to H_2.

6.24. By consulting Figure 6.9, from the stable forms of the *d*-block elements, identify (a) all oxidizing agents that should rapidly oxidize water to O_2; (b) all oxidizing agents that should slowly oxidize water to O_2; (c) all reducing agents that should rapidly reduce water to H_2; (d) all reducing agents that should slowly reduce water to H_2.

6.25. *By consulting Figure 6.10, from the stable forms of the *f*-block elements, identify (a) all oxidizing agents that should rapidly oxidize water to O_2; (b) all oxidizing agents that should slowly oxidize water to O_2; (c) all reducing agents that should rapidly reduce water to H_2; (d) all reducing agents that should slowly reduce water to H_2.

6.26. (a) Based on their electronegativities, arrange the following metals in an activity series: K, Au, Cu, Mg, Mn, Zn, Fe, and Ni. (b) Similarly list the following metals in order of decreasing activity: Ca, Cd, Ce, Cf, Co, Cr, Cs, Cu. (c) In each series, which of the metals should visibly react with (cold or hot) water to liberate hydrogen? What GHS pictogram are these likely to carry? (d) In each series, which of the metals should fail even to react with hot HCl to liberate hydrogen? What pictogram are these likely to carry?

6.27. *Based on their electronegativities, arrange in an activity series all the metals the names of which begin with (a) the letter T and (b) the letter P.

6.28. From the following lists of metals, select those that would (1) react with water; (2) react with HCl but not with water; (3) not react with either HCl or water: (a) Au, Be, Bi, Ce, Fe, Ga, Ir, Pu. (b) Rb, Sb, Pt, Y, Sm, Mn, Rh, Sn; and (c) Li, Bi, Sc, Hg, Tm, Os, Zn, Au.

6.29. *Describe the activity of each of the following elements with (1) cold water; (2) hot HCl solution; (3) concentrated HNO_3: (a) La; (b) Pt; (c) Co; (d) Sc; (e) Os; (f) Sr; (g) Cr.

6.30. Consider the metals Ba, Lu, Os, Ti, Bi, Sn, Cu, Au, and Pu. (a) Which of these will react with water? (b) Which of these will react with HCl but not with water? (c) Which of these will dissolve only in HNO_3, or will not dissolve in any of the above?

6.31. *Describe the activity of each of the following elements with (1) cold water; (2) hot HCl solution; (3) concentrated HNO_3: (a) In; (b) Ir; (c) Nd; (d) Cu; (e) Zn; (f) Sn; and (g) Ba.

6.32. If you did not perform Experiment 7, go back to it now. Describe what would have happened during the experiment, and answer the questions in part E.

6.33. *Gold will not dissolve in concentrated sulfuric acid but will dissolve in concentrated selenic acid. Explain why. Which acid is more deserving of the "Oxidizer" GHS pictograph?

6.34. Bromine is made from bromide ion in seawater: $2Br^- + Cl_2 \rightarrow Br_2 + 2Cl^-$. Explain why this works. What term could be applied to this type of reaction? (It is the term used for the corresponding reactions among metals and metal ions.)

6.35. *Based on periodic trends, which element in each set is least likely to oxidize something? Which is/are most likely to come in a container with an "Oxidizer" GHS label? (a) F_2, I_2, Br_2, Cl_2 and (b) F_2, C, O_2.

6.36. Based on the electronegativities of Cl and Sb, predict the following. Also identify which is more likely to get a GHS pictogram. (a) Which element will be the stronger oxidizing agent, chlorine or antimony? (b) Which oxo acid will be the stronger oxidizing agent, perchloric acid or antimonic acid? (c) Which hydride will be the better reducing agent, HCl or SbH_3?

6.37. *Use periodic trends to answer the following questions about the anions CH_3^-, Ge^{4-}, O^{2-}, Se^{2-}, F^-, and H^-. (a) Which two of these six anions would be the weakest reducing agents? (b) The salts of which three anions are most likely to be labeled with the "Flammable" pictogram?

6.38. Consider the four anions C^{4-}, Sn^{4-}, F^-, and I^-. Based on periodic trends, (a) which one is the strongest reducing agent? (b) Which one is the weakest reducing agent? (c) The salts of which two anions are most likely to be labeled with the "Flammable" pictogram?

6.39. *Consider the six anions C^{4-}, Ge^{4-}, O^{2-}, Se^{2-}, F^-, and Br^-. Based on periodic trends, (a) which one is the strongest reducing agent? (b) Which one is the weakest reducing agent? (c) The salts of which two anions are most likely to be labeled with the "Flammable" pictogram?

6.40. Using redox predominance (or Pourbaix) diagrams for the appropriate elements, tell which of these compounds or mixtures are likely to be thermodynamically stable. If they are not stable compounds or mixtures, which are likely to be potentially explosive? (a) $Ni^{2+} + NO_3^-$; (b) $CH_4 + F_2$; (c) $PH_3 + ClO_4^-$; (d) $NH_4^+ + Br^-$; (e) $Mn^{2+} + H_4XeO_6$.

6.41. *For each of the salts (a)–(h), tell whether (1) it will be thermodynamically stable in the sense that its cations and anions will not undergo redox reactions with each other; (2) it will be so unstable that, if it can be made at all, it will be potentially explosive; or (3) its ions may react with each other, but probably not explosively. (a) $CrFeO_4$; (b) $CsMnO_4$; (c) NH_4MnO_4; (d) $EuMnO_4$; (e) CaC_2; (f) Ag_2C_2; (g) $Cr(BrO_3)_2$; and (h) $Tl(N_3)_3$.

6.42. Using redox predominance (or Pourbaix) diagrams for the appropriate elements, tell which of these compounds or mixtures are likely to be thermodynamically stable. If they are not stable compounds or combinations, which are likely to be potentially explosive? (a) $NH_4^+ + H_3PO_4$; (b) $Ag^+ + N_3^-$; (c) $CH_4 + H_4XeO_6$; (d) $Se + O_2$; and (e) $Hg^{2+} + ClO_4^-$.

6.43. *Are silver salts (a)–(c) likely to be stable? If they exist, are they likely to be explosive? (a) $AgBrO_4$; (b) AgH; (c) $AgCl_2$.

6.44. Refer to the redox predominance diagrams for N, I, Co, and Cr. (a) Which of the following salts are expected to be nonexplosive and thermodynamically stable to redox reactions: $Co(NO_3)_3$; NH_4I; $(NH_4)_2Cr_2O_7$; and $Cr(IO_3)_2$? (b) Which of the following mixtures are potentially explosive: HNO_3 and I_2; Cr and I_2; $K_2Cr_2O_7$ and Cr; and NH_4I and CrI_2?

6.45. *(a) Ammonium permanganate (NH_4MnO_4) is a known salt. Comment on its expected solubility or insolubility, and on its expected stability upon being struck with a hammer. (b) Ammonium manganate [$(NH_4)_2MnO_4$] is not a well-known salt. Comment on its expected solubility or insolubility, and on its expected stability upon being struck with a hammer. (c) Comment on the likely safety aspects of grinding together, in a mortar and pestle, finely divided potassium nitrate with manganese metal; finely divided potassium nitrate with manganese dioxide.

6.46. Use redox predominance or Pourbaix diagrams to answer the following questions: (a) Which are expected to be nonexplosive and thermodynamically stable to redox reactions: $AgIO_3$, NH_4Br, NH_4Br_3, $TlReO_4$, and $FeBr_3$? (b) Which mixtures are potentially explosive: HNO_3 and BiH_3; NH_3 and H_2S; $HClO_4$ and SiO_2; Pu^{4+} and H_2Te; Yb^{2+} and H_2Te? (c) Write the formula of the salt containing Se and Fe that is most likely to be unstable (explosive or nonexistent).

6.47. *Use the redox predominance or Pourbaix diagrams of the three nonmetals Br, N, and Sb to answer the following questions: (a) List the anions and hydrides of these elements (NH_4^+, Br^-, SbH_3) in order of increasing activity as reducing agents. (b) List the highest oxo acids or anions of these elements (NO_3^-, BrO_3^-, H_7SbO_6) in order of increasing activity as oxidizing agents. (c) List the elements themselves (Br_2, N_2, and Sb) in order of increasing activity as oxidizing agents. (d) Of the nine species mentioned, the redox reaction of which two species would likely be most exothermic, and therefore might present the greatest risk of an explosion?

6.48. Which would most likely be an explosive compound: (a) methyl cyanate, CH_3OCN; (b) methyl isocyanate, CH_3NCO; or (c) methyl fulminate, CH_3ONC?

6.49. *Given guanidine [$HN=C(NH_2)_2$], nitroguanidine, [$O_2N—N=C(NH_2)_2$], and hydrazine ($H_2N–NH_2$), (a) which is least likely to be an explosive compound? (b) Which is most likely to explode?

6.50. Between methyl hypochlorite ($CH_3–O–Cl$) and trichloromethyl hypochlorite ($CCl_3–O–Cl$), which is more likely to be an explosive compound?

6.51. *(a)Which organic derivative of iodine is more likely to be explosive, $(CH_3)_5IO_6$ or CH_3I? (b) Which is the more dangerously unstable compound, H–O–O–H, H–S–S–H, or H–Se–Se–H? (c) On which astatine compound would it be most dangerous to pour liquid oxygen, $(CH_3)_5AtO_6$, or CH_3At?

6.52. Which of the following 2,4,6-trinitrophenyl (R) compounds is most likely explosive and why? (a) 2,4,6-trinitrophenyl nitrile (R–C≡N) or (b) 2,4,6-trinitrophenyl isonitrile (R–N≡C:) or (c) 2,4,6-trinitrophenylmethane (R–CH_3) or (d) 2,4,6-trinitrophenylgermane (R–GeH_3).

6.53. *Which is most likely explosive and why: (a) $H–C(NH_2)_3$, (b) $H–C(NO_2)_3$, or (c) $H–C(H_2PO_4)_3$?

6.54. According to redox predominance diagrams, which of the following insoluble iodides would be unstable (over the long term) in water if they had solubilities of 1 M, and why? (a) AgI; (b) CuI; (c) AuI; (d) Hg_2I_2; (e) PbI_2.

6.55. *According to redox predominance diagrams, which of the following insoluble chlorides would be unstable (over the long term) in water if they had solubilities of 1 M, and why? (a) AgCl; (b) CuCl; (c) AuCl; (d) Hg_2Cl_2; (e) $PbCl_2$.

6.56. Redraw the predominance diagram for thallium, Tl (Figure 6.8), to account for a change in total thallium ion concentration from 1 M to (a) 10^{-3} M; (b) 10^{-6} M.

6.57. *You are a metallurgist employed in industry, and you need to dissolve some samples of metal in acid to get solutions for analysis by atomic absorption spectroscopy. You are having difficulty in dissolving samples of certain metals in any single acid, even the good oxidizing agent nitric acid. Which acid, HF or HCl, would better help dissolve the indicated metals in nitric acid? How would you explain how it helped? (Include, if possible, a plausible formula for the product.) Metals: (a) tantalum (Ta) and (b) platinum (Pt).

6.58. (a) List the following elements in an activity series for their reaction with water or aqueous acids: Na, Pt, Co, Bi. (b) Might the position of Pt in this activity series be altered if the reaction were run in liquid hydrogen cyanide (HCN) as a solvent, instead of water? If so, how would its position be altered and why?

6.59. *Which ligand would best stabilize the +6 oxidation state of Pt: F^-, H_2O, NH_3, or CO?

6.60. In each set of compounds, all of which exist, select the compound that likely shows the most intense color: (a) CF_4, CCl_4, CBr_4, CI_4; (b) CrF_3, $CrCl_3$, $CrBr_3$, CrI_3; (c) TlF_6^{3-}, $TlCl_4^-$, $TlBr_4^-$, TlI_4^-; (d) CaS, HgS, CdS, ZnS; and (e) PI_5, SiI_4, AlI_3, MgI_2, NaI.

6.61. *In each of the following sets of compounds, identify the one compound that is most probably either nonexistent or very intensely colored, and the one compound that not only exists, but is least intensely colored or even colorless: (a) BiF_3, $BiCl_3$, $BiBr_3$, BiI_3; (b) InI_3, SnI_4, SbI_5; (c) PF_5, PBr_5, PI_5, PAt_5; and (d) FO_4^-, MnO_4^-, ReO_4^-. Give the name for the type of light absorption responsible for intense color among some of the above compounds. Explain it briefly.

6.62. Given ZnO, ZnTe, ZnSe, and ZnS, which two are colorless, which one is yellow-red, and which one is red?

6.63. *The heavier uranium(III) halides—UCl_3, UBr_3, and UI_3—are all intensely colored. Which one is black, which one is red, and which one is green?

6.64. The heavier vanadium chlorides—VCl_4, VCl_3, and VCl_2—have different colors. Which one is green, which one is red-violet, and which one is red-brown?

6.65. *A new element, wattsium (Wa), has just been discovered. Careful electrochemical measurements have established the Pourbaix diagram shown below. The thermodynamically stable species involved are Wa^{3+}, Wa^{2+}, $Wa(s)$, $Wa(OH)_3$, $Wa(OH)_2$, WaO_4^{3-}, and WaO_4^{2-}. (a) Locate each species in the appropriate predominance region in the diagram. (b) Which species is the best reducing agent? (c) Which species is the best oxidizing agent? (d) Which oxidation state disproportionates in acidic or neutral solution?

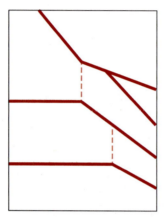

6.66. A blank Pourbaix diagram for the element ruthenium, Ru, is shown below. Without reference to the text, fill in the following chemical forms of Ru in the predominance areas of this diagram: Ru, RuO_4, RuO_4^-, RuO_4^{2-}, $Ru(OH)_3$, $Ru(OH)_4$, and RuO_5^{2-}.

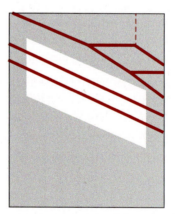

6.67. *Below is a Pourbaix diagram for the imaginary element Branchizium (Bz), which has six predominance areas labeled **A** through **F**. Branchizium has six stable forms: BzO_2, BzO_3, Bz^{4+}, BzO_2^{2+}, Bz, and BzO_4^{2-}. (a) Assign one of these forms to each of the areas **A**–**F**. (b) Crossing the boundaries (lines) separating which pairs of areas correspond to (pure) acid–base reactions? (c) Crossing the boundaries (lines) separating which pairs of areas correspond to (pure) redox reactions? (d) Crossing the boundaries (lines) separating which pairs of areas correspond to mixed acid–base/redox reactions?

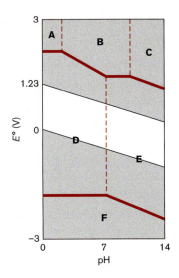

6.68. Below is a Pourbaix diagram for an imaginary element Yatesium (Yt). Ignore the two black lines (which are the window of stability for water; Section 6.5B). The five chemical forms of yatesium are $YtO_3(s)$, $Yt(s)$, $Yt^{3+}(aq)$, $Yt(OH)_3(s)$, and $YtO_2(s)$. Tell where each goes on the diagram.

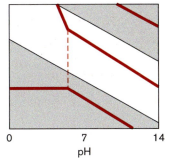

6.69. Figure 6.27 shows the $E°$/pH (Pourbaix) diagram for uranium. You are attempting to study the fate of waste uranium (from nuclear fuel reprocessing) in a natural lake. (a) If the lake is well aerated and not polluted, in what form will the uranium be found? Will it be present in solution or found in the sludge at the bottom of the lake? (b) If the lake is well aerated but is highly contaminated with acid rain (pH 3), in what form will the uranium be found? Will it be present in solution or found in the sludge at the bottom of the lake? (c) Suppose some metallic uranium were dropped into this lake (with pH 3), but it fell to the bottom of the lake, where oxygen was absent and reducing impurities (from decaying organic matter) were present. Would it remain as elemental uranium?

6.70. Figure 6.25 shows the Pourbaix diagram for manganese. People often find that clear well water that they draw from wells will deposit a black manganese-containing solid on standing in their toilet bowls. Explain what that solid is and why it forms in the toilet bowl and not underground.

6.71. *Figure 6.27 shows the Pourbaix diagram for plutonium. You are attempting to study the fate of waste plutonium (from atomic bomb assembly) in a lake. (a) If the lake is well aerated and of normal pH, in what chemical form will the plutonium be found? Will it be present in solution or found in the sludge at the bottom of the lake? (b) Will metallic plutonium dumped into the lake remain in the metallic form, or will it react with the water? If the latter, write a balanced chemical equation showing the reaction with the water. (c) If the lake is anaerobic and highly polluted with acid rain, in what chemical form will the plutonium be found? Will it be present in solution or found in the sludge at the bottom of the lake?

6.72. Identify the chemical form of (a) Mn and (b) N that is predominant in well-aerated lakes of normal pH, and in highly anaerobic lakes that are strongly contaminated with acid rain.

6.73. *Examine the Pourbaix diagrams for the elements Am, Cr, Mn, Bi, and I. If one exists, find a chemical form of each of these elements that, at some pH, would (a) disproportionate; (b) oxidize water; and (c) reduce water.

6.74. Consult the Pourbaix diagram for ruthenium, Ru, to answer the following questions: (a) Which chemical form of Ru listed is the best oxidizing agent at pH 14? (b) Which chemical form of Ru listed is the best reducing agent at pH 0? (c) Will any chemical form of Ru listed release H_2 in contact with water? If so, which form? (d) Which chemical form of Ru is most likely to be found in well-aerated natural waters of pH 7? (e) Which chemical form of Ru is most likely to be found in the anaerobic waters of a swamp at pH 7?

6.75. *In what part(s) of Earth are the elements (a) Ir, (b) Mo, (c) Au, (d) Co, and (e) Os classified as siderophiles? If two parts of Earth are mentioned, <u>underline</u> the part in which the element should have the higher concentration.

6.76. Problems (a) and (b) pertain to each of the four ions Ag^+, Eu^{3+}, Eu^{2+}, and Bi^{3+}. (a) Classify each as a lithophile, a chalcophile, a siderophile found in the core of Earth only, or a siderophile also found at the surface of Earth. (b) Write the formula of a plausible mineral source of that ion.

6.77. *Bacteria became involved in the geochemistry of the elements at which stage of differentiation and when? (a) The primary differentiation of the elements, at the time of the coalescence of the planet Earth; (b) the secondary differentiation of the elements, at the time of the coalescence of the planet Earth; (c) the secondary differentiation of the elements, after the coalescence of the planet Earth; (d) the primary differentiation of the elements, after the coalescence of the planet Earth.

6.78. Which produced the mineral ores that we now mine and when? (a) The primary differentiation of the elements, at the time of the coalescence of the planet Earth; (b) the secondary differentiation of the elements, at the time of the coalescence of the planet Earth; (c) the secondary differentiation of the elements, after the coalescence of the planet Earth; (d) the primary differentiation of the elements, after the coalescence of the planet Earth.

6.79. *Calculate the lattice energy of $CaCl_2$, given its enthalpy of formation of –798 kJ mol^{-1}.

6.80. We want to calculate the lattice energy of $AlCl_3$, given its enthalpy of formation of –698 kJ mol^{-1}. (a) Fill in the empty squares in Figure 6.33 (see Example 6.17) with the symbols and states of the chemical species involved in this Born–Haber cycle. (b) Adding the needed information from Tables 6.4–6.8, compute $U(AlCl_3)$.

6.81. *Set up a Born–Haber cycle for each of the following reactions, and calculate the experimental lattice energies of the products.

(a) $Ca(s) + F_2(g) \rightarrow CaF_2(s)$; $\Delta H_f = -1215$ kJ mol^{-1}

(b) $Ca(s) + 0.5O_2(g) \rightarrow CaO(s)$; $\Delta H_f = -636$ kJ mol^{-1}

(c) Calculate the theoretical lattice energies for these two compounds and compare them to your experimental values.

6.82. The standard enthalpies of formation (in kilojoules per mole) of the metal(II) oxides of the fourth-period elements of the *d* block are as follows (source: M. C. Ball and A. H. Norbury, *Physical Data for Inorganic Chemists*, Longman: London, 1974, pp. 62–64):

TiO	VO	MnO	FeO	CoO	NiO	CuO	ZnO
−518	−431	−385	−264	−239	−241	−155	−348

(a) Write a general equation suitable for calculating the lattice energy of any of these oxides, based on thermochemical data. (b) Find the lattice energies of TiO, CuO, and ZnO. (c) Overall the most dissimilar ΔH_f values are those of TiO and CuO; the most dissimilar ΔH_f values of neighboring oxides are those of CuO and ZnO. Which thermodynamic factor is apparently responsible for each of these dissimilarities? (d) What factor in the theoretical lattice-energy calculations (as done in Section 4.2B) is most responsible for the substantial change in lattice energies between TiO and CuO?

6.83. *The formula of the superconductor $YBa_2Cu_3O_7$ suggests that one-third of its copper ions are in the unusual +3 oxidation state, but some research suggested the surprising alternative that, instead of a structural subunit involving Cu^{3+} and O^{2-} ions, the superconductor may involve Cu^{2+} and singly charged O^- ions. (a) Sketch a thermochemical cycle to estimate the energetics of the transformation $(Cu^{3+})_2(O^{2-})_3(s) \rightarrow (Cu^{2+})_2(O^-)_2(O^{2-})(s)$. (b) Supply as many numbers as you can to complete the thermochemical analysis, but you will not be able to compute the lattice energies of the reactant and the product. (c) Which of the two lattice energies should be larger? Which energy terms in this thermochemical cycle favor the $Cu^{3+}O^{2-}$ structure? Which favor the $Cu^{2+}O^-$ structure?

6.84. We want to calculate the activity of calcium in the reaction

$$Ca(s) + 2H^+(aq) \rightarrow H_2(g) + Ca^{2+}(aq)$$

(a) Fill in the empty squares in the thermchemical cycle shown below with the symbols and states of the chemical species involved in this Born–Haber cycle. (For example, the reactants and products of this reaction go in the boxes on the bottom line of this drawing.)

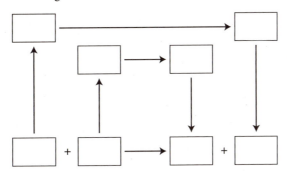

(b) The energy values for the hydrogen total −439 kJ mol^{-1} of H$^+$ that reacts. Use this total and the appropriate values for Ca to compute ΔH for the above reaction.

(c) Neglecting entropy, estimate the voltage produced by an electrochemical cell in which this reaction is carried out under standard conditions.

6.85. *Consider the possibility of a reaction between tin metal and aqueous strontium ion:

$$Sr^{2+}(aq) + Sn(s) \rightarrow Sn^{2+}(aq) + Sr(s)$$

(a) Draw a diagram of the thermochemical cycle for this reaction, and (using appropriate data) calculate ΔH for this reaction.

(b) Does this reaction go as written to products? Neglecting entropy effects, calculate the cell potential $E°$ for the reaction as written.

6.86. (a) Set up and evaluate a thermochemical cycle to evaluate the enthalpy change for the following reaction: $2Cr^{3+}(aq) + H_2 \rightarrow 2Cr^{2+}(aq) + 2H^+(aq)$. (b) Could H_2 be used as a reducing agent to prepare $Cr^{2+}(aq)$? (c) Ignoring entropy effects, calculate $E°$ for this reaction.

6.87. *The new transplutonium element qinjium (Qj) has just been discovered. Like Pu, in water it forms a +3 ion of radius 114 pm and hydration energy ΔH_{hyd} of –3441 kJ mol^{-1}. Other properties are compared in the following table:

Property	Qj	Pu
Atomization energy ΔH_{atom}	440 kJ mol^{-1}	352 kJ mol^{-1}
First ionization energy IE(1)	499 kJ mol^{-1}	585 kJ mol^{-1}
Second ionization energy IE(2)	1480 kJ mol^{-1}	1129 kJ mol^{-1}
Third ionization energy IE(3)	2114 kJ mol^{-1}	2103 kJ mol^{-1}

You want to determine which metal will be higher in the activity series: Qj or Pu.
(a) Draw a thermochemical cycle for the reaction: $Qj(s) + Pu^{3+}(aq) \rightarrow Pu(s) + Qj^{3+}(aq)$. Label the energy terms involved. (b) Calculate ΔH for the reaction in (a). Which metal is more active? (c) Estimate the voltage generated by the above reaction.

6.88. The first and second ionization energies of Mg and of Pb do not differ greatly, but their activities (in reacting with acid to form hydrated +2 ions) do. (a) Calculate ΔH for the reaction of Mg metal with 1 M hydrogen ion to produce hydrated Mg^{2+} ions and H_2. (b) Make the same energy calculation for Pb metal. (c) Which metal is more active? Calculate the approximate standard reduction potential $E°$ for each metal from your ΔH value. (d) Fundamentally, what is the cause of the difference in activity between Mg and Pb?

6.89. *Consider the possibility of a reaction between gaseous lead metal (from a metal-atom reactor apparatus) and aqueous magnesium ion: $Pb(g) + Mg^{2+}(aq) \rightarrow Pb^{2+}(aq) + Mg(s)$. (a) Draw a diagram of the thermochemical cycle for this reaction, and calculate ΔH for this reaction. (b) Does this reaction go as written to products? Neglecting entropy effects, calculate the cell potential $E°$ for the reaction as written.

6.90. Consider the possibility of a reaction between magnesium metal and aqueous lead ion:

$$Pb^{2+}(aq) + Mg(s) \rightarrow Mg^{2+}(aq) + Pb(s)$$

(a) Draw a diagram of the thermochemical cycle for this reaction, and (using appropriate data from the tables in the text) calculate ΔH for this reaction.

(b) Does this reaction go as written to products? Which metal is more active? Neglecting entropy effects, calculate the cell potential $E°$ for the reaction as written.

6.91. *Consider the possibility of a (balanced) reaction between Pb metal and aqueous Ag^+ to give Ag metal and Pb^{2+} ion. (a) Draw a diagram of the thermochemical cycle for this (balanced) reaction. (b) Calculate ΔH for this reaction. (c) Does this reaction go as written to products? Which metal is more active? Neglecting entropy effects, calculate the cell potential $E°$ for the reaction as correctly balanced.

6.92. Suggest a better way to attempt to correlate activities of metals (Experiment 7) with ionization energy than simply correlating with the first ionization energies of metals. (Hint: Look at Eq. 6.35.) Try this correlation (on graph paper) with the $E°$ values from Figure 6.14. Do you get a more satisfactory correlation with activity using your ionization-energy function than the correlation of activity with electronegativity (Figure 6.16)? If not, why not?

6.93. *Sketch thermochemical cycles for these reactions, showing the products formed after each step. Use standard abbreviations to indicate the energy change expected for each step, and show how these should be added or subtracted to obtain ΔH for the reaction.

(a) $U(s) + 2F_2(g) \rightarrow UF_4(s)$

(b) $2Cr^{3+}(aq) + Zn(s) \rightarrow Zn^{2+}(aq) + 2Cr^{2+}(aq)$

(c) $Co^{3+}(aq) + Cr^{2+}(aq) \rightarrow Co^{2+}(aq) + Cr^{3+}(aq)$

(d) Although you are not doing actual calculations, from the principles in the text identify which one energy term is likely responsible for the fact that the formation of UF_4 from the elements is exothermic.

ALFRED WERNER (1866–1919) was born in Mulhouse, Alsace. At the time, it was part of France, but it was annexed by Germany in 1871. He received his doctorate from the University of Zurich in 1890 for a study of the spatial arrangements of atoms in molecules containing nitrogen. Werner proposed that many trivalent compounds of nitrogen were tetrahedral, with the nitrogen at one corner of the tetrahedron and the three bonds to the nitrogen pointing at the other three corners. In 1893, he theorized that inorganic compounds, which we now recognize as complex ions, consisted of a central transition metal atom around which were arranged a set number of neutral or anionic species, which we now call ligands, in a simple geometrical pattern. The number of ligands, which Werner called the coordination number, was typically 4, 6, or 8. For example, Werner proposed that the structure of $CoCl_3 \cdot 6NH_3$ consisted of a central cobalt surrounded by six NH_3 ligands at the vertices of an octahedron. He was able to show, moreover, that the three chlorine atoms were dissociable as Cl^- ions. Werner distinguished between the "primary" valence (the Co–Cl bonds) and the "secondary" valence (the Co–NH_3 bonds). Up to that point, chemists defined the valence of a compound as its total number of bonds, without distinguishing between different kinds of bonds or interactions. Werner was made a professor at the University of Zurich in 1895 and he used his time there to synthesize many more inorganic compounds, including the geometrical isomers and optical isomers that were suggested by his ideas about the structures of coordination compounds. Werner was awarded the 1913 Nobel Prize in Chemistry "in recognition of his work on the linkage of atoms in molecules by which he has thrown new light on earlier investigations and opened up new fields of research especially in inorganic chemistry." He was the first inorganic chemist to win the award and he was the last until 1973.

CHAPTER 7

Introduction to Transition Metal Complexes

We discussed the overall chemistry of complex ions in Chapters 3 and 5. Here in Chapter 7, we emphasize some special features present in the complex ions of the *d*-block metals. First, we should stress that, in most respects, the complex ions of the *d*-block metals are like those found in the rest of the periodic table. Their stability results mainly from the donation of electrons from the Lewis base ligands to (empty orbitals in) the Lewis acid metal ion (Section 3.3). Their complexes with chelating ligands are more stable than with monodentate ligands (Section 5.1C). The hard and soft acid–base (HSAB) principle applies (Section 5.4).

However, there are certain features about complexes formed from metal ions having incomplete filling of the *d* (and *f*) orbitals that are not observed in other complexes. The first and most obvious of these features is the great variety of colors observed in these complexes—even in the simplest complex ions, the hydrated ions (Sections 1.2C and Tables 1.3 and 1.4). As if the colors of the $3d^n$ hydrated ions were not attractive enough, it turns out that addition of different ligands to the hydrated ions produces yet other colors, with each ligand seeming to produce a subtly or dramatically different color (Section 7.4; Experiment 8). With Ni^{2+} in particular, practically every color of the rainbow can be produced. A second important feature (especially in the *d* block) is the large number of possible ion charges and oxidation states (Table C and Section 1.4B), which (in contrast with elements in the *p* block) may differ by only one electron; even zero or negative oxidation states are possible. A third unique feature is the magnetic properties of these ions—they may have unpaired electrons, which results in the magnetic property called *paramagnetism* (Section 7.3). A fourth feature is that the *d*-block metal ions may bond well with certain types of ligands such as carbon monoxide (the carbonyl ligand) that very seldom bond to *p*-block elements (Sections 7.4B and 7.8B).

Particularly importantly, although the predictions of VSEPR theory for geometries and bond angles of complexes (Section 5.1A) usually work out for complexes of the *d*-block elements, some geometries such as the important square planar geometry show up unexpectedly in the *d* block (Section 7.6B). Secondly, VSEPR procedures are given in Section 5.1A for compounds with unshared central-atom electron pairs for *p*-block elements. Now we need to understand the role of central-atom electron pairs for *d*-block elements. Thirdly, the formation of the complexes of most of these metal ions are more exothermic than we can account for simply using bond energy trends in Section 3.3C or lattice energies in Section 4.2.

In this chapter, we develop two special theories used specifically to interpret the special properties of *d*-block complexes. The oldest of these is the **crystal field theory**, which was developed by physicists around 1930[1] to explain similar special properties that result when ionic crystals (such as NaCl) are irradiated; its existence and relevance to *d*-block complex-ion chemistry was unknown to chemists until the 1950s. As befits its origin, it treats bonding from an electrostatic point of view: Ligands are modeled as point negative charges. This obvious oversimplification works surprisingly well. Many of the terms and quantitative parameters used even today with the more sophisticated bonding theories have their origins in the crystal field theory.

For some purposes, however, it becomes imperative to consider the fact that the metal–ligand interaction involves covalent overlap. When this adjustment is made to the crystal field theory we have the **ligand field theory**, simple elements of which are also introduced in this chapter. In Chapter 11 we introduce the more powerful **molecular orbital theory**, applying it to some *d*-block complexes in Section 11.5.

7.1. Crystal Field Theory

OVERVIEW. Valence electrons in different *d* orbitals are repelled differently with the six (donated) electron pairs of six ligands in an octahedral complex, making them unequal in energy. You should be able to identify and draw the *d* orbitals within an octahedron (Figure 7.1) and classify them as t_{2g} or e_g in an octahedral complex. You may practice these concepts of Section 7.1A by trying Exercises 7.1–7.4. The energy difference between t_{2g} and e_g sets of *d* electrons in an octahedral complex is designated as Δ_o and can be used to calculate the crystal field stabilization energy (CFSE) that results from unequal occupation of the *d* orbitals in such a complex. You should be able to write the t_{2g} and e_g electron configuration and calculate the ligand field stabilization energy for an octahedral complex in which the metal ion has 1, 2, 3, 8, 9, or 10 *d* electrons. You may practice these concepts of Section 7.1B by trying Exercises 7.5–7.8.

The concepts of Section 7.1 draw on valence electron configurations from Section 1.3C, the shapes and labeling of *d* orbitals from Figure 3.7, and the methods of assembly of complex ions from Section 3.4. The concepts of this section will be developed further in Sections 9.4 and 11.5.

7.1A. Orientation of *d* Orbitals in Complexes—Repulsion of t_{2g} and e_g Sets of Orbitals.
Many of the special properties of the *d*- (and *f*-) block complexes result from the fact that the central metal ion in the complex is generally *nonspherical* due to *partial occupancy* of the valence (*d* or *f*) orbitals. Metal ions having partially filled electron configurations from d^1 to d^9 and from f^1 through f^{13} have irregular orbital surfaces that interact differently with incoming ligands, depending on the direction from which the ligands approach the metal. This is true for all electron configurations except those in

which the *d* and *f* orbitals are equally occupied. The d^0 and f^0 configurations are spherical. It can be shown from the wave functions, moreover, that the fully occupied d^{10} and f^{14} configurations are also spherical, as are the half-filled electron configurations d^5 and f^7, provided each of the *d* or *f* orbitals is occupied by one electron. Many of the special properties that we will discuss are absent in these cases. Some of these special properties are found among all complexes having incompletely filled *d* and *f* orbitals. However, many are *commonly* found only in the *d*-block complexes of the *fourth* period in which the *d* orbitals are incompletely filled—that is, the complexes of the eight metals Ti through Cu. The complexes of these metal ions will be emphasized.

Because the maximum total coordination number among the *d*-block metal ions of the fourth period is six, let us begin by discussing octahedral complexes of these ions, in which the ligands come in from the +x, –x, +y, –y, +z, and –z directions (Figure 7.1). Each of these ligands has a pair of electrons to be donated to the metal ion; in the crystal field theory model, these are treated as point charges. If the metal ion has a d^0 electron configuration, a complex is formed with no special properties. But if the metal ion has a d^1 electron configuration (e.g., Ti^{3+}), the properties will vary depending on which *d* orbital holds the electron. In Figure 7.1, we take the drawings of the shapes of the *d* orbitals from Figure 3.7, and locate them inside an octahedron of ligands (point negative charges). Two of the *d* orbitals, the d_{z^2} and $d_{x^2-y^2}$, have lobes that are oriented along the *x*, *y*, and *z* axes. If the electron is in one of these orbitals, it will repel the incoming point charge, producing a less stable complex. On the other hand, the other three orbitals, the d_{xy}, d_{xz}, and d_{yz} orbitals, have lobes that fall between the *x*, *y*, and *z* axes, so there is much less repulsion. It will be preferable in an octahedral complex, therefore, for the electron to occupy the d_{xy}, d_{xz}, or d_{yz} orbital (labeled the **t_{2g} set of orbitals**), rather than the d_{z^2} or $d_{x^2-y^2}$ orbital (the **e_g set of orbitals**). (For the origin of the labels t_{2g} and e_g, see Chapter 10.)

We want to compare this actual situation with the hypothetical one in which the consequences of the nonspherical electron distribution is removed. For a d^1 electron configuration, this could be achieved, for example, by putting 0.2 electrons in each of the five *d* orbitals. Because this is somewhat hard to envision, the equivalent hypothetical complex ion is conventionally taken in which the nonspherical shape is removed from the set of ligands instead, by distributing the electrons of the six octahedral donor atoms evenly over a sphere at a common distance, as sketched in **brown** in Figure 7.2b and d.

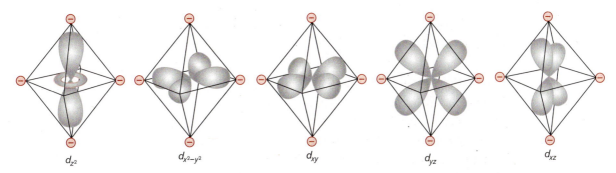

Figure 7.1. Orientations of the *d* orbitals inside an octahedron of ligands or point negative charges. [Adapted from G. Zumdahl, *Chemistry*, 3rd ed., D. C. Heath and Company: Lexington, MA, 1993.]

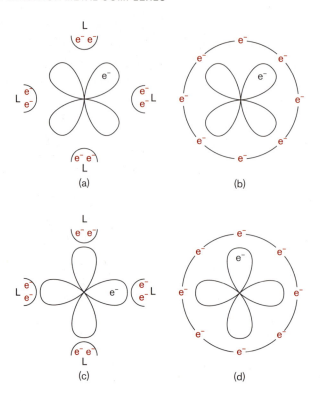

Figure 7.2. The *xy*-plane representation of the interaction of two metal *d* orbitals with the electrons (in **brown**) of four donor atoms in a complex. (a) and (c) The *real* octahedral set of donor atoms, with eight electrons being donated. (b) and (d) The *hypothetical* spherical complex, with eight electrons being donated equally from all positions on the sphere. (a) and (b) One metal *d* electron in the d_{xy} orbital. (c) and (d) One metal *d* electron in the $d_{x^2-y^2}$ orbital.

The complex ion with the one electron in the t_{2g} orbital experiences *less* repulsion of the ligand in the *real* octahedral complex (Figure 7.2a) than in the hypothetical spherical complex ion (Figure 7.2b, in **brown**). The ligands are able to move closer to the metal ion along the *x*, *y*, and *z* axes, thereby donating their electrons more effectively to the metal ion and producing a *more* stable complex. On the other hand, if the one electron of the metal ion is in the e_g set of orbitals, it experiences *more* repulsion in the real octahedral complex ion (Figure 7.2c) than in the hypothetical spherical complex ion (Figure 7.2d).

EXAMPLE 7.1

Reason whether each of the five *d* orbitals will be weakly or strongly repelled if a metal ion containing one *d* electron is placed in a two-coordinate linear complex ion. (Note that this complex, by convention, is considered to lie along the *z* axis.)

SOLUTION: The two ligands (or point negative charges) are those at the top and the bottom of Figure 7.1. The d_{z^2} orbital will be repelled strongly; the d_{xz} and d_{yz} orbitals will be weakly repelled; and the d_{xy} and $d_{x^2-y^2}$ orbitals will not be repelled at all in a linear complex (ligands at ends of the *z* axis).

Repulsion of *f* Orbitals in Octahedral Complexes

Figure 7.3 shows the orientations of the seven *f* orbitals as they appear in an octahedral complex (with ligands centered in each face of the drawing).

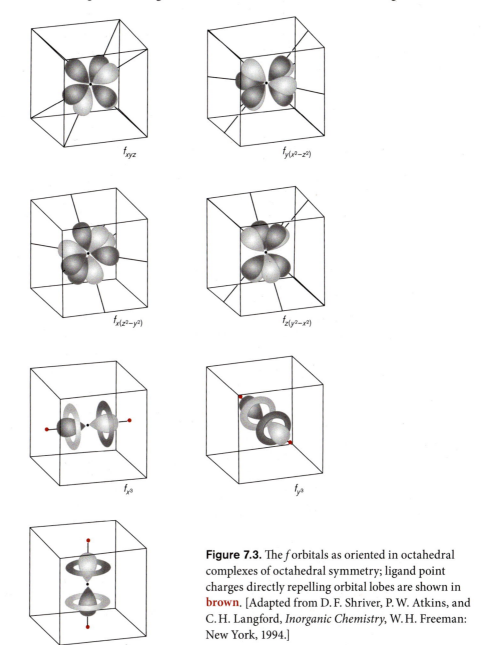

Figure 7.3. The *f* orbitals as oriented in octahedral complexes of octahedral symmetry; ligand point charges directly repelling orbital lobes are shown in **brown**. [Adapted from D. F. Shriver, P. W. Atkins, and C. H. Langford, *Inorganic Chemistry*, W. H. Freeman: New York, 1994.]

It is also true that the energies of the seven *f* orbitals of a given *f*-block metal ion are split in energy in an octahedral complex ion (as in Exercise 7.3). However, because the valence $(n-2)f$ orbitals have smaller sizes than the valence *ns* [or even the core $(n-1)p$ orbitals, the ligands cannot in fact get close enough to the *f* orbitals to cause much crystal field splitting at all, particularly for metals with 4*f* valence electrons. (This is discussed further in Section 9.4.)

7.1B. Crystal Field Splitting and Stabilization Energies. The extra energy released as a result of the nonspherical metal orbital–ligand interactions is called the **crystal field stabilization energy (CFSE)** of the complex ion. (This energy term also arises when considerations of covalency in the ligand–d-orbital interaction are introduced as part of the ligand field theory, so this energy term is also called the ligand field stabilization energy, LFSE.)

In Figure 7.4 we show an **energy-level diagram** comparing the relative energies of orbitals that the d electron could occupy in these two geometries of complexes: Any d orbital in the spherical complex (on the left) is taken as the zero of energy, while in the octahedral complex (on the right) the t_{2g}^1 set of orbitals is lower in energy and the e_g^1 set of orbitals is higher. The difference in energy between the two sets of orbitals (t_{2g} and e_g) is called the (octahedral) **crystal field splitting** (or ligand field splitting) and is symbolized as Δ_o (in which the subscript "o" stands for "octahedral"). (In some sources the crystal field splitting is symbolized by $10Dq$.)

The CFSE of t_{2g} orbitals and the complementary crystal field destabilization energy of e_g orbitals are set so that the net result, if all of the d orbitals are equally occupied, is a CFSE of zero:

$$\frac{-0.4\Delta_o(3\text{ orbitals}) + 0.6\Delta_o(2\text{ orbitals})}{5\text{ orbitals}} = 0.0\Delta_o \tag{7.1}$$

Returning to the case of the d^1 metal ion, this ion has a CFSE of $-0.4\Delta_o$ in the preferred t_{2g}^1 configuration.

The behavior of a complexed d^2 metal ion such as V^{3+} is quite similar: In such an ion the electrons preferentially occupy two of the three t_{2g} orbitals, with both spins aligned the same way (e.g., both spins up—Hund's rule). This gives twice the crystal field stabilization energy of the d^1 case, $-0.8\Delta_o$.

In a complexed d^8 metal ion such as Ni^{2+}, the eight electrons completely fill the three t_{2g} orbitals (with one electron spin up and one electron spin down in each orbital), and the remaining two electrons will be found in the e_g pair of orbitals. Both spins must be the same, by Hund's rule. The CFSE is then $6(-0.4\Delta_o) + 2(+0.6\Delta_o) = -1.2\Delta_o$.

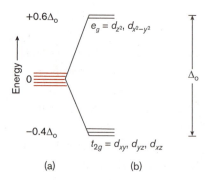

Figure 7.4. Energy-level diagram showing relative energies of the metal d orbitals in (a) a hypothetical spherical set of donor atoms, taken as the zero of energy, and (b) a real octahedral set of donor atoms. The Δ_o parameter is the crystal field splitting.

EXAMPLE 7.2

(a) Draw an energy-level diagram to indicate how, in an octahedral complex $[ML_6]^{n+}$, the energy levels of the d orbitals are split as a result of the repulsion of d and ligand electrons. (b) Fill in three electrons as arrows (↑,↓) correctly aligned to represent a d^3 metal ion in an octahedral complex. (c) Compute the CFSE in units of Δ_o for this d^3 complex. (d) Which (if any) of the complexes—$[Cr(H_2O)_6]^{3+}$, $[Cr(H_2O)_6]^{2+}$, or $[CrF_6]^{3-}$—would be considered to contain a d^3 metal ion?

SOLUTION: (a) and (b) Three energy levels are stabilized as a result of weaker repulsion, while two energy levels are destabilized because they have stronger repulsion than in a comparable spherical field. The three electrons all enter the lower energy level with the same spin (either all up or all down) as a consequence of Hund's rule.

   (c) The CFSE is $-0.4\Delta_o$ for each of the three electrons, for a total of $-1.2\Delta_o$. (d) The first two ions are hydrated Cr^{3+} and Cr^{2+} ions. As was covered in Section 1.3C, these ions have their valence electrons in d orbitals, so they are d^3 and d^4 ions, respectively. The third complex ion is assembled as indicated in Section 3.4 from a Cr^{3+} central ion (d^3) and six F^- ligands.

7.2. High-Spin and Low-Spin Electron Configurations: Magnetic Properties of Metal Ions

OVERVIEW. At first valence electrons fill a set of orbitals all with the same spin, resulting in unpaired electrons. When enough valence electrons are present, some of them must pair in orbitals with electrons of the opposite spin. These two paired electrons repel each other, however, so a pairing energy P must be provided. In octahedral complexes, when four to seven d electrons are present, the electrons can either pair (if Δ_o is larger than P) or occupy the higher-energy e_g orbitals (if P is larger than Δ_o). You should be able to write and draw the electron configuration, calculate the ligand field stabilization energy, and compute the number of unpaired electrons for the octahedral complex of a transition metal ion for which both weak-field and strong-field electron configurations are possible. You may practice this concept of Section 7.2A by trying Exercises 7.9–7.14. Unpaired electrons cause an ion or complex to be paramagnetic so that it is attracted to a magnetic field, and has a molar magnetic susceptibility χ_m^{corr}. You should know the terms describing magnetic properties of transition metals and their complexes, and know how to predict magnetic moments and magnetic susceptibilities using Equations 7.2 and 7.3. You may practice this concept of Section 7.2B by trying Exercises 7.15–7.18.

   This section contains a Connection of magnetic properties of complexes to organic and medicinal chemistry. In Chapter 8 you will extend your understanding of the magnetic properties of ions and complexes to solid-state materials, in which other magnetic properties are possible—namely, ferromagnetism, antiferromagnetism, and superconductivity.

7.2A. Paired and Unpaired Electrons—High-Spin and Low-Spin Electron Configurations.

Paired electrons are those for which each electron with a spin up ($\uparrow$) has a corresponding electron with the spin down ($\downarrow$). For example, two electrons in the same orbital must be paired, and all core electrons in an atom or ion are paired. The **number of unpaired electrons** in an atom or ion is the excess of electrons of one spin over electrons of the other spin. In a free metal ion, or a metal ion in a spherical complex, all d or f orbitals are of equal energy (as in the left side of Figure 7.4). In such a species one electron goes into each d or f orbital before any electron occupies one of these orbitals with the other spin. Thus, there can be up to *five* unpaired electrons in an uncomplexed or spherically complexed d-block metal ion, and up to *seven* in the f block.

Beyond these numbers, additional electrons must be paired: In a d^6 or an f^8 ion, in one valence orbital there is one set of paired electrons ($\uparrow\downarrow$). The energy required to force these two electrons into the same orbital is called the **pairing energy**, **P** (Table 7.1). From Table 7.1, it may be seen that pairing is more difficult in +3 ions (which are smaller) than in the larger +2 ions. The values in Table 7.1 apply to ligand-free gaseous metal ions; it is likely to be 15–30% more difficult to pair electrons in the more compressed or covalent orbitals in complexes.

The fourth electron of an octahedral complex of a d^4 metal ion such as Cr^{2+} must make a choice between two evils. If it goes into one of the three t_{2g} orbitals, thereby reducing the repulsion of the ligand electrons, it is repelled by the other electron already in that orbital. Therefore, an endothermic term, the pairing energy P, must be applied. If the fourth electron is to avoid paying the price of this pairing energy, it must instead go into one of the e_g orbitals, which are higher in energy by the crystal field splitting, Δ_o. Two possible electron configurations are possible for a d^4 metal ion.

Similarly, there are two possibilities in an octahedral complex of a d^5 metal ion such as Mn^{2+} (Figure 7.5). If the crystal field of the ligands is relatively weak, resulting in P being larger than Δ_o, the electron configuration $t_{2g}^3 e_g^2$ results, as shown in Figure 7.5b. Such a complex has a CFSE of $3(-0.4\Delta_o) + 2(0.6\Delta_o) = 0.0\Delta_o$. Complexes in which the metal d electrons can more readily overcome the crystal field of the ligands than pair are known as **weak-field complexes**. A weak-field complex is expected for water as a ligand and for all complexes of f-block metal ions. Such a complex has a higher number of unpaired electrons (five in this case) and is also known as a **high-spin complex**.

TABLE 7.1. Pairing Energies *(P)* for Some Gaseous 3d Metal Ions

Configuration	M²⁺	P (kJ mol⁻¹)	P (cm⁻¹)	M³⁺	P (kJ mol⁻¹)	P (cm⁻¹)
d^4	Cr^{2+}	281	23,500	Mn^{3+}	335	28,000
d^5	Mn^{2+}	305	25,500	Fe^{3+}	359	30,000
d^6	Fe^{2+}	211	17,600	Co^{3+}	251	21,000
d^7	Co^{2+}	269	22,500	Ni^{3+}	323	27,000

Source: D. S. McClure, "The Effects of Inner Orbitals on Thermodynamic Properties," in *Some Aspects of Crystal Field Theory*, T. M. Dunn, D. S. McClure, and R. G. Pearson, Eds., Harper and Row: New York, 1965, p. 82.

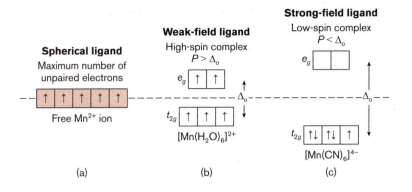

Figure 7.5. Octahedral complexes of Mn^{2+} with (a) a hypothetical spherical ligand, (b) a weak-field ligand such as H_2O, and (c) a strong-field ligand such as CN^-.

If Δ_o is larger than P for the metal ion in question, a **strong-field complex** is formed. The second electron configuration, t_{2g}^5, is selected, as shown in Figure 7.5c. Such a complex has a CFSE of $-2.0\Delta_o + 2P$. It has a lower number of unpaired electrons (one in this case) and is also known as a **low-spin complex**. Strong-field complexes are expected for the cyanide ligand (CN^-) and for nearly all complexes of metals with electrons in $4d$ and $5d$ valence orbitals.

Both types of complexes are common for metals with $3d$ valence electrons (top period of the d block)—thus, for Cr^{2+} an example of a strong-field complex is the cyanide complex $[Cr(CN)_6]^{4-}$, while the hydrated ion $[Cr(H_2O)_6]^{2+}$ is a weak-field complex.

The same choices exist among the complexes formed by metal ions having six and seven valence d electrons (e.g., Fe^{2+} and Co^{2+}, respectively), as illustrated in Figure 7.6. Note that the pairing energy term P in the CFSE is based on, not the total number of sets of paired electrons, but the number of sets of paired electrons *in excess of those found in the hypothetical spherical ion* (shown in Figure 7.6). Thus, for the strong-field d^6 complex $[Fe(CN)_6]^{4-}$, the CFSE is $6(-0.4\Delta_o + 2P)$, because the strong-field complex has three sets of paired electrons while the spherical-field complex has one set.

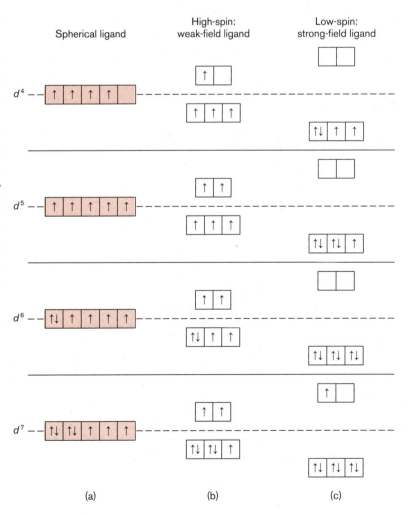

Figure 7.6. Energy-level diagrams showing the weak-field and strong-field electron configurations for d^4, d^5, d^6, and d^7 metal ions.

When we add one more electron to either the weak- or strong-field d^7 electron configuration to arrive at the d^8 electron configuration (as found in Ni^{2+}), we arrive at only one possible electron configuration: $t_{2g}^6 e_g^2$, with a CFSE of $-1.2\Delta_o$. All octahedral d^8 complexes have the same electron configuration, so there is no longer any need of the weak-field/strong-field distinction. Similarly, d^9 ions (e.g., Cu^{2+}) adopt the $t_{2g}^6 e_g^3$ electron configuration, with a CFSE of $-0.6\Delta_o$. The d^{10} ions are spherical, with zero CFSE and the $t_{2g}^6 e_g^4$ electron configuration.

EXAMPLE 7.3

In this example we contrast $[Fe(H_2O)_6]^{2+}$, which has a weak crystal field, with $[Fe(CN)_6]^{4-}$, which has a strong crystal field. (a) Draw energy-level diagrams for the two complexes so as to suggest the difference in the crystal field between the two complexes. (b) Fill in the electrons as $\downarrow$ and $\uparrow$, and express the electron configuration of each in terms of t_{2g} and e_g. (c) How many unpaired electrons are there in each complex? (d) In terms of Δ_o and P, what is the CFSE of each complex?

SOLUTION: (a) Each complex contains the Fe^{2+} central ion, with a d^6 electron configuration. The drawings should have energy-level splitting Δ_o that is larger for the strong-field complex than for the weak-field complex. (The pairing energies should theoretically be the same in each drawing, because the same central metal ion is involved.)

(b) When the electrons are filled in, the drawings will look like those found in Figure 7.6 in the third column. The electron configuration of $[Fe(H_2O)_6]^{2+}$ is $t_{2g}^4 e_g^2$, while that of $[Fe(CN)_6]^{4-}$ is t_{2g}^6.

(c) The number of unpaired electrons in $[Fe(H_2O)_6]^{2+}$ is four, while there are none in $[Fe(CN)_6]^{4-}$.

(d) The CFSE of $[Fe(H_2O)_6]^{2+}$ is $-0.4\Delta_o$, while that of $[Fe(CN)_6]^{4-}$ is $-2.4\Delta_o + 2P$ (because two more pairs of electrons are paired in this strong-field complex than are paired in spherical-field Fe^{2+}).

7.2B. Magnetism, Magnetic Moments, and Magnetic Susceptibility. Any ion, molecule, or atom in which all of the electrons are paired (e.g., most species in the p-block except those known as *free radicals*) exhibits a magnetic property known as **diamagnetism**—that is, the substance is weakly *repelled* by a magnetic field. Any ion, molecule, or atom that has unpaired electrons exhibits the stronger magnetic property known as **paramagnetism**. Because the magnetic effects of all the electrons with spin "up" are not cancelled by the opposed effects of an equal number of electrons with spin "down," the species is *attracted* to a magnetic field. The magnitude of this effect for the species is known as its **magnetic moment** (μ). Its theoretical value is most simply related to the number of unpaired electrons n by Equation 7.2:

$$\mu = \sqrt{n(n+2)} \text{ Bohr magnetons (BM)} \qquad (7.2)$$

Calculated and commonly observed values of μ are given in Table 7.2.

The experimental magnetic moment is calculated from the observed force with which 1 mol of sample is pulled into a magnetic field. This force, after corrections for the diamagnetism of the paired electrons in the metal ion, ligands, and counterions, is known as the corrected molar **magnetic susceptibility** of the substance, χ_m^{corr}. By *Curie's law* the magnetic susceptibility of a substance is also inversely proportional to the absolute temperature of the substance.

$$\chi_m^{corr} = (1/T)(\mu/2.83)^2 \text{ cgs units (units of cm g s)} \tag{7.3}$$

Measuring the magnetic susceptibility of a complex is therefore the most direct way of determining whether a given complex is a high-spin complex or a low-spin complex.

TABLE 7.2. Calculated and Typical Observed Magnetic Moments μ^a

n^b	μ_{calc}	μ_{obs}
1	1.73	1.7–2.2 (in Cu^{2+} complexes)
2	2.83	2.8–3.3 (in Ni^{2+} complexes)
3	3.87	4.5–5.2 (in Co^{2+} complexes)
4	4.90	5.1–5.7 (in Fe^{2+} complexes)
5	5.92	5.6–6.1 (in Mn^{2+} complexes)

[a] In units of Bohr magnetons (BM). The deviation of the observed from the calculated values is due to a contribution from the *orbital motion* (not just the spin) of the unpaired electrons. Observed values are for octahedral complexes.

[b] Number of unpaired electrons.

EXAMPLE 7.4

Predict (a) the number of unpaired electrons, (b) the magnetic moment, and (c) the molar magnetic susceptibility at 25°C of each of the following ions in a spherical electric field: Fe^{2+}; Ru^{3+}; Cr^{2+}; Eu^{2+}.

SOLUTION: (a) The d^6 Fe^{2+} has two paired and therefore four unpaired electrons, the d^5 Ru^{3+} ion has five unpaired electrons, the d^4 Cr^{2+} ion has four unpaired electrons, and the f^7 Eu^{2+} ion has seven unpaired electrons.

(b) Using Equation 7.2, the theoretical values of the magnetic moment μ are 4.90 BM, 5.92 BM, 4.90 BM, and 7.94 BM, respectively.

(c) The predicted molar magnetic susceptibilities, χ_m^{corr} at 298 K, are obtained from Equation 7.3: 0.0101 cgs units, 0.0147 cgs units, 0.0101 cgs units, and 0.0264 cgs units per mole, respectively.

EXAMPLE 7.5

Predict the number of unpaired electrons, the magnetic moment, and the molar magnetic susceptibility at 25°C for each of the following complex ions: (a) $[Fe(CN)_6]^{4-}$; (b) $[Ru(NH_3)_6]^{3+}$; (c) $[Cr(NH_3)_6]^{2+}$; and (d) $[EuCl_6]^{4-}$. If the ion has the possibility of being either high spin or low spin, make predictions for both cases, but indicate the more likely possibility.

SOLUTION: First, in order to determine the d^n electron configuration in each complex ion, it is necessary to determine the oxidation number of each central metal atom (Sections 1.4A and 3.4B). In (a), we have a complex of CN^- with the d^6 ion Fe^{2+}, which can be either high spin with four unpaired electrons, or low spin with zero unpaired electrons. With CN^- as a ligand, low spin is more probable. In (b), we have a complex of NH_3 with the d^5 Ru^{3+} ion, which can be either high spin with five unpaired electrons or low spin with one unpaired electron. In the fifth period (to which Ru belongs), low spin is more likely. In (c), we have a complex of NH_3 with the d^4 Cr^{2+} ion; in the fourth period, either the high-spin configuration with four unpaired electrons or the low-spin configuration with two unpaired electrons is possible. In (d), we have a complex of Cl^- ions with the Eu^{2+} ion, which has a f^7 electronic configuration that will certainly be high spin with seven unpaired electrons, because f orbitals have so little overlap and interaction with ligand orbitals. (See the Amplification entitled "Repulsion of f Orbitals" in Section 7.1A.)

Using Equation 7.2, the theoretical values of the magnetic moment μ are found: (a) maybe 4.90 but probably 0 BM; (b) maybe 5.92 but probably 1.73 BM; (c) either 4.90 or 2.83 BM; (d) 7.94 BM.

The predicted molar magnetic susceptibilities, χ_m^{corr} at 298 K, are obtained from Equation 7.3: (a) maybe 0.0101 cgs units but probably 0 cgs units; (b) maybe 0.0147 but probably 0.0013 cgs units; (c) either 0.0101 or 0.0034 cgs units; and (d) 0.0264 cgs units per mole.

CONNECTIONS TO ORGANIC CHEMISTRY AND MEDICINAL CHEMISTRY

Lanthanide Shift and MRI Contrast Agents

The slight transfer of paramagnetism of f-block complexes to nearby organic molecules has proven useful in simplifying the NMR (nuclear magnetic resonance) spectra of large organic molecules and enhancing the contrast in magnetic resonance imaging (MRI) of cerebral tumors.[2] NMR shifts can be simplified by adding the phenanthroline adduct of the europium(III) complex shown in Figure 7.7a, while medical imaging uses gadolinium(III) complexes such as that shown in Figure 7.7b. Eu(III) has six unpaired electrons, while Gd(III) has seven unpaired electrons.

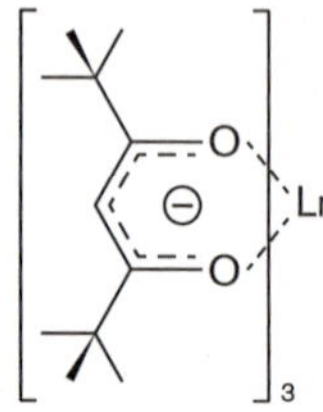

(a) dpm

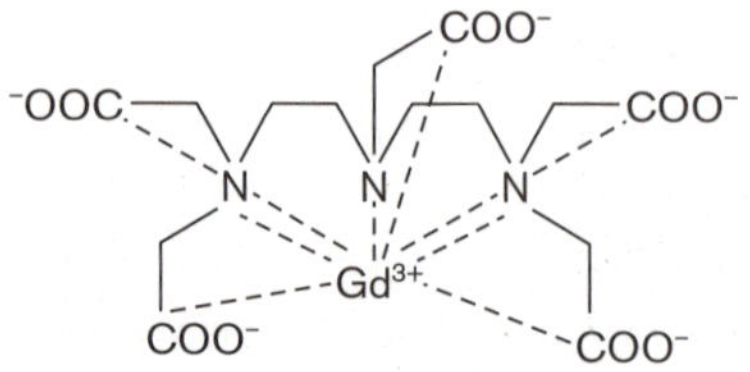

(b) $[Gd(dtpa)]^{2-}$
Magnevist®

Figure 7.7. (a) An NMR magnetic shift reagent, a lanthanide complex of dipivaloylmethane (dpm), and (b) an MRI contrast agent, the gadolinium (III) complex of the chelating ligand triethylenetetraminepentaacetic acid. [Adapted from J.-C. G. Bünzli et al., *J. Alloys Compd.* 303–4, 66 (2000).]

7.3. Electronic Absorption Spectra of *d*-Block and *f*-Block Complex Ions

OVERVIEW. The complexes of *d*-block metal ions absorb light in the near-infrared, visible, and ultraviolet portions of the electronic spectrum. If the metal ion has more than one *d* electron, electron–electron repulsion effects can produce more than one absorption band, but in octahedral complexes the lowest-energy band normally gives us the value of Δ_o. You should be able to use the customary and spectroscopic values of energy, and of light frequency and wavelength, in interpreting spectra and obtaining values of Δ_o. The visible light transmitted by the complex is complementary to that absorbed and gives the complex the color we see. With the aid of Table 7.3, from the color of light absorbed by a complex or from a drawing of its spectrum, we can obtain an approximate color of the complex. The values of Δ_o for hydrated ions characteristically increase from +2 to +3 to +4 ions, and from 3*d* to 4*d* and 5*d* metal ions. The intensities (molar absorptivities, ε) of these complexes are moderate, and are much lower than typically found in charge-transfer absorptions (Section 6.4B). You can practice these concepts by trying Exercises 7.19–7.24.

7.3A. Spectra of Complexes of d^1 Metal Ions. Complexes of *d*-block metals with partially filled *d* subshells absorb light in the visible (vis), ultraviolet (UV), and/or near-infrared (near-IR) regions of the electromagnetic spectrum (Figure 7.8). It is the portion of the absorption spectrum that occurs in the visible (middle) region that is responsible for the beautiful and varied colors such as those of the hydrated *d*- and *f*-block ions (Tables 1.3 and 1.4). Taking the first spectrum in this figure, that of the octahedral d^1 ion $[\mathrm{Ti(H_2O)_6}]^{3+}$ as an example, the one *d* electron occupies a t_{2g} orbital. If the ion captures a photon (particle of light) of the appropriate energy, that *d* electron can be promoted to the higher energy e_g orbital. The energy required to do this is just Δ_o. Hence, electronic, UV–vis, and near-IR spectroscopy is a convenient way of measuring the magnitude of the crystal field splitting (Δ_o) of a given complex.

We can express this energy in kilojoules per mole, as shown at the bottom of Figure 7.8, but spectroscopists customarily express the energy as the wavenumbers of the light absorbed, which has units of reciprocal centimeters (cm^{-1}) or kiloKaysers (kK). The relationships between the energy units are shown in Equation 7.4:

$$1 \text{ kK} = 1000 \text{ cm}^{-1} = 12.0 \text{ kJ mol}^{-1} \tag{7.4}$$

The instruments themselves are often calibrated in terms of the wavelength of the light, λ:

$$\text{Wavelength } \lambda \text{ in nm} = 10^7/\text{wavenumbers in cm}^{-1} \tag{7.5}$$

The absorption peak for $[\mathrm{Ti(H_2O)_6}]^{3+}$ occurs at 493 nm = 20,300 cm^{-1} = 243 kJ mol^{-1}, right in the middle of the visible region. Referring to the top of Figure 7.8, this means that blue and green light are absorbed out of the visible light; the other colors of light (particularly red and violet) pass through the solution of this ion and give it the red-violet color we observe.

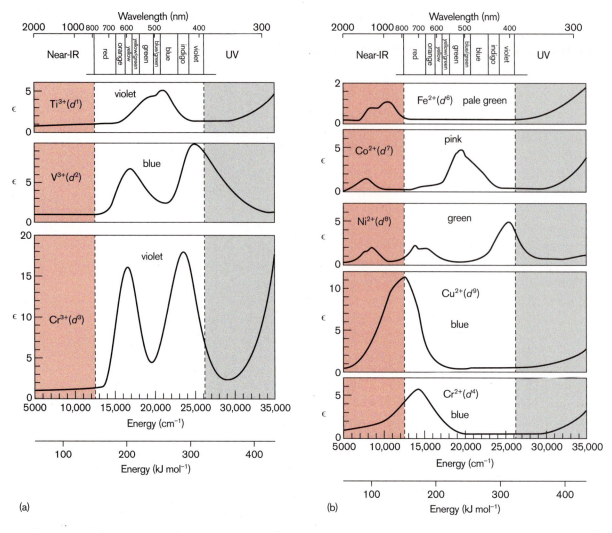

Figure 7.8. Electronic absorption spectra of some $3d^n$ hydrated metal ions in the ultraviolet, visible, and near-infrared regions. (a) Spectra of some +3-charged ions; (b) spectra of some +2-charged ions. [Adapted from B. N. Figgis and M. A. Hitchman, *Ligand Field Theory and Its Applications*, Wiley-VCH: New York, 2000, pp. 205–206.]

TABLE 7.3. The Colors of the Visible Spectrum and the Corresponding Wavelength and Wavenumber Ranges

Color of Light	Approximate Wavelength Ranges (nm)	Corresponding Wavenumbers (Approximate Values) (cm^{-1})	Color of Light *Transmitted* (i.e., Complementary Color of Absorbed Light)[a]
Red	700–620	14,300–16,100	Green
Orange	620–580	16,100–17,200	Blue
Yellow	580–560	17,200–17,900	Violet
Green	560–490	17,900–20,400	Red
Blue	490–430	20,400–23,250	Orange
Violet	430–380	23,250–26,300	Yellow

[a] These are the colors transmitted when (only one) color is absorbed.

7.3B. Spectra of Complexes of Metal Ions with More Than One *d* Electron. In the case of ions having more than one *d* electron in the t_{2g} orbitals (or more than one vacancy in the e_g orbitals), two or three absorption bands are present due to electron–electron repulsion effects (see the Amplification in Section 7.3C). Thus, the t_{2g}^2 V^{3+} ion shown in Figure 7.8 has an additional $d{\to}d$ absorption band in its spectrum. The *lowest-energy* absorption band (the one closest to the near-IR) observed in the typical octahedral complex occurs at the energy Δ_o. In the case of the hydrated V^{3+} ion, the absorption is at 18,600 cm^{-1}, which is its Δ_o. In the case of the hydrated Cr^{3+} ion, it is at about 17,000 cm^{-1}, and so on. These and other values for Δ_o are listed in Table 7.4.

If we add (measured or indirectly computed) crystal field splittings for hydrated metal ions from the fifth and sixth periods (Figure 7.9), we note a marked trend for crystal field splittings Δ_o to *increase* in the lower periods. As an important consequence, nearly all complexes (hydrated or otherwise) of fifth and sixth period metal ions are *strong-field* complexes.

TABLE 7.4. Crystal Field Splittings Δ_o of Hydrated Metal Ions of the Fourth Period

+2-Charged Ions	Δ_o (cm^{-1})	+3-Charged Ions	Δ_o (cm^{-1})
V^{2+}	12,840	Ti^{3+}	20,300
Cr^{2+}	14,000* (9000)	V^{3+}	18,600
Mn^{2+}	8500	Cr^{3+}	17,000
Fe^{2+}	10,000	Mn^{3+}	21,000
Co^{2+}	9200	Fe^{3+}	14,000
Ni^{2+}	8700	Co^{3+}	18,200
Cu^{2+}	12,000* (8500)		

Sources: B. N. Figgis and M. A. Hitchman, *Ligand Field Theory and Its Applications*, Wiley-VCH: New York, 2000, pp. 204–210; H. B. Gray, *Electrons and Chemical Bonding*, W. A. Benjamin, Inc.: New York, 1965, p. 196.

* Starred values fit the observed spectrum (Figure 7.8), which has a distorted geometry (Section 7.6B); the value in parentheses is estimated for an undistorted octahedral ion.

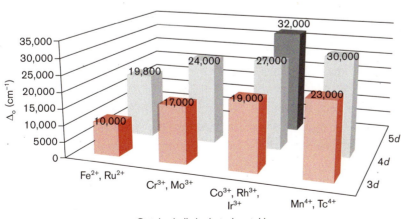

Figure 7.9. Crystal field splittings Δ_o of comparable octahedrally coordinated metal ions as a function of period. [Sources: B. N. Figgis and M. A. Hitchman, *Ligand Field Theory and Its Applications*, Wiley-VCH: New York, 2000, p. 219; J. E. Huheey, E. A. Keiter, and R. L. Keiter, *Inorganic Chemistry: Principles of Structure and Reactivity*, 4th ed., Harper-Collins: New York, 1993, p. 408.]

7.3C. Colors of Hydrated Ions. The colors of hydrated ions (or other species) are due, not to the color of light that is *absorbed*, but rather to the color(s) of light that are *transmitted* (pass through) the crystal or solution; the transmitted light shows up in *valleys* of absorption spectra such as those of Figure 7.8. We first consider spectra with a single range of transmitted light energy [such as $Ni^{2+}(aq)$]. The valley in its spectrum is centered at roughly 20,000 cm^{-1}, which (according to Table 7.4) corresponds to green light. This is the color of hydrated Ni^{2+} ion. Next we consider spectra with a single range of absorbed light [such as $Ti^{3+}(aq)$]. Its peak is at about 18,000–20,000 cm^{-1}, which by Table 7.4 is yellow-green. The light transmitted by the solution is the complementary color to yellow-green, which is red-violet (right side of Table 7.4).

AN AMPLIFICATION

Why Are There Additional Absorption Bands in Multielectron Ions?

After the photon is absorbed, the excited V^{3+} ion has an electron configuration of $t_{2g}^1 e_g^1$, but the two electrons could be in several pairs of orbitals. If the configuration is actually $d_{xy}^1 (d_{x^2-y^2})^1$, there will be a greater degree of repulsion between the two d electrons than if the configuration is actually $d_{xy}^1 (d_{z^2})^1$, and so on. The bottom line is that there will be more than one absorption band, because there will be different magnitudes of electron–electron repulsion energy for different combinations of orbitals involved. (When two or three absorption peaks have been recorded, it is possible to use all of the values to refine the value of Δ_o.)

The values of Δ_o in Table 7.4 correspond to a wide range of lowest-energy absorptions.

1. Those which are less than 14,300 cm^{-1} are located in the near-infrared part of the electromagnetic spectrum. These tend to come from hydrated ions of the +2-charged metal ions of the $3d$ metals. In theory these ions would be colorless, but often the absorption band has a shoulder in the visible spectrum (pale-green Fe^{2+} and blue Cu^{2+}; Figure 7.8), or else there are additional $d{\rightarrow}d$ absorbance bands in the visible spectrum (pink Cu^{2+} and green Ni^{2+}).

2. Δ_o values in the visible part of the spectrum tend to come from +3-charged ions of the $3d$ metals (Figure 7.8a).

3. Δ_o values in the ultraviolet range of the electromagnetic spectrum (>26,300 cm^{-1}) tend to come from hydrated +3 ions of $4d$ and $5d$ metals, or from +4 ions from any part of the d block. Most of these hydrated-ion absorption bands have shoulders reaching just into the violet region of the visible spectrum, so these ions tend to have the complementary color—namely, yellow.

7.3D. Intensity of Spectra. The intensities of absorption spectra are measured by their molar absorptivities (ε), as shown in Figure 7.8. If you have ever measured the UV spectra of organic compounds such as dyes, or of many analytically detected species, you will have been dealing with molar absorptivities on the order of perhaps 10,000 or 100,000. In theory, for a wave of light to be absorbed as it passes through a chemical species, there must be certain relationships between the orbital from which the electron originates and the orbital into which it is excited. The electronic transition is, in principle, forbidden if the two orbitals are on the same atom and are of the same type. Since both the t_{2g} and the e_g sets are of *d* orbitals, this "$d \to d$" transition is theoretically (but not completely in practice) forbidden. Hence, thousands of photons will pass through the complex ion unabsorbed before one is finally absorbed.

The problem is much more severe for the *f*-block ions in the sixth period. For example, these hydrated ions (Table 1.3) have molar absorptivities (ε) on the order of 0.01 to 0.1, and have very pale, "pastel" colors.

In contrast, charge-transfer spectra (Section 6.4B) involve electronic transitions between different atoms, so they feature much more intense absorptions, with typical ε values on the order of 2000 to 20,000. When soft-base ligands rather than water are present, charge-transfer absorptions may appear in the same spectrum with the $d \to d$ transitions. Often the charge-transfer absorptions are at higher energies, but not always. If the two types of transitions overlap, then the weaker $d \to d$ transitions are masked. As an example, the red-orange salts of the $[Fe(CN)_6]^{3-}$ ion (Section 5.3A) owe their color to a charge-transfer transition at about 24,000 cm^{-1} ($\varepsilon \approx 1000$); this band is so strong that the expected $d \to d$ transitions cannot be observed.

EXAMPLE 7.6

A hypothetical metal ion with one valence electron absorbs light with a wavelength of 640 nm; the absorption band has an ε of 8. (a) Is this metal ion a d^1 or an f^1 ion? Is this absorption a charge-transfer absorption? (b) What color of light is being absorbed? What color of light will the solution of the ion have? Will the color be intense, of moderate intensity, or weak (pastel)? (c) What is the energy of this absorption in kK? In cm^{-1}? In kJ mol^{-1}?

SOLUTION: (a) A molar absorptivity (ε) of 8 is characteristic of a $d \to d$ absorption. It is too high to be an $f \to f$ absorption and too low to be a charge-transfer absorption. (b) The color will be of moderate intensity, because of the intermediate value of ε. (c) By examination of Table 7.3, we can see that 640-nm light is orange light. This is the color of light absorbed, not the light that is transmitted by the ion, which is what gives the ion its color. From Table 7.3, we see that the transmitted color is the complementary color of orange, which is blue. The wavelength of 640 nm is equivalent to 15,600 cm^{-1}, by application of Equation 7.5. This is 15.6 kK and is equivalent to $15,600 \times 12/1000 = 188$ kJ mol^{-1}.

AN AMPLIFICATION

Spin-Forbidden Transitions

The hydrated Mn^{2+} ion has a pale pink color (Table 1.4) and very low ε values (0.01 to 0.03). It has the high-spin d^5 electron configuration, with each orbital having one electron aligned in the same direction. An electronic $d{\to}d$ transition is impossible unless the t_{2g} electron being excited simultaneously flips its spin so that it can pair with the electron already in the e_g orbital; this event is extremely improbable.

7.4. The Spectrochemical Series of Ligands and the Effects of Covalency

OVERVIEW. The spectrochemical series arranges ligands in order from those producing the smallest Δ_o to the largest Δ_o for a given metal ion. Given spectra or tabulated values of Δ_o for complexes of a given metal ion, you can arrange the ligands in a spectrochemical series. The series is largely the same for different metal ions. The spectrochemical series arranges ligands by the bonding properties of their donor atoms: π donors to metals before σ-only donors before π-acceptor ligands. You may practice applying these concepts by trying Exercises 7.25–7.34.

As Experiment 8 shows, the colors of the $3d^n$ hydrated ions (Table 1.4) are changed to other hues when the water molecules are replaced with other ligands. With Ni^{2+} in particular, practically every color of the rainbow can be produced! [Some of the colors result from geometries other than octahedral (Section 7.6).] If (as in Experiment 8) you arrange the complexes of Ni^{2+} in a "rainbow" order of colors as a function of the ligand added, then do the same for the corresponding complexes of Cu^{2+} and of Co^{2+}; you find that the "rainbow" orders of ligands are approximately the same for each metal. In Table 7.5 we add data for complexes of three other metal ions that were not practical to include in Experiment 8. You should find that the same "rainbow" ordering of ligands applies to the complexes of these metals. This ordering of ligands is known as the **spectrochemical series of ligands** (Eq. 7.6). For a given metal ion, this series arranges ligands from that giving the smallest value of Δ_o to the largest value. For some ligands in Equation 7.6, such as SCN^- and CO, the boldfaced letter indicates which atom bonds to the metal.

EXPERIMENT 8 MAY BE ASSIGNED NOW.
(See www.uscibooks.com/foundations.htm.)

WARNING: The author was hopelessly "hooked" on inorganic chemistry as a high school student when making multi-hued crystals of beautiful d-block complexes such as these!

$$I^- < Br^- < \mathbf{S}CN^- < F^- < OH^- < RCOO^- < C_2O_4^{2-} < \mathbf{O}NO^- < H_2O <$$
$$N\mathbf{C}S^- < EDTA^{4-} < CH_3CN < C_5H_5N < NH_3 < NH_2CH_2CH_2NH_2 <$$
$$\text{bipy} < \text{phen} < \mathbf{N}O_2^- \ll \mathbf{C}N^- < \mathbf{C}O \tag{7.6}$$

TABLE 7.5. Values of Δ_o for Some Octahedral Complexes of +3-Charged Ions[a]

Ion	Cl⁻	F⁻	H₂O	NH₃	en	CN⁻
Cr^{3+}	13,000	15,000	17,400	21,600	22,300	26,600
Co^{3+}		13,100	**18,200**	**22,900**	**23,200**	**34,800**
Rh^{3+}	**20,400**		**27,200**	**34,100**	**35,000**	

Sources: B. N. Figgis and M. A. Hitchman, *Ligand Field Theory and Its Applications,* Wiley-VCH: New York, 2000, p. 215; J. E. Huheey, E. A. Keiter, and R. L. Keiter, *Inorganic Chemistry: Principles of Structure and Reactivity,* 4th ed., Harper-Collins: New York, 1993, p. 406; H. B. Gray, *Electrons and Chemical Bonding,* W. A. Benjamin, Inc.: New York, 1965, p. 196.

[a] Values in **boldface** are for strong-field complexes. Units are cm⁻¹ for all values.

Somewhere along the spectrochemical series for most d^4–d^7 metal ions, Δ_o becomes large enough to exceed the pairing energy P for the metal ion in its complexes, and the complexes from there on will no longer be weak field and high spin, but will be strong field and low spin. As indicated in Table 7.5, the only weak-field complex of octahedral Co^{3+} is $[CoF_6]^{3-}$. As suggested by the data for Rh^{3+}, almost all complexes of fourth-and fifth-period d-block metal ions are strong-field.

EXAMPLE 7.7

The following table lists values of Δ_o taken from different sources for complexes consisting of the ligands shown across the top and the metal ions on the left.

Ion	F⁻	Cl⁻	Br⁻	H₂O	C₂O₄²⁻	NH₃	en
V^{3+}	16,000	13,000		17,850	18,000		
Rh^{3+}		20,300	19,000	27,200	26,000	34,000	35,000
Ir^{3+}		25,000	23,000			40,000	41,000
Fe^{3+}	14,000			13,700	14,100		

Can a common spectrochemical series be derived for these complexes? If so, what is it? Do these show expected vertical trends?

SOLUTION: Br⁻ < Cl⁻ in the Rh and Ir series. Cl⁻ < F⁻ in the V series. F⁻ < H₂O in the V series, but the reverse is true (barely) in the Fe series. H₂O is barely < $C_2O_4^{2-}$ in the Fe and V series, but the reverse is true in the Rh series and in Equation 7.6. NH₃ < en in the Rh and Ir series. Vertical trends are as expected: $3d$ (Co^{3+}; Table 7.5) $\ll$ $4d$ (Rh^{3+}) $\ll$ $5d$ (Ir^{3+}) among NH₃ and en complexes. (Note that some of these values have two significant figures and others have three, so differences between such close values may not be significant.)

Next, we look for *periodic trends* in the spectrochemical series of ligands. Given the crystal field theory's model of a ligand as a point source of charge, we might expect that the more negatively charged (or the more basic) a ligand is, the stronger the crystal field splitting it should produce (the higher it should fall in the spectrochemical series). This is not the case, however, because the two multiply charged ligands, $C_2O_4^{2-}$ and $EDTA^{4-}$, fall in the lower half of the series, and charged OH^- falls below uncharged H_2O.

Thinking back to Chapter 5, we might look to see if the softness of the donor atom (in boldface in Eq. 7.6) is the determining factor. Indeed, at the bottom of the spectrochemical series there are many soft donor atoms: I, Br, (Cl), and S. Next, we find ligands with hard donor atoms, F and O, followed with those with a borderline donor atom, N, followed by ligands with softer (C) donor atoms!

Instead, the spectrochemical series arranges ligands approximately by the group of the periodic table in which the donor atom is found:

$$\text{Group 17/VII donors} \leq \text{Group 16/VI donors} <$$
$$\text{Group 15/V donors} < \text{Group 14/IV donors} \qquad (7.7)$$

This is in part an order of electronic structure of the donor atom, because Group 17/VII and 16/VI donor atoms generally have two or more unshared electron pairs, while Group 15/V and 14/IV donor atoms generally have one unshared electron pair. The importance of having two or more unshared electron pairs on the donor atom is that it enables the ligand to act as a **π-donor ligand** and to attempt to form a π bond by donating the second unshared electron pair to the t_{2g} set of metal d orbitals (Figure 7.10a). However, if the metal ion already has an electron in that d orbital, the interaction must be one of repulsion. In this case, the repulsion raises the energies of the

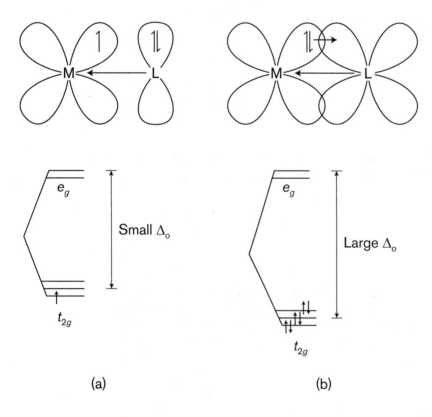

(a)   (b)

Figure 7.10. The π-type overlap of ligands with metal d orbitals in octahedral complexes and its effect on the magnitude of the crystal field splitting Δ_o. (a) The π-donor ligand has two electrons in a π-type orbital, but the metal has (at least) one electron in a t_{2g} orbital, resulting in repulsion that raises the energy of the t_{2g} orbitals relative to the e_g orbitals. (b) The π-acceptor ligand has an empty π-type orbital (postvalence d or antibonding π orbital), and the metal has (at least one, normally two) electrons in a t_{2g} orbital, resulting in π bonding that lowers the energy of the t_{2g} orbitals relative to the e_g orbitals.

TABLE 7.6. Examples of Important π-Acceptor Ligands (R = Organic groups)[a,b]

Acceptor Orbital Type: Antibonding π (Figure 3.8f)		Postvalence d (Section 3.3B)	
:N≡O⁺	Nitric oxide (nitrosyl cation)	:PF$_3$	Phosphorus trifluoride
:C≡S	Thiocarbonyl	P(OR)$_3$	Phosphites
:C≡O	Carbon monoxide (carbonyl)	:PAr$_3$	Triarylphosphines
:C≡Se	Selenocarbonyl	:PR$_3$	Trialkylphosphines
:C≡N–R	Isonitriles	:AsR$_3$	Trialkylarsines
:N≡N:	Dinitrogen	:SbR$_3$	Trialkylstibines

[a] The triple-bonded ligands on the left and the phosphorus compounds on the right are listed in order of decreasing π-acceptor strength (strongest at top). See R. C. Bush and R. J. Angelici, *Inorg. Chem.* 27, 681 (1988); R. C. Bush and R. J. Angelici, *Inorg. Chem.* 27, 681 (1988).

[b] The π-acceptor orbitals for the Group 15/V ligands on the right side of Table 7.6 are commonly identified as *postvalence* $3d$, $4d$, or $5d$ orbitals on the P, As, or Sb donor atoms (see the Amplification in Section 3.3B). The π-acceptor orbitals for the triple-bonded ligands on the left side are antibonding π orbitals (orbitals with negative overlap between adjacent p_x or p_y atomic orbitals, as in Figure 3.8f).

electrons in the t_{2g} orbitals, resulting in smaller crystal field splittings than are found in similar complexes in which either the metal has no t_{2g} electrons or the ligand has no π-donor electrons. Consequently, π-donor ligands usually fall at the low end of the spectrochemical series.

Nitrogen-donor ligands with only one unshared electron pair are normally just **σ-donor ligands**. We can also include :CH$_3^-$, which has spectrochemical properties very similar to those of :NH$_3$,[3] and :H⁻, which falls between bipyridyl and cyanide in the spectrochemical series.[4] Phosphorus- and carbon-donor ligands often have empty orbitals of π-type symmetry, such as the postvalence $3d$ orbitals on phosphorus. Such ligands (Table 7.6) act as π-acceptor ligands and form π bonds with the metal by accepting electrons from the π-symmetry t_{2g} set of metal d orbitals (Figure 7.10b). The formation of this additional π bond lowers the energy level of the t_{2g} set of electrons, increasing Δ_o (Figure 7.10b) so that the ligands moves up the spectrochemical series. Their large values of Δ_o result in the large majority of these complexes being low-spin (strong-field) complexes.

Some of the π-acceptor ligands (most notably carbon monoxide) have such strong fields that they not only render their complexes low spin, but they insure that they have the specific t_{2g}^6 configuration that maximizes the π-accepting interaction. These include some carbonyl anions and cations—[Ti(CO)$_6$]$^{2-}$, [V(CO)$_6$]⁻, [Cr(CO)$_6$], and [Mn(CO)$_6$]⁺—and similar compounds from the fifth and sixth periods (Section 11.5D). Other ligands have a lesser effect, and although still generally giving a low-spin complex, they will tolerate a greater variety of t_{2g}^n configurations. Cyanide is such a ligand: d-Block metals from Ti through Co form low-spin hexacyanometallates [M(CN)$_6$]$^{n-}$ with up to six t_{2g} electrons.

Magnetic and Spectral Properties of Prussian Blue and Other Hexacyanometallates

The low-spin hexacyanometallate ions $[M(CN)_6]^{n-}$ with t_{2g}^6 metal ions, such as $M = Fe^{2+}$, Ru^{2+}, and Co^{3+}, are diamagnetic. They form a number of Prussian-blue-type salts with other d-block metal ions, in which the magnetic moments are due to the other d-block metal ions. These are of interest in part because of their intense metal-ion–metal-ion charge-transfer absorptions. A classic example of such a charge-transfer transition is the intense 680-nm transition in Prussian blue ($KFeFeCN_6$; Section 5.3A and Figure 5.18). In the structure shown in Figure 5.18, the softer Fe^{2+} ions are coordinated by the softer, stronger-field carbon end of the bridging cyanide ion, while the harder Fe^{3+} ions are surrounded by the harder, weaker-field nitrogen ends of the cyanides. A light wave can easily transfer one $3d$ electron from the Fe^{2+} ion to the Fe^{3+} ion, giving an excited state in which the iron at the center of the carbon donor atoms is now Fe^{3+}, while the iron at the center of the nitrogen donor atoms is now Fe^{2+}. The electron is transferred fairly readily through the π system of the bridging cyanide ions, allowing the transfer to occur using visible light.

The intense colors and magnetic susceptibilities of a number of other d-block hexacyanometallates are listed in Table 7.7.

TABLE 7.7. Colors and Magnetic Susceptibility of Some Hexacyanometallates

Chemical Formula (water not included)	Color	μ_{BM}
$(Fe^{3+})_4[Fe^{II}(CN)_6]_3$	Prussian blue	5.7
$(Fe^{3+})_4[Ru^{II}(CN)_6]_3$	Prussian blue	6.0
$(Cr^{3+})_4[Fe^{II}(CN)_6]_3$	Dark green	3.9
$(Cu^{2+})_2[Fe^{II}(CN)_6]$	Brown	2.1
$(Co^{2+})_2[Fe^{II}(CN)_6]$	Deep blue-green	4.6
$(Mn^{2+})_2[Fe^{II}(CN)_6]$	Pale blue	6.0
$(Fe^{2+})_3[Co^{III}(CN)_6]_2$	Green-yellow	5.7
$(Co^{2+})_3[Fe^{III}(CN)_6]_2$	Deep red-brown	7.4
$(Fe^{2+})_3[Cr^{III}(CN)_6]_2$	Brick red	
$(Fe^{2+})_3[Mn^{III}(CN)_6]_2$	Blue	

Sources: P. G. Rasmussen and E. A. Meyers, *Polyhedron* 3, 183 (1984); E. Reguera, J. F. Bertrán, and L. Nuñez, *Polyhedron* 13, 1619 (1994).

The first seven examples listed all contain diamagnetic d^6 hexacyanometallate complex ions, so their magnetic susceptibilities should match the μ_{obs} values given in Table 7.2 for the weak-field, high-spin cation of the complex (which is coordinated by the nitrogen atoms of the bridging cyanides). The eighth example has a low-spin but still paramagnetic hexacyanometallate ion and a paramagnetic cation, so its magnetic properties are an additive effect from both ions.[5] The final two complexes undergo ready isomerism in which the cyanide ligands apparently turn around to coordinate the metal ions with exchanged nitrogen and carbon donor atoms, giving linkage isomers from the ambidentate cyanide ion.[6] In Section 8.4 we will find cases in which the unpaired electrons from the two metal ions interact strongly with each other, giving rise to the magnetic properties of ferromagnetism, antiferromagnetism, and superconductivity.

The Necessity of Considering Covalent Bonding in *d*-Block Complexes

We see a major conceptual defect of the crystal field theory: The important spectroscopic series of ligands depends on the covalent-bonding abilities of the ligands, not their point charges! **Ligand field theory**[7] modifies the crystal field theory to take the σ- and π-covalent bonding of metals and ligands into account.

The first effect of covalency is that pairing energies are likely to be 15–30% lower in complex ions than in the free ions (Table 7.1). Covalency allows the metal-ion d electrons to spend part of their time on the ligand, thereby reducing the mutual repulsion of the d electrons. This is known as the **nephelauxetic** (cloud-expanding) **effect,** and becomes more important to the right in the nephelauxetic series of ligands:

$$F^- < H_2O < NH_3 < NH_2CH_2CH_2NH_2 < Cl^- < CN^- < Br^- < I^-$$

For theoretical predictions of bonding and energetics in d-block complexes the molecular orbital theory is employed (Section 11.5). It is often desired to interpret the results in terms of bonding interactions of specific ligands and metal orbitals, as in crystal field theory and ligand field theory. For this purpose the **angular overlap** method is often employed.[8] This analyzes the ligand field effect of each ligand in terms of e_σ (its sigma-bonding interaction with the metal) and e_π (its pi-bonding interaction). In an octahedral complex the sigma-bonding interaction is with the e_g metal orbitals, raising them in energy, while the pi-bonding interaction is with the t_{2g} orbitals, lowering them in energy. In a simple case such as $M(halide)_6{}^{n-}$, Δ_o is then equal to $3e_\sigma - 4e_\pi$.

7.5. Thermodynamic and Structural Consequences of Crystal Field Effects

OVERVIEW. The hydration energies (Section 2.3A) of d-block metal ions are enhanced by up to about 10% by CFSEs. Because the CFSEs are greatest for high-spin d^3 and d^8 ions, plots of hydration energies versus group numbers show two "humps" at these electron configurations. Lattice energies are similarly enhanced by CFSEs, as are stability constants for forming complex ions. You may practice these concepts of Section 7.5A by trying Exercises 7.35–7.40. Radii of d-block metal ions decrease especially rapidly when nonrepelling t_{2g} orbitals are being filled, but increase when ligand-repelling e_g orbitals are being filled, giving two humps in high-spin complexes of Period 4 but only one hump for low-spin complexes of Periods 4, 5, and 6. Similar patterns of "humps" are found in the plots of Pauling electronegativities across the d block. You may practice these concepts of Section 7.5B by trying Exercises 7.41–7.44.

The properties of d-block complex ions are not as different from those found in the rest of the periodic table as we often suppose.[9] These properties are still mainly determined by the same fundamental atomic properties such as size, electronegativity, and charge or oxidation state. In Chapter 1, we noted the overall periodic trends and certain anomalies in the horizontal periodic trends in these atomic and ionic properties. Now, we will see that many of the anomalies across the d block are related to the presence of crystal field stabilization of certain d^n electron configurations.

7.5A Thermodynamic Effects. Hydration Energies. The CFSEs computed in Section 7.1A add to the stability of complex ions in a number of ways and in characteristic periodic patterns. Let us begin by examining the hydration energies of the +2 ions of the $3d$ set of metals, beginning with the $3d^0$ ion Ca^{2+} (hydration energy = –1592 kJ mol^{-1}; Table 2.3), and ending with the $3d^{10}$ ion Zn^{2+} (hydration energy = –2044 kJ mol^{-1}). If $3d$ metal ions were our hypothetical spherical ions, we would expect each successive ion between Ca^{2+} and Zn^{2+} to become smaller and therefore to have a greater (more negative) hydration energy (Section 2.2A). In fact, a plot of hydration energies across the $3d^n$ metal ions of +2 charge results in a characteristic *double-humped* curve (the solid line in Figure 7.11a).

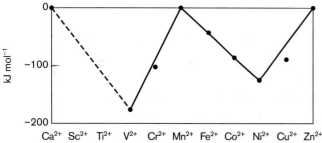

Figure 7.11. (a) Hydration energies of the +2-charged $3d^n$ ions (solid circles); hydration energies with the CFSEs of these ions subtracted (open **brown** circles). (b) CFSEs of the +2-charged ions in (a).

The two humps in such a curve are due to addition of the contribution of the weak-field, high-spin crystal CFSEs to the hydration energies expected for hypothetical spherical metal ions of steadily decreasing radii. The Δ_o values in the hydrated ions (Table 7.4) can be used to compute the CFSEs in kJ mol^{-1}; these values are plotted in Figure 7.11b. When these CFSEs are subtracted from the measured lattice energies, a reasonably straight line remains (the dashed line in Figure 7.11a), which represents the hydration energies of the hypothetical spherical hydrated ions. The CFSEs are of the order of 5–10% of the hydration energies. This reminds us that the main sources of hydration energies (and of the other energies discussed in this section) are those discussed in Chapters 2 and 3 (e.g., small size and high charge).

Lattice Energies. Similarly, CFSEs contribute to the stability of solid compounds of the d-block elements, as measured by the lattice energies of these compounds. As seen in Figure 7.12, plotting lattice energies of the $3d^n$-metal dihalides versus group numbers also produces double-humped curves. CFSEs are also responsible for about 5–10% of the lattice energies of these compounds.

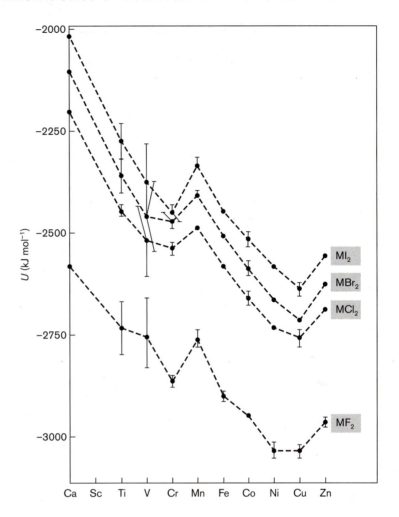

Figure 7.12. Lattice energies of the $3d^n$ halides MX$_2$ versus group number. Vertical lines indicate uncertainties in experimental values. [Adapted from J. E. Huheey, E. A. Keiter, and R. L. Keiter, *Inorganic Chemistry: Principles of Structure and Reactivity*, 4th ed., Harper-Collins: New York, 1993, p. 409.]

Stability Constants. The CFSEs resulting from the nonspherical nature of the d-block metal ions also affect their relative tendencies to form complexes with a ligand such as NH$_3$ or NH$_2$CH$_2$CH$_2$NH$_2$:

$$[M(H_2O)_6]^{2+} + NH_3 \rightleftharpoons [M(H_2O)_5(NH_3)]^{2+} + H_2O \qquad (7.8)$$

$$[M(H_2O)_6]^{2+} + NH_2CH_2CH_2NH_2 \rightleftharpoons$$
$$[M(H_2O)_4(NH_2CH_2CH_2NH_2)]^{2+} + 2H_2O \qquad (7.9)$$

These reactions represent the first steps in forming the fully substituted nitrogen–ligand complexes. Their first stepwise stability constants K_1 should also give a double-humped curve (Figure 7.13). For any weak-field ligand, the stability constants follow

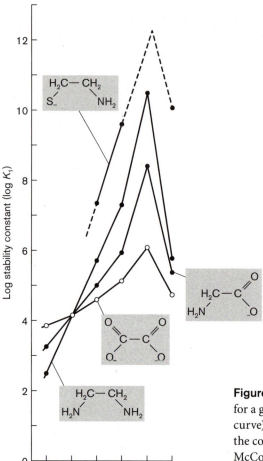

Figure 7.13. Logarithms of stability constants for a given bidentate ligand (shown next to each curve) versus the +2-charged metal ion forming the complex. [Adapted from H. Sigen and D. B. McCormick, *Acc. Chem. Res.* 3, 201 (1970).]

the order $Mn^{2+} < Fe^{2+} < Co^{2+} < Ni^{2+} < Cu^{2+} < Zn^{2+}$. This order is sometimes called the **Irving–Williams series**, and it is often used in discussing metalloenzyme stabilities (e.g., in bioinorganic chemistry). As illustrated in Example 7.8, the order in this series can be explained except for the peak position of Cu^{2+}, which is due to a tendency to form non-octahedral complexes (Section 7.6B).

EXAMPLE 7.8

Consider the five +3-charged metal ions from Co^{3+} through Ga^{3+}, including the hypothetical Zn^{3+}. List these in periodic-table order and insert the symbols "<" or ">" to indicate the expected order in the (absolute values of the) properties: (a) hydration energies; (b) lattice energies of trifluorides; (c) stabilities of weak-field complexes.

SOLUTION: These metal ions are d^6 through d^{10}, respectively. For weak-field complexes ($\Delta_o < P$) we expect the CFSEs of these ions to be $-0.4\Delta_o$, $-0.8\Delta_o$, $-1.2\Delta_o$, $-0.6\Delta_o$, and 0, respectively. Thus, the absolute values should peak at d^8. In practice, the Irving–Williams series shows that they peak at d^9; the reasons are covered in Section 7.6B. Hydrated ions and fluoride complexes are expected to be weak-field, so that the orders for (a), (b), and (c) should be $Co^{3+} < Ni^{3+} < Cu^{3+} < Zn^{3+}$ (according to Irving–Williams series) $< Ga^{3+}$.

7.5B. Effects on Fundamental Properties of *d*-Block Atoms and Ions. Radii.

In Figure 7.14, we show the horizontal periodic variation in the radii of octahedrally coordinated $3d^n$ ions of +2 charge (Figure 7.14a) and of +3 charge (Figure 7.14b). At first, the ions decrease in size to the right in the period, which is the normal trend. The $3d^4$ through $3d^7$ ions can take either high-spin or low-spin electron configurations; the radii for the high-spin ions are from 7–17 pm larger than the radii of the low-spin ions. Electrons in e_g orbitals repel the ligands, resulting in *an increase in radius each time an electron is added to an e_g orbital*. Since the high-spin ions occupy the e_g set of orbitals first at d^4, this is the point at which radii first begin to increase among the high-spin ions (open circles in Figure 7.14); the familiar double-humped curve results. The radii of low-spin ions (closed circles) decrease across the period until the e_g orbitals are first occupied at the d^7 electron configuration; the "anomalous" increases in radius continue until the e_g orbitals are filled at Zn^{2+}. Consequently, plots of periodic properties of low-spin complex ions should show a single-humped curve, with the extreme anomaly at d^6.

When we take into account these true radii of the 3*d* metal ions, the humps tend to disappear. For example, when hydration energies of these ions are plotted as a function of the *high-spin* radii (Figure 7.15) of the ions, no humps remain, and the 3*d* ions fit the trend represented by the Latimer equation (Eq. 2.15).

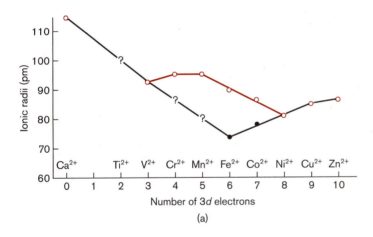

(a)

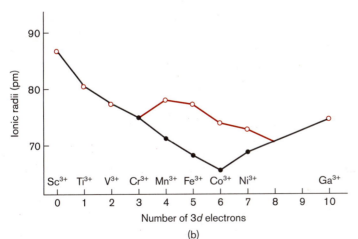

(b)

Figure 7.14. Radii of (a) the +2 ions and (b) the +3 ions as a function of the $3d^n$ electron configuration. From d^4 through d^7, open **brown** circles represent high-spin (weak-field) ions; closed circles represent low-spin (strong-field) ions. [Adapted from R. D. Shannon and C. T. Prewitt, *Acta Crystallogr., Sect. B* 26, 1076 (1970).]

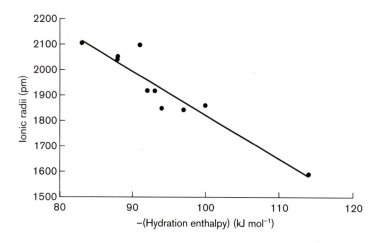

Figure 7.15. Negative of hydration enthalpies (kJ mol^{-1}) of Period 4 +2-charged ions from Ca^{2+} to Zn^{2+} versus their high-spin octahedral ionic radii. Leftmost point = Ni^{2+}; rightmost point = Ca^{2+}. (Data from Tables 2.3 and 7.8.)

Pauling Electronegativities. Pauling electronegativities (χ_P) in the d block are obtained[10] from the enthalpies of formation ΔH_f of salts MX of the metal (especially oxides and halides) in the +2 oxidation states, because bond energies (Section 3.3D) are not generally available in the d block:

$$\Delta H_f = -96.5n[\chi_P(M) - \chi_P(X)]^2 \tag{7.10}$$

These halides and oxides have lattices in which the metal ions generally have octahedral coordination and are surrounded by weak-field halide and oxide ligands. Their ΔH_f values include contributions from the CFSEs that peak at d^3 and d^8 [and also from Jahn–Teller distortions (Section 7.6B), which peak at d^4 and d^9].

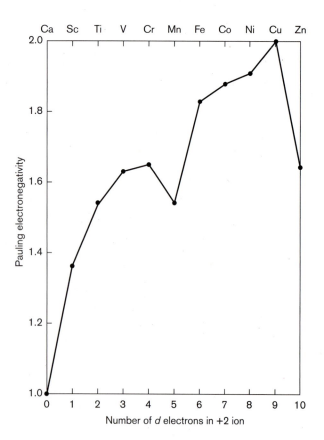

Pauling electronegativities tend to decrease when ionic radii increase (Section 1.6A). A plot of the Pauling electronegativities of the $3d$ metals is an (inverted) double-humped plot (Figure 7.16).

It is likely that a different series of Pauling electronegativities would be more appropriate among strong-field octahedral complexes, such as are predominant in Period 5. Because the CFSE is at a maximum at the d^6 electron configuration, we generally find the Period 5 χ_P values peaking at the d^6 Rh^{2+} ion.

In summary, one of the advantages of using appropriate ionic radii and Pauling electronegativities is that they build in some of the features of crystal field effects.

Figure 7.16. Pauling electronegativities χ_P of the $3d$ metals as a function of the $3d^n$ electron configurations. The electronegativities are computed for the +2-charged ions using Equation 7.10. The exception is for Sc ($3d^1$), which is for the +3-charged ion.

EXAMPLE 7.9

In the graph provided, four curves have been drawn connecting the fundamental atomic properties of d-block elements: Curve 1 connects the ionic radii from Table C for Period 4 +2-charged ions. Curve 2 connects the ionic radii for Period 5 +4-charged ions. Curve 3 connects the Pauling electronegativities from Table A for the Period 4 elements, and curve 4 connects the Pauling electronegativities for the Period 5 elements.

Tell how and why the brown curves 1 and 3 for Period 4 differ from the black curves 2 and 4 for Period 5 in the central region of the d block.

SOLUTION: The Period 4 curves show weak-field characteristics. They are double-humped (humps at about d^3 and d^8), showing normal periodic

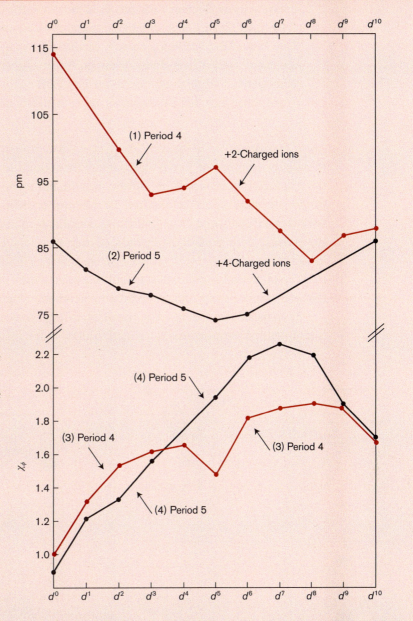

trends up to these electron configurations. After these electron configurations, electrons in weak-field complexes are entering repulsive e_g orbitals, causing anomalous *increases in radius* (curve 1) and *decreases in electronegativity* (curve 3). The Period 5 curves show strong-field characteristics (as expected for Period 5). Normal periodic trends apply through the d^6 electron configuration, at which point the single hump occurs. After that, repulsive e_g orbitals are being filled, causing anomalous *increases in radius* (curve 2) and *decreases in electronegativity* (curve 4).

7.6. Crystal Field Splitting in Complexes of Other Geometries

OVERVIEW. The CFSE in tetrahedral complexes is only about 4/9 that in octahedral complexes, but spectral bands are more intense. You may practice applying these Section 7.6A concepts to tetrahedral complexes in Exercises 7.45–7.48. CFSEs in other geometries such as square planar can be evaluated from Table 7.8. Octahedral complexes with e_g^1 and e_g^3 electron configurations are subject to the Jahn–Teller distortion. You may practice crystal field effects in other geometries as discussed in Section 7.6B by trying Exercises 7.49–7.60. This section includes Connections to biochemistry and to materials science.

7.6A. Tetrahedral and Cubic Complexes.

Because octahedral complexes have the maximum number of Lewis base–acid interactions feasible in the fourth period, the octahedral geometry is the most common, but by no means all complexes in this or the later periods are octahedral. Tetrahedral coordination results when a fourth-period metal ion adopts its penultimate coordination number, 4, as a consequence of the unusual steric bulk of the ligands, or their possession of good π-donor abilities (as in oxo anions; Section 3.5B). To see how the energies of the d orbitals split in a tetrahedral crystal field, it is useful first to look at a related but far more uncommon geometry, that of an eight-ligand cubic coordination geometry (Figure 7.17).

As suggested by Figure 7.17a, the outstanding characteristic of cubic coordination is that *none* of the d orbitals is pointed directly at the ligands. Consequently, there are

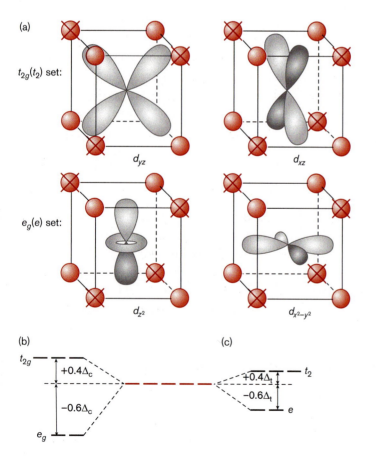

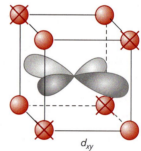

Figure 7.17. (a) Interaction of the d orbitals with a cubic set of ligands, or, (when the ligands indicated by Xs are omitted), a tetrahedral set of ligands. (b) Energy-level diagram for the cubic set of ligands. (c) Energy-level diagram for the tetrahedral set of ligands. [Adapted from I. S. Butler and J. F. Harrod, *Inorganic Chemistry: Principles and Applications*, Benjamin/Cummings: Redwood City, CA, 1989, p. 394.]

no strong repulsions between the *d* orbitals and the ligands. The ligands still fall into the same two sets as in octahedral geometry—t_{2g} and e_g—but now it is the t_{2g} set that experiences the greater repulsion. The reason is that the lobes of this set of orbitals fall in the edge centers, closer to the ligands than the lobes of the e_g set, which fall in the face centers. Consequently, the energy-level diagram (Figure 7.17b) is reversed from that for octahedral complexes. The total crystal field splitting, Δ_c (c for cubic), can be shown to be approximately eight-ninths of that present in an octahedral complex, Δ_o.

The more important case of tetrahedral complexation is simply generated by deleting the one-half of the cubic set of ligands crossed over with an X in Figure 7.17a. The pattern of interactions (Figure 7.17c) remains the same, but the total splitting is now halved. The total splitting Δ_t (t for tetrahedral) is now only four-ninths of Δ_o. Consequently, Δ_t is never greater than the pairing energy (P) and all tetrahedral complexes are weak-field (high spin).

Spectra of Tetrahedral Complexes. In a geometry such as tetrahedral, where the total crystal field splitting Δ_t is small, the frequency of the first absorption band will fall at a low energy—Δ_t ranges from 3200 to 5000 cm^{-1} in tetrahedral MCl_4^{2-}.[11]

EXAMPLE 7.10

Estimate the position of the lowest-energy absorption bands and the colors of (a) the hypothetical $Cu(H_2O)_4^{2+}$ cation and (b) the real $CuCl_4^{2-}$ anion, both of which are tetrahedral ions.

SOLUTION: (a) We expect the crystal field splitting and the energy of the lowest-energy absorption band of the tetrahedral hydrated Cu^{2+} ion to be approximately four-ninths of that of the octahedral hydrated ion, 13,000 cm^{-1} (Table 7.4). $\Delta_t = 4/9\Delta_o = (0.445)(13,000) = 5800$ cm^{-1}. From Figure 7.8 we see that this band would be in the near-IR, so the ion would be colorless (in the absence of charge-transfer bands). (b) Chloride is below water in the spectrochemical series, so its absorption band should fall below 5800 cm^{-1}. In fact, this absorption band occurs at 4500 cm^{-1}, well into the near-IR region, so this complex ion would be predicted to be colorless. In fact, tetrahedral $[CuCl_4]^{2-}$ is green due to a charge-transfer absorption.

For reasons related to symmetry, the intensities of tetrahedral absorption bands exceed the intensities of corresponding octahedral absorption bands (and we remove the subscripted "*g*" from the symmetry labels for the orbitals). Often a change in color intensity signals a change in coordination geometry. Indicating Drierite^{TM} is mainly anhydrous $CaSO_4$ impregnated with anhydrous $CoCl_2$. Once the Drierite^{TM} has absorbed its capacity of water from the solvent, the following changes in geometry, color, and color intensity occur about the Co^{2+} ion:

$$[CoCl_4]^{2-} \text{ (tetrahedral)} + 6H_2O \rightleftharpoons [Co(H_2O)_6]^{2+} \text{ (octahedral)} + 4Cl^- \quad (7.11)$$

$$\text{Bright blue } (\varepsilon \approx 60) \qquad\qquad\qquad \text{Pink } (\varepsilon \approx 5)$$

The blue color is not due to the lowest-energy absorption of $CoCl_4^{2-}$, which is well into the near-IR spectral region. The color is mainly due to an absorption at 15,000 cm^{-1} ($\varepsilon \approx 600$), due to one of the other absorption bands that result from electron–electron repulsion (Section 7.3).

7.6B. Other Geometries. Square planar and other geometries also have importance in d-block coordination chemistry. The crystal field splitting patterns for these can also be worked out and are tabulated in Table 7.8 in common units of Δ_o, the crystal field splitting in an octahedral field.

TABLE 7.8. Energy Levels of d Orbitals in Complexes of Various Geometries[a]

CN[b]	Geometry	d_{z^2}	$d_{x^2-y^2}$	d_{xy}	d_{xz}	d_{yz}	Largest Splitting
2	Linear[c]	1.028	−0.628	−0.628	0.114	0.114	0.914
3	Trigonal[d]	−0.321	0.546	0.546	−0.386	−0.386	0.867
4	Tetrahedral	−0.267	−0.267	0.178	0.178	0.178	0.445
4	Square planar[d]	−0.428	1.228	0.228	−0.514	−0.514	1.000
5	Trigonal bipyramid[e]	0.707	−0.082	−0.082	−0.272	−0.272	0.789
5	Square pyramid[e]	0.086	0.914	−0.086	−0.457	−0.457	0.828
6	Octahedron	0.600	0.600	−0.400	−0.400	−0.400	1.000
6	Trigonal prism	0.096	−0.584	−0.584	0.536	0.536	0.680
7	Pentagonal bipyramid[e]	0.493	0.282	0.282	−0.528	−0.528	0.810
8	Cube	−0.534	−0.534	0.356	0.356	0.356	0.890
8	Square antiprism	−0.534	−0.089	−0.089	0.356	0.356	0.445
9	Tricapped trigonal prism	−0.225	−0.038	−0.038	0.151	0.151	0.189
12	Icosahedron	0.000	0.000	0.000	0.000	0.000	0.000

Source: Adapted from J. E. Huheey, E. A. Keiter, and R. L. Keiter, *Inorganic Chemistry: Principles of Structure and Reactivity,* 4th ed., Harper-Collins: New York, 1993, p. 405.

[a] Units of Δ_o, the octahedral crystal field splitting, assuming the same overall charge density and distance. Geometries are illustrated in Figures 5.2 and 5.3.

[b] Coordination number = CN.

[c] Ligands lie along z axis.

[d] Ligands lie in xy plane.

[e] Pyramid base in xy plane.

EXAMPLE 7.11

Calculate the CFSE of the d^1 through d^5 ions in the (a) tetrahedral and (b) square pyramidal geometries. In the tetrahedral geometry, express your answers both in terms of Δ_t and Δ_o; in the square pyramidal geometry, express your answer in terms of Δ_o.

SOLUTION: (a) The energies in terms of Δ_t can be derived from the energy-level diagram in Figure 7.17c. They can then be expressed in terms of Δ_o by multiplying by four-ninths—or they can be obtained from Table 7.8. The electron configurations and energies are as follows:

Free-ion configuration	d^1	d^2	d^3	d^4	d^5
Tetrahedral configuration	e^1	e^2	$e^2 t_2^{\,1}$	$e^2 t_2^{\,2}$	$e^2 t_2^{\,3}$
CFSE (units of Δ_t)	−0.6	−1.2	−0.8	−0.4	0
CFSE (units of Δ_o)	−0.267	−0.533	−0.356	−0.178	0

(b) Table 7.8 gives the energies for the square pyramidal geometry in terms of Δ_o. The electron configurations are built up by filling the lowest orbital first; there is a fairly large energy gap between $d_{x^2-y^2}$ and d_{z^2}, which could perhaps result in spin pairing (a low-spin complex). Because this gap is less than Δ_o, let us assume that pairing does not result and that we have a high-spin complex after this gap is encountered. The electron configurations and energies are as follows:

Free-ion configuration	d^1	d^2	d^3	d^4	d^5
Square pyramidal configuration	$d_{xz,yz}^{\,1}$	$d_{xz,yz}^{\,2}$	$d_{xz,yz}^{\,2}\, d_{xy}^{\,1}$	$d_{xz,yz}^{\,2}\, d_{xy}^{\,1}\, (d_{z^2})^1$	$d_{xz,yz}^{\,2}\, d_{xy}^{\,1}\, (d_{z^2})^1\, (d_{x^2-y^2})^1$
CFSE	−0.457	−0.914	−1.000	−0.914	0

Radii in Other Geometries. In an Amplification in Section 4.1B we mentioned that the radius of a cation is *not* an invariant property, but depends on its coordination number. As fewer ligands are placed around a cation, their electrons repel each other less, and the measured cationic radius decreases. As seen in Figure 7.14 and tabulated in Table 7.9, the radii of d-block cations also depend on whether they are high spin or low spin.

TABLE 7.9. Radii of Selected *d*-Block Metal Ions as a Function of Geometry and Spin

Ion	Electron Configuration	CN = 4 Square Planar	CN = 4 Tetrahedral	CN = 6 (Octahedral) Low Spin		High Spin
+2-Charged Ions of Period 4						
Cr^{2+}	d^4			87		94
Mn^{2+}	d^5		80	81		97
Fe^{2+}	d^6	78	77	75		92
Co^{2+}	d^7		72	79		89
Ni^{2+}	d^8	63	69		83	
Cu^{2+}	d^9	71	71		87	
Zn^{2+}	d^{10}		74		88	
+3-Charged Ions of Period 4						
Mn^{3+}	d^4			72		79
Fe^{3+}	d^5		63	69		79
Co^{3+}	d^6			69		75
Ni^{3+}	d^7			70		74
+4-Charged Ions of Period 4						
Ti^{4+}	d^0		56		75	
Cr^{4+}	d^2		55		69	
Mn^{4+}	d^3		53		67	
Co^{4+}	d^5		54			67
Selected Ions of Periods 5 and 6						
Pd^{2+}	d^8	78			100	
Pt^{2+}	d^8	74			94	
Ag^{3+}	d^8	81			89	
Au^{3+}	d^8	82			99	
Ag^{2+}	d^9	93			108	
Cd^{2+}	d^{10}		92		109	
Hg^{2+}	d^{10}		110		116	
Ag^+	d^{10}	116	114		129	

Source: Data from R. D. Shannon, *Acta Crystallogr. Sect. A* 32, 751 (1976).

A CONNECTION TO BIOCHEMISTRY

Spectra of Complexes in Other Geometries

The electronic absorption spectrum of a *d*-block (or *f*-block) ion is very sensitive to and carries important information about the geometry of the coordination about that ion. The useful spectroscopic properties of *d*-block (and to some extent of *f*-block) elements are often exploited in biochemistry. Many metalloenzymes contain colorless metal ions such as Zn^{2+} and Ca^{2+} that have no accessible electronic absorption spectra, or any other useful properties such as those being appropriate for nuclear magnetic resonance (NMR) studies. If these metal ions can be replaced by very similar ions from the *d* or *f* blocks that are colored or are suitable for NMR or other studies, the metalloenzyme will hopefully retain its structure and perhaps even its activity. Measurement of the spectrum of the reporter metal ion will then tell us about the coordination geometry about the metal ion.

This approach worked very successfully in the study of the zinc-containing enzyme carboxypeptidase A (Section 5.4C; Figures 5.24 and 5.25). When Zn^{2+} is removed and replaced by Co^{2+} (with identical radius and charge and not too different an electronegativity), the enzyme not only retains its activity, but the activity is enhanced! The spectrum of Co^{2+}-substituted carboxypeptidase A was recorded[12] (peaks at 18,800 cm^{-1}, $\varepsilon = 210$; 17,500 cm^{-1}, $\varepsilon = 190$; 10,600 cm^{-1}, $\varepsilon = 20$; 6400 cm^{-1}, $\varepsilon = 17$) and compared with the spectra of model Co^{2+} complexes of octahedral, tetrahedral, and trigonal bipyramidal geometry, shown in Figure 7.18. The spectrum of the metalloenzyme was not like that of the octahedral model compound, but bore resemblance both to that of the tetrahedral and the trigonal bipyramidal complexes. When the crystal structure of the zinc-containing enzyme was determined (Figures 5.24 and 5.25), it was found that the coordination at Zn^{2+} is distorted from tetrahedral: Three bond angles at Zn are roughly near 90° and three are roughly near 120°, so the structure resembles a trigonal bipyramidal structure with one ligand missing.

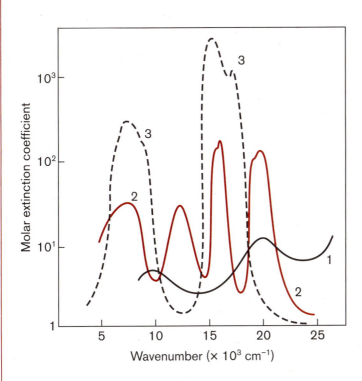

Figure 7.18. Electronic absorption spectra of model complexes of Co^{2+}. 1. (——) Octahedral Co^{2+} in $[Co(NH_2CH_2CH_2NH_2)_3]^{2+}$; 2. (in **brown**) trigonal bipyramidal Co^{2+} in $[CoBr(hexamethyl-diethylenetriamine)]^+$; and 3. (----) tetrahedral Co^{2+} in $[Co(NCS)_4]^{2-}$. [Adapted from E.-I. Ochiai, *Bioinorganic Chemistry: An Introduction*, Allyn and Bacon: Boston, 1977, p. 374.]

An appropriate substitution for Ca^{2+} in enzymes is more difficult to find, because all d-block +2-charged ions are too small, and all f-block +2-charged ions are very sensitive to air. Often it has proved possible to use f-block +3-charged ions, which are only slightly more acidic than Ca^{2+} and have useful colors (Table 1.4). Crystal field splittings in f-block ions are very small, but they have proved sufficient for this substitution to be useful for probing active-site geometry. (Usually other optical properties such as fluorescence are monitored rather than absorption.)

Square Planar Complexes. This geometry is seldom found in complexes and molecules of the p block, but it is found frequently for d-block complexes, even though VSEPR theory would predict another geometry. This geometry is especially favorable for d^8 complex ions with strong-field ligands and/or $4d$ and $5d$ metal ions. Ligands with strong enough fields to do this to the Ni^{2+} ion are infrequent: CN^- produces orange-yellow diamagnetic, square planar $[Ni(CN)_4]^{2-}$, and the chelate ligand dimethylglyoxime ($HON=CMe-CMe=NOH$) produces the characteristic red test species for Ni^{2+}, $[Ni(dimethylglyoximate)_2]$. As a result of the larger crystal field splitting and lower pairing energies for fifth- and sixth-period metal ions, the square planar geometry is normal for the d^8 Pd^{2+} and Pt^{2+} ions.

A square planar complex can be compared to an octahedral complex with two opposite ligands removed along the z axis. This relieves most of the repulsion of the ligands for the d_{z^2} orbital, which drops in energy from $+0.6\Delta_o$ to $-0.428\Delta_o$ (Table 7.8). Because the coordination number has now dropped from six to four, the ionic radius of the metal ion has decreased (Table 7.9). This brings the four ligands closer to the $d_{x^2-y^2}$ orbital, which rises in energy from $+0.6\Delta_o$ to $+1.228\Delta_o$. The difference in energy between this orbital and the next-highest orbital, d_{xy} at $+0.228\Delta_o$, is now equal to Δ_o. If $\Delta_o > P$, the one electron found in the $d_{x^2-y^2}$ orbital in an octahedral d^8 complex then drops down into the lower-energy d_{z^2} orbital, pairing with it. This converts the paramagnetic d^8 octahedral complex (two unpaired electrons) to a diamagnetic square planar complex (no unpaired electrons).

The spectra of yellow to red diamagnetic square planar Ni^{2+} complexes show an absorption at 20,000 to 24,000 cm^{-1} ($\varepsilon \approx 50$); this is approximately the value of Δ_o for these complexes. The corresponding absorptions for square planar Pd^{2+} and Pt^{2+} complexes are in the violet or ultraviolet region, at 25,000 to 35,000 cm^{-1}; these complexes are red to colorless.

Distorted Geometry and the Jahn–Teller Effect. Let us begin by considering a d^9 ion (e.g., Cu^{2+}) in an octahedral complex. In such an ion, there is an unequal population of the two e_g orbitals, $d_{x^2-y^2}$ and d_{z^2}. The electrons in the $d_{x^2-y^2}$ orbital repel (and are repelled by) the electrons donated by the four ligands located along the x and y axes; the electrons in the d_{z^2} orbital repel the electrons donated by the two ligands located along the z axis. If the two orbitals have equal populations, there are equal repulsions along all axes. However, if the two orbitals are unequally populated, the repulsions cannot be the same along the z axis as along the x and y axes. It is to be expected, and indeed is found, that the ligands along the more-populated axis will move further

Complexes That Switch Spins

Occasionally, a complex is discovered in which the crystal field splitting Δ_o is almost exactly equal to the pairing energy P. In such cases, a change in temperature can cause a partial change from high-spin to low-spin behavior, with unusual magnetic behavior resulting. For example, some monothiocarbamate complexes of iron(III) were discovered[13] in which the magnetic moment μ dropped from 5.7–5.8 BM at 300 K, to 4.7–5.0 BM at 150 K, to 3.6–4 BM at 78 K, as the proportion of molecules in the high-spin and low-spin states changed with temperature. Such complexes provide very small domains in which magnetic information can be stored, as for example in magnetic hard drives of computers.[14]

In cases in which the ligands are barely strong enough to stabilize square planar Ni^{2+} complexes, the tetrahedral complex may also be observed. This results not only in a change in color, but a change in magnetic properties between paramagnetic and diamagnetic. As an example, one isomer of the complex $[NiBr_2(PEtPh_2)_2]$ is square planar, diamagnetic, and brown, while the other isomer is tetrahedral, paramagnetic (μ = 3.20 BM at 300 K), and green.[15] In solution this compound exists as a mixture of isomers, with values of μ ranging from 0 to 3.20 BM, depending on the temperature and the solvent. Many similar changes occur in salts of the $[CuCl_4]^{2-}$ ion with alkylammonium cations, $[R_nNH_{4-n}]^+$. At low temperatures, hydrogen bonding of the cations to the chlorines of the anion stabilizes the square planar geometry; in this geometry the anion is yellow. Higher temperatures cause motion in the cation that breaks up the hydrogen bonding, allowing the $[CuCl_4]^{2-}$ anion to adopt a tetrahedral geometry, in which it has a deep green color.[16]

Recently a Ni^{2+} complex of a substituted porphyrin ligand has been prepared in which the porphyrin ligand has a side group that can reversibly have its geometry changed by light irradiation (Figure 7.19).[17] While the side chain is in its trans configuration, the nickel complex is square planar, low-spin, and diamagnetic. After irradiation with blue-green light, the side chain takes a cis conformation. Another donor atom is coordinated to the nickel, changing it to square pyramidal, high-spin, and paramagnetic. If the complex is then irradiated with blue-violet light, the ligand reverts to the trans configuration and the complex again becomes diamagnetic. This switching of spin offers promise that the complex could be used as a contrast agent in magnetic resonance imaging (MRI).

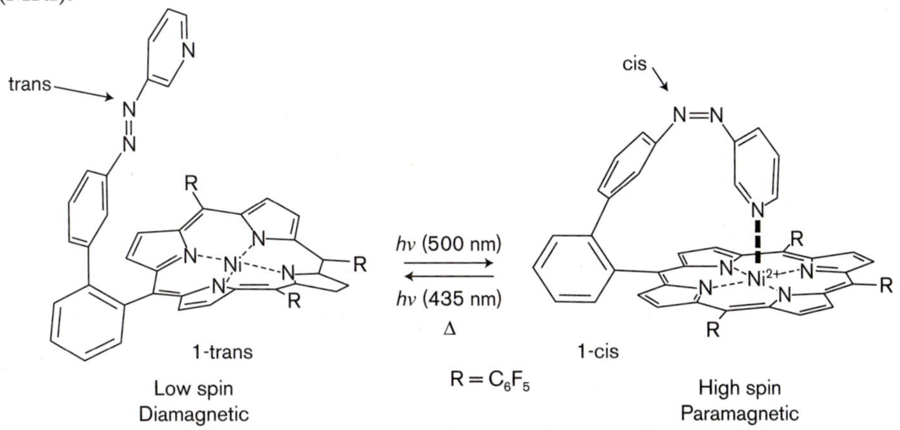

Figure 7.19. Complex with a reversible conformational change in its side chain that changes its Ni^{2+} from low-spin to high-spin. [Adapted from S. Venkataramani, U. Jana, M. Dommaschk, F. D. Sönnischen, F. Tuczek, and R. Herges, *Science* 331, 445 (2011).]

away, while the ligands along the less-populated axis will move closer; this motion produces a distorted octahedral geometry. This effect is known as a **Jahn–Teller distortion**. Experimentally, we find the lengthening occurs along the d_{z^2} orbital, as in Figure 7.20a. We deduce that the greatest repulsion is along the z axis; hence, the electron configuration of Cu^{2+} is $t_{2g}^6(d_{z^2})^2(d_{x^2-y^2})^1$.

The other important situation in which the e_g orbitals are unequally populated, and therefore Jahn–Teller distortion of the octahedron is observed, is in the electronic configuration e_g^1, as in the weak-field, high-spin case for d^4 ions such as Cr^{2+}.[18] The Jahn–Teller distortion has a definite effect on the stability and the spectra of ions having e_g^1 and e_g^3 electron configurations. As diagrammed at the bottom of Figure 7.20, one effect of the Jahn–Teller distortion is to increase the energy difference between the t_{2g} electrons and the unoccupied "hole" in the e_g electrons; consequently, higher-energy light must be absorbed, and the measured crystal field splitting is increased. Referring to Figure 7.8b, we see that among +2 ions, only in the hydrated Cr^{2+} and Cu^{2+} ions is the lowest-energy absorption shifted into the visible region of the spectrum. (Green Ag^{2+} in $Ag[AuF_4]_2$ is a related case.[19]) This shift accounts for the blue color of these ions and the larger crystal field splittings of these ions (Table 7.4). The effect has real thermodynamic consequences, because during the distortion two electrons are lowered in energy while only one is raised an equal amount in energy. This accounts for the fact that Cu^{2+} outranks Ni^{2+} in the Irving–Williams series (illustrated in Figure 7.13), in electronegativity (Figure 7.16), and to a lesser extent in other atomic and thermodynamic properties (Section 7.5).

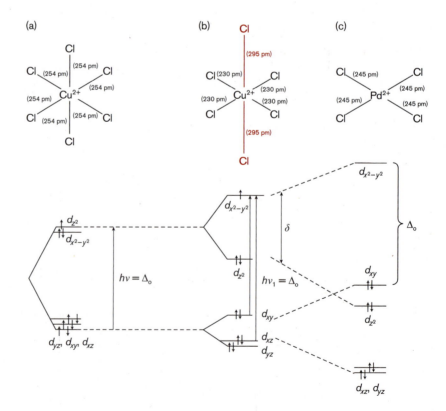

Figure 7.20. Effects of tetragonal distortion (bond lengthening) of an octahedral complex of a d^9 ion. (a) Cu^{2+} ion octahedrally coordinated by six Cl^- ligands in a hypothetical undistorted complex, showing the normal octahedral energy-level diagram and (hypothetical) bond lengths in pm. (b) The same ion after the distorted octahedral geometry is adapted (the lengthened bonds are shown in **brown**). (c) In a strong-field d^8 complex, the axial ligands are completely removed, producing a square planar complex.

AN AMPLIFICATION

Jahn–Teller Theorem

In 1937, Jahn and Teller proved a theorem stating that a nonlinear molecule cannot be stable in a degenerate electronic state, but must become distorted in such a way as to break down the degeneracy. The term "degenerate electronic state" refers to the situation in which the electrons can be arranged in either of two energy-equivalent ways. For the octahedrally coordinated Cu^{2+} ion, the e_g^3 electrons may be either $(d_{z^2})^2(d_{x^2-y^2})^1$ or $(d_{z^2})^1(d_{x^2-y^2})^2$. The distortion removes the degeneracy by making the former electronic configuration preferable.

7.7. Geometric Preferences; Rates and Mechanisms of Ligand Exchange

OVERVIEW. The geometric preference of a given d^n is influenced strongly by the CFSEs for different geometries. Computing differences in CFSEs of alternate geometries can help predict relative preferences, particularly as the value of n (the number of d electrons) varies. You can calculate or predict the relative preference of different d-block metal ions for one geometry of complex or another as described in Section 7.7A by trying Exercises 7.61–7.64. The difference between the octahedral and the square pyramidal CFSEs is related (along with the ion's charge) to the relative rates of ligand substitution in complexes. These rates can vary enormously, which allows us to identify inert and labile complexes. You can practice this Section 7.7B concept by trying Exercises 7.65–7.70.

7.7A. Geometric Preferences. The actual geometry chosen by the complex of a given d^n metal ion with a given set of ligands is the result of a balance of several factors: (1) The CFSEs for a given d^n ion can vary quite substantially from one geometry to another, as suggested by Example 7.6B. (2) More energy is released as a result of ligand–metal-ion electron-pair donation when the maximum coordination number is reached. (3) Contrariwise, larger destabilizing ligand–ligand repulsive energies can result from the crowding of bulky ligands that occurs with high coordination numbers. The latter two energies are important, and cannot be calculated using the crystal field theory. But they should not vary much across a series of $3d^n$ metal ions, while the CFSEs vary dramatically and predictably. Hence, it is possible, using the crystal field theory, to predict which $3d^n$ electron configurations show the greatest preference for a given geometry over some other geometry.

The two most common geometries for d-block ions are tetrahedral and octahedral; their energy-level diagrams are compared in Figure 7.21a. We can predict the relative preferences of different high-spin d^n ions for these two geometries by computing CFSEs for each value of n in the two geometries, then taking the energy difference

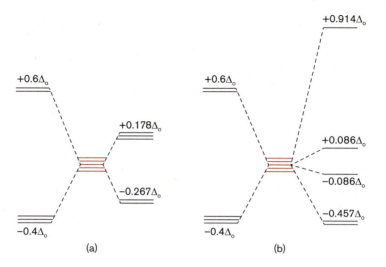

Figure 7.21. Comparative energy-level diagrams for different pairs of geometries: (a) octahedral ML_6 (on the left) versus tetrahedral ML_4 (on the right); (b) octahedral ML_6 (on the left) versus square pyramidal ML_5 (on the right).

CFSE (octahedral) – CFSE (tetrahedral). This **octahedral site preference energy** (Table 7.10) varies with n. First we computed CFSE (octahedral) for d^4–d^7 ions based on Figure 7.6, and we computed CFSE (tetrahedral) in Example 7.6B for the first five d^n ions. In order to make the subtraction we must use CFSE (tetrahedral) in units of Δ_o, as shown in Figure 7.17a. Taking these differences we obtain the data in Table 7.10.

These calculations show that ions with the d^3 electron configuration have the greatest preference for octahedral coordination; an extension of this calculation would also add ions with the d^8 electron configuration. The d^3 ion Cr^{3+} virtually never forms tetrahedral complex ions, and the d^8 ion Ni^{2+} does not form very many, even with fairly bulky ligands. In contrast, the d^5, d^6, and d^7 ions Mn^{2+}, Fe^{2+}, and Co^{2+} frequently form tetrahedral complexes, especially with bulky ligands or ligands that fall fairly low in the spectrochemical series.

TABLE 7.10. Octahedral Site Preference Energies for d^1–d^5 Ions

Free-Ion Configuration	d^1	d^2	d^3	d^4	d^5
Octahedral CFSE	−0.4	−0.8	−1.2	−0.6	0
Tetrahedral CFSE	−0.267	−0.533	−0.356	−0.178	0
Octahedral preference (units of Δ_o)	−0.133	−0.267	0.845	−0.422	0

EXAMPLE 7.12

Another useful choice of geometries is between high-spin octahedral and square pyramidal geometries (Figure 7.21b). Which electron configuration (d^1–d^5) most prefers octahedral over square pyramidal coordination?

SOLUTION: The high-spin square pyramidal complexes place one electron in each orbital before pairing any. The orbitals fill from lowest energy up, as determined from Table 7.8. Thus, the first two electrons have energies of $-0.457\Delta_o$, followed by one at $-0.086\Delta_o$, followed by one at $+0.086\Delta_o$, followed by one at high energy, $+0.914\Delta_o$. The totals are then shown in the third column of the following table, followed by the difference to indicate the preference for octahedral coordination when the term is negative.

Free-Ion Configuration	d^1	d^2	d^3	d^4	d^5
Octahedral CFSE	−0.4	−0.8	−1.2	−0.6	0
Square pyramidal CFSE	−0.457	−0.914	−1.000	−0.914	0
Octahedral site preference energy (units of Δ_o)	+0.057	+0.114	−0.2	+0.314	0

The preference for octahedral coordination is strongest for the d^3 (and d^8) configurations, while square pyramidal is seemingly preferred by the d^4 (and d^9) configurations. (This calculation ignores the loss in bond energy when the sixth ligand is removed, however.)

7.7B. Rates and Mechanisms of Ligand Exchange in Complexes.

Formation of a square pyramidal five-coordinate complex is presumed to be a necessary step in the substitution reactions of metal ions that have maximum coordination numbers of six. Such ions would be expected to show mechanisms in which, before a new ligand can come in, one of the existing ligands must first break its coordinate covalent bond to the metal, leaving a square pyramidal intermediate. The simplest possible ligand-exchange reaction is that of solvent water for coordinated water in hydrated ions. Figure 7.22 shows some measurements of rate constants for the exchange of water molecules by hydrated cations.

We would expect the relative rates of this reaction to depend on at least two factors: (1) the hydration energy of the metal ion, which is a function of its acidity classification; and (2) for d-block metal ions, the octahedral versus square pyramidal preference energies. As illustrations of the effects of cation acidity, we can see:

1a. The nonacidic and feebly acidic cations exchange solvent and coordinated water molecules with rate constants of 10^8–10^9 s^{-1} (rows 1 and 2 of Figure 7.22). That is, a given water molecule will stay in the hydration sphere of such a cation for a period of nanoseconds. This rate is almost as fast as the limiting factor, the rate of diffusion of liquid H_2O molecules in solution at room temperature.

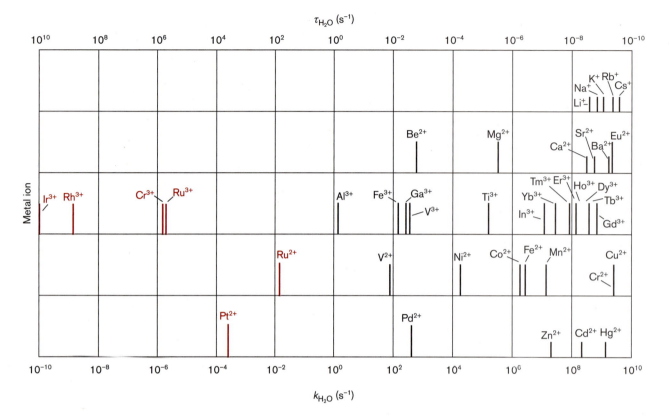

Figure 7.22. Mean lifetimes of a water molecule in the primary coordination sphere, τ_{H_2O}, and rate constants for exchange of water, k_{H_2O}, for solvated cations as measured by NMR (long lines) or as derived indirectly (short lines). [Adapted from S. F. Lincoln, D. T. Richens, and A. G. Sykes, "Metal Aqua Ions," in *Comprehensive Coordination Chemistry II*, J. A. McCleverty and T. J. Meyer, Eds., Elsevier: Amsterdam, 2004, vol. 1, pp. 515–555.]

1b. Most weakly acidic cations have smaller rate constants of 10^4–10^7 s⁻¹; most moderately acidic cations have still smaller rate constants of 10^0–10^3 s⁻¹. Water molecules in these ions are much more strongly bound to the cation, and it takes them periods ranging from microseconds to seconds to break away and be replaced by solvent water molecules.

2. Some *d*-block weakly and moderately acidic cations (shown in **brown** in Figure 7.22) are much slower to react than other metal ions of similar charge and size. These deviations can be correlated with octahedral versus square pyramidal preference energies (Example 7.12). The rates of exchange of coordinated and free water at d^3 and d^8 weak-field ions are in fact strikingly low (Figure 7.23), whereas the rates of ligand exchange for d^4 and d^9 ions are unusually high. A similar calculation for strong-field ions shows that there is a strong maximum in preference for octahedral coordination at the d^6 electron configuration. Because most Co^{3+} complexes and fifth- and sixth-period *d*-block complexes are strong-field, they also strongly resist ligand exchange and are considered **inert**.

Most metal complexes, provided that they are not immobilized by excessive chelation or macrocyclic–ligand coordination, are **labile**: Their ligands exchange with other ligands or with solvent water in less than a minute. The complexes of Cr^{3+}, low-spin Co^{3+}, and the d^6 complexes of the fifth and sixth periods last for well over 1 min before reacting with other ligands and the solvent. These are termed inert and are more easily isolated than labile complexes. In extreme cases, inert ions can be very slow to react indeed! The mean residence time of a given water molecule in the primary hydration

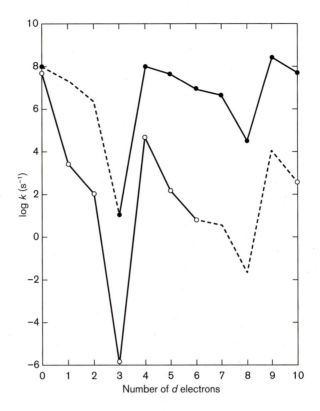

Figure 7.23. Rates of exchange of free and coordinated water for weak-field (high-spin) hydrated $3d^n$ metal ions. Solid circles are for +2-charged ions; open circles are for +3-charged ions. [Data from R. P. Hanzlik, *Inorganic Aspects of Biological and Organic Chemistry*, Academic Press: New York, 1976, p. 107; and J. Burgess, *Metal Ions in Solution*, Horwood Publishing: Chichester, UK, 1978, pp. 316–331.]

sphere of a $[Cr(H_2O)_6]^{3+}$ ion at 25°C is over 100 hours; for fifth-period d^6 $[Rh(H_2O)_6]^{3+}$, the time is nearly 1.5 years; for sixth-period d^6 $[Ir(H_2O)_6]^{3+}$, it is nearly 300 years![20] Werner's work in developing the original theory of coordination chemistry (see the Amplification in Section 3.2B) depended on being able to isolate and count all possible isomers of various octahedral Co^{3+}, Cr^{3+}, and Pt^{4+} complexes; this would have been impossible had these been labile complexes.

Synthesis of complexes by ligand substitution reactions of extremely inert d-block metal ions is therefore often impractically slow. However, in some cases the same metal with one more electron may be very fast to react. Although complexes of the kinetically inert d^3 Cr^{3+} ion react slowly, Cr^{2+} complexes have the labile electronic configuration d^4. Hence, substitution reactions of Cr^{3+} can often be speeded enormously by adding a small amount of a reducing agent to convert some of the chromium reactant to chromium(II). This reacts very rapidly, and it can then be reoxidized back to chromium(III).

A variation on this strategy is usually used to synthesize low-spin complexes of the Co^{3+} ion, which is an inert d^6 ion. Co^{2+} complexes are generally high spin and labile, and $[Co(H_2O)_6]Cl_2$ is a commonly available starting material. Then reaction with the ligand generates the labile cobalt(II) analogue of the desired complex. A suitable oxidizing agent (air, O_2, or H_2O_2) oxidizes it to the corresponding cobalt(III) complex:

$$[Co(H_2O)_6]Cl_2 + 6NH_3 \rightarrow [Co(NH_3)_6]Cl_2 + 6H_2O \tag{7.12}$$

Pink, weak-field, labile   Rose, weak-field, labile

$$4[Co(NH_3)_6]Cl_2 + 4NH_4Cl + O_2 \rightarrow 4[Co(NH_3)_6]Cl_3 + 4NH_3 + 2H_2O \tag{7.13}$$

Orange, strong-field, inert

Square planar complexes, which typically consist of d^7, d^8, and d^9 metal ions with strong-field ligands, are likely to exchange ligands by a different mechanism. The incoming ligand adds to the complex to form a square pyramidal transition state complex before the outgoing ligand is lost. Given the strong crystal field splitting in the square planar geometry, it is not surprising that square planar d^8 complexes of such ions as Pd^{2+} and Pt^{2+} are also generally inert. This inertness also applies to isomerization reactions: The medicinal square planar complex *cis*-platin (Section 5.8B) is slow to isomerize to the medicinally ineffective complex *trans*-platin.

EXAMPLE 7.13

In the synthesis of which of the following complexes (by a substitution reaction) would there be a significant advantage in adding some reducing agent, or in using a starting metal compound in which the metal ion had an oxidation state one lower than in the product? (a) $[Fe(NH_2CH_2CH_2NH_2)_3](NO_3)_3$; (b) $[Ir(NH_2CH_2CH_2NH_2)_3](NO_3)_3$; (c) $K_3[Co(C_2O_4)_3]$ (contains low-spin cobalt); and (d) $[Cr(NH_3)_6]Cl_2$.

SOLUTION: There would be a significant advantage if the metal ion in the complex is inert—that is, weak-field d^3 or d^8 octahedral or strong-field d^6, and can be reduced to a labile configuration—that is, weak-field d^4 or d^9 or strong-field d^7: This applies to (b), the d^6 Ir^{3+} complex, and to (c), the d^6 Co^{3+} complex.

7.8. Heme and Hemoglobin: A Connection to Biochemistry

> **OVERVIEW.** This section introduces no new concepts, but shows important connections to bioinorganic chemistry. It also reviews and applies several important concepts from this and earlier chapters, such as complexation by macrocyclic ligands (Section 5.2) and applications of the HSAB theory (Chapter 5). Exercises 7.71–7.76 can help you review and apply these concepts.

7.8A. Heme and Hemoglobin in Oxygen Transport.

Two metal complexes of vast importance in biochemistry are chlorophyll (Figure 5.1a) and heme (Figures 5.1b and 7.25a). Both are complexes of (substituted) tetradentate macrocyclic porphyrin^{2-} ligands. Because these ligands have a planar, conjugated system of π bonds around their perimeters, they are much more rigid ligands than are the crown ethers (Figure 5.10). Consequently, porphyrins are even more selective for a certain metal–donor atom distance (of close to 200 pm) than are the crown ethers. Their enforced square planar geometry of donor atoms gives them a strong preference for the d^8 Ni^{2+} ion ($r = 63$ pm; Table 7.9). But other coordinated metal ions may add fifth and maybe sixth ligands above or below the square plane.

The function of heme in hemoglobin is to bond to the weak ligand O_2 and carry it from the lungs to the cells, where it is released to another heme-containing protein, myoglobin. Hemoglobin itself (Figure 7.24b) is composed of a protein, globin, which contains (among many other amino acids) four units of histidine (Figure 5.23), which contain five-membered imidazole rings with nitrogen-donor atoms. These nitrogen-donor atoms bond four separate heme molecules to the globin through the Fe^{2+}, causing Fe^{2+} to be five-coordinate (Figure 7.24c). The vacant sixth site of the potential octahedron about Fe^{2+} is reserved for O_2 as a ligand. Myoglobin, in contrast, contains only one heme molecule. The binding ability of blood hemoglobin for O_2 must be delicately balanced with that of cellular myoglobin. Hemoglobin needs to bind O_2 efficiently in the lungs. However, the heme in hemoglobin must become inefficient at the blood–cell interface, so that the O_2 is released by the hemoglobin and picked up instead by the myoglobin.

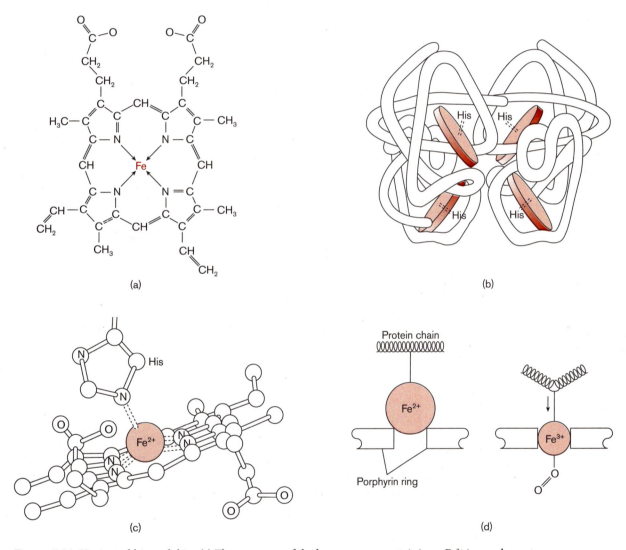

Figure 7.24. Heme and hemoglobin. (a) The structure of the heme group, containing a Fe^{2+} ion and a (macrocyclic) porphyrin^{2-} ligand. (b) The tetrameric hemoglobin molecule, containing four heme groups (represented by **brown** disks) bound to the protein by the amino acid histidine. (c) Details of the bonding of the heme group to histidine in myoglobin and hemoglobin, in the absence of bound O_2. Note that the Fe^{2+} is above the plane of the porphyrin ligand. (d) Sketch of the effects of coordination of O_2: The high-spin Fe^{2+} becomes low-spin Fe^{3+}, shrinks in size, falls into the plane of the porphyrin ligand, and pulls on the protein chain via the coordinated histidine. [Adapted from J. E. Huheey, *Inorganic Chemistry: Principles of Structure and Reactivity*, 3rd ed., Harper and Row: New York, 1983.]

Myoglobin exhibits normal (noncooperative) ligand bonding—its fractional saturation falls as the partial pressure of O_2 [$p(O_2)$] falls (dashed curves in Figure 7.25). Thus, it does not easily saturate with O_2 in the cells, where the pressure of O_2 is low. When blood hemoglobin reaches the cells, hemoglobin must have an even lower fractional saturation than myoglobin has—otherwise the O_2 would not transfer from the blood to the cells where it is needed. Therefore, the fractional saturation of hemoglobin must fall more rapidly with decreasing $p(O_2)$ than does the fractional saturation of myoglobin. As can be seen by comparing the solid hemoglobin and dashed myoglobin oxygen saturation lines in Figure 7.25, this is the case.

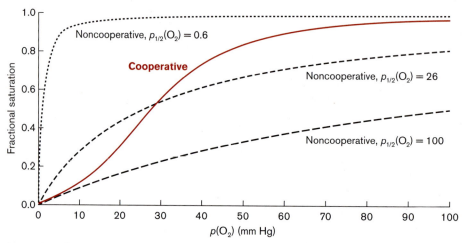

Figure 7.25. Noncooperative and cooperative binding of O_2. The curves show the fractional saturation of the hemoglobin with O_2 as a function of the partial pressure of O_2. The value of $p_{\frac{1}{2}}$ gives the partial pressure of O_2 when half of the heme units carry O_2; a low value of $p_{\frac{1}{2}}$ corresponds to a high equilibrium constant K_P for binding O_2 (the binding is highly efficient). [Adapted from I. Bertini, H. B. Gray, E. I. Steifel, and J. S. Valentine, *Biological Inorganic Chemistry: Structure & Reactivity*, University Science Books: Mill Valley, CA, 2007.]

This is achieved through the biochemical phenomenon known as **cooperative binding**, which means that the ability of a particular heme group to bind O_2 is strongly influenced by whether or not the other three heme groups in hemoglobin already possess O_2 ligands (i.e., information is passed from one heme group to the others). Consequently, in the lungs, where the pressure of O_2 is high (~100 mmHg) and much O_2 is bound, the affinity of heme in hemoglobin for O_2 is high, so that it is very efficiently loads up with as much O_2 as possible—it has a high fraction of saturation (solid curve in Figure 7.25). But near the cells, when the O_2 saturation is low, the whole structure of hemoglobin changes so that all of the heme units have low affinities for O_2, allowing it to be released to myoglobin.

The Perutz[21] mechanism by which this cooperative binding is achieved is a beautiful example of design employing both the principles of biochemistry and of inorganic chemistry (specifically, ligand or crystal field theory). In the absence of the O_2, the ligand field is (just barely) weak, so the Fe^{2+} is high spin, with the $t_{2g}^4 e_g^2$ electron configuration. In this spin state Fe^{2+} has too large a radius to fit in the hole of the porphyrin ligand, so the iron sits an average of about 36 pm above the plane of the porphyrin ring in the direction of the histidine (Figure 7.24c), achieving a distorted square pyramidal geometry.

When O_2 bonds to high-spin Fe^{2+} of heme in hemoglobin, it becomes, in effect, an O_2^- (superoxide) ion coordinated to a low-spin Fe^{3+} ion of radius 69 pm. This ion is small enough for Fe^{3+} to fit in the hole of the porphyrin ring, into which it now falls (Figure 7.24d). Because the iron is attached to the globin protein by the histidine amino acid, this causes various sorts of motion in the globin itself, which are passed along to the regions of the other three heme groups, thus transmitting the necessary message that an O_2 has been attached and that the other heme groups should go and do likewise.

There are some other necessary features of the chemistry of heme in hemoglobin. The O_2 molecule is the classical oxidizing agent, and Fe^{2+} is a good reducing agent, but if Fe^{2+} in heme is first oxidized to Fe^{3+} (hematin), it will not carry O_2! This oxidation to hematin happens readily in free heme in the presence of O_2 *and* a water molecule. Part of the function of the globin protein is to prevent access of H_2O to the heme while the O_2 is coming in.

7.8B. Binding of Other Ligands to Heme. Because O_2 is neither a strong-field nor a soft-base ligand, it is not easy to get it to coordinate to the fairly soft iron ion if soft-base ligands such as CO, NO, and CN^- are present. These ligands form much more stable complexes with iron in heme groups in hemoglobin, myoglobin, or other species containing heme; this complexation is the basis of the toxicity of these soft bases. Globin acts to bend and destabilize the Fe–CO bond enough to improve the chances of forming iron bonds to O_2; hydrogen bonding and electrostatic interactions between coordinated O_2 and glutamine and tyrosine side groups in the globin help stabilize the Fe–O_2 bond.[22]

In biochemistry, important molecules such as heme are often put to use in multiple ways. In the past two decades, Nobel-Prize-winning research has established the importance of three toxic ligands that bond to heme: hydrogen sulfide,[23] carbon monoxide (CO),[24] and especially nitric oxide (NO),[25,26] *Science* Magazine's 1992 Molecule of the Year.[27,28] Nitric oxide is generated deliberately in the body, where it bonds to metal-bearing proteins containing heme units. This bonding is done for a variety of purposes:

1. Both NO and CO act as an additional type of neurotransmitter, carrying nerve messages from heme groups in the enzyme guanylyl cyclase in certain types of neurons across synapses to other neurons. In the other neuron these again coordinate, causing a change of spin state and consequent conformational changes in this enzyme.

2. Both NO and CO are involved in long-term learning and in storing long-term memories in the brain.

3. NO acts to affect the contractions and relaxations of smooth muscles in blood vessels: In the penis, erections result; in response to septic bacterial infection, NO lowers blood pressure so dramatically as to cause deadly septic shock.

4. NO acts to suppress pathogens in the body, and pathogens use heme to sense the presence of NO and move away from it.

5. Unfortunately, during strokes, massive release of NO causes the death of nerve and brain cells.

7.9. Background Reading for Chapter 7

Relationship of d-Block Chemistry to Chemistry in the Whole Periodic Table

M. Gerloch, "The Roles of *d* Electrons in Transition Metal Chemistry: A New Emphasis," *Coord. Chem. Rev.* 99, 117–136 (1990). Identifies and distinguishes seven ways in which *d*-block chemistry resembles or does not resemble chemistry in the remainder of the periodic table.

Transition Metal Coordination Chemistry Summary

J. A. Roe, B. F. Shaw, and J. S. Valentine, "Fundamentals of Coordination Chemistry: Tutorial II," in *Biological Inorganic Chemistry: Structure and Reactivity*, I. Bertini, H. B. Gray, E. I. Steifel, and J. S. Valentine, Eds., University Science Books: Mill Valley, CA, 2007, pp. 700–709. Ligand field theory useful for bioinorganic chemistry.

Jack Barrett, *Inorganic Chemistry in Aqueous Solution*, Cambridge: Royal Society of Chemistry: Cambridge, UK, 2003, pp. 128–144 (Chapter 7: Periodicity of Aqueous Chemistry II: *d*-Block Chemistry). Enthalpies of hydration and redox potentials (metal activities) as functions of *d*-electron count and ligand field effects.

Rates of Water Exchange by Hydrated *d*-Block Metal Ions

John Burgess, *Ions in Solution: Basic Principles of Chemical Interactions*, 2nd ed., Horwood Publishing: Chichester, UK, 1999, pp. 111–124 (Chapter 9: Kinetics and Mechanisms: Solvent Exchange).

Heme, Hemoglobin, and Its Ligands

G. B. Jameson and J. A. Ibers, "Dioxygen Carriers," in *Biological Inorganic Chemistry: Structure and Reactivity*, I. Bertini, H. B. Gray, E. I. Steifel, and J. S. Valentine, Eds., University Science Books: Mill Valley, CA, 2007, pp. 354–388. A more detailed description of heme and related dioxygen carriers.

A. Butler and R. Nicholson, *Life, Death, and Nitric Oxide*, RSC Paperbacks: Cambridge, UK, 2003. Some undergraduates may have a special interest in Chapter 16, "The Truth About Viagra."

Numbered references from this chapter may be viewed online at www.uscibooks.com /foundations.htm.

7.10. Exercises

7.1. *Drawn here are some d orbitals in their xyz axes systems.

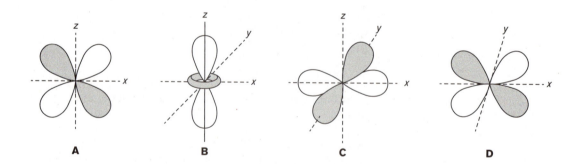

(a) List the names of these orbitals (d_{xy}, etc.). (b) Tell whether each belongs to the t_{2g} or the e_g set in an octahedral complex.

7.2. Drawn here are some d orbitals in their xyz axes systems. One of these is not a d orbital: Do not answer the questions about it!

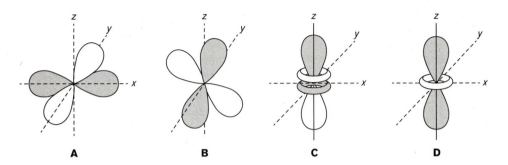

(a) List the names of these orbitals (d_{xy}, etc.). (b) Tell whether each belongs to the t_{2g} or the e_g set in an octahedral complex.

7.3. *The f orbitals are drawn and labeled in Figure 7.3 in forms that are appropriate for complexes of octahedral symmetry. (a) Group these into sets that will show equivalent interactions with external octahedral point charges or ligands. (b) Which of these sets will experience the greatest degree of repulsion? Which will experience the greatest stabilization?

7.4. If applicable, write the valence electron configurations of the central ions of each of the following metal complex ions in terms of t_{2g} and e_g populations: (a) $Na_3[TiF_6]$; (b) $[Ni(NH_3)_6]Cl_2$; (c) $K_2[SnCl_4]$; (d) $K_2[TiCl_6]$.

7.5. *Write the expected electron configuration in terms of t_{2g} and e_g and compute the CFSE in units of Δ_o for the central metal ion of each of the following complexes: (a) $[Cr(H_2O)_6]^{3+}$; (b) $[CuH_2O)_6]^{2+}$; (c) $[TiF_6]^{3-}$.

7.6. Which of these electron configurations give CFSE values of $-1.2\Delta_o$? (a) d^1, d^2, and d^3; (b) d^1, d^6, and d^9, (c) d^3 and d^8; (d) d^2 and d^7; (e) d^2 and d^8.

7.7. *(a) Draw an energy-level diagram that indicates qualitatively how the energy levels of the d orbitals are altered as a result of varying degrees of interaction of d and ligand electrons in an octahedral complex $[ML_6]^{n+}$. Show which d orbitals occur at which energy levels. (b) Show how the orbitals will be occupied by electrons for each of the following electron configurations: d^1, d^2, and d^3. (c) Write the expected electron configuration in terms of t_{2g} and e_g. (d) For each electron configuration, compute the CFSE in units of Δ_o. (e) From Table C, select and list all d-block ions having each of these electron configurations.

7.8. Repeat Exercise 7.7 for the d^8, d^9, and d^{10} electron configurations.

7.9. *Consider the following electron configurations in octahedral complexes: $t_{2g}^2e_g^0$; $t_{2g}^4e_g^0$; $t_{2g}^6e_g^2$; $t_{2g}^4e_g^2$; $t_{2g}^6e_g^0$. (a) Which of these are found *regardless* of whether the complex is weak-field or strong-field? (b) Which are found *only* in strong-field complexes? (c) Which result in two unpaired electrons?

7.10. The electron configuration of a d^4 metal ion such as Cr^{2+} in an octahedral complex is best written as which of the following? (a) t_{2g}^4, no matter what the ligands are; (b) e_g, no matter what the ligands are; (c) t_{2g}^4 with weak-field ligands but $t_{2g}^3e_g^1$ with strong-field ligands; (d) t_{2g}^4 with strong-field ligands but $t_{2g}^3e_g^1$ with weak-field ligands; (e) $t_{2g}^3e_g^1$, no matter what the ligands are.

7.11. *Consider each of the following metal d-electron configurations and ligand field splitting patterns in octahedral complexes: (a) d^6 strong field; (b) d^5 weak field; (c) d^6 weak field; and (d) d^5 strong field. For each, (1) write the expected electron configuration in terms of t_{2g} and e_g, (2) compute the ligand field stabilization energy as a multiple of Δ_o and P, (3) list the number of unpaired electrons expected, and (4) choose a likely example of this metal d-electron configuration and ligand field splitting from the following choices: $[Co(CN)_6]^{3-}$; $[Fe(H_2O)_6]^{3+}$; $[Fe(CN)_6]^{3-}$; and $[Co(H_2O)_6]^{3+}$.

7.12. Consider each of the following metal d-electron configurations and ligand field splitting patterns in octahedral complexes: (a) d^4 strong field; (b) d^7 weak field; (c) d^4 weak field; and (d) d^7 strong field. For each, (1) write the expected electron configuration in terms of t_{2g} and e_g, (2) compute the ligand field stabilization energy as a multiple of Δ_o and P, (3) list the number of unpaired electrons expected, and (4) choose a likely example of this metal d-electron configuration and ligand field splitting from the following choices: $[Co(CN)_6]^{4-}$; $[Mn(H_2O)_6]^{3+}$; $[Mn(CN)_6]^{3-}$; and $[Co(H_2O)_6]^{2+}$.

7.13. *All f-block complexes are weak-field. Which of the following would have the greatest number of unpaired electrons? (a) $[Gd(H_2O)_6]^{3+}$; (b) $Gd^{3+}(g)$; (c) $[Tb(H_2O)_6]^{3+}$; (d) $Tb^{3+}(g)$.

7.14. In the fifth and sixth periods, strong-field complexes are very predominant in the d block. Find all examples listed in Table C of fifth- and sixth-period d-block metal ions having each of the following electron configurations: (a) d^4; (b) d^5; (c) d^6; and (d) d^7. (e) Given that Ru^{2+} and Os^{2+} are also well known in complexes, which of these four electron configurations seem to have the most examples in the fifth and sixth periods, and why?

7.15. *A nickel(2+) salt is found to have an effective magnetic moment of 2.83 (i.e., the square root of 8) Bohr magnetons. (a) How many unpaired electrons are present on the Ni^{2+} ion? (b) Is the complex paramagnetic or diamagnetic? (c) Compute the corrected (molar) magnetic susceptibility χ_m^{corr} that you would expect to measure for this complex at 298 K and at 100 K.

7.16. (a) Write the expected electron configuration of a d^5 ion in terms of t_{2g} and e_g for a weak-field complex and for a strong-field complex. (b) Tell the number of unpaired electrons found in the weak-field complex and in the strong-field complex. (c) Calculate the expected magnetic moment μ in the case of the weak-field complex and the strong-field complex. (d) Calculate the magnetic susceptibility expected at 100°C for the weak-field complex and for the strong-field complex.

7.17. *Predict the magnetic moments and the corrected molar susceptibilities at 273 K of (a) $[Fe(H_2O)_6]^{3+}$; (b) $[Fe(CN)_6]^{4-}$; (c) $[Fe(H_2O)_6]^{2+}$; (d) $[Fe(CN)_6]^{3-}$; and (e) $[HoF_6]^{3-}$.

7.18. Predict the magnetic moments and the corrected molar susceptibilities at 273 K of (a) $[Ni(H_2O)_6]^{2+}$; (b) $[Co(CN)_6]^{3-}$; (c) $[Co(CN)_6]^{4-}$; and (d) $[Gd(H_2O)_6]^{3+}$.

7.19. *Below are shown two spectra: The first is of a Cr^{3+} octahedral complex ion, while the second is of a Ti^{3+} octahedral complex ion.

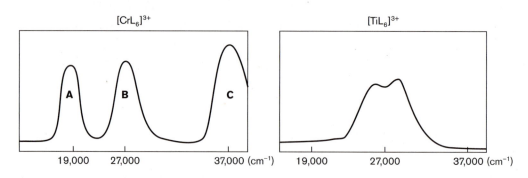

(a) Briefly explain why there is more than one absorption band in the spectrum of the chromium ion but not the titanium ion. (b) Which absorption band (**A**, **B**, or **C**) would you use to determine the value of Δ_o in the chromium complex ion? What is the value of Δ_o in this complex ion? (c) With the help of Table 7.4, describe approximately the colors of the titanium complex ion.

7.20. Below are shown three spectra of three octahedral complex ions of the same metal.

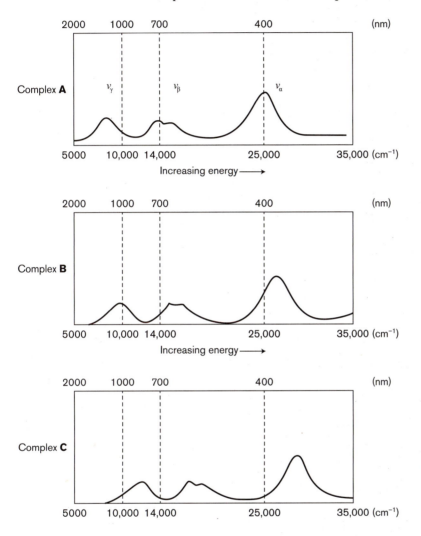

(a) Which of the following metal ions are almost certainly not the metal in these complex ions: Ti^{3+}, Ni^{2+}, Pt^{4+}, and Cu^{2+}. Briefly explain why you can rule out those metal ions.
(b) Which absorption band would you use to determine the value of Δ_o in each of these complex ions? (c) What are the approximate values of Δ_o in each of these complex ions?
(d) With the help of Table 7.3, describe the approximate colors of these complex ions.

7.21. *Below are shown the spectra of three octahedral complexes, **A**, **B**, and **C**.

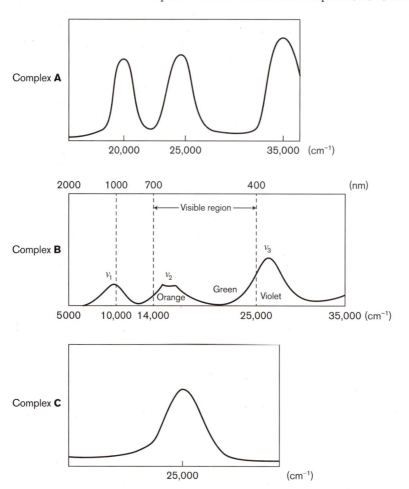

(a) Which complex is more likely that of a d^1 metal ion? Why did you choose that answer? (b) Is the color of complex **B** orange, green, or violet? (c) What is the value of Δ_o for complex **B** in cm^{-1}? In $kJ\ mol^{-1}$? (d) Which complex has the highest value of Δ_o?

7.22. Below are shown two spectra of two octahedral complexes, MA_6 and NB_6. (a) What is the value of Δ_o (in cm^{-1}) for the complex MA_6? What is the value of Δ_o (in cm^{-1}) for the complex NB_6? (b) What is the value of Δ_o (in $kJ\ mol^{-1}$) for the complex MA_6? What is the value of Δ_o (in $kJ\ mol^{-1}$) for the complex NB_6? (c) Which octahedral complex is more likely that of a d^1 metal ion? Why do you choose that answer? (d) What frequencies of light are being absorbed by each complex? What colors of light are being absorbed? (e) What color is likely for the NB_6 complex?

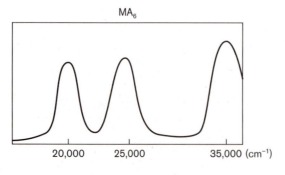

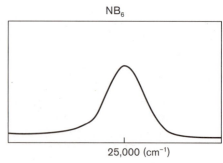

7.23. *Below we list some imaginary octahedral hydrated ions, along with their absorption energies or frequencies, and their classification or ε values. Assume that each has only one visible absorption band. Predict the color of each (including intensity) and their approximate ε value or classification (whichever is not given): (a) $A^{3+}(aq)$: $\nu = 550$ nm, an f–f transition; (b) $M^{5+}(aq)$: $\nu = 400$ nm, $\varepsilon = 15{,}000$; (c) $Sc^{2+}(aq)$: absorption at 9000 cm^{-1}, $\varepsilon = 5$.

7.24. (a) With reference to the known spectrum of $[Ti(H_2O)_6]^{3+}$, describe how you would expect the spectrum of $[Zr(H_2O)_6]^{3+}$ to appear. (b) With reference to the known spectrum of $[Cu(H_2O)_6]^{2+}$, describe how you would expect the spectrum of $[Zn(H_2O)_6]^{3+}$ to appear.

7.25. *Using the Cr^{3+} ion and the chelating glycinate ion ($NH_2CH_2CO_2^-$), you prepare the octahedral complex $[Cr(NH_2CH_2CO_2)_3]$. (a) Based on your results from Experiment 8, or based on periodic trends in the spectrochemical series, estimate the crystal field splitting Δ_o you expect to find in this complex. (b) Compute the CFSE you expect to find in this complex. (c) If Cr^{3+} were replaced by Mo^{3+} in the above complex, is it likely that Δ_o would be larger or smaller than in the above complex? (d) If, instead of using glycinate ion, you used the hypothetical phosphinoglycinate ion ($PH_2CH_2CO_2^-$), would the ligand field splitting increase or decrease? (e) Which one of the five spectra shown below would you expect to observe for the original octahedral Cr^{3+} complex $[Cr(NH_2CH_2CO_2)_3]$? (Assume that it is strictly octahedral.)

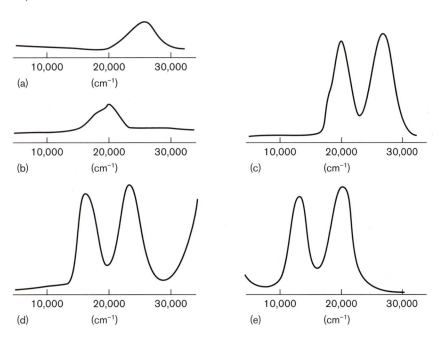

7.26. Referring to Equation 7.6, predict which complex in each set is most likely to be strong field, and predict which is most likely to be weak field: (a) $[CrBr_6]^{4-}$ or $[Cr(CN)_6]^{4-}$; (b) $[Co(NCS)_6]^{3-}$, $[Co(SCN)_6]^{3-}$, or $[CoF_6]^{3-}$.

7.27. *Exercise 7.20 shows the spectra of complexes **A**, **B**, and **C** with the same metal ion and three unidentified ligands. Which spectrum is most likely that of a weak-field ligand? Which is most likely that of a strong-field ligand?

7.28. Exercise 7.21 shows the spectra of complexes **A** and **B** with the same metal ion and three unidentified ligands. Which spectrum is most likely that of a weak-field ligand? Which is most likely that of a strong-field ligand?

7.29. *The following table gives the electronic spectra of some octahedral Ni^{2+} complexes.

Complex Ion	ν_1 (cm^{-1})	ν_2 (cm^{-1})	ν_3 (cm^{-1})
$[Ni(H_2O)_6]^{2+}$	8130	13,500	25,100
$[NiA_6]^{2+}$	11,500	18,500	30,000
$[NiB_6]^{2+}$	12,700	19,300	30,000
$[NiC_6]^{4-}$	6460	9860	17,000

(a) What is Δ_o for the hydrated nickel ion? (b) If the Ni^{2+} ions in the above complexes were replaced by Ni^{3+} ions, would the values of Δ_o increase or decrease? (c) If the Ni^{2+} ions in the above complexes were replaced by Pt^{2+} ions, would the values of Δ_o increase or decrease? (d) The three ligands A, B, and C include the following: iodide ion, ammonia, and carbon monoxide. Identify each ligand and explain your reasoning.

7.30. Among the bridging ligands found in Figure 5.4 are some that are ambidentate and are (at least in theory) capable of forming different linkage isomers. Identify these ambidentate ligands and tell in which isomer they should fall higher in the spectrochemical series.

7.31. *Based on their Lewis structures, predict whether each of the following ligands would be in the lower part, the middle, or the upper part of the spectrochemical series, and explain why: (a) CCl_2 (dichlorocarbene, with a C-donor atom); (b) NH_2^- (amide ion); and (c) the following hypothetical O donor ligand:

7.32. The three different classes of ligands have different consequences of overlap with the t_{2g} set of orbitals in octahedral complexes. Complete each statement about each class of ligands by filling in one of the following choices in each of the blanks: lower in energy, unchanged in energy; higher in energy; strong or large; moderate, weak, or small; Cl^-, NH_3, or CO.

(a) Pi-donor ligands make the t_{2g} electrons ____ and result in a ____ ligand field splitting Δ_o; an example of a π-donor ligand is ____.

(b) Pi-acceptor ligands make the t_{2g} orbitals ____ and result in a ____ ligand field splitting Δ_o; an example of a π-acceptor ligand is ____.

(c) Sigma-donor ligands make the t_{2g} electrons ____ and result in a ____ ligand field splitting Δ_o; an example of a σ-donor ligand is ____.

7.33. *Consider the ambidentate ligands fulminate ($C\equiv N-O^-$) and cyanate ($N\equiv C-O^-$), each of which can bond to a metal ion either via the oxygen donor atom or via the donor atom at the other end. List these four possible ligands in the order you would expect to find them in the spectrochemical series, starting at the weak-field end. Underline the donor atom you are using in each case. Which two of these cases should have approximately equivalent positions in the spectrochemical series?

7.34. Consider the ligands acetonitrile (in the spectrochemical series, Eq. 7.6) and methyl isonitrile (in Table 7.6 when R = CH_3). Which ligand is likely lower in the spectrochemical series?

7.35. *Compute the actual CFSEs (in kJ mol^{-1}) for the fourth-period +2-charged hydrated ions, and express each CFSE as a percentage of the total hydration energy for that ion.

7.36. Compute the actual CFSEs (in kJ mol^{-1}) for the high-spin fourth-period +3-charged hydrated ions, and express each CFSE as a percentage of the total hydration energy for that ion.

7.37. *Explain the overall trends and the reversals in trends that occur in the following series of oxide lattice energies (in kJ mol^{-1}), all of which have the NaCl lattice structure: CaO (–3461), TiO (–3879), VO (–3912), MnO (–3808), FeO (–3921), CoO (–3988), and NiO (–4071). (Data from D. F. Shriver, P. W. Atkins, and C. H. Langford, *Inorganic Chemistry*, W. H. Freeman: New York, 1990; p. 226.)

7.38. Sketch plots of lattice energies versus d^n electron configuration for octahedrally coordinated fourth-period +2-charged ions in the following lattices: (a) lattice with a weak-field anion [e.g., $MCl_2(s)$]; and (b) lattice with a strong-field anion [e.g., $M(CN)_2(s)$]. For simplicity, assume all compounds use the same lattice type.

7.39. *Describe the order of stabilities of metal complexes of a given ligand as the metal is varied in the second half of the d block in the fourth period among ions (a) of +2 charge in strong-field complexes and (b) of +3 charge in weak-field complexes.

7.40. Sketch plots of how you expect ΔH and the stability constants to vary as a function of the d^n electron configuration for fourth-period +2-charged ions in the following equilibria:

(a) $[M(H_2O)_6]^{2+} + 6CH_3NH_2 \rightleftharpoons [M(CH_3NH_2)_6]^{2+} + 6H_2O$

(b) $[M(H_2O)_6]^{2+} + 6CN^- \rightleftharpoons [M(CN)_6]^{4-} + 6H_2O$

(c) Explain why the equilibrium constants for these reactions would be smaller, on the average, on the left side of the d block than on the right side.

7.41. *Sketch a plot of radii of fourth-period d-block metal ions in octahedral complexes, showing one plot for weak-field ligands and one for strong-field ligands, showing clearly where the "humps" occur. Explain why there are "humps" in these plots and why they occur at the specific d^n electron configurations at which they do occur.

7.42. Plots of ionic radius versus the number of d electrons should show which of the following kinds of plot? (a) Single-humped (minimum at d^6) with weak-field ligands and double-humped (minima at d^3 and d^8) with strong-field ligands; (b) single-humped (minimum at d^5) with weak-field ligands and double-humped (minima at d^3 and d^8) with strong-field ligands; (c) double-humped (minima at d^3 and d^8) with both kinds of ligands; (d) a smooth straight line with no humps, because complexes are not camels; or (e) double-humped (minima at d^3 and d^8) with weak-field ligands and single-humped (minimum at d^6) with strong-field ligands.

7.43. *Plot radii from Table C for a series of like-charged ions (your choice of charges, but there should be several examples) in the fifth (or sixth) period as a function of the d^n electron configuration of the ion. Tell whether this plot resembles one expected for weak-field or for strong-field complexes; indicate where you expect the maxima and the minima in this graph to fall.

7.44. Plot the trend you would expect for the Pauling electronegativities of the fourth-period elements when they have the +2 oxidation state and are in strong-field (low-spin) complexes. For which elements would new electronegativities result? Would these be higher or lower than those given in Table A?

7.45. *Using Figure 7.17, explain what would have to be done to the ligands in an eight-coordinate cubic complex to convert their positions to those found in an eight-coordinate square antiprismatic complex. Would this raise or lower the ligand–ligand electronic repulsions described by VSEPR theory? Would this raise or lower the ligand–d-electron repulsions described by crystal field theory for the following d orbitals: d_{xy}; $d_{x^2-y^2}$; and d_{z^2}?

7.46. Cubic coordination is known in ionic lattices. (a) What are two lattice types in which the metal ion shows cubic coordination? (b) These lattice types do not feature d-block cations with d^1–d^9 electron configurations that would be colored or show crystal field effects. Invent an unknown d^1–d^9 d-block metal ion that, if it existed, would likely take each of these two lattice types.

7.47. *What factors tend to favor the loss of two ligands by an octahedral complex to form a tetrahedral complex? What factors tend to oppose this?

7.48. An octahedral hydrated d^1 metal ion has Δ_o = 22,000 cm^{-1}. Two tetrahedral complexes of the same metal ion have been isolated, (a) with Δ_t = 18,600 cm^{-1} and (b) with Δ_t = 8400 cm^{-1}. Are their ligands likely π-donor ligands, π-acceptor ligands, or σ-donor ligands? Predict the colors of these two complex ions (assume no charge-transfer bands interfere).

7.49. *The color changes possible (and the range of v_1 absorptions) in the spectrochemical series of ligands are more spectacular for the complexes of nickel(II) ion than for any other metal ion. (a) The chloro complex of Ni^{2+} is yellow, with Δ = 3850 cm^{-1}. What is its likely geometry? Estimate the Δ_o expected for octahedral [NiCl$_6$]$^{4-}$. (b) The cyano complex [Ni(CN)$_4$]$^{2-}$ is red, with v_1 at 31,100 cm^{-1}. What is its likely geometry? Estimate the Δ_t expected for tetrahedral [Ni(CN)$_4$]$^{2-}$, and give a plausible interpretation of the discrepancy. (c) The four-coordinate dimethylglyoxime complex of Ni^{2+} is bright pink and is used in lipstick and in the qualitative detection of Ni^{2+}; it has four nitrogen donor atoms attached to Ni and has no absorption in the near-IR region. What is its geometry? (d) Which of these geometries of nickel(II) complexes could you identify on the basis of magnetic susceptibility measurements alone?

7.50. When we were studying the formation of oxo anions in Chapter 3, we emphasized the role of sp hybrid orbitals in occupying space that would otherwise be used to bind ligands, but de-emphasized the roles of d electrons, stating that these tend to fall between the ligands. (a) For which of the following types of d electrons is this presumption basically true and for which is it basically false? The t_{2g} electrons in octahedral complexes; e_g electrons in octahedral complexes; t_2 electrons in tetrahedral complexes; e electrons in tetrahedral complexes; d_{z^2} electrons in square planar d^8 complexes; $d_{x^2-y^2}$ electrons in square planar d^8 complexes. (b) What geometry does valence-shell electron-pair repulsion (VSEPR) predict for four-coordinate d^8 complexes if the d electrons are presumed to be stereochemically inactive (i.e., to not block incoming ligands)? What geometry does VSEPR predict if allowance is made for the type of d electrons that do block electrons? Is either prediction correct?

7.51. *The d orbitals in a particular geometry of complex are found to be split by the crystal field of that geometry of ligands into the following energy levels: $d_{xz} = d_{yz} = -0.457\Delta_o$; $d_{xy} = -0.086\Delta_o$; $d_{z^2} = +0.086\Delta_o$; $d_{x^2-y^2} = +0.914\Delta_o$. (a) Which geometry could produce this particular splitting: octahedral, tetrahedral, square pyramidal (pyramid base in xy plane), cube, or linear (ligands along z axis)? (b) Assuming that your ligands were strong-field ligands, what metal ion electron configuration (d^3, etc.) would be most likely to favor this geometry? Draw and fill in the energy-level diagram with arrows representing electrons to support your answer. (c) Give the symbol of a common transition metal ion that actually shows this electron configuration. (d) Calculate the CFSE for a Fe^{3+} ion in this geometry, with strong-field ligands being present.

7.52. Platinum has two common positive ions, Pt^{2+} and Pt^{4+}. One of these strongly prefers to form octahedral complexes, and the other strongly prefers to form square planar complexes. Which ion prefers octahedral? Which ion prefers square planar? Briefly explain your choices.

7.53. *A real example of a linear two-coordinate complex ion is NiO_2^{3-} [A. Möller, M. A. Hitchman, E. Krausz, and R. Hoppe, *Inorg. Chem.* 34, 2684 (1995)]. (a) Look up the spectrum of this ion, and determine whether the authors agree with the order of d orbitals deduced in Example 7.1. How do you explain the number of bands observed? (b) Estimate the value of Δ_o for this ion. (c) Compare this value with those found for hydrated ions in Table 7.4. Comment on the reasonableness of this value.

7.54. In recent years several more linear two-coordinate complex ions have been discovered [P. P. Power, *Chem. Rev.* 113, 3482 (2012)]. (a) What types of ligands are used to stabilize these complexes? (b) What metal ions are characteristically found for the metal ions? (c) Tables 2–6 in the Power review give the magnetic moments μ. Comparing with μ_{calc} and μ_{obs} in Table 7.2, deduce the typical numbers of unpaired electrons for each metal's complexes. (Note that the first metal, in Table 2, can have either the +2 or +1 oxidation state). (d) In Table 7.8 a largest splitting of $0.914\Delta_o$ is listed for linear two-coordinate complexes. Perhaps this is large enough to produce some low-spin complexes. Which metals might have high-spin and low-spin alternatives? Which spin is in fact observed?

7.55. *(a) Which of the following octahedral complex metal-ion electron configurations are expected to undergo a strong Jahn–Teller distortion: t_{2g}^1, t_{2g}^3, $t_{2g}^3 e_g^1$, $t_{2g}^3 e_g^2$, t_{2g}^6, $t_{2g}^6 e_g^1$, $t_{2g}^6 e_g^3$, and $t_{2g}^6 e_g^4$? (b) Briefly describe how the shape of an octahedral complex changes when a Jahn–Teller distortion occurs.

7.56. Using the chelating glycinate ion ($NH_2CH_2CO_2^-$), you prepare two octahedral (weak-field) complexes, $[Cr(NH_2CH_2CO_2)_3]$ and $[Cr(NH_2CH_2CO_2)_3]^-$. Would either one be subject to the Jahn–Teller (tetragonal) distortion? If so, which one? Would either likely become square planar? If so, which one?

7.57. *The formation of an octahedral complex of the bidentate chelating ligand ethylenediamine (en, $NH_2CH_2CH_2NH_2$) proceeds in three steps; for each step an equilibrium constant (the stepwise formation constant) can be measured. When en coordinates to Ni^{2+}, the three stepwise formation constants are each about 10^{10}. When en coordinates to Cu^{2+}, the first two steps also proceed with similar stepwise formation constants, but great difficulty is experienced in attempting to chelate a third en ligand (the third stepwise formation constant is only 0.1). What difficulty might there be in the third chelation of Cu^{2+} that is not present for Ni^{2+}?

7.58. What strong-field d^n octahedral complexes are strongly subject to the Jahn–Teller distortion?

7.59. *Predict the two crystal lattice types likely to be found for solid fluorides of the fourth-period +2-charged d-block metal ions. Which two of these fluorides would be expected to show substantial differences from either possible predicted lattice type? What would be the nature of these differences? Would you expect the Madelung constants for these two lattice types to apply accurately to these two fluorides?

7.60. Which of the following octahedral-complex metal-ion electron configurations are expected to undergo a strong Jahn–Teller distortion? (Some of these are excited-state configurations that could only be found after absorption of light or other energy, but ignore this factor and answer for these as well.) (a) t_{2g}^2; (b) t_{2g}^4; (c) t_{2g}^6; (d) $t_{2g}^4 e_g^2$; (e) e_g^1; (f) e_g^2; (g) $t_{2g}^3 e_g^3$; and (h) e_g^3.

7.61. *Consider the possibility of Cr^{3+} forming a tetrahedral complex with the ligand glycine, $[Cr(NH_2CH_2CO_2)_2]^+$. (a) Using the answer given for Exercise 7.25a, $\Delta_o \approx 19{,}000$ cm^{-1}, estimate the expected tetrahedral ligand field splitting Δ_t in this complex ion. (b) Compute the expected tetrahedral CFSE in reciprocal centimeters (cm^{-1}). (c) Compare this result with the octahedral CFSE computed in Exercise 7.25b to determine which geometry of complex is favored—octahedral or tetrahedral. By how many reciprocal centimeters is it preferred?

7.62. Six-coordinate complexes are normally octahedral, but there is an alternate geometry, trigonal prismatic (as seen in Figure 4.6 for CdI_2, for example). From Table 7.8, the energy levels of the d orbitals in a trigonal prismatic complex are d_{z^2}, $+0.096\Delta_o$; $d_{x^2-y^2}$, $-0.584\Delta_o$; d_{xy}, $-0.584\Delta_o$; d_{xz}, $+0.536\Delta_o$; and d_{yz}, $+0.536\Delta_o$. (a) Draw an energy-level diagram comparing (on the left side) octahedral-complex energy levels with (on the right side) energy levels for a trigonal prismatic complex. (b) For a d^2 metal ion, the CFSE is of greater magnitude for which geometry? By how much energy (in units of Δ_o)? (c) For a d^3 metal ion, the CFSE is of greater magnitude for which geometry? By how much energy (in units of Δ_o)? (d) Which geometry is favored by VSEPR theory?

7.63. *Give some reasons why biochemists, in their research, might want to replace the Zn^{2+} ion in carboxypeptidase A with another d-block metal ion, and why they would end up replacing the Zn^{2+} ion with Co^{2+} rather than Ni^{2+} or Cu^{2+}.

7.64. In a Connection in Section 7.6B we mentioned that the complex $[NiBr_2(PEtPh_2)_2]$ crystallizes in both the tetrahedral and low-spin square planar isomers. This implies that there is little total preference energy between the two geometries. What is the difference in CFSEs for the square planar and tetrahedral geometries for a d^8 metal ion?

7.65. *Although the low position of the heavier halide ions (Cl$^-$, Br$^-$, I$^-$) in the spectrochemical series (and their fairly large size) allows most halo ions of the fourth-period d-block metal ions to be tetrahedral, this tendency is not pronounced for Ni^{2+} and is completely absent for Cr^{3+}, which only forms octahedral halo anions. Explain. Also explain why the tendency is more completely absent for Cr^{3+}.

7.66. (a) If you were to react the $[Cr(glycinate)_3]$ complex with another ligand to displace one of the glycinate ions, how fast would you expect the reaction to proceed at room temperature: in nanoseconds (10^{-9} s); in 10^{-8} s; in 10^{-4} s; in minutes or even years? (Choose the most plausible answer.) (b) The $[Cr(glycinate)_3]$ complex is best described inert or labile?

7.67. *The substitution reactions of octahedral d^3 and d^8 metal complex ions are very slow. (a) Why are the reactions of $[NiL_6]^{2+}$ ions not nearly as slow as the reactions of $[CrL_6]^{3+}$ ions, which may require days or weeks at room temperature? (b) The reactions of octahedral $[CrL_6]^{3+}$ can be speeded up enormously by adding some reducing agent (such as Zn) that will reduce Cr^{3+} to Cr^{2+}, then letting the Cr^{2+} complex be reoxidized to Cr^{3+}. Explain why this works.

7.68. Why is it possible to separate and isolate the isomers of square planar cis-$[Pt(NH_3)_2Cl_2]$ and $trans$-$[Pt(NH_3)_2Cl_2]$? Why is it likely to be impossible to separate and isolate the isomers of square planar cis-$[Cu(NH_3)_2Cl_2]$ and $trans$-$[Cu(NH_3)_2Cl_2]$?

7.69. *A ligand substitution reaction with the corresponding hydrated metal ion is likely to be unsatisfactory to produce which of the following complexes? (Hint: Consider all types of reactivity of hydrated metal ions.) (a) $[Co(NH_3)_6]Cl_3$; (b) $[Fe(C_5H_5N)_6]Cl_3$; (c) $[Ni(C_5H_5N)_6]Cl_2$; and (d) $[Tc(NH_3)_6](PF_6)_4$.

7.70. Identify the six most inert, slowly exchanging hydrated ions in Figure 7.22. Discuss the role of CFSEs and cation acidities in placing these ions among the "slowest six."

7.71. *Explain how it could be that the act of coordinating an O_2 molecule to the Fe^{2+} ion in hemoglobin could cause the Fe^{2+} ion to move a significant distance from its original position out of the plane of the porphyrin ligand to a new position in the plane of the porphyrin ligand. With what unusual aspect of the O_2 binding ability of hemoglobin is this connected?

7.72. Multiple choice: When O_2 bonds to hemoglobin, (a) the O_2 oxidizes Fe^{2+} to Fe^{3+}; (b) the O_2 oxidizes Fe^{2+} to Fe^{3+} and coordinates to it as superoxide ion; (c) the O_2 converts weak-field Fe^{2+} to strong-field Fe^{2+}; or (d) the O_2 converts weak-field Fe^{3+} to strong-field Fe^{3+}.

7.73. *The complex $[Ni^{2+}(porphyrin^{2-})]$ is planar. What difficulties would be experienced in trying to make square planar porphyrin complexes by replacing Ni^{2+} with (a) Fe^{2+}; (b) Pd^{2+}; (c) Al^{3+}?

7.74. Cu^+ has virtually the same radius as high-spin Fe^{2+}, and it is also involved in some biological redox reactions. Suppose that it were placed in a porphyrin complex. What difficulties might prevent it from coordinating O_2 to form a Cu^{2+}–O_2^- complex?

7.75. *Why would the porphyrin ring of heme not function properly if all its double bonds were hydrogenated, so that the ring only contained single bonds?

7.76. What are two other diatomic molecules that also bond to the iron in hemoglobins and related complexes, functioning as neurotransmitters? Do they bond more effectively than oxygen? Why? How does this affect their toxicity?

K. ALEX MÜLLER (1927–) was born in Basel, Switzerland. His mother died when he was 11, at which time he enrolled in the Evangelical College in Schiers (eastern Switzerland). After graduating and then fulfilling his obligation to the Swiss military, he attended the Swiss Federal Institute of Technology (ETH) in Zurich, where Wolfgang Pauli was one of his teachers. He received both his undergraduate degree and his doctorate (in 1958) from ETH. In 1963, Müller accepted a position at the IBM Zurich Research Laboratory, where he remained until his retirement. For 15 years, the main focus of his research was on the ceramic $SrTiO_3$ and related perovskite compounds.

J. GEORG BEDNORZ (1950–) was born in Neuenkirchen, North-Rhine Westphalia, in what was then West Germany. Although he developed an interest in chemistry in high school, he ended up taking his undergraduate degree in crystallography at the University of Münster. In 1972 Bednorz spent the summer at the IBM Zurich Research Laboratory, where he first met K. Alex Müller. After subsequent stints at IBM, where Bednorz worked with $SrTiO_3$, he began his doctorate research at ETH Zurich under the supervision of Heini Gränicher and Müller.

   After obtaining his Ph.D. in 1982, Benorz joined the IBM lab and worked with Müller on the electrical properties of ceramics made from transition metal oxides. In 1986 they prepared a lanthanum barium copper oxide whose critical temperature (the temperature at which it becomes superconducting) was 35 K. This was 12 K higher than the previous record and stimulated much additional research into high-temperature superconductivity, yielding materials with higher and higher critical temperatures. Müller and Bednorz were awarded the 1987 Nobel Prize in Physics for their work.

CHAPTER 8

Oxides and Silicates as Materials

This chapter applies concepts from many of the previous chapters to a major area of current research: materials science. Because this area is very broad, we focus on one type of material: the oxides of the elements (including the geochemically important silicates). Sections 8.1 and 8.2 set the basis for the study of materials science, using oxides as examples. Section 8.1 sets the background by asking, "What is a material? What are the physical properties we look for in a material? What kinds of structures can have these desirable properties? Why are some substances solids that are considered materials, while others are gases that are not of interest to materials science?" Section 8.2 then applies the background questions specifically to oxides. We extend the process of drawing Lewis structures (Section 3.1) to allow for the possibility that the ionic or covalent formula unit may polymerize, and derive principles that allow us to predict possible structure types for oxides of the elements. Based on the structure types, we predict periodic trends in useful physical properties of oxides.

Sections 8.3 and 8.4 zero in on some properties of ionic solids that contribute to the properties of solid oxides as materials. In Section 8.3 we examine the ability of more than one kind of metal ion to fit in oxide lattices to give mixed-metal oxides, and we look at important types of mixed-metal oxide lattices. These concepts are applied to radiochemistry, the doping of crystals in physics, physical chemistry and bioinorganic chemistry, and the technologies of microphones and loudspeakers. Section 8.4 looks at the electrical and magnetic properties of metal oxides, which are extremely important in modern-day technology. What are high-temperature superconductors? How can electronic production of sound work? The principles of this section are applied to conduction in lithium-ion batteries and solid oxide-ion conductors in the control of pollution from automobile exhausts and the building of maglev trains.

Although the emphasis here in Chapter 8 is on physical properties of oxides, the chemical reactivity of oxides must be taken into account. Section 8.5 discusses the acid–base reactivity, solubility properties, and contributions to the atmospheric chemistry of oxides of the elements, including cement chemistry and the failure of Biosphere II. In reading Section 8.5D, note what nonchemical considerations are also very important in successful pollution control.

Sections 8.6 and 8.7 roll back the calendar a few billion years and talk about the materials that are most important in the formation of the solid Earth: the simple, oligomeric, and polymeric silicates (Section 8.6) and aluminosilicates (Section 8.7). What silicate "lasts forever"? In Section 8.6 we address questions about the following applications: Which silicates were manufactured many thousands of years ago for use as pigments? How might the transformation of silicates be related to deep earthquakes? For what chemical reason do people choose to live on the slopes of very dangerous volcanoes? What minerals are being found on Mars, and what do they tell us about the early history of Mars? In Section 8.7 we discuss the chemistry of water softening, soil fertility, converting petroleum to gasoline, and try to answer questions, such as could silicates have had any role in the origin of life on Earth? What type of silicate is valued as much for what is not there as for what is present in the silicate?

8.1. Materials and Their Physical Properties

OVERVIEW. In modern materials science we look for specific physical properties in chemical substances such as oxides. Predicting substances likely to show such properties begins with an understanding of the relationship between physical properties and the structure type of a substance. The structure type is shown pictorially in Figure 8.1 and can be indicated in formulas using dimensionality prefixes such as $\frac{2}{\infty}$. You may practice applying these classifications by reading Section 8.1A and trying the "(a)" parts of Exercises 8.1, 8.2, and 8.4. Selected physical properties of oxides are discussed in Section 8.1B and summarized in Tables 8.1 and 8.2. In this subsection you can learn to predict likely structure types and physical properties for a given oxide by trying Exercises 8.1–8.7. Materials scientists can achieve unusual physical properties for a structure type by making unusual changes in the basic structural unit, as in the example of ionic liquids (Section 8.1C; Exercises 8.8–8.9).

In Section 8.1C you again employ the concept of lattice energies from Section 4.2 and the concept of speciation (Section 3.8) as applied to structural units. The concepts of Section 8.1 involving physical properties, structural types, linkage, and physical properties will be used again in extensively in Chapter 12.

In the remainder of this chapter we will be concerned with the physical properties of compounds such as oxides. Many of these are of interest to **materials science**. Substances considered to be materials have more complex or extended structures than simple gaseous molecules or ion pairs or multiplets, or than simple molecular liquids, and are valued principally for their physical (rather than chemical) properties. Most materials of interest are solids, but these also include some liquids with unusual structures (e.g., liquid crystals). In this section we discuss some general principles that affect the physical properties of inorganic materials. The earth we stand on and many of the materials out of which we build the edifices of modern civilization consist of inorganic materials. For many modern purposes (in electronics, spacecraft, and so on) the properties must be such as to withstand extreme conditions, or to show great strength and lightness, or to have various unusual combinations of properties. The principles in this section would be of use in suggesting the kinds of new materials that you would want to synthesize if you were seeking new materials with certain unusual physical properties.

8.1A. Structural Types. Fundamentally, many of the physical properties of materials depend on how extensively its **structural units** (MO_x in an oxide; M in a metal) are **linked** together by intermolecular forces or chemical bonds. In Figure 8.1, we categorize eight structural types (a) through (h) in which the same structural unit (shown as a sphere) is linked in different ways and to increasing extents.

At one extreme, the bulk material (e.g., a crystal) may consist of numerous, separate, much smaller molecules linked only very loosely together by **intermolecular attractions** (Section 8.1B). In these **monomeric molecular** substances (Figure 8.1a), the structural unit is a (at least fairly) simple molecule.

At the other extreme, the entire piece (e.g., the crystal) may consist of structural units that are linked into one giant three-dimensional **network polymer** (Figure 8.1g) or **ionic lattice** (Figure 8.1h) by strong chemical-bonding forces.

Between these extremes there are several intermediate possibilities, as illustrated in Figure 8.1. Next in complexity after the monomers are the **oligomeric** molecules or ions (Figure 8.1b, c, and d), which contain a few structural units linked together. In **chain oligomeric** molecules or ions (Figure 8.1b), each central structural unit is linked to *two* others, but after a definite number of units have been linked, the chain is terminated by a slightly different unit with only *one* link. (Examples include the homopolyatomic ions discussed in Section 3.1A.) In **cyclic oligomeric** molecules or ions (Figure 8.1c), all units are linked to *two* others and form a ring, so there are no terminal

I. MOLECULAR STRUCTURE TYPES

IA monomers

(a) Molecular

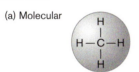

IB oligomers

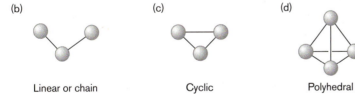

(b)   (c)   (d)

Linear or chain   Cyclic   Polyhedral

Figure 8.1. Main structure categories of materials. Structural units are represented by spheres and linking forces (covalent bonds or interionic attractions) are represented by lines.

II. POLYMERS (Macromolecular and ionic)

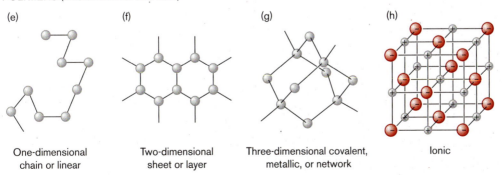

(e)   (f)   (g)   (h)

One-dimensional chain or linear   Two-dimensional sheet or layer   Three-dimensional covalent, metallic, or network   Ionic

units. In **polyhedral oligomeric** molecules or ions (Figure 8.1d) the units are linked in a small three-dimensional assembly to *three* or more other units.

Next in complexity are **polymeric** substances (Figure 8.1e–h). These can be divided into four categories. In **linear** or **chain** or **one-dimensional polymers** (Figure 8.1e), each structural unit is linked to *two* others to give long-chain molecules or ions. Their one-dimensional polymerization is often denoted with $_\infty^1$ in front of the formula of the structural or simplest-formula unit—for example, $_\infty^1$ [CH$_2$CH$_2$] for polyethylene. Sometimes a subscript *n* is used instead—for example, (CH$_2$CH$_2$)$_n$. In **layer** or **sheet** (or **two-dimensional**) **polymers** (Figure 8.1f) (denoted with $_\infty^2$, such as $_\infty^2$[C] for graphite), many or all structural units are linked to *three* other units, so the molecule or ion extends indefinitely in two dimensions. In either **ionic** (Figure 8.1h) **network** (or **three-dimensional**) **polymers** (Figure 8.1g) (denoted with $_\infty^3$, such as $_\infty^3$[SiO$_2$] for silicon dioxide), the individual structural units have enough links with orientations allowing the molecule or ionic lattice to extend indefinitely in three dimensions.

EXAMPLE 8.1

Which pairs of structure types in Figure 8.1 represent polymorphs of each other (Section 4.1C)?

SOLUTION: In Section 4.1C polymorphs are defined as different atomic arrangements (of a given structural unit) involving the same coordination numbers (of the unit's central atom). Let us take for granted that each structure type in Figure 8.1 has the same structural unit, with the same number *x* of terminal atoms bonded to the central atom. In addition, the central atom contains a certain number ℓ of links. (As a result, the coordination number of the central atom of each structural unit equals $x + \ell$.) Polymorphs have the same number of links ℓ. There are two pairs of polymorphs in Figure 8.1: cyclic molecules (Figure 8.1c) and linear polymers (Figure 8.1e), with two links to each central atom or structural unit, and polyhedral molecules (Figure 8.1d) and layer polymers (Figure 8.1f), with three links.

8.1B. Physical Properties. In Table 8.1 we summarize these eight categories of structure type, their structural units, links, what must happen to break the links, and then three important physical properties: hardness, electrical conductivity, and solubility.

As postulated in the ideal **kinetic-molecular theory** of gases, molecules capable of existing in the gaseous state exert no attractive forces on each other, move independently of each other, and remain relatively far apart. In practice, however, there are attractive forces between molecules (**intermolecular attractions**). The weakest of these is the van der Waals force, which occurs between molecules of all substances. The van der Waals force results when a momentary unsymmetrical distribution of electrons in one molecule (a temporary dipole) induces an opposed momentary unsymmetrical distribution of electrons in a neighboring molecule. The momentarily oppositely charged ends of the two molecules then attract each other. In small molecules the van

TABLE 8.1. Structure Types and Typical Properties of Solids

Material Type[a]	Structural Units	Held Together By	Vaporization Processes	Hardness	Electrical Conductivity	Solubility
Molecular (**a**)–(**d**)	Molecules	London forces, hydrogen bonds	Break these intermolecular forces	Soft	Poor	In appropriate solvents
Metallic (**g**) (Chapter 12)	Atoms (or ions)	Covalent bonds (or sea of electrons)	Break metallic bonds → M, M$_2$	Variable	Good	In other metals
Covalent network (**g**) (Chapter 12)	Atoms or molecular fragments (MO$_x$)	Covalent bonds	Break covalent bonds → monomers, oligomers	Hard	Usually poor	Insoluble in all solvents
Ionic (**h**) (Chapter 4)	Oppositely charge ions	Attractions of opposite charges	Lattices → ion pairs or multiples	Variable, brittle	Usually poor (good as liquid)	Sometimes in water
Chain (**e**)	Atoms or molecular fragments (MO$_x$)	Covalent bonds, van der Waals forces	Break covalent bonds → monomers, oligomers	Soft	Usually poor	In appropriate solvents
Layer (**f**)	Atoms or molecular fragments (MO$_x$)	Covalent bonds, van der Waals forces	Break covalent bonds → monomers, oligomers	Soft, slippery	Usually poor, can be good	Insoluble in all solvents

[a] The boldfaced letters in parentheses refer to the structure types in Figure 8.1.

der Waals force is very weak as compared to the average kinetic (thermal) energies that the molecules possess at room temperature. But in very large molecules with numerous highly polarizable (relatively loosely bound) electrons, the van der Waals forces may be so substantial that the kinetic energy available to molecules at room temperature is not enough to overcome such an attraction. Hence monomeric or oligomeric substances consisting of sufficiently large molecules are not gaseous at room temperature and atmospheric pressure.

The kinetic-molecular picture of the liquid state postulates that molecules or ions are held close by forces that are strong compared with the thermal energy at that temperature. But the molecules or ions are free to flow around each other, which implies that their intermolecular attractive forces can be stretched and modified. The molecules or ions must also either be relatively small, or if they are chain polymers, the chains must not be rigid; they cannot be layer or network polymers. Medium-sized molecular substances, or smaller molecules linked by somewhat stronger forces such as dipole–dipole forces and hydrogen bonding (i.e., water and alcohols), are most likely to be liquids at room temperature.

In the solid state molecules or ions are packed closely together and exert attractive forces on one another that are strong enough to hold the units in position in a lattice despite their vibrational energies. Large covalent molecules often have strong enough van der Waals forces to be solids at room temperature, but these forces seldom rival in strength the inter-unit covalent bonds or Coulombic forces found in the high-melting solid two- or three-dimensional polymeric substances.

Thus, *molecular substances* (Figure 8.1a–d), and especially monomeric molecular substances (Figure 8.1a), tend to be gases, liquids, or low-melting (i.e., below 300°C) solids at room temperature, and they have relatively low heats of fusion and heats of vaporization. As specific examples (Table 8.2) we may cite F_2 (in which the weak intermolecular attractions are due to van der Waals forces) and glycerol [$C_3H_5(OH)_3$], in which the stronger intermolecular forces are due to hydrogen bonding. Other properties generally follow from the absence of strong intermolecular attractions. Molecular substances tend to have higher solubilities and may even be soluble in nonpolar solvents. Like a brick wall with no mortar, these substances are lacking in structural strength, so they are deformable (soft in a mechanical sense, such as wax).

At the other extreme, the structural units of *three-dimensional network* or *ionic* substances (Figure 8.1g, h) are held together by strong chemical forces—namely, covalent or ionic bonding, which extends throughout the grain or crystallite of the

TABLE 8.2. Examples of Different Classes of Solids and Thermal Data Relating to Their Ease of Vaporization

Class[a]	Example	ΔH_{fus} (kJ mol^{-1})	Melting Point (°C)	ΔH_{vap} (kJ mol^{-1})	Boiling Point (°C)
Molecular (**a**)	F_2	0.5	~218	6.6	~188
	$C_3H_5(OH)_3$	18.3	18	61	290
Chain (**e**)	Selenium	6.69	220	96	685
Layer (**f**)	CdI_2	15.3	387	115	742
	Graphite	117	4489		
Metallic (**g**)	Na	2.6	98	99	1052
	W	52.3	3422	824	5500
Covalent Network (**g**)	Diamond			~713	4350
	Cristobalite	9.6	1722	516	3270
Ionic (**h**)	NaCl	28.2	801	170	1465
	SrF_2	28.5	1477	297	2477
	ZrO_2	87	2709		4300

Source of thermal data: D. R. Lide, Ed., *CRC Handbook of Chemistry and Physics*, 84th ed., CRC Press: Boca Raton, FL, 2003, pp. 6-109–6-139.

[a] The boldfaced letters in parentheses refer to the structure types in Figure 8.1.

substance. It is very difficult for the structural units to acquire mobility with respect to each other. At room temperature these substances are high-melting solids with high heats of fusion and vaporization (Table 8.2).

The subcategories of polymers have some characteristic variations on the typical properties of polymers. The one-dimensional chain polymers (Figure 8.1e) typically differ from other polymers in that the bonding usually does not give rigidity to the material. Consequently, chain polymers can be mechanically soft and have low melting points. As liquids they are viscous, since the long, spaghetti-like molecules can easily become entangled and hence resistant to flow. (The most familiar examples of these properties include thermoplastic organic polymers such as polyethylene.)

The two-dimensional layer polymers (Figure 8.1f) tend to be rigid in those two dimensions, but not in the third. These often consist of flat sheets that readily slide over each other (some are used as lubricants). They are mechanically soft in the third dimension. Examples include graphite, $CdCl_2$, and CdI_2 (Figure 4.5c; Section 4.1c).

The process of melting network polymeric (Figure 8.1g) and ionic (Figure 8.1h) solids requires stretching and bending strong inter-unit attractions, so it occurs at high temperatures (Table 8.2). This can happen for ionic solids, because Coulombic attractions are omnidirectional, and another attraction can replace the one lost when an ion becomes mobile. When *covalent* layer or network solids are melted, however, numerous covalent bonds must be ruptured. At this high temperature, any remaining linking covalent bonds may also be ruptured, so gaseous small molecules are formed. Hence, covalent layer or network solids frequently **sublime** (pass directly from the solid to the gaseous state) or decompose at high temperatures, rather than truly melting and boiling.

Densities. The distances between the separate molecules (Figure 8.1a–d) in the solid (or liquid) phase are governed by the relatively long van der Waals radii of whatever atoms are on the outside of the units, so molecular substances tend to have relatively low densities. Among the subcategories of polymers, the more-polymerized form tends to have the higher density and greater structural strength and mechanical hardness (Section 8.2B). Because the structural units are strongly bonded together at distances that are the sums of relatively short covalent or ionic radii, these materials have relatively high densities. As examples of the density trend involving the same structural unit, we may compare diamond, $^3_\infty[C]$ (density 3.51 g cm^{-3}), with graphite, $^2_\infty[C]$ (2.25 g cm^{-3}), and polyhedral molecular C_{60}, (1.65 g cm^{-3}).

EXAMPLE 8.2

Molybdenum(IV) sulfide is a solid (mp 1750°C) that is slippery; it is used as a lubricant in products such as WD40™. (a) To which category in Figure 8.1 does this solid likely belong? (b) Which would be the best concise way of indicating this category of solid: MoS_2, Mo_4S_8, $^3_\infty[MoS_2]$, or $^2_\infty[MoS_2]$?

SOLUTION: The very high melting point suggests a polymeric structure (but probably not one-dimensional). The property of being slippery suggests a two-dimensional layer structure (Figure 8.1f) with weak intermolecular forces between the layers—hence, $^2_\infty[MoS_2]$.

8.1C. New Physical Properties for Old Structure Types: Ionic Liquids.

The properties typically associated with a given structural type can often be modified considerably by changing details of the structures, or by making *composites* of two structure types. For example, ionic liquids[1] comprise an important group of exceptions to the general trends discussed above. When (nonacidic) cations and (nonbasic) anions are both *very* large, the lattice energies of their salts can become low enough to allow melting at less than 100°C or even below room temperature.[2] In practice, ionic liquids contain large nonacidic organic cations such as R_4N^+ (tetraalkylammonium) (Figure 8.2a), R_4P^+ (tetraalkylphosphonium) (Figure 8.2b), $C_5H_5NR^+$ (alkylpyridinium) (Figure 8.2c), and most frequently $C_3H_3(NCH_3)(NR)^+$ (1-alkyl-3-methylimidizolium) (Figure 8.2d), along with large nonbasic inorganic or fluorinated organic anions such as $Al_2Cl_7^-$ (Figure 8.2e), $AlCl_4^-$, Cl^-, Br^-, NO_3^-, SO_4^{2-}, $CF_3CO_2^-$ (Figure 8.2f), $CF_3SO_3^-$ (Figure 8.2g), BF_4^-, and PF_6^-.

The lattice energy of CsCl, with a large cation–anion distance of 348 pm, is −670 kJ mol^{-1} and its melting point is 645°C (918 K). 1-Ethyl-3-methylimidizolium (Figure 8.2d) hexafluorophosphate (PF_6^- anion) is anticipated to have a similar "radius" ratio to CsCl, but a much longer cation–anion distance of around 510 pm.[3] Proportionately its lattice energy might be expected to be in the neighborhood of −450 kJ mol^{-1};[4] its melting point is 60°C (333 K). However, the lattice energy is still much larger than the London, dipole–dipole, or hydrogen-bonding forces that hold molecules in the liquid state, so ionic liquids have essentially zero vapor pressure. They therefore do not contribute to air pollution and are considered to be *green solvents*. Their major disadvantage is that they are quite viscous due to the cation–anion attractions that are present.

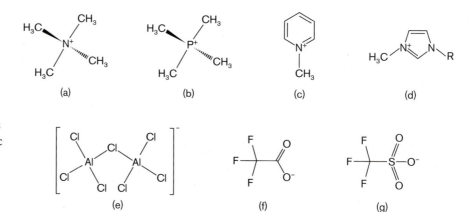

Figure 8.2. (a)–(d) Large nonacidic organic cations and (e)–(g) large nonbasic inorganic or fluorinated organic anions found in ionic liquids.

EXAMPLE 8.3

Let us contrast the properties of the two ionic compounds, BaI_2 and $[Ba(18\text{-crown-}6)(H_2O)_2](AlI_4)_2$. Compared to BaI_2, $[Ba(18\text{-crown-}6)(H_2O)_2](AlI_4)_2$ should have (a) a higher or lower lattice energy, (b) a higher or lower melting point, and (c) a higher or lower vapor pressure?

SOLUTION: With the larger cation and the larger anions, $[Ba(18\text{-crown-}6)(H_2O)_2](AlI_4)_2$ is more likely to be an ionic liquid. Therefore, it should have (a) a lower lattice energy, (b) a lower melting point, but (c) a faintly higher but still negligible vapor pressure.

AN AMPLIFICATION

Solubility and Ionic Liquids

Ionic liquids tend to be immiscible with nonpolar organic solvents. They somewhat resemble moderately polar organic solvents in being good solvents for many organic molecules.[5] Their miscibility with water and their ability to dissolve inorganic salts can be modified by the choice of the anion. With the Lewis base Cl^- as the anion, the ionic liquid is miscible with water. It is Lewis basic and can coordinate to acidic cations, dissolving (for example) acidic metal halides by forming halometallate anions:

$$AlCl_3(s) + (EMIM)^+Cl^-(l) \rightarrow (EMIM)^+[AlCl_4]^- \qquad (8.1)$$

[$(EMIM)^+$ is the cation shown in Figure 8.2d when R is the ethyl group, $-CH_2CH_3$.] With excess $AlCl_3$, the anion $[Al_2Cl_7]^-$ (Figure 8.2e) is formed, which tends to dissociate in solution to $AlCl_4^-$ and $AlCl_3$. The latter is a strong Lewis acid, so the ionic liquid is a Lewis acid. With large nonbasic anions such as $[PF_6]^-$, the ionic liquids are neither acidic nor basic; these salts are water insoluble by Solubility Tendency III (Section 4.3).

Given this variability of acid–base properties and miscibilities, it is often possible to find an ionic liquid to dissolve nonpolar organic substrates and ionic metal salts (as reactants or catalysts) in the same (ionic liquid) phase. After the reaction is complete, a nonpolar organic product can be extracted by an immiscible nonpolar liquid, such as hexane or supercritical CO_2, and recovered by evaporation. The remaining inorganic salt can be extracted with immiscible water, leaving the pure ionic liquid for reuse. If supercritical CO_2 has been used, there has been no air pollution.

8.2. The Structure Types and Physical Properties of the Oxides of the Elements

OVERVIEW. There are dramatic changes in structure types on crossing certain boundaries in the periodic table of the highest oxides of the elements (Table 8.3; Section 8.2A). Possible structure types can be predicted using an expansion of the procedure for drawing Lewis dot structures, which tends to assume a total coordination number of four. Higher total coordination numbers can be predicted in later periods of the periodic table using either the concept of changes in maximum/penultimate total coordination numbers or radius ratios. Equation 8.2 also gives us the number of coordinate-covalent links units form to each other via bridging fluorine or oxygen atoms. Hence, we can predict periodic trends in structure types. Section 8.2 also gives you a good chance to review the interpretation of observed bond lengths and bond angles. You may review these concepts by working Exercises 8.10–8.19.

You can then predict periodic trends in physical states, melting points and boiling points, enthalpies of fusion and vaporization, and usefulness as ceramics among fluorides or oxides as a function of their degrees of polymerization. Details of the effects of pressure on polymerization isomerism, and properties of ceramics such as brittleness and preparation by sintering, can be understood (Section 8.2B). You may try predicting these trends with Exercises 8.20–8.24.

Section 8.2 builds upon earlier concepts of covalent radii (Section 1.5B), electronegativity differences (Section 1.6), drawing Lewis structures (Section 3.1), total coordination numbers (Sections 3.4 and 3.5), and radius ratios and polymerization isomers (Section 4.1). Periodic trends in structure types will appear again in Section 12.1.

8.2A. Predicting Trends in Structure Types of Oxides. The overall goal in this section is to see (by example) *why* some oxides are materials and others are molecular. There are too many oxides of elements to enumerate or describe them all; Table 8.3 summarizes structural features of the oxides of elements in their highest oxidation states. Oxides in taller columns have greater molecularity.

The oxides in the shaded boxes have molecular structures, as also indicated at the bottom of the tallest columns. The large majority of the highest oxides are in short columns, which signifies polymeric structures. At the bottom of each box the type of polymeric structure of the major polymerization isomer is indicated. In this subset of oxides of the elements we may note three trends:

1. There are relatively *few molecular oxides* (taller columns in Table 8.3); these tend to be associated with relatively high oxidation states.

TABLE 8.3. Highest Oxide of Each Element and Its Structure Type[a,b]

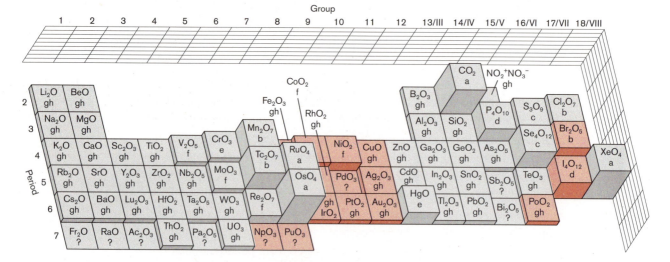

Source of most data: A. F. Wells, *Structural Inorganic Chemistry*, 5th ed., Oxford University Press: Oxford, UK, 1983.

[a] Solid-state structure types are as indicated in Figure 8.1 except as follows: **?** = unknown structure; **gh** = network (**g**) or ionic (**h**) structure.

[b] Entries in **brown** are for oxides in which the oxidation number is below the group number.

2. There are *numerous polymeric oxides* (shorter columns in Table 8.3), which might capture the interest of materials scientists. Among the polymeric oxides, the large majority have three-dimensional network covalent (Figure 8.1g) or ionic (Figure 8.1h) structures.

3. The transitions from the molecular structure types in Table 8.3 to ionic/network covalent structure types are often abrupt and result mainly from changes in oxidation number (and therefore structural unit).

Table 8.3 suggests that the overwhelming majority of oxides involve extensive bridging of structural units by oxygen atoms or ions to give *polymeric solid materials.* The O^{2-} ion is a very strongly basic anion, and it very frequently acts as a bridging ligand. Every time it forms a bridge to the central atom of another MO_x structural unit, it does so via a coordinate covalent bond, donating an electron pair to an empty orbital (a **vacant coordination site**, VCS) on another M oxide structural unit.

In a few cases (Table 8.3) oxides form chain polymers (Figure 8.1e). Examples include $^1_\infty[SO_3]$, $^1_\infty[CrO_3]$, and $^1_\infty[HgO]$. In a few cases oxides form layer polymers (Figure 8.1f); examples include $^2_\infty[Re_2O_7]$, $^2_\infty[V_2O_5]$, $^2_\infty[P_2O_5]$, and $^2_\infty[P_2O_3]$. (Notice that these are close to the network covalent or ionic structures in the periodic table of Table 8.3.)

Because Table 8.3 only shows a subset of oxides, we need a procedure to make predictions that apply to any binary compound. In many ways the procedure is similar to that for drawing Lewis structures (Section 1.4), but it adds three options. First, it is applicable both to ionic and covalent compounds. Second, it clarifies when the central atom may have more than an octet of electrons. Third, it allows a molecule/structural unit to undergo polymerization. Here we illustrate and apply the abstract principles of Section 8.1 to the prediction and understanding of structure types and the resulting trends in physical properties of the oxides of the elements.

A procedure for generating plausible structures for oxides (or fluorides) is outlined in Figure 8.3. The procedure may generate more than one plausible structure/polymerization isomer, but these should not be wildly different (e.g., not the structure types in Figures 8.1a and 8.1g), and should guide us to reasonable expectations of properties that are observed (Section 8.2B).

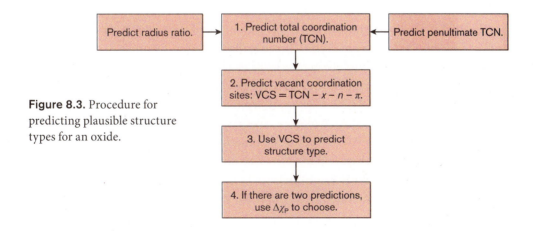

Figure 8.3. Procedure for predicting plausible structure types for an oxide.

Step 1. Start with the structural unit $(:)_n MO_x$ with only one M atom and x oxygen atoms (x can be a fraction). Generate one or two predictions of the total coordination number (TCN) of the central (normally the metal) atom or ion. One prediction is based on the radius ratio (Section 4.1). The other prediction is based on the expected penultimate TCN (Section 3.5).

The radius-ratio prediction is based on a presumption of ionic bonding. Because the ionic radius of oxide ion is 126 pm (Table C), use of the radius-ratio principle predicts eight-coordination for metal ions that are larger than 0.714×126 pm = 92 pm. Six-coordinate metal ions are predicted to be those with radii between 0.414×126 pm = 52 pm and 92 pm. Four-coordinate metal ions are predicted for ions with radii between 0.255×126 pm = 28 pm and 52 pm. If a metal ion is close to one of these boundaries, it is best to carry forth *both* predicted coordination numbers, because the radius ratio does not give precise results, and there may be polymerization isomers.

Step 2. The number of vacant coordination sites (VCS) of the central atom in the structural unit $(:)_n MO_x$ is then obtained by subtracting the number of oxygen (or other outer) atoms x, the number of central-atom unshared sp electron pairs n, and the number of M–O π bonds in the structural unit from the expected total coordination number obtained in Step 1:

$$VCS = \text{Links per } (:)_n MO_x = TCN - x - n - \pi \qquad (8.2)$$

Step 3. Select the structure type based on the number of VCS calculated from Equation 8.2: If VCS = 0, then the predicted structure type is a monomeric molecule (Figure 8.1a). If VCS is a fraction, then it is a chain oligomer (Figure 8.1b). If VCS = 1, then we predict a cyclic oligomer (Figure 8.1c) or a chain polymer (Figure 8.1e). If VCS = 1.5 or more, then we predict a polymer: a polyhedral oligomer (Figure 8.1d), a layer polymer (Figure 8.1f), a network polymer (Figure 8.1g), or an ionic solid (Figure 8.1h). Higher VCS numbers suggest higher degrees of polymerization. (However, a distinction between Figure 8.1g and Figure 8.1h is not expected yet.) Rarely, a *negative* VCS number may be computed, in which case one of the outer atoms in the structural unit is *expelled* and the compound is ionic—for example, NF_5 has TCN = 4, so VCS = –1. The compound is $[NF_4^+](F^-)$.

Step 4. If different predictions (e.g., Figure 8.1g, h) have resulted starting from the ionic-based radius-ratio rule and the covalent-based penultimate TCN, the electronegativity difference may now help narrow the choice of the more likely structure type and (if desired) the lattice type. This choice is illustrated in Example 8.5, in which the large Mg–O electronegativity difference leads us to favor the results of the radius-ratio calculation, the six-coordinate NaCl lattice type, over the four-coordinate prediction based on the penultimate TCN. Also, recall from Section 4.1C how, for a metal TCN of six in MX_2 or MX_3, increasing covalency favors layer structures (Figure 4.6b) over the ionic TiO_2 (Figure 4.2) or AlF_3 structures. Sometimes the same structure type is chosen regardless of the covalency: One of the ZnS structures (Figure 4.2) may be chosen for TCN = 4 regardless of the electronegativity difference (it is found both for oxides and sulfides). Step 4 is illustrated in Example 8.5 and Figure 8.4.

EXAMPLE 8.4

Without using radius ratios, calculate the numbers of VCS or links per structural unit for the (real or imaginary) oxides of third-period elements: MgO, SiO_2, SO_3, $ClO_{3.5}$, and "ArO_4."

SOLUTION: If we assume a constant penultimate TCN of four for these central atoms (Step 1), we see that this is achieved in the structural unit only in the hypothetical ArO_4; this is the only one of these oxides with no VCS or links and a monomeric molecular structure (Figure 8.1a). In all other cases the central atom falls short of its expected (penultimate) coordination number (corresponding to an octet of electrons). Thus, the central atoms of these oxide structural units have 3, 2, 1, and 0.5 VCS, respectively (Step 2). We can represent these potential Lewis-acid sites by empty parentheses: $(\,)_3MgO$, $(\,)_2SiO_2$, $(\,)SO_3$, and $(\,)_{0.5}ClO_{3.5}$. If π bonding is excluded, then the only way of achieving octets is by linking with neighboring structural units through bridging oxygen atoms or oxide ions.

EXAMPLE 8.5

Extend the results of Example 8.4 to predict the likely structure types for the remaining oxides: $ClO_{3.5}$, SO_3, SiO_2, and MgO.

SOLUTION: By Step 3, $(\,)_{0.5}ClO_3$ forms one oxygen link between two structural units, which produces an $O_3Cl–O–ClO_3$ (Cl_2O_7) "dimer" (Figure 8.4a). This conclusion is supported because we can draw a good Lewis structure for Cl_2O_7.

$(\,)SO_3$ has one VCS per structural unit. Each $(\,)SO_3$ structural unit donates an electron pair from one of its oxygen atoms to form an oxygen link and accepts an electron pair from an oxygen atom of another $(\,)SO_3$ structural unit to form another link, resulting in doubly linked structural units (Figure 8.4b). As shown there, at least two polymorphs can result: cyclic S_3O_9 and chain $^1_\infty[SO_3]$. Both conclusions are supported, because we can draw good Lewis structures for both.

$(\,)_2SiO_2$ (Figure 8.4c) has *two* VCS, so we anticipate correctly that it is likely to polymerize to a greater extent than does $(\,)SO_3$, using every oxygen atom and every VCS in quadruply linking the structural units, and giving each bridging oxygen atom a coordination number of *two*. Figure 8.4c shows the first step in polymerization to form large hexameric rings, $(SiO_2)_6$, of varying conformations. Formation of the six links to give this ring takes care of one VCS and one oxygen atom per SiO_2 unit. The other vacant site and oxygen atom per unit leads to three-dimensional polymerization to give the SiO_2 polymorphs shown in Figure 4.7a–c.

The $(\,)_3MgO$ structural unit has three VCS if its TCN is four. But (Step 4) radius-ratio calculations predict a TCN of six. In either case we predict polymerization in three dimensions to give a network or an ionic solid. Because the electronegativity difference of Mg and O is large, the radius-ratio calculations are more likely to be correct. In practice, MgO exhibits six-coordination for both Mg and O in a NaCl ionic lattice.

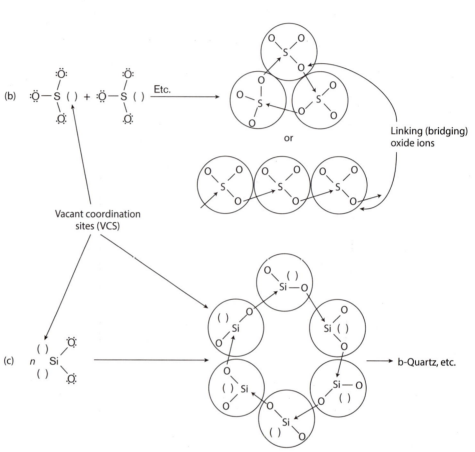

Figure 8.4. Lewis acid–base reactions between structural units (enclosed in circles in products) having vacant coordinate sites (represented by parentheses) to generate different polymerization isomers. (a) Dimerization of the $ClO_{3.5}$ structural unit to give Cl_2O_7. (b) Oligomerization or polymerization of S_3O_9 to give S_3O_9 or $^1_\infty[SO_3]$. (c) Cyclooligomerization of SiO_2 structural units to give intermediate *cyclo*-Si_6O_{12} units. These units still have VCS and polymerize further to give the polymorphs shown in Figure 4.7a–c.

EXAMPLE 8.6

Use radius ratios and the ions in Table C to predict the identities of the monomeric hexafluorides of the elements. Compare your predictions with the predictions you would obtain using maximum total coordination numbers.

SOLUTION: To be monomeric molecular, the coordination number of M in the structural unit must equal the number of F atoms in the formula. To obtain an MF_6, the radius ratio should be more than 0.414 but less than 0.732, so the radius of M^{6+} should be between 49 pm and 87 pm. There are a number of qualifying M^{6+} ions in Table C, including all of those of Groups 6, 16/VI, and the seventh-period elements of Groups 6F–8F. In practice, over a dozen monomeric MF_6 molecules are known.[6] This is more than would be expected based on maximum coordination numbers exceeding six for fifth-period elements or below. (The very electronegative fluorine atom of covalent fluorides is a very poor donor atom and thus is a very poor bridging atom.)

8.2B. Oxides and Their Physical Properties. In Figure 8.5 we show examples of structurally well-characterized molecular oxides of the elements. In Table 8.4 we tabulate ranges of important physical properties (melting points and boiling points, and heats of fusion and vaporization) of molecular and polymeric oxides.

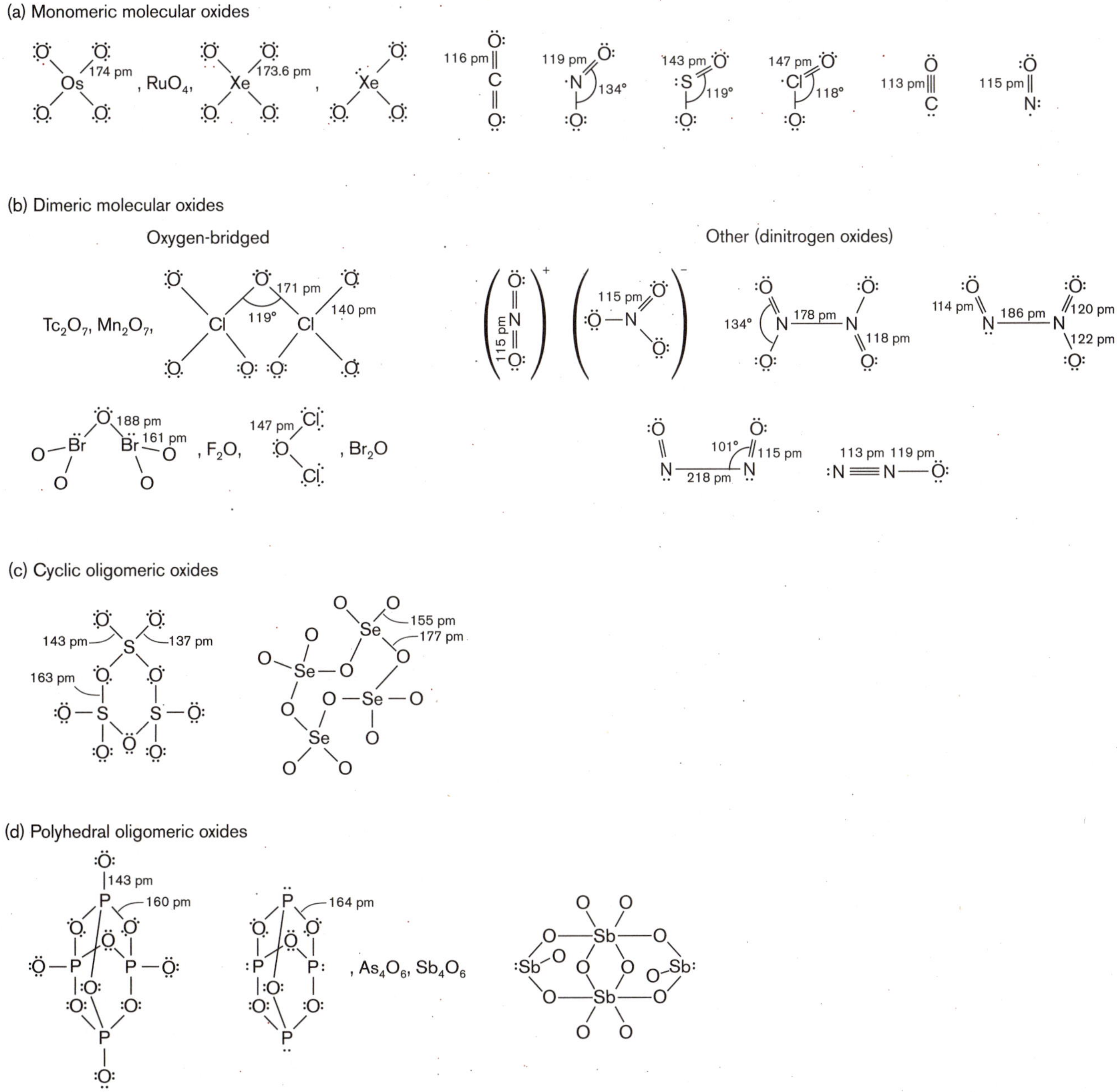

(a) Monomeric molecular oxides

(b) Dimeric molecular oxides

Oxygen-bridged   Other (dinitrogen oxides)

(c) Cyclic oligomeric oxides

(d) Polyhedral oligomeric oxides

Figure 8.5. Examples of monomeric, dimeric, cyclic oligomeric, and polyhedral oligomeric oxides. The letters in parentheses refer to the structure types in Figure 8.1. Where known, bond lengths and angles are also shown.

TABLE 8.4 Trends in Selected Physical Properties of Oxides by Structure Type

Structure Type[a]	Range of Oxides	ΔH_{fus} (kJ mol^{-1})	Melting Point (K)	ΔH_{vap} (kJ mol^{-1})	Boiling Point (K)
Monomeric (**a**)	CO to OsO$_4$	0.8–9.8	68–313	6–39.5	82–303
Dimeric (**b**)	F$_2$O to Tc$_2$O$_7$		39–392	26–59	128–485
Cyclic or Cluster (**c** and **d**)	S$_3$O$_9$ to Sb$_4$O$_6$			40–109	318–633
Ionic M^{2+}O^{2-} (**h**)	BaO to BeO	46–86	2193–2803	151–309	3023–4123
Ionic (M^{3+})$_2$(O^{2-})$_3$ (**h**)	Ga$_2$O$_3$ to Y$_2$O$_3$	105–111	2019–2711	290–402	2723
Ionic (M^{4+})(O^{2-})$_2$ (**h**)	ThO$_2$ to ZrO$_2$	87	2992–3763	514–567	4673–5123

[a] The boldfaced letters in parentheses refer to the structural types in Figure 8.1.

Physical Properties of Monomeric Molecular Oxides. The simplest monomers are XeO$_4$, RuO$_4$, OsO$_4$, IrO$_4$,[7] and :XeO$_3$. Because oxygen is good at forming π bonds, there are also doubly bonded and triply bonded monomeric molecular oxides, such as SO$_3$ (one polymerization isomer), :SO$_2$, CO$_2$, and :CO (Figure 8.5a). Also there are some simple oxides that not only involve π bonds but also are stable free *radicals* with odd numbers of valence electrons—namely, NO$_2$, ClO$_2$, and NO. For the monomeric molecular oxides, the structural unit is the molecule, so (except for the free radicals) it should be possible to draw a good Lewis structure for the structural unit.

The monomeric oxides have relatively low melting points, boiling points, enthalpies of fusion, and enthalpies of vaporization (Table 8.5)—it requires but little energy to overcome the weak intermolecular forces holding these molecules together in the liquid or solid.

Physical Properties of Dimeric Molecular Oxides. When the element of the oxide has an odd oxidation state, this requires the presence of two atoms of the element, and of one bridging oxygen atom to link the two element atoms in the dielement hepta–, penta–, and monoxides M$_2$O$_7$ (M = Cl, Mn, Tc), Br$_2$O$_5$, and M$_2$O (M = F, Cl, Br), respectively. The structures of the dinitrogen oxides differ, however, in having nitrogen–nitrogen bonds as the linking element instead of bridging oxygen atoms (Figure 8.5b). Again we can draw good Lewis structures for these dimeric molecular oxides M$_2$O$_{2x}$. Because the dimeric molecules are somewhat larger than the monomeric oxides, they have somewhat higher melting points and boiling points, and enthalpies of fusion and vaporization.

Physical Properties of Cyclic and Polyhedral Oligomeric Molecular Oxides. Some nonmetal oxides have cyclic oligomeric polymerization isomers, such as S$_3$O$_9$, and Se$_4$O$_{12}$. Others have polyhedral polymerization isomers, such as P$_4$O$_{10}$, :P$_4$O$_6$, :As$_4$O$_6$, :Sb$_4$O$_6$, and I$_4$O$_{16}$. In the first four of these M has a TCN of four, and good Lewis structures can be drawn (Figure 8.5c, d). These molecules are large enough to have substantial intermolecular attractions, so they are low-melting solids with medium enthalpies of vaporization. (The trends in the apparent melting points and enthalpies of fusion are complicated by the tendency of these oxides to convert to other polymerization isomers rather than to melt.)

Physical Properties of Network and Ionic Oxides—Heat Resistance and Mechanical Hardness. As seen in Table 8.4, network (Figure 8.1g) and ionic (Figure 8.1h) oxides have extraordinarily high melting points, boiling points, heats of fusion, and heats of vaporization. **Ceramics** are compounds of metals and metalloids with small nonmetal atoms such as boron, carbon, nitrogen, or oxygen. Nowadays ceramics are used for making far more than pottery and china: Advanced ceramics are used in abrasives, cutting tools, electrical insulators, heat shields, nuclear fuels, bone implants, and lasing crystals. The short, strong bonds in such materials give them not only very high melting points, but also the ability to keep their strength and mechanical hardness even when very hot (in contrast to most organic polymers). Hardness in this sense is the resistance of one body to indentation, scratching, cutting, abrasion, or wear by another body. The hardest materials are considered to be diamond, $_\infty^3[C]$, followed by the cubic polymerization isomer of boron nitride, $_\infty^3[BN]$. Other very hard carbide ceramics include B_4C, SiC, TiC, Be_2, ZrC, TaC, and WC. Very hard borides include AlB and ZrB_2; very hard nitrides include not only BN, but also TiN. Now, however, we emphasize very hard oxide ceramics.

Perhaps the hardest oxide is *alumina* (Al_2O_3), with *beryllia* (BeO) and *zirconia* (ZrO_2) not far behind. (Note the common industrial nomenclature by which the most common oxide of a metallic element is named by substituting "–ia" for "–ium" in the name of the metallic element.) Among the best known of these is silica.

The silica polymorphs quartz, tridymite, and cristobalite are not as hard because their structures (Figure 4.6) contain a good deal of empty space within the $(SiO_2)_6$ rings. The polymerization isomer stishovite, with six-coordinate silicon in a denser rutile lattice, has been found to be comparable in hardness to cubic boron nitride.[8] By the application of ultra-high pressure, solid molecular CO_2, a very soft material, is reduced in volume by 15% and is converted (reversibly) to $_\infty^3[CO_2]$, a much harder polymerization isomer of carbon dioxide having the tridymite structure with four-coordinate carbon atoms.[9]

The very high melting points and low volatility of these oxides make some of them (such as MgO) useful as **refractories** for providing surfaces capable of withstanding very high temperatures. MgO is used not only to line furnaces but also to cover the heating elements of electric ranges, because it conducts heat much more readily than it conducts electricity. A related use is that of *thoria* (actually 99% ThO_2 + 1% CeO_2) to provide luminosity to gas flames for lighting purposes: The oxides become white-hot without melting.

Brittleness is an unfortunate characteristic physical property found in ionic solid oxides. The mechanical strength of ionic materials stems from the large amount of work that has to be done to increase their cation–anion distances. However, if the mechanical stress is sufficient to overcome these attractions, the layers of ions then may slip to an arrangement in which cations are next to cations and anions are next to anions. Suddenly the attractions are replaced by repulsions, and the material shatters.

Difficulties of Synthesis

Because ceramics are hard and brittle, they cannot be formed into useful objects by some of the methods used with organic polymers, such as extrusion through holes or into molds; unlike metals, they are difficult to machine into shape. Because their melting points are so extraordinarily high, they are difficult to prepare from molten materials. So, often they are synthesized by **sintering**: heating finely divided powders in furnaces for long periods of time at very high temperatures (up to 2500°C), which are nonetheless well below the melting point of the oxide. Under these conditions the ions from different granules gradually diffuse and "weld" together the granules having dimensions of 100–10,000 nm to give a coherent solid. Between the grains impurities tend to congregate as defects that readily undergo failure when subjected to a stress.

Much modern research in ceramics deals with ways to generate fine, homogeneous *nanoparticles* of 1–100 nm dimensions of starting oxides that will sinter together (at lower temperatures) to give dense, strong ceramics with fewer defects. Alternatively, some of these problems can be overcome by creating **composites** of ceramics with either metals or organic or other polymers that are not brittle.

EXAMPLE 8.7

Predict the coordination numbers, the physical states, and the relative melting points and boiling points of the oxides of the second-period elements in their maximum oxidation state. State whether each should be monomeric, oligomeric, or polymeric or ionic.

SOLUTION: From Table 8.3 we see that these oxides are Li_2O, BeO, B_2O_3, CO_2, and N_2O_5. Six-coordinate metal ions are predicted to be those with radii between 52 pm and 92 pm; from Table C, Li^+ and Be^{2+} qualify (for Li_2O, this number would apply to the oxide ion, while Li^+ would have three-coordination). Four-coordinate metal ions are predicted for ions with radii between 28 pm and 52 pm; B^{3+} and C^{4+} qualify. Three-coordination is expected for N^{5+}. In contrast, the penultimate TCN for Period 2 is three.

We now compute the number of VCS per structural unit (Step 2 in Figure 8.3): 2.5 for Li in $LiO_{0.5}$; 5 or 2 for Be in BeO; 2.5 or 1.5 for B in $BO_{1.5}$. For CO_2 and N_2O_5 we should consider the number of π bonds in the structural unit—namely, two in CO_2 and one per N in N_2O_5. The resulting numbers of VCS by Equation 8.2 are 0 for CO_2 and 0.5 for $NO_{2.5}$.

Step 3 of Figure 8.3 is next. The oxides of Li, Be, and B have enough VCS for polymeric (ionic) structures to be likely, so we predict that these will be solids at room temperature, with high melting points and high boiling points. Our first three predictions are verified for these solids: Li_2O has a melting point of 1427°C, BeO of 2530°C, and B_2O_3 of 450°C (but with a boiling point of 1860°C). CO_2 is actually a gaseous monomeric molecule at room temperature, which sublimes at –79°C at atmospheric pressure. For dinitrogen pentoxide, the prediction is partially correct: In the gas phase one of the five oxygens bridges or links the two nitrogens, while each of the other oxygen atoms is bonded to only one nitrogen atom; each nitrogen has a coordination number of three. The solid form of this compound consists of NO_2^+ and NO_3^- ions, with π bonding in each ion.

EXAMPLE 8.8

Predict the coordination numbers, the physical states, and the trends in structures and melting points of the oxides of vanadium in its different oxidation states (II through V). Supplement this list by adding the oxides of Cr(VI) and Mn(VII).

SOLUTION: All of these would have the same penultimate coordination number, four, whereas the radii of the ions do decrease as the charge increases. For this reason, a more subtle prediction is possible using the trends in radius ratio. Of the ions involved, only V^{2+} might possibly be large enough for eight-coordination (six-coordination is actually found). The Cr^{6+} and Mn^{7+} ions are only a little larger than 52 pm, so they might be and are four-coordinate. The remainder should be six-coordinate. For most of these oxide structural units, the numbers of VCS are predicted to be greater than two, so that three-dimensional ionic or network covalent lattices are predicted. With only 1.0 and 0.5 VCS, the CrO_3 and $MnO_{3.5}$ structural units are expected to give either *cyclo*-$(CrO_3)_n$ or $_\infty^1[CrO_3]$ and Mn_2O_7, respectively. Any of these latter structures should have greatly reduced melting points—even a chain polymer such as $_\infty^1[CrO_3]$ could readily become mobile by twisting, in much the same way that the chain polymer, spaghetti, easily becomes fluid upon stirring.

The observed melting points are very high—1789° C and above for the first three oxides, VO, V_2O_3, and VO_2. The layer polymer $_\infty^2[V_2O_5]$ has a lower melting point (670°C). The chain polymer $_\infty^1[CrO_3]$ has a much lower melting point of 197°C, followed by dimeric molecular Mn_2O_7 at 6°C.

8.3. Close Packing of Anions: Isomorphous Substitution in Mixed-Metal Oxides

OVERVIEW. The structures of many oxides can be described as based on a lattice of close-packed large oxide ions (Section 8.3A). Smaller cations can fit in some or all of the small tetrahedral holes between anions in two layers; there are two tetrahedral holes per anion. Larger cations can fit in the larger octahedral holes; there is one octahedral hole per anion. You may try such identifications with Exercises 8.27–8.31.

Mixed-metal oxides have more than one kind of metal ion in different hole types. Two common lattice types for mixed-metal oxides are the spinel structure (AB_2O_4) and the perovskite structure (ABO_3). Nonstoichiometric oxides can arise when a given metal ion (usually from the *d* block) has two different adjacent oxidation states; then a smaller number of the higher-charged ions, plus a *defect* (missing ion) can replace a larger number of the lower-charged ions. Ions (usually cations) in a lattice can sometimes move into holes (such as vacant tetrahedral or octahedral sites) in the solid state, conducting electricity even in the solid state. You may practice identifying these structure types in Section 8.3B with Exercises 8.32–8.38.

Oxides often employ substitution of one cation for another of similar size according to the first and second principles of isomorphous substitution (Section 8.3C). With the first principle, the two cations have the same charge. The second principle substitutes pairs of ions for other pairs of ions of the same *total* charge. You may try predicting cases of isomorphous substitutions with Exercises 8.39–8.50.

This section includes Connections to biochemistry, physical chemistry, radiochemistry, and three Connections to materials science. This section uses the deduction of ionic charges from Section 1.4, ionic radii from Section 1.5, the lattice types and unit cells introduced in Section 4.1, oxidation–reduction half-reactions from Section 6.1, and charge-transfer absorptions from Section 6.4. The concepts of defects and holes in lattices will appear again in Chapter 12.

8.3A. Close Packing of Anions. Before we look at some additional types of ionic oxides of metals and their physical properties, we need to mention an alternative way of looking at crystal lattices that is useful for many ionic compounds. Anions are normally larger than cations and are often in contact with each other in layers in which each anion touches six other anions (Figure 8.6a.) Each anion of a second layer of anions then can contact three anions in the layer below it, falling in the center of each triangle of the first layer (Figure 8.6b).

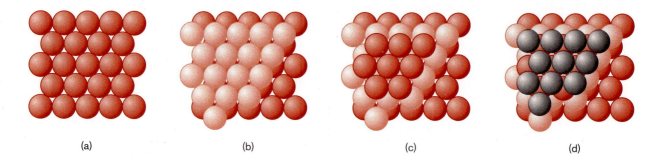

(a)   (b)   (c)   (d)

Figure 8.6. Close packing of atoms or (an)ions. (a) First layer. (b) Second layer (in **light brown**), with each atom of the second layer centered on an indentation between three atoms in the first layer. (c) Third layer (in **brown**) over indentations of the second layer, such that atoms of the third layer are directly over atoms of the first layer. This is the pattern of **hexagonal close packing**. (d) Third layer (in **gray**) also over indentations of the second layer, such that atoms of the third layer are also directly over indentations of the first layer. This is the pattern of **cubic close packing**. [Adapted from D. W. Oxtoby, H. P. Gillis, and A. Campion, *Principles of Modern Chemistry*, 7th ed., Brooks/Cole Cengage Learning: Belmont, CA, 2011, p. 1045.]

EXPERIMENT 9 MAY BE ASSIGNED NOW.
(See www.uscibooks.com/foundations.htm.)

Viewing a three-dimensional model built with the Polyhedral Model Kit described in Section 4.1 should improve your ability to understand how the layers relate to each other.

There are two possible patterns of close packing of a third layer of anions over the first and second layers. With one possibility, called **hexagonal close packing** (hcp) (Figure 8.6c), the third (top, **brown**) layer is matched up exactly with the first (bottom, **brown**) layer. This pattern is then normally continued throughout the whole crystal, with layers of spheres alternately in the A position (**brown**) and B position (light brown). The packing sequence of layers can be described as ".... ABAB....."

In the alternate pattern of close packing, called **cubic close packing** (ccp) (Figure 8.6d), the third (top, gray) layer does not directly coincide with either the bottom or the middle layer. Instead, the anions of the third layer fall directly over the three-sided cavities of both the lower and the middle layers (Figure 8.6d). Finally the fourth layer directly matches the first layer, so that the packing sequence of layers can be described as "...ABCABC...."

Lattices of anions only would be unstable, having only repulsions; an equal total charge of cations must be included. The much smaller cations can then fill some of the **holes** or interstices between the anions. Thus, if we were packing basketballs (e.g., anions) and baseballs and golf balls (e.g., cations) for shipment in the same large box, we would figure out first how to pack the larger basketballs (anions) in the most efficient way possible, with confidence that the baseballs and golf balls (cations) would fit in the spaces between the larger spheres.

There are two types of holes in the hcp and ccp lattices: smaller tetrahedral holes and larger octahedral holes. The locations of these may be seen using Figure 8.6b. Between three anions in the first layer drawn in **brown** and *directly below* the anion in the second neighboring layer drawn in light brown, there is a small open space. It is called a **tetrahedral hole**, because it is surrounded by four large anions. A small cation (e.g., a golf ball) put in here has a coordination number of four.

As can be seen from Figure 8.6b, there is a white channel that falls between three ions in the first (**brown**) layer of anions *and* between three ions in the second (light brown) layer of anions. A cation centered in this cavity has a coordination number of six and an octahedral geometry. This **octahedral hole** is larger (suitable for a baseball); this is easier to see with the Polyhedral Model Kit.

For each sphere in a given layer, there are *two* tetrahedral holes but only one octahedral hole. The stoichiometry of the salt does not normally allow all of the holes to be filled with cations, but (if close packing is utilized) the structure can be described in terms of the type and fraction of holes occupied.

EXAMPLE 8.9

Which of the following structural descriptions are inconsistent with the stoichiometry of the salt being described? (a) Al_2O_3 adopts an hcp lattice of oxide ions in which two-thirds of the octahedral holes are occupied by aluminum ions. (b) HgI_2 adopts a ccp lattice of iodide ions in which half of the tetrahedral holes are occupied by Hg^{2+} ions. (c) $SnBr_4$ adopts an hcp lattice of bromide ions in which one-eighth of the tetrahedral holes are occupied by Sn^{4+} ions.

SOLUTION: Recall that there are two tetrahedral and one octahedral holes per anion. (a) Two-thirds of an Al^{3+} ion has two positive charges, which match the two negative charges of the oxide ion. This description is *consistent* with the stoichiometry of the salt. (b) Filling one-half of the two tetrahedral holes per anion with Hg^{2+} ions gives two positive charges, which is *inconsistent* with the charge of the iodide ion, –1. (c) Filling one-eighth of two holes with Sn^{4+} ions generates one positive charge, which is *consistent* with the one negative charge of the bromide ion. (If these holes are filled properly, a molecular structure results, as is observed.)

8.3B. Mixed-Metal Oxides—Spinels, Perovskites, and Nonstoichiometric Oxides.

There are a number of technologically important oxides, the simplest formulas of which look like oxo salts (e.g., $BaTiO_3$), but which do not involve identifiable oxo anions. Such compounds are called **mixed-metal oxides**. The mixed-metal oxides are best regarded as consisting of lattices of oxide ions together with two (or more) different types of metal ions. In many of these, the oxide ions are close packed, and one kind of metal ion may occupy tetrahedral holes and the other kind, octahedral holes in the close-packed structure. An important class of over 100 mixed-metal oxides (and 30 mixed-metal sulfides) is known as the **spinels***, AB_2O_4. (Structures indicated with asterisks, *, can be built using the Polyhedral Model Kit.) Spinels are very important in the solid-state electronics industry for their electric and magnetic properties (Section 8.4). In spinels the oxide ions are cubic close packed. Normally the A metal ions are +2-charged ions of radius between 80 and 110 pm, which occupy one-eighth of the tetrahedral holes in the oxide-ion lattice. Spinels are known in which the A metal ions are the +2 ions of Mg, Cr, Mn, Fe, Co, Ni, Cu, Zn, Cd, and Sn. Normally the B metal ions are +3-charged ions of radius between 75 and 90 pm, which occupy one-half of the octahedral holes; these include the +3 ions of Ti, V, Cr, Mn, Fe, Co, Ni, Rh, Al, Ga, and In.

Of particular interest are spinels in which both A and B are the same element: They form compounds of stoichiometry $M^{2+}(M^{3+})_2(O^{2-})_4$ or M_3O_4, which have fractional average oxidation numbers because of the presence of two different oxidation states in the same compound. Mn_3O_4 (hausmannite), Fe_3O_4 (magnetite*), and Co_3O_4 are all intensely black, and much darker than the simple oxides of these metals, due to charge-transfer absorptions (Section 6.4B).

Another important class of mixed-metal oxides is the **perovskites*** (ABO_3), of which the prototype is $CaTiO_3$ (perovskite). It has an unusual cubic close-packed lattice of oxide *and* calcium ions (Figure 8.7), in the octahedral holes of which the much smaller Ti^{4+} ions can "rattle around." In this lattice the very large A ion ideally has a coordination number of 12.

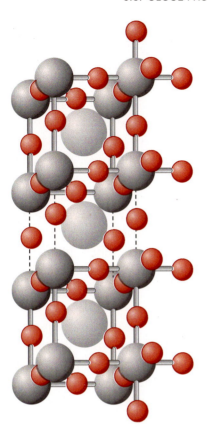

Figure 8.7. The structure of perovskites, ABO_3 (e.g., A = Ca^{2+} and B = Ti^{4+} in $CaTiO_3$). Oxide ions are shown in **brown**. Larger A ions (Ca^{2+}) are shown in light gray, at the center of dodecahedral (12) coordination by oxide ions. Smaller B ions (Ti^{4+}) are shown in **dark gray**, at the center of the octahedral coordination by oxide ions. [Adapted from D. W. Oxtoby, H. P. Gillis, and A. Campion, *Principles of Modern Chemistry*, 7th ed., Brooks/Cole Cengage Learning: Belmont, CA, 2011.]

EXAMPLE 8.10

Choose the appropriate classification—perovskite, spinel, or neither—for each of the following metal oxides: (a) $BaCO_3$; (b) $RaTiO_3$; (c) $BaThO_3$; (d) $TiBa_2O_4$; (e) hausmannite, Mn_3O_4; (d) Tl_2SiO_4.

SOLUTION: (a), (b), and (c) all have the stoichiometry to be perovskites, but the cation sizes may not be suitable. The Ba^{2+} and Ra^{2+} ions are very large +2 ions, so they are acceptable. There should also be a much smaller +4-charged metal ion. C^{4+} is not a realistic metal ion, however, and we may recognize that CO_3^{2-} is the polyatomic carbonate ion, so (a) is *neither*. The Ti^{4+} is indeed a much smaller metal ion, so (b) is a *perovskite*. It has an acceptable radius for a perovskite. The Th^{4+} ion has a quite large radius of 108 pm, on the other hand, so (c) is *neither*. (d), (e), and (f) have the correct overall stoichiometry to be spinels. In (d), however, the A ion is Ti^{4+}, which is not a +2 ion between 80 and 110 pm, and Ba^{2+} is not a +3 ion between 75 and 90 pm, so this salt is *neither*. In hausmannite the A ion is Mn^{2+}, which has the correct size and charge, and the B_2 ions are Mn^{3+}, which have the correct charge and size, so (e) is a *spinel*. In (f) the B_2 ions are Tl^+, which are too low in charge and too large in size. The A ion is Si^{4+}, which is too high in charge and too small in size. Indeed, you may recognize this as the *silicate* ion. So this is *neither*—it is the salt barium silicate.

Microphones and Loudspeakers

If the temperature is not too high, the Ti^{4+} ions of perovskite itself tend to be off the center of the lattice unit cell, giving rise to an electric charge separation or dipole; such materials are known as **ferroelectrics**. Application of mechanical pressure to one side of a perovskite crystal causes the Ti^{4+} ions to migrate, generating an electrical current; application of an electric current causes mechanical motion of the ions. The pressure effect, known as the **piezoelectric effect**,[10] makes perovskites useful in converting mechanical energy to electric energy, as in microphones, sonar, and vibration sensors, or vice versa. The conversion of electrical energy to mechanical energy is useful in sonic and ultrasonic transducers and in headphones and loudspeakers.

Perhaps the easiest mixed-metal-ion oxides to prepare are the **nonstoichiometric oxides** of metals that have cations of more than one oxidation state. Iron(II) oxide as normally prepared gives an actual elemental analysis corresponding approximately to $Fe_{0.95}O$; there is a **defect** consisting of missing iron atoms. The compound, however, must still be electrically neutral; this is accomplished by the replacement of three Fe^{2+} ions by two Fe^{3+} ions, leaving a hole or vacancy but keeping the overall electroneutrality (Figure 8.8). Because most of the d-block metals have cations differing by only one unit of charge, there are many nonstoichiometric d-block metal oxides, including (for example) those of ideal composition TiO, VO, MnO, FeO, CoO, and NiO. If the +2 and +3 oxidation states are of comparable stability, a wide range of nonstoichiometry can exist: Vanadium(II) oxide can range from an actual composition of $V_{0.77}O$ to $V_{1.27}O$, while nickel(II) oxide ranges only from $Ni_{0.999}O$ to $Ni_{1.000}O$. You should be able to verify that the structural segment shown in Figure 8.8 is neutral, and that its stoichiometry is $M_{0.917}O$.

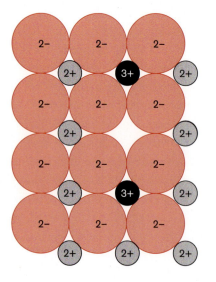

Figure 8.8. Planar segment of the structure of a nonstoichiometric oxide $M_{0.917}O$. **Brown** = oxide ions, gray = 2+ ions, and black = 3+ ions.

Electrical Conductivity of Solid Oxides. An electrical current consists of moving charged particles. In molten or dissolved ionic salts, the ions are free to move and to conduct an electrical current. In most *solid* metal oxides (and other salts) the ions are immobilized by the crystal lattice, so solid ionic compounds are generally not good conductors of electricity. Some metal oxides,[11] such as the nonstoichiometric early *d*-block metal ions of the fourth period, Ti_xO and V_xO ($x \approx 1.0$), have metal-like conductivity due to the easy transfer of *electrons* between titanium or vanadium ions differing by one unit of charge.

In certain types of lattices and under certain conditions, *ions* can move in solids, hence conduct an electric current: There are materials in which the smaller type of ion can move readily, while the larger counterions maintain the rigidity of the solid-state lattice as a whole. Such materials are variously known as solid electrolytes, super-ionic conductors, and fast-ion conductors.[12] They are important for their potential uses in high-energy-density batteries, fuel cells, and lasers.

Because anions are commonly larger than cations, most solid electrolytes involve rapid cation motion. For example, some solid *iodides* allow super-ionic conduction of electronegative cations into tetrahedral holes in the iodide lattice. The structures of copper(I) and silver(I) tetraiodomercurates, Cu_2HgI_4 and Ag_2HgI_4, feature four-coordinate metal ions but only three-coordinate iodide ions; one-fourth of the tetrahedral holes about the large (approximately cubic close-packed) iodide ions are vacant. At low temperatures the vacant holes are ordered and the cations cannot move into them; the compounds have low conductivity and light colors (red for the Cu^+ and yellow for the Ag^+ salt). At a specific temperature (50°C for the Ag^+ salt), a phase transition occurs in which the lattice opens up enough for the cations to move between the iodide ions. The cations and vacant holes are then disordered, the motion increases the electrical conductivity dramatically, and the color darkens (to black in the Cu^+ salt and orange in the Ag^+ salt). Similar types of phenomena are found in solid AgI. At room temperature AgI adopts the zinc blende structure (Figure 4.2), with Ag^+ ions filling half the tetrahedral holes in an orderly pattern. Above 145.8°C the Ag^+ ions become mobilized to move easily between the easily deformed (soft) I^- ions.

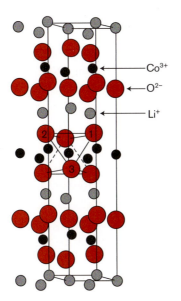

Figure 8.9. Structure of the $LiCoO_2$ solid-state electrolyte. Small black circles = Co^{3+} ions, small gray circles = Li^+ ions, and large brown circles = O^{2-} ions. [Adapted from H. J. Orman and P. J. Wiseman, *Acta Crystallogr.* C40, 12 (1984).]

Conduction in Lithium-Ion Batteries

Of particular importance for modern lightweight electronic devices such as laptop computers and cell phones are the rechargeable lithium-ion batteries (Sections 1.2C and 6.3A), in which the very light, relatively small Li^+ ion is mobile. The most common solid electrolyte[13] in these batteries is the mixed-metal oxide $LiCoO_2$, which has a distorted NaCl type of structure in which layers of Li^+ and Co^{3+} cations alternate.[14] The lithium ions are free to move within the two dimensions of its layer (Figure 8.9). When these batteries are charged, Li^+ ions are reduced to Li^0, which migrates out of the lattice (for safety reasons, Li^0 is absorbed between graphite layers as the compound of approximate composition LiC_6), and the cobalt is oxidized to Co(IV) oxide. During discharge the Li^0 releases its electron (producing a voltage of about 3.6 V) and migrates back into the solid, regenerating $LiCoO_2$.

Two of the technologies for renewable energy generation, capturing wind power and solar energy, suffer from intermittency: The wind does not always blow and the sun does not always shine. Hence, if these are to be relied on, the energy must be stored during times of peak production and released during times of excess demand; much larger lithium-ion batteries are being investigated for this use.

Oxide-Ion Conductors in Automobile Catalytic Converters

Oxide ions can be the mobile phase if the cations are large enough that oxide ions can move between them. This is the case between 500°C and 1000°C in zirconia (ZrO_2) doped with CaO or with Y_2O_3. When a Ca^{2+} ion or an Y^{3+} ion replaces a Zr^{4+} ion in the zirconia lattice, charge neutrality of the lattice is maintained by deleting oxide ions as well. The resulting vacancies in the fluorite-type lattice act as holes into which other oxide ions can move (and hence carry current) if enough thermal energy is available. Oxygen detectors for pollution control in automobile exhaust employ solid oxide-ion conductors.

8.3C. The First and Second Principles of Isomorphous Substitution. Often in nature or the laboratory we have more than one metal ion present, and when we form a metal oxide or other salt, some of the metal ions of the second metal replace the same number of the first metal ion with the same positive charge, without changing the lattice type. This is **isomorphous substitution** (Figure 8.10) for the first ion and gives a type of solid solution. Dissolving a small percentage of Cr_2O_3 into the lattice of

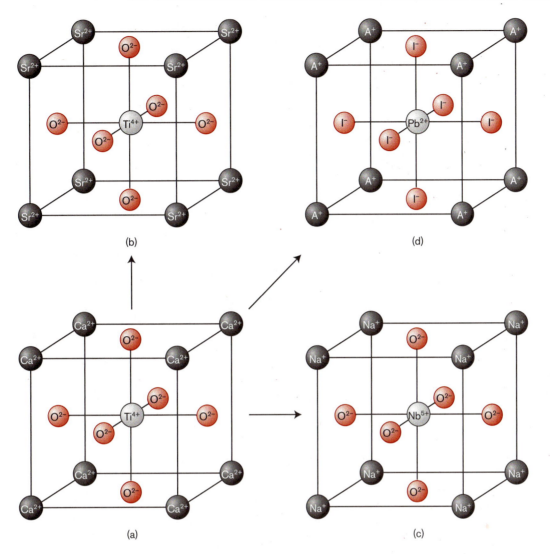

Figure 8.10. Isomorphous substitution in the perovskite unit cell ABO_3. Oxide (or halide) ions are shown encircled in **brown**, the large A cations are shown in black, and the smaller B cations are shown in gray. (a) The structure of $CaTiO_3$. (b) Isomorphous substitution of Sr^{2+} for Ca^{2+}. (c) Isomorphous substitution of Na^+ for Ca^{2+} and of Nb^{5+} for Ti^{4+}. (d) Isomorphous substitution of all ions: A^+ (= RNH_3^+) for Ca^{2+}, Pb^{2+} for Ti^{4+}, and I^- for O^{2+}.

colorless alumina (Al_2O_3) puts some colored Cr^{3+} ions in sites normally occupied by Al^{3+} ions. This gives the pink gemstone *ruby*. Similarly, dissolving some Ti_2O_3 (with colored Ti^{3+} ions) in Al_2O_3 results in blue *sapphires*.

The type of cations that can substitute isomorphously for other cations in a metal oxide (or other type of salt) depends on (1) the size of those cations and (2) the charge of those cations. The **first principle of isomorphous substitution** states that *one ion may substitute for another in a lattice if the two ions have identical charges and differ in radii by not more than 10 to 20%.*

Examination of Table C shows that quite a few sets of cations can be found that have the same charge and very similar ionic radii. If we limit ourselves to ions commonly found in Earth's crust, the sets found in the columns of Table 8.5 result. As a salt or mineral is formed by crystallization or the cooling of molten magma, there is little reason for one of these matched types of ions to be preferred over another. Naturally occurring minerals often have a mixture of cations present, which vary depending on the composition of the melt or solution from which the mineral grew. For example, Mg^{2+} and Fe^{2+} not only have identical charges but also have very similar radii (86 and 92 pm, respectively). The mineral olivine, with an ideal composition of Mg_2SiO_4, is often "impure" and can contain varying percentages of the Fe^{2+} ion in place of an equal number of Mg^{2+} ions. Thus, the formula of olivine is often written $(Mg,Fe)_2SiO_4$ to indicate that there are two magnesium or iron(II) cations present per mole of silicate ion, although there is no definite relationship between the number of magnesium and the number of iron(II) cations. Note how different this process is from what normally happens in synthesizing molecular (e.g., organic) compounds!

The substitution may be done more than once and may also be of anions for anions. In the latter case, polyatomic anions are likely involved, because there are more of them than there are simple anions. For example, the most important mineral in bones and teeth is *hydroxyapatite*, $Ca_5(OH)(PO_4)_3$. Because this compound contains a very strongly basic hydroxide anion, it is prone to reacting with acids generated in the mouth by bacterial action. The utility of adding trace amounts of fluoride to drinking water or toothpaste occurs when F^- (radius 119 pm) substitutes isomorphously for OH^- (no precise radius, but dominated by oxygen of ionic radius 126 pm) to give the mineral *fluorapatite*. Another set of hexagonal (six-sided) minerals arises by isomorphous substitution of polyatomic anions in minerals having the same stoichiometry as apatite. These minerals are lead minerals of stoichiometry $Pb_5Cl(EO_4)_3$. When the EO_4 polyatomic anion is vanadate, the mineral is known as vanadinite; when the anion is arsenate, the mineral is called mimetite; when it is phosphate, the mineral is pyromorphite.

TABLE 8.5 Sets of Common Ions Suitable for Isomorphous Substitution

Range of Radii (pm)	+1 Charge[a]	+2 Charge[a]	+3 Charge[a,b]	+4 Charge[a]
54–67			Al^{3+} (67)	Si^{4+} (54)
74–92	Li^+ (90)	Mg^{2+} (86)	Fe^{3+} (78)	Ti^{4+} (74)
		Fe^{2+} (92)		
100–117	Na^+ (116)	Ca^{2+} (114)	Ln^{3+} (100–117)	
149–152	K^+ (152)	Ba^{2+} (149)		

[a] Radii listed in parentheses are in picometers (pm).

[b] Lanthanide ions = Ln^{3+}.

A CONNECTION TO RADIOCHEMISTRY

Co-Precipitation

Isomorphous substitution is often deliberately used in radiochemistry to isolate exceedingly tiny quantities of radioactive elements. Thus, to isolate a tiny amount of radium ion from a large amount of uranium ore, we could (using the solubility principles of Section 4.3) add sulfate ion to precipitate radium sulfate while leaving uranium and most other cations in solution. But this would pose two problems: (1) only a tiny amount of precipitate would be formed, which would be difficult to handle without losing it; and (2) there might be so much solution present that even the low solubility product of radium sulfate would not be exceeded, so no precipitate would form. Marie Curie overcame these problems by adding not only sulfate ion but also barium ion. A large amount of barium sulfate then precipitated, in which the radium substituted isomorphously for the barium. This technique is known as **co-precipitation**; the barium ion is said to act as a *carrier* for the Ra^{2+}. Curie was then confronted with the formidable problem of separating the very similar barium and radium ions, but at least she was working with a much smaller volume of material.

A CONNECTION TO PHYSICAL CHEMISTRY

Doping Crystals by Isomorphous Substitution

Another illustration of this process occurs in a familiar experiment of growing crystals of ionic compounds. For example, large, beautiful octahedral crystals of *alum*, $KAl(SO_4)_2 \cdot 12H_2O$, are readily grown from solution. But there exist a whole series of similar compounds that also form large, beautiful octahedral crystals having the same lattice types and the same shapes (these compounds are said to be isomorphous). In the formula for alum, the K^+ ion can be replaced by other +1 cations of similar radius, such as Rb^+ and NH_4^+. The Al^{3+} ion can be replaced by numerous other +3 ions of similar radius, such as Cr^{3+} (giving purple crystals of *chrome alum*) or Fe^{3+} (giving pale violet crystals of *ferric alum*). The sulfate anion can even be replaced by the selenate anion. If nearly any combination of these three ingredients is mixed and crystallized, large crystals of an alum are formed; if a mixture, say, including both Al^{3+} and Cr^{3+} is used, crystals can be grown containing both ions, having whatever shade of light purple you desire! Such a crystal is sometimes said to be *doped* with a certain percentage of the less abundant ion.

The separation of the *f*-block cations involves quite a complex process of ion exchange. The most troublesome case of isomorphous substitution, however, is that of the elements Zr (Pauling electronegativity 1.33; ionic radius 86 pm) and Hf (Pauling electronegativity 1.3; ionic radius 85 pm). Hf occurs isomorphously substituted in all zirconium compounds to the same extent (about 2%), so there were no chemical discrepancies in the "pure" samples of zirconium prepared in 1825 and thereafter. As a result, the presence of Hf went undetected for a whole century!

CHEMICAL TRIVIA

What City Has the Greatest Number of Elements Named After It?

Perhaps the most extensive case of isomorphous substitution occurs in the minerals monazite and xenotime (MPO_4) and bastnaesite ($MFCO_3$). In 1794 J. Gadolin investigated a mineral obtained from the village of Ytterby, Sweden; from this mineral he extracted a metal oxide that he named *yttria*. But other chemists, in working with this material, kept getting slightly different properties; eventually it was realized that this oxide was a mixture. So, the mixture was separated, but the "pure" components also turned out to be mixtures. Eventually all of the *f*-block elements of the sixth period (except Pm), plus La and Y, turned out to be present in these minerals, isomorphously substituted for each other. (Refer to Tables A through C to note the extreme similarities of these elements to each other.) As element after element was discovered, the chemists were harder and harder pressed to come up with new names for them. Thus, it came to pass that the humble village of Ytterby has more elements named after it than any of the great cities of the world: yttrium, ytterbium, terbium, and erbium.

A CONNECTION TO BIOCHEMISTRY

Substituting NMR-Active Nuclei in Enzymes

There are often very practical reasons for substituting one metal ion for another in an ionic lattice (or even in compounds such as metalloenzymes). The *s*- and *p*-block metal ions are all colorless. If they can be isomorphously substituted with *d*- and *f*-block ions, the spectra of these ions will give us information on the crystal field environment of the ions in the lattice or enzyme (Section 7.3). Metal ions also have other properties that tell us about their environment: Some are fluorescent; some have unpaired electrons with magnetic properties that can be studied; others with appropriate nuclei can be studied by nuclear magnetic resonance (NMR) or Mössbauer spectroscopy. Some of the geologically and biochemically most important metal ions, such as K^+ and Zn^{2+}, however, are "silent metals" that lack most or all of these properties. Silent K^+ (radius = 152 pm) can often usefully be substituted with fluorescent and NMR-active Tl^+ (radius = 164 pm). Colorless Zn^{2+} (radius = 88 pm) is usefully replaced by colored Co^{2+} (radius = 88 pm). Silent Ca^{2+} (radius = 114 pm) can be replaced by NMR-active Eu^{2+} (radius = 131 pm), with seven unpaired electrons.

The Second Principle of Isomorphous Substitution. There are a number of other perovskites that can be generated from the structure of $CaTiO_3$ through the application of either the first or the **second principle of isomorphous substitution**. The second principle allows more versatility: Although substituting ions must still be about the same size as the ions replaced in order not to change the lattice type, within certain strict limits the *charge* of the entering ion need not be identical to the charge of the departing ion. The basic principle is that the *total* charge of the pair of replacing ions must equal the total charge of the pair of replaced ions. This means that isomorphous substitution can occur even if the new ions C have a charge one greater than the old ions A, if there is *simultaneous* substitution by new ions D with a charge one *less* than the old ions B. This conserves the electroneutrality of the salt, because the sum of charges of the new ions C and D equals the sum of charges of the old ions A and B. Table 8.5 shows sets of common ions that frequently substitute for one another; to use the second principle, the substituting ions should be from the same row of this table and from adjacent columns.

We illustrate this process starting with the prototype perovskite, $(Ca^{2+})(Ti^{4+})(O^{2-})_3$, in Figure 8.7. The perovskite shown in Figure 8.7c is $(Na^+)(Ta^{5+})(O^{2-})_3$, obtained by the simultaneous substitution of Na^+ ($r = 116$ pm) for Ca^{2+} ($r = 114$ pm) and of Ta^{5+} ($r = 78$ pm) for Ti^{4+} ($r = 74$ pm). The substituted ions have nearly the same size as those in the prototype perovskite, so radius ratios in the lattice are not appreciably altered. The

---

EXAMPLE 8.11

Which of the following minerals could arise by isomorphous substitution processes in leucite, $KAlSi_2O_6$? (a) $KYSi_2O_6$; (b) $RbAlSi_2O_6$; (c) $BaBeSi_2O_6$; (d) $BaAlSi_2O_6$.

SOLUTION: First, note that *the total charge of all the cations must equal the total charge of all the anions.* This requirement is implicit in the two principles of isomorphous substitution, but also gives a separate simple test in this example: If the total charge of all cations going into the structure is not the same as the total charge of all cations coming out, the new compound cannot exist, let alone be isomorphous. In (a) Y^{3+} replaces Al^{3+}; in (b) Rb^+ replaces K^+; in (c) Ba^{2+} and Be^{2+} replace K^+ and Al^{3+}; in (d) Ba^{2+} replaces K^+. Substitution product (d) cannot exist.

Note, too, that in order for the substitution to be isomorphous, *the total number of cations going in must be nearly equal to the total number coming out.* (If the two numbers are slightly unequal, a nonstoichiometric compound is formed, as was discussed on page 468.) All four possible substitution products given above obey this principle.

Once these two principles are satisfied, the two principles of isomorphous substitution are satisfied if the cations going into the replacement structures are within 10–20% of the radii of the cations coming out. In (a) Y^{3+} has a radius of 104 pm, which is too much bigger than Al^{3+} (67 pm) for the substitution to be isomorphous. However, (b) and (c) are satisfactory, since in (b) Rb^+ (166 pm) is close in size to K^+ (152 pm), and in (c) Ba^{2+} (149 pm) is close to K^+ (152 pm), and Be^{2+} (59 pm) is close to Al^{3+} (67 pm).

new ions have the same total charge (+6) as the old ions, and the opposite total charge to the three oxide ions, so the stoichiometry keeps the crystal neutral overall, as is also required. Among the suitable large A ions are Ca^{2+} ($r = 114$ pm), Y^{3+} ($r = 104$ pm), Ba^{2+} ($r = 149$ pm), and K^+ ($r = 152$ pm). Among the suitable small B ions are Cu^{3+} ($r < 87$ pm), Ti^{4+} ($r = 74$ pm), and Nb^{5+} ($r = 78$ pm).

It is even possible to substitute *all* of the ions, provided that the sizes are proportional and the total charges add up to zero for the neutral salt. In a Connection in Section 6.4B, we mentioned how the ability of white TiO_2 to absorb sunlight is enhanced by absorption of "perovskites" $(RNH_3^+)(Pb^{2+})(I^-)_3$. These also have the ABO_3 stoichiometry with charges adding up to zero. But because all ions have half the charge of the charges in $CaTiO_3$, the total lattice energy is *much* lower, so the physical properties of $(RNH_3^+)(Pb^{2+})(I^-)_3$ are quite different than those of $CaTiO_3$.

8.4. Ferromagnetic, Antiferromagnetic, and Superconducting Solid Ionic Oxides

> **OVERVIEW.** Paramagnetism found in simple ions (Section 7.2) can be modified by direct or indirect interaction with neighboring ions in solids such as some oxides to produce more complex magnetic properties: ferromagnetism, antiferromagnetism, ferrimagnetism, and superconductivity. The uses and difficulties of working with high-temperature superconducting oxides illustrate the challenges and potentials of materials science. Superconductors are identified and characterized in Exercises 8.51–8.55.
>
> Section 8.4 includes two Connections to materials science. It builds upon the fundamental types of magnetic properties from Section 7.2, valence electron configurations from Section 1.3, and overlap of orbitals from Section 3.3. Ferromagnetism and superconductivity will appear again in elements and alloys in Sections 12.3B and 12.4C.

In materials such as oxides in which paramagnetic ions (those of the *d*- and *f*-block metals) lie close together, their magnetic moments can align in cooperation with each other throughout large domains of the solid sample. In **ferromagnets** all of the individual magnets (metal ions) in the domain are aligned in the same direction; in **antiferromagnets** every other metal atom or ion has its magnetic field oriented in the opposite direction (Figure 8.11).

Ferromagnetism is found in materials in which unpaired electrons in *d* or *f* orbitals interact with unpaired electrons in similar orbitals in neighboring atoms or ions, so as to align their spins parallel to each other (Figure 8.11a). This interaction must be weak, however; otherwise the electrons pair up in the same bonding orbital, and no magnetism results. **Antiferromagnetism** is also found in oxides and other compounds when the two metal ions interact indirectly, through a ligand whose paired electrons couple weakly and in opposite directions with the unpaired electrons of neighboring metal ions (Figure 8.11b, c), so as to align their spins antiparallel to each other. **Ferrimagnetism** is a property found in some mixed-metal oxides in which the two types of ions are each ferromagnetic—but one type of ferromagnetism is stronger than the other. Although the spins for each set of ions are opposed, the effects do not cancel out.

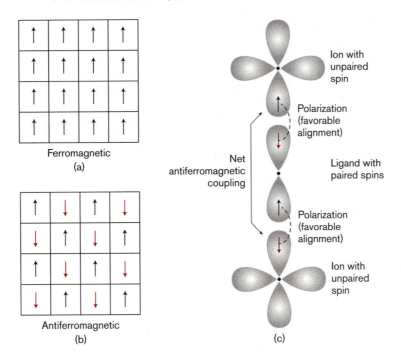

Ferromagnetic
(a)

Antiferromagnetic
(b)

Ion with
unpaired
spin

Polarization
(favorable
alignment)

Net
antiferromagnetic
coupling

Ligand with
paired spins

Polarization
(favorable
alignment)

Ion with
unpaired
spin

(c)

Figure 8.11. Schematic two-dimensional representation of the parallel and antiparallel orientations of the magnetic fields and electron spins of neighboring metal ions in (a) ferromagnets and (b) antiferromagnets, respectively. (c) The antiferromagnetic coupling of electron spins between the paired electrons of a ligand and the unpaired electrons of two neighboring metal ions, resulting in the two metal ions having antiparallel spins. [(a) and (b) Adapted from I. S. Butler and J. F. Harrod, *Inorganic Chemistry: Principles and Applications*, Benjamin/Cummings: Redwood City, CA, 1989, p. 434. (c) Adapted from D. F. Shriver, P. W. Atkins, and C. H. Langford, *Inorganic Chemistry*, W. H. Freeman: New York, 1990, p. 586.]

Figure 8.12a first shows the case of simple *paramagnetism*: The magnetic susceptibility χ_M as a function of temperature is as predicted by Equation 7.3. Figure 8.12b shows the enhanced magnetic susceptibility of *ferromagnets* at low temperature. The weak tendency of ferromagnetism to order electrons is readily overcome by the entropic tendency to disorder. Ferromagnetism fades out at higher temperatures and finally disappears at the **Curie temperature** T_{Curie}, above which paramagnetic behavior prevails. At low temperatures, *antiferromagnetic* materials show very low magnetic susceptibility, which increases (Figure 8.12c) with temperature up to the **Néel temperature** T_N, above which entropy prevails and normal paramagnetic behavior is found.

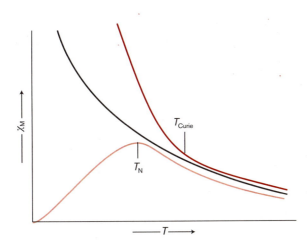

Figure 8.12. Plots of the temperature dependence of magnetic susceptibility χ_M of an idealized substance acting (a) as a paramagnetic substance (——); (b) as a ferromagnetic substance (——) below its Curie temperature, T_{Curie}; and (c) as an antiferromagnetic substance (——) below its Néel temperature, T_N.

Figure 8.13. Levitation of a magnet above a pellet of a high-temperature superconductor cooled below its T_C.

Among the mixed-oxidation state spinels of Mn, Fe, and Co, Mn_3O_4 and Co_3O_4 are antiferromagnetic, while Fe_3O_4 (the mineral *magnetite* or *lodestone*) is ferrimagnetic. Microscopic crystals of magnetite have been found in a number of different living organisms, including bacteria, pigeons, salmon, and (perhaps) in human brain tissue. These crystals might confer sensitivity to magnetic fields similar to Earth's.[15] Similarly, among the simple *d*-block dioxides, VO_2 and MnO_2 are antiferromagnetic, while ferromagnetic CrO_2 is used in magnetic storage devices.

Superconductivity was first observed in metals (Section 12.4C). Superconductors have two outstanding characteristics below their **critical temperature**, T_C, which is the highest temperature at which the superconducting state can be maintained. First, they are capable of excluding a magnetic field, a type of perfect diamagnetism known as the **Meissner effect**. This effect allows a superconductor to levitate a magnet above it (Figure 8.13). Second, the electrical resistance of superconductors drops to zero.

Molecular Magnetism in Prussian Blues

Prussian blues (Section 5.3) can be thought of as resembling oxides in which oxide-ion bridges are replaced with cyanide-ion bridges. The cyanide ion coordinated via its C donor atom is high on the spectrochemical series (Eq. 7.6), so the metal ion coordinated to the C donor atom is low spin. Furthermore, the e_g^* orbitals of the metal ion are so high in energy that they are not occupied. Hence, this metal ion must have six or fewer valence *d* electrons. In a d^6 complex such as $[Fe(CN)_6]^{4-}$ there are no unpaired electrons and the ion is diamagnetic, giving no possibility for ferromagnetism in salts of this ion. But earlier 4*d*-block complex cyanides such as $[Cr(CN)_6]^{3-}$ can be paramagnetic and participate in ferromagnetic interactions as illustrated in Figure 8.11.

The N donor atom of the bridging CN^- ion has a moderate position in the spectrochemical series, so the metal ion(s) coordinated to it can be low spin, but are more often high spin and can go beyond a d^6 electron configuration. This metal ion usually has unpaired electrons, is paramagnetic, and *may* have a ferromagnetic interaction with the metal ion coordinated to the C donor atom. As an example, $_\infty^3\{(Fe^{3+})[Fe(CN)_6]^{3-}\}$ has unpaired electrons on each type of Fe^{3+} ion and is ferromagnetic, but has a low Curie temperature.

Electrons in the *d* orbitals of early *d*-block metal ions extend further from the metal-ion nucleus (Section 9.4), so they can interact more strongly with the electrons in *d* orbitals of a $[M(CN)_6]^{n-}$ ion . With early *d*-block ions such as V^{2+}, Cr^{2+}, and Cr^{3+}, ferromagnetism may persist at or near room temperature.[16] A T_{Curie} of $-3°C$ is found in a mixed-valence chromium(II,III) hexacyanochromate(II,III), $_\infty^3[(Cr)_{2.12}(CN)_6]$.[17] An even higher T_{Curie} of $42°C$ is found when vanadium is incorporated in $_\infty^3\{(V^{2+})_{0.49}(V^{3+})_{0.67}[Cr^{III}(CN)_6]\}$.[18] Such materials are of interest as *molecular magnets*.

In 1987[19] the scientific world was very much surprised to learn that mixed-metal oxides can show the property of superconductivity. The first of these oxides, $La_{1.85}Sr_{0.15}CuO_4$, showed a higher T_C, 30 K, than any of the metals or alloys (Section 12.4C). Very soon a mixed-metal oxide with a T_C of about 90 K was discovered, $YBa_2Cu_3O_7$ (which is familiarly known as "123" for the ratios of the metal ions in the formula). With this high a T_C, the cheap and abundant coolant, liquid nitrogen, can be used to cool the superconductor below T_C. (Above T_C superconductors show the antiferromagnetic properties that are common in metal oxides.)

The structure of $YBa_2Cu_3O_7$ (Figure 8.14) is related to that of perovskite (Figure 8.7). This relationship can be seen if we take three adjacent A-centered unit cells (Figure 8.14) and allow Y^{3+} to occupy the A position in the central cell and Ba^{2+} to occupy the A positions in the outer cells. The other alteration is that three perovskite unit cells would contain a total of nine, not seven, oxide ions, and would require an impossibly high oxidation number for at least one of the metals. The $YBa_2Cu_3O_7$ structure deletes oxide ions from each of the three cells: along the front and back outer edges in the two Ba-centered cells, and along the middle edges in the Y-centered cell. In addition, the superconductor is slightly nonstoichiometric, so it has a formula of $YBa_2Cu_3O_{7-\delta}$.

Removing oxide ions changes the polyhedra of the copper ions. The d^8 Cu^{3+} ions found on the top and bottom surfaces of the new, larger unit cell of Figure 8.14 (those closer to Ba than to Y) show square planar coordination; bridging oxide ions link these squares into polymeric *chains*, as shown by the arrows in Figure 8.14. In contrast, the d^9 Cu^{2+} ions found in the center of the new unit cell (sandwiched between Ba and Y) lose only one oxide neighbor, so they show square pyramidal coordination; the square pyramids are also linked by bridging oxide ions, but this time in two dimensions to form *planes* of Cu^{2+} polyhedra.

Experiments in isomorphous substitution have helped decide which parts of the structure in Figure 8.14 are crucial for superconductivity. The diamagnetic Y^{3+} ions can be replaced with most of the lanthanide ions without altering the superconductivity (even though these ions are mostly paramagnetic!). Slight replacement (doping) of the Cu^{3+} ions from the chains of CuO_4 square planes with other +3-charged ions such as Ga^{3+} lowers T_C, but not as severely as does slight replacement of the Cu^{2+} ions from the planes of CuO_5 square pyramids with divalent ions such as Zn^{2+}.

$YBa_2Cu_3O_{7-\delta}$ is prepared by sintering the solid oxides Y_2O_3, CuO, and the oxide precursor $BaCO_3$ at 950°C for an extended time, followed by a slow cooling and exposure to O_2 at 500–600°C, whereupon one-third of the Cu^{2+} ions are oxidized to Cu^{3+} ions. Materials scientists continue to be challenged by the basic physical properties of such oxides. Synthesis by sintering requires high temperatures and long reaction times. It is difficult to insure that only one product is produced when so many mixed-metal oxides are possible; with such nonvolatile, insoluble products, normal methods of purification after synthesis are impossible. The brittleness of the products makes it very difficult to draw the superconductor into wires or to deposit it as films. In addition, with the presence of the high Cu^{3+} oxidation state, $YBa_2Cu_3O_7$ is air- and moisture-sensitive.

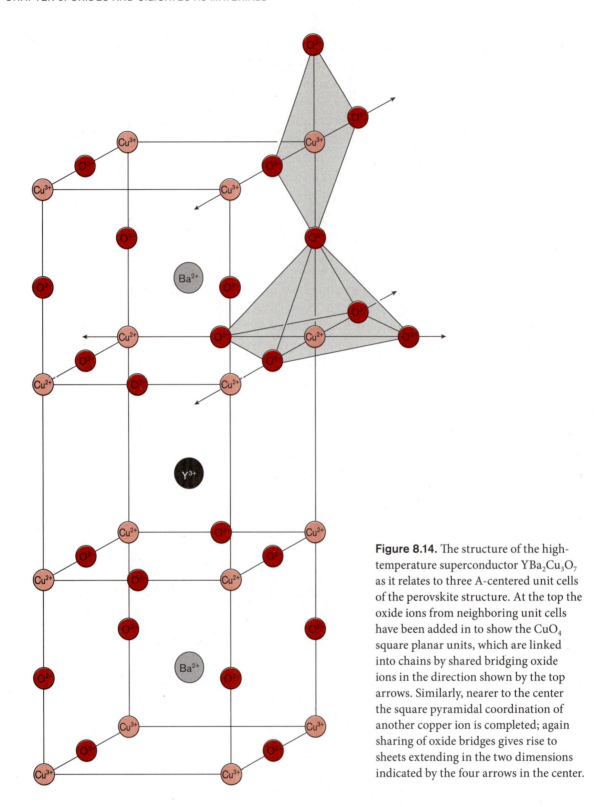

Figure 8.14. The structure of the high-temperature superconductor $YBa_2Cu_3O_7$ as it relates to three A-centered unit cells of the perovskite structure. At the top the oxide ions from neighboring unit cells have been added in to show the CuO_4 square planar units, which are linked into chains by shared bridging oxide ions in the direction shown by the top arrows. Similarly, nearer to the center the square pyramidal coordination of another copper ion is completed; again sharing of oxide bridges gives rise to sheets extending in the two dimensions indicated by the four arrows in the center.

A CONNECTION TO MATERIALS SCIENCE

Newer High-Temperature Superconductors

A variety of related "cuprate" high-temperature superconductors have since been discovered. $YBa_2Cu_4O_8$ and $Y_2Ba_4Cu_7O_{15}$ add additional chains of CuO_4 squares, but the most dramatic improvements in T_C result from incorporating heavy p-block metal ions of mercury, thallium, and bismuth, resulting in idealized formulas including $HgBa_2CuO_4$, $Bi_2Sr_2CuO_6$, $Tl_2Ba_2CuO_6$, and $TlBa_2Cu_2O_5$. The highest T_C yet found, 125 K and 133 K, belong to superconductors having three layers of Cu ions, $Tl_2Ba_2CaCu_3O_{10}$ and $HgBa_2CaCu_3O_8$, respectively. These structures lack chains of CuO_4 squares but still have the planes of cuprate units, confirming their importance in superconductivity. The production of the thallium- and mercury-containing cuprates is limited by the toxicity and high-temperature volatility of Tl_2O and HgO, which have low lattice energies and hence relatively high vapor pressures at temperatures at which the other oxides react to give superconductors.

Since these discoveries, superconductivity has been found in other classes of materials. One recent new class of high-temperature superconductors are based on a more complex lattice that has alternating oxide and arsenide layers, $^2_\infty[LaO]^+\ ^2_\infty[FeAs]^-$.[20,21] With about 10% of the oxide ions replaced by fluoride ions, the material showed a critical temperature of 26 K. By utilizing a high-pressure synthesis, the temperature was raised to 43 K.[22] Still more recently, a record-high T_C of 203 K was achieved in H_2S compressed to such high pressure that it decomposes to other species—perhaps H_3S.[23]

CONNECTIONS TO MATERIALS SCIENCE

SQUIDs and Maglev Trains

Limited commercialization of high-temperature superconductors has occurred.[24,25] A superconducting tape in which filaments of $Bi_2Sr_2Ca_2Cu_3O_{10}$ are sheathed with metallic silver is now produced, and superconducting power cables are now available, but they are expensive. Some applications involving low magnetic fields and electric current density and using liquid nitrogen cooling include microwave filters and superconducting quantum interference devices (SQUIDs). The application that most catches the public fancy is the construction of **maglev trains**, in which the train floats 8 mm above the track; despite formidable costs and difficulties, such a train is now operating on a 19-mile track between Shanghai and its international airport. With zero friction between the train and the tracks, speeds of 260 miles per hour have been reached.

8.5. Acidity, Solubility, and Atmospheric Chemistry of Oxides

OVERVIEW. Oxides of elements show periodic gradations in acid–base properties. They range from the soluble basic oxides containing the nonacidic and feebly acidic cations, through the insoluble basic oxides of other metals in moderate oxidation states, to the insoluble acidic oxides of metals in moderate oxidation states and metalloids, to the soluble acidic oxides of metals in high oxidation states and most nonmetals (Sections 8.5A). You can practice applying these classification schemes by trying Exercises 8.56–8.63. Acidic and basic oxides can react with each other without the involvement of water, in what is called a Lux–Flood acid–base reaction (Section 8.5B). You can practice predicting products of such reactants by trying Exercises 8.64–8.69. The acidic volatile oxides of the nonmetals, most of which are acidic or can be oxidized to forms that are acidic, are involved in environmental problems such as the formation of acid rain (Section 8.5C). This subsection does not introduce new principles, but Exercises 8.70–8.73 develop your practical knowledge of the atmospheric chemistry of these oxides, while applying principles from this text to understanding this chemistry.

Section 8.5 includes Connections to agricultural chemistry, atmospheric chemistry, environmental chemistry, and industrial chemistry. It will be helpful in applying these concepts to review the acidity classifications of cations (Section 2.2), the solubilities of metal oxides and hydroxides (Section 2.5), the drawing of Lewis structures of molecules (Section 3.1), and the formulas, acid–base properties, and speciation of oxo anions and acids (Sections 3.5–3.8).

8.5A Acidic and Basic Oxides. To a first approximation, the acid–base properties of oxides of the elements can be thought of as a composite of the very strongly basic properties of the oxide ion and the acidic properties of the element "cation," which can range from nonacidic to very strongly acidic. Many metal oxides are **basic oxides**: the basic properties of the oxide ions, which are very strong, prevail over the lesser acidic properties of most metal cations. We may further subdivide the basic oxides into those that are soluble and those that are insoluble.

The **soluble basic oxides** are the oxides of the nonacidic and feebly acidic metal cations in the lower left part of the periodic table: They dissolve sufficiently in water to give hydroxides of these metal ions in solutions of high pH.

$$BaO(s) + H_2O \rightarrow Ba^{2+}(aq) + 2OH^-(aq) \tag{8.3}$$

The **insoluble basic oxides** are oxides of the weakly acidic and yet more acidic metal cations. By Solubility Rule I these are insoluble in water, so they do not give solutions of elevated pH values. Nonetheless, they do dissolve in, and partially neutralize, strong acids, so they show basic properties.

$$Fe_2O_3(s) + 6H^+(aq) \rightarrow 2Fe^{3+}(aq) + 3H_2O \tag{8.4}$$

Often, however, the acidity of a very strongly acidic "cation" can prevail over the basicity of the oxide ion, so many oxides of the nonmetals at the right of the periodic table, or of metals in very high oxidation states, are **acidic oxides**. They can be categorized as soluble and insoluble, too.

Acidic Oxides. The process of dissolving an acidic oxide in water can be envisioned as involving at least two equilibria that influence each other. We may suppose that a soluble acidic oxide first reacts reversibly with water to give an oxo acid:

$$\tfrac{1}{\infty}[SeO_2] + H_2O \rightleftharpoons H_2SeO_3(aq) \tag{8.5}$$

If the resulting oxo acid is very strongly acidic, strongly acidic, or moderately acidic (Section 3.7), a second equilibrium then proceeds appreciably to the right:

$$H_2SeO_3(aq) + H_2O(aq) \rightleftharpoons H_3O^+(aq) + HSeO_3^-(aq) \tag{8.6}$$

For these three categories of oxo acids this equilibrium then shifts the position of the previous equilibrium far enough to the right to allow the original metal oxide to dissolve in neutral water and to lower its pH. Oxides that can dissolve to give very strongly acidic, strongly acidic, or moderately acidic oxo acids generally are **soluble acidic oxides**.

As an example, the soluble acidic oxide P_4O_{10} reacts very completely with water and hence is used as a drying agent. But it is dangerous to use it to dry organic liquids, because it releases so much heat via Equation 8.7 that it can set the organic liquid on fire:

$$P_4O_{10}(s) + 6H_2O(l) \rightarrow 4H_3PO_4(aq) \qquad \Delta H = -428.5 \text{ kJ} \tag{8.7}$$

The reaction with water is involved in the production of high-purity "syrupy" (85% or 15 M concentrated) phosphoric acid, which is used in making detergents, toothpaste, and in foods such as colas (about 0.05% H_3PO_4, pH = 2.3).

Most other nonmetal oxides, and a significant number of metal oxides, do not undergo sufficient ionization analogous to Equation 8.6 to allow them to dissolve in neutral water, but they dissolve in strong bases, partially neutralizing them. These may be termed **insoluble acidic oxides**. Major soluble and insoluble acidic oxides, and the products they form on dissolving in water, are indicated in Table 8.6.

Table 8.7 summarizes, in a general way, the locations in the periodic table of the four principal acid–base classes of oxides. Soluble basic oxides and soluble acidic oxides occupy opposite edges of the periodic table, except that metals in very high oxidation states can also give rise to soluble acidic oxides. In between we find insoluble basic oxides (more to the left) and insoluble acidic oxides (more to the right).

Amphoteric and Neutral Oxides. The realms of insoluble acidic oxides and insoluble basic oxides overlap extensively, however. There are many insoluble oxides such as Al_2O_3 that do not dissolve in water, but dissolve in strong acids *and* (in separate experiments) in strong bases:

$$Al_2O_3(s) + 6H^+ \rightleftharpoons 2Al^{3+}(aq) + 3H_2O \tag{8.8}$$

$$Al_2O_3(s) + 2OH^- + 3H_2O \rightleftharpoons 2[Al(OH)_4]^- \tag{8.9}$$

TABLE 8.6 Major Acidic Oxides of the _p_- and _d_-Block Elements[a–c]

$B_2O_3 \rightarrow H_3BO_3$(w)	$CO_2 \rightarrow H_2CO_3$(m)	$N_2O_5 \rightarrow HNO_3$(s)			
		$N_2O_3 \rightarrow HNO_2$(m)			
$Al_2O_3 \rightarrow AlO_4^{5-}$	$SiO_2 \rightarrow SiO_4^{4-}$	$P_4O_{10} \rightarrow H_3PO_4$(m)	$SO_3 \rightarrow H_2SO_4$(s)	$Cl_2O_7 \rightarrow HClO_4$(vs)	
		$P_4O_6 \rightarrow H_3PO_3$(m)	$SO_2 \rightarrow H_2SO_3$(m)	$Cl_2O \rightarrow HClO$(w)	
$Ga_2O_3 \rightarrow GaO_4^{5-}$	$GeO_2 \rightarrow GeO_4^{4-}$	$As_2O_5 \rightarrow H_3AsO_4$(m)	$SeO_3 \rightarrow H_2SeO_4$(s)		
		$As_4O_6 \rightarrow AsO_3^{3-}$	$SeO_2 \rightarrow H_2SeO_3$(m)	$Br_2O \rightarrow HBrO$(w)	
$SnO_2 \rightarrow SnO_6^{8-}$	$Sb_2O_5 \rightarrow SbO_6^{7-}$	$TeO_3 \rightarrow TeO_6^{6-}$			$XeO_4 \rightarrow H_4XeO_6$(s)
$SnO \rightarrow SnO_3^{4-}$	$Sb_2O_3 \rightarrow SbO_3^{3-}$	$TeO_2 \rightarrow TeO_3^{2-}$		$I_2O_5 \rightarrow HIO_3$(s)	$XeO_3 \rightarrow H_2XeO_4$(s)
$PbO_2 \rightarrow PbO_6^{4-}$					
	$V_2O_5 \rightarrow VO_4^{3-}$	$CrO_3 \rightarrow H_2CrO_4$(s)	$Mn_2O_7 \rightarrow HMnO_4$(vs)		
		$MoO_3 \rightarrow MoO_4^{2-}$	$Tc_2O_7 \rightarrow HTcO_4$(vs)	RuO_4	
		$WO_3 \rightarrow WO_4^{2-}$	$Re_2O_7 \rightarrow HReO_4$(vs)	OsO_4	

[a] Oxides above and to the right of the lines (in **brown** boxes) are _soluble_ acidic oxides; the acids they form on dissolving, and their strengths, are also shown.

[b] Oxides within the lines (in **light brown** boxes) are _insoluble_ acidic oxides; the corresponding nonprotonated oxo anions are also shown. (Normally the anions formed are then partially protonated—for example, $Sb_2O_5(s) + 5H_2O(s) + 2OH^-(aq) \rightleftharpoons 2[Sb(OH)_6]^-(aq)$.

[c] _Amphoteric_ oxides and their oxo anions are shown in the white region (for the _p_-block elements only).

TABLE 8.7. Soluble and Insoluble Acidic and Basic Oxides[a]

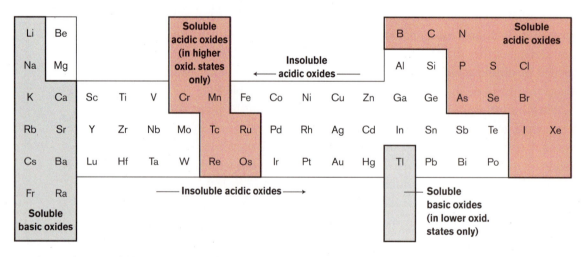

[a] Elements forming soluble acidic oxides are shaded in **brown**. Elements forming soluble basic oxides are shaded in gray. The elements in the unshaded region vary from insoluble basic oxides on the left to insoluble acidic oxides on the right, with no strong boundary between them.

Such oxides are known as **amphoteric oxides**. These are scattered somewhat unsystematically in the center part of the periodic table (Table 8.7). Some insoluble basic oxides derived from even weakly acidic cations (e.g., ZnO) do dissolve in strong bases, so they are amphoteric; some derived from moderately or even strongly acidic cations (e.g., Fe_2O_3 and TiO_2) do not, so they are insoluble basic oxides.

Because the acidity of a cation rises rapidly with its charge, there are several *d*-block elements possessing several oxidation states (such as chromium) that have one or more oxides that show only basic properties [e.g., chromium(II) oxide, CrO], one or more oxides that are amphoteric [e.g., chromium(III) oxide, Cr_2O_3], and one or more oxides that possess only acidic properties [chromium(VI) oxide, CrO_3]. Thus, the higher the oxidation number of a given element, the more acidic the corresponding oxide is.

EXAMPLE 8.12

Consider the following oxo acids: (a) H_3VO_4; (b) H_2XeO_4; (c) $HTcO_4$; (d) H_4SnO_3; (e) H_6MoO_6. For each oxo acid, give (1) its category of acidity; (2) its name; (3) the formula of the acidic oxide from which it might be prepared by the addition of water only (i.e., no redox chemistry); (4) whether it will be soluble or insoluble in water.

SOLUTION: Acidity categories of oxo acids are covered in Section 3.7, and they are based on the number of oxo groups. The nomenclature of oxo acids is covered in Section 3.9 and is based on the oxidation state of the central atom.

(a) H_3VO_4 has one oxo group and the vanadium has an oxidation number equal to the group number of +5, so this is vanadic acid, which is moderately acidic. The oxide of V^{5+} is V_2O_5, which should be soluble because it gives a moderately acidic oxo acid. (In fact, it is only slightly soluble.)

(b) H_2XeO_4 has two oxo groups and the xenon has an oxidation number equal to the group number $8 - 2 = +6$, so this is xenic acid, which is strongly acidic. The oxide of Xe^{6+} is XeO_3, which should be soluble because it gives a strongly acidic oxo acid.

(c) $HTcO_4$ has three oxo groups and the technetium has an oxidation number equal to the group number of +7, so this is pertechnetic acid, which is very strongly acidic. The oxide of Tc^{7+} is Tc_2O_7, which should be soluble because it gives a very strongly acidic oxo acid.

(d) H_4SnO_3 has no oxo groups (arguably it has –1 oxo group!) and the tin has an oxidation number equal to the group number $4 - 2 = +2$, so this is stannous acid, which is weakly acidic at best. The oxide of Sn^{2+} is SnO, which should be insoluble because it gives (at best) a weakly acidic oxo acid.

(e) Although the oxo anion of Mo is MoO_4^{2-}, the oxo acid, anomalously, is H_6MoO_6, which has no oxo groups and is weakly acidic. Hence, MoO_3 is relatively insoluble.

EXAMPLE 8.13

Classify each of the following oxides as acidic, basic, or amphoteric or neutral, and decide whether it will be soluble in water: (a) P_4O_{10}; (b) RaO; (c) Tl_2O; (d) SO_2; (e) Al_2O_3.

SOLUTION: (a) P_4O_{10} is a nonmetal oxide (so it is likely acidic) built from a P^{5+} "cation" in Period 3. Therefore, the oxo anion and oxo acid should show the penultimate TCN of four: PO_4^{3-} and H_3PO_4, a moderately acidic oxo acid. P_4O_{10} is therefore soluble. (b) RaO is a metal oxide built from a feebly acidic Ra^{2+} cation, so it is basic and soluble. (c) Tl_2O is a metal oxide built from a nonacidic or feebly acidic Tl^+ cation, so it is basic and soluble. (d) SO_2 is a nonmetal oxide (so it is likely acidic) built from a $:S^{4+}$ "cation" in Period 3. Therefore, the oxo anion and oxo acid hould show the penultimate TCN of four: $:SO_3^{2-}$ and H_2SO_3, a moderately acidic oxo acid. SO_2 is therefore soluble. (e) Al_2O_3 is the oxide of moderately acidic Al^{3+}, so this oxide will be insoluble. It could be an insoluble basic oxide or an insoluble acidic oxide, or both, in which case it is amphoteric. (This is actually the case.)

Oddball Oxides

Three monomeric nonmetal oxides from the upper-right portion of the p block have such low oxidation numbers for the nonmetal atom that they lack any acidic properties—namely, CO, N_2O, and NO; because these certainly don't have basic properties, they are neutral oxides. In addition, there are a few oxides, such as the free radicals NO_2 and ClO_2, which do not correspond in oxidation state to stable or known oxo acids or anions. These can give rise to a mixture of oxo acids or anions by disproportionation:

$$2NO_2 + 2OH^- \rightarrow NO_2^- + NO_3^- + H_2O \qquad (8.10)$$

8.5B. Lux–Flood Acid–Base Reactions. Because basic oxides can react with the hydronium ion, a strong aqueous acid, and acidic oxides can react with the hydroxide ion, a strong aqueous base, basic oxides such as SrO and acidic oxides such as SiO_2 can react directly with each other:

$$2SrO(s) + SiO_2(s) \rightarrow Sr_2SiO_4(s) \qquad \Delta H = -209.8 \text{ kJ} \qquad (8.11)$$

$$ZrO_2(s) + SiO_2(s) \rightarrow ZrSiO_4(s) \qquad \Delta H = -32.1 \text{ kJ} \qquad (8.12)$$

In such a reaction there is a transfer of *oxide ions*. In the Lux–Flood acid–base classification system an oxide-ion donor is a base and an oxide-ion acceptor is an acid. The products of these reactions are salts of oxo acids and, because water is not involved in the reaction, can be prepared even though oxides such as ZrO_2 and SiO_2 are insoluble in water, and the oxo anion SiO_4^{4-} is too basic to exist in water. Equation 8.12 produces the mineral zircon ($ZrSiO_4$) and can be carried out by sintering the two oxides at 1500°C for 4–6 hours. Direct reactions of acidic and basic oxides are important in such areas as control of pollution by gaseous acidic oxides and in the production of materials such as cement, glass, and ceramics. The ΔH of a Lux–Flood acid–base reaction can be approximately predicted from the Z^2/r ratios of the elements in the reacting acidic and basic oxides.[26]

EXAMPLE 8.14

Identify the Lux–Flood acid and base, then complete and balance the equations for their reactions, if any: (a) $N_2O_3 + Tl_2O \rightarrow$; (b) $BaO + OsO_4 \rightarrow$; (c) $SrO + ZrO_2 \rightarrow$; (d) $Fe_3O_4 + H^+ \rightarrow$.

SOLUTION: (a) N_2O_3 is a nonmetal oxide, so it is most likely an acidic oxide. With its oxidation state of +3 and penultimate TCN of +3, its corresponding oxo anion is $:NO_2^-$. Tl_2O is a basic oxide, so the two can react to give $TlNO_2$, the salt of Tl^+ with NO_2^-:

$$N_2O_3 + Tl_2O \rightarrow 2TlNO_2$$

(b) Because Ba^{2+} is a feebly acidic cation, BaO is a basic oxide. Because the hypothetical Os^{8+} cation is very strongly acidic, OsO_4 is an acidic oxide. Os would probably show its penultimate TCN of six and form the OsO_6^{4-} anion (Table 3.5). Balancing the –4 charge of this anion with two Ba^{2+} cations, we predict the following reaction:

$$2BaO + OsO_4 \rightarrow Ba_2OsO_6$$

(c) Because Sr^{2+} is a feebly acidic cation, its oxide SrO is a basic oxide. Zr^{4+} in ZrO_2 is strongly acidic, so its oxide is (unpredictably) either basic or amphoteric. If ZrO_2 is also a basic oxide (as it is, in fact), there can be no acid–base reaction between these oxides. (They might form a mixed-metal oxide.)

(d) Fe_3O_4 is a mixed-metal oxide in which the average oxidation state of iron is +2.67, which means that two-thirds of the iron ions are moderately acidic Fe^{3+} and one-third are weakly acidic Fe^{2+} ions. The oxides of such ions are insoluble in water, but they do dissolve in acids. Therefore, Fe_3O_4 is a basic oxide:

$$Fe_3O_4 + 8H^+ \rightarrow Fe^{2+} + 2Fe^{3+} + 4H_2O$$

A CONNECTION TO AGRICULTURAL CHEMISTRY

Fertilizer Composition

The three main nutrient elements provided in fertilizers are K, P, and N; the potassium and phosphorus are provided in the form of K^+ and PO_4^{3-} and the nitrogen either as the NH_4^+ ion or the NO_3^- ion. Interestingly, the K and P compositions of fertilizers are often expressed in terms of the masses of K_2O and P_4O_{10} or P_2O_5 needed to give the 100 g of the fertilizer via Lux–Flood acid–base reactions. Because the nitrogen may be present as NH_4^+, which cannot arise from a Lux–Flood acid–base reaction, its content is given simply as %N. A fertilizer labeled 16-48-0 contains 16% N, 48% P_2O_5, and 0% K_2O.

Concrete and Cement Chemistry

Concrete has been made since Roman times; it physically mixes sand with $Ca(OH)_2$ (hydrated or slaked lime). To make calcium hydroxide, $CaCO_3$ (limestone) is first heated above 800°C:

$$CaCO_3(s) \rightarrow CaO(s) + CO_2(g) \qquad \Delta H = +178.3 \text{ kJ} \qquad (8.13)$$

This is followed by the exothermic reaction of the basic oxide CaO with water to generate $Ca(OH)_2$. The strongly basic $Ca(OH)_2$ supported on the grains of sand reacts slowly with the acidic oxide CO_2 from the air to form interlocking strong crystals of $CaCO_3$.

Discovered in the early 1800s, Portland cement is generated by the high-temperature reaction of limestone ($CaCO_3$) with aluminosilicate clays. The firing (thermal decomposition) generates the basic oxide CaO (and some MgO from the clay), and the acidic oxides $CO_2(g)$, SiO_2, and Al_2O_3 (also from the clay). After firing, the dry cement contains principally the minerals $(Ca^{2+})_3(SiO_4^{4-})(O^{2-})$, Ca_2SiO_4, calcium aluminate, and some free CaO and MgO (from the clay). When water is added to Portland cement to cause it to set, reactions such as the following occur:

$$4(Ca^{2+})_3(SiO_4^{4-})(O^{2-}) + 12H_2O \rightarrow Ca_6(OH)_8[Si_4O_{10}] \cdot 2H_2O + 6Ca(OH)_2 \quad (8.14)$$

As a result, both concrete and Portland cement are strongly basic and much more reactive than people commonly realize, continuing to react with acidic CO_2 from the air for years to form $CaCO_3$. The reactivity of the cement used in its construction is what caused the failure of Biosphere II—a self-contained, sealed-off ecosystem—in Arizona in the early 1990s. CO_2 in its atmosphere was supposed to undergo photosynthesis to generate food for the inhabitants to eat and O_2 for them to breathe, but the CO_2 ended up reacting with the cement instead. As a result, the O_2 content of the atmosphere of Biosphere II fell from the normal 21% down to 14%, barely enough to sustain human life. Most species in the Biosphere became "extinct," with the few remaining species (cockroaches, ants, katydids, and morning glories) overrunning the environment.[27]

Recent research in cement chemistry has focused on its environmental impact: Equation 8.13 in the manufacture of Portland cement produces 5% of the total atmospheric load of the greenhouse gas CO_2. Among the changes being tried are to replace up to half of the Portland cement with blast-furnace slag and fly ash left over from coal-fired power plants.[28]

8.5C. Production, Uses, and Atmospheric Chemistry of Volatile Acidic Oxides— Connections with Environmental Chemistry. A number of the volatile (and usually acidic) oxides, and some of the oxo acids formed from them, are of immense significance in our technological civilization. These include some of the chemicals produced in the greatest total tonnage in the world. We can anticipate that the large quantities of these oxides in the atmosphere will result in air-pollution problems such as the production of **acid rain**. In the remainder of this section we deal with the uses and environmental chemistry of the most important of these oxides; there are many useful sources of this information that should be consulted for more details.[29,30] We should also be able to predict some uses and environmental chemistry for other volatile oxides, should they come into widespread use.

Carbon monoxide is produced industrially on a huge scale in the process of **coal gasification** as a mixture with H_2, by the reaction of steam with hot coal:

$$C(s) + H_2O(g) \rightarrow CO(g) + H_2(g) \qquad \Delta H = +131.3 \text{ kJ} \qquad (8.15)$$

This mixture of carbon monoxide and hydrogen is known as **water gas**; after adjustment of its hydrogen content it is known as **synthesis gas**, because it is used in the industrial production of a number of important organic chemicals. Intensive research is currently focused on the chemistry of carbon monoxide, because this mixture could be a source of organic chemicals from coal when our present source of many organic chemicals, petroleum, is exhausted. For example, synthesis gas can be converted to hydrocarbons via the Fischer–Tropsch process:

$$n\text{CO} + (2n + 1)\text{H}_2 \rightarrow \text{C}_n\text{H}_{2n+2} + n\text{H}_2\text{O} \qquad (8.16)$$

Environmentally, the main anthropogenic (human-created) source of carbon monoxide is the incomplete combustion of fuel in automobile engines (197 million tons per year). In urban environments this is a problem due to the high toxicity of this odorless soft base (Section 5.7B). In terms of the global atmospheric environment, however, carbon monoxide is not considered a serious pollutant, because natural production of CO [i.e., by the oxidation of CH_4 produced in swamps and in the tropics (Section 6.5B)] far outweighs human production, and because natural processes continuously remove CO from the atmosphere. These processes include microbial degradation in the soil and reactions in the atmosphere with reactive free radicals such as hydroxyl ($\cdot$OH) and hydroperoxyl ($HO_2\cdot$).

Carbon dioxide is produced industrially by a number of reactions, such as the combustion of carbonaceous fuels, and by the **water-gas shift reaction**:

$$CO(g) + H_2O(g) \rightleftharpoons CO_2(g) + H_2(g) \qquad \Delta H = -41.2 \text{ kJ} \qquad (8.17)$$

Because of its physical properties, CO_2 is widely used as a refrigerant. The solid form, dry ice, sublimes at $-78\,^\circ$C, thus cooling its environment without generating any messy liquids or toxic gases. It is also used on a large scale in fire extinguishers and to carbonate beverages.

Although of very low toxicity, carbon dioxide *is* considered to be a potentially serious global atmospheric pollutant. It is removed from the atmosphere by plant photosynthesis and put into the atmosphere by plant and animal respiration; presumably these processes have reached a balance over time. However, the atmospheric concentration of CO_2, as measured in such remote locations as Antarctica and the top of Mauna Loa, Hawaii, has been increasing 0.2% a year since about 1870. This is thought to be the result of the combustion of fossil fuels and the cutting and burning of many of the tropical forests of the world. This is of concern because of the vital role of CO_2 in the atmosphere: It absorbs infrared light (heat) emitted by Earth and returns some of it to the surface of Earth, thereby warming it; this process is popularly known as the **greenhouse effect**. An increase of only 2°C or 3°C would have profound effects on the climate of Earth; between 1880 and 1940 the mean temperature of Earth rose 0.4°C. Warmer air stores more water vapor, and it is predicted to result in a new climate with greater extreme variation between periods of drought and quick deluges of rain.*

At low temperature and high pressures (conditions under which CO_2 is a liquid), CO_2 absorbs water to form a clathrate compound, $CO_2 \cdot 5.75H_2O$ (clathrates were mentioned in Section 5.3B). This process has been filmed at a depth of 3600 meters;[31] a lake of liquid CO_2 with associated microbes has been seen at a hydrothermal vent.[32] Methods of disposing of CO_2 that has been removed from stationary sources such as smokestacks or cement kilns include injection in the deep ocean as well as in depleted oil and gas fields.

Nitrogen Oxides. Only three nitrogen oxides are stable enough to be of practical use or environmental importance. One of these, N_2O (commonly called **nitrous oxide**), is used as an anesthetic and as a propellant (to provide pressure to expel ingredients) in aerosol cans. It is involved in reactions in the upper atmosphere that could deplete the ozone layer.

Nitric oxide (NO) has two common sources. One is the catalytic oxidation of ammonia (which comes ultimately from petroleum and air):

$$4NH_3 + 5O_2 \rightleftharpoons 4NO + 6H_2O \tag{8.18}$$

The other is the direct combination of nitrogen and oxygen of the air in an electrical discharge (lightning) or at around 2000°C (e.g., in a power plant during the burning of coal or in an automobile engine):

$$N_2 + O_2 \rightleftharpoons 2NO \tag{8.19}$$

Nitric oxide is known to catalyze the destruction of vital ozone (O_3; Section 12.1) in the upper atmosphere:

$$NO + O_3 \rightarrow NO_2 + O_2 \tag{8.20}$$

$$O_3 \rightleftharpoons O_2 + O \tag{8.21}$$

$$O + NO_2 \rightarrow NO + O_2 \tag{8.22}$$

Overall, the reactions add up to

$$2O_3 \rightarrow 3O_2$$

Nitrogen Dioxide. After a few days in the atmosphere, NO is oxidized by oxygen to NO_2; hence in air-pollution work these two are often collectively referred to as NO_x. **Nitrogen dioxide** and its dimer, **dinitrogen tetroxide**, are readily interconverted in an equilibrium that is visible due to the brown color of nitrogen dioxide:

$$N_2O_4 \rightleftharpoons 2NO_2 \qquad \Delta H = +57 \text{ kJ mol}^{-1} \tag{8.23}$$

* The author is personally sensitive to this: His hometown, Cedar Rapids, IA, was massively flooded after an unprecedented rainfall of 15 inches in 2008, and he barely escaped a similar event in Nashville, TN, in 2010. Each event resulted in $2 billion in damages.

In the solid state this system is colorless, because it is completely in the form of N_2O_4. At its boiling point (21°C), the liquid is deep brown due to a 0.1% content of NO_2. The vapor becomes steadily darker with increasing temperature due to the increasing dissociation of the dimer, which is nearly complete at 140°C.

NO_2 reacts with water by disproportionation to produce nitric acid:

$$3NO_2 + H_2O \rightarrow 2HNO_3 + NO \tag{8.24}$$

This reaction occurs in the atmosphere and is one of the sources of acid rain and of nitrate as a plant nutrient. It is also carried out in industry in the manufacture of nitric acid, which is used on a large scale in the manufacture of ammonium nitrate fertilizer, nylon, steel, and in rockets (as the oxidizer of the rocket fuel).

In the lower atmosphere, NO_2 is involved in a complex series of **photochemical** reactions in air that is also contaminated with unburned hydrocarbons (from automobile exhaust) and in the presence of the ultraviolet component of bright sunlight (e.g., in Los Angeles). These reactions produce ozone, aldehydes, and organic nitrates such as peroxyacetyl nitrate (PAN) and peroxybenzoyl nitrate (PBN), which are powerful eye irritants and are quite damaging to vegetation.

Sulfur dioxide is made commercially by the combustion of sulfur, H_2S, or sulfide ores such as FeS_2. It is a colorless, poisonous gas with a choking odor and a relatively high boiling point (–10°C). Sulfur dioxide is useful as a solvent, refrigerant, food preservative, and (mainly) in the manufacture of sulfuric acid. In this process the SO_2 must first be oxidized by air to SO_3, which is a kinetically slow process, so a catalyst of platinum sponge, V_2O_5, or NO is required. The **sulfur trioxide** resulting from this oxidation reacts very exothermically with water to give H_2SO_4.

Sulfuric acid is the leading industrial chemical in terms of the number of tons produced per year. Concentrated sulfuric acid is 98% H_2SO_4 by weight, making it about 18 M. It boils at 338°C. It has a very strong affinity with water and releases a great deal of heat on absorbing water. Contact with the skin causes dehydration and chemical burns; should a spill occur, the acid should immediately be flushed away with large quantities of water for *at least 15 minutes*. The dilution of concentrated sulfuric acid should be carried out *cautiously*—the acid should be poured slowly into water with good stirring to dissipate the heat. Adding water to the acid can cause dangerous spattering of concentrated acid. Sulfuric acid even removes water from many organic molecules, converting carbohydrates to carbon, for example.

The uses of sulfuric acid are so many and varied that the figures for production of sulfuric acid in a given country have been used as a reliable indicator of that country's industrial capacity. The largest usage is in the production of fertilizer (Section 4.3); other major uses are in the refining of petroleum, in metallurgy, and in the manufacture of chemicals. For example, nitration of organic molecules to produce explosives (Section 6.3B) typically use a mixture of HNO_3 and H_2SO_4, the latter to remove water. This reaction not only generates toxic nitrogen oxides, but also produces waste dilute sulfuric acids. Clean H_2SO_4-free nitrating reagents are now being advocated, using the oxide N_2O_5 by itself.

Natural sources produce large amounts of SO_2 via decay of organic matter to H_2S, which is rapidly oxidized to SO_2 in the atmosphere. Anthropogenic SO_2 is produced in comparable quantities during the roasting of sulfide ores and the burning of oil and coal (which often contains substantial amounts of FeS_2). In the atmosphere, SO_2 is also oxidized to H_2SO_4. The resulting acid mist, which once prevailed in London fog (as opposed to Los Angeles or photochemical smog), has been found in many cities of the world in which homes were heated by burning soft coal. Breathing in droplets of sulfuric acid strains the lungs and heart, and has shortened the lives of many people during such episodes of smog.

In the vicinity of smelters in which sulfide ores are oxidized (roasted), the concentration of sulfuric acid and sulfur oxides has been so great that artificial deserts have been created. This problem has been alleviated by the construction of very high smokestacks ("the solution to pollution is dilution"). Unfortunately this spreads the sulfur oxides and sulfuric acid over whole continents and has helped cause the current problem of acid rain (Section 3.8B). This acid rain is corroding away many historic monuments and statues made of susceptible salts of basic oxo anions (marble and limestone are largely $CaCO_3$). Lakes that are in contact with limestone deposits are protected by the same reaction, which neutralizes the acid rain. Lakes that are not so fortunate become quite acidic, with harmful consequences already discussed in Section 3.8B. Effects on trees and vegetation are now also being discovered.

Due to these problems, a considerable amount of research has been done on methods of control of sulfur dioxide emissions from smelters and power plants. This would seem to be a simple matter, because acidic oxides such as SO_2 and SO_3 would be expected to react readily with inexpensive basic oxides or hydroxides. Thus, a solution of $Ca(OH)_2$ can be sprayed down the smokestack of the plant in a *scrubber* to react with the sulfur oxides according to the reaction in Equation 8.25:

$$Ca(OH)_2 + SO_2 + (1/2)O_2 \rightarrow CaSO_4 + H_2O \tag{8.25}$$

Finding a suitable chemical reaction is only the first step. A coal-fired power plant emits much more CO_2 than SO_2; CO_2 is also a weakly acidic oxide that would use up much of the $Ca(OH)_2$. As a consequence, for each ton of coal burned, up to 0.2 tons of limestone would be required, and an enormous quantity of wet $CaSO_4$ would be generated. This would convert an air-pollution problem into a solid-waste problem, because there are not enough uses of $CaSO_4$ to be able to market such quantities. A second problem is that the lime solution cools the exhaust gases so much that they no longer rise up out of the smokestack, so the stack gases have to be reheated!

A number of other alternative processes have been studied. For example, using $Mg(OH)_2$ or MgO instead of $Ca(OH)_2$ has advantages. The adsorption of SO_2 (as well as other acidic and even organic pollutants) by $MgO(s)$ is greatly improved if the MgO consists of tiny nanocrystals, because the nanocrystals have high surface areas with more coordinatively unsaturated surface sites.[33] After the SO_2 reacts with the slurry of $Mg(OH)_2$, the resulting $MgSO_3$ can be heated (in another location) to regenerate the SO_2:

$$MgSO_3 \rightleftharpoons MgO + SO_2 \tag{8.26}$$

The MgO can be recycled to form $Mg(OH)_2$; thus, there would be no solid-waste problem and no great investment in $Mg(OH)_2$. The SO_2 is formed at a concentration great enough to allow the manufacture of sulfuric acid, which could be sold. However, sulfuric acid is the cheapest acid, and the sale of this acid would not pay for the heat used to decompose the $MgSO_3$.

Although many other alternatives exist, each has its own drawbacks. It was estimated that equipping the power plants of the United States with devices to remove most of the SO_2 would cost about $32 billion. Unless subsidized by the government, this amount would be added to consumers' electric power bills.

EXAMPLE 8.15

As you read Section 8.5C, (a) what are the seven volatile oxides discussed? (b) Which of these are acidic? (c) Which of these seven oxides are represented in the redox predominance diagram of Figure 6.7? How are they represented? (d) Of the oxides represented in Figure 6.7, which should be oxidized by O_2? (e) The presence of these oxides in the atmosphere suggests that their oxidation may not be rapid. Why might this be?

SOLUTION: (a) CO, CO_2, N_2O, NO, NO_2, SO_2, and SO_3.

(b) CO_2, SO_2, SO_3, and NO_2 (by disproportionation).

(c) CO as unstable (to the right of the diagram), CO_2 as stable, NO as unstable, NO_2 as unstable, SO_2 as unstable H_2SO_3, and SO_3 as stable HSO_4^-.

(d) There are positive $E°$ gaps between O_2 at +1.23 V and the following oxides: CO at −0.11 V, NO at +1.03 V, NO_2 at +0.80 V, and H_2SO_3 at −0.16 V, so these oxides should be oxidized.

(e) One factor is that redox predominance diagrams are for 1 M aqueous solutions, not for gaseous oxides; water can play an important role in stabilizing reactants or products. Another factor is that redox predominance diagrams do not predict kinetics—whether the reaction will be fast or slow. (The two half-reactions of the oxide with O_2 would generally involve different numbers of electrons.)

8.6. Polysilicates: Basic Structural Types, Uses, and Chemistry

OVERVIEW. The various types of silicate and polysilicate ions in nature illustrate the different degrees of polymerization of oxo anions. The complex structures of many of the polysilicate ions are often represented with polyhedra, which you should be able to interpret. Some of the properties of these polysilicates follow from the degree of anion polymerization. You may practice taking the formula of a silicate mineral, determining which class of silicate ion it contains, and analyzing the structural unit from which it is built. From the charge densities per structural unit of different polysilicate ions, you can estimate their relative basicities (Section 8.6A; Exercises 8.74−8.80). Silicate minerals are subject to changes in polymerization isomerism with pressure, as they move to deeper layers of Earth and as simple silicates are subject to weathering to forms such as oxides and clays; you may apply the concepts of relative weathering rates (Section 8.6B) in Exercises 8.81−8.86.

Section 8.6 includes a Connection to the chemistry of art and five Connections to geochemistry. In Sections 8.6 and 8.7 you will use charge balance from Sections 1.4 and 3.2 to deduce the charges of silicate ions, the principles of Section 3.1 for assembling/disassembling the Lewis structure of a polyatomic anion, and the basicity classification of oxo anions from Section 3.7. The effects of pressure on the degree of polymerization was covered earlier in Section 4.1C.

8.6A. Classes of Silicate Anions. Of overwhelming importance in the geochemistry of Earth's crust are the silicates and polysilicates. These contain definite silicon oxo anions, with covalent Si–O single bonds and a TCN = 4 for Si. Most of them do not have the simple silicate ion SiO_4^{4-} that we discussed in Chapter 3, but rather have two-coordinate oxygen atoms covalently linking different silicon atoms into oligomeric or one-, two-, or three-dimensional polysilicate ions. We will see differences in physical properties caused by the different degrees of polymerization of the covalent bonding in the polysilicate ions. (Regardless of the degree of polymerization of the polysilicate ions, these are finally linked into three-dimensional ionic lattices by counterions.)

Simple Silicates (Nesosilicates). The simple silicate ion SiO_4^{4-} (also called *orthosilicate* ion, to distinguish it from the polysilicate ions, or called *nesosilicate* by geochemists) is not found in a wide variety of minerals: It is a very strong base that cannot persist in aqueous solution, but it does occur in nature as insoluble salts of acidic cations. Some of the mineral forms containing the orthosilicate ion are phenacite, Be_2SiO_4; willemite, Zn_2SiO_4; zircon, $ZrSiO_4$; thorite, $ThSiO_4$; the garnets, $(M^{2+})_3(M^{3+})_2(SiO_4)_3$ (M^{2+} = Ca, Mg, Fe, Mn; M^{3+} = Al, Cr, Fe); and olivine, $(M^{2+})_2SiO_4$ (M^{2+} = Mg, Fe)*. Although with these relatively acidic cations there is doubtless some covalent character to the M–O bonds, we will treat these as salts. Isomorphous substitution is very common in minerals such as these, so the formula of olivine, for example, is often written as $(Fe,Mg)_2SiO_4$, with commas separating ions that can and do substitute for each other in any proportion.

Oligosilicates: Disilicates (Sorosilicates). The basic structural unit of the disilicate ion, $[O_3Si–O–SiO_3]^{6-}$ or $[Si_2O_7]^{6-}$, is $()_{0.5}SiO_{3.5}^{3-}$, with one-half VCS per structural unit. The disilicate ion is found in hemimorphite, $Zn_4(OH)_2(Si_2O_7)\cdot H_2O$, and the rare mineral thortveitite, $Sc_2Si_2O_7$.

As we start building more polymeric silicate ions, their structural formulas become larger. In Figure 8.15 we show a more compact representation of these structures, in which tetrahedra are drawn to symbolize the silicon atom and its four coordinated oxides.

 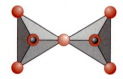

Figure 8.15. Tetrahedral models of the simple silicate and disilicate ions. The small black circles represent Si, the large **brown** circles represent terminal O, and the large **light brown** circles represent bridging O atoms. The point at the center of each tetrahedron is a terminal oxygen atom (**brown**) directly above a silicon atom (black); the point of view is directly down this O–Si bond.

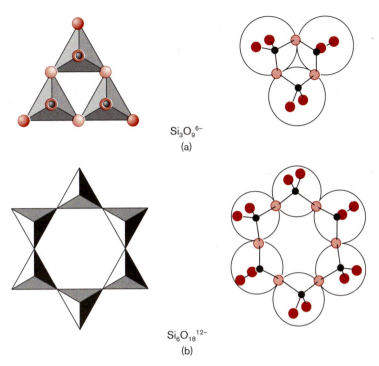

$Si_3O_9^{6-}$
(a)

$Si_6O_{18}^{12-}$
(b)

Figure 8.16. Tetrahedral and ball-and-stick models of two cyclic silicate ions: (a) $Si_3O_9^{6-}$; (b) $Si_6O_{18}^{12-}$. The drawings are color-coded as described for Figure 8.15. In the ball-and-stick models on the right, the structural units, $[(\ )SiO_3]^{2-}$, are enclosed in circles.

Conceptually, we can suppose that this ion is constructed by linking simple silicate ions in an acid–base neutralization reaction. To make room for a link to a bridging oxygen from one SiO_4^{4-}, an oxide ion must be removed from another:

$$SiO_4^{4-} + 2H^+ \rightarrow [(\ )SiO_3]^{2-} + H_2O$$
$$[(\ )SiO_3]^{2-} + SiO_4^{4-} \rightarrow [O_3Si-O-SiO_3]^{6-} \tag{8.27}$$

The disilicate ion has a lower *charge density* (three negative charges per Si structural unit or "nucleus") than does the orthosilicate ion; it is consequently less basic than SiO_4^{4-}.

Oligosilicates: Cyclic Silicates (Cyclosilicates). The structural unit of the cyclic silicate ions is $(\ )_1SiO_3^{2-}$. Cyclic trimers, $[SiO_3]_3^{6-}$, and hexamers, $[SiO_3]_6^{12-}$ (Figure 8.16) are most common and are found in such minerals as benitoite, $BaTi(Si_3O_9)$; beryl/ emerald, $Be_3Al_2(Si_6O_{18})$*; and the more complex tourmaline (which also contains borate anions). Cyclic silicates have a still lower charge density of –2 per silicon nucleus, so they are less basic than disilicates.

Chain Polysilicates: Single-Chain Polysilicates (Inosilicates). These chain polysilicates have the same structural unit as the cyclic silicates, $[(\ )SiO_3]^{2-}$. In chain silicates, the bridging oxygens are linked instead to form chain polymers, $\frac{1}{\infty}[SiO_3^{2-}]$ or $[SiO_3]_n^{2n-}$, rather than oligomers (Figure 8.17). Chain polymers are more common than cyclic oligosilicates and result in the important class of minerals known as the pyroxenes, which includes enstatite, $MgSiO_3$; diopside, $CaMgSi_2O_6$*; spudomene, $LiAlSi_2O_6$; and pollucite, $CsAlSi_2O_6$. The negative charge remains at –2 per structural unit.

EXAMPLE 8.16

(a) The tetrahedral drawing of the $Si_6O_{18}^{12-}$ ion in Figure 8.16b has not been colorized to distinguish atoms (including bridging and terminal oxygen atoms); do this. (b) Write the formula of the structural unit in this ion. Tell how many charges and how many VCS it has.

SOLUTION: (a) The colorized version of this ion is:

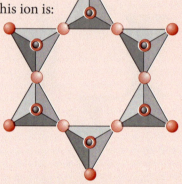

(b) The structural unit of this ion includes only one tetrahedron, and it looks like the following:

It has one silicon atom, two terminal oxygen atom (one at the peak of the tetrahedron and one at the top) and two bridging oxygen atoms (shown cut in half). So, the formula of the tetrahedral unit can be written as $[SiO_2(O_{1/2})_2]$. We can rewrite this as $[(\ )SiO_3]$ (showing one VCS to give silicon a coordination number of four). Its charge can be calculated from oxidation numbers: +4 for Si + 3(–2) for O = –2. Therefore, the structural unit is $[(\ )SiO_3]^{2-}$.

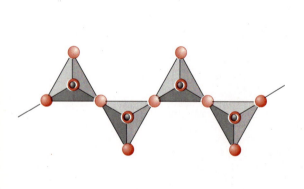

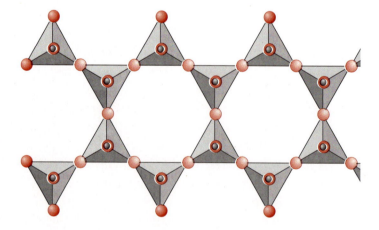

Figure 8.17. Tetrahedral models of chain and double chain polysilicate ions. The drawings are color-coded as described for Figure 8.15.

Chain Polysilicates: Double-Chain Polysilicates. Further reduction in anion basicity is achieved by expulsion of one oxide ion from every fourth SiO_3^{2-} structural unit found in chains. Then each chain is linked to its neighbor chain, giving rise to the *double-chain polysilicates* such as tremolite, $Ca_2Fe_5(OH)_2[Si_4O_{11}]_2$*. These contain the polymerized $\frac{1}{\infty}[Si_4O_{11}^{6-}]$ or $[Si_4O_{11}]_n^{6n-}$ anion (Figure 8.17). We can envision the structural unit as $(\ )_{1.25}SiO_{2.75}^{1.5-}$. Figure 8.18 represents the steps by which we can imagine the $(\ )_{1.25}SiO_{2.75}^{1.5-}$ structural unit polymerizing to give the final double chain.

The double-chain amphibole asbestos minerals such as crocidolite, $Na_2Fe_5(OH)_2$-$[Si_4O_{11}]_2$, and amosite, $(Mg,Fe)_7(OH)_2[Si_4O_{11}]_2$, have long been prized for their fire- and heat-resistance and for their fibrous nature (undoubtedly rooted in the long chain structure of the anion), which allows the weaving of insulating, nonflammable garments, as well as the fabrication of more than 3000 other products. Now these minerals are feared, though, because it has been found that the inhalation of the tiny fibers of asbestos often leads, after 20 or 30 years, to asbestosis (nonmalignant scarring of the lungs) or rare cancers such as mesothelioma. Finding materials to

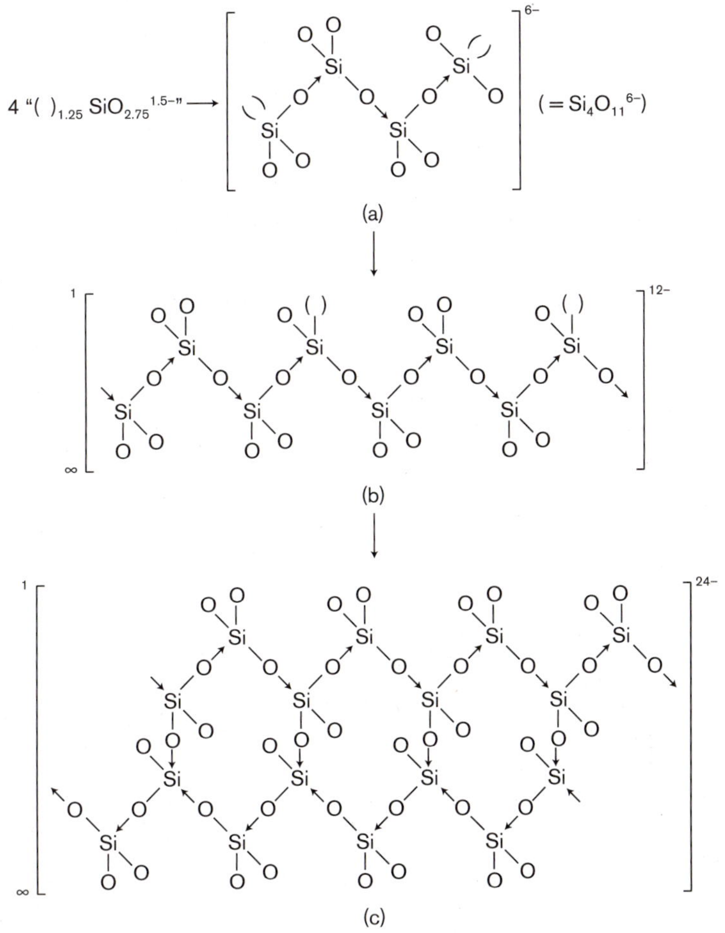

Figure 8.18. Hypothetical steps in the polymerization of an $(\ )_{1.25}SiO_{2.75}^{1.5-}$ structural unit to give: (a) a $[(\ )_2Si_4O_{11}^{6-}]$ simplest formula unit; (b) chain polymerization of the simplest-formula unit to give chains of $\frac{1}{\infty}[(\ )Si_4O_{11}^{6-}]$; (c) side-to-side crosslinking of two chains to eliminate the final VCS and give the double-chain structure, $\frac{1}{\infty}[Si_4O_{11}^{6-}]$.

Figure 8.19. A sample of asbestos, showing its fibrous nature.

replace asbestos in its 3000 uses poses quite a challenge to the industrial inorganic chemist.

Layer Polysilicates (Phyllosilicates). If the side-to-side linking of chains is continued indefinitely, still further oxide ions are eliminated to give a structural unit of $(\ )_{1.5}SiO_{2.5}{}^{0.5-}$ of still lower basicity with more VCS per Si nucleus (1.5) and a greater ability to link as in Figure 8.20 to give a layer silicate, $_{\infty}^{2}[Si_4O_{10}{}^{4-}]$ or $[Si_4O_{10}]_n{}^{4n-}$. The layer polymeric silicates have been extremely important to humans for millennia, due in part to their ready cleavage into thin sheets and other properties that can be related to their layer structures. These minerals include clay minerals such as kaolinite or china clay, $Al_4(OH)_8(Si_4O_{10})$*; pyrophyllite*, $Al_2(OH)_2(Si_4O_{10})$; talc or soapstone, $Mg_3(OH)_2(Si_4O_{10})$*; and serpentine, $Mg_6(OH)_8(Si_4O_{10})$. The contribution of the aluminum or magnesium hydroxide layers to the properties of these minerals is discussed in Section 8.7A. For now we note that the $Mg_6(OH)_8$ layers in serpentine impart a curvature to the sheets of this mineral, which therefore wrap into fibrous tubules. In this form they are the chrysotile form of asbestos, which was 95% of the asbestos used in the United States and which is not as dangerous as the double-chain forms.

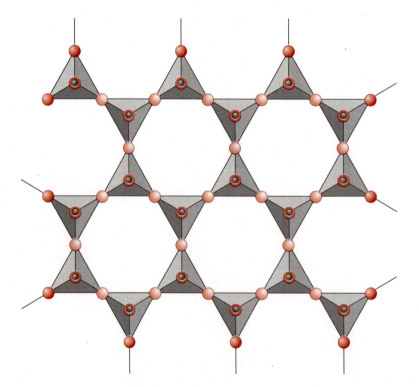

Figure 8.20. Tetrahedral model layer polysilicate ions. The drawings are color-coded as described for Figure 8.15.

Structure	Formula	Name	Total O/Si ratio	Vacant cordination sites per Si	Charges per Si
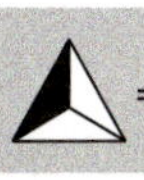 $= SiO_4$	SiO_4^{4-}	Orthosilicate	4	0	−4
	$Si_2O_7^{6-}$	Disilicate	3.5	0.5	−3
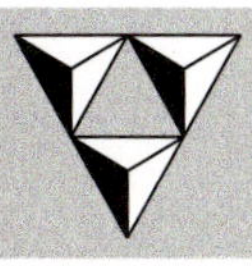	$Si_3O_9^{6-}$	Cyclic silicate	3	1	−2
	$Si_6O_{18}^{12-}$	Cyclic silicate	3	1	−2
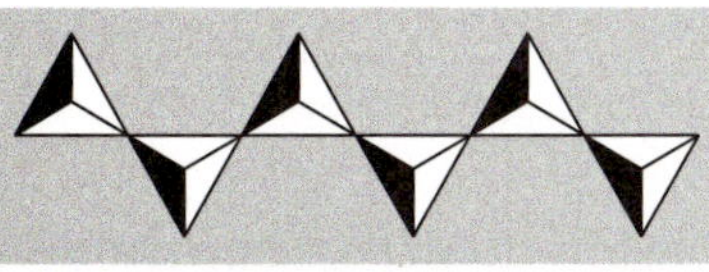	$_\infty^1[SiO_3^{2-}]$	Pyroxene (chain silicate)	3	1	−2
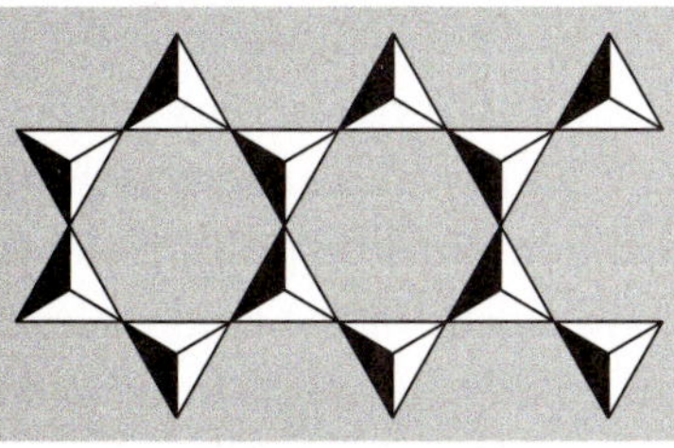	$_\infty^1[Si_4O_{11}^{6-}]$	Amphibole (double chain)	2.75	1.25	−1.5
	$_\infty^2[Si_4O_{10}^{4-}]$	Infinite sheet silicate	2.5	1.5	−1
	$_\infty^3[SiO_2]$	Silica	2	2	0

Figure 8.21. Structures of some polysilicates and the resulting ratios of oxygen atoms and charge to the number of silicon-atom structural units present. [Adapted from J. E. Ferguson, *Inorganic Chemistry and the Earth: Chemical Resources, Their Extraction, Use, and Environmental Impact*, Pergamon: Oxford, UK, 1982.]

A CONNECTION TO THE CHEMISTRY OF ART

Ancient Pigments

Early artists such as those working in the Lascaux cave or in South Africa 100,000 years ago[34] used naturally occurring inorganic pigments such as yellow hydrated iron(III) oxide (ocher) and red HgS (cinnabar), but stable blue pigments were unavailable in nature. Blue pigments based on layer polymeric polysilicate ions were invented in antiquity, however:[35] Egyptian blue, $CaCu[Si_4O_{10}]$, was first produced around 3600 BC by the heating to 800–900°C of limestone, sand, some Na_2CO_3, and a copper mineral such as malachite:

$$Cu_2(CO_3)(OH)_2 + 8SiO_2 + 2CaCO_3 \rightarrow 2CaCu[Si_4O_{10}] + 3CO_2 + H_2O \qquad (8.28)$$

Later, in China, the related pigment Han blue, $BaCu[Si_4O_{10}]$, was produced at a slightly higher temperature by using $BaCO_3$ (witherite) in place of the $CaCO_3$.

Other important pigments for paints include, above all, the extremely white TiO_2. Colored pigments include green Cr_2O_3, chrome yellow ($PbCrO_4$), molybdate red [$Pb(Cr,Mo,S)O_4$], and cobalt blue ($CoAl_2O_4$).

Classifying Structures and Relative Basicities of Silicates. As summarized in Figure 8.18, the successive steps of polymerization of the simple silicate ion have four results. (1) There is a successive reduction of the overall ratio of oxygen atoms to silicon atoms, from 4:1 in the orthosilicate ion to 2:1 in silica. (2) There is an increase in the number of VCS per silicon nucleus, which increases the number of possibilities for linkage. (3) There is a decrease in the charge per silicon nucleus in the anion. Hence, (4) the charge density per silicon tetrahedron decreases, and there is a decrease in basicity of the anion.

These changes are mathematically related, because (at atmospheric pressure) the silicon atoms always obey the octet rule (have a TCN of four). Thus, the total of (1) the number of oxygen atoms per silicon atom and (2) the number of VCS, must be *four*. Also, (3) the charge per silicon nucleus must be –2 for each oxygen atom in excess of two (because SiO_2 is uncharged). Hence, the charge density per silicon structural unit (charges per Si) is as shown in the right column of Figure 8.21. It decreases going down the figure, except that the charges per Si are the same, –2, for the third, fourth, and fifth entries. Hence, with the exception of these three polysilicates, the basicity decreases from top to bottom,

The chemical formulas of minerals are often written to set off the polysilicate ions from other anions, such as hydroxide, that may also be present. If this is done it is relatively easy to interpret the formula to tell what kind of polysilicate ion is present.

Framework Silicates (Tectosilicates). If the elimination of oxide ions is continued and sheets are linked into three-dimensional polymers, *all* of the oxide ions are ultimately eliminated, *all* of the remaining oxygens are converted into bridging oxygens, and the uncharged oxide silica, $^3_\infty[SiO_2]$, is produced (Figure 4.7). Having no negative charge, it is no longer basic at all; it is an acidic oxide.

EXAMPLE 8.17

Identify the degree of polymerization of the polysilicate ions found in the following minerals: (a) pyrophyllite, $Al_2(Si_4O_{10})(OH)_2$; (b) grunerite, $Fe_7(Si_4O_{11})_2(OH)_2$; (c) spessartite, $Mn_3Al_2(SiO_4)_3$; and (d) bustamite, $CaMn(SiO_3)_2$.

SOLUTION: Calculation of the ratio of oxygen to silicon atoms allows a choice among the structures shown in Figure 8.21 (or more generally in Figure 8.1), with the exception of the choice between chain and cyclic structures. The O:Si ratio in these minerals is 10:4 or 2.5 in pyrophyllite, 22:8 or 2.75 in grunerite, 4.0 in spessartite, and 3.0 in bustamite. Hence, the number of VCS per silicon nucleus are 1.5, 1.25, 0, and 1.0, respectively. Thus, the simple orthosilicate ion is present in spessartite (a type of garnet), and the degree of polymerization increases in the sequence bustamite < grunerite < pyrophyllite. Reference to Figure 8.21 identifies grunerite as a double-chain polysilicate and pyrophyllite as a layer silicate. Bustamite could be either a ring or a chain silicate; it happens to be the latter.

Aluminophosphates

Isomorphous substitution of two Si^{4+} in an Si_2O_4 fragment by one Al^{3+} ion and one P^{5+} ion gives rise to the first of a family of *aluminophosphates*, $^3_\infty[AlPO_4]$. Polymorphs of $^3_\infty[AlPO_4]$ are known with structures isomorphous to quartz (the mineral berlinite), tridymite, and cristobalite (Figure 4.7). Because Al^{3+} readily exceeds coordination number four, however, most aluminophosphates have a more open-chain structure analogous to the zeolites discussed in Section 8.7B.[36]

8.6B. Planetary Transformations of Silicates. As noted above, increasingly polymerized polysilicate ions have decreasing charges per silicon nucleus, so these ions are less basic. This fact has important consequences in soil chemistry: The more basic the polysilicate anion of a mineral, the more readily it reacts with weak acids, changing the polysilicate structure in the process of **weathering**. Rainwater is acidic even in the absence of sulfur and nitrogen oxides (Section 8.5C) due to dissolved carbon dioxide. This solution of carbonic acid is weakly acidic, and over the ages it reacts with the less polymerized silicate anions to remove oxide ions (as water). This results in replacement of the oxide ion with bridging oxygen to produce a more highly polymerized silicate:

$$Mg_2SiO_4 + 2H^+ \rightarrow Mg^{2+}(aq) + H_2O + Mg(SiO_3) \tag{8.29}$$

$$4Mg_2SiO_4 + 6H_2O \rightarrow Mg_6(OH)_8[Si_4O_{10}] + 2Mg(OH)_2 \tag{8.30}$$
$$\text{Serpentine}$$

Soils containing large amounts of simple silicates such as olivine are characterized as "youthful" soils. These soils may have crystallized from volcanic magma recently. Or they may be present in a desert region or the "dry valleys" of Antarctica, in which the liquid water necessary to weather the soil is absent. Volcanic soils, and desert soils when first irrigated, are often very fertile, because olivine has high concentrations of the nutrient ions Fe^{2+} and Mg^{2+}.

The cyclic and chain polysilicates weather somewhat more slowly; these are followed by the double-chain silicates, while layer polysilicates are quite persistent. Even SiO_2 may dissolve: SiO_2 is substantially more soluble in water than we might have expected. Its total solubility at pH values up to 9 is about 2×10^{-3} M. (It dissolves as $H_3SiO_4^-$ and as oligomeric silicic acids such as $H_6Si_2O_7$.) At the intermediate stage of weathering, such as is found in the temperate regions under a cover of grass or trees, layer silicates such as serpentine tend to predominate, along with some quartz. As can be seen from Equations 8.29 and 8.30, this weathering process is accompanied by a loss of soil cations, which is especially prominent for nonacidic and feebly acidic cations. Thus, the soil is less fertile than it was, due to the loss of the nonacidic plant nutrient potassium ion. The layer silicates present in the intermediate soils can still hold cations on their negatively charged surfaces, however. These cations can readily be exchanged for other ions (such as H^+), and are thus released as plants need them. Such soils are found in the still-quite-fertile corn and wheat belts of the world.

In the tropics, however, when the trees are cut down and frequent rain and heat speed up the weathering process, the aging process becomes quite advanced. Such soils have high levels of oxides of the most acidic cations, such as anatase and rutile (TiO_2), zirconia (ZrO_2), hematite (Fe_2O_3), and gibbsite [$Al(OH)_3$]. These soils can no longer hold the less acidic nutrient metal ions and are quite infertile. When tropical rain forests are removed in "slash and burn" agriculture, the soil can be used for agriculture for only a few years; after that it becomes infertile (and rock-hard).

EXAMPLE 8.18

Rank the order of weathering (from most quickly weathered to most persistent) of the silicate structure types shown in Figure 8.21.

SOLUTION: Generally, the more basic the silicate anion, the more rapidly its salts will be weathered (assuming constant acidities of the cation). As discussed above, the basicities fall in the order oligosilicates > disilicates > cyclic silicates = chain silicates > double-chain silicates > sheet silicates > silica. Therefore, this is the order of weathering as well. The exception, however, is that silica (SiO_2) is more soluble than expected, so it weathers before sheet silicates. Sheet silicates (clays, etc.) are the most long-lived in aqueous environments.

A CONNECTION TO GEOCHEMISTRY

The Chemistry of Mars

The concepts of weathering have been put to use in exploring the surface of Mars for evidence of its history. Mars has long been noted to have channels that suggest water erosion. The data from probes of Mars, including the two roving robots named *Spirit* and *Opportunity*, have provided a wealth of clues. The abundant presence of olivine was noted, suggesting rock that had not been weathered by contact with water. However, further sampling also revealed a number of minerals that are only formed on Earth in the presence of water. These include hydrated hydroxide-containing clays, magnetite (Fe_3O_4), goethite (FeOOH), and most remarkably the sulfate-containing mineral jarosite [$KFe_3(SO_4)_2(OH)_6$]. The current interpretation of the clues is that Mars may have had oceans for perhaps its first billion years of history (so may have had life!). After this period, water was largely absent, so more recent volcanically produced olivine would not have weathered.[37] Unlike Earth, Mars does not currently have a magnetic field, so it is bombarded heavily by cosmic rays, which can break water down to H_2 (which easily escapes the atmosphere) and O_2. It is speculated that Mars' core, being smaller than Earth's, may have solidified and lost its magnetism, with disastrous impacts on Mars' environment.

More recently the rover *Curiosity* landed in the ancient 96-mile-wide Gale Crater. At the center of this crater is a three-mile-high mountain, Aeolis Mons, which *Curiosity* ascended very slowly. Aeolis Mons is composed of layer after layer of sedimentary rock—layer silicates (clays) that can form only in water of near-neutral pH! This suggested a long history of (continuous or intermittent) immersion in water.[38]

In view of the low temperature and low atmospheric pressures on Mars, it was very unexpected when transient liquid water was detected flowing down the sides of Gale crater.[39] In contrast to Earth, the nonbasic but strongly oxidizing perchlorate ion (Section 6.3A) was detected on Mars by the *Phoenix* landing craft,[40] and is widespread. Although perchlorates of large nonacidic cations such as K^+ are insoluble (Section 4.3), those of feebly or weakly acidic cations are soluble and are deliquescent, so they can retain water even under the conditions on Mars. The spectra obtained from the *Mars Reconnaissance Orbiter* show clear evidence for the presence of hydrated magnesium perchlorate, magnesium chlorate, and sodium perchlorate.[41]

The Mantle of Earth

The mantle of Earth constitutes 85% of Earth's volume, and consists mainly of silicon, oxygen, magnesium, and iron. Deeper in the mantle, however, pressures increase and different silicate minerals predominate. The upper mantle, at typical depths of 35 to 660 km, has as a dominant mineral olivine, $(Mg,Fe)_2SiO_4$ (surface density 3.55 g cm^{-3}). Below a depth of about 410 km, the olivine then undergoes two phase transitions, to polymerization isomers that have close-packed spinel structures with higher densities and about 6% lower volume. At these depths the mineral ringwoodite has also been detected, which incorporates up to ~2.6% water as hydroxide ions. Ringwoodite would have formed during the planet's formation, so it could be the source of Earth's oceans.[42]

Below about 660 km, the lower mantle begins. The $(Mg,Fe)_2SiO_4$ decomposes to MgO (density 3.6 g cm^{-3}) and a form of $(Mg,Fe)SiO_3$ that has a perovskite structure ABO_3 with six-coordinate silicon atoms in the B sites and 12-coordinate Mg atoms in the A sites; this transformation results in a further loss of volume of about 8%.

Recently it has been discovered that below 2600 km (at a pressure exceeding 3 million atmospheres) the perovskite structure of $(Mg,Fe)SiO_3$ is transformed into an even more compressed structure.[43,44] This structure has silicate layers, but these differ from all previously discussed silicate structures in that each silicon is *six*-coordinated (and each oxygen is two-coordinate) in $_\infty^2[SiO_3^{2-}]$ anions. Between these layers lie eight-coordinated Mg^{2+} and Fe^{2+} ions. In this structure the Fe^{2+} ions can be so close to each other as to result in greatly enhanced electrical conductivity and conduction of heat from the core. Reaction of this *post-perovskite* mineral with iron in the core below 2900 km generates stishovite (SiO_2, Table 4.2), the iron silicide FeSi, and FeO.[45] The heat generated by this violent oxidation–reduction reaction may be involved in the thermal currents that drive continental drift.

The Chemistry of Earthquakes

Olivine, as noted above, is relatively reactive, and does not persist for long periods in contact with water. It is the most common mineral, however, in the upper mantle of Earth. Where it is exposed to water near the crust, it is readily transformed to the less-basic mineral serpentine by Equation 8.30. As serpentine-containing crust is subducted into the mantle at higher pressures as a result of plate tectonics where two plates are colliding, the serpentine $[Mg_6(Si_4O_{10})(OH)_8$; density 2.55 g cm$^{-3}]$ decomposes to olivine (Mg_2SiO_4; density 3.21 g cm^{-3}).

As the subduction continues (below a depth of about 410 km), the olivine then undergoes the changes discussed in the previous Connection to even more dense perovskite, then post-perovskite $(Mg,Fe)SiO_3$. The losses of volumes can, at least under some circumstances, result in the sudden shifting of layers to fill the voids, leading to deep earthquakes.[46]

The Most Stable Mineral

In contrast to olivine, zircon ($ZrSiO_4$) may be the most stable mineral on Earth. As the salt of a strongly acidic cation and a very strongly basic anion, it is very insoluble and undergoes no phase transitions to other polymorphs at temperatures found on the crust of Earth or during plate-tectonic subduction, so zircon crystals can last for the entire history of Earth. Furthermore, isomorphous substitution of U^{4+} for Zr^{4+} occurs when the zircon crystallizes initially from molten material. This U^{4+} is then essentially trapped forever, until it undergoes radioactive decay to isotopes of lead. Recent dating of tiny zircon crystals from Western Australia indicate that they crystallized 4.4 billion years ago—that is, only 100 million years after the formation of Earth.[47,48] However, these zircons may not have formed in rocks at the surface. Surface rock from 4.4 billion years ago may or may not have been found at Nuvvuagittuq in northern Canada.[49]

8.7. Aluminosilicates

OVERVIEW. The aluminum ion can readily replace the silicon ion in polysilicate ions to produce aluminosilicate ions. When this substitution occurs in layer polysilicates, aluminosilicate clays are produced (Sec. 8.7A). The charges of the aluminosilicate layer anions are neutralized in clays by, first of all, additional polymeric layer cations containing Al^{3+} and Mg^{2+} ions bridged by hydroxide ions. In some clays negative charge still remains, and is charge-balanced with layers of Group 1 or 2 cations or even hydronium ions that easily engage in ion exchange with hydrated cations in water. You may practice identifying and classifying the acid–base properties of layer aluminosilicate clays by trying Exercises 8.87–8.90.

When Al^{3+} replaces Si^{4+} in silica itself, framework aluminosilicates result (Section 8.7B). Negative charge is produced by this substitution, so the silica structure must open up to make room for additional cations (this substitution is non-isomorphous). This results in the feldspars and the more open-structure zeolites, which are of use in catalysis. Zeolites occur with a wide variety of structures and open connected channels. Because the anion charge is altered by this substitution, so is the anion basicity. You may practice identifying layer and framework aluminosilicates and their properties such as resistance to weathering by trying Exercises 8.91–8.96.

Section 8.7 contains Connections to agricultural chemistry, environmental chemistry, industrial chemistry, and organic chemistry.

8.7A. Aluminosilicate Clays. Aluminosilicates can be imagined as arising from the *non-isomorphous* replacement of part of the Si^{4+} by Al^{3+} *within* the structure of a polysilicate polymer such as a sheet polysilicate. To maintain charge neutrality we must also add a +1 ion that is *not* part of the polysilicate layer. This substitution is best indicated by writing the formula of the aluminosilicate ion in brackets, enclosing the framework aluminum ions inside the brackets but keeping the +1 or other counterions outside of the brackets.

Weathering reactions such as that of Equation 8.30 give rise not only to layer polysilicates such as serpentine but also layer aluminosilicates. Clays are characteristic of moderately weathered soils in the temperate zones of Earth.

Besides having layer polysilicate or polyaluminosilicate ions such as $_\infty^2[Si_4O_{10}{}^{4-}]$ containing *tetrahedrally coordinated* Al^{3+} or Mg^{2+} ions (shown in Figure 8.22 in **brown**), clays are characterized by having additional parallel layers of *octahedrally coordinated* Al^{3+} or Mg^{2+} ions such as $_\infty^2[Al_4(OH)_8{}^{4+}]$ bridged by OH^- ions (shown in Figure 8.22 in **light brown**). These are the only two layers in the simplest clay, kaolinite (Figure 8.22a); the bilayer "sandwiches" are bound to neighboring sandwiches only by comparatively weak hydrogen bonds. When kaolinite, $_\infty^2([Al_4(OH)_8][Si_4O_{10}])$, is heated in an oven, its hydroxide groups are dehydrated, and it is converted to a mixture including (the otherwise-rare mineral) mullite ($Al_2Si_6O_{13}$). This mixture is a chemical precursor to the ceramic that is the basis of brick, pottery, china, porcelain, and enamels.

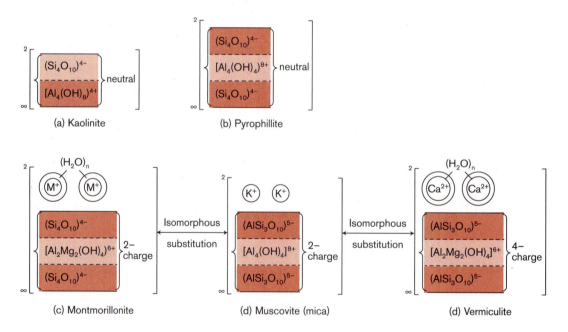

(a) Kaolinite   (b) Pyrophillite

(c) Montmorillonite   (d) Muscovite (mica)   (d) Vermiculite

Figure 8.22. Layer structures of selected clays and their interrelationships via isomorphous or non-isomorphous substitution. (a) Kaolinite, $_\infty^2([Al_4(OH)_8][Si_4O_{10}])^\star$; (b) pyrophyllite, $_\infty^2([Al_4(OH)_4][Si_4O_{10}]_2)^\star$; (c) montmorillonite, $_\infty^2(M_\delta[Mg_\delta Al_{(4-\delta)}(OH)_4][Si_4O_{10}]_2)^\star$ (illustrated for $\delta = 2$); (d) muscovite, $_\infty^2(K_2[Al_4(OH)_4][AlSi_3O_{10}]_2)^\star$; and (e) vermiculite, $_\infty^2\{(Ca,Mg)_2[Al_2Mg_2(OH)_4][AlSi_3O_{10}]_2\}^\star$. Aluminosilicate layers with tetrahedral coordination are shown in **brown**; hydroxide-bridged Al^{3+}–Mg^{2+} layers with octahedral coordination are shown in **light brown**.

Other important clays involve three layers, two of which are anionic tetrahedral polysilicate or polyaluminosilicate layers (shown in **brown**) sandwiching a central cationic octahedral aluminum hydroxide or magnesium aluminum hydroxide layer (shown in light brown). In pyrophyllite (Figure 8.22b), the central $^2_\infty[Al_4(OH)_4{}^{8+}]$ layer exactly neutralizes the two outer layers. Only hydrogen bonds link one triple layer to another, so cleavage into layers is very easy.

In muscovite (mica; Figure 8.22d), the two outer layers are each –5-charged aluminosilicate layers, so their total charge exceeds that of the central octahedral layer. Charge balance thus requires a fourth layer of cations such as K^+ (shown uncolored). Two other clays, montmorillonite (Figure 8.22c) and vermiculite (the familiar packing material; Figure 8.22e) are derived from the structure of muscovite by different isomorphous substitution processes of the second kind.

A shorthand way of writing the simplest formula of the mineral muscovite, $^2_\infty(K_2[Al_4(OH)_4][AlSi_3O_{10}]_2)$, is $KAl_2[AlSi_3O_{10}](OH)_2$, showing that only one-third of the Al^{3+} ions enter into the layer polysilicate ions. To determine the layer structure of this material from its formula, we can calculate the ratio of oxygens to silicon plus *nuclear aluminum atoms*: $10/(3 + 1) = 2.5$; thus, it is a layer structure with 1.5 VCS in the structural unit.

EXAMPLE 8.19

(a) Predict the formula of the anion of montmorillonite, $^2_\infty\{M_\delta[Mg_\delta Al_{(4-\delta)}(OH)_4][Si_4O_{10}]_2\}$, if only one-fourth of the cations in the central layer are magnesium ions and three-fourths are aluminum ions (i.e., if $\delta = 1$). (b) Predict the pK_b and the basicity classification of the anion, based on the (untested) hypothesis that this can be estimated from the formula of the mononuclear structural unit obtained by depolymerizing the anion, then counting the number of oxo groups and units of negative charge in the structural unit. (c) Predict the acidity classification of the acid form of this montmorillonite.

SOLUTION: (a) In Figure 8.22c, one-half of the cations in the central layer are Mg^{2+} and one-half are Al^{3+}. Making an isomorphous substitution of another Al^{3+}, the anion formula becomes $MgAl_3(OH)_4[Si_4O_{10}]_2{}^-$, containing 12 metal cations (which we combine as E) and one fewer negative charges. Dividing by 12, the structural unit is $EO_{1.67}(OH)_{0.33}{}^{0.083-}$.

(b) We can then apply Equation 3.14 to obtain $pK_{b1} = 10.0 + 5.7(1.67) - 10.2(0.083) = 18.4$. Therefore, this anion is presumably nonbasic.

(c) Neutralizing the negative charge of the structural unit by adding 0.083 H^+, the oxo acid structural unit becomes $(HO)_{0.413}EO_{1.587}$. Applying Equation 3.18, the presumed pK_{a1} of this acid is $8.5 - 5.7(1.587) = -0.55$, so the acidic form of montmorillonite is presumably strongly acidic.

Montmorillonite Chemistry

Montmorillonite, $^{2}_{\infty}\{M_{\delta}[Mg_{\delta}Al_{(4-\delta)}(OH)_4][Si_4O_{10}]_2\}$, is formed in large quantities on Earth from the weathering of volcanic ash, and it has been detected on Mars. As calculated in Example 8.19, its layer anion is nonbasic. The cations absorbed on the surfaces of the charged clays (muscovite, vermiculite, montmorillonite) are weakly attracted to the clay and are easily exchanged with ions from solution. Clays perform this function naturally in the soil. Enormous quantities of hydrated cations can be held to the negatively charged surfaces of clays; these cations include many nutrient ions such as Ca^{2+}, Mg^{2+}, K^+, and Na^+. As slightly acidic rainwater percolates over these clays, these nutrient ions are slowly released by ion exchange with the hydronium ion in the water and made available for plants to use (Figure 8.23). The largest and the least charged of these hydrated ions are held most loosely and are released first (and become depleted first). Recalling that smaller ions have larger secondary hydration spheres (Section 4.5B), we find that the order of release is Na^+ before K^+ before Mg^{2+} before Ca^{2+}. Because plants make little use of Na^+, the first deficiency that occurs among cationic nutrients is of K^+. Anionic nutrients such as NO_3^- and PO_4^{3-} cannot be bound to clays at all, so they are rapidly lost from soils. Thus, we see the need of fertilizers containing (especially) potassium, nitrogen, and phosphorus.

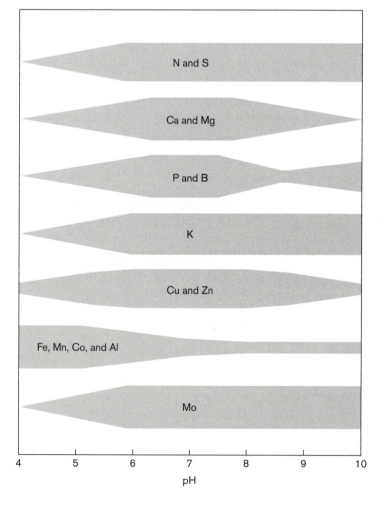

Figure 8.23. Relative availability of nutrient elements in soil as a function of soil pH. (The broader the bar, the more available is the nutrient.) [Adapted from J. W. Moore and E. A. Moore, *Environmental Chemistry*, Academic Press: New York, 1976.]

The cations on the surface of montmorillonite readily adsorb additional water (to form secondary hydration spheres) when montmorillonite is placed in water; these hydration spheres then hydrogen-bond to each other, thickening the suspension in water—at least until the suspension is stirred, which breaks up the hydrogen bonds and thins the suspension. Montmorillonite is thus a useful *thixotropic agent* and is used in paints to keep them from flowing while wet, and in drilling muds in the oil industry.

As also calculated in Example 8.19, the acidic form of montmorillonite is very strongly acidic (calculated $pK_a = -0.55$; observed pH after acid washing = –5.6 to –8). As an *insoluble strong acid*, it is much more easily handled than sulfuric or other strong acids. Montmorillonite is finding use in catalyzing typical organic reactions requiring acid catalysis, such as the Friedel–Crafts reaction.[50] Montmorillonite absorbs RNA nucleotide monomers by van der Waals attractions of the montmorillonite silicate layers to the purine and pyrimidine bases of the nucleotide. It then efficiently and (somewhat) selectively catalyzes the condensation of the bases into larger oligomers resembling RNA and containing 30–50 monomer units.

8.7B. Framework Aluminosilicates. To date we have discussed only one three-dimensional (framework) silicate, $^3_\infty[SiO_2]$ itself (Section 4.1C; Table 4.2). **Framework aluminosilicates** can be imagined as arising from the *non-isomorphous* replacement of part of the Si^{4+} by Al^{3+} *within* the $^3_\infty[SiO_2]$ network *and* by a +1 ion that is *not* part of the network. The extra +1 ion then goes into holes in expanded $^3_\infty[Si_xAl_{2-x}O_2]$ structures having larger ring sizes. Notice in Figure 8.24 that quartz (on the left) has smaller rings than the framework aluminosilicate albite, $Na[AlSi_3O_8]$, on the right.

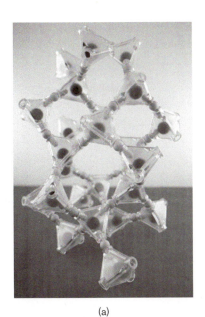

(a)

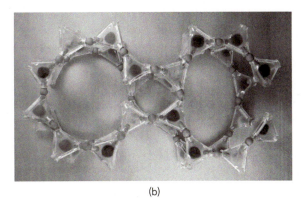

(b)

Figure 8.24. Structure of (a) quartz compared with the structure of (b) albite. Clear = oxygen, black = aluminum or silicon. Sodium ions are not shown in (b); they sit at the centers of the large rings. [Based upon materials developed by the Materials Research Science and Engineering Center on Structured Interfaces at the University of Wisconsin-Madison with funding from the National Science Foundation under award numbers DMR-1121288, DMR-0520527, DMR-0079983, and EEC-0908782.]

Feldspars. The most abundant of all minerals (about 60% of Earth's crust) are the **feldspars**, which include $K[AlSi_3O_8]$ (orthoclase), $Ba[Al_2Si_2O_8]$ (celsian), $Na[AlSi_3O_8]$ (albite)*, and $Ca[Al_2Si_2O_8]$ (anorthite). The feldspars themselves fall into two groups. One group includes orthoclase and celsian, and incorporates 150-pm ions such as K^+ and Ba^{2+}. Note that orthoclase and celsian are related to each other by isomorphous substitution of the second type. The other group, the plagioclase feldspars, involve the smaller 115-pm Na^+ and Ca^{2+} ions; compositions range continuously from $Na[AlSi_3O_8]$ (albite) through $Na_{0.33}Ca_{0.67}[Al_{1.67}Si_{2.33}O_8]$ (labradorite) to $Na_0Ca_1[Al_2Si_2O_8]$ (anorthite).

Zeolites. Perhaps the best known of the framework aluminosilicates are the **zeolites**,[51] with anions of general formula $[Al_xSi_yO_{2x+2y}]^{x-}$ with $x \leq y$ and an oxygen-to-(aluminum plus silicon) ratio of 2. The fundamental structural unit of zeolite structures is the **sodalite cage** $[(Si,Al)_{24}O_{48}]$ (Figure 8.25a). In Figure 8.25b the sodalite cage is drawn omitting all oxygen atoms and showing the 24 (Si,Al) nuclei at the intersections of the lines (which represent bridging oxygen atoms). Figure 8.25b shows how the zeolite cage structure consists of interlinked six-nucleus $[(Si,Al)O_2]_6$ rings (as in SiO_2 itself) and four-nucleus $[(Si,Al)O_2]_4$ rings.

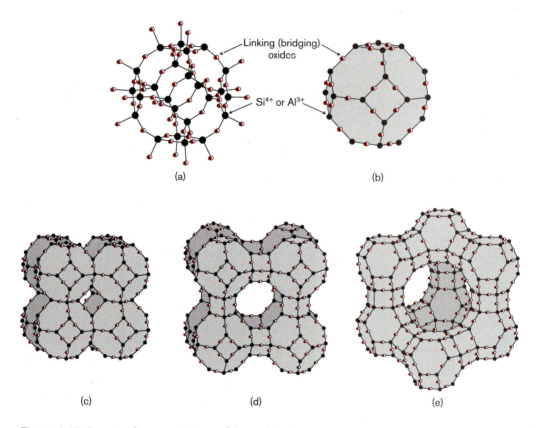

Figure 8.25. Structural representations of the zeolite structure. (a) Connection of aluminosilicate nuclei to form an open basket-like cluster known as the sodalite cage $[(Si,Al)_{24}O_{48}]$ (**brown** = oxygen; black circles = silicon or aluminum); (b) line representation of the sodalite cage in (a), in which black vertices represent silicon/aluminum nuclei and **brown** bridging oxygens lie along the edges; (c–e) connection of sodalite cages to give (c) sodalite, $\frac{3}{\infty}(Na_6[Al_6Si_6O_{24}] \cdot 2H_2O)$*; (d) Linde A, $\frac{3}{\infty}(Na_8[Al_8Si_{40}O_{96}] \cdot 24H_2O)$; and (e) faujasite, $\frac{3}{\infty}(Na_{58}[Al_{58}Si_{134}O_{384}] \cdot 240H_2O)$*. [Adapted from G. T. Kerr, *Sci. Amer.* **261**, 100 (1989).]

In zeolites, sodalite cages are linked to each other to give rise to large, three-dimensional structures that feature open channels (Figure 8.26). For example, in Figure 8.25c each square face of each sodalite cage is used twice (shared) between two adjacent sodalite cages to give the smallest possible network polymer, the mineral sodalite, $^3_\infty(Na_6[Al_6Si_6O_{24}]\cdot2H_2O)^*$, which has a central four-sided channel of 220 pm diameter.

In Figures 8.25d and 8.25e, the sodalite cages are instead linked to adjacent complete cages by converting terminal oxygen atoms of the original sodalite cage (Figure 8.25a) to bridging oxygens. In Figure 8.25d, square faces are linked to give an eight-sided channel with a diameter of 410 pm in the zeolite Linde A, $^3_\infty(Na_8Al_8Si_{40}O_{96}]\cdot24H_2O)$.

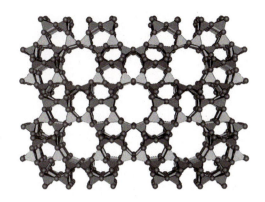

Figure 8.26. A larger view of the three-dimensional channels in the zeolite ZSM-11.

In Figure 8.25e, hexagonal faces are linked to give a 12-sided channel with a diameter of about 850 pm in the zeolite faujasite, $^3_\infty(Na_{58}[Al_{58}Si_{134}O_{384}]\cdot240H_2O)^*$. Many other natural and artificial zeolites are known involving other channel sizes, which can have up to 14 sides.[52]

AN AMPLIFICATION

Syntheses of Zeolites

The syntheses[53] of zeolites is a craft that involves the reaction of sodium silicate, sodium aluminate, and aluminum sulfate solutions in various ratios for extended times (e.g., 20–30 hours) at carefully controlled temperatures (e.g., 92 ± 2°C). Reactions are carried out in the presence of appropriate quaternary ammonium (R_4N^+) or other organic cations that act as templates to help determine the shape and size of the holes and channels in the aluminosilicate structures. After the synthesis is completed, the product is heated to a high temperature to vaporize or burn off the organic template, leaving behind voids where the organic cations had been.

In nature, the cations in the holes and cavities are normally hydrated nonacidic and feebly acidic cations; the water of hydration can be driven off by heating to leave a structure with enlarged channels. If the cation of the zeolite is the relatively large K^+ (in Zeolite 3A), a small channel of 300-pm (3-Å) diameter is available. If the cation is Na^+ (in Zeolite 4A), a larger channel of 400-pm diameter is present. If the cation is Ca^{2+} (in Zeolite 5A), a still larger channel of 500-pm diameter is present, because only half as many cations are present alongside the channel. The resulting zeolites have cavities of tailored sizes that are used as Molecular Sieves™ to adsorb molecules of different sizes from liquids. Among the small molecules most readily adsorbed are water molecules, which re-form the hydrated metal ions. Thus, these zeolites are very effective drying agents.

Water Softening and Decontaminating Radioactive Waste

Hydrated metal cations in such a structure, being large cations in a lattice with very large insoluble anions, are loosely bound and can readily be exchanged with other cations. Zeolites are used for *water softening*: the removal of ions of +2 charge found in tap water that would precipitate the anions used in detergents. A concentrated solution of NaCl is first percolated through the solid (insoluble) zeolite ion-exchanger, replacing whatever hydrated ions are present with hydrated Na^+ ions. Then the tap water is run through the zeolite; Ca^{2+} and other +2 ions become associated with the anion in the solid phase, while Na^+ ions go into solution. When the Na^+ ions in the zeolite are depleted, the solid is again "recharged" by running concentrated NaCl solution through it. With appropriate-sized cavities, zeolites can be made that selectively remove Cs^+ from water. Because $^{134}Cs^+$ and $^{137}Cs^+$ are among the most dangerous byproducts of nuclear power generation, bags of zeolites were immersed in the sea offshore from the earthquake- and tsunami-damaged Fukushima Daiichi power plant in April 2011.

Gasoline from Petroleum

Channels may penetrate through the zeolite in one dimension, or they may intersect other channels that go off in the other two directions. The channels thus may be tailored to be of such a size as to admit only molecules below a certain size, or only straight-chain but not branched-chain hydrocarbons, and so on. In combination with these possibilities of selectivity, the dehydrated zeolites have semi-bare dehydrated metal cations with Lewis acid properties. In addition, the metal ions can be ion-exchanged for hydrogen ions, which, if driven off by heat, leave behind positively charged silicon atoms with VCS that are powerfully acidic sites (Figure 8.27). The combination of powerful Lewis acidity with site selectivity makes these of great importance as catalysts. For example, ZSM-5* is used to catalyze the rearrangement of two moles of toluene to give one of benzene and one of *para*-xylene (*ortho*-xylene does not fit the cavity), to catalyze the alkylation of benzene with ethylene to give ethyl benzene, to catalyze the conversion of methanol (which can be made from coal and water) into gasoline plus water, and in the **cracking** of petroleum to form gasoline. Analogies have been drawn between the action of zeolites and the actions of enzymes, because both depend on the availability of reactive (e.g., Lewis-acidic) sites along with shape selectivity. And the easily recovered and regenerated zeolite and other solid acid catalysts are far more "environmentally friendly" than concentrated sulfuric acid.

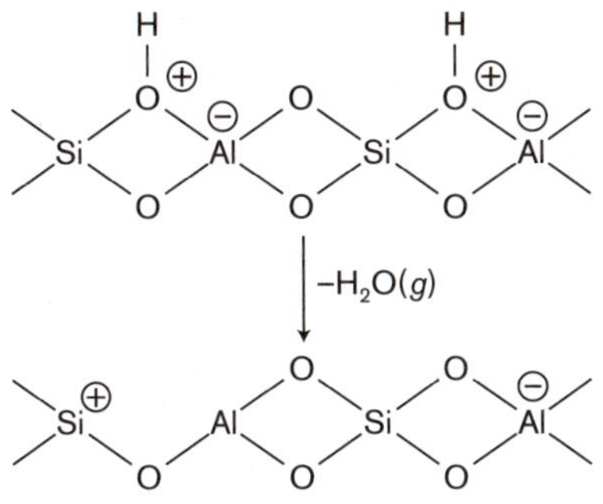

Figure 8.27. Dehydration of the surface of an acid-exchanged zeolite to leave coordinately unsaturated, hence strongly acidic, silicon cationic sites. Formal charges of atoms are shown in circles.

The cavities within zeolites also provide environments in which unstable chemical species can be protected. An ancient naturally occurring blue pigment is *lapis lazuli* or lazurite, a zeolite mineral the blue color of which is due to unstable polyatomic sulfur *radical anions* such as S_3^-. Correspondingly, zeolites such as anhydrous sodium zeolite X, $Na_{92}[Si_{100}Al_{92}O_{384}]$, in vacuum absorb 14 additional sodium atoms to form unstable polyatomic Na_3^{2+} cations,[54] and the all-silica zeolite ITQ-4 (Figure 4.7g) can encapsulate Cs^+ cations and *electride* ions (Section 6.4A).[55]

EXAMPLE 8.20

Consider the following mineral pairs. Classify each as related to the other by (1) isomorphous substitution of the first kind; (2) isomorphous substitution of the second kind; (3) non-isomorphous substitution; or (4) a pair that is completely unrelated in structure. (a) $Rb[AlSi_3O_8]$ and $K[AlSi_3O_8]$; (b) $Rb[AlSi_3O_8]$ and $Li[AlSi_3O_8]$; (c) $Na[AlSi_3O_8]$ and $Na_{12}[Al_{12}Si_{12}O_{48}]$; (d) $Rb[AlSi_3O_8]$ and $Ba[Al_2Si_2O_8]$; (e) $^3_\infty\{[SiO_2]_{24}\}$ and $^3_\infty\{Na_{12}[Al_{12}Si_{12}O_{48}]\}$

SOLUTION: All of these have ratios of oxygen to (silicon + aluminum) = 2:1, so they are framework silicates or aluminosilicates. Zeolites, however, have much larger ring sizes than feldspars, and the plagioclase feldspars have the larger cations. Hence, in (a) both are plagioclase feldspars with large cations, which differ by isomorphous substitution of the first kind. In (b), Li^+ is much smaller than Rb^+, so these two feldspars are different types—they are related by non-isomorphous substitution. In (c), the first salt is a feldspar and the second, with the larger rings, is a zeolite, so these are unrelated structures. In (d), we substitute Ba^{2+} and Al^{3+} for Rb^+ and Si^{4+}, which is isomorphous substitution of the second kind. In (e), the subscript 24 for both structures implies large rings of zeolite size, but replacement of 12 Si^{4+} with 12 Al^{3+} *and* 12 Na^+ is a non-isomorphous substitution.

8.8. Background Reading for Chapter 8

Physical Properties of Materials and Oxides (Periodic Trends)

R. J. Gillespie, "Covalent and Ionic Molecules: Why are BeF_2 and AlF_3 High Melting Point Solids whereas BF_3 and SiF_4 are Gases?" *J. Chem. Educ.* 75, 923–925 (1998).

A. Muñoz-Páez, "Transition Metal Oxides: Geometric and Electronic Structures," *J. Chem. Educ.* 71, 381–388 (1994).

J. D. Holbrey and R. D. Rogers, "Physicochemical Properties of Ionic Liquids: Melting Points and Phase Diagrams," in *Ionic Liquids in Synthesis*, 2nd ed., P. Wasserscheid and T. Welton, Eds., Wiley-VCH Verlag: Weinheim, 2008, Chapter 3.1, pp. 57–72.

V. A. Cocallia, A. E. Visser, R. D. Rogers, and J. D. Holbrey, "Solubility and Solvation in Ionic Liquids," in *Ionic Liquids in Synthesis*, 2nd ed., P. Wasserscheid and T. Welton, Eds., Wiley-VCH Verlag: Weinheim, 2008, Chapter 3.3, pp. 89–102.

High-Temperature Superconductivity in Oxides

A. Bussman-Holder and K. A. Müller, "The Route to High Temperature Superconductivity in Transition Metal Oxides," in *100 Years of Superconductivity*, H. Rogalla and P. H. Kes, Eds., CRC Press: Boca Raton, FL, 2012, Chapter 4.2, pp. 233–239.

D. Johnson, "Superconductivity above 10 K in Non-Cuprate Oxides," in *100 Years of Superconductivity*, H. Rogalla and P. H. Kes, Eds., CRC Press: Boca Raton, FL, 2012, Chapter 4.3, pp. 239–244.

C. W. Chu, "Cuprates—Superconductors with a T_c up to 164 K," in *100 Years of Superconductivity*, H. Rogalla and P. H. Kes, Eds., CRC Press: Boca Raton, FL, 2012, Chapter 4.4, pp. 244–255.

Chemistry of Cement and Concrete

F. A. Rodrigues and I. Joekes, "Cement Industry: Sustainability, Challenges, and Perspectives," *Environ. Chem. Lett.* 9, 151–166 (2011).

Environmental Chemistry of Volatile Acidic Oxides

S. Manahan, *Environmental Chemistry*, 9th ed., CRC Press: Boca Raton, FL, 2010, pp. 285–304 (Chapter 11: Gaseous Inorganic Air Pollutants).

Chemistry of Mars

P. R. Christensen, "The Many Faces of Mars," *Sci. Amer.* 293(1), 32–39 (2005).

J. Bell, "The Red Planet's Watery Past," *Sci. Amer.* 295(6), 62–69 (2006).

Chemistry of the Early and Deep Earth

K. Hirose, "The Earth's Missing Ingredient," *Sci. Amer.* 302(6), 76–83 (2010). A new form of $MgSiO_3$ at the border of the mantle and the outer core.

J. W. Valley, "A Cool Early Earth?" *Sci. Amer.* 293(4), 58–65 (2005).

C. Zimmer, "The Oldest Rocks on Earth," *Sci. Amer.* 310(3), 59–63 (2014).

Numbered references from this chapter may be viewed online at www.uscibooks.com/foundations.htm.

8.9. Exercises

8.1. *Sulfur trioxide exists in three forms, which have the molecular formulas SO_3, S_3O_9, and $_\infty^1[SO_3]$. (a) Classify each of these into one of the structural classifications given in Figure 8.1. (b) Assuming (contrary to actual fact) that the three forms are not readily interconverted, rank these three forms in order of increasing melting point. Which of the three is most likely to be a gas at room temperature?

8.2. The oxide of a certain element E in its group oxidation state has the polymorphs and/or structural isomers **(1)**, **(2)**, and **(3)**:

(a) Give the best description of each structure (e.g., as a monomer, a polyhedral oligomer, a layer polymer, etc.). (b) Which one of these three polymorphs and structural isomers would likely be most volatile? Which would most likely be a high-boiling liquid? (c) Which one of the elements C, Si, Pb, O, Se, Po, or Xe is element E most likely to be?

8.3. *Below are sketched some structures of four structural types **A**, **B**, **C**, and **D** of a particular element or element oxide, in which the structural unit is represented by an "•" and the links by lines "—."

Which of these (**A**, **B**, **C**, **D**, or none) is (a) a viscous liquid, (b) a high-melting solid, (c) a monomer, and (d) a substance about as hard as diamond?

8.4. Below are sketched some structures of four structural types **A**, **B**, **C**, and **D** of a particular element or element oxide, in which the structural unit is represented by an "•" and the links by lines "—."

Which of these (**A**, **B**, **C**, **D**, or none) is (a) a layer polymer, (b) an oligomer, (c) a low-boiling gas, and (d) a chain polymer?

8.5. *Select one of the materials C_{60}, He, $_\infty^3[(Th^{4+})(O^{2-})_2]$, $_\infty^3[C]$, $_\infty^2[C]$, or $_\infty^1[S]$ as an example of a material likely to exhibit the following physical property at room temperature: (a) viscous liquid; (b) slippery solid; (c) low-boiling gas; (d) soft solid; (e) hard solid; (f) low-density solid; (g) high-density solid.

8.6. For each of the room-temperature physical properties on the left, pick the one structure type on the right that is most likely to be associated with it.

Viscous liquid	Oligomer
Slippery solid	Molecular monomer
Low-boiling gas	Chain polymer
Soft solid	Layer polymer
Hard solid	Network polymer
Low-density solid	Ionic salt
High-density solid	

8.7. *Boron nitride has two polymerization isomers, and both have melting points in the vicinity of 3000°C. Polymerization isomer A is almost as hard as diamond and has a density of 3.47 g cm^{-3}; polymerization isomer B is slippery and has a density of 2.25 g cm^{-3}. (a) Which formula is best applied to polymerization isomer A: $B_{30}N_{30}$, $^2_\infty[BN]$, or $^3_\infty[BN]$? (b) Which formula is best applied to polymerization isomer B? (c) Consider the formula you did *not* choose for isomer A or isomer B. Should this third polymerization isomer have a density that is greater than 3.47 g cm^{-3} or less than 2.25 g cm^{-3}? Should its melting point be less than 3000°C, equal to 3000°C, or greater than 3000°C?

8.8. Which of the following salts or complexes are good candidates to be ionic liquids: FeO, $KClO_4$, $(C_3H_7)_4NCl$, and/or $(C_3H_7)_4PF_6$?

8.9. *Which of the following salts or complexes are good candidates to be ionic liquids: $C_3H_3(NCH_3)(NR)^+[Al_2Cl_7^-]$, $Ba(NO_3)_2$, $[Cr(NH_3)_3Cl_3]$, and/or $[Cr(NH_2C_8H_{17})_3]Cl_3]$?

8.10. (a) Consult Table C to find the eight cations with radii most nearly appropriate to having a TCN of 4 or less. (b) Considering only the *p* block, which ions expected to have penultimate total coordination numbers of 4 or less based on their period in the periodic table (i.e., on the ideas of Section 3.5) are *absent* from the list generated from Table C?

8.11. *The "cations" Be^{2+}, Si^{4+}, and Se^{6+} each have a radius of about 55 pm. Assuming this radius, predict the number of VCS in the structural units of the fluoride of each element.

8.12. Because the cations in all of the following structural units have radii $\le$ 54 pm, it is reasonable to expect total coordination numbers of four for SiO_2, $PO_{2.5}$, SO_3, and $ClO_{3.5}$. (a) How many VCS are there for each structural unit? (b) Which one will form a dimeric molecule? What is its formula? (c) Which one is most likely to form two polymorphs, one of which is a cyclic oligomer? What would its other polymorph be? (d) In which one will *all* of the oxygen atoms donate electron pairs to VCS on other structural units?

8.13. *Suppose that we have two new *p*-block cations with no tendency toward π bonding: *sixium* is found in Group 16/VI and forms the Sx^{6+} cation, whereas *eightium* is found in Group 18/VIII and forms the Eg^{8+} cation. Both cations have cationic radii of 52 pm, right on the boundary for different total coordination numbers. (a) What are the formulas of the structural units of these oxides? (b) Based on radius ratios, what are the two most likely total coordination numbers for each of these cations? (c) How many VCS would be found for the two possible structural units of sixium oxide? Of eightium oxide? (d) Write the formula (including prescripts such as $^1_\infty[\ ]$ if appropriate) of the oxide that is most likely to be a gas. (e) Write the formula (including prescripts such as $^1_\infty[\ ]$ if appropriate) of the oxide that is most likely to be a chain polymer.

8.14. If possible, draw Lewis structures for the following (presumably covalent) molecules or formula units. Use your results to help classify the oxides as (1) monomeric or dimeric molecular, (2) oligomeric molecular, or (3) ionic or macromolecular substances. (a) Na_2O; (b) Cr_2O_3; (c) CO_2; (d) SiO_2; (e) NO_2.

8.15. *If possible, draw Lewis structures for the following (presumably covalent) molecules or formula units. Use your results to help classify the oxides as (1) monomeric or dimeric molecular, (2) oligomeric molecular, or (3) ionic or macromolecular substances. (a) P_4O_{10}; (b) BaO; (c) Tl_2O; (d) SO_2; (e) Al_2O_3.

8.16. Consider the following dioxides of fifth-period elements: SnO_2, TeO_2, and XeO_2. (a) Use their common penultimate TCN to determine the number of vacant coordination numbers for each oxide. (b) Which oxide should be the most highly polymerized? (c) Which oxide should be the least polymerized? (d) Which oxide should have the highest melting point?

8.17. *Most of the oxygen-bridged dimeric and the polyhedral oxides in Figure 8.5 have great differences among the observed element–oxygen bond lengths. Compare these observed bond lengths with the sums of covalent radii (Table 1.11). Suggest one or more possible reasons for the differences.

8.18. Figure 8.5 compares the observed element–oxygen bond lengths for OsO_4 and XeO_4 with the sums of covalent radii (Table 1.11). If the bonds are much shorter than the sums of single-bond covalent radii, suggest an explanation.

8.19. *Assuming that each of the compounds is stable, predict whether melting points will increase or decrease in each of the following series of oxides. (If radius ratios are unavailable, use periodic trends in total coordination numbers.) (a) Cl_2O_5, Br_2O_5, I_2O_5; (b) MnO, Mn_2O_3, MnO_2, Mn_2O_7.

8.20. The two-coordinate bridging oxygen atoms in the rhenium oxide of Figure 4.4 and in mercury(II) oxide have widely differing metal–oxygen–metal bond angles: In the rhenium oxide this angle is 180°, while in mercury(II) oxide this angle is 109°. (a) What might this difference in the bond angles signify? (b) What arguments can you offer against this interpretation of the bond angles?

8.21. *(a) Predict the number of VCS for each of the following structural units: BaO, HfO_2, WO_3, and OsO_4. (b) Which one of these oxides is least likely to have an ionic or network covalent structure? (c) Which one of these oxides is least likely to be a ceramic?

8.22. Which one of the following—an element tetroxide MO_4, an element trioxide MO_3, and an element dioxide MO_2—is most likely to (a) be volatile, (b) form a chain polymer, (c) form a monomeric molecule, and (d) be of use as a ceramic?

8.23. *Put the following series of oxides in order of increasing melting points, and decide which are gases at room temperature: (a) Na_2O, Cr_2O_3, CO_2, SiO_2, P_4O_{10}; (b) ZrO_2, CO_2, SrO, Rb_2O, Y_2O_3.

8.24. An element forms three dioxide polymerization isomers, which do not readily convert into the other isomers. Isomer A is monomeric molecular, isomer B is dimeric molecular, and isomer C is cyclic molecular. (a) Enthalpies of fusion (in kJ mol^{-1}) for the three isomers are 20, 12, and 5. Which enthalpy of fusion is likely to belong to which isomer? (b) Boiling points for the three isomers are 90 K, 400 K, and 220 K. Which boiling point is likely to belong to which isomer?

8.25. *Which is more likely to have a higher enthalpy of fusion, and why? Is it (a) a Group 16/VI nonmetal trioxide or a Group 16/VI nonmetal hexafluoride, (b) a d block metal trioxide or a d block metal trifluoride, and (c) a Period 4 nonmetal dioxide or a Period 4 nonmetal tetrafluoride?

8.26. Why is it that many oxides but few if any fluorides can be considered to be ceramics?

8.27. *Which of the following structural descriptions is/are inconsistent with the stoichiometry of the salt being described? (a) $CdCl_2$ adopts an hcp lattice of chloride ions in which all of the octahedral holes are occupied by cadmium ions. (b) $CdCl_2$ adopts a ccp lattice of chloride ions in which half of the octahedral holes are occupied by cadmium ions. (c) Li_2SO_4 adopts an hcp lattice of sulfate ions in which all of the tetrahedral holes are occupied by lithium ions. (d) $(CH_3)_4NF$ adopts a ccp lattice of tetramethylammonium ions in which all of the octahedral holes are occupied by fluoride ions.

8.28. Which of the following are true statements about octahedral holes in a (cubic or hexagonal) close-packed lattice of anions? (a) There is one octahedral hole per anion in the lattice. (b) All of the octahedral holes are occupied in the lattice of CdI_2. (c) A cation in an octahedral hole is surrounded by eight anions. (d) An octahedral hole is larger than a tetrahedral hole. (e) All of the octahedral holes are occupied in the lattice of NaCl.

8.29. *Which of the following structural descriptions is/are inconsistent with the stoichiometry of the salt being described? (a) BiI_3 adopts an hcp lattice of iodide ions in which one-third of the octahedral holes are occupied by bismuth ions. (b) Ga_2S_3 adopts an hcp lattice of sulfide ions in which two-thirds of the tetrahedral holes are occupied by gallium ions. (c) Li_3N adopts a close-packed lattice in which all holes are occupied with Li^+ ions.

8.30. Which of the following are true statements about tetrahedral holes in a (cubic or hexagonal) close-packed lattice of anions? (a) There is one tetrahedral hole per anion in the lattice. (b) All of the tetrahedral holes are occupied in the lattice of ZnS. (c) A cation in a tetrahedral hole is surrounded by four anions. (d) A tetrahedral hole is larger than an octahedral hole; (e) All of the tetrahedral holes are occupied in the lattice of NaCl.

8.31. *The C_{60}^{3-} anion is nearly spherical, with a radius of roughly 500 pm. (a) Can it form a salt in which *all* holes are filled? Why? (b) If so, which salt is more likely: Rb_2CsC_{60} or $RbCs_2C_{60}$? Why?

8.32. In the following list, (a) which oxides are likely to be spinels, and (b) which oxides could be perovskites? $TiCo_2O_4$, $NaTaO_3$, $SrTiO_3$, Zn_2SiO_4, $ZnSeO_3$, $CoFe_2O_4$, and $FeCo_2O_4$.

8.33. *In the following list, (a) which oxides are likely to be spinels, and (b) which oxides could be perovskites? $NiFe_2O_4$, $BaFe_2O_4$, $BaTiO_3$, $BeTiO_3$, $BaSO_3$, $TiZn_2O_4$, Ni_3O_4, Pb_3O_4, and $NaTaO_3$.

8.34. There are also sulfide-ion-based spinels, where the anions are sulfide ions rather than oxide ions. Write the probable formulas of the following spinels: (a) greigite, containing only Fe; (b) linnaeite, containing only Co; and (c) daubreeite, containing Fe and Cr.

8.35. *Which of the following fluorides are likely to be perovskites? $KMgF_3$, $CaTiF_3$, $LiBaF_3$, $LiCF_3$, $TlCoF_3$, Cu_2F_3, or none of these.

8.36. Which of the following formulas correspond to possible nonstoichiometric oxides? $Sr_{0.95}O$, $Fe_{0.95}O$, $Ni_{0.95}O$, $C_{0.95}O$, $Cr_{0.95}O$, and $N_{0.95}O$.

8.37. *Which of the following formulas correspond to possible nonstoichiometric oxides? $Ca_{0.95}O$, $Fe_{0.95}O$, $Co_{0.95}O$, $C_{0.95}O$, $Cr_{0.95}O_3$, and $Eu_{0.95}O$.

8.38. Why would you not expect very high-oxidation-number oxides such as Mn_2O_7 and OsO_4 to be nonstoichiometric (i.e., $Mn_{2.2}O_7$ or $Os_{1.13}O_4$)?

8.39. *Which of the following are likely formulas for an isomorphously substituted salt: $(Be,Ra)SO_4$, $(Li,Mg)O$, and $(Zn,Fe)_2SiO_4$?

8.40. The formula of the mineral tremolite is $Ca_2(Mg,Fe)_5(OH)_2(Si_4O_{11})_2$. Interpret the part of the formula which is written $(Fe,Mg)_5$. What is the term for this phenomenon?

8.41. *You are studying the mineral hornblende, $Ca_2Mg_5(OH)_2(Si_4O_{11})_2$, and find samples in which isomorphous substitution of the magnesium and the calcium has occurred. Which of the following are possible minerals that could result from isomorphous substitution processes in hornblende? (a) $Y_2Mg_5(OH)_2(Si_4O_{11})_2$; (b) $Na_2Mg_5(OH)_2(Si_4O_{11})_2$; (c) $Na_2Mg_3(Fe^{III})_2(OH)_2(Si_4O_{11})_2$; (d) $Y_2Mg_3Li_2(OH)_2(Si_4O_{11})_2$; (e) $Sr_2Mg_5(OH)_2(Si_4O_{11})_2$.

8.42. The formula of the common form of the mineral garnet, andradite, can be written as $Ca_3Fe_2Si_3O_{12}$; there is a family of garnets, including (a)–(c). For each member of the family, tell whether the first or the second principle of isomorphous substitution (or both) is (or are) used to generate that member from common garnet (and confirm that it applies). Identify the oxidation state of any d-block metal ions. (a) Almandite, $Fe_3Al_2Si_3O_{12}$; (b) grossularite, $Ca_3Al_2Si_3O_{12}$; (c) spessartite, $Mn_3Al_2Si_3O_{12}$.

8.43. *The formula of the common form of the mineral garnet, andradite, can be written as $Ca_3Fe_2Si_3O_{12}$; there is a family of garnets, including (a)–(c). For each member of the family, tell whether the first or the second principle of isomorphous substitution (or both) is (or are) used to generate that member from common garnet (and confirm that it applies). Identify the oxidation state of any d-block metal ions. (a) Uvarovite, $Ca_3Cr_2Si_3O_{12}$; (b) the synthetic "YAG," used in lasers: $Y_3Al_5O_{12}$; (c) the synthetic $Y_3Fe_5O_{12}$, of importance for its magnetic properties.

8.44. The formula of one common form of feldspar, anorthite, can be written as $Ca[Al_2Si_2O_8]$. For the real or possible feldspars (a) and (b), tell which principle of isomorphous substitution—first or second—is used to generate this feldspar from anorthite. (a) Albite, $Na[AlSi_3O_8]$; (b) the synthetic $Sr[Ga_2Si_2O_8]$.

8.45. *Which principle of isomorphous substitution—the first or the second—is involved in converting (a) $^3_\infty[Si_2O_4]$ to $^3_\infty[AlPO_4]$ with the same structure; (b) $^3_\infty[LaPO_4]$ to $^3_\infty[CePO_4]$ with the same structure?

8.46. Which one of the following could arise by isomorphous substitution of ions in leucite, $KAlSi_2O_6$? $KYSi_2O_6$, $BaBeSi_2O_6$, $BaAlSi_2O_6$.

8.47. *Some semiconductors are soft-base salts that can be generated by isomorphous substitution of the second type, starting from the four-coordinate semiconducting elements, silicon and germanium. (a) Thinking of elemental silicon as $^3_\infty[Si^{4+}Si^{4-}]$ and estimating a radius for Si^{4-}, give a product that is obtained by carrying out the substitution first once and then twice. (b) Thinking of elemental germanium as $^3_\infty[Ge^{4+}Ge^{4-}]$ and estimating a radius for Ge^{4-}, give a product that is obtained by carrying out the substitution first once and then twice.

8.48. The mineral jarosite, $KFe_3(SO_4)_2(OH)_6$, was recently in the news when it was found on Mars. Consider the possibility of isomorphous substitution for the metal ion(s) in jarosite, with the following cations being candidates for substituting into jarosite: Li^+, Rb^+, Mg^{2+}, Ba^{2+}, Fe^{2+}, Fe^{3+}, and Cr^{3+}. (a) Which charge does Fe have in jarosite? (b) Which of the cations is the best candidate for replacing the K^+ in jarosite by isomorphous substitution of the first kind? Write the formula of the resulting jarosite in which all of the K^+ has been replaced.

(c) Which of the cations is the best candidate for replacing the iron in jarosite by isomorphous substitution of the first kind? Write the formula of the resulting jarosite in which all of the iron has been replaced. (d) Which two of the cations are the best candidates for replacing the K^+ and one of the iron ions in jarosite by isomorphous substitution of the *second* kind? Write the formula of the resulting jarosite in which this has been done.

8.49. *Consider mineral pairs (a)–(e). Classify each as related to the other by (1) isomorphous substitution of the first kind; (2) isomorphous substitution of the second kind; (3) non-isomorphous substitution; or (4) a pair that is completely unrelated in structure. (a) $KCr(SO_4)_2 \cdot 12H_2O$ and $RbAl(SO_4)_2 \cdot 12H_2O$; (b) $KCr(SO_4)_2 \cdot 12H_2O$ and $KCr(SO_4)_2$; (c) $Pb_5Cl(PO_4)_3$ and $Pb_5Cl(VO_4)_3$; (d) CO_2 and SiO_2; (e) xenotime (YPO_4) and zircon ($ZrSiO_4$).

8.50. Use the specified principle of isomorphous substitution and Table C to convert (a) wollastonite, $Ca(SiO_3)$, to another isomorphous mineral (first principle, replacing Ca); (b) magnesioriebeckite, $Na_2Mg_3Fe_2[Si_4O_{11}]_2(OH)_2$, second principle, replacing Na and Fe.

8.51. *(a) Verify that the large unit cell of $YBa_2Cu_3O_7$ shown in Figure 8.14 indeed shows the correct stoichiometry of ions. (b) If there were no mixing of Cu^{3+} and Cu^{2+} ions between the chains and sheets of copper ions, which type of copper ion would reside in the chains and which in the sheets in order to give the proper stoichiometry?

8.52. Choose the appropriate classifications(s)—perovskite, spinel, nonstoichiometric, defect perovskite, superconductor—for each of the following mixed-metal oxides: (a) $BaTiO_3$; (b) $La_{1.85}Sr_{0.15}CuO_4$; (c) $Fe_{0.95}O$; (d) Fe_3O_4; (e) $YBa_2Cu_3O_{6.9}$.

8.53. *Choose the appropriate classifications(s)—perovskite, spinel, nonstoichiometric, superconductor, impossible—for each of the following mixed-metal oxides: (a) $YBa_2Cu_3O_7$; (b) Co_3O_4; (c) $RbNbO_3$; (d) $Os_{0.95}O_4$.

8.54. Choose the appropriate classifications(s)—perovskite, spinel, nonstoichiometric, superconductor, or none of the above—for each of the following mixed-metal oxides: (a) $CsNbO_3$; (b) $LiPaO_3$; (c) $Mn_3Al_2(SiO_4)_3$; (d) $MgAl_2O_4$; (e) Ba_7SbO_6; (e) ThO_2.

8.55. *Determine the coordination number and list the nearest neighbors of the oxide ions in (a) perovskite and (b) $YBa_2Cu_3O_7$. (c) Confirm that Equation 4.1 applies to these mixed-metal oxides if it is modified to read: $\sum$(Coord. No. of M)(No. of M in formula) = (Coord. No. of Anion)(No. of Anions in formula).

8.56. Classify each oxide as acidic, basic, or amphoteric or neutral, and decide whether it will be soluble in water: (a) Na_2O; (b) Cr_2O_3; (c) CO_2; (d) SiO_2.

8.57. *Arrange the following oxides in order of decreasing acidity/increasing basicity: (a) Na_2O, Cr_2O_3, SiO_2, P_4O_{10}; (b) ZrO_2, CO_2, SrO, Rb_2O, Y_2O_3; (c) MnO, MnO_2, Mn_2O_3, Mn_2O_7; (d) TiO_2, TeO_2, SO_2, ThO_2.

8.58. Consider oxo acids (a)–(e): (a) H_2SeO_3; (b) H_6TeO_6; (c) $HMnO_4$; (d) H_3PO_4; (e) H_4XeO_6. For each oxo acid, (1) determine whether it is soluble or insoluble in water, and (2) give the formula of the corresponding acidic oxide from which it might be prepared by the addition of water only (assuming solubility).

8.59. *Suppose that five oxides of new elements have been discovered: hoffmon tetroxide, HmO_4 (Hm = fourth-period element); dimongium heptoxide, Mi_2O_7 (Mi = sixth-period element); yangium dioxide, YaO_2 (Ya = second-period element with no unshared *sp* electron pairs); vermillium dioxide, VeO_2 (Ve = fifth-period element with one unshared *sp* electron pair); and luium dioxide, LmO_2 (Lm = fifth-period element with no unshared *sp* electron pair). (a) Which two oxides are most likely to be gases at room temperature? (b) Which oxide is most likely to be strongly acidic? (c) Which oxide is most likely to be basic? Is it likely to be water soluble? (d) Which oxide is most likely to be useful for making ceramic plates to shield space shuttles from the heat of re-entry into the atmosphere?

8.60. Consider the following set of oxides: MnO_2, OsO_2, SnO_2, CO_2, OsO_4, XeO_4, and FeO_4. (a) Which three of these oxides are high-melting solids? (b) Which four of these oxides are easily vaporized (are gases or at least volatile solids)? (c) Three of the four easily vaporized solids are in the same class of acid–base reactivity. Are these acidic or basic oxides? The fourth of these easily vaporized oxides is, in contrast to the others, a neutral oxide. Which one is this? Briefly explain why.

8.61. *Consider the following set of oxides: SrO, ZrO_2, TeO_3, and XeO_4. (a) Which one of these oxides is a solid, insoluble in water, which reacts with the acidic oxide SiO_2 to give a simple silicate salt? Write an equation for this reaction. (b) Which one of these oxides is a solid that dissolves in water to give a basic solution? Write an equation for this process. (c) Which one of these oxides is a gas that dissolves in water to give an acidic solution? Write an equation for this process.

8.62. The elements V, Cr, Tc, and Os all form soluble acidic oxides. For each element, write the simplest formula of its soluble acidic oxide, then write the balanced chemical equation showing its reaction with water. (a) V; (b) Cr; (c) Tc; (d) Os.

8.63. *You have produced minute quantities of radioactive element number 109, meitnerium, and are investigating its positive oxidation states. (a) Suppose that you have oxidized it very strongly in a hot acidic solution, and find the radioactivity coming from the vapors above the solution; you conclude that you have a volatile oxide. What oxidation state do you probably have? Give arguments why you chose that oxidation state and not some other. Draw a likely structure of the oxide. (b) Suppose that you subsequently obtain meitnerium in the +6 oxidation state in basic solution, and find that it gives a precipitate not only with Ba^{2+} but also with most acidic cations. What does this lead you to suspect about the formula of the +6 species? (c) You have only a trace of Mt left, in the +7 oxidation state in basic solution. Outline a plan for precipitating it from solution.

8.64. Complete and balance the following chemical equations (or tell if no reaction will occur):

(a) $N_2O_5 + H_2O \rightarrow$
(b) $Cl_2O_7 + H_2O \rightarrow$
(c) $Na_2O + H_2O \rightarrow$
(d) $Cr_2O_3 + H^+(aq) \rightarrow$
(e) $Cr_2O_3 + H_2O \rightarrow$
(f) $SiO_2 + OH^- \rightarrow$

8.65. *Complete and balance the following chemical equations (or tell if no reaction will occur):

(a) $Tl_2O + H_2O \rightarrow$
(b) $I_2O_5 + H_2O \rightarrow$
(c) $Cl_2O(g) + H_2O \rightarrow$
(d) $La_2O_3 + H^+ \rightarrow$
(e) $B_2O_3 + OH^- + H_2O \rightarrow$
(f) $MnO + H^+(aq) \rightarrow$

8.66. Consider the oxides SrO, ZrO_2, MoO_3, and RuO_4. If needed, take the radius of Ru^{8+} to be 52 pm. (a) Which of these oxides will be soluble in water to give a basic solution? Write a chemical equation for this process. (b) Which of these oxides will be soluble in water to give an acidic solution? Write an equation for this process. (c) Which of these oxides (if any) will be monomeric molecular substances? (d) Which of these oxides (if any) will most easily become a gas? (e) Which one of these oxides will most likely be a basic oxide that will be *insoluble* in water? Write an example of an equation for a chemical reaction that would justify classifying it as a basic oxide.

8.67. *Classify each oxide below as acidic, basic, etc. Then complete and balance the following equations for Lux–Flood acid–base reactions, or note if no such reaction is to be expected.

(a) $BaO(s) + P_4O_{10}(s) \rightarrow$
(b) $TeO_3(s) + I_2O_5(s) \rightarrow$
(c) $SrO(s) + MoO_3(s) \rightarrow$
(d) $CaO(s) + TeO_3(s) \rightarrow$
(e) $CaO(s) + MnO(s) \rightarrow$

8.68. Classify each oxide below as acidic, basic, etc. Then complete and balance the following equations for Lux–Flood acid–base reactions, or note if no such reaction is to be expected.

(a) $Y_2O_3(s) + P_4O_{10}(s) \rightarrow$
(b) $Na_2O(s) + V_2O_5(s) \rightarrow$
(c) $Na_2O(s) + BaO(s) \rightarrow$
(d) $Na_2O(s) + P_4O_{10}(s) \rightarrow$
(e) $Fe_2O_3(s) + BaO(s) \rightarrow$
(f) $FeO + P_4O_{10} \rightarrow$

8.69. *(a) Would you expect a solid-state reaction between BaO and Sb_2O_5? (b) If so, which would be the product: $BaSbO_3$, $Ba_3(SbO_4)_2$, $Ba_7(SbO_6)_2$, or $BaSb_2O_6$?

8.70. Write three balanced chemical equations showing the three steps by which ammonia is converted industrially to nitric acid.

8.71. *(a) Write three balanced chemical equations showing the three steps by which elemental sulfur in coal is converted to sulfuric acid in acid rain. (b) Sulfur dioxide can be removed from smokestack gases by reaction with (by "scrubbing" with) magnesium oxide. Write a chemical equation for this process. (c) Calculate the number of grams of magnesium oxide that would be needed to clean the smokestack gases from burning 1,000,000 g of coal that is 3.2% S by weight.

8.72. Which volatile oxides are associated with each of the environmental problems given in (a)–(d)? Why are the other volatile oxides discussed in this section not involved?
(a) Acid rain; (b) the greenhouse effect; (c) the acidification of seawater; (d) London smog.

8.73. *What methods are discussed for disposing of volatile acidic oxides from smokestacks?

8.74. Classify each of the silicates (a)–(e) as (1) a layer polysilicate, (2) a chain or cyclic polysilicate, (3) a simple silicate, or (4) a double-chain polysilicate: (a) bustamite = $CaMn(SiO_3)_2$; (b) spudomene = $LiAl(SiO_3)_2$; (c) tremolite = $Ca_2Mg_5(OH)_2[Si_4O_{11}]_2$; (d) coffinite = $U(SiO_4)$; (e) kaolinite = $Al_2(OH)_4(Si_2O_5)$.

8.75. *Select one mineral—(1) wollastonite = $CaSiO_3$; (2) talc = $Mg_3(OH)_2[Si_4O_{10}]$; (3) grunerite = $Fe_7(OH)_2[Si_4O_{11}]_2$; (4) monticellite = $CaMgSiO_4$; (5) stishovite = SiO_2—as an example of each of the following: (a) contains a monomeric silicate ion; (b) contains a chain polysilicate ion; (c) contains a double-chain polysilicate ion; (d) contains a layer polysilicate ion.

8.76. Classify silicates (a)–(f) as (1) a simple silicate, (2) a disilicate, (3) a chain polysilicate or cyclic oligosilicate, (4) a layer polysilicate, or (5) a double-chain polysilicate: (a) acmite = $NaFe(SiO_3)_2$; (b) akermanite = $Ca_2Mg(Si_2O_7)$; (c) anthophyllite = $Mg_7(OH)_2(Si_4O_{11})_2$; (d) fayalite = $Fe_2(SiO_4)$; (e) pyrophyllite = $Al_2(OH)_2(Si_4O_{10})$; (f) zircon = $Zr(SiO_4)$; (g) talc = $Mg_3(OH)_2(Si_4O_{10})$.

8.77. *Classify silicates (a)–(d) as (1) a simple silicate, (2) a disilicate, (3) a chain polysilicate or cyclic oligosilicate, (4) a layer polysilicate, or (5) a double-chain polysilicate: (a) crocidolite, $Na_2Fe_5(OH)_2[Si_4O_{11}]_2$; (b) enstasite, $Mg(SiO_3)$; (c) olivine, $(Mg,Fe)_2(SiO_4)$; (d) diopside, $CaMg(SiO_3)_2$.

8.78. Solutions of tetraalkylammonium silicates have been found to contain an anion of formula $Si_8O_{20}^{8-}$: M. Wiebke, M. Grube, H. Koller, G. Engelhardt, and J. Felsche, *Microporous Mater.* 2, 55 (1993). (a) Which class of polysilicate would this appear to be from its formula? (b) Check the literature article to see whether this is indeed the proper class. If not, would it be a member of a new class? In this case, suggest a new class based on the terminology of Figure 8.21.

8.79. *Show that you understand the condensed drawings of the fragments of polysilicate structures shown in Figure 8.21 by redrawing them, using closed circles for Si atoms and open circles for O atoms. Redraw the following: (a) $^1_\infty[SiO_3^{2-}]$, (b) $^1_\infty[Si_4O_{11}^{6-}]$, and (c) $^2_\infty[Si_4O_{10}^{4-}]$.

8.80. A mineral proposed as an asbestos replacement called xonotlite, $Ca_6(OH)_2[Si_6O_{17}]$, is also a double-chain polysilicate, but it differs slightly in structure. Propose a structure for xonotlite.

8.81. *Predict the basicity classification of the anion in polysilicates (a)–(d), based on the (untested) hypothesis that their pK_b values can be estimated from the formula of the structural unit of the polysilicate using Equation 3.14. (a) $Si_2O_7^{6-}$; (b) $Si_3O_9^{6-}$; (c) $Si_4O_{11}^{6-}$; (d) $Si_4O_{10}^{4-}$. (e) Estimate the pK_a values of the corresponding conjugate acids of these silicate anions. If the hypothesis is valid, which of these polysilicates, when ion-exchanged into the H^+ forms, would be very strongly acidic?

8.82. Select one of the minerals mentioned in Exercise 8.75 as (a) the mineral that would weather the most rapidly; (b) the mineral that would weather the most slowly.

8.83. *Select one of the minerals mentioned in Exercise 8.76 as (a) the mineral that would weather the most rapidly; (b) the mineral that would weather the most slowly.

8.84. Consider the following minerals and oxides: lime = CaO; ureyite = $NaCr(SiO_3)_2$; chabazite = $Ca_6[Al_{12}Si_{24}O_{72}]$; zirconia = ZrO_2; tremolite = $Ca_2(Mg,Fe)_5(OH)_2[Si_4O_{11}]_2$; olivine = $(Fe,Mg)_2(SiO_4)$; and kaolinite = $Al_4(OH)_8[Si_4O_{10}]$. (a) Which one would weather most quickly? (b) Which one would most characteristically be found on Earth's surface only in the youthful soil of a desert? (c) Which would be the last to weather away? (d) Which is abundant in the mantle and is converted to a spinel polymerization isomer deep in the mantle?

8.85. *Consider the following types of silicates or related soil minerals: layer aluminosilicates, layer polysilicates, chain or cyclic polysilicates, simple silicates, three-dimensional polymeric metal oxides (e.g., TiO_2), and double-chain polysilicates. (a) Which one would weather most quickly? (b) Which one would be most characteristically found in a desert soil? (c) Which one would be most characteristically found in the soil of a tropical region in which the forest has been cut down?

8.86. The atmosphere of Venus is very hot (~800°C) and dense, is composed mainly of CO_2 with some N_2, and supports opaque clouds made of sulfuric acid. (a) How would you compare the acidity of this atmosphere to that of Earth? (b) How would you expect the rate of weathering of silicate minerals to be on Venus as compared to on Earth?

8.87. *Which of the following minerals are layer aluminosilicates? (a) Hedenbergite = $CaFe(SiO_3)_2$; (b) lepidolite = $KAl_2[AlSi_3O_{10}](OH,F)_2$; (c) phenacite = $Be_2(SiO_4)$; (d) petalite = $Li[AlSi_4O_{10}]$; (e) mesolite = $Na_2Ca_2[Al_2Si_3O_{10}]\cdot8H_2O$.

8.88. Which of minerals (a)–(d) is/are layer aluminosilicates? Which are clays? (a) Ureyite = $NaCr(SiO_3)_2$; (b) natrolite = $Na_2[Al_2Si_3O_{10}]\cdot2H_2O$; (c) biotite = $K(Mg,Fe)_3[AlSi_3O_{10}](OH)_2$; (d) cummingtonite = $Mg_7(OH)_2[Si_4O_{11}]_2$.

8.89. *Devise a possible asbestos substitute that would be the aluminosilicate obtained from chrysotile (serpentine) asbestos $Mg_6(OH)_8(Si_4O_{10})$ by isomorphous substitution of the second kind.

8.90. Figure 8.23 shows the relative availability of nutrient elements in soil as a function of soil pH. Using the principles in the book to date, explain why each element is available and unavailable at the pH values given.

8.91. You are studying the feldspar $K[AlSi_3O_8]$ and find samples in which isomorphous substitution has occurred. Which of the following are possible minerals that could result from isomorphous substitution processes in this feldspar? (a) $Ba[Al_2Si_2O_8]$; (b) $Rb[AlSi_3O_8]$; (c) $Na[AlSi_3O_8]$; (d) $Ba[AlSi_3O_8]$.

8.92. Select the proper structural type—framework aluminosilicate, layer polysilicate, chain or cyclic polysilicate, simple silicate, or double-chain polysilicate—for each of the following silicates: (a) ureyite = $NaCr(SiO_3)_2$; (b) natrolite = $Na_2[Al_2Si_3O_{10}]\cdot2H_2O$; (c) cummingtonite = $Mg_7(OH)_2[Si_4O_{11}]_2$.

8.93. *The formula of one common form of the mineral feldspar, anorthite, can be written as $Ca[Al_2Si_2O_8]$. Which principle(s) of isomorphous substitution (first and/or second) is(are) used to generate feldspars (a) and (b) from anorthite? (a) Albite, $Na[AlSi_3O_8]$; (b) the synthetic $Sr[Ga_2Si_2O_8]$.

8.94. Select the proper structural type—framework aluminosilicate, layer polysilicate, chain or cyclic polysilicate, simple silicate, double-chain polysilicate, or layer aluminosilicate—for each of the following silicates: (a) willemite = $Zn_2(SiO_4)$; (b) celsian = $Ba[Al_2Si_2O_8]$; (c) chabazite = $Ca_6[Al_{12}Si_{24}O_{72}]$; (d) tremolite = $Ca_2(Mg,Fe)_5(OH)_2[Si_4O_{11}]_2$; (e) phenacite = $Be_2(SiO_4)$.

8.95. *Consider the following silicates and aluminosilicates: tremolite = $Ca_2(Mg,Fe)_5(OH)_2[Si_4O_{11}]_2$; ureyite = $NaCr(SiO_3)_2$; chabazite = $Ca_6[Al_{12}Si_{24}O_{72}]$; kaolinite = $Al_4(OH)_8[Si_4O_{10}]$; and phenacite = $Be_2(SiO_4)$. (a) Draw the structure of the simplest formula unit of each silicate or aluminosilicate anion except that of chabazite. (b) Write the formula of the (mononuclear) structural unit of each anion, along with its (possibly fractional) charge. (c) List the anions in order of increasing predicted basicity. (d) Briefly explain why aluminum is written inside the brackets in the formula of chabazite but outside the brackets in the formula of kaolinite, $Al_4(OH)_8[Si_4O_{10}]$.

8.96. Predict the basicity classification of the anion in each of the following polyaluminosilicates, based on the (untested) hypothesis that these can be estimated from the formula of the mononuclear structural unit obtained by depolymerizing the anion, then counting the number of oxo groups and units of negative charge in the structural unit: (a) anion of orthoclase; (b) anion of celsian; (c) anion of sodalite; (d) anion of ZSM-5. (e) Estimate the pK_a values of the corresponding conjugate acids of these aluminosilicate anions, assuming that the hypothesis in it has some validity. If so, which of these aluminosilicates, when ion-exchanged into the H^+ forms, would be very strongly acidic?

ALBERT EINSTEIN (1879–1955) was born in Ulm, in Württemberg, Germany. In 1896 he entered the Swiss Federal Polytechnic School in Zurich to be trained as a physics and mathematics teacher. Graduating in 1901, he was unable to find a teaching position, so he took a job as an assistant examiner in the patent office in Bern, Switzerland. In 1905, Einstein received his doctorate from the University of Zurich, and he published separate papers on the photoelectric effect, Brownian motion, special relativity, and mass–energy equivalence. Based on Einstein's theory of general relativity, the sun's gravity should bend light from another star. Sir Arthur Eddington confirmed this prediction during a solar eclipse in 1919 and the resulting notoriety made Einstein internationally famous. Einstein held various academic positions in Switzerland, Germany, and what is now the Czech Republic, but he was unable to return to the one he held at the Humboldt University of Berlin in 1933 when the Nazis came to power. He took a position as a resident scholar at the Princeton Institute for Advanced Study, but not before trying to get as many Jewish scientists and their families safely out of Germany as possible. While at Princeton, he worked on a unified field theory, but his efforts were unsuccessful. Einstein received the 1921 Nobel Prize in Physics "for his services to Theoretical Physics, and especially for his discovery of the law of the photoelectric effect."

CHAPTER 9

The Underlying Reasons for Periodic Trends

We have now completed the foundations of inorganic chemistry that apply to aqueous and bioinorganic chemistry. In Chapter 9 we develop deeper reasons for these foundations; in Chapters 10–12 we develop some foundations for more advanced study.

This chapter focuses on the underlying reasons for the periodic trends that are covered in Chapter 1 (periodic table shape and ion charges, Section 1.2; core and valence electron configurations, Section 1.3; oxidation states, Section 1.4; radii of atoms and ions, Section 1.5; and electronegativities, Section 1.6). Explanations are given for orbital shapes and bond types (Section 3.3), and for ionization energies and electron affinities (Section 6.6). Your instructor may assign the relevant explanations in conjunction with the above sections. Or, because the explanations flow from the concepts of quantum mechanics as applied to atoms (Section 9.1) through Slater's rules (Section 9.2) to the reasons, your instructor may want to consider these reasons together at this time.

9.1. Quantum Mechanics: Wave and Particle Properties of Electrons

OVERVIEW. Section 9.1A is a review of the principles of quantum mechanics from general chemistry and your Foundations of Physical Chemistry course. Exercises 9.1–9.4 should help you recall those principles. In Section 9.1B we focus on orbitals for electrons in atoms. The complex wave equation for electron orbitals in atoms contain three types of quantum numbers–namely, n, l, and m_l. It can be separated into a radial part and an angular part. In Section 3.3A the drawing of the angular part of the wave function for s, p, d, and f orbitals was covered; at this time we discuss the relationship of the drawings to the values of the quantum numbers l and m_l. Understanding the radial part of these wave functions includes identifying or drawing nodal planes and surfaces and the signs of the wave functions. You may review these concepts in Section 9.1B and Section 3.3A, and by trying Exercises 3.31, 3.32, and 9.5–9.14.

9.1A. Waves and Particles (Review). Since the late 1920s and early 1930s we have realized that the atomic basis for the organization of the periodic table lies in the electrons of the atoms and in the types of paths those electrons follow as they move around the nuclei of atoms (i.e., in the types of **orbitals** occupied by electrons). Chemical change occurs among the electrons of atoms; the theoretical study of chemical change focuses on the study of electrons in orbitals. Just before this time, Bohr had developed a model of the atom in which the electrons were particles that followed orbits around the nucleus, in close analogy to the orbits of planets, asteroids, and comets around the sun. However, by the known laws of physics, negatively charged particles should not have remained in orbit about the positively charged nucleus, but should have spiraled inward, leading to collapse of the atom. Bohr avoided this catastrophe by using the assumption that only certain well-defined, "quantized" orbits were allowed for electrons. However, he was unable to explain why this should be so.

The rationale for this assumption came with the development of **quantum mechanics**. The formulation of quantum mechanics began with the discarding by de Broglie of the earlier view that things in the universe were either waves or particles. De Broglie suggested that this is a false dichotomy: All things have properties of waves, such as characteristic **wavelengths** λ, and have properties of particles, such as **momentum** (mass m times velocity v). The de Broglie relationship connects these two properties of a wave/particle:

$$mv = h/\lambda \tag{9.1}$$

The constant h, **Planck's constant**, has the value 6.626×10^{-34} J s. For calculations using Equation 9.1, one should express the mass m in kg, the velocity v in m s^{-1}, and make use of the definition that 1 joule is 1 kg m^2 s^{-2}.

A typical wave property is that of **diffraction**, in which a front of waves strikes a barrier or grating that has openings or slits spaced by distances comparable to the wavelengths of the waves. An example is ocean waves passing through a seawall (Figure 9.1a). As the waves pass through the openings, new wave fronts develop at each opening. These then spread out to strike other wave fronts. At some points there is **constructive interference** of the wave fronts, creating higher amplitude crests and troughs. At other points there is **destructive interference**, where the crest of one wave front coincides with the trough of another, and the two waves annihilate each other.

Electromagnetic radiation, which had always been thought of as waves, shows this property. X rays have very short wavelengths (3 pm and below), which are comparable in magnitude to the spacing between atoms and ions in crystals (~100 pm). The experiment of **X-ray diffraction** involves passing X rays through the gaps between the nuclei of atoms or ions in a crystal. This produces characteristic diffraction patterns that can be analyzed to determine the arrangement and spacing of the atoms or ions in a crystal. Electron diffraction can also be performed on crystals using the wave property of electrons, if they are accelerated by a potential difference of about 20,000 V, so that they have a wavelength of about 10 pm.

We are thus justified in thinking of an electron in an atom as a wave located in a particular wave pattern around the atomic nucleus. The electron wave has its crests and troughs (points of maximum amplitude) and its level or calm points (nodes, at

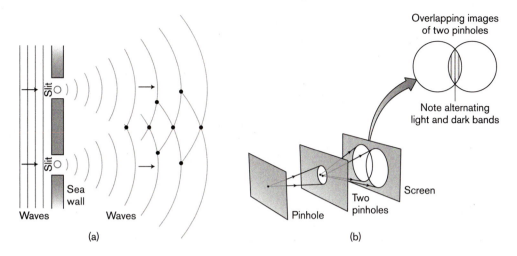

Figure 9.1. (a) Water waves (crests shown as lines, troughs between lines) passing through two gaps in a sea wall, being diffracted, and engaging in interference on the other side. Constructive interference of crests with crests is shown with heavy dots. Destructive interference of crests with troughs occurs along waves halfway between the dots. (b) The corresponding experiment of passing light through two pinholes, giving an interference pattern on the other side. [Adapted from W. J. Lehmann, *Atomic and Molecular Structure: The Development of Our Concepts*, John Wiley and Sons: New York, 1972, pp. 178 and 180.]

which the wave has zero amplitude). However, macroscopic objects of much larger mass (say 1 mg) have such short wavelengths that there are no physically real gratings with slits spaced closely enough to produce diffraction or other observable wave properties. Therefore, we neglect the wave properties of macroscopic objects.

EXAMPLE 9.1

(a) What is the momentum of the standard 1-kg platinum bar if it is moving at 60 mph (88 m s^{-1})? (b) What is its wavelength? (c) What (if anything) could diffract this platinum bar? Is it useful to regard it as a wave? As a particle?

SOLUTION: (a) Its momentum $mv = (1 \text{ kg})(88 \text{ m s}^{-1}) = 88 \text{ kg m s}^{-1}$. (b) We will employ Equation 9.1, which includes the unit of joules (J). We can first convert the momentum to units including joules:

$$88 \, \frac{\text{kg m}}{\text{s}} \cdot \frac{1 \text{ J}}{\text{kg m}^2 \text{ s}^{-2}} = 88 \text{ J s m}^{-1}$$

Then we substitute this into the rearranged form of Equation 9.1 to get

$$\lambda = \frac{6.626 \times 10^{-34} \text{ J s}}{88 \text{ J s m}^{-1}} = 7.5 \times 10^{-36} \text{ m}$$

(c) Diffraction could only occur with something having particles separated by around 10^{-36} m. This is far below interatomic distances, so diffraction of the Pt bar is unlikely to be observed. Hence, the bar can be regarded as a particle, but it is not useful to regard it as a wave.

As suggested earlier by Bohr, particles such as electrons that are confined within atoms or other regions of space cannot take just any path. Their paths (in atoms, orbitals) are described by equations called **wave functions** (ψ). Wave functions contain constants called **quantum numbers** that cannot have just any value, but must take values that are integers (or half-integers in some cases).

Simpler than the case of an electron in a three-dimensional atom is the case of an electron confined to a one-dimensional path of length L by two barriers at either end (Figure 9.2). The wave function for this electron is a sine function:

$$\psi = \sin\left(\frac{2\pi x}{\lambda}\right) \tag{9.2}$$

For a wave pattern of arbitrary wavelength λ, it is likely that the wave will reflect off of one barrier out of phase with itself, so that crests and troughs do not match up with each other after reflection; destructive interference then results. A stable standing wave pattern can only result if the wave has a wavelength that is integrally related to the distance L between the walls:

$$\lambda = \frac{2L}{n} \qquad n = 1, 2, 3, \ldots \tag{9.3}$$

The number n is a quantum number, which must be a whole positive number. For each quantum number there is a corresponding standing wave that can persist inside the walls of this one-dimensional box. Combining these two equations, we obtain the wave functions ψ for possible standing waves:

$$\psi = \sin\left(\frac{n\pi x}{L}\right) \qquad n = 1, 2, 3, \ldots \tag{9.4}$$

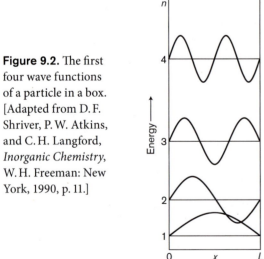

Figure 9.2. The first four wave functions of a particle in a box. [Adapted from D. F. Shriver, P. W. Atkins, and C. H. Langford, *Inorganic Chemistry*, W. H. Freeman: New York, 1990, p. 11.]

The standing waves drawn in Figure 9.2 may be characterized by the number of **nodes** present. The standing wave drawn at the bottom of Figure 9.2 has no nodes (points of zero amplitude) within the box itself; each successive wave ($n = 2, 3, 4, \ldots$) has one more node than the one before. These wave patterns are also related to those obtained by plucking a stringed musical instrument: In music theory, the zero-node wave is known as the fundamental, the one-node mode is called the first overtone, and so on.

The energy of the particle in this one-dimensional box can also be calculated, and it depends strongly on the value chosen for n. The particle has only kinetic energy:

$$E = \tfrac{1}{2}mv^2 \tag{9.5}$$

Substitution in this equation of the de Broglie relationship (Eq. 9.1) gives us Equation 9.6:

$$E = \frac{1}{2}\left(\frac{h^2}{m\lambda^2}\right) \tag{9.6}$$

We now introduce the wavelength of the electron in the box, from Equation 9.3:

$$E = \frac{1}{2}\left(\frac{h^2}{m}\right)\left(\frac{n}{2L}\right)^2 = \frac{n^2h^2}{8mL^2} \tag{9.7}$$

The kinetic energy of the electron in the box depends strongly on the value chosen for the quantum number n, as is also suggested by Figure 9.2. Note that its kinetic energy also increases as the number of nodes in the standing wave increases.

9.1B. Electrons in Atoms—Angular and Radial Parts of the Electronic Wave Function.
An electron in an atom is in a more complex environment. It is not confined in a box, but can (in principle) be found at any distance from its origin at the nucleus of the atom. It moves in three dimensions, not one, so *three* quantum numbers are required to define its path, not one. It is influenced by a potential derived from its attraction to the oppositely charged nucleus and modified by the repulsions between it and any other electrons present. The equation that summarizes this principle is known as the **Schrödinger wave equation**:

$$\frac{-h^2}{8\pi^2m}\left(\frac{\partial^2\psi}{\partial x^2} + \frac{\partial^2\psi}{\partial y^2} + \frac{\partial^2\psi}{\partial z^2}\right) + V\psi = E\psi \tag{9.8}$$

The first term of this partial differential equation is essentially the kinetic energy of the electron. The potential V in the second term includes the attractive potential of the nucleus for the electron and any electron–electron repulsions with other electrons. The parameter E then gives the total energy of the electron in the atom (or molecule). This equation is so formidable that it can be solved in an algebraic form only if just two particles are present: the nucleus and one electron. In this case, the potential function V is just the Coulombic force of attraction between the nucleus and the electron:

$$V = \frac{(\text{charge of nucleus})(\text{charge of electron})}{(\text{distance between})} = \frac{-Ze2}{r} \tag{9.9}$$

In this equation, e is the charge of the electron, $-Ze$ is the charge of the nucleus, and r is the distance between them.

To solve this equation, the wave function ψ (path of the electron) is written in terms of spherical polar coordinates (r,θ,ϕ):

$$\Psi = R_{n,l}(r) \cdot Y_{l,m_l}(\theta,\phi) \tag{9.10}$$

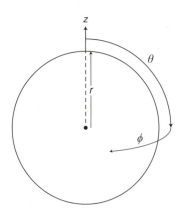

Figure 9.3. Spherical polar coordinates, where *r* is the distance from the nucleus, θ is the angular distance from the *z* axis (the "north pole"), and ϕ is the azimuthal angular distance from the reference vertical plane (the "Greenwich meridian").

In spherical polar coordinates, *r* is the distance from the center of the atom (the nucleus), while θ and ϕ give the angles of the point in question away from the polar axis through the nucleus and a reference vertical plane, respectively (Figure 9.3). The great advantage of this system of coordinates is that it is possible to factor the wave function ψ into two parts:

1. a radial wave function *R*, a description of the distance of the electron (or of its crest, trough, etc.) from the nucleus, and

2. an angular wave function *Y*, a description of the angular distance of the electron from the "north pole" and the reference vertical plane of the atom.

Let us first illustrate this distinction using the orbits of the planets as illustrations. The *angular part* of any planet's orbit is substantially the same. All move roughly in the equatorial plane of the solar system, so that $\theta = 90° = \pi/2$ rad; the orbit is substantially unaltered as the planet sweeps the angle ϕ around the sun. These angular parts of the planetary orbits define the shape and the orientation of the planetary orbits: All are more or less circular and in the equatorial plane of the solar system. The planetary orbits differ predominantly in their *radial parts*: Each planet moves in an orbit of a different radius.

The general form of the solution of the Schrödinger wave equation is too complex to make much sense on casual inspection. Because the atom is a three-dimensional "box," the general (nonrelativistic) form of this solution for ψ contains three quantum numbers, *n*, *l*, and m_l, to be discussed below. When specific values of these three quantum numbers are substituted in the general form of the equation, the specific solutions obtained are much simpler (Table 1.2). We note that the solutions have been separated into radial and angular parts. The overall wave function ψ is obtained by multiplying the two parts together.

The **principal quantum number *n*** occurs only in the radial part of the wave function and therefore is important in determining the *size* of the orbital. It is the first number shown in the usual representation of orbitals. It can take any integral value from 1 upward, indicating increasing size. For example, the 1*s* orbital is smaller than the 2*s* orbital.

TABLE 9.1. Orbitals for Hydrogen-Like Atoms or Ions

A. Radial Wave Functions $$R_{n,l}(r) = f(r)\left(\dfrac{Z}{a_0}\right)^{3/2} e^{-\rho/2}$$ with a_0 the Bohr radius (0.53 Å) and $\rho = \dfrac{2Zr}{na_0}$				B. Angular Wave Functions $$Y_{l,m_l}(\theta,\phi) = \left(\dfrac{1}{4\pi}\right)^{1/2} y(\theta,\phi)$$			
n	l	$f(r)$	Orbital Designation	l	m_l	$y(\theta,\phi)$	Orbital Designation
1	0	2	$1s$	0	0	1	s
2	0	$\left(\frac{1}{2}\sqrt{2}\right)(2-\rho)$	$2s$	1	0	$3^{1/2}\cos\theta$	p_z
2	1	$\left(\frac{1}{2}\sqrt{6}\right)\rho$	$2p$	1	± 1	$\mp\left(\frac{3}{2}\right)^{1/2}\sin\theta\, e^{\pm i\phi}$	p_x, p_y
3	0	$\left(\frac{1}{9}\sqrt{3}\right)(6-6\rho+\rho^2)$	$3s$	2	0	$\left(\frac{5}{4}\right)^{1/2}(3\cos^2\theta-1)$	d_{z^2}
3	1	$\left(\frac{1}{9}\sqrt{6}\right)(4-\rho)\rho$	$3p$	2	± 1	$\mp\left(\frac{15}{4}\right)^{1/2}\cos\theta\sin\theta\, e^{\pm i\phi}$	d_{xz}, d_{yz}
3	2	$\left(\frac{1}{9}\sqrt{30}\right)\rho^2$	$3d$	2	± 2	$\left(\frac{15}{8}\right)^{1/2}\sin^2\theta\, e^{\pm 2i\phi}$	$d_{xy}, d_{x^2-y^2}$

Source: Adapted from D. F. Shriver, P. W. Atkins, and C. H. Langford, *Inorganic Chemistry*, W. H. Freeman: New York, 1990, p. 11.

The **secondary** or **orbital angular momentum quantum number** l occurs in both parts of the wave function; it determines the *shape* of an orbital. The shapes of orbitals are described in Section 3.3A and illustrated in Figure 3.7. Visual representation of shapes of orbitals is of paramount importance at our level of inorganic chemistry, but these shapes of orbitals originate from mathematical expressions such as those found in Table 9.1. This quantum number can take any integral value from zero up to one less than the principal quantum number. Note that the number of **nodal planes** passing through the nucleus (Figure 3.7) is equal to the secondary quantum number. In designating orbitals, it is common to replace the numerical value of this quantum number with the corresponding letter designation shown in Table 9.2.

The **magnetic quantum number** m_l occurs only in the angular part of the wave function, and it determines the *orientation* of the orbital of a given (nonspherical) shape. This quantum number can take any positive or negative whole-number value, including zero, provided it falls between l and $-l$ inclusive. By convention, $m_l = 0$ corresponds

TABLE 9.2. Relationship of Secondary Quantum Numbers to Orbital Designations

Secondary quantum number l	0	1	2	3	4	5
Orbital shape designation	s	p	d	f	g	h

to an orbital oriented *along the z axis.* As shown in Table 9.1B, these include orbitals such as the *s*, p_z, and d_{z^2} orbitals. After some mathematical manipulation to remove imaginary numbers, we obtain equations for ψ. Then ψ^2 is the *probability density* from which we derive the shape of the orbital (Figure 3.7).

For a given orbital size and shape (a subshell of orbitals, such as the 2*p* orbitals) there are $2l + 1$ possible orientations. There are $2l + 1$ orbitals in a given subshell (Section 1.3A).

Although not part of the orbital wave function, it also turns out that the electron, as it occupies a given orbital with a unique set of values of (n, l, m_l), can either go in with its spin up or down, depending on the value of its **spin angular momentum quantum number m_s**, which can be either $+\frac{1}{2}$ or $-\frac{1}{2}$. A filled orbital has two electrons, one with each value of m_s. Consequently, an *s* subshell is filled by two electrons, a *p* subshell by six electrons, a *d* subshell by 10 electrons, and an *f* subshell by 14 electrons. On building up the periodic table atom by atom across a period, it takes this many electrons, hence this many atoms, to complete the corresponding block of the periodic table.

Radial Part of the Electronic Wave Function. The relative size of an orbital (i.e., the most probable single radius at which the electron will be found) is a function of its *principal quantum number, n.* In addition to this major feature, orbitals also have structure (lobes and nodal surfaces) outward from the nucleus. This structure can be seen in the radial portion of the wave function for an electron, which passes from crests through zero levels to troughs and back again for most types of orbitals. There are many ways that we can plot the radial wave function. In Figure 9.4, for example, we plot what is known as the radial probability function. This involves squaring the original wave function, so that negative parts of the wave function (troughs) become positive wave amplitudes or probabilities of finding the electrons. We also multiply the squared function by $4\pi r^2$. The result gives the probability of finding the electron at *any* point (not just one point) in the spherical shell that is at a distance r from the nucleus.

Figure 9.4 shows that, in a given atom, the most probable distance from the nucleus for the electron increases with increasing values of the principal quantum number n. But it also shows that there are some differences for different shapes of orbitals. The 1*s*, 2*p*, 3*d*, 4*f*, … orbitals have the simplest radial probability functions, with one maximum and no radius above $r = 0$ at which there is zero probability of finding the electron. But 2*s*, 3*p*, 4*d*, … orbitals all have one such distance, which corresponds to a **nodal sphere**: a nodal surface that does *not* pass through the nucleus. On one side of this surface, the wave function has a positive sign (crest); on the other side, it has a negative sign (trough). The 3*s* and other orbitals have two nodal spheres; the 4*s* and others have three nodal spheres. In summary (Table 9.3), an orbital with a given principal quantum number n and a secondary quantum number l:

- has a total of $n - l - 1$ *nodal spheres* in its radial wave function, and

- has l *nodal planes* through the nucleus in its angular wave function.

- This gives an orbital a total of $n - 1$ **nodal surfaces**, which is the same number found earlier for the electron in the one-dimensional box.

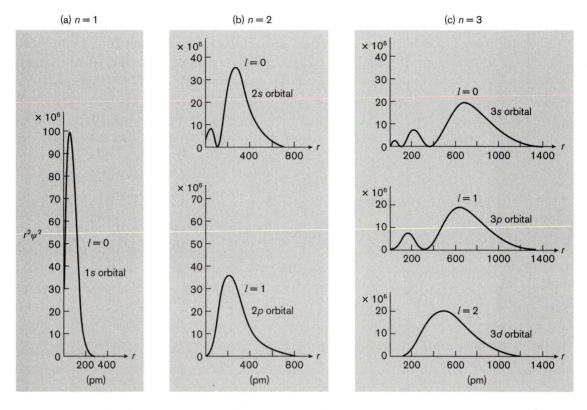

Figure 9.4. Radial probability functions (ψ^2) plotted against distance from the nucleus for $n = 1$, 2, and 3 for the hydrogen atom.

TABLE 9.3. Relationship of Numbers of Nodal Planes, Spheres, and Surfaces to Principal and Secondary Quantum Numbers

Number of nodal planes	l
(+) Number of nodal spheres	$n - l - 1$
(=) Number of nodal surfaces	$n - 1$

It is possible and useful to combine the radial and angular wave function probability plots to give a total picture of where in the atom the electron in a given orbital is likely to be found. Without a laser holograph it is difficult to do this in three dimensions, but we can draw cross-sections of orbitals, which is done for selected orbitals in Figure 9.5. These cross-sections also contain contours of equal probability of finding the electron, much like geological topographic maps. Note that nodal spheres are indicated by dashed circles, while nodal planes are indicated by dashed lines.

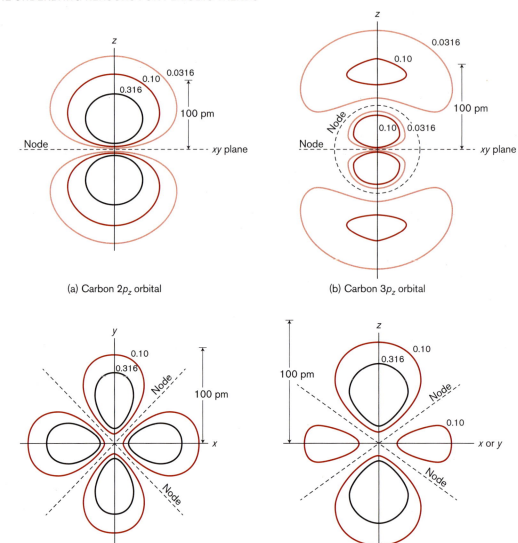

Figure 9.5. Electron-density contour maps for various orbitals. [Adapted from E. A. Ogryzlo and G. B. Porter, *J. Chem. Educ.* 40, 256 (1963).]

(a) Carbon $2p_z$ orbital

(b) Carbon $3p_z$ orbital

(c) Titanium $3d_{x^2-y^2}$ orbital

(d) Titanium $3d_{z^2}$ orbital

EXAMPLE 9.2

Sketch cross-sections in the *xy* plane (without density contours) of the following orbitals: (a) $4d_{x^2-y^2}$, (b) $5d_{xy}$, and (c) $5s$. Indicate nodal planes and spheres by dashed lines; show positive and negative signs of the wave function in the different lobes.

SOLUTION: It is best to start by counting up the number of nodal spheres (equal to $n - l - 1$) and nodal surfaces through the nucleus (equal to l). Then, draw these before attempting to fill in the actual lobes of the orbital. For the $4d$ orbital, there will be $4 - 2 - 1 = 1$ nodal sphere; draw a dashed circle out a distance from the nucleus. For the $5d$ orbital, there will be two concentric nodal spheres. The $5s$ orbital will have $5 - 0 - 1 = 4$ nodal spheres to be drawn concentrically

Example 9.2 continues on next page ▶

EXAMPLE 9.2 *(cont.)*

about the nucleus. There will be two nodal planes through the nucleus for the two *d* orbitals; draw these as dashed lines through the nucleus. For the $d_{x^2-y^2}$ orbital, the lobes fall along the *x* and *y* axes, so draw the nodal planes (lines) midway between the axes; the reverse is true for the d_{xy} orbital.

Now shade in the lobes, which will fall in between the nodal lines and circles; the last lobe(s) will fall outside the outer nodal sphere. The inner lobes in a cross-section will look approximately like circles; the outer ones can be drawn as shields or curved disks, or concentric circles.

Finally, fill in plus and minus signs to indicate the signs of the wave function. Start anywhere with either a + or − sign (the first choice is completely arbitrary), but thereafter any time you cross a nodal surface you must change signs. If this procedure results in any contradictions, a mistake has been made. Cross-section sketches in the *xy* plane of (a) the $4d_{x^2-y^2}$ orbital, (b) the $5d_{xy}$ orbital, and (c) the 5*s* orbital are shown below.

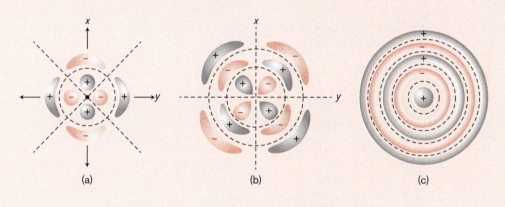

(a)　　　　　　　　　　　(b)　　　　　　　　　　　(c)

9.2. Shielding, Slater's Rules, and Core and Valence Electrons

OVERVIEW. In this section you build upon Section 1.3 to justify the distinction of core and valence electrons. You begin by considering the concept of penetration or shielding of two different orbitals on the same atom. You will find that this gives rise to an effective nuclear charge that differs from the actual nuclear charge (Section 9.2A; Exercises 9.15–9.20). You can use Slater's rules to calculate Z^* for specified orbitals of a given atom (Section 9.2B; Exercises 9.21–9.26). On going to the right across a period, **Z^*** for the valence electrons increases by +0.65 unit of charge per element in the *s* and *p* blocks, by +0.15 unit of charge per element in the *d* block, and is unchanged on crossing the *f* block. You will use Z^* values for different electrons of an atom to help justify their classification as core or valence electrons (Section 9.2C; Exercises 9.27–9.30).

Slater's rules are one of the few cases in which you make use of full electron configurations (Section 1.3A).

9.2A. Shielding and Penetration of Orbitals.

In the hydrogen atom (or other one-electron species), the energy with which the one electron in the orbital is held depends only on the principal quantum number:

$$E = -\frac{2\pi m Z^2 e^4}{n^2 h^2} = -\frac{1312 Z^2}{n^2} \text{ kJ mol}^{-1} \tag{9.11}$$

[This expression is somewhat more complex in Systeme International d'Unites (SI) units.] Although the one electron of such an atom or ion can be excited to any other orbital, in the normal *ground state* it is in the 1s orbital.

Equation 9.11 indicates that the energies of electrons in orbitals quickly become more negative with increasing nuclear charge (E is proportional to Z^2). This finding would appear to predict, for example, that the Li atom should be much harder to ionize than the H atom, whereas the converse is true. Evidently, the outermost (2s) electron of the Li atom does not experience the full attractive power of all three Li nuclear protons. Equation 9.11 does not consider the repulsion of the inner ($1s^2$) electrons on the outer ($2s^1$) Li electron, which makes that electron much easier to ionize.

According to classical electrostatics, if the 2s electron of Li is completely outside both the +3-charged nucleus and the two 1s electrons (–2 charge), it is the same as if it were outside a nucleus of +3 – 2 = +1 charge. Thus, we may say that the 2s Li electron feels an **effective nuclear charge** that is less than the true nuclear charge due to the **shielding** ability of the core 1s electrons. Therefore, in Equation 9.11, we might want to replace Z, the true nuclear charge, with an effective nuclear charge, Z^*, that is less than Z by the sum of the *screening constants S* of the inner orbitals:

$$Z^* = Z - \Sigma S \tag{9.12}$$

But the 2s electron of Li does not behave as if it were experiencing exactly a +1 charge either: Its properties are more consistent with those of an electron feeling a Z^* of +1.279. This difference can be explained by the fact that the 2s orbital has two maxima in its radial probability function (Figure 9.4) and by the fact that the lesser of these **penetrates** within the maximum of the inner 1s orbital. Hence, although most of the time the 2s electron is in the outer lobe of that orbital, feeling a Z^* of +1, for some of the time it is inside of the 1s orbital, experiencing the full nuclear charge of +3. This penetration of the inner orbital affects the chemical properties of that valence Li electron: It will affect the ease of ionization of that electron, the size of its orbital, the electronegativity of the atom, and so on.

Equation 9.11 contains only the *principal* quantum number as a variable, and thus implies that the third electron of Li is equally stable in the 2s and the 2p orbitals. But, in fact, when two or more electrons are present, the electrons repel each other, and they repel each other differently, depending on the structures of the orbitals in which they are located. The 2p orbitals of Li lack the penetrating inner lobes that the 2s orbital has. Consequently, the valence electron of Li (and other atoms) prefers the 2s orbital, and fills it first. Likewise, we can see (from Figure 9.4c) the superior penetrating power of the 3s orbital as compared to the 3p orbital, which is better than the 3d. Thus, electrons in orbitals of common n but increasing l have progressively poorer *penetration* of core electrons, so the orbitals of increasing l are filled later.

Once the 3s, 3p, and 3d orbitals are filled, they have differing abilities to shield outer electrons (such as 4s) from the charge of the nucleus. It is easier for these outer electrons to penetrate the electrons of the 3d orbitals, which have little likelihood of being near the nucleus, than to penetrate the 3p or especially the 3s orbitals. Consequently, electrons in orbitals of common n but increasing l exhibit progressively poorer *shielding* of outer electrons and should have smaller shielding constants for use in Equation 9.12.

We can rationalize the relative penetrating and shielding properties of orbitals of the *same shape* (same value of l) but *different sizes* (different values of n) by referring to Figure 9.4. Viewing this figure across the top (so as to see only s orbitals), note that the 1s orbital is much closer to the nucleus than the 2s, so the 1s shields other orbitals (such as 3s) from the nucleus more effectively than does the 2s. Shielding and penetration are better (for a common shape of orbital) for lower values of the principal quantum number.

EXAMPLE 9.3

Consider two orbitals, A and B, in an atom. Orbital A has one nodal plane and two nodal spheres, whereas orbital B has two nodal planes and one nodal sphere. (a) Identify the orbital types of A and B. (b) Which one is more effective at shielding electrons with higher principal quantum numbers from the nucleus? (c) Orbital A is better at shielding which kind of electrons from the nucleus: 5s or 5f?

SOLUTION: (a) Each orbital has three nodal surfaces, so the principal quantum number of each is 4. Orbital A has one nodal plane, so it is a 4p orbital. Orbital B has two nodal planes, so it is a 4d orbital. (b) Orbital A has more inner lobes (two inside the two nodal spheres) than orbital B (one inside the one nodal sphere), so it is more effective at shielding outer orbitals. (c) 5s orbitals have more inner lobes (four) than 5f (one), so they are harder to shield from the nucleus (they penetrate better). 5f is more easily shielded.

9.2B. Slater's Rules. It is useful to have some simple rules for estimating the degree to which electrons in the various types of orbitals shield other electrons from the nucleus, and hence for estimating the Z^* values experienced by other electrons. Based on calculations done in 1930 and earlier, Slater[1] proposed some simplified rules for making such estimations. Although more modern calculations, such as those of Clementi and Raimondi,[2] give more accurate values of Z^*, Slater's rules are adequate for predicting most periodic *trends* in Z^*. These allow us to predict periodic *trends* in the properties we use most to predict reaction chemistry in this book—namely, electron configurations, radii, electronegativities, and so on. The fact that Slater's calculations can be stated as rules allows us to see *why* these trends exist, in terms of the different penetrating and shielding properties of different kinds of orbitals. Slater's seven rules are outlined in Figure 9.6.

Slater's rules give different results for electrons in different types of orbitals. Therefore, we need to note which orbital type (subshell) is specified. (When we want the Z^* felt by the *valence s* and *p* electrons of an atom, we will put the symbol Z^* in **boldface**.) Rules 2–5 compute the shielding constants of the other electrons for the particular electron in question; Rule 6 simply applies Equation 9.12 to compute the Z^* for the specified electron. Rule 7 is a completely separate calculation of an effective principal quantum number n^* for the electron in question, which is needed for some applications in Sections 9.3 and 9.4.

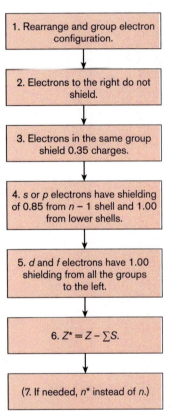

1. Rearrange and group electron configuration.

2. Electrons to the right do not shield.

3. Electrons in the same group shield 0.35 charges.

4. *s* or *p* electrons have shielding of 0.85 from *n* − 1 shell and 1.00 from lower shells.

5. *d* and *f* electrons have 1.00 shielding from all the groups to the left.

6. $Z^* = Z - \sum S$.

(7. If needed, n^* instead of n.)

Figure 9.6. Flow chart summarizing Slater's rules.

1. After writing out the full electron configuration in the way you learned in Section 1.3A, rearrange the electron configuration of the element in the following groupings and order. Group *ns* and *np* orbitals together (their shielding properties are roughly equal); group all other types of orbitals separately (their shielding properties are dissimilar). The order is numerical:

$$(1s)(2s,2p)(3s,3p)(3d)(4s,4p)(4d)(4f)(5s,5p)\dots$$

2. Electrons to the right of the group of electrons in question contribute (in theory) nothing to the shielding of that group of electrons.

3. All other electrons in the same group (enclosed in the same parentheses) as the electron in question shield that electron to an extent of 0.35 units of nuclear charge each.

4. If the electron in question is an *s* or *p* electron: (a) All electrons with principal quantum number one less than the electron in question shield it to an extent of 0.85 units of nuclear charge each. (b) All electrons with principal quantum number two or more less than the electron in question (in theory) shield it completely (i.e., to an extent of 1.00 unit of nuclear charge each).

5. If the electron in question is a *d* or *f* electron: All electrons to the *left* of the group of the electron in question shield the *d* or *f* electron completely (i.e., to an extent of 1.00 unit of nuclear charge each). This is a manifestation of the poor penetrating power of the *d* or *f* electron.

6. Sum the shielding constants from Steps 2–5, and subtract them from the true nuclear charge Z of the atom in question (Eq. 9.12) to obtain the Z^* felt by the electron in question.

7. Values of n^* are not the same as the true principal quantum number for higher values of n, as shown in Table 9.4.

TABLE 9.4. Values of the Effective Quantum Number n^*

n	1	2	3	4	5	6
n^*	1	2	3	3.7	4.0	4.2

EXAMPLE 9.4

Calculate Z^* for an electron in each type of occupied orbital in Sc ($Z = 21$).

SOLUTION: In Step 1, we rewrite the full electron configuration of Sc in the order $(1s)^2(2s,2p)^8(3s,3p)^8(3d)^1(4s)^2$. There are five different groups of occupied orbitals, so the question calls for five different values of Z^* to be calculated.

(a) For any $1s$ electron, the *only* shielding is from the one other $1s$ electron, which by Rule 3 shields out 0.35 unit of nuclear charge. Hence, by Rule 6 (Eq. 9.12), $Z^*_{1s} = 21 - 0.35 = +20.65$.

(b) For any $2s$ and $2p$ electron (treated the same by Slater's rules), we have two sources of shielding. The seven other $2s$ and $2p$ orbitals contribute shielding of $7 \times 0.35 = 2.45$ units of nuclear charge (Rule 3); the two $1s$ electrons contribute shielding of $2 \times 0.85 = 1.70$ units of nuclear charge (Rule 4a). By Rule 6, $Z^*_{2s,2p} = 21 - (2.45 + 1.70) = +16.85$.

(c) For each $3s$ and $3p$ electron, there are three sources of shielding: the seven other $3s$ and $3p$ electrons, which contribute shielding of 2.45 units (Rule 3); the eight $2s$ and $2p$ electrons, which contribute $8 \times 0.85 = 6.80$ units (Rule 4a); and the two $1s$ electrons, which shield completely—that is, 2.00 units of nuclear charge (Rule 4b). By Rule 6, $Z^*_{3s,3p} = 21 - (2.45 + 6.80 + 2.00) = +9.75$.

(d) For the $3d$ electron, which has no other $3d$ companions to which Rule 3 would apply, only Rule 5 need be applied: all 18 electrons inside of it shield by 1.00 unit each, so $Z^*_{3d} = 21 - 18.00 = +3.00$.

(e) For each $4s$ electron, Rule 3 shows one companion $4s$ electron shielding 0.35 units of nuclear charge, Rule 4a shows nine of principal quantum number 3 ($3s$, $3p$, $3d$) shielding $9 \times 0.85 = 7.65$ units, and Rule 4b shows 10 electrons of principal quantum number 2 or below ($1s$, $2s$, $2p$) shielding 10.00 units of nuclear charge. Hence, $\mathbf{Z^*_{4s}} = 21 - (0.35 + 7.65 + 10.00) = +3.00$.

9.2C. Periodic Trends in Z^*; Core and Valence Electrons. The calculations of Example 9.4 indicate that two types of orbitals in Sc, the $4s$ and $3d$, can be distinguished from all of the other types by the relatively low Z^* that they feel. When appropriate calculations of the type suggested by Equation 9.11 are made, it is confirmed that the $4s$ and $3d$ orbitals are nowhere near as stable as the $1s$ through $3p$. This justifies our dividing the electrons and orbitals into two classes:

1. The low-energy, more stable, strongly bound (high Z^*) **core electrons** in core orbitals are too strongly bound to be affected by the kinds of energy exchanged in chemical reactions.

2. The higher energy, loosely bound (low Z^*) **valence electrons** (and orbitals) are chemically active. **Postvalence orbitals** are normally too high in energy to hold electrons. *Chemical change is concentrated in the electrons occupying the valence orbitals.*

TABLE 9.5. Z^* **Values for Different Electrons in Sc and Ga**

	Z	1s	2s,2p	3s,3p	3d	4s,4p
Sc	21	+20.65	+16.85	+9.75	+3.00	+3.00
Ga	31	+30.65	+26.85	+19.75	+9.85	+5.00

It is instructive to repeat these calculations for the element with 10 more units of nuclear charge than Sc (e.g., Ga). The electron configuration written in the style of Rule 1 is $(1s)^2(2s,2p)^8(3s,3p)^8(3d)^{10}(4s,4p)^3$. Repetition of the calculations for each type of electron gives the following results (Table 9.5):

- For 1s, the shielding is unchanged from Sc, so Z^* increases by the 10 additional protons (Z^*_{1s} = +30.65).

- The shielding is also unchanged for (2s,2p) and (3s,3p) electrons, so Z^*_{2sp} = +26.85 and $Z^*_{3s,3p}$ = +19.75.

- However, the shielding does change for the remaining electrons. Each addition of a 3d electron is accompanied by the addition of another unit of nuclear charge, from which the first 3d electron is 65% shielded (Rule 3). After adding 10 protons in going from Sc to Ga, Z^*_{3d} increases to +9.85.

- For (4s,4p) valence electrons, the shielding has increased quite a bit, because the (4s,4p) electrons are 85% shielded by each of the 3d electrons that have been added (Rule 4a). $Z^*_{4s,4p}$ for Ga is +5.00, only two units higher than in Sc.

The 4s and 4p electrons now stand alone at relatively low Z^* values; these are the only valence electrons in Ga. The 3d electrons, which were valence electrons with low Z^* in Sc in the d block, now have relatively high Z^* and are core electrons in the p block in Ga.

The gradual plunge of the 3d orbitals into the (pseudo-noble-gas) core can be seen from the results of calculations across the d block[3] (Figure 9.7). This is because the 3d orbitals that fill in the d block do not do as good a job of shielding each other (Rule 3) as they do of shielding outer 4s or 4p electrons (Rule 4a). Eventually the Z^* values of 3d orbitals become too large for them to function as valence orbitals. The better-shielded 4s and 4p orbitals remain valence orbitals through the rest of the period.

This conclusion matches the statement from Section 1.3B that only the last-filled s orbital remains a valence orbital outside of its block, the s block—it remains a valence orbital across the entire period before becoming a core orbital in the next period. All other less-penetrating types of orbitals are shielded out by the end of the block in which they fill, so they remain valence orbitals only in their characteristic block of the periodic table. Hence, as introduced in Section 1.3B, a given element has *two* types of valence electrons: those in the *last-filled ns orbital* and those in the *last-filled orbital characteristic of the block*. These consist of *np* in the p block, $(n-1)d$ in the d block, and $(n-2)f$ in the f block (Table 1.7).

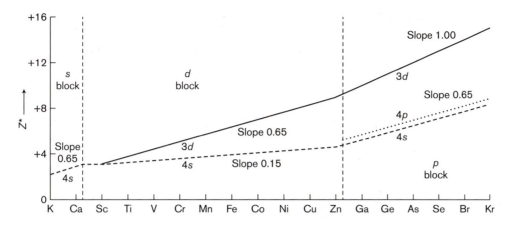

Figure 9.7.
The calculated effective nuclear charges (Z^*) for the 3d, 4s, and 4p electrons of the early d-block elements.

Horizontal Periodic Trends in Z of the Outermost Valence Electrons. Many applications of these calculations specifically require the Z^* value for the valence electron of highest principal quantum number—that is, for the ns or np valence electron. Because we will be interested in periodic trends in this property, it is useful to calculate variations in Z^* across the different blocks and down the different groups of the periodic table.

EXAMPLE 9.5

Calculate Z^* for the last valence electron in each atom of the second period.

SOLUTION: In Step 1, we can write the electron configurations of these atoms in the general form $(1s)^2(2s,2p)^n$, in which n is the number of valence electrons (and also the group number). Step 2 does not apply. In Step 3, we obtain a total shielding constant of $(n - 1) \times 0.35 = -0.35 + 0.35n$ for the 2s and 2p electrons. In Step 4, we obtain $2 \times 0.85 = 1.70$ for shielding by the two 1s electrons. Step 5 does not apply. In Step 6, we add the shielding constants to obtain $1.35 + 0.35n$. This is subtracted from the atomic number, $Z = (2 + n)$, to obtain $Z^*_{2s,2p} = 0.65 + 0.65n$.

From Figure 9.8, we can see that Z^* *increases on going to the right*:

- rather rapidly across the *s and p blocks*, by +0.65 unit of charge per element.

- more slowly on crossing the *d block* (0.15 units of nuclear charge per element). This occurs because the $(n - 1)d$ electrons being added across the d block are basically inside the ns electrons, and hence do a better job of shielding the outer ns electrons than can the np electrons in the p block.

- The computation of Z^* for the outermost s orbital of the f-block elements produces a very interesting result: The value of Z^* is 2.85 regardless of how many f electrons are present in the element! This result is a consequence of the fact that the $(n - 2)f$ orbitals are *two* shells below the s electrons. Hence, according to Rule 4b, the deeply buried $(n - 2)f$ electrons completely shield the ns electrons.

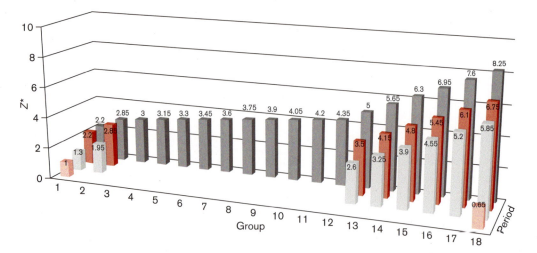

Figure 9.8. Values of Z^* for the last valence (ns or np) electron for atoms of the first four periods of the periodic table. Period 1 is in light brown, Period 2 is in gray, Period 3 is in **brown**, and Period 4 is in black. (Values in Periods 5, 6, and 7 match those in Period 4.)

Vertical Periodic Trends in Z^*. The calculation of Z^* for the outermost s electrons in a vertical group of atoms by Slater's rules also gives interesting results (Figure 9.8). For Group 2, we obtain Z^*(Be) = 1.95, Z^*(Mg) = 2.85, Z^*(Ca) = 2.85, Z^*(Sr) = 2.85, and Z^*(Ba) = 2.85. We see that, in the s block, the Z^* values are unchanged from the third period down, which contributes to the similarity in chemical properties of the members of a group of elements.

If we carry out the same calculations for Group 13/III, the trends are somewhat different. Here we obtain Z^*(B) = 2.60, Z^*(Al) = 3.50, Z^*(Ga) = 5.00, Z^*(In) = 5.00, and Z^*(Tl) = 5.00. The values for B and Al have increased by +0.65 from those of Be and Mg in Group 2, the normal trend in the p block. However, the values for Ga, In, and Tl

---

EXAMPLE 9.6

Oganesson (Og, element 118) presumably has the electron configuration of the preceding noble gas, Rn, plus the following: $7s^2 5f^{14} 6d^{10} 7p^6$. (a)–(d) Use Slater's rules to calculate Z^* for each of these four types of electrons. (e) Do your results justify the classification of two of these subshells of electrons as core electrons and two as valence electrons? If so, which two are core?

SOLUTION: The rearranged electron configuration of Og is
$1s^2 (2s,2p)^8 (3s,3p)^8 3d^{10} (4s,4p)^8 4d^{10} 4f^{14} (5s,5p)^8 5d^{10} 5f^{14} (6s,6p)^8 6d^{10} (7s,7p)^8$.
(a) The shielding for $7s^2$ is $(7 \times 0.35) + (18 \times 0.85) + (92 \times 1.00) = 109.75$, so Z^*_{7s} is $118 - 109.75 =$ +8.25. (d) The shielding and Z^* for $7p^6$ are the same as for $7s^2$. (b) The shielding for $5f^{14}$ is (Rule 2) 26×0 + (Rule 3) 13×0.35 + (Rule 5) $78 \times 1 = +82.55$. So, $Z^* = 118 - 82.55 = +35.45$.
(c) The shielding for $6d^{10}$ is (Rule 2) 8×0 + (Rule 3) 9×0.35 + (Rule 5) $100 \times 1 = 103.15$.
 So $Z^* = 118 - 103.15 = +14.85$. (e) The two electron types of lowest $Z^* = +8.25$ are the $7s$ and $7p$, so these are the valence orbitals. The other two electron types, $6d$ and $5f$, have much higher Z^* values of +14.85 and +35.45, respectively, so these are the core electrons.

have increased by +2.15 from those of their Group 2 analogues. The difference is that, in the meanwhile, d orbitals have been filling. Even though $Z^\star$ for outer s electrons increases only slowly across the d block, it does increase 10 times. Consequently, the $Z^\star$ values of the elements beyond the d block—Ga, In, Tl—are 0.65 *plus* 1.50 higher than the corresponding elements in Group 2. This produces an extra jump in $Z^\star$ between Al and Ga, which was not present in Group 2 between Mg and Ca.

This increase in $Z^\star$ occurs among the elements Ga, Ge, As, Se, Br, and Kr, which immediately follow the first filling of the d orbitals. Likewise, the fact that $Z^\star$ of Mg is higher than that of Be is due to Mg following the first filling of the p orbitals. And if Rule 4 were less approximate, we would also find an increase in $Z^\star$ in Tl, which follows the first filling of the f orbitals. We shall return to this point in the next sections.

Order of Energies of Electrons in Orbitals

Equation 9.11 was derived for a one-electron atom and shows that, even in this simple case, the energy of the electron becomes more negative (lower) with an increase in (effective) nuclear charge, and becomes less negative (increases) with increases in the principal quantum number. As a result of poorer penetration, the energies of electrons in multielectron atoms also become less negative (i.e., increase) with increases in their secondary quantum numbers. Consequently, there can be a complex analysis of the familiar but seemingly irregular *aufbau order of filling* of the orbitals in *neutral* atoms, which you will recall is the order in which the orbitals are encountered in the periodic table as one scans it from low atomic number to high atomic number (Table 1.7).

We saw in Example 9.4 and Figure 9.7 that $3d$ and $4s$ electrons have equal $Z^\star$ in Sc (+3.00), but $n^\star$ values are not the same (Table 9.4, 3.0 vs. 3.7). The preferred ground state electron configuration is the one in which the energies of *all* electrons are minimized, not just the energies of the valence electrons.[4] Detailed calculations are necessary to determine the ground state valence electron configurations in cases such as this. These detailed calculations explain why the $4s$ orbital of neutral K fills first even though the $3d$ orbital is lower in energy.[5]

In Section 1.3A we mentioned that many (21) atoms of the d and f blocks have *exceptional electron configurations*. (For example, many of the neutral atoms early in the f block have one or two electrons in the subsequent d orbital.) There is little predictability to these anomalies; fortunately, they apply only to isolated gaseous atoms, which we seldom encounter.

There is a similar change in electron configurations when an atom ceases being neutral and becomes positively charged: the late-filling $(n-1)d$ or $(n-2)f$ orbitals, under the influence of the extra positive charge, drop in energy relative to the ns or np orbital. Consequently, the order of filling in positive ions of the d and f blocks is more hydrogen-like, in that the $(n-1)d$ or $(n-2)f$ orbitals are occupied in preference to the ns orbital (Section 1.3C). There are therefore no anomalous electron configurations among the chemically important positive ions.

9.3. Periodic Trends in Ionization Energies and Electron Affinities

OVERVIEW. In Section 6.6 and Tables 6.5–6.8 you noted the main horizontal and vertical periodic trends in ionization energies and electron affinities. There are situations in which the main trend is not followed—for example, after a subshell has been half-filled. You may practice predicting first ionization energies from Slater's rules, and explaining these trends and exceptions, by trying Exercises 9.31–9.36.

In Section 6.6 we introduced first, second, and third ionization energies (Tables 6.6–6.8) and electron affinities/zeroth ionization energies (Table 6.9), which were useful in predicting and explaining the enthalpies of formation of ionic compounds (Section 6.6A) and the activity series of elements (Section 6.6B). There are attractive features about relating chemical reactivity trends to ionization energies. (1) Ionization energies can be measured with good precision. (2) They can be calculated from Z^* values, such as those developed from Slater's rules, or more accurately by other modern computational methods. (3) If derived from Slater's rules, they can give us insight into the reasons for their periodic trends and the anomalies in the trends.

The basic equation for the energy of an electron in an atom or ion is Equation 9.11. This equation gives generally very poor results, especially for elements of high Z and n. Slater's rules replace Z with Z^*, giving improvement. For further improvement, Slater's Rule 7 (Table 9.5) replaces n with n^* for higher values of n:

$$E_n = -1312(Z^*/n^*)^2 \text{ kJ mol}^{-1} \tag{9.13}$$

AN AMPLIFICATION

Calculations of Ionization Energies

To obtain an ionization energy in a multielectron atom, it is necessary to apply Equation 9.13 to *each* electron in the atom being ionized, then to *each* of the n electrons in the ion produced, and then take the difference:[6]

$$\text{IE} = \sum_1^n En(\text{ionized species}) - \sum_1^n En(\text{un-ionized species}) \tag{9.14}$$

In practice, for *most* cases in which only valence electrons are being ionized, Z^* values for core electrons are the same in the atom and the ion, so only valence electron energies need to be considered. Even so, the calculations are detailed. For example, we can calculate the first ionization energies of (a) Li, (b) Be, and (c) F.

(a) The electron configuration of Li is $1s^2 2s^1$; that of the product of ionization, Li$^+$, is $1s^2$. The Z^* value for each $1s$ electron in either Li or Li$^+$ is $3 - (0.35) = +2.65$, so their energy term, $-1312(2.65/1.00)^2$, cancels out on applying Equation 9.14. The Z^* value for the $2s$ electron of Li is $+1.30$ (Example 9.6), so the ionization energy of Li is equal to $1312(1.30/2)^2$ = 554 kJ mol^{-1}, as compared to the experimental value, 520 kJ mol^{-1}, in Table 6.5.

(b) The electron configuration of Be is $1s^2 2s^2$; that of the product of ionization, Be$^+$, is $1s^2 2s^1$. When one valence $2s$ electron of Be is ionized, this changes the Z^* value for the other $2s$ electron. We must therefore do separate calculations for the valence electrons of

Be $(2s^2)$ and Be$^+(2s^1)$. The Z^* value for the 2s electron of Be$^+$ is $4 - 2(0.85) = +2.30$, while the value for each of the two 2s electrons of Be is $4 - [0.35 + 2(0.85)] = +1.95$. The ionization energy of Be is equal to $-1312(2.30/2)^2 + (1312)(2)(1.95/2)^2 = 1312[(-1.3225 + (2)(0.9506)] = 1312(0.5787) = 759$ kJ mol^{-1}, as compared to 899 kJ mol^{-1} in Table 6.5.

(c) The electron configuration of F is $1s^2(2s,2p)^7$; that of its ionization product, F$^+$, is $1s^2(2s,2p)^6$. Again ionization results in no change in the Z^* felt by the core 1s electron, so the ionization energy depends only on the changes felt by the valence electrons. The Z^* parameter for each of the seven $(2s,2p)$ electrons in F is $+5.20$, while Z^* for each of the six $(2s,2p)$ electrons in F$^+$ is $+5.55$. Applying Equation 9.14 to these electrons and subtracting the energies for F from the energies for F$^+$, we obtain: IE(1) $= -1312[6(5.55/2)^2 - 7(5.2/2)^2] = -60,600 + 62,100 = 1500$ kJ mol^{-1}, as compared to the experimental value of 1681 kJ mol^{-1}.

The results using Equation 9.14 (and in particular using n^*) still leave much to be desired.[7] However, Equation 9.13 suggests: (1) *Ionization energies should increase as $(Z^*)^2$ increases,* as is found in the Coulombic relationship (Section 2.3A). (2) *Ionization energies should decrease as the principal quantum number of the orbital increases.*

9.3A. Horizontal Periodic Trends. We generally find that a given ionization energy (including the zeroth) becomes more difficult across a block of a period as the numbers of valence electrons of the block type increase, and as Z or Z^* increases. There is a drop on crossing from the s block to the p block, because (contrary to the simplifications of Slater's rules) the Z^* values felt by ns and np electrons are not precisely the same. In the p block of Period 2, the first ionization energies (Table 6.5 and Figure 9.9) increase from B ($2p^1$, 801 kJ mol^{-1}) to Ne ($2p^6$, 2081 kJ mol^{-1}).

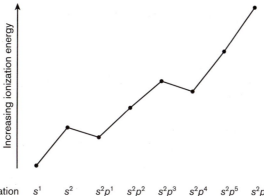

Valance electron configuration	s^1	s^2	s^2p^1	s^2p^2	s^2p^3	s^2p^4	s^2p^5	s^2p^6
Examples	He$^-$	Li$^-$	Be$^-$	B$^-$	C$^-$	N$^-$	O$^-$	F$^-$
	Li	Be	B	C	N	O	F	Ne
	Be$^+$	B$^+$	C$^+$	N$^+$	O$^+$	F$^+$	Ne$^+$	Na$^+$
	B^{2+}	C^{2+}	N^{2+}	O^{2+}	F^{2+}	Ne^{2+}	Na^{2+}	Mg^{2+}

Figure 9.9. Horizontal periodic trends in ionization energies. The examples listed are across Period 2, among the zeroth, first, second, and third ionization energies, respectively.

In the *d* and *f* blocks, *d* and *f* electrons are ionized third, after the two valence *s* electrons. Hence, to see their general trends, we look at third ionization energies (Table 6.7). We note the increases from Sc^{2+} ($3d^1$, 2389 kJ mol^{-1}) to Zn^{2+} ($3d^{10}$, 3833 kJ mol^{-1}), and from La^{2+} ($4f^1$, 1850 kJ mol^{-1}) to Yb^{2+} ($4f^{14}$, 2415 kJ mol^{-1}).

When the electron being removed first comes from a more than half-filled sub-shell, ionization is also somewhat easier, because an electron in a filled orbital is re-pelled by the other electron occupying the same orbital (this is related to the pairing energy; Section 7.2A). Thus, the first IE of O ($2s^2 2p^4$, with one paired-up *p* electron) is 1314 kJ mol^{-1}, somewhat closer to zero than the first IE of N ($2s^2 2p^3$, 1402 kJ mol^{-1}, with each *p* electron in its own orbital). Likewise in the *d* and *f* blocks, the third ionization energies drop rather than rise on going from d^5 Mn^{2+} to d^6 Fe^{2+}, and on going from f^7 Eu^{2+} to f^8 Gd^{2+}.

The same effects can be found for other ionization energies of, for example, ions using 2*s* and 2*p* valence electrons. Thus, in general, in the second period ionization is easiest for the $2s^1$ electron configuration [IE(1) of Li, IE(2) of Be$^+$, IE(3) of B^{2+}, IE(0) of He$^-$] (Figure 9.9). In general, the nth ionization becomes more difficult farther to the right for atoms or ions of the same charge. The highest ionization energy is required for atoms or ions of the $2s^2 2p^6$ electron configuration [IE(1) of Ne, IE(2) of Na$^+$, IE(3) of Mg^{2+}, IE(0) of F$^-$]. Again, there are drops on going from the *s* block to the *p* block, and going from the $2s^2 2p^3$ to the $2s^2 2p^4$ electron configuration.

9.3B. Vertical Periodic Trends. Generally speaking, we find that *the nth ionization energy or electron affinity decreases down a group* (Tables 6.5–6.8): Toward the bottom of a group, Z^* is unchanged, but n^* increases. Thus, the trend is because the electron being ionized is in a larger orbital of higher principal quantum number, and hence is not attracted as strongly to the nucleus.

AN AMPLIFICATION

Irregularities in Vertical Trends

(1) The electron affinities of the *p*-block elements of the second period (Table 6.8) are unexpectedly less than those of the third (or later) periods because of the greater electron–electron repulsions found in such small anions as F$^-$ and O$^-$ (Table 3.2 and the Amplification in Section 3.3C). (2) The ionization energies and electron affinities of some of the later *d*-block elements of the sixth period and early *f*-block elements of the seventh period are higher than expected, because of the scandide and lanthanide contractions, and the relativistic effects discussed in Section 9.6A.

9.3C. Successive Ionization Energies. Not all electrons from a set of orbitals ionize at the same energy. As detailed in the Amplification earlier in this section, the energy of ionization depends on the number of other electrons ionized from the same subshell and the number of pairing energies. Hence, for a given element, successive ionizations generally become more difficult in a relatively regular pattern. Equations 9.13 and 9.14 indicate that ionization energies are proportional to $(Z^*)^2$, so they rise very rapidly.

Taking the square root of successive ionization energies gives a more nearly linear relationship, as seen in Figure 9.10.[8] (In this figure the first ionization energy is lowest on the graph, with the second ionization energy next above the first, and so on.)

If, however, the nth ionization is one that begins removing core electrons from an atom, that ionization is much more difficult and gives (1) the very large gaps in Figure 9.10 between $1s$ and $2s$ orbitals, or between $2p$ and $3s$ orbitals, and (2) the smaller gaps between pseudo-noble-gas core $3d$ and valence $4s$ and $4p$ electrons in elements 31–36 (in the p block). (3) In addition, there are smaller gaps between the energies of ionizations of different subshells of the same principal quantum number (e.g., $3s < 3p < 3d$).

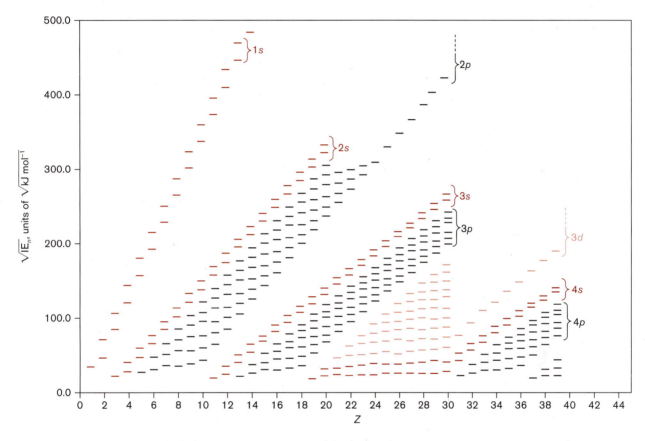

Figure 9.10. Square roots of the known ionization energies of the first 39 elements. First ionization energies are shown at the bottom for each element of atomic number Z, followed by second ionization energies, and so on. **Brown** lines represent ionization from s orbitals, black lines represent ionization from p orbitals, and light brown lines represent ionization from d oribitals. [Adapted from P. Mirone, *J. Chem. Educ.* 68, 132 (1991).]

EXAMPLE 9.7

(a) Which elements have (the square roots of) their seventh ionization energies shown in Figure 9.10?
(b) Indicate how these seventh ionization energies change from one element to the next through the end of Period 3 by inserting connectors between the element symbols: ≈ for about the same as, < for less than, << for a lot less than, > for greater than, or >> for a lot greater than. (c) Whenever you use ≈, <<, or >>, explain why this is so, in terms of the orbital types being ionized.

Example 9.7 continues on next page ▶

EXAMPLE 9.7 *(cont.)*

SOLUTION: (a) The first element to have seven electrons is N ($Z = 7$). Seven ionization energies are indicated for all subsequent elements through Zn ($Z = 30$). (There are also seven values indicated for elements 34 through 39, but we will ignore those.)

(b) For each element we must count up to the seventh level (as shown below). For N the level is at about $250\sqrt{kJ\ mol^{-1}}$, and for O it is at about $265\sqrt{kJ\ mol^{-1}}$. Then it drops precipitously to about $140\sqrt{kJ\ mol^{-1}}$ for F. So, the series reads N < O >> F < Ne ≈ Na < Mg < Al ≈ Si < P < S >> Cl < Ar.

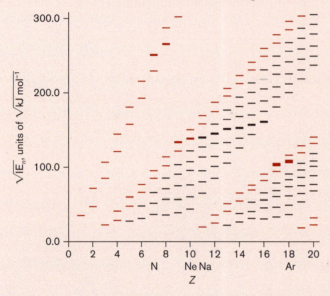

(c) Between O and F the ionization changes from the ionization of the low-energy $1s$ electron to the higher-energy $2s$ electron, so the IE required is much less. The change between S and Cl is similar, involving $2p$ versus $3s$ electrons. Between Ne and Na the ionization changes from that of slightly lower-energy $2s$ to slightly higher-energy $2p$ electrons, which roughly cancels out the expected increase in IE due to the increase in atomic number. Between Al and Si the ionization becomes that of a paired $2p$ electron, so the increase in IE expected due to the increase in Z is roughly cancelled out by loss due to relief of pairing energy upon ionization (this is the small change in Figure 9.9 between the s^2p^3 and s^2p^4 electron configurations).

9.4. Calculations of Atomic Radii and Reasons for Their Periodic Trends

OVERVIEW. You can use Z^* to calculate the radius $<r_{max}>$ for a given atom. You can use trends in $<r_{max}>$ values to explain periodic trends in the experimental r_{cov} atomic radii from Section 1.5B and Table 1.13. With these values of $<r_{max}>$ and with the more refined $<r_{max}>$ values of Desclaux (which you will be given as needed), you can explain the lanthanide and scandide contractions discussed earlier in Section 1.5B. You may practice these concepts by trying Exercises 9.37–9.44.

In Table 1.13 we presented the covalent radii of the elements; we discussed their main periodic trends and the lanthanide and scandide contraction countertrends in Section 1.5B. To find underlying reasons for these periodic trends, we need a model from which to calculate radii that resemble covalent radii. The old Bohr model of the atom, which puts electrons in orbits, allows an easy estimation of $<r_{max}>$, the expected distance from the nucleus to the maximum in the radial probability function for the atom (Figure 9.4):

$$<r_{max}> = a_0 \frac{(n^*)^2}{Z^*} \tag{9.15}$$

In this equation $a_0 = 52.9$ pm is the radius of the Bohr hydrogen atom. We could, for example, apply Equation 9.15 to estimate that a hydrogen atom with its one electron in a 252s orbital should have a radius of 0.00339 mm and be visible under an ordinary microscope. Absurd as this prediction sounds, radio astronomers have detected, in interstellar gas regions, hydrogen atoms with this excited-state electron configuration![9]

The calculation for $<r_{max}>$ of Li, for example, gives $<r_{max}> = (52.9)(2^2/1.30) = 162$ pm. Because $<r_{max}>$ is not exactly the same as any of the types of experimental radii we have examined, it is not surprising that our result differs from the covalent radius of Li, 134 pm (Table 1.13). The usefulness of $<r_{max}>$ calculations lies in seeing whether we can duplicate the periodic *trends* in covalent radii, then explain those trends in terms of n^* and Z^*.

9.4A. Horizontal Periodic Trends (Table 9.6).

Atoms of the second period are calculated to contract across the s and p blocks at an average rate of about 10% per element, which is approximately what r_{cov} does. According to Equation 9.15, the radius *decreases* because of the *increase* in Z^*. This suggests that atoms decrease in size across the s and p blocks because additional s and p electrons of the same principal quantum number do a poor job of shielding each other (Rule 3). The additional protons in each nucleus across the period pull in *all* the valence electrons more strongly.

The Period 4 series of Table 9.6 shows a calculation across the d block. Atoms are calculated to decrease in size here by an average of 3% per element, which is what r_{cov} values do. The smaller change is due to the smaller increases per element in Z^* across the d than the p block.

TABLE 9.6. Horizontal Periodic Trends in Atomic Radii[a]

Period 2 Element	Li	Be	B	C	N	O	F	Ne		
$<r_{max}>$	162	108	81	63	54	46	41	36		
r_{cov}	128	96	84	76	71	66	57			
Period 4 Element	Sc	Ti	V	Cr	Mn	Fe	Co	Ni	Cu	Zn
$<r_{max}>$	241	230	219	210	201	193	186	179	172	166
r_{cov}	170	160	153	139	139	132	126	124	132	122

[a] Covalent radii are taken from Table 1.13.

A calculation not shown in Table 9.6 is that for crossing the *f* block, for the simple reason that neither $n^\star$ nor $Z^\star$ changes, so $\langle r_{max} \rangle$ is 327 pm for the whole series of elements La through Yb. In fact, the covalent radii of these elements (Table 1.13) decrease slowly from 207 pm for La to 190 pm at Tm (a rate of about 0.6% per element). This finding suggests that there really is some increase in $Z^\star$ across this series, and that, contrary to Rule 4b of Slater's rules, the *6s* orbital really does penetrate the *4f* orbital to some extent.

9.4B. Vertical Periodic Trends (Table 9.7). Atoms of Group 2 are calculated to *increase in size down the group*, which is in fact observed (Figure 9.11). From the calculations, this trend must simply be a consequence of increasing the (effective) principal quantum number of the valence orbital: *higher n or n* means a larger orbital.*

Near the beginning of the *p* block the atoms again increase in size down a group, but there is a smaller-than-expected increase in $\langle r_{max} \rangle$ between Si and Ge (Table 9.7; Figure 9.12) (or, in Group 13/III, between Al and Ga). This smaller-than-expected increase, the *scandide contraction* (or nonexpansion), is found even more emphatically in r_{cov} (Table 1.13; Section 1.5B). Between these two periods, $n^\star$ is increasing as usual, but $Z^\star$ partly counteracts this effect by increasing 1.50 units due to the first insertion of the *d*-block elements (Section 9.2C). There is normal expansion between Ge and Sn because both have *d* orbitals and equally increased $Z^\star$ values. In general, for properties depending on size, we may anticipate an unusual resemblance of the corresponding *third- and fourth-period elements beyond the d block* because of the scandide contraction. (Recall the consequence from Section 2.3C that cations from these two periods share the same maximum coordination number of four.)

The last calculation shown in Table 9.7 is for the Group 12 elements. These are calculated to expand normally on going down from Zn to Cd to Hg, but r_{cov} values in fact *contract* from Cd to Hg. This is (in part) the *lanthanide contraction* occurring in the sixth-period elements after the first filling of *f* orbitals (Section 1.5B). For properties of the elements that depend on size, we expect an unusual resemblance of the corresponding fifth- and sixth-period elements (beyond the *f* block). Thus, cations of Periods 5 and 6 tend to have the same maximum coordination number (Section 2.3C).

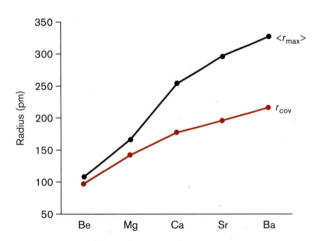

Figure 9.11. Group 2 trends in atomic radii: $\langle r_{max} \rangle$ in black, and r_{cov} in **brown**. [The data come from Table 9.7.]

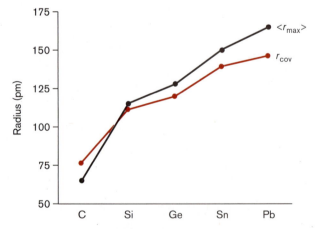

Figure 9.12. Group 14 trends in atomic radii: $\langle r_{max} \rangle$ in black, and r_{cov} in **brown**. [The data come from Table 9.7.]

TABLE 9.7. Vertical Periodic Trends in Atomic Radii[a]

Group 2 Element	Be	Mg	Ca	Sr	Ba
$<r_{max}>$	108	167	254	297	327
r_{cov}	96	141	176	195	215
Group 14 Element	C	Si	Ge	Sn	Pb
$<r_{max}>$	65	115	128	150	165
r_{cov}	76	111	120	139	146
Group 12 Element	Zn	Cd	Hg		
$<r_{max}>$	166	195	214		
r_{cov}	122	144	132		

[a] Covalent radii are taken from Table 1.13.

EXAMPLE 9.8

In Exercise 9.25, you compared Z^* for the outermost s electron of the following sixth-period elements: Ba, Yb, Hg, and Rn. Use these numbers to (a) calculate $<r_{max}>$ for these four elements; (b) explain what happens to these radii of atoms as we cross the s and p blocks of elements and why (in terms of Z^* and of shielding or penetration); (c) explain what happens to $<r_{max}>$ on crossing the d block of elements and how the magnitude of this change compares with that in the s and p blocks and why; (d) explain what happens to $<r_{max}>$ on crossing the f block of elements. (e) If your calculations fail to capture the observed trend in any of these cases, tell what the true trend is, and (if possible) why your calculations do not show it.

SOLUTION: From Figure 9.8 or the answers at the back of the book for Exercise 9.25, we find the following Z^* values: Ba +2.85, Yb +2.85, Hg +4.35, and Rn +8.25.

(a) Ba $<r_{max}> = 52.9(4.2)^2/2.85 = 327$ pm, Yb $<r_{max}> = 52.9(4.2)^2/2.85 = 327$ pm, Hg $<r_{max}> = 52.9(4.2)^2/4.35 = 214$ pm, and Rn $<r_{max}> = 52.9(4.2)^2/8.25 = 113$ pm.

(b) Atoms shrink from left to right across the s and p blocks of elements. This is caused by the increase in Z^*, due to the increased population of the $6s$ and $6p$ orbitals, which shield each other poorly from the additional protons. (While the shielding from Rule 3 increases from 0.35 to 2.45, the nuclear charge increases much more, from 80 to 86.

(c) Atoms shrink across the d block of elements, but by less than on crossing the p block, because the increase in Z^* is less (from 2.85 in Lu to 4.35 in Hg) than in the p block. The d electrons are being added to orbitals with a smaller radius ($n = 5$) than the $6s$ orbitals, where they are more effective (85%) at shielding each additional proton.

(d) According to the computations, radii are unchanged across the f block, because Z^* is unchanged; this is because electrons are being added to the $4f$ orbitals, which are so far inside the atom as compared to the $6s$ orbitals that (according to Slater's rules) they shield the additional protons completely.

(e) Atoms actually do contract slightly across the f block. This is because f orbitals have no inner lobes and are not really 100% effective at shielding the highly penetrating $6s$ orbitals, in contradiction of Slater's Rule 4b.

9.4C. Some Consequences of Improved Calculations of $<r_{max}>$. More modern methods of calculation can overcome many of the limitations of Slater's rules, at a cost of making it more difficult to see the reasons why the periodic trends are as they are. One such set of calculations is that of Desclaux,[10] who obtained more refined values of $<r_{max}>$. For the most part these values show similar trends. However, there are some exceptions.

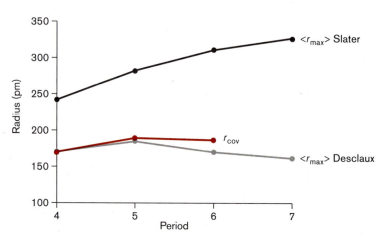

Figure 9.13. Radii of Group 3 elements (in pm) as a function of period of the periodic table. [Sources of data: $<r_{max}>$ Slater (black) are calculated from Slater's rules; $<r_{max}>$ Desclaux (gray) are calculated from data given in J. P. Desclaux, *Atom. Data Nucl. Data Tables* 12, 311 (1973) after multiplying by 52.93 to convert from a.u. to pm; and r_{cov} values (**brown**) are from Table 1.13.]

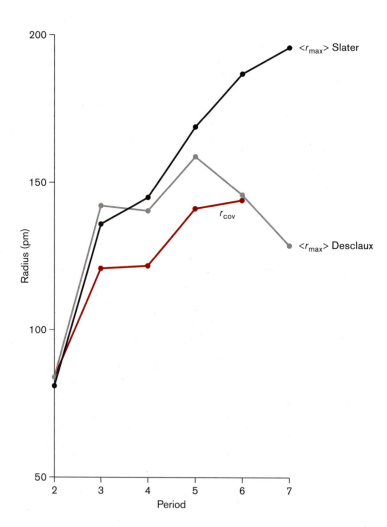

Figure 9.14. Radii of Group 13/III atoms (in pm) as a function of period of the periodic table. [Sources of data: $<r_{max}>$ Slater (black) are calculated from Slater's rules; $<r_{max}>$ Desclaux (gray) are calculated from data given in J. P. Desclaux, *Atom. Data Nucl. Data Tables* 12, 311 (1973) after multiplying by 52.93 to convert from a.u. to pm; and r_{cov} values (**brown**) are from Table 1.13.]

Most notably, Desclaux's calculations show the effects of the lanthanide contraction. Calculations of $Z^\star$ for the first group in which this should emerge, Group 3, give the same $Z^\star$, 3.00, for Sc, Y, and Lu. Consequently, as $n^\star$ grows, the Slater-calculated values of $<r_{max}>$ increase from 241 pm to 282 pm to 311 pm. However, the experimental covalent radii decrease, as do the $<r_{max}>$ values of Desclaux (Figure 9.13). The Desclaux $<r_{max}>$ values (but not those of Slater) show the scandide contraction from Al to Ga *and* the lanthanide contraction from In to Tl. [The contraction from Tl to Nh (nihonium, element 113) remains to be explained in Section 9.6A.]

AN AMPLIFICATION

The Diminished Role of *sp* Hybridization in Inorganic Chemistry

Desclaux obtained separate values of $<r_{max}>$ for *s* and *p* orbitals. Normally $<r_{max}>$ values for valence *p* orbitals exceed $<r_{max}>$ values for the valence *s* orbital of the same atom. The exception is in Period 2, in which the two types of valence orbitals have virtually the same $<r_{max}>$ values (top row of Table 9.8).

TABLE 9.8. Differences in $<r_{max}>$ Values of Valence *s* and *p* Orbitals for Lighter *p*-Block Elements

Period 2 Element	B	C	N	O
$2p <r_{max}> - 2s <r_{max}>$ (pm)	3	0	−2	−2
Period 3 Element	Al	Si	P	S
$3p <r_{max}> - 3s <r_{max}>$ (pm)	31	21	14	11
Period 4 Element	Ga	Ge	As	Se
$4p <r_{max}> - 4s <r_{max}>$ (pm)	35	25	19	16
Period 5 Element	In	Sn	Sb	Te
$5p <r_{max}> - 5s <r_{max}>$ (pm)	39	29	23	19

Kutzelnigg[11] attributes the unusual second-period trend to the fact that, among *p* orbitals, only 2*p* orbitals do not have $(n - 1)p$ orbitals inside them to expand them by repulsion. This has a couple of consequences: (1) Incorporation of both types of valence orbitals into covalent bonding via *sp* hybridization (Section 3.3B) is commonly invoked in organic chemistry for carbon and other Period 2 elements. However, lower in the periodic table the *s* and *p* valence orbitals differ in size, so it is harder to get them to overlap to form sp^n hybrid orbitals. Hence, the idea of hybridization is much less useful in inorganic chemistry than it is in organic chemistry. (2) As the valence *ns* orbitals become relatively more difficult to ionize, there is an increased tendency for them not to ionize, allowing the heavier elements to form ions with ns^2 valence electron configurations and oxidation numbers two less than the group numbers (Table 1.5).

Relative Radii of Core *p* and Valence *d* Electrons in an Atom

Although the orbitals of valence electrons normally are larger than those of core orbitals, this is not necessarily the case. In the *d* block the valence $(n-1)d$ valence orbitals may have larger $<r_{max}>$ values than the outer core $(n-1)s$ and $(n-1)p$ orbitals. Figure 9.15 illustrates by how much Desclaux's $<r_{max}>$ values for valence 3*d*, 4*d*, 5*d*, and 6*d* orbitals exceed the $<r_{max}>$ values for the corresponding core 3*p*, 4*p*, 5*p*, and 6*p* orbitals; Figure 9.16 illustrates the differences for all of the *d* block. Valence *d* orbitals are generally larger than the corresponding core *p* orbitals. The great exception is in Period 4 (the bottom line in Figure 9.16). Except to a slight extent at

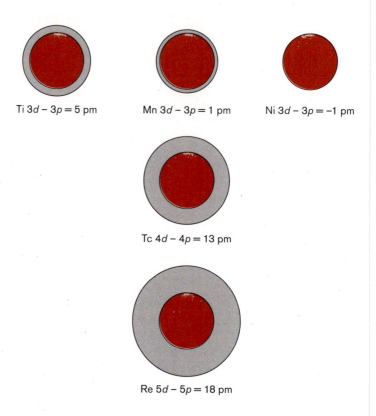

Ti 3*d* – 3*p* = 5 pm   Mn 3*d* – 3*p* = 1 pm   Ni 3*d* – 3*p* = −1 pm

Tc 4*d* – 4*p* = 13 pm

Re 5*d* – 5*p* = 18 pm

Figure 9.15. Qualitative differences in radii of core *p* electrons (in **brown**) and valence *d* electrons of the same principal quantum number (in gray).

the left of Period 4, the valence 3*d* orbitals are no larger than the corresponding core *p* orbitals. Kutzelnigg[11] attributes the exceptionally small size of the 3*d* orbitals to the fact that the core contains no *smaller* type of *d* orbitals repelling the 3*d* orbitals and pushing them outward. All other types of *nd* orbitals are repelled by the corresponding core $(n-1)d$ orbitals.

The conclusion that valence *nd* orbitals have larger $<r_{max}>$ values than core electrons (except late in Period 4) allows us to explain some consequences seen in previous chapters.

1. Valence *d* electrons extending beyond the core are expected to be more strongly repelled by ligand electrons (Chapter 7). Therefore, fifth-period (and below) *d*-block metal ions have larger crystal field splittings and tend to form more strong-field complexes.

2. Valence *d* orbitals extending beyond the core are more likely to give covalent overlap (Sections 3.3B and 3.3C) with other *d* orbitals or *p* orbitals on neighboring oxygen and nitrogen atoms. This enables π bonding with π-donor ligands at the top of the spectrochemical series (Section 7.4), and the formation of multiply bonded *d*-block oxo and nitrido anions such as $[V=O]^{3+}$ and $[WO_4]^{2-}$ showing the antepenultimate total coordination number (Section 3.5).

3. This allows π and δ multiple bonding to other d-block atoms (Section 3.3B) and the formation of metal–metal quadruple and quintuple bonds.

4. Similar results are found in the f block; in Period 6 the valence $4f$ electrons do not extend beyond the core $5s$ and $5p$ orbitals, so they show virtually no crystal field effects. The relatively larger $5f$ orbitals in Period 7 form oxo cations such as the very stable $[O=U=O]^{2+}$ cation.

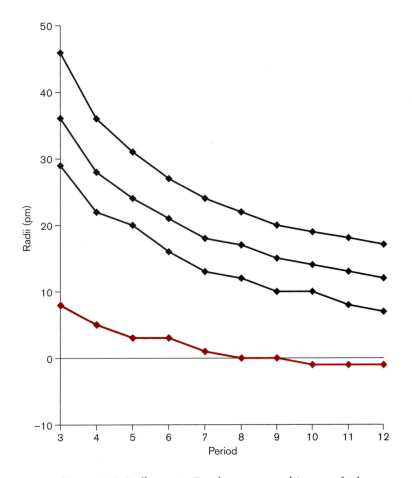

Figure 9.16. Difference in Desclaux $<r_{max}>$ radii $r_d - r_p$ of valence d and core p orbitals (in pm). Bottom set of data points (in **brown**) is for Period 4; next is Period 5; next is Period 6; top set of data points is for Period 7. Differences $<r_{max}>$ of valence d minus $<r_{max}>$ of outermost core p electrons are calculated from data given in J. P. Desclaux, *Atom. Data Nucl. Data Tables* 12, 311 (1973) after multiplying by 52.93 to convert from a.u. to pm.

9.5. Periodic Trends in Electronegativities of Atoms: Explanation and Theoretical Prediction

OVERVIEW. Given the covalent radius of an element, you can use Z^* to calculate the Allred–Rochow electronegativity of an element. You can explain periodic trends in electronegativities in terms of periodic trends in Z^*. You can also explain discrepancies between Allred–Rochow and Pauling electronegativity values for heavier d- and p-block elements. You may practice applying these ideas by working Exercises 9.45–9.52.

Pauling described an atom's electronegativity as its power to attract shared electrons to itself. The term "power" does not have a chemical definition; it was not clear how such a property of atoms was to be measured. Pauling's electronegativity values were presented in Section 1.6 and derived from experimental data (i.e., bond energies) in Section 3.3D. Many other scientists have suggested other interpretations of how to define "power" and to assign it, resulting in quite a few different measures of electronegativity.

Allred and Rochow[12] attempted to produce a more theoretically based electronegativity scale, the **Allred–Rochow electronegativity** scale. To do this, they interpreted Pauling's concept of electronegativity as being the force exerted by an atom of an element on its (s and p) valence electrons:

$$F = \frac{e^2 Z^*}{r^2} \tag{9.16}$$

For the charge, they used the Z^* of the atom from Slater's rules, and for the radius they used the covalent radius of the atom. They added empirical parameters so that the range of their Allred–Rochow electronegativity values χ_{AR} would correspond to the Pauling scale of electronegativities and obtained the following equation:

$$\chi_{AR} = \frac{(3590)(Z^* - 0.35)}{(r_{cov})^2} + 0.74 \tag{9.17}$$

The resulting Allred–Rochow electronegativities, χ_{AR}, of the elements are shown in Table 9.9. Pauling and Allred–Rochow electronegativities are compared graphically in Figure 9.17.

The computed Allred–Rochow and experimental Pauling electronegativities of the elements are in excellent agreement in the s block and in most of the p block. Hence, the Allred–Rochow electronegativity calculations provide us with explanations of the main periodic trends in Pauling electronegativities.

1. Electronegativities increase from left to right because Z^* increases. As we add additional electrons on crossing a period, we also add additional protons, which are only 35% shielded by the additional electrons (Rule 3 of Slater's rules). Hence, all valence electrons (including those in the bond) are more strongly attracted to the nucleus.

2. Down Groups 1 and 2, Z^* often remains constant, so the small decline in electronegativity is due to the larger size: The bond electrons are farther from the nucleus and, hence, somewhat more weakly attracted to it.

TABLE 9.9. Allred–Rochow Electronegativities of the Elements

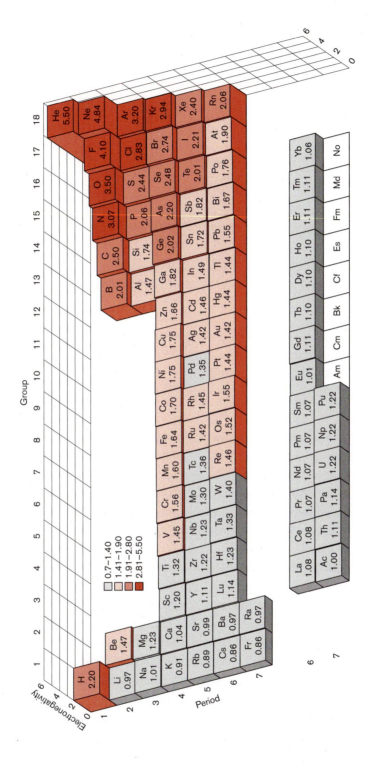

Sources: Values from A. L. Allfred and E. G. Rochow, *J. Inorg. Nucl. Chem.* 5, 264 (1958); E. J. Little and M. M. Jones, *J. Chem. Educ.* 37, 231 (1960); and J. E. Huheey, *J. Inorg. Nucl. Chem.* 42, 1523 (1980).

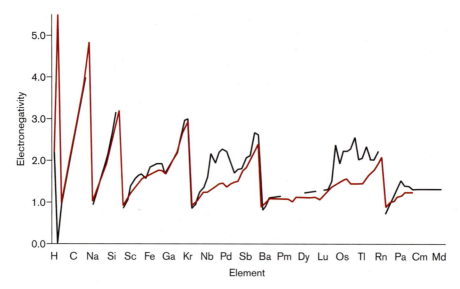

Figure 9.17. Plot of Pauling (black) vs. Allred–Rochow (**brown**) electronegativities of the elements. [Courtesy of Gabriel Gillette.]

3. The Allred–Rochow electronegativities also duplicate the anomalously high Pauling electronegativities of Ga (>Al) and Ge (>Si). This trend is due to the increase in Z^*, which results from the first filling of the poorly shielding d orbitals—the same factor responsible for the scandide contraction discussed earlier.

4. Even though Slater's rules (and hence the Allred–Rochow electronegativities) do not predict the lanthanide contraction, we know it is there and expect that it should cause the Pauling electronegativities of the post-f-block sixth-period elements to be higher than those of the corresponding elements of the fifth period. Thus, we find the electronegativity relationships Au > Ag, Hg > Cd, Tl > In, Pb > Sn in the Pauling but not the Allred–Rochow scales.

Discrepancies among electronegativity scales, and difficulties in obtaining values, are common in the d block, in which all electronegativity scales show some difficulties in implementation. In Figure 9.17 we see two major discrepancies in the d block between the Pauling and the Allred–Rochow graphs. (1) Double humps appear in the Pauling but not the Allred–Rochow electronegativities in the d block. We can recognize these as the results of crystal field effects (Section 7.5), which affect Pauling values but are not incorporated in Slater's rules. (2) In the Pauling scale, the fifth- and sixth-period elements of the d block are strikingly more electronegative than the corresponding fourth-period elements. The lanthanide contraction cannot explain the discrepancy for the fifth-period elements; we suspect that the Pauling values are affected by partial multiple bonding in the metal–halogen and metal–metal bonds.

Other Electronegativity Scales

Mulliken electronegativities[13] are based on the average of the ionization energies and electron affinities of atoms in appropriate hybridization scales—unfortunately hybridization schemes cannot be deduced clearly when *s*, *p*, and *d* orbitals are all in play in the *d* block. Allen electronegativities[14] are based on weighted averages of ionization energies of all the valence-shell electrons. However, his analysis for *d*-block elements seems to depend on an asssumption that no metal can have a higher electronegativity than that of silicon, the lowest metalloid.[15]

AN AMPLIFICATION

EXAMPLE 9.9

All electronegativity scales are not the same! The greatest discrepancy between the Pauling and Allred–Rochow electronegativity scales occurs at gold. Electronegativity is one of our three fundamental atomic variables for predicting chemical reactivity. We use Pauling electronegativities (Table A), but if we were to use the Allred–Rochow value for gold (Table 9.9), how would this change our predictions of the chemical reactivity of Au^+ in (a) cation acidity (Section 2.2B); (b) softness (Section 5.4); and (c) activity of gold metal (Section 6.2B)?

SOLUTION: $\chi_P(Au) = 2.54$, but $\chi_{AR}(Au) = 1.42$, a change of 1.12 units. Using an electronegativity of 1.42 for Au^+ would:

(a) After applying Equation 2.14, give Au^+ a pK_a of 15.14 – 88.16[1/151 + 0.096(1.41 – 1.50)]. But Equation 2.14 cannot be used for an electronegativity of <1.50; instead Equation 2.13 is used, which lacks the last term. Then the pK_a is 14.5, and Au^+ is nonacidic. Using Equation 2.14 with the Pauling electronegativity of 2.54 gives us a pK_a that is diminished by a term of $88.16 \times 0.096(2.54 – 1.50) = 8.80$ units, so it is 14.5 – 8.8 = 5.7, which makes Au^+ moderately acidic (Figure 2.2).

(b) After using Table 5.3, change Au^+ from a soft acid to a hard acid!

(c) After using Figure 6.16, change Au metal from an electronegative metal with a positive standard reduction potential (very inactive in the activity series) to an electropositive (almost very electropositive) metal with a standard reduction potential around –1.6 V (similar to Al). If you are married, you would have to be careful about putting the fourth finger of your left hand in vinegar!

9.6. Relativistic Effects on Orbitals

OVERVIEW. Relativistic effects cause contraction of *s* and *p* orbitals and expansion of *d* and *f* orbitals in very heavy atoms. In Section 9.6A, you learn how to predict how this affects their ionization energies, electron affinities, electronegativities, and likely oxidation numbers (Exercises 9.53–9.56). You also learn what spin-orbit coupling is, which subshells of orbitals are most affected, and how this varies with the atomic number of an element in a given group. In Section 9.6B you learn how spin-orbit coupling affects the ionization energies, electron affinities, electronegativities, and likely oxidation numbers of heavy atoms (Exercises 9.57–9.62).

9.6A. Relativistic Expansion and Contraction. The orbital wave functions we discussed in Section 9.1 do not incorporate any effects of relativity theory. But solutions of the Schrödinger wave equation that do incorporate relativity theory show that both the radial and the angular parts of the wave functions for the heavier atoms in the periodic table are appreciably altered by relativistic effects.[16,17,18,19] These relativistic effects increase approximately in proportion to the square of the atomic number.

The radial effect is known as **relativistic contraction**. We can think of the electron in an orbital as a particle that is accelerated to a certain radial velocity by the attraction of the nucleus. As the attractive nuclear charge builds up, this radial velocity builds up, too, approaching the speed of light as the nuclear charge approaches $Z = 137.036$. According to Einstein's theory of relativity, the mass (m) of a particle increases over its rest mass (m_0) when its velocity (v) approaches the speed of light (c):

$$\frac{m_0}{m} = \sqrt{1 - \frac{v^2}{c^2}} \tag{9.18}$$

Hence, the mass of a 1*s* electron in mercury is about 1.2 times its rest mass, while that of fermium is 1.46 times its rest mass.

The radius of the Bohr orbit of an electron is inversely proportional to the mass of the electron:

$$r = a_0 \sqrt{1 - \frac{v^2}{c^2}} \tag{9.19}$$

Therefore, we expect the radius of the 1*s* orbital in mercury to be about 80% of the nonrelativistic value, while the radius for fermium is 68% of the nonrelativistic value. This value is the relativistic contraction. It also affects the 2*s* and higher *s* orbitals roughly as much because of their inner lobes, which are close to the highly charged nucleus; the radius of the 7*s* orbital of Cn is expected to be 75% of the nonrelativistic value.[20]

Because these valence orbitals are drawn closer to the nuclei, their energies are lower and they are harder to ionize. Thus, the relativistically calculated energy for the 7*s* orbital of Cn is double the nonrelativistically calculated value (about 600 kJ mol⁻¹ lower).[21]

The effect is present to a lesser extent among *p* orbitals and is nearly absent among *d* and *f* orbitals, which have fewer lobes near the nucleus. Indeed, the *d* and *f* orbitals are more effectively screened out by the contracted *s* (and *p*) orbitals, so they undergo **relativistic expansion**. Thus, in uranium the relativistically calculated radius of the 7*s* orbital is 86.1% of the nonrelativistic radius (it is contracted), while the corresponding values for the 5*f* and 6*d* orbitals are 113% and 112%, respectively (they are expanded).[16] This expansion of the 6*d* orbitals increases their $<r_{max}>$ to become comparable with the $<r_{max}>$ of the 7*s* electrons. This increases the ability of the 5*d* orbitals to form additional and stronger covalent bonds in the sixth period.

The average energies of 6*d* orbitals also rise (they become easier to ionize, with lower ionization energies). Because the opposite trends affect the 7*s* orbital, this causes their energy levels to switch places in an element such as Sg (Figure 9.18).

The higher orbital energies and lower ionization energies of the 6*d* orbitals make it easier for sixth-period *d*-block elements to form compounds in high oxidation states. For similar reasons affecting the 5*f* orbitals, uranium characteristically is found in higher oxidation states (+6) than neodymium (+3).

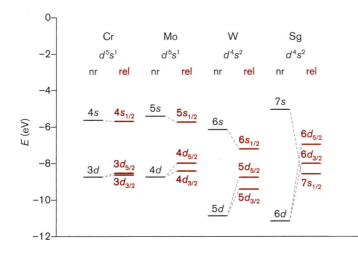

Figure 9.18. Relativistic (rel, in **brown**) and nonrelativistic (nr, in black) energy levels of the Group 6 valence *ns* and $(n-1)d$ electrons. (To convert the left scale to kJ mol⁻¹, multiply by 96.5.) [Adapted from M. Schädel, *Angew. Chem. Intl. Ed.* 45, 368 (2006).]

EXAMPLE 9.10

The $<r_{max}>$ values calculated by Desclaux[10] include relativistic effects. For the lower part of Group 1 the $<r_{max}>$ values (in pm) are Rb 245 pm, Cs 272 pm, Fr 264 pm, and element 119 230 pm. (a) Explain the change in the periodic trend below Cs. (b) Would you expect the periodic trend in first ionization energies to change at Cs? If so, how and why? (c) Would you expect the periodic trend in electronegativities to change at Cs? If so, how and why?

SOLUTION: (a) As we go down the group, the inner lobes of the *ns* orbitals are close to nuclei of increasing charge (closer to $Z = 137$), so they are more massive and more strongly attracted; they experience relativistic contraction. (b) Yes. As the $<r_{max}>$ values decrease, it should become harder to ionize the electrons, and the ionization energies should increase. In fact, in Table 6.5 we see that IE(1) for Fr (400 kJ mol⁻¹) exceeds that for Cs (376 kJ mol⁻¹). It is possible to calculate an IE(1) for Element 119 of at least 417 kJ mol⁻¹. (c) Based on Equation 9.17, as the radius decreases below Cs, the electronegativities should increase.

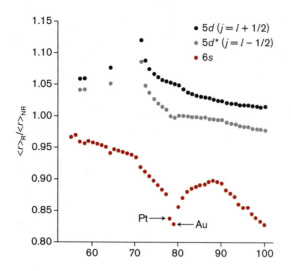

Figure 9.19. The relativistic contraction of the 6s orbital (in **brown**) and the relativistic expansion of 5d orbitals (in black and gray) in the elements Cs ($Z = 55$) to Fm ($Z = 100$). The left axis gives the ratio of the relativistic to the nonrelativistic radii for these elements calculated in J. P. Desclaux, *Atom. Data Nucl. Data Tables* 12, 311 (1973). [Adapted from M. Jansen, *Solid State Sci. 7,* 1464 (2005); and P. Pyykkö and J.-P. Desclaux, *Acc. Chem. Res.* 12, 276 (1979).]

While the 5d orbitals are filling and expanding across the sixth period, they are doing their usual poor job of shielding the 6s orbital from the nucleus. Once the 5d orbitals are filled, they are no longer considered valence orbitals, and in Groups 11 and 12 the electronegativity depends on the size of the valence 6s orbitals. With their inner lobes, these orbitals are contracted strongly in gold and mercury (Figure 9.19).[17] Beyond mercury, the less penetrating p orbitals are the largest valence orbitals, so $<r_{max}>$ increases for a few elements. Eventually, however, the increase in nuclear charge again causes atomic contraction.

Relativistic Effects on the Chemistry of Gold, Mercury, and Nearby Elements. The relativistic contraction is calculated to be especially substantial for gold and mercury, but also significant for the elements preceding them.[22,23] In support of this, we find in Table 1.13 that the covalent radii of Os through Hg are all *less* than the covalent radii of the elements just above them. Although Z^* calculations do not incorporate relativistic effects, we expect (due to its inner lobes) that when the 6s orbital is a valence orbital it will experience an anomalously high Z^*. Allred–Rochow electronegativity calculations also do not incorporate relativistic effects, but we would expect unusually high Pauling electronegativities for these elements. The experimentally based Pauling values of Pt, Au, and Hg are higher than the elements above them. The electron affinities (Table 6.8) of Os through Au are higher than the elements above them. Gold has the highest first ionization energy of any metal (the only one over 1000 kJ mol^{-1}; Table 6.5). These anomalies have chemical consequences.

One consequence is the formation of the only known d-block metal anions, Au$^-$ (the auride anion) and Pt^{2-} (the platinide anion). Mercury, with a $5d^{10}6s^2$ valence electron configuration, has essentially filled all of its valence orbitals; this makes it very near to being a noble gas (or rather a noble liquid). Gold may then be thought of as one electron short of a noble liquid electron configuration and thus might be expected to show some resemblance to a halogen such as iodine. Gold forms a stable Au$^-$ ion, similar in size to the Br$^-$ ion, in the compounds RbAu, CsAu, (Me$_4$N)Au, [Cs(18-crown-6)] Au·8NH$_3$, and (Cs$^+$)$_3$(Au$^-$)(O^{2-}), and in liquid ammonia solution.[24] There are other resemblances to halogens—namely, Au has a first ionization energy similar to that of I, and the Au$_2$ molecule is known and has an Au–Au bond energy of 221 kJ mol^{-1}, similar to that of the Br_2 molecule.[24] Platinum can be thought of as being two electrons short of a noble liquid electron configuration, and has been shown to form an ionic compound, Cs_2Pt.[25]

Beyond gold and mercury, the $6s$ electrons are still difficult to ionize. The metallic elements among these retain their two $6s$ electrons in their most common ions and oxidation states (Table 1.11). Thus, Tl normally occurs as Tl^+, Pb as Pb^{2+}, and Bi as Bi^{3+}, while the most common oxidation state of Po is Po(IV). This preference for the oxidation state that is two less than the group number is sometimes known as the **inert pair** (of $6s$ electrons) **effect**. It is also advisable to use separate electronegativity values for Tl^+ and Tl^{3+}, and for Pb^{2+} and Pb^{4+} (Table A), because the strongly contracted $6s$ orbital participates directly in the covalent bonding in Tl^{3+} and Pb^{4+} but not in Tl^+ and Pb^{2+}.

A CONNECTION TO INDUSTRIAL CHEMISTRY

Relativity and the Lead-Acid Battery

The lead-acid battery, although it is 150 years old, still dominates in automotive uses. The reaction in this battery is

$$Pb(s) + PbO_2(s) + 2H_2SO_4(aq) \rightarrow 2PbSO_4(s) + 2H_2O(l) \qquad E° = 2.11 \text{ V}$$

Calculations on a similar model system excluding relativistic effects gave a much-diminished $E° = 0.39$ V. Most of the driving force of this reaction is due to relativistic effects, particularly in destabilizing the +4 oxidation state in $PbO_2(s)$.[26]

9.6B. Spin-Orbit Coupling. In addition to the relativistic contraction of s orbitals and the relativistic expansion of d and f orbitals, there is another relativistic effect in heavy atoms known as **spin-orbit coupling**. The relativistic treatment of the Schrödinger wave equation was the first to reveal that electrons in atoms also have spin, and need a fourth quantum number, m_s, which can either have the value of $+\frac{1}{2}$ or $-\frac{1}{2}$. This number corresponds to angular rotation of the electron in either a clockwise or a counterclockwise direction. This angular rotation occurs even in the lightest atoms. But in heavy atoms the angular motion due to the spin of the electron about its own axis begins to mix with the angular motion of the electron among the lobes of the nonspherical orbitals. These two types of motion may either reinforce or cancel each other out in part. In a given group, spin-orbit coupling effects increase roughly in proportion to Z^2.

Consequently, in atoms heavier than bromine, we observe that the p, d, or f electrons ionize at either of two energies, depending on whether the two types of angular motion are in harmony or are opposed. In the heaviest atoms, the difference between these two energies may be hundreds of kilojoules per mole, which is as large as or larger than their covalent bond energies (Table 9.10). As a result, their bonding abilities are affected, and it becomes necessary to speak not just of a $6p$ electron but to distinguish which type of $6p$ electron is meant. The distinction is made by specifying an additional quantum number j, which can have either of two values: $j = l + \frac{1}{2}$ or $j = l - \frac{1}{2}$. This number is attached as a subscript on the right of the orbital designation.

TABLE 9.10. Periodic Trends in Orbital Energy Differences (kJ mol⁻¹) Due to Spin-Orbital Coupling

Period	*f*-Block Element	ΔEnergy[a]	*d*-Block Element	ΔEnergy[b]	*p*-Block Element	ΔEnergy[c]
2					N	2
3					P	6
4			Mn	12	As	28
5			Tc	27	Sb	62
6	Sm	52	Re	82	Bi	203
7	U	100	Hs	168	Mc	607

Source: J. P. Desclaux, *Atom. Data Nucl. Data Tables* 12, 311 (1973). Converted from a.u. (hartrees) to kJ mol⁻¹ by multiplying by 2625.45 kJ mol⁻¹ hartree⁻¹.

[a] Orbital energy differences between $f_{7/2}$ and $f_{5/2}$ orbitals.

[b] Orbital energy differences between $d_{5/2}$ and $d_{3/2}$ orbitals.

[c] Orbital energy differences between $p_{3/2}$ and $p_{1/2}$ orbitals.

Thus, we find two types of $6p$ orbitals, designated the $6p_{3/2}$ and the $6p_{1/2}$ orbitals. There are also two types of nd orbitals, the $nd_{5/2}$ and $nd_{3/2}$ orbitals, and two types of nf orbitals, the $nf_{7/2}$ and $nf_{5/2}$ orbitals. Each set of orbitals can hold $2j + 1$ electrons; thus, there can be four $6p_{3/2}$ electrons and two $6p_{1/2}$ electrons. The $6p_{1/2}$ orbitals penetrate the core better and experience relativistic contraction; hence, they are filled first and are more stable than the $6p_{3/2}$ orbitals by many kilojoules per mole (Figure 9.20).

Consequently, in the sixth period of the *p* block the $6p_{1/2}$ orbital is a valence orbital, while the $6p_{3/2}$ orbitals are higher enough in energy that they can be left unoccupied in stable compounds. Thus, we may expect to find atoms or ions of valence electron configuration $6s^2 6p_{1/2}^2$. This plays a role in the low reactivity of lead metal (oxidation number zero), stabilizes the ion Bi⁺ [known in a solid compound $(Bi^+)(Bi_9^{5+})(HfCl_6^{2-})_3$],[27] and allows Po^{2+} to be found more frequently.[18]

This tendency is expected to be much more pronounced in the seventh period, in which the computed stable oxidation states in the gas phase are +1 for Nh (nihonium,

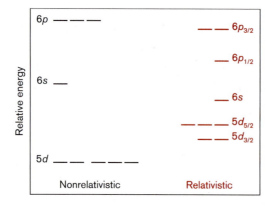

Figure 9.20 Sketch of nonrelativistic (in black) and relativistic (in **brown**) energy level diagrams applicable to a *p*-block element late in the sixth period. [Adapted from J. S. Thayer, *J. Chem. Educ.* 82, 1721 (2005).]

$Z = 113$), +2 for Fl (flerovium, $Z = 114$), +1 for Mc (moscovium, $Z = 115$), +2 and +4 for Lv (livermorium, $Z = 116$), +3 for Ts (tennessine, $Z = 117$), and +2 and +4 (but *not* zero) for Og (oganesson, $Z = 118$).[20] In contrast with all other Group 18/VIII elements, relativistic calculations for Og give it a nonzero electron affinity (5 kJ mol^{-1}),[28] indicating that the 8*s* orbital comes into the bonding.[29]

EXAMPLE 9.11

The calculated first ionization energies are 1126 kJ mol^{-1} for Cn, 823 kJ mol^{-1} for Fl, and 834 kJ mol^{-1} for Og.[30] (a) Compare these values to those of the elements just above them (Table 6.5). Why are the trends different for these three elements? (b) All three of these elements have been suggested as possible "noble liquids." Do these values suggest that these elements should be less active chemically than the elements above them? (c) Figure 9.9 shows the normal trend for first ionization energies across the second period in the *p* block. Would you expect a similar trend across the seventh period? If not, why not?

SOLUTION: (a) Cn is more difficult to ionize than Hg; Fl is more difficult to ionize than Pb; but Og is considerably easier to ionize than Rn. The first ionizations of Cn and Fl are of $7s^2$ and $7p_{1/2}$ electrons, which experience enhanced relativistic contractions. But the first ionization of Og is of one of the $7p_{3/2}$ electrons, which are raised in energy by spin-orbit coupling.

(b) Cn and Fl should be less active chemically than Hg and Pb, but Og should be more active chemically than Rn.

(c) Not completely, because the dip in IE(1) between N and O is due to repulsion from the first filling of a 2*p* orbital in O. However, in Period 7 the break would be between filling the $7p_{1/2}$ orbital and the $7p_{3/2}$ orbitals, between Fl (Group 14/IVA) and Mc (Group 15/VA).

The half-lives of even the most stable isotopes of the elements in the *d* block in Period 7 (the *transactinides*) become shorter and shorter, eventually becoming less than one second. The nuclear reactions producing them by bombardment of available isotopes with fully ionized nuclei of other atoms give very low yields. Hence, the chemical experiments that have been done on these elements involve only a few *atoms* at a time and must be done extremely rapidly.[31,32,33] Aqueous experiments are limited to the longer-lived Rf, Db, and Sg.[34] It is helpful if the product molecules are gaseous and can rapidly be removed from the reactor. This is helpful especially for the element Hs, which forms a volatile oxide HsO_4 similar to the known OsO_4, and for the element Cn, which is a volatile "noble liquid" like Hg.

9.7. Background Reading for Chapter 9

Relative Radii of Core *p* and Valence *d* Electrons: An Amplification

W. Kutzelnigg, "Chemical Bonding in Higher Main Group Elements," *Angew. Chem. Intl. Ed. Engl.* 23, 272–295 (1984). Contrast between bonding concepts used in second-row chemistry (such as organic chemistry) and concepts that apply to heavier *p*-block elements.

Chemical Periodicity Changes at the Bottom of the Periodic Table

E. Scerri, "Cracks in the Periodic Table," *Sci. Amer.* 308(6), 68–73 (2013).

A. Turler, "Chemical Experiments with Superheavy Elements," *Chimia* 64, 293–298 (2010). Chemistry of elements as heavy as Fl.

M. Schädel, "Chemistry of Superheavy Elements," *Radiochim. Acta* 100, 579–604 (2012). Includes summary of isotopes and half-lives.

P. Pyykkö, "A Suggested Periodic Table up to $Z \leq 172$, based on Dirac-Fock Calculations on Atoms and Ions," *Phys. Chem. Chem. Phys.* 13, 161–168 (2011). New orders of filling orbitals in the eighth (and ninth) periods, with the possibilities of oxidation states as high as +12.

Relativistic Effects

J. S. Thayer, "Relativistic Effects and the Chemistry of the Heaviest Main-Group Elements," *J. Chem. Educ.* 82, 1721–1727 (2005).

P. Pyykkö, "Relativistic Effects in Chemistry: More Common Than You Thought," *Annu. Rev. Phys. Chem.* 63, 45–64 (2012).

V. Pershina, "Relativistic Electronic Structure Studies on the Heaviest Elements," *Radiochim. Acta* 98, 459–476 (2011).

Numbered references from this chapter may be viewed online at www.uscibooks.com /foundations.htm.

9.8. Exercises

9.1. *Calculate your wavelength if you are striding at 4 m s^{-1}. Are you likely to be diffracted by a grating? Is it more useful to regard yourself as a wave or as a particle?

9.2. (a) Calculate the wavelength of a neutron (mass of 1.67×10^{-27} kg) if it is moving at 1000 m s^{-1}. (b) Could it conceivably be diffracted? (c) Is it useful to regard a neutron as a wave? As a particle?

9.3. *Do all objects in the solar system have circular paths around the sun? If not, would the equations describing their paths be more or less complex than those of the planets?

9.4. (a) Which of 3, π, –3, ½, and 7774 can be values of the principal quantum number n (for either a particle in a box or an electron in an atom)? (b) For which quantum number does the particle in the box have the highest energy?

9.5. *Which of the orbitals drawn in Figure 3.7 have values of (a) the secondary quantum number equal to zero; (b) the secondary quantum number equal to one; (c) the magnetic quantum number equal to zero?

9.6. An f_{xyz} orbital is shown in Figure 3.7. (a) What is the value of its secondary quantum number? (b) Is its magnetic quantum number equal to zero?

9.7. *How many nodal planes do each of the following orbitals have? How many nodal spheres? Does the orbital have any inner lobes? (a) 13*s*; (b) 11*d*; (c) 5*g*; (d) 6*h*; (e) 9*g*; and (f) an orbital for which $n = 232$ and $l = 117$.

9.8. How would we designate (i.e., as 6*f*) an orbital having the following numbers of nodal planes and nodal spheres, respectively? (a) 2 and 1; (b) 0 and 4; (c) 2 and 5; (d) 4 and 0; and (e) 5 and 3.

9.9. *Sketch xy-plane cross-section diagrams for the orbitals listed below. Indicate nodal planes and spheres by dashed lines, and regions of high electron probability by shading. Show the x and y axes. Indicate the positive and the negative parts of the wave function with plus and minus signs. (a) $4p_x$; (b) $3s$; (c) $4d_{xy}$; (d) $3d_{x^2-y^2}$; and (e) $4s$.

9.10. Some orbitals are drawn in Figure 9.21a, along with their axis systems. (a) Tell how many nodal planes are present in each orbital; sketch these in. (b) Tell how many nodal spheres are present in each orbital; sketch these in. (c) Fill in the signs of the wave function on each lobe of each orbital. (d) Identify each orbital (e.g., $4f_{xyz}$).

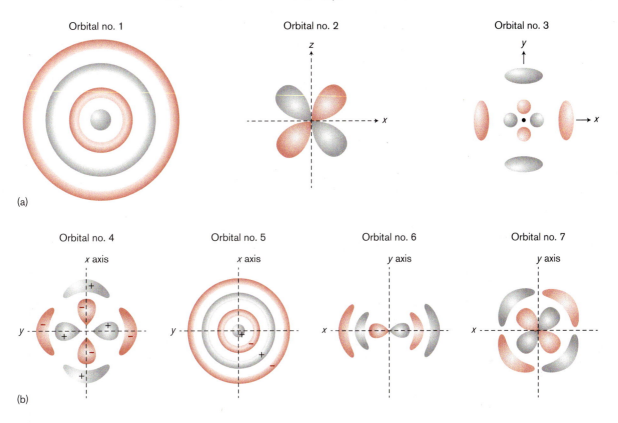

Figure 9.21. (a) Drawings of three orbitals for use in Exercises 9.10. (b) Drawings of four orbitals for use in Exercises 9.11 and 9.16.

9.11. *(a) Tell how many nodal planes are present in each orbital sketched in Figure 9.21b. (b) Tell how many nodal spheres are present. (c) Identify each orbital as specifically as possible (i.e., $4f_{xyz}$).

9.12. (a) How many nodal planes are present in each orbital sketched below? (b) How many nodal spheres are present in each? (c) Identify each orbital as specifically as possible.

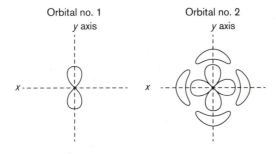

9.13. *(a) How many nodal planes are present in each orbital sketched below? (b) How many nodal spheres are present in each? (c) Identify each orbital as specifically as possible.

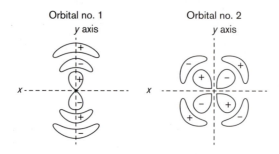

9.14. (a) Sketch in the nodal planes as dashed lines on the orbitals shown below. (b) Sketch in the nodal spheres as dashed circles on these orbitals.

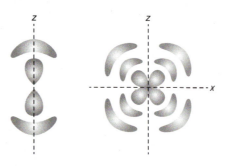

9.15. *Rank the $5s$, $5p$, $5f$, $5d$, and $5g$ orbitals in order of increasing (a) penetrating ability, (b) shielding ability, and (c) degree to which they are shielded by inner electrons.

9.16. Of the orbitals drawn in Figure 9.21(b), (a) which two will be poorest at penetrating the nucleus? (b) Which one will best shield other outer orbitals from the nucleus?

9.17. *In Exercise 9.13, (a) which orbital is poorest at penetrating inner orbitals? (b) Which orbital is best at shielding other outer orbitals from the nucleus?

9.18. In Exercise 9.12, (a) which orbital is poorest at penetrating inner orbitals? (b) Which orbital is best at shielding other outer orbitals from the nucleus?.

9.19. *You are considering the attraction of a Pr nucleus ($Z = 59$) for its $4d$ electrons. (a) Which of these other Pr electrons is most effective at shielding the Pr $4d$ electrons from the nucleus: $1s$, $4s$, other $4d$, $4f$, $5f$. (b) Which of these other Pr electrons is the second most effective at shielding the Pr $4d$ electrons from the nucleus: $1s$, $4s$, other $4d$, $4f$, $5f$. (c) Which of these other Pr electrons is least effective at shielding the Pr $4d$ electrons from the nucleus: $1s$, $4s$, other $4d$, $4f$, $5f$. (d) Which of these other Pr electrons is second least effective at shielding the Pr $4d$ electrons from the nucleus: $1s$, $4s$, other $4d$, $4f$, $5f$.

9.20. You are considering the attraction of a Y nucleus ($Z = 39$) for its $3d$ electrons. (a) Which of these other Y electrons is most effective at shielding the Y $3d$ electrons from the nucleus: $3s$, $4s$, other $3d$, $4d$. (b) Which of these other Y electrons is least effective at shielding the Y $3d$ electrons from the nucleus: $3s$, $4s$, other $3d$, $4d$.

9.21. *Write the full electron configurations of the following atoms, then rearrange them in the manner required for the application of Slater's rules: (a) Mg; (b) Si; (c) V; (d) Pm; (e) I; (f) Ir; and (g) No.

9.22. Tungsten has an atomic number of 74 and a characteristic electron configuration of $1s^2 2s^2 2p^6 3s^2 3p^6 4s^2 3d^{10} 4p^6 5s^2 4d^{10} 5p^6 6s^2 4f^{14} 5d^4$. (a) Rewrite this electron configuration in the order and groupings required for use with Slater's rules. (b) Compute **Z*** for a $6s$ electron in this atom. (c) Compute Z^* for a $5d$ electron in this atom.

9.23. *Calculate Z^* for (a) the $6s$ and (b) the $4f$ valence electrons of Ce.

9.24. You have discovered a new element, which has an atomic number of 125 and the following electron configuration:

$1s^2 2s^2 2p^6 3s^2 3p^6 4s^2 3d^{10} 4p^6 5s^2 4d^{10} 5p^6 6s^2 4f^{14} 5d^{10} 6p^6 7s^2 5f^{14} 6d^{10} 7p^6 8s^2 5g^5$

(a) Rewrite this electron configuration in the order and groupings required for use with Slater's rules. (b) Compute Z^* for a $5s$ electron in this atom. (c) Compute Z^* for a $5f$ electron in this atom.

9.25. *(a) From Figure 9.8, deduce **Z*** for the outermost s electron of the following sixth-period elements: Ba, Yb, Hg, and Rn. (b) Across which block of the periodic table does **Z*** of the outermost valence s electron vary most rapidly? Least rapidly? (c) Explain why these answers are so, in terms of shielding.

9.26. (a) Calculate Z^* for electrons in the following types of orbitals found in the Rn atom: $3p$, $3d$, $4p$, $4d$; $5p$, and $5d$. (b) Show by calculation for which principal quantum number ($n = 3$, $n = 4$, or $n = 5$) the np and nd orbitals have the most nearly equal Z^* values. Would it be true to say that the deeper in the core one of these sets of orbitals is, the more hydrogen-like is the ordering of their energies?

9.27. *Both Sb and Nb have five valence electrons (Section 1.3B). Let us compare the Z^* values of the $5s$ electrons of Sb (+6.30), the $4d$ electrons of Sb (+11.85), and the $4d$ electrons of Nb (+4.30). Which one of these electron sets is more likely to be a set of core electrons? Why?

9.28. Both Ca and Zn have two valence electrons (Section 1.3B). Let us compare the Z^* values of the $4s$ electrons of Ca (+2.85), the $4s$ electrons of Zn (+4.35), and the $3d$ electrons of Zn (+8.85). Which of these three types of electrons is more likely to be a set of core electrons? Why?

9.29. *A new type of orbital is expected to be occupied in the eighth period. (a) What orbital is that? (b) In which blocks of Period 8 is this orbital likely to be a valence orbital? (c) In which blocks is it likely to be a core electron?

9.30. The valence electron configurations of the Group 3 elements are Sc, $4s^2 3d^1$; Y, $5s^2 4d^1$; and Lu, $6s^2 5d^1$. Unlike the other two Group 3 elements, Lu has 17 electrons more than the previous noble gas. What are the additional electrons, and why are they not valence electrons?

9.31. *Below are listed some important periodic trends in the properties of elements. For each property, a selection of elements from a given group or period are listed. Choose the element from the set at which that property reaches a maximum, or indicate if this property is the same for all elements of the set.

(a) Zeroth ionization energy	Na	Mg	Al	P	Cl	Ar
(b) First ionization energy	Na	Mg	Al	P	Cl	Ar
(c) Second ionization energy	Na	Mg	Al	P	Cl	Ar

9.32. (a) Which atom of the sixth period has the highest IE(0) (least exothermic electron affinity)? Why? (b) A secondary maximum value is reached at which atom in the middle of the period? Why?

9.33. *It was reported [K. Moock and K. Seppelt, *Angew. Chem. Intl. Ed. Engl.* 28, 1676 (1989)] that the common cesium ion Cs^+ can be oxidized in acetonitrile solution to give the Cs^{3+} ion, at an $E°$ of +3.0 V. This finding was subsequently disputed [C. Jehoulet and A. J. Bard, *Angew. Chem. Intl. Ed.* 30, 836 (1991)]. Perhaps thermochemical computations could illuminate this disagreement. Let us assume for this problem that the reaction can be done in aqueous solution. (a) Explain why there is such a huge jump between IE(1) and IE(2) for Cs as compared to the smaller jump between IE(2) and IE(3). If this report is confirmed, what would be unique in the periodic table about the product? (b) Draw a thermochemical cycle showing the steps involved in the oxidation: $Cs^+(aq) \rightarrow Cs^{3+}(aq) + 2e^-$. Describe briefly what happens in each step and show what energy terms are required in a thermochemical analysis of this process. (c) One energy term you need for your thermochemical cycle is not known or listed either, but if you have identified it correctly you should be able to identify about how many times larger it should be than another term. (d) Do IE(2) and IE(3) appear to be prohibitively high for Cs for this reaction to occur? In discussing this, compare them with IE(2) and IE(3) for other elements in the periodic table that do form compounds in the +3 oxidation state. Also, complete your calculation of the thermochemical cycle using your estimate from part (c).

9.34. No table of fourth ionization potentials is included in this book. Nonetheless, you should be able to predict some trends involving them. (a) What general trend in the values of the fourth ionization energies is to be expected in crossing the *d* block from Sc to Zn? (b) Some anomalies in this trend are to be expected. At what elements should these countertrends show up? Why?

9.35. *Table 6.9 shows the electron affinities (zeroth ionization energies) of the elements. The fourth-period *d*-block elements show generally small positive values, except for Mn and Zn. Why do these elements have zero electron affinities?

9.36. Which of the following sets of ionization energies show normal vertical trends? If anomalies are found, explain them if possible. (a) IE(1) of Group 3; (b) IE(2) of Group 14/IV; (c) IE(1) of Group 18/VIII; (d) IE(1) of Group 12.

9.37. *(a) Using Slater's rules or Figure 9.8, calculate or find Z^* for the 4s, 3d, and 3p electrons in the Fe atom. (b) Calculate $<r_{max}>$ for the 4s, 3d, and 3p electrons in the Fe atom.

9.38. In a certain atom, an electron in orbital A experiences a Z^* of +4.00, while an electron in orbital B experiences a Z^* of +21.85. Suppose also that the outermost s orbital of this atom is a 6s orbital ($n = 6$, $n^* = 4.2$), which experiences a Z^* of +2.60. (a) Is orbital A or orbital B more likely a valence orbital? (b) Compute $<r_{max}>$ for this atom.

9.39. *(a) From Figure 9.8, find Z^*_{4s} for Zn. (b) Using Slater's rules, calculate Z^*_{3d} for Zn. (c) Calculate $<r_{max}>$ for these two orbitals of zinc.

9.40. In a certain atom, an electron in orbital C experiences a Z^* of +14.00, while an electron in orbital D experiences a Z^* of +4.85. Suppose also that the outermost s orbital of this atom is a 6s orbital ($n = 6$, $n^* = 4.2$), which experiences a Z^* of +3.60. (a) Is orbital C or orbital D more likely a valence orbital? (b) Compute $<r_{max}>$ for this atom.

9.41. *Using your results from Exercise 9.37, tell which orbital of Fe (4s or 3d) is likely to give better overlap with the orbital of another atom bonded to Fe? Briefly explain why.

9.42. Application of Slater's rules leads you to calculate the following Z^* values for Ti and Zn: 3.15 for the $4s$ orbital of Ti, 3.65 for the $3d$ orbital of Ti, and 4.35 for the $4s$ orbital of Zn. (a) Use Slater's rules to calculate Z^* for the $3d$ orbital of Zn. (b) Now calculate $<r_{max}>$ for the $3d$ and $4s$ orbitals in Ti; for the $3d$ and $4s$ orbitals in Zn. (c) In the text, it was noted that $3d$ orbitals are not as good at covalent overlap with ligand orbitals as $4s$ orbitals. Is this equally true for Zn and Ti? If not, for which element (Ti or Zn) is the statement more strongly true? (d) Explain your calculated trend in terms of penetration or shielding of orbitals. Explain how your results are relevant to the fact that Ti and its neighbors form more stable oxo cations than do Zn and its neighbors.

9.43. *Using characteristic electron configurations, compute Z^* and $<r_{max}>$ for electrons in the valence d and s orbitals of (a) Cr, (b) Mo, and (c) W. Explain whether or not these calculations show the lesser covalent-bond overlapping ability of the Cr $3d$ orbitals as compared to the Mo $4d$ and the W $5d$ orbitals. What factor in Slater's rules is responsible?

9.44. Two of the elements in Exercise 9.43, Cr and Mo, actually have the "anomalous" electron configurations $3d^54s^1$ and $4d^55s^1$, respectively, while W has the predicted electron configuration. Repeat the calculations of Exercise 9.43 using the actual electron configurations, and determine whether the conclusions of Exercise 9.43 are altered.

9.45. *The covalent radius of zinc is 121 pm; calculate the Allred–Rochow electronegativity of Zn. How do you think this value compares with the Pauling value for Zn? If you think that it is either substantially (>0.5 units) higher or substantially lower, explain why this is the case.

9.46. The Allred–Rochow electronegativity of Rn is not known; assuming a covalent radius for Rn of 130 pm, calculate it.

9.47. *In a certain atom, an electron in orbital A experiences a Z^* of +4.00, while an electron in orbital B experiences a Z^* of +21.85. Suppose also that the outermost s orbital of this atom is a $6s$ orbital ($n = 6$, $n^* = 4.2$), which experiences a Z^* of +2.60, and suppose that the covalent radius of this atom is 170 pm. Compute χ_{AR} for this atom.

9.48. In a certain atom, an electron in orbital C experiences a Z^* of +14.00, while an electron in orbital D experiences a Z^* of +4.85. Suppose also that the outermost s orbital of this atom is a $6s$ orbital ($n = 6$, $n^* = 4.2$), which experiences a Z^* of +3.60, and suppose that the covalent radius of this atom is 190 pm. Compute χ_{AR} for this atom.

9.49. *Find Z^* for the valence s orbital, then compute $<r_{max}>$ and χ_{AR} (a) for Na and Cu; (b) for W, Nd, and U. Before World War II periodic tables had U listed below W, not Nd. From these numbers, does U appear to be more closely related to Nd or to W?

9.50. Identify all elements in the periodic table for which the Pauling and Allred–Rochow electronegativities differ by more than 0.5 units. In what parts of the table do these occur?

9.51. *Briefly explain (in terms of the atomic numbers Z and the principal quantum numbers n, or in terms of the shielding and penetrating powers of the orbitals why (a) the Mg atom is more electronegative than the Ca atom; (b) the Si atom is more electronegative than the Al atom.

9.52. The standard reduction potential for Rg^+ has been estimated to be +3.2 V [R. Hancock, L. J. Bartolotti, J. Libero, and N. Kaltsoyannis, *Inorg. Chem.* 45, 10780 (2006)]. (a) Use the empirical relationship found in Chapter 6 to estimate χ_P for Rg. (b) Use this value and the charge and size of Rg^+ given to estimate the pK_a of Rg^+. (c) Now classify the acidity of Rg^+. Does it work out to be a "very strong acid" as claimed?

9.53. *Explain why relativistic effects can cause gold to have both (a) a lower oxidation number than either Ag or Cu (–1 in CsAu) and (b) a higher oxidation number than either Ag or Cu (+5 in AuF_5).

9.54. Gold shows the greatest discrepancy between its Allred–Rochow electronegativity, 1.42, and its Pauling electronegativity, 2.54. Two theoretical effects are responsible for the discrepancy. (a) Name these two effects and describe them in terms of penetration or shielding. (b) For each of the two causes, tell whether the discrepancy should be the same, worse, or less for Rg, the element below gold. Overall, would you expect the electronegativity of Rg to be greater than, less than, or equal to 2.54? (c) Name an element in an entirely different group of the periodic table to which the chemistry of Rg might be more related (than it would be to the chemistry of copper in Group 11).

9.55. *Which element is most likely to have the indicated property: (a) the higher electronegativity: Ba or Ra; (b) the larger covalent radius: Hg or Cn; (c) the higher value of Z^*: Au or Rg?

9.56. Choose one of the given causes as the best explanation of each of the effects listed later. CAUSES: (1) Scandide contraction. (2) Lanthanide contraction. (3) Relativistic increase in mass of a valence s electron penetrating a heavy nucleus. (4) Relativistic shielding of a valence d or f electron by massive s electrons. (5) No special cause needed; this is a normal periodic trend. EFFECTS: (a) Gold has the highest electron affinity of any element other than a halogen. (b) Gold has a +5 oxidation state, higher than is known for Cu or Ag. (c) It took 100 years to discover that the element hafnium is always present in samples of the element zirconium. (d) Gallium is more electronegative than aluminum. (e) Indium is less electronegative than gallium.

9.57. *Add the calculated stable oxidation states for elements 111–117 to Table 1.11. (a) Which of these oxidation states are not found for the elements above them? (b) What relativistic properties might make these new common oxidation states stable?

9.58. Exercise 3.89 gave the reference to calculations that suggested that iridium might exhibit an oxidation state of +9; this prediction was subsequently confirmed. These calculations suggested that the formation of the cation IrO_4^+ (as a salt with a large nonbasic anion) by oxidation of IrO_2 with the salt of the known O_2^+ cation could be exothermic, and its decomposition to $IrO_2^+ + O_2$ could be endothermic. They also calculated the possibilities for the corresponding meitnerium cation MtO_4^+. (a) For which of the two reactions—the formation or the decomposition—did the authors construct and evaluate a thermochemical cycle? (b) Did the authors conclude that the MtO_4^+ cation could be formed exothermically? (c) The valence d orbitals of which element—Ir or Mt—would be more affected by spin-orbit coupling? (d) Is this spin-orbit coupling more relevant to the stability of the MO_4^+ cation or its MO_2^+ decomposition product? Why? (e) The authors concluded that which cation, IrO_4^+ or MtO_4^+, should be more stable?

9.59. *Spin-orbit coupling effects are likely to be the largest in (a) the $7s$ subshell or the $7p$ subshell; (b) the $5d$ subshell or the $6p$ subshell; (c) Sb, Bi, or Mc (moscovium, element 115)?

9.60. Write valence electron configurations reflecting spin-orbit coupling for the (ions of) elements Pb, Bi, Po, At, or Rn in the following compounds: (a) "BiX"; (b) Po^{2+}; (c) At^+; (d) RnF_2; (e) RnF_4.

9.61. *Pershina[33] in her Table 17 and Figures 16–18 gives estimated $E°$ values (in volts) for the first transactinide aqueous ions.

$Lr^{3+} + e^- \rightarrow Lr^{2+}$; $E° = -2.6$ V; $Lr^{3+} + 3e^- \rightarrow Lr$; $E° = -2.06$ V (from Figure 6.14)

$Rf^{4+} + e^- \rightarrow Rf^{3+}$; $E° = -1.5$ V; $Rf^{3+} + e^- \rightarrow Rf^{2+}$; $E° = -1.7$ V

$Db_2O_5 + 5e^- \rightarrow Db$; $E° = -0.81$ V

$SgO_4^{2-} + e^- \rightarrow SgO_4^{3-}$; $E° = -1.34$ V.

(a) Draw redox predominance diagrams for these elements. (b) Compare these with the lighter elements in their groups. Do these four transactinide elements show the expected periodic trends in the stability of the group oxidation state?

9.62. Pyykkö's calculations[35] suggest the following order of filling of orbital groupings in atoms of the new eighth period of the periodic table: $8s$ before $5g$ before $8p_{1/2}$ before $6f$ before $7d$ before $9s$ before $9p_{1/2}$ before $8p_{3/2}$ (which is filled in the ninth period). (a) The placement of which of these orbital groupings violates the pattern established in the rest of the periodic table? (b) What relativistic effects are responsible for the changes for these orbital groupings?

MAURITS CORNELIS ESCHER (1898–1972) was
born in Leeuwarden, the Netherlands. In primary and
secondary school, Escher excelled at drawing, but his other
grades tended to be poor. He failed his high school exams,
but he eventually enrolled in the School for Architecture
and Decorative Arts in Haarlem. After graduating in 1922,
he traveled extensively in Italy and Spain. The intricate
designs and geometrical patterns of the Alhambra in
Granada inspired his interest in the mathematics of
tessellation—that is, the tiling of a flat surface using one
or more geometric shapes with no overlaps and no gaps.
Escher lived in Rome until 1935, when the political climate
under Mussolini became unacceptable. Escher then moved
his wife and children first to Switzerland, then to Belgium,
and finally to Baarn, the Netherlands, in 1941. Escher's
love of symmetry and interlocking patterns permeates his
many lithographs, his woodcuts and wood engravings, and
his drawings and sketches. Besides his work as a graphic
artist, Escher illustrated books and designed tapestries,
postage stamps, and murals.

CHAPTER 10

Symmetry

We begin our study of more advanced bonding concepts with a consideration of a topic that would seem to belong to the realm of aesthetics. One of the properties that goes into making a work of art or nature (Figure 10.1) or a person, beautiful, is the presence of elements of **symmetry**. Molecules, too, may have greater or lesser degrees of symmetry. One of the things that surely has contributed to the interest in the discovery of one of the new forms of the element carbon, buckminsterfullerene (C_{60}, Figure 10.1h), is the fact that it is more symmetrical than almost any other nonlinear polyatomic molecule that has ever been discovered. Although buckminsterfullerene has so beautiful a structure that we feel compelled to humble it by calling it "buckyballs," such beauty (rooted in symmetry) has some real chemical consequences that we shall see here in Chapter 10 and in Chapter 11. There may even be evolutionary reasons why people and animals prefer symmetrical (beautiful) mates.[1]

This brief study of symmetry and group theory is a useful foundation for the in-depth study of many topics in inorganic chemistry, physical chemistry, and organic chemistry:

- The presence or absence of symmetry operations and elements for a molecule or polyatomic ion (Section 10.1) are useful in determining whether the species is chiral or polar (Section 10.3).

- The symmetry point group of the molecule or polyatomic ion (Section 10.2) is useful in constructing molecular orbitals for the species (Section 10.4 and Chapter 11).

- The presence or absence of the inversion center (Section 10.1) and other symmetry properties of the molecular orbitals determine important spectral properties, such as the higher intensity of charge-transfer transitions (Section 6.4C) as compared to electronic transitions between d orbitals (Section 7.3). Symmetry labels such as e_g and t_{2g} (Chapter 7) come from Section 10.3, so your instructor may choose to cover this chapter before Chapter 7.

Figure 10.1. Symmetry in art and nature. (a) The Eiffel Tower. (b) The "Star of Bethlehem," *Campanula isophylla* "*Mayi.*" (c) Three-blade propeller. (d) Children's toy pinwheel. (e) Radiolarian. (f) Radiolarian. (g) Japanese crest. (h) Buckminsterfullerene, C_{60}.

10.1. Symmetry Operations and Elements

OVERVIEW. There are five basic types of symmetry elements (and of operations around those elements), as summarized in Table 10.1: (rotation around an) n-fold proper rotation axis C_n, (reflection through a) mirror plane σ, (inversion through an) inversion center i, (rotation then reflection through a perpendicular plane around an) improper rotation axis S_n, and (doing nothing via an) identity E. You may practice identifying symmetry elements in a molecule, ion, or object by trying Exercises 10.1–10.6.

If necessary, you may want again to review the drawing of Lewis structures (Section 3.1A) and the predictions of molecular geometries using VSEPR theory (Section 5.1A). For square planar complexes in the d block, you may want to revisit Section 7.6B.

Symmetry is defined in Webster's dictionary as "similarity of form or arrangement on either side of a dividing line or plane," with "correspondence of opposite parts in size, shape, and position." Operationally, we can state that if a molecule or ion has two or more orientations that are indistinguishable, that molecule or ion possesses symmetry.

There are only a few ways in which we can reorient molecules or ions to see whether all parts coincide after the operation. These reorientations or **symmetry operations** take place about the five kinds of points, lines (axes), or planes listed in Table 10.1, which are the possible **symmetry elements** of the molecule or ion being tested. (The effects of these operations are best appreciated by working with three-dimensional models of the molecules,[2] because most of the operations are difficult to envision with two-dimensional drawings of molecules on paper.)

Of these elements and operations, the most trivial-seeming is that of the **identity** operation, which consists of leaving the molecule or ion alone while your eyes are closed; all molecules remain unchanged upon carrying out this operation. Its listing is required by the mathematical properties of group theory; the symbol for the identity element and operation is E.

The nature of the other elements and operations are best illustrated with concrete examples; we begin with H_2O. The most prominent symmetry element in most molecules is that of the **n-fold proper rotation axis** (symbol C_n). The corresponding operation is one of rotation of the molecule by an angle of $360°/n$ about this axis. In water

TABLE 10.1. Elements and Operations of Molecular Symmetry

Symmetry Element	Symbol	Operation	Example
Identity	E	Do nothing: leave all parts in place	
n-Fold proper axis	C_n	Rotation by $360°/n$ about this axis	Figure 10.2(a)
Mirror plane	σ	Reflection through this plane	Figure 10.2(b)
Inversion center	i	Inversion through this point	Figure 10.4
n-Fold improper axis	S_n	Rotation by $360°/n$ followed by reflection through a plane ⊥ this axis	Figure 10.5

this axis passes through the oxygen atom and midway between the hydrogen atoms. The two hydrogen atoms are interchanged by rotating by 180° about this C_2 (*twofold rotation*) axis (Figure 10.2a). One rotation of 180° about this axis generates a new but indistinguishable orientation of the molecule. (In Figure 10.2a the brown and white atoms have been interchanged, but the real atoms are indistinguishable because they are neither brown nor white). Carrying out the rotation operation twice, however, puts each H atom back in its original position (brown to brown and white to white), so it is the same as the identity operation. (We can write this symbolically as $C_3^3 \equiv E$.)

Molecules with more symmetry than H_2O may have more than one C_n axis, and they may have axes in which n exceeds 2. The rotation axis with the highest value of n is designated the *principal rotation axis*. Its direction is taken as the z (vertical) direction in the molecule for purposes of labeling in symmetry.

The second most prominent type of symmetry element is the **mirror plane**(s) (symbol σ) that may be present in a molecule or ion. As shown in Figure 10.2b, H_2O has two mirror planes. Both of these include the vertical C_2 axis, so both are **vertical** mirror planes, and are designated by the symbol $σ_v$. The mirror plane including all

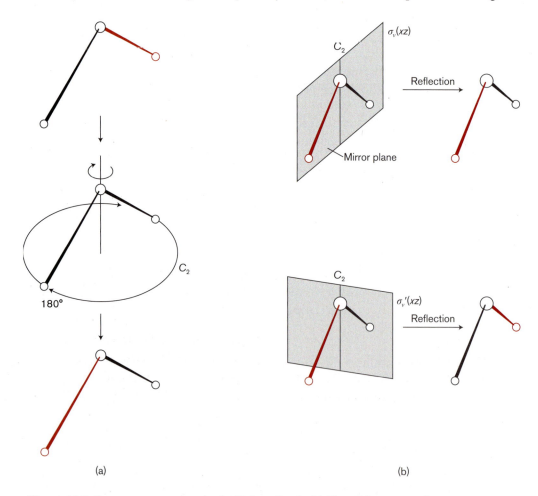

(a)   (b)

Figure 10.2. Symmetry operations in the H_2O molecule. (a) The 180° rotation of a water molecule about its C_{2v} axis, an operation that leaves it apparently unchanged but actually interchanges the white and brown hydrogen atoms; (b) the two vertical mirror planes in the water molecule, $σ_v(xz)$ and $σ_v'(yz)$, and the corresponding reflection operations.

EXAMPLE 10.1

(a) What are the rotation axes C_n with the highest value of n in tetrahedral CCl_4 and square planar $[AuCl_4]^-$? (b) How many times do the corresponding symmetry operations have to be carried out on each to generate an identity element? (c) Do the repetitions of any of these rotation operations generate a distinct rotation axis?

SOLUTION: (a) In Figure 10.3 we draw structures of these species, with the chlorine atoms labeled with letters A, B, C, and D. CCl_4 is three-dimensional; the highest rotation axes are C_3 axes, coinciding with any of the C–Cl bonds. $AuCl_4^-$ is planar, and its highest rotation axis is a C_4 axis, at the Au atom but perpendicular to the plane of the species.

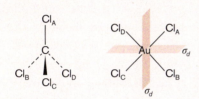

Figure 10.3. Structures of tetrahedral CCl_4 and square planar $[AuCl_4]^-$, with the outer chlorine atoms individually lettered to aid in visualizing the effects of carrying out symmetry operations. Dihedral mirror planes of $[AuCl_4]^-$ (σ_d) are shown in brown.

(b) As shown below, carrying out the threefold rotation once around the C–Cl_A axis of CCl_4 generates an equivalent but nonidentical orientation in which Cl_B, Cl_C, and Cl_D have been interchanged. Carrying out this operation again further interchanges these three atoms. But carrying out the operation a third time brings each Cl atom to the identical position at which it started (Cl_B at Cl_B, etc.), so an identity element E is generated.

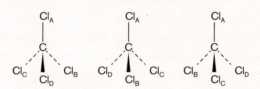

(b) and (c) In $AuCl_4^-$ the fourfold rotation has to be carried out four times to generate an identical orientation and the identity element. However, carrying it out twice is identical to carrying out a C_2 rotation about the same axis, so a C_2 rotation axis (collinear with the C_4 axis) is generated. Note that both the C_2 rotation and the C_4 rotation carried out twice interchange Cl_A with Cl_C and Cl_B with Cl_D.

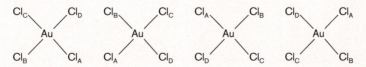

three atoms of this planar molecule is finally labeled the $\sigma_v(xz)$ mirror plane; the plane bisecting the molecule is labeled the $\sigma_v'(yz)$ plane. There is one independent operation—reflection one time—that can be carried out about each of these planes.

There are two other types of mirror planes that can be found in $[AuCl_4]^-$ (Figure 10.3b). **Horizontal** mirror planes, σ_h, are perpendicular to ($\perp$) the (vertically oriented) principal rotation axis. In $[AuCl_4]^-$ the plane of the species is a horizontal mirror plane. **Dihedral** mirror planes, σ_d, are similar to vertical mirror planes in that they also include the principal rotation axis, but they are found to fall between adjacent bonds in the molecule, thus bisecting these bond angles. In $[AuCl_4]^-$ there are two such planes that include the Au atom but not any of the Au–Cl bonds); these are shown in **brown**. One cuts from top to bottom, interchanging Cl_A with Cl_D and interchanging Cl_B with Cl_C. The other cuts from left to right, interchanging Cl_A with Cl_B and interchanging Cl_C with Cl_D.

Because there are no other elements of symmetry present in the water molecule, there are four (independent) symmetry operations possible: E, C_2, $\sigma_v(xz)$, and $\sigma_v(yz)$. These labels also identify the four symmetry elements of the water molecule.

To illustrate the other two types of symmetry elements, we select examples in which each type is especially prominent, beginning with the staggered conformation of *meso*-1,2-dibromo-1,2-dichloroethane, BrClHC–CHClBr, shown in Figure 10.4. The only symmetry operation and element (other than identity) present in this conformation is **inversion** about an **inversion center** (symbol i). This operation consists of taking every atom with atomic coordinates (x,y,z) through the center of the molecule and out the other side an equal distance until the position $(-x,-y,-z)$ is reached. In this process each Br passes through the center of the molecule (the midpoint of the C–C bond) and is exchanged with the other Br; the same occurs for the H, Cl, and C atoms.

The *n*-**fold improper rotation** (or **rotation–reflection**) axis is especially prominent in the organic molecule allene (propadiene) (Figure 10.5). In this molecule the two π bonds are cumulated (they both involve the same carbon atom), so one must be in the plane of the paper while the other is above and below the paper. This forces the CH_2 group at one end of the paper to be perpendicular to the plane of the paper while the other CH_2 group is in the plane of the paper. In this molecule the long molecular axis is a fourfold improper rotation axis, S_4. A S_n improper rotation consists of

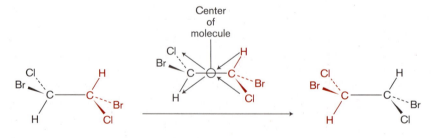

Figure 10.4. One conformation of *meso*-CHBrCl–CHBrCl. The inversion operation of the pairs of identical substituents (black and **brown**) is partially illustrated in the middle of the figure by showing the motion of **Cl** and **H** through the center of the molecule and out the same distance in the same direction to the positions of Cl and H, respectively. The complete results of the inversion operation are shown at the right.

rotation by 360°/*n* about the axis, followed by a *reflection* through a plane perpendicular to the rotation axis. In the case of allene, neither the rotation nor the reflection by itself is a symmetry element. The fourfold rotation takes H_A 90° into the plane of the paper, where no H atom previously was located. The reflection then takes it across the central mirror plane to the position where H_D was previously located, meanwhile placing H_D where H_A had been. Simultaneously, H_B and H_C are interchanged.

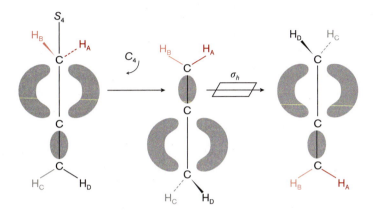

Figure 10.5.
Improper rotation about the S_4 axis in allene, C_3H_4.

EXAMPLE 10.2

Identify and count all symmetry elements and operations in the trigonal bipyramidal molecule PCl_5 (Figure 10.6a).

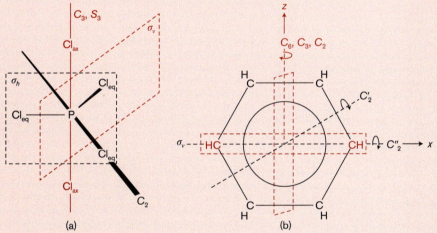

(a)   (b)

Figure 10.6.
(a) Phosphorus pentachloride, showing its C_3 axis, its σ_h mirror plane, one of three σ_v mirror planes, and one of three C_2 axes. (b) Benzene, showing its C_6, C_3, and C_2 axes. Axes, atoms, and bonds that are above and below the equatorial plane (the plane of the paper) are shown in **brown**.

SOLUTION: This molecule, like all others, possesses the identity element, *E*. Its most obvious element of symmetry is the threefold rotation axis, C_3, which passes through P and the two axial Cl atoms. Two independent operations can be carried out about this axis: rotation one time, designated C_3, and rotation twice, designated C_3^2. Each generates a new (although indistinguishable) orientation of the molecule.

The C_3 axis is the principal rotation axis of this molecule, but it is not its only rotation axis. PCl_5 can also be rotated by 180° about any of the three equatorial P–Cl bonds; as a result, the molecule also has three C_2 axes and three C_2 operations.

Example 10.2 continues on next page ▶

EXAMPLE 10.2 *(cont.)*

Among mirror planes, the most prominent is the equatorial plane of the molecule. With the principal C_3 axis being vertical, this plane is placed horizontally ($\perp C_3$), so it is designated as a horizontal mirror plane, σ_h. But this is not the only mirror plane: There are three vertical mirror planes, σ_v, each including the vertical C_3 axis and one of the three equatorial P–Cl bonds.

PCl$_5$ has a threefold improper rotation axis, S_3; it coincides with the C_3 axis. The rotation part of this operation does nothing to the axial chlorines but does interchange the equatorial ones. The reflection part of the operation through the (horizontal, in this case) reflection (mirror) plane does nothing to the equatorial chlorine atoms in this plane, but does interchange the axial chlorines. Overall, every chlorine undergoes an interchange upon carrying out the S_3 operation.

There is no inversion center in this molecule, because inversion carries an equatorial chlorine into a position that was vacant before.

In summary, the PCl$_5$ molecule has the following symmetry elements: E, C_3, three different C_2's, σ_h, three different σ_v's, and S_3, for a total of 10. Each of these elements has one (independent) operation associated with it, except that each threefold axis—C_3 and S_3—has two operations associated with each: C_3, C_3^2; and S_3, S_3^2. Hence, there are 12 symmetry operations.

EXAMPLE 10.3

Identify and count all symmetry elements and operations in the hexagonal planar molecule benzene, C$_6$H$_6$ (Figure 10.6b).

SOLUTION: C$_6$H$_6$ has the E element. The highest rotation axis in this molecule is the sixfold one, C_6. The operations involving this axis include rotations by 60° (C_6), 120° ($C_6^2 \equiv C_3$), 180° ($C_6^3 \equiv C_2$), 240° ($C_6^4 \equiv C_3^2$), 300° (C_6^5), and 360° ($C_6^6 \equiv E$). This sixfold axis is also a threefold axis (the operations C_3 and C_3^2 result from rotation about a C_3 axis) and a twofold axis. There are two other types of C_2 axes, however, running across the plane of the benzene ring. Three of them cut between carbon atoms and three of them include pairs of C–H bonds; these are labeled C_2' and C_2'' axes to distinguish them from the C_2 axis that coincides with the C_6 axis.

Benzene is also rich in mirror planes. The plane of the molecule is perpendicular to the (vertical) principal (C_6) axis, so it is a σ_h (horizontal mirror plane). There are also six vertical mirror planes: Three of them include pairs of C–H bonds and are labeled σ_v (vertical mirror planes), whereas the other three of them cut between the C atoms and C–H bonds and are labeled σ_d (dihedral mirror planes).

The principal rotation axis of benzene coincides with an improper rotation axis: It is an S_6 axis (with two operations) and an S_3 axis (with two operations). The principal rotation axis is also an S_2 axis; however, this symmetry element is identically equal to the remaining type of symmetry element, the inversion center.

The complete list of symmetry elements in as symmetric a molecule as benzene is substantial: E, C_6, C_3, C_2, three C_2', three C_2'', σ_h, three σ_v, three σ_d, S_6, S_3, and i, for a total of 20 elements of symmetry of 12 types. Because the C_6, C_3, S_6, and S_3 elements each have two operations, the total number of symmetry operations is 24.

10.2. Molecular Point Groups, Polarity, and Chirality

OVERVIEW. Many molecules and ions can have a common set of symmetry elements and operations and belong to the same molecular point group. We identify 10 families of common point groups. The flow chart of Figure 10.9 can be used to identify the point group to which most molecules and ions belong. With the help of this flow chart, you can use the symmetry elements of a molecule or ion to identify its symmetry point group. You may practice this Section 10.2A concept by trying Exercises 10.7–10.15. In Section 10.2B we find that polar molecules belong to one of to the C_n, C_{nv}, S_n, and C_s point groups. Chiral molecules belong to one of the C_n or D_n point groups or the low-symmetry C_1 group. You can practice applying this concept by trying Exercises 10.16–10.22.

10.2A. Molecular Point Groups. It is fortunate that there is no need to go through as detailed an analysis and as careful a count for every molecule as we did for benzene! First, many molecules and ions [e.g., C_6F_6, C_6Cl_6, and $C_6(CH_3)_6$, if the methyl groups are treated as freely rotating symmetric tops] have exactly the same symmetry operations and elements as benzene itself does. Such molecules or ions with the same symmetry operations and elements are said to belong to the same molecular **point group**. There are symbols for each point group: The point group including benzene and its relatives is labeled the D_{6h} point group. Similarly, the point group of water is labeled C_{2v}, and it includes not only very similar molecules such as H_2S, but also H_2CCl_2 and $COCl_2$ (phosgene). The point group of PCl_5 is labeled D_{3h}, and it includes such diverse molecules as the trigonal planar BF_3, the planar benzene derivative 1,3,5-tribromobenzene, and the nine-coordinate tricapped trigonal planar anion $[ReH_9]^{2-}$ (Figure 5.2).

The chemically most important point groups can be grouped into several sets, A–J; we list them in the order that they will be encountered in the procedure for assigning point groups.

A. The **linear** point groups $D_{\infty h}$ and $C_{\infty v}$ apply to homonuclear diatomic molecules, M–M, and heteronuclear diatomic molecules, M–N, respectively. Their symmetry is preserved upon rotation about an infinitesimally small angle; this operation can be carried out an infinite number of times.

B. The **higher-order (high-symmetry)** groups have more than one three- or higher-fold rotation axes. These groups include the tetrahedral point group T_d, which has four C_3 axes and includes CH_4; the octahedral point group O_h, which has three C_4 axes, four C_3 axes, and i, and includes SF_6; and the icosahedral point group I_h, which includes six C_5 axes and 120 symmetry operations, and includes buckminsterfullerene (Figure 10.1h) and methane hydrate (Figure 5.21).

C. The **nonaxial** groups have no rotational axes and can be characterized as **low-symmetry** groups. These include three point groups: C_1, which has only the identity element, E (e.g., CFBrClI in Figure 10.7); C_s, which has only E and a mirror plane, σ (e.g., $H_2C=CFCl$, $SOCl_2$, and the external human body); and C_i, which has only E and an inversion center, i (e.g., BrClHC–CHClBr in Figure 10.4).

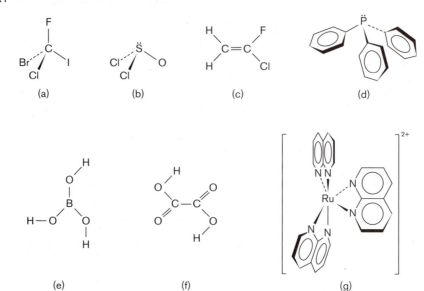

Figure 10.7.
Examples of molecules or conformations of molecules for the nonaxial, C_n, C_{nh}, and D_n point groups.

(a)   (b)   (c)   (d)

(e)   (f)   (g)

D. The D_{nh} point groups characteristically contain the prominent σ_h. A square planar complex such as $[AuCl_4]^-$ falls in the D_{4h} point group.

E. The D_{nd} point groups have the C_n axis and n perpendicular C_2 axes characteristic of all the D groups, and characteristically add n σ_d planes. The allene molecule, $CH_2=C=CH_2$ (Figure 10.5), falls in the D_{2d} point group.

F. The D_n groups have a C_n axis and n C_2 axes perpendicular to this principal axis. An example of an ion in the D_3 point group is the tris-chelated complex ion $[Ru(1,10\text{-phenanthroline})_3]^{2+}$ (Figure 10.7g).

G. The C_{nh} groups characteristically have a C_n axis and σ_h (other elements and operations are present in particular point groups). An example in the C_{3h} point group is $B(OH)_3$, which has a propeller-like arrangement of OH groups around the trigonal planar B atom. $H_2C_2O_4$ (oxalic acid) is similarly an example of the C_{2h} point group (Figure 10.7e, f).

H. The C_{nv} groups have a C_n axis and n vertical and/or dihedral mirror planes. We have seen H_2O and other examples of the C_{2v} point group; pyramidal NH_3 is an example of the C_{3v} point group.

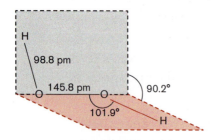

Figure 10.8. The structure of H_2O_2 in the solid state. [Adapted from N. N. Greenwood and A. Earnshaw, *Chemistry of the Elements*, 2nd ed., Butterworth-Heinemann: Oxford, UK, 1997, p. 635.]

I. The S_n groups (n = an even number ≥ 4) have E plus S_n symmetry elements, and no other elements except the $C_{n/2}$ elements that arise from such identities as $C_2 \equiv S_4^2$, and so on. An example of a molecule in the S_2 point group is H_2O_2 (Figure 10.8).

J. The C_n groups have a C_n axis. An example of a molecule in the C_3 point group is triphenylphosphine, $:P(C_6H_5)_3$, in which the three phenyl groups have a propeller-like arrangement about the trigonal pyramidal P atom (Figure 10.7d).

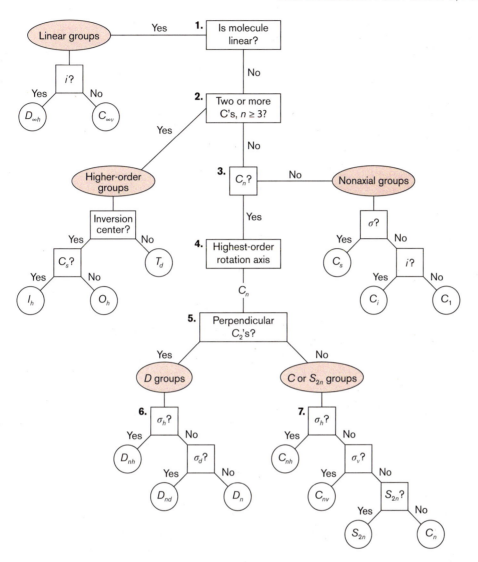

Figure 10.9. Diagram of procedure for assigning point groups. **Brown** ovals enclose the five larger categories of point groups. Rectangular boxes numbered 1 through 7 correspond to the seven classification questions or steps given in the text. Circles enclose specific sets of symmetry point groups. [Adapted from G. L. Miessler and D. A. Tarr, *Inorganic Chemistry*, Prentice Hall: Englewood Cliffs, NJ, 1991, p. 97; and P. Atkins and J. de Paula, *Physical Chemistry*, 7th ed., W. H. Freeman: New York, 2002, p. 460.]

The use of the applications of symmetry and group theory depend on the assignment of molecules and ions to their appropriate symmetry point groups (which in turn must follow the determination of their shape using VSEPR theory; Section 3.4). We suggest the following procedure, which is diagrammed as a flow chart in Figure 10.9. Steps 1, 2, and 3 of the procedure pull out the first three special classes of molecules—namely, *linear* molecules, *high-symmetry* molecules, and *nonaxial* molecules. The next four steps are then used to classify the many remaining molecules.

1. If the molecule is *linear* (A, above), then look for an inversion center. If i is found, the molecule belongs to the point group $D_{\infty h}$; if there is no inversion center, it belongs to $C_{\infty v}$.

2. If the molecule is *not* linear, determine whether the molecule belongs to one of the *high-symmetry* groups (B, above). These symmetry groups are identified by the presence of two or more threefold or higher rotation axes.

   Then, if no inversion center is found, the molecule is tetrahedral (T_d). If an inversion center and a fivefold axis are found, the molecule is icosahedral (I_h). Otherwise it is octahedral (O_h).

3. Is there an axis of rotation? If no, then the molecule belongs to one of the *non-axial* point groups (C, above), depending on what symmetry elements it has.

4. If the molecule has an axis of rotation, then it belongs to one of the medium-symmetry point groups found at the center and base of Figure 10.9. Next, find the highest-order rotation axis (C_n with highest n) in the molecule.

5. Does the molecule have any C_2 axes perpendicular to the highest-order C_n axis? If the answer is yes, then the molecule falls in a D point group). If the answer is no, then the molecule falls in a C or S_{2n} point group.

6. Does the D-group molecule have a horizontal mirror plane (σ_h) perpendicular to the highest-order rotation axis? If the answer is yes, then it belongs in a D_{nh} point group (D, above).

   If no, see whether it has a dihedral mirror plane (σ_d). If yes, it belongs to a D_{nd} point group (E, above). If no, it belongs to a D_n point group (F, above)

7. Does the C- or S_{2n}-group molecule have a horizontal mirror plane (σ_h) perpendicular to the highest-order rotation axis? If yes, then it is classified in a C_{nh} point group (G, above).

   If the answer is no, does the molecule have any vertical mirror planes? If the answer is yes, the molecule is classified in a C_{nv} point group (H, above).

   If the answer is no, is there an S_{2n} axis collinear with the C_n axis? If this answer is yes, then the molecule is finally classified in an S_{2n} point group (I, above); if this answer is no, the molecule is classified in a C_n point group (J, above).

EXAMPLE 10.4

Identify the point groups for the molecules PCl_5 (Example 10.2) and benzene (Example 10.3).

SOLUTION: If possible, build models of the molecules in question so you can visualize their structures. In any event, look for symmetry elements, as was already done in Examples 10.1 and 10.2. The flow chart (Figure 10.9) suggests the order in which to look for the symmetry elements, and then gives the final point group classification.

First we apply Step 1 to PCl_5: The molecule is *not* linear. Steps 2 and 3: The molecule has one and only one threefold rotation axis, so it does not belong to either the high-symmetry or the nonaxial point groups. Step 4: The rotation axis is a C_3 axis. Step 5: The molecule has twofold rotation axes perpendicular to the C_3 axis (Figure 10.6a). Therefore, it is in a D group. Step 6: There is a σ_h—namely, the plane including the P atom and three of the Cl's. Therefore, we can conclude that the molecule belongs to D_{3h}.

We apply Step 1 to C_6H_6: The molecule is *not* linear. Steps 2 and 3: The molecule has one sixfold axis but no other threefold or higher rotation axes, so it does not belong to either the high-symmetry or the nonaxial point groups. Step 4: The rotation axis is a C_6 axis. Step 5: The molecule has twofold rotation axes perpendicular to the C_6 axis (Figure 10.6b). Therefore, it is in a D group. Step 6: There is a σ_h—namely, the plane including the entire planar molecule. Therefore, we can conclude that the molecule belongs to D_{6h}.

EXAMPLE 10.5

Assign the complexes, ions, and molecules shown below to the appropriate symmetry point groups: (a) octahedral *cis*-$[CrL_2(CO)_4]$; (b) octahedral *trans*-$[CrL_2(CO)_4]$; (c) a free SO_4^{2-} anion; (d) a SO_4^{2-} ion coordinated to a metal in a linear fashion through one O donor atom; (e) a SO_4^{2-} ion bridging two metal ions; (f) S_2Cl_2; (g) $C_8H_4Cl_4$; (h) planar $C_8H_8^{2-}$; (i) $B_{12}H_{12}^{2-}$.

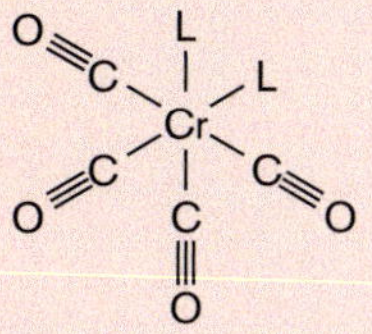

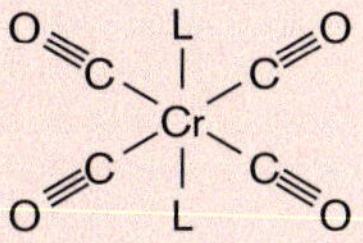

SOLUTION: Step 1. None of the molecules is linear.

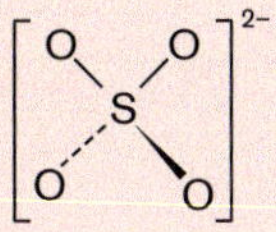

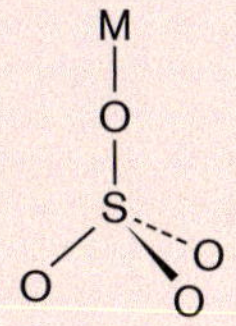

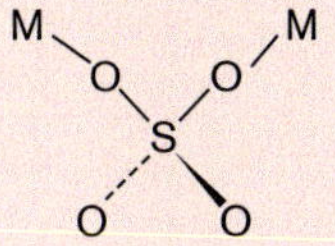

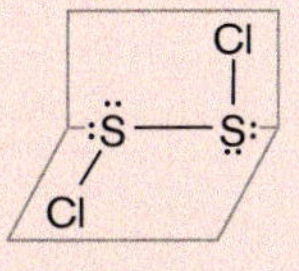

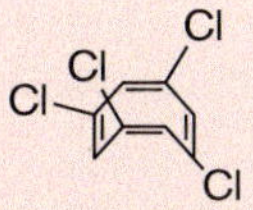

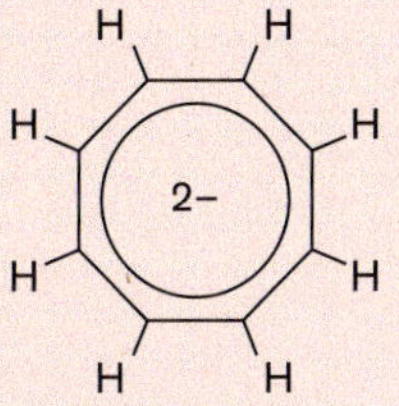

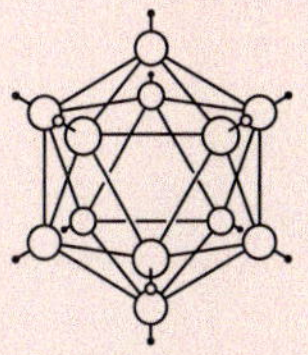

In Step 2, we note the presence of more than one of the highest-order C_3 axis in the tetrahedral SO_4^{2-} ion (c, above), which has no inversion center and therefore belongs to the high-symmetry T_d point group. We note the presence of more than one of the highest-order C_5 axes in the $B_{12}H_{12}^{2-}$ ion (i, above), and the presence of an inversion center. This ion therefore belongs to the high-symmetry I_h (icosahedral) point group.

Step 3. All of the remaining molecules have C_n axes, so none is in a nonaxial point group. [This is most difficult to see in molecule (f), but is illustrated below—note that this axis is not in either of the planes shown for this molecule, but falls halfway between them.]

Step 4. The highest-order C_n axis in each of the remaining molecules is as follows (as illustrated below): (a) a C_2 axis bisecting the angle between the two Cr–L bonds; (b) a C_4 axis including both Cr–L bonds; (d) a C_3 axis along the M–O–S axis; (e) a C_2 axis bisecting the angle between the two donor O–S bonds; (f) a C_2 axis located halfway between the two S–Cl bond planes and passing through the middle of the S–S bond; (g) a C_2 axis running top to bottom through the center of the molecule; (h) a C_8 axis running top to bottom through the center of the ion.

Example 10.5 continues on next page ▶

EXAMPLE 10.5 *(cont.)*

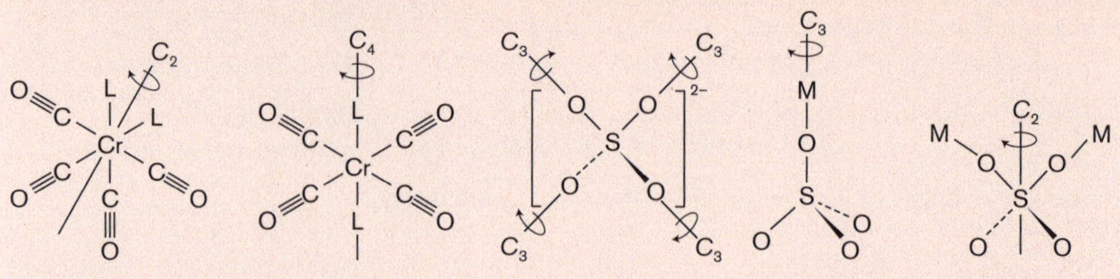

Step 5. The following cases have C_2 axes $\perp$ the C_n axis: (b) along the opposite OC–Cr–CO bonds; (h) along the opposite pairs of C–H bonds. Species (b) and (h) therefore belong to D point groups; (a), (d), (e), (f), and (g) belong to C or S point groups.

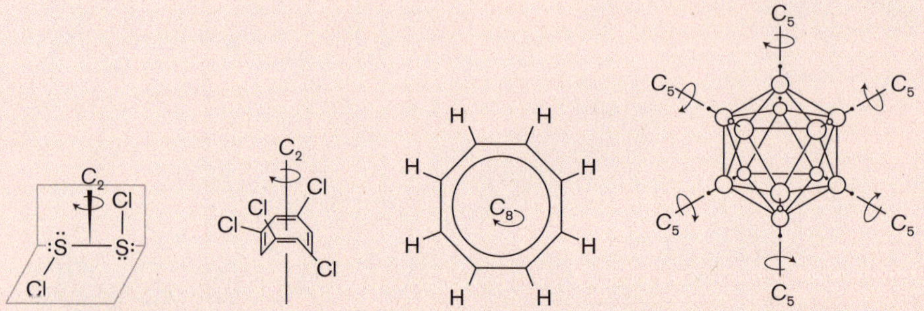

Step 6. Horizontal mirror planes (σ_h, $\perp$ the highest-order C_n axis) are found in (b), which therefore belongs to point group D_{4h}, and (h), which belongs to point group D_{8h}.

Step 7. Vertical or dihedral mirror planes (those containing the highest-order axis) are found in (a) and (e), both of which belong to point group C_{2v}, and (d), which belongs to point group C_{3v}. An S_4 axis (collinear with the C_2 axis) is found in (g), which therefore falls in point group S_4. No such axis is found in (f), which therefore falls in point group C_2.

In addition to molecules, polyatomic ions, and works of art, many macroscopic objects[3] have symmetry elements and can be assigned to point groups, including tennis balls,[4] hubcaps,[5] and tires.[6]

10.2B. Chirality and Polarity. Polarity of Molecules. One of the properties of molecules determined by their symmetry is their **polarity**. A polar molecule is a molecule with a permanent electric dipole moment: The molecule has positive and negative ends. (This dipole moment results mainly from an unbalanced arrangement of the polar bonds in the molecule, but has other contributions, such as from the polarity of any unshared electron pairs.) To be polar, the molecule *cannot* have symmetry operations interchanging *all* of the positive and negative ends of the molecule. In practice, a molecule cannot be polar if it has a principal C_n axis (to interchange ends of the x and y axes) *and* either a C_2 axis $\perp$ the principal axis, or a σ_h plane (to interchange ends of the z axis). Only molecules belonging to the C_n, C_{nv}, S_n, and C_s point groups can be polar (types H, I, J, and A (if in $C_{\infty v}$) in our listing in Section 10.2A).

Chiral molecules are those that are not superimposable on their mirror images in the same way that your right and left hands cannot be superimposed on each other. Such molecules rotate the plane of polarized light and hence are optically active. A chiral molecule can have no elements of symmetry other than proper rotation axes (and E). Chiral molecules, in practice, *must* belong to one of the C_n or D_n point groups (types F and J in our listing in Section 10.2A), or the low-symmetry C_1 group. (A chiral molecule and its mirror-image isomer are called *enantiomers*.)

Chirality and Polarity among Chelated Complexes and Geometric Isomers. Chirality occurs more frequently in complexes of chelating ligands than complexes of monodentate ligands, due to their reduced symmetry. As an example, the octahedral complex ion $[Co(NH_3)_6]^{3+}$ is of (for practical purpose) O_h symmetry (because the NH_3 ligands rotate very rapidly about the Co–N bonds). This complex can neither be polar nor chiral. However, the tris-chelate complex $[Co(\eta^2-NH_2CH_2CH_2NH_2)_3]^{3+}$ (similar to Figure 5.6a; see Section 5.1B on η designations of polydentate ligands) falls in the less-symmetric point group D_3, which allows chirality. (Unlike monodentate ligands, chelated ligands are not free to rotate around their donor atoms.)

Geometric isomerism (introduced earlier in Figure 3.6) also alters the symmetry point groups of complexes, so it can affect their polarity and chirality. Among disubstituted octahedral complexes, MA_4B_2, the two B groups may either be cis (adjacent) to each other (Figure 10.10a), or trans to (across from) each other (Figure 10.10b); these two geometric isomers fall in different point groups. In trisubstituted octahedral complexes, MA_3B_3, there are also two geometric isomers possible: The three ligands may all be on the same triangular face of the octahedron (the facial or fac isomer; Figure 10.10c), or they may lie in the same equatorial plane of the complex (the meridional or mer isomer; Figure 10.10d).

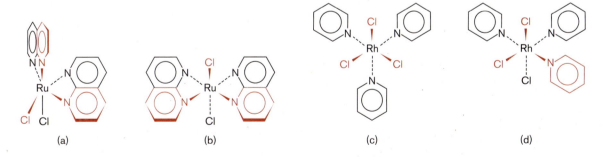

(a)   (b)   (c)   (d)

Figure 10.10. Examples of geometric isomers: (a) *cis-* and (b) *trans-*[Ru(1,10-phenanthroline)$_2$Cl$_2$]; (c) *fac-* and (d) *mer-*[RhCl$_3$(pyridine)$_3$]. The portions of each complex that are above the plane of the paper are shown in brown.

EXAMPLE 10.6

Determine whether the following isomers are polar, and whether they possess optical activity: (a) *cis*-[RuCl$_2$(1,10-phenanthroline)$_2$]; (b) *trans*-[RuCl$_2$(1,10-phenanthroline)$_2$]; (c) *fac*-[RhCl$_3$(pyridine)$_3$]; (d) *mer*-[RhCl$_3$(pyridine)$_3$]. The structures of these complexes are shown in Figure 10.10.

SOLUTION: We begin by assigning the molecules to point groups, from which the identification of the two properties follows readily. Isomer (a) possesses, as its only nontrivial element of symmetry, a C_2 axis that bisects the Cl–Ru–Cl angle. It thus falls into the C_2 point group and, with this low symmetry, can be (and is) polar and chiral. In isomer (b), the two phenanthroline rings are coplanar. The molecule possesses a principal C_2 axis (the Cl–Ru–Cl bonds), two other C_2 axes $\perp$ the principal axis, a σ_h plane, and so on. It falls in the D_{2h} point group, so it cannot be either polar or chiral.

Isomer (c) possesses a C_3 axis as viewed in Figure 10.10c; there are also three mirror planes including the C_3 axis. This molecule falls in point group C_{3v}. It can be polar, but it cannot be chiral. In isomer (d), the three N atoms form a coplanar arc, as do the three Cl atoms; the central-N–Rh–central-Cl axis is a C_2 axis. There are two vertical mirror planes: one including the three Cl and the central N donor atoms; the other including the three N and the central Cl donor atoms. The molecule is a member of the C_{2v} point group. It can be polar, but it cannot be chiral.

AN AMPLIFICATION

Can Chirality Only Be Produced by Living Organisms?

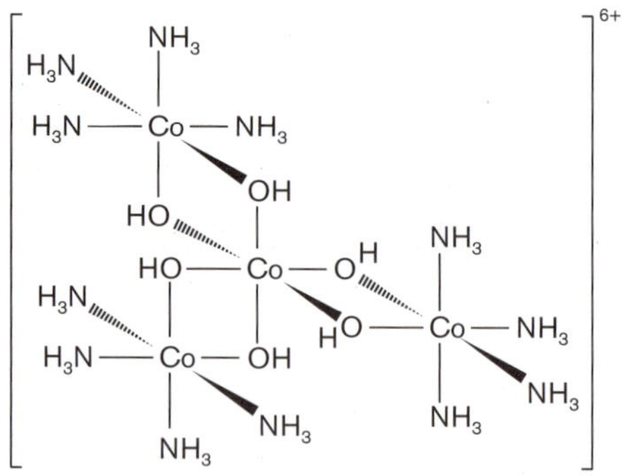

Over a century ago the founder of the systematic study of coordination compounds, Alfred Werner, relied heavily on the isolation and counting of isomers—both geometric and optical—to establish the characteristic geometry about a given metal ion. (He was the first inorganic chemist to win the Nobel Prize.) For example, if [Ru(η^2-1,10-phenanthroline)$_3$]$^{2+}$ were based on a trigonal prismatic rather than an octahedral geometry (Figure 10.7g), it would have two geometric isomers, but neither of these would be chiral. The detection of optical activity in a D_3 inorganic complex such as [Ru(1,10-phenanthroline)$_3$]$^{2+}$ was one more step in destroying the idea that the characteristic properties of bioorganic compounds such as chirality could be produced only by living organisms and hence found only in organic compounds. Some skeptics still pointed out that the chelate ligands were organic; perhaps the optical activity somehow resided in these (optically inactive) ligands. So, as a final proof, Werner prepared and resolved the carbon-free complex cation shown here.

10.3. Character Tables for Symmetry Point Groups

OVERVIEW. Character tables for symmetry point groups provide a wealth of information. The column headings identify the independent symmetry elements and operations in the point group. The left side of a table shows the molecule's irreducible representations (symmetry labels). Atomic and molecular orbitals generally lack the full symmetry of the molecule, but have partial symmetry corresponding to one of the irreducible representations. The body of a character table gives numbers (characters) indicating the results of applying the symmetry operations of the point group to the atomic or molecular orbital. Degenerate orbitals are pairs or sets of orbitals that only have the symmetry of an irreducible representation when they are considered together. You may practice reading character tables and identifying degenerate sets of orbitals by trying Exercises 10.23–10.32.

10.3A. Character Tables. Many of the applications of symmetry take advantages of some of the consequences of the fact that symmetric molecules are covered by the mathematical theory known as **group theory**. The mathematics of this theory is beyond the scope of this book, but we shall make particular use of the fact that, for each symmetry point group, there is a corresponding **character table**. Nine pages of character tables for the major chemically important point groups are available online at www.uscibooks.com/foundations.htm. For convenience we reproduce some examples in Table 10.2, beginning with the character table for the C_{2v} point group of water, our first molecular example.

Many of the components of a character table are already familiar. In the upper left corner the point group C_{2v} is listed. Over the main body of the table the classes of symmetry operations of the point group are summarized. Closely related operations are

TABLE 10.2. Selected Character Tables

C_{2v}	E	C_2	$\sigma_v(xz)$	$\sigma_v'(yz)$		
A_1	1	1	1	1	z	x^2, y^2, z^2
A_2	1	1	−1	−1	R_z	xy
B_1	1	−1	1	−1	x, R_y	xz
B_2	1	−1	−1	1	y, R_z	yz

C_{3v}	E	$2C_3$	$3\sigma_v$		
A_1	1	1	1	z	$x^2 + y^2, z^2$
A_2	1	1	−1	R_z	
E	2	−1	0	$(x, y), (R_x, R_y)$	$(x^2 - y^2, xy), (xz, yz)$

Table 10.2 continues on next page ▶

TABLE 10.2. **Selected Character Tables** *(cont.)*

D_{4h}	E	$2C_4$	C_2	$2C_2'$	$2C_2''$	i	$2S_4$	σ_h	$2\sigma_v$	$2\sigma_d$	(x axis coincident with C_2')		
A_{1g}	1	1	1	1	1	1	1	1	1	1		x^2+y^2, z^2	
A_{2g}	1	1	1	-1	-1	1	1	1	-1	-1	R_z		
B_{1g}	1	-1	1	1	-1	1	-1	1	1	-1		x^2-y^2	
B_{2g}	1	-1	1	-1	1	1	-1	1	-1	1		xy	
E_g	2	0	-2	0	0	2	0	-2	0	0	(R_x, R_y)	(xz, yz)	
A_{1u}	1	1	1	1	1	-1	-1	-1	-1	-1			
A_{2u}	1	1	1	-1	-1	-1	-1	-1	1	1	z		
B_{1u}	1	-1	1	1	-1	-1	1	-1	-1	1			
B_{2u}	1	-1	1	-1	1	-1	1	-1	1	-1			
E_u	2	0	-2	0	0	-2	0	2	0	0	(x, y)		

summarized together. For example, in the point group C_{3v} the two operations C_3 and C_3^2 are listed as "$2C_3$." It may be seen from Table 10.2 that the C_{2v} point group has four classes of operations: E, C_2, $\sigma_v(xz)$, and $\sigma_v(yz)$.

Each *row* of the character table corresponds to an **irreducible representation (symmetry label)** in that point group, labeled with a **symmetry species** at the left—namely, A_1, A_2, B_1, and B_2 in the C_{2v} character table. Atomic and molecular orbitals generally lack the full symmetry of the molecule, but must have the reduced symmetry corresponding to one of the irreducible representations of that molecule's point group. The heart of the character table is a square array containing numbers called **characters**.

If the atom or orbital in an irreducible representation is left in an undistinguishable position by a symmetry operation at the top, a **character** of "1" is shown in the table under that operation. If the orbital is left in the identical position, but its positive and negative lobes are interchanged, a character of "–1" is shown in the table. When all of the operations are carried out, a set of characters are produced. If these match the characters in a particular row, the orbital belongs to that irreducible representation, and the symmetry label (perhaps using a lowercase letter) can be applied to it.

For example, H_2O falls in the C_{2v} point group. An *s* orbital on the oxygen atom is left in the same position with the same sign by all operations of the point group. Therefore, it has characters of "1" under each operation. This matches the A_1 symmetry label, so the orbital can be labeled as belonging to the A_1 irreducible operation and symmetry species.

To evaluate the other oxygen orbitals (or hydrogen orbitals), we need to see how the *x*, *y*, and *z* axes are defined. The *z* axis is taken as the *highest-order rotation axis* of the molecule. As shown in Figure 10.11, the C_2 axis becomes the *z* axis. The *x* axis is the axis $\perp$ the *z* axis that is in the plane of the molecule (Figure 10.11a), while the *y* axis is

EXAMPLE 10.7

In Examples 10.2 and 10.3, we discovered the symmetry elements and operations for PCl_5 and for C_6H_6; in Example 10.4, we classified these molecules into the symmetry point groups D_{3h} and D_{6h}, respectively. Verify from the character tables for these two point groups (Appendix C) that we have all of the correct symmetry operations.

SOLUTION: (a) The D_{3h} character table has the following headings: $D_{3h} \mid E$ $2C_3$ $3C_2$ σ_h $2S_3$ $3\sigma_v \mid$ (x axis coincident with C_2). The first entry is simply the point group; the last entry indicates that sometimes we need a clear convention on labeling the x and y axes. Let us compare the central column headings (symmetry operations) with the list of 12 symmetry operations we deduced in Example 10.2: E, C_3, C_3^2, three different C_2's, σ_h, three different σ_v's, S_3, and S_3^2. The headings match our deductions (C_3 and C_3^2 are consolidated as $2C_3$; S_3 and S_3^2 are consolidated as $2S_3$).

(b) The D_{6h} character table has the following headings: $D_{6h} \mid E$ $2C_6$ $2C_3$ C_2 $3C_2'$ $3C_2'' $ i $2S_3$ $2S_6$ σ_h $3\sigma_v$ $3\sigma_d \mid$ (x axis coincident with C_2). The list of symmetry operations we deduced in Example 10.3 is substantial: E, two C_6, two C_3, C_2, three C_2', three C_2'', σ_h, three σ_v, three σ_d, two S_6, two S_3, and i, for a total of 24 operations about the 12 types of symmetry elements. (Note that the sum of coefficients of the headings is $1 + 2 + 2 + 1 + 3 + 3 + 1 + 2 + 2 + 1 + 3 + 3 = 24$ operations). Our examples are in accord with the character table.

the axis that is $\perp$ the plane of the molecule (not shown). After the axes are defined, we can see which mirror plane is which. The mirror plane $\sigma_v(yz)$ is the yz plane ($\perp$ the x axis); the $\sigma_v(xz)$ is in the xz plane (Figure 10.2b).

Now let us consider the p_x orbital of the oxygen atom of the water molecule. Referring to Figure 10.11a, this orbital lies in the xz plane (the plane of the H_2O molecule), along the x axis. Let us consider what each operation does in turn to this orbital.

- The identity operation E leaves the orbital unchanged, so the character of this operation is +1.

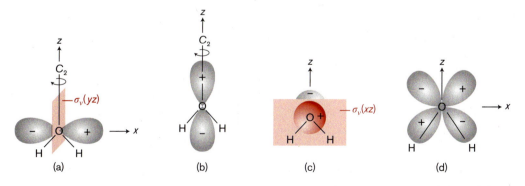

Figure 10.11. Drawings of axes, atoms, and selected oxygen atomic orbitals in the water molecule—to be used in determining the irreducible representations of these orbitals. (a) p_x orbital; (b) p_z orbital; (c) p_y orbital; (d) d_{xz} orbital.

- The C_2 operation interchanges the positive and the negative lobes of the p_x orbital, so the character for this operation interchanging positive and negative parts is –1.

- The $\sigma_v(xz)$ reflection leaves the p_x orbital unchanged, so the character of this operation is +1.

- Finally, the $\sigma_v(yz)$ operation once again interchanges the + and – lobes of the p_x orbital, so it has a character of –1.

- In the order in which the operations are listed in the character table, the characters of the operations are 1, –1, 1, and 1. This corresponds to the third row of characters found in the character table, following the irreducible representation label B_1. We say that the p_x orbital transforms under the symmetry operations of the C_{2v} point group according to the irreducible representation B_1.

Not only orbitals, but many other things associated with a molecule transform according to one or another of the irreducible representations of that group. For example, a vector along the x axis, with its head on one side of the oxygen atom and its tail on

EXAMPLE 10.8

Determine the irreducible representations and symmetry labels of the (a) p_z, (b) p_y, and (c) d_{xz} orbitals on the oxygen atom of water.

SOLUTION: (a) The positive and negative lobes of the p_z orbital (Figure 10.11b) are never interchanged by any of the four symmetry operations of this point group, so the p_z orbital also transforms in the totally symmetric A_1 irreducible representation of this point group. So does a vector centered on oxygen and pointing in the z direction; the z shown at the right of the A_1 row indicates both that p_z orbital and a z vector transform in this manner.

(b) The p_y orbital (Figure 10.11c) is affected differently by the four symmetry operations. The identity operation leaves it unchanged and has a character of 1. The C_2 operation interchanges + and – lobes and has a character of –1. The $\sigma_v(xz)$ operation does the same and has a –1 character. The $\sigma_v(yz)$ leaves the orbital unchanged, so it has a character of +1. This set of characters matches that of the B_2 irreducible representation; y is listed on the right side of this representation for a convenient reminder of this.

(c) The d_{xz} orbital (Figure 10.11d) is unchanged by E (automatic character of +1). All of the + lobes and – lobes of the d_{xz} orbital are interchanged by the C_2 and the $\sigma_v(yz)$ operations, so there are characters of –1 under each. The lobes of the d_{xz} orbital are unchanged by the $\sigma_v(xz)$ operation (character of +1). This set of characters matches that of the B_1 irreducible representation, which is also indicated by the xz entry at the far right of the row. The irreducible representations of the other d orbitals (and of the f orbitals in Appendix C) are also indicated by the entries at the far right of the table.

the other side, transforms identically to the p_x orbital. For convenient reference, an x is listed at the right side of the row; this indicates that anything with symmetry like a vector in the x direction, such as a p_x orbital, transforms in this irreducible representation.

In general, the symbols used for the irreducible representations also carry some of this information. Of particular importance in many groups are the subscripts g (from the German *gerade*, meaning even), meaning that the representation is symmetric to inversion, and u (from the German *ungerade*, meaning odd), meaning that the representation changes sign upon inversion. When an inversion center is present, gerade irreducible representations are shown in the top half of the table, and ungerade representations are shown in the lower half. By the nature of the patterns of their wave function signs (Figure 3.7), central-atom s, d, and g orbitals are gerade, while p and f orbitals are ungerade. In classifying an object or orbital, it is often prudent to look first for an inversion center in order to determine whether the object or orbital falls in a gerade or an ungerade irreducible representation.

Details in Character Tables

Sometimes the entries at the right of the irreducible representation show orbitals indirectly: Since both x^2 and y^2 are listed with the A_1 irreducible representation, this implies that $d_{x^2-y^2}$ transforms in this irreducible representation as well. Among symmetry labels, the letter A indicates that representation is symmetric to the principal rotation operation (has a character of +1), while B indicates that it is antisymmetric to that operation (has a character of –1). The subscript 1 indicates that the representation is symmetric either to the perpendicular C_2 axis (if present), or to a vertical mirror plane; the subscript 2 indicates antisymmetry.

AN AMPLIFICATION

10.3B. Orbital Degeneracy.

Additional symbols appear in groups of higher order: those with C_3 or higher axes. Let us consider as an example the XeF_4 molecule, which has a C_4 axis and falls in the D_{4h} point group (see Table 10.2).

The sign of the wave function of the s orbital of the Xe atom is unchanged by any symmetry operation, so the s orbital falls in the totally symmetric A_{1g} irreducible representation. The Xe p_z orbital has the signs of the wave function at its two lobes interchanged when the operation of inversion is carried out, so it must fall in one of the ungerade irreducible representations found in the lower half of the character table. The p_z orbital is unchanged in sign on carrying out rotation about the principal (C_4) axis, so it falls in an A irreducible representation. Reference to the character table for D_{4h} shows an entry "z" at the right side of the row for the A_{2u} irreducible representation, so the p_z orbital is classified here.

When we carry out a C_4 rotation on the p_x axis, however, we find that it neither keeps nor changes its sign, but rather is superimposed on the original position of the p_y orbital. Simultaneously the p_y orbital is rotated to coincide with the p_x orbital (but with

EXAMPLE 10.9

Assign the s and p orbitals of the NH_3 molecule (Figure 10.12) to the proper irreducible representations. (NH_3 is in point group C_{3v}; its character table is in Table 10.2.)

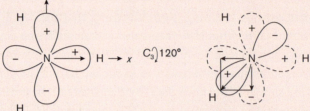

Figure 10.12. View down the C_3 axis and the p_x orbital of the NH_3 molecule: (a) before rotation; (b) after rotation, with the former orbital positions shown by dotted lines. The vector representing the positive lobe of the p_x orbital is also shown as a vector sum of the negative lobes of the former p_x and p_y orbitals.

SOLUTION: NH_3 is drawn in Figure 10.12a in a view down its threefold (z) axis, with its valence p_x and p_y orbitals shown explicitly. Without further thought we can assign the s orbital of the central atom to the totally symmetric A_1 irreducible representation. With a little thought, it will be seen that the p_z orbital in ammonia is very similar in symmetry to the p_z orbital in water: Neither is altered by any symmetry operation, so p_z also transforms in the A_1 irreducible representation. The situation with regard to the p_x and p_y orbitals is more complex, however. A threefold rotation does not take the positive lobe of p_x (shown with a vector inside it) all the way to its negative lobe, but instead leaves it in between the negative p_x and the negative p_y lobes. By vector analysis we see that it can be expressed as a resultant of the two. A somewhat similar situation applies to the p_y orbital upon threefold rotation. These two orbitals must therefore fall into the doubly degenerate irreducible E representation in the C_{3v} point group. The effects of the C_3 operation are difficult to envision, because they intermix the two orbitals. The positive lobe of the p_x orbital is partially converted to the negative lobe; vector analysis shows that the partial character of this part of the operation is $-1/2$. Because the same result holds for the lobe of the p_y orbital, the overall character of this operation is $(-1/2) + (-1/2) = -1$. Fortunately for us, the character table spares us the necessity of doing this vector analysis, because it shows (p_x, p_y) at the right side of the E irreducible representation. The calculations of the characters of the other operations also take into account the fact that the operations act on more than one orbital. Thus, the identity operation E leaves both the p_x and the p_y components of the set unchanged, in which case the character of this operation is 2. Let us next consider any one of the σ_v operations, say the one shown in the center of Figure 10.12. This mirror reflection interchanges + and − lobes of the p_x orbital for a partial character of −1, but it also leaves the p_y orbital unchanged, for a partial character of +1. The overall character for the operation on both parts of the set is therefore zero. Finally, we see that the three characters, in order, are 2, −1, and 0. This set of characters is found under the irreducible representation labeled E. This is confirmed by the appearance of the (x, y) set at the right of this row.

a reversal in the sign of the wave function). Neither the p_x nor the p_y orbital considered alone show even the reduced symmetry necessary to transform according to any of the irreducible representations of this point group, but together, considered as a pair, they do interchange lobes and signs with each other. The p_x and p_y orbitals thus must be considered together as a **doubly degenerate** set of orbitals. Doubly degenerate pairs of orbitals fall in irreducible representations beginning with the letter E. Since all p orbitals change sign upon inversion, this pair of orbitals falls in the doubly degenerate irreducible representation E_u, as indicated by the appearance of the pair (x, y) on the right side of the character table. The character "2" appears under the identity operation (also coincidentally labeled E), because two orbitals are unchanged in sign upon this operation. The character "0" appears under the C_4 operation, because this operation retains the sign of the p_x orbital (partial character of +1), while at the same time inverting the sign of the p_y orbital (partial character of –1). The sum of these two contributions to the character is zero.

In the higher-order symmetry point groups, even greater degeneracy of orbitals can occur. For example, the p_x, p_y, and p_z orbitals are completely equivalent to each other in strictly tetrahedral (point group T_d) and octahedral (point group O_h) molecules and are interchanged with each other during symmetry operations; such **triply degenerate** irreducible representations use the letter T (or F) in their labels. Only in the I_h (icosahedral) point group [e.g., buckminsterfullerene (C_{60}) and dodecahedrane ($C_{20}H_{20}$)] are higher degeneracies possible: Quadruply degenerate representations include the letter G, while quintuply degenerate representations include the letter H.

The labels of irreducible representations have a number of uses in inorganic chemistry. We have already seen one use in Section 7.1A, in which the d orbitals in an octahedral (O_h) complex were grouped in symmetry-equivalent sets (d_{xy}, d_{xz}, d_{yz}) and (d_{z^2}, $d_{x^2-y^2}$), and labeled as t_{2g} and e_g sets, respectively. Consulting the character table for the O_h point group shows that these are the (uncapitalized) labels of the irreducible representations of these sets of orbitals. In Section 7.6A we considered tetrahedral (T_d) complexes. This point group lacks an inversion center, so these sets of d orbitals were labeled without g or u subscripts, as t_2 and e.

10.4. Symmetry Labels for Bonding and Antibonding Orbitals in Diatomic Molecules

OVERVIEW. Recall from Section 3.3B and Figure 3.10 that two atoms can form σ, π, δ, and φ bonds. These [and the corresponding negative-overlap (*antibonding*) combinations] can be classified into their symmetry types (irreducible representations) in the $D_{\infty h}$ point group if the two atoms are identical (M₂ molecules) and in the $C_{\infty v}$ point group if the atoms differ (MN molecules). The most important difference between the two point groups is the presence of the inversion operation in the $D_{\infty h}$ point group, so $D_{\infty h}$ symmetry labels contain g or u subscripts. You may practice assigning bonds and antibonding combinations to irreducible representations by trying Exercises 10.33–10.38.

In *homoatomic* diatomic molecules M₂, both atoms are identical; they fall in the $D_{\infty h}$ point group. *Heteroatomic* diatomic molecules contain two different types of atoms MN; they fall in the $C_{\infty v}$ point group. Their covalent bonding and antibonding orbitals

TABLE 10.3. Character Tables for Linear Molecules

$C_{\infty v}$	E	$2C_\infty^\phi$	$\dots$	$\infty\sigma_v$			
$A_1 \equiv \Sigma^+$	1	1	$\dots$	1	z	x^2+y^2, z^2	z^3
$A_2 \equiv \Sigma^-$	1	1	$\dots$	-1	R_z		
$E_1 \equiv \Pi$	2	$2\cos\phi$	$\dots$	0	$(x,y); (R_x, R_y)$	(xz, yz)	(xz^2, yz^2)
$E_2 \equiv \Delta$	2	$2\cos 2\phi$	$\dots$	0		(x^2-y^2, xy)	$[xyz, z(x^2-y^2)]$
$E_3 \equiv \Phi$	2	$2\cos 3\phi$	$\dots$	0			$[x(x^2-3y^2), y(3x^2-y^2)]$
$\dots$	$\dots$	$\dots$	$\dots$	$\dots$			

$D_{\infty h}$	E	$2C_\infty^\phi$	$\dots$	$\infty\sigma_v$	i	$2S_\infty^\phi$	$\dots$	∞C_2			
$A_{1g} \equiv \Sigma_g^+$	1	1	$\dots$	1	1	1	$\dots$	1		x^2+y^2, z^2	
$A_{2g} \equiv \Sigma_g^-$	1	1	$\dots$	-1	1	1	$\dots$	-1	R_z		
$E_{1g} \equiv \Pi_g$	2	$2\cos\phi$	$\dots$	0	2	$-2\cos\phi$	$\dots$	0	(R_x, R_y)	(xz, yz)	
$E_{2g} \equiv \Delta_g$	2	$2\cos 2\phi$	$\dots$	0	2	$2\cos 2\phi$	$\dots$	0		(x^2-y^2, xy)	
$\dots$	$\dots$	$\dots$	$\dots$	$\dots$	$\dots$	$\dots$	$\dots$	$\dots$			
$A_{1u} \equiv \Sigma_u^+$	1	1	$\dots$	1	-1	-1	$\dots$	-1			
$A_{2u} \equiv \Sigma_u^-$	1	1	$\dots$	-1	-1	-1	$\dots$	1	z		z^3
$E_{1u} \equiv \Pi_u$	2	$2\cos\phi$	$\dots$	0	-2	$2\cos\phi$	$\dots$	0	(x,y)		(xz^2, yz^2)
$E_{2u} \equiv \Delta_u$	2	$2\cos 2\phi$	$\dots$	0	-2	$-2\cos 2\phi$	$\dots$	0			$[xyz, z(x^2-y^2)]$
$E_{3u} \equiv \Phi_u$	2	$2\cos 3\phi$	$\dots$	0	-2	$2\cos 3\phi$	$\dots$	0			$[x(x^2-3y^2), y(3x^2-y^2)]$
$\dots$	$\dots$	$\dots$	$\dots$	$\dots$	$\dots$	$\dots$	$\dots$	$\dots$			

(illustrated in Figure 3.10 and discussed in Section 3.3B) can be classified according to their irreducible representations in these point groups.

The three σ bonds on the left side of Figure 3.10 are formed by *positive* overlap of (at the top) two s orbitals on adjacent atoms A and B; (in the center) two p_z orbitals along the principal (z or internuclear) axis of the molecule; and (at the bottom) of two d_{z^2} orbitals. Each of these σ bonds was characterized in Section 3.3B as having *no* nodal planes along the internuclear (principal) axis. In our current language, this means that, upon rotation about all angles ϕ around the internuclear C_∞ axis, the sign of the orbital never changes; its character remains 1.

Referring to the character tables, (Table 10.3), these σ bonds for MN molecules fall in the A_1 irreducible representation of the $C_{\infty v}$ point group. In the $D_{\infty h}$ point group for M_2 molecules, these orbitals are gerade, so their irreducible representations are A_{1g}. When we apply these labels to the orbitals themselves, we refer to them using lower cases: a_1 and a_{1g}. Finally, as indicated in Table 10.3, we can change these symmetry labels to the common σ–π type of notation, as σ^+ or σ_g^+. (The "+" is often omitted.) To distinguish the three types of σ bonds, we can label them as $\sigma_g^+[s(A) + s(B)]$, $\sigma_g^+[p_z(A) - p_z(B)]$, and $\sigma_g^+[d_{z^2}(A) + d_{z^2}(B)]$, respectively.

The π bonds in the center of Figure 3.10 are formed by positive overlap of: (a) two p_x orbitals on adjacent atoms A and B $[p_x(A) + p_x(B)]$, and (b) two d_{xz} orbitals along the principal axis of the molecule $[d_{xz}(A) - d_{xz}(B)]$, because they have their two positive lobes oriented in different directions. Each of these π bonds was characterized in Section 3.3B as having *one* nodal plane along the internuclear (principal) axis. In our current language, this means that, upon rotation about angles ϕ around the internuclear C_∞ axis, the sign of the orbital does change, from +1 at $\phi = 0°$ to 0 at 90° (in the nodal plane) to –1 at 180° to 0 at 270° to +1 at 360°. Its character is therefore proportional to $\cos \phi$.

However, there is another π bond that is paired with each of these π bonds, so these are doubly degenerate (as shown in Figure 10.13e). Referring to the character tables for the $D_{\infty h}$ and $C_{\infty v}$ point groups under C_∞ (Table 10.3), they fall in the E_1 or Π irreducible representation (e_1 or π labels for orbitals). In the $D_{\infty h}$ point group these orbitals are ungerade, and the irreducible representation is E_{1u} or Π_u (e_{1u} or π_u as orbitals).

The combinations of orbitals on atoms A and B having *negative* overlap, although ignored in valence bond theory, have a good deal of importance in molecular orbital theory (see Chapter 11), in which we refer to them as **antibonding** orbitals. Thus, the combination of s orbitals on atoms A and B having *opposite* signs of their wave functions is shown in Figure 10.13b. In the $D_{\infty h}$ point group this combination is ungerade, but it otherwise has the same symmetry properties as the positive combination (σ bond). Hence this orbital can be labeled as σ_u^+ in the $D_{\infty h}$ point group. We often add a star after the label to indicate antibonding, giving us a σ_u^{+*} orbital label.

There is also a corresponding degenerate antibonding set of orbitals resulting from negative overlap of p_x and p_y orbitals (Figure 10.13f). These have the symmetry of the Π_g irreducible representation in the $D_{\infty h}$ point group, so they are properly labeled as π_g^* antibonding orbitals. (The bonding and the antibonding combinations in M_2 have opposing g and u subscripts.)

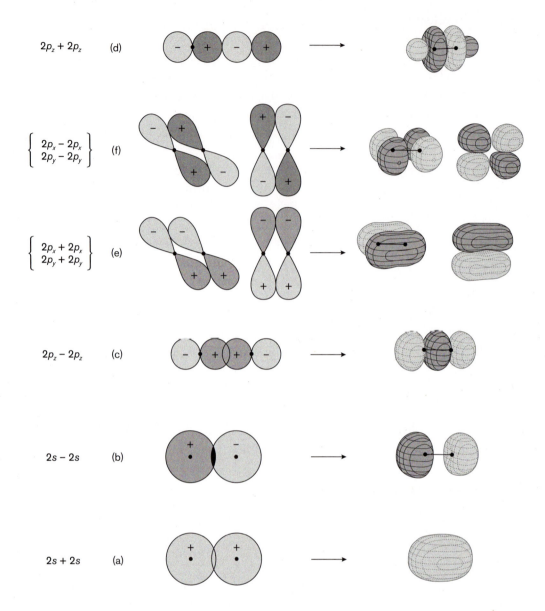

Figure 10.13. Positive overlap of valence orbitals in M_2 to give bonds [(a), (c), (e), and (f)] and negative overlap to give antibonding combinations [(b) and (d)]. At the right, the contour diagrams are the results of calculations for N_2. [Adapted from W. L. Jorgensen and L. Salem, *The Organic Chemist's Book of Orbitals*, Academic Press: New York, 1973, p. 79.]

EXAMPLE 10.10

The lower right bond shown in Figure 3.10 is a δ bond, formed by positive overlap of $d_{x^2-y^2}$(A) + $d_{x^2-y^2}$(B). (a) What other pair of d orbitals gives a bonding combination that falls in the same irreducible representation? (b) Justify the fact that these two δ bonds fall in the Δ_g irreducible representation in the $D_{\infty h}$ point group. (c) The negative-overlap (antibonding) combinations of these orbitals fall in what irreducible representation? (d) What would the views of these orbitals be like from the end of the principal (z) axis?

SOLUTION: (a) The same type of overlap is achieved by d_{xy}(A) + d_{xy}(B). This combination is rotated by 45° from that shown in Figure 3.10. (b) The quick* answer is that, at the right of the Δ_g irreducible representation of the $D_{\infty h}$ character table, we find the entry ($x^2 - y^2$, xy). (c) If we change the signs on each lobe of the right orbital shown at the bottom right of Figure 3.10, the overlap becomes negative and antibonding results. This means that the combination is ungerade. All other symmetry operations give the same results, so this combination (and the corresponding combination of d_{xy} orbitals) falls in the Δ_u irreducible representation. (d) The end view of these orbitals of diatomic molecules would be the same as a view along the z axis of the parent d orbital. The nodal planes would show especially clearly in this view. Thus, the end views of either the bonding or antibonding combinations involving d_{xy} would look like d_{xy}; the end view for the combinations involving $d_{x^2-y^2}$ would look like $d_{x^2-y^2}$.

* The more thorough answer is (1) these combinations are *gerade*; (2) since this pair of d orbitals is doubly degenerate, it has a character of 2 under the symmetry operation E and falls in an E irreducible representation; (3) on rotation about the principal axis (C_∞ operation), the original +2 character is again duplicated at the angles $\phi = 180°$ *and* 360°, so it is proportional to cos 2ϕ.

10.5. Background Reading for Chapter 10

Molecular and Non-chemical Examples of Symmetry Elements and Operations

M. Hargittai and I. Hargittai, *Symmetry through the Eyes of a Chemist*, 3rd ed. Springer: New York, 2009, pp. 1–95 (Chapters 1 and 2). Numerous examples using many photographs and drawings.

E. B. Flint, "Teaching Point-Group Symmetry with Three-Dimensional Models," *J. Chem. Educ.* 88, 907–909 (2011). Includes schemes for using models for teaching some of the difficult-to-visualize symmetry elements (rotation axes perpendicular to the principal axis, inversion centers, mirror planes, and improper rotation).

C. J. Luxford, M. W. Crowder, and S. L. Bretz, "A Symmetry POGIL Activity for Inorganic Chemistry," *J. Chem. Educ.* 89, 211–214 (2012). Student learning-cycle activities for discovering the difficult-to-visualize elements (rotation axes perpendicular to the principal axis, inversion centers, mirror planes, and improper rotation). Points out that students can be confused because mirror planes are applied differently in organic chemistry (for finding enantiomers) and inorganic chemistry (for classifying symmetry).

Point Groups and Character Tables

M. Hargittai and I. Hargittai, *Symmetry through the Eyes of a Chemist*, 3rd ed. Springer: New York, 2009, pp. 169–176 and 191–204 (Sections 4.1 and 4.5–4.6).

Numbered references from this chapter may be viewed online at www.uscibooks.com /foundations.htm.

10.6. Exercises

10.1. *Figure 10.1 shows a number of three-dimensional natural and artificial objects. For each object, identify the highest-order rotation axis present; tell whether an inversion center is present; tell whether any mirror planes are present; and tell whether any improper rotation axes are present.

10.2. Figure 10.14 shows a number of two-dimensional representations of natural organisms, objects, or works of art. For each object, identify the highest-order rotation axis present; tell whether an inversion center is present; and tell whether any mirror planes (other than the plane of the object itself) are present.

10.3. *What kinds of symmetry elements (aside from E) are present in the molecules shown in Figure 5.2 as examples for VSEPR predictions?

10.4. What kinds of symmetry elements (aside from E) are present in the molecules shown in Figure 5.3 as examples for VSEPR predictions?

10.5. *Below are shown several molecules or ligands, along with their basic geometric shapes as predicted by VSEPR theory. For each molecule or ligand, identify the highest-order rotation axis present; tell whether there is a C2 rotation axis perpendicular to this highest-order axis; tell whether an inversion center is present; and tell whether any mirror planes are present. (a) pentagonal pyramidal TlC5H5; (b) pentagonal planar C5O52−; (c) bidentate chelated C5O5M; (d) chelated tetrahedral Si(CH2CH2CH2CH2)2.

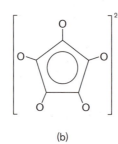
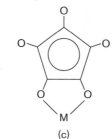
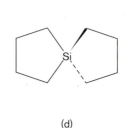

(a)   (b)   (c)   (d)

10.6. Below we draw four simple molecules—tetrahedral CH_4, square planar $[PtCl_4]^{2-}$, trigonal planar boric acid, and nonplanar H_2O_2 (see also Figure 10.8). (a) What is the highest-order rotation axis in each one? (b) Are any mirror planes present? (c) Is an inversion center present? (d) Is there a C_2 axis perpendicular to the highest-order rotation axis?

10.7. *Classify the objects you analyzed in Exercise 10.1 into symmetry point groups.

10.8. Classify *cis*-[PtCl₂(NH₃)₂] and *trans*-[PtCl₂(NH₃)₂] into symmetry point groups. Assume that the NH₃ groups rotate freely so that the hydrogens do not lower the symmetry of these complexes.

10.9. *Classify the following derivatives of the perxenate ion into symmetry point groups: (a) the free XeO_6^{4-} ion itself; (b) the free oxo acid H_4XeO_6, in which the two oxo groups lie in trans positions and the four OH groups are bent at oxygen in a conformation resembling a swastika; (c) the free oxo acid H_4XeO_6, in which the two oxo groups lie in cis positions and the OH groups are all bent away from the oxo groups; (d) a hypothetical organic derivative [(*ortho*-$C_6H_4O_2$)₃Xe]²⁺, in which three benzenedioxy groups each chelate the Xe⁸⁺ "cation."

Figure 10.14. Two-dimensional artistic or natural patterns for use in the Exercises.

(a) American Indian decoration.
(b) Quilt pattern with oak leaf wreath.
(c)–(h) Pueblo Indian pottery designs.
(i) M. C. Escher's "Whirlpools."

(j) M. C. Escher's "Circle Limit IV."
(k) Stalked jellyfish.
(l) Cross-section of protein disk of tobacco mosaic virus.
(m) Starfish and other organisms.
(n) Color-stained snowflake crystal.

10.10. For each of the following species, list the highest-order rotation axis present, and give the symmetry point group to which the molecule belongs: (a) $CHCl_3$; (b) the aromatic hydrocarbon $C_5H_5^-$, as a free ion; (c) the aromatic hydrocarbon $C_5H_5^-$, coordinated to a metal ion on one face at the center of the five carbon atoms; (d) the anion $Cl_3CSO_3^-$ with an eclipsed conformation (Cl's directly above O's); (e) $Cl_3CSO_3^-$ with a staggered conformation (Cl's falling midway between the O's).

10.11. *Classify into a symmetry point group the crown ether 18-crown-6 (Figure 5.10a): (a) assuming that the crown ether is completely planar; (b) assuming (more realistically) that the CH_2CH_2 groups alternately lie above and below the plane of the six O atoms.

10.12. The rhodizonate ion, $C_6O_6^{2-}$, is a planar hexagonal organic anion in which (due to resonance) all oxygen atoms are equivalent. Give the symmetry point groups of the following rhodizonate derivatives: (a) the free rhodizonate anion; (b) a bidentate chelate complex $[C_6O_6M]$; and (c) $[C_6O_6M_3]$, a complex in which the rhodizonate anion bridges three different metal atoms of the same type, chelating each one of them.

10.13. *Classify into symmetry point groups the examples of the basic VSEPR geometry types shown in Figure 5.2.

10.14. Classify the examples of the basic VSEPR geometry types having unshared electron pairs shown in Figure 5.3 into symmetry point groups.

10.15. *Consider ICl_4^-, SbFClBr, TeO_6^{6-}, and XeO_4^{2-}. (a) What is the highest-order rotation axis present in each molecule or ion? (b) Which of the molecules or ions have centers of inversion? (c) Which of the molecules or ions have mirror planes? (d) Give the symmetry point groups of each of these molecules or ions.

10.16. (a) Molecules in which point groups—D_{3d}, T_d, C_s, O_h, $D_{\infty h}$, $C_{\infty v}$, C_{17}, and D_2—can be polar? (b) What symmetry elements in the other point groups keep the molecules from being polar? (c) Molecules in which point groups—D_{3d}, T_d, C_s, O_h, $D_{\infty h}$, $C_{\infty v}$, C_{17}, and D_2—are chiral?

10.17. *Which of the objects shown in Figure 10.1 are chiral, and which would be polar (if they were molecules containing polar covalent bonds)? (Recall that symmetry point groups for these objects were assigned in Exercise 10.7.)

10.18. (a) Classify *mer*-$[CoCl_3(NH_3)_3]$ and *fac*-$[CoCl_3(NH_3)_3]$ into symmetry point groups. Assume that the NH_3 groups rotate freely. (b) Then tell which (if any) are chiral and which (if any) must be nonpolar.

10.19. *The dianion of 1,2-benzenediol (catechol), *ortho*-$[O_2C_6H_4]^{2-}$, readily forms chelated complexes with many metal and nonmetal cations. Draw structures of the chelate complexes that this dianion would form if it displaced all of the fluoro, oxo, etc., ligands already present in the selected structures drawn out in Figures 5.1 and 5.2. (Where isomeric choices are possible, use the benzenediolato dianion to span neighboring axial and equatorial positions, or top and bottom planes, and so forth.) Then tell whether the resulting complex is polar and whether it is chiral. (a) $[C(O_2C_6H_4)_2]$, formed from CF_4; (b) $[S(O_2C_6H_4)_3]$, formed from SF; (c) $[I(O_2C_6H_4)_4]^-$, formed from IF_8^-; (d) $[S(O_2C_6H_4)_2]$, formed from SF_4.

10.20. Which (if any) of the isomers of (a) $[PtCl_2(NH_3)_2]$ (Exercise 10.8) or (b) $[CoCl_2(NH_3)_4]^+$ or (c) $[CoCl_2(NH_2CH_2CH_2NH_2)_2]^+$ are polar? Which (if any) are chiral?

10.21. *Each of the following pairs of trisubstituted octahedral complexes of Cr can exist as two geometric isomers: facial and meridional. (a) Is *fac*-$[Cr(NH_3)_3(CH_3CO_2)_3]$ chiral? (b) Is *mer*-$[Cr(NH_3)_3(CH_3CO_2)_3]$ chiral? (c) Draw all four possible isomers of the octahedral complex $[Cr(NH_2CH_2CO_2)_3]$ in which $NH_2CH_2CO_2^-$ is the anion of the simplest amino acid, glycine. (Hint: Two will be mer and two will be fac isomers.) How many pairs of enantiomers are included among the isomers?

10.22. Consider complexes of monodentate ligands and complexes of chelating ligands having the same donor atoms. (a) Which (if either) is more likely to be chiral? (b) Which (if either) is more likely to be polar? (c) Which (if either) is likely to be higher on the spectrochemical series?

10.23. *Figure 7.3 shows the appearances and names of the *f* orbitals in forms appropriate for a central atom in an octahedral (O_h) complex ion. (a) Similarly to what was done in Exercise 7.3a, group the seven *f* orbitals into three smaller sets of orbitals that are likely interchanged with each other during the symmetry operations of the O_h point group. (b) The irreducible representation labels for each set of *f* orbitals would begin with which of the following letters: *A*, *B*, *E*, *T*, *G*, or *H*? (c) The labels for each set would include which subscript: *g* or *u*?

10.24. In the complex $[Ti@C_{60}]$ the Ti atom could sit at the center of a complex of icosahedral symmetry. Table 7.7 indicates that all five *d* orbitals of the Ti atom experience the same crystal field splitting in an icosahedral complex—namely, none, because they are equivalent by symmetry. In the I_h point group, will the irreducible representation for the five *d* orbitals be labeled with a *g* subscript or a *u* subscript? Which of the following letters—*A*, *B*, *E*, *G*, *H*, or *T*—will begin the label of this irreducible representation?

10.25. *In the complex $[La@C_{60}]$ the La atom could sit at the center of a complex of icosahedral symmetry. In this symmetry the *f* orbitals are grouped into two energy levels, a set of three *f* orbitals and a different set of four *f* orbitals. (a) In the I_h point group, is the irreducible representations for the *f* orbitals labeled with *g* subscripts or *u* subscripts, or neither? (I_h contains the inversion center *i*.) (b) Which two of the following letters—*A*, *B*, *E*, *G*, *H*, or *T*—will begin the labels used for the irreducible representations to which the *f* orbitals belong?

10.26. For a trigonal prismatic complex, $[TiL_6]$, Table 7.7 indicates that the d_{z^2} orbital is at an energy level of $0.096\Delta_o$, $d_{x^2-y^2}$ and d_{xy} are at an energy level of $-0.584\Delta_o$, and d_{xz} and d_{yz} fall at an energy level of $+0.536\Delta_o$. The pairings of orbitals are due to their being equivalent by symmetry (falling in the same irreducible representations). Which of these *d* orbitals will be classified in irreducible representations beginning with the capital letter *A*? Which beginning with capital letter *E*? Which beginning with capital letter *T*?

10.27. *There are suggestions that plutonium might show the +8 oxidation state in PuO_4 [although calculations indicate that, if PuO_4 exists, it would actually be a superoxide complex of plutonium(V): W. Huang, W-H. Xu, J. Su, W. H. E. Schwarz, and J. Li, *Inorg. Chem.* 52, 14237 (2013).] Suppose that PuO_4 would fall in the T_d point group. (a) Give the symmetry labels (irreducible representations) for each of the *p* orbitals and for each of the *d* orbitals of the plutonium ion at the center of the molecule. (b) Referring to the drawing of the plutonium f_{xyz} orbital in Figure 7.3, mentally orient this orbital inside the PuO_4 tetrahedron such that each negative lobe coincides with a Pu–O bond, and decide in which irreducible representation the f_{xyz} orbital belongs.

10.28. The UF_6 molecule is octahedral and falls in the O_h point group; you may consult the right side of the O_h character table to answer the following questions: (a) In which irreducible representation does the uranium f_{xyz} orbital fall? Explain what the capital letter in the label for this irreducible representation means. Explain what the lowercase letter in the label for this irreducible representation means. (b) In which irreducible representation does the uranium f_z^3 orbital fall? Explain what the capital letter in the label for this irreducible representation means. Symmetry operations of this point group interchange the f_z^3 orbital with which other f orbital(s)?

10.29. *Table 7.8 shows the energy levels of the central-atom d orbitals in complexes of various geometries. Except for the rare accidental degeneracy, all orbitals at the same energy level are interconverted by symmetry operations of the point group, so they belong to the same irreducible representation. Assume that all ligands for the complexes in question are identical, so that these geometries lose none of their symmetry. List the irreducible-representation labels that would be used for the sets of d orbitals in complexes of the following geometries: (a) square planar; (b) trigonal prismatic; (c) icosahedral.

10.30. The commonly used labels for the sets of metal d electrons in octahedral complexes, t_{2g} and e_g, are labels of irreducible representations if the complex belongs to the O_h point group (e.g., $[Co(NH_3)_6]^{3+}$ with free rotation of the NH_3 groups), but they are not strictly applicable if the complex belongs to a lower-symmetry point group. Referring to the character tables, pick the proper irreducible-representation labels for each metal d orbital in each of the following complex ions: (a) cis-$[CoCl_2(NH_3)_4]^+$; (b) trans-$[CoCl_2(NH_3)_4]^+$. (c) Would you expect the visible spectra of these complex ions to be more or less complex than that of $[Co(NH_3)_6]^{3+}$?

10.31. *The crown ether complex $[K(C_{12}H_{24}O_6)]^+$ was classified in an appropriate point group in Exercise 10.11b. Referring to the character table for this point group, give the symmetry labels (irreducible representations) for each of the p orbitals and for each of the d orbitals of the potassium ion at the center of the complex ion.

10.32. Assume that the gaseous mercury(II) hydroxide molecule has a planar structure linear next to Hg but bent at both ends, so that it falls in the C_{2h} point group. Using the symmetry operations from the character table for C_{2h} but without consulting the right side of the table, determine in which irreducible representation each of the following orbitals on the Hg atom fall: (a) s; (b) the p orbital that lies along the O–Hg–O direction; (c) p_z; (d) p_x; (e) $d_{x^2-y^2}$.

10.33. *Consult the $D_{\infty h}$ character table to find the bonding and antibonding possibilities for the f orbitals.

10.34. In Example 10.10 we found that the d_{xy} and $d_{x^2-y^2}$ orbitals can combine to form Δ_g bonding and Δ_u^* antibonding orbitals. Consult the $D_{\infty h}$ character table to find the possible combinations for the other d orbitals in M_2 molecules.

10.35. *Drawn below are three combinations of orbitals located on the two adjacent atoms of a homonuclear diatomic molecule. (a) Tell whether each of the combinations of orbitals has positive, negative, or zero overlap. (b) Describe each of the combinations as bonding,

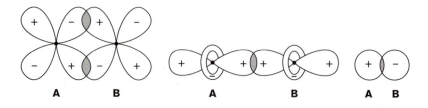

antibonding, or neither. (c) Classify each combination as gerade, ungerade, or neither. (d) After looking for orbital degeneracy, classify each bonding or antibonding combination using appropriate symmetry labels (for a homonuclear diatomic molecule M_2).

10.36. Drawn below are four combinations of orbitals located on the two adjacent atoms of a homonuclear diatomic molecule. (a) Tell whether each of the combinations of orbitals has positive, negative, or zero overlap. (b) Describe each of the combinations with nonzero overlap as bonding or antibonding, or neither. (c) Classify each combination as gerade, ungerade, or neither. (d) After looking for orbital degeneracy, classify each bonding or antibonding combination using appropriate symmetry labels (for a homonuclear diatomic molecule M_2).

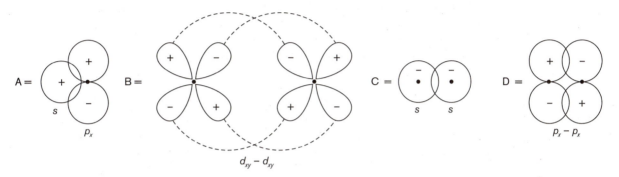

10.37. *In Figure 3.8g–l, combinations of orbitals are drawn, some of which gave positive overlap, some negative, and some zero (see Exercise 3.34). Omitting those giving zero overlap, find their irreducible representations (a) in the $D_{\infty h}$ point group; (b) in the $C_{\infty v}$ point group. Some do not fall in any irreducible representation. Note if a combination is not in an irreducible representation, but can be paired with another combination in a doubly degenerate irreducible representation.

10.38. In Figure 3.8a–f, combinations of orbitals are drawn, some of which gave positive overlap, some negative, and some zero (see Exercise 3.33). Omitting those giving zero overlap, find their irreducible representations (a) in the $D_{\infty h}$ point group; (b) in the $C_{\infty v}$ point group. Some do not fall in any irreducible representation. Note if a combination is not in an irreducible representation, but can be paired with another combination in a doubly degenerate irreducible representation.

GEOFFREY WILKINSON (1921–1996) was born near Todmorden, in west Yorkshire, United Kingdom, to a working-class family. After attending local schools, he obtained a Royal Scholarship to Imperial College, London, where he graduated in 1941. From 1943 to 1946 he did nuclear fission research in Canada, followed by four years at the University of California, Berkeley, where he worked with Glenn Seaborg making artificial isotopes that were neutron-deficient. In 1950, Wilkinson began working on transition metal complexes containing carbonyl and olefin ligands as a Research Associate at MIT. In 1951, though, he was hired as an Assistant Professor at Harvard University on the strength of his nuclear chemistry background. Still, he continued to pursue his interests in transition metal chemistry, including elucidating the structure of dicyclopentadienyl iron (ferrocene) after it was reported in the literature in 1952. He was appointed the chair of Inorganic Chemistry at Imperial College, London, in 1955, where his research focused on complexes of ruthenium, rhodium, and rhenium that contained unsaturated hydrocarbons and metal–hydrogen bonds. As part of this research, he prepared $RhCl(PPh_3)_3$, now known as Wilkinson's catalyst, which is used industrially to hydrogenate alkenes to alkanes. In 1973, Wilkinson and Ernst Otto Fischer were awarded the Nobel Prize in Chemistry "for their pioneering work, performed independently, on the chemistry of the organometallic, so called sandwich compounds." They were the first inorganic chemists since Alfred Werner in 1913, and the first organometallic chemists ever, to win the prize.

CHAPTER 11

Molecular Orbital Theory
A Bridge Between Foundational and Advanced Inorganic Chemistry

Overview of the Chapter

In previous chapters the bonding theory we used to describe molecules and ions was the **valence bond (VB) theory**, which is the bonding theory behind the drawing of Lewis structures and the octet "rule" (Chapter 3). Valence bond theory treats chemical bonds as the result of sharing electron pairs between *two* atoms. In **molecular orbital (MO) theory** a chemical bond can involve several, or even an infinite number of atoms. Valence bond theory is adequate for describing most properties of the vast majority of molecules and ions in common experience; it is relied upon in general chemistry and much of organic chemistry. For work in advanced inorganic, organic, and physical chemistry, however, we need the molecular orbital theory.

Even in homonuclear diatomic molecules (Section 11.1), MO theory introduces the concept of *antibonding orbitals*, which are useful (in Section 11.3) in explaining the Lewis acidities of molecule such as I_2. MO theory also allows bonding orbitals to involve just one electron, which can result in nonintegral bond orders. Occasionally, there are discrepancies that arise even when a "good" Lewis structure can be drawn. The classic example of this is the simple O_2 molecule, for which the Lewis structure shows an O=O double bond and two sets of (paired) unshared electrons on each O atom. It turns out, however, that the O_2 molecule is *paramagnetic* (Section 7.2B), with two unpaired electrons, and in liquid form it can be seen to be attracted to the poles of a strong magnet. This property emerges immediately from MO theory in Section 11.1.

In Section 11.2 we use the concept of *hybridization* (mentioned in an Amplification in Section 3.3) in a different sense, to allow the mixing of different molecular orbitals of the same type and similar energy. Section 11.2 also allows us to compare the bonding descriptions produced by VB theory with those from MO theory, which usually are quite compatible with each other.

MO theory is particularly advantageous in interpreting the various types of *electronic spectroscopy* of polyatomic ions and compounds. In spectroscopic transitions, electrons often are promoted from orbitals that are readily described using valence bond theory into other orbitals such as antibonding orbitals (Section 11.3). When the electrons are completely ionized away from the molecule, as in the method of *photoelectron spectroscopy* described in this section, the spectra match up well with the expectations from MO theory.

Section 11.4 can usefully be introduced with Experiment 10, in which computations are carried out on related *heteroatomic* diatomic molecules and ions. The quantitative results from these computations are useful for explaining periodic trends in the properties of these molecules.

In Section 11.5 we apply MO theory to selected classes of polyatomic molecules, in which the bonds involve more than two atoms. MO theory is more satisfactory in explaining the bonding in *electron-deficient molecules* (Section 11.5B) such as B_2H_6 that do not have enough electrons in them to complete a Lewis structure, but which exist and may be relatively stable nonetheless. In this section MO theory gives us a better way of explaining the results of crystal and ligand field theories (Chapter 7), particularly as they apply to the complexes of π-acceptor ligands at the top of the spectrochemical series of ligands (Section 7.4). We also develop *d*-block analogues to the octet rule, the 16- and 18-electron rules, and learn when these can be successfully applied.

In Section 11.6 we introduce the chemistry and bonding in *d*-block *organometallic compounds* and *d*-block *metal carbonyls*, and we see how well the 16- and 18-electron rules apply to these important compounds.

MO theory is even more extensively applied in more advanced topics in inorganic chemistry. Chapter 12 introduces one important application, to the properties of elemental substances themselves, which are usually polymeric materials. In such materials, the valence bond theory is quite inadequate to clearly explain some of their important properties, such as semiconductivity and metallic bonding.

11.1. Molecular Orbital Theory for Homonuclear Diatomic Molecules

OVERVIEW. Molecular orbital theory allows not only for σ, π, δ, and φ bonding as described in Sections 3.3B and 10.4, but also for the destabilizing effects of "antibonding" σ*, π*, δ*, and φ* orbitals in molecules. Using a given type of atomic orbital, the effectiveness of orbital overlap of two like atoms occurs in the order σ > π > δ > φ. It is possible, therefore, to construct an energy-level diagram for homonuclear diatomic molecules. Given such a diagram, you can usually write the electron configuration of a given molecule, using the proper symmetry labels, and determine its bond order, its total number of unpaired electrons, and its paramagnetism or diamagnetism. You may practice these concepts by trying Exercises 11.1–11.8.

In molecular orbital theory, molecules, by analogy with atoms, have orbitals called **molecular orbitals (MOs)**. MOs involve several atoms or the entire molecule, just as atomic orbitals (AOs) spread over (more or less) the entire atom. Like AOs, each MO has a definite energy. There can be ionization energies and electron affinities for MOs. The electron configuration of a molecule is obtained by filling electrons into MOs starting with those at the lowest energy and in accord with the Pauli exclusion principle, and so on.

However, the computational problems become formidable when we try to apply the Schrödinger wave equation (Section 9.1B) to electrons moving around several nuclei that attract them. Therefore, it is common to assume that, when the electrons are close to a given atomic orbital, their wave function closely resembles an atomic orbital of that atom. Consequently, MOs are commonly generated as a *linear combination of atomic orbitals* on the nuclei in question, which are allowed to overlap with each other.

In Section 3.3B, we saw that the overlap of two atomic orbitals can be of three types: positive, negative, and zero. Valence bond theory emphasizes only the result of positive overlap: When two atomic orbitals (e.g., two $1s$ orbitals on two H atoms) have positive overlap, the (one or two) electrons that can reside in that orbital can spend a lot of time in the region of positive overlap, which is between the two nuclei. In this region, the electrons are attracted to the positive charges of the two nuclei. This attraction is a major source of the bond energy that makes H_2 a more stable species than two separate H atoms. This orbital, therefore, is a **bonding orbital**. In molecular orbital theory, however, overlap is not restricted to the region between two nuclei—it can fall between three, four, or any number of nuclei.

In contrast to valence bond theory, MO theory also stresses the importance of the case of negative overlap of two atomic orbitals (as in Figure 3.6b, d, f, and j–l). When two atomic orbitals have negative overlap, they act as waves that are out of phase with one another. The crest of one coincides with the trough of the other in the region of overlap, canceling out any wave amplitude. The electrons then must spend their time where the wave still has amplitude, which is in the nonoverlapping parts outside of the two nuclei. This leaves the two positive nuclei repelling each other with little or no electron density in between to help pull them together. Putting electrons into a MO built upon negative overlap not only does not contribute to bonding, but also allows internuclear repulsion to proceed unchecked. If only such an orbital is involved, the atoms will fly apart; this is an **antibonding** interaction.

Alternatively, if a molecule has two electron pairs, and one goes into a bonding orbital composed of (say) $1s$ orbitals on each atom, while the other goes into an antibonding overlap composed of $1s$ orbitals, the two effects cancel out: There is no bonding, and the electron pairs behave as unshared, **nonbonding** electron pairs do in Lewis structures. Actually, when the calculations are done, it turns out that the antibonding effect is somewhat larger than the bonding, so the two atoms not only do not bond, they also fly apart: Two He atoms do not form a stable molecule.

Antibonding orbitals are important in MO theory in part because, in assembling MOs as linear combinations of atomic orbitals, there is a type of *conservation of orbitals*. If we combine N atomic orbitals to make molecular orbitals, we obtain N molecular orbitals; no orbitals are "lost." Thus, if $1s$ orbitals are combined in a diatomic molecule to give a MO featuring positive overlap, there must also exist a corresponding MO featuring negative overlap. The existence of a MO with negative overlap does not mean, however, that we must put electrons into such an orbital: Commonly observed molecules have some or all of their antibonding orbitals unoccupied.

In Figures 3.10 and 10.13a we drew the σ-bonding molecular orbital formed by positive overlap of two $1s$ orbitals on adjacent atoms A and B; in Section 10.4 we determined that it was gerade. Mathematically, we can describe this MO as a linear combination of the two $1s$ atomic orbitals:

$$\sigma_g(1s) = c_A 1s(A) + c_B 1s(B) \tag{11.1}$$

In a homonuclear diatomic molecule (but not a heteronuclear one), the two atoms are equally involved in the overlap, so the coefficients c_A and c_B are equal. (Here, for simplicity, we assume them to be one.)

Because this MO was obtained by combining two atomic orbitals, there must be a second MO based on negative overlap, which is produced by taking one of the atomic $1s$ orbitals (say on atom B) as a trough rather than a crest, with a negative sign to its contribution to the wave function. This combination was drawn in Figure 10.13b, and was found in Section 10.4 to be ungerade, so it is labeled as a σ_u^* orbital. Often the asterisk (*) is added to indicate that this is an antibonding orbital:

$$\sigma_u^*(1s) = c_A 1s(A) - c_B 1s(B) \tag{11.2}$$

We write electron configurations of molecules just as we do those of atoms. Molecular and atomic orbitals can take, at most, two electrons each, and the lowest-energy one (in this case, the one with positive overlap) is filled first. We can write electron configurations for molecules or ions using the above two orbitals to hold from one to four electrons:

One electron (e.g., H_2^+)	$\sigma_g(1s)^1$ or just σ_g^1
Two electrons (e.g., H_2)	σ_g^2
Three electrons (e.g., He_2^+)	$\sigma_g^2 \sigma_u^{*1}$
Four electrons (e.g., He_2)	$\sigma_g^2 \sigma_u^{*2}$ (11.3)

In valence bond theory, an important classification of bonds is by their **bond order**: single, double, triple, or quadruple. In MO theory these bonds can occur along with nonintegral bond orders, so a formal definition of bond order is needed: Bond order is *one-half the difference between the number of electrons in bonding orbitals, n_b, and the number in antibonding orbitals, n^*.*

$$\text{Bond order} = \tfrac{1}{2}(n_b - n^*) \tag{11.4}$$

Stability?

AN AMPLIFICATION

It is worthwhile to put the notion of "stability" in its proper context here. We shall see that H_2^+ and He_2^+ have substantial bond energies and are stable compared to the separate atoms and ions (e.g., H and H^+). However, as we proceed through the rest of the text, we will not mention these species, or most of the diatomic molecules that we may treat by MO theory in this chapter, any further. The reason is that most of them are unstable to polymerization or to reaction with other molecules. We study them now because they are the simplest species with which we can investigate bonding; knowledge of them will also be of value to those who become radio astronomers or spectroscopists, or who study the chemistry of the upper atmosphere.

Thus, the bond order in H_2 is ½(2 – 0) = 1. In agreement with its Lewis structure from valence bond theory, it has a single bond.

The bond order in He_2 is ½(2 – 2) = 0. In agreement with our inability to draw a Lewis structure for such a molecule, He_2 does not exist.

Although we cannot draw likely looking Lewis structures for them, the other two ions, which each have nonintegral bond orders of ½, can exist in high vacuum; a salt of the green Xe_2^+ cation has been isolated.[1]

In Chapter 7 we indicated the relative energies of different atomic orbitals in an energy-level diagram to illustrate the effects of the crystal field theory. Similar diagrams can be drawn for molecules (Figure 11.1). These diagrams are in fact even more valuable for molecules because they can express, not only the fact that the $\sigma_u^*(1s)$ MO is higher in energy (is less stable) than the $\sigma_g(1s)$ MO, but it can also indicate how each stands in relationship to its parent (1s) atomic orbitals.

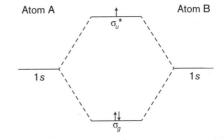

Figure 11.1. Energy-level diagram for a homonuclear diatomic molecule or ion using 1s atomic orbitals (at the relative energies indicated on the left or the right) to form the molecular orbitals at the relative energies indicated at the center. Illustrated is the electron configuration of the ion He_2^+.

To draw MO energy-level diagrams, we first draw lines on either side of the diagram at the energy levels of the parent atoms (in polyatomics, central atoms are placed on one side and outer atoms on the other). *For our purposes of approximation*, these energy levels may be taken as the atomic **valence orbital potential energies (VOPEs)** defined and tabulated in Table 11.1. Note that here in Chapter 11, the standard unit of energy is the electron volt (eV), where 1 eV = 96.5 kJ mol^{-1}.

Next, we draw in bars in the center of the diagram to represent the energies of the MOs. Although computations are necessary to locate these energy levels quantitatively, we know that the energy level of a bonding MO is located below the levels of its parent atomic orbitals, while the energy level of an antibonding MO is located (approximately) the corresponding distance above that of its parent atomic orbitals.

Finally, we draw in lighter or dashed lines to connect the energy levels of the MOs to those of their parent atomic orbitals; the MOs are labeled by symmetry type, and arrows can be filled in to represent occupancy of the orbitals. In Figure 11.1 we show the three electrons of He_2^+.

When we go beyond four electrons, we must occupy a third MO, which requires the use of additional parent atomic orbitals. The next type of atomic orbital available is the 2s. It would seem possible to have positive overlap between a 1s orbital on one atom and a 2s orbital on the other atom, but there are two criteria that rule this out in a homoatomic molecule. The first criterion is that *strong covalent bonding interactions are possible only between atomic orbitals that are fairly close in energy*. In an atom in which 2s orbitals are valence orbitals, the 1s orbitals are core electrons with much higher ionization energies (Section 9.3); the dissimilarities in energy levels make the overlap

TABLE 11.1. Valence Orbital Potential Energies[a]

Atomic Number	Element	1s	2s	2p	3s	3p	3d	4s	4p
1	H	−13.6							
2	He	−24.6							
3	Li		−5.4						
4	Be		−9.3						
5	B		−14.0	−8.3					
6	C		−19.4	−10.6					
7	N		−25.6	−13.2					
8	O		−32.3	−15.8					
9	F		−40.2	−18.6					
10	Ne		−48.5	−21.6					
11	Na				−5.1				
12	Mg				−7.6				
13	Al				−11.3	−5.9			
14	Si				−14.9	−7.7			
15	P				−18.8	−10.1			
16	S				−20.7	−11.6			
17	Cl				−25.3	−13.7			
18	Ar				−29.2	−15.8			
19	K							−4.3	
20	Ca							−6.1	
21	Sc						−4.7	−5.7	−3.2
22	Ti						−5.6	−6.1	−3.3
23	V						−6.3	−6.3	−3.5
24	Cr						−7.2	−6.6	−3.5
25	Mn						−7.9	−6.8	−3.6
26	Fe						−8.7	−7.1	−3.7
27	Co						−9.4	−7.3	−3.8
28	Ni						−10.0	−7.6	−3.8
29	Cu						−10.7	−7.7	−4.0
30	Zn							−9.4	
31	Ga							−12.6	−6.0
32	Ge							−15.6	−7.6
33	As							−17.6	−9.1
34	Se							−20.8	−10.8
35	Br							−24.1	−12.5
36	Kr							−27.5	−14.3

Source: From R. L. DeKock and H. B. Gray, *Chemical Structure and Bonding*, University Science Books: Mill Valley, CA, 1989, pp. 227 and 330.

[a]All energies are the negatives of ionization energies (for convenience of use in energy-level diagrams) and represent averages for all terms of the specified orbitals. The *d*-block energies in the last three columns represent the ionization $3d^{n-1}4s \to 3d^{n-2}4s$, $3d^{n-1}4s \to 3d^{n-1}$, and $3d^{n-1}4p \to 3d^{n-1}$, respectively.

ineffective. The second criterion is that *molecular orbitals must have the symmetry of one of the irreducible representations of the point group* (Sections 10.3 and 10.4). A combination of $1s(A) + 2s(B)$ is neither gerade nor ungerade, so it is not allowed in $D_{\infty h}$.

The combinations we need to consider are

$$\sigma_g(2s) = 2s(A) + 2s(B) \tag{11.5}$$

and

$$\sigma_u^*(2s) = 2s(A) - 2s(B) \tag{11.6}$$

Now we can write electron configurations for molecules or ions with from five to eight electrons:

Five electrons (e.g., Li_2^+)	$\sigma_g^2\sigma_u^{*2}\sigma_g(2s)^1$ or (showing only valence electrons) σ_g^1
Six electrons (e.g., Li_2)	$\sigma_g^2\sigma_u^{*2}\sigma_g^2$ or σ_g^2
Seven electrons (e.g., Be_2^+)	$\sigma_g^2\sigma_u^{*2}\sigma_g^2\sigma_u^{*1}$ or $\sigma_g^2\sigma_u^{*1}$
Eight electrons (e.g., Be_2)	$\sigma_g^2\sigma_u^{*2}\sigma_g^2\sigma_u^{*2}$ or $\sigma_g^2\sigma_u^{*2}$ (11.7)

The latter versions of the electron configurations are really *valence* electron configurations of the molecules—because the two lower energy MOs are constructed from core atomic orbitals, they are chemically inactive as well. In fact, these core MOs are virtually indistinguishable from atomic $1s$ orbitals on the two atoms, because a third criterion is that *in order to form a MO, there must be significant overlap of the parent atomic orbitals.* As suggested in Figure 1.4, core orbitals have substantially smaller radii than valence orbitals and do not overlap well with core orbitals on adjacent atoms.

When we go to molecules or ions possessing more than eight electrons, we need to incorporate atomic $2p$ orbitals among the parent atomic orbitals. However, the $2p_z$ orbital has different symmetry properties than the doubly degenerate $(2p_x, 2p_y)$ set in the $D_{\infty h}$ point group. Taking linear combinations of the $2p_z$ orbitals as shown in Figure 10.13c, d on the two atoms gives two new σ_g and σ_u MOs:

$$\sigma_g(2p_z) = 2p_z(A) - 2p_z(B) \tag{11.8}$$

$$\sigma_u^*(2p_z) = 2p_z(A) + 2p_z(B) \tag{11.9}$$

The doubly degenerate $(2p_x, 2p_y)$ set of orbitals give positive overlap as seen in Figure 10.13e to produce the familiar π_u bonding MOs (i.e., the degenerate pair of π_x and π_y bonds). There is also a corresponding degenerate antibonding set of orbitals resulting from negative overlap, the π_g^* antibonding MOs.

An energy-level diagram including these MOs having parentage in p orbitals is shown in Figure 11.2. Note that in the isolated parent atoms, electrons in all p orbitals are at the same energy level. However, as we previously noted in Section 3.3B, σ-type overlap is more effective than, and gives more bonding than, π-type overlap. Sigma-type negative overlap is also more antibonding than π-type negative overlap. Consequently, the bonding MOs resulting from σ-type overlap drop in energy from

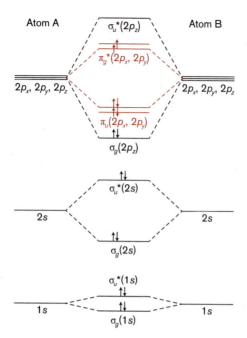

Figure 11.2. Energy-level diagram for a homonuclear diatomic molecule or ion of the second period, with valence 2s and 2p atomic orbitals. In contrast to Figure 11.7 in Section 11.2, *no* interaction is shown of the σ_g or of the σ_u^* orbitals arising from valence 2s and 2p atomic orbitals. The electron configuration of O_2 is illustrated.

the parent atomic orbitals more than the bonding MOs resulting from π-type overlap, and the antibonding MOs resulting from σ-type overlap rise in energy from the parent atomic orbitals more than the antibonding MOs resulting from π-type overlap. The order of filling of these MOs is $\sigma_g(2p_z) < \pi_u(2p_x, 2p_y) < \pi_g^*(2p_x, 2p_y) < \sigma_u^*(2p_z)$. The valence electron configuration of the 12-valence-electron molecule O_2 is therefore $\sigma_g(2s)^2\sigma_u^*(2s)^2\sigma_g(2p_z)^2\pi_u(2p_x, 2p_y)^4\pi_g^*(2p_x, 2p_y)^2$. The bond order can always be calculated from valence electron configurations only. In this case, it is ½(2 – 2 + 2 + 4 – 2) = 2 and the O_2 molecule has the expected double bond. However, by Hund's rule, the two electrons in the degenerate π_g^* pair of orbitals must each occupy one of the orbitals with parallel spins; this accounts for the presence of two unpaired electrons and the paramagnetism in O_2.

Homonuclear diatomic molecules may also involve *d*- and *f*-block atoms and *d* and *f* orbitals. At high temperatures, enough energy is available to vaporize *d*- and *f*-block metals. Although these metals mostly vaporize as atoms (Table 6.4; Section 11.8), it is possible to generate a sufficient concentration of gaseous diatomic molecules to allow their study. As was discussed in Sections 3.3B and 10.4, the overlap of *d* or *f* orbitals from the two metals gives rise not only to the possibility of forming σ and π bonds and the corresponding antibonding orbitals, but also to the possibility of forming δ and ϕ bonding and antibonding orbitals (Figure 11.3).

For like atoms in a diatomic molecule, the effectiveness of the covalent overlap, and hence the degree to which the bonding MO is more stable than the *d* or *f* orbitals of the parent atoms, falls in the order $\sigma > \pi > \delta > \phi$ (Section 3.3B). This is therefore also the order in which the corresponding antibonding MO is less stable than (is raised in energy above) the *d* or *f* orbitals of the parent atoms. This order results in an energy-level diagram such as that shown in Figure 11.4 for homoatomic diatomic molecules of the *d* block.

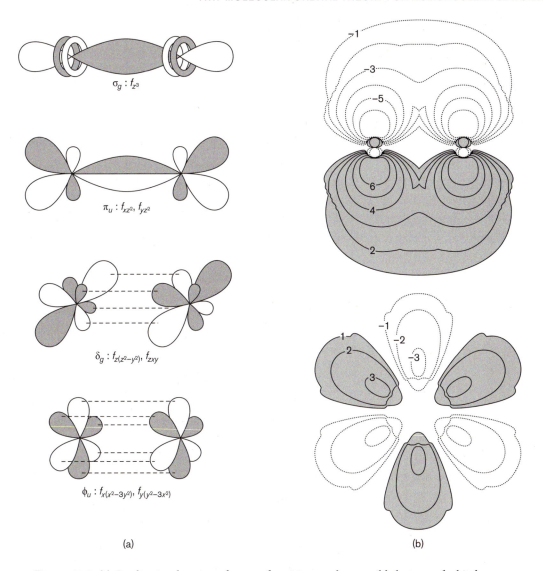

(a)

(b)

Figure 11.3. (a) Qualitative drawing of types of positive overlap possible between f orbitals on two adjacent atoms; positive and negative lobes are shown in different colors. (The parent f orbitals are drawn and named differently than in Figure 7.3, because the symmetry of the molecule is different: f orbitals are normally drawn in different manners for molecules in higher-order point groups than in others.) (b) Quantitative contour diagrams of the ϕ_u bonding orbital from a side view (top) and from an end view (bottom). [Adapted from B. E. Bursten and G. A. Ozin, *Inorg. Chem.* 23, 2910 (1984).]

Figure 11.4. Energy-level diagram for a homonuclear diatomic molecule or ion of the fifth period of the d block, with valence $5s$ and $4d$ orbitals as parent atomic orbitals.

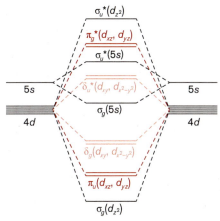

One complication that comes up in the d and f blocks is that the parent atoms have two types of valence orbitals. The s orbitals are also valence orbitals, and give rise to two additional MOs: a σ_g and a σ_u^*. Where these fall in the energy-level diagram depends on how much higher in energy the parent s orbitals are than the parent d orbitals, and on the relative magnitudes of the two types of overlap. These factors cannot be predicted qualitatively and vary across a period and down a group (Section 3.3C). For the purpose of Example 11.1, we assume the energy-level diagram has the order of orbitals shown in Figure 11.4.

Another complication is that the energy differences between adjacent levels in an energy-level diagram such as Figure 11.4 may be smaller than the pairing energies of the electrons (Section 7.2A), so the diatomic molecule may adopt a high-spin electron configuration with a large number of unpaired electrons.

AN AMPLIFICATION

Relativistic Effects

There are additional complications for heavy elements that arise from relativistic effects (Section 9.6). These include spin-orbit coupling effects (Section 9.6B), which result in the splitting of the d and f orbitals of elements heavier than Br into two subsets: the $nd_{5/2}$ and the $nd_{3/2}$ orbitals, and two types of nf orbitals, the $nf_{7/2}$ and the $nf_{5/2}$ orbitals. These are affected differently by relativistic expansion (Section 9.6A); the $5f_{7/2}$ orbitals are so expanded that they cannot overlap other orbitals, while $5f_{5/2}$ orbitals have a better chance of forming (quite weak) ϕ bonds. Because the valence s and $p_{1/2}$ orbitals are affected in the opposite direction by relativistic contraction, the predictions we make for (especially) seventh-period elements are likely to be quite different from what is observed.

Recent relativistic calculations[2] suggest, as a result, that the highest bond order that can actually be achieved between two atoms is six, in Mo_2 and W_2, which would have *two* σ bonds, two π bonds, and two δ bonds. In the f-block elements of Period 7, Th_2 can have a quadruple bond, and Pa_2 can have a quintuple bond. The situation for U_2 is much more complicated in view of the relativistic effects and the presence of 16 valence or post-valence orbitals in close energy proximity to each other (these include the $5f$, $6d$, $7s$, and $7p$ orbitals).[3] More recent calculations[4] show the presence of a clear triple bond $\sigma_g^2 \pi_u^4$ based on the strongly overlapping $7s$ and $6d$ orbitals. Beyond that, things are murkier: An oversimplified description has six singly occupied molecular orbitals based on $6d$ and $5f$ atomic orbitals, with the singly occupied ϕ_u and ϕ_g^* orbitals being very close in energy due to poor overlap of the contracted $5f$ orbitals. Consequently, this molecule is expected to have six unpaired electrons and a bond order according to Equation 11.4 of $n = \frac{1}{2}(11 - 1) = 5$. By the time that Pu_2 is reached, no bonding involving $5f$ orbitals is expected, so the bond order is predicted to be low.[5]

EXAMPLE 11.1

Write the expected electron configuration and predict the bond order and the number of unpaired electrons for the homonuclear diatomic molecules (a) Mo_2 and (b) U_2.

SOLUTION: (a) This diatomic molecule has 12 valence electrons, enough to fill the six lowest orbitals in Figure 11.4 (because there are two sets of doubly degenerate energy levels, this means the lowest four distinct energy levels). Consequently, the electron configuration is expected to be $\sigma_g^2\pi_u^4\delta_g^4\sigma_g(5s)^2$. Because all of these orbitals are bonding, the predicted bond order is six; this molecule should have a sextuple bond! (Actual calculations have supported this prediction.[6]) It should have no unpaired electrons and hence be diamagnetic.

(b) The diatomic U_2 molecule has 12 valence electrons. If the energy-level diagram is similar to Figure 11.4, in the sense of having the orbitals derived from the $7s$ above all of the bonding orbitals derived from the $5f$ orbitals, then the predicted electron configuration is $\sigma_g^2\pi_u^4\delta_g^4\phi_u^2$ and the bond order should be six, but with two unpaired electrons. As described in the previous Amplification, the actual bond order is likely to be five, with six unpaired electrons![7]

11.2. Orbital Mixing in Homonuclear Diatomic Molecules: Bond Energies and Bond Lengths

OVERVIEW. MOs in a homonuclear diatomic molecule can undergo mixing or "hybridization" if the two MOs fall in the same irreducible representation and are of similar energies. Similarity in MO energies is more likely if the atomic orbital valence orbital potential energies (Table 11.1) are close. From the bond orders of diatomic molecules, we can predict trends in bond energies and bond lengths of closely related species. It is also possible to interpret the electron configurations of the molecules in terms of equivalent valence bond structures or sets of resonance hybrid structures by applying three guidelines of translation. You can practice applying these concepts by trying Exercises 11.9–11.18.

In Section 11.1 we selected our linear combinations of atomic orbitals for construction of MOs based on three criteria, which included that the atomic orbitals be of comparable energy, and that the resulting MOs have the symmetry of one of the irreducible representations. These criteria do not actually exclude some possibilities that we have ignored so far.

Atoms from the p, d, and f blocks have more than one type of valence orbital. The energies of these different types of valence orbitals are often rather close. As indicated by Table 11.1 and in the graph of valence orbital energies (Figure 11.5), the parent atoms' valence s and p orbitals are closest in energy in the early elements of the p block, B and C. Consequently, B and C form some unusual diatomic molecules and ions (Example 11.3). We may also note that s–d hybridization is also favored for the same reason early in the d block (see the Amplification in Section 9.4 entitled "The Diminished Role of sp Hybridization in Inorganic Chemistry"). Hybridization is also favored by spatial overlap of the s and p orbitals, which is best if they are both small (have low values of $<r_{max}>$; Section 9.4). Hence, s–p hybridization overall is most likely

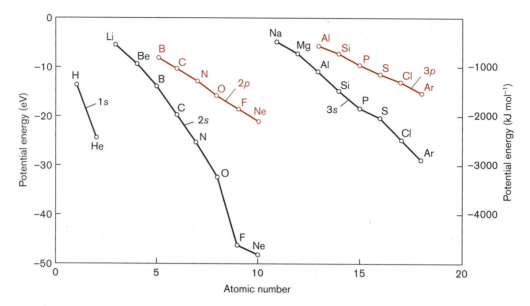

Figure 11.5. Valence orbital potential energies. [Adapted from G. L. Miessler and D. A. Tarr, *Inorganic Chemistry*, Prentice Hall: Englewood Cliffs, NJ, 1991, p. 137.]

for elements close to carbon, which is why this concept is employed so extensively in organic chemistry. When we formed the $\sigma_g(2s)$ orbital for a second-period diatomic (Figure 11.2), we did not consider the possibility that not only the $2s$ but also the $2p_z$ orbitals of the two atoms would contribute to this σ_g molecular orbital.

A corollary of these criteria is that if two MOs falling in the same irreducible representation have energies close to each other (i.e., are formed from atomic orbitals with energies close to each other), they interact with each other—they undergo a type of electron configuration interaction or orbital mixing, which can be called the **hybridization of molecular orbitals**. The $2s$ orbitals are the parents of the $\sigma_g(2s)$ molecular orbital, and the $2p$ orbitals are the parents of the $\sigma_g(2p)$ molecular orbital. However, these MOs are close in energy to each other, and they both belong to the same symmetry type, σ_g. Consequently, the actual MOs are each partly derived from the $2s$ and partly derived from the $2p$ atomic orbitals:

$$\sigma_g(2s, 2p_z) = c_A 2s(A) + c_B 2s(B) + d_A 2p_z(A) + d_B 2p_z(B) \tag{11.10}$$

In this "s–p hybridized" wave function, although $c_A = c_B$ and $d_A = d_B$ by the requirements of symmetry, $c_A \neq d_A$. The s orbitals are somewhat emphasized in the lower-energy MOs and the p orbitals, being higher in energy to start with, are somewhat more highly emphasized in the higher-energy MOs.

It is perhaps easiest to view this "hybridization" process as one of mixing the two σ_g molecular orbitals with each other. In combining two MOs, which must be of the same irreducible representation, we can make two new hybridized MOs (Figure 11.6). The lower energy σ_g orbitals result from positive overlap of the two original MOs and have more positive overlap than either one; hence they are lower in energy than either starting MO (Figure 11.6a). The other, higher energy σ_g orbital (Figure 11.6b) is made with negative overlap of the two starting orbitals. The new negative overlap just about cancels out the original positive overlap present in the original MOs, so this σ_g orbital

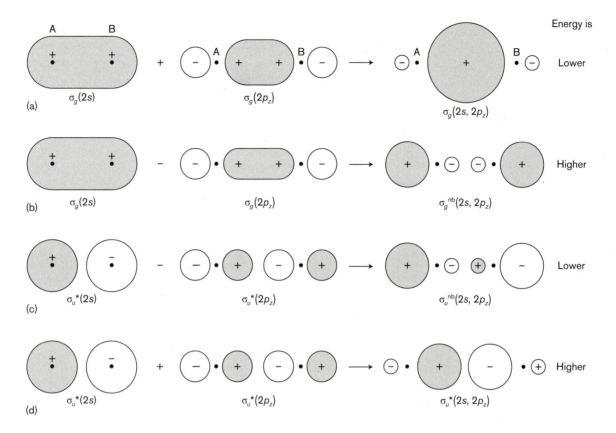

Figure 11.6. Mixing or hybridization of molecular orbitals of the same symmetry type and similar energies. Top, positive (a) and negative (b) combinations of $\sigma_g(2s)$ with $\sigma_g(2p_z)$ molecular orbitals. Bottom, negative (c) and positive (d) combinations of $\sigma_u^*(2s)$ with $\sigma_u^*(2p_z)$ molecular orbitals.

has close to zero overlap (it is only slightly bonding). It is higher in energy than either of the two starting MOs, and because of its low overlap of atomic orbitals, it is an approximately nonbonding MO, σ_g^{nb}.

If we examine the energy-level diagram for second-row diatomic molecules carefully (Figure 11.2), we see that there is only one other pair of MOs that share the same irreducible representation and that can therefore hybridize if they are close in energy: $\sigma_u^*(2s)$ and $\sigma_u^*(2p_z)$. Upon combining these two antibonding orbitals, they give rise to one stabilized, approximately nonbonding (actually slightly antibonding) MO, σ_u^{nb} (Figure 11.6c), and one very antibonding MO, σ_u^* (Figure 11.6d). When we incorporate the changes in energies of these four orbitals into the energy-level diagram for second-row diatomics, we get a new diagram that may have one significant change in order: σ_g^{nb} may rise above π_u (Figure 11.7).

As the bond order in a diatomic molecule or ion increases, there are two important practical consequences: (1) the bond energy increases (the bond becomes stronger) and (2) the bond becomes shorter. These phenomena

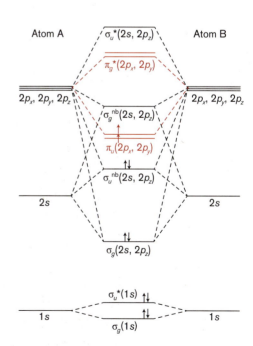

Figure 11.7. Energy-level diagram for a homonuclear diatomic molecule with s–p hybridization or mixing. The electron configuration of B_2 is illustrated.

are readily illustrated in the first period by the data in Table 11.2 for the species with 1–3 electrons, H_2^+ through He_2^+. The maximum bond energy and the minimum bond lengths are found for H_2, the species with the maximum bond order. Species such as H_2^+ and He_2^+, which have bond orders of ½, have bond energies roughly one-half that of H_2, and bond lengths substantially longer than H_2.

TABLE 11.2. Bond Energies and Lengths of Some Homonuclear Diatomic Molecules and Ions[a]

Molecule	Bond Energy (kJ mol⁻¹)	Bond Length (pm)	Molecule	Bond Energy (kJ mol⁻¹)	Bond Length (pm)
H_2^+	255	106	N_2^+	842	112
H_2	432	74	N_2	942	110
He_2^+	230	108	P_2	477	189
O_2^+		112	As_2	382	229
O_2	494	121	Sb_2	298	221
O_2^-	393	126	O_2	494	121
O_2^{2-}		149	S_2	421	189
B_2	274	159	Se_2	325	217
C_2	602	124	Te_2	261	256
Si_2	314	225	Cu_2	198	222
Ge_2	272		Ag_2	162	
Sn_2	192		Au_2	226	247
Pb_2	96				

Source: Data from R. L. DeKock and H. B. Gray, *Chemical Structure and Bonding*, University Science Books: Mill Valley, CA, 1989, p. 229.

[a]Bond energies for the diatomic Group 1 and 17/VII molecules are found in Table 3.2.

EXAMPLE 11.2

Partial bond energy and bond length data for the series of diatomic oxygen ions and the O_2 molecule are given in Table 11.2. Explain these data, and fill in the gaps, including predicting the data for O_2^{2+}.

SOLUTION: The electron configurations of these diatomics can be predicted as in Example 11.1. These electron configurations lead to predictions of bond orders as follows: O_2^{2+}, 3.0; O_2^+, 2.5; O_2, 2.0; O_2^-, 1.5; and O_2^{2-}, 1.0. Bond energies should decrease and bond lengths should increase along this sequence. The strongest and shortest bond should therefore be that of O_2^{2+} (bond energy about 846 kJ mol⁻¹ from data in Tables 3.2 and 3.3 for one σ plus two π bonds; bond length < 112 pm for O_2^{2+}). The unknown bond energy for O_2^+ should be about 670 kJ mol⁻¹ and for O_2^{2-} it should be about 142 kJ mol⁻¹ for a single σ bond.

EXAMPLE 11.3

Write the electron configuration and describe the bonding in the B_2 molecule, using (a) the energy-level diagram appropriate for no s–p mixing and (b) the energy-level diagram appropriate for s–p mixing (hybridization).

SOLUTION: The B_2 molecule has six valence electrons. (a) Using the energy-level diagram of Figure 11.2, we obtain the following valence electron configuration: $\sigma_g(2s)^2\sigma_u^*(2s)^2\sigma_g(2p_z)^2$. The bond order is $\frac{1}{2}(4 - 2) = 1$. (b) Using the energy-level diagram of Figure 11.7, we obtain the following valence electron configuration: $\sigma_g^2(\sigma_u^{nb})^2\pi_u^2$. The bond order is $\frac{1}{2}(2 - {\sim}0 + 2) =$ somewhat less than 2. The bond consists of one π bond, containing two unpaired electrons, plus most of a σ bond (because σ_u^{nb} is actually slightly antibonding).

B_2 is a strange species! However, it does exist and has a bond energy of 274 kJ mol^{-1} and a bond length of 159 pm. This is somewhat weaker than the 293 kJ mol^{-1} B–B single bond energy given in Table 3.2, and it is somewhat shorter than twice the 84-pm covalent radius of B (Table 1.13). However, it is longer than the 145-pm B≡B bond found in a boryne complex with two bulky axial ligands.[8]

AN AMPLIFICATION

Designation of Nonbonding and Antibonding Orbitals in the Literature

When MOs involve the mixing of different parent orbitals, or when they involve several atoms, the clear distinction of bonding, nonbonding, and antibonding can be lost. Some orbitals may be "approximately nonbonding" or may be more bonding than other bonding MOs. Consequently, the literature often omits the suffixes "nb" and "*" due to these ambiguities, and does not show the parent atomic orbitals, which may be of several types. To distinguish the several σ molecular orbitals that may be present in a molecule, they are often listed numerically: "1σ" is the lowest energy σ molecular orbital in the molecule, "2σ" is the second lowest (regardless of whether it is bonding or antibonding), and so on. We have elected not to enumerate the MOs in this text, however, because this requires keeping track of all core MOs, which is otherwise fairly pointless. We also assign approximate nonbonding status because it allows us to compute approximate bond orders; hence it helps us to visualize the overall bonding.

You are probably more comfortable with valence bond theory than with MO theory at this point from having used it more in your organic chemistry course. Therefore, it may be useful to translate some of the MO bonding pictures into their nearest valence bond equivalents (Lewis dot structures or resonance structures). The "guidelines of translation" are as follows:

1. Electron pairs in bonding MOs in diatomic molecules correspond to individual bonds of the indicated type (σ, π, and δ).

2. When both a bonding MO and an antibonding MO of the same symmetry type (e.g., σ and σ^*) and atomic parentage are occupied by two electrons,

their combined effect is to cancel each other's bonding contributions, and to act as two unshared electron pairs, one on each atom.

3. When an electron pair occupies a nonbonding MO in a homonuclear diatomic molecule, it corresponds to one unshared electron pair located equally on each atom, a situation that can only be represented in valence bond theory by drawing resonance structures.

For example, we know that F_2 has the MO valence electron configuration $\sigma_g(2s)^2\sigma_u*(2s)^2\sigma_g(2p_z)^2\pi_u(2p_x, 2p_y)^4\pi_g*(2p_x, 2p_y)^4$. Does this or does this not disagree with the Lewis dot structure of F_2, which shows a single bond and six unshared electron pairs on the two F atoms? The F_2 molecule has four electron pairs in bonding MOs and three in antibonding MOs, which results in a bond order of one: a single bond. The three pairs in antibonding MOs cancel out three pairs found in the corresponding type of bonding MO, so these account for the six unshared electron pairs in the Lewis structure of F_2. There is no discrepancy between the MO and valence bond pictures of the bonding in this molecule.

There is no way that we can draw a Lewis structure for B_2 that obeys the octet rule. What Lewis structure is suggested by its electron configuration from MO theory? In Example 11.3, we obtained electron configurations of $\sigma_g(2s)^2\sigma_u*(2s)^2\sigma_g(2p_z)^2$ (with no s–p mixing) and $\sigma_g^2(\sigma_u^{nb})^2\pi_u^2$ (with s–p mixing). Considering the first electron configuration, we can apply Guideline 2 to translate $\sigma_g(2s)^2\sigma_u*(2s)^2$ into two unshared electron pairs, one on each boron. By Guideline 1, the $\sigma_g(2p_z)^2$ translates into a σ bond, so the Lewis structure would be :B–B:. The second electron configuration includes one σ bond, one π bond, and one unshared electron pair located equally on *both* boron atoms. This situation can only be represented in valence bond theory by using resonance structures—namely, :B=B ⟷ B=B:.

It is also possible to carry out the reverse process, starting with a Lewis structure and adding electrons to get an indication as to whether the first unoccupied molecular orbital is bonding (it is if the bond order increases) or antibonding (it is if the bond order decreases).[9] For example, we can see that the first unoccupied MO of O_2 is antibonding (π*) by adding two electrons to the Lewis structure of :Ö=Ö: to give $[:Ö–Ö:]^{2-}$, which has a lower bond order (one less π bond) than O_2.

By subtracting electrons from the Lewis structure and then rearranging to give the best possible structure of the electron-depleted substance, we can also surmise whether the highest-energy occupied MO is bonding or antibonding. The highest-energy occupied MO of O_2 is π*, because removing its two electrons generates $[:O=O:]^{2+}$, with one more π bond than O_2.

11.3. Electronic Spectroscopy and Molecular Orbital Theory

OVERVIEW. On an energy-level diagram of a molecule or ion, you can identify the HOMO and LUMO, and sketch lines that would correspond to bands that might be observed in the UV or visible spectrum of the molecule or ion. Note whether the spectrum should contain intense absorption bands. You may practice these concepts of Section 11.3A by trying Exercises 11.19–11.22. You can interpret a photoelectron spectrum to determine whether a given MO is nonbonding, bonding, or antibonding. You may practice these concepts of Section 11.3B by trying Exercises 11.23–11.24 and (after reading Section 11.4) Exercises 11.31, 11.33, and 11.34. Trends in their ionization energies and electron affinities (or orbital energies) give the relative softnesses of a series of molecules. You can identify examples of secondary or halogen bonding. You may practice these concepts of Section 11.3C by trying Exercises 11.25–11.30 and (after reading Sections 11.4 and 11.5) Exercises 11.35, 11.37, 11.38, and 11.50.

11.3A. Frontier Orbitals and Ultraviolet–Visible Spectra. Electrons can be promoted from full (or partially full) orbitals to vacant (or partially vacant) orbitals by photons of light, as previously discussed in Section 7.3. The lowest-energy absorption usually occurs in the visible or ultraviolet (UV) region of the spectrum. F_2 is virtually colorless because this absorption occurs in the UV region of the spectrum. Cl_2 is yellow because the absorption band spans some of the UV and some of the adjacent visible region. Br_2 is deep red because the band is clearly in the visible. I_2 is purple because the band is at even lower energy in the visible region. The absorption usually results in the promotion of an electron from the **highest occupied molecular orbital (HOMO)** to the **lowest unoccupied molecular orbital (LUMO)** of the molecule or ion. Collectively, these orbitals are the focus of much chemical reactivity and are known as the **frontier orbitals** of the molecule or ion. Paramagnetic molecules such as O_2 also have **singly occupied molecular orbital(s)** or **SOMOs**, which can be involved in the spectra and in free radical reactions.

After consulting Figure 11.2, we can identify the lowest-energy electronic transition in the dihalogen molecules as $\pi_g^* \rightarrow \sigma_u^*$. This electronic transition is found at 19,230 cm^{-1} = 230 kJ mol^{-1} = 2.39 eV in I_2. Based on the colors of the halogens and using Table 7.3, we can see that the trend in the HOMO–LUMO separation is to decrease from top (>26,000 cm^{-1} = >312 kJ mol^{-1} = >3.2 eV in F_2) to bottom of the group of elements.

Two factors affect this shift to lower energy as we go down the group. First, the entire MO diagrams (analogous to Figure 11.2), including the HOMOs and LUMOs, are shifted to less negative energies for heavier halogens by the upward trend in halogen valence np orbital potential energies from F (–21.6 eV) through Br (–12.5 eV) and on to I (Table 11.1). Second, overlap and bond strength decrease down a group of the periodic table as the orbitals become more diffuse (Section 3.3C; Table 3.2), so the overall separations of the energy levels (e.g., of the HOMO and the LUMO) decrease from F_2 through I_2.

Intensity of Spectra. If the molecule has an inversion center, the intensity of the light absorption is restricted by the irreducible representations of the molecular orbitals involved. In order for the wave of light to be absorbed with high intensity, *either* the HOMO or the LUMO (but not both) must be ungerade. In the case of the diatomic halogen molecules, the absorptions (and the colors) are intense because the HOMO is gerade while the LUMO is ungerade. In *d*-block octahedral (O_h) transition metal complexes, the absorptions (Section 7.3) are not intense because the transitions are between two gerade orbitals, t_{2g} and e_g. In contrast, *d*-block tetrahedral (T_d) complexes do not have an inversion center, so their $e \rightarrow t_2$ electronic transitions are allowed, and they are more intense than those in the corresponding octahedral complexes.

EXAMPLE 11.4

As discussed in Section 7.6A, Indicating Drierite™ is mainly anhydrous $CaSO_4$ impregnated with anhydrous $CoCl_2$. Once the Drierite™ has absorbed its capacity of water from the solvent, the following changes in geometry and color occur about the Co^{2+} ion:

$$[CoCl_4]^{2-} \text{ (tetrahedral)} + 6H_2O \rightleftharpoons [Co(H_2O)_6]^{2+} \text{ (octahedral)} + 4Cl^-$$
Bright blue   Pink, not intense color

Which of the three spectra shown in Figure 7.18 could belong to $[CoCl_4]^{2-}$? Which could belong to $[Co(H_2O)_6]^{2+}$? Using intensities, explain.

SOLUTION: The lowest-energy absorption for octahedral $[Co(H_2O)_6]^{2+}$ is $t_{2g} \rightarrow e_g$, so it would be of low intensity, as is observed in spectrum (1). The lowest-energy absorption for tetrahedral $[CoCl_4]^{2-}$ would be $e \rightarrow t_2$. No inversion center is involved, so this can have a higher molar extinction coefficient, as is observed in spectra (2) or (3). Spectrum (3) is the more likely, because it has fewer absorption bands than spectrum (2).

11.3B. Photoelectron Spectroscopy and Ionization Energies of Molecules.

Photoelectron spectroscopy uses photons of high but constant energy to strike gaseous molecules and eject photoelectrons from the different occupied MOs in the molecules. This allows us to determine the ionization energies of molecules. *Koopman's theorem* posits that the ionization energy of the molecule is equal to the negative of the orbital energy of the electron in the molecule, provided that, on ionizing that electron, the other electrons do not reorganize themselves and thereby change energy. From its photoelectron spectrum, we find that the ionization energy of the H_2 molecule is 15.426 eV. This reflects the fact that the σ_g electrons in the H_2 molecule are more firmly bonded to the H nuclei than is the 1s electron in the single H atom (ionization energy = 13.60 eV). Using Koopman's theorem we deduce that the *approximate* orbital energy of the σ_g HOMO in H_2 is –15.45 eV.

In practice, it is often observed that photoelectron spectra mainly contain, not single absorptions at precise energies such as 15.45 eV, but multiplets with even spacing between several lines. As explained in the following Amplification, if a photoelectron band has *vibrational fine structure*, the ionization producing it is ionization from a *bonding or antibonding MO*. If the photoelectron band has *no fine structure*, the ionization is from a *nonbonding molecular orbital*.

In Figure 11.8, we show the photoelectron spectra of the three most stable diatomic molecules of the second period. Note that sharp, single absorption bands that indicate nonbonding MOs are found only in the spectrum of N_2. Approximately nonbonding orbitals occur in homonuclear diatomic molecules only with s–p mixing, which therefore applies to N_2. By referring to Figure 11.7 and adding enough electrons to give the 14-electron configuration of N_2, we see that the HOMO predicted by this energy-level diagram is σ_g^{nb}, which is approximately nonbonding. The only other nonbonding energy level in Figure 11.7 is the third one down, σ_u^{nb}. In accord with this, we observe in the photoelectron spectrum that the first and third bands from the top lack vibrational structure. So the photoelectron spectrum not only confirms these assignments, it suggests quantitatively at what energies the different levels in the energy-level diagram should go. If we have done quantitative MO calculations, the photoelectron spectrum can support or contradict our calculations.

AN AMPLIFICATION

Fine Structure in Photoelectron Spectra

In H_2 the absorption at 15.45 eV is followed by several other evenly spaced lines at higher energies. The extra energies of the higher absorptions are used to excite the vibrations of the molecule. The energy spacing corresponds to the H–H stretching frequency of the molecule. This phenomenon occurs when the electron being ejected comes from a bonding or antibonding MO. In such cases the product ion (e.g., H_2^+) has a substantially shorter or longer bond at equilibrium. It is not formed with this equilibrium bond length, but rather with a nonequilibrium one (in a vibrationally excited state), and it may use some of the energy of the photon to vibrate toward the equilibrium bond length. However, if the electron is ejected from a nonbonding MO, the bond order and the bond length in the product ion is unchanged from that in the molecule, so no vibrations need be excited to reach an equilibrium bond length.

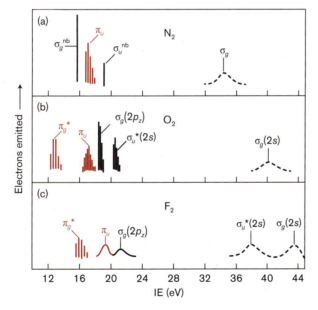

Figure 11.8. Schematic photoelectron spectra of (a) N_2, (b) O_2, and (c) F_2 molecules. [Adapted from W. C. Price, "Ultraviolet Photoelectron Spectroscopy: Basic Concepts and the Spectra of Small Molecules," in *Electron Spectroscopy: Theory, Techniques, and Applications*, vol. 1, C. R. Brundle and A. D. Baker, Eds., Academic Press: New York, 1977, p. 151.]

EXAMPLE 11.5

The halogen molecules do not involve *s*–*p* hybridization. Figure 11.9 shows their photoelectron spectra. With reference to the appropriate MO diagram: (a) Assign the different photoelectron peaks to the MOs from which the ejected electrons originate; read off the ionization potential of the dihalogen molecules. (b) Now consider the visible and UV spectra of these molecules, with electronic transitions from these same orbitals to the LUMO of the molecule. Which of these transitions should be intense?

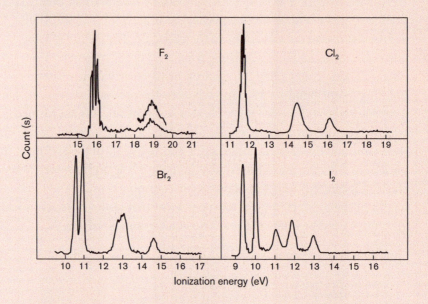

Figure 11.9. Photoelectron spectra of the halogens. Relativistic effects mask the fine structure expected for the bands in Cl_2, Br_2, and I_2. [Adapted from A. W. Potts and W. C. Price, *Trans. Faraday Soc.* 67, 1242 (1971).]

SOLUTION: The appropriate MO diagram is Figure 11.2, in which the HOMO is π_g^*. This is the lowest-energy band in each spectrum; its energy should be the ionization energy of the molecule (accepted values are 15.697 eV in F_2, 11.480 eV in Cl_2, 10.516 eV in Br_2, and 9.31 eV in I_2).[10] The penultimate energy band in each spectrum should be ionization from the π_u MO. In Cl_2 and Br_2, third-lowest energy bands can also be seen, which should originate from the $\sigma_g(2p_z)$ MO. (b) The LUMO of these molecules is $\sigma_u^*(2p_z)$. The intense absorptions will be those from gerade MOs—namely, those from the π_g^* and $\sigma_g(2p_z)$ MOs.

11.3C. Electron Affinities, Frontier Orbitals, and Soft Lewis Acid–Base Chemistry of Molecules.

LUMOs and Electron Affinities of Molecules. Electron affinities (EAs) of molecules can now be determined by the method of laser photo-detachment of electrons from negative ions. According to Koopman's theorem, orbital energies of LUMOs should be *approximately* equal to the negative of the electron affinities.

Frontier Orbitals and Lewis Acid–Base Chemistry. *Molecular soft acids and soft bases are those in which the frontier orbitals are close in energy.*[11,12] The softness of a molecular acid or base can be expressed as

$$\text{Softness parameter} = \frac{2}{(\text{IE} - \text{EA})} = \frac{2}{(\text{orbital energy of LUMO} - \text{orbital energy of HOMO})} \quad (11.11)$$

This closeness in energy means that the valence electrons in a molecular soft acid or soft base can easily be promoted into another orbital, the LUMO. Hence, these species are described as "polarizable" or "soft." For example, the relative softnesses of the dihalogen molecules increase from $2/(15.70 - 3.08) = 2/12.62 = 0.158$ for F_2 up to $2/(9.31 - 2.55) = 2/6.76 = 0.296$ for I_2.

With this model, we can further illuminate the nature of the soft-acid–soft-base interaction. The key element for a soft base is its HOMO, which contains its most reactive (most easily donated) electron pair. The key element for a soft acid is its LUMO. Soft bases tend to have high-energy HOMOs, while soft acids tend to have low-energy LUMOs. Such a combination means that the HOMO of the acid has good overlap with the LUMO of the base. This favors the formation of a fairly stable chemical bond.

EXAMPLE 11.6

The electron affinity of N_2 has been estimated at -2.2 eV. Adding data from Table 11.3, illustrate and explain the trend in electron affinities across Period 2.

SOLUTION: The available electron affinities (in eV) across Period 2 are C_2, 3.269; N_2, -2.2; O_2, 0.451; and F_2, 3.08. The LUMOs of these molecules are C_2, σ_g^{nb}; N_2 and O_2, π_g^*; and F_2, σ_u^*. Based on the MO energy-level diagram of Figure 11.7, the σ_g^{nb} energy level is much lower than the others, so an electron will be most strongly attracted to fill it in C_2. The other antibonding MOs are π_g^* orbitals, which decrease in energy from N_2 to F_2 as the $2p$ valence orbital potential energies decrease (Table 11.1). Because electron affinities are taken as the negatives of the orbital energies, the electron affinities increase from N_2 to F_2.

TABLE 11.3. Electron Affinities of Some Homonuclear Diatomic Molecules[a]

Group\Period	1		11		14		15		16		17	
2					C_2	3.269			O_2	0.451	F_2	3.08
3	Na_2	0.430			Si_2	2.201	P_2	0.589	S_2	1.670	Cl_2	2.38
4	K_2	0.497	Cu_2	0.836	Ge_2	2.035	As_2	0.739	Se_2	1.94	Br_2	2.55
5	Rb_2	0.498	Ag_2	1.023	Sn_2	1.962	Sb_2	1.282	Te_2	1.92	I_2	2.55
6	Cs_2	0.469	Au_2	1.938	Pb_2	1.366						

Source: D. R. Lide, Ed., *CRC Handbook of Chemistry and Physics*, 84th ed., CRC Press: Boca Raton, FL, 2003, pp. 10-149 through 10-150.

[a] All electron affinity values are in eV.

Lewis Acid–Base Chemistry of I$_2$. The –2.55 eV LUMO of I$_2$ (a $\sigma_u{}^*$ orbital) is fairly low in energy, allowing I$_2$ to have some capacity to act as a Lewis acid. The iodine acceptor atom sits along the z axis of the molecule, and allows Lewis bases to form weak **secondary** coordinate covalent bonds to the I$_2$. **Secondary bonding** is characterized by "intramolecular distances that are much longer than normal bonds and intermolecular distances that are much shorter than van der Waals distances," and focuses on approximately linear arrangements, Y–X–X.[13] More recently, the related term **halogen bonding** has been defined as "a net attractive interaction between an electrophilic region of a halogen atom on a molecule or molecular fragment and a nucleophilic region of another molecule or molecular fragment" and applied to this interaction.[14]

This weak bonding interaction with the LUMO changes the energy of the LUMO and changes the color of the I$_2$ species. I$_2$ in nonbasic solvents such as CCl$_4$ is bright violet. In weakly basic aromatic hydrocarbons, the color changes to pink or reddish brown; in stronger donors such as alcohols or ethers, the color becomes deep brown.

In larger molecules containing terminal iodine atoms and electronegative functional groups such as CF$_3$, there is a similar $\sigma_u{}^*$ acceptor LUMO orbital on iodine (at –1.57 eV in CF$_3$I). In these cases the seemingly paradoxical result is that the electronegative iodine atom actually has a partial *positive charge* along the molecule's C–I bond axis, and it can be described as having a shorter van der Waals radius in this direction than perpendicular to the C–I axis. Such molecules halogen-bond exothermically with donor molecules. The interaction energies can vary from 10–200 kJ mol^{-1},[15] and they are comparable to or greater than the strength of the other quite strong intermolecular attraction, *hydrogen bonding* (Section 8.1B and Table 8.1). Halogen bonding is emerging as an important type of intermolecular bonding in biochemistry and structural engineering, alongside the more familiar hydrogen bonding.[16,17]

In Section 3.1B we encountered the product when the Lewis base is the I$^-$ ion itself: the yellow *triiodide* ion, I$_3{}^-$. In triiodide ion salts, the bonding energy is at a maximum, and the secondary I–I bond is 20–32 pm longer than the covalent I–I bond in the ammonium and cesium salts. However, in the potassium and tetraphenylarsonium salts, there is no difference:[18] The triiodide ion becomes a linear triatomic ion such as we discussed in Section 3.1B.

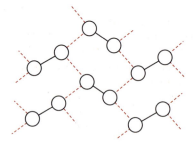

Figure 11.10. Molecular arrangement of I$_2$ molecules in the solid state. Normal 272-pm covalent bonds are drawn with solid lines (——); weaker 350-pm secondary bonds are drawn with broken lines (-----). [After A. F. Wells, *Structural Inorganic Chemistry*, 5th ed., 1988, p. 388.]

In I$_2$ itself, the HOMO at –9.31 eV (a $\pi_g{}^*$ orbital) is fairly high in energy, allowing I$_2$ to have some capacity to act as a Lewis base, donating a π electron pair along the x or y axis of the molecule. In the solid state, I$_2$ molecules can form secondary or halogen bonds to other I$_2$ molecules in the same plane (Figure 11.10). In order to have the proper orientation, the two molecules must be roughly perpendicular to each other so that the π electron pair on one iodine molecule overlaps with the σ^* orbital on another iodine molecule. This weak secondary bonding interaction results in a contact distance between individual I$_2$ molecules that is longer than the sum of covalent radii, but is shorter than the sum of the van der Waals radii. The phenomenon is not observed in the second period, but it is found in the later periods not only among the halogens, but also the Group 15/V and 16/VI elements (Table 11.4).

TABLE 11.4. Interatomic Distances in Heavier Group 16/VI and Group 17/VII Elements[a]

Element	$2r_{cov}$	d_{cov}	$2r_{vdw}$	$d_{secondary}$
Cl_2	198	198	360	332
Br_2	228	227	380	331
I_2	266	272	408	350
S_8	204	204	360	337
Se_8	234	234	380	335
$Se_{metallic}$	234	237	380	344
$Te_{metallic}$	270	284	420	350

Source: Covalent bond and secondary bond distances *d* are from A. F. Wells, *Structural Inorganic Chemistry*, 5th ed., 1988, p. 286. Covalent and van der Waals radii are from Tables 1.13 and 1.14.

[a] All values are in pm.

11.4. Heteroatomic Diatomic Molecules and Ions

OVERVIEW. When we change a homonuclear molecule such as N_2 to an isoelectronic heteronuclear molecule such as CO or BF, the symmetry point group and labels for the MO change. More significantly, the energy levels and wave functions for the MO become asymmetrical, having larger amplitudes on the atom in the MO that is closest to it in energy, so the bonding tends to become more polar. These changes may increase the Lewis acidity and the Lewis basicity of the molecule, making it more likely to engage in π backbonding. You may compute some of these properties by carrying out Experiment 10 and practice applying these concepts by trying Exercises 11.31–11.36 and (if you did Experiment 10) Exercises 11.37 and 11.38.

The construction of MOs for heteronuclear diatomic molecules and ions (those composed of atoms of two different elements), while fundamentally no different than above, does involve two changes. First, heteroatomic diatomic molecules fall in a different point group, $C_{\infty v}$, which does not have an inversion center, so the MOs are no longer gerade or ungerade. Second, the orbitals of the two parent atoms are no longer at the same energies. It then becomes more important to know their actual energies, because the further apart in energy atomic orbitals are on two parent atoms, the less effectively they can overlap.

AT THIS POINT, YOUR INSTRUCTOR MAY ASK YOU TO PERFORM A COMPUTATIONAL LAB MODELED ON EXPERIMENT 10.

This experiment is found at www.uscibooks.com/Foundations.htm

Let us begin by considering the diatomic molecule HF. Hydrogen has only one valence orbital, $1s(H)$, while F has one core orbital, $1s(F)$, and four valence orbitals, $2s(F)$, $2p_z(F)$, $2p_x(F)$, and $2p_y(F)$. From considerations of the signs of the atomic wave functions (as in Figure 3.6) we quickly determine that the only hydrogen orbital, $1s(H)$, must have *zero* overlap with two fluorine orbitals, $2p_x(F)$ and $2p_y(F)$. Consequently, the

$2p_x$(F) and $2p_y$(F) orbitals must be *strictly nonbonding*. The π_x and π_y orbitals are degenerate in the $C_{\infty v}$ point group, so they are at one energy level, which we label π^{nb}. There is substantially no change in the energy level of this MO from that of the parent $2p_x$(F) and $2p_y$(F) atomic orbitals.

Symmetry alone does not exclude overlap between $1s$(H) and any of the remaining three orbitals on fluorine, so we need to see how close in energy these orbitals are. From Table 11.1, we see that the orbital energy of $1s$(H) is –13.6 eV; the orbital energies for the three fluorine orbitals are $1s$(F) << –40.2 eV (not listed); $2s$(F) = –40.2 eV; $2p_z$(F) = –18.6 eV. These are the levels at which we draw the atomic orbitals on the sides of the energy-level diagram for HF. There must be an extreme mismatch of energies between the core $1s$(F) orbital and the valence $1s$(H) orbital, so the core $1s$(F) orbital is also a nonbonding orbital of unaltered energy. The main (positive and negative) overlap is between the $1s$(H) and the $2p_z$(F) orbitals. The positive combination of these orbitals gives a σ bonding MO, while the negative combination gives a σ^* antibonding combination.

The energy-level diagram and the MOs that result are shown in Figure 11.11. We can write the wave function of the σ bonding orbital as a combination of the two parent orbitals:

$$\sigma = c_H 1s(H) + c_F 2p_z(F) \tag{11.12}$$

However, because the two parent orbitals are at dissimilar energies, we do not have the case that $c_H = c_F$. Instead, *a given MO has a larger contribution (coefficient) from the atomic orbital that is closest to it in energy.* In the bonding MO represented by Equation 11.12, c_F is substantially greater than c_H for this reason. Consequently, electrons in this orbital spend the greatest part of their time in the vicinity of the fluorine atom, which we have attempted to represent pictorially on the left side of Figure 11.11. Occupation of this bonding orbital results in a *polar bond* polarized toward fluorine.

The opposite weighting holds for the antibonding σ^* orbital, which is closer in energy to its hydrogen parent, $1s$(H). Electrons in this orbital spend more time closer to the hydrogen atom. However, this orbital is unoccupied in HF, which has the valence electron configuration $(\sigma^{nb})^2\sigma^2(\pi^{nb})^4$ with a bond order of 1.0 (as expected from the Lewis structure of the molecule). The photoelectron spectrum of HF shows a sharp band without vibrational structure at 16.0 eV assignable to the HOMO, π^{nb}, a broadened band at about 20 eV assignable to σ, and a band assignable to σ^{nb} at 40 eV.

The energy-level diagrams of the heavier Group 17/VII hydrides (HX) are qualitatively similar. The energies of the HOMOs, which in all cases are π^{nb} molecular orbitals, are available from photoelectron spectra: –16.0 eV for HF, –12.7 eV for HCl, and –10.5 eV for HI. The energies of the σ^* LUMOs have been computed as +6.0 eV for HF, +3.3 eV for HCl, and +0.0 eV for HI.[11] The LUMO energies drop through this series because there is less overlap between the hydrogen $1s$ orbital and the halogen np orbitals as n increases and the orbital becomes more diffuse. Consequently, the antibonding LUMOs are less antibonding in an absolute sense.

Carbon monoxide (CO) is an important ligand molecule that is isoelectronic with N_2, so it has the same number of valence electrons and an energy-level diagram that is related to that of N_2 (Figure 11.7), but with the significant difference that the parent atoms are not identical. Therefore, they have different valence orbital energies. The

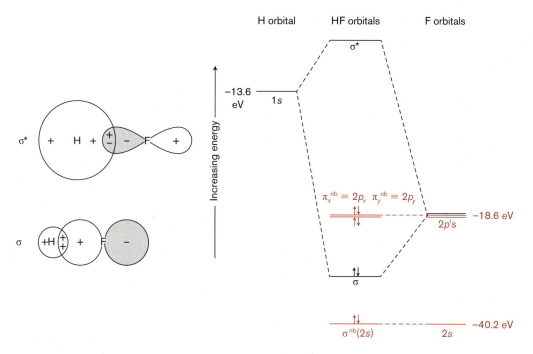

Figure 11.11. Energy-level diagram for HF. The energies of the $2s$ atomic orbital and $\sigma^{nb}(2s)$ molecular orbital (enclosed in the dashed box) are off the scale to the bottom. [Adapted from R. L. DeKock and H. B. Gray, *Chemical Structure and Bonding*, University Science Books: Mill Valley, CA, 1989, p. 253.]

eight valence $2s$ and $2p$ atomic orbitals once again give rise to eight valence MOs, as shown in Figure 11.12a. These include a doubly degenerate π set that show up in the photoelectron spectrum (Figure 11.12c) at around –17 eV, and the π^{*} LUMO at a positive value of +1.8 eV. Note that this pair of MOs has a greater amplitude on carbon (left atom in Figure 11.12) than on oxygen. This leads us to expect CO to act as a soft acid. Because the LUMO energy level is so high, CO is a very weak Lewis acid.

Once again, s–p hybridization of the four σ molecular orbitals derived from the four atomic orbitals—$2s(C)$, $2p_z(C)$, $2s(O)$, and $2p_z(O)$—occurs. In Figure 11.12a, these MOs are at the lowest, the second-lowest, the fourth-lowest, and the highest energy levels. The HOMO is at the fourth-lowest energy level, with an ionization energy of 14.014 eV (Figure 11.12c). The drawing of the MO (fourth from the bottom in Figure 11.12a) shows that it has a greater amplitude on carbon than on oxygen. Therefore, the donor atom of CO is carbon rather than oxygen, and CO acts as a soft Lewis base. Its low HOMO energy of –14.014 eV makes CO a weak Lewis base. As such, it shows little or no tendency to bond to s- and p-block atoms. However, CO can donate electrons to the e_g pair of d-block metal orbitals if these are empty.

With this model, we can further illuminate the nature of the soft-acid–soft-base interaction. The key element for a soft base is its HOMO, which contains its most reactive (most easily donated) electron pair, whereas the key element for a soft acid is its LUMO. Soft bases tend to have high-energy HOMOs, while soft acids tend to have low-energy LUMOs. The HOMO of the acid therefore is close in energy to the LUMO of the base, which favors the formation of a fairly stable coordinate covalent chemical bond between the two.

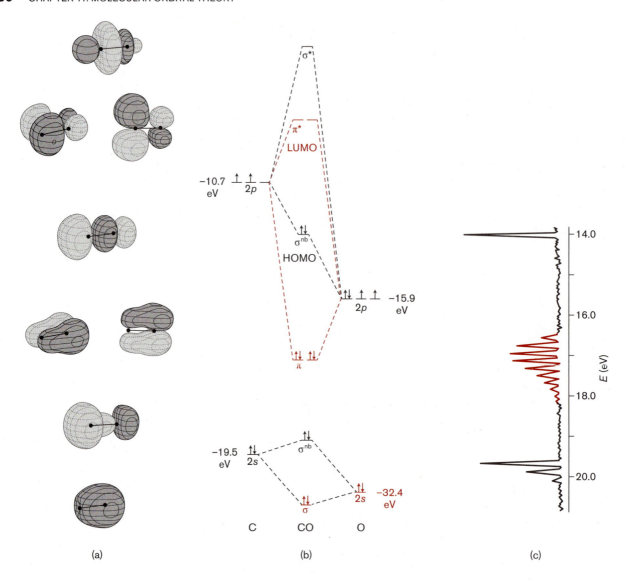

Figure 11.12. Carbon monoxide (CO): (a) contours of MOs (carbon atom on left, oxygen on right); (b) energy-level diagram; (c) photoelectron spectrum. The energies of the 2s atomic orbital of oxygen and the $\sigma^{nb}(2s)$ molecular orbitals (enclosed in the box) are off the scale to the bottom. [Credits: (a) Reprinted with permission from W. L. Jorgensen and L. Salem, *The Organic Chemist's Book of Orbitals*, Academic Press: San Diego, 1973, p. 78; (c) adapted from L. Gardner and J. A. R. Samson, *J. Chem. Phys.* 62, 1447 (1975).]

Boron monofluoride (BF) is isoelectronic with CO and N_2, but the valence orbital energies of its two parent atoms are quite different from each other, because fluorine is much more electronegative than boron. The only good matchup of valence orbital potential energies between B and F is of the boron 2s (−14.0 eV) with the fluorine 2p (−18.7 eV). The fluorine 2s orbital lies far below the energy of any boron orbital, and the matchup of the fluorine 2p with the boron 2p (−8.3 eV) is much poorer in BF than was the corresponding matchup in CO or N_2.

Considering symmetry only, the energy-level diagram (Figure 11.13) of BF has the same sequence of MOs as CO, so the electron configuration comes out the same, and the HOMO is also σ^{nb}, with an ionization energy of 11.12 eV. Because effective overlap also requires that the parent atomic orbitals be similar in energy, however, some of the MOs that are bonding in N_2 and CO may be more or less nonbonding in BF. Qualitatively we can see that there is such an energy level mismatch between the parent orbitals of the lowest-energy σ molecular orbital that this is likely to be nonbonding in BF even though it was not in N_2.

Similarly, the 10.3-eV energy gap between the B and F $2p$ parent orbitals of the π molecular orbital makes this π orbital less bonding in BF than it is in N_2 or CO. Consequently, the B–F π bond strength is about 80% lower than the C–O or N–N bond strengths.[19] If we assume that there is indeed π overlap in BF, this molecule has a valence electron configuration of $(\sigma^{nb})^2\sigma^2\pi^4(\sigma^{nb})^2$, for a bond order of 3. If we assume that there is no π overlap, the valence electron configuration is $(\sigma^{nb})^2\sigma^2(\pi^{nb})^4(\sigma^{nb})^2$, for a bond order of 1. The possible resonance structures for this π bond are $\overset{2-}{:}\overset{2+}{B}\!\!\equiv\!\!\overset{}{F}\!: \longleftrightarrow \overset{1-}{:}B\!\!=\!\!\overset{1+}{F}: \longleftrightarrow :B\!\!-\!\!\ddot{F}:$. Experimentally, the bond order of BF has been computed as 1.4.[19] The bond energy of BF, 548 kJ mol[-1], is considerably less than that of CO, 1070 kJ mol[-1], and in fact is close to the B–F single bond energy, 613 kJ mol[-1] (Table 3.4). Although it is the only resonance structure to obey the octet rule, the triple-bonded Lewis dot structure :B≡F: is the most unimportant, due to the low B–F π bond strength. Moreover, the computed ionization energy of the π orbital, –19.09 eV, is nearly unchanged from the –18.6 eV VOPE of the F $2p$ orbital. This suggests that the π orbital is essentially localized on the F $2p$ orbital.

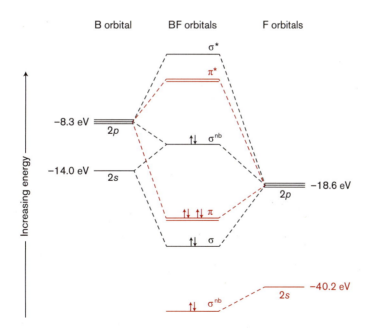

Figure 11.13. Energy-level diagram for BF. The energies of the $2s$ atomic orbital on F and the $\sigma^{nb}(2s)$ molecular orbital are off the scale to the bottom. [Adapted from R. L. DeKock and H. B. Gray, *Chemical Structure and Bonding*, University Science Books: Mill Valley, CA, 1989, p. 258.]

AN AMPLIFICATION

The Chemistry of BF

The principal resonance structure for BF is the third, singly bonded one. (The triple-bonded resonance structure B≡F: shows a formal charge of –2 on boron and +2 on fluorine and is not a good resonance structure, especially because the multiple positive charge is on the more electronegative atom). In stark contrast to the results for CO and N_2, BF does not come close to obeying the "octet rule." As you may find if you do Exercise 11.38, BF has a low-energy LUMO and is thus a strong Lewis acid, and it has a high-energy HOMO, and is thus a strong Lewis acid. Although it can be prepared in high yield from $BF_3(g)$ and $B(s)$ at high temperature and low pressure, at lower temperatures it reacts with itself (by Lewis acid–Lewis base bond formation?) to form a green polymer of unknown structure.[20]

EXAMPLE 11.7

The series of diatomic molecules C_2, BN, BeO, and LiF are isoelectronic, but because the sets of parent atoms have increasingly diverse valence orbital potential energies (Table 11.1), their energy-level diagrams progressively diverge in character. (a) Set up the left and right sides of the energy-level diagrams for each of these four molecules. How do the two sides change through this series? (b) Qualitatively, what effects does this have on the shapes of the MOs and their energy level through this series? (c) The bond lengths in the gaseous diatomic molecules in this series are 124 pm in C_2, 128 pm in BN, 133 pm in BeO, and 156 pm in LiF. The bond energies (in kJ mol⁻¹) are 602 in C_2, 385 in BN, 444 in BeO, and 568 in LiF.[21] Discuss these trends.

SOLUTION: (a) The left and right sides of the energy-level diagrams are constructed from the data in Table 11.1 and are shown in Figure 11.14. As we progress through the series there is less and less matching of energy levels between the atoms on either side of the diagram. (b) This means that the degree of overlap of parent atomic orbitals decreases from C_2 to LiF, which results in less splitting of corresponding bonding and antibonding MOs. Gradually, the low-energy occupied MOs more and more come to resemble the atomic orbitals of the more electronegative element on the right. Consequently, the polarity of the bonds increases, so that the bonding in LiF comes about as close as possible to being completely ionic. An approximate energy-level diagram for LiF is also shown in Figure 11.14. Note that all electrons are in orbitals that are much like fluorine orbitals in energy (and appearance)—this is really a gaseous Li^+F^- ion pair. (c) The bond lengths grow somewhat because the double bonding of C_2 is lost later in the series, then the covalent single bonding. However, the bond energies show no clear trends: The ionic (electrostatic) attractions of Li^+ and F^- in the gaseous ion pair give similar stabilizations to the covalent interactions in a molecule such as C_2.

Example 11.7 continues on next page ▶

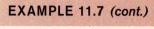

EXAMPLE 11.7 *(cont.)*

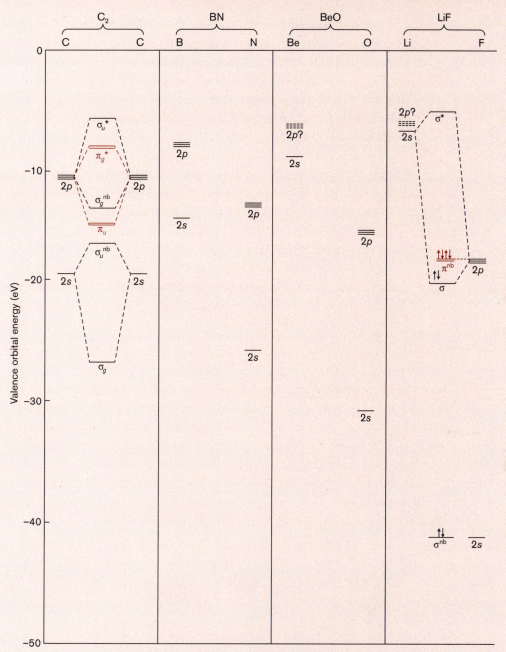

Figure 11.14. Partial energy-level diagrams, showing orbital energies of parent orbitals only, for the series of isoelectronic molecules C_2, BN, BeO, and LiF; the approximate molecular orbital energy levels for LiF are also included to complete the last energy-level diagram.

11.5. Molecular Orbitals for Selected Two- and Three-Dimensional High-Symmetry (D_{nh}, T_d, and O_h) Complexes and Molecules

OVERVIEW. Molecular orbitals for polyatomic species with a central atom are most usefully constructed using, not orbitals of individual terminal atoms, but terminal-atom symmetry-adapted orbitals (TASOs). You can construct qualitative pictures of the MOs for planar polyatomic species AB_n as described in Section 11.5A. These MOs can be given appropriate symmetry labels and be classified as being overall bonding, nonbonding, or antibonding. You can draw energy-level diagrams and write electron configurations for such molecules, and identify their HOMOs and LUMOs (or SOMOs). You can practice by trying Exercises 11.39–11.44.

Similar principles and objectives are applied to tetrahedral molecules in Section 11.5B and to octahedral molecules in Section 11.5C. You will be able to identify, in the energy-level diagram of a tetrahedral or octahedral complex of a d-block metal ion, the part that is emphasized in crystal field theory (Chapter 7). Exercises 11.45–11.47 give you practice in applying the concepts to tetrahedral molecules. You can explain how, in octahedral complexes, the presence of π-type orbitals on the ligands changes the nonbonding status of the t_{2g} metal orbitals, and how this alters the position of the ligand in the spectrochemical series. You may practice these concepts by trying Exercises 11.50–11.52.

11.5A. Borane (BH_3) and Related Planar Complexes. We begin with the study of two-dimensional (planar) metal–ligand complexes. These fall in the D_{nh} point groups, in which the x and y directions are indistinguishable. Therefore, many of the MOs fall in doubly degenerate pairs of equal energy. We start with transient BH_3 (borane), which has only six valence electrons, enough to fill three bonding molecular orbitals. These MOs are shown in Figure 11.15 and are labeled with the appropriate irreducible representations (symmetry labels) from the D_{3h} character table (which your instructor can obtain for you at www.uscibooks.com/foundations.htm). Because there are seven atomic orbitals used, there are seven resulting MOs, including the boron $2p_z$ orbital. A sketch of the energy-level diagram for BH_3 is shown in Figure 11.16.

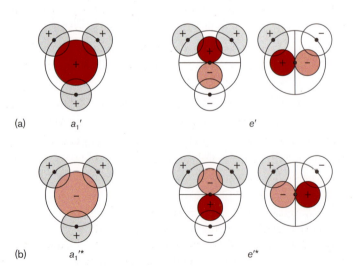

(a) a_1' e'

(b) $a_1'^*$ e'^*

Figure 11.15.
(a) Bonding MOs and (b) antibonding MOs of BH_3 in the molecular (xy) plane. (left) MOs involving the boron $2s$ orbital; (center and right) orbitals involving the boron $2p_x$ and $2p_y$ orbitals. (The nonbonding boron $2p_z$ orbital is not shown. It lies above and below the molecular plane and is in the a_2'' irreducible representation.)

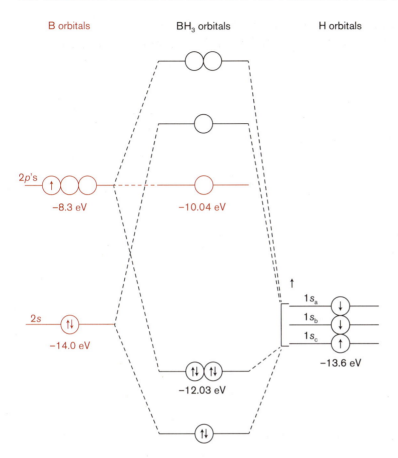

Figure 11.16. An energy-level diagram for BH_3 with VOPEs for the B and H atoms from Table 11.1 and estimates of the energies of the HOMO and LUMO (shown below the energy levels) from the ionization energy and electron affinity of the molecule.

The MOs involve H $1s$ orbitals not individually, but in combinations of them that match the irreducible-representation symmetry of the central atom orbital with which they must overlap: **terminal-atom symmetry-adapted orbitals** (TASOs) (also known as SALCs (symmetry-adapted linear combinations). TASOs can have positive overlap with, and bond to, real boron central-atom orbitals only if both belong to the same irreducible representation. Hence, the bonding a_1' molecular orbital (left side of Figure 11.15a) is formed by positive overlap between the a_1' TASO $[1s_A + 1s_B + 1s_C]$ of the three hydrogen ligands A, B, and C, and the valence boron orbital in the A_1' irreducible representation, the $2s$ orbital. Likewise, a bonding pair of e' molecular orbitals (right side of Figure 11.15a) are formed by positive overlap between the e' TASOs shown and the pair of valence boron orbitals in the E' irreducible representation: the $2p_x$ and $2p_y$ orbitals. These three bonding MOs hold the six valence electrons of BH_3. The electron configuration of BH_3 is $(a_1')^2(e')^4$.

Borane has a very high electron affinity (10.04 eV), so it is a strong Lewis acid. It has a relatively low ionization energy (12.03 eV), so it also has Lewis-base properties—even though it has no unshared electron pairs! The HOMO–LUMO gap is extremely small, so this species is extremely soft and very reactive—it dimerizes to give diborane (Figure 11.17), B_2H_6, which is discussed further in Section 11.5B.

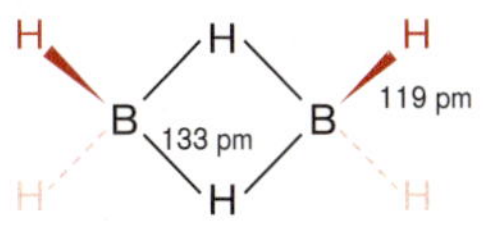

Figure 11.17. Molecular structure of B_2H_6. Bridging hydrogen atoms (shown in black) are in the plane of the paper, while terminal hydrogen atoms (shown in **brown** and light brown) are directly above and below the plane of the paper, respectively.

Of more real interest is boron trifluoride (BF_3). The σ bonding in this molecule is quite like that in BH_3, except that the circles shown for the TASOs in Figure 11.15 can be taken to be the inner lobes of F $2p$ orbitals directed at boron. Each terminal F ligand also has a filled $2p_z$ orbital extending along the z axis (the threefold axis) of the molecule. The three $2p_z$ orbitals of the three terminal F atoms generate three additional TASOs, one of which [$2p_z(A) + 2p_z(B) + 2p_z(C)$] has positive overlap everywhere with the a_2'' valence orbital on boron, $2p_z$. In principle this gives us one π-bonding MO per molecule. This π bonding is limited by the poor energy match between the boron (VOPE –8.3 eV) and fluorine (VOPE –18.6 eV) orbitals (Table 11.1). BF_3 is still a strong Lewis acid, but it has no Lewis-base properties, is very hard, and does not dimerize, in contrast to BH_3.

If we try to draw Lewis structures showing such a π bond, we are forced to draw three resonance structures, each showing a π bond between boron and a different fluorine atom. Averaging the resonance structures, we can say that the B–F π bond order is ⅓. From the point of view of molecular orbital theory, there is one π-bonding MO, but it is **multicentered**, involving the boron atom and all three fluorine atoms. We can reconcile these two descriptions by describing a **bond order per link**:

$$\text{Bond order per link} = \frac{(\text{Total bond order from Eq. 11.4})}{(\text{Number of interatomic links})} \tag{11.13}$$

In the case of BH_3 the result is unsurprising: There are three bonding MOs and three B–H links, so the bond order per link is one—that is, each B–H link is a single bond. We come to the same answer for the σ bonding in BF_3: Each B–F link involves one σ bond. In the case of the π-bond order per link of BF_3, we divide the total number of π-bonding MOs, one, by the three B–F links, to get a π-bond order of ⅓ π bond per B–F link. The total bond order is then 1⅓ bonds per B–F link.

EXAMPLE 11.8

Based on periodic trends in π-bonding ability discussed earlier, discuss how the extent of π bonding in BCl_3 should compare with that in BF_3.

SOLUTION: The VOPE of the Cl $3p$ orbitals, –13.7 eV, are closer than that of the F $2p$ orbital, –18.6 eV, to the VOPE of the B $2p$ orbital, –8.3 eV (Table 11.1). This factor would favor stronger π bonding in BCl_3. But the greater length of the B–Cl bond over the B–F bond (Section 3.3C) favors stronger π bonding in BF_3. Hence, the trend is not readily predicted. ([35]Cl nuclear quadrupole resonance data on BCl_3 suggests that there is appreciable π bonding in BCl_3.[22])

11.5B. Tetrahedral Molecules—The T_d Point Group. The simplest tetrahedral molecule is methane, CH_4, which has eight valence orbitals from its C and H atoms, so it therefore has eight MOs. Its central carbon atom has four valence atomic orbitals, $2s$ (A_1 irreducible representation) and the three degenerate $2p$ orbitals (T_2 irreducible representation), oriented with respect to the tetrahedron as shown in Figure 11.18. The four H atoms can also give rise to four TASOs, the irreducible representations of which match those of the central atom. Positive overlap of the carbon valence orbitals and the H_4 TASOs of the same irreducible representation produce the two levels of bonding MOs that are fully occupied in the electron configuration of methane, $(a_1)^2(t_2)^6$. Negative overlap of the carbon valence orbitals and the H_4 TASOs of the same irreducible representation produce the two levels of antibonding MOs that are vacant (and not shown in the figure): a_1^* and t_2^*. There are no orbitals left over to be nonbonding. The energy-level diagram of methane is shown in Figure 11.19.

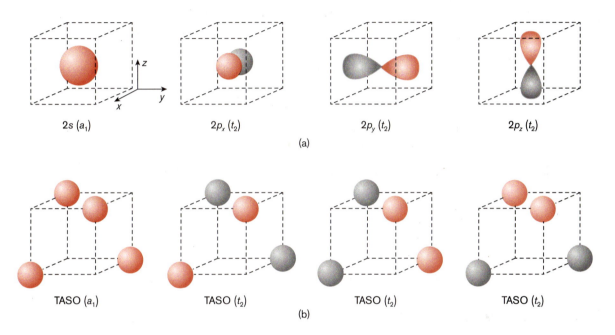

$2s\ (a_1)$ $2p_x\ (t_2)$ $2p_y\ (t_2)$ $2p_z\ (t_2)$

(a)

TASO (a_1) TASO (t_2) TASO (t_2) TASO (t_2)

(b)

Figure 11.18. Orbitals to use in constructing the MOs of CH_4. (a) The four valence orbitals of carbon; (b) the four TASOs involving the four hydrogen $1s$ orbitals.

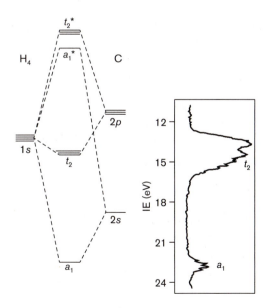

Figure 11.19. Energy-level diagram and photoelectron spectrum of the CH_4 molecule. [Adapted from R. L. DeKock and H. B. Gray, *Chemical Structure and Bonding*, University Science Books: Mill Valley, CA, 1989, p. 283.]

The photoelectron spectrum of methane (right side in Figure 11.19) shows two ionizations: one from the bonding t_2 HOMO, at about 13 eV, and one from the bonding a_1 MO, at about 23 eV. This seems not to support our expectations from valence bond theory, which describes methane as having sp^3-hybridized carbon, so that all H–C bonding orbitals are equivalent. However, each set of MO bonding orbitals involves all hydrogens equally, so that MO theory, just like valence bond theory, results in all H–C links being equivalent.

The next alkane, C_2H_6, has 14 valence electrons and therefore 14 valence MOs, which give a more complex MO energy-level diagram and photoelectron spectrum. However, the energy levels and photoelectron spectral peaks are at virtually the same energies as in CH_4. With seven bonding MOs and seven C–H and C–C links, the bond order per link is one, in agreement with the Lewis structure of ethane.

Of more interest is ethane's boron analogue B_2H_6, *diborane*. Its structure is different from that of ethane: It has two BH_2 units linked by two *bridging hydrogen* atoms (Figure 11.17).

This molecule has eight B–H links, but only 12 valence electrons, which is not enough to put an electron pair into each link. Diborane is an example of an **electron-deficient molecule**; a Lewis structures cannot be drawn for it. In diborane, each boron-terminal hydrogen bond is a "normal" two-electron bond shared by two atoms. The remaining four electrons are used in two longer *three-center, two-electron* B–H–B bonds; each B–H$_{bridge}$ link has a bond order of ½ per B–H link. Electron-deficient molecules such as diborane are common in boron chemistry and can be remarkably stable.

EXAMPLE 11.9

How would you expect the energy-level diagram for methane (Figure 11.19) to change as the central atom is changed from carbon (in methane) to silicon (in *silane*), germanium (in *germane*), and tin (in *stannane*)?

SOLUTION: Two effects are involved here. First, the VOPEs shift upward from carbon (–19.4 and –10.6 eV) to silicon (–14.9 and –7.7 eV). Then, because fourth-period germanium is very similar to third-period silicon but has a higher electronegativity, its VOPEs are about the same (–15.6 and –7.6 eV). Although the VOPEs of tin are not tabulated in Table 11.1, they presumably resume the normal periodic trend and are upwards of those of Si and Ge. This change on the right side of Figure 11.19 should generally shift all MOs upwards in the diagram.

The second trend is that bonds lengthen down the group (except from Si to Ge), which should reduce the covalent overlap between the central atoms and their TASOs. This should have the effect of reducing the spreads of the energy-level diagrams, so that, for example, the HOMO–LUMO gaps decrease and the molecules become softer and less stable.

The loss of stability makes these hydrides thermodynamically unstable in the lower part of Group 14/IV. To obtain actual data, we can replace the hydrogens with methyl groups (as we did in going from methane to ethane) with little change in the energy levels. The resulting molecules such as *tetramethylsilane* are far more stable and easy to characterize. The experimental LUMO energies (EAs) therefore decrease (C +6.1 > Si +3.9 ≅ Ge +3.7 > Sn +2.9 eV). The experimental HOMO energies (–IEs) increase (C –11.4 < Si –10.5 ≅ Ge –10.2 < Sn –9.7 eV).[23] As a result, the HOMO–LUMO gap decreases quite substantially as we go down the group.

11.5C. Octahedral Molecules—The O_h Point Group. We begin by considering the bonding in the extremely stable *sulfur hexafluoride*, SF_6. The central sulfur atom has four valence orbitals, $3s$ (A_{1g} irreducible representation, VOPE = –20.7 eV) and the three $3p$ orbitals (T_{1u} irreducible representation, VOPE = –11.6 eV). It may also be able to bond with its postvalence $3d$ orbitals (Section 3.3B), especially because the oxidation state of sulfur is high in SF_6 and these d orbitals are likely to be contracted. However, let us show how the bonding can be accommodated without using the $3d$ orbitals.

The six terminal F atoms have 24 atomic orbitals. To make the treatment of these manageable, we will treat the six $2s$ orbitals as nonbonding, because their VOPEs are so much lower than those of any sulfur orbitals. We divide the eighteen $2p$ orbitals of the six fluorines (VOPE = –18.6 eV) into two sets. The *radial* set is the six $2p$ orbitals (one per atom) directed toward the sulfur at the center of the octahedron; these are oriented to be able to form σ bonds with sulfur. The *tangential* set is the 12 $2p$ orbitals that lie more or less along the outer surface of the octahedron and is destined to hold nonbonding fluorine lone electron pairs, unless overlap with an appropriate central-atom set of orbitals can result in π bonding.

We first consider the MOs that result from overlap of the radial set of six F_6 TASOs with the four sulfur valence orbitals. Because there are two more F_6 TASOs than there are sulfur valence orbitals, two F_6 TASOs are destined to be nonbonding. Otherwise, we can anticipate an equal number of bonding and antibonding MOs (four each).

Figure 11.20 shows the four bonding (a_{1g}, no nodal planes, and t_{1u}, one nodal plane) and two nonbonding MOs that result for SF_6 when no d orbitals are used in bonding. (The corresponding unoccupied sets of antibonding MOs are not shown.) Finally, there is a doubly degenerate e_g set of TASOs.

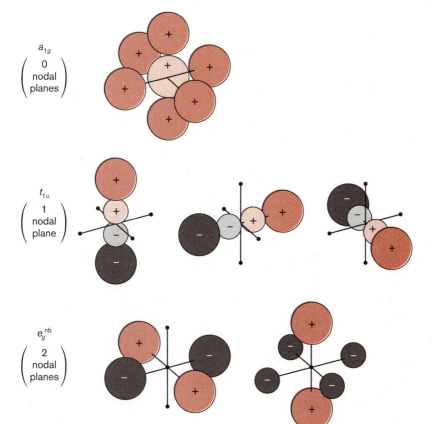

a_{1g}
$\begin{pmatrix} 0 \\ \text{nodal} \\ \text{planes} \end{pmatrix}$

t_{1u}
$\begin{pmatrix} 1 \\ \text{nodal} \\ \text{plane} \end{pmatrix}$

e_g^{nb}
$\begin{pmatrix} 2 \\ \text{nodal} \\ \text{planes} \end{pmatrix}$

Figure 11.20. σ-Bonding and nonbonding MOs for an octahedral (O_h) complex in which the central atom cannot use d orbitals. (Orbitals of the central atom are shaded.) [Adapted from D. F. Shriver, P. W. Atkins, and C. H. Langford, *Inorganic Chemistry*, W. H. Freeman: New York, 1990, p. 679.]

Octahedral (O_h) Transition Metal Complex Ions. In d-block complexes, the central-atom d orbitals are not only involved in the bonding, they are lower in energy than the valence s and p orbitals. Consequently, not only the a_{1g} and the t_{1u} TASOs but also the e_g pair of TASOs give rise to bonding and antibonding molecular orbitals. If the ligands in these complexes are strictly σ-donor ligands (Section 7.4B), then there are no TASOs to match the t_{2g} set of central-atom d orbitals, which therefore is nonbonding. The six σ-donor ligands contribute six pairs of electrons, enough to fill the six bonding MOs (Figure 11.21). If the central metal ion has a d^1 through a d^6 electron configuration, these electrons occupy the t_{2g}^{nb} MOs. These MOs are nonbonding, so their occupancy or nonoccupancy does not alter the total bond order of the complex. Hence, there is no strong preference for d^0 or d^6 over the intermediate electron configurations, all of which are found in d-block complexes.

For these complexes, the HOMOs (or SOMOs) are therefore t_{2g}^{nb}, and the LUMOs are the first antibonding MOs, e_g^*. These are the orbitals emphasized in crystal and ligand field theory (Chapter 7). In crystal field theory, the t_{2g}-orbital electrons were described as avoiding repulsion by the ligands; in MO theory they are described as "nonbonding." In crystal field theory, the e_g-orbital electrons were described as being repelled by the ligands; in MO theory they are described as "antibonding." These orbitals are enclosed in the box of Figure 11.21. The octahedral crystal field splitting energy, Δ_o, in crystal field theory is equivalent to the separation of the HOMO and the LUMO in MO theory.

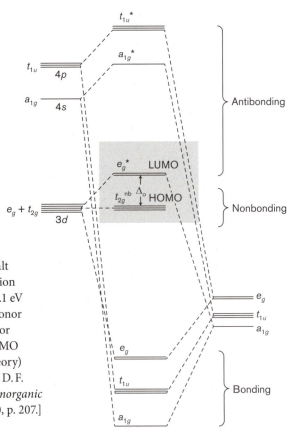

Figure 11.21. Energy-level diagram for a salt of a complex ion of a typical d-block metal ion (e.g., Co^{3+} with VOPEs of –8.7 eV for $3d$, –7.1 eV for $4s$, and –3.7 eV for $4p$ orbitals) with σ-donor ligands (such as H^- with VOPE of –13.6 eV or N of NH_3 with VOPE of –13.2 eV). The HOMO and LUMO (emphasized in crystal field theory) are enclosed inside the box. [Adapted from D. F. Shriver, P. W. Atkins, and C. H. Langford, *Inorganic Chemistry*, W. H. Freeman: New York, 1990, p. 207.]

Complexes with π Bonding. If the octahedral transition metal complex involves π-donor or π-acceptor ligands (Section 7.4B), such as halide ions or carbon monoxide, respectively, then we must consider how the tangential π_x and π_y or the π^* orbitals of these ligands interact with the metal d orbitals. Six of these ligands have 12 such orbitals; these must be combined into 12 tangential TASOs, which fall in four triply degenerate sets. The π-bonding set is the t_{2g} set (Figure 11.22), which overlaps with the metal t_{2g} set of d_{xy}, d_{xz}, and d_{yz} atomic orbitals. Hence, in a metal complex with π-donor or π-acceptor ligands, the metal-ion t_{2g} orbitals are no longer nonbonding, but become either bonding or antibonding MOs.

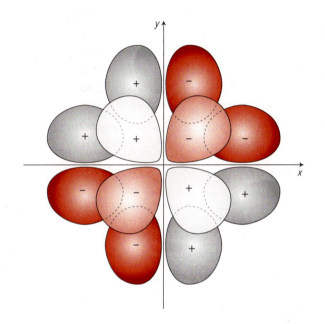

Figure 11.22. One of the three degenerate π-bonding t_{2g} set of TASOs (shaded in **brown** and gray), behind the corresponding one of the metal t_{2g} set of d_{xy}, d_{xz}, and d_{yz} atomic orbitals (in foreground, shaded in **light brown** and white). Note the positive overlap.

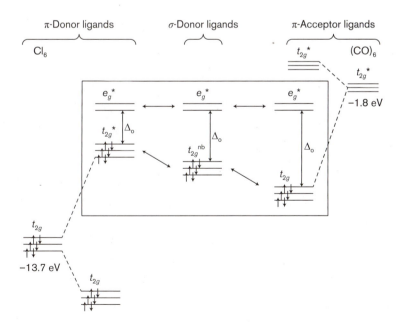

Figure 11.23. Comparative energy levels of the metal t_{2g} and e_g sets of orbitals in complexes of: (center) σ-donor ligands (as in Figure 11.21); (left) π-donor ligands such as chloride; (right) π-acceptor ligands such as carbon monoxide, with the resulting trend in the magnitude of Δ_o. The d^6 metal-ion electron configuration is illustrated in each case.

With π-donor ligands such as the halide ions, the tangential t_{2g} set of TASOs is filled and is relatively low in energy (left side of Figure 11.23). These TASOs interact with higher-energy filled metal t_{2g} orbitals to give a low-energy bonding t_{2g} MO (located primarily on the ligands) and a higher-energy antibonding $t_{2g}{}^{*}$ MO, located primarily on the metal ion. This elevation of the energy of $t_{2g}{}^{*}$ then reduces the magnitude of Δ_o in such π-donor complexes as compared to the case of σ-donor ligands (center of Figure 11.23). Consequently π-donor ligands fall at the low end of the spectrochemical series of ligands (Section 7.4).

With π-acceptor ligands such as carbon monoxide having high-energy empty tangential orbitals, and with metals having electrons in the t_{2g} set, two bonding interactions can occur. While the ligand is donating its σ electron pair to the e_g orbital of the metal ion, the metal can partially donate its t_{2g} electrons to the ligand tangential t_{2g} TASOs. This "backbonding" from the metal ion as Lewis base to the ligand as Lewis acid (Figure 7.10b) gives some double-bond character to the metal–carbon bond. This changes the metal t_{2g} set of orbitals from nonbonding to bonding and lowers their energy (right side of Figure 11.23), thus increasing the magnitude of Δ_o for such π-acceptor ligands. Consequently, π-acceptor ligands fall at the high end of the spectrochemical series of ligands (Section 7.4).

EXAMPLE 11.10

Contrast our expectations for valence electron configurations of the four octahedral tungsten compounds $W(CH_3)_6$, WCl_6, $W(CO)_6$, and hypothetical WXe_6 with Xe as a weak-field ligand (Section 7.4).

SOLUTION: First identify the ligands and their classifications as σ donors, π donors, or π acceptors, then identify the charges of the tungsten ions and their d^n electron configurations. Using the procedure outlined in Section 3.2B, the ligands must be $CH_3{}^{-}$, Cl^{-}, and neutral CO and Xe, respectively. These are a σ donor, a π donor, a π acceptor, and a σ donor, respectively. For charge neutrality of the four compounds, the charge on tungsten must be $+6$, $+6$, 0, and 0, respectively.

The W^{6+} ions have d^0 electron configurations, while the neutral W atoms in $W(CO)_6$ and WXe_6 can be expected to have a d^6 electron configuration. Each set of ligands contributes two valence electrons to the σ bonding, so the first two compounds will have 12 σ valence electrons in an energy-level diagram similar to Figure 11.21. Their σ valence electron configurations can be described as $a_{1g}{}^{2}t_{1u}{}^{6}e_g{}^{4}$.

$W(CO)_6$ and WXe_6 have 18 valence electrons, with the additional six electrons being in t_{2g}. With the π-acceptor ligand CO, the t_{2g} orbital is bonding (Figure 11.23), so its complete valence electron configuration is $a_{1g}{}^{2}t_{1u}{}^{6}e_g{}^{4}t_{2g}{}^{6}$. With the π-donor ligand Xe, the t_{2g} orbital is antibonding. Additionally, if it is weak-field, then this complex will be high spin (Section 7.2A), so the complete valence electron configuration is $a_{1g}{}^{2}t_{1u}{}^{6}e_g{}^{4}t_{2g}{}^{*4}e_g{}^{*2}$.

Recall that we listed π-acceptor ligands in Table 7.4. The ligands on the left side of Table 7.4 are similar electronically to CO, with LUMOs consisting of ligand π* orbitals (Section 11.4B). The ligands on the right side of Table 7.4 are Group 15/V soft bases with P, As, or Sb donor atoms. Example 11.9 discusses the decrease in the σ* LUMO energies in the series of Group 14/IV tetramethyl derivatives. Back in Section 7.4B we said that the empty ligand π-acceptor orbitals might be P, As, or Sb postvalence *d* orbitals. Now we add that the σ* LUMOs of the Group 15/V ligands may also contribute to the π-acceptor properties of the ligands on the right side of Table 7.4.

11.6. *d*-Block Organometallic Compounds: The Role of σ-Donor, π-Donor, and π-Acceptor Ligands

OVERVIEW. In Section 11.6A, we compute the total valence electron counts (VECs) of strong-field *d*-block carbonyl or related complexes to check for compliance with the 18-electron rule and to assign metal-atom oxidation numbers. Alternately, we can use the 18-electron rule to predict the formulas of these complexes. You may practice this concept by trying Exercises 11.53–11.58.

In Section 11.6B, we construct qualitative pictures of the molecular orbitals for cyclic polyatomic molecules (especially conjugated cyclic *polyenes* $C_nH_n^{z\pm}$), showing nodal planes; give them appropriate symmetry labels; give the number of positive and negative overlaps in each MO, and classify each as being overall bonding, nonbonding, or antibonding; and draw the energy-level diagram. For such a molecule, we write its electron configuration and give its total bond order, bond order per atom–atom link, and its total number of unpaired electrons; we also identify the HOMO and LUMO (or SOMO). You may practice these concepts by trying Exercises 11.59–11.61.

In Section 11.6C, you learn how to tell which orbitals of a cyclic molecule are suitable for overlap with which orbitals of an atom M in forming an "open-faced sandwich" molecule C_nH_nM. You can check the formulas of *d*-block metallocene or sandwich compounds for compliance with the 18-electron rule. These do not as readily obey the 18-electron rule, but show surprising stability. You can compute VECs and metal oxidation numbers for these complexes. You may practice these concepts by trying Exercises 11.62–11.71.

11.6A. *d*-Block Metal Carbonyls and the 18-Electron Rule. The π-acceptor ligand CO falls at the top of the spectrochemical series (Section 7.4; Eq. 7.6) and results in low-spin, high-field complexes. In contrast to most *d*-block complexes, the complexes of π-acceptor ligands such as CO have regular and predictable electron configurations, but often have unusually low metal oxidation numbers, such as +1, 0, −1, or as low as −4.[24] Table 11.5 shows most of the *d*-block metal carbonyls, carbonylate anions (lower left), and carbonyl cations (upper right). The arrangement of complexes in Table 11.5 is designed to suggest the electronic similarity of the carbonyls, carbonyl cations, and carbonylate anions of a given geometry. The chemistry of carbonyl cations is relatively recent.[25] The most recently discovered neutral carbonyl is seaborgium hexacarbonyl.[26]

TABLE 11.5. Typical Carbonyls of the d-Block Elements[a]

Octahedral d^6	$[Ti(CO)_6]^{2-}$	$[V(CO)_6]^-$	$Cr(CO)_6$	$[Mn(CO)_6]^+$	$[Ru(CO)_6]^{2+}$			
Trigonal bipyramidal d^8		$[V(CO)_5]^{3-}$	$[Cr(CO)_5]^{2-}$	$[Mn(CO)_5]^-$	$Fe(CO)_5$	$[Co(CO)_4]^+$		
Tetrahedral d^{10}			$[Cr(CO)_4]^{4-}$	$[Mn(CO)_4]^{3-}$	$[Fe(CO)_4]^{2-}$	$[Co(CO)_4]^-$	$Ni(CO)_4$	
Other		$V(CO)_6$				$[Co(CO)_3]^{3-}$		
						$[Rh(CO)_4]^+$	$[Pt(CO)_4]^{2+}$	$[Au(CO)_2]^+$ $[Hg(CO)_2]^{2+}$

[a] Neutral carbonyls are enclosed in shaded boxes. Formulas of carbonylate anions are shown in black; those of carbonyl cations are shown in **brown**. The carbonyls in the last row are either linear or square planar. In most cases the remaining elements in each group also form the corresponding carbonyl.

When we add the number (2n) of electrons donated to the metal atom by the n carbonyl ligands to the number of d electrons contributed by the metal, we obtain the metal's total **valence electron count (VEC)**. This VEC is characteristically the same number for a given geometry. We use the octahedral carbonyls in Example 11.11.

EXAMPLE 11.11

What are the metal oxidation numbers and the valence electron counts in (a) $[Ti(CO)_6]^{2-}$, (b) $[Ir(CO)_6]^{3+}$, and (c) $V(CO)_6$? What are the metal electron configurations in terms of bonding, nonbonding, and antibonding t_{2g} and e_g orbitals?

SOLUTION: Recalling the procedure for assigning central-atom charges/oxidation numbers given in Section 3.2B, assigning metal oxidation numbers in these simple carbonyl species is pretty straightforward, because the carbonyl ligands are neutral. Thus, in (a) the Ti has an oxidation number of –2! In (b) the Ir has an oxidation number of +3, and in (c) the V has an oxidation number of 0. From this, the number of d electrons assigned to the metal atom or ion can be deduced as described previously in Section 1.3C, taking note that, in complexes such as carbonyls, all of the valence electrons are assigned to d orbitals. Thus, in (a), a neutral Ti atom has four valence electrons, so a Ti^{2-} species has two more (six), which are all assigned to d orbitals to give a valence electron configuration of $3d^6$. In (b) the Group 9 Ir atom has lost three valence electrons in its Ir^{3+} cation, which has the valence electron configuration of $5d^6$. In (c) the Group 5 V atom has five valence electrons, which are all assigned to the 3d orbitals.

By extension, each of the metal atoms in the octahedral carbonyls in Table 11.5 except $V(CO)_6$ has a d^6 valence electron configuration, which can be labeled as shown on the right side of Figure 11.23 as (bonding) t_{2g}^6. Each contributes six valence electrons to the VEC. [The V in $V(CO)_6$ anomalously contributes only five electrons.] Six carbonyl ligands donate 12 valence electrons to the metal atom or ion, so they contribute 12 to the VEC.

This characteristic VEC of 18 in octahedral carbonyls fills nine MOs, one per valence *s*, *p*, and *d* orbital of the metal ion. Eighteen electrons fill all of the bonding orbitals and leave all of the antibonding orbitals vacant in an octahedral complex (Figure 11.23). Hence, we often predict the electron configurations of *d*-block carbonyls (and related strong-field complexes) using the **18-electron rule**. Corresponding to the frequent occurrence of *octet*s in compounds of *p*-block elements, stable carbonyls and other complexes of π-acceptor ligands of the *d*-block elements tend to have 18 valence electrons about the metal atom. Thus, the *VEC in octahedral metal carbonyls is 18 electrons*, except in $V(CO)_6$, in which it is 17 electrons.

The 18-electron rule emerges from the energy-level diagram of octahedral complexes (Figure 11.21). We point out three regions of orbitals in this diagram.

- Below the boxed region are six bonding MOs. These orbitals are closer in energy to the original ligand orbitals (right side of the diagram) than the metal atomic orbitals (left side), so they are still mainly ligand-like orbitals; there is one per electron pair donated by the set of ligands.

- Within the boxed region are the five orbitals emphasized by the crystal field theory, which are either equal in energy or not too far removed from the energies of the parent metal *d* atomic orbitals; the HOMO and LUMO are in this region.

- Of less importance, because they are neither occupied nor are they the LUMOs, are the four antibonding orbitals above the boxed region; they are closest in energy to, and more or less correspond to, the metal ion's valence *s* and *p* orbitals.

For other geometries, we refer back to the crystal field splittings given in Table 7.8, keeping in mind that these correspond to the boxed region of the corresponding MO energy-level diagrams. Thus, for a trigonal bipyramidal five-coordinate organometallic, first we need to count the 10 electrons donated by the five ligands. Next, we consider the five *d* orbitals whose energies are given in Table 7.8. Four of the five have negative (low) crystal field splitting energies; in the presence of π-acceptor ligands, we presume that MO calculations would find these to be bonding (or perhaps nonbonding) MOs. These should logically be filled to give 18 valence electrons. The high-energy d_{z^2} orbital likely would turn out to be antibonding and would remain empty. You can readily verify that the trigonal bipyramidal carbonyls shown in Table 11.5 also obey the 18-electron rule.

For a tetrahedral complex, this procedure does not appear to work. Four ligands donate eight electrons, and only two *d* orbitals ($e = d_{z^2}$ and $d_{x^2-y^2}$) have negative crystal field splitting energies. The 18-electron rule, however, is also obeyed in tetrahedral complexes of π-acceptor ligands such as CO, because the splitting of *e* and t_2 in tetrahedral complexes is small, and both the *e* and the t_2 sets of *d* orbitals donate electrons to the CO π-accepting orbitals of the CO ligands, so that both interactions become bonding.[27]

One major geometry definitely does not obey the 18-electron rule, however: the square planar geometry, including the square planar tetracarbonyl cations $[Rh(CO)_4]^+$, $[Pd(CO)_4]^{2+}$, and $[Pt(CO)_4]^{2+}$. In this four-coordinate geometry, the ligands donate eight electrons, and four of the five metal d orbitals either have negative or only small positive crystal field splitting energies. The $d_{x^2-y^2}$ orbital has a very large positive crystal field splitting energy, so it normally remains unoccupied; in MO terms it is definitely antibonding. In this case, the VEC is $8 + 8 = 16$. Hence, square planar complexes of π-acceptor ligands tend to have 16 valence electrons. Linear two-coordinate carbonyl cations $[Au(CO)_2]^+$ and $[Hg(CO)_2]^{2+}$ owe their low coordination numbers and VECs to relativistic effects (Section 9.6).

The neutral metal carbonyls are colorless or lightly colored volatile liquids or solids. The M–C bonds in these compounds are not especially strong; the bond strengths are around 160 kJ mol^{-1} and are highest in the carbonylate anions and lowest in the carbonyl cations.[28] Metal carbonyls readily lose CO on heating or upon irradiation with ultraviolet (UV) light. Thus, these compounds should be handled in hoods to avoid CO poisoning. Some of them are even more toxic than CO. Nickel carbonyl, for example, is 100 times more toxic than CO; it apparently decomposes in the lungs, in effect nickel-plating the lungs.

EXAMPLE 11.12

Use the 18-electron rule to predict the formulas of the metal carbonyls with 0, +1, and –1 overall charges.

SOLUTION: We will not consider any coordination numbers above six, because this is the maximum total coordination number for the fourth period. Octahedral, trigonal bipyramidal, and tetrahedral carbonyls should give a total count of 18 valence electrons. The six CO atoms of a hexacarbonyl contribute 12 valence electrons, so the parent metal ion or atom should therefore have six valence electrons. Therefore, only Group 6 metals are expected to form neutral $M(CO)_6$; in fact, Group 6 metals as well as vanadium do. The possible M^+ species in a carbonyl cation $[M(CO)_6]^+$ must be Group 7 metals in order to have six valence electrons; this is confirmed in Table 11.5. The possible M^- species in a $[M(CO)_6]^-$ carbonyl anion should be and are Group 5 metals.

For trigonal bipyramidal pentacarbonyl species, the ligands donate 10 electrons, so it is up to M, M^+, or M^- to provide eight valence electrons. Hence, M in $M(CO)_5$ should be and is a Group 8 metal, M^+ should be and is a Group 9 metal ion, and M^- should be and is a Group 7 metal.

For tetrahedral tetracarbonyl species, in which the ligands donate eight electrons, M or M^+ should donate 10 valence electrons. Hence, neutral M should be a Group 10 metal atom; in practice, it can be Ni. Anionic M^- should be a Group 9 metal; in practice, it is Co. Cationic M^+ should be a Group 11 metal; in practice, it can be Cu$^+$.

Related Complexes of Other π-Acceptor Ligands (Table 7.6). Full substitution of all CO groups by incoming ligands is usually not practical, but fully substituted **homoleptic** (having only one type of ligand) products are known in several cases—for example, $[V(CNR)_6]^+$, $Ni(PF_3)_4$, $[Co\{P(OMe)_3\}_5]^+$, $Pt(PPh_3)_4$, and $Cr(NO)_4$.

The ligands from Table 7.6 have diverse σ-donor and π-acceptor abilities. For example, the phosphine and related ligands show the following order of basicity (σ-donor ability):[29]

$$:PR_3 > :PAr_3 > :P(OR)_3 > :P(OAr)_3 \gg :PF_3 \tag{11.14}$$

where Ar = aryl and R = alkyl. Their order of π-accepting ability is thought to be the reverse:

$$PR_3 < PAr_3 < P(OR)_3 < P(OAr)_3 < PF_3 \tag{11.15}$$

Triple-bonded ligands can also be rated in order of π-acceptor abilities:[30]

$$:C\equiv N^- < :N\equiv N < :C\equiv NR < :C\equiv O < :C\equiv S < :N\equiv O^+ \tag{11.16}$$

The two charged end members of this series differ substantially from the other members: Electron-rich cyanide ion is not only a poor π acceptor, but is also the best σ donor of the series. It is compatible with higher oxidation states in complexes than the other ligands, and its complexes (Section 5.3A) are sufficiently lower on the spectrochemical series (Section 7.4) to no longer be considered organometallic compounds, despite the presence of M–C bonds. At the other end of the series, NO^+ is strongly electron withdrawing.

Some of the substituted complexes are of importance for what they do to the stability and reactivity of the coordinated ligand. Both CS and CSe are stable in complexes, but not as free ligands. Coordinated N_2, in which π backbonding has reduced the N≡N triple bond closer to a double bond, is sometimes thereby activated toward protonation[31] and further reduction by H_2. This could be an important step in improved methods for the manufacture of ammonia fertilizers.

11.6B. Donation of Shared π Electrons by Alkenes and Aromatic Hydrocarbons.

Alkene Complexes. Classically, only unshared electron pairs can be donated by ligands to metals. However, many double- or triple-bonded organic compounds can act as soft bases by donating *shared π electron pairs* from adjacent carbon atoms with metal ions. (Note how this differs from the π donation of *unshared* electron pairs from ligands such as Cl^- at the low end of the spectrochemical series; Section 7.4B.)

The earliest organometallic compound of all, the anion of *Zeise's salt* (Figure 11.24a), was prepared in 1827 by the reaction of $K_2[PtCl_4]$ with ethylene:

$$K_2[PtCl_4] + CH_2=CH_2 \rightarrow KCl + K[PtCl_3(CH_2=CH_2)] \tag{11.17}$$

In these complexes there is also π backbonding of appropriate metal *d* electrons to the empty π* orbitals of the ligand (Figure 11.24b), so organic π-bonded ligands are also *π-acceptor ligands*. The π backbonding to the π* orbital of the ligands reduces the bond order in the ligand. If the ligand is a strong enough π acceptor, such as tetracyanoethylene $[(NC)_2C=C(CN)_2]$, the alkene bond order can be reduced to one, generating a saturated three-membered *metallacyclopropane* ring (Figure 11.24c). Dienes can form chelated complexes, as shown in Figure 11.24d–e.

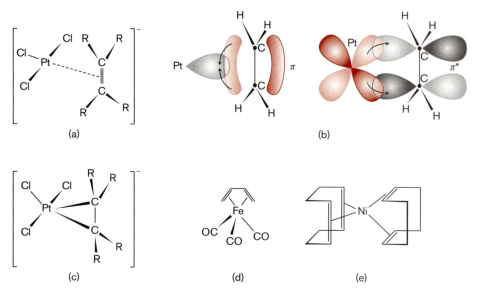

Figure 11.24. Alkene complexes. (a) The anion of Zeise's salt, $K[PtCl_3(CH_2=CH_2)]$, emphasizing its role as a π donor; (b) (left) electron donation from the π HOMO of an alkene to an empty orbital on Pt; (right) back donation from an occupied Pt d orbital to the π^* LUMO of the alkene; (c) an analogue of Zeise's salt with strong back donation, producing a metallacyclopropane structure; (d) (butadiene)iron tricarbonyl; (e) bis(1,5-cyclooctadiene)nickel(0). [Adapted from D. F. Shriver, P. Atkins, and C. H. Langford, *Inorganic Chemistry*, 2nd ed., W. H. Freeman: New York, 1994, pp. 685–687.]

The counting of coordination numbers and the description of geometries (Section 5.1) becomes more complex with π-donor ligands, because each π bond involves two carbon donor atoms but only one electron pair being donated. If we focus on the number of electron pairs being donated rather than the number of donor atoms, the complexes of these strong-field nonclassical ligands show a strong tendency to obey the 18- and 16-electron rule. As examples:

- The VEC of (butadiene)iron tricarbonyl (Figure 11.24d) includes eight from the Fe atom, six from the three CO ligands, and four from the two butadiene π bonds to give 18 electrons, which is expected for a five-coordinate trigonal bipyramidal complex.

- The VEC of bis(1,5-cyclooctadiene)nickel(0) includes 10 from the Ni atom and eight from the four π bonds of the two 1,5-cyclooctadiene ligands. Because this total is 18 electrons, the complex is tetrahedral rather than square planar.

- In contrast, the total in the anion of Zeise's salt (Figure 11.24a) is 16 electrons (eight for Pt^{2+} plus six for three Cl^- ligands plus two for ethylene), so Zeise's salt is square planar.

Pi MOs of Conjugated Cyclic Polyenes—Review from Organic Chemistry. The *conjugated cyclic polyenes* $C_nH_n^{z\pm}$ (including the *aromatic hydrocarbons*), in which carbon p_z atoms overlap to form a ring structure, are the most important organic ligands that are both π donors and π acceptors. In these compounds, n can be 3 to 8 and z can be 0, 1, or 2 (Table 11.6). We assume initially that these have the geometries of regular n-sided polygons.

TABLE 11.6. π-Electron Counting for Conjugated Cyclic Polyenes

Conjugated Cyclic Polyene	Formula	π-Electron Count
Cyclopropenium ion	$\eta^3\text{-}C_3H_3^+$	2
Cyclobutadiene	$\eta^4\text{-}C_4H_4$	4
Cyclopentadienide ion	$\eta^5\text{-}C_5H_5^-$	6
Benzene	$\eta^6\text{-}C_6H_6$	6
Tropylium ion	$\eta^7\text{-}C_7H_7^+$	6
Cyclooctatetraenide ion	$\eta^8\text{-}C_8H_8^{2-}$	10

Before drawing the MOs it is helpful to draw the vertical nodal planes in the polygon. The π-type MOs of conjugated cyclic polyenes also each have *one horizontal nodal plane* that passes through all of the nuclei of the ring; this is the same nodal plane that separates the positive and negative lobes of the carbon $2p_z$ orbitals that go into forming the π MOs.

The lowest-energy *a* MO has *no* vertical nodal planes. After that there are some *e* pairs of MOs that have 1, 2, 3, . . . vertical MOs. Within each e_n level, the vertical nodal planes are rotated as far away from each other as possible. For example, referring to the TASOs of Figure 11.15a, we see that the lowest-energy a_1' MO has *no* vertical nodal planes. In the e' set one MO has a vertical nodal plane separating the left and right halves of the molecule, while the other has a vertical nodal plane separating the upper part of the MO from the lower part.

If there is just one MO left at the top of the polygon, the location of the vertical nodal planes cannot run through all of the atoms (this would obliterate the MO), but must be oriented to fall between all of the atoms.

As the number of vertical nodal planes increases, there are fewer *net positive overlaps*, and the energy of the orbital increases.

Net positive overlaps = Number of positive overlaps of adjacent orbitals –
Number of negative overlaps of adjacent orbitals   (11.18)

You may recall from organic chemistry that there is a simple way of drawing the energy-level diagrams for conjugated cyclic polyenes: inscribe the polygon with the proper number of carbon atoms inside a circle, with one apex of the polygon pointed straight down. The energy-level diagram for that polyene has an energy level at the energy corresponding to each apex of the polygon. This automatically generates at lowest energy a unique bonding MO (symmetry label starting with *a*) which has *no* vertical nodal plane interrupting the π overlap of adjacent p_z orbitals). This MO is followed by as many doubly degenerate pairs of MOs as possible (symmetry labels involving *e* followed by the subscript "1" if the pair of MOs have one vertical nodal plane each, "2" if they have two vertical nodal planes, etc.). Those energy levels falling in the lower half of the circle are bonding MOs; those in the upper half are antibonding; those at the halfway point are nonbonding. Filling the bonding energy levels only requires 2, 6, 10, 14, . . . electrons, depending on the size of the polygon; these are the number of π electrons required to impart aromatic stability to a conjugated cyclic polyene (Table 11.6).

EXAMPLE 11.13

Draw and label the six π MOs of benzene. Compute the net positive overlaps of each, and draw the energy-level diagram.

SOLUTION: With six carbon atoms utilizing six p_z orbitals, this molecule has six π MOs. The lowest-energy MO has no nodal planes and six positive overlaps (it is an a_{2u} bonding orbital). The next two MOs each have one vertical nodal plane. As drawn in Figure 11.25a, the two MOs have slightly different appearances, because in one of them two atoms do not participate—but each has two net positive overlaps. (These are labeled e_{1g}.) These are followed by two slightly antibonding MOs with two vertical nodal planes each and a net of two negative overlaps (e_{2u}^*). The sixth and highest-energy MO positions the three nodal planes so that they fall *between* orbitals. This one MO is not degenerate, so it is not labeled e_{3g} (it turns out to be b_{2g}^*).

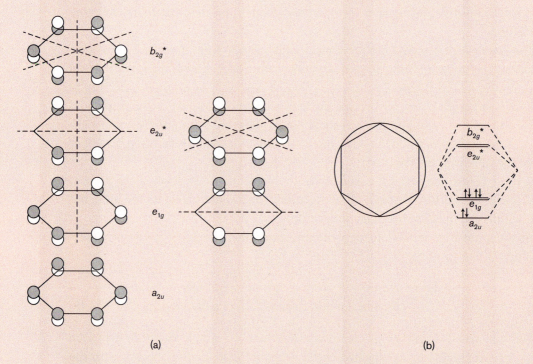

Figure 11.25. (a) Pi (π) molecular orbitals of benzene; (b) energy-level diagram of benzene. [Adapted from R. L. DeKock and H. B. Gray, *Chemical Structure and Bonding,* University Science Books: Mill Valley, CA, 1989, p. 314.]

The energy-level diagram of C_6H_6 can be generated from a hexagon inscribed point down in a circle (as shown in Figure 11.25b). The most stable form of C_6H_6 fills the three bonding MOs with six carbon valence p_z electrons. Hence, C_6H_6 is an uncharged molecule.

EXAMPLE 11.14

Draw and label the MOs and the energy-level diagram of the conjugated cyclic polyene, C_4H_4 (cyclobutadiene). Predict its number of unpaired electrons.

SOLUTION: This four-carbon-atom polyene has four π MOs, which utilize the zero-nodal plane pattern, the two one-nodal-plane patterns, and the two-nodal-plane pattern of Figure 11.26. The first MO (a_{2u}) has four positive overlaps, so it is bonding and is lowest in the energy-level diagram. The second and third MOs are a degenerate pair with two positive and two negative overlaps, so they have zero net positive overlaps and are nonbonding (e_g^{nb}). The top MO has four negative overlaps and is antibonding; it is not degenerate and cannot be labeled e_{2u} (it is b_{1u}^*).

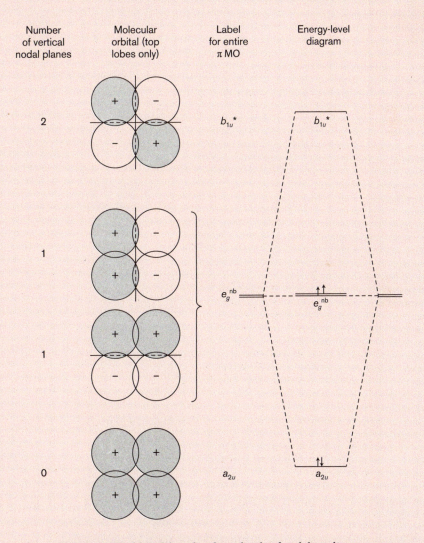

| Number of vertical nodal planes | Molecular orbital (top lobes only) | Label for entire π MO | Energy-level diagram |

Figure 11.26. (a) Pi (π) molecular orbitals of cyclobutadiene; (b) energy-level diagram of cyclobutadiene.

The cyclobutadiene molecule is predicted to have two unpaired electrons in a degenerate pair of molecular orbitals, so its π bond order is only 1, which is 1/4 per C–C link. It is held together, not only by π, but also by σ bonds that we have not discussed, so it does not fly apart into two C_2H_2 molecules, but it retains some other peculiarities. By a variation of the Jahn–Teller theorem (Section 8.7) it undergoes a distortion to relieve the double orbital degeneracy of its partially filled e_g^{nb} orbitals: It elongates to become a *rectangular* molecule, which resembles two ethylene molecules loosely attached by very long σ bonds. With a rectangular geometry, the energy-level diagram of Figure 11.26 no longer applies. Two MOs are bonding and two are antibonding; all π electrons are paired in the bonding MOs, so the molecule is now diamagnetic. But with such a strained geometry, it is not at all stable: It can be isolated at –78°C, but it rapidly decomposes at room temperature. In contrast to the familiar case of the cyclic aromatic hydrocarbon benzene, C_6H_6, cyclobutadiene has been *destabilized* by the conjugation of its π electrons in a ring; it is an example of an *antiaromatic* molecule.

11.6C. Metallocenes and Sandwich Molecules.

The 1951 discovery of *ferrocene*, $[Fe(C_5H_5)_2]$, was a revolutionary discovery in the history of inorganic chemistry. In 1951, Kealy and Pauson[32] were attempting a synthesis of the elusive organic compound, fulvalene, which contains two cyclopentadienyl (Cp) rings joined by a double bond, by the oxidation of the σ-bonded Grignard reagent C_5H_5MgX with Fe^{3+}. To their surprise, they obtained instead an orange solid, stable to 400°C, which analyzed as $C_{10}H_{10}Fe$ and which they reported with the suggestion that the Cp^- ligands were each σ bonded to two-coordinate iron (Figure 11.27a). Almost simultaneously, Miller et al.[33] reported another failed attempt at an organic synthesis of an amine by heating cyclopentadiene and N_2 over an iron catalyst, which also gave $C_{10}H_{10}Fe$ and for which they proposed the same structure.

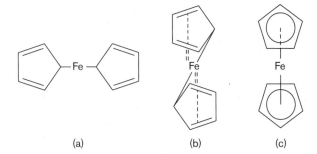

Figure 11.27. Ferrocene. (a) The structure originally proposed for ferrocene, with just two coordinate covalent bonds to iron (the pentagonal ring represents $C_5H_5^-$). (b) The structure originally conceived by Woodward, with two classical and four nonclassical coordinate covalent bonds to iron. (c) The structure now given for ferrocene, in which all 10 carbon atoms bond equally to iron.

Wilkinson[34] read the first of these two papers, saw the proposed structure of this unprecedentedly stable *d*-block organometallic compound, said to himself, "[Expletive deleted], it can't be that!" and, having read about Zeise's salt (Figure 11.24a), thought that the π bonds of the Cp^- groups ought also to be coordinated (Figure 11.27b). Then, recalling lectures on the aromaticity of six π-electron systems such as Cp^-, he conceived of the structure in Figure 11.27c, with all carbon and hydrogen atoms being equivalent, and thought, "It's a sandwich." The next day Wilkinson and Woodward conceived of the proof of this equivalence: The infrared (IR) spectrum of the com-

TABLE 11.7. Simple Metallocenes and Metallocenium Ions[a]

Neutral Metallocene Temperature (°C) Color	VCp_2 167°C Purple 3 unpaired electrons M–C 228	$CrCp_2$ 172°C Scarlet 2 unpaired electrons M–C 217	$Mn(C_5Me_5)_2$ Orange 1 unpaired electron M–C 211	$FeCp_2$ 173°C Orange 0 unpaired electrons M–C 206	$CoCp_2$ 173°C Purple 1 unpaired electron M–C 212	$NiCp_2$ 173°C Green 2 unpaired electrons M–C 220
Neutral Metallocene Temperature (°C) Color			$MnCp_2$ 172°C Pink 5 unpaired electrons M–C 238	$RuCp_2$ 199°C Yellow 0 unpaired electrons		
Neutral Metallocene Temperature (°C) Color			$Re(C_5Me_5)_2$ Purple 1 unpaired electron	$OsCp_2$ 230°C Colorless 0 unpaired electrons		
Metallocenium Ions Color		$[CrCp_2]^+$ Black 3 unpaired electrons	$[Re(C_5Me_5)_2]^-$ Orange 0 unpaired electrons	$[FeCp_2]^+$ Blue 1 unpaired electron	$[CoCp_2]^+$ Yellow 0 unpaired electrons	$[NiCp_2]^+$ Yellow 1 unpaired electron
Metallocenium Ions Color		$[CrCp_2]^-$		$[RuCp_2]^+$ Yellow 1 unpaired electron	$[RhCp_2]^+$ Colorless 0 unpaired electrons	$[NiCp_2]^+$ Yellow 0 unpaired electrons
Metallocenium Ions Color					$[IrCp_2]^+$ Yellow 0 unpaired electrons	$[Pd(C_5Me_5)_2]^{2+}$
Metallocenium Ions						$[Pt(C_5Me_5)_2]^{2+}$

[a]Entries in the table include melting point (°C), color, numbers of unpaired electrons, and M–C bond distances (pm).

pound ought to be especially simple (due to its high symmetry). This was confirmed in another day or so. This compound is commonly known as *ferrocene*; related neutral compounds or cations containing other metals are known as *metallocenes* and *metallocenium ions* (Table 11.7).

The Cp⁻ anion forms sandwich molecules not only with Fe, Rh, and Os, but also with the *d*-block metals V, Cr, Mn, Co, and Ni. (Using Ti resulted in a product that was so reactive that it dismembered the Cp ring in various ways. Beginning in 1998, substituting the Cp ring with groups even more sterically demanding and harder to dismember finally resulted in extremely reactive green or brown-pink Ti metallocenes with two unpaired electrons.[35]) Metallocenium cations are formed with Fe, Ru, V, Co,

Ru, Ir, Cr, Mn, and Ni (Table 11.7). Remarkably, the fourth-period neutral metallocenes (e.g., vanadocene, chromocene, manganocene, ferrocene, cobaltocene, and nickelocene) all have melting points very close to 173°C: The slightly differing metal atoms are evidently so buried within the sandwiches that they have no effect on the intermolecular forces that determine physical properties such as melting points.

After confirming the structure of ferrocene by X-ray crystallography, E. O. Fischer[36] synthesized bis(benzene)chromium and bis(benzene)molybdenum. Reduction of $CrCl_3$ in the presence of benzene and $AlCl_3$ gave a yellow salt $[Cr(C_6H_6)_2][AlCl_4]$.

$$CrCl_3 + \tfrac{2}{3}Al + AlCl_3 + 2C_6H_6 \rightarrow [Cr(C_6H_6)_2]AlCl_4 + \tfrac{2}{3}AlCl_3 \qquad (11.19)$$

This cation was then reduced in aqueous solution to give the product $[Cr(C_6H_6)_2]$, a black solid. Fischer shared the 1973 Nobel Prize in Chemistry with Wilkinson.

Among the other polyenes, only the eight-membered ring has to date given sandwich compounds involving two rings of the same size; these are made by the Lewis acid–base reactions in THF of $C_8H_8^{2-}$ (generated from neutral C_8H_8 and 2 mol K metal) and f-block metal halides:

$$UCl_4 + 2K_2(C_8H_8) \rightarrow [U(C_8H_8)_2] + 4KCl \qquad (11.20)$$

Covalent Overlap between Conjugated Cyclic Polyenes and d-Block Metal Ions. We expect the metal ion to bond above one face of the polyene, with overlap occurring between the lobes of the π bonds on that face of the polyene and the lobes of the metal orbitals on the side of the metal ion adjacent to the polyene. (Hence it is not necessary to worry about the signs on the orbitals on the other half of the polyene.) The main criterion for overlap is that *the metal ion and the ligand MO share the same number of vertical nodal planes.* This is equivalent to saying that the metal orbitals and polyene orbitals must fall in the same irreducible representation, if we take into account the fact that we have now eliminated the inversion center i and lowered the symmetry from D_{nh} to C_{nv}.

Let us take the common π ligand, benzene (C_6H_6), as an example; its MOs are shown in Figure 11.25. To find the metal-ion orbitals of appropriate symmetry to interact with these TASOs, we can either recall those having the same number of vertical nodal planes, or else consult the C_{6v} character table.

The π MOs of benzene, when lowered to C_{6v} symmetry, have the g or u removed from their labels and may have other changes: The MOs in Figure 11.25, from bottom to top, in the C_{6v} point group have symmetry labels a_1, e_1, e_2, and b_2, with 0, 1, 2, and 3 vertical nodal planes, respectively. If the metal atom sitting above the ligand face is a d-block metal ion, the metal orbitals eligible for bonding with these ligand MOs are:

- s, p_z, and d_{z^2} with zero vertical nodal planes (a_1 irreducible representation);

- (p_x, p_y) and (d_{xz}, d_{yz}) with one vertical nodal plane (e_1 irreducible representation);

- $(d_{x^2-y^2}, d_{xy})$ with two vertical nodal planes (e_2 irreducible representation); or

- no d orbitals with three vertical nodal planes (b_2 irreducible representation).

EXAMPLE 11.15

Consider the planar $C_8H_8^{2-}$ (cyclooctatetraenide dianion) ligand. In Exercise 11.60 you drew facial views of each of the π molecular orbitals of this ligand. Now identify the metal-ion orbitals (if any) that can overlap with each of these ligand orbitals if the metal ion is a *d*-block metal ion.

SOLUTION: (See the illustrations at the end of this example.) The lowest-energy orbital of $C_8H_8^{2-}$ (a_{2u}) has zero vertical nodal planes. When we bring a metal ion above its center along the *z* axis, for positive overlap the metal ion must have zero nodal planes along its *z* axis. This ligand orbital gives positive overlap with the metal *s*, p_z, and d_{z^2} orbitals.

Next, the ligand has a pair of e_{1g} orbitals with one vertical nodal plane each—one along the *x* axis and one along the *y* axis. These orbitals overlap metal orbitals with one nodal plane each: the metal (p_x, p_y) and (d_{xz}, d_{yz}) orbital pairs.

Next, the ligand has a pair of e_{2u}^{nb} orbitals with two vertical nodal planes each—one pair is oriented along the *x* and *y* axes and the other is oriented between the *x* and *y* axes. These orbitals overlap the metal (d_{xy}, $d_{x^2-y^2}$) orbital pairs.

Next, the ligand has a pair of e_{3g}^* orbitals with three vertical nodal planes each. However, no metal *s*, *p*, or *d* orbitals have this many vertical nodal planes—although a metal employing *f* valence orbitals would.

Last, the ligand has a unique MO with four vertical nodal planes. No metal orbitals of any kind have four vertical nodal planes, so there can be no metal–ligand overlap involving this MO.

(As described in our 2000 text, it is possible to consult the D_{8h} character table and find the overlapping metal orbitals listed on the right side of the table.)

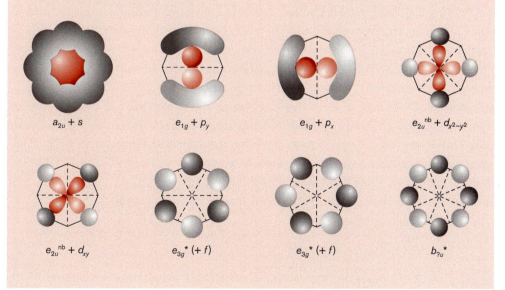

$a_{2u} + s$	$e_{1g} + p_y$	$e_{1g} + p_x$	$e_{2u}^{nb} + d_{x^2-y^2}$
$e_{2u}^{nb} + d_{xy}$	$e_{3g}^* (+ f)$	$e_{3g}^* (+ f)$	$b_{?u}^*$

MOs for Sandwich Compounds; Consequences

Now suppose that we place another benzene ring on the opposite side of the metal ion to give a sandwich molecule, $M(C_6H_6)_2$. Depending on the orientation of this second ring, this addition can restore the molecule to D_{6h} symmetry and return the inversion center i and the u and g labels. As the second benzene ring comes in, it can either come in with the same lobe signs pointing to the metal ion as the first ring had, which generates a gerade MO, or it can come in with the reverse-sign face towards the metal, which generates an ungerade MO. Thus, 12 TASOs will be generated for the two benzene rings. Some of these, however, will find no d-block metal-ion orbitals of matching symmetry, because d-block metal ions have only nine valence orbitals (including their np orbitals). Some TASOs must therefore be nonbonding to the metal ion. In Figure 11.28a, we draw the eight benzene MOs that do find metal-ion orbital partners; to save space we do not show the corresponding antibonding MOs. The ungerade p orbitals on the metal ions end up paired with the ungerade benzene TASOs, while the gerade s and d orbitals overlap with the gerade TASOs.

In Figure 11.28b we draw the lower part only of the energy-level diagram for dibenzenechromium, which shows the energy levels of the eight bonding MOs shown in Figure Figure 11.25. It shows the energy level of only the lowest of the antibonding MOs (the LUMO). One metal-ion orbital remains approximately nonbonding in dibenzene chromium: the d_{z^2}. (This is because both d_{z^2} and s overlap with the ligand a_{1g} MO; consequently, the metal uses s–d hybridization for this MO, with one metal orbital becoming approximately nonbonding, as in Section 11.2. Also, the ring orbitals tend to fall in the conic nodes of d_{z^2}, further reducing overlap.)

This molecule, composed of atoms of intermediate electronegativity, is expected to fill its bonding MOs. It may or may not fill its nonbonding MO. The eight bonding MOs require 16 electrons, 12 of which come from the π systems of the two benzene rings, and four of which come from Cr. If the nonbonding MO is also filled (to give a total of 18 valence electrons), Cr must provide the two extra electrons: Cr with six valence electrons is neutral Cr^0, so the resulting complex is neutral dibenzenechromium, $[Cr(C_6H_6)_2]$. If the nonbonding MO takes only one electron, the other known derivative results: $[Cr(C_6H_6)_2]^+$, the dibenzenechromium(I) ion.

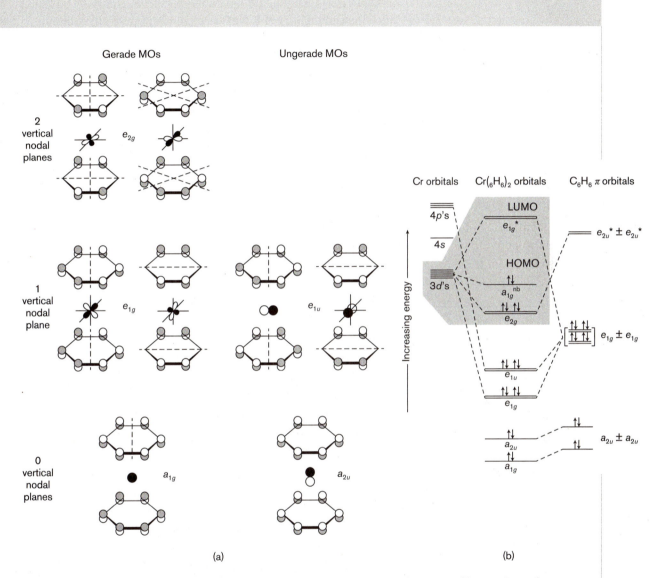

Figure 11.28. Selected MOs for bis(benzene)chromium. (a) Bonding MOs only, arranged by numbers of vertical nodal planes, with gerade MOs on the left and ungerade MOs on the right; (b) Energy-level diagram, showing only one antibonding MO, the LUMO, and enclosing in the box the region emphasized in crystal field theory for the corresponding geometry, the trigonal prism. (In order to simplify the diagram, not all of the dashed correlation lines have been drawn.) [Adapted from R. L. DeKock and H. B. Gray, *Chemical Structure and Bonding*, University Science Books, Mill Valley, CA, 1989; pp. 389–390.]

The 18-Electron Rule and Valence Electron Counts (VEC) in Sandwich Compounds. We illustrate the application of these concepts by checking the electron count in our two prototypic sandwich compounds—$FeCp_2$ and $Cr(C_6H_6)_2$.

1. The polyene ligands are identified in Table 11.6 and the numbers of electrons each donates are noted. Both C_6H_6 and Cp^- are six-electron donors.

2. The charges on the central metal atoms are assigned as in Section 3.2B and Equation 3.3, and the metal's number of d (or f) electrons is noted. Because each Cp^- is an anion, to conserve the electroneutrality of $FeCp_2$, the iron must be counted as the cation Fe^{2+} with six valence electrons (d^6 electron configuration). $Cr(C_6H_6)_2$ is a neutral species, and each benzene ring is a neutral ligand. Therefore, the Cr atom is a neutral d^6-electron species.

3. The ligand and metal-ion valence electrons are then totaled to give the VEC, which is 18 if the 18-electron rule is obeyed. In ferrocene the VEC is six from each of two Cp^- and six from Fe^{2+}, for a total of 18, so ferrocene is an 18-electron species. In $Cr(C_6H_6)_2$ the VEC is six from each of two benzene rings and six from the Cr atom, for a total of 18, so bis(benzene)chromium obeys the 18-electron rule, too.

Although the largest polyene, $C_8H_8^{2-}$, is too large to form a good sandwich with d-block metal ions, it does form the most important f-block organometallic compounds, such as $U(C_8H_8)_2$, commonly called *uranocene*. Because uranium is not a d-block metal, the 18-electron rule does not apply to its compounds. The π-donor and π^*-acceptor orbitals of $C_8H_8^{2-}$ interact both with metal valence $(n-2)f$ orbitals and postvalence $(n-1)d$ orbitals. These are among the few f-block metal compounds with a pronounced degree of covalent bonding.

EXAMPLE 11.16

Calculate the metal VEC in each of the following metallocenes or metallocenium ions: (a) $NiCp_2$; (b) $Ni(C_4H_4)_2$; (c) $TiCp(C_7H_7)$; (d) $U(C_8H_8)_2$; (e) $[CoCp(C_3H_3)]^+$.

SOLUTION: 1. Each ligand is recognized as the corresponding species from Table 11.6 with the indicated charge and donating the cited number of electrons: (a) two –1 ions, each donating six electrons; (b) two neutral molecules, each donating four electrons; (c) a +1 ion and a –1 ion, each donating six electrons; (d) two –2 ions, each donating 10 electrons; (e) a +1 ion donating two electrons and a –1 ion donating six electrons.

2. The charge on the metal ion is then deduced, and its number of d (or f) electrons is noted: (a) Ni^{2+}, d^8; (b) Ni^0, d^{10}; (c) Ti^0, d^4; (d) U^{4+}, f^2; and (e) including the metallocenium ion charge, Co^+, d^8.

3. The ligand and metal-ion valence electrons are then totaled to give the VEC: (a) 20; (b) 18; (c) 16; (d) 22; and (e) 16 electrons.

As a consequence of the presence of closely spaced (including nonbonding) MOs, non-18-electron metallocenes exist and show a variety of magnetic properties. According to Table 11.7, manganocene has an anomalously large number of unpaired electrons. This anomaly disappears on substituting the C_5 ring with methyl groups. Evidently, the small electronic change is sufficient to cause Δ_o to exceed P, the pairing energy (Section 8.3), in $Mn(C_5Me_5)_2$ and $Mn(C_5H_4Me)_2$, which are low-spin complexes (Chapter 8), but not in $MnCp_2$, which is a high-spin complex.

CONNECTIONS TO MEDICINAL CHEMISTRY AND INDUSTRIAL CHEMISTRY

Titanocene and Zirconocene Dichlorides

The successful use of cisplatin (Section 5.8B) to treat cancer prompted Köpf and Köpf-Maier to test other *d*-block complexes with ligands in cis positions that can readily leave. The common organometallic compound titanocene dichloride or dichlorobis(η⁵-cyclopentadienyl)titanium(IV) (Figure 11.29) quickly showed antitumor properties when implanted in mice. However, the chloride ions were lost so much more rapidly than in the case of cisplatin that the drug could not be administered in solution. Even one Cp⁻ was eventually lost. The second-generation drug budotitane or diethoxybis(1-phenylbutane-1,3-dionato) titanium(IV) was given clinical trials, but ultimately also proved too unstable for use.

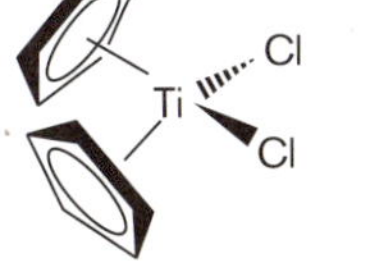

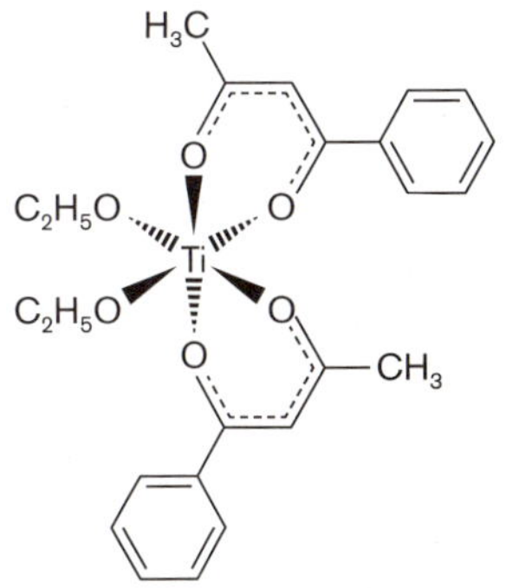

$[(\eta^5\text{-Cp})_2\text{TiCl}_2]$
Titanocene dichloride

Figure 11.29. Structures of titanocene dichloride and one isomer of budotitane. [Adapted from J.C. Dabrowiak, *Metals in Medicine*, John Wiley and Sons Ltd.: Chichester, UK, 2009, p. 168.]

However, titanocene dichloride or zirconocene dichloride in combination with trimethylaluminum and water [or their reaction product, methylaluminoxane (MAO)] have become commercially important catalysts. They polymerize alkenes such as propylene ($CH_3CH=CH_2$) stereospecifically to give isotactic or syndiotactic polypropylene or related polymers.[37]

The non-18-electron metallocenes are much more reactive than ferrocene, however. Thus, the 19-electron species cobaltocene is about as good a reducing agent as sodium metal, and it is oxidized readily to the stable 18-electron cobaltocenium ion, $CoCp_2^+$. The 20-electron nickelocene is readily oxidized to the 19-electron $NiCp_2^+$ ($E° = +0.18$ V), and even to the 18-electron dication $NiCp_2^{2+}$ ($E° = +1.11$ V). This formally contains nickel in the unusually high +4 oxidation state. (In contrast to CO, Cp^- as a ligand is compatible with relatively high metal oxidation states.)

11.7. Background Reading for Chapter 11

General Background on Applied Molecular Orbital Theory

R. Hoffman, S. Shaik, and P.C. Hiberty, "A Conversation on VB vs MO Theory: A Never-Ending Rivalry," *Acc. Chem. Res.* 36, 750–756 (2003).

R.L. DeKock and H.B. Gray, *Chemical Structure and Bonding*, University Science Books: Mill Valley, CA, 1989, pp. 183–418 (Chapter 4: The Molecular Orbital Theory of Electronic Structure and the Spectroscopic Properties of Diatomic Molecules, Chapter 5: Electronic Structures, Photoelectron Spectroscopy, and the Frontier Orbital Theory of Reactions of Polyatomic Molecules, and Chapter 6: Transition-Metal Complexes).

Multiple Bonding in Diatomic Metallic Molecules

B.O. Roos, A.C. Borin, and L. Gagliardi, "Reaching the Maximum Multiplicity of the Covalent Chemical Bond," *Angew. Chem. Intl. Ed.* 46, 1469–1471 (2007).

Frontier Orbitals and Hardness and Softness of Molecules

R.G. Pearson, "Absolute Electronegativity and Hardness," *Inorg. Chem.* 27, 734–740 (1988).

Secondary and Halogen Bonding

P. Politzer, J.S. Murray, and T. Clark, "Halogen Bonding and Other σ-Hole Interactions: A Perspective," *Phys. Chem. Chem. Phys.* 15, 11178–11189 (2013).

Metallocene Catalysis

W. Kaminsky, "The Discovery of Metallocene Catalysts and Their Present State of the Art," *J. Polym. Sci., Part A: Polym. Chem.* 42, 3911–3921 (2004).

Numbered references from this chapter may be viewed online at www.uscibooks.com /foundations.htm.

11.8. Exercises

11.1. *A hypothetical and incompletely labeled energy-level diagram is shown below for a diatomic molecule M_2, which is composed of two transition metal atoms bonding to each other with their valence *s* and *d* orbitals. (a) Label each of the eight MO energy levels using the appropriate combinations of the following symbols: σ, *g*, *u*, δ, π, and *. (b) Three of these MOs were drawn in Exercise 10.35. Identify which energy level in the diagram below belongs to each of the MOs drawn in Exercise 10.35.

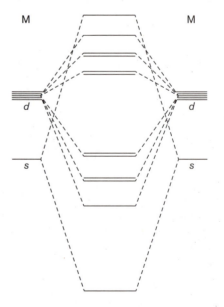

11.2. Below is an energy-level diagram for a diatomic molecule M_2 composed of two transition metal atoms bonding to each other with their valence s and d orbitals. The MOs and their energy levels have been labeled alphabetically starting from the lowest energy levels. (a) Orbitals A and E are connected by dashed lines to single lines at the left and right of the diagram. These single lines represent the energies of which type of orbitals of the individual M atoms (s or d)? (b) Orbital A is which type of MO (bonding or antibonding)? (c) Orbital E is which type of MO? (d) Draw a picture of the appearance of orbital E, showing signs of the wave function. (e) Which lettered MO, if any, should really be labeled σ_u^*? Which should really be labeled δ_g? Which should really be labeled ϕ_g^*? (f) The δ_u^* MO results from which type of overlap (positive or negative) of which atomic orbitals of the two M atoms? Is the energy level of the δ_u^* molecular orbital singly or doubly degenerate? What does this mean?

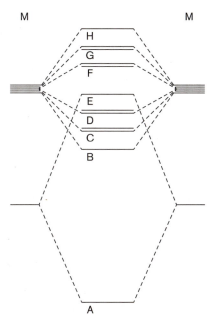

11.3. *Write the (valence) electron configurations of the following diatomic molecules or ions (assuming no s–p mixing; Section 11.2). For each molecule give the bond order and the number of unpaired electrons. (a) B_2^{2-}; (b) N_2^+; (c) O_2^-; (d) F_2; and (e) H_2.

11.4. Write the (valence) electron configurations of the following diatomic molecules or ions (assuming no s–p mixing; Section 11.2). For each molecule give the bond order and the number of unpaired electrons. (a) B_2; (b) C_2^{2-}; (c) N_2^{2+}; (d) Be_2; and (e) C_2.

11.5. *Using Figure 11.4, write the valence electron configurations of, compute the bond orders of, and tell the number of unpaired electrons in each of the following molecules. For f-block elements, assume that molecular orbitals based on $7s$ atomic orbitals are higher in energy than all of the MOs based on atomic f orbitals. (a) Ti_2; (b) V_2; (c) Pd_2; (d) Am_2; (e) Th_2; and (f) Bk_2.

11.6. Using the same assumptions as in Exercise 11.5, write the expected valence electron configurations of, compute the bond orders of, and tell the number of unpaired electrons in each of the following ions: (a) Mo_2^{4+}; (b) Np_2^{2+}; (c) Hg_2^{2+} (treat d electrons as core electrons); and (d) Pt_2^{2-}.

11.7. *Using the energy-level diagram and lettering scheme in Exercise 11.2, assume normal orders of filling for each of the following diatomic molecules or ions: Hf_2, Mo_2^{2-}, and Pd_2^{4+}. (a) Write valence electron configurations, adding stars (*) to indicate antibonding MOs (i.e., A^3B^{*7}); (b) indicate the number of unpaired electrons in the molecule or ion; and (c) indicate its bond order.

11.8. The fourth-period diatomic molecule V_2 is calculated to have the following electron configuration: $\sigma_g^2\pi_u^2\delta_g^2\sigma_g(s)^1\delta_u^{*2}\pi_g^{*1}$. (a) What is the bond order for this electron configuration? (b) Why is this so dissimilar from the electron configuration calculated for Nb_2, which resembles that of Mo_2 as given in the text example? (c) How can you explain the fact that the lowest energy orbitals are not always filled before the highest energy orbitals? (d) How many unpaired electrons is this configuration likely to represent? (Apply Hund's rule to the maximum extent.)

11.9. *(a) Give the symmetry labels of the orbitals in Exercise 11.1 that may be altered in energy and appearance as a result of s–d mixing or hybridization. What condition is necessary for this to occur? On which side of the d block are these conditions most likely to be met? (b) In the f block, hybridization or mixing is possible involving the $(n-2)f$, the $(n-1)d$, and the ns orbitals. Molecular orbitals based on which f orbital types (Figure 11.3) might be affected by hybridization involving MOs based on which other atomic orbital types if the parent atomic orbitals are of appropriate relative energies?

11.10. (a) Draw an energy-level diagram for a second-row homonuclear diatomic molecule with no s–p mixing or hybridization. Label each energy level with the appropriate symmetry label from the $D_{\infty h}$ point group, and use the superscripts b, nb, and * to indicate whether the orbital is bonding, nonbonding, or antibonding, respectively. (b) Give the symmetry labels of the orbitals in your energy-level diagram that are altered as a result of s–p mixing or hybridization.

11.11. *Cite data from Table 11.1 to justify *horizontal* trends in the likelihood of s–p hybridization across the p-block elements of (a) Period 2, (b) Period 3, and (c) Period 4. (d) If you took data from Table 11.1 for a vertical trend in Group 14/IV, you would incorrectly assume that s–p hybridization increased from C to Pb. What other factor negates the effects of the energy differences from Table 11.1? (See the first Amplification in Section 9.4.)

11.12. Which is the best Lewis dot structure for (a) B_2 with no s–p hybridization (choose one): :B–B: or :B≡B: or B≡B or the resonance structures :B=B ⟷ B=B:; (b) C_2 with no s–p hybridization (choose one): :C=C: or C≡C or :C–C: or the resonance structures :C≡C ⟷ C≡C:; and (c) C_2 with s–p hybridization (choose one): :C=C: or C≡C or the resonance structures :C≡C ⟷ C≡C:?

11.13. *Explain the following trends in bond energies and bond lengths, based on data from Tables 3.2, 3.3, and 11.2: (a) The bond in N_2^+ is longer and weaker than the bond in N_2, but the bond in Cl_2^+ is shorter (189 pm) and stronger (415 kJ mol^{-1}) than the bond in Cl_2. (b) Bond energies fall among diatomic molecules going down Groups 14/IV, 15/V, and 16/VI, but increase going down Group 11. (c) Bond energies decrease much more rapidly going down Group 14/IV than going down Group 17/VII.

11.14. Predict trends in bond energies and bond lengths in the following series: (a) C_2^{2+}, C_2, C_2^{2-}; (b) Mo_2, Mo_2^{2+}, Mo_2^{4+}; (c) Cr_2, Mo_2; and (d) Nb_2, Mo_2, Tc_2, Ru_2.

11.15. *For each of the following species, write the electron configuration (assuming no s–p hybridization) and compute the bond order. Then tell: (a) Which should have the longer bond, O_2 or O_2^{2-}? (b) Which should have the stronger bond, B_2 or B_2^{2-}? (c) Which should have the weaker bond, C_2 or C_2^{2-}? (d) Which should have the shorter bond, O_2^+ or O_2^{2-}?

11.16. Referring to the electron configurations in Exercise 11.8 and the energy-level diagram in Exercise 11.2, (a) which should have the longer bond, Mo_2 or Mo_2^{2-}? (b) Which should have the stronger bond, Hf_2 or Hf_2^{2-}? (c) Which should have the weaker bond, Pd_2^{2+} or Pd_2^{4+}?

11.17. *The computed bond lengths for diatomic molecules of the early actinides are as follows: Ac_2, 363 pm; Th_2, 276 pm; Pa_2, 237 pm; U_2, 243 pm.[4] Compare these with the single-bond lengths you would predict for these molecules from data in Table 1.13. Are these single bonds? Where along the series are bond orders increasing left to right and where are they decreasing?

11.18. The computed bond dissociation energies for diatomic molecules of the early actinides are as follows: Ac_2, 1.2 eV; Th_2, 3.3 eV; Pa_2, 4.0 eV; U_2, 1.2 eV (experimental 52 kcal mol^{-1}); Pu_2, 0.30 eV.[4] Convert these to common units of kJ mol^{-1}. Where along the series are bond orders increasing left to right, and where are they decreasing?

11.19. *Choose the (vis, UV, or near-IR) electronic transition(s) from the following list that should be intense: σ_g to σ_u^*, σ_g to δ_g, π_u to δ_g, δ_u^* to σ_u^*.

11.20. After adjusting Figure 11.2 to reflect the valence electron configuration of the N_2 molecule (with no s–p hybridization), draw in arrows (↑) to represent all of the intense UV absorptions to be extended. (These arrows should start at an occupied orbital and point to an unoccupied orbital.)

11.21. *The halogen–molecule HOMO–LUMO electronic transition discussed in Section 11.3, $\pi_g^* \rightarrow \sigma_u^*$, is intense. List all other electronic transitions among valence MOs in halogen molecules that should also be intense.

11.22. Although the acetylide ion C_2^{2-} is colorless, the C_2 molecule absorbs visible light of frequency 19,300 cm^{-1}. (a) Referring to Section 7.3, convert this frequency to energy units of kilojoules per mole and electron volts. What color is C_2 likely to be? (b) Is C_2 likely to show s–p hybridization or to be unhybridized? Explain, with the aid of its most likely electron configuration, why C_2 absorbs relatively low-energy visible light (note that C_2^{2-}, O_2, and N_2 absorb mainly or exclusively in the UV or vacuum UV regions). (c) Is the color of C_2 likely to be intense? Why or why not?

11.23. *Sketch the expected photoelectron spectrum of (a) B_2 and (b) C_2. Will any of the bands be lacking fine structure?

11.24. Diatomic molecules having hybridization have a type of MO that is not present in unhybridized diatomic molecules. Describe briefly how the photoelectron spectrum should consequently differ in appearance from that of an otherwise-identical but unhybridized molecule.

11.25. *Show by calculation which of each of the following pairs of molecules is the softest: (a) N_2 or P_2; (b) O_2 or S_2. The ionization energies are 15.581 eV for N_2, 12.070 eV for O_2, 10.487 eV for P_2, and 9.356 eV for S_2; the electron affinities are –2.2 eV for N_2, 0.451 eV for O_2, 0.589 eV for P_2, and 1.670 eV for S_2.

11.26. Based on qualitative principles, would you expect the diatomic molecules of the lower part of the d block (such as Mo_2) to be harder or softer than the diatomic molecules of the top of the p block (such as N_2)? Why?

11.27. *Explain why the electron affinity of C_2 is much greater (3.269 eV) than that of O_2 (0.451 eV) or N_2 (about –2.2 eV).

11.28. Below are the energy-level diagrams of diatomic molecules of two hypothetical atoms, sengsavangium (Ss) and herrium (Hr). The numbers beside the levels represent the actual energies of the orbitals in electron volts, relative to zero for an electron at an infinite distance from the nucleus. (a) Draw an arrow on each diagram to indicate the process responsible for the first ionization energy of each molecule. (b) Draw an arrow on each diagram to illustrate the process responsible for the electron affinity of each molecule. (c) Which molecule is softer, Ss_2 or Hr_2? Calculate the softness of each molecule.

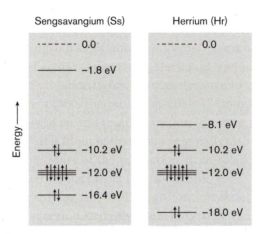

11.29. *Given the interatomic distances in Table 11.4, argue in which case the secondary bonding is most pronounced: (a) Cl_2 or Br_2 or I_2; (b) S_8 or Se_8 or $Se_{metallic}$ or $Te_{metallic}$.

11.30. Using data from Table 11.3, illustrate and explain the trend in electron affinities across Period 4.

11.31. *The series of diatomic molecules I_2, IBr, and ICl are isoelectronic, but because the sets of parent atoms have increasingly diverse valence orbital potential energies (Table 11.1) and electronegativities, their energy-level diagrams progressively diverge in character. Their photoelectron spectra[38] yield the following ionization energies:

I_2	IBr	ICl
9.22 eV	9.85 eV	10.10 eV
10.74 eV	11.99 eV	12.88 eV
12.66 eV	13.70 eV	14.26 eV

(a) Set up the left and right sides of the energy-level diagrams for each of these three molecules. Fill in the irreducible-representation symmetry labels on each energy level. How and why do the labels change as the molecules change across this series? (b) Assign the tabulated ionization energies to the energy levels in your drawing. Discuss the periodic trends in these ionization energies and energy levels in going from I_2 to ICl. (c) Qualitatively, how do shapes of the MOs change in going from I_2 to ICl? (d) The ionization energy of IF is 10.54 eV; the electron affinities of I_2 and IBr are both 2.55 eV. Characterize the trends in softness of these four molecules.

11.32. Suppose that you changed a homonuclear diatomic transition metal molecule M_2 to a heteroatomic molecule MM′, composed of two different transition metals of differing electronegativities. Briefly describe (a) one type of change that you would make on the left and right of the energy-level diagram of M_2 to make it appropriate for MM′, and (b) one type of change that you would make in the irreducible-representation symmetry labels.

11.33. *(a) Using Figure 11.12 as a starting point, write the electron configuration of the molecule NO and calculate its bond order. (b) Is the bonding in NO weaker or stronger than that in CO? (c) Is the NO bond longer or shorter than the CO bond? (d) Suggest how the photoelectron spectrum of NO should compare with that of CO.

11.34. (a) Which of the following three should have the highest bond order: NO, NO^+, or NO^-? (b) Which of these three should have the weakest bond? (c) Which of these three should have the shortest bond? (d) Which of these should have the photoelectron spectrum with the fewest bands? (e) Which (if any) of these should be paramagnetic, with how many unpaired electrons?

11.35. *The ionization energy of the N_2 molecule is 15.58 eV, and its electron affinity is −2.2 eV. Based only on this information and the corresponding data for CO given in the text, (a) which is the softer ligand, N_2 or CO? (b) Which should be the better σ-donor ligand? (c) Which should be the better π-acceptor ligand? (d) Which of the two ligands should have MOs with a shape or contour better suited to act as a σ donor? As a π acceptor?

11.36. Sketch lines in the energy-level diagram of LiF (Figure 11.14) to represent the following energies or processes: (a) the electron affinity of LiF; (b) the ionization energy of LiF; and (c) the softness of LiF.

11.37. *If your class carried out Exercise II of the computational Experiment 10 on the molecules and gaseous ions N_2, NO^+, and CN^-, (a) take the symmetry-labeled MO energy-level diagrams that your class produced in part (d) and arrange them in logical order for these isoelectronic neutral molecules and charged ions. (b) Compare your results for CN^- with the calculations and photoelectron spectra given by M. Considine, J. A. Connor, and I. H. Hillier, *Inorg. Chem.* 16, 1392 (1977) for Na^+CN^-. Why did they adjust their calculations by 5.5 eV? (c) Compare and explain how the energies of the LUMO, HOMO, (HOMO-1 and HOMO-2), and HOMO-3 vary across this series. (d) Compare how the softnesses of these molecules (or molecules and ions) vary across each series. For each diatomic species, identify (e) the likely Lewis-base donor atom and (f) the likely Lewis-acid acceptor atom.

11.38. If your class carried out Exercise II of the computational Experiment 10, gather your results on the molecules N_2, CO, and BF. If you did not do Experiment 10, you can obtain partial results for the HOMOs and LUMOs from U. Radius, F. M. Bickelhaupt, A. W. Ehlers, N. Goldberg, and R. Hoffman, *Inorg. Chem.* 37, 1080 (1998). (a) To the extent possible, assign the photoelectron bands shown in Figure 11.8a and Figure 11.12c to the energy levels of your MO diagrams. You may add the following ionization energy from the photoelectron spectrum of BF: 11.12 eV. (b) Compute and compare the softnesses of these molecules (or molecules and ions) across the series. For each diatomic species, identify (c) the likely Lewis-base donor atom and (d) the likely Lewis-acid acceptor atom.

11.39. *Consider the hypothetical square planar molecule XeH_4. (a) Draw the H_4 TASOs. (b) Which (if any) Xe valence atomic orbitals are nonbonding? Which (if any) symmetry-adapted sets of H_4 orbitals are nonbonding? What is the maximum degeneracy found in any MO of XeH_4? (c) Draw in the Xe valence orbitals in the middle of the above sets of H_4 TASOs so as to produce the $a_{1g}*$ molecular orbital of XeH_4; the e_u set of bonding MOs of XeH_4.

(d) The energy-level diagram for XeH_4 is shown below. Fill in the appropriate symmetry labels in the diagram, and show the proper number of valence electrons as arrows on the diagram. (e) Compute the total bond order of XeH_4 and the bond order per Xe–H link.

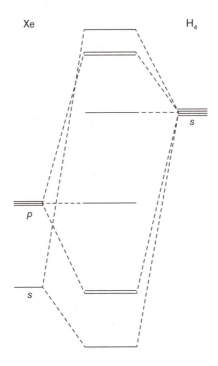

11.40. Consider the hypothetical square planar complex ion $[PtH_4^{2-}]$, in which the Pt^{2+} ion uses some of its $5d$ orbitals to σ-bond TASOs constructed from four H^- ligands located at the ends of the x and y axes. (a) Which $5d$ orbitals can form bonding and antibonding MOs with the ligands? (b) Draw the bonding MOs of this complex ion. (c) Give the symmetry labels (irreducible representations) of these MOs.

11.41. *(a) Draw and label (using proper symmetry labels) the possible σ-type TASOs in the pentagonal planar XeF_5^- ion. (b) How many unshared electron pairs are there on Xe in this ion, and where are they located? If appropriately hybridized, Xe atomic orbitals are used to hold these unshared electron pairs, and if Xe cannot use d orbitals in this bonding, what Xe atomic orbitals remain to bond with these TASOs? Which TASOs must remain nonbonding? (c) With these assumptions, what is the bond order per Xe–F link in this ion?

11.42. An important species in the study of organic chemistry is the carbonium ion, CH_3^+. (a) How would you have to adjust the energy levels on the left side of Figure 11.21 to make it apply to CH_3^+ instead of BH_3? (b) This change would change the Lewis acidity of CH_3^+ in comparison to BH_3. Describe this change. (c) What other difference between the two species would increase the Lewis acidity of CH_3^+ in comparison to BH_3?

11.43. *Compare the π bonding in BF_3 and the three important trigonal planar oxo anions listed in Table 3.6. Based on VOPEs from Table 11.1, list the four species in order of decreasing π bonding and justify your choice.

11.44. Compare the π bonding in the $[C(NH_2)_3]^+$ (guanidinium ion), CO_3^{2-}, and SiO_3^{2-} ions. (a) Based on VOPEs from Table 11.1, list the three species in order of decreasing π bonding and justify your choice. (b) Does this affect the stability of any of these species? (c) Can you generalize the comparison of CO_3^{2-} and SiO_3^{2-} ions to the general vertical trend in total coordination numbers found in oxo anions?

11.45. Consider a tetrahedral (T_d) transition metal complex MH_4^{n-}. (a) Draw the four TASOs of the H_4 ligand set, giving each its proper symmetry label. (b) Find the valence orbitals of the transition metal that match each of the TASOs of the four hydride ions in symmetry. (c) Draw the energy-level diagram for this tetrahedral complex, and label the orbitals with their irreducible representations and with nb or * if appropriate. Include only metal valence s and d orbitals. (In many cases, the exact ordering of energy levels is uncertain without calculations, so put down a reasonable order.) Circle the part of the energy-level diagram that is emphasized by the crystal field theory (CFT), and show where the crystal field splitting ($10\Delta_o$ or Δ_t) is located.

11.46. Consider the tetrahedral molecule CH_4 (T_d point group); its energy-level diagram is shown below. (a) Place the appropriate irreducible-representation symmetry labels next to each MO energy level. Also label each carbon atomic orbital energy level at the left (as being s, π_x, etc.). (b) Why would someone say that CH_4 is not sp^3 hybridized? What type of experimental evidence could be cited to support this view? (c) How could you argue the opposing view, that methane is sp^3 hybridized? (d) Suppose the molecule were now changed to tetrahedral TiH_4. What new orbitals would have to be added to the left side of the energy-level diagram? In what irreducible representations do these orbitals fall? Is there now any possibility of hybridization in the TiH_4 molecule? If so, which atomic orbitals on Ti would be involved?

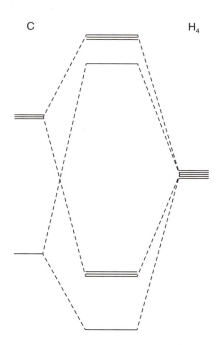

11.47. *The π bonding changes in extent across the series of third-period tetrahedral oxo anions EO_4^{x-} (E = Si, P, and Cl). (a) Taking into account VOPEs and changes in radius, how would you predict the degrees of π bonding to vary? (b) The formal charges (see the Amplification in Section 3.1B) also change across this series, growing larger than the preferred +1 maximum. How can you draw resonance structures to avoid this problem? Do any of the resonance structures have bond orders per E–O link that do not match those found in the tetrahedral MO picture? (c) Carry out computer calculations on these three oxo anions. What additional type of central-atom atomic orbital is employed in the calculations to improve the results? Which central atom (Si, P, or C) is most able to use this central-atom orbital?

S F₆

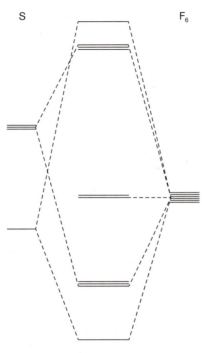

11.48. A possible energy-level diagram for SF₆ that uses no post-valence *d* orbitals is shown at left. The six orbitals shown on the right are the six fluorine tangential 2*p* orbitals (VOPE = –18.6 eV). (a) Label the sulfur valence atomic orbitals on the left with their identities (as being s, p_x, etc.), with their irreducible-representation symmetry labels, and with their VOPEs. (b) Identify each MO in the center of the diagram with its irreducible-representation symmetry label and its nonbonding or antibonding character (if appropriate). (c) Fill in the valence electrons and label the HOMO with its energy (IE = 15.32 eV) and label the LUMO with its energy (EA = 1.05 eV).

S H₆

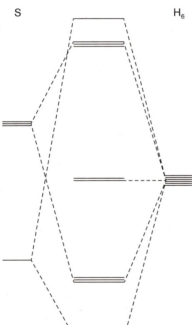

11.49. *Consider the hypothetical octahedral molecule SH₆ (O_h point group); its energy-level diagram is shown at left. (a) Place the appropriate irreducible-representation symmetry labels next to each MO energy level. Also label each sulfur valence atomic orbital energy level on the left (as being *s*, p_x, etc.) (b) Suppose the molecule were now changed to octahedral WH₆. What new valence orbitals would have to be added to the left side of the energy-level diagram? In what irreducible representations do these valence atomic orbitals fall? (c) After adding these new valence AOs, are there now any nonbonding MOs in the WH₆ molecule? If so, give their symmetry label. (d) If the H₆ set of ligands were replaced by a set of six π-acceptor ligands, would there now be any nonbonding MOs?

11.50. Basing your answers on the valence electron configurations derived in Example 11.10, predict how the photoelectron spectra of the following four octahedral tungsten compounds would compare: (a) W(CH₃)₆; (b) WCl₆; (c) W(CO)₆; (d) hypothetical WXe₆ with Xe as a weak-field ligand (Section 7.4). (Omit consideration of spectral peaks arising from electrons in orbitals that are exclusively on the ligands.)

11.51. *Compute the VEC in each of the following four-coordinate organometallic species, and tell which are tetrahedral and which are square planar: (a) Pd(CO)₄; (b) Ni(PPh₃)₄; (c) [Pd(CO)₄²⁺]; (d) [Fe(CO)₄]²⁻; and (e) [Cr(CO)₄]⁴⁻.

11.52. Predict the formulas (including charges) of the following organometallics: (a) the neutral carbonyl of Pd; (b) the least charged carbonyl anion of Cr; (c) the least charged carbonyl anion of Zr; (d) the cationic cobalt complex of MeNC; and (e) a square planar cationic complex of triphenylarsine with iridium.

11.53. *Predict the formulas (including charges) of the following carbonyls: (a) the neutral carbonyl of Os; (b) the least-charged carbonyl anion of Fe; and (c) the least-charged anionic tantalum complex of PF_3.

11.54. Which d-block elements would be expected to form species of the following stoichiometries? (a) $[M(CO)_5]^{2-}$; (b) $[M(CO)_6]^{2-}$; (c) $[M(CO)_4]^{2-}$; (d) $[M(CO)_5]^{3-}$; (e) $[M(CO)_6]^{3-}$; and (f) $[M(CO)_4]^{3-}$.

11.55. *Ellis has reported the synthesis of the diphosphine-substituted carbonyl $[Ti(CO)_5(Me_2PCH_2CH_2PMe_2)]$. (a) Classify the diphosphine ligand into one of the categories discussed in Section 5.1. (b) With this information, what is the likely coordination number of Ti in this complex? Discuss this in view of the maximum coordination number expected in the fourth period and the maximum observed in neutral carbonyls. (c) Does this compound obey the 18-electron rule?

11.56. Recently a transition metal *heptacarbonyl* cation, $[M(CO)_7]^+$, has been found that also obeys the 18-electron rule.[39] (a) In which group of the periodic table should M be located? (b) In fact, the fourth-period element in this group is *unable* to form such a heptacarbonyl cation. Suggest a reason why heptacarbonyl cations were found in Periods 5 and 6, but not in Period 4.

11.57. *Unlike other neutral ligands isoelectronic to CO, CN^- is not high enough in the spectrochemical series to be classified reliably as a π-acceptor ligand (Table 7.6), the complexes of which almost always obey the 18-electron rule. What would the VEC be for the following d-block cyano anions (Section 5.3A)?[40] $[M(CN)_6]^{4-}$ for M = (a) V^{2+} and (b) Mn^{2+}; $[M(CN)_6]^{3-}$ for M = (c) Ti^{3+}, (d) Cr^{3+}, (e) Mn^{3+}, and (f) Co^{3+}; and (g) $[Mn(CN)_6]^{2-}$.

11.58. If you performed Experiment 11B and answered Exercise 11.37, use your answers to suggest why CN^- is not included among the π-acceptor ligands, N_2 is included, and NO^+ is listed at the top of the list (Table 7.6).

11.59. *Consider a ring of six hydrogen atoms, *cyclo*-H_6. Draw the six molecular orbitals of this molecule. Indicate nodes (nodal lines or planes) with dashed lines. Indicate + and – signs of the wave function within the molecular orbitals. Give the number of positive overlaps, the number of negative overlaps, and the net number of positive overlaps for each M.O. Classify each as bonding, antibonding, or nonbonding. Draw the energy-level diagram for *cyclo*-H_6.

11.60. Consider the (hypothetically) planar ring, cyclooctatetraene (*cyclo*-C_8H_8). (a) Draw the eight π molecular orbitals of this molecule (from a top view). Indicate nodes (nodal lines or planes) with dashed lines. Indicate + and – signs of the wave function within the molecular orbitals. Give the number of positive overlaps, the number of negative overlaps, and the net number of positive overlaps for each MO. (b) Draw the energy-level diagram for *cyclo*-C_8H_8 and use the appropriate number of arrows to show electrons occupying the orbitals. How many unpaired electrons would the planar molecule cyclooctatetraene have? How many ultraviolet–visible absorptions would it have? Would this be classified as an aromatic or an antiaromatic molecule? Would it be expected to adapt a distorted geometry?

11.61. *Draw and label the molecular orbitals and the energy-level diagram, and determine the expected charges, for the following cyclic species: (a) *cyclo*-H_5; (b) *cyclo*-C_5H_5 (π electrons only); (c) *cyclo*-H_7; and (d) *cyclo*-C_7H_7 (π electrons only).

11.62. Suppose that cyclobutadiene were to form a "half-sandwich" molecule by coordinating a transition metal ion above its top face (i.e., forming a complex $C_4H_4Mn^+$.) Identify which s, p, and d orbitals of the transition metal would have the appropriate symmetry to bond with each of the four p molecular orbitals of cyclobutadiene.

11.63. *Consider a cyclopentadienide ion ($C_5H_5^-$) ring coordinated to a d-block or an f-block metal ion. Identify which s, p, and d orbitals of the transition metal ion would have the appropriate symmetry to bond with each of the π molecular orbitals of $C_5H_5^-$. (These are drawn in the answer to Exercise 11.61.) Tell in which resulting molecular orbitals the $C_5H_5^-$ ring is acting as a π-donor ligand and in which it can be accepting electrons from the metal ion.

11.64. Below are drawn one student's idea of the top views of the five π-type molecular orbitals to be found in $C_5H_5^-$. (For purposes of answering some of the questions, each molecular orbital has been given a personal "name," such as Harry.) (a) Draw in the vertical nodal planes of these five MOs. In doing this, you should find that one of these MOs is improper and cannot be a true MO of this molecule. Choose the irreducible representation label (a_2'', e_1'', or e_2''*) for each proper orbital. (b) Suppose that a d-block metal ion were to coordinate the $C_5H_5^-$ ring from above, just above the center of the ring. Identify the specific s, p, and d orbitals of the metal ion that can bond to each proper named MO. (c) Draw the expected energy-level diagram for the p orbitals of $C_5H_5^-$, and place the names of the four proper orbitals at the appropriate energy levels.

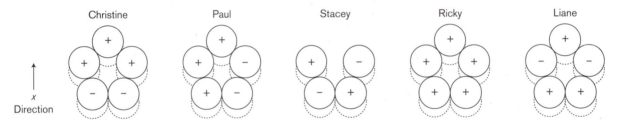

11.65. *Below are drawn one student's idea of the p-type molecular orbitals to be found in H_4. (For purposes of answering some of the questions, each molecular orbital has been given a personal "name.") (a) Draw in the nodal planes of these five MOs. In doing this, you should find that one of these MOs is improper and cannot be a true MO of this molecule. Choose the irreducible representation (e_u^{nb}, a_{1g}, b_{1g}*) of each proper MO. (b) Suppose that a d-block metal ion were to coordinate the H_4 ring from just above the center of the ring. Identify the specific s, p, and d orbitals of the metal ion that can bond to each named proper H_4 MO. (c) Draw the expected energy-level diagram for the orbitals of H_4, and place the names of the four proper orbitals at the appropriate energy levels.

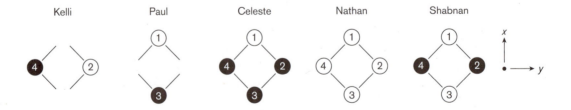

11.66. Compute the metal oxidation numbers and VECs in the following compounds or ions: (a) $CpMn(CO)_3$; (b) $C_6H_6Cr(CO)_3$; (c) $[Cr(C_6H_6)_2]^+$; (d) $[FeCp_2]^+$; (e) $[V(C_6H_6)_2]^-$; (f) $K_2[Ti(C_6H_6)_2]$; and (g) $[NdCp_3(OMe_2)]$.

11.67. *Compute the VEC in each of the following species: (a) NiCp$_2$; (b) Fe(C$_6$H$_6$)$_2$; (c) [NpCp$_3$Cl]; and (d) [CpFe(CO)$_3$]$^+$.

11.68. Predict the formulas of (a) the cyclopentadienyl carbonyl of Co, [CpCo(CO)$_n$]; (b) the most stable sandwich species (Cp rings) for Ir; (c) the most stable sandwich species (benzene rings) for Mn; and (d) a plausible mixed-(cyclic polyene) neutral sandwich species for Mn.

11.69. *In principle, what metals should form the most stable neutral sandwich molecules employing (a) two benzene rings; (b) two cyclopentadienyl rings; (c) one benzene and one cyclopentadienyl ring; (d) two cyclobutadiene rings; (e) one tropylium and one benzene ring; and (f) one cyclopropenium and one cyclopentadienide ring?

11.70. Show calculations to determine what fourth-period d-block metal should form the most stable −1-charged anionic sandwiches employing: (a) one benzene and one cyclopentadienyl ring; (b) one tropylium and one cyclopentadienyl ring; and (c) one cyclobutadiene ring and three CO's.

11.71. *The energy-level diagram of metallocenes in the d-orbital region is similar to the corresponding region of bis(benzene)chromium (Figure 11.28): electrons may occupy a_2'', e_1'', and e_2'' MOs. The spacing between these three orbitals can vary so as to give low-spin and high-spin compounds; in some cases intermediate-spin compounds may also be possible. (a) Write two possible valence d-electron configurations (in terms of the a_2'', e_1'', and e_2'' MOs) for each of the metallocenes listed in the first row of Table 11.7. (b) Use these to explain the numbers of unpaired electrons given in the table for these metallocenes and also MnCp$_2$.

11.72. The pK_a values of the titanocene dichloride (Figure 11.29) diaquation product are pK_{a1} 3.51 and pK_{a2} 4.35. (a) Classify the acidity of this cation. (b) Based on its charge, size, and electronegativity, predict the pK_{a1} value of [Ti(OH$_2$)$_6$]$^{2+}$ and its acidity classification. (c) The pK_{a1} value of the simple hydrated titanium(IV) ion [Ti(OH$_2$)$_6$]$^{4+}$ is −4.0, and that of [Ti(OH$_2$)$_6$]$^{3+}$ is 2.2 (Table 2.1). Comment on the relative pK_{a1} values of these four ions in light of their charges and oxidation numbers, (c) Would you expect the titanium atoms of these anticancer drugs to bind to the same site on DNA as cisplatin? Why or why not?

Harold W. Kroto

Richard E. Smalley

Robert F. Curl, Jr.

HAROLD W. KROTO (1939–2016) was born in Wisbech, Cambridgeshire, United Kingdom, but he grew up in Bolton, Lancashire, UK. Early experiences that developed his interests and skills in science included having a Meccano set (Erector set in the USA) and working in his father's balloon factory. He obtained an undergraduate degree in chemistry in 1961 and a Ph.D. in molecular spectroscopy in 1964, both from the University of Sheffield. Kroto took a job at the University of Sussex in 1967, where his research focused on spectroscopic studies of unstable and semi-stable species, such as carbon multiply bonded to S, Se, and P.

RICHARD E. SMALLEY (1943–2005) was born in Akron, Ohio, but he grew up in Kansas City, Missouri. The successful launching of *Sputnik* in 1957 made him determined to pursue a career in science. He graduated from the University of Michigan in 1965 with a degree in chemistry, but he took a job in the chemical industry instead of going immediately to graduate school. In 1969, though, he began graduate studies at Princeton University, where he used optical and microwave spectroscopy to study molecular single crystals cooled to liquid helium temperatures. Graduating in 1973, he spent three years as a postdoctoral fellow at the University of Chicago, where he collaborated with Donald Levy and Lennard Wharton to develop supersonic beam laser spectroscopy. In 1976, he took a job at Rice University, in part because he wanted to collaborate with Robert F. Curl, Jr.

ROBERT F. CURL, JR. (1933–) was born in Alice, Texas. A lifelong love of chemistry began for him at age nine when his parents gave him a chemistry set. He attended Rice Institute (now Rice University) because it was free at the time and it had a good football team. He earned his B.S. in chemistry in 1954 and then his Ph.D. in physical chemistry in Kenneth Pitzer's lab at the University of California, Berkeley, in 1957. For his doctoral research, Curl used infrared spectroscopy to determine the Si−O−Si bond angle in disiloxane. After a year at Harvard University using microwave spectroscopy to determine the bond rotation barriers of molecules, he returned to Rice as an Assistant Professor.

In 1985, Kroto convinced Curl and Smalley to use a laser beam apparatus built by Smalley to investigate the formation of carbon chains in the atmospheres of red giant stars. In amongst the chains they were looking for were unknown compounds of pure carbon containing 60 and 70 atoms. They named the C_{60} allotrope buckminsterfullerene for its resemblance to Buckminster Fuller's geodesic domes. The three of them were awarded the 1996 Nobel Prize in Chemistry "for their discovery of fullerenes."

CHAPTER 12

The Elements as Molecules and Materials

This chapter can serve as a timely summary of the Foundations of Inorganic Chemistry as we draw together numerous references to previous chapters (especially Chapters 4 and 8) and as we conclude our study of one of the liveliest areas of inorganic chemistry, inorganic materials science. Materials science comes into play when a liquid or powdered chemical is converted into a coating, a film, a fiber, or a functional solid object.[1] The properties of a useful material originate in its chemical bonding, but are also strongly affected by its structure type, secondary bonding, and levels and types of impurities. The interplay of these is complex enough that materials that we expect to be useful for a particular purpose may fail because of secondary effects—but the material may turn out to have uses for other purposes we had not anticipated. In a foundations textbook such as this, we can only go so far in studying advanced materials; more in-depth study (and some understanding of band theory and solid-state physics) will be required for success in this field. Research in materials science involves collaboration with scientists and engineers having many different backgrounds. Those scientists and engineers may depend on you to provide the key chemical insights for their materials, however! This chapter includes work that was rewarded with four recent Nobel Prizes and includes some of the most highly publicized advances that involve inorganic substances.

Here in Chapter 12, we focus on **elemental substances**. (We often add "substances" to make it clear that we are not referring to properties of single atoms of elements.) This chapter introduces another major type of bonding—metallic bonding. The study of the electrical and magnetic properties of materials was first of interest to solid-state physicists, who used their own language of *band theory* to derive and describe their results. We are *not* introducing this language, because it is not generally familiar to chemists in most fields of chemistry, and it is not introduced in courses that are usually prerequisite to a Foundations of Inorganic Chemistry course.[2] Hawkes[3] and Edwards[4] have pointed out that there are some important conductivity properties of the elemental substances that are better explained without band theory; we will go into these.

This textbook has emphasized the usefulness of atomic properties (charge, size, and electronegativity, in particular) in predicting the properties of inorganic substances. How well can we predict the varied properties of elemental materials using these atomic properties? Attempts to answer this question have a venerable history, having first been considered by van Arkel around 75 years ago.[5] We will come to this question in the final section. Seventy-five years later, how well do charge, size, and electronegativity serve with this most challenging and important group of inorganic substances?

12.1. Elemental Substances: Structures and Physical Properties

OVERVIEW. Prepared-for-bonding electron configurations maximize the number of valence atomic orbitals occupied with one electron each. This allows us to determine the number of covalent bonds an atom can form, and therefore allows us to propose different structure types, coordination numbers, and polymerization isomers (allotropes) for molecules of the elements. The structure types can be classified as monomeric, oligomeric, or polymeric, just as we did with oxides in Section 8.1, Table 8.1, and Figure 8.1. You may practice applying these concepts of Section 12.1A by trying Exercises 12.1–12.4. In Section 12.1B we see that σ- and π-bond energies also affect the relative stabilities of allotropes of elements that can employ π bonding. You may practice this concept by trying Exercises 12.5–12.13. As we saw earlier in Section 8.2B for oxides, the structure type strongly influences many physical properties of the allotrope, such as its physical state (solid, liquid, or gas), its relative heats of atomization and vaporization, and its density, boiling point, and hardness. You may practice these concepts of Section 12.1C by trying Exercises 12.14–12.18.

12.1A. Structure Types and Prepared-for-Bonding Electron Configurations of Metallic and Nonmetallic Atoms.

Elemental substances include many examples of metallic bonding, secondary bonding, covalent bonding, and (surprisingly) even an example of ionic bonding. Elemental substances show the whole range of structure types that we introduced in Section 8.1, Table 8.1, and Figure 8.1, which you may want to review. There we summarized the main structure types of solids (including oxides and elemental substances), and identified their *structural units* and how these units are held together—how they are *linked*. We now use this same classification scheme to classify the main structure types of the elemental substances.

We saw in Figure 4.7 that there are many ways of assembling the structural units of SiO_2 into its material forms. These we called the *polymorphs* and *structural isomers* of SiO_2; they had different physical properties. The structural units of elemental substances are atoms. Often the atoms of one element can be assembled in different ways, which are commonly known as **allotropes** of the elements. Molecular structures of major allotropes of sulfur, phosphorus, and carbon are illustrated in Figure 12.1.

When we assemble the structures of nonmetallic elemental substances from their constituent atoms, it is usually best to think of the structural units as atoms, which are linked by *nonpolar covalent bonds* involving two-centered *shared electron pairs*, with

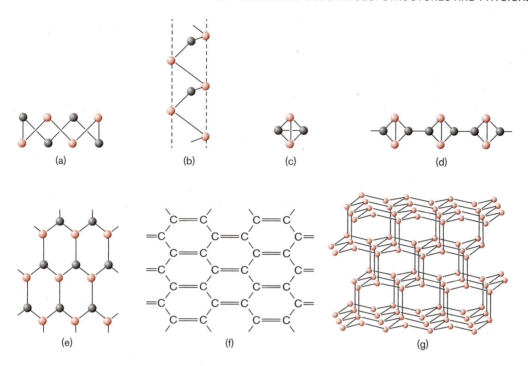

Figure 12.1. The structures of some of the allotropes of sulfur, phosphorus, and carbon. Black circles represent atoms below the plane of the paper, whereas **brown** circles represent those above the plane of the paper. (a) S_8; (b) the helical structures of $^1_\infty[S]$ and gray $^1_\infty[Se]$; (c) P_4 (white phosphorus); (d) red phosphorus $^1_\infty[P_4]$ (in one possible conformation); (e) black phosphorus $^2_\infty[P]$ (rhombohedral form); (f) graphite $^2_\infty[C]$ (one resonance structure); (g) diamond $^3_\infty[C]$. [Adapted in part from F. A. Cotton, G. Wilkinson, and P. Gaus, *Basic Inorganic Chemistry*, 3rd ed., John Wiley and Sons: New York, 1995.]

each atom contributing one of the two electrons in the pair. Because each *p*-block atom has four valence orbitals (one *s* and three *p*), the maximum number of two-centered bonds can be achieved in Group 14/IV, when each atom has four unpaired valence electrons—one per orbital. This electron configuration is called the **prepared-for-bonding electron configuration** (Table 12.1). In other *p*-block groups, lower numbers of unpaired electrons can be prepared for bonding. This reduces the number of covalent bonds that can be formed, because filled orbitals cannot effectively overlap half-filled or filled orbitals on other atoms. Thus, Group 13/III and 15/V atoms can participate in only three two-centered covalent bonds, Group 2 and 16/VI atoms can form only two, Group 1 and 17/VII atoms can form only one, and Group 18/VIII atoms can form no two-centered covalent bonds. Similarly, *d*-block atoms can share the greatest number of electrons and form the greatest number of covalent links, if they rearrange their valence *s* and *d* electrons as shown at the bottom of Table 12.1.

Some known ways of linking these *p*-block atomic structural units are shown in Figure 12.2. We expect the Group 18/VIII elements to exist only as monoatomic substances (Figure 8.1a). Atoms of Group 17/VII, with one prepared-for-bonding unpaired electron each, are expected to give single-bonded dimeric molecules X_2 (also Figure 8.1a).

The Group 16/VI atoms have two prepared-for-bonding unpaired electrons. Therefore, they can link to form either chain polymers such as $^1_\infty[S]$ (Figures 8.1e and 12.1b)

TABLE 12.1. Horizontal Periodic Trends in Numbers of Unpaired Electrons Available for Bonding

Lewis structure of atom	Na·	·Mg·	·Äl·	·S̈i·	·P̈·	:S̈·	:C̈l·
Valence electron configuration	s^1	s^2	s^2p^1	s^2p^2	s^2p^3	s^2p^4	s^2p^5
Prepared-for-bonding configuration	s^1	s^1p^1	s^1p^2	s^1p^3	s^2p^3	s^2p^4	s^2p^5
Number of unpaired electrons available	1	2	3	4	3	2	1
Lewis structure of atom	Cs·	·Ba·	·Lu·	·Hf·	·Ta·	·W·	:Re:
Valence electron configuration	s^1	s^2	s^2d^1	s^2d^2	s^2d^3	s^2d^4	s^2d^5
Prepared-for-bonding configuration	s^1	s^1d^1	s^1d^2	s^1d^3	s^1d^4	s^1d^5	s^1d^6
Number of unpaired electrons available	1	2	3	4	5	6	5

or cyclic oligomers such as S_8 (Figures 8.1c and 12.1a). The possibility also exists that these two unpaired electrons can be shared between just two atoms to form diatomic double-bonded molecules such as O=O (Figure 8.1a).

The Group 15/V elements, with three unpaired electrons in the prepared-for-bonding electron configuration, also have a variety of possibilities. At one extreme, each atom could serve as a link to three other atoms, producing a layer polymeric structure such as $^2_\infty$[P] (Figures 8.1f and 12.1e). At the other extreme, two atoms could pair these electrons by forming one triple bond, once again producing diatomic molecules such as :N≡N: (Figure 8.1a); intermediate possibilities, such as shown in Figure 12.1c and d, can also be imagined.

The Group 14/IV elements have four unpaired electrons in the prepared-for-bonding electron configuration, but a quadruple bond is not possible in the p block, so diatomic molecules are less likely. Each Group 14/IV atom could link to four other atoms by single bonds, producing a network polymer $^3_\infty$[C] (Figures 8.1g and 12.1g). Each atom could use two unpaired electrons to form a double bond and the other two electrons to link to other atoms to give layer polymers $^2_\infty$[C] (Figures 8.1f and 12.1f) or oligomeric or chain polymeric molecules. Each atom could use three unpaired electrons to form a triple bond and one unpaired electron to link to another atom to give a chain polymer $^1_\infty$[C] (Figure 8.1e).

Figure 12.3 summarizes the degree to which the p-block elements (in their major allotropes) form metallic, polymeric, or monomeric molecules or individual atoms.

The p-block elements do not all have the properties of nonmetallic substances. As indicated in Figure 12.3, all elements of Group 13/III except B, and also Pb, Po, and one allotrope of Sn, are classified as **metals**. In contrast to the total coordination numbers of four found in nonmetallic structures, *metal* atoms link to eight or 12 neighbors in the solid state. The metallic bonding in these elements is described in Section 12.4A, and their characteristic physical properties are described in Section 12.4C.

Between the metallic and the nonmetallic substances lies a group of p-block elemental substances having some characteristics of each, but some important characteristics that contrast with each. Because not all of these elemental substances have

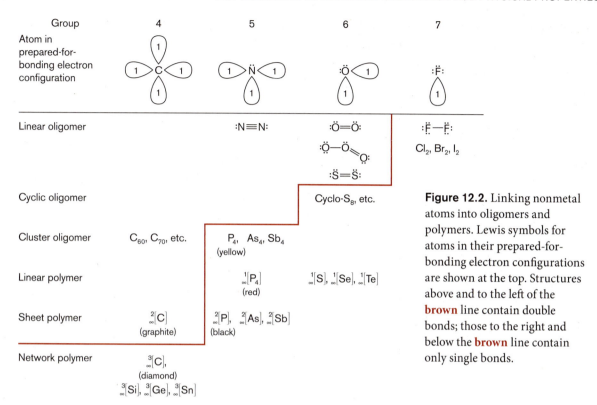

Figure 12.2. Linking nonmetal atoms into oligomers and polymers. Lewis symbols for atoms in their prepared-for-bonding electron configurations are shown at the top. Structures above and to the left of the **brown** line contain double bonds; those to the right and below the **brown** line contain only single bonds.

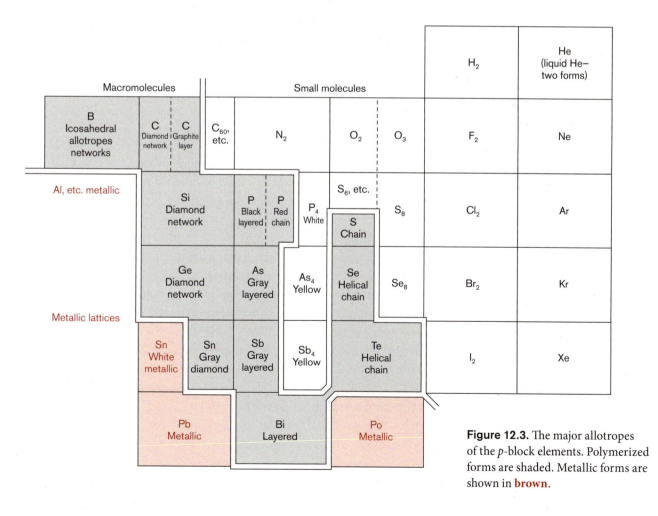

Figure 12.3. The major allotropes of the *p*-block elements. Polymerized forms are shaded. Metallic forms are shown in **brown**.

all of these characteristics, this group, which is often called the **metalloids**, cannot be uniquely defined or identified. Members of the group lie along a diagonal, and include boron from Group 13/III; silicon and germanium from Group 14/IV, arsenic and antimony from Group 15/V, and tellurium from Group 16/VI. Other elements are sometimes included—namely, Bi from Group 15/V, Se from Group 16/VI, and At from Group 17/VII.

EXAMPLE 12.1

Based on the given number of prepared-for-bonding electrons and observed coordination numbers, tell whether the elemental substance is more likely to have a structure that is monomeric, oligomeric, linear polymeric, sheet polymeric, network polymeric, or metallic: (a) a Group 17/VII atom with one prepared-for-bonding electron and a coordination number of one; (b) a Group 16/VI atom with two prepared-for-bonding electrons and a coordination number of one; (c) a Group 16/VI atom with two prepared-for-bonding electrons and a coordination number of two; (d) a Group 16/VI atom with two prepared-for-bonding electrons and a coordination number of eight.

SOLUTION: (a) This atom Xa can and does form one covalent bond, to form Xa_2, a dimeric (oligomeric) molecule. (b) This atom Xb can form two covalent bonds, but with a coordination number of one, it is forming Xb_2, a dimeric (oligomeric) molecule with a double bond. (c) This atom Xc can form two covalent bonds, and with a coordination number of two, these are to different atoms. Therefore, among the choices given the best is linear polymeric $_\infty^1[Xc]$. (d) This atom Xd has a coordination number of eight, well in excess of its number of prepared-for-bonding electrons, so the structure is expected to be metallic.

12.1B. The Role of π-Bond Energies in Allotropy. The choice among possible oligomeric versus polymeric allotropes for a given nonmetal depends in part on which type of covalent bond is stronger—σ or π. Hence, we need also to consider not only the number of inter-unit links, but also the relative strength of each inter-unit link.

Let us consider, as an example, the limiting possibilities for the atoms of Group 15/V: three σ bonds to three different atoms in a polymeric structure, or one σ and two π bonds (a triple bond) to one other atom in a diatomic molecule. We can predict the most stable allotrope at low temperatures (when entropy is a negligible factor) and ambient pressure by comparing the sums of bond energies present in each allotrope. These bond energies were tabulated in Table 3.2.

This example can be generalized: Multiply bonded allotropes (graphite, N_2, O_2, and O_3) are favored for p-block elements in the second period. These allotropes have lower degrees of polymerization than singly bonded allotropes, which are favored for p-block elements in the third period and below.

EXAMPLE 12.2

Compute the enthalpy changes for forming (from gaseous atoms) triple-bonded versus single-bonded allotropes of nitrogen and of phosphorus.

SOLUTION: Each Group 15/V atom contributes three prepared-for-bonding electrons to form three covalent bonds, so each atom releases three-halves times its σ-bond energy upon forming an allotrope involving only σ bonds. The same atom releases one-half of the σ-bond energy plus two-halves times the π-bond energy upon forming a diatomic allotrope having one triple bond. For nitrogen, the formation of three single bonds releases $\frac{3}{2} \times 167$ kJ mol^{-1} = 250 kJ mol^{-1} of nitrogen atoms, while the formation of one triple bond releases $\frac{1}{2} \times 167 + 387$ = 470 kJ mol^{-1} of nitrogen atoms. The latter is the preferable alternative for nitrogen. However, for phosphorus, the single-bonded alternative releases more energy (301 kJ mol^{-1} of atoms) than does the triple-bonded alternative (240 kJ mol^{-1} of atoms); hence, a single-bonded allotrope is preferred.

12.1C. Physical Properties of Elemental Substances. Table 8.1 also listed some characteristic physical properties that vary among solids: vaporization processes, mechanical hardness, electrical conductivity, and solubility. In Section 8.2B and in Table 8.4, we saw that the oxides of the elements show widely varying physical properties. The nonmetals themselves also show an enormous diversity even in their physical states—many are solids at room temperature and atmospheric pressure, but 11 elemental substances or allotropes are gases (the Group 18/VIII gases, H_2, F_2, Cl_2, O_2, O_3, and N_2), and only Br_2 is a liquid. Therefore, these 12 are unlikely to be of interest in materials science, at least at room temperature. The vast majority of metals are solids at room temperature.

The *atomization energies* of elemental substances (Table 6.4) depend in large degree on the number of covalent bond pairs that must be broken to form atoms. This is a function of the number of prepared-for-bonding electrons—with classical two-centered, two-electron bonds, breaking one bond produces a prepared-for-bonding electron on each atom. Atomization energies show an enormous range, from 860 kJ mol^{-1} for W (three bond pairs/atom) and 717 kJ mol^{-1} for C (two bond pairs/atom) to zero for the Group 18/VIII gases.

Two other related physical properties of elemental substances are their **enthalpies of vaporization** and *boiling points*, which also vary widely. The heats of vaporization of the elemental substances vary from 824 kJ mol^{-1} for W to 0.08 kJ mol^{-1} for He. The boiling points at 1-atm pressure of elemental substances also vary widely. At one extreme, helium is a gas (at 1-atm pressure) at any temperature in excess of 4 K. At the other extreme, tungsten metal does not become a gas until a temperature of 5936 K is reached. It is said that tungsten is so nonvolatile at ordinary temperatures that, if the entire universe consisted of tungsten at room temperature, one atom of tungsten would be in the vapor phase.

We can see periodic trends in boiling points and enthalpies of vaporization and atomization of the most important allotropes of the *p*-block elements (Figure 12.4). The enthalpies of vaporization and atomization are approximately equal if the element vaporizes as atoms rather than as diatomic or other molecules, which is the case in Groups 14/IV and 18/VIII. These are not even approximately equal for most nonmetals of Groups 15/V, 16/VI, and 17/VII, which form diatomic or tetratomic molecules in the gas phase. The trend in boiling points then is parallel to the trend in the enthalpies of vaporization, not the enthalpies of atomization. Note that boiling points and enthalpies of vaporization and atomization tend to be highest in Group 14/IV, in which there are the greatest number of prepared-for-bonding electrons and therefore the greatest degree of polymerization, and decrease in groups either to the left or the right.

Down groups of molecular nonmetals (Groups 18/VIII and 17/VII), boiling points and enthalpies of vaporization and atomization tend to *increase* down a group, as van der Waals and secondary bonding (Section 11.3C) forces increase. But in Group 14/IV, when all elemental substances have network polymeric structures, these enthalpies and boiling points *decrease* down the group as covalent bond energies decrease. However, enthalpies of vaporization and atomization and boiling points tend to *increase* at the bottom of the *d* block (i.e., tungsten in Group 6).

The *densities* of the elemental substances (tabulated in Lide, Ed., *CRC Handbook of Chemisty and Physics*, 84th ed., CRC Press: Boca Raton, FL, 2003, pp. 4-137 to 4-140) also show enormous variability and unusual periodicity, ranging from 0.000090 g cm^{-3} for $H_2(g)$ or (if we exclude gases) 0.071 g cm^{-3} for $H_2(l)$ (Figure 12.5) to 22.5 g cm^{-3} for Os(*s*).

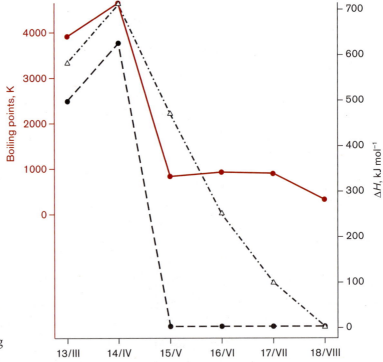

Figure 12.4. Elemental substances of the second period: Graph of boiling points in K ● (use scale at left), enthalpies of atomization at room temperature Δ (use scale at right), and enthalpies of vaporization at the boiling point ● (use scale at right). [Data from M. C. Ball and A. H. Norbury, *Physical Data for Inorganic Chemists*, Longman: London, 1974, pp. 21–30.]

EXAMPLE 12.3

Without referring to Table 7.1, choose the elemental substance from each set that should have the highest atomization energy and the elemental substance that should be easiest to atomize. Referring also to Section 3.3, page 113 to obtain trends in the d block, explain why you made the choice you did: (a) C, Si, Ge, Sn, Pb; (b) V, Nb, Ta; (c) Li, Be, B, C, N, O, F, Ne; and (d) Sr, Zr, Mo, Ru, Pd, Cd, Sn, Te, Xe.

SOLUTION: (a) Highest is C, lowest is Pb. Smaller atoms give the strongest covalent bonds, which must be broken in order to atomize an elemental substance. (b) Highest is Ta, lowest is V. The $5d$ orbitals overlap each other better than $3d$ orbitals do. (c) Highest is C, lowest is Ne. Carbon can participate in four covalent bonds in its prepared-for-bonding electron configuration of $4s^1 4p^3$; Ne can form none because all its valence orbitals are full. (d) Highest is Mo, lowest is Xe [for the same reason as Ne in (c)]. Molybdenum can participate in as many as six covalent bonds in its prepared-for-bonding electron configuration of $5s^1 5d^5$.

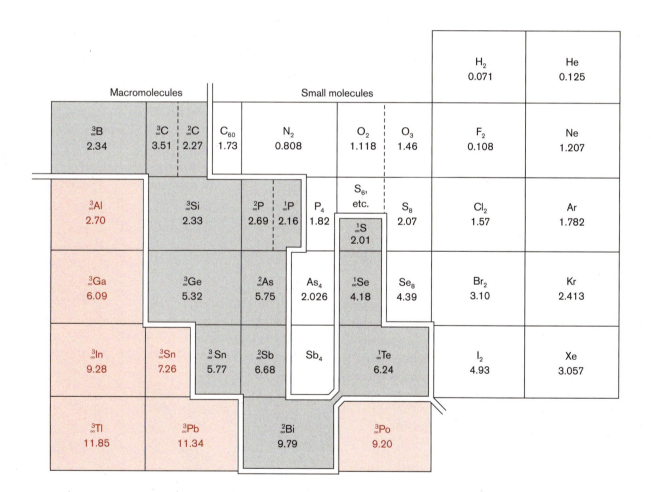

Figure 12.5. The densities (in g/cm³) of liquid and solid p-block elemental substances, including some major allotropes. Data are for ambient pressure and at room temperature or the boiling point (if lower). Polymeric forms are shaded in gray. Metallic forms are shown in brown.

Part of this variation in density is simply due to the fact that the masses of individual atoms increase going down the periodic table. It is more instructive to graph the *atomic densities* of the elemental substances—that is, the number of moles of atoms of each element present in a fixed volume (Figure 12.6). We calculate this by dividing the densities by the atomic weights of the element:

$$\text{Atomic density} = \text{density (in g cm}^{-1}) \times \frac{(1000\ \text{cm}^3/\text{dm}^3)}{(\text{atomic weight (in g/mol)}}\tag{12.1}$$

Atomic densities also reach a maximum near the *middle* of a block of elements, where the bond orders per link are highest.

The physical (Mohs) hardnesses (Sections 8.1B and 8.2C) of the solid nonmetals also strongly depend on the bond order per link and the degree of polymerization. Three-dimensional $_\infty^3[\text{C}]$ (diamond) sets the upper limit for the Mohs scale at 10.0; diamonds will cut almost any other substance. Boron is also very hard (9.0), and silicon has a hardness of 7.0. Other nonmetal compounds of these elements can also be very hard: Carborundum $_\infty^3[\text{SiC}]$ has a hardness of 9.3, and boron carbide $_\infty^3[\text{B}_4\text{C}]$ is of similar hardness. Compounds of these nonmetals with metals having many prepared-for-bonding electrons also give very hard materials, such as the tungsten carbide $_\infty^3[\text{WC}]$ used for the tips of saw blades. Much lower hardnesses are found for the two-dimensional polymers $_\infty^2[\text{C}]$ graphite, 0.5; $_\infty^2[\text{As}]$, 3.5; $_\infty^2[\text{Sb}]$, 3.0–3.3; and $_\infty^2[\text{Bi}]$, 2.5, while the chain polymers are still softer: $_\infty^1[\text{Se}]$, 2.0, and $_\infty^1[\text{Te}]$, 2.3. The molecular P_4 solid is very soft, with a hardness of 0.5. Metals also increase in hardness to the right of the Group 1 metals, which have such low bond orders per link and consequently are so soft that many of them can be cut with a (dull) table knife or spatula.

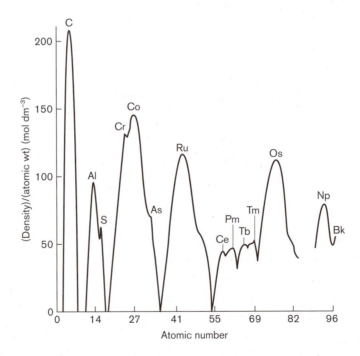

Figure 12.6.
The atomic densities (numbers of moles of atoms of the elements present in 1 L or 1 dm³) of elements under standard conditions of temperature and pressure.

12.2. Molecular Orbital Theory and Conductivity Properties of Linear Oligomers and Polymers of Elements

OVERVIEW. For linear oligomeric and polymeric molecules and ions (H_n, etc.), you can construct qualitative pictures of the MOs, showing nodal planes, give them appropriate symmetry labels, give the number of positive and negative overlaps in each MO, and classify each as being overall bonding, nonbonding, or antibonding. You can draw an energy-level diagram for such a molecule; writing the electron configuration of the molecule, giving its total bond order, bond order per atom–atom link, and its number of unpaired electrons. You can identify the HOMO and LUMO (or SOMO). You may practice these concepts of Section 12.2A by trying Exercises 12.19–12.23. In Section 12.2B, we examine properties of linear chain polymeric molecules and ions related to their classification as insulators, semiconductors, and conductors: electrical conductivity, bandwidths, energy gaps and band gaps, and Peirls distortions. You may apply these concepts by trying Exercises 12.24–12.27.

12.2A. Molecular Orbitals for Linear Oligomers and Polymers of Elements. In this section we outline how MO theory can be extended to linear (one-dimensional) polymeric materials. For simplicity, we will focus on polymers constructed from one type of atom, even though there are enormous numbers of polymers that are heteroatomic. The process is outlined in Figure 12.7.

Ideally linear polymers belong to the point group $D_{\infty h}$. The simplest possible oligomeric and polymeric molecules or ions would be those constructed from atoms using only valence s orbitals. Examples could be constructed from sodium and hydrogen atoms, $(Na_n)^{z\pm}$ and $(H_n)^{z\pm}$.

The MOs for linear molecules are wave functions and show one-dimensional standing wave patterns characteristic of vibrating stringed musical instruments or of particles in one-dimensional boxes (Section 9.1A, Figure 9.2). Such wave functions are sine waves (Eq. 9.2) that can be differentiated by the numbers of nodal planes present that cut across the z direction of the string. The first wave pattern has no nodes, the second has one, and the nth has $(n - 1)$, much as in an atom.

1. Because our $(Na_n)^{z\pm}$ or $(H_n)^{z\pm}$ species has n valence s atomic orbitals, we can construct n molecular orbitals (Section 11.1). Draw strings of n empty circles to represent the chain of atomic ns orbitals. A total of n such chains should be drawn, one for each MO.

2. As for atoms in Section 9.1B, the wave functions are most conveniently generated by first locating the nodal planes of the MOs. No nodal planes should cut the z axis of the first chain; one nodal plane should cut through the center of the second chain, two nodal planes should be located one-third of the way and two-thirds of the way along the second chain, and so on. Now draw in sine waves that start at zero to the left of the

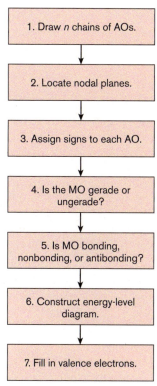

Figure 12.7. Flow chart for constructing MOs of $(Na_n)^{z\pm}$ or $(H_n)^{z\pm}$.

leftmost atom and return to zero every time a nodal plane is reached, as well as to the right of the rightmost atom. The first four waves should look like those illustrated in Figure 9.2.

3. The signs of all atoms located beneath *crests* of the sine wave are now filled in as positive; the signs of all atoms located above *troughs* of the sine wave are now filled in as negative. Any atom that is cut in half by a nodal plane is omitted from the wave function (is neither + nor –), because an *s* orbital cannot be positive on one side and negative on the other.

4. The $D_{\infty h}$ irreducible-representation labels can now be assigned to each MO. Because they are constructed from *s* orbitals, all are necessarily σ orbitals. The first MO is positive everywhere, so it is gerade (Section 10.3B). The next MO has one nodal plane, so it changes sign once and is consequently ungerade. Because each successive MO adds one more nodal plane and sign change, the MOs alternate between gerade and ungerade.

5. The bonding, antibonding, or nonbonding (Section 11.1) nature of each orbital is now assessed. (a) Count the number of positive overlaps between adjacent orbitals. (b) Count the number of negative overlaps between adjacent orbitals. Note that if an atomic orbital is deleted from a particular MO, its neighbors on either side do not overlap each other spatially, and the situation is neither positive nor negative overlap. (c) Compute the net number of positive overlaps, which equals the number of positive overlaps minus the number of negative overlaps. If this number comes out *positive*, the MO is a *bonding* MO. If it comes out *zero*, the MO is a *nonbonding* MO and can be labeled "nb." If this number is *negative*, the MO is *antibonding* and can be labeled with a " *."

6. Construct an energy-level diagram (Section 11.1) for the species (with the levels of the atomic valence *s* orbitals on the left and right sides of the diagram). The relative positions of the different energy levels are approximately those of the net number of positive overlaps, with the most positive (most bonding) located at the bottom. Each successive MO of a homoatomic linear molecule has two fewer positive overlaps, so the MOs will be spaced evenly up the diagram, with all bonding MOs located below the level of the valence atomic orbitals, all antibonding MOs located above it, and any nonbonding MO located at the same level. There will be no degeneracy of orbitals.

7. After determining the number of valence electrons in the species, fill in these valence electrons in the energy-level diagram.

8. If desired, compute the total bond order (Eq. 11.4, Section 11.1) and the bond order per link (Eq. 11.13, Section 11.5A).

9. If desired, electronic transitions from occupied to unoccupied MOs can be sketched in, and their intensity described: intense if it is a gerade → ungerade or ungerade → gerade transition; weak and probably hidden ("forbidden") if it is a gerade → gerade or ungerade → ungerade transition (Section 7.3).

10. If desired, you may compare the shapes of the MOs you have drawn with those obtained from molecular orbital calculations such as those of Experiment 10 (parts of which you may have done in Chapter 11). You will find that the computer calculations add some refinements. (a) Orbitals that are partially cut by the sine wave are diminished in size until the sine wave no longer cuts through them, and those that do not fill the space up to the sine wave are enlarged. This refinement is a consequence of the fact that not all atomic orbitals contribute equally to each MO. (b) With quantitative results, you can calculate the energy of the HOMO–LUMO gap, which will shortly prove important in determining the electrical conductivity of the species.

In Figure 12.8, we draw the three MOs of a triatomic H_3 species. Superimposed on each qualitative MO drawing we show the corresponding standing wave pattern of that MO. Mathematically, we can express the three MOs of the H_3 species as linear combinations of atomic orbitals:

$$\sigma_g^* = 1s(H_A) - 1s(H_B) + 1s(H_C) \tag{12.2}$$

$$\sigma_u^{nb} = 1s(H_A) - 1s(H_C) \tag{12.3}$$

$$\sigma_g = 1s(H_A) + 1s(H_B) + 1s(H_C) \tag{12.4}$$

Figure 12.8 also shows the energy-level diagram of an H_3 species. A species consisting of atoms of moderate electronegativity (such as H_n or hydrocarbons) would normally be expected to fill all of its bonding MOs, which are of low energy, and none of its antibonding MOs, which are of high energy. (Stable species involving highly electronegative atoms such as halogens commonly fill up all orbitals below about -7 to -10 eV in energy.) The nonbonding MOs may or may not be filled because they are of intermediate energy and contribute nothing to the stabilization of the molecule.

We can conceive of three linear H_3 species that might be reasonably stable: H_3^+, with a σ_g^2 electron configuration; H_3, with a $\sigma_g^2(\sigma_u^{nb})^1$ electron configuration, and H_3^-, with a $\sigma_g^2(\sigma_u^{nb})^2$ electron configuration. Each of these populates only one bonding MO,

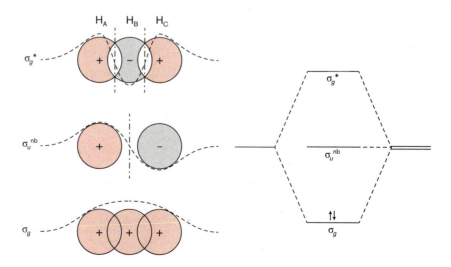

Figure 12.8. Molecular orbitals and energy-level diagram for a homonuclear linear triatomic molecule or ion using s atomic orbitals. The wave pattern of each molecular orbital is sketched as a dashed line (----); the nodal planes for each molecular orbital are indicated by dash-dotted lines ($-\cdot-\cdot-$). Illustrated is the electron configuration of the ion H_3^+.

so according to Equation 11.4, each has a total bond order of one. The bond order per link (Eq. 11.13) of each is 0.5. However, this one bonding MO uses two electrons to bond together three positively charged nuclei. This type of bonding is known as three-center, two-electron bonding (3c–2e), and it was introduced in Section 11.5B for the bridging hydrogen atoms of diborane.

The bonding in H_3 cannot be as strong an interaction as the bonding in H_2, where the bond order per link is 1.0. Indeed, the neutral species H_3 is very unstable and, if ever synthesized, would decompose quite exothermically:

$$2H_3 \rightarrow 3H_2 \tag{12.5}$$

This is because the products of this reaction have a total bond order of three in all molecules, whereas the reactants have a total bond order of two.

Although it is not known in liquids or solids, $H_3^+(g)$ is formed exothermically from $H_2(g)$ and $H^+(g)$ ($\Delta H = -424$ kJ mol^{-1}) and is thought to be important in the chemistry of interstellar clouds, where it has been detected.[6] (However, the H_3^+ species turns out to be cyclic, not linear.)

EXAMPLE 12.4

Draw and label the MOs of a chain of 10 hydrogen atoms, H_{10}. Draw the energy-level diagram of this species, and predict what charge (if any) it would have. Compute its total bond order and bond order per link. Identify the intense electronic transitions in H_{10}.

SOLUTION: The strings of atomic orbitals and nodal planes are shown in Figure 12.9.

Step 1: Orbitals were originally drawn centered on dots 10 mm apart on graph paper.

Step 2: Nodal planes were drawn 10×10 mm apart, $10 \times 10/2$ mm apart, $10 \times 10/3$ mm apart, and so on, then the sine waves were drawn in as shown.

Step 3: The orbitals were labeled with positive signs if they are in a crest and negative signs if they are in a trough; they were omitted if a nodal plane passed through or very close to their center.

Step 4 (shown in Figure 12.9c): The gerade and ungerade nature of each MO can be assigned visually by comparing signs on the left and right halves of the MOs.

Step 5: The numbers of positive and negative overlaps were counted up as shown to give the tabulated net numbers of positive overlaps (Figure 12.9b), which identified the top five orbitals as antibonding and the bottom five as bonding.

Step 6 (Figure 12.9d): Because the net numbers of positive overlaps increase by two for each MO, the energy-level diagram is drawn with equal spacing of energy levels as shown at the right.

Step 7: Filling up the five bonding MOs requires 10 electrons, which is just the number brought in by 10 hydrogen atoms, so the H_{10} species is predicted to be neutral.

Step 8: Applying Equation 11.4, we have 10 electrons in bonding orbitals and none in nonbonding orbitals, so the total bond order is ½(10 – 0) = 5. Because the H_{10} chain has nine H–H links, the bond order per link is 5/9.

Step 9: The intense electronic transitions are from filled orbitals to empty orbitals of the opposite inversion symmetry, so there are a number of these. The lowest-energy one of these is from the fourth to the fifth MO in Figure 12.9d—that is, from the third σ_g (often labeled as $3\sigma_g$) to the first antibonding orbital, the third σ_u (often labeled as $3\sigma_u$).

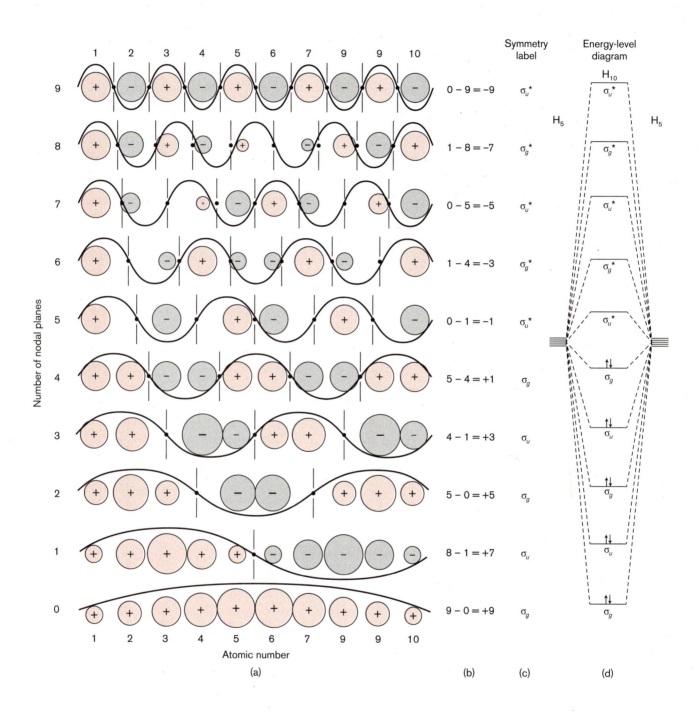

Figure 12.9. (a) Drawings of the 10 MOs of H_{10}. The wave pattern of each molecular orbital is sketched as a solid curve; the nodal planes for each molecular orbital are indicated by vertical dash-dotted lines (——•——). (b) Calculations of net positive overlaps in each MO. (c) Symmetry labels (irreducible representations) for each MO. (d) Energy-level diagram for H_{10}.

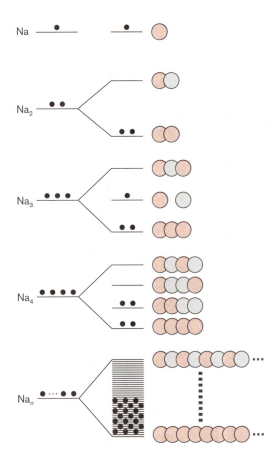

Figure 12.10. Predictions for small molecules $H_1(g)$ or $Na_1(g)$ through $H_4(g)$ or $Na_4(g)$ and on to the chain polymer $^1_\infty[H](g)$ or $^1_\infty[Na](g)$. The left column shows the energy level of the s valence electron of a gaseous Na or H atom; the center column shows the energy levels of the gaseous molecules; and the right column sketches the appearances of the corresponding molecular orbitals. [Adapted from A. B. Ellis, M. J. Geselbracht, B. J. Johnson, G. C. Lisensky, and W. R. Robinson, *Teaching General Chemistry: A Materials Science Companion*, American Chemical Society: Washington, DC, 1993, p. 190.]

Figure 12.10 illustrates the process of Example 12.4 being applied to longer chains, culminating in an $^1_\infty[H]$ or $^1_\infty[Na]$ chain polymer built from an infinite number of atoms and producing the same number of molecular orbitals and energy levels.

12.2B. Conductivity and Related Properties of Linear Polymeric Homoatomic Molecules and Ions.

The H_n, Na_n, $^1_\infty[H]$, and $^1_\infty[Na]$ species are of no practical interest because they are unstable. Linear conjugated π-bonding systems, as in polyalkenes in organic chemistry, or (by using both $2p_x$ and $2p_y$ orbitals on each carbon) **polyynes**, are more stable because they are supported by strong C–C σ-bond chains. Although polyenes are more stable and abundant, let us examine instead the strictly linear polyynes $R-[C\equiv C]_n-R$. The MOs derived in Example 12.4 and shown in Figure 12.9 can be modified to become MOs of linear π-conjugated chains if the empty circles drawn in Step 1 to represent s orbitals are replaced by p-orbital dumbbells drawn above and below the chain z axis. In polyynes there would also be p-orbital dumbbells in front of and behind the chain z axis, producing doubly degenerate π MOs. In Figure 12.11 we show student computations of one each of the first five doubly degenerate π-symmetry MOs of hexatriyne, $H-[C\equiv C]_3-H$.

Band Widths. As we work our way towards a true linear polymer of hydrogen atoms, the drawing of Figure 12.9 seems to suggest that the spread of energy levels increases by the same amount for each lengthening of the chain. However, this is not the case. The procedure of Figure 12.7 counts numbers of overlaps, not how strong each overlap is. The bond order per link decreases as the chain lengthens, so the overlap of neighboring AOs decreases, and the energy separation of neighboring MOs also decreases (Figure 12.12). It turns out that there is a limiting value to the total spread of energy levels—the **band width**—as the chain lengthens.

If we were to apply the process of Example 12.4 to the hypothetical Na_{10} molecule, we would obtain the results similar to those in Figure 12.9. However, as we saw earlier in Example 11.9, the energy gaps between HOMOs and LUMOs, and between other MOs as well, tend to decrease as we go down a group. Thus, an actual MO calculation would show that the overlap between the more diffuse Na $2s$ orbitals is less than the overlap between the H $1s$ orbitals. Counteracting this, there is increased overlap between nonadjacent atoms in the case of a heavier atom such as Na.[7] Thus, the band widths are roughly similar in Na_{10} and H_{10}.

It was originally concluded that strictly linear or planar polymeric crystals were thermodynamically unstable and could not exist.[8] This is due to the process of **Peirls distortion**, which is related to the Jahn–Teller distortion in (complex) molecules

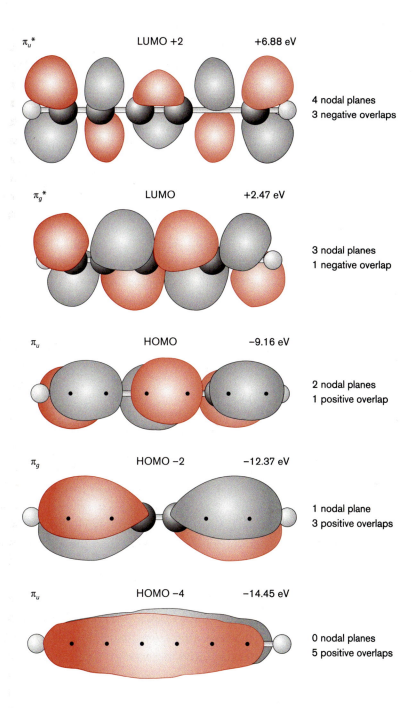

Figure 12.11. Student computations of the five lowest-energy π MOs of hexatriyne, their numbers of nodal planes, their net numbers of positive overlaps, their symmetry labels, their position relative to the HOMO and LUMO, and their energies in eV. Light brown-shaded regions represent positive lobes, whereas gray-shaded regions represent negative lobes. Not shown are the degenerate mates of each MO, at right angles to each MO shown. Also not shown is the highest-energy π* MO, which is above some σ* orbitals in energy.

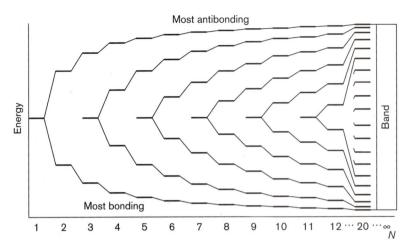

Figure 12.12. The change in MO energy levels of a chain of N atoms as the chain length increases. [Adapted from D. F. Shriver, P. W. Atkins, and C. H. Langford, *Inorganic Chemistry*, 2nd ed., W. H. Freeman: New York, 1994, p. 92.]

discussed in Section 7.6B. A Peirls distortion in a half-filled band of molecular orbitals, such as in H_{10} in Figure 12.12, lowers the total energy by distorting the structure of the polymer (here, doubling the distance required to repeat the structural unit). The H_{10} molecule (with 10 structural H units linked as suggested on the left side of Eq. 12.6) is not known—it decomposes to five H_2 molecules (with structural units H–H). The five H_2 molecules have a lower total energy.

$$H-H-H-H-H-H-H-H-H-H \rightarrow H-H \ \ H-H \ \ H-H \ \ H-H \ \ H-H \qquad (12.6)$$

In practice, one-dimensional homoatomic oligomers (and perhaps polymers) of carbon atoms do exist, anchored by the presence of strong C–C σ bonding. Linear or chain structures in *polyenes* and *polyynes* can be stabilized through the use of extremely bulky R groups at the ends of the chain (to prevent polymerization). If the chain is long enough, the possible carbon allotrope *carbyne* $^1_\infty[-C\equiv C-C\equiv C-]$ might result. It may even have been synthesized—this is very controversial, because even the definition of carbyne is in dispute.[9] In carbyne the Peirls distortion favors the structure with alternating single and triple bonds rather than the alternative $^1_\infty[=C=C=]$ structure with all double bonds.

HOMO–LUMO and Band Gaps. Recently polyynes R–[C≡C–]$_n$–R*, with R* = *tris*-[3,5-di(*t*-butyl)phenyl]methyl and n up to 22, have been synthesized and studied.[10,11] Their UV–visible spectra are shown in Figure 12.13. The lowest-energy electronic absorption λ_{max} (the transition from the HOMO to the LUMO) shifts further into the visible (to lower energy) with increasing chain length, but it does *not* approach zero energy at infinite length. It was deduced that a limiting value of λ_{max} = 485 nm (a HOMO–LUMO energy gap of 2.56 eV) is reached at or beyond n = 48. Such a finite HOMO–LUMO gap or **band gap** at infinite chain length is not what we might have expected by examination of Figures 12.9 or 12.12. Notice in Figure 12.11 that the computed energy difference between the HOMO and the LUMO (the band gap) is *much* greater than the energy differences between other adjacent MOs (HOMO and HOMO-2, etc.).

Electrical conductivity is a property of paramount interest in materials. Conductivity is the reciprocal of the resistivity of a material, which is directly proportional, in turn, to its electrical resistance in ohms (Ω). As a result, the units of conductivity are

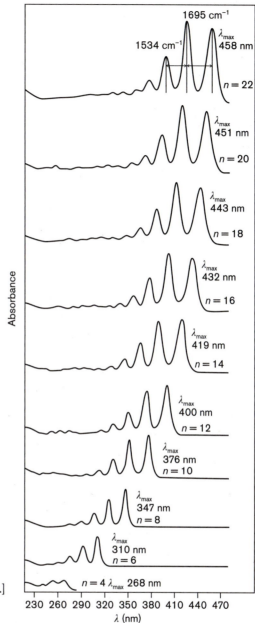

Figure 12.13. UV–visible spectra of polyynes Tr*[–C≡C–]$_n$–Tr* as measured in polyynes. Tr* = *tris*-[3,5-di(*tert*-butyl)phenyl]methyl. [Adapted from W. A. Chalifoux and R. R. Tykwinski, *Nat. Chem.* 2, 967 (2010). Copyright 2010 Macmillan Publishers Limited.]

commonly expressed in reciprocal ohm-centimeters (Ω^{-1} cm^{-1}) or Siemens per centimeter (S cm^{-1}).

Insulators *have large band gaps and zero conductivity at 0 K.* At room temperature they have very low conductivities, from about 10^{-18} Ω^{-1} cm^{-1} for liquid Group 18/VIII elements (or covalent oxides such as SiO$_2$) to about 10^{-7} Ω^{-1} cm^{-1}. Insulators such as diamond and most solid ionic salts have (when pure and at absolute zero) complete occupation of the valence band of bonding MOs and no electron occupation of the *conduction band* of antibonding MOs. Thermal energies are insufficient to promote electrons across the large band gap into the LUMO, where they could conduct an electric current.

Conductors *have zero band gaps and high conductivities at 0 K.* Their conductivity *decreases with increasing temperature,* as atoms in the lattice increasingly vibrate and resist electron flow. At room temperature, their conductivities range from about 7×10^3 Ω^{-1} cm^{-1} for Mn up to 10^6 Ω^{-1} cm^{-1} for Cu. Conductors include not only metallic solids, but also some oxides such as $NaWO_3$, ReO_3 (Figure 4.3), and CrO_2 (Section 8.4), and sulfides (Section 5.6B) such as TaS_2.[4]

In a conducting solid, one band of MOs is only partially filled. With no band gap, the MOs are so close together that the LUMO must lie only infinitesimally higher in energy than the HOMO. Thermal energies and entropic tendencies cause some of the higher-energy filled MOs to be depopulated in favor of the lower-energy vacant MOs. Some of these accessible orbitals bring that electron closer to a positive electrode at one edge of a crystal—this movement of charged particles constitutes conduction of electricity.

Semiconductors have modest band gaps and intermediate conductivities. Their electrical conductivity *increases with increasing temperature* (among metalloids, from 10^{-6} Ω^{-1} cm^{-1} for B or ~10^{-9} Ω^{-1} cm^{-1} for Se to ~10^4 Ω^{-1} cm^{-1} for As). At higher temperatures more electrons can readily be promoted into the empty MOs of the conduction bands of semiconductors, so the material conducts electricity weakly and is an *intrinsic semiconductor*. The conductivity also increases greatly upon irradiation with light, which also promotes electrons to the conduction band.

Superconductors such as $YBa_2Cu_3O_7$ (Section 8.4) have *infinite* conductivity at 0 K and at temperatures below T_c. The metals showing low-temperature superconductivity at ambient pressure are shown in Figure 12.14.

Both metals and metalloids have the property of **metallic luster**, or silvery appearance. This property is also found in many conducting or semiconducting compounds, such as $NaWO_3$, ReO_3, and many sulfides, selenides, and arsenides, such as FeS_2 ("fool's gold"). For the electrons in the higher-energy occupied MOs, there are available (at only a slightly higher energy) alternative, unoccupied MOs. Any frequency of light will suffice to promote an electron from some occupied to some unoccupied orbital. So, metals readily absorb all frequencies of visible light. However, they then re-radiate light at these same frequencies, so they are very reflective (as in silvered mirrors).

Figure 12.14.
Elements exhibiting ferromagnetism (shaded in gray) and low-temperature superconductivity (unshaded) at ambient pressure.

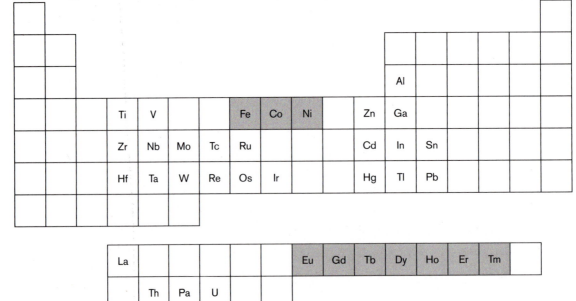

EXAMPLE 12.5

A series of linear chain polycations of mercury are known as salts of large nonbasic fluoro anions.[12] These cations are colorless Hg_2^{2+} (so-called mercurous ion), metallic lustrous golden $Hg_{2.86}^{2+}$, bright yellow Hg_3^{2+}, and black Hg_4^{2+}. (Non-whole-number formulas such as $Hg_{2.86}^{2+}$ result when these complexes have stacks of cations and anions that repeat in the lattice at different chain lengths.) Assume that mercury uses only its $6s$ valence electrons and orbitals in bonding. (a) What are the number of $6s$ valence electrons in each of these ions? (b) What are the valence electron configurations of the ions having integral numbers of mercury atoms? What are their bond orders per link? (c) Based on Figure 12.10, describe the electron configuration for $Hg_{2.86}^{2+}$. (d) Based on (c), explain the metallic lustrous appearance of $Hg_{2.86}^{2+}$. Compare the expected electrical conductivity of each of these ions. (e) Hg–Hg bond lengths (not in order) are 245 pm, 255 pm, and 264 pm; one of the ions has two different Hg–Hg bond lengths of 257 and 270 pm. Based on the bond orders per link and symmetry, assign the bond lengths to the appropriate ions.

SOLUTION: (a) These cations can be considered as being assembled from one $6d^{10}7s^0$ Hg^{2+} ion (with no valence $6s$ electrons) and the remainder $6d^{10}7s^2$ Hg atoms (contributing two valence $6s$ electrons to the bonding). The total numbers of valence electrons in the Hg chains are 2 in Hg_2^{2+}, 3.72 in $Hg_{2.86}^{2+}$, 4 in Hg_3^{2+}, and 6 in Hg_4^{2+}.

(b) Based on Equation 11.13, Figure 12.8, and Figure 12.10, the electron configurations would be σ_g^2 in Hg_2^{2+} with bond order per link = 1; $\sigma_g^2(\sigma_u^{nb})^2$ in Hg_3^{2+} with bond order per link = ½(2/2) = ½; and $\sigma_g^2\sigma_u^2(\sigma_g^*)^2$ in Hg_4^{2+} with bond order per link = ½(4 – 2)/2 = ½.

(c) The electrons of a hypothetical neutral $_\infty^1[Hg_{2.86}]$ would completely fill the band of bonding and antibonding MOs with 5.72 electrons per $Hg_{2.86}$ structural unit, giving a bond order of zero. Giving this polymer a +2 charge per structural unit means that we fill in 3.72 electrons per structural unit—enough to fill in all bonding and some of the antibonding MOs.

(d) Because the band of antibonding MOs (conduction band) is only partially filled, this ion has metallic conductivity and luster. The other ions are not polymeric and do not have conduction bands, so they are insulators.

(e) Hg_2^{2+} has the highest bond order per link, 1.0, and has the shortest Hg–Hg bond length, 245 pm. By symmetry, Hg_4^{2+} has two kinds of Hg–Hg bonds, one involving two bridging Hg atoms, and the other two involving bonds between bridging and terminal atoms, so it has the two distinct bond lengths, 257 and 270 pm. The other two ions have (in the absence of Peirls distortion) all equivalent Hg–Hg bond lengths, so we cannot easily assign the two remaining bond lengths, 255 and 264 pm.

12.3. Allotropes of the Nonmetals

OVERVIEW. Allotropy is common among *p*-block elemental substances, especially sulfur and carbon. Allotropes further down a group are more likely to show enhanced conductivity and reduced HOMO–LUMO band gaps. You should know the structures and likely physical properties of common allotropes of nonmetallic elements at normal pressure (Figure 12.1). You may practice these Section 12.3A concepts by trying Exercises 12.28–12.32. The classically known network allotrope of carbon, diamond, and layer allotrope, graphite, have been added to in recent years by some allotropes with extremely important physical properties: the polyhedral oligomeric fullerenes (including fullerene anion salts and endohedral metallofullerenes); the carbon nanotubes; and graphene (Section 12.3B). You should be able to give the important physical properties for these new allotropes, as suggested in Exercises 12.33–12.43.

12.3A. Allotropes of Nonmetals Other Than Carbon. We first encounter the phenomenon of allotropy in Group 16/VI, as suggested in Figure 12.2 by the diverse structure types listed. The double-bonded diatomic allotrope X_2 is stable only for oxygen, although violet S_2 occurs at high temperatures in the vapor phase. Although these molecules possess double bonds, they are not diamagnetic: Each has two unpaired electrons in a set of $\pi^\star$ orbitals (Section 11.2). Oxygen also has a less stable allotrope ozone (O_3), which is diamagnetic and involves a single and a double bond; sulfur also forms a cherry-red vapor-phase S_3 molecule. The O_2 molecule is now often called *dioxygen* by inorganic chemists to distinguish it from ozone and from the element in general. Atoms of oxygen, O, which are the principal form in the atmosphere at altitudes between 180 and 650 km,[13] are called *atomic oxygen*.

The stable allotropes of the heavier Group 16/VI elements involve two single bonds to each atom (Figure 12.1a, b). Therefore, these allotropes include cyclic oligomers and

---

A CONNECTION TO ENVIRONMENTAL CHEMISTRY

Ozone

The allotrope ozone is produced in low yield by high-energy processes acting upon O_2 (electrical discharges in the laboratory; sunlight in the upper atmosphere). Ozone in the upper atmosphere absorbs high-energy ultraviolet light that would likely cause skin cancer and increase mutation rates at Earth's surface. There is concern that its concentration may be adversely affected by pollutants such as nitrogen oxides, Freons, or nuclear warfare. As indicated in Figures 6.2 and 6.19, ozone is a much more powerful oxidizing agent than O_2. It is an undesirable air pollutant in the lower atmosphere, where it is irritating, attacks materials such as rubber, and reacts with other air pollutants such as unburned hydrocarbons to generate irritating pollutants.

linear polymers. Sulfur is reported to have more allotropes than any other element (except perhaps carbon).[14] The stable allotrope of sulfur is a yellow solid that contains the cyclic S_8 molecule. There are a number of additional cyclic allotropes, such as S_6 through S_{15}, S_{18}, and S_{20}. Above 160°C, liquid S_8 shows some remarkable changes. The S–S bonds in the rings begin to break, giving rise to open chains with unpaired electrons at each end, which join together to give longer and longer one-dimensional polymeric μ-sulfur. The chains may exceed 200,000 sulfur atoms in length at 180°C. Consequently, the color darkens and the liquid becomes quite viscous, because the long chains cannot flow effectively. Above 195°C, entropy effects favor shorter chains, and the liquid decreases in viscosity up to its boiling point of 444°C. If the hot liquid is rapidly cooled, various polymeric allotropes can be frozen out; these allotropes involve helical chains of sulfur atoms (Figure 12.1b). If gaseous sulfur is heated sufficiently, entropy effects prevail and the small molecules S_3 and S_2 are produced.

Selenium also shows several allotropes. Red selenium contains Se_8 rings like those of S_8; *cyclo*-Se_6 and *cyclo*-Se_7 are also known. The stable gray "metallic" form of selenium contains one-dimensional polymeric helical chains. Black (commercial) Se contains various sizes of large rings up to about Se_{1000}. Tellurium has only one allotrope, analogous to gray selenium. Polonium adopts a metallic lattice. Thus, the Group 16/VI elements show almost the whole range of allotropic forms to be found among all the nonmetals.

These elemental substances also span the range of conductivity properties. The band gaps decrease and the room-temperature conductivities increase down the group. The stable allotropes of oxygen and sulfur are insulators. $_\infty^1[Se]$ has a HOMO–LUMO gap of 1.84 eV and a semiconducting conductivity of ~10^{-9} Ω^{-1} cm^{-1}. $_\infty^1[Te]$ has a HOMO–LUMO gap of 0.33 eV and a semiconducting conductivity of 1 Ω^{-1} cm^{-1}. Po has a metallic conductivity of 2.4×10^4 Ω^{-1} cm^{-1}.

In Group 15/V, for nitrogen only the triple-bonded diatomic molecule N_2 is stable. In contrast, the diatomic form X_2 for the heavier Group 15/V elements is found only at high temperatures and low pressures in the vapor phase; allotropes having larger single-bonded molecules are favored under normal conditions.

The simplest phosphorus allotrope is known as white phosphorus, and it consists of tetrahedral P_4 molecules (Figure 12.1c). The bond angles in this tetrahedron are abnormally small (60°) and strained, which in part accounts for the fact that this allotrope is thermodynamically unstable and extremely more reactive than the others (e.g., it ignites spontaneously in air). The two other major allotropes are known as red phosphorus and black phosphorus. Red phosphorus is a linear polymer in which one of the tetrahedral P–P bonds has opened up and is replaced by a P–P bond that is external to the tetrahedron (Figure 12.1d). This relieves strain and imparts greater stability and resistance to oxidation than is found in white phosphorus. As we may expect, it also increases the melting point and lowers its solubility in nonpolar solvents. It also greatly decreases its toxicity: Red phosphorus is essentially nontoxic, whereas white phosphorus is dangerously toxic, even by absorption through the skin. The two main allotropes called black phosphorus contain puckered sheets of phosphorus atoms (Figure 12.1e); consistent with such structures, black phosphorus is flaky, with a graphite-like appearance.

Allotropy is less extensive in As, Sb, and Bi than in P. Tetrahedral As_4 and Sb_4 exist but are not thermodynamically stable at room temperature; the stable modifications of these three elemental substances involve two-dimensional layer structures related to that of black phosphorus (Figure 12.1e). These stable allotropes show conductivities of $3 \times 10^5 \ \Omega^{-1} \ cm^{-1}$, $2 \times 10^5 \ \Omega^{-1} \ cm^{-1}$, and $8 \times 10^4 \ \Omega^{-1} \ cm^{-1}$, respectively.

Group 14/IV Elements Other Than Carbon. At normal pressures, silicon and germanium essentially have only one allotrope each, with the diamond structure. These have lower melting points than diamond (1420°C for Si and 945°C for Ge). Tin, on the other hand, has two allotropes: Its gray form, which has the diamond structure, is stable below 13°C, while its white form, stable above that temperature, has a metallic lattice. The much less dense gray allotrope forms and crumbles on prolonged exposure of very pure tin to temperatures below 13°C; this "tin disease" has been a very troublesome phenomenon in the tin pipes of old organs in European cathedrals. Continuing the type of periodic trend we have seen before (Figure 12.3), lead shows only a metallic form. As we would expect, conductivities increase and band gaps decrease down Group 14/IV: Si, $0.021 \ \Omega^{-1} \ cm^{-1}$ and 1.107 eV; Ge, $0.021 \ \Omega^{-1} \ cm^{-1}$ and 0.67 eV; gray Sn, $9 \times 10^4 \ \Omega^{-1} \ cm^{-1}$ and 0.08 eV; and Pb, $5 \times 10^5 \ \Omega^{-1} \ cm^{-1}$ and 0.0 eV.

Boron. Boron has only three valence electrons and, even in the prepared-for-bonding electron configuration of $2s^1 2p^2$, cannot form ordinary two-centered covalent bonds to each of its nearest neighbors either in many of its compounds such as diborane, B_2H_6 (Section 11.5B), or within the icosahedron and any nearest neighbors outside the icosahedron in the allotropes of elemental boron. Hence, the best treatment of the bonding in elemental boron is the multicentered approach to covalent bonding embodied in the molecular orbital theory.

Elemental boron is extraordinarily difficult to purify. There appear to be four[15] allotropes of pure boron: $^3_\infty[B_{12}]$, $^3_\infty[B_{28}]$ (Figure 12.15), $^3_\infty[B_{106}]$, and $^3_\infty[B_{192}]$. All feature a structural unit very characteristic of boron chemistry: the icosahedron of 12 boron atoms (Figure 10.1h). Within the icosahedron, each boron has five nearest neighbor boron atoms. These icosahedra are linked to each other in different ways via two- or three-centered bonds to form three-dimensional polymers. All allotropes would be expected to have high melting points (e.g., 2300°C), high atomization and covalent bond energies, and to be very hard and quite resistant to chemical attack.

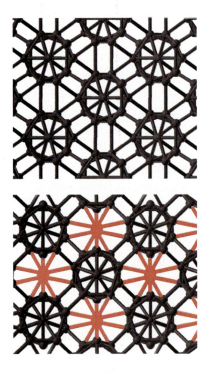

Figure 12.15. Structures of α-B_{12} (top) and γ-B_{28} (bottom). In each case the B_{12} icosahedra are shown in black as two (staggered) pentagons with the top and bottom boron atoms at the center. In γ-B_{28} space between these icosahedra are filled by linear B_2 units (viewed endwise, in **brown**). [Adapted from A. R. Oganov et al., *Nature* 457, 863 (2009).]

AN AMPLIFICATION

An Ionic Allotrope of Boron

The γ-B$_{28}$ allotrope (Figure 12.15b) is calculated to be very unusual for an element in that its internal bonding is significantly *ionic*:[15] The B$_2$ group (B–B distance 173 pm) is *cationic* and the B$_{12}$ icosahedra are *anionic*. In accord with these ionic properties, the γ-B$_{28}$ allotrope is an insulator.

EXAMPLE 12.6

Many allotropes of the nonmetals do not melt and/or boil, but convert to other allotropes on heating (or cooling). Pretending for the moment that this does not happen, classify each allotrope in Figure 12.3 as one of the following: (a) likely to have very high melting and boiling points; (b) likely to be a gas or a liquid at room temperature; or (c) likely to be a solid with a relatively low melting point. Explain, in general, why classes (a), (b), and (c) contain the allotropes they do.

SOLUTION: (a) All layer and network allotropes are likely to have high melting and boiling points—B, diamond, graphite, Si, Ge, black P, gray As, gray Sb, Bi, all Group 14/IV forms). Reason: To melt and boil, these must be converted to small molecules by rupturing covalent bonds. (b) The smallest of the small molecules: He, Ne, Ar, Kr, Xe, H$_2$, F$_2$, Cl$_2$, Br$_2$, O$_2$, O$_3$, N$_2$. Reason: These molecules are held together in the liquid or solid state only by very weak van der Waals forces. (c) The larger of the small molecules: I$_2$, S$_8$, S$_6$ and so on, Se$_8$, P$_4$, As$_4$, Sb$_4$, C$_{60}$, and the chain polymers: S, Se, Te, and red P. Reason: Having more electrons, these molecules are held together in the liquid or solid state by stronger van der Waals forces.

12.3B. Allotropes of Carbon. The stable form of carbon is graphite, in which the atoms participate in both single and double bonds, and are linked into a two-dimensional polymeric sheet structure based on six-membered rings (Figure 12.1f). This structure gives graphite its soft, flaky nature, and makes it useful as a lubricant due to the absence of bonding between the layers. Graphite is almost metallic, with a band gap of about 0 eV and a conductivity that varies with direction: In the plane of the sheet structure it is much higher (about 10^4 Ω^{-1} cm^{-1}) than it is perpendicular to the sheets (about 1 Ω^{-1} cm^{-1}).

Another allotrope, diamond (Figure 12.1g), is a three-dimensional polymer involving only single bonds, and it is only slightly less stable thermodynamically than graphite (ΔH_f = +1.9 kJ mol^{-1}). Because diamond is bonded in all dimensions, it is the denser allotrope (Figure 12.5). Hence, although graphite is the more stable allotrope at normal pressures, under very high pressures the more-dense form, diamond, is favored and formed. These high pressures occur some 100–300 km deep in the mantle of Earth, so diamonds seldom reach the surface (in formations known as kimberlite pipes). Diamond is a classic insulator, with a large band gap of 5.4 eV and a low conductivity of 10^{-14} to 10^{-16} Ω^{-1} cm^{-1}.

Fullerenes, Fullerene Anions, and Endohedral Metallofullerenes. In recent decades the number of allotropes of carbon has expanded dramatically.[16] The first of the new allotropes to be discovered were the polyhedral oligomeric fullerenes (also known as buckminsterfullerenes or "buckyballs") such as C_{60} (Figures 10.1h and 12.16). The discovery[17] of these earned Richard Smalley, Robert Curl, and Harold Kroto the 1996 Nobel Prize in Chemistry. These were first made by the high-energy evaporation of graphite using a laser, and later[18] in an electric arc in a low-pressure atmosphere of helium gas. They are now made commercially from the soot obtained by burning benzene under special conditions.[19] The product of a well-annealed laser-arc reaction mixture contains, most prominently, C_{60} and C_{70} (ΔH_f = +2327 kJ mol^{-1} and +2655 kJ mol^{-1}, respectively). However, the mass spectrum shows peaks for other fullerenes containing even numbers of carbon atoms—up to 350 or more. Fullerenes have been found in deposits that resulted from meteor impacts with Earth, in regions of lightning strikes, and in young planetary nebula.[20]

The physical properties of the fullerenes contrast notably with those of graphite and diamond. In contrast to the insoluble black graphite and the insoluble colorless diamond, the lower-density C_{60} dissolves in toluene to give a purple solution and gives only a single ^{13}C NMR peak. The second most abundant fullerene, C_{70}, is orange-red and has the shape of a rugby ball, with D_{5h} symmetry (Figure 12.16) and five ^{13}C NMR peaks. The next-higher fullerene, C_{76}, is yellow-green, belongs to the chiral D_2 point group, and has 19 ^{13}C NMR peaks. The C_{78} molecule has been isolated in two isomeric forms, a chestnut-brown allotrope of C_{2v} symmetry with 21 ^{13}C NMR peaks, and a golden-yellow isomer of chiral D_3 symmetry with 13 ^{13}C NMR peaks. Two C_{84} molecules, belonging to point groups D_2 and D_{2d}, have been characterized crystallographically,[21] and others have been characterized by their ^{13}C NMR spectra.[22]

Some fullerenes act as semiconducting electron-poor molecules. C_{60} has a HOMO–LUMO gap of about 1.62 eV between the h_u HOMO and the t_{1u} LUMO; C_{70} also has a relatively large HOMO–LUMO gap.[23] Gas-phase C_{60} has a substantial electron affinity of 2.668 eV (257 kJ mol^{-1}) because it forms its –1-charged anion. In the solid state $E°$ for the reduction of C_{60} to C_{60}^- is –0.61 V. In contrast to oxygen atoms or most molecules, fullerenes also accept electrons in the gas phase to form –2-charged anions,[24] the photoelectron spectra of which can be measured.[25] The cation C_{60}^+ has been identified by astronomers in diffuse interstellar space.[26]

Group 1 salts of C_{60} have stoichiometries of MC_{60}, M_2C_{60}, M_3C_{60}, M_4C_{60}, and M_6C_{60}. (In many of these salts, C_{60} anions act as very large spherical anions with 500-pm radii.) Group 1 MC_{60} salts crystallize in two phases: a conducting $^1_\infty[MC_{60}]$ and an insulating dimeric form with single C–C bonds between C_{60} icosahedra.[27] The M_3C_{60} lattice has two tetrahedral and one octahedral site per C_{60}^{3-} anion (Section 8.3A); these are completely occupied in Rb_2CsC_{60}. This salt is superconducting to a relatively high temperature of 33 K. Conduction pathways are provided by the π system of the anions, which are essentially in contact with each other in three dimensions in the solid state. The ultimate reduction product, M_6C_{60}, has too many cations to allow contact between the anions, so it is an insulator.

Fullerenes larger than C_{70} have smaller HOMO–LUMO gaps. For example, C_{74} has a very small HOMO–LUMO gap of 0.05 eV. Such small-gap fullerenes are generally

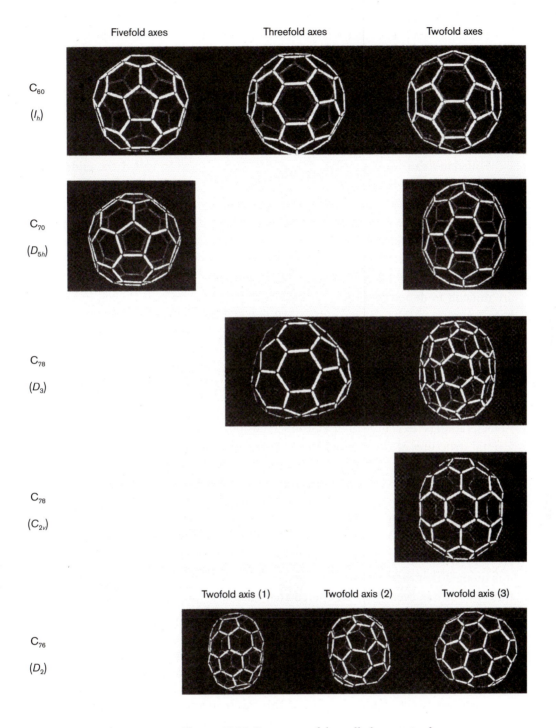

Fivefold axes   Threefold axes   Twofold axes

C$_{60}$
(I_h)

C$_{70}$
(D_{5h})

C$_{78}$
(D_3)

C$_{78}$
(C_{2v})

Twofold axis (1)   Twofold axis (2)   Twofold axis (3)

C$_{76}$
(D_2)

Figure 12.16. Structures of the well-characterized
higher fullerenes, viewed along their fivefold, threefold,
and/or twofold rotation axes as indicated. [Reprinted with
permission from W. O. J. Boo, *J. Chem. Educ.* 69, 605 (1992);
copyright ©1992, Division of Chemical Education, Inc.]

insoluble—with such a small gap, the fullerenes can easily have unpaired electrons that form bonds between adjacent molecules.[23]

Fullerenes such as C_{60} have large internal cavities ($r \approx 350$ pm). Noble gases from helium through xenon can be incorporated by heating these fullerenes in noble gases at high pressures,[28] and are found in meteorites and deposits from the time of the extinction of the asteroids.[29] There is no obvious chemical interaction between the noble gas and the fullerene. Incorporation of H_2 inside C_{60} was achieved by organic synthesis: closure of the C_{60} molecule around a pre-inserted H_2 molecule. The properties of the H_2 and the C_{60} are essentially unchanged.[30]

Fullerenes can form another class of metal compounds, called the **endohedral metallofullerenes**,[31,32] in which the metal cation (or cations) fits inside the large cavity of the fullerene anion. Their chemical formulas (e.g., La@C_{60}) are written using the symbol @ to indicate this unusual intercalation of a metal cation inside a nonmetal anion. These are made by high-energy evaporation of graphite rods that were previously impregnated or packed with metal powders, metal oxides, or metal oxides of the Groups 2–4 and the *f*-block elements. They are produced in up to 1% yield and are difficult to dissolve and purify. The structure of the cationic [Li@C_{60}]($SbCl_6$) has been determined.[33] Its one-peak ^{13}C NMR and UV–visible electronic spectra are very similar to that of C_{60} itself, but differ from the other metallofullerenes, in which the C_{60} behaves as an anion. The Li^+ cation sits off the center of the icosahedron, however, which suggests the presence of a Jahn–Teller distortion (Section 7.6B).

The compounds M@C_{60} are less commonly formed than endohedral metallofullerenes involving larger fullerenes such as C_{80} and C_{82}. M is more commonly a Group 3 or *f*-block cation of +3 charge than a Group 2 or *f*-block +2-charged cation such as Eu^{2+}.[32] Crystal structures of La@C_{82} and Y@C_{82} show that the metal ions do not reside at the center of the fullerene, but reside close to the carbon cage.[34] Some of the larger fullerenes, such as C_{80} and C_{82}, can accommodate two *f*-block +3-charged cations, with the fullerene anion thereby acquiring a –6 charge. Finally, two or three cations and a nonmetal anion can together be intercalated in products such as Sc_3N@C_{80} and Sc_2C_2@C_{84}.

Some very large tetravalent metal atoms (U, Ti, Zr, and Hf) form stable gas-phase endofullerenes M@C_{28} with a carbon cage that is exceptionally small.[35] In these cases, it is thought that there is direct covalent bonding between these metal atoms and the fullerene. It has even been computed that such endofullerenes involve 16 covalent bonds between Th^{4+} or U^{6+} and the tetrahedral $C_{28}{}^{4-}$ carbon cage. Such bonding would involve 16 valence or post-valence *s*, *p*, *d*, and *f* orbitals.[36]

Carbon Nanotubes. Vaporizing a graphite rod that also includes cobalt and nickel promotes the curling of the growing graphite sheets in one dimension only. This process gives rise to high yields of another series of carbon allotropes, the **nanotubes** (Figure 12.17). Depending on the conditions under which they are formed and annealed, the edges of the hexagonal layers may meet each other upon one rolling to give *single-walled nanotubes*[37] or curl over each other to produce *multi-walled nanotubes*[38] two or more layers thick. The diameters of single-walled carbon nanotubes are 0.6–2.4 nm, and their lengths are a thousand or more times their diameters. The ends of the tubes are capped with fullerene-like hemispheres (Figure 12.18). Smaller-diameter single-walled nanotubes of only 0.4 nm diameter have been produced inside the cavities of zeolites (Section 8.7B); these show superconductivity at temperatures up to 15 K.[39]

Armchair or zigzag single-walled nanotubes (Figure 12.17) show near-zero HOMO–LUMO gaps and metallic conductivity, whereas about two-thirds of chiral nanotubes have larger HOMO–LUMO gaps and are semiconducting.[40] Syntheses normally produce mixtures of different nanotubes. Because the properties of different nanotubes can differ so much, separating them is quite important. However, they tend to be of low solubility and are thus quite difficult to disperse in solution and then separate both from residues of the nickel or cobalt, fullerenes, and from each other. Chemical and physical methods of attempted purification have been reviewed.[41] More recently, two methods have been discovered for generating single types of single-walled nanotubes with metallic conductivity and high purity;[42,43] the yields, however, leave much to be desired.

Physically, nanotubes resemble long needles and are of such small diameters that they can pierce cells. Recall that this was also a property of asbestos fibers (Section 8.6A), and carbon nanotubes of length greater than 20,000 nm also have cytotoxic

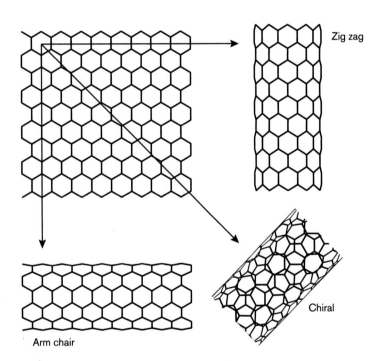

Zig zag

Arm chair

Chiral

Figure 12.17. By rolling a graphite sheet in three directions, different nanotubes can be obtained: zig zag, arm chair, and chiral.

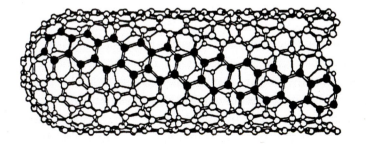

Figure 12.18. Hypothetical structure of a single-layer nanotube with fullerene-like caps.

properties,[44] such as when inhaled.[45] It has been found that agglomerations of nanotubes show more persistent toxicity than dispersed single-walled nanotubes, which can be cleared by macrophages over time.[46]

Graphene. The discrete layers of graphite are held together by intermolecular attractions involving the π electrons. These attractions are weak enough to be broken rather readily, as when a "lead" (actually graphite) pencil is used to write on paper, or when graphite is used as a slippery solid-state lubricant. However, because the π-electron molecular orbitals are very extensive and polarizable, the attractions are by no means negligible. Overcoming these attractions to produce very thin layers of graphite was expected to result in a material with significantly altered properties. Scraping graphite crystals led to thin films having fewer than 100 layers. The method found in 2004[47] for producing films that were only one layer thick was amazingly simple: Stick the graphite flake on plastic adhesive tape, fold the tape over onto the other side, then pull the tape apart—the flakes are then pulled into thinner layers. Repeating this process a number of times eventually produces one-atom-thick layers of **graphene**—the thinnest of all possible materials. For their experiments with plastic tape and graphite, Andre Geim and Konstantin Novoselov won the 2010 Nobel Prize in Physics.[48]

These atom-thick sheets of graphene have amazing properties.[49] Because they are held together by extremely strong C–C and C=C bonds, they do not undergo Peirls distortion, and they are (unexpectedly) stable at room temperature. The crystals are of extraordinarily high quality, being without any defects such as the holes normally found in ionic materials (Section 8.3B). The electrons are not slowed down by collisions with defects and are slowed very little by motions of the carbon atoms, which are held rigidly in place by the strong C–C bonds, even at room temperature. As a result, the electrons in this conducting material move at a faster rate than in any other substance, even metals: at about 1/300 the speed of light. Consequently, the electrons

---

EXAMPLE 12.7

Consider the following four allotropes of carbon: graphite, diamond, C_{60}, and C_{70}. (a) Assign the structures of each of these to the appropriate structural category: monomer or oligomer, one-dimensional polymer, two-dimensional polymer, and three-dimensional polymer. (b) Only two of these allotropes will dissolve in organic solvents. Which two? Why? (c) One of these was suspected of being present in interstellar space, while another is found in the cones of extinct volcanos. Identify the allotropes, and explain why each is found where it is. (d) Another possible allotrope is the hypothetical diatomic molecule C_2. If this existed, would you expect to find it in interstellar space, in volcanic cones, or on Earth's surface? What does MO theory predict for its bond order? How does this explain its instability?

SOLUTION: (a) Graphite is a two-dimensional polymer; diamond is a three-dimensional polymer; C_{60} and C_{70} are oligomers. (b) C_{60} and C_{70}, because no strong forces (chemical bonds) need be broken to separate the polyhedral molecules. (c) Interstellar space: C_{60} (low pressure should favor the smaller molecules, which are volatile); volcano cones: diamond (high pressures favor the most dense, most polymerized form). (d) Interstellar space; bond order = 2, or one per C atom, which is lower than all other allotropes, so less bond energy is released in forming this allotrope.

acquire very unusual relativistic properties (see Section 9.5). The electrons are always accompanied by pairs of electrons and antielectrons (positrons), which are created out of empty space, then rapidly disappear. These particles can, for example, "tunnel" through dauntingly high energy barriers with 100% efficiency.

Production of graphene with plastic tape is not an efficient method of producing more than miniscule quantities of graphene. More efficient methods of separating virtually defect-free graphene sheets from graphite have been found, such as by microwaving a suspension of graphite in an oligomeric ionic liquid.[50] This allows physicists to examine such peculiar quantum effects in a room-temperature material produced cheaply.

Increasing Conductivity via Nonstoichiometric Oxidation or Reduction. The organic polymer *polyacetylene* or all-*trans*- $\frac{1}{\infty}$[–HC=CH–HC=CH–] has a calculated band gap (HOMO–LUMO gap) of about 2 eV, so it should be a semiconductor. The conductivity of polyacetylene can be increased to 10^5 Ω^{-1} cm^{-1} by *doping* it with up to 10% of an electron acceptor such as I_2 or AsF_5, or an electron donor such as Li or NH_3.[51] Electron acceptors serve to carry out a nonstoichiometric oxidation of the polymer by removing some electrons from the band of MOs (the valence band) below the 2-eV gap. As a result, the gap between the new HOMO and LUMO (or SOMOs) is very small, and conductivity increases dramatically. Likewise, electron donors add electrons above the 2-eV gap, into the conduction band. Again, the gap between the new HOMO and LUMO (or SOMOs) is very small, so conductivity increases dramatically.

Graphite intercalation compounds of increased conductivity are formed by intercalating (inserting) strongly reducing or strongly oxidizing elements between the layers of graphite, which are held together only loosely by van der Waals forces. Strongly reducing elemental substances such as Group 1 metals then insert electrons into the band of LUMOs (conduction band), thereby eliminating the HOMO–LUMO gap and improving the conductivity; some of these are superconductors. Group 1 metal atom intercalation gives lamellar compounds such as $\frac{2}{\infty}$[LiC_6], $\frac{2}{\infty}$[NaC_{64}], and $\frac{2}{\infty}$[MC_8] for M = K, Rb, and Cs. These compounds ignite in air and react explosively with water, but the lithium-graphite lamellar compound is still preferred to lithium metal as an electrode in lithium-ion batteries.

Similar results can be obtained by intercalating halogen molecules to partially oxidize the graphite layer to give $\frac{2}{\infty}$[C_8^+] alternating with layers of, for example, Br^-; some products of this type have conductivities exceeding that of copper metal in that plane.

One-dimensional chains with metal–metal bonding can be constructed from colorless, yellow, or green salts such as $K_2[Pt(CN)_4]Cl_{0.3} \cdot 3H_2O$, which have stacked square planar complexes of d^8 metal cations such as Pt^{2+}.[52] Along the z axes of such complexes, overlap can occur between filled valence $5d_{z^2}$ orbitals of the Pt ions. Overlap of filled $5d_{z^2}$ orbitals produces no net bonding, so the Pt–Pt distances are 309–371 pm —much longer than twice the covalent radius of Pt (Table 1.13). Partial oxidation of compounds such as this (e.g., with Br_2 or Cl_2) results in removal of some electrons from the $5d_{z^2}$ orbitals, so there is net Pt–Pt bonding, with a Pt–Pt distance of 287 pm (about twice the covalent radius of Pt) in the bronze-colored $K_2[Pt(CN)_4]Cl_{0.3} \cdot 3H_2O$. These chain polymers show metallic conductivity at room temperature, but become semiconductors at low temperatures.

12.4. Metals and Alloys

OVERVIEW. In Section 12.4A you see that metals commonly occur in three lattice types: the more dense 12-coordinate hexagonal close-packed and cubic close-packed lattices, or the less dense eight-coordinate body-centered cubic lattice. You can also interpret the three indices of electron density in a solid: its flatness, its molecularity, and its degree of charge transfer. You can practice applying these concepts by trying Exercises 12.44 and 12.45. You can recognize the order of structural and conductivity changes that are likely to occur in an elementary substance as pressure increases. From an element's ionization energy and electron affinity, you can estimate whether it should be compressible to become a metallic conductor under high pressure. You may practice these concepts of Section 12.4B by trying Exercises 12.46–12.49. In a horizontal or vertical series of metals, you can find the one that should show the maximum in atomization energies or ferromagnetism. You can give examples of elemental substances that will be most useful (1) as structural metals, (2) for applications requiring malleability or ductility, or (3) as ferromagnets. For an alloy of two metals, tell whether it is likely to be substitutional or an intermetallic compound, and whether the alloy will have a high heat of formation. You may practice these concepts of Section 12.4C by trying Exercises 12.50–12.56. You can explain how and why nanoparticles of metals have different properties than bulk samples of the same metal or than gaseous metal atoms. You can identify giant cluster numbers of metal atoms from their numbers of shells or vice versa. You may practice these concepts of Section 12.4D by trying Exercises 12.57–12.60.

12.4A. Metallic Structures and Bonding. Metal atoms link to eight or 12 neighbors in solid metallic substances. In all cases this exceeds the number of prepared-for-bonding electrons present for the structural units, so there are not enough electrons present for each link to be a classical two-centered two-electron covalent bond. If we apply Equation 11.13, we find that the bond orders per link in metals can be very low indeed—as low as 1/8 for Group 1 solids, and no higher than 6/12 in Group 6 metals (six bonding electrons for 12 links). Therefore, the bonding in metals is often described using multicentered molecular orbitals.

At ambient pressure, most metals have one of three types of metallic lattice. Two of these are the cubic close-packed [ccp; also known as face-centered cubic (fcc)] and hexagonal close-packed (hcp) lattices. (Close packing of anions was illustrated in Figure 8.6.) The third common form of metallic lattice is the body-centered cubic (bcc) lattice, in which each metal atom is eight-coordinate and sits at the center of a cube of nearest-neighbor metal atoms. [The body-centered cubic lattice was previously encountered in the structure of CsCl (Figure 4.2).] The bcc structure is not close packed and is somewhat less dense than the fcc (ccp) structure.[53] It tends to be adopted by the metals at the left of the periodic table, which have the fewest prepared-for-bonding electrons to bond atoms closely, and by other metals at higher temperatures, when greater amplitudes of atomic vibration favor more open structures.

Metals placed under high pressure respond to this stress by adopting a higher density structure, which relieves some of the stress of the high pressure (Le Châtelier's principle). Iron, for example, which is normally bcc, can adopt the close-packed (dense) hcp structure under high pressures, such as those found in the inner core of Earth.

EXAMPLE 12.8

Interpret the following sets of solid-state electron density indices: (a) Flatness low, molecularity high, and charge transfer medium; (b) flatness low, molecularity low, and charge transfer high.

SOLUTION: (a) The low flatness and high molecularity suggest a covalent molecular solid; the medium charge transfer suggests that it is a polar covalent solid. (b) The low flatness and low molecularity suggest a network solid; the high charge transfer suggests that it is an ionic solid.

A CONNECTION TO GEOCHEMISTRY

The Largest Crystal?

It has been proposed that Earth's entire inner core may be one single crystal of hcp iron, which has grown slowly over Earth's lifetime.[54] The hcp structure is of lower symmetry than the cubic bcc and ccp alternatives, so it has a threefold z axis that is distinct from its x and y axes. The hcp structure is thought to explain the fact that earthquake waves pass through Earth 3–4% faster when traveling from one pole to the other than when traveling across the equatorial plane. Because a metal crystal is a macromolecule, this Fe crystal could be the largest molecule on (or in) Earth—the size of Earth's moon!

Modern methods of crystal structure determination can, in favorable cases, locate the total electron density in the crystal; this property can also be computed.[55] The remarkable result for metals is that the electron density *is not confined to the regions between neighboring atoms.* This was illustrated in Figure 1.4b for a four-atom fragment of a metallic structure. As expected, the atom at the upper left of Figure 14.b shows orbital overlap with its nearest neighbors, the atoms at the upper right and the lower left. However, in **metallic bonding** there is also *overlap of orbitals on nonadjacent atoms*—that is, the atom in the lower right. There is significant electron density in the region that is farthest from the four nuclei, indicated in Figure 1.4b by the blackened region. This is an extreme type of delocalization of the electron density, which results in the characteristic physical properties of metals (Section 12.4C). The electrons are located in orbitals with an unusual degree of **flatness** or even density throughout the crystal. Therefore, in freshman chemistry these valence electrons are often described in a simplified manner as being in a "sea of electrons." The Group 1 metals show the greatest degree of flatness, followed by the Group 2 metals, followed by most other metals and alloys.

The same calculations[55] also give an index of **molecularity**, which is a depletion of charge density between discrete covalent molecules (Figure 1.4a). This index is large for discrete covalent molecules (top of Figure 8.1), but is small for covalent network molecules (bottom of Figure 8.1, excepting ionic compounds).

The third index is of *charge transfer,* which measures the degree to which charge density is confined to separate ions and approaches the value of the nominal oxidation state. This is high for ionic compounds, but it is *not* zero for metals, which have some negative charge density in the region between nuclei (as in the black area of Figure 1.1b) and can be thought of as *metal electrides* (H in Section 12.4B).

Atomic Factors That Favor Metallic Bonding

Metallic bonding overlap is favored if the nuclei are close together, if the valence orbitals extend well beyond the filled core orbitals, and if these valence electrons have low enough ionization energies that they are not bound tightly to their original nucleus.[56] The degree to which the valence orbitals extend beyond the core orbitals can be examined by comparing $<r_{max}>$ values, as was done in the Amplification in Section 9.4 entitled "Relative Radii of Core p and Valence d Electrons in an Atom." It may be seen that the extension of the valence electrons beyond the core is greatest among the Group 1 metallic atoms, so that Group 1 metallic substances show the greatest flatness of their electron density and show the most pronounced metallic properties, such as malleability and ductility (Section 12.4C).

TABLE 12.2. Differences in $<r_{max}>$ Between Smallest Valence and Largest Core Orbitals[a,b] of s- and p-Block Atoms

ns-Core	ns-Core	np-Core	np-Core	np-Core	np-Core	np-Core	np-Core
145 (Li)	94 (Be)	73 (B)	55 (C)	44 (N)	37 (O)	32 (F)	28 (Ne)
147 (Na)	108 (Mg)	116 (Al)	91 (Si)	76 (P)	65 (S)	57 (Cl)	51 (Ar)
168 (K)	128 (Ca)	109 (Ga)	90 (Ge)	78 (As)	70 (Se)	63 (Br)	58 (Kr)
169 (Rb)	130 (Sr)	108 (In)	91 (Sn)	80 (Sb)	72 (Te)	66 (I)	62 (Xe)
176 (Rb)	137 (Ba)	100 (Tl)	85 (Pb)	76 (Bi)	69 (Po)	63 (At)	59 (Rn)
158 (Fr)	125 (Ra)	113* (Nh)	96* (Fl)	86*(Mc)	78* (Lv)	72* (Ts)	67* (Og)

[a] Radii differences are in pm, and are computed from $<r_{max}>$ values given by J. P. Desclaux, *Atom. Data Nucl. Data* 12, 311 (1973).

[b] The valence $<r_{max}>$ values used for the atoms identified with an asterisk, *, are for the $7p_{3/2}$ orbitals, rather than for the filled $7p_{1/2}$ orbitals.

12.4B. Effects of High Pressure on Structures, Types of Bonding, and Conductivity. As we previously saw for the example of SiO_2 in Section 4.1C and Figure 4.7, *pressure* also affects the types of structures and allotropes to be found. Under very low pressures (and with high temperatures), metals can vaporize into gaseous small covalent molecules or individual atoms. Figure 12.19 summarizes some examples of structural changes in various materials resulting from applications of very high pressures—up to 380 GPa = 3,800,000 atm = the pressure of Earth's core. Grochala et al[57] suggest 10 types of changes that occur with increasing pressure; we summarize the most significant of these.

1. The first change for molecular substances is that they are compressed to occupy smaller volumes and have higher densities. This does not alter their structure types, however.

2. Higher pressures cause changes in structures and bonding. As pressures increase and molecules are pressed closer together, weak van der Waals interactions may become stronger secondary bonding (Section 11.3C) interactions, then become multicentered bonding (Section 11.5A), as higher coordination numbers and densities result. This often results in increasing the degree of polymerization of covalent materials. An example is the polymerization of CO_2 to superhard $_\infty^3[CO_2]$ at $P > 40$ GPa.[58]

As the density of an elemental substance increases with increasing pressure, the atom's valence electrons become progressively less tightly bonded to the nucleus of the atom. At first we can envision the electrons being transferred from some atoms of the elemental substance to other atoms, endothermically producing ion pairs M^+M^-. The **Mott energy** U for this process is given by the difference of the element's first ionization energy (Table 6.5) and its electron affinity (Table 6.8).[4]

$$U = IE_1 - EA_1 \qquad (12.7)$$

As the pressure and density increase, atomic orbitals overlap to give molecular orbitals, and the process becomes exothermic. A bandwidth develops. When the bandwidths exceed the Mott energy, the HOMO–LUMO gap vanishes and the elemental substance becomes a metallic conductor. This is likely to happen when the Mott energy is less than about 8–10 eV or about 770–965 kJ mol^{-1}. Conversely, the HOMO–LUMO gap is likely to remain when the Mott energy is more than about 9 eV or about 870 kJ mol^{-1}.[4]

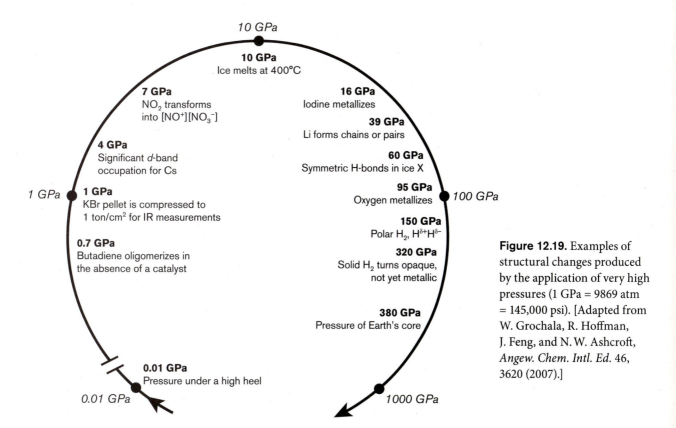

10 GPa

10 GPa
Ice melts at 400°C

7 GPa
NO$_2$ transforms into [NO$^+$][NO$_3^-$]

16 GPa
Iodine metallizes

4 GPa
Significant *d*-band occupation for Cs

39 GPa
Li forms chains or pairs

60 GPa
Symmetric H-bonds in ice X

1 GPa

1 GPa
KBr pellet is compressed to 1 ton/cm^2 for IR measurements

95 GPa
Oxygen metallizes

100 GPa

0.7 GPa
Butadiene oligomerizes in the absence of a catalyst

150 GPa
Polar H$_2$, H$^{\delta+}$H$^{\delta-}$

320 GPa
Solid H$_2$ turns opaque, not yet metallic

380 GPa
Pressure of Earth's core

0.01 GPa
Pressure under a high heel

0.01 GPa

1000 GPa

Figure 12.19. Examples of structural changes produced by the application of very high pressures (1 GPa = 9869 atm = 145,000 psi). [Adapted from W. Grochala, R. Hoffman, J. Feng, and N. W. Ashcroft, *Angew. Chem. Intl. Ed.* **46**, 3620 (2007).]

3. Therefore, "all materials become metallic under sufficiently high pressure,"[57,59] as coordination numbers grow. Among elements substances we can cite the conversion of monatomic Xe to a metallic form[60] with a melting point of about 3300 K, and the polymerization of semiconducting secondary-bonded diatomic I_2 with a 1.3-eV band gap in the plane of the secondary bonding to a ccp metallic structure with a zero band gap at $P > 21$ GPa.[61]

Other changes include the polymerization of N_2 to $_\infty^3[N_2]$ with three-coordinate pyramidal nitrogen atoms at $P > 110$ GPa,[62] and the polymerization of O_2[63] and S_8.[57] The greatest challenge has been to metallize hydrogen—data from satellites flying by the planets Jupiter and Saturn suggest that the cores of these very massive planets may contain liquid metallic hydrogen.[64] In 2011 the laboratory metallization of hydrogen at very high pressure and room temperature was claimed.[65]

4. Pressure may cause the occupation of *post-valence* orbitals (Section 3.3B) that are close in energy to valence orbitals normally used. This can influence the ability of the (metal) atom to form alloys (Section 12.4B).

5. Under extremely high pressure, electrons in metals may move completely away from nuclei into the holes between cations,[66] forming ionic **electrides** such as those formed when cations are dissolved in liquid ammonia or encrypted in "crypt" ligands (Section 6.4A). This destroys the metallic bonding and causes the metals to lose their metallic conductivity and luster, becoming colorless.[67]

EXAMPLE 12.9

Calculate the Mott energy U required to be overcome for the compression of hydrogen to H^+H^-. Does this suggest that the compression of hydrogen to a metallic form is likely to be achievable under pressure?

SOLUTION: Applying Equation 12.7 to data for hydrogen from Tables 6.6 and 6.9, we find that $U = 1312$ kJ mol^{-1} – 73 kJ mol^{-1} = +1239 kJ mol^{-1}. This HOMO–LUMO gap exceeds the 870 kJ mol^{-1} that is likely to be overcome by increasing the pressure. (However, Eq. 12.7 is an approximation that does not specify how *much* pressure increase is involved.)

Novel Ionic Salts at High Pressure

Redox chemistry is also altered at high pressures. The reaction of Na and Cl_2 under very high pressures gives, as alternatives to NaCl(s), the otherwise-unknown salt $Na^+(Cl_3)^-$, with a trichloride ion related to I_3^-, and Na_3Cl, which has a layer structure of alternating ionic and metallic parts, $_\infty^2[(NaCl)(Na_2)]$.[68] Under very high pressure, Hg and F_2 react to give HgF_4, a square planar d^8 derivative of Hg^{4+}.[69] Computations predict that, under high pressure, Cs may employ *core* 6p electrons to form linear molecular CsF_2, ionic $[CsF_2]^+F^-$, and pentagonal planar CsF_5.[70]

AN AMPLIFICATION

12.4C. Physical Properties of Metals and Alloys. Although metals and some nonmetallic substances such as graphite can be electrical conductors, the *mechanical* properties of metals tend to be quite different than those of nonmetallic substances. Gilman[71] argues that the characteristic physical property of metals is **ductility** (the abilty to be drawn into thin wires). Closely related is **malleability** (the ability to be hammered into thin sheets). When metals are deformed under stress and their atoms change positions, the flatness of their electron density makes it easy for stressed atoms to move to new positions. In contrast, even conducting nonmetallic substances such as graphite are **brittle**, and break readily under stress. The mobile conduction electrons also make metals good **conductors of heat**, because moving electrons can collide with vibrating atoms, and acquire and transfer some of that energy to atoms elsewhere in the crystal.

In contrast with the nonmetals, which have widely varying structures, the metals have very few variations in structure type. Nonetheless, the metals show substantial variety in many physical properties, mainly because of the great differences in the bond order per M–M link (Eq. 11.13), which depend, in turn, on the great variety of numbers of prepared-for-bonding electrons that the metal atoms can furnish (Table 12.1).

Periodic Trends in Atomization Energies of Metals. Because some metals come from the *p* block of elements, but even more come from the *d* and *f* blocks, we need to note whether the same horizontal periodic trend in atomization energies (Table 6.5) found among the *p*-block nonmetals (Section 12.1A) are also found among the metals of the *d* and *f* blocks. In the *d* block, the maximum atomization energies seem to correspond to (at most) half-filling of the one valence *s* and five valence *d* orbitals, as in W in Group 6 (Table 12.1).

However, in the fourth period of the *d* block and especially in both periods of the *f* block, the maximum in atomization energies and bonding occurs well before the valence orbitals can be half-filled in the middle of the block. Recall from Section 9.4 that in the fourth period of the *d* block and in the seventh and especially the sixth periods of the *f*-block, the radii of valence *d* and *f* orbitals tend to become less than the radii of the other valence orbitals of these atoms, resulting in poor overlap with like orbitals of neighboring atoms. Thus, the valence *d* and *f* electrons in them are not so readily available for metallic bonding.

The **strengths** of metals are also related to the number of valence electrons available for metallic bonding. Group 1 metals, Group 11 metals (copper, silver, and gold), and some of the subsequent base metals such as thallium and lead have few electrons prepared for bonding. Consequently, they lack structural strength. On the other hand, metals with many available valence electrons (from the center of the *d* block) have much greater strength and are more useful for structural purposes, although they often tend to break upon hammering (they may be brittle). The *d*-block metals in Groups 4 and 5 (except vanadium), with their high atomization energies, have high melting points and structural strength and resist corrosion well, so they are useful for structural purposes. Titanium is also light, so it finds use in jet engines and aircraft and bicycle frames, where light weight and strength are both valuable.

Tensile strength is the ability to resist being stretched and to bear weight, and it can be found both in metallic and in nonmetallic substances. It is quite low in rubber bands, but is quite high in piano wire and spider web strands. Although quantitative data are scarce, we would expect tensile strength to be low for elemental substances with weak, long bonds, such as $\frac{1}{\infty}[Te]$. Because of the strong single and triple bonds in polyynes or carbyne, it has been estimated that these could be the strongest materials, with tensile stiffnesses and strengths well above those of other very strong materials—namely, carbon nanotubes, graphene, diamond, and steel piano wire.[72]

Alloys. We can improve the strength and related properties of pure metals by preparing alloys, because incorporating other metals alters the average number of valence electrons available for bonding. Most of the metals in the middle of the *d* block are used mainly for making alloys of iron (steels), to which they impart various properties of toughness, hardness, and corrosion resistance. Pure iron itself is fairly soft, silvery, and very reactive, so it cannot be used for the numerous structural purposes that steel is used for. The metals used to form stronger alloys with iron include Mo, W, and the fourth-period metals from V through Cu.

Substitutional alloys can be thought of as solid solutions of one metal in another. For this type of substitution to occur, the two atoms should have covalent radii within 15% of each other, the two metals should show the same type of crystal lattice, and the electronegativities of the two elements should be similar. Thus, Na does not dissolve in K, although both have bcc lattices, because Na is 18% smaller than K. The neighbors of Cu (ccp), Ni and Zn, both have similar radii and electronegativities, but only Ni is also ccp, so only Ni alloys with Cu in any proportion; Zn (hcp) forms solid solutions over only a limited range of composition.

Other alloys form solids with their own distinct lattice types, which are often unrelated to those of their parent metals. These alloys have approximately constant compositions and are known as **intermetallic compounds**. Intermetallic compounds between metals from the left side of the *d* block with metals of much higher electronegativity from the right side, such as $ZrPt_3$, may have quite substantial heats of formation (ΔH_f) (up to ~330 kJ mol^{-1}). Miedema's[73] interpretation of these heats of formation of alloys[74] includes: (1) Exothermicity is favored by a large electronegativity difference between the two metals, and (2) exothermicity is hindered by a large difference in charge densities (therefore, of atomic densities, Figure 12.6) of the two metals.

Group 1 metals are large because they use large valence *s* orbitals, so their charge density is quite different from the charge densities of *d*-block metal atoms, and they do not form compounds with most of them. Under very high pressure, K, Rb, and Cs move 0.5 or more electrons per atom into smaller postvalence *d* orbitals, so they can form compounds with nickel and silver such as K_2Ag and K_3Ag.[75]

As another major exception, **amalgams** are easily prepared alloys of metals with the liquid metal mercury. Metals dissolve to widely varying extents in mercury, with the highest solubilities belonging to the metals of Groups 1, 2, Eu, and Yb, and to Au, Zn, Cd, Ga, In, and Tl. Amalgams have been employed for a long time in dentistry (the silver-mercury alloy is soft at first and can easily be inserted into a tooth cavity, where it soon hardens), as reducing agents (sodium amalgam), and in the extraction of gold from ores.[76] Using elemental mercury to extract the gold as a gold/mercury alloy, then boiling off the mercury, gives terrible pollution problems.[77]

EXAMPLE 12.10

Palladium (χ_P = 2.20, density = 12.0 g cm^{-1}) should form an alloy most exothermically with which metal of χ_P about 1.5: (a) Ti (density = 4.5 g cm^{-1}), Mn (density = 7.4 g cm^{-1}), Pa (density = 15.4 g cm^{-1}), or Ta (density = 12.0 g cm^{-1})?

SOLUTION: Alloys should form most exothermically, for two metals of a given χ_P, with the metals with the closest atomic densities. Applying Equation 12.1, we find the following atomic densities, in mol dm^{-3}: Pd, 12,000/106.4 = 112.8; Ti, 4500/47.9 = 93.9; Mn, 7400/54.94 = 134.7; Pa, 15,400/231.0 = 66.7; and Ta, 16,600/180.9 = 91.7. The atomic density of Pd is most nearly matched by that of Ti, although Ta comes close. (Miedema[73] does not mention this particular alloy, but gives a calculated ΔH_f for ZrPd$_3$ of –113 kJ mol^{-1}.)

A CONNECTION TO THE CHEMISTRY OF ART

Memory Metal

A particularly interesting intermetallic compound is Memory Metal[78] or Nitinol (NiTi, ΔH_f = –34 kJ mol^{-1}), which is able to recall its previous shape after it undergoes a phase transition. The sculptor Olivier Deschamps used it to make a statue of a man whose hands point downward in cold weather and upward in warm weather. There is even a beautiful purple alloy, AuAl$_2$,[79] that is exothermically formed (ΔH_f = –49 kJ mol^{-1}).

AN AMPLIFICATION

Quasicrystals

Unusual forms of elemental substances or alloys can sometimes be produced by cooling molten elemental substances at a rate (e.g., 1,000,000°C s^{-1}) that is too fast to allow crystallization, thus giving amorphous materials. Amorphous silicon has become of interest for its use in solar energy conversion. Rapid cooling of a mixture of molten aluminum and manganese produces an alloy having grains in the shapes of dodecahedra—that is, with a type of icosahedral symmetry. Although these materials appear crystalline, it is impossible for the unit cell of the lattice of a solid to have the fivefold symmetry elements found in the I_h point group. These grains cannot be crystals composed of identical unit cells. They are known as **quasicrystals**, and they have unusual properties, such as low density and high strength.[80] Most quasicrystalline materials are alloys of three elements with disordered structure. However, the crystal structure of the quasicrystalline alloy YbCd$_{5.7}$ was determined and established how this impossible symmetry of crystal is actually possible.[81] For his work in establishing that the impossible was possible, Dan Shechtman won the 2011 Nobel Prize in Chemistry.

Ferromagnetic Metals and Alloys. In materials in which paramagnetic atoms (those of the d- and f-block metals) are close together, the magnetic moments can align in cooperation with each other throughout large domains of the solid sample (Figure 8.11). In ferromagnets, all of the individual magnets (metal atoms) in the domain are aligned in the same direction. This interaction must be weak, otherwise electrons on neighboring atoms form metal–metal bonds and pair up. Thus, ferromagnetism is not seen in the fifth-, sixth-, or early fourth-period d-block metals, in which the d orbitals extend outward far enough for covalent overlap to occur (Figure 9.16). It is found in late fourth-period d-block and sixth-period f-block metals (Table 12.2), in which overlap is insufficient for pairing, but sufficient for information exchange.

The magnetic field of Earth is generated by the liquid iron in the core, which rises, cools off, and sinks. Earth's rotation twists the iron stream and generates the magnetic field and Earth's magnetic poles. The process is chaotic, however, and the poles can wander and occasionally weaken and reverse position (the magnetic north pole is now moving towards Siberia).[82] It is thought that Mars originally also had a liquid iron core, but that it froze, depriving Mars of the protection from solar radiation, which then destroyed most of the water on the planet (Section 8.6B).

Alloys can also be ferromagnetic: Although Sm in not ferromagnetic, the alloy $SmCo_5$ is strongly ferromagnetic. Currently, the strongest permanent ferromagnets are made from the alloy $Nd_2Fe_{14}B$,[83] which is used in computer disk drives, audio speakers, and many types of motors. Alloys can also be superconducting, although none superconducts to as high a temperature as the oxides discussed in Section 8.4. Superconducting magnets for high-field NMR spectrometers are still made from the alloy Nb_3Sn: Although it is only superconducting to about 18 K and therefore must always be cooled with expensive liquid helium, it maintains its superconductivity in the presence of higher current densities.

Superconductivity in metals and alloys is attributed by physicists to the presence of **Cooper pairs** of electrons that are capable of moving together through a metallic lattice in such a way that, when one electron is deflected from its path by collision with a nucleus (i.e., experiences nonzero electrical resistance), it is attracted back onto the path by the other electron. This phenomenon is limited in metals to temperatures below about 20 K, because at higher temperatures vibrations of nuclei disrupt the Cooper pairs. This means that superconductivity in metals and alloys can only be achieved practically using the very expensive coolant, liquid helium. The alloys NbTi and Nb_3Sn are used to construct superconducting magnets for the NMR spectrometers used in chemistry research and medically in MRI (magnetic resonance imaging).

12.4D. Nanoparticles and Giant Clusters of Metallic Elements. The properties of materials depend not only on the atoms present in them, but also on the size of the particles of the material. You may recall (Section 6.6B) that the standard reduction potential and activity of a metal are drastically altered if it is *atomic*. **Nanoparticles** are larger than individual atoms, but they have enough fewer atoms (tens to thousands) and enough fewer links per atom that their properties are very significantly different than the bulk elemental substances. Metallic nanoparticles have standard reduction potentials between those of bulk metals (Section 6.2B) and metal atoms (Section 6.6B).

For example, the activity of aluminum was enhanced by producing 80-nm nanoparticles of aluminum. In a paste with ice, these reacted vigorously enough to send a rocket aloft.[84]

A CONNECTION TO TOXICOLOGY

Health Effects of Nanoparticles

Nanoparticles of silver are used as disinfectants in many consumer products. Although it had been hypothesized that the silver nanoparticles disinfected by entering cells through pores, another explanation has now been established. Bulk silver is not active enough to be attacked by oxygen, but the surface silver atoms of nanoparticles are active enough to be oxidized to produce silver ions, which are the germ-killing agents.[85]

The author suspects that nanoparticles are behind a current medical problem: the high rate of failure of artificial hips that have metal-on-metal joints.[86] One type is made of chromium-cobalt alloy. Neither metal reacts with neutral water in the activity series of metals, so it was thought that such artificial hip joints would last longer than the older metal-on-plastic joints, which eventually abraded. However, in too many cases patients with metal-on-metal joints show an unexpected buildup of metal ions [e.g., $Cr^{3+}(aq)$ and $Co^{2+}(aq)$ in their blood]. The author hypothesizes that the wear of these metal joints on each other can cause abrasion to produce metal nanoparticles, the activity of which would be enhanced sufficiently to allow them to react with the nearly neutral water in blood. However, the author may be wrong: Recent studies suggest cell-induced corrosion of the metal surface.[87]

Nanoscience has been with us for hundreds or thousands of years, although we did not know it. Medieval artisans who made red stained glass windows produced their color by incorporating gold chloride in molten glass, in which it was reduced to particles of elemental gold of around 25 nm size. Gold particles twice this size are green, and particles four times this size are orange. Similar variation in size (and shape) of silver nanoparticles produces a different range of colors: dark blue for 40-nm spheres, gold for 100-nm spheres, and red for 100-nm prisms. Besides color, many other properties are changed by changing size: electrical conductivity, HOMO–LUMO gap, catalytic ability, toxicity, catalytic activity, and even polarity![88] Indeed, some properties such as toxicity may depend more on size than chemical composition—the nanoscale-diameter fibers of asbestos and carbon nanotubes derive their cytotoxicity from their ability to pierce cell walls, rather than from chemical reactions of their component atoms.

Clusters of Metal Atoms. When it has been possible to determine the structures of pure nanoparticles, they have often turned out to consist of **clusters** of specific numbers of atoms. Clusters of gold atoms have been most extensively studied. Some of these clusters resemble fullerenes and are called **golden fullerenes**. Others resemble carbon nanotubes[89,90]; still others resemble small fragments of a bulk metallic ccp or hcp lattice. These have normally been characterized first by computation, then experimentally in the gas phase.

A small golden fullerene of particular stability is Au_{20},[91] which is calculated to have T_d symmetry with all 20 gold atoms on the surface in face-centered cubic positions. The HOMO–LUMO gap is calculated to be 1.8 eV, larger than the 1.63 eV gap in C_{60}. Its electron affinity is 2.75 eV, which is again slightly higher than that of C_{60} (2.69 eV). To condense them into the solution or solid state, many of the empty coordination sites of the outer atoms must be occupied by adding external ligands. The clusters $Au_{20}(PPh_3)_8$ and $Au_{20}(PPh_3)_4$ have been prepared and characterized in the condensed state.[92]

Earlier endohedral icosahedral clusters $Mo@Au_{12}$ and $W@Au_{12}$ were computed[93] to be stabilized both by the 18-electron rule (Section 11.5D) and by relativistic secondary gold–gold bonding interactions known as **aurophilic bonding**.[94] These involve 20–40 kJ mol^{-1} interactions between filled $5d$ orbitals on one gold atom and empty post-valence $6s$ and $6p$ orbitals on neighboring gold atoms.[95] $Mo@Au_{12}$ and $W@Au_{12}$ were characterized by photoelectron spectroscopy[96] and have large energy gaps (1.48–1.68 eV) between the h_g HOMOs and the h_g LUMOs; the electron affinities of the two clusters are 2.25 and 2.02 eV. Additional properties of $W@Au_{12}$ have been computed.[97] An all-gold endohedral icosahedral cluster $[Au@Au_{12}(Ph_2P-CH_2-CH_2-PPh_2)_6](NO_3)_2$ has also been characterized.

Giant Clusters. How big would a metallic cluster have to become in order to show the properties of metals? Some features of metals and metalloids are not reproduced in the above cluster compounds: They are not close packed and they do not have a band structure to their energy levels. Two non-icosahedral geometries for 12 or (including an endohedral metal atom) 13 cluster atoms are fragments of close-packed metallic structures: The D_{3h} truncated hexagonal bipyramid, which is a fragment of an hcp lattice, and the O_h cuboctahedron, which is a fragment of a ccp lattice (Figure 12.20). The 13-atom $M@M_{12}$ fragments give the central metal atom the characteristic coordination number of 12 found in these close-packed lattices. Note that the new ($l = 1$) layer of metal atoms wrapped around the endohedral ($l = 0$) atom adds $10l^2 + 2$ (e.g., 12) atoms to the cluster.

If we were now to add a second layer of close-packed metal atoms around the 13-atom cluster, this $l = 2$ layer would contain $10l^2 + 2$ (e.g., 42) additional atoms, to give an $M@M_{12}@M_{42}$ cluster containing 55 metal atoms. Such giant clusters can be formed (for M = Rh, Ru, Pt, and Au) by the reduction of appropriate metal salts by B_2H_6 in organic solvents in the presence of the ligands L = R_3P and R_3As.[98] The ^{197}Au Mössbauer spectrum of $Au_{55}(PPh_3)_{12}Cl_6$ shows the presence of four types of Au atoms: 12 Au atoms coordinated to PPh_3, six Au atoms coordinated to Cl, and 24 "bare" Au atoms, all on the surface, and 13 inner atoms; the Mössbauer parameters of only the inner atoms are very similar to those of gold metal itself. These 1.4-nm clusters are unusually cytotoxic, evidently because they are the right size to interact with the major grooves of DNA (Figure 12.21).

Figure 12.22 shows the continuous addition of layers to these clusters to form the 147-atom (four-shell) and 309-atom (five-shell) clusters. Five-shell, seven-shell, and even eight-shell Pd clusters are known.[99] A large five-shell cluster, $Pt_{309}L_{36}O_{30}$, shows an NMR Knight shift, which is a property very characteristic of metals; the Mössbauer parameters of the 147 inner atoms are similar to those of bulk metal.[100]

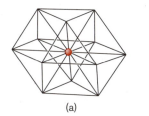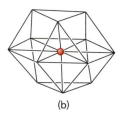

Figure 12.20. Close-packed cluster structures.
(a) The cuboctahedron (O_h point group), which is cubic close packed if an (interstitial) metal atom is included at the center (shown in **brown**). (b) The truncated hexagonal bipyramid (D_{3h}), which under the same conditions is hexagonal close packed. [Adapted from J. W. Lauher, *J. Am. Chem. Soc.* 100, 5305 (1978).]

(a)   (b)

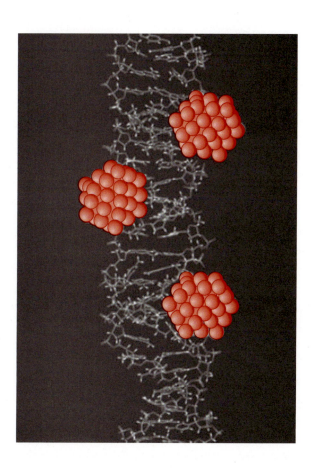

Figure 12.21. Modeling of the interaction of bare Au_{55} clusters with the major grooves of B-DNA. [Reprinted with permission from Y. Liu, W. Meyer-Zaika, S. Franzka, G. Schmid, M. Tsoli, and H. Kuhn, *Angew. Chem. Intl. Ed.* 42, 2853 (2003). Copyright 2003, Wiley-VCH.]

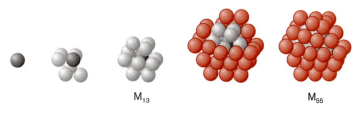

M_{13} M_{55}

Figure 12.22. Building up multilayer full-shell clusters starting with a single atom (black) followed by 12 more (**gray**), then 42 more (**brown**), then a shell of **light brown** atoms, and finally another shell of black atoms. [Adapted from G. Schmid, *Chem. Soc. Rev.* 37, 1909 (2008).]

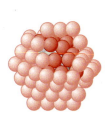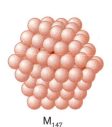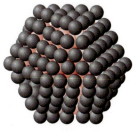

M_{147} M_{309}

12.5. Inorganic Materials That Are Structurally Related to Carbon Allotropes

> **OVERVIEW.** You can distinguish intrinsic and extrinsic semiconductors among materials related to diamond, silicon, and carbon. Using Figure 12.23, you can estimate the color of light emitted by an intrinsic binary semiconductor when employed in an LED (a light-emitting diode). You can describe how to use doping to produce extrinsic semiconductors. Be able to identify semiconducting layer materials that are related to graphite and graphene. You can practice these concepts by trying Exercises 12.61–12.66.

When disulfur dinitride is allowed to sit at 0°C for several days, it polymerizes to give **polythiazyl**, $_\infty^1[SN]$.[101] Polythiazyl is a bronze-colored solid with metallic luster that conducts electricity about as well as mercury. Its structure involves nonlinear chains of alternating sulfur and nitrogen atoms as in μ-sulfur (Figure 12.1b), but with one less electron per nitrogen atom. The best resonance structures are shown in Figure 12.23. (The sulfur atoms are involved in secondary bonding to sulfur atoms in neighboring chains.)

Considering molecular orbitals, the chains of μ-sulfur completely occupy π* MOs. In polythiazyl there are too few electrons to fill the π* MOs, so it has a half-filled HOMO (conduction band). Consistent with this, the conductivity of $_\infty^1[SN]$ is in one direction only.

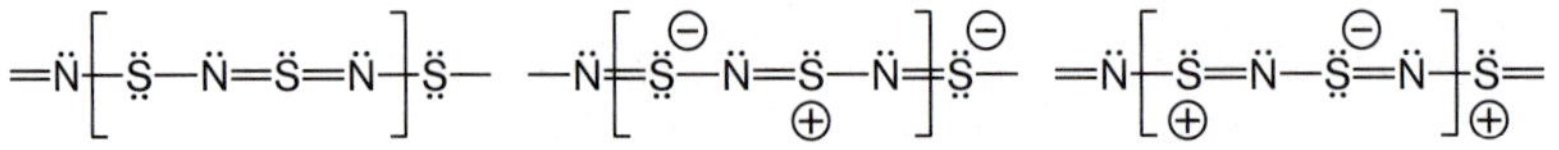

Figure 12.23. Resonance structures for polythiazyl.

12.5A. Materials Related to Diamond. III–V, II–VI, and I–VII Intrinsic Semiconductors.
Intrinsic semiconductors are semiconducting when pure. In Sections 12.3A and 12.3B, we examined the semiconducting properties of the intrinsic semiconductors silicon and germanium. These have band gaps of 1.107 eV and 0.67 eV, respectively. For many applications in materials science, we would like to have the strongly bonded network solid structure of the Group 14/IV elemental substances but with a different selection of band gaps.

We can produce a broader range of band gaps by applying the second principle of isomorphous substitution (Section 8.3C and Table 8.5). III–V compounds substitute one Group 13/III and one Group 15/V atom for two Group 14/IV atoms. II–VI compounds substitute one Group 12 and one Group 16/VI atom for two Group 14/IV atoms, and so on. If both atoms retain tetrahedral coordination, the resulting network covalent compounds are isomorphous and isoelectronic to diamond, silicon, and germanium. Table 12.3 lists the band gaps of these compounds.

An important application of these materials is in the fabrication of light-emitting diodes (LEDs),[102] in which electrical energy is converted to light. These are used in pocket calculators, wristwatches, scientific instrument displays, and the like. The color of light emitted depends on the energy gap.

TABLE 12.3. Band Gaps (eV) in Binary Compounds Related to Diamond, Silicon, and Germanium

Period (average)[a]	IV–IV	III–V	II–VI	I–VI
2	Diamond 5.4	BN 4.6		
2.5	SiC 2.3	AlN 6.02		
3	Si 1.107	AlP 2.45 BAs 1.5 GaN 3.34	ZnO 3.2	
3.5		AlAs 2.16 GaP 2.24 InN 2.0	ZnS 3.54	CuCl 3.17
4	Ge 0.67	AlSb 1.60 GaAs 1.35 InP 1.27	ZnSe 2.58 CdS 2.42	CuBr 2.91
4.5		GaSb 0.67 InAs 0.36	ZnTe 2.26 CdSe 1.74	CuI 2.95 AgBr 2.50
5	Gray Sn 0.8	InSb 0.163	CdTe 1.44 HgSe 2.10	AgI 2.22
5.5			HgTe 0.06	

Source: D. L. Lide, Ed., *CRC Handbook of Chemistry and Physics*, 84th ed., CRC Press: Boca Raton, FL, 2003, pp. 12-101 to 12-102. There is some variation between zinc blende and wurtzite lattice types of the same compound.

[a] Average of the period numbers of the two elements in a binary semiconductor; this gives a rough indication of the average size of the two types of atom.

Not all of the binary materials listed in Example 12.11 have the needed physical properties to be useful in LEDs. GaAs emits IR light, and it saw early use in remote controls for many kinds of consumer electronics devises (such as television sets). GaP emits green light, AlSb emits red light, and AlAs emits orange light (there are some slight discrepancies with the expectations from Example 12.11). The range of LEDs can be increased by creating solid solutions of one binary semiconductor in another, thus giving *ternary* semiconductors such as indium gallium nitride.

Potentially the most useful LED of all would be one that produced white light. One-fourth of the world's electrical energy is used for illumination, and LED bulbs are four times as efficient as fluorescent lights and 20 times as efficient as incandescent bulbs. Furthermore, they last much longer than competing bulbs. In order to produce white LEDs, red, green, and blue LEDs must be combined. As can be seen from Example 12.11, blue LEDs are the most difficult to find and produce. In 1986 a way was found to grow high-quality crystals of gallium nitride and treat them to produce blue light. This opened the way to produce white LED bulbs, an achievement which was rewarded with the 2014 Nobel Prize in Physics for Isamu Akasaki, Hiroshi Amano, and Shuji Nakamura.

EXAMPLE 12.11

Table 7.3 gives colors of light emitted on the left and the corresponding energies in wavenumbers. However, Table 12.3 gives the band-gap energies in electron volts. Using the conversion 1 eV = 8065.54 cm⁻¹, (a) modify Table 7.3 to show the colors in eV. (b) Using your color ranges, match the colors with the binary compounds in Table 12.3. (c) Derive the conversion given.

SOLUTION: (a) The left side of Table 7.3 as re-expressed in eV is as follows:

Color of Light	Corresponding eV
Red	1.77–2.00
Orange	2.00–2.13
Yellow	2.13–2.22
Green	2.22–2.53
Blue	2.53–2.88
Violet	2.88–3.26

(b) The last three columns of Table 12.3 give the following expected colors:

III–V	II–VI	I–VI
BN 4.6 ultraviolet		
AlN 6.02 ultraviolet		
AlP 2.45 green BAs 1.5 infrared GaN 3.34 ultraviolet	ZnO 3.2 violet	
AlAs 2.16 yellow GaP 2.24 green InN 2.0 red-orange	ZnS 3.54 ultraviolet	CuCl 3.17 violet
AlSb 1.60 infrared GaAs 1.35 infrared InP 1.27 infrared	ZnSe 2.58 blue CdS 2.42 green	CuBr 2.91 violet
GaSb 0.67 infrared InAs 0.36 infrared	ZnTe 2.26 green CdSe 1.74 infrared	CuI 2.95 violet AgBr 2.50 green
InSb 0.163 infrared	CdTe 1.44 infrared HgSe 2.10 orange	AgI 2.22 yellow-green
	HgTe 0.06 infrared	

(c) The conversion between eV and kJ mol⁻¹ is the *Faraday*, 1 eV = 96.5 kJ mol⁻¹, and was introduced with Equation 6.5. Equation 7.4 gives the conversion 12.0 kJ mol⁻¹ = 1000 cm⁻¹. Combining these, 1 eV = 96,500/12 = 8040 cm⁻¹. (This calculation has only three significant figures, so within those limits it agrees with 1 eV = 8065.54 cm⁻¹.)

EXAMPLE 12.12

From the several semiconductors listed in Table 12.3, select the one in each column that most closely matches Ge in bond length. Use the band gap of these semiconductors to determine, when structure types and the bond lengths are controlled, how the band gap depends on the electronegativity differences of the two elements.

SOLUTION: Consulting the covalent radii in Table 1.13, we find that the Ge–Ge bond length should be 240 pm. Among the III–V semiconductors listed, the bond lengths should be AlSb, 260 pm; GaAs, 241 pm; and InP, 249 pm, so GaAs is the best match for Ge in size. Among the II–VI semiconductors, the bond lengths should be ZnSe, 242 pm, and CdS, 249 pm. The I–VII semiconductor CuBr should have a 252-pm bond length. For the semiconductors with ~240-pm bond lengths, we find the following electronegativity differences and band gaps: Ge, 0.00 and 0.67 eV; GaAs, 0.37 and 1.35 eV; ZnSe, 0.90 and 2.58 eV; and CuBr, 1.06 and 2.91 eV. We deduce that when structure type and bond length are both controlled, the band gap increases with increasing electronegativity difference (ionic character).

As indicated in Example 12.12, when we control for structure type, increasing the electronegativity difference of the two elements in a binary semiconductor increases the band gap and decreases the conductivity. In accord with these relationships, for a given type of semiconductor (e.g., boron–Group V semiconductors), the bond flatness increases and the charge transfer decreases as the Group V element is varied from N to As.[55]

Doping to Form Extrinsic Semiconductors. Silicon and germanium are extremely important for their semiconducting properties in modern solid-state electronic devices (transistors, etc.). Their conductivities can be improved dramatically by inclusion of (often very tiny levels of) impurities. To begin, Si and Ge are prepared in an extraordinary state of purity (at least 99.9999999% pure) for the electronics industry. Impurities of similar atomic sizes and electronegativities such as Ga or As are then deliberately introduced (doped) at controlled levels (as low as 1 atom in 10^9) into ultrapure Si or Ge. The doped silicon, germanium, and diamond are known as **extrinsic semiconductors**.

Doping a small level of As in a Ge semiconductor introduces one extra electron per As atom. Because the occupied MOs of Ge are filled, these electrons must go up into the formerly unoccupied band of MOs. Now a new HOMO–LUMO band gap of virtually zero is created, so the conductivity of the doped material is vastly improved. Such a semiconductor is known as an **n-type (negative) semiconductor**, because the negative electrons are free to move among many MOs, some of which bring them closer to an electrode. Similarly, impurities of Ga (with three valence electrons) in a Ge semiconductor lead to a vacancy or **hole** in the band of occupied MOs. Upon the input of a relatively small amount of energy, another electron can jump into this positive hole, also increasing the electrical conductivity in this **p-type (positive) semiconductor**. When an n-type and a p-type semiconductor are put in contact, the resulting diode

Doping Diamonds

Because the insulator diamond is so hard and high-melting, it cannot be doped by conventional methods. By depositing diamond films and crystals by decomposing vapors of organic compounds mixed with diborane (B_2H_6), blue diamonds are produced that are *p*-type semiconductors. It is more difficult to produce a useful *n*-type semiconductor: If N is doped in, it is so much more electronegative than C that the excess electrons are localized, and hence are relatively immobile, while P is too much larger than C for effective substitution in the diamond lattice.[103]

will conduct an alternating current better when it is flowing from the *n*-type to the *p*-type semiconductor than in the other direction. Hence, an alternating current (ac) can be converted to direct current (dc).

Materials Related to Graphite, Graphene, Carbon Nanotubes, and Fullerenes. Boron nitride (BN) exists in diamond-like and graphite-like forms. The polarity of the B–N bonds makes these materials less inert than carbon, however: BN is slowly hydrolyzed by water. The reaction of BCl_3 with excess NH_3 at 750°C produces slippery white hexagonal boron nitride, $^2_\infty[BN]$, with a layer structure much like that of graphite. At high temperature and pressure, hexagonal $^2_\infty[BN]$ is converted to cubic $^3_\infty[BN]$ with the diamond structure, a material so hard that it will scratch diamond. The nitrides of Al, Ga, and In are similar to cubic $^3_\infty[BN]$. GaN has been of great interest because it can be used to make blue LEDs.

Thinner layers of $^2_\infty[BN]$ can be produced by the use of adhesive tape, but single layers (analogous to graphene) require more elaborate methods.[104] Boron nitride nanotubes with few walls and large (~5.5-eV) band gaps can now be produced in quantity.[105]

Carbon nitride (C_3N_4) would not appear to resemble boron nitride or graphite. Graphitic carbon nitride has been synthesized by heating dicyandiamide ($C_{2N^{4H}}$) to produce ammonia and the multilayer polymeric carbon nitride (Figure 12.24), a semiconductor with a band gap between 1.6 eV and 2.0 eV.[106]

Metal dichalcogenides such as MoS_2 also adopt layer structures (Figure 5.27) and may show metallic luster (Section 5.6B). Multilayer MoS_2 has long been used as a solid-state lubricant with appearance and properties resembling those of graphite. Crystallites of monolayer MoS_2 and $NbSe_2$ were produced by the adhesive-tape method; however, they are admixed with much larger amounts of multilayer graphite-like materials.[107] A direct solution-phase synthesis of monolayer WSe_2 nanocrystals has been developed; it has a band gap of ~1.2 eV and conductivity up to 92 S cm^{-1}.[108] Several methods of generating single-layer and few-layer MoS_2 sheets have now been found.[109] Multi-walled nanotubes and multi-shell fullerene-like nanoparticles of dichalcogenides such as TiS_2, NbS_2, MoS_2, $MoSe_2$, and WS_2 have been generated by vapor-phase reactions of metal halides or carbonyls with hydrogen chalcogenides (H_2S and H_2Se).[110]

Dimetal trichalcogenides can adopt another type of layer structure and often are semiconductors with low band gaps (e.g., Ga_2Te_3, In_2Te_3, Sb_2Se_3, Bi_2Se_3, and Bi_2Te_3).[111]

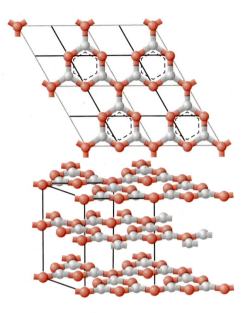

Figure 12.24. Structure of graphitic carbon nitride. C atoms are shown in gray, whereas N atoms are shown in brown. [Adapted from G. Algara-Siller et al., *Angew. Chem. Intl. Ed. 33*, 750 (2014).]

Crystal and film growth of such insoluble (Section 5.6A) chalcogenides has been difficult, but a mixture of the basic chelating ethylenediamine and the acidic chelating 1,2-ethanedithiol dissolves most of these dimetal trichalcogenides quite well.[112]

Magnesium Diboride (MgB₂). This compound had been known for decades and was cheaply available in large quantities, but was not checked for superconducting properties until 2001, when it was found to be superconducting up to the relatively high temperature of 39 K.[113] Its structure resembles that of an intercalated graphite, with the graphite layers being constructed of B⁻ structural units that are isoelectronic to carbon atoms and are connected in a two-dimensional network of σ bonds. Intercalated between the layers of poly(diboride) ions are layers of Mg^{2+} ions, which do not fully separate the diboride layers: The π-bonding MO networks of each diboride layer also overlap the π-bonding MO networks of neighboring layers. The Mg^{2+} ions induce charge transfer from the σ-bond MOs (creating a hole in them) to the π MO networks. Unlike previously known superconductors, MgB₂ has *two* band gaps: one within the σ-bonding MOs and one within the π-bonding MOs.

Although MgB₂ is difficult to dope, its low cost and high superconducting temperature make a derivative of it promising for replacing the alloy Nb₃Sn in coils for NMR (MRI) imaging, which must be able to carry a high current without losing their superconductivity. MgB₂ could be used at higher temperatures than Nb₃Sn and would lose much less conductivity if accidentally warmed above the superconducting temperature, thus more readily avoiding expensive failure ("quenching").[114]

12.6. Summary Overview of Relationships Among Ionic, Covalent, and Metallic Structures and Bonding

OVERVIEW. This section is presented, not as new materials to learn, but as a chance to review concepts on structures and bonding from several chapters in the book. Exercises 12.67–12.71 are suggested as opportunities for you to critique some earlier proposals for classifying substances as ionic, covalent, and metallic, and thus to tie together and review concepts from this chapter and before. (These Summary Exercises are not answered at the end of the chapter.)

One of the earliest suggestions that substances could be divided into three fundamental classes was made by van Arkel,[5] when he drew a triangle representing structure and bonding types, with metallic, covalent, and ionic structures/bonding at the vertices, as exemplified by Cs, F_2, and CsF as extreme examples. More recently, Sproul[115] and Jensen[116] kept the shape of the "van Arkel triangle," but rendered the threefold division quantitative by using both electronegativity differences $\Delta\chi$ and average electronegativities χ_{av} to classify binary compounds (Figure 12.25). In our previous book we discussed and evaluated these triangles from the point of view of what was known in the twentieth century.[117] However, as you may have noticed, much has been learned since then.

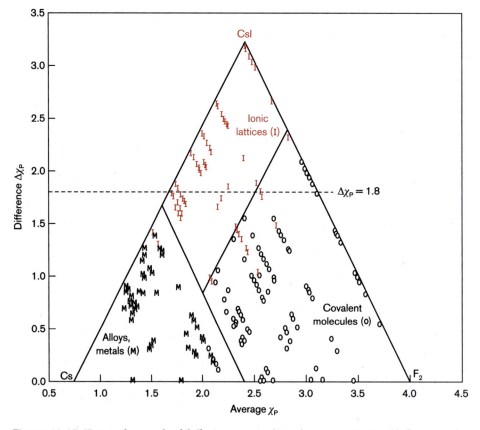

Figure 12.25. Triangular graph of differences in Pauling electronegativities (difference χ_P) versus average Pauling electronegativities (average $\Delta\chi_P$), showing Sproul's best division of the binary compounds involved into three regions. Compounds are symbolized by M if classified structurally as metallic, O if classified as covalent, and **I** if **ionic**. The dashed line represents the traditional bipartite division of ionic and covalent compounds based on a Pauling electronegativity difference of 1.8. [Adapted from G. D. Sproul, *J. Phys. Chem.* **98**, 6699 (1994).]

Since then, Sproul[118] has simplified the axis system by plotting the lower electronegativity versus the higher electronegativity (Figure 12.26). He also suggested the following:

1. Metallic substances (metals and alloys) occur when the more electronegative substance has an electronegativity less than about 2.2 (left half of Figure 12.26). This electronegativity value of 2.2 corresponds in some definitions of electronegativity to a valence electron energy of about 13.5 eV. This low electronegativity or ionization energy allows the valence electrons to show the property of flatness characteristic of metallic bonding (Section 12.4A).

2. When the more electronegative substance has an electronegativity above 2.2 and the less electronegative substance has an electronegativity less than about 1.7, the substance is ionic (bottom right of Figure 12.26). The electronegativity value of 1.7 corresponds in some definitions of electronegativity to a valence electron energy of about 9.5 eV. This electronegativity difference of from 0.5 to 3.5 allows (in most cases) enough charge transfer (Section 12.4) to occur to stabilize an ionic lattice.

3. When the more electronegative substance has an electronegativity above 2.2 and the less electronegative substance has an electronegativity greater than about 1.7, the substance is covalent molecular or covalent network (top right of Figure 12.26). In this case, the electronegativities are all too large to allow either flatness of the electron density or charge transfer (Section 12.4).

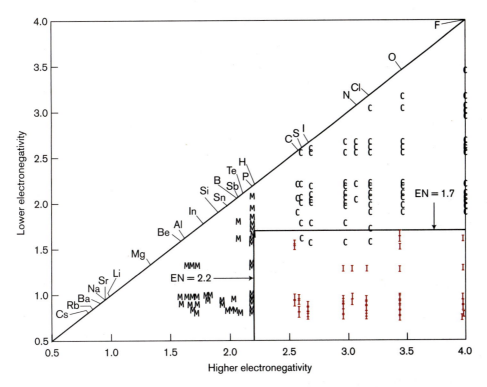

Figure 12.26. Structure type versus electronegativities of elements and binary compounds of *s*- and *p*-block elements. M = metallic alloys (left); **I = ionic compounds** (lower right); C = covalent molecular or network compounds (upper right); elements occur along diagonal and are individually identified. [Adapted from G. Sproul, *J. Chem. Ed.* 78, 387 (2001).]

The homoatomic elemental substances then occur on the diagonal boundary of this right triangle. These are individually labeled along the diagonal of Figure 12.26.

Summary Exercises 12.67–12.71 can be a good review of the interrelationships of concepts in this section with other concepts that are discussed in this chapter and earlier in the book. The materials science of elemental substances and their binary analogues is developing very rapidly, so keep your eyes open to new developments and concepts. This book is by no means the last word on the Foundations of Inorganic Chemistry!

12.7. Background Reading for Chapter 12

The Elements as Materials

P. Calvert, "Advanced Materials," In *The New Chemistry*, N. Hall, Ed., Cambridge University Press: Cambridge, UK, 2000, pp. 352–374; P. P. Edwards, "What, Why, and When is a Metal," ibid., pp. 85–114.

R. Hoffman, *Solids and Surfaces: A Chemist's View of Bonding in Extended Structures*, VCH Publishers: New York, 1988. A translation of the language of the band theory used by solid-state physicists into the language used by chemists.

Allotropes of Boron

A. R. Oganov, J. Chen, C. Gatti, Y. Ma, Y. Ma, C. W. Glass, Z. Liu, T. Yu, O. O. Kurakeyvch, and V. L. Solozhenko, "Ionic High-Pressure Form of Elemental Boron," *Nature* 457, 863–867 (2009).

Allotropes of Carbon

The entire March 1992 issue of *Accounts of Chemical Research* (vol. 25, no. 3) was devoted to reviews of fullerene chemistry.

H. Cong, B. Yu, T. Akasaka, and X. Lu, "Endohedral Metallofullerenes: An Unconventional Core-Shell Coordination Union," *Coord. Chem. Rev.* 257, 2880–2898 (2013).

P. J. F. Harris, *Carbon Nanotube Science: Synthesis, Properties, and Applications*, Cambridge University Press: Cambridge, UK, 2009.

A. K. Geim and K. S. Novoselov, "The Rise of Graphene," *Nat. Mater.* 6, 183–191 (2007); A. K. Geim and P. Kim, "Carbon Wonderland," *Sci. Amer.* 298(6), 90–97 (2008).

Metallic Structure and Bonding

P. Mori-Sànchez, A. M. Pendàs, and V. Luaña, "A Classification of Covalent, Ionic, and Metallic Solids Based on the Electron Density," *J. Am. Chem. Soc.* 124, 14721–14723 (2002).

Effects of High Pressure on Structures, Types of Bonding, and Conductivity

W. Grochala, R. Hoffman, J. Feng, and N. W. Ashcroft, "The Chemical Imagination at Work in *Very* Tight Places," *Angew. Chem. Intl. Ed.* 46, 3620–3642 (2007).

Memory Metal and the Purple Plague: Connections to Art

A. B. Ellis, M. J. Geselbracht, B. J. Johnson, G. C. Lisensky, and W. R. Robinson, *Teaching General Chemistry: A Materials Science Companion*, American Chemical Society: Washington, DC, 1993, pp. 423–426 ("A Shape Memory Alloy, NiTi") and pp. 119 and 147 ("Synthesis and Reactivity of $AuAl_2$").

Quasicrystals: An Amplification

P. W. Stephens and A. I. Goldman, "The Structure of Quasicrystals," *Sci. Amer.* 264(4), 44–53 (1991).

A. I. Goldman, J. W. Anderegg, M. F. Besser, S.-L. Chang, D. W. Delaney, C. J. Jenks,

M. J. Kramer, T. A. Lograsso, D. W. Lynch, R. W. McCallum, J. E. Shield, D. J. Sordelet, and P. A. Thiel, "Quasicrystalline Materials," *Am. Sci.* 84, 230–241 (1996).

Gold Clusters and Aurophilicity

H. Schmidbaur, "Ludwig Mond Lecture: High-carat Gold Compounds," *Chem. Soc. Rev.* 24, 391–400 (1995).

M. J. Katz, K. Sakai, and D. B. Leznoff, "The Use of Aurophilic and Other Metal-Metal Interactions as Crystal Engineering Design Elements to Increase Structural Dimensionality," *Chem. Soc. Rev.* 37, 1884–1895 (2008).

G. Schmid, "The Relevance of Shape and Size of Au_{55} Clusters," *Chem. Soc. Rev.* 37, 1909–1930 (2008).

Inorganic Materials Related to Carbon Allotropes

R. Tenne, "Inorganic Nanotubes and Fullerene-Like Nanoparticles," *Nat. Nanotechnol.* 1, 103–111 (2006).

Numbered references from this chapter may be viewed online at www.uscibooks.com /foundations.htm.

12.8. Exercises

12.1. *Below are listed some important periodic trends in the properties of atoms or elements. For each property, a selection of elements from a given group or period are listed. Choose the element from the set for which that property reaches a maximum, or indicate if this property is the same for all elements of the set. (a) Number of prepared-to-bond electrons: Sr, Zr, Mo, Ru, Pd, Cd. (b) Assuming all have the same coordination number of 12, the bond order per link: Sr, Zr, Mo, Ru, Pd, Cd. (c) Number of prepared-to-bond electrons: C, Si, Ge, Sn, Pb. (d) Coordination number: elemental phosphorus (P) at high temperature and low pressure or elemental phosphorus (P) at extremely high pressure.

12.2. Below are shown the structures (or parts of the structures in the case of polymers) of allotropes of some of the elements. Assume that none of the lines represents multiple bonds.

(a) Describe each allotropic structure as linear oligomeric, chain (one-dimensional) polymeric, sheet (two-dimensional) polymeric, or network (three-dimensional) polymeric. (b) Give the coordination numbers of the atoms in each allotrope. (c) If the atoms obey the octet rule, give their total coordination numbers and numbers of unshared electron pairs. (d) If the atoms obey the octet rule, identify the group in which each atom is located.

12.3. *Below are drawn some structures of possible allotropes of elements. Assume that none

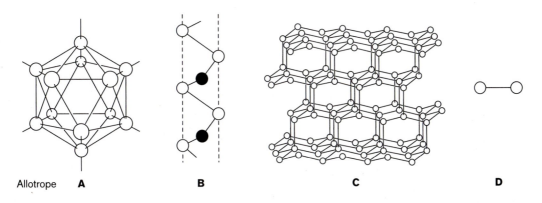

Allotrope   **A**   **B**   **C**   **D**

of the lines represents multiple bonds. (a) Describe each allotropic structure as monomeric molecular, oligomeric, chain (one-dimensional) polymeric, sheet (two-dimensional) polymeric, or network (three-dimensional) polymeric. (b) Give the coordination numbers of the atoms in each allotrope. (c) If the atoms obey the octet rule, give their total coordination numbers and numbers of unshared electron pairs. (d) If the atoms obey the octet rule, identify the group in which each atom is located.

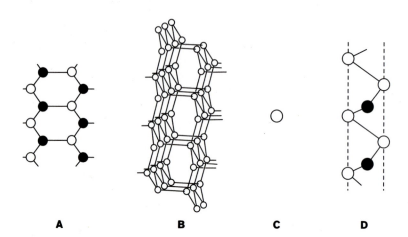

A   B   C   D

12.4. Below are shown the structures (or parts of the structures in the case of polymers) of possible allotropes of some of the elements. Assume that none of the lines represents multiple bonds.

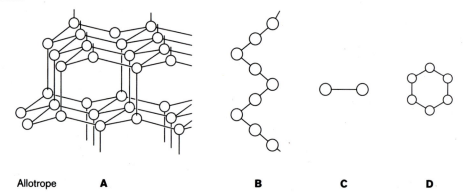

Allotrope   **A**   **B**   **C**   **D**

(a) Describe each allotropic structure as monomeric molecular, linear oligomeric, cyclic oligomeric, chain (one-dimensional) polymeric, sheet (two-dimensional) polymeric, or network (three-dimensional) polymeric. (b) Give the coordination numbers of the atoms in each allotrope. (c) If the atoms obey the octet rule, give their total coordination numbers and numbers of unshared electron pairs. (d) If the atoms obey the octet rule, identify the group in which each atom is located.

12.5. *From the σ and π O–O bond energies, calculate ΔH for the conversion of 3 mol of O_2 to 2 mol of O_3. Which is more stable?

12.6. For each of the Group 15/V and Group 16/VI elements except Bi and Po, write balanced chemical equations illustrating the following process:

$$\text{multiply bonded allotrope} \rightarrow \text{singly bonded allotrope}$$

(One of these allotropes may have to be an imaginary analogue of a known allotrope of another element in the group.) Using the average σ- and π-bond energies from Tables 3.2 and 3.3, calculate ΔH for this process in each case. What is the general trend in the stability of multiply bonded allotropes? Can you explain this in terms of overlap of orbitals and bond lengths?

12.7. *Calculate ΔH for the conversion of 6 mol of sulfur atoms from the allotropic form S_2 to the allotropic form S_3 (with a structure like that of ozone). Is this reaction favored?

12.8. Given that the Se–Se σ-bond energy is 172 kJ mol^{-1} and the Se–Se π-bond energy is 100 kJ mol^{-1}, calculate ΔH for the conversion of 2 mol of Se_3, with a Lewis structure like that of ozone, to 1 mol of the ring compound Se_6. Is this reaction favored by ΔH?

12.9. *For Group 15/V elements, there are two general types of allotropic structures: (1) those with three single bonds to three different atoms (as in P_4) and (2) those with one triple bond to one other atom (N_2). Try to imagine an allotrope with the intermediate situation, in which each Group 15/V atom has one single bond to one neighbor and one double bond to another neighbor. (a) Draw a small-molecule nitrogen allotrope with this structure. Predict some physical properties and compare it with the physical properties of existing allotropes for N and P. (b) Estimate ΔH for this new allotrope of N decomposing to the more stable allotrope of N. (c) Estimate ΔH for this new allotrope of P decomposing to the more stable allotrope of P.

12.10. Suppose that a new cyclic allotrope of arsenic, cyclo-As_6, were to be discovered. (a) Draw a reasonable Lewis structure (one obeying octet rules) for this allotrope, and determine the number of σ and π bonds present. (b) Estimate ΔH for the conversion of 1 mol of the cyclo-As_6 allotrope to the normal allotrope of arsenic.

12.11. *The diatomic molecules C_2 and Zr_2 are known in the gas phase. If you follow the normal rules for drawing Lewis structures, you will come up with a quadruple bond in each of these compounds. For which one of these two is a quadruple bond impossible? For which one is it plausible? Explain why.

12.12. The multiply bonded allotropes of the second period are thermodynamically favored over the singly bonded allotropes. Is this tendency as strong for carbon as for nitrogen and oxygen? Why or why not?

12.13. *What conditions of temperature and pressure should favor the formation of multiply bonded allotropes for Groups 15/V and 16/VI, and why?

12.14. (a) Which elemental substance of Group 13/III is the hardest to atomize? Explain why. (b) Which elemental substance of Group 7 is the easiest to atomize? Explain why.

12.15. *Suppose that valence f orbitals and valence s and p orbitals all had similar values of Z^* and $<r_{max}>$ in a given atom, but that its valence d orbitals were spatially and energetically unavailable. At approximately which atom would you expect to find the largest heat of atomization among the f-block elemental substances of the seventh period?

12.16. (a) The highest boiling point for a p-block elemental substance is about 4623 K. Which elemental substance should this be for? (b) The lowest boiling point for a p-block elemental substance is 4.22 K. Which elemental substance should this be for? (c) The highest boiling point for a d-block elemental substance is about 5923 K. Which elemental substance should this be for?

12.17. *Which elemental substance of the second period is the hardest to atomize and to boil? In one sentence, explain why.

12.18. Below are listed some important periodic trends in the properties of elemental substances. For each property, a selection of elements from a given group or period are listed. Choose the elemental substance from the set for which that property reaches a maximum, or indicate if this property is the same for all elemental substances of the set. (a) Highest boiling point: C, Si, Ge, Sn, Pb. (b) Highest boiling point: Sr, Zr, Mo, Ru, Pd, Cd, Sn, Te, Xe. (c) Strongest element–element single bond: C, Si, Ge, Sn, Pb. (d) Highest bond order per atom in an elemental substance: C, Si, Ge, Sn, Pb. (e) Highest bond order per atom in an elemental substance: C, N, O, F, Ne.

12.19. *The molecular orbitals of the linear H_3 molecule and their symmetry labels (irreducible representations) are which? (Vertical lines | represent nodal planes.)

(a) ① ② ③ = σ_g; ① | ❸ = σ_u^{nb}; and ① | ❷ | ③ = σ_g^*

(b) ① ② | = σ_g; and | ❷ ❸ = σ_u^*

(c) ① ② ③ = σ_g; ① ② | ❸ = σ_u^{nb}; and ① | ❷ | ③ = σ_g^*

(d) ① ② ③ = σ_g; ❶ ❷ ❸ = σ_g; and ① | ❸ = σ_u^*

(e) None of the above is correct.

12.20. Consider a linear string of four hydrogen atoms, H_4. Draw the four MOs of this molecule. Indicate nodes (nodal lines or planes) with dashed lines; give the total number of nodal planes. Indicate positive and negative signs of the wave function within the MOs. Give the number of positive overlaps, the number of negative overlaps, and the net number of positive overlaps for each MO. Give the symmetry label for each MO. Draw the energy level diagram for linear H_4.

12.21. *Consider a linear string of five hydrogen atoms, H_5. Draw the five MOs of this molecule. Indicate nodes (nodal lines or planes) with dashed lines; give the total number of nodal planes. Indicate positive and negative signs of the wave function within the MOs. Give the number of positive overlaps, the number of negative overlaps, and the net number of positive overlaps for each MO. Give the symmetry label for each MO. Draw the energy level diagram for linear H_5.

12.22. Consider a linear string of six hydrogen atoms, H_6. (a) Draw the six MOs of this molecule. Indicate nodes (nodal lines or planes) with dashed lines; give the total number of nodal planes. Indicate positive and negative signs of the wave function within the MOs. Give the number of positive overlaps, the number of negative overlaps, and the net number of positive overlaps for each MO. Give the symmetry label for each MO (σ_g^b, etc.). (b) Draw the energy-level diagram for H_6 and fill in arrows for the electrons occupying those orbitals.

12.23. *Which (if any) of the photoelectron spectra sketched at right most nearly fits with the expected bonding in the H_3 molecule?

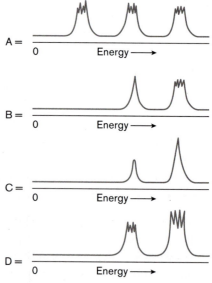

A = 0   Energy ⟶

B = 0   Energy ⟶

C = 0   Energy ⟶

D = 0   Energy ⟶

E =   None of the above

12.24. The 12 π molecular orbitals of (linear) hexatriyne, C_6H_2, can be generated from the σ molecular orbitals of H_6 (Exercise 12.22.). (a) Why is it really only necessary to draw six of these MOs? (b) Draw the six needed π MOs of this molecule. Indicate nodes (nodal lines or planes) with dashed lines; give the total number of nodal planes. Indicate positive and negative signs of the wave function within the MOs. Give the number of positive overlaps, the number of negative overlaps, and the net number of positive overlaps for each MO. Give the symmetry label for each MO ($σ_g^b$, etc.). (c) Draw the energy-level diagram for hexatriyne and fill in arrows for the electrons occupying those orbitals. (d) Compare your results with the results of computations illustrated in Figure 12.11. Between which two orbitals is the energy gap the largest?

12.25. *The six π molecular orbitals of (linear) hexatriene, C_6H_8, can be generated from the σ molecular orbitals of H_6 (Exercise 12.22). Describe the changes that must be made to the MOs of H_6 and their symmetry labels in order for these to apply to hexatriene.

12.26. If your class carried out Exercise III of the computational Experiment 10, gather your results on the π-type MOs of the molecules ethylene, *trans*-butadiene, all-*trans*-hexatriene, and their silicon and tin analogues. (If you did not do the experiment, your instructor can provide you with printouts of (incomplete) student results from 2006.) On a sheet of graph paper: (a) plot the energies of the π-type MOs of these (poly)alkenes of these three elements as a function of their chain lengths, and (b) plot the wavelengths of the electronic absorption spectra of these alkenes. (c) Do you expect the colors of polyalkenes to depend on the length of the alkene chains? (d) Do you expect the colors of polyenes to depend on the Group 14/IV element used in its backbone?

12.27. *B. A. Anderson [*J. Chem. Educ.* 74, 985 (1997)] has suggested the use of $(C_6H_5)–(CH=CH)_x–(C_6H_5)$ as a model series of one-dimensional oligomers that have easy-to-measure UV spectra and therefore HOMO–LUMO gaps. Furthermore, he suggests a particle-in-a-box calculation, based on Equation 9.7, to predict these spectra. Note that on the right side of Equation 9.7 half of the parameters are constants that do not change as the length of the oligomer changes. (a) When the length of the chain increases (i.e., when x doubles from 2 to 3, then to 4), what happens to the values of each of the variable parameters? (b) When these increases occur, what happens to $ΔE$ for the HOMO–LUMO gap? (c) According to these calculations, what would happen to $ΔE$ for the HOMO–LUMO gap as the chain length approaches infinity? (d) Is this realistic? Why or why not?

12.28. In each part, give the symbols of two elements that: (a) occur in the diamond structure; (b) occur in polymeric helical chains; (c) occur in a black, sheet-like allotrope; (d) occur as diatomic molecules; (e) occur as tetrahedral X_4 molecules; (f) occur as X_8 molecules; (g) occur at room temperature in allotropes with multiple bonds; and (h) occur as monoatomic gases.

12.29. *Draw the structure of each of the following elemental substances, and characterize each as either a gas, liquid, easily vaporized solid, or hard-to-vaporize solid: (a) boron, (b) white phosphorus, (c) black phosphorus, and (d) ozone.

12.30. (a) Draw the structures of the three major allotropes of phosphorus. (b) Which of the three forms should have the highest vapor pressure at room temperature? (c) Which form is chemically the most reactive? (d) Which has the least stable σ bonds and why? (e) Which has no counterpart in arsenic or antimony chemistry?

12.31. *The structures (or parts of the structures in the case of polymers) of allotropes of some of the elements were drawn in Exercise 12.2. (a) For each structure, list one element that actually has an allotrope (maybe not the stable one) with that structure. (b) List the structures **A**, **B**, **C**, and **D** in order of expected increasing melting points. (Assume the equal strength of all covalent bonds that consist of electron pairs shared between two atoms.)

12.32. Some structures of possible allotropes of elements were drawn in Exercise 12.3. For each structure, select the best answer. (a) An allotrope with this structure is likely to be which of the following: a gas; a soft low-melting solid; a slippery high-melting solid; or a hard, dense high-melting solid? (b) An elemental substance that actually has a structure similar to the one shown is which of the following: Se, Xe, P, Si?

12.33. *Structures (or parts of the structures in the case of polymers) of possible allotropes of some of the elements were drawn in Exercise 12.4. Assume that (somehow) each of these allotropes were possible for one element. Which of the allotropes (**A**, **B**, **C**, or **D**) is most likely to (a) show the lowest boiling point, (b) show the greatest mechanical strength (hardness), (c) be the densest, and (d) be a gas at room temperature? (e) List one element that actually has an allotrope (maybe not a stable one) with a structure corresponding to **A**, to **B**, to **C**, and to **D**.

12.34. Below are listed some important periodic trends in the properties of elemental substances. For each property a selection of elements from a given group or period are listed. Choose the element from the set for which that property reaches a maximum, or indicate if this property is the same for all elements of the set.

(a) Predominance of polymeric forms of element: Si, P, S, Cl, Ar.

(b) Tendency to form π bonds: C, Si, Ge, Sn, Pb.

(c) Electrical conductivity: O, S, Se, Te, Po.

(d) Band gap: S, Se, Te, Po.

12.35. *A possible new allotrope of carbon that would be of interest for its electrical properties is carbyne, $\frac{1}{\infty}[-C\equiv C-C\equiv C-]$, which contains alternating single and triple bonds. (a) There are how many σ and how many π bonds per carbon atom in this allotrope? (b) What is the stable allotrope of carbon? There are how many σ and how many π bonds per carbon atom in the stable allotrope of carbon? (c) Calculate ΔH for the conversion of 6 mol of C atoms from the new hypothetical allotropic form to the stable allotropic form. Is this reaction favored?

12.36. Which should be the highest density allotrope of carbon: C_{60}, diamond, graphite, or graphene?

12.37. *Three of the rarely seen allotropes of carbon are C_2, carbyne, and graphene. (a) Classify each of these into one of the structural classifications given in Figure 8.1. (b) Assuming (contrary to actual fact) that the three forms are not readily interconverted, rank these three forms in order of increasing melting point. Which of the three is most likely to be a gas at room temperature?

12.38. Would you expect to be able to separate the fullerenes from the other allotropes of carbon by vacuum sublimation? Why or why not?

12.39. *Given that the radii of the nearly spherical C_{60} and its anions are about 500 pm, and with the help of Figure 2.16, predict the pK_b values and the basicity classifications of the anions C_{60} through C_{60}^{6-}. Which of these would you predict could persist in water?

12.40. The C–C bond length of graphite is 141.5 pm. Would you expect the C–C bond lengths in the intercalates of graphite with halogens (such as C_8Br) to be shorter or longer than 141.5 pm? Why?

12.41. *The C–C bond length of graphite is 141.5 pm. Would you expect the C–C bond lengths in the intercalates of graphite with Group 1 metals (such as RbC_8) to be shorter or longer than 141.5 pm? Why?

12.42. Consider the conducting polymer formed by oxidizing $[Pt(CN)_4]^{2-}$. Suggest a reason why the very similar square planar complex $[Pt(CNMe)_4]^{2+}$ might be more likely to oxidize to form a metallic polymer than the similar square planar complex $[Pt(CNTr^*)_4]^{2+}$, where $Tr^* = tris$-[3,5-di(*tert*-butyl)phenyl]methyl.

12.43. *The ultimate polyyne is carbyne, $\frac{1}{\infty}[-C\equiv C-C\equiv C-]$, which has alternating single and triple bonds. (a) Based on the discussion of Figure 12.13, would you expect this to be a conductor, insulator, or semiconductor? (b) If carbyne were oxidized with a small (nonstoichiometric) amount of iodine, what would happen to its conductivity? (c) If carbyne were reduced with a small (nonstoichiometric) amount of lithium, what would happen to its conductivity?

12.44. Describe in detail the changes you would have to make to the lattice structures of Figure 4.2 to make them apply to ccp, hcp, and bcc metals.

12.45. *Interpret the following sets of solid-state electron density indices: (a) flatness high, molecularity low, and charge transfer small but not zero; (b) flatness low, molecularity high, and charge transfer zero; (c) flatness low, molecularity low, and charge transfer zero.

12.46. Calculate the Mott energy U required to be overcome for the compression of the metalloids (a) Si and (b) Se to E^+E^-. Does this suggest that the compression of these semiconducting metalloids to metallic forms are likely to be achievable under pressure?

12.47. *Calculate the Mott energy U required to be overcome for the compression of the nonmetals (a) S and (b) Rn to E^+E^-. Does this suggest that the compression of these insulators is likely to be achievable under pressure?

12.48. Critique the following statement: "Because there is no electronegativity difference between identical atoms, elemental substances cannot have ionic structures."

12.49. *The following structural changes can happen to elemental substances under pressure. Arrange these in increasing order of pressure likely needed to achieve them: **A** = formation of ionic electrides; **B** = formation of secondary bonds; **C** = reduction of van der Waals radii of atoms; **D** = development of bands that overlap in energy (zero HOMO–LUMO energy gap); and **E** = promotion of some valence electrons to post-valence orbitals that are close in energy.

12.50. Below are listed some important periodic trends in the properties of elemental substances. For each property, a selection of elements from a given group or period are listed. Choose the element from the set for which that property reaches a maximum, or indicate if this property is the same for all elements of the set. (a) Ferromagnetism: Gd, Cm, eka-Cm (element below Cm); (b) ferromagnetism: Co, Rh, Ir; (c) mechanical hardness of material: pure Rb, Rb/Sr alloy, pure Mo; (d) highest atomization energy: B, Al, Ga, In, Tl; (e) ductility: Hf, W, Os, Au.

12.51. *Below are listed some important periodic trends in the properties of elemental substances. For each property, a selection of elements from a given group or period are listed. Choose the element from the set for which that property reaches a maximum, or indicate if this property is the same for all elements of the set. (a) Highest bond order per link in elemental substance: Sr, Zr, Mo, Ru, Pd, Cd; (b) greatest variability in composition: Na/K alloy, Rb/K alloy, Cu/Zn alloy; (c) ferromagnetism of elemental substance: Cs, Gd, Os, Pb, Rn; (d) strongest metal or alloy: pure Sr, Sr/Pd alloy, pure Au, Au/Ag alloy, pure Fe, Fe/V alloy; (e) most ductile metal or alloy: pure Sr, Sr/Pd alloy, pure Au, Au/Ag alloy, pure Fe, Fe/V alloy.

12.52. Zirconium (χ_P = 1.33, density = 6.56 g cm^{-1}) should form an alloy most exothermically with which metal of χ_P about 1.9? (a) Fe (density = 7.87 g cm^{-1}) or Re (density = 21.04 g cm^{-1}) or Ag (density = 10.49 g cm^{-1})?

12.53. *Uranium (χ_P = 1.38, density = 19.0 g cm^{-1}) forms the following alloys exothermically: UFe$_2$, ΔH_f = –11 kJ mol^{-1}; URh$_3$, ΔH_f = –64 kJ mol^{-1}; and UIr$_2$, ΔH_f = –71 kJ mol^{-1}.[73] Explain the relative values. Except for Fe, the alloying metals all have χ_P values of about 2.3; their densities are Fe, 7.87 g cm^{-1}; Rh, 12.4 g cm^{-1}; and Ir, 22.5 g cm^{-1}.

12.54. The metals showing the highest solubility in liquid mercury to form amalgams are the metals of Groups 1, 2, Eu, Yb, Au, Zn, Cd, Ga, In, and Tl. (a) Group these metals into two groups based on the type of amalgam that is probably formed (substitutional alloys or intermetallic compounds). (b) Based qualitatively on Miedema's interpretation, which of these two groups of metals should form amalgams most exothermically?

12.55. *The standard reduction potentials of the Group 1 metals given in Figures 6.7 and 6.8 are altered considerably when the Group 1 metals are employed as amalgams: the $E°$ values for the amalgams are –2.00 V for Li amalgam, –1.84 V for Na amalgam, –1.81 V for Rb amalgam, and –1.78 V for Cs amalgam.[119] (a) Does amalgamation move these metals higher or lower in the activity series of metals (Section 6.2B)? (b) Offer two possible explanations of why the activities of these metals change so much. (c) Would you expect as great a change in the activity of silver when it is amalgamated to fill a cavity in your tooth? Why or why not?

12.56. Brittleness is characteristic of two types of solids, while malleability is characteristic of two others. Identify these and explain why.

12.57. *(a) How many shells are present in the close-packed giant cluster Pt$_{309}$L$_{36}$O$_{30}$? (b) How many of the Pt atoms in this giant cluster should show metallic properties? (c) How many metal atoms should be present in the giant cluster having one more shell than the above?

12.58. Suggest the formulas (including charges) of other endohedral icosahedral clusters isoelectronic to W@Au$_{12}$ involving sixth-period metals, including gold.

12.59. *Give a qualitative explanation in terms of molecular orbitals for why the Ag and Au clusters described in Section 12.4D differ in color as the nanoparticle size changes.

12.60. Why are the cluster geometries shown in Figure 12.20 for M@M$_{12}$ more relevant than the icosahedral geometry for studying the formation of giant metal clusters leading to bulk metal structures?

12.61. *Select three of the most likely elemental substances or compounds from Table 12.3 to be able to emit ultraviolet light in a laser or LED. What would have to be done chemically to them to make them act as semiconductors?

12.62. From the III–V semiconductors listed in Table 12.3, select those that have similar electronegativity differences of 0.24–0.40. Use the band gap of these semiconductors to discuss, when structure types and electronegativity differences are controlled, how the band gap depends on the element–element bond lengths.

12.63. *How would you dope MgB$_2$ to make it an *n*-type semiconductor? A *p*-type semiconductor? (Don't worry about complications that might arise from the presence of two band gaps in MgB$_2$.)

12.64. How could you dope GaAs to make it an *n*-type semiconductor? A *p*-type semiconductor?

12.65. *(a) Which structural types of non-carbon-based inorganic polymers or macromolecules would most likely have their conductivities and band gaps altered by intercalation of an oxidizing or reducing agent? (b) Would you expect greater changes upon redox intercalation in the forms of BN or the forms of MoS_2? Why?

12.66. Compare and contrast the different forms of boron nitride to each other and to the corresponding carbon allotropes. In your answer, discuss (a) how the structures compare, (b) how the chemical reactivity with water compares, and (c) how the band gaps compare.

12.67 (Summary Exercise 1). Do the metalloids (Section 12.1A) fall between the metallic and covalent substances in Figure 12.25 and Figure 12.26?

12.68 (Summary Exercise 2). Are d-block elements included in Figure 12.25 and Figure 12.26? The different electronegativity scales generally agree for s- and p-block elements, except for the handful that show relativistic effects. Do the different scales agree for d-block elements? If not, which electronegativity scale works best? (Try adding some d-block elements to the graph to see.)

12.69 (Summary Exercise 3). Some electronegativity scales do not take into account the bonding properties that are unique to d-block elements, so they do not work so well. (Hence, they were not included in Sproul's or other triangles.) What are some of these bonding properties that might alter covalent bonding energies in ways not caught by the electronegativity scale?

12.70 (Summary Exercise 4). d-Block elements vary more in oxidation number than do most p-block elements. Is there evidence from the structure types of d-block metal oxides that structure types change with oxidation number? If so, one electronegativity element per element might not fit all predictions of bonding type.

12.71 (Summary Exercise 5). What other factors besides electronegativity strongly influence the ionic, molecular covalent, network covalent, or metallic structure and bonding type of an elemental substance or a binary substance?

APPENDIX

Answers to Odd-Numbered Exercises

Chapter 1

1.1. a. I^-; **b.** I_2; **c.** $I + e^- \rightarrow I^-$; **d.** $I_2 + 2e^- \rightarrow 2I^-$.

1.3. a. Pale green; **b.** silvery.

1.5. $Pb^{2+} \rightarrow Pb^{4+} + 2e^-$.

1.7. a. K^+; **b.** either Fe^{2+} or Fe^{3+}.

1.9. a. Chromium(II) ion, **b.** chromium(III) ion; **c.** selenide ion; **d.** $(Cr^{2+})(Se^{2-})$ = chromium(II) selenide; **e.** $(Cr^{3+})(Cl^-)_3$ = chromium(III) chloride; **f.** $(Cr^{3+})(N^{3-})$ = chromium(III) nitride.

1.11. a. Fe(II) set: Fe^{2+} = iron(II) ion, FeS = iron(II) sulfide; Fe(III) set: Fe^{3+} = iron(III) ion, $FeCl_3$ = iron(III) chloride, FeN = iron(III) nitride; **b.** Sn(II) set: Sn^{2+} = tin(II) ion, $SnCl_2$ = tin(II) chloride, Sn_3N_2 = tin(II) nitride; Sn(IV) set: Sn^{4+} = tin(IV) ion, SnO_2 = tin(IV) oxide, Sn_3N_4 = tin(IV) nitride; **c.** Tl(I) set: Tl^+ = thallium(I) ion, TlCl = thallium(I) chloride, Tl_2O = thallium(I) oxide; Tl(III) set: Tl^{3+} = thallium(III) ion, TlN = thallium(III) nitride, Tl_2O_3 = thallium(III) oxide.

1.13. a. Cu^+, Cu^{2+}, CuO, CuCl, Cu_3N_2; **b.** Cr^{3+}, Cr_2O_3, CrF_3, Cr^{6+}, CrO_3, CrF_6.

1.15. a. $1s^2 2s^2 2p^6 3s^2 3p^6 4s^2 3d^{10} 4p^6 5s^2 4d^{10} 5p^6 6s^2 4f^{14} 5d^{10} 6p^6 7s^1$; **b.** $1s^2 2s^2 2p^6 3s^2 3p^6 4s^2 3d^{10} 4p^3$; **c.** $1s^2 2s^2 2p^6 3s^2 3p^6 4s^2 3d^{10} 4p^6 5s^2 4d^{10} 5p^6 6s^2 4f^{14} 5d^8$; **d.** $1s^2 2s^2 2p^6 3s^2 3p^6 4s^2 3d^{10} 4p^6 5s^2 4d^{10} 5p^6 6s^2 4f^{10}$; **e.** $1s^2 2s^2 2p^6 3s^2 3p^6 4s^2 3d^{10} 4p^2$.

1.17. a. $7s^1$ Fr·; **b.** $4s^2 4p^3$:As·:; **c.** $6s^2 5d^8$; **d.** $6s^2 4f^{10}$; **e.** $4s^2 4p^2$:Ge·.

1.19. a. $5s^2$:Sr; **b.** $6s^2 6p^5$.:At::; **c.** $6s^2 5d^4$; **d.** $7s^2 5f^9$; **e.** $4s^2 3d^5$; **f.** $7s^2 5f^3$.

1.21. a. $6s^2 6p^6 = (::Bi::)^{3-}$; **b.** $6s^2 =$:Bi^{3+}; **c.** $6s^0 = Ba^{2+}$; **d.** $5d^8$; **e.** $4f^9$; **f.** $4s^2 =$:Ge^{2+}; **g.** $4s^2 4p^6 = (::Ge::)^{4-}$.

1.23. a. $3s^2 3p^6 = (::Cl::)^-$; **b.** $6s^2 6p^2 = (:At:)^{3+}$; **c.** $5d^6$; **d.** $4d^2$; **e.** $4s^2 =$:Se^{4+}; **f.** $4s^2 4p^6 = (::Se::)^{2-}$.

1.25. a. Xe and Te^{2-}; **b.** Mn; **c.** Pr^{3+}; **d.** Hg and Au^-.

1.27. a. $6s^25d^{10}$; **b.** $6s^2$; **c.** $6s^25d^{10}$; **d.** $6s^2$.

1.29. a. O = –2, C = +4; **b.** O = –2, N = +3; **c.** O = –2, N = +3; **d.** H = +1, O = –1; **e.** H = +1, C = –4; **f.** F = –1, I = +5; **g.** O = –2, Os = +8.

1.31. a. O = –2, S = 2.5; **b.** S = –2, P = 3.5.

1.33. a. +3; **b.** +5.

1.35. a. H = +1, O = –1; **b.** O = –2, +3⅓.

1.37. a. –4/9; **b.** +3; **c.** +3.

1.39. a. $3d^6$; **b.** $5s^05p^0$ Sn^{4+}; **c.** $3s^2$:P^{3+}; **d.** $5s^2$:Sb^{3+}.

1.41. O, S, Se, Te, Po, Cr, Mo, W, Nd, U, Sg. 124 (in g block), 142 (in f block), 156 (in d block), 166 (in p block).

1.43. S, Ba, Ta, U.

1.45. a. 226 pm; **b.** 106 pm; **c.** 226 pm.

1.47. a. Predicted 146 pm, measured 139 pm; **b.** predicted 86 pm, measured 69 pm; **c.** predicted 206 pm, measured 198 pm.

1.49. a. Pr, by about 60 pm; **b.** As^{3-}, by about 120 pm.

1.51. a. Ionic = 234 pm, covalent = 223 pm; **b.** ionic = 308 pm, covalent = 284 pm; **c.** ionic = 235 pm, covalent = 223 pm; the ionic bonds would appear to be longer.

1.53. a. Ne < F < C < Li; **b.** Be < Ca < Ba < Ra; **c.** B < Al ≈ Ga < In ≈ Tl; **d.** V < Nb ≈ Ta < Db.

1.55. a. Cr^{6+} < Cr^{4+} < Cr^{3+} < Cr^{2+}; **b.** Be^{2+} < Mg^{2+} < Sr^{2+} < Ra^{2+}; **c.** Re^{7+} < W^{6+} < Ta^{5+} < Hf^{4+} < Lu^{3+} < Ba^{2+} < Cs$^+$; **d.** No^{3+} < ... < Ac^{3+}.

1.57. a. F$^+$ < Cl$^+$ < Br$^+$ < I$^+$ < At$^+$; **b.** Au^{7+} < Au^{5+} < Au^{3+} < Au$^+$; **c.** Bi^{3+} < Pb^{2+} < Tl$^+$ < Au$^-$; **d.** H$^+$ < Li$^+$ < Rb$^+$ < Fr$^+$; **e.** Te^{4+} < Sb^{3+} < Sn^{2+} < In$^+$ (from expected trends, but Table C indicates Te^{4+} > Sb^{3+}).

1.59. a. Nd and Pm; **b.** Al and Ga; **c.** Zr and Hf; **d.** In and Tl.

1.61. a. Increase; **b.** decrease; **c.** less rapidly; **d.** decrease, decrease, less rapidly; **e.** increase, increase, less rapidly.

1.63. a. Decreases; **b.** increases; **c.** unaltered.

1.65. a. Ba < Sr < Ca < Mg < Be; **b.** Na < Al < P < Cl; **c.** Ge ≈ Si < C; without taking anomalies into account, Ge < Si < C; **d.** without taking anomalies into account, Rg < Au < Ag. Considering only the lanthanide contraction (and assuming it is still important at the right side of the d block): Rg < Au ≈ Ag.

1.67. a. Mg < Si < S; **b.** Mg > Sr > Ra; **c.** B > Al ≈ Ga; **d.** > and >.

1.69. Main trends: **a.** increases left to right; **b.** decreases left to right; **c.** increase left to right. Countertrends: **a.** in parts of the d block; **b.** in parts of the d block; **c.** at the right sides of the f, d, and p blocks.

1.71. a. K; **b.** C; **c.** Li; **d.** we predict Cu, but (due to lanthanide or relativistic effects, Chapter 9) in fact Au or even Rg; **e.** Si and Ge; **f.** Zr and Hf.

1.73. a. Pm ≈ Sm; **b.** Be; **c.** O; **d.** Zr ≈ Hf.

Chapter 2

Note: The classifications of several metal ions, particularly those that are nonacidic, feebly acidic, or weakly acidic, depend on whether Equation 2.14 or the calculations summarized in Table 2.2 are used for classification.

2.1. a. $K_a = \dfrac{[H^+][ObOH^+]}{[Ob^{2+}]} = 10^{-3}$ **b.** $K_a = \dfrac{[H^+][CtOH^{3+}]}{[Ct^{4+}]} = 3.2 \times 10^5$.

2.3. Selected from Table 2.1: **a.** Ba^{2+} and Sr^{2+}; **b.** Fe^{2+} and La^{3+}; **c.** Sc^{3+}, In^{3+}, and Cr^{3+}; **d.** Au^{3+} and Ce^{4+}.

2.5. a. Nonacidic cations; **b.** feebly or weakly acidic cations; **c.** moderately acidic cations; **d.** strongly acidic cations; **e.** very strongly acidic cations.

2.7. a. Nogium (Ng); **b.** cochranium (Cc); **c.** Cc feebly acidic, Ng very strongly acidic; Ro moderately acidic.

2.9. a. Weakly acidic and $pK_a = 6$ to 11.5; **b.** strongly acidic and pK_a between -4 and 1; **c.** nonacidic and pK_a over 14.

2.11. a. Approximately 4; **b.** 10^{-4}; **c.** moderately acidic cation.

2.13. a. Approximately 9; **b.** 10^{-9}; **c.** weakly acidic cation.

2.15. a. Cr^{2+} weakly acidic, Cr^{3+} moderately acidic, Cr^{4+} very strongly acidic, Cr^{6+} very strongly acidic; **b.** Np^{3+} weakly acidic, Np^{4+} moderately acidic, Np^{5+} very strongly acidic, Np^{6+} very strongly acidic, Np^{7+} very strongly acidic; **c.** assuming that they exist, CrF_4 and CrF_6; **d.** assuming that they exist, NpF_5, NpF_6, and NpF_7.

2.17. a. U^{3+} weakly acidic, Ag^+ feebly acidic, Pa^{5+} very strongly acidic, C^{4+} very strongly acidic, As^{3+} strongly acidic, Tl^+ nonacidic, Th^{4+} moderately acidic; **b.** Pa^{5+}, As^{3+}, possibly with Th^{4+}; this could be cleared up by adding excess acid for As^{3+} and Th^{4+}, but not for Pa^{5+}.

2.19. a. 0.219, 0.027, 0.008, 0.148, and 0.417, respectively; **b.** strongly acidic, feebly acidic, nonacidic, moderately acidic, and very strongly acidic; **c.** very strongly acidic, feebly acidic, feebly acidic, moderately acidic, and very strongly acidic; **d.** Ba^{2+} and Ag^+; **e.** B^{3+} and As^{5+}.

2.21. a. Nonacidic, weakly acidic, very strongly acidic; **b.** Rb^+ 14.6, La^{3+} 8.5, P^{5+} -33.1; **c.** Rb^+ 6.5, La^{3+} 5.5, P^{5+} 0; **d.** Rb^+ hydration only, La^{3+} perhaps very slight cloudiness (mainly just hydration), P^{5+} violent reaction with hissing HCl and heat evolution.

2.23. a. Na^+, K^+, Rb^+, Cs^+, Fr^+, Tl^+; **b.** Li^+, Ra^{2+}, Ba^{2+}, Sr^{2+}, Ca^{2+}, Eu^{2+}, Dy^{2+}, Tm^{2+}, Yb^{2+}, No^{2+}, Ag^+, Au^+.

2.25. b. $AlCl_3$ versus $PaCl_3$ represents the best test because the charges of both cations are $+3$ and the electronegativities are also close, while the radii differ considerably.

2.27. a. NaCl versus $CaCl_2$ versus $LaCl_3$—all have radii close to 115 pm and electronegativities close to 1.0, while the charge varies from $+1$ to $+2$ to $+3$.

2.29. a. For Na^+, $\Delta H_{hyd} = -367$ kJ mol^{-1}; for Ca^{2+}, $\Delta H_{hyd} = -1485$ kJ mol^{-1}; for La^{3+}, $\Delta H = -3282$ kJ mol^{-1}; for Hg^{2+}, $\Delta H_{hyd} = -1467$ kJ mol^{-1}; **b.** the tabulated values are more negative than the calculated ones by 38 kJ mol^{-1} for Na^+, 107 kJ mol^{-1} for Ca^{2+}, 1 kJ mol^{-1} for La^{3+}, and 357 kJ mol^{-1} for Hg^{2+}; **c.** the discrepancy is greatest for Hg^{2+}, due to its much higher electronegativity, which adds covalent bonding effects to the electrostatic effects incorporated in the Latimer equation.

2.31. Pa^{5+}; $\Delta H_{hyd} = -10,700$ kJ mol^{-1}; no.

2.33. Doubling the charge should quadruple the hydration energies and acidities. Doubling the radius should halve the hydration energies and acidities, but this is the lesser effect. Grecium ion should have the higher hydration energy and K_a.

2.35. a. KCl dissolves smoothly in water; it will not fume in air; **b.** $NbCl_5$ fumes in air and reacts violently and irreversibly with water to give HCl and $Nb(OH)_5$ or Nb_2O_5; **c.** $AlBr_3$ might fume in air some; it will react with moderate heat with water; **d.** CBr_4 will not fume, nor will it react with or dissolve in water; **e.** BaI_2 will not fume; it will dissolve smoothly in water; **f.** IF_7 fumes in air and reacts violently and irreversibly with water; fuming occurs if the cation of the halide has high enough acidity and has not yet reached its maximum coordination number.

2.37. a. UF_6; **b.** $COCl_2$; **c.** SO_2F_2. Maximum coordination number. The number of halogen atoms around the central atom adds up to the maximum coordination number in CCl_4 and SF_6, so there is no room for a water molecule to enter and begin the hydrolysis. In the others, the number of groups attached is less than the maximum coordination number, so there is room for an H_2O molecule to attack. They fume if the acidity is high enough and if the maximum coordination number has not been met.

2.39. a. $H_2Te(aq)$; **b.** $HTe^-(aq)$; **c.** $Te^{2-}(aq)$.

2.41. a. See below; **b.** for H_3Hm, $pK_{a1} = 2$; for H_2Hm^-, $pK_{a2} = 7$; for HHm^{2-}, $pK_{a3} = 12$; **c.** H_3Hm is moderately acidic, H_2Hm^- is weakly acidic, HHm^{2-} is feebly acidic; **d.** the left species is hydrogen hoffmide, and the right species is the hoffmide ion. (Although not covered in the text, the middle two would commonly be named dihydrogen hoffmide ion and hydrogen hoffmide ion.)

H_3Hm	H_2Hm^-	HHm^{2-}	Hm^{3-}

pH 0   2   4   6   8   10   12   14

2.43. a. HS^- and HTe^-; **b.** Cl^- and either H_2S or HS^-, depending on the final pH; **c.** HSe^- and Te^{2-}.

2.45. a. $HBy^- + H_2Kl$; **b.** not expected to react; **c.** not expected to react.

2.47. a. Sulfide ion, amide ion, hydrogen sulfide ion, nitride ion; **b.** La_2S_3, $La(NH_2)_3$, $La(SH)_3$, LaN.

2.49. a. $Ca(NH_2)_2$, $Ca(SeH)_2$, Ca_3P_2, CaI_2; **b.** Iron(III) amide, iron(III) hydrogen selenide, iron(III) phosphide, iron(III) iodide.

2.51. a. Feebly acidic, $pK_a = 12.5$; **b.** moderately acidic, 3.0; **c.** nonacidic, > 16; **d.** very strongly acidic, < −4.

2.53. a. Pb^{4+} third diagram, Pb^{2+} second diagram, Yb^{2+} first diagram; **b.** Pb^{4+}.

2.55. Top diagram: As^{3+}. Second diagram: Rb^+. Third diagram: N^{3+}. Fourth diagram: Pb^{2+}.

2.57. a. First diagram is Ro cation, second diagram is Cc cation, third diagram is none of these cations, fourth diagram is for Ng cation; **b.** add acid to lower the pH of the solution to below 3; **c.** cannot be done.

2.59. Oxides of nonacidic cations are soluble (as hydrated cations and OH^- ions) at pH values of 0, 7, and 14. Oxides of feebly acidic cations are soluble at pH values of 0 and 7, but insoluble at pH 14. Oxides of weakly acidic cations are soluble at pH 0, insoluble at 14, and perhaps insoluble at 7. Oxides of moderately acidic cations are soluble at pH 0, insoluble at 7, and insoluble at 14. Oxides of strongly acidic cations are probably insoluble at pH 0, insoluble at 7, and insoluble at 14. Oxides of very strongly acidic cations are insoluble at pH values of 0, 7, and 14.

2.61. $Bb^- < Dd^- < Cc^{2-} < Ee^{3-} < Aa^{3-}$.

2.63. a. I^-; **b.** C^{4-}.

2.65. Feeble acid–base reactivity implies a pK_a or pK_b between 11.5 and 14; it arises from a relatively small charge and/or a relatively large radius; the reactivity is detectable in water only under special conditions; the predominance range of the species is very broad but does not quite cover all pH ranges.

2.67. a. Br^-; **b.** C^{4-}, Ge^{4-}, O^{2-}; **c.** OH^- and SeH^-.

2.69. a. Vo, Gr, and Da form 3– anions; Su, Go, and Ch form 2– anions; Sh, Gb, and Bw form 1– anions; **b.** Vo^{3-}; **c.** Bw^-.

2.71. a. Beckwithide, udezide; **b.** very strongly basic, weakly basic, strongly basic, and nonbasic, respectively; **c.** beckwithide, udezide; **d.** beckwithide, udezide; **e.** beckwithide fourth diagram, campbellide third diagram, gallianide first diagram, udezide second diagram.

2.73. a: Weakly basic Ff^{2-} would have a pK_b between 6 and 11.5, say at 9; then the pH boundary would be at $14 - 9 = 5$.

b. GgH^-, being feebly basic, will have a left boundary to its predominance range of about 2; being feebly acidic, it will have a right boundary of perhaps 12.

c. Jj^{3-}, being very strongly basic, will have no predominance range; based on the precedent of the Group 15/V anions, HJj^{2-} and H_2Jj^- may not either, so the entire diagram may feature only H_3Jj.

d. Because Kk^- is moderately basic, its pK_b is between 1 and 6, say about 3; then its left pH boundary is $14 - 3 = 11$.

	HKk	Kk⁻(aq)
pH 0		11 14

Chapter 3

3.1. a. O_2^-; **b.** O_2^{2-}; **c.** probably O^{2-} in titanium(IV) oxide but hypothetically this could be O_2^{2-} in titanium(II) peroxide.

3.3. +13, which is impossible because Cr does not have 13 valence electrons. The anion could have four peroxide ions bonded to a central Cr^{5+} ion and thus be a $[Cr(O_2)_4]^{3-}$ ion.

3.5. Advantages: Using parentheses clarifies that a polyatomic anion is present, and showing charges of ions makes identifying ions much easier. Disadvantages: Using parentheses and showing charges requires more typesetting (which costs money), keystrokes, or blackboard space. Furthermore, it isn't done that way in the past literature.

3.7. a. $H-\ddot{N}=N=\ddot{N}-H$ **b.** $(:\ddot{N}=N=\ddot{N}:)^-$ **c.** $\left(:\ddot{F}-\overset{\overset{:\ddot{F}:}{|}}{\underset{\underset{:\ddot{F}:}{|}}{N}}-\ddot{F}:\right)^+$ **d.** $:\ddot{O}=C=\ddot{S}:$

3.9. a. $H-\overset{\overset{:\ddot{C}l:}{|}}{\underset{\underset{:\ddot{C}l:}{|}}{C}}-\ddot{C}l:$ **b.** $:N\equiv\ddot{S}-\ddot{F}:$ **c.** $H-\overset{\overset{H}{|}}{\underset{\underset{H}{|}}{C}}-\overset{\overset{H}{|}}{\underset{\underset{H}{|}}{C}}-H$ **d.** $H-\overset{\overset{H}{|}}{\underset{\underset{H}{|}}{C}}-\overset{\overset{H}{|}}{\underset{\underset{H}{|}}{C}}-\overset{\overset{H}{|}}{\underset{\underset{H}{|}}{C}}-\ddot{C}l:$

e. Five ($-C\equiv O:$) and four unshared electron pairs around Fe.

f. $:\ddot{C}l-\overset{\overset{H}{|}}{\underset{\underset{H}{|}}{C}}-\overset{\overset{H}{|}}{\underset{\underset{H}{|}}{Si}}-\overset{\overset{H}{|}}{\underset{\underset{H}{|}}{Si}}-\overset{\overset{H}{|}}{\underset{\underset{H}{|}}{C}}-\ddot{C}l:$

3.11. a. $\overset{H-\ddot{O}:}{\underset{\underset{:\ddot{F}:}{|}}{}}$ **b.** $H-\overset{\overset{H}{|}}{\underset{\underset{H}{|}}{Si}}-\overset{\overset{H}{|}}{\underset{\underset{H}{|}}{Si}}-H$ **c.** $H-\overset{\overset{H}{|}}{\underset{\underset{H}{|}}{Si}}-\overset{\overset{H}{|}}{\underset{\underset{H}{|}}{Si}}-\overset{\overset{H}{|}}{\underset{\underset{H}{|}}{Si}}-H$ **d.** $:\ddot{Se}=C=\ddot{Se}:$

e.

f. $:\ddot{C}l-\overset{\overset{H}{|}}{\underset{\underset{H}{|}}{C}}-\overset{\overset{H}{|}}{\underset{\underset{H}{|}}{C}}-\ddot{C}l:$

3.13. a. $H-\overset{\overset{H}{|}}{\underset{\underset{H}{|}}{\ddot{N}}}-H$ N donor atom

b. $H-\overset{\overset{H}{|}}{\underset{\underset{H}{|}}{\ddot{N}}}-\overset{\overset{:O:}{\|}}{C}-\ddot{O}-H$ N and two O donor atoms

c. $\left(:\ddot{O}-\overset{\overset{:\ddot{O}:}{|}}{\underset{\underset{:O:}{|}}{S}}-\ddot{O}:\right)^{2-}$ Four O donor atoms

d. $\left(:\ddot{O}-\overset{\overset{:\ddot{O}:}{\|}}{\underset{\underset{:O:}{|}}{S}}-\ddot{O}:\right)^{2-}$ S and three O donor atoms

3.15. a. Cl; **b.** N, S, and F; **c.** none; **d.** Cl; **e.** (Fe and) O; **f.** Cl.

3.17. a. O and F atoms; **b.** none; **c.** none; **d.** Se atoms; **e.** (V and O) atoms; **f.** Cl atoms.

3.19. a. $As^{5+} + 5 F^- \rightarrow AsF_5$, so As acts as Lewis acid; **b.** $(CH_3)_2O: + AsCl_3 \rightarrow (CH_3)_2O:AsCl_3$, so As acts as Lewis acid; **c.** $Cl:^- + AsCl_3 \rightarrow [Cl:AsF_3]^-$, so As acts as Lewis acid; **d.** $Ni^{2+} + 4:AsCl_3 \rightarrow [Ni(:AsCl_3)_4]^{2+}$, so As acts as Lewis base.

3.21. a. Ligands or Lewis bases or nucleophiles or donors; **b.** N and O, respectively; **c.** Lewis acid or acceptor or central atom or electrophile.

3.23. a. Zero; **b.** zero; **c.** –1.

3.25. a. +4; **b.** +5; **c.** +6; **d.** the Al is Al^{3+}.

3.27. a. $Cs[AlCl_4]$; **b.** $[Si(C_5H_5N)_4Cl_2]Cl_2$; **c.** $[Pb(C_5H_5N)_2Cl_4]$.

3.29. a. $[Pd((CH_3)_3P)_4]Cl_2$, $[Pd((CH_3)_3P)_3Cl]Cl$, $[Pd((CH_3)_3P)_2Cl_2]$, $K[Pd((CH_3)_3P)Cl_3]$, $K_2[PdCl_4]$; **b.** the Cl^- ions in the first two compounds and the K^+ ions in the last two compounds; **c.** 4.

3.31. Find the drawings in Figure 3.7.

3.33. **a.** p_x, p_y, zero; **b.** s, s, negative; **c.** s, p_z, positive, σ; **d.** p_z, p_z, negative; **e.** s, p_x, zero; **f.** p_x, p_x, negative.

3.35. **a.** Three; **b.** π.

3.37. In the order shown: δ, π, σ. Bond dissociation energies: $\delta < \pi < \sigma$. For bonding interactions, signs of wave functions should match in regions of overlap (while changing on crossing nodal planes).

3.39. Probably; **b.** is better, because although π bonds are indeed especially stable in the second period, so are σ bonds, except among N, O, and F.

3.41. **a.** Pb–Pb < Sn–Sn < Ge–Ge < Si–Si < C–C; **b.** Li–Li < B–B < C–C; **c.** $\delta < \pi < \sigma$.

3.43. **a.** O_2; **b.** N_2; **c.** C_2^{2-}.

3.45. 319 kJ mol^{-1} in CCl_4; 366 kJ mol^{-1} in $(CH_3)_3CCl$.

3.47. 2.03.

3.49. χ_P for Ge(II) = 1.83. Yes, the electronegativity of an element in a lower oxidation state should be lower.

3.51. Calculated oxidation states are: **a.** +8 for Os; **b.** +5 for N; **c.** +4 for Xe; **d.** +4 for Si; **e.** +3 for I.

3.53. Calculated oxidation states are: **a.** +5 for N; **b.** +4 for Xe; **c.** +5 for Sb; **d.** +5 for I; **e.** +3 for N.

3.55. Calculated oxidation states are: **a.** +4 for Xe; **b.** +3 for Sb; **c.** +8 for Xe; **d.** +4 for Xe; **e.** +4 for C.

3.57. **a.** MoF_8^{3-}; **b.** SeF_5^{-}; **c.** XeF_7^{-}; **d.** XeF_8.

3.59. **a.** TeF_8^{2-}; **b.** $:XeF_7^{-}$; **c.** WF_8^{2-}; **d.** OsF_8^{2-}.

3.61. **a.** CF_4; **b.** $[SiF_6]^{2-}$; **c.** $[ClF_4]^{-}$; **d.** $[ClF_6]^{+}$; **e.** $[IF_8]^{-}$.

3.63. **a.** BO_3^{3-}; **b.** $:BrO_2^{-}$; **c.** AsO_4^{3-}; **d.** $:SO_3^{2-}$.

3.65. **a.** BO_4^{5-}; **b.** BO_3^{3-}; **c.** AsO_4^{5-}; **d.** RuO_5^{2-}; **e.** IO_4^{3-}.

3.67.

3.69. Oxides are neutral compounds; two oxide ions neutralize the +4 charge of the Group 14/IV cation. Oxo anions add additional oxide ions, up to the penultimate total coordination number, which varies down the group.

3.71. a. ReO_4^- (or perhaps ReO_6^{5-}), AsO_3^{3-}, SO_2^{2-}, ClO_3^-, SbO_6^{7-}, PO_4^{3-}; **b.** ReO_4^- 22.6 (or perhaps ReO_6^{5-} –6.8), AsO_3^{3-} –3.5, SO_2^{2-} 1.0, ClO_3^- 16.9, SbO_6^{7-} –27.2, PO_4^{3-} 2.2; **c.** ReO_4^- nonbasic (or ReO_6^{5-} very strongly basic), AsO_3^{3-} (very) strongly basic, SO_2^{2-} moderately or strongly basic, ClO_3^- nonbasic, SbO_6^{7-} very strongly basic, PO_4^{3-} moderately basic.

3.73. a. $pK_b = 18.3$, so nonbasic; **b.** $pK_b = -11.6$, so very strongly basic; **c.** $pK_b = -11.3$, so very strongly basic.

3.75. a. Nonbasic, very strongly basic, moderately basic; **b.** ReO_4^- 22.6, AlO_4^{5-} –18.2, AsO_4^{3-} 2.2; **c.** (in the same order) 6.5, 14.0, 12.0; **d.** AlO_4^{5-} cannot exist as such in water.

3.77. a. –3.5; **b.** strongly or very strongly basic; **c.** $Ca_3(AsO_3)_2$; **d.** $\left(\begin{matrix}\ddot{\text{O}}\text{—}\overset{..}{\text{As}}\text{—}\ddot{\text{O}}:\\|\\:\ddot{\text{O}}:\end{matrix}\right)^{3-}$

3.79. (One way of reasoning:) Increasing negative charge increases the attraction for the H of water, and hence increases basicity. More oxo groups allow the dispersal of the negative charge, and hence reduce basicity. PO_4^{3-} is moderately basic, with three negative charges dispersed over four oxo groups. In the pyrophosphate ion, four negative charges are dispersed over six oxo groups; in the tripolyphosphate ion, there are five negative charges dispersed over eight oxo groups. If we consider the number of oxo groups to be proportional to size, we can calculate the ratio $Z^2/r \approx Z^2/$(number of oxo groups). For PO_4^{3-} this ratio is 9/4 = 2.25; for $P_2O_7^{4-}$ this is 16/6 = 2.67; for $P_3O_{10}^{10-}$ this is 25/8 = 3.12. Therefore, the basicities may decrease slightly along this series. (Experimentally the pK_b values are 2.0, 4.9, and 3.5, respectively. The differences are small; all are experimentally moderately basic anions.)

3.81 a. IF_4^-; **b.** WF_8^{2-}; **c.** TeF_7^-.

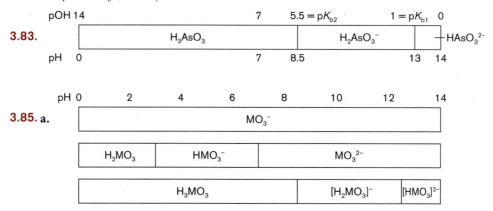

b. MO_3^-, MO_3^{2-}, HMO_3^{2-}, and none, respectively.

3.87. a. The second diagram is for arsenic(III), the first is for arsenic(V); **b.** reading from the left side of the first diagram, the forms of arsenic(V) are H_3AsO_4, $H_2AsO_4^-$, $HAsO_4^{2-}$, and AsO_4^{3-}; in the second diagram, the forms are H_3AsO_3, $H_2AsO_3^-$, $HAsO_3^{2-}$, and AsO_3^{3-}; **c.** H_3AsO_3, weakly acidic; **d.** H_3AsO_4, moderately acidic; **e.** $H_2AsO_4^-$ and H_3AsO_3; **f.** 3.

3.89. a. IrO_4^+ or IrO_6^{3-}

b.

pH –4	–2	0	2	4	6	8
$HIrO_6^{2-}$		IrO_6^{3-}				

c. IrO_6^{3-}; **d.** IrO_4^+.

3.91. a. Strongly acidic; **b.** weakly acidic; **c.** H_6TeO_5 weakly acidic; **d.** H_3IrO_6 would be very strongly acidic.

3.93. a. Three, very strong acid; **b.** the three oxo groups are found on three different phosphorus central atoms; each has one oxo group, as in H_3PO_4; the steps in Section 3.7 should be applied to the number of oxo groups *per central atom*.

3.95. a. Na, Sr, Ca, Mg, and Y; **b.** Na, Sr, Ca, Mg, Y, Th, Al, and Zr (as the oxo cation ZrO^{2+}); **c.** Na (as cation), Al (as protonated oxo anion), Sn (as protonated oxo anion).

3.97. a. Cl as ClO_4^-, S as HSO_4^-, P as $H_3PO_4(aq)$, and Al as Al^{3+}; **b.** Cl as ClO_4^-, S as SO_4^{2-}, P as an equimolar solution of HPO_4^{2-} and PO_4^{3-}, and Ge as $H_3GeO_4^-$.

3.99. a. Nonacidic; **b.** feebly acidic.

3.101. a. +1, $Li^+(aq)$; **b.** +3, $Al(OH)_3$ or Al_2O_3; **c.** +6, WO_4^{2-}; **d.** HsO_2 or $Hs(OH)_4$.

3.103. PuO_2; perhaps Pu^{4+} in the case of acid-rain contamination.

3.105. a. Cation; **b.** PoO_5^{6-}, not present; or PoO_3^{2-}, might be present.

3.107. a. Titanium(III) chloride; **b.** iron(III) phosphate; **c.** silver(I) periodate; **d.** mercury(II) tellurate.

3.109. a. Strontium bromate; **b.** gold(III) selenate; **c.** potassium stannite; **d.** mercury(II) periodate; **e.** thallium(I) iodate; **f.** zinc(II) perneptunate.

3.111. a. $HMnO_4$, very strongly acidic; **b.** H_2SeO_4, strongly acidic; **c.** H_3AsO_3, weakly acidic; **d.** H_2SeO_3, moderately acidic; **e.** H_6TeO_6, weak acid.

3.113. a. SiO_4^{4-}, very strongly basic; **b.** TeO_6^{6-}, very strongly basic; **c.** BrO_4^-, nonbasic; **d.** SO_3^{2-}, moderately basic.

3.115. a. OCl^-, moderately basic; **b.** NpO_6^{5-}, very strongly basic; **c.** NO_2^-, (feebly) basic; **d.** FeO_4^{2-}, feebly basic.

3.117. a. $Ca_3(BO_3)_2$; **b.** $Sr(ClO_4)_2$; **c.** $EuSO_3$; **d.** $Fe_3(PO_4)_2$; **e.** $CrCO_3$.

3.119. a. N 3, Br 4, P 4, Sb 6 (or 4), Np 6 (or 4); **b.** NO_3^-, BrO_3^-, PO_4^{3-}, SbO_6^{7-}, NpO_6^{7-}; **c.** nitrate, bromate, phosphate, antimonate, neptunate; **d.** nonbasic, nonbasic, moderately basic, very strongly basic, very strongly basic; **e.** dissolved, dissolved, dissolved, sediments, sediments; **f.** phosphate ion, as HPO_4^{2-} and $H_2PO_4^-$.

3.121. a. Permanganate, perneptunate, octafluorotungstate(VI), hypoiodite; **b.** nonbasic, very strongly basic, nonbasic, moderately basic; **c.** 22.6, −6.8, (not an oxo anion), 5.5; **d.** 11.2; **e.** MnO_4^-, WF_8^{2-}, IO^-, $H_4NpO_6^-$.

3.123. If your class did not do Experiments 3 and 4, ask your instructor to download and provide you with the results and answers for this question.

Chapter 4

4.1. Fluorite stoichiometry MX_2: **a, b, f**. Anti-fluorite stoichiometry M_2X: **d, e**.

4.3. a. 8/3 or 2.67; two of the Li^+ ions have coordination number 3 and one has coordination number 2; **b.** no; the stoichiometry is not 1:1.

4.5. a. Four cations and four anions; **b.** one cation and one anion.

4.7. a. Radius ratio = 0.51, coordination numbers 6 and 6, NaCl lattice; **b.** radius ratio = 0.42, coordination numbers of 6 and 4, Al_2O_3 (corundum) lattice; **c.** radius ratio = 0.39, coordination numbers of 4 and 2, SiO_2 (β-quartz) lattice.

4.9. a. NaAt = 0.513 NaCl, KAt = 0.672 NaCl, RbAt = 0.735 CsCl, CsAt = 0.800 CsCl; **b.** $MgAt_2$ = 0.38 SiO_2, $CaAt_2$ = 0.504 TiO_2, $SrAt_2$ = 0.584 TiO_2, $BaAt_2$ = 0.659 TiO_2; **c.** MgPo = 0.38 ZnS, CaPo = 0.504 NaCl, SrPo = 0.584 NaCl, BaPo = 0.659 NaCl.

4.11. a. 0.802; coordination number of Ce = 8, of O = 4; fluorite (CaF_2); **b.** 0.261; coordination number of Si = 4, of Te = 2; SiO_2; **c.** 0.807; coordination number of Tl = 8, of O = 5.33; lattice type not listed; **d.** 0.310; coordination number of B = 4, of N = 4; ZnS.

4.13. Rutile (TiO_2) lattice type with coordination numbers of 6 for Ra^{2+} and 3 for At^-.

4.15. a. 8 for La, 8 for N, CsCl; **b.** 6 for Sc, 6 for N, NaCl.

4.17. No. The CdI_2 and $CdCl_2$ lattice types would be appropriate only for layer lattices with about 90° Si–O–Si bond angles, which is characteristic of quite covalent bonding. This degree of covalent bonding is not expected for elements with as large an electronegativity difference as Si and O.

4.19. a. Not for AgF; **b.** for AgF(s), U = –864 kJ mol^{-1}; for AgBr(s), U = –702 kJ mol^{-1}; **c.** both values are much further from the measured values than in the case of NaCl(s); **d.** the Ag–Br covalent bond energy is more of a factor than the Ag–F covalent bond energy.

4.21. a. –2628 kJ mol^{-1}; **b.** –11,034 kJ mol^{-1}.

4.23. Born exponent for $Th^{4+} \approx$ 14; U = –8788 kJ mol^{-1}.

4.25. a. $U \approx$ –1647 kJ mol^{-1}; **b.** $U \approx$ –1601 kJ mol^{-1}.

4.27. U = –7918 kJ mol^{-1}.

4.29. a. U = –7841 kJ mol^{-1}; **b.** yes; **c.** $ThTe_2$.

4.31. $CsBrO_4$, $CePO_4$, $BaSeO_4$, $Cd_5(IO_6)_2$, TiO_2.

4.33. $CsReO_4$, Al_2TeO_6, Ga_2O_3. EuO may be insoluble or at least less soluble than 1 M.

4.35. a. U^{3+} weakly acidic, Rb^+ nonacidic, Pb^{4+} very strongly acidic, Al^{3+} moderately acidic; **b.** Rb^+; **c.** U^{3+}, Pb^{4+}, and Al^{3+}.

4.37. $Pu_3(PO_4)_4$, $(CH_3)_4N^+PF_6^-$, and $EuSO_4$; $Eu_3(PO_4)_2$ is probably insoluble; $Pu(SO_4)_2$ is probably soluble.

4.39. a. All chromates are soluble except those of Ca^{2+}, Sr^{2+}, Ba^{2+}, Ra^{2+}, Tl^+, Eu^{2+}, Tm^{2+}, and Yb^{2+}; **b.** ferrates would be predicted in the same manner as chromates; **c.** all pertechnetates are soluble except those of (Na^+?), K^+, Rb^+, Cs^+, and Fr^+; **d.** all silicates are insoluble except those of Na^+, K^+, Rb^+, Cs^+, and Fr^+; **e.** all tellurates are insoluble except those of Na^+, K^+, Rb^+, Cs^+, and Fr^+.

4.41. a. Because a solution of very strongly basic SiO_4^{4-} also contains OH^-, Hg_2SiO_4, and $Hg(OH)_2$ or HgO; **b.** mix Hg^{2+} with OH^- only to see the color of the precipitate; **c.** PO_4^{3-} is not very strongly basic, so it does not produce as high a concentration of OH^-.

4.43. a. Add a solution of a large nonbasic anion such as nitrate; **b.** add a solution of a large, feebly acidic cation such as Ba^{2+} (or perhaps nonacidic Cs^+); **c.** add a solution of a large +1-charged cation such as Cs^+ or Tl^+.

4.45. LiF, $U = -1033$ kJ mol^{-1}; NaCl, $U = -779$ kJ mol^{-1}; AgCl, $U = -910$ kJ mol^{-1}; CaCl$_2$, $U = -2249$ kJ mol^{-1}.

4.47. a. ΔH_{hyd} for NO$_2^-$ = -353 kJ mol^{-1}; **b.** ΔH_{hyd} for NO$_3^-$ = -314 kJ mol^{-1} from Table 2.4; the smaller NO$_2^-$ ion would be expected to have a ΔH_{hyd} that is greater in magnitude.

4.49. a. Soluble in neither; **b.** soluble in neither; **c.** probably soluble in both; **d.** nonpolar solvents only; **e.** soluble in water only; **f.** soluble in neither; **g.** soluble in nonpolar solvent only.

4.51. a. Positive, favorable; **b.** negative, unfavorable; **c.** positive, favorable; **d.** negative, unfavorable; **e.** positive, favorable.

4.53. a. More negative; **b.** more negative; **c.** decrease; **d.** Solubility Rule I and Solubility Tendency IV.

4.55. a. (C$_4$H$_9$)$_4$N$^+$; **b.** Al^{3+}; **c.** Rb$^+$.

4.57. a. Al^{3+}; **b.** Zn^{2+} and Al^{3+}; **c.** K$^+$; **d.** Zn^{2+} and Al^{3+}; **e.** Al$_4$(SiO$_4$)$_3$ and Zn$_2$SiO$_4$; **f.** for Zn(ClO$_4$)$_2$, Al(ClO$_4$)$_3$, K$_4$SiO$_4$, K$_2$SeO$_4$, ZnSeO$_4$, and Al$_2$(SeO$_4$)$_3$, lattice energies are low due to the great difference in size of the cation and anion, so hydration energies predominate and the salt is soluble.

4.59. a. Precipitates, $-T\Delta S$; **b** soluble salts, ΔH, cross combinations have mismatched radii, so they have poor lattice energies compared to hydration energies.

Chapter 5

5.1. a. ~109° at N donor atom; **b.** ~109° at N donor atom and the O donor atom of the –OH group, ~120° at the O donor atom of the C=O group; **c.** ~109° at O donor atoms; **d.** ~109° at S donor atom and O donor atoms.

5.3. a. Linear and 180°; **b.** octahedral and 90°; **c.** square antiprismatic or dodecahedral; are not given in Figure 5.1.

5.5. b.

5.7. Only **e**.

5.9. a. C and D; **b.** C is sexidentate (hexadentate), D is bidentate.

5.11. a. "A" 90° at Co, 109.5° at O, S, and O; "B" 90° at Co, 109.5° at N, C, C, and N; "C" 90° at Co, 109.5° at N, C, C, C, and N; "D" 90° at Co, 120° at O, C, C, C, and O; **b.** "B" is closest to ideal ($n = 5$), followed closely by "D" ($n = 6$); **c.** $n = 6$, as in complex "C," in a boat or chair conformation.

5.13. a. Products; **b.** reactants.

5.15. a. Donor atoms are N's, S's, and S's, respectively; **b.** *cyclo*-(SCH$_2$CH$_2$)$_3$ and CH$_3$SCH$_2$CH$_2$SCH$_2$CH$_2$SCH$_3$; **c.** *cyclo*-(SCH$_2$CH$_2$)$_3$; **d.** *cyclo*-(SCH$_2$CH$_2$)$_3$ is tridentate, CH$_3$SCH$_2$CH$_2$SCH$_2$CH$_2$SCH$_3$ is tridentate.

5.17. a. N, S, and two O in "A"; S, four F in "B"; S, N, three O in "C"; **b.** "A" and "C" can be chelating; "C" can be macrocyclic; **c.** "C."

5.19. a. B < A < C ; **b.** A and C; **c.** C.

5.21. a. Left side (reactants); **b.** left side (reactants); **c.** left side (reactants); **d.** right side (products).

5.23. a. $CH_3-\underline{O}-CH_2-CH_2-\underline{O}-CH_2-CH_2-\underline{O}-CH_3$ (let us call this "A"), $\underline{C}H_3^-$, $\underline{N}_3^-$, $\underline{Cl}^-$, $H_2\underline{O}$, NH_3, $C_6H_5-CH(\underline{N}H_2)-C(=\underline{O})-\underline{O}^-$ (let us call this "B"), $\underline{S}_2O_3^{2-}$; **b.** "A" is tridentate, "B" is bidentate; **c.** N_3^-, Cl^-, $S_2O_3^{2-}$; **d.** each Mo has a coordination number of 7.

5.25. 1, 2, and **9** can be converted to chelating ligands by moving the nitrogen donor atoms from the *para* to the *ortho* positions in the pyridine rings. Moving the carboxylate (COO^-) functional groups to *ortho* positions in **5** and **6** does *not* generate chelating ligands, however, because if we number a carboxylate oxygen atom as #1 in one carboxylate functional group, an oxygen atom in the next functional group is #6, which is too far away to form a good five- or six-membered chelate ring. Similarly, for **3, 4, 7,** and **8**, moving the functional groups as close as possible would put them too close (for **3**) or too far apart (for **4, 7,** and **8**).

5.27. a. 1, 5, and **9** with **b.** the octahedrally coordinated $[M(OH_2)_6]^{2+}$ ions; **c.** Prussian blue.

5.29. a. Structure **8** of Figure 5.17; **b.** Structure **16**; **c.** Structure **10**; **d.** the structure based on the center TPDC ligand (structure **16**); **e.** Structure **8** with the 2,6-NDC ligand.

5.31. a. Even if MOF-5 and ITQ-4 had the same pore volume, they would have different densities, because their frameworks (SiO_2's or MOF's) have different masses; **b.** all polymorphs of SiO_2 have the same framework unit, SiO_2; **c.** the storage device (the gas cylinder or MOF or SiO_2 polymorph) adds to the weight of the vehicle, which reduces the distance the vehicle can travel per unit of fuel; this is a major problem for gas cylinders, but would not be negligible even for a low-density MOF, the mass of which is mostly *not* fuel.

5.33. a. Yes, but they would not resemble Prussian blue by having cubic unit cells; **b.** ligands 2, 6, and 7 involve 120° angles, so with secondary building units also having 120° angles, you might expect six-sided pores; ligands 3, 4, and 8 involve 109.5° angles, so with secondary building units also having 109.5° angles, you expect tetrahedral pores (holes).

5.35. b, c, and **e.**

5.37. a. Products; **b.** products; **c.** reactants; **d.** products; **e.** products.

5.39. a. Products; **b.** products; **c.** reactants.

5.41. a. Pd^{2+} is the soft acid, and SCN^- is a soft base; **b.** Co^{3+} is the hard acid, NH_3 is a borderline base, and SeO_4^{2-} is a hard base.

5.43. a. $K^+ < Cu^+ < Ag^+ < Au^+$; **b.** $F^- < Cl^- < Br^- < I^-$; **c.** $Mo^{6+} < Mo^{4+} < Mo^{2+}$; **d.** $BF_3 \lesssim B(OCH_3)_3 < B(CH_3)_3$; **e.** $Fe^{3+} < Fe^{2+} < FHg^+ < CH_3Hg^+$; **f.** $(CH_3)_2O < (CH_3)_2S < (CH_3)_2Se$.

5.45. a. $F^- < Cl^- < Br^- < I^-$; **b.** $Ir^{4+} < Ir^{3+} < Ir^{2+}$; **c.** $Pa^{5+} < Po^{4+} < Po^{2+} < Pd^{2+} < Pt^{2+}$; **d.** $Ac^{3+} < As^{3+} < Ag^{2+} < Ag^+ < Au^+$; **e.** $BF_3 < InF_3 < InI_3 < TlI_3$; **f.** $[Co(H_2O)_5]^{3+} < [Co(NH_3)_5]^{3+} < [Co(NH_3)_5]^{2+} < [Co(PH_3)_5]^+$.

5.47. a. Reactants; **b.** products; **c.** products.

5.49. a. Reactants; **b.** reactants; **c.** reactants; **d.** products.

5.51. a. N donor atom is borderline; **b.** N donor atom is borderline and O donor atoms are hard; **c.** O donor atoms are hard; **d.** O donor atoms are hard and S donor atom is soft.

5.53. a. Reactants; **b.** products; **c.** products; **d.** products.

5.55. The monatomic anion or atom. Adding hard-base oxo groups increases the oxidation number of the central atom and diminishes the softness associated with it.

5.57. a. $Hg(CN)_2$; **b.** $Ag_2(S-SO_3)$; **c.** $Mg(O_3S-S)$; **d.** $Hg(NCO)_2$; **e.** $Hg(CNO)_2$.

5.59. Very small size (and high charge in the case of B^{3+}).

5.61. a. $U = -1850$ kJ mol^{-1}; **b.** the experimental lattice energy of CaI_2 is 237 kJ mol^{-1}, which is more exothermic than expected, because there is some covalency involved; **c.** the discrepancy for CdI_2 itself is 567 kJ mol^{-1} more; it is greater for CdI_2 because this is a borderline acid–soft-base salt, which would have more covalent character than CaI_2, a hard-acid–soft-base salt.

5.63. a. As_2O_5; **b.** SrF_2; **c.** $La_2(CO_3)_3$; **d.** MgF_2; **e.** LiF.

5.65. a. Softness , chelation, and strength; **b.** strength and chelation; **c.** strength.

5.67. CdTe, AgI, TiO_2, $PtAs_2$.

5.69. a. Tl_2S; **b.** Cu_2S; **c.** none; **d.** in theory, $Cr_2(CO_3)_3$—but in practice, $Cr(OH)_3$; **e.** SnS.

5.71. a. Zr soluble, Ag insoluble, Sb soluble; **b.** Zr insoluble, Ag insoluble, Sb insoluble.

5.73. a. Sr^{2+} soluble, Bi^{3+} soluble, Eu^{2+} soluble, Co^{2+} soluble; **b.** Sr^{2+} soluble (but reacts with water), Bi^{3+} insoluble, Eu^{2+}soluble (but reacts with water), Co^{2+} insoluble; **c.** Sr^{2+} lithophile, Bi^{3+} chalcophile, Eu^{2+} lithophile, Co^{2+}chalcophile; **d.** Sr^{2+} silicate, Bi^{3+} sulfide, Eu^{2+} silicate, Co^{2+} sulfide.

5.75. a. Zr is a lithophile, Ag is *not* a lithophile, Sb is *not* a lithophile; **b.** Zr silicate, Ag sulfide, Sb sulfide.

5.77. a. Sr^{2+} lithophile, Bi^{3+} chalcophile, Eu^{2+} lithophile, Co^{2+} chalcophile; **b.** Sr^{2+} silicate, Bi^{3+} sulfide, Eu^{2+} silicate, Co^{2+} sulfide.

5.79. a. Lithophile, Be_2SiO_4; **b.** lithophile, CaF_2; **c.** chalcophile/siderophile, $CoAs_2$; **d.** lithophile, ThO_2.

5.81. a. Fe^{4+} is listed in Table C, so it might be possible; **b.** Pt^{6+} is not listed in Table C, but is the maximum oxidation number, so it might be plausible; **c.** Pd^{6+} is not listed in Table C and also exceeds the maximum oxidation number of Pd, so it is *not* plausible.

5.83. All show toxicity pictograms. The Poison/Toxic pictogram is found for nickel(II) sulfate and copper(II) sulfate. The Health Hazard pictogram is shown for cobalt(II) sulfate and nickel(II) sulfate. The Environmental Hazard (fish) pictogram is shown for all of them except iron(II) sulfate, which is probably the least toxic.

5.85. Soft-acid metal ions; cysteine groups.

5.87. c.

5.89. a. Ac^{3+}; **b.** Mt^{2+}; **c.** Po^{2-}; **d.** Fr^+.

5.91. a. H_2Te is expected to be very toxic; **b.** ZrO_2 is insoluble and unavailable in natural waters, so it is probably nontoxic; **c.** $CoCl_2$ contains the borderline-acid cation, Co^{2+}; such cations tend to be essential in small quantities but toxic in larger quantities; **d.** Shahrokhi Semiconductor Co., the company that discharges H_2Te.

5.93. a. Mt^{2+}; **b.** Pt^{2+}; **c.** Pb^{2+}.

5.95. a. HgS(*s*); **b.** $HgCl_2$(*aq*); **c.** $Hg(CH_3)_2$; **d.** $Hg(CH_3)_2$.

5.97. a. This contains As^{5+}, a hard acid due to its high charge. **b.** This contains :As^{3+}, which could be a soft base due to the unshared electron pair on arsenic, but its acidity is likely more dominant. Its charge of +3 and small size favor hardness, while its high electronegativity favors softness. **c.** The As atom has an unshared electron pair; this is a soft base. **d.** The unshared electron pairs are on oxygen; this is a hard base. **e.** The unshared electron pair on As could make this a soft base, while the unshared electron pairs on oxygen make this ambidentate ligand also a hard base. **f.** This insoluble salt is probably not an acid or a base. **g.** Only $FeAsO_4$ might be nontoxic due to its great insolubility. $AsCl_3$, containing an evidently soft As^{3+} cation, is among the most toxic. AsO_3^{3-} would react with stomach acid to generate As^{3+}, so it would also be very toxic. :$As(C_6H_5)_3$, as a soft base, could be quite toxic, attacking essential borderline acids.

5.99. The last ligand.

5.101. (From left to right:) Useless; Be^{2+}; useless; Hg^{2+}; useless; Cu^{2+}.

5.103. a. G; **b.** F.

5.105. a. The isonitrile ligand is neutral; the Tc is Tc^+. **b.** Each methylenediphosphonate ligand has a –2 charge; the central metal ion is Tc^{4+}. **c.** Tc^{5+}. **d.** Both Tc^{4+} and Tc^{5+} are expected to be hard acids; there is no way to rate which is harder or softer, but Tc^+ is definitely a soft(er) acid. The MDP^{2-} and OH^- ligands are both hard bases; the O^{2-} ligand is a hard base; $HMPAO^{2-}$ is a borderline base; $R–N{\equiv}C:$ is a soft base. The softest Tc ion, Tc^+, is associated with the softest ligand, $R–N{\equiv}C:$. But since the hardest acid cannot be identified, we cannot say whether it is associated with the hardest ligand set.

5.107. No; there is no reason to expect a 17-membered chelate ring to be stable. Pt^{2+} is a soft acid and would be expected to coordinate to soft S- or Se-donor atoms in the body in preference to the borderline-base N atoms of guanine.

Chapter 6

6.1. a. $ClO_2 + 5e^- + 4H^+ \rightarrow Cl^- + 2H_2O$, $Br_2 + 6H_2O \rightarrow 2BrO_3^- + 10e^- + 12H^+$;
b. $V_2O_5 + 2e^- + 6H^+ \rightarrow 2VO^{2+} + 3H_2O$, $NO + 2H_2O \rightarrow NO_3^- + 3e^- + 4H^+$;
c. $O_2 + 4e^- + 4H^+ \rightarrow 2H_2O$, $2HF + H_2O \rightarrow OF_2 + 4e^- + 4H^+$.

6.3. a. $2MnO_4^- + 3U^{4+} + 2H_2O \rightarrow 2MnO_2 + 3UO_2^{2+} + 4H^+$; **b.** $E°_{cell} = 1.37$ V.

6.5. a. WO_3 +6, WO_2 +4, H_2MoO_4 +6, MoO_2 +4, $Cr_2O_7^{2-}$ +6, Cr^{3+} +3;
b. A = WO_3, B = WO_2, C = H_2MoO_4, D = MoO_2, E = $Cr_2O_7^{2-}$, F = Cr^{3+};
c. $WO_3 + 2H^+ + 2e^- \rightarrow WO_2 + H_2O$, $H_2MoO_4 + 2H^+ + 2e^- \rightarrow MoO_2 + 2H_2O$, $Cr_2O_7^{2-} + 14H^+ + 6e^- \rightarrow 2Cr^{3+} + 7H_2O$.

6.7. a. CoO_2; **b.** H_2Se; **c.** Eu^{2+}; **d.** IrO_4^{2-}; **e.** AmO_2^{2+}.

6.9. a. Ta; **b.** Pr has no +5 oxidation state—otherwise Bi as Bi_2O_5.

6.11. Br as BrO_4^-.

6.13. a. MnO_4^- is the most oxidizing *stable* form, while MnO_4^{2-} is the most oxidizing of all forms; **b.** Mn(s); **c.** $Mn^{2+} + 2e^- = Mn$; **d.** $MnO_4^- + 3e^- + 4H^+ = MnO_2 + 2H_2O$.

6.15. a. $E°_{cell} = +0.96$ V, reaction goes to products; **b.** $E°_{cell} = +0.93$ V, reaction goes to products; **c.** $E°_{cell} = -0.81$ V, reaction does not go; **d.** $E°_{cell} = +0.70$ V, reaction goes to products.

6.17. a. Only Cl_2; **b.** Zn, Bi, Sc, Tm, Os; **c.** none.

6.19. a. NO_3^-, N_2, (N_2O_4, NO_2, NH_3OH^+, $N_2H_5^+$); **b.** for example,
$2NO_3^- + 5Mn + 12H^+ \rightarrow N_2 + 6H_2O + 5Mn^{2+}$; **c.** MnO_4^-, MnO_2, (MnO_4^{2-}); **d.** for example,
$2MnO_4^- + 2NH_4^+ \rightarrow N_2 + 2MnO_2 + 4H_2O$, $3MnO_2 + 2NH_4^+ + 4H^+ \rightarrow N_2 + 3Mn^{2+} + 6H_2O$.

6.21. a. All Group 1 and 2 elements except hydrogen; **b.** B, Al, F, Si, P, Cl, Ga, Ge, In, Tl, and the Group 18/VIII noble gases; **c.** Sc, Y, Lu, Zr, Hf, Nb, Ta, Co, Ni, Zn, Cd, and Au; **d.** all *f*-block elements except U, Np, Pu.

6.23. a. BrO_4^-, BrO_3^-, H_4XeO_6, and XeO_3; **b.** H_5IO_6, H_5AtO_6, AtO_3^-, PoO_3, Bi_2O_5, PbO_2, and Tl^{3+}; **c.** K, Ca, GeH_4, Rb, Sr, H_2Te, Cs, Ba, BiH_3, and H_2Po; **d.** Ga, Ge, AsH_3, H_2Se, In, Sn, SbH_3, Tl, and Pb.

6.25. a. None; **b.** Ce^{4+}, NpO_2^{2+}, AmO_2^{2+}, Bk^{4+}, and No^{3+}; **c.** *all* of the *f*-block metals, and also Yb^{2+}, Cf^{2+}, Es^{2+}, and Fm^{2+}; **d.** Eu^{2+}, Pa^{4+}, U^{3+}, and Md^{2+}.

6.27. a. Tb, V, Tm > Th > Ta, V, Ti > Tl > Sn > Tc > W; **b.** K > Pr, V, Pm > Pu > Pa > Pd > Pt.

6.29. a. Reacts with all; **b.** unreactive with 1, 2, perhaps 3; **c.** reacts with 2 and 3; **d.** reacts with all; **e.** unreactive with 1, 2, perhaps 3; **f.** reacts with all; **g.** reacts with 2 and 3.

6.31. a. Reacts with 2, 3; **b.** unreactive with all; **c.** reactive with all; **d.** reacts with 3; **e.** reacts with 2, 3; **f.** reacts with 2 and 3; **g.** reacts with all.

6.33. Selenate is a stronger oxidizing agent than sulfate. $HSeO_4^-$ has a predominance region that is higher than that of HSO_4^-, so it would be more deserving of the "Oxidizer" symbol.

6.35. a. I_2 is least likely to oxidize, while (F_2 and) Cl_2 are most likely to be labeled oxidizers; **b.** C is least likely to oxidize, while (F_2 and) O_2 are most likely to be labeled oxidizers.

6.37. a. F^- and O^{2-}; **b.** salts of CH_3^-, Ge^{4-}, and H^-.

6.39. a. Ge^{4-}; **b.** F^-; **c.** salts of the C^{4-} and Sn^{4-} anions.

6.41. a. Potentially explosive; **b.** stable; **c.** potentially explosive; **d.** potentially explosive; **e.** probably stable; **f.** probably unstable; **g.** potentially explosive; **h.** potentially explosive.

6.43. a. Neither stable nor explosive; **b.** explosive; **c.** neither stable nor explosive.

6.45. a. Insoluble; unstable and is likely to be explosive; **b.** soluble; unstable and likely to be explosive; **c1.** could be explosive; **c2.** it should be possible to grind KNO_3 and MnO_2 together safely.

6.47. a. $Br^- < NH_4^+ < SbH_3$; **b.** $Sb_2O_5 < NO_3^- < BrO_4^-$; **c.** $Sb < N_2 < Br_2$; **d.** SbH_3 with BrO_4^-.

6.49. a. Guanidine; **b.** nitroguanidine.

6.51. a. $(CH_3)_5IO_6$; **b.** H–O–O–H; **c.** CH_3At.

6.53. b. Trinitromethane is explosive, because it contains both oxidizing and reducing functional groups.

6.55. b; CuCl would be unstable because $Cu^+(aq)$ is unstable; **c.** AuCl would be unstable because $Au^+(aq)$ is unstable *and* because there is a gap of 0.47 V between the cation and anion predominance regions.

6.57. a. The F^- ion in HF would help complex Ta^{5+}; **b.** the Cl^- ion in HCl would help complex Pt^{4+}.

6.59. F^-.

6.61. a. BiI_3 and BiF_3; **b.** SbI_5 and InI_3; **c.** PAt_5 and PF_5; **d.** FO_4^- and ReO_4^-. Charge-transfer; see Section 6.4B.

6.63. UI_3 is black; UBr_3 is red; UCl_3 is green.

6.65. a. On the left, acidic side of the diagram: Wa^{3+} above Wa^{2+} above $Wa(s)$; on the right, basic side of the diagram: WaO_4^{2-} above WaO_4^{3-} above $Wa(OH)_3$ above $Wa(OH)_2$ [above $Wa(s)$]; **b.** Wa; **c.** WaO_4^{2-}; **d.** WaO_4^{3-}.

6.67. a. $A = BzO_2^{2+}$, $B = BzO_3$, $C = BzO_4^{2-}$, $D = Bz^{4+}$, $E = BzO_2$, $F = Bz$; **b.** $A|B$, $B|C$, $D|E$;
c. $D|F$; **d.** A/D, $B\backslash D$; B/E, $C\backslash E$; $E\backslash F$.

6.69. a. $UO_2^{2+}(aq)$ in solution or $UO_3(s)$ or perhaps $U_3O_8(s)$ in the sludge;
b. $UO_2^{2+}(aq)$ in solution; **c.** no.

6.71. a. PuO_2, in sludge; **b.** react; the probable overall reaction is $Pu + 2H_2O \rightarrow PuO_2 + 2H_2$,
although the first step could be $2Pu + 6H^+ \rightarrow 2Pu^{3+} + 3H_2$; **c.** Pu^{3+} in solution.

6.73. a. $Am(OH)_4$ or $Am_2O_5 \cdot H_2O$; CrO_2; Mn_2O_3, Mn_3O_4, or MnO_4^{2-}; Bi_4O_7 or BiO_2; I_3^- or I_2;
b. AmO_3, no Cr species, MnO_4^- in acid solution, $Bi_2O_5(?)$, no I species(?); **c.** Am metal,
Cr metal, Mn metal (not shown), BiH_3, none for N or I (assuming 0.5-V overvoltage).

6.75. a. <u>Core</u> and crust; **b.** core only; **c.** <u>core</u> and crust; **d.** core only; **e.** <u>core</u> and crust.

6.77. c. The secondary differentiation of the elements, after the coalescence of Earth.

6.79. -2257 kJ mol^{-1}.

6.81. a. $\Delta H_f = \Delta H_{atom}(Ca) + 2\Delta H_{atom}(F) + IE(1)Ca + IE(2)Ca - 2EA(F) + U - 1215$,
$U = -2630$ kJ mol^{-1}; **b.** $\Delta H_f = \Delta H_{atom}(Ca) + \Delta H_{atom}(O) + IE(1)Ca + IE(2)Ca - EA(1)(O) +$
$EA(2)(O) + U - 636$, $U = -3401$ kJ mol^{-1}; **c.** CaF_2 -2628 kJ mol^{-1}. CaO: Predict CsCl lattice
type and -3571 kJ mol^{-1}; actually has NaCl lattice type and -3540 kJ mol^{-1}.

6.83. a. $\Delta H = -U(Cu_2O_3) - 2IE(3) + 2EA(2) + U(Cu^{2+}oxide)$;
b. $-2IE(3) + 2EA(2) = -2(3554) + 2(-532)$; **c.** the lattice energy of $(Cu^{3+})_2(O^{2-})_3(s)$ will be
larger than that of $(Cu^{2+})_2(O^-)_2(O^{2-})(s)$ due to the larger ionic charges and will favor the Cu^{3+}
structure; the high third ionization energy favors the Cu^{2+} structure.

6.85.

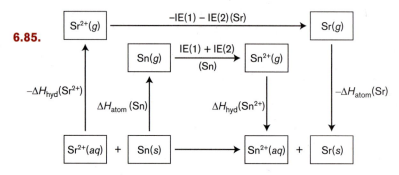

a. $+535$ kJ mol^{-1}; **b.** no, $E° = -2.77$ V.

6.87. a. 3 steps for Qj are $Qj(s) \rightarrow Qj(g) \rightarrow Qj^{3+} \rightarrow Qj^{3+}(aq)$;
3 steps for $Pu^{3+}(aq) \rightarrow Pu^{3+}(g) \rightarrow Pu(g) \rightarrow Pu(s)$;
sum: $Qj(s) + Pu^{3+}(aq) \rightarrow Qj^{3+}(aq) + Pu(s)$.
For Qj: $\Delta H_{atom} + IE(1) + IE(2) + IE(3) + \Delta H_{hyd}$;
for Pu: $-\Delta H_{hyd} - IE(1) - IE(2) - IE(3) - \Delta H_{atom}$; **b.** $+364$ kJ, Pu is more active; **c.** $E° \approx -1.26$ V.

6.89. a. For Pb: $Pb(g) \rightarrow Pb^{2+}(g) \rightarrow Pb^{2+}(aq)$, and $\Delta H = IE(1) + IE(2) + \Delta H_{hyd}$;
for Mg: $Mg^{2+}(aq) \rightarrow Mg^{2+}(g) \rightarrow Mg(g) \rightarrow Mg(s)$, and $\Delta H = -\Delta H_{hyd} - IE(1) - IE(2) - \Delta H_{atom}$;
the overall $\Delta H = +273$ kJ; **b.** no, -1.42 V.

6.91. a. For Pb: $Pb(s) \rightarrow Pb(g) \rightarrow Pb^{2+}(g) \rightarrow Pb^{2+}(aq)$; $\Delta H = +\Delta H_{atom} + IE(1) + IE(2) + \Delta H_{hyd}$;
for Ag: $2Ag^+(aq) \rightarrow 2Ag^+(g) \rightarrow 2Ag(g) \rightarrow 2Ag(s)$; $\Delta H = -2\Delta H_{hyd} - 2IE(1) - 2\Delta H_{atom}$;
b. $\Delta H = -201$ kJ; **c.** yes, Pb; $+1.04$ V.

6.93. a. For U: $U(s) \rightarrow U(g) \rightarrow U^{4+}(g)$;
for F: $2F_2(g) \rightarrow 4F(g) \rightarrow 4F^-(g)$;

and, forming the lattice: $U^{4+}(g) + 4F^-(g) \rightarrow UF_4(s)$;

$\Delta H_f(UF_4) = \Delta H_{atom}(U) + IE(1)U + IE(2)U + IE(3)U + IE(4)U + 4\Delta H_{atom}(F) - 4EA(F) + U(UF_4)$.
b. For Cr: $2Cr^{3+}(aq) \rightarrow 2Cr^{3+}(g) \rightarrow 2Cr^{2+}(g) \rightarrow 2Cr^{2+}(aq)$;
for Zn: $Zn(s) \rightarrow Zn(g) \rightarrow Zn^{2+}(g) \rightarrow Zn^{2+}(aq)$;
$\Delta H = \Delta H_{atom}(Zn) + IE(1)Zn + IE(2)Zn + \Delta H_{hyd}(Zn^{2+}) - 2\Delta H_{hyd}(Cr^{3+}) - 2IE(3)Cr + 2\Delta H_{hyd}(Cr^{2+})$.
c. For Co: $Co^{3+}(aq) \rightarrow Co^{3+}(g) \rightarrow Co^{2+}(g) \rightarrow Co^{2+}(aq)$;
for Cr: $Cr^{2+}(aq) \rightarrow Cr^{2+}(g) \rightarrow Cr^{3+}(g) \rightarrow Cr^{3+}(aq)$;
$\Delta H = -\Delta H_{hyd}(Co^{3+}) - IE(3)Co + \Delta H_{hyd}(Co^{2+}) - \Delta H_{hyd}(Cr^{2+}) + IE(3)Cr + \Delta H_{hyd}(Cr^{3+})$.
d. U is so large for the salt UF_4.

Chapter 7

7.1. a. $A = d_{xz}$, $B = d_{z^2}$, $C = d_{x^2-y^2}$, $D = d_{xy}$; **b.** $A = t_{2g}$, $B = e_g$, $C = e_g$, $D = t_{2g}$.

7.3. a. f_{xyz}, $(f_{x^3}, f_{y^3}, f_{z^3})$, $(f_{y(x^2-z^2)}, f_{x(z^2-y^2)}, f_{z(y^2-x^2)})$; **b.** greatest repulsion: $(f_{x^3}, f_{y^3}, f_{z^3})$, greatest stabilization: f_{xyz}.

7.5. a. Cr^{3+} is t_{2g}^3 with a CFSE of $-1.2\Delta_o$; **b.** Cu^{2+} $t_{2g}^6 e_g^2$ with a CFSE of $-1.2\Delta_o$; **c.** Ti^{3+} is t_{2g}^1 with a CFSE of $-0.4\Delta_o$.

7.7. a. **b.**

c. $d^1 = t_{2g}^1$, $d^2 = t_{2g}^2$, $d^3 = t_{2g}^3$; **d.** $d^1 = -0.4\Delta_o$, $d^2 = -0.8\Delta_o$, $d^3 = -1.2\Delta_o$; **e.** $d^1 = V^{4+}, Nb^{4+}, Ta^{4+}, Mo^{5+}, W^{5+}, Os^{7+}$; $d^2 = Ti^{2+}, V^{3+}, Nb^{3+}, Ta^{3+}, Cr^{4+}, Mo^{4+}, W^{4+}, Tc^{5+}, Re^{5+}$; $d^3 = V^{2+}, Cr^{3+}, Mn^{4+}, Tc^{4+}, Re^{4+}, Ru^{5+}, Os^{5+}$.

7.9. a. $t_{2g}^2 e_g^0 = d^2$ and $t_{2g}^6 e_g^2 = d^8$; **b.** $t_{2g}^4 e_g^0$ and $t_{2g}^6 e_g^0$; **c.** $t_{2g}^2 e_g^0$, $t_{2g}^6 e_g^2$, and $t_{2g}^4 e_g^0$.

7.11. a. t_{2g}^6, $-2.4\Delta_o + 2P$, 0 unpaired e⁻, $[Co(CN)_6]^{3-}$; **b.** $t_{2g}^3 e_g^2$, $0\Delta_o$, five unpaired e⁻, $[Fe(H_2O)_6]^{3+}$; **c.** $t_{2g}^4 e_g^2$, $-0.4\Delta_o$, four unpaired e⁻, $[Co(H_2O)_6]^{3+}$; **d.** t_{2g}^5, $-2.0\Delta_o + 2P$, one unpaired e⁻, $[Fe(CN)_6]^{3-}$.

7.13. a. $[Gd(H_2O)_6]^{3+}$; **b.** $Gd^{3+}(g)$.

7.15. a. Two; **b.** paramagnetic; **c.** $\chi_m^{corr} = 0.0034$ when $T = 273$ K and 0.0100 when $T = 100$ K.

7.17. a. $\mu = 5.92$ BM and $\chi_m^{corr} = 0.0159$ cgs units; **b.** $\mu = 0$ BM and $\chi_m^{corr} = 0$ cgs units; **c.** $\mu = 4.90$ BM and $\chi_m^{corr} = 0.0109$ cgs units; **d.** $\mu = 1.73$ BM and $\chi_m^{corr} = 0.00136$ cgs units; **e.** $\mu = 4.90$ BM and $\chi_m^{corr} = 0.0109$ cgs units.

7.19. a. Electron–electron repulsions differ among different pairs of d orbitals if more than one d electron are present; **b.** A, 19,000 cm⁻¹; **c.** yellow.

7.21. a. C, because it has only one peak (there is no electron–electron repulsion); **b.** green; **c.** 10,000 cm⁻¹ or 120 kJ mol⁻¹; **d.** complex C.

7.23. a. Pale or pastel violet, with $\varepsilon \approx 0.01$; **b.** intensely yellow; **c.** colorless, d–d transition.

7.25. a. Between hydrated Cr^{3+} (17,000 cm⁻¹) and $[Cr(NH_3)_6]^{3+}$ (21,600 cm⁻¹), Δ_o may be about 19,000 cm⁻¹; **b.** $\approx -23,000$ cm⁻¹; **c.** larger; **d.** increases Δ_o; **e.** spectrum (c).

7.27. A most likely has a weak-field ligand; C most likely has a strong-field ligand.

7.29. a. 8130 cm^{-1}; **b.** increase; **c.** increase; **d.** A is ammonia, B is CO, and C is iodide; the order of the Δ_o values should match the approximate order of the ligands in the spectrochemical series.

7.31. a. Upper part, because the donor C atom has an empty p orbital in the Lewis structure, so the ligand can be a π acceptor; **b.** lower part, because the donor N atom has two unshared electron pairs, so the ligand can be a π donor; **c.** middle, because the donor O atom has only one unshared electron pair, so the ligand is thus only a σ donor.

7.33. $N\equiv C-O^- = C\equiv N-O^-$ (O donors) $< N\equiv C-O^-$ (N donor) $< C\equiv N-O^-$ (C donor).

7.35. V^{2+} CFSE = -170 kJ mol^{-1}, which is 8.9% of the hydration energy. $Cr^{2+} = -101$ kJ mol^{-1} = 5.6%. $Mn^{2+} = 0$. $Fe^{2+} = -48$ kJ mol^{-1} = 2.5%. $Co^{2+} = -86$ kJ mol^{-1} = 4.2%. $Ni^{2+} = -124$ kJ mol^{-1} = 5.9%. $Cu^{2+} = -94$ kJ mol^{-1} = 4.5%.

7.37. Overall, lattice energies become more negative in this series because the radii of the metal ions decrease, and U is proportional to Z^+Z^-/r. But we expect reversals when the e_g orbitals are occupied, at d^4, d^5, d^9, and d^{10} for weak-field oxides.

7.39. a. Peak at d^6: $Mn^{2+} < Fe^{2+} > Co^{2+} > Ni^{2+} > (?) Cu^{2+} > Zn^{2+}$;
b. peak at d^8 or (in practice) d^9: $Fe^{3+} < Co^{3+} < Ni^{3+} < Cu^{3+} < (?) (Zn^{3+}) > Ga^{3+}$.

7.41.

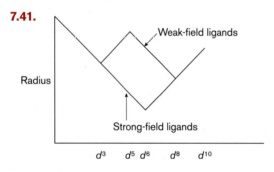

The radii increase when the e_g orbitals are occupied, because these repel the ligands. This occurs at d^7–d^{10} for strong-field complexes and at d^4, d^5, d^9, and d^{10} for weak-field complexes.

7.43. The most extensive set of radii (in pm) to plot would be one of the following series: Period 5 +4 ions, Period 6 +4 ions, or Period 5 +3 ions.

	d^0	d^1	d^2	d^3	d^4	d^5	d^6	d^7	d^8	d^9	d^{10}
Per. 5 +4	Zr	Nb	Mo	Tc	Ru	Rh	Pd				Sn
	86	82	79	78	76	74	75				83
Per. 6 +4	Hf	Ta	W	Re	Os	Ir	Pt				Pb
	85	82	80	77	77	76	76				91
Per. 5 +3	Y		Nb			Ru	Rh		Ag		In
	104		86			82	80		89		94

In any of the three series, the radii drop from d^0 to a minimum at d^6 and then rise to a maximum at d^{10}.

7.45. Twist the top square by 45° with respect to the bottom square of ligands. This reduces the repulsion between the two squares of ligands. If we take the z axis as the one connecting the two squares, this twisting will reduce the repulsions between the ligands and the d_{xy} orbital, raise the repulsions involving the $d_{x^2-y^2}$ ligand, and have no effect on the repulsions involving the d_{z^2} ligand.

7.47. Favoring loss of ligands: steric bulk of large ligands. Favoring retention of ligands: loss of crystal field stabilization energy in tetrahedral complexes; loss of (coordinate covalent) bond energies when metal does not achieve its maximum (total) coordination number.

7.49. a. Tetrahedral, 8660 cm^{-1}; **b.** square planar, 13,800 cm^{-1}; **c.** square planar; **d.** only the square planar complexes are diamagnetic.

7.51. a. Square pyramidal. **b.** d^7.

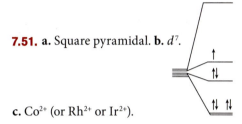

c. Co^{2+} (or Rh^{2+} or Ir^{2+}).

d. Strong-field d^5 would *probably* have the electron configuration $(d_{xz}, d_{yz})^3(d_{xy})^1(d_{z^2})^1$, for which the CFSE $= -1.371\Delta_o$ $(+ P)$, but in an *extremely* strong field it could be $(d_{xz}, d_{yz})^4(d_{xy})^1$, for which the CFSE $= -1.914\Delta_o$ $(+ 2P)$.

7.53. a. Authors agree (their Fig. 7); the spectrum has two bands, as expected for a d^9 Ni^{1+} ion with electronic transitions from $(d_{xy}, d_{x^2-y^2}) \rightarrow d_{z^2}$ and from $(d_{xz}, d_{yz}) \rightarrow d_{z^2}$; **b.** $\Delta_o \approx 7100$ cm^{-1} to 7500 cm^{-1}; **c.** this Δ_o is below the Δ_o of Ni$^{2+}(aq)$, 8600 cm^{-1}, which is reasonable for a complex of Ni$^+$, which has a lower charge.

7.55. a. $t_{2g}^6 e_g^1$, $t_{2g}^3 e_g^1$, $t_{2g}^6 e_g^3$; **b.** the ligands along the z direction move away from the metal ion (while the others move in), giving a distorted octahedron.

7.57. The Jahn–Teller distortion greatly lengthens the two axial bonds to Cu, which labilizes ligands in those positions. Additionally, in a tris-chelate complex of Cu^{2+}, a ligand is required to stretch too far to bond both to this position and to an equatorial position.

7.59. TiO$_2$ and CaF$_2$; because F$^-$ is a weak-field ligand, Jahn–Teller distortions are expected mainly at d^4 CrF$_2$ and d^9 CuF$_2$; no.

7.61. a. $\Delta_t = 8400$ cm^{-1}; **b.** (–)6800 cm^{-1}; **c.** the octahedral complex is strongly favored, by 14,600 cm^{-1}.

7.63. Zn^{2+} (d^{10}) is colorless and gives no d–d absorption spectra (or fluorescence or NMR spectra, etc.); the d^1 through d^9 ions are spectroscopically superior. Co^{2+} has the same radius as Zn^{2+} (88 pm), so it is most likely to fit in the enzymes without loss of activity, especially because its d^7 electron configuration, like that of Zn^{2+} but in contrast to those of Ni^{2+} and Cu^{2+}, results in no strong preference for a predetermined geometry, such as octahedral.

7.65. These two ions are d^8 and d^3, respectively, which have strong stabilization energies for octahedral complexes over tetrahedral complexes; this is especially strong for the more highly charged (more acidic) Cr^{3+} ion.

7.67. a. Cr^{3+} is more acidic than Ni^{2+}, so the ligands that must dissociate are held more strongly. **b.** For d^3 Cr^{3+}, the energetic preference for octahedral *versus* square pyramidal coordination is $-0.200\Delta_o$, so the octahedral geometry is favored and dissociation is slow. But for d^4 Cr^{2+}, the energy difference is $+0.314\Delta_o$, so the square pyramidal geometry is favored, and dissociation is rapid. Also, Cr^{3+} is more acidic than Cr^{2+}, so it holds ligands more strongly.

7.69. Complexes **a**, **b**, and **d**.

7.71. The weak-field, high-spin $t_{2g}{}^4 e_g{}^2$ ion Fe^{2+} is too large to fit in the plane of a porphyrin ligand, but if it becomes strong-field, low-spin Fe^{3+} ($t_{2g}{}^5$), it shrinks in size and it can fit into the ring. This is connected to the cooperative bonding of O_2.

7.73. a. Square planar Fe^{2+} is larger than Ni^{2+}, and it does not have the d^8 electron configuration that gives CFSEs favorable for square planar coordination; **b.** Pd^{2+} is a much larger ion; **c.** although Al^{3+} is the right size to fit in the porphyrin ring, its enhanced charge and acidity would result in the coordination of additional ligand(s) to give a coordination number of 5 or 6.

7.75. The ring would no longer be inflexible enough to keep the Fe^{2+} from falling into the plane of the four N donor atoms. Then cooperative bonding would not work.

Chapter 8

8.1. a. SO_3 is monomeric (molecular), S_3O_9 is oligomeric, $_\infty^1[SO_3]$ is chain polymeric; **b.** $SO_3 < S_3O_9 < {}_\infty^1[SO_3]$, SO_3 is most likely a gas.

8.3. a. D; **b.** A; **c.** B; **d.** none.

8.5. a. $_\infty^1[S]$; **b.** $_\infty^2[C]$; **c.** He; **d.** C_{60} or $_\infty^2[C]$ (or $_\infty^1[S]$); **e.** $_\infty^3[C]$ or $_\infty^3[(Th^{4+})(O^{2-})_2]$; **f.** C_{60}; **g.** $_\infty^3[C]$ or $_\infty^3[(Th^{4+})(O^{2-})_2]$.

8.7. a. $_\infty^3[BN]$; **b.** $_\infty^2[BN]$; **c.** < 2.25 g cm^{-3}, $< 3000°C$.

8.9. $C_3H_3(NCH_3)(NR)^+[Al_2Cl_7^-]$.

8.11. 4 for BeF_2, 2 for SiF_4, and 0 for SeF_6.

8.13. a. SxO_3 and EgO_4; **b.** 4 and 6; **c.** SxO_3 has 1 VCS with TCN = 4, or 3 VCS with TCN = 6; EgO_4 has 0 VCS with TCN = 4, or 2 VCS with TCN = 6; **d.** EgO_4 (if 0 VCS); **e.** $_\infty^1[SxO_3]$ (if 1 VCS).

8.15. a. Lewis structure is in Figure 8.5(d), oligomeric molecule; **b.** ionic or macromolecular; **c.** ionic or macromolecular; **d.** Lewis structure is in Figure 8.5(a), monomeric molecule; **e.** ionic or macromolecular.

8.17. Cl_2O_7 has bond lengths of 140 and 171 pm versus the calculated 172 pm. Br_2O_5 has 161 pm and 188 pm versus 187 pm. S_3O_9 has 137, 143, and 163 pm versus 175 pm. Se_4O_{12} has 155 and 177 pm versus 190 pm. P_4O_{10} has 143 and 160 pm versus 183 pm. In all cases the terminal bond is shorter, or the most seriously shorter, than calculated, probably due to its greater involvement of π bonding.

8.19. a. $Cl_2O_5 < Br_2O_5 < I_2O_5$; **b.** $MnO < Mn_2O_3 < MnO_2 > Mn_2O_7$.

8.21. a. 5 or 7 for BaO, 4 for HfO_2, 3 for WO_3, and 2 for OsO_4; **b.** OsO_4; **c.** OsO_4.

8.23. a. $CO_2(g) < P_4O_{10} \ll Na_2O$, Cr_2O_3, and SiO_2; **b.** $CO_2(g) \ll Rb_2O$, SrO, Y_2O_3, and ZrO_2.

8.25. a. A Group 16/VI nonmetal trioxide, because its coordination number can be less than the maximum, so it can link units via bridging oxygen atoms; **b.** the dioxide—although both involve M^{4+} ions; the (bridging) attraction to an O^{2-} ion is stronger than that to an F^- ion.

8.27. a.

8.29. b.

8.31. a. Yes, because the anion lattice has two tetrahedral holes and one octahedral hole per anion. These could be filled by three +1-charged cations. **b.** Because Rb^+ is smaller than Cs^+, the two tetrahedral holes are more likely to be filled by Rb^+ cations, giving Rb_2CsC_{60}.

8.33. a. $NiFe_2O_4$ and Ni_3O_4; **b.** $BaTiO_3$ and $NaTaO_3$.

8.35. $KMgF_3$, $LiBaF_3$, and $TlCoF_3$.

8.37. $FeO_{0.95}O$, $CoO_{0.95}O$, and $EuO_{0.95}O$.

8.39. Only $(Zn,Fe)_2SiO_4$.

8.41. c, **d**, and **e**.

8.43. a. First principle; **b.** first and second principles; **c.** second principle.

8.45. a. The second; **b.** the first.

8.47. a. First substitution, AlP; second substitution; **b.** first substitution, GaAs; second substitution, ZnSe.

8.49. a. (1); **b.** (4); **c.** (1); **d.** (4); **e.** (2).

8.51. a. Clearly shown are one Y and two Ba ions. Eight Cu ions on top and bottom corners give one Cu ion. Eight Cu ions along the edges give two more Cu ions. Six O ions along edges in the top cube and six more in the bottom cube give three O ions. Eight O ions within faces give four more O ions, for a grand total of seven O ions; hence, $YBa_2Cu_3O_7$. **b.** The stoichiometry works out properly for the composition $Y^{3+}(Ba^{2+})_2(Cu^{3+})(Cu^{2+})(O^{2-})_7$, with one-third of the Cu ions being Cu^{3+}. Hence, if these were ordered, these would be the ones in the chains.

8.53. a. Superconductor; **b.** spinel; **c.** perovskite; **d.** impossible.

8.55. a. CN(O) = 6; **b.** CN(O) in top and bottom cubes = 6; CN(O) in middle cube = 6; **c.** equation is valid for both structures.

8.57. a. (Most acidic) $P_4O_{10} > CO_2 > SiO_2 > Cr_2O_3 > Na_2O$ (least acidic); **b.** $CO_2 > ZrO_2 > Y_2O_3 > SrO > Rb_2O$; **c.** $Mn_2O_7 > MnO_2 > Mn_2O_3 > MnO$; **d.** $SO_2 > TeO_2 > TiO_2 > ThO_2$.

8.59. a. HmO_4 and YaO_2; **b.** Mi_2O_7; **c.** LmO_2, no; **d.** LmO_2.

8.61. a. ZrO_2; $ZrO_2 + SiO_2 \rightarrow ZrSiO_4$; **b.** SrO; $SrO + H_2O \rightarrow Sr^{2+}(aq) + 2OH^-(aq)$; **c.** XeO_4; $XeO_4 + 2H_2O \rightarrow 4H^+ + XeO_6^{4-}(aq)$; **d.** SrO.

8.63. a. A volatile oxide would have a high oxidation state so as to have a small central atom and few bridging oxides: MtO_4 or even Mt_2O_9; **b.** a partially protonated MtO_6^{6-}; **c.** a partially protonated MtO_6^{5-} ion, which would be precipitated by any acidic cation.

8.65. a. $Tl_2O + H_2O \rightarrow 2Tl^+(aq) + 2OH^-(aq)$; **b.** $I_2O_5 + H_2O \rightarrow 2H^+(aq) + 2IO_3^-$; **c.** $ClO_2 + 2OH^- \rightarrow ClO_3^-(aq) + ClO_2^-(aq) + H_2O$; **d.** $La_2O_3 + 6H^+ \rightarrow 2La^{3+}(aq) + 3H_2O$; **e.** $B_2O_3 + 2OH^- + 3H_2O \rightarrow 2[B(OH)_4]^-(aq)$; **f.** $MnO + 2H^+ \rightarrow Mn^{2+}(aq) + H_2O$.

8.67. a. $6BaO$ (basic) $+ P_4O_{10}$ (acidic) $\rightarrow 2Ba_3(PO_4)_2$; **b.** TeO_3 (acidic) $+ I_2O_5$ (acidic) $\rightarrow$ NR; **c.** SrO (basic) $+ MoO_3$ (acidic) $\rightarrow SrMoO_4(s)$; **d.** CaO (basic) $+ TeO_3$ (acidic) $\rightarrow CaTeO_4(s)$; or 3CaO (basic) $+ TeO_3$ (acidic) $\rightarrow Ca_3TeO_6(s)$ **e.** CaO (basic) $+ MnO$(basic) $\rightarrow$ NR.

8.69. a. Yes; **b.** $(Ba^{2+})_7(SbO_6^{7-})_2$.

8.71. a. $S + O_2 \rightarrow SO_2$, $SO_2 + \frac{1}{2}O_2 \rightarrow SO_3$, $SO_3 + H_2O \rightarrow H_2SO_4$;
b. $MgO(s) + SO_2(g) \rightarrow MgSO_3(s)$; **c.** 40,300 g MgO.

8.73. CO_2 could be injected underground in depleted gas or oil deposits, or in deep seawater where it would precipitate as a clathrate compound. SO_2 (and SO_3) could be diluted by emitting them from tall smokestacks, or by scrubbing the smokestack with a solution of $Ca(OH)_2$ or a suspension of $Mg(OH)_2$.

8.75. a. Monticellite; **b.** wollastonite; **c.** grunerite; **d.** talc.

8.77. a. (5), a double-chain polysilicate; **b.** (3), a chain polysilicate or cyclic oligosilicate; **c.** (1), a simple silicate; **d.** (3), a chain polysilicate or cyclic oligosilicate.

8.79.

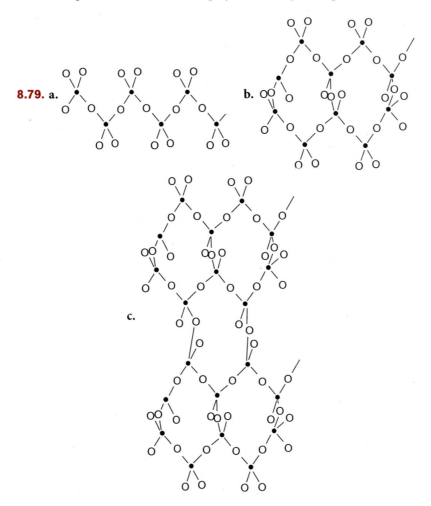

8.81. a. Very strongly basic; **b.** moderately basic; **c.** moderately basic; **d.** nonbasic;
e. pK_a of $(HO)_3SiO_{0.5} = 5.65$, pK_a of $(HO)_2SiO = 2.8$, pK_a of $(HO)_{1.5}SiO_{1.25} = 1.37$, pK_a of $(HO)SiO_{1.5} = 0.00$, so strongly acidic.

8.83. a. Fayalite; **b.** pyrophyllite or zircon.

8.85. a. Simple silicates; **b.** simple silicates; **c.** three-dimensional polymeric metal oxides.

8.87. Only **b.** lepidolite and **d.** petalite.

8.89. $(Mg_5Fe^{III})_2(OH)_8[AlSi_3O_{10}]$.

8.91. a and **b**.

8.93. a. The second principle; **b.** the first principle is used twice.

8.95. a.

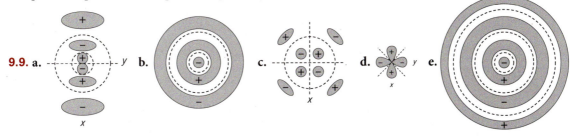

$$\tfrac{1}{\infty}(SiO_3)^{2-} = (O{-}SiO_2)_n; \ (SiO_4)^{4-} = \text{as written.}$$

b. Tremolite $(SiO_{2.75})^{1.5-}$, ureyite $(SiO_3)^{2-}$, chabazite $(EO_2)^{0.33-}$, kaolinite $(SiO_{2.5})^-$, phenacite $(SiO_4)^{4-}$. **c.** Chabazite anion $[Al_{12}Si_{24}O_{72}]^{12-}$ < kaolinite anion $(Si_4O_{10})^{4-}$ < tremolite anion $(Si_4O_{11})^{6-}$ < ureyite anion $(SiO_3)^{2-}$ < phenacite anion SiO_4^{4-}. **d.** Al is part of the aluminosilicate anion in chabazite, but is part of an external counterion in kaolinite.

Chapter 9

9.1. Assuming that you weigh 80 kg, your wavelength $= 2.07 \times 10^{-37}$ m. No; no grating has grates spaced this closely. A particle and not a wave.

9.3. No—comets have elliptical orbits with more complex wave equations.

9.5. a. s orbital; **b.** the p orbitals; **c.** s, p_z, and d_{z^2} orbitals.

9.7. a. No planes, 12 spheres, yes; **b.** 2 planes, 8 spheres, yes; **c.** 4 planes, 0 spheres, no; **d.** 5 planes, 0 spheres, no; **e.** 4 planes, 4 spheres, yes; **f.** 117 planes, 114 spheres, yes.

9.9.

9.11. a. 2, 0, 1, 2 nodal planes; **b.** 1, 3, 2, 1 nodal spheres; **c.** $4d_{x^2-y^2}$, $4s$, $4p_x$, $4\,d_{xy}$.

9.13. a. Orbital #1 has one nodal plane while orbital #2 has two; **b.** orbital #1 has two nodal spheres while orbital #2 has one; **c.** orbital #1 is $4p_x$ while orbital #2 is $4d_{xy}$.

9.15. a. $5g < 5f < 5d < 5p < 5s$; **b.** $5g < 5f < 5d < 5p < 5s$; **c.** $5s < 5p < 5d < 5f < 5g$.

9.17. a. Orbital #2; **b.** orbital #1.

9.19. a. The $1s$ orbital; **b.** the $4s$ orbital; **c.** the $5f$ orbital; **d.** the $4f$ orbital.

9.21. Rearranged full electron configurations are: **a.** $(1s)^2(2s,2p)^8(3s)^2$; **b.** $(1s)^2(2s,2p)^8(3s,3p)^4$; **c.** $(1s)^2(2s,2p)^8(3s,3p)^8(3d)^3(4s)^2$; **d.** $(1s)^2(2s,2p)^8(3s,3p)^8(3d)^{10}(4s,4p)^8(4d)^{10}(4f)^5(5s,5p)^8(6s)^2$; **e.** $(1s)^2(2s,2p)^8(3s,3p)^8(3d)^{10}(4s,4p)^8(4d)^{10}(5s,5p)^7$; **f.** $(1s)^2(2s,2p)^8(3s,3p)^8(3d)^{10}(4s,4p)^8(4d)^{10}(4f)^{14}(5s,5p)^8(5d)^7(6s)^2$; **g.** $(1s)^2(2s,2p)^8(3s,3p)^8(3d)^{10}(4s,4p)^8(4d)^{10}(4f)^{14}(5s,5p)^8(5d)^{10}(5f)^{14}(6s,6p)^8(7s)^2$.

9.23. a. $Z^* = 2.85$; **b.** $Z^* = 11.65$.

9.25. a. Z^* for Ba = 2.85, Z^* for Yb = 2.85, Z^* for Hg = 4.35, Z^* for Rn = 8.25. **b.** Most rapidly across the p block; least rapidly across the f block. **c.** The two types of valence orbitals have the same principal quantum number in the p block: Neither is inside the other very much, so they shield each other poorly, allowing Z^* to increase rapidly. In the f block, the f valence orbitals have principal quantum numbers two below those of the s valence orbitals. Hence, they lie quite far inside the s orbitals, and they are 100% effective at shielding out the s valence orbitals from increases in the true nuclear charge Z as the f block is crossed.

9.27. The $4d$ electrons of Sb.

9.29. a. The $5g$ orbital; **b.** only in the new g block; **c.** in the f block, the d block, and the p block.

9.31. a. Cl; **b.** Ar; **c.** Na.

9.33. a. The first ionization is of a *valence* electron, $6s^1$, but the second and third ionizations are of *core* electrons (from $5p^6$); **b.** $Cs^+(aq) \rightarrow Cs^+(g) \rightarrow Cs^{3+}(g) \rightarrow Cs^{3+}(aq)$
$\Delta H = -\Delta H_{hyd}(Cs^+) + (IE2 + IE3)(Cs) + \Delta H_{hyd}(Cs^{3+})$;
c. $\Delta H_{hyd}(Cs^{3+})$ is unknown, but should be at least nine times as large as $\Delta H_{hyd}(Cs^+)$ due to Z^2/r dependence;
d. $\Delta H = -(-263) + 2230 + 3400 - 9(263) = +3526$ kJ mol^{-1}, which is prohibitively endothermic due to IE(2) and IE(3). However, these values are not worse than the IE(2) and IE(3) values for boron and copper; boron readily forms compounds (but not ions) in the +3 oxidation state, while Cu apparently forms Cu^{3+} ions in the high-temperature superconductor.

9.35. Adding an electron to a Mn atom of valence electron configuration $4s^2 3d^5$ puts two electrons in the same d orbital, which repel each other. Adding an electron to a Zn atom of valence electron configuration $4s^2 3d^{10}$ puts one electron in a higher energy post-valence $4p$ orbital.

9.37. a. Z^* for $4s = 3.75$, Z^* for $3d = 6.25$, Z^* for $3p = 14.75$; **b.** $<r_{max}>$ for $4s = 193.1$ pm, $<r_{max}>$ for $3d = 76.2$ pm, $<r_{max}>$ for $3p = 32$ pm.

9.39. a. $Z^* = +4.35$; **b.** $Z^* = 8.85$; **c.** for $4s$: $<r_{max}> = 166$ pm, for $3d$: $<r_{max}> = 53.8$ pm.

9.41. $4s$: The inner $3d$ orbitals do not extend out as far, due to their lower principal quantum number and their higher Z^*.

9.43. a. $Z^* = 4.95$ for $3d$ and 3.45 for $4s$; $<r_{max}> = 96.2$ pm for $3d$ and 210 pm for $4s$.
b. $Z^* = 4.95$ for $4d$ and 3.45 for $5s$; $<r_{max}> = 146$ pm for $4d$ and 245 pm for $5s$.
c. $Z^* = 4.95$ for $5d$ and 3.45 for $6s$; $<r_{max}> = 171$ pm for $5d$ and 270 pm for $6s$. Ratio of $<r_{max}>$ values of d/s orbitals is 0.458 for Cr, 0.596 for Mo, and 0.632 for W. These calculations slightly show the effect. Due to the small increase in n^*, the $(n + 1)s$ orbital gains less in size relative to the nd orbital for higher values of n.

9.45. $\chi_{AR} = 1.72$. This is close to the Pauling value.

9.47. $\chi_{AR} = 1.02$.

9.49. a. Z^*: Na 2.20, Cu 4.20; $<r_{max}>$: Na, 216 pm, Cu 172 pm; χ_{AR}: Na 1.02, Cu 1.50.
b. Z^*: W 3.45, Nd 2.85, U 2.85; $<r_{max}>$: W 270 pm, Nd 327 pm, U 359 pm; χ_{AR}: W 1.40, Nd 1.07, U 1.18. U appears to be more closely related to Nd.

9.51. a. The Mg atom has a lower value for n, so its valence orbital is smaller and the nucleus can pull in the valence electrons more effectively. This results in Mg being more electronegative than Ca. **b.** The Si atom has a larger Z^*, and the nucleus can pull in the valence electrons more effectively. As a result, Si is more electronegative than Al.

9.53. a. Relativistic effects cause gold's $6s$ orbital to contract and have a greater attraction for electrons (electron affinity)—hence, gold can form an anion, Au^-. **b.** Relativistic effects cause gold's $5d$ orbitals to expand due to the increased shielding by electrons in the contracted s orbitals. Hence, they are better able to overlap other atoms' orbitals, allowing a higher valence and oxidation state.

9.55. a. Ra; **b.** Hg; **c.** Rg, due to greater relativistic contraction of the valence s orbital in Rg.

9.57. a. Elements 115–117 have predicted stable oxidation states = group number – 4; **b.** spin-orbit coupling.

9.59. a. The $7p$ subshell; **b.** the $6p$ subshell; **c.** Mc.

9.61. a.

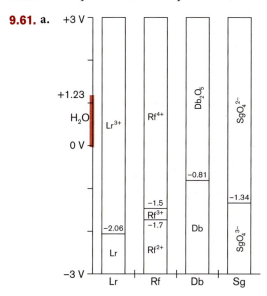

b. The redox predominance region for Lr^{3+} is larger than those for Sc^{3+}, Y^{3+}, or Lu^{3+}, in accord with the expected trend towards greater stability (reduced oxidizing ability) down the group. The data for Rf involves multiple stable cations, like Ti but unlike Zr or Hf; however, the predominance regions for Rf are shifted much lower than for Ti, in general accord with expectations of increased stability of the group oxidation state on going down the group. The predominance diagram for Db matches that for Ta, and differs from those of V, Nb, and Pa, which is a reasonable periodic trend. The Sg diagram includes fewer oxidation states than the diagrams of Cr, Mo, or W, but the predominance region for the group oxidation state is largest for Sg, which is an expected periodic trend.

Chapter 10

10.1. Object **a.** C_4, no i, yes σ, no S_n; **b.** C_5, no i, no σ, no S_n; **c.** C_3, no, no, no; **d.** C_4, yes, no, no S_4 (yes $S_2 = i$); **e.** C_3, yes, yes, yes; **f.** C_5, yes, yes, yes (S_{10}); **g.** C_5, no, no, no; **h.** C_5, yes, yes, yes (S_{10}).

10.3. Linear, C_∞, C_2, σ_v, σ_h, i, (S_∞); equilateral triangular, C_3, C_2, σ_h, σ_v, S_3; tetrahedral, C_3, C_2, S_4, σ_d; trigonal bipyramidal, C_3, C_2, σ_h, σ_v, S_3; octahedral, C_4, C_3, C_2, σ_h, σ_d, i, S_6, S_4; pentagonal bipyramidal, C_5, C_2, σ_h, σ_v, S_5; capped octahedron, C_3, σ_v; capped trigonal prism, C_2, σ_v; square antiprismatic, C_4, C_2, σ_d, S_8; dodecahedral, C_2, σ_d, S_4; tricapped trigonal prismatic, C_3, C_2, i, σ_d, S_6.

10.5. a. C_5, no C_2, no i, σ; **b.** C_5, C_2, no i, σ; **c.** C_2, no C_2, no i, σ; **d.** C_2, C_2, no i, σ.

10.7. a. C_{4v}; **b.** C_5; **c.** C_4; **d.** C_4; **e.** D_{3d}; **f.** I_h; **g.** D_5; **h.** I_h.

10.9. a. O_h; **b.** C_{4h}; **c.** C_{2v}; **d.** D_3.

10.11. a. D_{6h}; **b.** D_{3d}.

10.13. Linear, $D_{\infty h}$; equilateral triangular, D_{3h}; tetrahedral, T_d; trigonal bipyramidal, D_{3h}; octahedral, O_h; pentagonal bipyramidal, D_{5h}; capped octahedron, C_{3v}; capped trigonal prism, C_{2v}; square antiprismatic, D_{4d}; dodecahedral, D_{2d}; tricapped trigonal prismatic, D_{3d}.

10.15. a. C_4, C_1, C_4, and C_2, respectively; **b.** ICl_4^- and TeO_6^{6-}; **c.** ICl_4^-, TeO_6^{6-}, and XeO_4^{2-}; **d.** D_{4h}, C_1, O_h, C_{2v}.

10.17. a. Point group C_{4v} is no for chirality, yes for polarity; **b.** point group C_5 is yes for chirality, yes for polarity; **c.** point group C_4 is yes for chirality, yes for polarity; **d.** point group C_4 is yes for chirality, yes for polarity; **e.** point group D_{3d} is no for chirality, no for polarity; **f.** point group I_h is no for chirality, no for polarity; **g.** point group D_5 is yes for chirality, no for polarity; **h.** point group I_h is no for chirality, no for polarity.

10.19. a. (Point group D_{2d}) neither polar nor chiral;

b. (point group D_3) chiral but not polar;

c. (point group D_4) chiral but not polar;

d. (point group C_1) both polar and chiral.

10.21. With free rotation of monodentate ligands about their donor atoms, complex **a** can be treated as *fac*-CrN$_3$O$_3$ (point group C_{3v}) and complex **b** can be treated as *mer*-CrN$_3$O$_3$ (C_{2v} point group). **a.** No. **b.** No. **c.** The symmetry point group of the *fac* complex falls to C_3 and that of the *mer* isomer falls to C_2. Both can be chiral, so they have structures that are mirror images of each other (enantiomers):

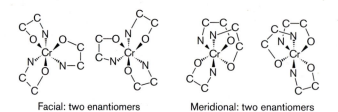

Facial: two enantiomers   Meridional: two enantiomers

10.23. a. f_{xyz}; $(f_{x^3}, f_{y^3}, f_{z^3})$; $f_y(x^2-z^2), f_x(z^2-y^2), f_z(y^2-x^2)$; **b.** f_{xyz}: A or B, because singly degenerate (consult character table operations to choose between these); others, T, because in triply degenerate sets; **c.** u.

10.25. a. The u subscript; **b.** G (for the quadruply degenerate set) and T (for the triply degenerate set).

10.27. a. $(p_x, p_y, p_z) = T_2$, $(d_{z^2}, d_{x^2-y^2}) = E$, $(d_{xz}, d_{yz}, d_{xy}) = T_2$; **b.** A_1.

10.29. a. In D_{4h} square planar complexes, and using character tables, we can identify: $d_{z^2} = A_{1g}$, $d_{x^2-y^2} = B_{1g}$, $d_{xy} = B_{2g}$, and $(d_{xz}, d_{yz}) = E_g$. Without using the tables, we know that all d orbitals are gerade, and we can readily tell that $(d_{xz}, d_{yz}) = E_g$ and that $d_{z^2} = A_{1g}$.
b. In D_{3h} trigonal prismatic complexes, and using character tables, we can identify: $d_{z^2} = A_1'$, $(d_{x^2-y^2}, d_{xy}) = E'$, and $(d_{xz}, d_{yz}) = E''$. Without using the tables, we can readily tell that $d_{z^2} = A'$ and the other two pairs of d orbitals are in E irreducible representations. **c.** In I_h icosahedral complexes, all d orbitals are in the same irreducible representation, which is therefore quintuply degenerate and must be labeled H_g.

10.31. In the D_{3d} point group, $p_z = A_{2u}$, $(p_x, p_y) = E_u$, $d_{z^2} = A_{1g}$, $(d_{x^2-y^2}, d_{xy}) = E_g$, and $(d_{xz}, d_{yz}) = E_g$.

10.33. In the last column on the right side of the character table we find the labels that apply to f orbitals, and which apply to either their bonding or antibonding combinations in diatomic molecules. So, $f_{z^3}(A) + f_{z^3}(B)$ gives us an ungerade σ_u^+ bonding orbital, which is shown below. The antibonding $f_{z^3}(A) - f_{z^3}(B)$ combination is gerade, so it is labeled σ_g^+. We find (xz^2, yz^2) listed in the Π_u irreducible representation. The bonding combination is labeled π_u, and is shown below; the antibonding combination is gerade, so it can be labeled π_g^*. We find $(xyz, z(x^2-y^2))$ listed in the Δ_u irreducible representation. The bonding combination is gerade, is labeled Δ_g, and is shown below; the antibonding combination can be labeled Δ_u^*. Finally, we find $(x(x^2-3y^2), y(3x^2-y^2))$ listed in the Φ_u irreducible representation. The bonding combination is ungerade, is labeled ϕ_u, and is shown below; the antibonding combination can be labeled ϕ_g^*.

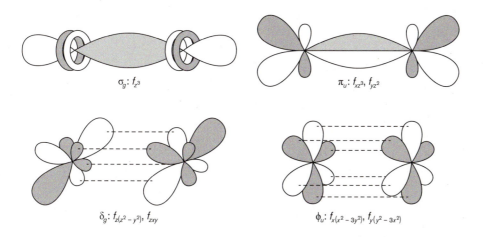

σ_g: f_{z^3}

π_u: f_{xz^3}, f_{yz^2}

δ_g: $f_{z(z^2-y^2)}$, f_{zxy}

ϕ_u: $f_{x(x^2-3y^2)}$, $f_{y(y^2-3x^2)}$

10.35. a. From the left: negative overlap, positive overlap, negative overlap; **b.** antibonding, bonding, antibonding; **c.** gerade, gerade, ungerade; **d.** from the left: π_g^*, σ_g, σ_u^*.

10.37. h. In the $D_{\infty h}$ point group, not in any irreducible representation. $A_1 = \Sigma^+$ in $C_{\infty v}$.
i. In $D_{\infty h}$, not in any irreducible representation. In $C_{\infty v}$, is doubly degenerate with the $p_y - d_{yz}$ combination, with the symmetry label $E_1 = \Pi$. **j.** In the $D_{\infty h}$ point group, doubly degenerate with the $d_{yz} - d_{yz}$ combination. $E_1 = \Pi$ in $C_{\infty v}$ and $E_{1g} = \Pi_g$ in $D_{\infty h}$. **k.** In $D_{\infty h}$, doubly degenerate with the $d_{xy} - d_{xy}$ combination. $E_2 = \Delta$ in $C_{\infty v}$ and $E_{2u} = \Delta_u$ in $D_{\infty h}$. **l.** In $D_{\infty h}$, not in any irreducible representation. $A_1 = \Sigma^+$ in $C_{\infty v}$.

Chapter 11

11.1. a. From the bottom: σ_g, σ_g, π_u, δ_g, δ_u^*, π_g^*, σ_u^*, σ_u^*; **b.** from the left: π_g^*, σ_g, σ_u^*.

11.3. a. $\sigma_g^2\sigma_u^{*2}\sigma_g^2\pi_u^2$, bond order 2, 2 unpaired e⁻; **b.** $\sigma_g^2\sigma_u^{*2}\sigma_g^2\pi_u^3$, bond order 2.5, 1 unpaired e⁻; **c.** $\sigma_g^2\sigma_u^{*2}\sigma_g^2\pi_u^4\pi_g^{*3}$, bond order 1.5, 1 unpaired e⁻; **d.** $\sigma_g^2\sigma_u^{*2}\sigma_g^2\pi_u^4\pi_g^{*4}$, bond order 1, 0 unpaired e⁻; **e.** σ_g^2, bond order 1, 0 unpaired e⁻.

11.5. a. $\sigma_g^2\pi_u^4\delta_g^2$, bond order 4, 2 unpaired e⁻; **b.** $\sigma_g^2\pi_u^4\delta_g^4$, bond order 5, 0 unpaired e⁻; **c.** $\sigma_g^2\pi_u^4\delta_g^4\sigma_g(5s)^2\delta_u^{*4}\sigma_u^*(5s)^2\pi_g^{*2}$, bond order 2, 2 unpaired e⁻; **d.** $\sigma_g^2\pi_u^4\delta_g^4\phi_u^4\phi_g^{*4}$, bond order 5, 0 unpaired e⁻; **e.** $\sigma_g^2\pi_u^4\delta_g^2$, bond order 4, 2 unpaired e⁻; **f.** $\sigma_g^2\pi_u^4\delta_g^4\phi_u^4\phi_g^{*4}\delta_u^{*4}$, bond order 3, 0 unpaired e⁻.

11.7. a. Hf_2 $A^2B^2C^4$, Mo_2^{2-} $A^2B^2C^4D^4E^{*2}$, Pd_2^{4+} $A^2B^2C^4D^4E^{*2}F^{*2}$; **b.** Hf_2 0, Mo_2^{2-} 0, Pd_2^{4+} 2; **c.** Hf_2 4, Mo_2^{2-} 5, Pd_2^{4+} 4.

11.9. a. Orbitals of the same irreducible representation: $\sigma_g(d_{z^2})$ with $\sigma_g(5s)$, also $\sigma_u^*(d_{z^2})$ with $\sigma_u^*(5s)$; the parent d_{z^2} and $5s$ orbitals should be of comparable energy; the left side of the d block; **b.** $\sigma_g(f_{z^3})$ with $\sigma_g(d_{z^2})$ and $\sigma_g(s)$, $\pi_u(f_{xz^2}, f_{yz^2})$ with $\pi_u(d_{xz}, d_{yz})$, $\delta_g(f_{xyz}, f_{z(x^2-y^2)})$ with $\delta_g(d_{xy}, d_{x^2-y^2})$.

11.11. s–p hybridization is least likely when the s–p energy separation is largest. **a.** In Period 2, with F; **b.** in Period 3, with Cl; **c.** in Period 4, with Br; **d.** the difference in $<r_{max}>$ between valence ns and np orbitals increases going down this group; this increasing mismatch in radii makes s–p hybridization more difficult.

11.13. a. N_2 has a bond order of 3, while N_2^+ has a bond order of only 2.5; Cl_2 has a bond order of 1, while Cl_2^+ loses an antibonding electron to give a bond order of 1.5. **b.** s and p orbitals decrease their overlap going down a group in the p block, while d orbitals increase their overlap going down a group in the d block. **c.** The Group 14/IV diatomic molecules involve π bonding, which decreases more rapidly down a group than does σ bonding.

11.15. a. O_2 $\sigma_g^2(\sigma_u^*)^2\sigma_g^2\pi_u^4(\pi_g^*)^2$ with bond order 2, O_2^{2-} $\sigma_g^2(\sigma_u^*)^2\sigma_g^2\pi_u^4(\pi_g^*)^4$ with bond order 1, so O_2^{2-}; **b.** B_2 $\sigma_g^2(\sigma_u^*)^2\sigma_g^2$ with bond order 1, B_2^{2-} $\sigma_g^2(\sigma_u^*)^2\sigma_g^2\pi_u^2$ with bond order 2, so B_2^{2-}; **c.** C_2 $\sigma_g^2(\sigma_u^*)^2\sigma_g^2\pi_u^2$ with bond order 2, C_2^{2-} $\sigma_g^2(\sigma_u^*)^2\sigma_g^2\pi_u^4$ with bond order 3, so C_2; **d.** O_2^+ $\sigma_g^2(\sigma_u^*)^2\sigma_g^2\pi_u^4(\pi_g^*)^1$ with bond order 2.5, O_2^{2-} $\sigma_g^2(\sigma_u^*)^2\sigma_g^2\pi_u^4(\pi_g^*)^4$ with bond order 1, so O_2^+.

11.17. We predict the following bond lengths: Ac_2 430 pm, Th_2 412 pm, Pa_2 400 pm, and U_2 392 pm. The bonds do not seem to be single bonds. The bond orders increase if the computed bond lengths shrink more than do the values derived from Table 1.13: Bond orders increase $Ac_2 < Th_2 < Pa_2 \leq U_2$.

11.19. $\sigma_g \rightarrow \sigma_u^*$ and $\pi_u \rightarrow \delta_g$.

11.21. $\sigma_g(2p) \rightarrow \sigma_u^*$ and $\sigma_g(2s) \rightarrow \sigma_u^*$.

11.23. a. ‖‖‖‖ ‖ ‖‖‖‖. **b.** ‖‖‖‖ ‖ ‖‖‖‖‖. The middle bands corresponding to $(\sigma_u^{nb})^2$ lack fine structure.

11.25. The softnesses are: **a.** N_2 0.112, P_2 0.202; **b.** O_2 0.172, S_2 0.260.

11.27. During the attachment of another electron, C_2 gains an electron in its relatively low-energy LUMO, σ_g^{nb}, while O_2 and N_2 each gain an electron in a relatively high-energy antibonding orbital, π_g^*.

11.29. a. $d_{primary} - 2r_{cov} = 0$ pm for $Cl_2 < 1$ pm for $Br_2 < 6$ pm for I_2; $|d_{secondary} - 2r_{vdW}| = 28$ pm

for Cl_2 < 49 pm for Br_2 < 58 pm for I_2. Secondary bonding is most prevalent in I_2.
b. $d_{primary} - 2r_{cov} = 0$ pm for S_8 and Se_8 < 3 pm for $Se_{metallic}$ < 14 pm for $Te_{metallic}$; $|d_{secondary} - 2r_{vdW}| = 23$ pm for S_8 < 36 pm for $Se_{metallic}$ < 45 pm for Se_8 < 70 pm for $Te_{metallic}$. Secondary bonding is most prevalent in $Te_{metallic}$.

11.31. a.

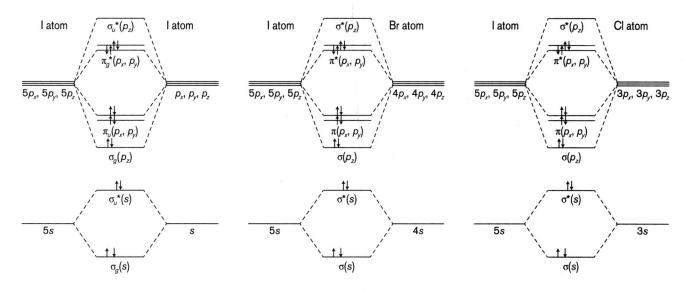

b. π_g^* or $\pi^* = -9.22$ eV in I_2, −9.85 eV in IBr, −10.10 eV in ICl. π_u or $\pi = -10.74$ eV in I_2, −11.99 eV in IBr, −12.88 eV in ICl. σ_g or $\sigma = -12.66$ eV in I_2, −13.70 eV in IBr, −14.26 eV in ICl. As the halogen on the right side becomes more electronegative, its energy level drops. This drops the energy levels of these three MOs (from I_2 to ICl). **c.** As the other halogen goes from I to Cl, the σ- and π-bonding MOs tend to localize on that other halogen atom. However, the π*-antibonding MO tends to localize more on the iodine atom. **d.** I_2 > IBr (> ICl > IF).

11.33. a. $\sigma^2(\sigma^{nb})^2\pi^4(\sigma^{nb})^2\pi^{*1}$, with bond order 2.5; **b.** weaker; **c.** longer; **d.** the photoelectron spectrum of NO should have an additional band with fine structure due to ionization of the π* electron.

11.35. a. Softness = 0.112 for N_2 < 0.127 for CO; **b.** CO; **c.** CO; **d.** CO, CO.

11.37. The numerical answers you obtain will depend on the computational program and basis sets you use. The following are (unchecked) answers obtained by our students. Energies are given in eV.

MO	NC⁻	N_2	NO⁺
LUMO (π*)	+14.29	+5.06	−7.36
HOMO (σ^{nb})	−4.11	−16.77	−31.14
HOMO-1, -2 (π)	−4.25	−16.86	−19.93
HOMO-3 (σ)	−8.16	−20.5	−22.92

a. The order of MTSU student-calculated orbital energies is NO⁺ < N_2 < NC⁻. **b.** The literature calculated ionization potentials for CN⁻ differ by roughly 5.5 eV from the student results for CN⁻. This difference was attributed to the effect of the cation on the energies. **c.** Energies decrease drastically as the positive charge increases, as expected from the repulsive or attractive Coulombic interaction of the entering electron with the ion. **d.** The softnesses decrease from cyanide (0.109) through N_2 (0.092) to nitrosyl cation (0.084). **e.** For N_2, N; for CN⁻, C; for NO⁺, N. **f.** For N_2, N; for CN⁻, C; for NO⁺, N.

11.39.

a.

b. P_z; b_{2g}; 2

c.

d. (Reading from the bottom of the figure) $a_{1g}(\uparrow\downarrow)e_u(\uparrow\downarrow\uparrow\downarrow)a_{2u}^{nb}(\uparrow\downarrow)b_{2g}^{nb}(\uparrow\downarrow)e_u^*(\uparrow\uparrow)a_{1g}^*$.

e. 2; ½.

11.41.

a.

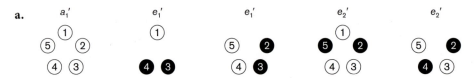

b. Two; one above and one below the plane of the pentagon, in sp_z-hybridized orbitals; $(p_x, p_y) = e_1'$ remain to bond with the TASOs; a_1' and e_2' are nonbonding. **c.** 0.4.

11.43. A closer match in energy levels (VOPEs) for the $2p$ orbitals of the two atoms involved should favor stronger π bonding. The $2p$ VOPE of oxygen, –15.8 eV, is closest to that of nitrogen, –13.2 eV, followed by carbon, –10.6, then that of boron, –8.3 eV. Lastly we find the VOPE for fluorine, –18.6 eV, to be quite distant from that of boron, –8.3 eV. So, the order should be $NO_3^- > CO_3^{2-} > BO_3^{3-} > BF_3$.

11.45.

a.

b. s matches the a_1 TASO; (p_x, p_y, p_z) and (d_{xy}, d_{xz}, d_{yz}) match the t_2 TASOs.

c.

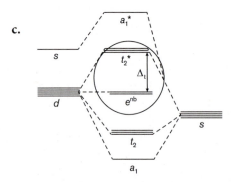

11.47. a. π overlap in oxo anions increases from Si to P to Cl. **b.** Formal charge of the central atom becomes quite high for Cl unless you draw resonance structures having two or three Cl–O π bonds. But these give higher bond orders per Cl–O link than the 1.0 value we have from our MO diagram for EO_4^{x-}. **c.** MO computations on the computer add post-valence $3d$ orbitals to the basis set, which can have π overlap with oxygen $2p$ orbitals. The computations you do likely show the highest such π overlap for Cl.

11.49. **a.** From the bottom: a_{1g}, t_{1u}, e_g^{nb}, t_{1u}^*, a_{1g}^*, and at left, s below (p_x, p_y, p_z); **b.** d orbitals, e_g and t_{2g}; **c.** yes, t_{2g}^{nb}; **d.** no.

11.51. **a.** 18, tetrahedral; **b.** 18, tetrahedral; **c.** 16, square planar; **d.** 18, tetrahedral; **e.** 18, tetrahedral.

11.53. **a.** $Os(CO)_5$; **b.** $[Fe(CO)_4]^{2-}$; **c.** $[Ta(PF_3)_6]^-$.

11.55. **a.** Bidentate chelating ligand; **b.** seven, which exceeds the maximum coordination number, 6; **c.** yes.

11.57. **a.** 15 electrons; **b.** 16 e⁻; **c.** 13 e⁻; **d.** 15 e⁻; **e.** 16 e⁻; **f.** 18 e⁻; **g.** 15 e⁻.

11.59.

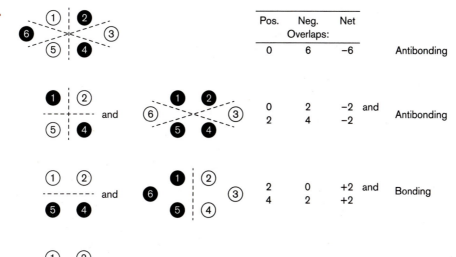

	Pos.	Neg.	Net	
		Overlaps:		
	0	6	−6	Antibonding
	0	2	−2	and
	2	4	−2	Antibonding
	2	0	+2	and
	4	2	+2	Bonding
	6	0	+6	Bonding

The energy level diagram has four equally spaced levels; the middle two levels are each doubly degenerate.

11.61.

a.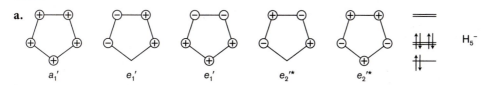

b. Similar to the above but with lobes of opposite sign below each lobe, and with the labels:

a_2'' $\quad$ e_1'' $\quad$ e_1'' $\quad$ $e_2''^*$ $\quad$ $e_2''^*$ $\quad$ $C_5H_5^-$

c.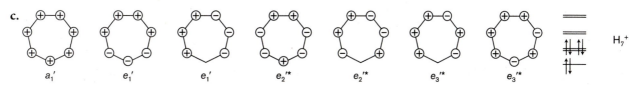

d. Similar to the above but with lobes of opposite sign below each lobe, and with the labels:

a_2'' $\quad$ e_1'' $\quad$ e_1'' $\quad$ $e_2''^*$ $\quad$ $e_2''^*$ $\quad$ $e_3''^*$ $\quad$ $e_3''^*$ $\quad$ $C_7H_7^+$

11.63. a_2'' orbital overlaps with metal s, p_z, d_{z^2}, and f_{z^3} orbitals; ligand acts as a π donor; e_1'' orbital overlaps with metal p_x, p_y, d_{xz}, d_{yz}, f_{xz^2}, and f_{yz^2} orbitals; ligand acts as a π donor; e_2'' overlaps with metal d_{xy}, $d_{x^2-y^2}$, f_{xyz}, and $f_{x(z^2-y^2)}$ orbitals; ligand acts as a π acceptor.

11.65. a. Kelli has an xz nodal plane e_u^{nb}, Paul has a yz nodal plane e_u^{nb}, Celeste is improper, Nathan has no nodal plane a_{1g}, Shabnan has two nodal planes between the x and y axes b_{1g}^*. **b.** Kelli p_y and d_{yz}; Paul p_x and d_{xz}; Nathan s, p_z, and d_{z^2}; Shabnan $d_{x^2-y^2}$. **c.** Nathan is at the lowest energy level, Kelli and Paul both are at the middle (zero) energy level, and Shabnan is at the upper energy level.

11.67. a. 20 electrons; **b.** 20 e⁻; **c.** 23 e⁻; **d.** 18 e⁻.

11.69. a. Cr, Mo, and W; **b.** Fe, Ru, Os; **c.** Mn, Tc, Re; **d.** Ni, Pd, Pt; **e.** V, Nb, Ta; **f.** Ni, Pd, Pt.

11.71. a. As an example, for Mn^{2+} these are $e_2''^4 a_2''^{nb1}$ (low spin) or $e_2''^2 a_2''^{nb1} e_1''^{*2}$ (high spin). **b.** For Mn^{2+} the low-spin configuration gives the 1 unpaired electron found in $Mn(C_5Me_5)_2$ while the high-spin configuration gives the 5 unpaired electrons found in $Mn(C_5H_5)_2$.

Chapter 12

12.1. a. Mo; **b.** Mo; **c.** all the same; **d.** elemental phosphorus at extremely high pressure.

12.3. a. **A** sheet polymeric, **B** network polymeric, **C** monomeric molecular, **D** chain polymeric; **b.** **A** 3, **B** 4, **C** 1, and **D** 2; **c.** **A** has one unshared electron pair and TCN = 4, **B** has no unshared electron pairs and TCN = 4, **C** has four unshared electron pairs and TCN = 4, **D** has two unshared electron pairs and TCN = 4; **d.** **A** Group 15/V, **B** Group 14/IV, **C** Group 18/VIII, **D** Group 16/VI.

12.5. +210 kJ; O_2.

12.7. $\Delta H = -27$ kJ. The reaction is slightly favored by the ΔH term (although not by the entropy term).

12.9. a.

This is isoelectronic to benzene and of a similar molecular weight, so it would probably have similar physical properties: It might be a liquid of b.p. about 80°C! It surely would have a higher boiling point than $N_2(g)$ and a lower boiling point than black P or red P. **b.** $\Delta H = -660$ kJ (or −110 kJ mol⁻¹ of N atoms). **c.** $\Delta H = -183$ kJ (or −60.5 kJ mol⁻¹ of P atoms).

12.11. Impossible for C_2; plausible for Zr_2. A quadruple bond must include a δ bond, which requires overlap of d orbitals.

12.13. High temperature and low pressure. Both increase the importance of the $T\Delta S$ term in $\Delta G = \Delta H - T\Delta S$, as larger numbers of smaller multiply bonded molecules are favored.

12.15. Bk.

12.17. Carbon. In its prepared-for-bonding electron configuration $2s^1 2p_x^1 2p_y^1 2p_z^1$, carbon can form more pure covalent bonds to its neighbors (4) than any other second-period atom.

12.19. (a).

12.21.

Orbital	+ Overlaps	− Overlaps	Net overlaps	
①│**②**│③│**④**│⑤	0	4	−4	σ_g^*
①│**②**│ │④│**⑤**	0	2	−2	σ_u^*
① │ **③** │ ⑤	0	0	0	σ_g^{nb}
①│② │ **④**│**⑤**	2	0	+2	σ_u
①│②│③│④│⑤	4	0	+4	σ_g

The energy-level diagram has five equally spaced levels.

12.23. B.

12.25. Add lobes of opposite sign below each lobe shown for H_6; convert all σ_g labels to π_u; convert all σ_u labels to π_g.

12.27. a. The variable parameter L of the box increases as do the energy levels of the π HOMOs and π^* LUMOs. **b.** For each increase in x, the value of ΔE is halved. **c.** The HOMO–LUMO gap approaches zero. **d.** This is unrealistic: A Peirls distortion would open up a nonzero HOMO–LUMO gap.

12.29. For structures, see Figures 12.1 and 12.2. **a.** Hard-to-vaporize solid; **b.** easily vaporized solid; **c.** hard-to-vaporize solid; **d.** gas.

12.31. a. A: boron, **B:** S or Se or Te, **C:** C or Si or Ge or Sn, **D:** N or O or H or halogens or P or As; **b. D < B < A ≈ C.**

12.33. a. C; **b.** A; **c.** A; **d.** C; **e. A:** C, Si, Ge, Sn; **B:** S, Se, Te; **C:** H, N, O, halogens, P, As. **D:** S.

12.35. a. One σ and one π; **b.** graphite: 1.5 σ and 0.5 π; **c.** for six atoms $\Delta H = -270$ kJ; yes.

12.37. a. C_2 monomeric (molecular), graphene is (sheet or layer) polymeric, carbyne is (chain or linear) polymeric; **b.** C_2 < carbyne < graphene; C_2 is most likely a gas.

12.39. C_{60}^- $pK_b \approx 28$, nonbasic, it could persist in water. C_{60}^{2-} $pK_b \approx 14$, nonbasic, it could persist in water. C_{60}^{3-} $pK_b \approx 0$, strongly basic, it could persist only at high pH values in water. The remaining fullerene anions would be very strongly basic and could not persist in water: for example, C_{60}^{4-} would have a $pK_b \approx -18$.

12.41. Longer, because the electrons from the Group 1 metals enter the LUMOs, which are antibonding (π^*) orbitals.

12.43. a. A semiconductor; **b.** its conductivity would increase into the metallic range; **c.** its conductivity would increase into the metallic range.

12.45. a. The solid is metallic; **b.** the solid is nonpolar covalent; **c.** the solid is (nonpolar) network covalent.

12.47. a. For S, $U = 800$ kJ mol^{-1}, so probably yes; **b.** for Rn, the electron affinity is not listed, but $U \approx 1037$ kJ mol^{-1} or even more, so no.

12.49. C < B < D (becomes metallic) < E for some elements < A for some elements.

12.51. a. Mo; **b.** Rb/K alloy; **c.** Gd; **d.** Fe/V alloy; **e.** pure Au.

12.53. Because Fe differs least from U in electronegativity, and differs the most from U in atomic density, its alloy with U should and does have the smallest enthalpy of formation. Rh and Ir differ little in atomic density; the enthalpies of formation of their alloys should be and are similar to each other.

12.55. **a.** Lower. **b.** One explanation would be based on the Nernst equation: The activities and concentrations of the metals are higher in the pure metals than in the amalgams. The second explanation would be that, in exothermically forming intermetallic compounds such as MHg_{11}, their reactions with water would necessarily become less exothermic. **c.** Less; the activity of Ag is only changed by virtue of its reduced concentration in silver amalgam.

12.57. **a.** Four shells plus the central atom; **b.** 55; **c.** 561.

12.59. Since the bandwidths and HOMO–LUMO gaps change as the length of one-dimensional polymers change, we would also expect this to be true for clusters with three dimensions.

12.61. Diamond, boron nitride, aluminum nitride. Their band gaps are so large that they act as insulators, so they would need to be doped.

12.63. For a *p*-type semiconductor, you want an atom from one group to the left in the periodic table that can substitute isomorphously. For an *n*-type semiconductor, you want an atom from one group to the right. C could perhaps substitute for B in the diboride layer.

12.65. **a.** Layer structures or perhaps fullerene-type clusters; **b.** MoS_2, because Mo^{4+} is more susceptible to one-electron oxidation or reduction than B^{3+}; probably S^{2-} is also more susceptible than N^{3-}.

PHOTO AND FIGURE CREDITS

INDEX

Page numbers for definitions of terms are in **boldface**.